THE OXFORD
COMPANION TO
AUSTRALIAN
MILITARY
HISTORY

THE OXFORD COMPANION TO AUSTRALIAN MILITARY HISTORY

PETER DENNIS
JEFFREY GREY
EWAN MORRIS
ROBIN PRIOR
with
JOHN CONNOR

Melbourne
OXFORD UNIVERSITY PRESS
Oxford Auckland New York

OXFORD UNIVERSITY PRESS AUSTRALIA

Oxford   New York
Athens   Auckland   Bangkok   Bombay
Calcutta   Cape Town   Dar es Salaam   Delhi
Florence   Hong Kong   Istanbul   Karachi
Kuala Lumpur   Madras   Madrid   Melbourne
Mexico City   Nairobi   Paris   Singapore
Taipei   Tokyo   Toronto

and associated companies in
Berlin   Ibadan

OXFORD is a trade mark of Oxford University Press

National Library of Australia
Cataloguing-in-Publication data:

The Oxford companion to Australian military history.

ISBN 0 19 553227 9.

1. Australia — History, Military — Encyclopedias.
2. Australia — Armed Forces — Biography.
I. Dennis, Peter, 1945– .
II. Title: Companion to Australian military history.

355.00994

Edited by Katherine Steward with Venetia Somerset
Maps drawn by Paul Ballard
Text designed by Steve Randles
Cover design by Cora Lansdell
Typeset by Desktop Concepts P/L, Melbourne
Printed by Australian Print Group
Published by Oxford University Press,
253 Normanby Road, South Melbourne, Australia

CONTENTS

PHOTOGRAPHS

All photographs, with the exception of that mentioned below, have been reproduced by kind permission of the Australian War Memorial. The Australian War Memorial negative number for each photograph appears in brackets after the caption.

Permission to reproduce the photograph of the Australian Vietnam Veteran's Memorial was kindly granted by the photographer, Deborah Jenkin.

MAPS

PREFACE

War has been one of the defining forces in Australian history. Recent scholarship has established what was previously unknown or denied, that the European settlement of Australia was not the peaceful spread of White civilisation but a story of violence that fully deserves the name of war. Participation in war, whether actively in the armed forces or on the home front, or in opposition to a particular war, has shaped the lives of successive generations of Australians in the twentieth century. Today, more than 20 years after Australia withdrew its forces from Vietnam, questions relating to defence do not usually seize the attention of the media, but the Department of Defence remains one of the largest single consumers of the annual federal budget, and Anzac Day continues to hold a special place in the Australian calendar.

The Oxford Companion to Australian Military History seeks to explore the richness and diversity of Australia's military history. We have conceived of military history as widely as possible. The result, we trust, is not a compendium of great battles and great captains, but a reference work that will explain how military questions have affected Australian history. While we have sought to make our coverage as comprehensive as we can, the result to some extent reflects our own interests and expertise, and in some cases the state of current scholarship. We have sought to combine information with comment and analysis, and while our focus is Australian, it is essential that Australian military history be seen in a wider context. Word length is not necessarily an index of the importance we attach to a subject. Not all entries are as long as their subjects deserve, none more so than in the case of entries on individual women. For many of these only the barest information is available, and we hope that by identifying important women whose lives need further exploration we might stimulate research on them.

Inevitably a work on military history contains a great deal of technical information. We hope to convey this accurately, but also in a manner that does not exclude readers who are not specialists or who have no knowledge of military matters. We seek the indulgence of those who might feel that some detail or other offends their understanding of a particular subject: much of the literature is at odds on certain facts, and styles (for example on rank and the names of service departments and sections) have changed over time. We have sought to be consistent rather than pedantic. Despite our best efforts, there are no doubt errors of fact and questions of inclusion or omission that will occur to readers: we welcome any suggestions for changes or additions that we could make to future editions of this *Companion*. In the course of writing this book we have drawn extensively on the works of others, none more so than in the case of biographical entries. The *Australian Dictionary of Biography* is one of the crowning achievements of Australian scholarship, and it would be foolish to deny that when writing about individuals we have not in the first instance consulted the *ADB*. Our entries, however, have a somewhat different, and more specifically military, focus from the *ADB*'s, and we have made it a point of policy not to ask authors who have written on an individual for the *ADB* to write an entry on the same person for the *Companion*.

In the process of completing this work we have incurred many debts. We are grateful to Mr A. J. Hill and Dr A. B. Lodge who first conceived this project and who, when circumstances made it impossible for them to continue, entrusted it to us. The successive Rectors of the University College at the Australian Defence Force Academy, Professor G. V. H. Wilson and Professor H. P. Heseltine, have been generous supporters, and have enabled us to appoint a research assistant. In the first instance this was Ewan Morris, whose contribution is acknowledged by his inclusion as a main author. He was succeeded by John Connor, who has not only kept us to our schedule by a combination of discipline and humour but has been entirely responsible for the selection of photographs in the book. We are immensely grateful to John for his willingness and patience — with us and with the project. As is always the case in the Department of History, we have been helped considerably by Helen Boxall, Elizabeth Greenhalgh, Christine Kendrick, Barbie Robinson and Elsa Selleck, whose continuing contribution to our work we value highly. A number of entries have been written by those with special knowledge in certain areas: such articles are identified as the work of an individual author, whose names appear collectively in the list of contributors. We are indebted to them all, not only for the expertise they have brought to the *Companion* but also for their willingness and cooperation. Paul Ballard of the Department of Geography and Oceanography at the University College has drawn the maps with his customary skill, and has shown a degree of tolerance that we had no right to expect.

Peter Rose of Oxford University Press has been everything that we could have asked for in a publisher. His encouragement, forbearance and wise advice have been invaluable to us, and we have found working with him a wonderful and enjoyable experience. Our editors, Katherine Steward and Venetia Somerset, have brought keen eyes and incisive minds to our manuscript, and we have benefited greatly from their ability and willingness to seek clarification of what we had thought was patently obvious, only to discover on careful rereading that it was far from that. We have tried to make the *Companion* approachable and non-exclusive, and if we have succeeded in that aim, much of the credit must go to them. We are deeply grateful to Professor R. J. O'Neill of All Souls College, Oxford, for his very extensive and helpful comments on the manuscript. We are also indebted to Mr A. J. Hill, who gave us the benefit of his wide knowledge of the field, and who saved us from error on a number of occasions. The responsibility for any errors that remain, however, is entirely ours.

DIRECTORY OF CONTRIBUTORS

Eric Andrews
Associate Professor of History
University of Newcastle

Paul Ballard
Department of Geography and
  Oceanography
University College, University of
  New South Wales
Australian Defence Force Academy

Joan Beaumont
*Professor of Australian and International
  Studies*
Deakin University

Coral Bell
Visiting Senior Fellow
Strategic and Defence Studies Centre
Australian National University

Peter Burness
Senior Curator, Military Heraldry
Australian War Memorial

Enrico Casagrande
Wing Commander, Legal Branch
Royal Australian Air Force

John Coates
Lieutenant-General (rtd), Honorary Fellow
Department of History
University College, University of
  New South Wales
Australian Defence Force Academy

John Connor
Department of History
University College, University of
  New South Wales
Australian Defence Force Academy

Chris Coulthard–Clark
Military and Defence historian
Canberra

Peter Dennis
Professor of History
University College, University of
  New South Wales
Australian Defence Force Academy

Jeff Doyle
Department of English
University College, University of
  New South Wales
Australian Defence Force Academy

T. R. Frame
Naval historian
Wagga Wagga
New South Wales

Anna Gray
Senior Curator, Art Section
Australian War Memorial

Jeffrey Grey
Senior Lecturer in History
University College, University of
  New South Wales
Australian Defence Force Academy

Bob Hall
Executive Director
Australian Defence Studies Centre
University College, University of
  New South Wales
Australian Defence Force Academy

Deborah Jenkin
Lecturer in Communications
University of Canberra

Ian Kuring
Warrant Officer Class Two
Army History Section
Department of Defence

Peter Londey
Historical Research Section
Australian War Memorial

Michael McKernan
Deputy Director
Australian War Memorial

Bruce Moore
*Director, Australian National Dictionary
   Centre*
Australian National University

Ewan Morris
Department of History
University of Sydney

Peter Pierce
Senior Lecturer
Centre for Australian Studies
Monash University

Robin Prior
Senior Lecturer in History

University College, University of
   New South Wales
Australian Defence Force Academy

Peter Stanley
Head, Historical Research Section
Australian War Memorial

Gerald Walsh
Senior Lecturer in History
University College, University of
   New South Wales
Australian Defence Force Academy

Craig Wilcox
Historical Research Section
Australian War Memorial

Craig Wood
Canberra

Directory of Thematic Entries

STYLE NOTES

In this *Companion* entries are arranged in alphabetical order up to the first punctuation in the headword. The surnames of people (e.g. **GORTON, John Grey**) are printed in bold capitals at the beginning of articles, with given names in bold upper and lower case letters. A given name by which a person is not generally known is presented in brackets and nicknames are in inverted commas, for example, **DUNLOP, Lieutenant-Colonel (Ernest) Edward 'Weary'**. Pseudonyms are printed in inverted commas within brackets, for example, **HOGUE, Major Oliver ('Trooper Bluegum')**. Titles (other than those indicating military rank), honours and awards have not been included in biographical headwords. Ranks are the highest attained by an individual. The names of other entities, such as departments, organisations and events, are printed in bold capitals (e.g. **ARMOUR**). Readers will note that the definite article has been deleted in these headwords.

Cross-references are indicated by 'q.v.' or by the use in brackets of 'see' followed by the entry title in small capitals. Occasionally, 'See also' is used to inform the reader of related subjects that may be of interest. The cross-reference indicator 'q.v.' has been used with some discretion. It has generally been included for names of individuals, organisations, institutions and less well-known elements of the defence force that may not obviously warrant an entry, but not for subjects for which one could reasonably expect to find an entry (e.g. Australian Imperial Force, Department of Defence, Gallipoli, Korean War, New Guinea campaign, RMC Duntroon). Readers are thus encouraged to use the *Companion* to see if an item not covered by a 'q.v.' is nevertheless subject to an entry.

Other aspects of style to be noted are that, wherever possible, place-names in the text and on the maps have been given their English names; brief quotes within the text have not been sourced but, where appropriate, references for further reading have been included at the end of entries; and some words that appear in the list of abbreviations (e.g. POW, RAR, RMC, UK, UN, US) may also be spelt out where the text demands it.

Readers who are not familiar with the military may find it useful to read the information on army organisation and military ranks that follows.

ARMY ORGANISATION

The army has always been made up of constituent parts ranging from the section to corps and beyond. In hierarchical terms, an army looks as follows:

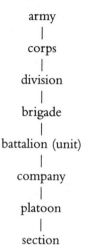

army
|
corps
|
division
|
brigade
|
battalion (unit)
|
company
|
platoon
|
section

The basic building block of an army is the section consisting of 10 men and an NCO. Three sections make a platoon, commanded by a lieutenant. Three platoons form a company, generally a major's command. Four companies make up a battalion, commanded by a lieutenant-colonel. This is a unit; its component parts are sub-units. A brigade consists of three battalions, with attached supporting arms, commanded by a brigadier. A division (the lowest level of formation) comprises three brigades, and is commanded by a major-general. A corps consists of two or more divisions, and is a lieutenant-general's command. Two corps, at least, may be grouped to form an army.

All of the above has varied enormously during the twentieth century, and between arms. During the First World War, brigades consisted of four battalions. Initially, in 1914, a brigade was commanded by a colonel, and the latter were promoted to the rank of brigadier-general only some months after the outbreak of war and in line with British practice. Battalions went from an eight company to a four company organisation in the same time span. The divisions of the 2nd AIF had divisional cavalry squadrons attached in 1939–40, only to lose these in the army reorganisation of 1942. During the Pentropic (q.v.) experiment in the early 1960s, divisions of the Australian Army looked utterly different from the above model. Mounted and armoured divisions were organised differently, and used different nomenclature: the equivalent of a platoon was a troop, that of a company a squadron. Before the First World War the basic unit of the field artillery was the artillery brigade, which was later designated a field regiment. Neither corresponded to a brigade or regiment as those terms were understood in the infantry. A division has always incorporated various other arms in support: engineers (of several kinds), ordnance, artillery, medical, veterinary (where appropriate), provost, service, et cetera. The term 'regiment' has always enjoyed a precise organisational definition in the American and similar armies, which it lacks entirely in British pattern ones: a regiment may comprise three battalions or nine, as the Royal Australian Regiment (q.v.) has done, or may be used to encompass the whole of the Royal Australian Artillery. In American usage a regiment is the equivalent of a brigade or, if designated a Regimental Combat Team as in Vietnam or Korea, a brigade group. There were numerous other variations to army organisation in the twentieth century, including those involving machine gun battalions, mounted and armoured and army tank regiments, and the introduction and demise of the task force in the 1960–70s.

MILITARY RANKS: SERVICE EQUIVALENTS

Commissioned officers

Royal Australian Navy	Army	Royal Australian Air Force
Admiral	General	Air Chief Marshal
Vice-Admiral	Lieutenant-General	Air Marshal
Rear-Admiral	Major-General	Air Vice-Marshal
Commodore	Brigadier	Air Commodore
Captain	Colonel	Group Captain
Commander	Lieutenant-Colonel	Wing-Commander
Lieutenant-Commander	Major	Squadron Leader
Lieutenant	Captain	Flight Lieutenant
Sub-Lieutenant	Lieutenant	Flying Officer
Midshipman	2nd Lieutenant	Pilot Officer

Warrant and non-commissioned officers

Royal Australian Navy	Army	Royal Australian Air Force
Warrant-Officer	Warrant-Officer Class 1	Warrant-Officer
Chief Petty Officer	Warrant-Officer Class 2	
	Staff Sergeant	Flight Sergeant
Petty Officer	Sergeant	Sergeant
Leading Seaman	Corporal or Bombardier	Corporal
	Lance-Corporal or Lance-Bombardier	

ABBREVIATIONS

ABC	Australian Broadcasting Commission
AC	Companion of the Order of Australia
ADC	aide-de-camp
ADF	Australian Defence Force
AFC	Air Force Cross
AIF	Australian Imperial Force
ALP	Australian Labor Party
AM	Member of the Order of Australia
AMF	Australian Military Forces
ANZUS	Australia, New Zealand and the United States
AO	Officer of the Order of Australia
AOC	Air Officer Commanding
ASEAN	Association of South-East Asian Nations
AWM	Australian War Memorial
BBC	British Broadcasting Corporation
BGS	Brigadier General Staff
BGGS	Brigadier-General, General Staff
BHP	Broken Hill Proprietary Company Limited
C-in-C	Commander-in-Chief
CAS	Chief of the Air Staff
CB	Companion of the Order of the Bath
CBE	Commander of the Order of the British Empire
CGM	Conspicuous Gallantry Medal
CGS	Chief of the General Staff
CMF	Citizen Military Forces
CMG	Companion of the Order of St Michael and St George
CNS	Chief of Naval Staff
CO	Commanding Officer
CRA	Commander Royal Artillery
CVO	Commander of the Royal Victorian Order
DCM	Distinguished Conduct Medal
DFC	Distinguished Flying Cross
DFM	Distinguished Flying Medal
DSC	Distinguished Service Cross
DSM	Distinguished Service Medal
DSO	Distinguished Service Order
GBE	(Knight or Dame) Grand Cross of the Order of the British Empire
GC	George Cross
GCB	(Knight or Dame) Grand Cross of the Order of the Bath
GCMG	(Knight or Dame) Grand Cross of the Order of St Michael and St George
GHQ	General Headquarters
GM	George Medal

GOC	General Officer Commanding
GOC-in-C	General Officer Commanding-in-Chief
GPMG	general purpose machine-gun
GSO	General Staff Officer
HMAS	His/Her Majesty's Australian Ship
HMCS	Her Majesty's Colonial Ship
HMS	His/Her Majesty's Ship
KBE	Knight Commander of the Order of the British Empire
KCB	Knight Commander of the Order of the Bath
KCMG	Knight Commander of the Order of St Michael and St George
KCVO	Knight Commander of the Royal Victorian Order
KG	Knight of the Order of the Garter
LMG	light machine-gun
LSW	light support weapon
MBE	Member of the Order of the British Empire
MC	Military Cross
MG	machine-gun
MGGS	Major-General, General Staff
MM	Military Medal
MMG	medium machine-gun
NATO	North Atlantic Treaty Organization
NCO	non-commissioned officer
OBE	Officer of the Order of the British Empire
PNG	Papua New Guinea
POW	prisoner of war
q.v.	quod vide (see this reference)
QC	Queen's Counsel
qq.v.	quae vide (see these references)
RAAF	Royal Australian Air Force
RAF	Royal Air Force
RAN	Royal Australian Navy
RANC	Royal Australian Naval College
RAR	Royal Australian Regiment
RFC	Royal Flying Corps
RMC	Royal Military College
RN	Royal Navy
RNZAF	Royal New Zealand Air Force
RRC	(Lady of the) Royal Red Cross
RSL	Returned and Services League
RSSILA	Returned Sailors' and Soldiers' Imperial League of Australia
SAS	Special Air Services
SEATO	South-East Asia Treaty Organization
SLR	self-loading rifle
SMG	sub-machine gun
UAP	United Australia Party
UK	United Kingdom
UN	United Nations
US	United States
USAF	United States Air Force
USS	United States Ship
VC	Victoria Cross
WRAAC	Women's Royal Australian Army Corps

A

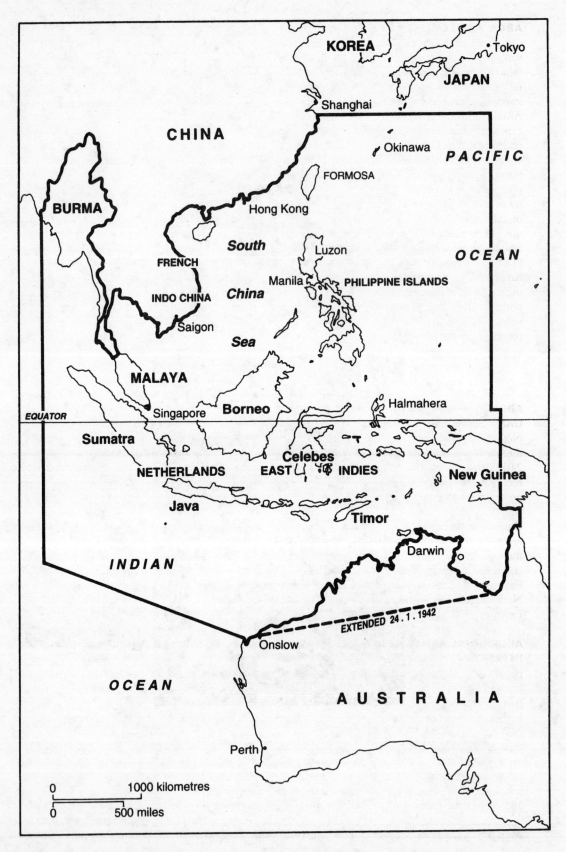

ABCA AGREEMENT The America–Britain–Canada–Australia Armies Standardisation Program was set up under an agreement signed in 1964, to which New Zealand adhered with observer status in 1965. It has its origins in the 1947 plan to effect standardisation between the British, Canadian and American armies, which arose out of problems experienced in north-west Europe during the Second World War. The program is designed to foster cooperation, collaboration and inter-operability through standardisation of equipment, doctrine and procedures. It emphasises inter-operability, the ability to operate jointly with the minimum of friction or difficulty, rather than commonality of equipment. The program is managed by an executive board which meets every 18 months at the TEAL (Tactics, Equipment And Logistics) conferences. It has been of particular benefit to Australia by allowing the army to remain in close and regular contact with the latest developments in equipment, technology and military thinking. There are equivalent programs for the navy and air force.

Thomas-Durrell Young, *Australian, New Zealand and United States Security Relations, 1951–1986* (Westview Press, Boulder, Colorado, 1992).

ABDA (AMERICAN–BRITISH–DUTCH–AUSTRALIAN) COMMAND, covering Burma, Malaya, the Netherlands East Indies and the Philippines (with the north-western part of Australia included from 24 January), was established on 15 January 1942 to reconcile the differing interests of the member powers, and to coordinate military resistance to Japan's advance in south-east Asia. Commanded by General Sir Archibald Wavell (q.v.), it proved powerless to prevent the collapse of Allied control throughout the region, and with nothing left to command following the fall of Singapore on 15 February and the defeat of the Dutch in the Netherlands East Indies, it was dissolved on 25 February 1942.

ABORIGINAL ARMED RESISTANCE TO WHITE INVASION It is impossible to know how many people were living in Australia at the start of the European invasion in 1788, but a figure of 750 000 is currently accepted as a reasonable estimate. Our knowledge about the nature of Aboriginal societies before the invasion is also sketchy, relying on a mixture of archaeological research, observations by early White settlers, and more recent anthropological studies. In addition, Aboriginal social and territorial organisation was extraordinarily complex, and varied from place to place. It seems, however,

Map 1 Outline of the ABDA Command area, 1942.

that Aboriginal people generally formed themselves into bands made up of several families which would travel around a known and owned territory carrying out subsistence activities of hunting, gathering and fishing. Such bands should be distinguished from land-owning groups, which for convenience may be termed 'clans'. Some writers have depicted clans as owning discrete territories with clear boundaries, while others have argued that clans, defined principally through descent, owned constellations of sacred sites which made up a territory without absolute boundaries. In the second model the territories of neighbouring clans might overlap or intersect; furthermore, all members of a clan may not have related to all sites within a territory. Aboriginal ownership of land involved strong spiritual and emotional ties to country, ties that were reflected in detailed knowledge of and feelings of responsibility for the land. Outsiders could be excluded from a clan's territory if they earnt the clan's enmity or if resources were scarce, but usually neighbouring clans ranged freely over each other's estates and strangers were allowed to pass through and use the land's resources. Although clans could on rare occasions assert exclusive rights to land, the ideas of individual ownership and of transfer of land ownership through sale or conquest were completely foreign to them. Above the clan level there were larger groupings (sometimes coinciding with divisions between the hundreds of indigenous Australian languages) which may be termed 'tribes', though they lacked the cohesiveness implied by this word. Trade in material items took place over vast distances and ceremonies were shared between groups, but these activities did not provide the basis for inter-tribal alliances. There was no sense of continent-wide, or even region-wide, Aboriginal identity before 1788, nor indeed for many years afterwards.

Aboriginal society was egalitarian rather than hierarchical in the sense that there were no privileges or handicaps based on birth (though the question of whether or not women constituted a subordinate group remains a vexed one) and that the power of individuals to give orders or to enforce their will on others was limited. Nevertheless, there were differences of authority based on degrees of religious knowledge, and therefore closely related to age and perhaps to gender. This reflected the fact that Aboriginal politics centred on control of land and that land ownership was justified in religious terms. Disputes over land do not, however, seem to have been the cause of violent conflicts either within or between groups, though bloodshed could result from unauthorised uses of the land's resources. Violence was also used for revenge or punishment in cases of sacrilege, disputes over women and murder (including deaths believed to have been caused by sorcery). In such cases, parties of men would often set out to ambush a particular victim or to creep into his camp at night and kill him. Individuals could also fight duels, which sometimes escalated into pitched battles between two bodies of men. Casualties were strictly limited, however, by the highly ritualised nature of Aboriginal fighting, and a halt was usually called once blood was drawn. Fighting was limited not only by convention but also by practical considerations. Aboriginal societies did not produce the economic surpluses necessary to sustain prolonged warfare or to support standing armies. James Morrell, a White man who lived for many years with Aborigines in the nineteenth century, wrote of Aboriginal 'wars' that 'they cannot keep it up for many hours as they are forced to go and get supplies … and they seldom renew the conflict'. There was no separate military class in Aboriginal society, so when fighting did escalate all able-bodied men were potentially involved. Weapons used in Aboriginal fighting included stabbing spears and sharpened fighting poles; throwing spears, which were often thrown with a spear-thrower; boomerangs and throwing sticks; clubs, including some in northern Queensland that were shaped like swords; fighting picks (stone points attached at right angles to long wooden handles); and knives. Shields were the only defensive items: thin hardwood shields were used for parrying blows in close combat, while broad lightwood shields were used both for deflection and to absorb the impact of weapons, particularly thrown spears. Not all of these weapons were used in all parts of Australia. They were made of wood, with points, blades and barbs made of stone or, more rarely, of animal materials such as sharks' teeth or stingray spines.

Britain claimed sovereignty over Australia in three stages. The initial annexation by Governor Phillip in 1788 only extended as far inland as the 135th degree of east longitude. This annexation was pushed back to the 129th meridian when a port was established at Melville Island in 1824, an extension that was confirmed in 1825. Finally, the entire continent became British in 1829 when the remaining third (present-day Western Australia) was claimed for the British Crown. European powers at this time agreed that countries with no recognisable form of law or political organisation could legitimately be annexed. Finding no chiefs or obvious form of government in Australia, British authorities never felt the need to establish sovereignty by negotiating a treaty with the indigenous inhabitants, as

happened in New Zealand. Australia was deemed to be a colony of settlement, not of conquest, so English law came immediately into force with the claim of sovereignty and Aboriginal people automatically became subjects of the British Crown.

The legal situation with regard to land is more complex. The justification for taking Aboriginal land was that Australia was *terra nullius*, a land belonging to no one. This doctrine has two meanings: first, a land without a sovereign recognised by European powers, and second, a country with no system of land tenure. When the colonisation of Australia began it was believed that Australia was *terra nullius* because Aborigines had no concept of land ownership and were therefore not in effective possession of the land. Historian Henry Reynolds has argued that as British governments gained greater knowledge of Aboriginal societies they recognised the existence of native title to land, but that this recognition was ignored by colonial authorities in Australia. As a result, *terra nullius* remained the legal justification for the dispossession of Aborigines until 1992, when it was struck down by the decision of the High Court of Australia in the case of *Mabo v. Queensland* (1992).

Because Aboriginal people became British subjects when Britain claimed sovereignty over Australia, the British government considered that a state of war could not exist with them. This is one reason why, unlike in other British colonies, warfare against Aborigines was not carried out primarily by the armed forces. Although there were instances in which troops were used against Aborigines, most notably during the period of martial law in Tasmania from 1828 to 1832, most of the fighting was done by civilians and police. There were practical as well as legal reasons for this: Aborigines never constituted a sufficient threat to justify the enormous expense of using the military to pursue them over Australia's vast distances and difficult terrain. As a result, this was a war in which there was no clear distinction between combatants and civilians.

If the White invasion and attendant Aboriginal dispossession provide the overall context for warfare between Aborigines and Whites, there were also more specific causes of conflict. Desecration of sacred sites by Whites, conflict over women and disputes over property all led to violence. Cross-cultural misunderstandings could also have tragic results; for example, Whites confronted by Aborigines shouting and waving weapons (a standard, ritualised approach to visitors) may sometimes have assumed they were hostile and opened fire. But above all it was the attempt by Whites to assert exclusive ownership of land that led to warfare. Whites took the most productive land and most of the available water, while their animals drove out native wildlife and destroyed vegetation. Aborigines were pushed into more marginal country, where they were forced to over-utilise local food resources. Competition for scarce resources produced increasing conflict between Aboriginal groups, as well as between Aborigines and Whites.

As conflict spread around the continent with the expansion of White settlement, it tended to follow predictable patterns. At first, Aborigines ignored, tolerated or even welcomed Whites, often believing them to be relatives returned from the dead. As this illusion faded, and as White behaviour became more intolerable, Aborigines began killing individual Whites in revenge for particular grievances. At the same time there may have been an increase in fighting between Aboriginal groups, not only because of competition for resources, but also because deaths from conflict with Whites and from introduced disease may have been blamed on sorcery practised by hostile Aborigines. Aboriginal revenge attacks against Whites were in turn usually met with reprisals designed to deter such attacks by inflicting large numbers of casualties. Whites saw such reprisals as a way of ending conflict by 'teaching the blacks a lesson', but to Aboriginal people they must have seemed out of all proportion to the original killings. The scale of death had now become too great for revenge attacks, and if Aborigines did not abandon armed conflict with Whites at this point then violence inevitably escalated. Aborigines came to regard all Whites as potential enemies and the conflict took on the character of a war between peoples rather than disputes between individuals.

Although there were regional variations, the tactics used in this war were generally quite similar across Australia. Aboriginal tactics were almost entirely based on their pre-existing hunting and fighting practices. This raises the question of why Aboriginal people, unlike indigenous people in New Zealand and North America, did not radically adapt their fighting techniques to changed circumstances. It is not enough to point to the absence in Aboriginal societies of the concept of war as understood by Europeans: in the nineteenth century the Zulus of South Africa were able to achieve a military revolution which took them in a short time from small-scale fighting like that of Australian Aborigines to mass warfare involving conquest. Nor is it sufficient to identify Aboriginal 'conservatism' as the reason why they did not adapt their military tactics. Aboriginal culture had changed before and would change again in many ways after the invasion. There are, however, several characteristics of Aboriginal societies which help explain their continued use of traditional fighting methods. First, the fact that fighting in Aboriginal societies was a highly ritualised, rule-governed activity may have inhibited innovation. Second, Aboriginal people lacked the economic base necessary for sustained warfare. Third, the egalitarian nature of Aboriginal societies meant that change could not be imposed from above (as happened in the Zulu case). This egalitarianism also made it more difficult for different Aboriginal groups to unite and fight the invaders together, as there was no mechanism for arranging such an unprecedented alliance.

That it was not lack of ability which stopped Aboriginal people from using new methods of fighting is shown by their use of firearms. From quite early on Aborigines displayed an interest in and desire for guns. Those who acquired them generally cleaned and maintained them carefully. Protector of Aborigines George Robinson discovered that Tasmanian Aborigines had caches of well-preserved guns hidden away, and wrote in 1831 that 'those of the natives who have learnt the use of firearms are excellent shots — they have excellent sight'. Yet Aborigines rarely used guns for anything other than hunting, perhaps because, as Victorian Aboriginal Protector William Thomas wrote in 1858, 'they consider guns a cowardly means of defence'. It is also worth noting that some Aborigines who had lived among Whites, such as Mosquito and Walyer in Tasmania and Jandamarra (Pigeon) in north-west Australia, employed non-traditional authoritarian leadership styles and used guns when they turned against their former employers.

From time to time large groups of Aboriginal people attacked Whites in relatively open country, and something like a conventional battle ensued. In such cases the Aborigines tried to take advantage of superior numbers: there are reports of them advancing in crescent formation, perhaps with the aim of outflanking and surrounding the Whites. Once they learnt not to break ranks in the face of gunfire, Aborigines attacking *en masse* could sometimes be quite effective, particularly before the improvements in firearms technology that occurred from about 1850. A disciplined group of Aboriginal fighters could wait for the first volley of shots then hurl their spears while the Whites were reloading. Many White observers were impressed by Aboriginal ability with the spear. Pastoralist G. S. Lang noted in 1865 that, 'A blackfellow, with some eight or ten spears in his hand and some paddy-melon sticks, will throw them all while a white man is reloading after firing two shots; and I have known one man to be pierced in the thigh by two spears successively, thrown at seventy yards off.'

Nevertheless, open battle was usually more costly to Blacks than to Whites, and as their numerical advantage disappeared Aborigines generally abandoned it. Perhaps the most famous battle was that at Battle Mountain near Cloncurry in Queensland in 1884, which resulted in an overwhelming defeat for the Kalkatungu (Kalkadoon) people and put an end to their six-year war against the invaders.

Because open battle usually put them at a disadvantage, Aborigines preferred to use tactics of stealth and ambush like those employed by revenge parties in pre-invasion times. This form of warfare was summed up by the Lieutenant-Governor of Van Diemen's Land, George Arthur, in 1829: 'they suddenly appear, commit some acts of outrage, and, then, as suddenly vanish — if pursued, it seems impossible to surround and capture them, and, if the Parties fire, the possibility is that Women and children are the victims'. The first step in such a surprise attack was the gathering of intelligence, and here Aborigines' ability to hide and move unseen in the bush was invaluable. Aborigines often carefully observed those they planned to attack and struck when Whites were most vulnerable. Attacks on settlers' huts frequently started with the Aborigines enticing Whites out into the open or forcing them out by setting fire to the roof. They would then attempt to spear the Whites or to lead them on a chase into the bush while other Aborigines plundered the hut and perhaps killed any Whites left inside. On other occasions Aborigines taunted besieged Whites, encouraging them to exhaust their supply of powder and shot, before attacking.

Most Aborigines did not subscribe to European-style ideals of heroism requiring a fight to the death. Instead, as Tasmanian settler J. E. Calder noted, they 'invariably retired directly when overmatched which was part of their system of warfare', and successful raids were likewise followed by a rapid retreat into the bush. Because of the superiority of Aboriginal bushcraft, attempts to pursue Aboriginal raiding parties were usually futile, which was one reason why White reprisals were carried out against the Aboriginal population in general. Their ability to make effective use of the natural environment was the greatest advantage Aborigines had over Whites in this war. Resistance lasted longest in rugged country and in areas of thick scrub or forest. In areas of dense vegetation Aborigines could hide among the trees, bushes or river reeds, while in rugged terrain they could hide in caves and retreat to narrow gorges or rocky hills where Whites could not follow on horseback. Aborigines covered their tracks by burning grass behind them and hidden Aboriginal scouts observed White movements so that they could warn their people of approaching reprisal parties.

Although Aboriginal fighting techniques remained largely unchanged from pre-invasion times they did adapt by engaging in various forms of economic warfare against Whites. At its most basic level this involved killing settlers' livestock in numbers far in excess of what they took for food. Fire was another weapon that Aborigines used very effectively against White property. Aborigines also attacked and pillaged the drays on which White settlers relied for their supplies, and on one occasion Aborigines barricaded the dray road from Ipswich to the Darling Downs with the declared intention of blocking off supply and communications to the stations. In north-west Australia Aborigines knocked down telegraph lines throughout the 1890s. Although they were motivated partly by a desire for fragments of the glass insulator caps which they used to make spear heads, the possibility that Aborigines understood the importance of the telegraph to White communications cannot be dismissed. These forms of economic warfare undoubtedly hurt the Whites, who on occasion were so economically devastated by the attacks that they were forced to abandon their properties, but they did not affect the balance of power in the long term.

Although the defeat of armed Aboriginal resistance eventually proved to be total and overwhelming, the degree of fear experienced by Whites on the frontiers of settlement can hardly be overemphasised. Writing in 1853, Victorian settler Alfred Thompson recalled that Aboriginal attacks had Whites living 'in a continual state of anxiety, apprehension, and alarm'. Many other writers reported equally intense and widespread fear at other times and places around the continent. Nor was such fear confined to those who lived on isolated stations: rumours of impending Aboriginal attack periodically caused panic among townsfolk as well. The result of this fear was a society in which White settlers routinely travelled armed and in groups if possible. Houses were well-stocked with guns or even swivel cannon in a few cases, and were built with loopholes through which to fire them. Telegraph stations in central Australia were built like fortresses, with thick stone walls, loopholes and barred windows. Such defensive measures had the support of colonial governments, which from the very early days of White settlement encouraged settlers to arm themselves and to drive Aborigines away from their farms by shooting at them.

Even 'defensive' violence occurred within the context of invasion and was designed to protect land that had been taken by force. In such a context most Whites were not content merely to ward off Aboriginal attacks but sought, by one means or another, a permanent end to 'the Aboriginal problem'. A minority of Whites sought some sort of peace through conciliation and accommodation with Aborigines: these were mostly people who lived away from the frontiers, but also included some more humane frontier settlers. Even among the substantial majority who supported violent action against Aborigines, attitudes and motivations varied. Clearly, though, the armed conflict had a brutalising effect, particularly on those Whites who experienced it at first hand. Visitors and newly-arrived immigrants from Britain were shocked to hear Australian settlers speaking openly and callously about killing Aborigines. Writing in 1846 Gippsland squatter Henry Meyrick admitted to being 'familiarized with scenes of horror from having murder made a topic of everyday conversation'. Four decades later the British colonial official Sir Arthur Gordon observed that otherwise humane and cultured Queenslanders could 'talk, not only of the *wholesale* butchery ... but of the *individual* murder of natives, exactly as they would talk of a day's sport, or of having had to kill some troublesome animal'. At its worst, this brutalisation could lead Whites to advocate a war of extirpation against Aborigines, and the many recorded instances of mass poisonings of Aborigines by Whites testify to the appeal of exterminationist thinking to some settlers. Characterisations of Aboriginal people as subhuman, aided in the late nineteenth century by the 'scientific' racism of Social Darwinism, helped provide the justification for holding Aboriginal lives cheap. On the whole, though, the fear and thirst for revenge which motivated Whites did not drive them to contemplate systematic genocide.

Whites represented their own violence as a necessary reaction to Aboriginal savagery. Aborigines were seen as bloodthirsty and treacherous, and many Whites believed that the only realistic response was to deter Aboriginal violence by striking terror into Aboriginal communities. The editor of Rockhampton's *Northern Argus* made his views brutally clear in 1864: there was 'but one law for them that they will ever respect — the bullet; the sole logic, the cock of the rifle'. However, this cruel but apparently calculating approach should not obscure the extent to which Whites were motivated by sheer racial hatred, the burning desire for revenge, and blood-lust, just as they claimed Aborigines were. In any event, whether they sought to take revenge or to terrorise Aboriginal communities, White reprisals were invariably out of all proportion to the number of Whites killed in Aboriginal attacks.

Central to the ability of Whites to inflict terror on Aborigines was the settlers' use of firearms. As one of the most brutal enforcers of White law in the Northern Territory, Constable W. H. Willshire, wrote with characteristic frankness: 'It's no use mincing matters — the Martini–Henry carbines ... were talking English.' However, before the mid-nineteenth century the use of guns did not always give Whites the upper hand. The guns in use before about 1850 were muzzle-loading, mainly smooth bore, single-shot weapons with flintlock firing mechanisms. In flintlock guns the trigger released a striker which hit a flint, igniting powder in a pan outside the barrel. The burning powder then entered the barrel through a tiny hole, detonating a secondary charge which propelled the bullet. Loading such a gun involved filling the pan with powder, inserting the propelling charge and holding wad into the barrel, then ramming the bullet on to the wad. These weapons had several significant deficiencies. First, they had a high rate of misfire (one in every 6.5 shots according to British army tests in 1834) due mainly to flints wearing down and powder becoming damp. Second, the 'hang-fire' or delay between the flash of powder igniting in the pan and the detonation of the propelling charge gave alert opponents the chance to avoid the shot. Third, loading them was time-consuming, so only two or at best three shots could be fired per minute. Fourth, they were only accurate at short range: long-arm muskets had good accuracy at 50 metres, but only 50 per cent accuracy at 150 metres. Accuracy could be increased by using bullets that fitted more tightly into the barrel, but this also increased the time spent ramming.

The deficiencies of these weapons were not so important when used by massed ranks of soldiers firing volleys at each other at close range, but their disadvantages when used in frontier warfare were great. The hang-fire made fast-moving Aborigines hard to hit, while the time-consuming loading process made Whites vulnerable to surprise attack. Flintlock firearms could be kept ready for instant firing, but for no longer than a day. The slowness of loading, the high rate of misfire and the fact that the guns had to be used at short range all combined to even the odds between Aborigines and Whites. A misfire or a missed shot gave Aborigines more than 20 seconds in which to shower an opponent with spears. However, the problems with pre-1850s firearms should not be exaggerated. They could still be devastating when large numbers were fired at Aborigines who could not escape, and a wound from a musket was much more likely to be lethal than one from a spear.

By the 1850s advances in firearms technology were beginning to give Whites the advantage over Aborigines. The improved six-shot Colt revolver appeared in 1849, while in 1862 the Snider single-shot breech-loading rifle was patented. In the hands of the native police from 1870, the Snider carbine was to become the single most effective means of breaking Aboriginal resistance in Queensland. With their self-contained cartridges and efficient breech-loading system, Sniders had a faster rate of fire than earlier guns. They also had a longer range, though they were still most effective at 50–60 metres, at which range a Snider bullet could inflict massive injury. The Martini-Henry rifle, introduced later, was even more powerful than the Snider and twice as accurate at 500 metres. Finally, rapid-fire rifles, such as the Winchester of 1866 which was to be adopted by the Western Australian police, allowed a magazine of rounds to be fired as fast as the lever under the gun could be pulled. None of these guns immediately came into widespread use in Australia. Revolvers and breech-loading rifles were available in the 1860s, but most settlers still used muzzle-loading carbines, rifles, shotguns and pistols. By about 1870, though, the new guns were relatively cheap (£5 for a Snider) and easy to obtain. From then on Aborigines armed with spears stood little chance against White firepower.

If Aborigines had to adjust to a completely new kind of fighting, Whites, too, had to learn how to fight fast-moving, often hidden enemies who generally operated in small groups. There were two situations in which Whites could gain a clear advantage. One was on open ground, where they could use the superior mobility provided by horses to surround Aborigines then decimate them with gunfire. Surprise attacks, when Whites chose the time and place to strike, were the other form of offensive action in which the advantage was clearly with the invaders. It might be difficult for Whites to track Aborigines down, but once they were found and surrounded Aborigines had little chance of escape. Usually attacking at night, Whites crept up with guns loaded then opened fire. Now the very geographical features which had hitherto offered protection from discovery became death traps, as Aborigines were forced off cliff ledges or died scrambling up the sheer faces of gorges. Often Aborigines would be surprised when camping by a river, and would retreat into the river while Whites fired from either bank. In such circumstances men, women and children were killed; resistance was difficult and generally futile.

It was not the Whites themselves who most successfully took the war into the enemy's camp, however; those Aborigines who fought for the Whites could be much more effective. Some pastoralists used their Aboriginal workers to hunt down 'wild' Aborigines (the Eastern and African Cold Storage Company employed gangs of Aborigines for this purpose on their stations near Roper River in the 1900s), but most Aborigines fighting on the White side did so as part of the police force. By the 1820s Aborigines were being used to track down White criminals, and the difficulties encountered in trying to capture Aborigines (exemplified by the failure of the 'Black Line' in Tasmania to push Aborigines into the south-east corner of the island) led to calls for the employment of Aboriginal skills in the suppression of Black resistance. The advantages of employing Aboriginal police troopers were that they were cheap (as they were paid only a fraction of the wages paid to White police), they were highly mobile (as they could live off the land, thus allowing them to travel light and stay in the bush for long periods at a time) and they were skilled at tracking. Some Whites also hoped that if Black police killed Aborigines the Aboriginal clans would not take revenge on Whites, while others with a more humanitarian bent thought that employing Aborigines in the police force was a good way of 'civilising' them.

After several unsuccessful earlier attempts, the first native police force was formed at Port Phillip in 1842. Reaching a peak size of 65, it operated throughout the colony of Victoria against both White and Black law-breakers before it was disbanded in 1852. A South Australian native police force, in existence from 1852 to 1856, was a failure. Much more successful was the force established in the northern rivers district of New South Wales (taking in what became southern Queensland) in 1848. Control of this force was transferred from the New South Wales Inspector-General of Police to the Government Resident in Brisbane in 1856, then to the government of Queensland when that colony was formed in 1859. The Police Act of 1863 confirmed the existence of the Queensland native police, who were brought under the control of the Commissioner of Police. The force's numbers varied greatly over the years in response to changing policies and financial circumstances, but at its highest points it numbered around 200 men.

Queensland native police detachments generally consisted of 6 to 10 Aboriginal troopers under the firm and often brutal command of a White officer. Their main duty was to patrol their district, visiting stations to find out if the settlers were having any problems with the local Aboriginal people. In theory such patrols were meant to act primarily as a deterrent against Aboriginal attacks, but in practice the main role of the native police was to carry out punitive expeditions at the behest of the settlers. Such activities were carried out in accordance with official instructions which made it the responsibility of native police 'to disperse any large assemblage of blacks', and 'dispersal' soon became a grim euphemism for slaughter. The tactics of the native police in punitive raids were similar to those of Whites, often involving sneaking up on Aboriginal camps and then opening fire, but because of their superior stealth and tracking skills the native police could be much more effective than Whites in such attacks. The purpose of the native police force, not acknowledged officially but widely recognised in practice, was to terrorise Aborigines, and the available evidence suggests that Aboriginal people did indeed live in dread of native police attacks. The details of native police operations were not usually made known to the public, and little interest was taken in the legality of their actions. Public opinion was generally in favour of the force, which was given a fairly free hand to operate as its officers saw fit. There were some humanitarian critics of the native police, however, and criticism increased from the 1870s, especially in the southern colonies and in those parts of Queensland that were away from the frontier.

Once the native police were deemed to have broken the back of Aboriginal resistance in a particular district they were moved to other 'unsettled' districts further to the north and west and replaced in the 'settled' areas by White police and Black trackers. By the 1880s official policy was to replace the native police force completely with ordinary police, but this process was carried out very slowly. There was also an attempt at this time to use the native police to conciliate Aborigines, but this was largely unsuccessful. By 1896, when an inquiry recommended the disbanding of the native police, the force's days were truly numbered. The number of troopers declined throughout the 1890s, and though the government did not abolish the force it did gradually merge it into the regular police. Now that Aboriginal armed resistance in Queensland had all but ended they also had more success in using the native police to pursue a conciliatory policy. As late as 1913 there was still one native police detachment operating on Cape York Peninsula, but most Aborigines employed by the police were working as trackers in the regular force. Armed Aboriginal resistance was now largely restricted to the Northern Territory and the northern part of Western Australia, and though native police forces were never formed in these two regions, Black trackers employed by the police forces were used to hunt down and kill Aboriginal people there.

There is little direct evidence about why Aborigines joined police forces, but a number of possible reasons can be advanced. First, Aboriginal troopers gained access to the goods that Whites possessed. Unlike many other Aboriginal workers they were generally well fed and clothed, and the Queensland native police were even paid, albeit at a much lower rate than Whites. They were given guns and horses, and although they received little training, they developed their horse-riding, shooting and other skills through their work. Furthermore, their skills were recognised by a White society whose usual attitude towards Aborigines was one of contempt. Most Aborigines who lived with or near Whites were forced to accept an outcast or servile status, but Black troopers, though they were clearly subordinate to White officers, had real power and importance. For young Aboriginal men, excluded by their age from power in Aboriginal society and by their colour from power in White society, this may have been the greatest attraction of all. In particular, because they were feared by Aboriginal communities they readily acquired Aboriginal women, while the troopers' status within White society meant that these women were generally left alone by Whites. Finally, it is important to understand that native police would not have seen themselves as killing 'their own' people. In almost all cases, as a matter of deliberate policy, native police were recruited to serve far away from their homelands. The ruthlessness with which they carried out their duties is testimony to the fact that they felt no affinity with the people they were employed to kill.

Because of the nature of the war between Aboriginal people and the invaders it is very difficult to say when the war ended in any particular area. There were no formal declarations of peace, no treaties signed. Violence did not cease suddenly, but rather petered out; a climate of violence remained long afterwards, however. Whether they lived on cattle stations, on reserves or on the fringes of towns, Aboriginal people continued to experience violence at the hands of White overseers, police and vigilantes. Even after Aborigines had supposedly been 'quietened', White violence could sometimes extend beyond the usual whippings and beatings to murder. The casual manner in which some White Australians are still able to joke about killing Aboriginal people is a chilling reminder that attitudes which developed at the time of frontier conflict have not yet died away. For the most part, though, a sustained but low level of violence was enough to keep Aboriginal people in their place at the bottom of the social ladder. The memory of past violence combined with continuing brutality in the present was enough to prevent many Aboriginal people from directly challenging White power. Speaking in 1982, Riley Young Winpilin, from the Victoria

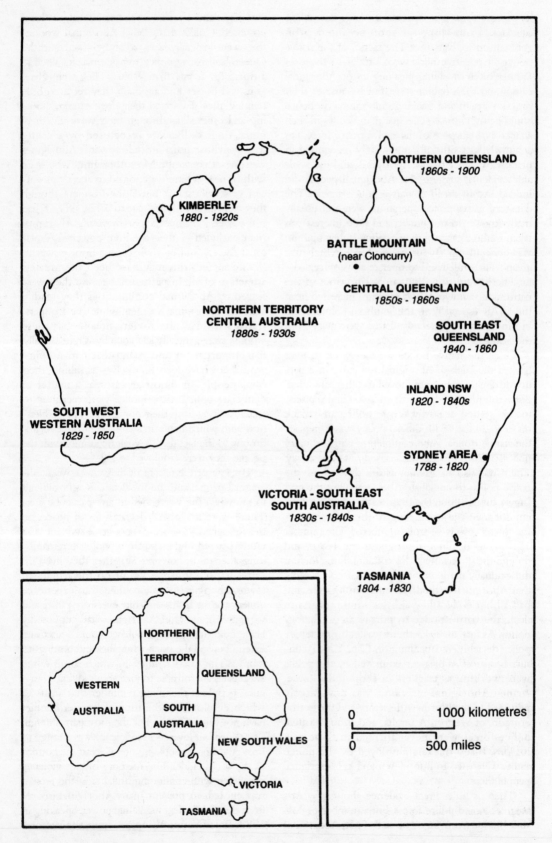

NORTHERN QUEENSLAND
1860s - 1900

KIMBERLEY
1880 - 1920s

BATTLE MOUNTAIN
(near Cloncurry)

CENTRAL QUEENSLAND
1850s - 1860s

NORTHERN TERRITORY
CENTRAL AUSTRALIA
1880s - 1930s

SOUTH EAST
QUEENSLAND
1840 - 1860

INLAND NSW
1820 - 1840s

SOUTH WEST
WESTERN AUSTRALIA
1829 - 1850

SYDNEY AREA
1788 - 1820

VICTORIA - SOUTH EAST
SOUTH AUSTRALIA
1830s - 1840s

TASMANIA
1804 - 1830

NORTHERN
TERRITORY

QUEENSLAND

WESTERN

AUSTRALIA

SOUTH

AUSTRALIA

NEW SOUTH WALES

VICTORIA

TASMANIA

0   1000 kilometres

0   500 miles

River district in the Northern Territory, recalled being told by his family not to get into a fight with White people: 'Because olden times, you know, you can get shot like a dog. They shoot you like a dog and just let it, let you, burn on the fire.'

Although no definite endpoint for frontier warfare can be established, it is possible to identify roughly the periods in which Aboriginal people were carrying out reasonably sustained campaigns of resistance to the settlers. The frontier moved around the continent as Whites encroached on more and more Aboriginal land, and fighting usually was occurring in several parts of the continent simultaneously. In New South Wales conflict began with the arrival of the first European settlers in 1788, and fighting continued in Sydney and surrounding areas until about 1820. The war then spread into outback New South Wales, and continued through the 1840s. In Tasmania Black–White violence occurred from 1804 to about 1830. Victoria and the southern part of South Australia experienced warfare in the 1830s and 1840s, while south-western Western Australia was the site of conflict from 1829 to about 1850. The war in Queensland started in Brisbane and nearby districts around 1840, continuing to about 1860. It moved into central Queensland in the 1850s and 1860s, and war raged in northern Queensland from the 1860s to the turn of the century. Meanwhile, in Western Australia, White settlement and the violence that accompanied it gradually moved north until by about 1880 violent clashes were occurring in the Kimberley region in the north-west. Conflict in the Kimberley continued until the 1920s, and was roughly contemporaneous with the war in the Northern Territory and central Australia, which lasted from the 1880s to the 1930s.

The death toll from almost 150 years of war around Australia is impossible to assess with any accuracy. Evidence about deaths in frontier conflict is widely scattered, and is in any case far from complete. Aboriginal deaths are particularly difficult to quantify, as in many cases the killers themselves were the only witnesses, and the bodies were burned after the killing was done. Despite these problems, however, a number of historians have attempted to calculate the death tolls in particular regions as accurately as possible. Drawing on this work, Henry Reynolds has estimated that around 2000 to 2500 non-Aboriginal people and at least 20 000 Aboriginal people were killed as a direct result of armed conflict; many more on both sides were physically or psychologically wounded. What-

ever the exact figures, it is clear that the war between Blacks and Whites was surpassed only by the First and Second World Wars in terms of the number of Australians killed. By far the largest number of deaths on both sides occurred in Queensland, but the proportional impact and the ratio of Aboriginal to White deaths varied widely around Australia. In some areas Aboriginal survival was aided by rugged terrain or by relatively close government supervision of settler activities; in other areas Aboriginal people suffered because settlement took place at a time of improving firearms technology or of government policies which gave settlers and native police a free hand. Nowhere in Australia, though, does the death toll from violence alone explain the dramatic drop in Aboriginal population to about 95 000 nationwide in 1901. Disease was undoubtedly the greatest killer of Aboriginal people, and also played a significant role in reducing the ability of Aborigines to resist the invasion. This fact is sometimes pointed to by those who wish to deny that Australia was conquered by force, as if the one fact cancels out the other. However, although the Aboriginal population decline was primarily due to disease, it remains true that Aboriginal resistance was suppressed by armed force; that had it not been suppressed Aborigines would have made it impossible for Whites to live in Australia; and that the non-Aboriginal presence in Australia therefore rests ultimately on conquest.

That Australia was a conquered territory was not news to those Whites who actually lived through the frontier wars, and many of them admitted as much at the time. Later generations, however, were more squeamish and chose to believe that Australia was 'the quiet continent', with a uniquely peaceful history. What anthropologist W. E. H. Stanner called 'the great Australian silence' descended, as Aborigines were simply written out of history. Traumatic memories of Black–White conflict were passed on within Aboriginal communities, but frontier warfare remained unstudied by non-Aboriginal historians until the 1970s. Then, prompted by a resurgent Aboriginal political activism, there was an explosion of interest in Aboriginal history generally, and in the armed conflict specifically. A number of local studies appeared which examined frontier violence in particular areas, while Henry Reynolds's work provided an Australia-wide overview. For the first time Aboriginal people were seen as historical actors, and historians began to realise that the invasion must have looked very different from 'the other side of the frontier'. Indeed, the term 'invasion' itself began to be used in relation to White settlement in Australia for the first time since the nineteenth century. Even

Map 2 Areas of sustained Aboriginal resistance from 1788 to the 1920s. *Inset* State boundaries.

those who jib at the word 'invasion' have been forced to acknowledge that at best the European settlement of Australia was accompanied by sustained violence against the indigenous population. Work on frontier warfare has continued, but much remains to be done. For example, regional differences need to be examined and explained more carefully; the position of women on both sides in the war deserves detailed consideration; and a substantial study of the Queensland native police is long overdue. There is still room, too, for further examination of the methods by which the war was carried out, particularly on the Aboriginal side. Perhaps most importantly of all, historians have become increasingly aware that relations on the frontier could involve a complex mixture of conflict and accommodation, and as this insight is pursued we will develop a much more well-rounded view of frontier life.

The work of historians has made the myth of peaceful settlement untenable, but some of the uses made of the new history at a popular level are in danger of introducing new distortions. Increasingly, frontier warfare is being written and spoken about entirely in terms of massacres, invariably perpetrated by Whites on Aborigines. If the term 'massacre' is understood to mean the mass slaughter of defenceless people, then the conflict between Aborigines and Whites involved much more than just massacre. It is certainly true that many massacres occurred and that most were carried out by Whites or native police. Many more Aborigines were massacred than Whites: whether this was because the tradition of warfare in Aboriginal societies did not encompass the idea of general slaughter or simply because Whites had the edge in weaponry and logistics is unclear. Undoubtedly, defenceless women, children and old people were killed, and armed men were killed when they were trapped and could not fight back effectively. Some cases of mass killing, however, were not massacres but military defeats in which fighting took place and where one side suffered far greater casualties than the other. Even in the case of surprise attacks on Aboriginal camps, Aboriginal men were able to fight back, though they were clearly at a disadvantage. It is also worth bearing in mind that punitive expeditions by Whites or native police were in reprisal for Aboriginal attacks. These reprisals were usually indiscriminate and disproportionately severe, but the fact that they followed Aboriginal attacks is a reminder that this was a two-sided war. To depict Aborigines as always defenceless victims may seem to give present-day Aboriginal people the moral high ground, but it does so at the expense of the bravery and fighting skills of their ancestors. Aboriginal people were not passive in the face of White invasion, as both the old myth of peaceful settlement and the new myth of universal massacre imply. Rather, they used every means at their disposal to defend their land, their resources and their independence.

On the other hand, there has been a contrary tendency to depict the Aboriginal armed resistance to White invasion as involving set-piece battles and heroic Aboriginal leaders. Clashes like that at Battle Mountain are highlighted rather than the more typical use by both sides of stealth and surprise attacks, while figures such as the Eora leader Pemulwuy are turned into guerrilla commanders on the model of the leaders of anti-colonial movements elsewhere in the world. This approach risks turning Aboriginal resistance into a cardboard-cutout story of heroic struggle every bit as one-dimensional as the Anzac legend (q.v.). Like all wars, this conflict was a brutal one, with little room for romantic notions of heroism. Because so much of the fighting took place at close quarters, thousands of people watched as friend and foe alike suffered cruel and sometimes lingering deaths. Many, both Black and White, remained haunted by the experience for years afterwards, and to turn this experience into a glorious epic struggle is surely inappropriate. Ironically, too, given that it has the laudable aim of countering White myths that Aborigines were passive or even cowardly, this approach is based on Eurocentric ideas about what constitutes a 'real' war. That this was not, by and large, a war fought by opposing ranks of fighting men, or one in which individual leaders played a prominent role, is no reflection on Aboriginal societies. It is no compliment to the Aboriginal people of the past to measure them against models of warfare that were foreign to them. A simple recognition that Aborigines fought with courage and skill, drawing on their traditions to meet an unprecedented threat, is tribute enough.

Henry Reynolds, *The Other Side of the Frontier: Aboriginal Resistance to the European Invasion of Australia* (Penguin Books, Melbourne, 1981); Richard Broome, 'The struggle for Australia: Aboriginal–European warfare, 1770–1930', in M. McKernan and M. Browne (eds), *Australia: Two Centuries of War and Peace* (Australian War Memorial, Canberra, 1988).

ABORIGINES AND TORRES STRAIT ISLANDERS IN THE ARMED FORCES Aborigines and Torres Strait Islanders have participated in each war Australia has fought since the end of the nineteenth century. Aboriginal trackers were employed by the Australian contingent in the Boer War, and approximately 400 to 500 Aborigines served as enlisted soldiers in the First World War. About one-third of those who served became casualties; one, Douglas

Men of the Torres Strait Light Infantry Battalion on parade, Thursday Island, Queensland, 29 October 1945. (AWM 119195)

Grant (q.v.), became a POW and several, including Albert Knight and William Rawlings, won medals for their outstanding courage during battle. On their return to Australia, however, Aboriginal soldiers found that their performance of the duties of citizenship did not win them citizens' rights. Throughout Australia, but particularly in the 'frontier states' of Queensland, the Northern Territory, Western Australia and South Australia, Aborigines and Torres Strait Islanders remained marginalised, without the vote in State or federal elections, and under the control of the so-called 'Protection Acts' — highly repressive and paternalistic acts imposing strict White control over almost every aspect of Aboriginal and Islander life.

Despite the failure of White Australia to respond to the sacrifice made by Aboriginal and Islander people during the First World War, it was in the Second World War, and particularly during the war in the Pacific, that Aborigines and Torres Strait Islanders made their greatest contribution to national defence in terms of raw numbers and in the range of duties they performed. Although the federal government had in 1939 replaced its 'Protection' policy with a policy of assimilation, which was aimed at assimilating detribalised Aborigines and Islanders into the broad stream of White Australian society, the army and navy initially rejected all applications from Aborigines and Islanders, and anyone else not 'substantially of European origin or descent'. Senior officers felt that White Australians would not serve satisfactorily with Black Australians and that the operational performance of the services would suffer as a result. Faced with the heavy manpower demands imposed on it by the Empire Air Training Scheme (q.v.), the RAAF could not afford to reject otherwise suitable recruits on racial grounds and so tended to admit non-Europeans, including Aborigines and Islanders, more freely than the other services. Despite the bans imposed on their enlistment, some Aboriginal and Islander recruits managed to join up very early in the war while recruiting officers were confused about enlistment policy. This contingent included men like Reg Saunders (q.v.) who was to become the only Aborigine to serve as a commissioned officer in the Australian forces during the war, and Charles Mene whose many years of exemplary service included winning the Military Medal in the Korean War. The service of these men and others who enlisted at this time proved that the concerns of senior officers were baseless.

Aboriginal political organisations such as the Aborigines' Progressive Association and the Aborigines' Advancement League saw the war as an opportunity to press for an extension of full citizens' rights to Aborigines. Although the powerful National Association for the Advancement of Colored Peoples had to struggle to achieve the desegregation of the military in the United States, in Australia, where the small number of Aborigines and Islanders available for military service all but eliminated the possibility of forming segregated units, Aboriginal political organisations urged just that. This was to ensure that the war effort of Aboriginal and Islander Australians would be highly visible and therefore form a more potent argument for improving the lot of Black Australians. Apart from some isolated attempts, the armed forces had no interest in enlisting Aborigines and Islanders or in forming segregated units until mid-1941 when the Japanese threat emerged. This caused the demand for military manpower to increase enormously; the services abandoned their earlier exclusive attitudes to the enlistment of Aborigines and Islanders and began to admit them in larger numbers. Most of the Aborigines who enlisted served in integrated units. The cohesive forces that weld small groups together in military organisations resulted in them finding levels of acceptance they had seldom experienced in their prewar lives. Military service in integrated units provided opportunities for personal advancement, and many Aborigines and Islanders rose to the rank of NCO where they had command over White Australians in battle — an unheard-of status for Aborigines in prewar Australia. Others used the war as a means of broadening their personal horizons through travel and the acquisition of skills. One man, Warrant Officer Leonard Waters, fulfilled his boyhood dreams of becoming a fighter pilot, an impossibility for an Aborigine before the war.

As well as abandoning their opposition to the enlistment of non-Europeans, the armed forces also overturned their resistance to the formation of segregated Aboriginal and Islander units. The Torres Strait Light Infantry Battalion, manned mainly by Islanders but also including some Cape York Aborigines, began recruiting in June 1941 and reached its peak strength of 745 Islanders and Aborigines in August 1943. Officers and senior NCOs were White Australians; segregation of the unit was accompanied by discrimination. The Black soldiers of the Light Infantry Battalion were illegally underpaid throughout the war, and the underpayment was not made good by the Commonwealth government until 1986. Elsewhere in northern Australia, small irregular units of Aborigines were formed to provide surveillance of isolated parts of the coast. In east Arnhem Land, an area unmapped and devoid of White population yet a potential site for a Japanese landing, the Northern Territory Special Reconnaissance Unit was formed. Commanded by Squadron Leader Donald Thomson, an anthropologist before the war, it consisted of 50 tribal Aboriginal warriors, six Solomon Islanders, a Torres Strait Islander and several White NCOs. The Aborigines patrolled east Arnhem Land searching for signs of Japanese landings. They were to use their traditional weapons to wage a guerrilla war against the Japanese while reporting the enemy's movements to Darwin. Similar irregular forces were raised at Bathurst and Melville Islands, on the Cox Peninsula across the harbour from Darwin, and at Groote Eylandt. In each case the armed forces sought to make use of the Aborigines' outstanding local knowledge and bushcraft skills for the defence of Australia. Not one of these irregular forces was formally enlisted or paid for their service. The Commonwealth government finally acknowledged the service of these groups in 1992, when they were paid and awarded their war service medals.

As Japanese forces advanced towards Australia, the northern coast was fortified for defence. Some White Australians believed that Aborigines, particularly those who had worked in the pre-war pearling industry which was dominated by the Japanese, would assist the enemy, but this proved false. In fact, Aborigines overwhelmingly gave their support to the defence effort. Across north Australia they laboured to build airstrips at isolated outposts for the RAAF; they worked to keep the supplies moving north to Darwin for the army, they rescued downed airmen, salvaged crashed aircraft, located unexploded bombs and performed a host of other tasks. Along the Stuart Highway from Alice Springs to Darwin, labourers and their families were moved into army-run labour settlements. These settlements set new standards for the accommodation, food rations and welfare of Aboriginal labour in northern Australia and had a considerable but indirect influence on the future of Aboriginal labour relations in this region.

It is impossible to make an accurate tally of the number of Aborigines and Islanders who served in the Second World War because the armed forces generally did not record the race of their recruits. An estimated 3000 Aborigines and Islanders served as formally enlisted soldiers, sailors or airmen. A further 150 served as irregular soldiers in northern Australia and about 3000 worked for at least part of the war as labourers directly employed by the armed forces. This level of effort was a remarkable achievement. Taken alone, the war effort of Torres Strait Islanders, which is much more accurately documented, is even more impressive. Virtually every able-bodied Islander male of military age —

about one in every four or five Islanders — served in one or other of the services. This compares very favourably with the figure for Australians as a whole, which was one in every seven or eight.

At the war's end, when the demand for military manpower had again subsided, the armed forces reintroduced their bans on the service of non-Europeans and, for a time, no Aborigines or Islanders were permitted to join the peacetime armed forces, although those who had joined during the war were allowed to remain in service. Aborigines and Islanders who had served in the Second World War were given the vote in Commonwealth elections, and therefore in State elections, through an amendment to the Commonwealth Electoral Act, but the general extension of full citizens' rights to all Aborigines — which Aboriginal servicemen and political organisations had been advocating — was not to eventuate until the 1960s.

Some Aboriginal and Islander servicemen remained in the forces following the end of the Second World War and became members of the British Commonwealth Occupation Force (q.v.) in Japan, and later served in the Korean War. Other Aborigines with a high public profile, like Reg Saunders, were able to re-enlist for the Korean War. Reg Saunders served with the rank of Captain during the Korean War and commanded a rifle company in 3RAR at the battle of Kapyong.

As late as 1954, and despite the Commonwealth government's assimilation policy, recruiting advertisements continued to warn that persons must be of primarily European descent before they could be eligible for military service. Aborigines were also excluded from the 1951 to 1959 National Service scheme (see CONSCRIPTION) which conscripted Australians for home defence only. The National Service scheme adopted between 1965 and 1972 conscripted Australians for overseas service but Aborigines were again excluded despite demands from the public for their inclusion. This raised the question of what constituted Aboriginality, and at least one person of 'part-Aboriginal' descent added to the anti-conscription debate by arguing that he should be exempt from conscription on the grounds that he was an Aborigine, although he lived a detribalised lifestyle. Numerous Aborigines served in the Vietnam War as volunteers.

In the post-Vietnam era the armed forces continue to be troubled by the problem of ensuring the security of remote parts of northern Australia. The army has created a series of regionally based Army Reserve surveillance units, namely the North West Mobile Force (NORFORCE), 51 Far North Queensland Regiment and the Pilbara Regiment. The first two draw heavily on the local Aboriginal and Islander population for their recruits and Aborigines and Islanders are heavily overrepresented in their ranks. In 1992 Aborigines and Islanders were represented within the armed forces as a whole in accordance with their representation within the community. However, they remain underrepresented in the officer corps, not least through inability to meet stringent educational requirements.

Robert A. Hall, *The Black Diggers: Aborigines and Torres Strait Islanders in the Second World War* (Allen & Unwin, Sydney, 1989).

ROBERT A. HALL

ACHERON, HMCS see **NEW SOUTH WALES NAVAL FORCES SHIPS**

ACUTE, HMAS see **ATTACK CLASS PATROL BOATS**

ADELAIDE (I), HMAS (Modified *Chatham* Class light cruiser). Laid down 1915, launched 1918; displacement 5560 tons; length 463 feet; beam 50 feet; 1922 armament 9 × 6-inch guns, 1 × 3-inch anti-aircraft gun; 1939 armament 8 × 6-inch guns, 3 × 4-inch guns, 4 × 3-pounder guns; 1942 armament 7 × 6-inch guns, 2 × 4-inch guns; speed 25 knots.

A three-funnel light cruiser built at Cockatoo Island (q.v.) in Sydney, HMAS *Adelaide* was not launched until a few months before the end of the First World War, and peace delayed the ship's completion until 1922. In 1927 HMAS *Adelaide* went to the Solomon Islands to help put down indigenous opposition to British colonial rule. She was laid up from 1928 to 1938, when the ship was modernised and converted from coal to oil-fired engines. During the Second World War she served off New Caledonia in September 1940 and was then used for convoy duties off the coast of Western Australia. Her only action of the war occurred in these waters when she sank the German blockade runner *Ramses* on 28 November 1942. HMAS *Adelaide* was decommissioned after the war and broken up for scrap in 1949.

ADELAIDE (II), HMAS see **OLIVER HAZARD PERRY CLASS GUIDED MISSILE FRIGATES**

ADMINISTRATIVE HEADQUARTERS, AIF In January 1915 Major-General William Throsby Bridges (q.v.) set up an Australian Intermediate Base under Colonel V. C. M. Sellheim to handle the administration of the AIF in Egypt and Gallipoli. At first this was set up as an Australian section of the British base under General Sir John Maxwell, but in early 1916 Maxwell separated out the Australian and New Zealand administrative bodies. Sellheim became

Australian soldiers on leave in Horseferry Road, London, during the First World War. The War Chest Club is on the left and AIF Administrative Headquarters is on the right. (AWM D00796)

commander of AIF Headquarters and was responsible to General Sir William Birdwood (q.v.), who in April 1916 decided to move the AIF Base from Egypt to the United Kingdom. In May Sellheim and his staff arrived in Britain, where they took over offices in Horseferry Road, central London, which had been occupied since October 1915 by a records office established to deal with AIF convalescents.

Sellheim's organisation was now known as Administrative Headquarters, AIF. As commandant Sellheim was not only responsible for administration but also for liaison with the War Office and command of AIF troops in Britain. Doubts about his ability to handle such great responsibilities led to his being relieved of these posts in August 1916. He was replaced as commandant of Administrative Headquarters by former businessman Colonel Robert Anderson, and as commander of all AIF depots in the United Kingdom by Brigadier-General Newton Moore. When Birdwood became GOC, Australian Imperial Force, in September 1916 he had to divide his attention between his responsibilities as controller of AIF Headquarters and as commander in France. As a result he had to rely on Anderson, who had overall responsibility

for the AIF training depots in England, to keep administration running smoothly. Unfortunately, while Anderson was an able administrator, he was tactless and distrusted the permanent soldiers, who returned his antipathy. Furthermore, though Anderson had an initial success in his important role as the Australian government's representative at the War Office when he arranged a complete readjustment of accounting procedures with that office, he soon antagonised the War Office as well. In April 1917 Anderson was replaced by the most successful commandant of the Administrative Headquarters, Colonel Thomas Griffiths, who remained in that position until May 1919. Under the direction of Birdwood and his Chief of Staff, Major-General Brudenell White (q.v.), Griffiths immediately drafted orders for the AIF's administration. His term as commandant earned him high praise from White, who said Griffiths was primarily responsible for the successful administration of the AIF, and from C. E. W. Bean (q.v.), who called him 'one of the great figures in the Australian Army'.

The day-to-day work of the Administrative Headquarters involved keeping records of AIF personnel, servicing divisions in the field, liaising with the Department of Defence in Australia, and administrative duties relating to AIF personnel in Britain. Its Horseferry Road offices, which within a year of the Headquarters' establishment had become too small and were being criticised for being located in what some saw as a dingy area, soon became well known to Australian soldiers. Soldiers newly arrived on leave in London would visit the offices to collect pay, tourist information and usually a new uniform. They might then cross the road to the War Chest Club, a club and hostel funded by the Australian Comforts Fund (q.v.), or get a meal at the nearby Anzac Buffet before heading out to see the city. W. H. Downing's *Digger Dialects* (1919) defines Horseferry Road as 'anathema to the fighting men', and the attitude of many front-line soldiers to those who worked in administration was expressed in the many versions of this popular Australian song of the First World War:

> *He went up to London and straight away strode*
> *To Army headquarters on Horseferry Road,*
> *To see all the bludgers who dodge all the strafe*
> *By getting soft jobs on the headquarters staff.*

ADROIT, HMAS see *ATTACK CLASS PATROL BOATS*

ADVANCE, HMAS see *ATTACK CLASS PATROL BOATS*

ADVANCED LAND HEADQUARTERS (LANDOPS) opened in Brisbane in the grounds of the University of Queensland on 1 August 1942 in response to the decision by General Douglas MacArthur (q.v.) to move his General Headquarters north from Melbourne. LANDOPS became General Thomas Blamey's (q.v.) operational headquarters and was responsible for the conduct of fighting during the Papuan campaign. Blamey's administrative headquarters, Allied Land Headquarters (q.v.), remained in Melbourne, and as a consequence Blamey spent a great deal of time moving between the two. While this dual structure, and Blamey's twin roles as C-in-C of the Australian Military Forces and Commander of Allied Land Forces, violated various command principles, it was made necessary by the politics of the alliance with the United States, the lack of military experience on the part of the Prime Minister, John Curtin (q.v.), and Blamey's belief, probably well founded at this stage of the war, that no other Australian general was equipped to deal with the variety of urgent operational, administrative and political tasks that confronted him.

ADVISORY WAR COUNCIL (AWC), established on 28 October 1940, was designed to enhance the national war effort by drawing all major political parties in federal parliament into the decision-making process. Prime Minister Menzies initially argued for a government of national unity, but with the balance of power almost evenly divided between the conservative parties and the Labor Party, the Labor Party under John Curtin was not prepared to join an all-party government. Conversely the war situation was so critical that a dissolution of parliament had to be considered a last resort. Rather than join an executive body by whose decisions it would be bound, thereby denying it the right to criticise the government in parliament, the Labor Party agreed to join an advisory body. The AWC consisted initially of four cabinet ministers from the War Cabinet, and three members of the Opposition, with the Secretary of the War Cabinet (and of the Department of Defence Coordination), F. G. Shedden (q.v.), acting as secretary. It met for the first time on 29 October 1940. The Labor government that came into office in October 1941 maintained the AWC, and adopted the principle that a recommendation made by it, with a majority of ministers supporting the recommendation, would normally be accepted by the War Cabinet. Following the landslide Labor win in the election of August 1943, there was not the political necessity to seek an accommodation with the Opposition. The changed political scene, as well as the markedly improved military situation, led to the decline of the AWC, although it was maintained as a courtesy to the Opposition. It ceased to exist on 30 August 1945 when the Opposition members suggested that it had achieved the purposes for which it had been formed.

AE1, AE2 see *'E' CLASS SUBMARINES*

AGENT ORANGE is the term used to describe a mixture of the herbicide chemicals 2,4D and 2,4,5T, used as a defoliant during the Vietnam War. More colloquially, it has come to stand for a range of issues arising from that war, and in particular to symbolise the grievances of a disaffected group of Vietnam veterans who claim that chemical agents (for Agent Orange was only one of a number used) are responsible for birth defects in their children and for post-traumatic stress disorders and various other health problems to which they themselves are subject. This became a serious social and political issue in the early 1980s, dissatisfaction with its handling having led to the formation of the Vietnam Veterans' Association (q.v.) in 1979–80. This body lobbied hard and sometimes effectively for government action, resulting in a number of studies and culminating in a Royal Commission under Mr Justice Phillip Evatt. The Royal Commission's report largely exonerated chemical agents from responsibility for health defects in Australian veterans and their children; this angered the VVA, who proceeded to attack both the report and the commission. The political fallout led the government to refer the report to a political adviser, Bob Hogg, whose findings, presented in October 1987, concluded that the report's findings were unassailable and should be accepted. His further suggestion, that the government sponsor additional studies into Vietnam veterans' health, was ignored. In May 1988 the government announced that it accepted Hogg's findings. After continued lobbying by veterans, however, the government was forced to set up an independent medical committee, whose report *Veterans and Agent Orange: Health Effects of Herbicides used in Vietnam* (September 1994) overturned the Evatt Royal Commission findings and accepted links between Agent Orange and diseases such as leukaemia, Hodgkin's disease and lung cancer. This finally enabled ex-servicemen to claim compensation.

F. B. Smith, 'Agent Orange: The Australian aftermath', in Brendan O'Keefe and F. B. Smith, *Medicine at War: Medical Aspects of Australia's Involvement in Southeast Asian Conflicts 1950–1972* (Allen & Unwin, Sydney, 1994).

AIF see **AUSTRALIAN IMPERIAL FORCE**

AIF NEWS see Service newspapers

AIR, DEPARTMENT OF, created on 13 November 1939, took the control of air force administration and finance away from the Department of Defence. At the time of its abolition on 30 November 1973 it was responsible for air defence as well as for organisation and control of the RAAF. These functions were taken over by the Air Office within the Department of Defence.

AIR BOARD OF ADMINISTRATION was established on 9 November 1920 to control and administer the air force according to policy determined by the Air Council. The Air Council, formed at the same time as the Air Board, consisted of the minister for Defence, the CNS, the CGS, two members of the Air Board, and the Controller of Civil Aviation. The purpose of making the Air Council superior to the Air Board was partly to provide guidance for the RAAF's relatively young and inexperienced senior officers, but it also enabled the army and navy to maintain a strong influence over the air force. The Air Council did not meet after 1925, and was abolished when the Defence

Council (see COUNCIL OF DEFENCE) was created on 8 March 1929. From then on the RAAF was on an equal footing with the other services, and the Air Board reported directly to the minister. The Air Board originally consisted of a Director of Intelligence and Organisation, a Director of Personnel and Training, a Director of Equipment and a Finance Member. In 1922 the membership was changed to comprise the First Air Member (CAS), the Second Air Member (Chief of Administrative Staff) and the Finance Member. In 1929 an Air Member for Supply was added and the Chief of Administrative Staff became the Air Member for Personnel. In 1940 the Air Member for Supply became the Director-General of Supply and Production, and an Air Member for Organisation and Equipment was added. A Business Manager joined the board between 1940 and 1948, and the titles of some other members were changed over the years. In 1954 the Secretary, Department of Air, replaced the Finance Member and became Secretary to the Air Board. By 1975 the Air Board was composed of the CAS, the Air Member for Personnel, the Air Member for Technical Services, the Air Member for Supply and Equipment, and the Special Deputy of the Permanent Head, Department of Defence. The Air Board was abolished on 9 February 1976 along with the Military and Naval Boards (qq.v.).

AIR FORCE LIST Compiled from air force personnel records, this provides a gradational list of seniority by branch (General List, General Duties, Engineering, Supply, Medical, Special Duties, Chaplains), and an alphabetical listing of all active duty officers in the RAAF. It is published annually as a Defence Instruction (Air Force).

AIR SUPPORT This term covers the range of tactical air operations that are used in support of a land battle. It includes the attainment and maintenance of air superiority, tactical air reconnaissance, battlefield interdiction (isolating the battlefield from enemy resupply, reinforcement, operational movement etc.), close air support, aerial resupply, and the provision of battlefield mobility through the use of tactical air transport. The provision of air support has often been a source of bitter contention between ground and air forces in the armed services of most countries, including Australia. Since 1990 the provision of air support using helicopters such as the Sikorsky Blackhawk (q.v.) has been the responsibility of the Australian Army Aviation Corps (q.v.). An early example of air support was the air drop of extra ammunition to forward troops during the battle of Hamel on the Western Front on 4 July 1918. During the Vietnam War air support to the 1st Australian Task Force was often provided by the Iroquois (q.v.) helicopters of No. 9 Squadron RAAF.

AIR TRAINING CORPS (ATC)

The Air Training Corps was a wartime creation designed to help prepare boys under 18 years of age for future aircrew training. Formed in 1941, by October 1943 it consisted of 97 squadrons with an establishment for 122, and during the course of the war 11989 members of the ATC joined the RAAF, 6704 as aircrew, the remainder in ground musterings. At the end of the war the corps was reorganised, with squadrons redesignated flights and wings becoming squadrons, one to each State. Each squadron was headed by a commandant, usually a reserve officer with a distinguished wartime record. The headquarters staff consisted of three permanent RAAF officers as commanding officer, adjutant and chief instructor. Other staff were drawn from permanent NCOs and reserve officers. Flights were divided into school flights (those based on a school) and town flights (which catered for those no longer at school, or whose school lacked an ATC link). Those between 14 and 18 years of age were eligible to join, and until May 1982 girls were excluded. In 1975 the Whitlam government decided to disband all cadet forces, but the following year the Fraser government established the Australian Services Cadet Scheme, into which the three existing service cadet organisations were subsumed.

AIRCRAFT, ROYAL AUSTRALIAN AIR FORCE

When the RAAF was formed in 1921 a numbering system was introduced whereby each aircraft type was assigned an 'A' prefix followed by a number. The first series of aircraft were all biplanes, and were numbered A1 to A12. A new sequence of aircraft, numbered A1 to A100, began in 1935 with the Hawker Demon. A third series was introduced in 1961 with the Bell Sioux helicopter. Many of the aircraft used for transport and communication during the Second World War were civilian aircraft impressed (taken into military use) by the RAAF. Table 1 includes all aircraft flown by the RAAF with their 'A' serial number if one was allocated. Aircraft operated by RAAF squadrons during the Second World War in the Mediterranean and Europe were not assigned 'A' serials. The table includes aircraft flown by the RAAF for the RAN before the formation of the Fleet Air Arm (q.v.). Also included in the 'A' series, but not included in this table, are aircraft operated by the army (see AUSTRALIAN ARMY AVIATION CORPS) and some aircraft types assigned 'A' prefixes but never actually used by the RAAF.

Stewart Wilson, *Military Aircraft of Australia* (Aerospace Publications, Canberra, 1994).

A line-up of RAAF aircraft from the late 1940s. From left to right: De Havilland Tiger Moth, CAC Wirraway, North American Mustang, Airspeed Oxford, De Havilland Mosquito, Douglas Dakota, Consolidated Liberator and GAF Lincoln. Central Flying School, Point Cook, Victoria, March 1947. (AWM P0448/212/200)

Table 1   Aircraft flown by the RAAF

Aircraft	Total aircraft	In service
Fighter aircraft		
Airacobra, Bell A53	22	1942–43
Temporarily loaned to the RAAF by the USAAF. Used in Australia by No. 23, 24, 82 and 83 Squadrons.		
Beaufighter, Bristol and DAP A8, A19 (q.v.)	581	1942–57
Boomerang, CAC A46 (q.v.)	250	1942–46
Buffalo, Brewster A51 (q.v.)	63	1941–43
Bulldog, Bristol A12	8	1930–40
Replaced the SE5A fighter and also used for aerobatic displays. Operated by No. 1 and 2 Squadrons in Australia.		
Defiant, Boulton Paul	18	1941
Used as night fighters in the UK by No. 456 Squadron.		
Demon, Hawker A1	64	1935–45
Last biplane fighter-bomber aircraft flown by the RAAF. Two used as trainers with No. 1 Flying Training School until 1945.		
Gauntlet, Gloster	6	1940
Used in Egypt by No. 3 Squadron.		
Gladiator, Gloster	12	1940–41
Used by No. 3 Squadron during the First Libyan campaign.		
F/A 18 Hornet, McDonnell Douglas A21 (q.v.)	75	1985–
Hurricane, Hawker	unknown	1941, 1942–46
Used in north Africa by No. 3 and 451 Squadrons in 1941. One RAF Hurricane that escaped Singapore transferred to the RAAF and served to 1946 with the Central Flying School and No. 2 Communications Flight.		
Kittyhawk, Curtiss A29 (q.v.)	848+	1942–47
Meteor, Gloster A77 (q.v.)	111	1946–47, 1951–63
Mirage, GAF A3 (q.v.)	116	1964–88
Mosquito, De Havilland A52 (q.v.)	285+	1942–54
Mustang, North American and CAC A68 (q.v.)	499+	1944–60
Sabre, CAC A94 (q.v.)	112	1954–71
SE5A, Royal Aircraft Factory A2	35	1921–28
Imperial Gift aircraft (q.v.). First fighter to be flown by the RAAF. One aircraft repainted in Australian Flying Corps colours is preserved in the collection of the Australian War Memorial.		
Spitfire, Supermarine A58 (q.v.)	656+	1941–45
Tomahawk, Curtiss	unknown	1941
An early version of the Kittyhawk (q.v.) used in the Syrian and 2nd Libyan campaigns by No. 3 Squadron.		
Vampire, De Havilland and De Havilland Australia A78, A79 (q.v.)	193	1949–70
Bomber aircraft		
Baltimore, Martin (q.v.)	unknown	1943–45
Beaufort, Bristol and DAP A9 (q.v.)	701	1941–46
Blenheim, Bristol	unknown	1942–44
Used in the Mediterranean by No. 454 Squadron as a bomber and briefly by No. 459 Squadron for maritime reconnaissance until replaced by Hudsons (q.v.).		
Boston, Douglas A28 (q.v.)	69	1942–45
Canberra, GAF A84 (q.v.)	55	1951–82
DH9 and DH9A, De Havilland A6, A1	58	1921–30
Imperial Gift aircraft (q.v.). Used in Australia by No. 1 and 3 Squadrons and No. 1 Flying Training School. The DH9A differed from the DH9 in having an American Liberty rather than a British Puma engine.		
F-111, General Dynamics A8 (q.v.)	43	1973–
Halifax, Handley Page (q.v.)	unknown	1942–45

Table 1   Aircraft flown by the RAAF *cont.*

Aircraft	Total aircraft	In service
Bomber aircraft *cont.*		
Hampden, Handley Page (q.v.)	unknown	1941–42
Hudson, Lockheed A16 (q.v.)	247+	1940–48
Lancaster, Avro (q.v. and see 'G FOR GEORGE')	unknown	1942–46
Liberator, Consolidated A72 (q.v.)	277	1944–48
Lincoln, GAF A73 (q.v.)	73	1946–61
Mitchell, North American A47 (q.v.)	50	1942–46
Phantom, McDonnell Douglas A69 (q.v.)	24	1970–73
Vengeance, Vultee A27 (q.v.)	342	1942–46
Ventura, Lockheed A59 (q.v.)	75+	1942–46
Wapiti, Westland A5	44	1929–43
Replaced the DH9 and DH9A and used as a bomber from 1929 to 1935 with No. 1 and 3 Squadrons. Used as a trainer and tug aircraft until 1943.		
Wellington, Vickers (q.v.)	unknown	1941–45
Helicopters		
Allouette, Sud A5	3	1964–66
Used at Woomera by the Long Range Weapons Research Establishment (q.v.).		
Chinook, Boeing Vertol A15 (q.v.)	12	1974–89
Iroquois, Bell A2 (q.v)	66	1962–90
S-51, Sikorsky A80	3	1947–64
First helicopter acquired by the RAAF. Used in Australia.		
Sioux, Bell A1 (q.v.)	65	1961–65
Squirrel, Aérospatiale A22 (q.v.)	18	1984–90
Sycamore, Bristol A91	2	1951–65
Used at Woomera by the Long Range Weapons Research Establishment (q.v.).		
Reconnaissance aircraft		
Auster AOP A11	62	1944–59
Small aircraft used by No. 16 and 17 Air Observation Post Flights during the Second World War in the South-West Pacific Area. Two used by RAAF Antarctic Flight during the 1953–54 and 1955–56 expeditions.		
Catalina, Consolidated A24 (q.v.)	168	1940–50
Cub, Piper	2?	1943–44
Small aircraft borrowed from the USAAF and used by No. 4 Squadron in New Guinea.		
Empire, Short A18	5	1939–43
Qantas flying boats taken into service on outbreak of war and used as coastal reconnaissance aircraft until replaced by Hudsons (q.v.). One was destroyed by the Japanese raid on Broome on 3 May 1942. The Sunderland (q.v.) was a military design based on the Empire.		
Fairey IIID A10	6	1921–28
Seaplane used for fleet cooperation with RAN. In 1924 a Fairey IIID flown by S. J. Goble (q.v.) and Ivor McIntyre became the first aircraft to fly around Australia.		
Kingfisher, Vought Sikorsky A48	18	1942–48
American seaplanes ordered by the Netherlands East Indies government and diverted to Australia in 1942 after the fall of the Netherlands East Indies. Used by No. 107 Squadron to patrol for Japanese submarines off the Australian coast during the Second World War. A Kingfisher was used in the 1947–48 Antarctic expedition.		
Lancer, Republic A56	8	1942–43
Fighter used by the RAAF for photographic reconnaissance in the South-West Pacific Area. An early version of the famous Thunderbolt fighter.		

cont. next page

Table 1 Aircraft flown by the RAAF *cont.*

Aircraft	Total aircraft	In service
Reconnaissance aircraft *cont.*		
Learjet, Gates	8	1982–87
Leased for use by Survey Flight of No. 6 Squadron.		
Lightning, Lockheed A55	3	1942–44
Fighter used by the RAAF for photographic reconnaissance in the South-West Pacific Area.		
Lysander, Westland	6	1940
Used in Egypt by No. 3 Squadron.		
Neptune, Lockheed A89 (q.v.)	24	1951–77
Orion, Lockheed A9 (q.v.)	31	1968–
Seagull III, Supermarine A9	9	1926–36
Seaplane used on seaplane carrier HMAS *Albatross* and later HMAS *Australia* (II) and HMAS *Canberra* (qq.v.) until replaced by the Seagull V (q.v.).		
Seagull V (Walrus), Supermarine A2 (q.v.)	61	1935–47
Southhampton, Supermarine A11	2	1928–39
Seaplane used in Australia by the Coastal Reconnaissance Flight and No. 1 Flying Training School.		
Sunderland, Short A26 (q.v.)	146	1939–46
Swordfish, Fairey	3	1942
Three aircraft in crates were on a ship diverted to Perth. They were assembled and used in Western Australia by No. 25 Squadron for anti-submarine patrols.		
Transport aircraft		
BAC-111 A12	2	1967–90
A British passenger jet used by No. 34 Squadron for VIP transport.		
Beaver, De Havilland Canada A95	5	1955–64
Small rugged transport used by RAAF Antarctic Flight.		
Boeing 707 A20	4	1979–
Former Qantas jets used as passenger transport and as inflight refuellers for F/A-18 Hornets (q.v.) by No. 33 Squadron.		
Bombay, Bristol	8	1942–44
Used in the Mediterranean by No. 1 Air Ambulance Unit.		
Caribou, De Havilland Canada A4 (q.v.)	28	1964–
Dakota, Douglas A65 (q.v.)	124	1939, 1943–
DC-2, Douglas A30	14	1940–47
Forerunner of the DC-3 Dakota (q.v.). The RAAF purchased 10 from Eastern Airlines in the USA in 1940. DC-2s were used by No. 34, 35 and 37 Squadrons and in training paratroops. One DC-2 was shot down by Japanese aircraft while flying between Java and Timor during the Netherlands East Indies campaign in 1942.		
Delta, Northrop A61	1	1942–44
Aircraft sold to the federal government by American Antarctic explorer Lincoln Ellsworth in February 1939. The Delta was flown by the Department of Civil Aviation until it was impressed by the RAAF in 1942, serving in succession with No. 35, 34 and 37 Squadrons until it was broken up after being damaged.		
DH86, De Havilland A31	8	1939–45
The DH86 was a four-engined passenger aircraft which entered service with Qantas in 1934. They were impressed into service in 1939 by the RAAF and used as transports and air ambulances in the Mediterranean and South-West Pacific Area.		
DHA-G2 glider, De Havilland Australia A57	8	1942–50
An Australian-designed glider never used in action.		

Table 1   Aircraft flown by the RAAF *cont.*

Aircraft	Total aircraft	In service
Transport aircraft *cont.*		
Do24K, Dornier A49	6	1942–44
Seaplane operated by the Netherlands East Indies Air Force. During the Netherlands East Indies campaign they were used to fly refugees to Australia and five were destroyed in Broome harbour during the Japanese raid on 3 May 1942. Surviving aircraft was used by No. 41 Squadron as transports between Australia and Papua.		
Dragon Rapide, De Havilland A3, A33	8	1935–38, 1940–44
British passenger aircraft used in the 1930s by the RAAF for aerial surveying. In 1940 seven aircraft were impressed from airlines and used as transports.		
Dragonfly, De Havilland A43	1	1942
Civilian aircraft impressed by the RAAF in 1942 and used by No. 2 Communications Flight and No. 34 Squadron. Used in New Guinea as an air ambulance.		
Falcon 900, Dassault A26	5	1989–
French jets leased for use as VIP transport with No. 34 Squadron. In September 1990 they were used to fly Australian citizens released from Iraq prior to the Gulf War.		
Fox Moth, De Havilland DH83 A41	4	1941–45
Two aircraft impressed from civilian owners and two from Qantas. Used as air ambulances and communications aircraft.		
Freighter, Bristol A81	4	1949–67
Large transport aircraft used at Woomera by the Long Range Weapons Establishment (q.v.).		
Gannet, Wackett A14	6	1935–46
Designed by Lawrence Wackett (q.v.) as a photographic survey aircraft but used 1942–45 as an air ambulance with No. 2 Air Ambulance Unit in South-West Pacific Area. In 1938 a Gannet flew to Singapore becoming the first Australian-built military aircraft to fly overseas.		
Goose, Grumman	1	1942
Used by No. 1 Air Ambulance Unit until it crashed into the Mediterranean.		
Hercules, Lockheed A97 (q.v.)	36	1958–
Lodestar, Lockheed A67	10	1943–47
American passenger aircraft similar in design to Lockheed Hudson and Ventura (qq.v.) bombers. Used by No. 37 Squadron in Australia, Papua, New Guinea and the Netherlands East Indies.		
Mariner, Martin A70	12	1943–46
Flying boat used in Australia and the South-West Pacific Area by No. 40 and 41 Squadrons.		
Metropolitan, Convair A96	2	1956–68
American passenger aircraft used as VIP transport by No. 34 Squadron.		
Mystere, Dassault A11	3	1967–89
French jet used as VIP transport by No. 34 Squadron.		
Nomad, GAF	3	1989–93
Australian designed and built transport used by No. 75 Squadron at Tindal (q.v.).		
Otter, De Havilland Canada A100	2	1961–67
Small Canadian passenger aircraft used at Woomera Long Range Weapons Project (q.v.).		
Proctor, Percival A75	1	1945–47
Small British passenger aircraft used by Governor-General's Flight during its brief existence. The civilian version, the Percival Vega Gull, was also used by the RAAF.		
Trimotor, Ford A45	2	1942–43
Flown by Guinea Airways and impressed by the RAAF in 1942. Used as air ambulances by No. 24 and 33 Squadrons in New Guinea.		

cont. next page

Table 1   Aircraft flown by the RAAF *cont.*

Aircraft	Total aircraft	In service
Transport aircraft *cont.*		
Viking, Vickers A59	1	1947–51
British passenger aircraft used at Woomera Long Range Weapons Establishment (q.v.).		
Viscount, Vickers A6	2	1964–69
British passenger aircraft used by No 34 Squadron for VIP transport.		
York, Avro A74	1	1945–47
A passenger version of the Lancaster (q.v.) bomber. Used by Governor-General's Flight during its brief existence and in the repatriation of Australian POWs from Singapore.		
Communications aircraft		
Airmaster, Cessna C34 A40	1	1941–45
Civilian aircraft impressed by the RAAF in 1941 and used by No. 2 Communications Flight.		
Audax, Hawker	unknown	1940–41
Similar to the Hawker Demon, unofficially used as a communications aircraft by Australian squadrons in north Africa.		
Beechcraft 17 A39	3	1941–47
Civilian aircraft impressed by the RAAF and used by No. 34 Squadron and No. 2, 3, and 4 Communications Flights.		
DH50A, De Havilland A8, A10	2	1926–29, 1943–45
In 1926 A8-1 became the first RAAF aircraft to fly beyond Australian territory on a trip to the Solomon Islands. A10-1 was a civilian aircraft impressed in 1942 and used as a communications aircraft.		
Fairchild 24 A36	4	1940–46
Civilian aircraft impressed by the RAAF and used by No. 36 Squadron and No. 1, 2, and 4 Communications Flights.		
Junkers aircraft A44	3	1942–43
Junkers G31, W34f and W34d passenger aircraft impressed from Guinea Airways in 1942.		
Miles aircraft A37	6	1940–45
Miles Hawk, Falcon and Merlin civilian aircraft impressed by the RAAF between 1940 and 1942.		
Norseman, Noorduyn A71	14	1943–46
Canadian aircraft used by No. 1, 3, 4, 5, and 7 Communications Flights.		
Prince, Percival A90	3	1952–57
British passenger aircraft used at Woomera Long Range Weapons Establishment (q.v.).		
Reliant, Stinson A38	1	1941–45
Civilian aircraft impressed in 1941 by the RAAF and used by No. 2 Communications Flight.		
Vega, Lockheed A42	1	1941–44
Civilian aircraft impressed in 1941 by the RAAF and used by No. 24 and 33 Squadrons and No. 3 Communications Flight.		
Vega Gull, Percival A32	2	1940–46
Civilian aircraft impressed in 1940 by the RAAF and used by No. 1 Communications Flight. A military version called the Proctor also served with the RAAF.		
YQC-6, Waco A5	1	1942–44
Civilian airliner impressed in 1942 by the RAAF and used by No. 3 Communications Flight.		
Training aircraft		
Airtrainer, New Zealand Aerospace Industries A19	51	1975–92
New Zealand-built aircraft used for basic air training after which pilots graduated to the Macchi (q.v.).		
Anson, Avro A4 (q.v.)	1020	1937–55

Table 1   Aircraft flown by the RAAF *cont.*

Aircraft	Total aircraft	In service
Training aircraft *cont.*		
Avro 504 A4	61	1922–28
The RAAF operated 20 former Australian Flying Corps (q.v.) aircraft, 35 Imperial Gift aircraft (q.v.) and six manufactured at Mascot, NSW, which were the first locally built aircraft flown by the RAAF.		
Battle, Fairey A22	366	1940–49
Obsolete bomber used as a trainer for the Empire Air Training Scheme (q.v.).		
Cadet, Avro A6	34	1935–45
British intermediate trainer used in conjunction with the Gipsy Moth basic trainer.		
Cirrus Moth, De Havilland A7	34	1926–35
Basic trainer that replaced the Avro 504. The Gipsy Moth was a development of the Cirrus Moth with a better engine.		
Dolphin, Douglas A35	4	1940–44
Amphibious civilian aircraft impressed in 1940 by the RAAF and used by the Seaplane Training Flight, No. 3 Operational Training Unit, No. 9 Squadron and No. 4 Communications Flight.		
Dragon, De Havilland A34	98	1940–45
Aircraft built by De Havilland at Bankstown in Sydney, and 11 civilian aircraft impressed in 1940 and 1941 and used by the RAAF for training and communciations.		
Gipsy Moth, De Havilland A7	98	1930–46
A British basic trainer used until replaced by the Tiger Moth (q.v.). Forty-eight were civilian aircraft impressed in 1939–40. Several were converted to floatplanes including one used in 1936 to search for the missing American Antarctic explorer Lincoln Ellsworth, whose Northrop Delta was later flown by the RAAF.		
HS 748 A10	10	1966–
British passenger aircraft. The RAAF uses eight as navigational trainers at the School of Air Navigation and two for VIP transport with No. 34 Squadron.		
Macchi, CAC A7 (q.v.)	87	1968–
Moth Minor, De Havilland A21	42	1940–45
Three impressed civilian aircraft and aircraft built by De Havilland Australia that were used as basic trainers with the Empire Air Training Scheme (q.v) until they were replaced by Tiger Moths (q.v.). They were then used as communications aircraft.		
Oxford, Airspeed A25	391	1940–53
Two-engined trainer used in the Empire Air Training Scheme (q.v.).		
PC-9, Pilatus A23 (q.v.)	67	1987–
Pup, Sopwith A4	11	1922–25
Used as fighter trainers by No. 1 Flying Training School at Point Cook.		
ST-M, Ryan A50	34	1942–45
Trainer aircraft flown by the Royal Netherlands East Indies Air Force and used in Australia after the Netherlands East Indies campaign.		
Tiger Moth, De Havilland A17 (q.v.)	885	1940–57
Wackett, CAC A3 (q.v.)	202	1941–46
Winjeel, CAC A85 (q.v.)	64	1951–94
Wirraway, CAC A20 (q.v.)	755	1939–58
Prototypes and trial aircraft		
A86 P1081, Hawker	1	1950–51
British swept-wing fighter. The RAAF ordered 75 in 1950 but cancelled the order in 1951 and instead purchased the Sabre (q.v.).		

cont. next page

Table 1 Aircraft flown by the RAAF *cont.*

Aircraft	Total aircraft	In service
Prototypes and trial aircraft *cont.*		
Avro 707A	1	1956
British experimental delta-wing aircraft used by the Aircraft Research and Development Unit for aerodynamic research. Preserved privately in Melbourne.		
CA-15, CAC A62 (q.v.)	1	1946–50
Jet Provost, Percival A99	1	1959
British trainer trialled by the RAAF.		
Magister, Miles A15	1	1938–40
British trainer acquired for trials. Magisters were also used unofficially by Australian squadrons in north Africa.		
Pika, GAF A93	2	1950–54
Piloted prototype of the Jindivik (q.v.). The first jet-powered plane designed and built in Australia.		
Sea Hornet, De Havilland A83	1	1948–50
Royal Navy twin-engined single-seat fighter trialled by the RAAF.		
Shrike, Curtiss A69		
USAAF name for the US Navy Helldiver dive bomber. The RAAF ordered 150 in 1943 but then decided dive bombers were not needed, withdrawing the Vultee Vengeance (q.v.) dive bomber from service and cancelling the Shrike order. Ten aircraft from the order which had already arrived were returned to the USAAF.		
Valiant, Vickers	2	1956–57
RAF jet bombers used at Woomera by the Long Range Weapons Establishment (q.v.). One dropped an atomic bomb at Maralinga (see ATOMIC TESTS IN AUSTRALIA).		
Warrigal, Wackett A12	2	1927–33
Designed by Lawrence Wackett (q.v.) at the RAAF experimental station as fighter and trainer prototypes (Marks I and II). Also known as Avro 598 and 599.		
Washington, Boeing A76	2	1952–56
British name for B-29 bomber. Used in weapons trials at Woomera by the Long Range Weapons Establishment (q.v.).		
Widgeon, Wackett A12	2	1927–33
General purpose amphibian prototype designed by Lawrence Wackett (q.v.) at the RAAF experimental station for use on the seaplane carrier HMAS *Albatross*.		
Woomera, CAC A23	2	1942–46
Australian designed bomber prototype not put into production.		

Captured aircraft

Captured German, Italian and Japanese aircraft flown by the RAAF during the Second World War:
  Breda 25
  Cant 100
  Caproni Ca-309 Ghibli
  Fiat CR-42
  Focke Wulf Fw-190
  Macchi MC-205
  Messerschmitt Bf-109
  Mitsubishi A6M5 'Zero'
  Mitsubishi Ki-21
  Mitsubishi Ki-51
  Tachikawa Ki-54

A captured Japanese Mitsubishi A6M5 'Zero' fighter with Australian markings, Labuan, Borneo, 1945. (AWM P0590/09/08)

AITAPE, HMAS see *ATTACK CLASS PATROL BOATS*

ALAMEIN, BATTLE OF EL In 1942 the pivotal battles of the Desert War were fought around the El Alamein area in Egypt. The German Commander Field Marshal Erwin Rommel had pushed the British Eighth Army back to this natural defensive position in June but had failed to smash his way through it. In July the British C-in-C, General Sir Claude Auchinleck, made three attempts to push the German and Italian forces back. The 9th Australian Division, which had reinforced Auchinleck on 4 July 1942, took part in all these actions, on 10, 21–22 and 26–27 July. The Australians were operating on the northern coastal strip of the battlefield against some of the most formidable Axis defences. The 9th Division broke into Rommel's defensive line at the first attempt but a forceful German response with armour and artillery meant that the early gains could not be exploited. Events further south were also inconclusive leading to a temporary stalemate.

Rommel's next attempt to break the British line was made on 30 August. By this time the British Command had been reorganised with General Sir Bernard Montgomery in charge of the Eighth Army and General Sir Harold Alexander in overall command. Rommel's aim was to outflank the Alamein position from the south but this attempt was easily beaten off. Meanwhile on 1 September in the north the 9th Division launched a small diversionary attack which was beaten back after heavy fighting.

The next few months saw a steady build up of British forces. By the third week in October Montgomery was ready. His plan was to send the infantry forward, protected by a mass of artillery, and clear paths through the minefields protecting the Axis forces. He anticipated that the enemy infantry divisions that tried to interfere with this process would be 'crumbled' away by throwing themselves against his forces which would be dug in on ground of their own choosing and protected by the artillery. At the appropriate moment he hoped to push his armour through the corridors to protect the infantry from anticipated counter-attacks from Rommel's armour. In this way too the enemy's mobile forces would be destroyed in a similar 'crumbling' manner to their infantry. The 9th Division was to participate by clearing a corridor in the north and then threatening the German forces between the coastal road and the sea.

Very few of these objectives went according to plan. The battle opened on the night of 23 October with a bombardment from Montgomery's 900 guns. The Australian infantry attacked at 10.00 p.m.

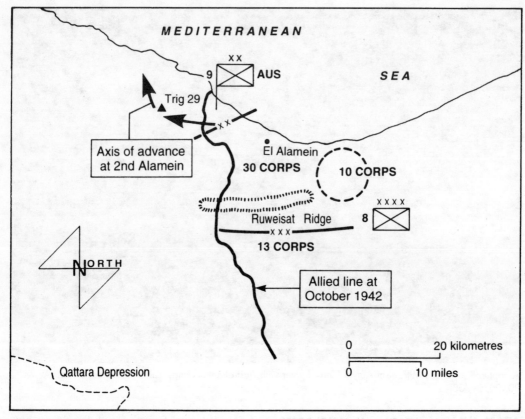

MEDITERRANEAN   SEA

9   X X   AUS

Trig 29

Axis of advance
at 2nd Alamein

El Alamein

30 CORPS   10 CORPS

Ruweisat Ridge   8   X X X X
X X X

13 CORPS

NORTH

Allied line at
October 1942

0   20 kilometres
0   10 miles

Qattara Depression

Map 3   El Alamein, 1942, showing the Eighth Army area of operations.

They fought their way into Rommel's defences but did not manage to clear all the minefields before dawn. The armoured sortie in this area was therefore cancelled, which was unfortunate because the German command was in chaos. The temporary commander, General Stumme, had died of a heart attack and Rommel was still back in Germany. By the time the corridor was cleared on the 24th the enemy had established a new front; on the 25th the Australians were counter-attacked by the German 15th Armoured Division which they drove off after heavy fighting.

Montgomery now changed his plan. He withdrew all his armour for regrouping and continued with his infantry/artillery policy to 'crumble' the German armour. The part given to the 9th Division was to attack between the northern flank and the sea. The operation opened on the 26th. During the next five days the division was involved in severe fighting around a high dune called Trig 29. This attack brought almost the entire Afrika Korps against the Australians and resulted in some of the most severe fighting they had encountered. With Rommel's key forces already engaged Montgomery pushed the Allied armour against enemy positions

to the south. Again all did not go according to plan and the armour was tardy in following through with its attacks. Nevertheless by this time attrition had taken its toll. Rommel had no option but to withdraw or be annihilated. He withdrew to the Tunisian border. Axis total casualties for this battle were 90 000, Allied 13 000. Of the latter the 9th Division suffered 2700. The losses of the 9th Division over the whole period of the El Alamein battles totalled 5800.

ALBATROSS, HMAS see **FLEET AIR ARM**

ALBERT, HMVS see **VICTORIAN NAVY SHIPS**

ALCOCK, A. U. see **SCIENCE AND TECHNOLOGY**

ALIENS, WARTIME TREATMENT OF Under the *War Precautions Act 1914* (q.v.) a total of 6890 'enemy aliens', almost all male and mostly classified as 'Germans' or 'Austro-Hungarians', were interned in Australia during the First World War. Only about 4500 had been living in Australia before August 1914; the rest were sailors of enemy nationality taken off ships while in Australian ports or were enemy residents of Britain's Asian and Pacific colonies transported to Australia at the British government's request. At first only subjects of nations at war with Britain were considered 'enemy aliens', but the concept was gradually widened until it included native-born Australians of 'enemy descent' and other 'persons of hostile origin or association'. Not all enemy aliens were interned, but all had to be registered with the police and were subject to surveillance and restrictions. Although this policy was principally motivated by British race patriotism and wartime paranoia directed against non-Britons in general, it also had more specific targets. Unsuccessful migrants dependent on government assistance, political radicals, leaders of the German-Australian community, and those who were seen as competing too successfully with British Australians in business or employment were singled out for internment and deportation. Internment was administered by the Defence Department and by local military authorities who established detention camps in all military districts, the largest being at Holsworthy (q.v.). Prisoners were not usually forced to work, and although they resented their imprisonment most internees did not protest. After the war 6150 'enemy aliens' were deported from Australia.

During the Second World War all aliens were required under the *National Security Act 1939* (q.v.) to register at police stations and were subject to various restrictions including limitations on movement. The minister for the Army administered these regulations until August 1942 when the Attorney-General took over this responsibility, though the Army remained responsible for the custody of internees. A total of 16798 aliens were interned under these regulations, of whom 8921 were local internees and the rest, received from overseas, were held on behalf of other Allied governments. Most were Italians (4754 local, 425 from overseas), Germans (2013 local, 3753 from overseas) and Japanese (1141 local, 3160 from overseas). Only adult male aliens who directly threatened security or public safety were supposed to be interned, and from November 1940 internees could appeal to the Aliens Tribunal, which could recommend their release if it believed they were not a threat. Nevertheless some notable injustices occurred, including the internment of 2542 predominantly Jewish refugees from Nazi Germany sent from Britain on HMT *Dunera* in 1940, several hundred Indonesian political prisoners sent from Dutch New Guinea, and many Australian residents who had no fascist associations. After Japan entered the war all Japanese, excepting consular officials but including Australian-born or naturalised British subjects of Japanese ori-

gin, were interned. In the opinion of the War Cabinet their 'fanaticism and devotion to their country' made all Japanese potential saboteurs. In February 1942 all aliens over 18 other than POWs and internees were required to register for the Citizen Forces or for non-military service, and in May 1943 the Civil Aliens Corps was formed to help relieve the manpower shortage. However, of about 16000 aliens registered between 1942 and 1945 only 1671 joined the Civil Aliens Corps, and no more than 4000 were employed under the Aliens Service Regulations at any time. Most of the rest were exempted on medical grounds or because they were already employed in essential industries. In May 1945 the Civil Aliens Corps merged with the Civil Constructional Corps (q.v.). After the war most internees were given the choice of returning to their country of origin or remaining in Australia, but almost all Japanese were deported. Of the latter, only Australian-born people of Japanese descent, Japanese married to British subjects, and those who were unfit to travel were excluded from forced deportation.

Gerhard Fischer, *Enemy Aliens: Internment and the Homefront Experience in Australia 1914–1920* (University of Queensland Press, Brisbane, 1989); Kay Saunders and Helen Taylor, 'The enemy within? The process of internment of enemy aliens in Queensland 1939–45', *Australian Journal of Politics and History*, vol. 34, no. 1 (1988), pp. 16–27; Margaret Bevege, *Behind Barbed Wire: Internment in Australia during World War II* (University of Queensland Press, Brisbane, 1993).

ALLEN, Major-General Arthur Samuel 'Tubby' (10 March 1894–25 January 1959). Allen served in two world wars and had the distinction of commanding in action at every level from platoon to division, a distinction of which he was justly proud. During the First World War he served as a lieutenant with the 13th Battalion, and a captain and major with the 45th Battalion, and he commanded successively the 45th, 48th and 13th Battalions. In the interwar period he served in the militia as a battalion and brigade commander. His command of the 16th Brigade in the first Australian actions of the Second World War at Bardia and Tobruk was highly creditable, while his personal courage and spontaneous manner greatly appealed to his troops. After serving in the campaign in Greece in 1941, he commanded the 7th Division in Syria and Papua. Allen was relieved of command on 29 October 1942 by General Thomas Blamey (q.v.) for alleged lack of progress against the Japanese. His removal, however, reflected an ignorance of local conditions on the part of General Douglas MacArthur (q.v.) and Blamey in Australia rather than any shortcomings on Allen's part, or that of his division, which the day before his removal had broken through the Japanese positions at Eora Creek after earlier halting their advance at Ioribaiwa. He then held inactive commands in New Guinea and the Northern Territory for the rest of the war. 'Honest, impetuous, brave', in Gavin Long's (q.v.) view, Allen was an excellent example of the best kind of militia officer at the unit and brigade levels, although he was less well prepared for the command of a division.

ALLIED GEOGRAPHIC SECTION (AGS)

ALLIED GEOGRAPHIC SECTION (AGS) was established between June and October 1942 because of the absence of useful maps of many parts of New Guinea. It collected and checked available geographic and hydrographic information, then made this information available in printed studies designed for operational use. With headquarters in Brisbane and an Australian director, Colonel W.V. Jardine-Blake, the AGS was nevertheless an inter-Allied organisation in terms of its personnel and its structure, with Australian, American and Dutch subsections. It was dissolved on 30 November 1945.

ALLIED INTELLIGENCE BUREAU (AIB)

ALLIED INTELLIGENCE BUREAU (AIB) was established on 6 July 1942 to control and coordinate the activities of various intelligence organisations that had been set up after the outbreak of war with Japan. Special Operations Australia (q.v.) was also known by the cover name of the Inter-Allied Services Department (ISD, founded in April 1942), which was headed by Colonel G. E. Mott of the British Special Operations Executive (SOE); it became Section A of the AIB, responsible for obtaining information about the enemy and carrying out acts of sabotage. AIF personnel serving in SOA were attached to Z Special Unit (q.v.). Section B consisted of Secret Intelligence Australia (SIA); it was in effect a branch of the British Secret Intelligence Service (SIS). The Combined Field Intelligence Service, Section C, was based on the coastwatchers' organisation (q.v.), and was divided into sub-units covering the North-East Area, the Philippines, and the Netherlands East Indies. Section D consisted of a propaganda organisation, the Far East Liaison Office (FELO), which had been set up on the orders of the Allied Land Commander, General Sir Thomas Blamey (q.v.), on 19 June 1942. Problems of control, partly arising out of regular forces' suspicions of irregular or special units, were deepened by the overlapping responsibilities of the various sections, and together with diverging allied interests in different areas as military action changed the strategic picture, led to a reorganisation of the structure. Section C was divided into three regional divisions (Philippines, NEI, and North-East); SIA was focused on the NEI; and SOA was given the new cover name of the Services Reconnaissance Department (SRD), and made answerable to Blamey. As Z Special Unit moved with SOA out of AIB, M Special Unit was formed as an administrative unit for AIF personnel who remained with AIB. Under the command of Lieutenant-Colonel P. J. F. Chapman Walker, SRD mounted the JAYWICK (q.v.) and RIMAU raids against Singapore Harbour in 1943 and 1944. Despite the tangled lines of authority and the continued duplication of function, the AIB was a significant step forward in that it constituted a break from Australia's previous reliance on British sources for military and political intelligence, enabling it to establish a foundation for more independent assessments on which to base future policy.

ALLIED LAND HEADQUARTERS

ALLIED LAND HEADQUARTERS was established by General Thomas Blamey (q.v.) on 9 April 1942 in anticipation of his assuming the command of Allied Land Forces in the South-West Pacific Area (q.v.). The Australian Army was reorganised into a field army and lines-of-communications areas, the Military Board (q.v.) was suspended and its members became Blamey's principal staff officers, while Blamey himself became C-in-C of the Australian Military Forces. Allied Land Headquarters was based in Melbourne and was subordinated to General Douglas MacArthur's (q.v.) General Headquarters, which was in Brisbane, as was Blamey's Advanced Land Headquarters (q.v.). Blamey was the only Australian to hold a senior command position within the SWPA structure, but the arrangements were never really honoured by MacArthur, who had no intention of allowing Blamey to command American forces and systematically cut him out of command of any but Australian formations.

ALLIED TRANSLATOR AND INTERPRETER SECTION (ATIS)

ALLIED TRANSLATOR AND INTERPRETER SECTION (ATIS) was established by General Headquarters, South-West Pacific Area (GHQ SWPA), on 19 September 1942 to translate seized Japanese documents and to provide interpreters for the interrogation of POWs. Under the command of US Army Colonel Sidney F. Mashbir, ATIS was Allied and inter-service in character and in operation. From an original staff of 25 officers and 10 enlisted men, its personnel expanded steadily until it reached a peak of 250 officers and 1700 enlisted men and women in 1945. As the Allies took the offensive against the Japanese from 1943, ATIS Advanced Echelons were formed to follow the combat forces, seizing and rapidly translating documents as they went. When GHQ SWPA moved from Melbourne to Hollandia in New Guinea, then to

Leyte and finally Manila in the Philippines, Base ATIS moved with it. In the course of the war ATIS screened some 350000 captured documents, fully translating 18000 of them, and published over 2800 interrogation reports. It also produced interpretive reports and summaries of Japanese activities. After the war Base ATIS moved to Tokyo, where it assisted Occupation authorities until it was dissolved on 30 April 1946 and replaced by a new Translator and Interpreter Service.

AMBERLEY, RAAF BASE, near Ipswich, Queensland, is Australia's second largest operational air force base. The first operational aircraft landed there on 4 July 1940 and it is currently home to No. 1 and 6 Squadrons, the RAAF's bomber squadrons, which are equipped with F-111 (q.v.) strike/ reconnaissance aircraft.

AMBON A small island at the eastern extreme of the Indonesian archipelago, Ambon had been strategically important for over four centuries because of its deep-water harbour which penetrates between the two peninsulas of the island. By 1941 Ambon had the additional asset of an airfield at Laha on the harbour side of the northern (Hitu) peninsula. Under an agreement made in early 1941 between the Australian and Netherlands East Indies governments, Gull Force (q.v.) was sent immediately after the Japanese attack on Pearl Harbor to assist the 2600-strong Dutch forces (commanded by Lieutenant-Colonel J. L. R. Kapitz) defending the island.

Like the comparable forces Sparrow and Lark (qq.v.), which were deployed on Timor and New Britain respectively, Gull Force faced a hopeless task. The defending forces were under-equipped; such limited air support as they had, including Hudsons (q.v.) of No. 13 Squadron RAAF, was withdrawn before the Japanese attack on the night of 30–31 January 1942. The efforts of the charismatic commander of Gull Force, Lieutenant-Colonel L. N. Roach, to convince Army Headquarters in Melbourne of the need to evacuate his force led to his replacement by Lieutenant-Colonel W. J. R. (Jack) Scott (q.v.).

The vastly superior Japanese forces, consisting of the three battalions of the 228th Regiment and approximately a battalion of naval troops, landed on the northern and southern coasts of the island to avoid the major fortification, the fixed battery at Benteng overlooking the harbour. Within 24 hours the Dutch forces, many of whom were indigenous, surrendered; the Japanese crossed the southern Laitimor Peninsula to capture the town of Ambon and isolate the Australian defensive position to its

south. At Kudamati, where the Australian, Driver Bill Doolan, ensured his immortality with the local Ambonese for his bravery, and on the commanding Mount Nona, the Australians of 5 Platoon managed to repulse Japanese attacks. However, believing the Australian position to be untenable, Scott evacuated the Amahusu line stretching up the side of Nona on the night of 1–2 February. Retreating to Eri at the tip of the peninsula, Scott surrendered on 3 February. On the other side of the bay at Laha, Australians under the command of Major H. Newbury resisted fiercely but were overwhelmed by 2 February. With the exception of a few Australians who managed to escape by island-hopping to Australia, all personnel at Laha were captured and murdered.

Gull Force's barracks at Tan Tui became the POW camp for the majority of Gull Force for the remainder of the war. The island was subjected to Allied air attacks from May 1942, two of which in February 1943 and August 1944 devastated the prison camp, killing a number of the more able Australian officers. Allied airmen who fell into Japanese hands were executed. Bypassed and blockaded by MacArthur's advance to the Philippines, the island was not liberated until after the Japanese armistice in August 1945. After an attempt to enter Ambon Harbour on 16 August was aborted for fear of Japanese retaliation, Australian naval forces rescued 302 prisoners on 10 September 1945. Four Japanese servicemen were executed for war crimes (q.v.) committed on Ambon. In recognition of the help given to Australian prisoners by the Ambonese population, the Gull Force Association and the Australian government have maintained a program of medical aid since 1967. Annual pilgrimages commemorate Anzac Day at the war graves cemetery on the site of the prison camp.

Joan Beaumont, *Gull Force: Survival and Leadership in Captivity 1941–1945* (Allen & Unwin, Sydney, 1988); Lionel Wigmore, *The Japanese Thrust*, vol. 4, Series 1 (Army), *Australia in the War of 1939–1945* (Australian War Memorial, Canberra, 1957).

JOAN BEAUMONT

AMF see **AUSTRALIAN REGULAR ARMY**

ANDERSON, Lieutenant-Colonel Charles Groves Wright (12 February 1897–11 November 1988). Born in Cape Town, Anderson saw regimental service with the King's African Rifles in East Africa in the First World War, during which he was awarded the MC. Migrating to Australia in 1934, he was commissioned into the CMF in 1939 and in July 1940 was appointed as second-in-command of the 2/19th Battalion, AIF; he succeeded to command in August. During operations against the Japanese in the Malayan campaign in January 1942, he was engaged at close quarters with the enemy in four days of desperate fighting against heavy odds in the Muar area. He successfully extricated his unit and surrounding troops, which sustained heavy casualties in the withdrawal to Singapore. For his leadership and personal gallantry he was awarded the VC, the only Australian unit commander in the Second World War so honoured. Taken prisoner at the campaign's end, he returned to Australia in August 1945 and served as a Liberal MP in federal parliament between 1949–51 and 1955–61. His actions at Muar provide one of the few positive examples of Australian command in Malaya, while the action itself was judged a minor epic in an otherwise disastrous campaign.

ANGLO-AUSTRALIAN NAVAL AGREEMENTS, 1887, 1903, defined the naval relationship between Britain and the Australian colonies and ultimately provided the spur for the development of an Australian navy. The 1887 agreement sought to regularise the growth of colonial naval forces and to define carefully the relationship between those forces and the RN. The *Colonial Naval Defence Act 1865* (q.v.) had authorised the colonies to develop their own naval forces, but required that such forces, whether deployed in home waters or beyond, remain under the control of the Admiralty, a provision to which the colony of Victoria in particular objected. The 1878 report by Major-General William Jervois (q.v.) had recognised that while the imperial government remained responsible for the external defence of the colonies, the colonies themselves had need of naval forces to protect their ports. That need was reinforced by the Russian war scare of 1885 (see COLONIAL WAR SCARES), which resulted in every station being placed on alert for several months. Underlying the defence question was the issue of cost, which combined with the matter of control to present the authorities, both imperial and colonial, with a difficult problem: the colonies argued that since they would bear the cost of developing naval forces, they should have authority over the deployment of those forces; the imperial government insisted that all naval vessels, whether deployed in colonial waters or on the high seas, should be under the control of the Admiralty. The 1887 agreement required the Admiralty to enlarge the Australian Squadron by the creation of the Auxiliary Squadron, Australia Station (q.v.), comprising five fast cruisers and two torpedo gunboats, and to bear the costs of construction, although the Australian colonies and New Zealand undertook to contribute towards the interest charges on the capital

cost and the annual maintenance bill. This was a significant concession by the Australian colonies, which had hitherto insisted that the cost of imperial defence should be borne by Britain; but in making that concession the colonies extracted from the Admiralty an undertaking that the ships could not be moved beyond Australian waters without the specific consent of the colonial governments.

That restriction, reluctantly given in 1887, had by the turn of the century become increasingly unacceptable to the Admiralty in the new strategic environment. The development of the German navy and the rise of Japan as a major naval power in the Pacific made the notion of an auxiliary squadron that was effectively dedicated to local Australian defence completely at odds with the Admiralty's 'blue-water' strategy, which focused on the offensive power of the RN to seek out and destroy the enemy fleet, wherever it might be, and thereby defend the territorial and trade interests of the whole Empire.

The 1887 agreement was due to run for 10 years from the time of arrival of the new ships on station, which was 1891. A new agreement was therefore negotiated in 1903, this time between the British government and the new Commonwealth government. It provided for a larger Australian squadron in return for increased financial contributions from Australia and New Zealand. The earlier restrictions on deployment were lifted to the extent that it was agreed that the ships could operate in the waters of the China, East Indies and Australian stations. In return for this concession to the notion of centralised control and a blue-water focus, the Admiralty agreed that three of the nine ships of the squadron would be drill ships (which had been provided for in the 1887 agreement but never honoured), manned as far as possible by Australian and New Zealand sailors. Although each party had won concessions to its position, neither was satisfied with the result. Increasingly the Australian government looked to the establishment of a local navy to defend local interests, while the Admiralty pressed for a global perspective in which there was no room for purely local naval forces.

ANIMALS have played important roles, both psychological and practical, throughout the history of the Australian armed forces, though more recently their practical utility has declined as a result of technological advances. A special affinity with animals, particularly horses, has often been attributed to Australian servicemen because of their supposed experience of bush life. This idea served useful propaganda and morale-boosting purposes, allowing Australians to compare themselves favourably not

An Australian soldier holding a pet koala, Cairo, Egypt, 1915. (AWM P0156/81/31)

only with inexperienced European city-dwellers, but also with enemies and 'natives', whose allegedly cruel treatment of animals was taken as further evidence of contempt for 'civilised' values. Whether or not Australians work better with animals than other people, it is clear that many Australian servicemen developed genuine affection for the animals they worked with. Caring for animals provided some relief from the stresses of service life, allowing servicemen to express their gentle, affectionate and sentimental sides.

This aspect of the role of animals within the armed forces is shown clearly in the tradition of animal mascots. Mascots are meant to bring luck, to boost morale and to provide fun. Originally they had no official status within the Australian Army, but were usually soldiers' pets which were unofficially adopted by a unit. During the First World War animals of all kinds became mascots, including Australian native species such as kangaroos, possums and koalas. The continuing popularity of mascots during the Second World War was revealed in a 1945 article in the army magazine *Salt*, which reported attempts by Australian servicemen to bring home 'at least 10 goats, 220 dogs, 170 cats, 150 birds, and 50 assorted monkeys, squirrels, etc.'

from the start of the Pacific War. Mascots have been more common in the army than in the other services, and only in the army have selected mascots been given official recognition. A small number of army mascots have been assigned regimental numbers, have official records, and are eligible for promotion. The navy and air force have also had unofficial mascots, though in the navy mascots are prohibited at sea.

Single-humped Arabian camels were used by Australian troops during the First World War because their ability to store water and to walk on sand made them superior to horses for some purposes during the Palestine campaign. The Imperial Camel Corps, formed in 1916, consisted of 18 companies, 10 of them Australian. Unlike the mounted infantry of the light horse (q.v.), troops mounted on camels were pure infantry, as they had to dismount at a distance from the enemy and could not keep their camels close at hand for a quick escape. The corps had some victories, but the Australian official historian H. S. Gullett (q.v.) considered that it contributed little to the desert campaign and that once the desert terrain was past its members would have been of greater value mounted on horses. Military authorities reached this conclusion in mid-1918 and the Camel

Milo the tracker dog with men of 2RAR on patrol near Nui Dat, South Vietnam, July 1970. (AWM WAR70/547/VN)

Corps was disbanded, its Australian members forming the 5th Australian Light Horse Brigade. Camel trains were used to transport supplies in the desert, but their slowness and the resources required to feed the camels and their drivers meant that, where possible, horse-drawn and motor transport was used in preference. Likewise, the wounded were transported on cacolets (chairs on stretchers slung on either side of a camel) only when this was unavoidable, as the jolting ride was torture for the patient.

Dogs are perhaps the most versatile animals used by the military, but the Australian Army has never used them on such a large scale as other armies and was comparatively late in making use of the full range of their abilities. During the First World War, and at some times subsequently, dogs were used to carry messages. These dogs were taught to run from a forward position to their handler at a rear station, or to run between two handlers. Their advantages over pigeons as message-carriers were that they could run at night and could make several runs during a tour of duty in the line, but they were also easier targets for enemy fire and their efficiency was often impaired by the development of bonds of affection with their handlers. Both the army and the RAAF used German Shepherds as guard dogs during and after the Second World War. Guard dogs are trained to protect important establish-

ments by detecting the presence of intruders, alerting their handlers and, if necessary, attacking on command. German Shepherds have also been used as infantry patrol dogs, trained to avoid ambushes. The patrol dog uses its senses of smell and hearing to detect hidden people and silently indicates their presence to its handler by 'pointing'. In contrast to patrol dogs which work on air scent, tracker dogs work on ground scent. Trained to follow human scent only, tracker dogs work with a harness and a long leash. They can track a scent for an average of about 5 to 6.5 kilometres when it is up to eight hours old. Mine-detection dogs are taught to recognise the scent and appearance of explosives. Generally working without a leash, they move back and forth within a limited area until they find an explosive, whereupon they sit down close by to indicate its presence to accompanying sappers. Mine-detectors can discover explosives in metallic containers, but mine-detection dogs are useful for finding non-metallic devices and explosives hidden in areas where other metal is present, such as railway lines. Australian Army training of dogs and dog-handlers began after the successful use of patrol dogs by Australians trained by British specialists during the Korean War. A dog-training depot was established at the School of Military Engineering, Casula, in 1954, and soon afterwards tracker dogs played an important part in Australian operations in the Malayan Emergency. This training was discontinued for a time, but in 1965 a Tracking Wing was established at the Infantry Centre, Ingleburn, and tracker dogs began to be used during the Vietnam War. This conflict also produced a resurgence of interest in mine-detection dogs. A Mine Dog Section, established at the School of Military Engineering in 1971, became the Military Dog Wing in 1974 and amalgamated with the Explosive Ordnance Section in 1986. The RAAF Police Dog Training Centre has been operating since 1954.

Donkeys and mules were used as beasts of burden in both world wars. Donkeys generally did lighter work such as carting water or meals and carrying wounded soldiers. Although they were less useful than mules their good tempers made them more popular with the troops. However, mules, despite their cantankerous nature, slowly earned grudging respect for their extraordinary strength and endurance. They made excellent pack animals, were agile even in mountainous terrain, were quiet, needed little attention and could eat almost anything. Horses were also sometimes used as pack animals in addition to being used to pull wheeled transport. The advent of motorised transport did not immediately make horses and mules obsolete, despite their comparative slowness and low carrying capacity. They remained useful for their flexibility and their ability to perform over rough or slushy ground which was difficult for motor transport to negotiate, particularly before the invention of the multi-axle drive.

Of all the animals used by Australian troops, it was probably riding-horses that were regarded with the greatest affection. The close bonds that developed between men and horses were revealed in the reaction of Australian light horsemen to the news that, because of cost and the danger of introducing diseases to Australia, their horses would not return with them. 'Trooper Bluegum' (Oliver Hogue [q.v.]) expressed the heartbreak of many in his poem 'The Horses Stay Behind':

I don't think I could stand the thought of my old
   fancy hack
Just crawling round old Cairo with a 'Gyppo' on his
   back.

'Trooper Bluegum' went on to claim that rather than let his horse meet such a fate he would shoot it, but most soldiers were happy to leave to the Remount and Veterinary Services the unpleasant task of destroying those horses that could not be transferred to other armies or sold. The only horse returned to Australia after the First World War was Sandy, the favourite horse of the 1st Division's commander General W. T. Bridges (q.v.). The head

and one hoof of this horse have been preserved and stored at the Australian War Memorial.

The performance of Australian horses in the Middle East during the First World War defied expectation when they proved capable of withstanding heavier loads, longer distances and harsher conditions than had previously been thought possible. Their endurance was often attributed to the hardiness of the waler (q.v.) stock, but was due at least as much to the work of the Remount Service. The Remount Branch had been formed in 1911–12 to train and care for military horses, and in 1915 two remount units were sent to Egypt where they were incorporated into the Imperial Remount Service. They originally consisted of a mixture of middle-aged men unfit for active service and younger 'rough-riders' needed for horse-breaking. As the units were consolidated and reduced in size in 1916, most of the older men were released and the proportion of rough-riders rose. In addition to the important work of taming and acclimatising new horses there were plenty of routine duties to keep the Remount Service busy: feeding, watering, grooming and exercising the horses and disposing of manure. Horses continued to be used by the Australian Army after the First World War, but horse transport was phased out from 1944 and the Remount Service was disbanded in 1946.

Pigeons have been useful to the military because of their homing instinct, which makes them return immediately to their home loft after being removed from there and then released. Troops could carry pigeons with them to a forward area, release them with messages in small cylindrical containers attached to the birds' legs, and be sure that each bird, unless it had a mishap in flight, would take the message to its home loft back at base. Pigeons were useful in isolated areas where other means of communication were not available and in situations in which the use of a wireless was considered inadvisable. With an average speed of 50 kilometres per hour, pigeons have a normal flying range of 60–100 kilometres if being used from a mobile loft or 200 kilometres from a stationary loft. They will not fly at night or in bad weather. Although there was some use of pigeons by the Australian Army in the early twentieth century it was not until the First World War that they were widely used. In France, pigeons were obtained by the AIF from the British Army and each infantry brigade's signallers set up a pigeon station. During the Second World War the vulnerability of communication by line and wireless meant that an alternative means of communication between coastal defences was needed in case of invasion.

This need prompted the formation in 1942 of the Corps of Signals Pigeon Service, which was made up largely of men who had been pigeon-fanciers as civilians. Pigeon-fanciers also donated thousands of birds to the army. In addition to forming an alternative communications network across the Australian continent, the Pigeon Service operated in New Guinea. Their birds were of particular use to raiding and reconnaissance parties, and to small boats which used them for ship-to-shore communication. So effective was the Australian Pigeon Service that from 1943 its personnel and pigeons were used by US forces in the South-West Pacific Area (q.v.). On many occasions pigeons carried information that was crucial to the success of particular operations or which saved the crews of small boats that were in trouble. The Australian Corps of Signals officially ended the use of pigeons in 1946.

Animals were also used to determine the poisonous effects of absorbing mustard gas through the skin in the initial stages of chemical warfare research in Australia (see GAS, SECOND WORLD WAR).

ANSON, AVRO (3 crew twin-engine maritime patrol bomber or trainer [Mark I]). Wingspan 56 feet 6 inches; length 42 feet 3 inches; armament 2 × 0.303-inch machine-guns, 360 pounds bombs or 500 pounds depth-charges; maximum speed 188 m.p.h.; range 790 miles; power 2 × Armstrong Siddeley Cheetah 350 h.p. engines.

The Anson entered service with the RAAF as a reconnaissance bomber and maritime patrol aircraft in 1936. It was replaced as a bomber in 1940 by the Hudson (q.v.) but was used between 1942 and 1945 by No. 66, 67, 71 and 73 Squadrons for anti-submarine patrols on the east coast of Australia. During the Second World War over 1000 Ansons were used by the RAAF as twin-engine training aircraft, mainly for the Empire Air Training Scheme (q.v.). After the Second World War several ex-RAAF Ansons were operated as civilian airliners.

ANTHROPOMETRY A term originating in the mid-nineteenth century, it describes the measurement of the human body with a view to determining its development at different stages and among different races and social classes. In the early twentieth century anthropometrical measurements were taken of schoolchildren, and for a time analysis of the data acquired appeared in the Commonwealth *Year Book*. Such measurements were also taken of all young men who presented for enlistment under the military training scheme (see CONSCRIPTION) in order 'to study the development of the Australian nation'. The practice was part of the fascination with questions of 'racial

hygiene' common to a number of European societies. It stopped in Australia in the interwar period.

ANTI-SUBMARINE WARFARE (ASW) Until recently this was the main postwar operational function of the RAN, although the navy had developed this specialisation in 1938 with the establishment of an anti-submarine branch at HMAS *Rushcutter* (q.v.). Australian ships were heavily involved in anti-submarine tasks during the Second World War, particularly in the Mediterranean with the RN. RAN ships assisted in the sinking of seven Italian, German and Japanese submarines including the Italian submarine *Gondar*, sunk by HMAS *Stuart* (q.v.) in the Mediterranean on 30 September 1940, the German U-boat *U 127*, sunk by the 'N' Class destroyer (q.v.) HMAS *Nestor* off Gibraltar on 15 December 1941, and the Japanese submarine *RO 33* sunk by the Tribal Class destroyer (q.v.) HMAS *Arunta* off Papua on 29 August 1942. In 1948 the anti-submarine and torpedo schools were combined as the Torpedo and Anti-Submarine Branch (TAS) and in 1956 moved to HMAS *Watson* (q.v.) in Sydney, now the Naval Warfare Centre. The emphasis on ASW after 1945 came about because of Australia's reliance on seaborne trade and American emphasis on the Soviet submarine threat. In the 1950s four 'Q' Class destroyers (q.v.) were converted to 'Q' Class anti-submarine frigates (q.v.) armed with Limbo ASW mortars (q.v.). Over its service life the aircraft-carrier HMAS *Melbourne* (q.v.) used Fairey Gannet and Grumman Tracker aircraft and Westland Wessex and Westland Sea King helicopters (qq.v.) in the ASW role. The development of the Ikara ASW missile, the Mulloka sonar and Barra sonobuoy (qq.v.), with US Navy assistance, enhanced Australian capabilities in this area, and sales of Ikara were made to a number of Western navies.

ANZAC is the acronym formed from the initial letters of the Australian and New Zealand Army Corps, the formation into which Australian and New Zealand soldiers were grouped in Egypt prior to the landing at Gallipoli in April 1915. (In many unit histories and on the battle honours of those units involved, the coming ashore at Gallipoli on the first day is referred to as 'The Landing'.) Allegedly devised by a New Zealand signaller, it was a convenient piece of telegraphese, devised for security purposes. It quickly gave its name to the cove where the men landed (Anzac Cove, or simply Anzac), and soon after to the men themselves (the Anzacs, an Anzac). Initially an Anzac was any man who had fought at Gallipoli, a distinction marked by a small brass 'A' attached to his colour patches (q.v.), but in time, although still during the war, it came to mean any Australian or New Zealand soldier. It has been suggested that the view which a man held of the war and his participation in it was connoted by whether he used the term 'Anzac', more overtly patriotic and hence identified with conservative forces, to describe himself, or the more demotic, and allegedly more democratic, 'digger' (q.v.), with its associations with manual work and the labour movement. As an adjective it appeared in a variety of forms. An Anzac button was a nail pressed into service on a pair of trousers; an Anzac overcoat was one worn with the collar turned up (presumably to ward off the elements). Anzac wafers were wartime issue hard tack (i.e. army issued biscuits), while Anzac biscuits (oats, flour, sugar, butter, golden syrup) were sufficiently imperishable to be capable of being sent to the soldiers from Australia by sea. In time, as the historian Ken Inglis has observed, they became a kind of folk food. This was especially the case after July 1916, when the term Anzac became protected by law under the War Precautions Act to prevent its exploitation for commercial purposes. There was even some suggestion during the First World War that the planned federal capital, Canberra, should be named Anzac. Anzac Day (25 April) is widely, though not universally, regarded as Australia's national day, albeit an unofficial one.

ANZAC (I), HMAS (*Marksman* Class destroyer leader). Laid down 1916, launched 1917; displacement 1310 tons; length 325 feet; beam 32 feet; armament 4 × 4-inch guns, 2 × 2-pounder pom-pom guns, 4 Lewis machine-guns, 4 × 21-inch torpedo tubes; speed 34 knots.

Transferred from the Royal Navy in 1919, *Anzac* was the first ship in the RAN to carry that name. During the Depression HMAS *Anzac* was one of only four ships, and the only destroyer, left in commission. She was paid off in 1933 and was sunk off Sydney for target practice in 1936.

ANZAC (II), HMAS see **BATTLE CLASS DESTROYERS**

ANZAC AREA see **AUSTRALIA–NEW ZEALAND AGREEMENT**

ANZAC CLASS FRIGATES (*Anzac, Arunta, Warumungu, Stuart, Parramatta, Ballarat, Toowoomba, Perth*) will replace the RAN's River Class anti-submarine frigates (q.v.) in the 1990s. Based on the German Meko 200 design, the frigates will be armed with one 127 mm gun, a Seasparrow air-defence missile system and torpedoes. A Seahawk (q.v.) helicopter will be carried for anti-submarine warfare (q.v.), and the frigates will be equipped with sonar and air-

surveillance radar. Eight frigates have been ordered and will be delivered between 1995 and 2004.

ANZAC DAY, a national public holiday to commemorate Australia's war dead and to honour its war veterans, is held on 25 April, the anniversary of the Gallipoli landing. The first anniversary of the landing in 1916 was observed very widely in Australia, where the mingled pride and grief produced by the Gallipoli campaign were still fresh. Acting Prime Minister George Pearce (q.v.) officially named 25 April Anzac Day, but took no other action. Queensland Premier T. J. Ryan took a more active role, trying unsuccessfully to organise a coordinated approach to the day by State governments, and requesting an Anzac Day message from the King. The King's message was duly sent and, despite the low-key approach to the commemoration taken by State and federal governments, there proved to be tremendous public support for remembering the Gallipoli campaign. Large crowds turned up for church services and public ceremonies, with between 60 000 and 100 000 people packing the Domain in Sydney. Recruiting officers took advantage of the crowds, and several recruiting rallies were held in Sydney during the day. The mood of the day mixed solemn commemoration with celebration, a mixture that made some people uncomfortable. Before the event, one writer in Sydney's *Truth* wondered whether it was appropriate 'To hold a picnic o'er Australia's dead'. In Egypt, however, Australian soldiers, including men who had been at Gallipoli, thought it quite appropriate to commemorate the event with both solemnity and light-heartedness. After 'a short but very dignified Service', reported Brigadier-General John Monash (q.v.), the troops spent the rest of the day in sports and entertainments, including what would be unthinkable later, 'a skit on the memorable landing'. Meanwhile in England 2000 Australian and New Zealand troops marched through London, watched by huge and enthusiastic crowds, to a service at Westminster Abbey attended by Lord Kitchener and the King and Queen. This imperial recognition of Australian troops seemed to set the seal on Anzac Day as the anniversary of Australia's 'coming of age'.

There was no royal message for Anzac Day 1917, and there would not be another one until 1921. Nevertheless, the commemoration was repeated, this time with the federal government taking a more active role. Defence minister Pearce approved parades of AIF troops in each capital city on the morning of 25 April, and similar parades were held in provincial centres. In both 1917 and 1918 Anzac Day was again used as an occasion for

Ex-servicemen of the Victorian State Police, equipped with overcoats, take part in a wet Anzac Day March in Melbourne, 25 April 1944. (AWM 140893)

recruiting and for patriotic rallying calls to the nation at large. In 1919, at W. M. Hughes's (q.v.) insistence, Australian soldiers led by Monash again marched through London on Anzac Day. Also in 1919 Australians, along with people throughout the British Empire, observed the two minutes' silence on Armistice Day (11 November), but Armistice Day and Remembrance Sunday would never challenge Anzac Day's pre-eminence as Australia's national day of commemoration. Nor was serious consideration given after the war to assessing which of the AIF's battles was most worthy of commemoration, though Pearce had suggested this in 1916. By 1919 the importance of Anzac Day was almost universally recognised, and pressure mounted for it to be made a public holiday. Holidays were essentially a State matter, with the Commonwealth's powers limited to declaring holidays for its own public servants and for the territories. Uniform national commemoration was thus held up by debates within each State about the most appropriate form of commemoration. Some people preferred an observance on the Sunday nearest to 25 April, fearing that declaring a public holiday invited frivolity rather than solemnity. Employers worried

about the disruption to business and that they might be made to provide a paid holiday for their employees; workers worried that they might lose a day's wages unless a paid holiday was provided for in their award.

The Returned Sailors' and Soldiers' Imperial League of Australia (see RSL) adopted a resolution in 1918 calling for 25 April to be made a public holiday throughout Australia, and it lobbied governments to enact the necessary legislation. The first to do so was the government of Western Australia, which in 1919 gazetted 25 April as a public service and bank holiday, and from 1920 closed shops on Anzac Day as well. In 1920 Anzac Day fell on a Sunday, but in 1921 all States except New South Wales declared a public holiday. Later in 1921 the Queensland government made 25 April a holiday, and went further than Western Australia by ordering that liquor stores be closed and racing banned on the day (Western Australia followed suit in 1923). At the 1921 Premiers' Conference, Prime Minister Hughes gained agreement for the observance of 25 April as a uniform national holiday. Anzac Day became a statutory holiday in South Australia in 1922, but despite continuing pressure from the RSL (which now wanted a ban on trading, industry and gambling on 25 April until one o'clock) it took several more years for the other States to come into line. The New South Wales government legislated in 1924 to make Anzac Day a bank holiday and in 1925 it followed the policy of Western Australia, South Australia and Queensland by providing for a holiday on the Monday when 25 April fell on a Sunday. The tenth anniversary of the landing, in 1925, saw the most impressive Anzac Day yet, as for the first time parades of returned servicemen were held in all capital cities. It was also the year in which the Victorian and Tasmanian governments, after extensive public debate, enacted Anzac Day legislation. Victoria's law provided for a public and bank holiday, with a Monday holiday in lieu of Sunday as in the other States. With newspapers, shops, factories, hotels, theatres, cinemas and racetracks closed, Victoria's was the strictest observance in the nation until 1930, when Queensland added the closure of shops, factories, theatres and newspapers to its existing restrictions. In Tasmania, a 1925 law provided for the closure of shops on Anzac Day, and this was followed in 1927 by the declaration of Anzac Day as a statutory holiday.

By 1926 General Monash could say with satisfaction that Anzac Day had 'grown year by year from small beginnings to a mighty solemnisation'. In 1926 Anzac Day fell on a Sunday, so 1927 was the first year in which all the States observed some form of public holiday together on Anzac Day. The national holiday proved to be a tremendous success, helped along by the presence of the Duke and Duchess of York, in Australia to open the new Parliament House in Canberra. They attended the Anzac Day ceremony in the old capital, Melbourne, where 28 000 men marched, compared to 15 000 the previous year and 7000 in 1925. Sydney also saw a larger parade than usual, but even so it could only muster some 4000 marchers; apparently Melbourne's prestige as national capital still counted for something. Sydney did, however, manage a brilliant innovation in 1927: the first dawn service was held at the cenotaph (q.v.), and was quickly added to the program around the nation. In the years that followed the power of Anzac Day was undiminished, and by 1938 over 500 000 people were watching more than 43 000 marchers in the Sydney parade. After the Second World War there was no thought of choosing a new commemorative day relating to that war. Instead, Anzac Day became a day for the commemoration of all wars in which Australia has taken part, and men who had fought in the Second World War joined in the Anzac Day parades. The same would happen after later wars.

By the 1930s the main components of the Anzac Day ritual were well established, and have remained largely unchanged up to the present. The rules of Anzac Day are unwritten, having developed by common agreement over time, with the RSL acting as the enforcer of correct procedure. Tom Griffiths wrote in 1982 of his participation in an Anzac Day ceremony in a small Victorian country town:

> Some Right Way of Doing Things seemed to hang over the R.S.L. member's confrontation with the ritual of Anzac Day. There was no definition of who decides the Right Way, but there was a definite groping towards its fulfilment, whatever it was. Questions were continually asked about how things should be done, and the answers precipitated through the assertion of a vague group memory.

As it was in 1916, Anzac Day is still observed with a mixture of solemnity, celebration and festiveness. The intensity of emotion that the day stirs up varies widely, from those who treat the day as simply another holiday, through the casual but sympathetic onlookers, the bereaved, and the veterans themselves, who have really made the day their own. For the veterans, the day begins before dawn, when they gather in small groups, often sharing a tot of rum before marching to the local war memorial (q.v.) for the dawn service. In the capital cities dawn services take place at many suburban memorials in addition to the main ceremony organised by the State or territory branch of the RSL at the city's main cenotaph or memorial. The service, its timing intended to recall the dawn landing at Gallipoli, is a solemn ceremony of mourning and reflection. It is first and foremost a time for veterans to remember their dead comrades, but is usually attended by other members of the community as well. Dawn services generally include prayers, hymns, a brief address, the playing of the Last Post, the observance of one minute's silence, and the recitation of the last verse of Laurence Binyon's 'For the Fallen', adopted by the RSL as its ode:

> They shall grow not old, as we that are left grow old:
> Age shall not weary them, nor the years contemn.
> At the going down of the sun and in the morning
> We will remember them.

The ceremony concludes with the playing of Reveille (q.v.), and is followed by the laying of wreaths against the memorial. In some places a Field of Remembrance is created by planting crosses, each representing a local person who died in war, in the ground near the memorial. After the service veterans, friends and relatives often share a breakfast as the mood lightens in preparation for the march.

After breakfast, veterans assemble at the starting point of the march, the men dressed in suits and wearing their medals, while spectators gather along the march route. In large towns and cities, veterans march with the other members of their former units, and units are preceded by identifying banners, decorated with the units' colour patches and battle honours (qq.v.). In addition to the Australian veterans themselves, other groups which march in the parades may include cadet units, serving members of the armed forces, bands, legatees (see LEGACY), boy scouts, girl guides, Red Cross members, and people from other countries who served in Allied forces during the wars. The ostensible purpose of the march is to bring veterans together for a commemorative ceremony at the war memorial, but for many veterans the march itself is the highlight of their day. It is their moment of glory, the crowds of onlookers an acknowledgment of the veterans' importance. It is a dignified, but not a solemn event, as its focus is not those who died but those who survived the wars. At the conclusion of the march some marchers and onlookers disperse, while others remain for the commemorative service. This is similar in content to the dawn service — hymns, prayers, one minute's silence, the ode, and usually a somewhat longer address. Speeches at Anzac Day ceremonies are generally bland and conservative in a very general sense, stressing the debt the nation owes to those who served and died in war, the virtues of the country for which they fought, and the need to inculcate the spirit of Anzac in the younger generation. At times, however, Anzac Day speeches have been used to make more specific political points about avoiding class conflict, protecting the nation from communism, preserving the flag and the monarchy, and maintaining defence preparedness.

Following the ceremony some veterans make their way to hotels, where they join those of their mates who left after the march for boisterous reunions. Here stories are swapped, dead friends recalled, and old songs sung. These afternoon sessions have also become associated with illegal gambling on games of 'two-up' or 'swy', rowdiness and public drunkenness, to all of which the police and public have tended to turn a blind eye. This aspect of the day was, however, an easy target for critics of Anzac Day in the 1950s and 1960s. It is seized on by the character Hughie in Alan Seymour's play *The One Day of the Year* (1960) as a symbol of the hypocrisy of Anzac Day: 'Yeah, it's a lot of old has-beens getting up in the local RSL and saying, "Well, boys, you all know what we're here for, we're here to honour our mates who didn't come back." And they all feel sad and have another six or seven beers.' Catharsis is perhaps a more apt description of this behaviour than hypocrisy, however. After a day in which their emotions have run the gamut from grief to elation, the afternoon's revel is probably for some veterans a final letting off of steam before they return, contented, to their homes and families.

Controversy over the appropriateness of drunkenness and larrikinism on Anzac Day is just one example of contestation about the form, meaning and purpose of the day. Many of the debates that have occurred over the years relate to tensions and contradictions that are inherent in the day. Is it primarily a day of mourning or of celebration? Does it belong to the veterans, to the bereaved, or to the nation as a whole? Drinking, gambling and general frivolity, for example, are perhaps out of place on a day of mourning; certainly, representatives of the bereaved have tended to think so. Soldiers' Mothers' Associations were at the forefront of the campaign for Anzac Day to be a 'closed' holiday, on which hotels, racetracks and other places of public entertainment would be closed. They got their way in Queensland, Victoria and Western Australia, so Anzac Day was less associated with public drunkenness in those states (though veterans could still organise private parties). This form of observance sat ill with many returned men, however, and RSL leaders who supported the closed holiday sometimes found themselves at odds with their members. Men like the 'mere digger' who wrote to the *Daily Telegraph* in 1922 'to register a kick against petticoat control of war anniversaries' resented being told what to do by church, temperance and women's organisations, and preferred 'to celebrate epic deeds' rather than holding 'solemn grief-reviving' ceremonies. In Victoria the controversy resurfaced in 1959, when the government decided to end decades of grumbling by polling RSL members to see if they wanted the day to remain closed (this was itself an interesting reflection of the status of the RSL). The War Widows and Widowed Mothers' Association resolved that Anzac Day should remain closed, but those RSL members who voted supported change by a wide margin, and in 1960 the government relaxed the restrictions in accordance with their wishes.

Religion has also been a regular subject of Anzac Day controversies. Many of these controversies have been the result of the uneasy position of the Catholic minority in Protestant-dominated Australia. Traditional Irish Anglophobia made some Australian Catholics sceptical about the war and the Anzac legend from the start, but probably more typical was the attitude of the Catholic *Freeman's Journal* which in 1916 welcomed Anzac Day as the start of a truly national tradition which had saved Australia from being merely 'a joint in the tail of a great Empire'. Catholic concerns about Anzac Day were, therefore, more of a religious than a political nature. Initially, Catholics faced the problem that their priests were forbidden to conduct a mass for the souls of the dead on April 25, as this date was already the feast day of St Mark, but this obstacle was overcome when the Pope gave permission for Anzac Day masses in time for Anzac Day 1923. A more lasting problem, however, was that Catholics were not allowed by their church to participate in any kind of religious service other than their own. As a result, many Catholics dropped out of Anzac Day parades before they reached the war memorials, so as to avoid participation in commemorative services which included prayers, hymns and addresses by Protestant clergy. Concerned to maximise participation in all parts of the Anzac Day commemoration, the RSL tried on several occasions, with varying degrees of success, to institute services that would be free of Christian references and clerical involvement. Not surprisingly, such attempts to secularise the public ceremonies met with approval from Catholics, but were strongly condemned by Protestant clergy. This conflict only began to be resolved in the 1960s, when the Archbishop of Sydney, Cardinal Gilroy, allowed Catholics to attend the main service on the understanding that any clergy who spoke at the ceremony would not say anything religious.

To many Protestant clergy, attempts to remove religion from Anzac Day were a disturbing sign that the RSL considered the Anzac tradition to be more important than religion, or perhaps even a substitute for religion. This concern emerged again in 1965, when Anzac Day fell on a Sunday. On previous occasions when this had happened, the RSL had held the Sydney march in the afternoon so that it would not compete with ordinary church services. In 1965, however, it decided, without consulting or informing the churches, to hold the march in the morning. Many Protestant church leaders were outraged, and called on Christian veterans not to march. The result was a heated debate in which leading roles were taken by the Reverend Alan Walker and Sir William Yeo, State president of the RSL. To Walker there was 'a sharp choice between the worship of the state and its traditions and the worship of God ... The churches will supply a more worthy remembrance of the day than the pagan atmosphere engendered by the R.S.L.' Yeo responded with the provocative statement that 'Anzac Day is bigger than churches and individuals'. As in the argument about whether Anzac Day should be a closed holiday, there was a clear differ-

ence of opinion about who owned Anzac Day: the RSL believed it was their day, and there was no reason to consult anyone else about it.

Because RSL leaders have tended to see Anzac Day as the exclusive property of their members, they have jealously guarded the right to take part in Anzac Day ceremonies. In the 1930s, for example, women were banned from attending the dawn service at Melbourne's Shrine of Remembrance (q.v.), on the grounds that 'This is purely a soldier's ceremony, reminiscent of the dawn vigils kept by the men before their hop-over', as the RSL State secretary put it. A similar ban was initially placed by the RSL in 1935 on women being present on the official stand during the Anzac Day march in Melbourne. This ban, imposed 'because of the general feeling among the men that it is their march', was immediately undermined by the presence on the stand of a woman who was there 'not as a woman, but as head of an army unit': Grace Wilson, head of the Australian Army Nursing Service (q.v.). Ten years later the ban on women at the dawn service was also breached by a detachment of the Women's Auxiliary Australian Air Force (q.v.). With the creation of the women's services the idea that it was for men to march and for women to mourn began to break down, but some parts of the day, particularly the dawn service and the boozy reunions of the afternoon, remained very much the province of men. Women were still more often found on the sidelines or behind the scenes (making and serving breakfasts, hot drinks and dinners for the men) than centre stage.

By the 1980s the scope of Anzac Day had expanded considerably to include not only ex-servicewomen but also, for example, former members of the Australian Women's Land Army (q.v.) and of the forces of Australia's wartime allies. Efforts have also been made to give greater prominence to Vietnam veterans, some of whom had felt marginalised at earlier Anzac Day ceremonies. One issue on which the RSL has remained firm, however, is that participants must march with their former units, not under banners that proclaim other identities. Attempts to assert another identity in addition to that of veteran were rare on Anzac Day before the 1980s, but following the development of 'identity politics' in the 1960s and 1970s groups such as the Gay Ex-Servicemen's Association and the National Aboriginal and Islander Ex-Service Association have sought to march in Anzac Day parades. Permission for such groups to march as separate contingents has been denied by the RSL, which claims that the march would be destroyed if 'splinter groups' marched under their own banners.

Another vexed question over the years has been whether Anzac Day glorifies war. In the 1920s, the labour movement charged Anzac Day organisers with promoting militarism and imperialism, and similar accusations were levelled by student activists in the 1960s. By the 1980s, feminists had expanded the agenda, not only condemning what they saw as Anzac Day's glorification of war, but also arguing that Anzac Day ceremonies neglected the suffering of women in war and promoted a national identity that excluded women. The basis for these attacks on Anzac Day is the confusion between the day's multiple purposes. If the day were simply an occasion for mourning the war dead, it would be hard to argue that it glorifies war. But because to many people Anzac Day is the anniversary of the day Australia became a nation, and is therefore as much a day for celebrating Australia's wartime achievements as for mourning the dead, it is not a day with which people who question Australia's involvement in wars can easily feel comfortable. In 1924 General Monash proclaimed that 'Mourning should not dominate the day; the keynote should be a nation's pride in the accomplishments of its sons. The day should be one of rejoicing'. Little wonder, then, that a 1927 Anzac Day poem in *Labour Call* railed against those who tell 'The strange, insidious lie, / That nationhood is born where men / In bitter warfare die.' The task of those who wish to criticise the Anzac legend's coupling of war, manhood and nationhood is, however, made more difficult by the nature of Anzac Day. Because the public ceremonies of Anzac Day are represented as expressions of people's private memories, any challenge to the day or to its associated myths can be seen as a violation of these very personal memories.

This conflation of the public and the private on Anzac Day has meant that protests on the day have been rare. When, in 1966, a group of women protested against conscription (q.v.) for Vietnam by making an unauthorised entry into the Anzac Day ceremony at Melbourne's Shrine of Remembrance to lay a wreath of mourning, this was seen as a desecration. This protest would, however, seem mild by comparison with the protests of women two decades later. Starting in the late 1970s, feminists began marching on Anzac Day in major Australian cities to highlight the suffering of women raped in war. These protests continued throughout most of the 1980s, and soon objections to the perceived militarism of Anzac Day and its celebration of military manhood were added to the protesters' initial focus on rape. Their tactics varied, from confrontation, to attempts to participate in the ceremonies, to the holding of separate, non-confrontational marches and ceremonies. These differing tactics reflected divisions within the feminist movement between those who wanted to reclaim Anzac Day

as a day of mourning and those who sought the abolition of Anzac Day. Whatever their tactics, these protests produced outrage among other Anzac Day marchers and spectators. The women were accused by the president of the Vietnam Veterans' Association of Australia (q.v.) of committing 'an act of symbolic rape' against Anzac Day, and were assailed with often viciously misogynistic comments from the crowds. By the end of the decade most feminists seem to have decided that Anzac Day protests were counter-productive, gaining them only negative publicity, and Anzac Day has been largely free of protest since then.

By the 1990s Anzac Day was looking popular and well attended. In the 1960s newspapers had pointed out that crowds were dwindling, but since then both Anzac Day and the Anzac legend have experienced a remarkable revival. By expanding the definition of those who can take part in the march, the RSL has ensured that Anzac Day parades remain large and has made them more representative of an increasingly diverse Australian society. At the same time, the crowds of spectators have been growing and young people are much more in evidence on Anzac Day than in the 1960s. The redefinition of the Anzac legend which took place in the 1980s has made Anzac Day more than ever an occasion for the expression of national pride, though it is a pride still mixed with deep sorrow. Periodic suggestions that 25 April replace 26 January as Australia's official national day have so far come to nothing, but for many people Anzac Day is, de facto, already the national day. A number of special events that took place in the 1980s and 1990s, including the Welcome Home march for Vietnam veterans in 1987, the pilgrimages to Gallipoli in 1985 and 1990, and the entombment of the Unknown Australian Soldier (q.v.) in 1993, have also helped to spark renewed interest in Anzac Day. Even as the men of the First World War, whose actions Anzac Day was originally intended to commemorate, pass from life into memory, others are taking their places, while more and more people are echoing Laurence Binyon's words: 'We will remember them'.

ANZAC LEGEND The terms Anzac legend, Anzac myth and Anzac tradition are widely used, not only in academic writing but also in journalistic and popular accounts of Australian history. Yet, as with the bush myth of Russel Ward's *The Australian Legend* (1958), the content and meaning of this legend are extremely difficult to pin down. It is a legend which deals with the qualities of Australian fighting men, but also one which is centrally concerned with Australian nationhood. The roots of the legend, then, lie in early twentieth-century ideas about Australian nationality and national character.

By the time of Federation Australia was already highly urbanised, but this did not stop many Australians from believing that the qualities of the 'typical Australian' were to be found in the bush. Bushmen were supposed to be tough, resourceful, practical, sceptical, anti-authoritarian and loyal to their mates, and many city-dwelling Australian men liked to imagine that these were qualities which they shared with their brothers in the bush. With the possible exception of anti-authoritarianism, they were also the characteristics that were thought to make good soldiers. C. E. W. Bean (q.v.), who was to become one of the most influential promoters of the idea that Australian soldiers owed their fighting prowess to their bush background, was writing as early as 1907 that, 'The Australian is always fighting something ... All this fighting with men and with nature, fierce as any warfare, has made of the Australian as fine a fighting man as exists.' Others, however, worried that the British 'racial stock' might have degenerated as a result of Australia's temperate climate and peaceful history (the memory of the war against the Aborigines having been by now almost erased). Whatever their beliefs about 'Australian character', however, most commentators agreed that the best test of this character would be war.

War was seen both as a proving-ground for national character and as an experience that all nations must go through to forge a true sense of national identity and to prove themselves in the eyes of the world. In 1896 the poet Henry Lawson, having abandoned his youthful enthusiasm for regeneration through socialist revolution, looked forward to the day when Australasia's star would rise 'in the lurid clouds of war . . . / For ever the nations rose in storm, to rot in a deadly peace'. Lawson's sentiments may have been influenced by the growing belief in the redemptive power of war among European intellectuals, who feared that a prolonged peace produced enervation and decadence, but it was also a view strongly influenced by Australian conditions. Half a world away from Britain, Australians lived in fear of invasion by European or Asian powers (see COLONIAL WAR SCARES; JAPANESE THREAT). Their sense of vulnerability was increased by Australia's small and militarily inexperienced population. Hence the desire for Australian men to gain experience in war, but hence also the wish to believe that Australians, through a combination of racial superiority and positive environmental influence, were naturally good fighters. There was also a strong element of Social Darwinism in these beliefs, a feeling that,

with no experience of war, Australia had no history worth speaking of, because it was widely thought that war was the forging ground of nationhood. If Australia was to mark its place in the world it could do so most simply and forcibly by its military achievements. As the French writer Alexis de Tocqueville had remarked many years before, military greatness was the most pleasing achievement to the mind of a democratic people, as it was 'a greatness of vivid and sudden lustre, obtained without toil, by nothing but risk of life'.

There was some hope that the Boer War, coinciding as it did with the movement towards Federation, would be Australia's 'baptism of fire', and, indeed, there was a certain amount of self-congratulation during and after the war about the performance of Australian troops. In particular, the war gave heart to those who believed that natural ability and skills acquired in the bush counted for more than formal military training. However, while the war in South Africa gave a foretaste of the spirit in which Australians would react to news of the Gallipoli landing, it did not adequately meet the perceived need for Australian character and nationality to be tried in war. Australian soldiers fought as part of larger British formations, so they were unable to prove themselves fighting on their own; furthermore, the nature of the war, in which the might of the British Empire was brought to bear on a relatively small Boer Army (and, increasingly, on civilians) was very different from the glorious struggle many Australians expected. With the start of a much larger war in 1914, many hoped that Australian soldiers would finally get a chance to show their mettle. The *Sydney Morning Herald* welcomed the war as 'our baptism of fire' in which 'the discipline will help us find ourselves. It will test our manhood and our womanhood.' Little would subsequently be heard about womanhood, however; it was men who were primarily responsible for defending nation and empire, and men's deeds in battle that brought glory to their nation. Whether or not Australian men would live up to the high expectations of them remained to be seen until finally, on 25 April 1915, the landing at Gallipoli tested them in battle.

Although no one person can be credited with the creation of the Anzac legend, one man clearly played a key role at the outset: the British war correspondent Ellis Ashmead-Bartlett. His account of the landing, the first to reach the world's press, was published in Australia on 8 May, and could hardly have been better designed to appeal to an Australian audience. It was a thrilling description, full of stirring phrases which told Australians that their men had performed just as they expected: 'Not waiting for orders, ... they sprang into the sea'; 'facing an almost perpendicular cliff ... those colonials, practical above all else, went about it in a practical way'; 'this race of athletes proceeded to scale the cliffs without responding to the enemy's fire'. Afterwards, the wounded Australians 'were happy because they knew that they had been tried for the first time and not found wanting', an assessment that Ashmead-Bartlett wholeheartedly endorsed. 'There has been no finer feat in this war', he concluded. 'These raw colonial troops in these desperate hours proved worthy to fight side by side with the heroes of Mons, the Aisne, Ypres, and Neuve Chapelle.' Not only was this just what Australians wanted to hear, but it came from the perfect source. As an Englishman, Ashmead-Bartlett could be assumed to be authoritative; but in addition, with his concluding remarks he conferred on Australians the British approval they sought. In the years that followed much more praise would be heaped on Australian soldiers by British journalists, generals and politicians, and all would be eagerly lapped up by Australians at home, but arguably no subsequent comment would have as much impact as the words of Ashmead-Bartlett's despatch. To one Australian soldier it was 'fairly certain that future historians will teach that Australia was discovered not by Captain Cook, but by Mr. Ashmead-Bartlett, war correspondent'.

Ironically, Ashmead-Bartlett became increasingly disillusioned with the Dardanelles campaign, but on his speaking tour of Australia in early 1916 he was careful once again to lavish praise on the Australian troops, for by then his interpretation of the landing had been taken up in Australia with enthusiasm. The soldiers at Gallipoli themselves commented on the importance of their achievement for Australia, and at the end of 1915 an Australian officer looked back on the past year as 'Australia's entry into the Company of nations — no finer entry in all history ... to have leapt into Nationhood, Brotherhood and Sacrifice at one bound ... what a year: — never can Australia see its like again.' By October 1916 the Sydney *Truth* could say with confidence that '"Brave as an Anzac" has become a household word'. Politicians and others with a cause or a product to promote rushed to associate themselves with the developing Anzac legend, though unauthorised use of the word 'Anzac' itself was prohibited by law. During the war the legend was promoted by pro-conscriptionists and recruiters, as the Deputy Chief Censor admitted in August 1915: 'There is no better way of stimulating recruiting than the publication of spirit-stirring stories, fresh and unconventional, of the gallant lads now fighting at Gallipoli.' Journalists

could generally be relied on to write such stories, as could popular fiction writers such as C. J. Dennis, whose best-selling book *The Moods of Ginger Mick* (1916) tells the story in verse of an urban larrikin who redeems himself by dying at Gallipoli, having first learnt that 'Pride o' Race' is more important than 'pride o' class'. Another important vehicle for disseminating the legend was *The Anzac Book*, a souvenir of the Gallipoli campaign which had sold 100 000 copies by September 1916. Written by Australian soldiers at Gallipoli, it was carefully edited by Bean to exclude references to danger and personal bereavement, as well as material that might show the Anzacs in a bad light. Some of these various types of war writing (though only those with the highest possible moral tone) also appeared in publications for schools, where children read Bean's and Ashmead-Bartlett's despatches from Gallipoli along with patriotic poems and stories of military heroism. Finally, from 1916 Anzac Day speeches played a crucial role in defining and reinforcing the Anzac legend.

As the war progressed the deeds of Australians on the Western Front were incorporated into the legend, though the Gallipoli landing retained its pre-eminence. By the end of the First World War the rhetoric of the Anzac legend was already highly conventionalised, and it remained so throughout the interwar period. Certainly there was room for differing interpretations, particularly between those who took a conventionally patriotic view of the cause for which Australian soldiers had died and those who preferred to believe that they had died to bring peace to the world. But in all the mass of writing and speeches about Australia's part in the war, two principal themes emerge. One concerns the character of Australian servicemen, the other deals with the importance of the war for Australia's sense of nationhood.

At the end of the war Bean remarked that 'the big thing in the war for Australia was the discovery of the character of Australian men', and one year before the start of the Second World War he wrote that one of the main questions he sought to answer in his near-completed official history was that of how the Australian character came through the test of war. Bean's answer, and the answer of almost everyone else who addressed this topic, was that Australian character had passed the test with flying colours. The most typical representative of Australian character, to this way of thinking, was the private soldier who had volunteered for the duration of the war only. He (for the representative of Australian character was most emphatically male) was less likely to be thought of as the sailor, the airman, the officer, the regular soldier (or, in later wars, the conscript), though such people could also be seen as displaying the characteristics of the 'typical' Anzac. The stereotype of the Australian soldier which developed during the war has been well outlined by the historian Lloyd Robson:

> These men reflected the egalitarian colonial origins of Australia and were direct and straightforward in their dealings with each other, and contemptuous of lesser breeds; they could and did fight like threshing machines when they had to; ... they showed up all other soldiers and especially the British to be lacking in initiative and go; they revealed that they were rather undisciplined when that discipline was merely a formality, but really needed no controlling when it came to the deadly business of battle — then they became highly effective, skilful and feared killers; they were a classless army; they stuck to their mates through thick and thin; their burden as soldiers was lightened by a sardonic sense of humour ... and in their ranks abounded many wags and tough nuts who made it a rule always to outwit the authorities; they did not give a damn for anyone on earth, in heaven or in hell. Their highly distinctive tunics and hats were perhaps never cleaned and brushed as they might have been ... ; they had a penchant for removing objects of value left in their way and were expert con men; their contempt for 'Gyppos' was notorious; though they at first hated the Turks ... very soon they developed a respect for Johnny Turk; their attitude to the German soldier was not one of hate but of respect ... The stereotypic Australian soldier was very tall and sinewy and hatchet-faced. He had a great respect for the institutions of the 'old country' and what he perceived as its quaintness, but little time for pommy officers and men as a rule, or until they proved themselves manly. He got on well with the Scots. The stereotype was not formally religious but had a lot of time for the Salvation Army and some of the 'fighting' padres.

These characteristics are strikingly similar to the stereotypical qualities of the Australian bushman and, not surprisingly, the bush influence was a common explanation for the supposedly special abilities of Australian men as fighters. This was a favourite explanation of Bean's, though as time went on he de-emphasised the bush element and stressed instead the relative egalitarianism of the social order in Australia, which he believed was responsible for Australian independence, resourcefulness and initiative. Others linked the Australians' success in battle to their racial heritage, Joseph Cook rejoicing in federal parliament that 'there is no indication of decadence' in these transplanted 'boys of the bull-dog breed'. But whatever their preferred explanation, most commentators looked to the character of Australian men, formed before they went to war, as the reason for their soldierly ability. There was a marked reluctance to look instead, or as well, at the importance of military discipline, training, experience and leadership.

The reason why there was so much interest in the character and performance of Australian soldiers is that they were seen as representing the nation as a whole. Defending the nation was the prime duty of the male citizen, just as giving birth to and nurturing the nation's children was the prime duty of the female citizen. But war, not childbirth, was the supreme test of citizenship, as it showed the willingness of the citizen to sacrifice himself for the nation. Thus, the soldier, the citizen in uniform, was not only proving his own manhood in battle but also showing that the nation as a whole had 'stepped into the world-wide arena in the full stature of great manhood' as the Melbourne *Argus* put it in 1915. Hence also the stress on the egalitarian nature of the AIF, and on the fact that it included Australian men of all classes and backgrounds, for only if there was equality of sacrifice could the army be seen as representing the nation. In this way, the concern with the character of Australian soldiers and the concern with proving Australian nationhood came together.

As soon as news of the Gallipoli landing reached Australia, people began saying that it had made Australia a nation. The rhetoric on this point was not always consistent, however. Sometimes it was said that Australia had literally become a nation on 25 April 1915; on other occasions it was rather that 'Australia had sprung into a realisation of nationhood' or that 'the consciousness of Australian nationhood was born', as Bean put it, suggesting that Australians simply became aware of an already existing nationhood. Sometimes the Gallipoli landing had given birth to the nation, at other times it was said to have brought it to maturity or manhood. To later generations there would also seem to be a contradiction between the Australian and the imperial elements of the legend, but few Australians would have perceived such a contradiction before the 1960s. Australians not only relied on Britain for their economic and military security, but also relied on notions of Britishness for their sense of who they were. To be Australian was to be a member of the British race, and because the Anzac tradition celebrated the participation of an Australian force in the larger British war effort it could simultaneously be a focus for Australian nationalism and for Empire loyalty. Thus, for example, the Melbourne *Age* could write in 1926 that Australia emerged 'into full bloom of nationhood' at Gallipoli where 'its men had proved themselves worthy of the highest traditions of the British race'. Some said that Australian history had started at Gallipoli, but it had done so because for the first time Australians had played an important role in the history of Europe and the British Empire. On 25 April, according to New South Wales Director of Education Peter Board, Australia 'became an active partner in a worldwide Empire … On 25 April history and Australia's history were fused, and fused at white heat. Never again can the history of this continent of ours stand detached from World history.'

Two motifs occur repeatedly in the rhetoric about Anzac and nationhood: the 'baptism of fire', or the test, and the 'baptism of blood', or the sacrifice. The baptism of fire motif appeared whenever people wrote or spoke of the achievements of Australian soldiers, of their having successfully passed through the ordeal of trial by battle. At Gallipoli, it was said, 'with the fierce search-light of every nation turned upon it, our representative manhood showed no faltering' (R. H. Knyvett, *'Over There' with the Australians*, 1918) or 'the courage, strength and endurance of our southern manhood was put to a supreme test and did not fail' (Brisbane *Courier*, 1916). As a result, Australians had taken their place among the nations, engraved their name on the map, written their name in blood, received their charter of nationality, and so on. The idea of the baptism of blood, on the other hand, relies on an analogy with the Christian doctrine of sacrifice, resurrection and redemption. Just as Christ died to redeem humanity, so Australian soldiers died to redeem or give life to the nation; and because they were all volunteers, they had truly *given* their lives. 'If the nation is to be born,' said the *Sydney Morning Herald* in 1922, 'if the nation is to live, someone must die for it.' The proximity of Anzac Day to Easter reinforced this theme, allowing churchmen like the Reverend Irving Benson to expound (as late as 1960) on the 'spiritual affinity' of the two commemorations, 'for the Crucifixion and the Resurrection speak to us of the success of failure'. The Anzacs' baptism of blood, he continued, 'was to weld us into a nation'. The analogy could be taken still further, since just as Christ rose again from the dead, so too Australia's war dead lived on, whether it be 'in the hearts of men', in 'the deathless name of Anzac', or in the continuing life of the nation they had died to save. With all the talk of birth and sacrifice, it is not surprising that some, like the Reverend Henry Glazier in 1921, also drew a comparison with the sacrifices made by women: 'Life becomes fruitful only as it becomes sacrificial … The highest dignity of womanhood is only reached through the sacrifices and dangers of motherhood … Nations cannot be saved without sacrifice.' He did not need to say that the duty of saving the nation was men's; and if women reached their highest dignity giving birth to humans who

were all too mortal, how much greater was the dignity of those men who gave birth to the immortal nation.

Between the two world wars the cult of Anzac became firmly entrenched as a national tradition, with Anzac Day as its focal point and the Returned Sailors' and Soldiers' Imperial League of Australia as its guardian. It was overwhelmingly a conservative tradition; though there were attempts by the left to claim the tradition (even the Communist *Worker's Voice*, during the popular front period, could declare that Australians fighting for the Spanish Republic had 'more than upheld the glorious traditions of the Anzacs at Gallipoli') these failed to break the association between Anzac and conservatism in the minds of most Australians. Where once labour propagandists had claimed that socialism meant being mates, now Anzac mateship meant being anti-socialist. Although many returned servicemen must have resumed their union membership and their involvement in labour politics, the organised ex-service movement was often hostile to class politics. Such a stance was supported by returned men like the one who wrote to the *Argus* in 1919 that 'we who fought, squatter and station hand, as privates shoulder to shoulder, think of each other in terms of digger and not in terms of classes'. Certainly there was an element of self-interest when returned soldier and successful grazier Frederic Hinton deplored in 1932 the 'ugly spectre of class hatred' which could be overcome by reviving the 'spirit of Anzac … the spirit of sacrifice'. But there was more to this philosophy than simply the defence of entrenched privilege, and such views were strongly supported by many returned servicemen whose stake in the status quo was less obvious.

There are three principal reasons for the alienation of a substantial body of returned men from labour politics, and the resulting inability of the left to make convincing use of the Anzac legend. First, many men who had been through the appalling experience of war were disgusted by what they saw as the pettiness and decadence of civilian life. To such men the way to make Australia great was not to encourage further divisiveness but to seek to bring about at a national level the unity they had found in the trenches. Second, these returned men found themselves in conflict with the unions over the issue of preference in employment for ex-servicemen versus preference for unionists, and with the ALP over the memory of the war. Having lost its most belligerent members in the wartime split, Labor was uncomfortable with the old patriotic justifications for the war which many returned soldiers relied on to give meaning to their wartime experiences. The conservatives, by contrast, were quite happy to indulge in rhetoric which ranged from conventional patriotism to aggressive hyperbole (an example of the latter coming in 1921 from journalist, war historian and anti-socialist polemicist F. M. Cutlack (q.v.), who wrote that the word Anzac 'conveys something savagely masculine, ruthless, resolute, clean driven home'). Third, there were many men for whom the identity of 'digger' (q.v.) overrode all other ties or allegiances. The shared experience of war made them feel they were a class apart from the rest of society, and lionisation on their return to Australia may even have made them feel they were a class above. Consequently, such men were more concerned to secure their rights and privileges as returned servicemen than to engage in what they saw as the sordid politics of economic class.

Thus, by the time of the Second World War, the Anzac legend had largely been captured by the political right. The memory of Anzac was frequently recalled to inspire the men of the 2nd AIF, who in 1940 were called 'proud bearers of the standard bequeathed to them by the original Anzacs' by the *Sydney Morning Herald*. The actions of the Australians in this war were assimilated into the existing legend, and it was agreed that they had shown the same fighting spirit as their predecessors. The experience of Australian prisoners of war (q.v.) proved difficult to integrate into the legend, however, and it would be some time before they, too, entered the national pantheon as exemplars of the Anzac virtues of stoicism, mateship, resourcefulness and anti-authoritarianism (in this case, subtly defying Japanese authority). There was some repetition of the rhetoric about war as a test of nationhood, but this test could not have as much meaning as the last one since, as the historian Ken Inglis has written, 'You cannot be baptised, or come of age, or become a nation, twice.' Instead, the Second World War was seen as confirming what Australians had learnt about nationality and national character in the First.

Because the ALP, and eventually even the Communist Party, supported the war effort, the grip of the conservative parties on the Anzac legend loosened somewhat during the Second World War. But, if anything, the fact that the ALP had been reconciled to it only provided a broader base of support for what was still essentially a conservative legend. The legend saw service again in Korea, where Australian troops were once again depicted as chips off the Anzac block, but by the time it was pressed into service for the Vietnam War the conventional phrases no longer went unchallenged. From the late 1950s the legend began to be questioned by

social critics who associated it with what they saw as the complacency and conformism of the Menzies era. Satirist Barry Humphries gently poked fun at the conventionality of the RSL culture through the character of Sandy Stone, the returned serviceman who left his house in Gallipoli Crescent every week for 'a really nice night's entertainment' at the local RSL club. Less gentle was Alan Seymour's play *The One Day of the Year* (1960) which, though it is more a dramatisation of the generation gap than a critique of the Anzac legend, puts some stinging attacks on the legend into the mouth of the university student, Hughie. In the same year that the play was first performed, real-life university students, writing in the Sydney University newspaper *Honi Soit*, called Anzac Day 'a glorification of war to create hysteria, maudlin sentiment and jingoism'. The start of the Vietnam War and the introduction of conscription (q.v.) turned more young people against the legend, and a spray-painted message on a Melbourne war memorial declared what many radicals believed: 'Vietnam war explodes Anzac myth'. The RSL and other custodians of the legend resisted such attacks, but the authority of the received version of the Anzac legend was gradually weakened.

Meanwhile, academic historians (inspired perhaps in part by the fiftieth anniversary of the First World War and in part by the new 'debunking' histories of the war emerging from Britain) began to examine the roots of the Anzac legend. With an article entitled 'The Anzac Tradition', published in 1965, Ken Inglis began a debate about the origins of the legend and, in particular, about the role of C. E. W. Bean. It is a debate which continues today, and which has two main strands. The first strand of the debate concerns the accuracy of the assessment of Australian soldiers by Bean and others. In the words of Tony Gough, the Anzacs have emerged from revisionist studies of this topic 'not so bronzed, not so democratic, not so courageous, not so physically superior nor so well behaved as has been popularly imagined'. Military historians have revised the notion of Australian uniqueness and natural fighting ability, highlighting the similarity of Australian soldiers to those of other armies and the fact that training, command, experience, logistics and equipment were often more important than 'character'. Such revisionism has not generally sought to denigrate Australian troops, only to reduce them from superhuman to human dimensions, and to place their achievements within an appropriate context.

The second strand of debate involves the ideological roots of the legend, and the role of Bean and others in creating it. There have been studies of

The Anzac Book, the impact of Ashmead-Bartlett's despatch, the history of Anzac Day, the 'big-noting' theme in war literature, the creation of the myth of Simpson and his donkey (q.v.), and, above all, endless re-examinations of Bean's output as war correspondent and official historian. Much of this analysis has focused on the conservatism of the legend, while feminist critique has centred on the ways in which the legend enshrined men as national heroes. Such studies are not concerned with the truth or falsehood of the legend so much as with the question of why particular events and ideas are elevated to the level of national myth.

The revisionist positions in both of these debates have been challenged by historians such as Bill Gammage, John Barrett and John Robertson, who have elaborated on Bean's interpretation and approach. That passions can still be roused in this debate is apparent in John Robertson's *Anzac and Empire* (1990), in which he responds to examination of Bean's role in making the Anzac legend with the claim that 'The creators of the Anzac legend were, of course, the men [i.e. the soldiers] themselves', and goes on to remark that 'One wonders what qualifies people who have never experienced the rigours of campaigning or the terrifying savagery of battle to belittle the valour of those who have.' It is beyond dispute, in Robertson's view, that Anzac Day has pride of place in the national tradition because of 'a highly meritorious military achievement' by 'a force genuinely representative of Australia'.

Despite the undermining of the traditional Anzac legend and despite confident predictions in the 1970s of the imminent demise of Anzac Day and its associated tradition, the 1980s and 1990s have seen a resurgence in popularity of the legend, albeit partly in a new form. Both the revival of the legend and its changed meaning can be explained by the rise of Australian nationalism since the 1970s. The breakup of the British Empire, the increasing involvement of the United Kingdom in European rather than Commonwealth affairs, the massive postwar migration to Australia which has significantly decreased the proportion of the population which is of British descent, all have made Australian nationalism increasingly incompatible with loyalty to Britain. The suspicion of overseas military entanglements engendered by the Vietnam War has also forced a reassessment of Australia's military myths. In searching for a national identity which is not defined in terms of Britishness, politicians, artists, historians and others have returned to Australia's existing myths to find the origins of a truly independent nationality. Thus, shorn of its imperial elements, the Anzac legend has emerged as a much more unambiguously nationalist story, which might be summarised as follows: Australia has a history of fighting other people's wars (initially Britain's wars, later those of the United States). These wars were not necessarily in Australia's interests but, spurred on by imperial ideas, Australians enthusiastically fought in them nonetheless. In the First World War Australians fought bravely, and displayed typically Australian characteristics such as mateship, initiative, ingenuity, larrikinism and egalitarianism (in this respect the new legend is much like the old). However, Australians were let down by the British military establishment who, contemptuous of the lives of mere colonials, sacrificed them in the futile Gallipoli campaign and as shock troops on the Western Front. The war achieved nothing and, far from proving Australian nationhood, actually demonstrated Australian subservience, though Australian soldiers did learn how different they were from the British. Australia's real baptism of fire came during the Second World War when, deserted by Britain, Australians found themselves for the first time fighting to defend their country. This war was worth fighting, but unfortunately Australia swapped its dependence on Britain for a reliance on the United States, and it took the disaster of Vietnam to shock Australians into giving political expression to the independent character their soldiers had displayed in battle.

This new Anzac legend has by no means gained universal acceptance, and indeed has been criticised as grotesque, even ahistorical. It has so far been associated more with the left (using that term very broadly) than with the right, and the assumption that Australia had no interest in the outcomes of wars fought far from its shores has been attacked by many, not all of them conservatives. Some leaders of the RSL still display an attachment to Britain, and there is no doubt that this reflects the views of a proportion of their members. At the same time, the new legend is clearly appealing to a section of those Australians who came of age during or after the Vietnam era. The new interpretation of Australia's military history was promoted in two of the flagship films of the Australian film revival which began in the 1970s, *Breaker Morant* and *Gallipoli*, both of which celebrate the Australian bushman hero while taking a dim view of British military officialdom (see FILM AND TELEVISION, WAR IN AUSTRALIAN). Renewed interest in the legend was evident at the Australian War Memorial, which recorded a 50 per cent increase in visitor numbers in September 1981, the year of *Gallipoli's* release, compared with the same month in 1980. Also becoming apparent at the memorial was the way in which the search for national identity had become intertwined with the search for personal 'roots'. As genealogical research has increased in popularity, more and more people have been using the Memorial's Research Centre to find ancestors or relatives who fought in the First World War. In so doing they gain satisfaction by linking themselves to events seen as historically significant, while at the same time acquiring new reasons for perpetuating the idea that war is central to national identity. Whether the Anzac legend will continue to be the subject of vigorous debate, or whether the renewal of faith symbolised by the entombment of the Unknown Australian Soldier (q.v.) will lead to a new conformity of opinion, remains to be seen; but it is clear that the Anzac legend will remain an important national myth for some time to come.

Alistair Thomson, *Anzac Memories: Living with the Legend* (Oxford University Press, Melbourne, 1994).

ANZAC MEMORIAL, SYDNEY see **WAR MEMORIALS**

ANZUS (AUSTRALIA, NEW ZEALAND AND UNITED STATES) SECURITY TREATY came into force on 29 April 1952, and remains a cornerstone of Australian security arrangements. From the outbreak of war in the Pacific in December 1941 the establishment of a defence treaty with the United States was a prime aim of Australian diplomacy. The attempts both during and immediately after the war of H. V. Evatt, Minister for External Affairs 1941–49, to win American agreement to such an arrangement came to nought, and with the American refusal to retain Manus Island (q.v.), north-east of New Guinea, as a postwar Pacific base, America's lack of interest in the south-west Pacific was made manifest. By 1950 conditions in the Pacific had changed markedly. The United States had yet to conclude a peace treaty with Japan, the Korean War had broken out, and China had intervened in Korea. Australia responded quickly to the United Nations' call for military assistance for South Korea, and used its show of support for the United States to push again for a security arrangement. The US Secretary of State, Dean Acheson, opposed any collective security agreement other than the North Atlantic Treaty Organization, which had been established in 1949, although the United States was prepared to discuss an 'island security chain' linking Japan, the Philippines, Australia, New Zealand and the western Pacific islands in a consultative arrangement rather than a formal treaty. The United Kingdom, however, opposed even this weak system, even though it would not necessarily exclude British participation. The British feared that any Pacific-centred defence arrangement would undermine the ability of Australia and New Zealand

to fulfil their primary obligations in the Middle East in the event of a major war, and they were also concerned that an arrangement which did not include the mainland Asian states, such as Malaya, might undermine those states' will to resist communist encroachment. In negotiations with the United States, the Australian Minister for External Affairs, P. C. Spender (q.v.), countered that it was precisely such a security arrangement with the United States that would enable Australia to fulfil its obligations outside its immediate region (i.e. in the Middle East). Taking a calculated risk, Spender also argued with the US negotiator John Foster Dulles that Australia would not sign the peace treaty with Japan without a tripartite security treaty between Australia, New Zealand and the United States. As Spender himself admitted, the force of this threat should not be exaggerated, as the changing circumstances in eastern Asia made the Americans willing to accept what they had formerly rejected. The Korean War was a telling demonstration of the possibility of the Cold War exploding into direct military confrontation between the major powers. Australian regional preoccupations overlapped with the global concerns of the United States.

In the event, the ANZUS Treaty was less binding and comprehensive than Spender had wanted. The three parties pledged merely to consult if one of them was attacked, which was a far cry from Spender's intention of securing an American guarantee to defend Australia. While a Council of Ministers was established to provide a forum for political exchanges, Australian hopes for access to the US Joint Chiefs of Staff were not realised; instead military consultation was to be between the Australian and New Zealand Chiefs of Staff and the C-in-C, Pacific. Australian–American security issues were also discussed in other forums, notably the South-East Asia Treaty Organization (SEATO) (q.v.). Within these limitations, close defence ties developed between the three signatories, with Australia and New Zealand being granted 'allied' status in the area of logistics support, priority in arms delivery and, perhaps most important, in access to American military intelligence. The establishment of Joint Defence Facilities (q.v.) in Australia has been the most public manifestation of this increasingly close link. The tripartite nature of the ANZUS Treaty was formally suspended in August 1986 by the US government in response to the decision of the New Zealand Labour government to ban the entry of American warships into New Zealand ports (see NEW ZEALAND) because the US Navy refused to confirm or deny that specific ships were nuclear-powered and/or armed. Since that time the US government has excluded New Zealand from the diplomatic and defence arrangement, although in 1994 steps were taken to restore the relationship.

ARARAT, HMAS see **BATHURST CLASS MINESWEEPERS (CORVETTES)**

ARCHER, HMAS see **ATTACK CLASS PATROL BOATS**

ARDENT, HMAS see **ATTACK CLASS PATROL BOATS**

ARMED MERCHANT CRUISERS were passenger ships requisitioned and refitted for naval use. During the Second World War the *Manoora* and *Westralia*, requisitioned by the RAN, were armed with seven 6-inch guns, two 3-inch anti-aircraft guns and one Seagull V (q.v.) aircraft each. The *Kanimbla*, requisitioned by the RN, but crewed mainly by members of the Royal Australian Naval Volunteer Reserve and commanded by Captain Frank Getting (q.v.), was similarly armed but carried no aircraft. Initially the vessels were used to patrol the China Station, the Indian Ocean and the Pacific. In June 1940 HMAS *Manoora* forced the surrender of the Italian merchantman *Romolo* in the Solomon Islands. The *Kanimbla* was transferred to the RAN and all three ships were converted to Landing Ships (Infantry) by 1943. In this capacity they were involved in most of the major assault landings in the south-west Pacific, including Arawe on Bougainville, Hollandia and Aitape in the New Guinea campaign, the landing at Moratai, the Australian landings on Borneo and American landings in the Philippines. The ships were paid off in 1946 and returned to passenger use.

ARMIDALE, HMAS see **BATHURST CLASS MINE-SWEEPERS (CORVETTES)**

ARMOUR Australian troops had their first experience with armoured warfare during the First World War. Some of their most significant victories (Hamel, Amiens) were achieved with the assistance of British tanks. As a result of this experience the Australian government decided to acquire tanks of its own. Due to a number of factors, including postwar economy measures, this did not occur until 1927 when four Vickers Medium Mark II tanks were purchased in Britain and the Australian Tank Corps was established. The tanks, which with ammunition and spares cost £72 000, arrived in 1929 and were formed into the 1st Tank Section, based at the Small Arms School (forerunner of the Infantry Centre) at Randwick in Sydney.

Development of armour came to a halt with the Depression, although a low level of familiarity with armoured warfare was maintained by the regular army tank cadre, which operated in an instruc-

tional capacity. The implementation of various proposals to mechanise the cavalry were deferred and it was not until 1937 that the ageing Medium IIs were replaced by 11 Mark VIA light tanks. These were formed into the 1st Light Tank Company based at Randwick. The 2nd Light Tank Company was formed in 1939 at Caulfield, Victoria.

At the outbreak of war Australian units sent to the Middle East in 1940 were quite unprepared for armoured warfare. The only formation with any claim to be armoured was the 6th Divisional Cavalry Regiment, one squadron of which was equipped with armoured machine-gun carriers and operated in a reconnaissance role. Meanwhile the victory of the Germans in western Europe had revived interest in tanks in Australia. A decision was made to raise an armoured division as part of the 2nd AIF and to set up an Armoured Fighting Vehicle (AFV) School to train personnel for the division. The AFV School was located at Puckapunyal (q.v.), Victoria, in 1941. Major-General John Northcott (q.v.) was placed in command of the embryo division which was formed by mechanising light horse units.

The first priority was to equip the new formation with tanks, the 11 Mark VIs being the only mobile tanks in the country. Orders were immediately placed in Britain and the United States and a decision taken to design and manufacture the Sentinel tank (q.v.) in Australia. Thirteen M3 light tanks arrived in September 1941, and were soon followed by 400 more, while 140 British Matildas began arriving in July 1942. By December there were no less than 1450 tanks in the country, the majority being Matildas and light and medium M3s. Other types were to arrive in Australia during the war and various types of armoured cars and scout cars were also used.

Meanwhile the training of the 1st Australian Armoured Division was proceeding under a new GOC, Major-General H. C. H. Robertson (q.v.). Manoeuvres were held and training courses attended, although the equipping of the division took much longer. With the rapid Japanese advance and the threat to Papua New Guinea, however, it became increasingly obvious that the exigencies of the situation and the nature of the territory under threat meant that 1st Armoured Division would be unlikely to fight as a formation. This is indeed what occurred. The 1st Armoured Division was broken up in September 1943, but small groups of regular officers were sent to the Mediterranean theatre to acquire armoured warfare experience with the British Army. During the course of 1942–43 Australia attempted to raise three armoured divisions, which proved to be well beyond the nation's capa-

The first tank type issued to the Australian Army, the Vickers Medium Tank (Mark II), shown here in use with the 1st Australian Armoured Training Regiment, Puckapunyal, Victoria, October 1941. (AWM P1440/11)

bilities. Cavalry units were sent to Papua and other areas as required, with M3 tanks in operation at Buna and Gona in 1942, Matildas in the Huon Campaign in 1943 and at Wewak and Bougainville in 1944 and 1945. The final Australian tank operations of the war took place in Borneo in 1945. In general the Matilda had proved the most successful tank for jungle fighting. It was manoeuvrable in close country and its 2-pounder gun was useful in destroying Japanese concreted bunkers. During the occupation of Japan after the war the 1st Australian Armoured Car Squadron provided a light armoured capability for the British Commonwealth Occupation Force (q.v.).

The postwar period and demobilisation led to a scaling down of Australian armoured forces, the main unit in the postwar army being the 1st Armoured Regiment, formed in 1949 as part of the new Australian Regular Army. It was equipped with Churchill tanks and trained at the old AFV school at Puckapunyal. The armoured regiment had been intended to form the nucleus for a much larger force, but the introduction of National Service (see CONSCRIPTION) in 1951 and the drain of regular personnel to train CMF armoured units led to the rescheduling of these plans. With the outbreak of war in Korea in June 1950 Australian ground forces were dispatched to the peninsula as part of a Commonwealth division. No armour was included in the original offer because the Churchill tanks were obsolete. By the time these were replaced by the Centurion tanks in 1952 the armoured component of the Commonwealth Division had been established with British and Canadian units, and there was no scope for Australian participation. As in the last stages of the Second World War, however, individual Australian officers were attached to British units.

The 1950s were a slow period in the development of Australian armoured capabilities. No new equipment was purchased after the Centurions in 1952, and the 1st Armoured Regiment remained isolated (at Puckapunyal) from the rest of the army, although much effort was directed into training the continuous intakes into the CMF. Nor did matters improve in the early 1960s. A positive step was taken in allowing armoured units to venture from Puckapunyal and take part in tank/infantry training exercises. However the adoption of the Pentropic (q.v.) organisation saw the reduction of the five medium tank regiments to two. The Pentropic idea was abandoned in 1965, but only one regular army unit was equipped with medium tanks, although the 2nd and 3rd Cavalry Regiments (used in a reconnaissance and armoured personnel carrier [APC] role respectively) were also part of the regular establishment, while the CMF maintained two medium tank regiments and five cavalry regiments.

During the Vietnam War the initial armoured component of the 1st Australian Task Force was a light one, provided by APC squadrons equipped with the American-produced M113A1 APCs. The initial force of eight APCs operated in support of infantry operations and were the only Australian armour in action during 1965–67. In late 1967, however, the decision was taken to send a troop of Centurion tanks. The first of these arrived in February 1968. Eventually the armoured contingent of the task force consisted of C Squadron, 1st Armoured Regiment (Centurions) and A Squadron, 3rd Cavalry Regiment (APCs). During the next few years B Squadron and A Squadron were rotated through the task force. The tanks undertook a range of tasks — fire support and perimeter defence as well as direct assault on enemy positions. The APCs played an important role in relieving D Company of 6RAR during the Battle of Long Tan in August 1966 and acted in a similar supporting role at Binh Ba in June 1969, while the Centurions played a significant supporting role in the series of actions associated with Fire Bases Coral and Balmoral in May 1968. The Centurions were replaced in Australian service by the German Leopard tank, produced by Krauss-Maffei and modified for Australian use, from 1974. Although upgraded again in the course of their service, they are now obsolete in terms of armoured technology outside Australia's own region. The M113 APC is also obsolete despite modification and upgrading, while the army has introduced the LAV 25 wheeled armoured fighting vehicle for use in the cavalry role; the majority of these are based in northern Australia. In recent years the future role of medium or heavy armour in the defence of Australia has been questioned, although a requirement for the combination of mobility, firepower and protection which light armoured vehicles provide seems unquestioned.

ARMS CONTROL see **DISARMAMENT**

ARMY, DEPARTMENT OF THE, created on 13 November 1939, took over control of army administration and finance from the Department of Defence. It was abolished on 30 November 1973 as part of the amalgamation of the armed services departments into the Department of Defence, and its functions were taken over by the Army Office within the Defence Department.

ARMY FARMS were established from 1940 as a way of providing fresh food and avoiding the cost of transport and storage of perishable commodities. Farm companies under the Australian Army Service Corps were set up in the Northern Territory, North Queensland and New Guinea. Men of 'B' medical grade were drafted to do the work; attempts to recruit experienced farmers and Aboriginal labourers were largely unsuccessful. The efficiency and enthusiasm of army farm workers impressed visiting food technologists, but they were less impressed with the planning of production. Other problems, including labour shortages, distribution problems and poor productivity in the wet season, meant that demand could never be fully satisfied. Nevertheless, army farms in the Northern Territory were remarkably successful by the standards of Territory agriculture, reaching a peak production level of 1.7 million kilograms of vegetables and tropical fruit in 1944. Tomatoes, cabbages, beans and melons were the main vegetable products; chickens, eggs, pigs and honey the chief animal products. The farms in New Guinea, established from 1943, achieved a total production of 6.3 million kilograms of fruit and vegetables plus small quantities of other products.

ARMY LIST The first *Military List of the Commonwealth of Australia* was published in March 1904, and set out the seniority of all officers on the active list of Australia's army. The Army List has gone through various modifications since, but remains an invaluable reference source for the career details of army officers. It contains a list which gives the ranking by seniority of all officers by corps or service, and a gradation list which contains a full career profile on each. It also provides details of appointments as colonels-in-chief, colonels-commandant, and honorary colonels of regiments and corps, the battle honours of the AMF, and the alliances between Australian and British units. During the First World War there was a separate list covering the AIF, while after the Second World War there were separate volumes covering the regulars and the CMF. Often referred to as the 'stud book', it was also known as the 'wine list' when the colour of the binding was changed from blue to red in the late 1960s. Cost factors led to its appearing in a much abbreviated paperback form in the mid-1970s, and it was discontinued altogether in the 1980s.

ARMY QUARTERLY, THE, which began publication in London in 1920, provided a forum for the discussion of army matters. It consistently attracted articles from junior and middle-ranking officers, as well as contributions by famous and influential military writers including T. E. Lawrence, B. H. Liddell Hart, J. F. C. Fuller, and Admiral Sir Herbert Richmond. In the absence of any comparable Australian publication, it provided an outlet for the views of Australian officers, all the more valuable because those views, partly influenced by study at the British Army Staff College at Camberley or the Imperial Defence College (q.v.) (where Richmond was Commandant, 1927–28), often ran counter to the official centralist policy on imperial defence. In 1927, for example, after study at the IDC, Lieutenant-Colonel H. D. Wynter (q.v.) published an article in the *AQ* arguing that imperial defence meant the defence of Britain rather than the Empire, and calling for greater Australian self-reliance, especially in military, as opposed to naval, matters. This theme was continued in articles in 1933 and 1934 by Colonel J. D. Lavarack (q.v.), Commandant of RMC Duntroon (q.v.) 1933–35, and Major H. C. H. Robertson (q.v.), but they had little influence, arguing as they were for a fundamental reorientation of Australian official thinking about defence.

ARMY RESERVE (ARES) replaced the CMF as a result of recommendations made in 1974 by the Millar Committee (q.v.), which enquired into the state of the CMF and the school cadets (see Aus-TRALIAN CADET CORPS). Millar's recommendation was for more than merely a change of name, although he saw this as important in helping to change attitudes towards part-time soldiering inside and outside the regular army. Central to his report was a 'total force' or 'one army' concept, in which regulars and reservists were central to the achievement of army goals and missions. The Defence White Paper of 1976 suggested an increase in Army Reserve numbers in the order of 25 per cent, to about 25 000 by 1981. This was in fact exceeded (a ceiling of 30 000 being imposed and reached), but was not sustained for more than a couple of years. Peaking in 1983 at over 33 000, ARES strength had declined by 1985 to under 24 000, and in the early 1990s had stabilised at around 25 000. More important in some respects than overall establishment has been wastage rate. Between 1986 and 1989 nearly 50 per cent of the force was aged 19 years or younger, that is to say, with less than two years' service. In the same period, 27 per cent of members left within 12 months of enlistment, and a further 25 per cent within the second year. Officers in the ARES, on the other hand, tend to be older than their regular counterparts.

In 1964 it was decided that pay for the CMF would be untaxed (in 1972 this was extended to allowances), but this principle was broken in 1992 when the government decided that 50 per cent of pay and allowances was declarable as income. (This had such a negative impact on ARES members that it was subsequently rescinded.) Together with perceived lack of support for ARES activities among many private businesses (those in government employ are granted leave to attend training periods which fall in the working week), many aspects of ARES training and conditions have tended to militate against strong retention rates. Training is often unimaginative, equipment is not as modern as in regular units, and many of the static facilities, especially depots, are old and in need of maintenance. On the other hand, the Army Reserve has a better rate of female participation (17 per cent) than does the regular army (9.5 per cent). Enlistment age is between 17 and 35, and the term of engagement is four years (although it is possible to opt out much earlier). Officers are commissioned through officer cadet training units, and through university regiments. The latter have been criticised for the high level of resources they consume relative to the numbers produced, and for being merely a source of finance for undergraduates whose commitment to ARES service after their graduation is doubted.

There are some strikingly successful ARES units, such as the commando companies, which enjoy much higher than average recruitment and retention of ex-regular personnel, and the Regional Force Surveillance Units (Far North Queensland Regiment and NorForce), which make extensive and intelligent use of local Aboriginal and Torres Strait Islander people, and which have markedly lower wastage rates. The establishment of the Ready Reserve (RRes), which came out of the Force Structure Review in 1991, was felt by some to pose a threat to the attraction and viability of the Army Reserve among educated and motivated young people (for whom the access to university or technical education places offered by the RRes scheme has obvious attraction), but it is too early to tell whether this has affected the Army Reserve to any significant extent. The Army Reserve's problems tend to be those which dogged the CMF, a patronising attitude on the part of some regular personnel not least amongst them. Its strengths, especially the dedication of its longer-serving members, are often those of the older force also. Given government unwillingness to increase personnel costs in the regular army, the Reserve is likely to be regarded as a key feature of the Australian Army, at least at the policy and planning levels, for the foreseeable future.

ARMY SHIPS see **SMALL SHIPS SQUADRON**

ARROW, HMAS see **ATTACK CLASS PATROL BOATS**

ARTILLERY The artillery was one of the earliest arms of the service to appear in Australia. British regiments brought cannon with them, both in mobile form and for coastal defence, although gunners of the Royal Artillery did not become a permanent feature of colonial garrisons until the 1850s. When the British units were replaced with colonial volunteers the British weapons were often handed over to the local forces. The most common artillery pieces were the 6-pounder bronze smooth-bore gun and the 12-pounder bronze howitzer. Most colonies possessed these guns and for years they formed the basis of the volunteer artillery units.

The major issue in gun design in the mid-nineteenth century was the development of rifled ordnance. (Rifling imparted spin to shells as they left the barrel, thus enabling a more accurate trajectory.) In Britain rifled guns began to replace smooth-bore weapons from 1859. The first colony to order such guns was Victoria, which in 1864 took delivery of a battery of Armstrong 6-pounders. Other colonies were to replace their smooth-bore guns with rifled ordnance over the next 20 years. The second major development, breech-loading, followed in the 1860s, and it was common in colonial inventories to find a mix of rifled breech-loading (RBL) and rifled muzzle-loading (RML) guns. While breech loading ordnance eventually superseded muzzle loaders, at the technical level then available the RML guns were superior in range, accuracy, cost and endurance.

There was little opportunity to use any of these guns on active service. In 1885 a contingent of artillerymen was despatched from New South Wales as part of the Sudan contingent. However before these men could familiarise themselves with the battery provided by the War Office (six 9-pounder guns), the chance for engagement had passed. The next opportunity for service arose during the Boer War. On 30 December 1899 'A' Battery of the Royal Australian Artillery sailed for Cape Town. The battery remained in South Africa for 18 months but was rarely in action as a single unit, on most occasions being split into sections of two and attached to various British units. In the event the guns hardly saw any action.

Despite the disparate nature of the guns in service in the various colonies, at least one unifying measure had taken place before Federation. In 1885 a School of Gunnery was established in Sydney. Initially the school trained only the New South Wales Artillery. From 1897, however, gunners from the other colonies attended the school and common procedures and techniques of gunnery were established. On 1 March 1901 the State military establishments were transferred to the Commonwealth. The artillery resources of the new nation comprised 32 field and heavy guns as well as a range of obsolete pieces.

In the first years of the new century the most significant artillery development was the standardisation and increased weight of ordnance. From 1905 the 12- and 15-pounder field guns began to be replaced with the new British 18-pounders (q.v.), while 5-inch howitzers began to replace the multiplicity of guns of that type. On the outbreak of war the AIF proceeded to the conflict with its main artillery component being 36 × 18-pounder field guns. This was actually below establishment, which called for batteries of six guns, not four, but reflected the shortage of modern artillery pieces then available. There was, however, limited scope for the use of artillery on Gallipoli, although the absence of howitzers in Australian batteries was keenly felt on occasions. The Western Front was a different matter. Eventually there were to be 60 field and a number of medium and heavy batteries manned by AIF personnel. The two guns most used in support of Australian operations were the 18-pounder and 4.5-inch howitzer (qq.v.).

There were a number of significant tactical developments in the use of artillery during the course of the war. The main tactic affecting infantry operations was the 'creeping barrage'. This consisted of a curtain of shells, often some hundreds of yards wide, behind which the infantry formed up for the attack. The barrage then moved forward (or crept) towards the enemy front line. This forced the hostile machine-gunners and riflemen to keep their heads down (and thus unable to fire their weapons) until the point that the attacking infantry arrived and was able to deal with them at short range. The other artillery development that aided the Australians in their later battles was the detection of enemy batteries by sound location. This method enabled the German batteries to be deluged with shells at zero hour, thus eliminating them as an infantry-destroyer. Australian heavy batteries used these methods throughout the climactic battles of 1918.

Between the wars there was a great reduction in the size of the artillery (along with other arms of the services). Ammunition for training was limited and the standard guns continued to be those used during the First World War: 4.5-inch howitzers, 18-pounders, 60-pounders and 6-inch howitzers. Consequently, at the outbreak of the Second World War, the Australians were equipped with obsolescent guns — at a relatively lesser capability than in 1914. The 2nd AIF sailed to the Middle East with 18-pounder and 4.5-inch howitzer batteries, and

25-pounder guns of the 2/7th Field Regiment and a 3.7-inch anti-aircraft gun of the 132nd Heavy Anti-Aircraft Battery, with Matilda Tanks, firing on Japanese positions, Tarakan, Borneo, 10 June 1945. (AWM T08886)

although these **guns** performed useful service in the early campaigns against the Italians in the Western Desert, they were of no use against the better-equipped Germans. Indeed, even during the first Libyan campaign Australian units had 'salvaged' captured Italian guns, and during the siege of Tobruk in 1941 these formed the 'bush artillery'. By 1942 the Australians finally began to be re-equipped with the new British 25-pounder guns (q.v.), the most versatile field gun of the war. Anti-tank guns, however, continued to be a problem. The Australian troops found the standard 2-pounder just as inadequate against medium armour as their British counterparts. The war in the desert provided the best opportunity for the use of artillery in support of Australian operations in the Second World War, culminating at Alamein where the massive British and Commonwealth superiority in artillery was the crucial factor in the crumbling and eventual destruction of the Afrika Korps.

Artillery was of less use in jungle warfare. The terrain was difficult, the targets were often not obvious, and the opportunity for the large-scale deployment of guns limited. Light pack-guns, including a version of the 25-pounder (the 'baby' 25), were, however, used to some effect, as was the 3.7-inch pack howitzer. Occasionally the guns could be decisive, as they were at Lababia Ridge

and Mount Tambu in New Guinea in 1943 and in some of the Bougainville operations in 1944.

At the end of the war there was another massive reduction in the size of the AMF, which included reductions in the artillery. In Korea no Australian guns were committed, although individual officers of the Royal Australian Artillery were attached to allied units. In the Malayan Emergency the 4.2-inch mortar was found to be more useful than standard field artillery in jungle warfare. This was less true in Borneo during Confrontation, where artillery was able to outrange mortars in providing fire support for Security Force patrols extricating themselves from the Indonesian side of the border.

It was not until 1959 that the 25-pounder was finally replaced as the standard gun of the army. In that year the American-designed 105 mm howitzer entered service, followed by a light 105 mm pack howitzer, the L5, designed in Italy. In September 1965 105 Field Battery was the first Australian artillery unit to be ordered to Vietnam, where its guns provided support for various American units. At the Battle of Long Tan in August 1966, the 1st Field Regiment fired in excess of 3000 rounds in support of D Company, 6RAR, and their efforts, together with those of the New Zealand guns, helped turn the battle in favour of the Australians. The general role played by the artillery in the Vietnam conflict was mainly that of providing fire support from forward bases. Guns were often moved from one firebase to another using Chinook helicopters. During this early phase it was discovered that the L5 was insufficiently robust for long-term operations in harsh conditions. The guns were easily damaged when towed over rough terrain and their elevation mechanism would not stand up to continuous firing. They were withdrawn in 1967 and replaced with 105 mm howitzers. The artillery again featured heavily in the defence of Fire Base Coral, to the north of Saigon, in May 1968. In the initial heavy assault on 12 May part of the perimeter was overrun and one of the guns lost. Later, patterned mortaring and artillery fire beat back the attack and stabilised the situation.

Following the Vietnam War the remaining L5s were replaced by 105 mm howitzers. During the 1970s the widespread introduction of computers to assist in the calculation of firing and ranging data gave an additional dimension to the rapidity of response and volume of fire capable of being brought to bear on a target or targets. In the late 1980s a new 105 mm light gun, designated the 'Hamel', came into service. Manufactured in Australia under·licence, it has also been sold to the New Zealand Army.

ARUNTA (I), HMAS see **TRIBAL CLASS DESTROYERS**

ARUNTA (II), HMAS see **ANZAC CLASS FRIGATES**

ASSAIL, HMAS see *ATTACK* **CLASS PATROL BOATS**

ATOMIC WEAPONS TESTS were conducted in Australia at the request of the British government, which sought to maintain Britain's great power status by developing an independent nuclear arsenal, with the enthusiastic agreement of the Australian government led by Prime Minister Robert Menzies. Twelve weapons, with yields of between 1 and 60 kilotons, were exploded in Australia, starting at the Monte Bello Islands off north-western Australia on 3 October 1952. The Monte Bello Islands were also used for two explosions in May–June 1956, but the rest of the trials were carried out in the central desert regions of South Australia: at Emu Field in October 1953 and at Maralinga in September–October 1956 and September–October 1957. In addition, hundreds of minor trials, mostly involving components of nuclear weapons, took place at Emu Field and Maralinga between 1953 and 1963, and these trials caused most of the lasting contamination by scattering plutonium around the field at Maralinga. The nature of radiation is such that the long-term effects, if any, of working at the

test sites, as well as the effects on the general population of the fallout that was dispersed on several occasions as the result of unexpected weather conditions, may never be known with any certainty. Those who suffered most as a result of the tests were the Aboriginal people from the Maralinga area, some of whom continued to inhabit the prohibited zone during and after the tests, and some of whom were almost certainly dusted with fallout.

Controversy about the tests resurfaced in the 1970s, and in 1984 the Australian government appointed a Royal Commission into the matter under Justice J. R. McClelland. The commission's report, which was extremely critical of the Australian and British governments, recommended that the British government should pay to clean up the test sites, while the Australian government should compensate the Aboriginal people of the Maralinga area. However, in 1993 the Australian government accepted a British offer of $A45.2 million, which is less than half the expected cost of rehabilitating Maralinga.

Robert Milliken, *No Conceivable Injury: The Story of Britain and Australia's Atomic Cover-up* (Penguin Books, Melbourne, 1986); Lorna Arnold, *A Very Special Relationship: British Atomic Weapon Trials in Australia* (Her Majesty's Stationery Office, London, 1987).

ATTACK CLASS PATROL BOATS Laid down and launched 1966–68; displacement 149 tonnes; length 32.6 m; beam 6.1 m; armament 1 × 40/60 mm Bofors gun, 1 machine-gun; speed 24 knots.

The 20 *Attack* Class patrol boats were ordered in 1965 as a response to the Indonesian Confrontation (q.v.) and entered service between 1967 and 1969. They were formed into three squadrons in 1972 based in Sydney, Cairns and Darwin. Equipped with radar and echo-sounders, their main task has been to patrol Australia's northern coasts to deter foreign fishing vessels from entering its territorial waters. Two *Attack* Class boats were used in the ABC television series 'Patrol Boat'. Many were transferred to Indonesia and Papua New Guinea in the 1970s and 1980s, with the remainder transferring to the RAN reserve as they were replaced by *Fremantle* Class patrol boats (q.v.) from 1980.

AUSSIE see **SERVICE NEWSPAPERS**

AUSTRALIA (I), HMAS (*Indefatigable* Class battle-cruiser). Laid down 1910, launched 1911; displacement 18800 tons; length 590 feet; beam 80 feet; armament 8 × 12-inch guns, 16 × 4 inch guns, 4 × 3-pounder guns, 2 × 18-inch torpedo tubes; speed 25 knots.

HMAS *Australia* was the first capital ship built for the Commonwealth. It was designed to be the flagship of the Australian Fleet Unit, which was to be integrated into the main British fleet during wartime. On the outbreak of war in 1914 the *Australia* was sent to escort the expedition which captured the colonies of German New Guinea and Samoa. In December 1914 with the demise of the German outlying possessions, HMAS *Australia* left to join the Battle Cruiser Fleet at Rosyth in Scotland. During this voyage, on 6 January 1915, she sank the German merchantman *Eleanore Woermann* off the Falkland Islands, the only time HMAS *Australia* would ever fire her guns in anger. With the Grand Fleet she became the flagship of the 2nd Battle Cruiser Squadron. The *Australia's* tour of duty in the North Sea was uneventful except for two naval collisions (q.v.). In April 1916 she hit HMS *New Zealand*, the resultant damage causing both to miss the Battle of Jutland. A later collision with HMS *Repulse* in 1917 caused more minor damage. HMAS *Australia* was part of the fleet that witnessed the German surrender at Scapa Flow in 1918. The ship returned home in June 1919 and was sunk off Sydney Heads in 1924 as part of the disarmament (q.v.) provisions of the Washington Naval Treaty.

AUSTRALIA (II), HMAS (*Kent* Class heavy cruiser). Laid down 1925, launched 1927; displacement 9850 tons; length 630 feet; beam 68 feet 6 inches; armament (1939) 8 × 8-inch guns, 8 × 4-inch guns, 4 × 3-pounder guns, 4 × 2-pounder guns, 12 machine-guns, 8 × 21-inch torpedo tubes; 1 Seagull V (q.v.) aircraft; speed 31.5 knots.

The second HMAS *Australia,* sister ship to the *Canberra* (q.v.), was built at Clydebank in Scotland. After six years in Australian waters, she spent 1935–36 on exchange duty with the RN, taking part in the Naval Review to commemorate George V's Jubilee in 1935. Paid off into reserve in 1938, HMAS *Australia* was recommissioned on the outbreak of the Second World War. Her first major action was at the French colonial port of Dakar in west Africa in September 1940, where, during an unsuccessful attempt to land General Charles de Gaulle, *Australia,* with two British destroyers, sank the Vichy French destroyer *L'Audacieux* and took part in the bombardment of Dakar harbour. During this action *Australia's* Seagull V seaplane was shot down. *Australia* was serving in Australian waters when Japan entered the war and formed part of Vice-Admiral John Crace's (q.v.) Task Force 44 during the Battle of the Coral Sea in May 1942. From 1943, *Australia* took part in bombardments prior to landings in New Guinea, the Netherlands

East Indies and the Philippines. On 21 October 1944, during the landings at Leyte Gulf in the Philippines, a Japanese aircraft hit the bridge in a suicide attack, killing the ship's Captain, Emile Frank Verlaine Dechaineux (q.v.), and injuring the commander of the Australian Squadron, Commodore John Collins (q.v.). During further landings in the Philippines in January 1945 *Australia* was again hit by kamikaze attacks. HMAS *Australia* was paid off a second time in 1954 and sold for scrap to a British company.

AUSTRALIA–NEW ZEALAND AGREEMENT (the 'Canberra Pact'), signed on 21 January 1943, asserted the right of the two dominions to participate in all decisions affecting the south and south-west Pacific. The two governments were concerned at their apparent exclusion from the councils of the great powers, most recently demonstrated by the Cairo Declaration of 1 December 1943, in which the leaders of Britain, the United States and China (Churchill, Roosevelt and Chiang Kai-shek) made fundamental decisions about the future of the post-war Asia–Pacific region without consulting Australia and New Zealand.

The reactions in Washington and London were critical; in a subsequent meeting with the New Zealand Prime Minister, Peter Fraser, the American Secretary of State, Cordell Hull, blamed the Australian Minister for External Affairs, H.V. Evatt, for the tone of the agreement and likened his behaviour to that of the Soviet government over Poland, while the British government expressed concern that the customary channels through which the two governments had conducted their external affairs, that is, London, had been flouted, and attempted to convince the Australians and New Zealanders that arrangements for the subsequent regional conference called for by the agreement should be postponed until after the 1944 Prime Ministers' Conference in London. At Wellington in November 1944 the Australian and New Zealand governments again conferred on the issue of representation in the processes that would arrive at armistices with enemy governments in the Pacific, and on the formation of a South Seas Regional Commission. Although both governments were accorded representation, belatedly, at the surrender ceremony in Tokyo Bay in September 1945, the major aims of the agreement were not realised because the Americans and British had no pressing reason to include either government in the surrender negotiations. The South Pacific Commission, however, was established following further meetings and a conference held in Canberra in January 1947, and this represented the only concrete outcome of the

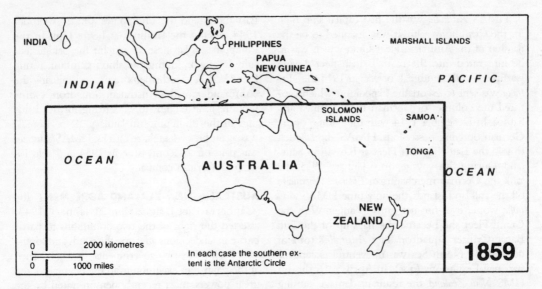

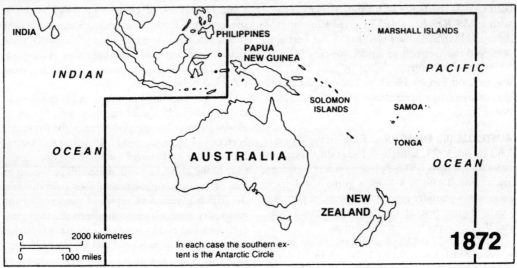

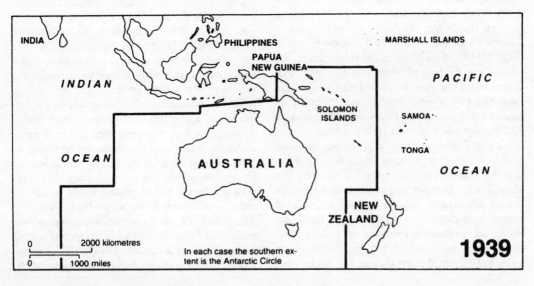

whole affair. The Australia–New Zealand Agreement was significant, however, as a sign of Australian assertiveness, especially in regional affairs, which was to become more marked as a consequence of the Pacific War, and as an attempt to remind the major powers of the obligation to involve smaller regional powers in matters which concerned them directly. This latter issue was to be one which Evatt pursued vigorously in the forum of the United Nations immediately after the war.

Robin Kay (ed.), *The Australian–New Zealand Agreement 1944* (Historical Documents Branch, Wellington, 1972); W. J. Hudson et al. (eds), *Documents on Australian Foreign Policy*, vol. 7 (Australian Government Publishing Service, Canberra, 1988).

AUSTRALIA STATION was the administrative term for the naval command based on Australian waters. In the early years of the colony of New South Wales ships based in these waters came under the control of the East Indies Station, but in 1859 the Admiralty delineated a separate station, under the command of a commodore. The decision was in part a recognition of the circumstances in which a good part of the East Indies Station had been detached for duty in Australian waters, and reflected concerns about the strategic situation in the western Pacific, especially over Tahiti, and the lead-up to the Maori Wars of the 1840s. There was also a need to police the trade in island labourers. In 1884 the station was upgraded to a rear-admiral's command. Modifications to the boundaries were made in 1864, 1872, 1893 and 1908, such that by the time responsibility for the station passed to the newly founded RAN in 1913 it covered the mainland of Australia and its island dependencies to the north and east, while touching no other shores. It was redrawn again in 1958, by which time it differed little from the original boundaries set almost exactly a century before.

John Bach, *The Australia Station: A History of the Royal Navy in the South West Pacific 1821–1913* (New South Wales University Press, Sydney, 1986).

AUSTRALIAN AIR CORPS (AAC) Formed on 1 January 1920 out of the Aviation Instructional Staff and disbanded AIF units at the army's Central Flying School at Point Cook, Victoria, the AAC was created as an interim arrangement pending the establishment of the (Royal) Australian Air Force as a new and permanent air defence service. With an initial strength of five officer pilots and 70 other ranks, filled mainly by former members of the wartime Australian Flying Corps (AFC), the AAC's role was to maintain equipment in good order and perform useful flying work such as preparing air routes. Initially it operated some 30 Avro 504 trainers and Sopwith Pup scouts received in Australia during the first half of 1919, along with a few obsolete types used at Point Cook for training.

The arrival in Australia from March 1920 of huge quantities of surplus war stores, provided by the British government to dominions wishing to start their own air force, caused a rapid expansion of the AAC to more than 175 all ranks. Included in the imperial gift (q.v.) were 128 aircraft, including SE5a fighters, DH9 and DH9a bombers; while a few were unpacked and put into use, the remainder were stored and formed the initial equipment of the RAAF.

The existence of the corps enabled a number of notable activities to be undertaken. During the visit of the Prince of Wales to Melbourne in May 1920, the AAC mounted aerial displays involving 15 to 20 aircraft; the corps' limited number of pilots meant some machines had to be flown by ex-AFC volunteers. Test flights of uncrated gift machines provided the occasion for an Australian altitude record (27 000 feet) set in June 1920, which stood for more than a decade. The first non-stop flight between Sydney and Melbourne was achieved the following month. Between July and November an Avro 504 floatplane and AAC personnel were also embarked in HMAS *Australia* (I) (q.v.), and later HMAS *Melbourne*, for sea trials.

In August nearly 20 AAC aircraft undertook flights in all States to promote the government's Second Peace Loan. Again, returned servicemen had to be engaged to fly many of the machines, four of which crashed in the course of the program of stunt flying, exhibitions and aerial derbies. Although not a great success for the purpose intended, the Peace Loan did much to raise the profile of aviation within Australia.

While assisting in the search for a missing schooner off the Tasmanian east coast in September, one of two AAC aircraft was lost with its two-man crew. These were the only fatalities in the corps' brief existence and raised important compensation issues affecting the new air service. Although purely a holding organisation while the future of defence aviation was being decided in the initial postwar period, the AAC achieved some significant milestones during its 15 month existence.

CHRIS COULTHARD-CLARK

Map 4 Boundaries of the Australia Station from 1859 to 1939.

AUSTRALIAN ARMOURED CORPS see **ROYAL AUSTRALIAN ARMOURED CORPS**

AUSTRALIAN ARMY AVIATION CORPS (AA AVN)

AUSTRALIAN ARMY AVIATION CORPS (AA AVN) was formed on 1 July 1968 after command of the 1st Aviation Regiment was transferred from the RAAF to the army in 1964. Its role is to transport troops and equipment in battle zones as well as to undertake aerial photography and radio relay work. Since 1990 the Aviation Corps has been responsible for the helicopter training of all three services, using Squirrel (q.v.) helicopters at the Australian Defence Force Helicopter School at Fairbairn. The Aviation Corps currently operates Blackhawk and Iroquois (qq.v.) helicopters with the 5th Aviation Regiment at Townsville for battlefield transport; Chinook (q.v.) helicopters for heavier transport; Kiowa (q.v.) helicopters with No. 161 and 162 Squadrons for observation; and Nomad aircraft with No. 173 Squadron for transport. Training is conducted at the School of Army Aviation at Oakey, Queensland.

AUSTRALIAN ARMY CATERING CORPS (AACC) was formed on 12 March 1943, when cooks from all units (except Australian Army Medical Corps units, which kept their own cooks for some time) were brought into a single corps. Corps members prepare, cook and serve meals in army establishments and in the field.

(See also RATIONS.)

AUSTRALIAN ARMY DENTAL CORPS see ROYAL AUSTRALIAN ARMY DENTAL CORPS

AUSTRALIAN ARMY EDUCATIONAL CORPS see ROYAL AUSTRALIAN ARMY EDUCATIONAL CORPS

AUSTRALIAN ARMY JOURNAL (AAJ) First published in June–July 1948, under the editorship of E. G. Keogh (q.v.), the *Australian Army Journal* was designed to continue the role of the wartime *Army Training Memoranda*. The latter had appeared monthly from the Directorate of Military Training and gave the latest information on enemy tactics, organisation and weapons, and on Australian training methods and techniques. The *ATMs* were in fact the first regular and consistent set of doctrinal statements produced by the Australian Army and intended for specifically Australian conditions. The *AAJ* encountered early problems, mostly because of attempts to publish through a seriously overworked Army Headquarters Printing Press, and in 1950 production was contracted out to a private firm and the journal began to appear monthly from June that year. Keogh retired as editor in February 1965, to be succeeded by A. J. Sweeting, long-serving research officer to the official historian Gavin Long (q.v.). He was succeeded in turn in September 1967 by C. F. Coady,

and in August 1968 the journal was redesignated the *Army Journal*, and the format was revised. It continued to function, as its subtitle proclaimed, as 'a periodical review of military literature', although during the 1960s and the Vietnam War it became an increasingly bland one. In November 1976 it was replaced altogether by a new *Defence Force Journal*, itself retitled in January 1991 the *Australian Defence Force Journal*. As with all its predecessors, it is well produced but the quality of the content varies considerably.

AUSTRALIAN ARMY LEGAL CORPS (AALC) was formed on 30 September 1943, replacing the Australian Army Legal Department which had been created in 1922. It consists of officers who have been admitted as barristers or solicitors of the High Court or a State or Territory Supreme Court. These officers provide legal advice to the army, give instruction in military law, and provide legal aid to soldiers on civil matters. Unlike in the British forces, they also act as defending officers and judges advocate at courts martial. At the end of the Second World War AALC members were heavily involved in preparation for and conduct of war crimes trials (q.v.) of alleged Japanese war criminals.

(See also LAW, MILITARY.)

AUSTRALIAN ARMY MEDICAL CORPS see ROYAL AUSTRALIAN ARMY MEDICAL CORPS

AUSTRALIAN ARMY MEDICAL WOMEN'S SERVICE (AAMWS) had its origins in the Voluntary Aid Detachments raised from the Australian Red Cross (q.v.) and the Order of St John during the First World War. When the Second World War began, labour shortages soon brought female voluntary aids (VAs) into the military hospital system to release men for other duties. They were employed largely as ward and theatre assistants and as medical technicians, and were not paid, even on full-time duty, until January 1940. In June 1941 approval was given for the employment of VAs on overseas service, and a party of 200 embarked for the Middle East in October. Others followed, serving in Cairo, Gaza, Ceylon and on hospital ships.

The Pacific War heightened the problem of labour shortages and created new opportunities for women in the services. In March 1942 the Military Board approved the call-up of VAs to replace men in new hospitals, and from this point on these Voluntary Aid Detachments were administered as a service within the Army Medical Service. In order to make clear the distinction between these women working full time within the army and those VAs still based in the Red Cross or Order of

St John, the former were redesignated as the Australian Army Medical Women's Service in December. Women in the AAMWS served in hospitals throughout northern Australia, often in isolated areas, and from 1943 some were deployed to New Guinea (q.v.), serving at Buna, Dobodura, Lae and Finschhafen. Towards the end of the war they were also to be found on Bougainville, Borneo (qq.v.) and Morotai. After hostilities ended they were employed in the care of recovered POWs and civilian internees, while in 1946 a party of volunteers was dispatched to Japan as part of the British Commonwealth Occupation Force (q.v.).

There was some friction between the AAMWS and the Australian Army Nursing Service (AANS), as it was then, at the administrative and command levels, over what the Matron-in-Chief of the AANS saw as 'duplication of administrative personnel' and conditions of service, which differed between the three female branches within the army. The use of AAMWS in lieu of trained nurses in the AANS also led to friction and some professional jealousy. The AAMWS was disbanded in February 1951 and its functions were incorporated into the new Royal Australian Army Nursing Corps (q.v.).

AUSTRALIAN ARMY NURSING SERVICE (AANS)

There were army nurses in some of the colonial military forces (q.v.), the first Army Nursing Service being set up in New South Wales in May 1899, and Australian nurses served with the colonial contingents sent to the Boer War (q.v.). The Australian Army Nursing Service Reserve was created in July 1902 to provide a body of trained personnel available for service in the event of war. A Lady Superintendent was responsible for nursing services in each military district, and by 1914 the AANS came under the control of the Director-General of Medical Services. Disagreements between the commanding officer of No. 1 Australian General Hospital, Colonel W. Ramsay Smith, and the Principal Matron, Jane Bell (q.v.), over who should be responsible for nurses led to both being recalled to Australia from Egypt and the establishment of the position of Matron-in-Chief in May 1916, first held by Evelyn Conyers (q.v.). The Matron-in-Chief administered the service and

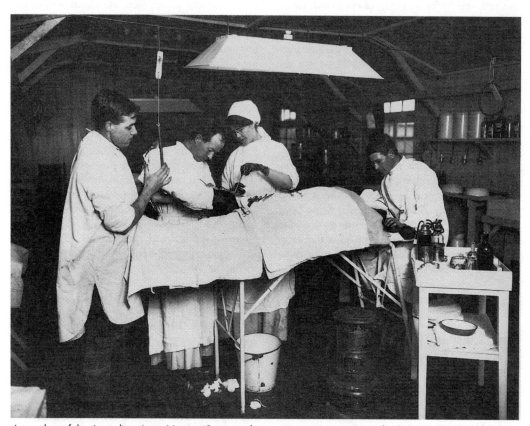

A member of the Australian Army Nursing Service taking part in an operation at the 1st Australian Casualty Clearing Station, Oultersteene, France, 23 November 1917. (AWM E01304)

acted as adviser to the Director-General Medical Services on nursing matters.

In the First World War nurses were recruited from both the nursing service and the civilian profession and served as an integral part of the AIF. They served in Egypt and Lemnos during the Gallipoli campaign, in England, France and Belgium in support of the fighting on the Western Front, and in Greece, Salonika, Palestine, Mesopotamia and India. At least 2139 nurses served abroad between 1914 and 1919, and a further 423 worked in military hospitals in Australia, while 29 died on active service. Between the wars the AANS reverted to a reserve status, and nursing services in the military districts were run by a Principal Matron, answerable to the Matron-in-Chief at Army Headquarters. On the outbreak of war in 1939 the service was placed on active duty, and nurses were again enlisted for service overseas with the 2nd AIF. For most of the war they were the only female service personnel to serve outside Australian territory. Once again they served in disparate parts of the world: England, Egypt, Palestine, Libya, Greece, Syria, Ceylon, Malaya, Singapore, Papua New Guinea, the Solomon Islands, throughout Australia and, with the war's end, as part of the British Commonwealth Occupation Force (q.v.) in Japan.

In the First World War nurses had enjoyed the privileges of commissioned rank but did not actually hold rank. During the Second World War this position was regularised, although some nurses disliked becoming part of the military structure, believing that an officer's status was incompatible with a nurse's calling. Casualties were heavier than in the previous war: 71 nurses were killed, 41 of them in Malaya and Singapore. In July 1947 members of the AANS were transferred to the Interim Army, and in November 1948 the service was designated a 'Royal' one. In July 1949 the RAANS became part of the Australian Regular Army (q.v.). In February 1951 the service became a corps, further emphasising its integration into the army mainstream, becoming known as the Royal Australian Army Nursing Corps (q.v.).

Jan Bassett, *Guns and Brooches: Australian Army Nursing from the Boer War to the Gulf War* (Oxford University Press, Melbourne, 1992).

AUSTRALIAN ARMY ORDNANCE DEPARTMENT see ROYAL AUSTRALIAN ARMY ORDNANCE CORPS

AUSTRALIAN ARMY ORDNANCE STORES CORPS see ROYAL AUSTRALIAN ARMY ORDNANCE CORPS

AUSTRALIAN ARMY PAY CORPS see ROYAL AUSTRALIAN ARMY PAY CORPS

AUSTRALIAN ARMY PROVOST CORPS see ROYAL AUSTRALIAN CORPS OF MILITARY POLICE

AUSTRALIAN ARMY PSYCHOLOGY CORPS (AA PSYCH) replaced the Australian Army Psychology Service, formed in 1945. When the corps was created on 22 October 1952 it was unique in the British Commonwealth. Its function is to provide psychological advice and practice so as to make effective use of those serving in the army. To this end, corps members are involved in such areas as personnel selection, assessment, allocation and training; personnel research and psychological testing of personnel; and individual counselling of personnel.

AUSTRALIAN ARMY SERVICE CORPS see ROYAL AUSTRALIAN ARMY SERVICE CORPS

AUSTRALIAN ARMY TRAINING TEAM UGANDA (AATTU) After the collapse of the tyrannical regime of Idi Amin in 1979, and the temporary resolution of the civil war occasioned by the return of Dr Milton Obote, Australia agreed to furnish a component to the Commonwealth Military Training Team Uganda, set up to help re-establish discipline and professional standards within the Ugandan Army. The Australian contribution consisted of five staff: a component commander with the rank of major, and an instructional team comprising one captain, one warrant officer and two senior NCOs. While in Uganda the Australians helped to run courses for officers, warrant officers and NCOs at the Ugandan School of Infantry at Jinja Barracks. They assisted in the re-establishment of the Ugandan Military Training School (MTS) at Kabamba and the conduct of basic skills courses at the MTS and monitored the activities of the Ugandan National Liberation Army (the Obote-controlled forces). Australians were committed to Uganda for these purposes from March 1982 until March 1984, and each component team undertook a tour of approximately six months' duration, during which time staff were considered to be on war service.

AUSTRALIAN ARMY TRAINING TEAM VIETNAM (AATTV) Following representations by the United States, the Australian government announced on 24 May 1962 the commitment of 30 military advisers to assist the Army of the Republic of

Warrant Officer 2nd Class 'Locky' Scowcroft of the Australian Army Training Team Vietnam training a Montagnard soldier in how to search a captured enemy soldier, Pleiku, South Vietnam, October 1969. (AWM BEL/69/696/VN)

Vietnam (ARVN). The Australians were deployed intially as instructors in jungle warfare techniques and technical areas such as signals and engineering, and were specifically forbidden to accompany the South Vietnamese into combat. The first commander of the AATTV was Colonel F. P. Serong (q.v.), who had instructed at the Jungle Training Centre at Canungra (q.v.) during the Second World War and who was regarded as an authority on counterinsurgency. Thirty-six advisers, all officers and senior NCOs with considerable operational experience, arrived in August 1962, marking the beginning of Australia's troop commitment to the Vietnam War. The unit was increased to 83 advisers in June 1964, and the limitations on operational deployment were lifted. The first adviser was killed in action in July.

Ultimately the AATTV grew to 100 strong by December 1964, and operated across the whole of South Vietnam in a variety of roles. Some Australians acted as advisers with regular ARVN units and formations, others worked with the Montagnard hill tribes as part of the US Special Forces network, and a few were involved in Project PHOENIX run by the Central Intelligence Agency, designed to destabilise the Vietcong infrastructure through infiltration, arrest and assassination. AATTV members won all four of the VCs awarded for Vietnam, as well as a US Presidential Unit Citation, and became the most highly decorated Australian unit to serve there. After the Australian Task Force was withdrawn between late 1971 and early 1972 the AATTV remained in Phuoc Tuy province in a training and advisory role, also assisting in the training of Cambodian soldiers of the Forces Armées Nationales Khmer. The last Australians were withdrawn in mid-December 1972 after the election of the Whitlam government.

Ian McNeill, *The Team: Australian Army Advisers in Vietnam 1962–1972* (University of Queensland Press, Brisbane, 1984).

AUSTRALIAN ARMY VETERINARY CORPS

(AAVC) was formed in 1909, replacing the veterinary department of the Commonwealth Military Forces, which had itself absorbed the colonial veterinary departments. With the establishment of permanent artillery batteries and of a remount department to supply these batteries with horses, it became necessary to establish a permanent section of the AAVC. The formation of the AAVC (Permanent Forces) was authorised in 1911, with its officers responsible not only for veterinary care of horses but also for training farriers and NCOs in such things as shoeing, care of horses and veterinary first aid. Initially administered by the Director of Remounts, the veterinary services got their own Director in 1914 and remained under the control of a veterinarian thereafter. During the First World War 120 AAVC officers served overseas with the AIF, but because of mechanisation there were only a few specialised roles that required the expertise of AAVC members during the Second World War. The corps was disbanded in 1946.

AUSTRALIAN CADET CORPS (ACC) School cadets, as they were commonly known, originated in the middle of the nineteenth century. The first cadet unit was raised at St Mark's Collegiate School in Sydney in 1866, and was maintained when that school merged with King's College three years later. The cadet unit at The King's School, Parramatta, is thus the oldest continuous cadet unit in Australia. Sydney Grammar formed a cadet unit in 1869, and state schools followed suit, Fort Street School setting up a unit in 1873. By 1883 there were 22 state school cadet units with a strength of over 1000 cadets, and by 1891 there were nine permanent full-time state officials ('cadet quartermasters') employed to oversee the scheme. By the time of Federation in 1901 there were 25 cadet units throughout Australia in all States except Queensland, with a total of 2600 cadets, the number of units having declined during the 1890s Depression.

The Commonwealth Cadet Corps was established in July 1906, as part of the revision of defence thinking that followed the Japanese victory over the Russians the previous year (see JAPANESE THREAT). In 1911 cadet service was made mandatory as part of the compulsory military training scheme (see CONSCRIPTION): boys aged 12 to 14 were enrolled as junior cadets, those between 15 and 18 as senior cadets. Rural youths were exempt, and many urban boys failed to comply. The compulsory scheme was abandoned in 1922 for juniors and in 1929 for seniors. It was replaced by a parallel system in 1932, with senior cadets organised either in schools or based on militia units; this latter, the Regimental Cadets system, was abandoned in turn in 1939.

School cadets were maintained during the Second World War, despite the competing demands for equipment and personnel occasioned by the war. By 1945 there were 143 cadet units, mostly in state schools, and by the war's end they had come under the Army's Directorate of Cadet Services and Reserves. The cadets, noted the *Official Yearbook of the Commonwealth of Australia*, occupied 'a foremost position in the scheme of national defence', and the support of cadet units by the regular army was to become a feature of cadet policy for several decades. In 1951 the Australian Cadet Corps was formed. It had its own Director (with the rank of colonel) with a headquarters staff, and each State was designated a cadet brigade with schools organised into battalion groups; a few schools, such as King's, were big enough to comprise a cadet battalion on their own. The administrative and instructional staff were officers and warrant officers of the regular army, while officers of each school unit were usually schoolteachers who held cadet commissions. By 1975 there were 330 regular staff and 35 civilians employed in running the ACC. The cadets reached a peak strength of 42 000 in 1966, but the Vietnam War and a brief surge of anti-military feeling in some quarters created pressures for its abolition, although in 1975 there were still 33 000 boys in cadet units, representing 6.36 per cent of eligible male students, and there were 79 schools in New South Wales and Queensland waiting to form cadet units.

As a consequence of the Millar Committee (q.v.) report into the CMF (q.v.) and ACC, the Whitlam government decided to abolish the school cadets despite Millar's noting that the scheme attracted broad community support. The Fraser government re-established the cadets after it came to power in December 1975, but on a fundamentally different basis from the previous postwar scheme. Cadets were now based on the Army Reserve, not the regulars, with attendant implications for resources. In 1983 the Hawke government announced that school-based units would no longer receive direct support from the army, meaning that henceforth only wealthy private schools would be in a position to maintain their units. Cadets were now to be community-based in regional units. In 1989 there were 130 such fully supported units, together with a few limited support units still based on schools and which paid for any army support received, covering some 16 000 cadets. These were administered by just 13 regular army personnel, and cost $2.26 million in the financial year 1988–89. In May 1982, as a result of the deliberations of a joint working party of the three services, eligibility was extended to females.

The abolition of cadets in 1975 generated considerable heat, and various arguments were advanced for the morally improving nature of cadet service and the importance of cadet service for recruitment, especially to RMC Duntroon and, later, to the Australian Defence Force Academy (qq.v.). From 1961 the *Official Yearbook* stated that 'the Australian Cadet Corps … does not form part of the Australian Military Forces', further emphasising the move away from direct military utility. It seems unlikely that a large school-based cadet scheme will be revived, despite occasional advocacy from various conservative interest groups.

AUSTRALIAN COMFORTS FUND (ACF) was established in August 1914 as an umbrella organisation for most of the voluntary bodies that had been set up soon after the outbreak of the war. Its purpose was to aid soldiers still under military command (i.e. in the field) as opposed to those under medical control, responsibility for whom became the preserve of the Australian Red Cross (q.v.), although in practice this division was not strictly observed. The ACF was active in two main areas: raising money to be sent overseas for the purchase of needed items (e.g. cigarettes, cocoa, and 'Tommy cookers' — crude but effective devices for heating food in the trenches), and making clothing for the troops (1.3 million pairs of socks, as well as many thousands of shirts, vests, and pyjamas). A secondary area of ACF activity was the establishment of a residential club in London for Australian soldiers on leave, and similar clubs in Sydney, Melbourne and Adelaide. The ACF was officially wound up on 16 April 1920. It was reformed in January 1940, with much the same aims as in the First World War: fund-raising (£7 million), clothing (3 million pairs of socks), and the provision of meals and accommodation, especially in Australia. An important feature of the ACF in both wars was its reliance on the voluntary efforts of women, which kept costs to a minimum. It was, however, slow to recognise the needs of servicewomen; not until 1943 did the ACF establish a hostel for servicewomen on leave in Sydney.

AUSTRALIAN COMMONWEALTH HORSE Formed in 1902 for service in the Boer War, the Australian Commonwealth Horse was the first military unit established by the Australian federal government. Squadrons or companies of mounted infantry were raised in each State and combined to form battalions. Eight battalions were formed, the first of which arrived in Durban, South Africa, in March 1902. The first two battalions of the Australian Commonwealth Horse took part in patrols against the Boers, but the war ended before the remaining battalions could see action. Approximately 4000 men enlisted in the Australian Commonwealth Horse, including Brudenell White (q.v.), later CGS, and they were the first Australian troops to wear the rising sun badge (q.v.), a design chosen especially for the unit by the GOC, Australian Military Forces, Major-General Edward Hutton (q.v.).

AUSTRALIAN CORPS OF SIGNALS see **ROYAL AUSTRALIAN CORPS OF SIGNALS**

AUSTRALIAN DEFENCE FORCE ACADEMY (ADFA) opened in 1986 on a 52-hectare site in the Can-

berra suburb of Campbell, adjacent to RMC Duntroon. It was established only after prolonged and
heated controversy within the three services and in
the wider community. Proposals in the 1920s to
amalgamate the army and naval colleges foundered
on the difficulties of combining two very different
systems; the establishment of the RAAF College at
Point Cook in 1947 and the resolve of each service
to defend its own position in the postwar period
thwarted any further attempts to establish a single
institution for some time. Several army committees
(the Vasey Committee in 1944 and the Rowell
Committee in 1946) emphasised the need for the
army to restore the Duntroon course to its traditional four-year duration, and to provide a university education for officer cadets. The Wade
Committee (1959) went further and argued for the
establishment of a tri-service college that would
offer a university education.

Each of the services supported the need for university education but jibbed at the proposal for a
tri-service institution, and each went its own way
in satisfying the perceived need for university qualifications: the RAAF by affiliating the RAAF College with the University of Melbourne (1961), the
army by entering into an agreement with the University of New South Wales (UNSW) to establish a
Faculty of Military Studies at RMC Duntroon in
1967, and the RAN by sending its midshipmen to
UNSW to complete their degrees after one year's
study at the RAN College, Jervis Bay. The government intended that these would be temporary
measures only, and a committee was established in
1967 under Sir Leslie Martin to recommend ways
of achieving the objective of a tri-service college.
When Malcolm Fraser, Minister for Defence, submitted the report to Cabinet in 1970, it was
rejected by the Prime Minister, John Gorton, and
the Treasurer, William McMahon. Four years later,
following the election of a Labor government that
was intent on establishing a single Department of
Defence and on improving the quality of advice
that the services proffered the government, the
Minister for Defence, Lance Barnard, announced
that a tri-service academy would be established.
Progress was stalled by the defeat of the government the following year, but Fraser, the new Liberal
prime minister, gave 'in principle' support for the
proposal in 1976.

The Academy Development Council, which
had been appointed by Barnard in early 1975,
announced its plans in 1978. It proposed that the
academy should also be an independent university,
which it was felt would establish the institution's
academic credentials. The government endorsed
this approach, and introduced a bill into parliament

to establish 'Casey University — Australian
Defence Force Academy'. When the building program was referred to the Parliamentary Standing
Committee on Public Works, critics of the broader
proposal found an avenue for expressing their concerns. Many officers opposed the tri-service concept as endangering the single-service ethos that
they held to be fundamental to the well-being of
the individual services; they questioned the desirability of a three-year academic program so early in
an officer cadet's career; they were concerned that
the academy would be 'too academic'; and the
RAAF and the RAN were suspicious that with the
academy in Canberra, on a site next to RMC
Duntroon, there would be an inevitable tendency
for army practices to predominate. Nor were some
members of the wider academic community any
less opposed. Many felt that a 'military university'
was a contradiction in terms, and that such an institution could not provide a liberal university education. Both groups of opponents were united in
criticising the cost of the project, which they felt
could be distributed better elsewhere.

After hearings throughout 1978–79, the Public
Works Committee unanimously rejected the proposed academy. Although the government itself was
not wholly behind the proposal, and although the
service chiefs had made it clear to the parliamentary inquiry that they regarded the academy as a
compromise at best, Fraser and his Defence minister, James Killen, were determined to push ahead
with a tri-service institution. Accordingly, the government announced in May 1980 that it would
not proceed with an independent university along
the 'Casey' lines, but would instead enter into an
arrangement with the University of New South
Wales, whereby UNSW would establish a University College within the academy in much the same
way as it had operated the Faculty of Military
Studies at RMC Duntroon since 1968. An interim
Academy Council, charged with developing the
academic and military curricula and planning and
constructing the campus, was established in 1982,
and its work continued after the election of the
Labor government the following year.

The role of the academy, as defined in the
agreement between the Commonwealth and the
University of New South Wales, is 'to provide military education and training for officer cadets ...
and by way of foundation for their careers as officers of the Defence Force a balanced and liberal
university education in a military environment'.
The academy is a tri-service institution, and the
undergraduate body has both male and female officer cadets, as well as cadets from New Zealand,
Singapore and Thailand. An exchange program has

operated with The Royal Military College of Canada since 1993, whereby each year the two institutions host three visiting students for a semester. In 1995 the undergraduate body, which is restricted to members of the Defence Force and foreign military forces, consisted of 1005 students, the breakdown being: Arts, 430 (297 males, 133 females); Science 252 (205 males, 47 females); and Engineering 323 (280 males, 43 females). The normal academic course is three years, with Honours students in Arts and Science and students in Engineering returning for a fourth year of study, in the case of army students after completing a pre-commissioning year at RMC Duntroon.

As well as the exclusively military undergraduate body, there is a proportionately large postgraduate population (397 in 1995), enrolled in research and course-work degrees at Masters and PhD level. A number of course-work programs have a specific defence orientation, such as the Australian Technical Staff Officers' Course and the Master of Defence Studies. Entry at postgraduate level is open to any qualified person, military or civilian. Each year the University College attracts postgraduate students from all over Australia, and from a number of foreign countries, many of them winning scholarships. The academic staff of the University College are very active in their respective fields of research, and consistently attract high levels of external research funding.

The common military program, which runs throughout the year, combines wide-ranging lecture and discussion courses on topics such as military law, defence studies and communications, with instructional periods in drill and ceremonial and weapons handling. Single-service training is conducted by each of the three services at the beginning and end of the academic part of the year, and enables officer cadets from each service to be exposed to the practicalities of life in the army, RAN and RAAF. All cadets participate in 'Cryptic Challenge' at the end of second year, a practical one week exercise in the demands of leadership in the field. Upon completing the three year academic and military program, cadets transfer to their respective single-service training establishments. By the end of 1994, a total of 1932 degrees had been awarded to graduates of the University College, and 1831 officer cadets had graduated from the academy.

AUSTRALIAN DICTIONARY OF BIOGRAPHY (ADB), which is the largest joint voluntary project in the humanities ever undertaken in Australia, records the lives of thousands of men and women from all walks of life who have contributed to the Australian experience. The first 12 volumes, published from 1966 to 1990, contain 7211 separate entries dealing with over 10 000 people and are divided into three periods: 1788–1850 (two volumes), 1851–90 (four volumes) and 1891–1939 (six volumes). The inclusion of a person in the appropriate period was generally determined by when he/she did his/her most important work (floruit). From 1940 the 'date of death' principle has replaced the floruit, so that the volumes now in preparation will include only people who died during the period 1940–80.

As well as the biographies of many British soldiers and marines stationed in Australia, the first 12 volumes of the *ADB* contain the lives of hundreds of Australian soldiers, sailors and airmen, including 491 soldiers (4.7 per cent of total names listed) who fought in the First World War. All the prominent commanders, Victoria Cross winners and other highly decorated personnel are included, along with 30 nurses from both world wars. With the exception of a few already included, Second World War, along with Korean, Malayan and Vietnam personnel are included in volumes 13–16. For example, volume 13 (A–De) contains the biography of Field Marshal Sir Thomas Blamey (q.v.) and VC winner Albert Chowne. An index volume published in 1991 lists air force officers, marines, naval officers, and soldiers; the last mentioned are divided into the following categories: British, colonial, Maori Wars, South African War, Sudan War, World War I and World War II. Material on many other members of Australia's defence forces, not included in the *ADB*, can be found in the 'Biographical Register' located at *ADB* headquarters in the Research School of Social Sciences at the Australian National University, Canberra.

AUSTRALIAN FLYING CORPS (AFC) Of all the dominions, Australia was best placed to enter the new dimension of aerial warfare. The Central Flying School had been set up at Point Cooke (now Point Cook) in Victoria in 1913 and on the eve of war was ready for its first intake of trainee pilots. Australia's early involvement in aerial warfare, however, was modest enough. One aircraft (a BE2c) accompanied the expedition against German New Guinea (q.v.) but was never unpacked from its crate. Then in February 1915 the Indian government requested trained pilots to assist in its Mesopotamian campaign against the Turks. In April four officers and support staff left Melbourne for Basra where they arrived at the end of May. The Mesopotamian Half Flight (MHF), as it was called, was thus the first Australian aerial unit to enter a war zone. The experience was not to be a happy

Ground crew turning the propeller to start an SE5A aircraft of No. 2 Squadron Australian Flying Corps, Palestine, 1917. (AWM P1034/81/61)

one. The aircraft available, Caudron, Maurice Farman Longhorn, and Martinsyde, were underpowered and mechanically unreliable. Despite reinforcements in August which converted the flight to a squadron (No. 30 RFC) the attrition rate was high. Of the nine pilots who flew with the squadron two were lost presumed killed and six were captured by the Turks. In November 1915 only one of the original contingent remained. The work of the squadron largely involved reconnaissance for the expeditionary force. Initially the aircraft were unarmed except for mortars and rifles carried by the crew. Later small 9-pound bombs were employed, and in October bombing by three Australian planes of an Arab tribe known to be hostile induced the tribe's sheik to surrender. During the siege of Kut the remaining pilot, H. A. Petrie, dropped supplies to the garrison, but to no avail. When Kut surrendered the Australian support crew still in the city were captured. So ended the brief and disastrous history of the MHF.

A more fruitful experience awaited Australian airmen in the Middle East. In late 1915 Australia was asked by Britain to form complete squadrons

for service with the RFC. Australia responded by despatching No. 1 Squadron AFC (28 officers, 195 other ranks) to Egypt. Twelve aircraft for the squadron were supplied by the British. During the next two years the squadron operated first from Heliopolis and later in Palestine and Syria. They supported the ground forces in all the major battles of the Palestine campaign against the Turks: Romani, Magdhaba, Gaza, Beersheba (q.v.), Amman and Nablus. The duties of the squadron consisted of aerial reconnaissance bombing missions against enemy positions and communications and observation for the artillery. In 1918 the squadron photographed an area of Turkish-held positions of approximately 1000 square kilometres to enable a new set of maps to be compiled. There were many aerial duels during the period and it was in this theatre that the Australians first encountered anti-aircraft fire. The only VC awarded to an Australian airman was given to Lieutenant F. H. McNamara for rescuing a fellow pilot shot down by anti-aircraft fire and forced to land near Turkish positions. A great variety of aircraft was flown by the Australians. At first BE2cs and Martinsyde

G100s were used. In 1917 the squadron was equipped with BE2as, RE8s and Martinsyde G102s. Most of these were replaced with Bristol Fighters in 1918.

Meanwhile more squadrons were being raised in Australia for the Western Front. No. 2, 3 and 4 Squadrons arrived in France during August, September and December 1917 respectively. The first major action involving the AFC was the Battle of Cambrai on 20 November 1917. No. 2 Squadron (flying DH5s) was attached to the Third Army and during the battle carried out patrol duties, ground strafing of enemy troops and bombing. Seven aircraft out of 18 were shot down during the battle and two pilots were lost. Meanwhile No. 3 Squadron (RE8s) was supporting the last phase of the Passchendaele campaign in Flanders. Their main duty was spotting for the artillery but the bad weather during the winter months severely limited their opportunities. The last Australian Squadron to arrive (No. 4 flying Sopwith Camels) took up its duties with the First Army in a quiet sector of the front centred on Lens. Its most memorable encounter took place in March 1918 when 11 Camels from the squadron were attacked by German aircraft commanded by Manfred von Richtofen, the 'Red Baron'. The Australians shot down two of the Red Albatrosses for the loss of one of their own planes. The situation for No. 4 Squadron changed dramatically when the Germans launched their major offensive on 21 March. Their activities then involved much low-level flying and bomb-dropping to mask the approach of the tanks supporting the 4th Australian Division.

Australian airmen played a limited role in the great advance from Amiens on 8 August 1918. No. 3 Squadron assisted the attack with reconnaissance work and by flying contact patrols to help identify the position of the advancing infantry. In the next phase of the advance they photographed the whole of the Australian Corps' front to aid remapping and locating of German artillery batteries. Meanwhile No. 2 and 4 Squadrons were operating in the Ypres–Arras area carrying out patrols and raids on the railways around Lille. From September the Germans were in retreat on this part of the front and the squadrons were hard put to keep up. The most spectacular actions in which the Australians were involved occurred in the last days of the war. On 29 October, 15 Snipes of No. 4 Squadron encountered 60 Fokkers. What ensued was one of the largest air battles of the war. At the end of the dogfight 10 Fokkers had been shot down for the loss of one Snipe and several more badly damaged.

When the war ended on 11 November 1918 No. 4 Squadron was the only Australian unit in the British Army of Occupation. It arrived in Germany in December and was based in Cologne. In February 1919 the three squadrons handed over their equipment to the RAF. They arrived back in Australia in May. The AFC was disbanded and replaced by the Australian Air Corps, which became the Royal Australian Air Force in 1921.

F. M. Cutlack, *The Australian Flying Corps,* vol. 8 of the *Official History of Australia in the War of 1914–18* (Angus & Robertson, Sydney, 1923).

AUSTRALIAN FORCES VIETNAM see **VIETNAM WAR**

AUSTRALIAN IMPERIAL FORCE (AIF) The Australian Imperial Force was the name given to the expeditionary forces fielded by Australia for overseas service in the two world wars. While in 1914 both the Canadian and New Zealand governments designated their forces 'expeditionary' (Canadian Expeditionary Force, New Zealand Expeditionary Force), the first commander of the AIF, Major-General William Throsby Bridges (q.v.), chose 'imperial' to signify the twin nature of the Australians' duty: nation and empire. In 1939 the 2nd AIF was naturally assumed to be heir to the name as well as the traditions of the earlier force.

The first units of the AIF were raised in August 1914 following the outbreak of war. The government had offered a force of 20000 men for service anywhere, and enlistment and organisation were based on mobilisation plans drawn up in 1912. Enlistment under the provisions of the *Defence Act 1903* (q.v.) was voluntary, and the AIF, with South Africa, was to remain the only army on either side that did not resort to conscription during the First World War. Recruiting began on 10 August, and was regionally based. The 1st Brigade (1st–4th Battalions) was drawn from New South Wales, while the 2nd (5th–8th Battalions) came from Victoria. The 3rd Brigade, the 'All Australian Brigade', was a mixed force drawing on enlistments from the 'outer states': the 9th Battalion came from Queensland, the 10th from South Australia, the 11th from Western Australia and the 12th from South Australia, Western Australia and Tasmania. The 1st Light Horse Brigade, comprising the 1st, 2nd and 3rd Regiments, was recruited from New South Wales, Queensland, and South Australia and Tasmania respectively. This policy of drawing units from particular States, and often from particular regions or districts within a State, was also to be a feature of the AIF throughout the war.

Officers in the early contingents were drawn overwhelmingly from those who had held some form of commission before the war, either in the very small regular Australian or British Army, as officers of the CMF (see CITIZEN MILITARY FORCES), or in the Australian Cadet Corps (q.v.). Specialist officers such as Regimental Medical Officers often came from the prewar Army Medical Corps or the reserve, but some civilian doctors who volunteered were appointed directly, with the rank of captain. From January 1915 the only method of commissioning was from the ranks, and the only point of entry became enlistment as a private soldier. There were some minor exceptions to this, chiefly newly graduated subalterns from RMC Duntroon (q.v.), but the numbers involved here were small. This policy undoubtedly added to the generally egalitarian ethos of the AIF. The other ranks came from the full range of social, religious and economic backgrounds represented in the wider Australian population, but the original intention that half the force be drawn from men with militia training and half without does not appear to have been realised even in the early contingents, and certainly as the war went on the majority of enlistments had no previous military experience (for what that would have been worth in any case). They were, however, the best paid soldiers of the war. Bridges's chief of staff, Colonel Brudenell White (q.v.), decided that the pay of a volunteer should be higher than that of the militia, which then stood at 4 shillings a day. A private in the AIF was paid 5 shillings a day with an additional shilling 'deferred', to be paid upon discharge. Married men were required to allot 2 shillings a day to dependants; a separation allowance came later, in 1915, but was only available to men in the ranks. Junior officers were also paid more highly than their British counterparts, but senior officers received considerably less than was paid in the imperial army. While generous by the standards of other First World War armies, such conditions were not necessarily all that favourable when compared with the average wage in Australia in 1914.

The 1st Division reached Egypt in December 1914, to be joined by a further brigade of infantry and two more of light horse early the following year. The Australians and New Zealanders trained in the country surrounding their camp at Mena, before embarking to make the assault at Anzac Cove on 25 April 1915 which began the Gallipoli campaign. The eight-month campaign was a disaster, costing Australian casualties of 1007 officers and 25104 other ranks, of whom 362 and 7779 respectively were killed or died from wounds or disease, and included the AIF's first commander, Bridges. The AIF returned to Egypt to absorb reinforcements and rebuild units which had been sorely depleted by months of privation, and in early 1916 the force underwent substantial expansion and

reorganisation. It was also split, and for the rest of the war the Australians would fight in two widely separated theatres: the Western Front and the Middle East.

The 1st Division and the 4th Brigade were split in two. Their constituent infantry battalions were paraded and divided in half, the two halves then being brought up to strength with recently arrived reinforcements from Australia. As a result, the 4th and 5th Divisions were formed from a strong nucleus of men with recent combat experience, commanded by young officers who had earned promotion through their performance in battle, but containing a leavening of men who had not yet begun to feel the enervating effects of prolonged exposure to combat. The 2nd Division, which had arrived in mid-1915 and seen service on Gallipoli from early September, was retained intact because it had not had the opportunity to complete its organisation or training before being committed to battle. The 4th Brigade was allotted to the 4th Division and the 8th Brigade to the 5th Division. The 3rd Division meanwhile was forming in Australia and would sail directly for England and, ultimately, France.

The 1st and 2nd Divisions and the New Zealand Division were formed into I Anzac Corps under the command of a British general of the Indian Army, W. R. Birdwood (q.v.), who had also succeeded to overall command of the AIF itself following Bridges's death on Gallipoli. II Anzac Corps, comprising the 4th and 5th Australian Divisions, was commanded by another British officer, Lieutenant-General Sir Alexander Godley (q.v.), who was also the commander of the New Zealand Expeditionary Force. Birdwood's corps sailed for France in March 1916, followed by Godley's some weeks later; both were still in the process of retraining and re-equipping for the much greater demands that the war in France would make upon them. The light horse units stayed in Egypt under the command of Major-General Harry Chauvel (q.v.), who had led the 1st Light Horse Brigade on Gallipoli, and with New Zealand, British and Indian units fought a long and difficult campaign through Sinai and Palestine against the Turks, who were only finally defeated a week or so before the end of the war in Europe in 1918. Although Birdwood retained overall responsibility for these men, in practice Chauvel had responsibility for the AIF in the Middle East as well as ultimately commanding a corps in the Egyptian Expeditionary Force.

The two Anzac corps formed part of the British Expeditionary Force in France and Flanders and fought through more than two years of costly campaigning before taking a leading role as part of the Fourth Army in the fighting that ultimately defeated the Germans in the field in the last months of 1918. In November 1917, as a result of manpower shortages occasioned by the heavy fighting that year and the decline in voluntary enlistments in Australia, the five Australian divisions were formed into a single Australian corps, with the 4th Division initially acting as a depot division for the others. In May command of the Australian Corps devolved upon an Australian, Lieutenant-General Sir John Monash (q.v.), and the Australians earned their most illustrious plaudits under his leadership. With the signing of an armistice on 11 November 1918, Monash went to London and supervised the repatriation of the AIF to Australia, a task completed by the end of the following year. On their return to Australia the men were demobilised and mustered out of the army, and on 1 April 1921 the AIF officially ceased to exist. On the recommendation of a committee of senior officers, which was convened in 1920 at the direction of the Minister for Defence, George Foster Pearce (q.v.), the numerical designations, traditions and battle honours (q.v.) of the infantry battalions and light horse regiments were transferred to the CMF, which was reorganised accordingly.

With the declaration of war against Germany on 3 September 1939, consideration was again given to raising an expeditionary force for service overseas, based on a mobilisation scheme, Plan 401, initially drawn up in 1922. The new brigades and divisions were to be numbered sequentially as in 1914, and as the militia disposed of five divisions and 15 brigades, the new AIF formation was designated the 6th Division, and its brigades the 16th, 17th and 18th. Battalions, however, used the original numbers with a '2nd' prefix, for example, 2/1st Battalion. The 16th Brigade recruited in New South Wales, the 17th in Victoria, while the 18th drew units from Queensland (2/9th), South Australia (2/10th), Western Australia (2/11th) and Tasmania (2/12th). Throughout 1940 the 7th, 8th and 9th Divisions were raised, but the intention of maintaining the neat symmetry of the original AIF organisation broke down and the result was an organisational mess.

The first of the major changes came about through a British decision to reduce brigades from four battalions to three. The 18th Brigade had been diverted to England in June 1940, and the 6th Division was completed by combining the 2/4th and 2/8th Battalions, now surplus to establishment, with the 2/11th, which had arrived in Egypt ahead of the rest of the 18th Brigade, to form a new brigade, the 19th. This decision, together with the needs of divisions committed to action, set off a

train of organisational movement which lasted over a year. The 7th Division was made up with the addition of a new brigade, the 26th, which consisted of the 2/23rd, 2/24th and 2/48th Battalions, and which joined the 20th and 21st Brigades. The 8th Division, which was to be lost at the fall of Singapore in February 1942, consisted originally of the 22nd, 23rd and 24th Brigades, but never took the field as a complete division because the battalions of the 23rd Brigade were spread thinly across Ambon, Timor and Rabaul (qq.v.), where they were quickly overwhelmed in the first Japanese attacks. The 9th Division was formed from those forces diverted to Britain and formed into the 25th Brigade, together with the 24th Brigade, which was transferred from the 8th Division and replaced by a new 27th Brigade, and the 20th Brigade, which was transferred from the 7th Division and replaced in that formation by the 18th Brigade. After the final reorganisation following the first Libyan campaign, the four divisions then overseas were organised as follows: 6th Division (16th, 17th, 19th Brigades); 7th Division (18th, 21st, 25th Brigades); 8th Division (22nd, 23rd, 27th Brigades); 9th Division (20th, 24th, 26th Brigades). As a result of the lessons of the early campaigns in Europe, the decision was taken to raise an armoured division in Australia at the beginning of 1941, and the 1st Armoured Division joined the AIF order of battle, although by the beginning of 1942 it was still well under strength, lacked much of its equipment, and was destined never to serve outside Australia as a formation.

The organisation of the 2nd AIF was complicated further by the decision that any militia unit in which three-quarters of the men volunteered to serve anywhere outside Australian territory became an AIF unit. By February 1943, 530 000 men out of a total of 820 000 were volunteers of this kind, and during the course of the war about 200 000 men transferred from the CMF to the AIF. The result was that the army now consisted of units that had been raised as part of the 2nd AIF, units that had been militia but had been given AIF status, and units that remained militia ones and hence could not serve outside a clearly circumscribed geographical area. The 'two armies policy' was unwise at a number of levels, and gave rise to considerable animosity between men on both sides. Early volunteers for the 2nd AIF had been dubbed 'five-bob-a-day murderers' and 'economic conscripts'; they got their own back by designating the militia 'chocos' (q.v.), short for chocolate soldiers, or 'koalas' ('not to be exported or shot'). The ill will was compounded at the beginning of the war by the government's decision to make an unmarried private soldier's pay in the 2nd AIF less than a volunteer militia soldier, and less even than the dole. By 1943, after the heroic performance of some militia battalions in the Papuan campaign, much of this animosity had dissipated, and besides, the Americans now provided a common focus for the resentment of Australian soldiers, AIF or CMF, over leave, pay, amenities, and women. The army was demobilised rapidly at the war's end, and while at the end of September 1945 there were 286000 men in the 2nd AIF, 12 months later the strength of the entire army was just 61 903. The 2nd AIF was formally disbanded on 30 June 1947.

AUSTRALIAN INSTRUCTIONAL CORPS (AIC) was formed on 14 April 1921 to replace the Instructional Staff created shortly after Federation. It was responsible for training all soldiers as well as men called up for compulsory military training (see CONSCRIPTION). The AIC was the administrative and training backbone of the CMF between the wars, and many of its members received commissions in the 2nd AIF in the early months of the Second World War. They filled specialist functions within CMF units such as regimental sergeant-major and regimental quartermaster sergeant, and as signals and ordnance NCOs ran qualification courses for CMF NCOs and specialist schools. Many also assisted in running school cadet units. The AIC was abolished on 19 May 1955, though for practical purposes it had ceased functioning several years earlier.

AUSTRALIAN INTELLIGENCE CORPS (AUS INT CORPS) was formed on 6 December 1907 for the purposes of training soldiers in intelligence (q.v.) work; collecting and recording information about the topography and military resources of Australia, its dependencies and foreign countries (especially those in the Pacific region); and preparing strategical and tactical maps and plans. The first Director of Military Intelligence was Lieutenant-Colonel James Whiteside McCay (q.v.). The corps was effectively disbanded on 30 September 1914 and replaced by Intelligence Sections of the General Staff in each military district. The corps was re-formed in 1939 with responsibilities including intelligence; security; passport control; rail, air and shipping security; censorship; and prisoner of war interrogation and data compilation. At present the corps' main task is to provide intelligence personnel in every formation headquarters from brigade to Army Office. The School of Military Intelligence has been located at Canungra (q.v.) since 1974.

AUSTRALIAN LEGION OF EX-SERVICEMEN AND WOMEN was formed in December 1944, absorbing several existing ex-service organisations. It is distinguished by the fact that, unlike the RSL, it admits as members all ex-service personnel, regardless of where they served. This includes British Commonwealth and Allied ex-service personnel, as well as former peacetime members of the Australian Regular Forces and Reservists. The legion is formed in each State as a non-profit company, and has sub-branches in all major cities and towns. Its national headquarters is in Melbourne, and the National Council, which meets annually, makes submissions to government on matters affecting ex-service personnel. The legion is involved in welfare and social activities, as well as in assisting with benefit and compensation claims by ex-service personnel.

AUSTRALIAN MILITARY FORCES see **AUSTRALIAN REGULAR ARMY**

AUSTRALIAN MILITARY REGULATIONS AND ORDERS (AMR&O) These formed the regulatory base for the discipline and administration of the AMF in peace and war whether in Australia or overseas, and operated together with the Defence Act (q.v.) of 1903 and the British Army Act. They covered a great variety of subjects, from the appointment of officers and enlistment of other ranks, the preferring of charges and conduct of courts martial, and promotion, resignation and discharge, to the powers of commanding officers, provision of leave, and the employment of soldiers as officers' servants. Like the Defence Act itself, they were regularly amended. They were replaced by the Defence Force Discipline Act of 1982 (which came into force in July 1985), and by the *Defence Force Discipline Regulations and Rules*.

AUSTRALIAN NATIONAL DEFENCE LEAGUE, formed on 5 September 1905 in Sydney, was modelled on Britain's National Service League, but differed from the British body in significant ways. It sought a system of universal training based on the Swiss militia system rather than full-time national service, it had the support of Labor and Liberal politicians (including W. M. Hughes, J. C. Watson (qq.v.), and Prime Minister Alfred Deakin) as well as conservatives, and it was motivated by fears of an Asian rather than a European threat. The league's founding came only months after the Japanese naval victory over the Russians at Tsushima on 28 May 1905, and its journal *The Call* promoted fears of a Japanese threat (q.v.). The league was active in lobbying and published a list of candidates for the 1910 elections who supported compulsory military training. Such training was introduced after this election (see CONSCRIPTION), so the league disbanded. At the time of its disbandment it had 21 branches. Its membership, which peaked at 1600, included political, academic, religious and business leaders.

AUSTRALIAN NATIONAL WAR MEMORIAL, at Villers-Bretonneux near Amiens in northern France, commemorates Australians who died in France during the First World War. The memorial is sited in a military cemetery, and consists of a three-sided court and a 105-foot tower, made of fine ashlar. On the three walls, which are faced with Portland stone, are the names of 10 885 Australians who were killed in France and who have no known grave. The tower contains an observation room with a circular stone table with bronze pointers indicating the battlefields in France and Belgium in which Australians fought and, far beyond, Gallipoli and Canberra. After the war an appeal in Australia raised £22 700, of which £12 500 came from Victorian schoolchildren, with the request that most of the funds be used to build a new school in Villers-Bretonneux. The boys' school opened in May 1927, and contains an inscription stating that the school was the gift of Victorian schoolchildren, 1200 of whose fathers are buried in the Villers-Bretonneux cemetery, with the names of many more recorded on the memorial. Villers-Bretonneux is now twinned with Robinvale, Victoria, the centre of a large soldier-settlement scheme, which has in its main square an impressive memorial to the links between the two towns.

JOHN CONNOR

AUSTRALIAN NAVAL INSTITUTE was incorporated in Canberra on 10 June 1975 with the aim of advancing knowledge and understanding of naval and maritime affairs among serving and retired members of the RAN. The institute publishes the *Journal of the Australian Naval Institute*, organises seminars on naval matters, and attempts to bring professional expertise to the public discussion of naval and defence questions. It has branches throughout Australia.

AUSTRALIAN NAVAL AND MILITARY EXPEDITIONARY FORCE (ANMEF) see **GERMAN NEW GUINEA**

AUSTRALIAN NAVAL WHITE ENSIGN was adopted on 1 March 1967. It replaced the British White Ensign, dating from 1864 as the official flag of British men-of-war, the use of which had been denied the various Australian naval forces from 1884 to 1911, when the Commonwealth Naval Forces were founded. The Australian White Ensign retained the Union Jack in the canton and the white field, but replaced the Cross of St George with the blue Southern Cross and the seven-point Federation Star.

AUSTRALIAN NEW GUINEA ADMINISTRATIVE UNIT (ANGAU) was created on 10 April 1942, bringing those sections of Papua and New Guinea not in Japanese hands under a single military administration subordinate to New Guinea Force (q.v.). Headquarters was established in Port Moresby, and in August Major-General B. M. Morris (q.v.) was appointed GOC, ANGAU. ANGAU was initially made up of men who had been members of the prewar civil administrations, but as its strength grew from 450 in December 1942 to 2026 in July 1945 it took on inexperienced and sometimes unsuitable staff. ANGAU's activities included recruitment and management of indigenous labourers (see 'FUZZY-WUZZY ANGELS'); carrying out propaganda campaigns among the indigenous people; gathering military intelligence; supervising the indigenous police force; maintaining law and order; providing health and education services; providing relief and rehabilitation of refugees and of devastated areas recaptured from the Japanese; participation in actual fighting when necessary; arranging recruitment and non-operational administration of indigenous battalions (see PACIFIC ISLANDS REGIMENT); and acting as guides for forward troops. It also controlled production of essential war materials, but in May 1943, after pressure from plantation-owners, responsibility for copra and rubber production was transferred to the quasi-civilian Production Control Board. ANGAU remained responsible, however, for recruitment, administration and the welfare of plantation labourers. In ANGAU's activities military requirements took precedence over other considerations, including indigenous people's welfare. As a result, its achievements in helping the Allies to attain victory in the Papuan and New Guinea campaigns were not matched by achievements in administration or welfare. Perhaps its most lasting legacy lies in the continued administration of Papua and New Guinea as a single unit after the provisional civil administration extended its control over the territories, a process which was completed on 24 June 1946.

AUSTRALIAN RED CROSS was established on 13 August 1914, one week after the outbreak of war. It quickly became the main aid society, with over 100000 members, of whom 80 per cent were women, directing its efforts towards the raising of funds to purchase comfort supplies for Australian

Margaret Young, an Australian Red Cross worker, giving a transistor radio to Private Colin McLeod, a patient at the 8th Field Ambulance, Nui Dat, South Vietnam, December 1968. (AWM COM/68/1048/VN)

servicemen overseas and at home, and to producing huge quantities of clothing, especially gloves, socks and shirts. Red Cross members worked in hospitals and convalescent homes as part of the Voluntary Aid Detachments, and from late 1915 male members of the VAD helped supervise the sick and wounded on train journeys. Red Cross motor transport sections were established in each State in 1915 to provide transport from ships to hospitals for the returning casualties. A POW section was formed in London to look after the interests of Australian POWs, and large numbers of parcels and sums of money were sent to prisoners. The POW section was the official link between prisoners and their families, and maintained full casualty lists for each State.

This pattern of concern for the welfare of Australian servicemen continued in the Second World War, and in every subsequent war in which Australian troops have been employed. The activities of the Red Cross have also extended to the problems of postwar reconstruction and rehabilitation, ranging from assistance in the liberation of the Belsen concentration camp in 1945 to the provision of medical clinics in the 'New Villages' established during the Malayan Emergency in the 1950s as part of the drive against communist insurgents.

AUSTRALIAN REGULAR ARMY In its current guise the regular army has had a relatively short history, being formed on 30 September 1947 from the postwar Permanent Military Forces (PMF), but the army has had a proportion of regular, or permanent, soldiers throughout its existence. Before the Second World War the PMF existed to staff, train and administer the CMF. As well as some artillery, engineer, ordnance and medical units, the PMF had three main constituent parts over the course of its operation. The Australian Staff Corps (q.v.), formed on 1 October 1920, contained all officers of the combatant permanent forces and not just graduates of RMC Duntroon (q.v.). The Administrative and Instructional (A&I) Staff, which had fulfilled the organisational and training functions in the pre-1914 army, was disbanded at that time. The Australian Instructional Corps (q.v.) was created in October 1921 and incorporated the majority of permanent warrant and non-commissioned officers. The strength of the PMF was always low, especially in the interwar years when career opportunities were few, pay was poor and promotion painfully slow. In both world wars the regulars provided the leavening of professional military knowledge necessary to raise, equip, train and deploy an army for war, and many shone in command positions when given the opportunity. It was recognition of the fact that reliance on part-time military institutions was an inadequate basis for national defence that led the government after the Second World War to authorise the creation of a regular army available to meet a growing number of commitments overseas, most immediately in the British Commonwealth Occupation Force (q.v.) for the postwar occupation of Japan. It was the regular army which fought the Korean War, the Malayan Emergency, Confrontation and the Vietnam War; and senior military professionals who became the principal source of advice to government on military and defence policy. At present it consists of 22 corps. The strength of the postwar regular army has fluctuated depending on the number and scale of its commitments and the budgetary decisions of government, and in the early 1990s as part of force 'downsizing' the regular establishment was reduced and some functions were located either with the Army Reserve or civilian contractors.

(See also CORPS; ROYAL AUSTRALIAN REGIMENT.)

AUSTRALIAN SQUADRON was created as a result of decisions made at the 1909 Imperial Conference (q.v.) at which the British government reversed its long-standing opposition to separate dominion navies and agreed to the creation of an Australian fleet unit, which was to become the Royal Australian Navy. Admiralty thinking originally intended the formation of an 'Eastern' or 'Pacific' fleet, to which contributions would be made by the RN and by the governments of the dominions of Australia, New Zealand and Canada. Internal politics and a marked lack of interest in the Pacific meant that the Canadian contribution was never likely, and indeed did not come about, while the government of New Zealand chose to underwrite the cost of a battlecruiser, HMS *New Zealand*, which before the First World War was to form part of the China Squadron. The significant features of the Australian Squadron were that it was to be manned as far as possible by Australians, that in peacetime it was firmly under the control of the Australian authorities, but that at the outbreak of war control reverted automatically to the Admiralty. As a consequence of this last provision, ships of the Australian navy were governed by regulations similar to those which pertained in the RN, that is, the King's Regulations. This was felt to be doubly important because a widespread program of interchange between the two services was contemplated from the outset, and conditions within each navy therefore needed to be in general accord. Rates of pay in the Australian service, for example, were identical to those in the RN. These provisions were also the basis of the markedly British flavour which the RAN demonstrated from the outset of its existence, and which it was to maintain until long after the end of the Second World War. The Australian Army, in contrast, was not brought under the provisions of the British Army Act, which meant,

among other things, that the automatic provision of the death penalty could not be extended to Australian soldiers serving overseas. In the Australian Squadron (and subsequently again in the RAN at the outbreak of war in 1939), disciplinary policy emanated from the Admiralty and not from the Naval Board. At the outbreak of war in August 1914 the Australian Squadron consisted of the battle cruiser HMAS *Australia*, the light cruisers HMA Ships *Melbourne*, *Sydney*, and *Brisbane* (then still building), the destroyers *Parramatta*, *Yarra*, and *Warrego*, and the submarines *AE 1* and *AE 2*.

AUSTRALIAN STAFF CORPS Visiting British Field Marshal Lord Kitchener had recommended in his report of 1910 that a staff corps of 350 be formed to fill the administrative, staff and instructional positions in the permanent forces, with RMC Duntroon as the sole source of officers for the corps. The college was established in 1911, but the Staff Corps was not created until 1 October 1920. It included all officers of the combatant permanent forces, and not just graduates of Duntroon. The Administrative and Instructional Staff, to which many non-Duntroon graduates had been previously assigned, was disbanded in October 1921 and replaced by the Australian Instructional Corps (q.v.), to which the majority of permanent warrant and non-commissioned officers were allotted. Relations between the Staff Corps and the citizen officers of the CMF were sometimes bad during the Second World War, a situation that arose from the markedly less favourable pay and conditions, and restricted professional opportunities, which were the lot of the permanent officers between the wars.

D. M. Horner, 'Staff Corps versus Militia: The Australian experience in World War II', *Defence Force Journal*, vol. 26 (January–February 1981), pp. 13–26.

AUSTRALIAN SURVEY CORPS see **ROYAL AUSTRALIAN SURVEY CORPS**

AUSTRALIAN VETERANS AND DEFENCE SERVICES COUNCIL (AVADSC), formed in 1971 at the suggestion of the then Minister for Repatriation, R. M. Holten, is an umbrella organisation of national ex-service associations. Each member association has two representatives on the council, originally called the Australian Services Council. Membership has grown from eight associations in 1971 to 28 in 1993, but the RSL is not a member and no longer even sends an observer to council meetings. The council restricts its activities to action on veterans' affairs and defence force benefits, taking no part in discussion of wider political issues.

The first meeting of the Australian War Cabinet, Melbourne, 27 September 1939. From left to right: Senator George McLeay, Minister for Commerce; Sir Henry Gullett, Minister for External Affairs and Information; Richard Casey, Minister for Supply and Development; Robert Menzies, Prime Minister and Treasurer; Geoffrey Street, Minister for Defence; and Frederick Shedden, Secretary of the Department of Defence and the War Cabinet. (AWM 042822)

AUSTRALIAN VIETNAM VETERANS' NATIONAL MEMORIAL see **VIETNAM VETERANS' NATIONAL MEMORIAL**

AUSTRALIAN WAR CABINET met for the first time on 27 September 1939, its formation having been foreshadowed in the War Book in the 1930s and formally announced on 15 September 1939. Membership consisted of the Prime Minister and Treasurer, and the Ministers of Defence, External Affairs, Supply and Development, and Commerce, with the Secretary of the Department of Defence acting as War Cabinet secretary. In June 1941, in line with developments that Menzies had observed in the United Kingdom, the Australian War Cabinet was expanded to include the service chiefs, ministers of departments involved in particular issues under discussion, and representatives of Allied governments, and it assumed responsibility for the general direction of war policy. The Secretariat headed by Mr (later Sir) Frederick Shedden (q.v.) marked the beginning of modern administrative methods in the cabinet, with the maintenance of systematic records of cabinet discussions and supporting papers, in place of the *ad hoc* and informal notes that had been kept previously.

Paul Hasluck, *The Government and the People*, 2 vols (Australian War Memorial, Canberra, 1952, 1970).

AUSTRALIAN WAR MEMORIAL was founded on a unique proposition: that the commemoration of Australian war dead in the Great War should involve knowledge and understanding in association with symbolism. Australia would create more than a cenotaph or shrine, constructing, in addition to commemorative features, a major museum, archive and library as a repository for an extraordinarily diverse collection.

In describing the origins of this idea, some have concentrated on the impact the remains of the former battlefield at Pozières had on Charles Bean (q.v.) and his small band of supporters in 1917. In this account, Bean, ever a romantic, studying that most disastrous battlefield for Australia, began to ponder the problem of the unknowability of war for those at home. Faced with the debris of battle and the evidence of the awful and savage loss of life, Bean looked for a way of helping Australians at home to know what their troops had endured and achieved. It was at Pozières that he decided that only in knowledge and truth could commemoration properly exist.

No doubt Pozières made a significant impact on Bean's thinking, but it should also be acknowledged that he urged the Australian government to consider a major war museum in response to the efforts of the Canadians and the British. Indeed, such were the plans for the Imperial War Museum in London that Bean feared that there would be few objects of significance left for Australia. Such was the scale of the First World War, such the terrible loss of life, that nations everywhere, in the century of democratic wars, determined to ennoble those of their citizens who had died. That Australia was to do so with a cultural institution of significance was remarkable and far-sighted and suggests the widely accepted view that the AIF had a profound influence on shaping the development of Australia's brief history.

Throughout its history the Australian War Memorial has suffered from a certain ignorance and misunderstanding of its purpose, and if its founders, so assiduous and determined in all matters relating to the Memorial, failed it in one thing, it was in explaining and convincing the nation that this institution would be unique and quite different. Too often those who have professed to love it most have understood it least. They have wanted a place of the heart while Bean also constructed a place for the mind.

The failure of the founders can be blamed in part at least on the extraordinary length of time between conception and birth. Charles Bean thought that he had secured government approval for his museum by late 1917, by which time he already had the outlines of a striking collection of objects and records relating to Australia's role in the war. In May 1917 he had secured the appointment of a young army officer, John Linton Treloar (q.v.), as head of the newly formed Australian War Records Section. Treloar became perhaps Australia's most passionate collector and certainly Australia's first great museum professional. He thoroughly organised the creation and collection of the official records of the Australian units at war, building on the copious war diaries that every unit of each arm was required to keep: a unique and precious record at first hand, often of platoon or company level engagements. Treloar also organised a methodical and rigorous accumulation of objects that would flesh out museum displays with a reality that would be the hallmark of his eventual galleries. Uniforms, water bottles, duckboards, evidence of humour and the day-to-day life of the Australians and those who opposed them, weapons of all kinds, and machinery constituted the material evidence that Treloar sought. His collection was returned to Australia in 1919 in staggering quantity and diversity. To this Bean added the works of war photographers and war artists, specifically recruited to the AIF, commissioned to create an Australian perspective on war, the hallmarks of which would be honesty and diversity.

Dogged by bad luck and government and possibly veterans' indifference, the Memorial did not open until November 1941, by which time Australia was at war again. Treloar had left the Memorial by 1940 and indeed had been so disheartened by the delays during the interwar years that he had actively sought to work elsewhere. The Memorial sorely missed Treloar's experience and vision during the installation of the galleries and in guiding its formative years in Canberra. No one knew the collection as he did and a quiescent Board of Management had largely left affairs entirely in his hands. Returning to the Memorial in 1946, Treloar faced an even bigger task in securing government support for extensions to the building to house the equally massive Second World War collection, and in integrating that collection with the earlier one. Despite extraordinary devotion to duty as shown in punishingly long hours, and with very little support or public sympathy, Treloar increasingly became less effective and on his death in 1952 the Memorial faced a parlous future. The board, led by Charles Bean, whose vast intellectual powers had now substantially declined, found it difficult to attract staff of dynamism and vision and the Memorial, the centre-piece of a largely remote Canberra, became something of a backwater. Bean undertook a survey of the 'surplus' First World War collection and, lacking experience and appropriate guidelines, broke up what Treloar had so diligently put together. The new Director, J. J. McGrath, removed items of utmost importance in the Second World War collection, arguing that governmental and public indifference made the storage of items impossible to envisage even in the long term.

A visitor to the Memorial in the 1960s would have been greeted on arrival by the sound of rather funereal music that was played constantly in the Memorial's public areas. The commemorative courtyard displayed the Roll of Honour that listed the name of every serviceman and woman to die for Australia, the merchant navy excepted. The names, on enormous bronze panels, were listed without rank or decoration and by ship's company, battalion or squadron. The cloisters housing the Roll open on to the Hall of Memory, envisaged by Bean as a place of reflection that would also house the AIF's most sacred objects. Even in the Hall of Memory commemoration was to be assisted by understanding. The Hall, with its massive stained glass windows, had the feel of a church but no trace of religious symbolism.

The museum galleries, surrounding and beneath the commemorative courtyard, focused almost exclusively on the First World War using Treloar's collection to tell a unified, chronological and coherent story. Photographs and paintings gave a sense of realism and that was assisted by the several dioramas over which skilled sculptors and craftsmen had laboured throughout the 1920s. A frozen moment in time, and designed to be absolutely accurate, the dioramas gave, to those who knew something of the story, a graphic account of some of the highlights of Australian participation in the First World War. Supporting them were what Bean called 'plan models': three-dimensional, free-standing, large-scale terrain models of significant battlefields. A father might show his sons under which bridge he had been sheltering at Villers-Bretonneux when the Germans began a gas attack. Text panels, largely drafted by John Treloar some 40 years earlier, gave the visitor the detail to make sense of the objects, paintings and photographs. Some few Second World War objects reminded the visitor that Australia had more recently fought for its survival as a nation.

It was in 1968 that the board secured government approval for the extension of the Memorial building. Sir Edmund Herring (q.v.), a distinguished civilian soldier in both the First and Second World Wars, was the chairman who would oversee this work and at 76 years of age might not have been best placed to understand the needs of the Memorial's changing visitors. Initially at least, the Memorial confronted the need for change reluctantly and each departure from the established way of doing things provoked criticism from sections of the public. However changing community expectations, and particularly the development of a mature museum profession in Australia, required that Memorial staff reconsider approaches within the galleries. Increasingly professionally qualified staff were appointed to positions at the Memorial and while they lacked direct experience of war they brought skills and training that their predecessors had lacked. In 1980 parliament significantly increased the Memorial's responsibilities, extending its coverage to all wars in which Australians had participated. Growing demands on research and education services and the need to match sophisticated and professional museums and galleries elsewhere in Canberra and in the other States placed additional strains on the Memorial's small staff, and change created difficulties and conflicts.

Nevertheless the Memorial demonstrated an increasing relevance to national life even as the experience of war receded from popular memory. The Memorial became recognised as an important custodian of crucial aspects of Australia's national collection, numbers visiting both the galleries and the research centre increased dramatically and the Memorial deliberately embraced Bean's challenge to it of commemoration through understanding.

The visitor in the 1990s has a variety of galleries to inspect, including displays on peace-keeping, the South African War, the Vietnam War, and particular themes such as women in war. The research collection is listed and described, making it far more accessible than ever before, and along with other collections, such as the the photographic collection, is now available on a worldwide basis using the most recent information technology.

Critics complained that the Memorial was departing from its original charter by introducing such change, while Memorial advocates suggested that the mission of commemoration through understanding was more essential than ever as generations came to maturity blessed with no direct experience of war. To reaffirm that commemoration remained the essence of the institution's purpose, the Memorial's council decided to mark the 75th anniversary of the end of the First World War with the return, for reburial, of an Unknown Australian Soldier (q.v.) from France. This matter had first been discussed in the 1920s but the difficulties involved and exaggerated imperial sentiment had deflected detailed consideration of the proposal. With the cooperation of the Commonwealth War Graves Commission and the Australian Defence Force, an Australian Unknown Soldier lying in the Adelaide Cemetery at Villers-Bretonneux was returned to the Memorial's Hall of Memory for reburial. The Unknown Soldier was laid to rest with appropriate military honours in a ceremony in which millions of Australians were able to feel involved. Many believed that the commemorative area for the first time truly reflected the aspirations of those who had laboured to create a unique Australian war memorial.

MICHAEL MCKERNAN

AUSTRALIAN WOMEN'S ARMY SERVICE (AWAS), originally the Australian Army Women's Service from October 1941, was authorised by a decision of the War Cabinet (see AUSTRALIAN WAR CABINET) on 13 August 1941 'to release men from certain military duties for employment with fighting units', and followed on from a proposal initiated by the army in May that year. The cabinet decision specified that no woman was to be sent overseas without cabinet approval, and that details governing conditions of service and rates of pay would be worked out by the Treasury Finance Committee.

Miss Sybil Howy Irving, daughter of Major-General G. G. Irving, was appointed the Controller, with the rank of lieutenant-colonel, in late September. A former secretary of the Girl Guides' Association in Victoria, she was assistant secretary of the Australian Red Cross (q.v.) in Victoria at the time of her appointment. She had two assistant controllers, with the rank of major; each of the four area commands (Northern, Central, Southern and Western) had an assistant controller responsible for the AWAS personnel within that command. Rates of pay were set at two-thirds of equivalent male rates. Women seeking to enlist had first to gain a clearance from the Directorate of Manpower, since certain categories of female labour came under protected employment provisions. War Cabinet had directed originally that the AWAS establishment should be restricted to 1600 women aged between 18 and 45, but by March 1942 pressure was being exerted to have this increased in the face of the labour needs occasioned by the outbreak of the Pacific War. Ultimately 24028 women served in the AWAS, and its establishment was lifted to 6000. This included 3618 who served with the Royal Australian Artillery manning fixed defences — usually radars and searchlights attached to anti-aircraft guns — and about 3600 who worked in the Australian Corps of Signals.

The AWAS did many and varied jobs, considerably beyond the original intention to restrict them largely to 'traditional' areas of female employment as clerks, typists, cooks, and motor transport drivers.

General T.A. Blamey requested that 500 be sent to the Middle East, but the advent of war with Japan stopped this. They served all over Australia, often in isolated areas, and worked as drivers, signallers and clerks, and in intelligence, chemical warfare units, watercraft workshops, electrical and mechanical engineers repair shops and in the Ordnance Corps. In 1945 a contingent was sent to Lae after advanced headquarters, Allied Land Forces, moved to Hollandia, and to make up shortfalls in personnel at the headquarters of First Australian Army. Another group attached to intelligence units spent three months in Hollandia in the first months of 1945. Others went to Lae and Rabaul after the end of hostilities. The AWAS was disbanded at the end of the war, final demobilisation being completed by 30 June 1947.

Jean Beveridge, *AWAS: Women Making History* (Boolarong, Brisbane, 1988); Ann Howard, *You'll Be Sorry: Reflections of the AWAS from 1941–1945* (Tarka, Sydney, 1990).

AUSTRALIAN WOMEN'S LAND ARMY (AWLA)

was established on a national basis following its approval by the Minister for Labour and National Service on 27 July 1942. The outbreak of war had

A member of the Australian Women's Land Army learning to milk on a papier mâché cow at a Land Army training farm, Darley, Victoria, 6 September 1944. (AWM 141684)

meant that while production of food and materials such as flax and cotton was needed more than ever to supply both Australia and its allies, many male agricultural workers were leaving the countryside to join the armed forces. Women's organisations responded by setting up 'land armies' in each State, and most of the women from these private groups were later absorbed into the AWLA. The AWLA, which was under the control of the Director-General of Manpower, was open to all women who were British subjects or 'friendly aliens' aged between 18 and 50 (or between 16 and 18 with parental approval) and not already engaged in rural work. As well as the full-time members, who enrolled for periods of not less than 12 months or for the duration of the war, there was an AWLA Auxiliary of women who worked for periods of not less than four weeks at nominated times of the year. Enrolment never reached the projected level of 6000; at its peak in October 1944 the AWLA had 2565 permanent and 503 auxiliary members.

AWLA women worked 48-hour weeks, performing almost every kind of agricultural labour as well as working in canneries and packing houses. The women were to be paid the award rate, the going rate for the district or the AWLA minimum wage (30 shillings a week plus keep or 50 shillings a week without keep), whichever was higher. In almost every case they were paid much less than men for doing exactly the same work. They had their own uniform and badge, and were trained at AWLA establishments set up in every State except New South Wales and South Australia. AWLA workers lived either in communal living quarters with a supervising matron or on the property where they were working. Many former AWLA members remember their time in the Land Army as an enjoyable experience which taught these mostly urban women a lot about rural life, but they express resentment about the government's failure to treat them as equal to the armed forces' women's auxiliaries. During the war the government gave the impression that the AWLA was an official fourth women's service, a position endorsed by Cabinet in January 1943, but a bill to this effect was not completed before the war's end. After the war, with the disbandment of the AWLA in December 1945 pending, the matter was simply dropped, and as a result AWLA members were not eligible for various benefits made available to ex-servicewomen. Since then, many former AWLA women have felt that their important wartime role has not received due recognition, with the RSL refusing to let them participate in Anzac Day marches until the 1980s. A Committee of Enquiry into defence and defence-related awards recommended in its March 1994 report the creation of a new Civilian Service Medal 1939–45 within the Australian honours and awards system to recognise wartime civilian service by groups such as the AWLA.

Sue Hardisty, *Thanks Girls and Goodbye! The Story of the Australian Women's Land Army 1942–45* (Viking O'Neil, Melbourne, 1990).

AUSTRALIANS IN THE SERVICE OF OTHER NATIONS Although Australia has a long tradition of sending official expeditions to fight overseas, there has not been a strong tradition of individual Australians joining the armed forces of foreign countries. The major exception to this rule concerns the armed forces of the United Kingdom, which was not strictly a foreign country as until recently Australians were British subjects. Many Australians have served in the British forces both in wartime and in peacetime, some choosing to spend their entire careers in British service and rising to quite senior ranks. Large numbers of Australians joined the British armed services during both the First and Second World Wars; of these, perhaps the best known are those who served in the RAF. Only a few hundred Australians flew in the British forces during the First World War, but they included some of Australia's most skilful flying aces as well as future aviation pioneers such as Charles Kingsford Smith. In 1939 there were more Australian aircrew in the RAF than in the RAAF, and throughout the Second World War thousands of Australians served in RAF units. Most had been provided under the Empire Air Training Scheme (q.v.) and they saw service in Europe, the Middle East and south-east Asia.

In 1870 the British Parliament passed the Foreign Enlistment Act, which applied to all British subjects, including those in the colonies and dominions. The act made it an offence to enlist or recruit someone to enlist in military service against a foreign State at peace with the Crown, or to leave Britain or the dominions in order so to enlist. The purpose of the act was clearly to prohibit British subjects from fighting in any war in which Britain was neutral, so 'foreign State' was defined so broadly that it could encompass both sides in a civil war. Australia did not pass its own legislation dealing with this area until federal parliament passed the *Crimes (Foreign Incursions and Recruitment) Act 1978*. Under section six of this act it is an offence punishable by 14 years' imprisonment for Australian citizens or residents to engage in hostile activity in a foreign country against the government of that country, or to enter a foreign country with the intention of engaging in such hostile activity.

Recruitment, financing, training or other activities carried out in Australia in preparation for the commission of an offence under section six are also prohibited. Finally, the act makes it illegal to recruit in Australia for foreign armed forces or to advertise for this purpose, but it does not seek to prevent Australian citizens from joining the armed forces of foreign governments recognised by Australia.

The Australians who have fought for other nations can be divided into mercenaries who are motivated primarily by desire for money and adventure, ideologues who are motivated by political commitment, and migrants who are motivated by attachment to their country of origin. The boundaries between these groups are blurred, however: most people who enlist for active service overseas are probably seeking adventure as much as anything else, while even the most hardened mercenary may have some political commitment (often of an anti-communist and/or White supremacist nature). It seems reasonable to assume that small numbers of Australians have made their living as mercenaries throughout the twentieth century. Eight Australians served with the French Foreign Legion during the First World War, while four joined the German 'British Free Corps' during the Second World War. These latter men were POWs in German camps; two joined in the hope of escaping, and when this was impossible, returned to camp, while the other two, a soldier and a merchant seaman, enlisted for the mercenary reason of gaining better food and conditions and remained with the unit until the end of the war. The soldier, Albert Stokes, was court-martialled and sentenced to a year's imprisonment for 'voluntarily aiding the enemy'. More recently, in Africa, Australian mercenaries have been reported serving in the Nigerian air force during the Biafran war, in the Rhodesian army during the war which led to the creation of Zimbabwe, and as officers of South Africa's No. 32 'Buffalo Battalion', made up of black Angolans.

The number of Australians who have fought for foreign countries because of ideological commitment is probably very small, though they are perhaps the best known group. One early example was Arthur Lynch, Australian-born of a radical Irish Catholic father, whose republican and anti-imperialist principles prompted him to lead a brigade for six months on the Afrikaner side in the Boer War. Elected as an Irish nationalist MP, he was arrested and tried for treason in 1902 when he arrived in London to take up his seat. He was found guilty, but his death sentence was immediately commuted to life imprisonment and following mass petitioning he was released after only a

year. Ironically, he spent the First World War recruiting in Ireland for the British Army.

A more famous instance of Australian radicals fighting for a foreign country was the involvement of at least 66 Australian men and women in the Spanish Civil War of 1936–39. Other countries contributed many more people to the war; the relatively small number of Australians can perhaps be explained by Australia's distance from Europe and by the strength in Australia of Catholic opposition to Spain's Republican government. Although the Australian government could have prosecuted those who fought in Spain under the Foreign Enlistment Act, it received no reports of Australians travelling to Spain to fight in the war and therefore took no action. Only one Australian, Nugent Bull, is known to have fought for General Franco's Nationalist forces; the rest fought on the Republican side. At least 14 were killed in the war, while Bull, who joined the RAF during the Second World War, was killed in Bomber Command. Because there were so few Australians in Spain they were assigned to battalions of other nationalities rather than forming their own unit.

Among the Australians who fought in Spain were some Spanish migrants to Australia, and in a sense migrants who return to fight in their country of origin are a subcategory of those who fight out of political commitment. It is impossible to know how many migrants have done this, though during the 1960s and 1970s a few Croatians engaged in military training in Australia for a guerrilla campaign against the Yugoslav government, and more recently there have been reports in the media of former Australian residents fighting in nationalist struggles from the former Yugoslavia to Kurdistan. In addition, in some cases citizens of foreign countries who are also Australian citizens, or who are Australian residents, can be made to undertake compulsory military service under the laws of their country of origin if they return there.

The largest known groups who have enlisted from Australia to fight for the country of their birth were in the First World War. A special force, the 'Jugo Slav Battalion', was raised in Australia for service with the Serbian Army; several hundred men enlisted and embarked from Australia in September 1917, accompanied by a small number of Australian officers who went only as far as Egypt. The men were promised either repatriation back to Australia at the end of the war, or a land grant in Serbia. The collapse of the Austro–Hungarian Empire in 1918 and the consequent turmoil in Serbia meant that neither of these promises could be kept by the Serbian government, and those who chose to return to Australia had to wait until 1920 before their repatriation could be arranged. A second, and much larger, special force drew on Italians living in Australia. Designated 'Italian Reservists', they left Australia in four contingents, in May, June, July and November 1918. As with the 'Jugo Slav Battalion', their story remains almost entirely unknown.

Amirah Inglis, *Australians in the Spanish Civil War* (Allen & Unwin, Sydney, 1987).

AUXILIARY MINESWEEPERS When the Second World War broke out in 1939 there was a perceived need for a series of craft to sweep mines in Australian waters. For this purpose a wide variety of ships including tankers, steamers and tugboats was requisitioned by the navy. In all, 35 vessels were taken over and stationed in groups in Adelaide, Brisbane, Darwin, Fremantle, Hobart, Melbourne, Newcastle and Sydney. One vessel, HMAS *Patricia Cam,* was sunk by Japanese aircraft off the Northern Territory coast on 22 January 1943.

AUXILIARY SQUADRON, AUSTRALIA STATION, was established under the provisions of the Australasian Naval Defence Act of 1887 (see ANGLO-AUSTRALIAN NAVAL AGREEMENTS). The Australian and New Zealand colonial governments were to pay 5 per cent of the original cost of the ships plus the cost of maintenance and manning; in return they gained an assurance that, although the squadron would be under the control of the British C-in-C of the Australia Station (q.v.), it would not be deployed outside the station without the colonies' consent. Seven vessels under construction for the Royal Navy were set aside for the squadron: the 2nd Class cruisers *Ringarooma*, *Tauranga*, *Mildura*, *Katoomba* and *Wallaroo*, and the torpedo gunboats *Boomerang* and *Karrakatta*. The cruisers were lightly armoured ships with powerful engines which made them faster than the other major warships on the station. The torpedo gunboats were unarmoured vessels designed to catch and destroy torpedo boats, a task for which they proved too

slow but one which they were not required to carry out in Australia. The ships arrived in Sydney in September 1891 and operated like the other British warships on the station, never engaging in joint exercises with the colonial navies. Usually two cruisers and one torpedo gunboat were kept in reserve. The only break from their routine duties came in 1900 when the *Wallaroo* was released with colonial approval for service in China during the Boxer Rebellion (q.v.). Dissatisfaction with the Auxiliary Squadron grew amongst both the British authorities, who resented having their ships restricted to one part of the Empire, and the Australians, who found the ships unsuitable for the rough local waters and objected to having them crewed exclusively by British sailors. A new Anglo-Australian naval agreement in 1903 replaced the Auxiliary Squadron with an improved British fleet crewed partly by Australians and New Zealanders but free to operate in the Australia, East Indies and China Stations. Between 1901 and 1906 the Auxiliary Squadron's ships returned to Britain, where all but one were sold in 1905–06.

Katoomba, Mildura, Ringarooma, Tauranga and *Wallaroo*: displacement 2575 tons; length 265 feet; beam 41 feet; speed 19 knots; armament 8 × 4.7-inch guns, 8 × 3-pounder guns, 4 × 14-inch torpedo tubes, 4 machine-guns.

Boomerang and *Karrakatta*: displacement 735 tons; length 230 feet; beam 27 feet; speed 21 knots; armament 2 × 4.7-inch guns, 4 × 3-pounder guns, 3 × 14-inch torpedo tubes.

AVERNUS, HMCS see **NEW SOUTH WALES NAVAL FORCES SHIPS**

AVIATION CORPS see **AUSTRALIAN ARMY AVIATION CORPS**

AWARDS see **DECORATIONS**

AWARE HMAS see **ATTACK CLASS PATROL BOATS**

B

BALLARAT (I), HMAS see **BATHURST CLASS MINESWEEPERS (CORVETTES)**

BALLARAT (II), HMAS see **ANZAC CLASS FRIGATES**

BALTIMORE, MARTIN (4-crew bomber [Mark IV]). Wingspan 61 feet 4 inches; length 48 feet 6 inches; armament 8–14 × 0.303-inch machine-guns, 2000 pounds bombs; maximum speed 308 m.p.h.; range 1082 miles; power 2 × Wright 1660 h.p. engines.

The Baltimore was an American-designed bomber ordered by the British government and used to equip squadrons serving in the Mediterranean. Two Australian squadrons, with crews trained through the Empire Air Training Scheme (q.v.), used the Baltimore during the Second World War. No. 454 Squadron replaced its Bristol Blenheim bombers with Baltimores in 1943 and No. 459 Squadron replaced its Lockheed Hudsons and Venturas (qq.v.) with Baltimores in 1944. Both squadrons used the Baltimore until 1945 for anti-shipping strikes off Italy and the Greek islands and for bombing raids against targets in Yugoslavia.

BANDOLIER, HMAS see **ATTACK CLASS PATROL BOATS**

BANKA ISLAND MASSACRE see **BULLWINKEL, Lieutenant Vivian**

BARBER, Major-General George Walter (20 November 1868–24 July 1951). Born in Lancashire, England, Barber studied medicine and migrated to Australia in 1895, working as a doctor in Kalgoorlie. He was commissioned in the West Australian (Volunteer) Medical Staff in 1900 (see COLONIAL MILITARY FORCES) and joined the Australian Medical Corps when it was formed. When the First World War began, Barber was made a major in the AIF. During the Gallipoli campaign he served as second in command and later commanding officer of No. 2 Australian Stationary Hospital, which worked from hospital ships anchored off the Greek island of Lemnos. Barber went to the Western Front as Assistant Director of Medical Services for the 4th Division. Taking account of the special circumstances of trench warfare, Barber moved regimental aid posts close to the lines, increased the number of stretcher-bearers and set out strict hygiene and sanitation rules in standing orders which were imitated by the other Australian divisions. In 1918 he became Deputy Director of Medical Services for the Australian Corps. In 1925 he joined the Permanent Military Forces as Director-General of Medical Services and was promoted to major-general in 1927. From 1927 to his retirement in 1934, Barber was also Director-General of Medical Services for the RAAF and the Department of Civil Aviation.

BARBETTE, HMAS see **ATTACK CLASS PATROL BOATS**

BARCOO, HMAS see **RIVER CLASS FRIGATES**

BARNARD, Lance Herbert (1 May 1919–). A schoolteacher by training, Barnard served overseas with the 9th Division of the 2nd AIF between 1940 and 1945. He held the federal seat of Bass for the ALP from 1954 until his retirement from the parliament in 1975, was deputy leader of the parliamentary party between 1967 and 1974, and minister for Defence in the Whitlam government from 1972 to 1975. In this position he presided as minister over the implementation of the thorough-going reform in the defence sector usually associated with the name of Sir Arthur Tange (q.v.), the secretary of the department. Following his retirement from politics he held an ambassadorial post and was Director of the Office of Australian War Graves within the Department of Veterans' Affairs.

BARRA is a sonobuoy developed at the Defence Research Centre at Salisbury, South Australia, during the late 1970s for the detection of submarines. Dropped into the sea by an Orion (q.v.) maritime patrol aircraft, the Barra sonobuoy is able to trace the sound of even the quietest submarine, and sends information of the submarine's location back to the aircraft. The Barra is manufactured by Plessey Australia and has also been sold to the United Kingdom.

BARRACKS began to be constructed extensively in Britain during the French Revolutionary and Napoleonic Wars, replacing the previous system of billeting troops with local inhabitants. Intended to make it easier to keep soldiers under control and to keep them from what the barracks department called 'the contaminating influence' of the lower orders, barracks were at first controversial because they symbolised the separation of the standing army from the rest of society. British troops in Australia were housed in barracks from the early days of settlement, though the temporary barracks built at new settlements were crude buildings of clay and bark. In time, however, more substantial stone barracks were built in major towns. They were generally built on elevated sites with good views of the surrounding district; the ability to observe people's movements was particularly important in penal settlements where troops might have to move quickly to quell convict rebellion. In general, major barracks in Australia seem to have been quite comfortable, perhaps more comfortable than those in Britain. Barracks were important social centres in colonial towns: they were the location for sporting events and other activities that were popular with the general public, as well as for events such as officers' balls attended by the élite. Some barracks built in the nineteenth century went on to be used by colonial military forces and are now used as military headquarters. Notable examples include Anglesea Barracks, Hobart (Australia's oldest military establishment still occupied by the military, built around 1814–18), Victoria Barracks, Sydney (occupied from 1848), Victoria Barracks, Melbourne (occupied from 1860 and used as the Commonwealth's defence headquarters for almost 60 years from 1901) and Victoria Barracks, Brisbane (occupied from 1864).

Barracks went on being constructed throughout the twentieth century. Often temporary structures, built at times when there was a sudden need to house large numbers of soldiers, such as during the Second World War and the second National Service scheme (see CONSCRIPTION), they became permanent or semi-permanent. By the late 1980s it had become clear that the neglect of on-base barracks accommodation, combined with the lack of privacy due to frequent sharing of rooms, was a major problem. Those who live in barracks are single members of the armed services, and though their accommodation costs are still partially subsidised, since the early 1970s they have paid a rations and quarters charge for living in barracks. (Members with families are provided with houses, which have also been scandalously neglected by governments.) The poor accommodation provided to single service personnel caused many to live off base, usually at their own expense, and even contributed to the decision of some to leave the Defence Force. Consequently, the government announced in 1987 that it planned to spend $240 million over five years on upgrading barracks accommodation. Since then there has been a concerted effort to improve living conditions in barracks and to provide private rooms for all service personnel (with the exception of those in certain training establishments where multi-occupant rooms are believed to help in developing a sense of group membership).

BARRICADE, HMAS see *ATTACK* **CLASS PATROL BOATS**

BARTOLOMEO COLLEONI, an Italian cruiser, was sunk off Crete on 19 July 1940 in the RAN's first major victory of the Second World War. The *Bartolomeo Colleoni* was a 5069-ton Condottieri Class

cruiser launched on 21 December 1930. It had sailed from Tripoli in Libya on 17 July 1940 to attack Allied shipping in the Aegean Sea. HMAS *Sydney* (II) (q.v.), under the command of Captain J. A. Collins (q.v.), came to the rescue of four British destroyers which were being attacked by two Italian cruisers, the *Bartolomeo Colleoni* and the *Giovanni Delle Bande Nere*. Surprising the cruisers, *Sydney* put *Bartolomeo Colleoni* out of action and then unsuccessfully pursued the other cruiser. British destroyers sank the *Bartolomeo Colleoni* and rescued its crew while under heavy aerial bombardment from Italian aircraft.

BARWON, HMAS see **RIVER CLASS FRIGATES**

BATAAN, HMAS see **TRIBAL CLASS DESTROYERS**

BATHURST CLASS MINESWEEPERS (CORVETTES)
Launched 1940–43 (56 ships); length 186 feet; beam 31 feet; displacement 650–790 tons; armament 1 × 4-inch and 1 × 40 mm gun, 5–6 machine-guns (with wide variations); speed 15 knots.

On the outbreak of the Second World War there was an immediate need for a warship of simple design that could be built in shipyards unaccustomed to naval construction and that could perform escort and minesweeping functions. In 1940 a modified version of the British *Bangor* Class minesweeper was selected. They were known in the RAN as corvettes. In addition to minesweeping duties around the Australian coast, *Bathurst* Class corvettes carried out a wide variety of tasks including convoy escort (q.v.), troop transport, hydrographic survey (q.v.) and on occasion bombardment in support of ground operations. They served in the Pacific and Indian Oceans, the Persian Gulf and the Mediterranean. HMA Ships *Deloraine, Katoomba* and *Lithgow* sank the Japanese submarine *I 124* in the Arafura Sea on 20 January 1942, while HMAS *Wollongong* helped sink the German U-boat *U 617* off Morocco on 11 September 1943. Corvettes assisted at American landings in the Philippines and in Australian landings during the New Guinea campaign and on Borneo. HMAS *Armidale* was sunk by Japanese aircraft while carrying Netherlands East Indies troops in the Arafura Sea on 1 December 1942. Eight corvettes, including HMAS *Bendigo*, were attached to the British Pacific Fleet as minesweepers and led the fleet into Hong Kong on its return in August 1945. Most corvettes were sold or scrapped in the 1950s and 1960s.

BATTLE CLASS DESTROYERS (*Anzac* [II] and *Tobruk* [I]). Laid down 1946, launched 1947–48; displacement 2400 tons; length 379 feet; beam 41 feet; armament 4 × 4.5-inch guns, 12 × 40 mm anti-aircraft guns, 10 × 21-inch torpedo tubes; speed 31 knots.

HMA Ships *Anzac* and *Tobruk* first saw service during the Korean War between 1951 and 1953 when they carried out escort duties and coastal bombardments. They later engaged in shore bombardments during the Malayan Emergency and exercised with the South-East Asia Treaty Organization (q.v.). HMAS *Tobruk* was put into reserve in 1960 and HMAS *Anzac* was converted to a training ship in 1963. Both were scrapped in the early 1970s.

BATTLE HONOURS are a form of public commemoration of a campaign, battle, action, or engagement and recognise the presence of a unit at and its contribution towards the outcome of a particular battle. Honours are not generally awarded for defeats, and a unit must have taken an active part in the action in order to qualify; merely being present is generally insufficient. The headquarters and at least 50 per cent of the posted strength must be involved. In the army, only infantry, armoured and cavalry regiments are entitled to battle honours, while in the navy individual ships bearing the same name carry the battle honours of all previous ships of that name.

The first battle honour awarded to an Australian unit was 'Suakin', earned by the New South Wales colonial contingent to the Sudan (q.v.) in 1885, and still carried by its lineal descendant, the Royal New South Wales Regiment. In the Boer War the general honour 'South Africa' was awarded to recognise the involvement of successive colonial contingents, and this honour too is now held by the Army Reserve (q.v.) regiment in each State. In the First World War militia units served only in Australia, and thus were ineligible for battle honours. As the units of the AIF did not have colours (q.v.) and were to be disbanded after the war, permission was given for the honours of the AIF to be transferred to the colours of militia counterparts. During the Second World War, militia units served in the South-West Pacific Area (q.v.) themselves, and thus were eligible for honours in their own right. The honours earned by the 2nd AIF were transferred after 1945 in the same manner as had occurred after the previous war. With the amalgamation of existing CMF units into State regiments in July 1960, all these honours likewise were transferred to the colours of the successor units.

Certain units of the Australian Regular Army (q.v.) have earned battle honours in the campaigns fought since 1945, principally for service in the Korean and Vietnam Wars. In 1994 a committee of inquiry into honours and awards recommended that battle honours be given in recognition of service during the Malayan Emergency and in Borneo during Confrontation, and that further honours be extended to include particular battles such as Maryang San in the Korean War.

BATTLE MOUNTAIN see **ABORIGINAL ARMED RESISTANCE TO WHITE INVASION**

BAY CLASS FRIGATES (*Condamine, Culgoa, Murchison, Shoalhaven*). Laid down 1943, completed 1946 (*Culgoa* 1947); displacement 1544 tons; length 302 feet; beam 36 feet 6 inches; armament 4 × 4-inch guns, 3 × 40 mm anti-aircraft guns, 4 × 20 mm guns, 1 × Hedgehog (an anti-submarine weapon), 4 depth-charge throwers; speed 19.5 knots.

Construction began on these anti-aircraft frigates during the Second World War, but the war ended before they could see service. *Culgoa*, *Murchison* and *Shoalhaven* formed part of the British Commonwealth Occupation Force (q.v.) in Japan. HMAS *Shoalhaven* was in Japanese waters with the Tribal Class destroyer (q.v.) HMAS *Bataan* when the Korean War began and was sent to Korea. All four ships made one tour of duty each during the war and were put into reserve in the 1950s. HMAS *Culgoa* was used as an accommodation ship at HMAS *Watson* (q.v.) from 1962 to 1972.

BAYONET, HMAS see **ATTACK CLASS PATROL BOATS**

BAYONET Bayonets have been in use since the sixteenth century, and in their original form were inserted into the muzzle of the musket to provide the soldier with an offensive weapon, a rudimentary form of pike. The bayonets that the British army in Australia (q.v.) used in the early colonial period were ring or socket bayonets, which fitted over the end of the muzzle, thus permitting the soldier to fire his weapon as well as wield the bayonet. A bewildering variety of bayonets was used by colonial military forces (q.v.) in the nineteenth century, reflecting the diversity of armaments carried by Australian colonial soldiers. The bayonet most commonly used by Australian soldiers in the twentieth century was the Pattern 1907 Sword Bayonet and its variations, for use with the 0.303 short-magazine Lee Enfield rifle (see SHORT-MAGAZINE LEE ENFIELD RIFLE), the standard-issue infantry weapon in the two world wars. Many of these bayonets were made at the small-arms factory at Lithgow in New South Wales. The jungle campaigns of the Second World War revealed the need for smaller, lighter bayonets for use in close terrain and thick vegetation, and various models were trialled and manufactured for use with the Owen gun (q.v.) and the Austen. With the decision to adopt the L1A1 Self-Loading Rifle (q.v.) in 1957, a committee of inquiry recommended that a short tropical bayonet be adopted as suitable for use in the terrain of south-east Asia. When the FN rifle (known colloquially in Australia as the SLR) was scheduled to be replaced in the mid-1980s, a replacement bayonet was also considered, although with the diminishing importance of this weapon it was very much a secondary consideration. The bayonet that accompanies the Steyr (q.v.) is more a general-purpose tool, with a saw-backed edge for part of its length, a wire-cutter formed by attaching the bayonet blade to a stud on the back of the scabbard, and a screwdriver point at the scabbard's tip. The bayonet has been of decreasing importance on the modern battlefield since the Second World War, although it is worth noting that at the battle of Fire Bases Coral and Balmoral in the Vietnam War in May 1968 Australian soldiers fixed bayonets in order to clear one of their own positions which had been temporarily overrun by the enemy.

Ian Skennerton, Australian service bayonets (unpublished, 1976).

BEAN, Charles Edwin Woodrow (18 November 1879–30 August 1968). No other Australian has been as influential as C. E. W. Bean in shaping the ways in which Australians see their military history, in particular their view of the First World War.

Bean was a product of empire and late Victorian values. Schooled in England at Clifton, alma mater of Generals Haig and Birdwood (q.v.) and Henry Newbolt, the poet of Empire, he read classics at Oxford and received a Bachelor of Civil Laws in 1903. He returned to Australia in 1904, was admitted to the New South Wales Bar and worked as an associate to Justice Sir William Owen on the New South Wales circuit. A growing interest in writing and a lack of briefs led him to take a job as a junior reporter with the *Sydney Morning Herald* in January 1908. In August that year he spent some time on board HMS *Powerful* observing the visit to antipodean waters of the American Great White Fleet, from which sprang his first, self-published, book, *With the Flagship in the South* (1909). In 1909 he was sent to western New South Wales to write about the wool industry, an assignment that initially held little attraction for him. The trip produced two further books, *On the Wool Track* (1910) and *The Dreadnought of the Darling* (1911), an increasing fascination with rural Australia, and the beginnings of an idea about the nature of Australia and the virtues of its society. Already enthusiastic about Australians as a superior form of overseas Briton, Bean now began to see the values of outback Australia as the dominant influence in the creation of 'outstanding national types'. Between 1910 and 1912 he lived in

England as the *Herald*'s London correspondent, writing a series of articles on the building of the battlecruiser *Australia* (q.v.), and the cruisers *Melbourne* and *Sydney* (q.v.) for the new Australian navy; these, together with a good deal of reworked material from his earlier book *With the Flagship*, were published in 1913 as *Flagships Three*. He returned to Australia at the beginning of 1913 as a leader-writer.

In September 1914 Bean was selected by a ballot of the Australian Journalists Association to become Australia's official correspondent with the AIF, narrowly beating Keith Murdoch (q.v.) of the Melbourne *Herald*. He was to be accorded the pay and privileges of a captain. Even at this early stage there seems to have been some suggestion that Bean might write the history of Australians in the war, and Bean's work habits throughout the war were clearly predicated on gathering material for such an undertaking. He sailed to Egypt with the first contingent, earned an early measure of unpopularity with the troops for writing an article that expressed a rather priggish moral disapproval of their rowdy behaviour, and early in April left with the main body for the landing at Gallipoli (q.v.).

His experiences at Gallipoli set the pattern for the rest of the war. He went ashore on the first morning, accompanied the 2nd Brigade in its futile attack at Krithia in May, was lightly wounded in August, and was among the last to leave in late December. During the campaign he was out and about in the front lines every day, talking to the soldiers, sharing their hardships and dangers, and writing it all up in his copious diaries and notebooks at night. His first despatch describing the landing was delayed by military censorship for several weeks, and the first account that reached Australian readers was by a British journalist, Ellis Ashmead-Bartlett, who wrote in more highly coloured and sensational terms than Bean ever felt comfortable with. While some editors complained that Bean's despatches were insufficiently graphic, his writing was sober and painstakingly accurate, and sought to convey, within the limitations imposed on him, something of the experience of Australians at the front. Already he was beginning to see in that experience the traits he had identified in outback Australia before the war: 'the big thing in the war for Australia', he was later to write, 'was the discovery of the character of Australian men. It was character which rushed the hills at Gallipoli and held on there'.

In March 1916 he went to France with the infantry divisions, and in the course of the next two and a half years was present at or visited shortly thereafter every significant feature and position held by the Australians on the Western Front. While in France he began to map out in his mind the shape that the history would take, and in early 1918 he spent a few weeks in London discussing the problems with Australian and British officials. It was at this stage, too, that he conceived of the Australian War Memorial (q.v.) as part of his desire to explain to Australians what their men had done and experienced in the war. To this end he set up the Australian War Records Section, modelled on its Canadian counterpart, and began to gather physical relics of the fighting. The Australian soldiers justified their reputation as scroungers by assisting the collection process, and huge quantities of material were shipped back to Australia at the war's end.

Few Australians can have had a better or closer knowledge of the war and the AIF than Bean. On only one occasion during the war did he clearly abuse his position, and that occurred in 1918 when the question of a replacement for Birdwood as commander of the Australian Corps arose. Bean and Murdoch attempted to intervene with Australian Prime Minister W. M. Hughes (q.v.) and others to block the appointment of Lieutenant-General John Monash (q.v.) and secure the command of the corps for Major-General C. B. B. White (q.v.), whom Bean hero-worshipped. White's role in the affair was entirely proper, dissociating himself from the manoeuvre and attempting to dissuade Bean from pursuing it. Bean's motivation is only partially clear; he did not like Monash and underrated him consistently as a commander both during the war and in the early volumes of the history. He admitted his error here many years later, and it is difficult to escape the conclusion that he was motivated in part by the mild anti-semitism often found in the public schools of the late Victorian era, as well as clashes between the two dating back to June 1915.

At the end of the war Bean returned to Australia via Gallipoli, to resolve some of the imponderables of the campaign, collect more relics, and report on the condition of the Australian graves there. Out of this came *Gallipoli Mission* (1948). His proposal for the appointment of an official historian and for the creation of the War Memorial were accepted by the government, and in October 1919 Bean and a small staff set up at Tuggeranong homestead, outside Canberra, to begin the great task. It was to take him 23 years.

Bean insisted that the history was to be free from censorship, and in the main this was what happened, although the naval volume by A. W. Jose (q.v.) had some technical and intelligence material removed, and Bean himself ameliorated some of Jose's judgements. Within his own volumes, Bean was prepared to be quite critical of British strategy and command, especially over the Dardanelles campaign or on the

Western Front in 1917. As well as writing six volumes himself, he edited a further five, compiled the volume of photographs, and assumed editorial responsibility for A. G. Butler's (q.v.) three-volume medical series when that commenced. Over and above this he, with John Treloar and Henry Gullett (qq.v.), steered the nascent War Memorial through the financial and political shoals of the 1920s and 1930s. It was an astonishing work-load. The history's theme was stated in an address Bean delivered in 1938: 'how did the Australian people — and the Australian character, if there is one — come through the universally recognised test of this, their first great war?' His approach from the beginning had been to describe the war experience as much from the view of the private in the forward trench as from that of the commanding general in the headquarters to the rear, and to assign responsibility for success or failure where it was due, which in Bean's view was frequently among the soldiers and junior officers at the front. The enormously detailed approach, with footnotes giving the personal particulars of every man mentioned in the text, from private to field marshal, was possible not least because of the limited scale with which Bean dealt; had he been writing about the army of one of the major combatants, several million strong, he could not possibly have adopted the same technique, as he himself fully realised.

The hallmarks of the official history were its democratic nature, its detailed approach, and the authority that came from Bean's close personal knowledge of events supplemented by the extraordinary volume of records that Treloar had assembled and the private papers that Bean himself had amassed. Despite some financial difficulties during the Depression, the histories sold solidly, on average about 10000 copies of each volume, and were generally well received both within Australia and by reviewers and critics abroad. A one-volume abridgement, *Anzac to Amiens*, appeared just after the Second World War in 1946. Bean's last major book, a study of Major-General W. T. Bridges (q.v.) and of his great friend and patron Brudenell White, appeared in 1957 as *Two Men I Knew*.

Bean was essentially a modest man. He declined a proffered knighthood on several occasions. He had been mentioned in despatches for gallantry at Gallipoli; in 1930 the Royal United Service Institution honoured him with its Chesney Gold Medal in recognition of the achievement that the official history represented. Honorary degrees from the University of Melbourne and the Australian National University followed. He was instrumental in the appointment of Gavin Long (q.v.) as official historian in the middle of the next world war, and played an important part in the establishment of the Australian Archives. He was chairman of the Trustees of the Australian War Memorial from 1952 to 1959, by which time his health had begun to fail. He was admitted to Concord Repatriation Hospital in early 1964, and died there four years later.

The influence of Bean's history on the way in which the study of Australian military history developed is undoubted, although difficult to quantify. The great, claret-coloured volumes of the official history, each on average about 300 000 words in length, grace many public and private libraries, but are more frequently referred, even deferred, to than actually read. His influence is seen less perhaps in what he wrote than in the approach he took: the concentration on individual and small group experience, and the elevation of the other ranks to an equal role in the telling, made sense in the context in which he developed it, but in time it came to exercise a somewhat negative hold on the ways in which military history was studied and written. His romantic belief in rural virtues and their direct relation to martial ones, although it diminished over time and in the course of successive volumes, was at best an inadequate explanation for the military prowess of the AIF which had developed by the final year of the war. Bean, superlative journalist that he was, had little real understanding of the military technicalities of doctrine, organisation, command, training and administration, and these hardly feature in his work. They were, however, the key to success on the battlefields of the First World War. The official history, as a result, has all the virtues, and the vices, of Bean's journalistic background.

Dudley McCarthy, *Gallipoli to the Somme: The Story of C. E. W. Bean* (John Ferguson, Sydney, 1983).

'BEAT HITLER FIRST' STRATEGY, the policy of considering Germany the primary enemy in the Second World War, and fighting a purely defensive war against Japan until Germany was defeated, was formally adopted at the Anglo-American 'Arcadia' conference in December 1941–January 1942. Although Australian Prime Minister John Curtin (q.v.) and External Affairs minister H.V. Evatt seem to have been aware that 'Beat Hitler First' was the guiding principle of Allied strategy, they did not learn of its ratification at 'Arcadia' until May 1942. They continued to advocate increased priority for the Pacific in Allied strategy, but it was the American reassessment of the altered strategic situation in all theatres of war rather than Australian government lobbying that led to the escalation of operations against Japan from early 1943.

John Robertson, 'Australia and the "Beat Hitler First" strategy, 1941–42: A problem in wartime consultation', *Journal of Imperial and Commonwealth History*, vol. 11, no. 3 (1983), pp. 300–21.

BEAUFIGHTER, BRISTOL AND DAP (2-seat fighter-bomber [DAP Mark 21]). Wingspan 57 feet 10 inches; length 41 feet 8 inches; armament 4 × 20 mm cannon, 4 × 0.5-inch and 1 × 0.303-inch machine-guns, 500 pounds bombs, 8 × 20-pound rockets; maximum speed 315 m.p.h.; range 1480 miles; power 2 × Bristol Hercules 1770 h.p. engines.

The Beaufighter was a Second World War fighter design based on the Beaufort (q.v.) bomber. In Europe two Australian squadrons, with crews from the Empire Air Training Scheme (q.v.), flew Bristol-built Beaufighters. No. 456 Squadron used Beaufighters as night fighters from 1941 until they were replaced by Mosquitos (q.v.) in 1942. No. 455 Squadron used Beaufighters against German shipping in 1944 and 1945. In the South-West Pacific Area (q.v.) five RAAF squadrons flew British-built Beaufighters from 1942 and Australian-built aircraft from 1944. The Australian Beaufighters, built by the Department of Aircraft Production, incorporated some improvements over the original design and had more powerful engines. In 1943 Beaufighters played an important role in the Battle of the Bismarck Sea. In the South-West Pacific theatre they were known as 'Whispering Death'. After the Second World War Beaufighters were used for target-towing until 1956.

BEAUFORT, BRISTOL AND DAP (4-crew light bomber [DAP Mark VIII]). Wingspan 57 feet 10 inches; length 44 feet 3 inches; armament 4–6 × 0.303-inch machine-guns, 2000 pounds bombs; maximum speed 267 m.p.h.; range 1060 miles; power 2 × Pratt & Whitney Twin Wasp (q.v.) 1200 h.p. engines.

The Beaufort was a British bomber, chosen in 1939 for manufacture by the Department of Aircraft Production, which entered service in 1941, just months before the outbreak of the Pacific War. The Beaufort was the main aircraft used by RAAF squadrons for bombing raids against Japanese land targets and shipping during the New Guinea campaigns of 1942 and 1943. The Beaufort design was developed into the Beaufighter (q.v.) fighter aircraft. They were also used for coastal reconnaissance and as transport aircraft (called 'Beaufreighters'). In 1946 some Beauforts were used to spray DDT on farm crops.

BEAVIS, Major-General Leslie Ellis (25 January 1895–27 September 1975). An early graduate of RMC Duntroon in 1915, Beavis served with the AIF in the Middle East and France as an artillery officer, winning the DSO as a battery commander in 1917. After the war he undertook a number of

A Beaufighter of No. 31 Squadron attacking Japanese seaplanes moored at Taberfane, Aru Islands, Netherlands East Indies, 12 June 1943. (AWM P1160/01)

artillery and ordnance staff courses in England, and it was these that in many respects laid the foundations of his future career. Like many regulars in the interwar period, he served in a succession of staff and CMF regimental postings, in his case a routine career enhanced by attendance at the Staff College, Camberley, in 1928–29 and a two-year attachment to the Imperial General Staff in London. In 1936 he was appointed chairman of the Defence Resources Board, from which position he argued for the development of defence-related capabilities in industry and a greater encouragement of private industry for defence purposes. This was at variance with prevailing advice, which argued that scarce resources should be allocated to government factories first, despite the fact that in 1937 there were very few of these in existence. The latter view prevailed.

Beavis served with the 2nd AIF in the Middle East, becoming Director of Ordnance Services on AIF headquarters in December 1940. Returning to Australia after the outbreak of the Pacific War, he was promoted to major-general and made Master-General of the Ordnance, the senior officer respon-

sible for army supply and logistic matters both in Australia and overseas. In this demanding job he enjoyed considerable success, the official historian Gavin Long (q.v.) noting that he had 'to organise and direct a very large army organisation, one particularly subject to ministerial (and Secretariat) criticism, and one which had to be built on very small foundations'. An able and firm administrator, he exercised great influence on the reform and development of army administration at a time of enormous growth and in the middle of a major war.

The Second World War marked the pinnacle of his career, but Beavis made a number of significant contributions to the development and acquisition of army weapons in the postwar period. He was centrally involved in the Long Range Weapons Establishment (q.v.) and the development of Woomera. He sat on or (more usually) chaired a range of key committees within the Department of Defence between 1946 and his retirement in 1952, including the Principal Administrative Officers Committee (Maintenance and Materials), the Joint War Production Committee (subsequently renamed the Defence [Industrial] Committee), and the Joint War Planning Staff. On retirement from the army he spent two years as high commissioner to Pakistan.

Beavis was a key figure in the regularising of army administration, especially during the Second World War, and his career illustrates the growth of professionalism within the army, particularly after 1942.

BEAZLEY, Kim Christian (14 December 1948–). From a family with strong Labor traditions, Beazley took a master's degree in strategic studies at Oxford while a Rhodes scholar, and taught politics at Murdoch University before winning the federal seat of Swan for the Labor Party in 1980. He moved into the ministry early and was appointed minister for Defence in 1984, having been minister assisting in the portfolio in 1983–84. He quickly stamped the direction of Labor's defence policy as his own, and presided over fundamental changes to Australia's defence thinking, characterised by the adoption of the Dibb Report (q.v.), a major re-equipment and acquisition program for all three services, and a program of privatisation and commercialisation of defence industrial concerns. He gave every sign of enjoying the portfolio at all levels, which led to jibes from the Left of the ALP about 'toys for the boys' and a process of alleged militarisation of Australian interests in the Asia–Pacific regions in particular. Unquestionably one of the most able and knowledgable defence ministers since 1945, he was also one of the most popular and highly-regarded within the ADF. He has subsequently held the portfolios of Transport and Communications, and Finance, and was appointed government leader in the House of Representatives in 1988 and deputy prime minister in June 1995.

BECHER, Rear-Admiral Otto Humphrey (13 September 1908–15 June 1977). After a good scholastic and sporting record at the RAN College Jervis Bay, which he entered in 1922, Becher served as a midshipman on HMAS *Adelaide* and HMAS *Brisbane* before proceeding to Britain where he served on HMS *Repulse* and specialised in gunnery. He returned to Australia in 1930 as a lieutenant and spent time on board the heavy cruisers *Australia* and *Canberra* before a further period in Britain attending the long course in naval gunnery between 1932 and 1934. At the outbreak of war he was squadron gunnery officer in HMS *Devonshire*, and took part in the Norwegian campaign, for which he was awarded the DSC. In November 1940 he went to the Mediterranean where he remained until April 1942. He returned to Australia and commanded the gunnery school at the Flinders Naval Depot until March 1944, when he

was given command of the destroyer *Quickmatch*; his powers of command during a naval attack against Japanese forces at Sabang near Sumatra led to a bar to his DSC and a mention in despatches. Following the war he spent a period ashore in Navy Office before returning to sea on HMAS *Sydney*, the first of the RAN's newly acquired aircraft carriers. He commanded HMAS *Warramunga* during its deployment to Korean waters in 1950–51, for which he received the DSO and the American Legion of Merit. For the next decade he held a succession of staff posts ashore, including Deputy CNS on two occasions, interspersed with attendance at the Imperial Defence College (q.v.) in London and ship commands aboard HMAS *Vengeance* between 1954 and 1955 and HMAS *Melbourne* between 1957 and 1958. He returned to London in 1962–63 as head of the Joint Services Staff. He was Flag Officer Commanding HM Australian Fleet when the *Voyager* (q.v.) disaster occurred in February 1964, although he had only taken up the posting the previous month. His evidence before the Royal Commission headed by Sir John Spicer may have led the latter to ascribe blame for the collision to the captain of the *Melbourne*, Captain R. J. Robertson, while the fact that he had discussed the events with Robertson prior to his own appearance became public knowledge and led to suggestions of collusion. He retired from the navy in March 1966, but spent a period between 1966 and 1969 as Director-General of Recruiting. In a job made sensitive by the fact of conscription for overseas service in Vietnam, he became known for his opposition to conscription on the grounds that it diluted professional standards within the services. A naval officer with very considerable operational experience who commanded wide respect, he was, like many Australian naval officers of his generation, highly professionally competent but perhaps not equally attuned to the political difficulties that attended the navy during the course of the 1960s.

BEERSHEBA, BATTLE OF Part of the first phase of the Third Battle of Gaza in the Palestine campaign, the action at Beersheba on 31 October 1917 is best known to Australians for the mounted charge of the 4th Light Horse Brigade late on the first day of the battle, but this was only one part, albeit a crucial one, of a much larger and more involved battle.

Gaza was the strategic key to the defences of southern Palestine, and the British made two costly and unsuccessful attempts to take it directly in March and April 1917. Thoroughly alerted by these attempts, the Turks fortified the town heavily and constructed a defensive line running in a south-easterly direction and anchored at the other end on the small town of Beersheba. Although the British forces outnumbered their opponents, they had to contend with supply difficulties, especially shortages of water, created by the harsh environment in which they fought.

The plan of the British GOC, Sir Edmund Allenby, called for a demonstration against Gaza itself in order to make the Turks think that the British were attempting another unimaginative frontal assault. At the same time, a wide flanking movement was to be mounted at the other end of the Turkish lines, at Beersheba, but this movement was to be presented to the enemy as the feint. Strategically complex, the plan was implemented successfully, not least through a brilliant deception on the part of an intelligence officer, Colonel Richard Meinertzhagen, and by virtue of the greatly increased presence of the RFC in the theatre, which attained air superiority and denied the Turks an aerial reconnaissance capability. The key to success then lay in the hands of the attacking troops, as it was vital that the town be taken on the first day in order to guarantee water supplies for the whole force.

On the night of 30–31 October about 40 000 troops were on the move towards Beersheba. The mounted units of the Desert Mounted Corps, two divisions of about 15 000 men in total, under the command of Lieutenant-General Sir Harry Chauvel (q.v.), covered 25–30 miles in a night approach march. Bombardment of the defences at Gaza began on 27 October, with an attack planned as soon as success at Beersheba had been signalled in order to facilitate the next phase of Allenby's plan: the rolling up of the Turkish defensive positions from east to west.

Early on the morning of 31 October, infantry of the British 60th Division attacked forward Turkish positions around Beersheba, and a general engagement developed which lasted most of the day. Progress in the infantry assault was steady but slow, and by late afternoon the decision still eluded the attacking force. Chauvel was conscious of the extreme importance of securing the wells in the town, particularly for his horses, many of whom had been without water for several days. The town was still strongly defended, and he decided to draw in the uncommitted 4th Light Horse Brigade (4th and 12th Light Horse Regiments) under Brigadier-General W. Grant (q.v.): 'Put Grant straight at it', he ordered. The decision to charge the defences cavalry-style appears to have been Grant's own; orthodox mounted infantry doctrine called for light horse to use their mounts for operational movement, but to attack dismounted. It was an inspired decision which may have owed some-

thing to a similar manoeuvre carried out by the 10th Light Horse Regiment at Magdhaba in December the previous year. The charge moved off at around 4.30 p.m. Many men carried their long sword bayonets in their hands, and were carried quickly through the zone of machine-gun and artillery fire by the impetus of their charge. Having breached the defences, they fell upon the defenders, inflicting casualties and taking more than 1000 prisoners for the loss of 31 killed and 32 wounded. Most of the wells were captured intact, a complete Turkish division, the 27th, was destroyed, and the way was thrown open to the capture of Gaza on 7 November and the advance into Palestine, although this was not achieved without more hard fighting.

Beersheba, and the capture of Gaza of which it formed a part, was one of the great feats of Australian arms in the Middle East during the Great War, and deserves the epic status it has acquired in Australia. Two feature films depict the event, Charles Chauvel's *Forty Thousand Horsemen* (1940) and Ian Grant's *The Light Horsemen* (1987), and Frank Dalby Davidson treated it memorably in his short novel *The Wells of Beersheba* (1933).

BELL, Flight Officer Mary see **WOMEN'S AUSTRALIAN AUXILIARY AIR FORCE**

BELL, Matron Jane (16 March 1873–6 August 1959). Migrating as an orphan to Sydney in 1886, Bell trained as a nurse at Royal Prince Alfred Hospital and by 1904 was matron at Brisbane General Hospital. She undertook midwifery courses at Queen Charlotte's Hospital in London in 1906 and worked at the Edinburgh Royal Infirmary between 1907 and 1910. She returned to Australia in the latter year. In June 1913 she was appointed Lady Superintendent of the 3rd Military District, and at the beginning of the war recruited the Victorian section of the 1st Australian General Hospital, which left for Egypt in December 1914. Army nurses occupied an indeterminate position in the military hierarchy at this time, accorded many of the privileges of officers but not holding commissions, and their position in the command structure likewise had yet to be settled. Bell waged an outspoken campaign seeking to clarify these and other issues with the Army Medical Service, which resisted attempts by the nursing service to acquire administration and disciplinary powers over its own members. When a series of recommendations on staffing and organisation were rejected, she requested to be relieved, and was returned to Australia in September 1915. An inquiry into the administration of the nursing service subsequently vindicated her position, but her appointment in the

AIF was terminated in October 1915. She took up a position at Melbourne Hospital, where she remained until retirement in 1934, and was instrumental in various reforms in civilian nursing and nursing training.

BENALLA, HMAS see **SURVEY SHIPS**

BENDIGO (I), HMAS see **BATHURST CLASS MINESWEEPERS (CORVETTES)**

BENDIGO (II), HMAS see **FREMANTLE CLASS PATROL BOATS**

BENNETT, Air Vice-Marshal Donald Clifford Tyndall (10 September 1910–17 September 1986). Born in Toowoomba, Queensland, the fourth son of a wealthy pastoral family, Bennett was educated at Brisbane Grammar School where his matriculation results were below the standard required to gain entry into medical school. Influenced by stories of pioneering flights, including the early bush pilots of Qantas, Bennett determined upon a career in the RAAF. He failed his first medical examination and it was not until July 1930 that he reported to the Air Board, one of an entry of 15. The original intention of the authorities was to send two or three out of the graduating class to serve with the RAF, but in the case of Bennett's class they were told that because of the financial exigencies arising from the Depression, all would have to agree to accept a transfer to the RAF as a condition of training with the RAAF. Bennett accepted this condition, and passed out second in his class, coming top in flying. He served in the RAF in a variety of positions, and in March 1934 gained the very rare qualification of a First Class Navigator's licence, but finding little challenge in the peacetime service resigned on 11 August 1935. For the next five years he was a pilot for Imperial Airways, and set a number of records for long-range flights. In 1940 he was invited by Lord Beaverbrook, Minister of Aircraft Production in the British War Cabinet, to establish a trans-Atlantic ferry route. Bennett led the first flight of Hudson aircraft from Gander airfield, Newfoundland, to Ireland on 11 November 1940, during which flight he acted as navigator, drawing on his outstanding navigational skills. He remained in command of the ferry service until July 1941, and on 9 August 1941 he rejoined the RAF with the rank of group captain (subsequently reduced to wing commander) and was posted to command the new elementary navigation school. After establishing the school and setting its curriculum in place, he moved in December 1941 to No. 77 Squadron where he flew many sorties until mid-April 1942, when he joined No. 10 Squadron. After only several days in his new position he joined a bomber force in an attack on the German battle-cruiser *Tirpitz*, was shot down and escaped into Sweden, and managed to return to England within a month; for this he was awarded the DSO. On 5 July 1942, he was instructed to form the Pathfinder Force, an élite group of aircraft equipped with the most advanced radar and blind bombing devices and manned by hand-picked crews, whose task it was to guide the massed formations of Bomber Command to their targets with the greatest possible accuracy. Bennett personally developed the marking techniques used by the Pathfinders, which became indispensable in the bombing campaign from 1943. Bennett was promoted to air vice-marshal that year and made CBE; the following year he became CB. He was widely regarded as the outstanding tactician of strategic bombing in the war and the most brilliant and technically able navigator, an ability that was the basis of his, and the Pathfinders', wartime success. He served as Liberal MP for Middlesborough West from 1945 to 1947, and continued his involvement in the development of civil aviation.

BENNETT, Lieutenant-General Henry Gordon (16 April 1887–1 August 1962). A prominent interwar citizen soldier, Bennett is one of the most controversial figures in Australia's military history. Commissioned into the CMF in August 1908, Bennett had reached the rank of major by the time the Great War broke out. An early volunteer for the AIF, he embarked for Egypt as major and second-in-command of the 6th Battalion. He was wounded in the attacks against Lone Pine on the day of the Gallipoli landing, returning to the fighting before he could be evacuated. By May he had been promoted to temporary lieutenant-colonel in command of his battalion, and rapid promotion and disregard for his own safety were to characterise his service in the First World War. Transferring to France after the evacuation, he was promoted to command of the 3rd Brigade in December 1916, at the age of 29 the youngest brigadier-general in the British Empire armies. He was mentioned in despatches eight times during the course of the war, was decorated on several occasions and earned a considerable reputation as a front-line commander. He became well known to his superiors for a prickly temperament, argumentative nature and proneness to quarrel which prompted his divisional commander, Major-General Sir William Glasgow (q.v.), to label him 'a pest'.

Returning to civilian life, Bennett combined a business career, citizen soldiering and a taste for

extra-parliamentary conservative politics, becoming prominent in the All for Australia League and the Defence of Australia League. Despite this his first love remained soldiering. He was promoted to major-general in 1930 in command of the 2nd Division, and took an active, indeed activist, interest in citizen soldier matters throughout the 1930s. He was the leading advocate of the alleged superiority of citizen officers over their regular, Staff Corps colleagues, and generated much ill feeling with an unwise set of newspaper articles in 1937 which contained intemperate criticism of the regulars and which led to his censure by the Military Board.

Notwithstanding this, at the outbreak of war in 1939 he was the senior citizen soldier on the Army List, and believed that the top command in any expeditionary force would be his by right. Not only was he ignored for this appointment, the command of the 2nd AIF going to Major-General Sir Thomas Blamey (q.v.), but the commands of the 7th, 8th and 9th Divisions, as each was raised, went to others. It was only after the death of the CGS, Brudenell White (q.v.), and the appointment of Major-General V. A. H. Sturdee (q.v.) as CGS from command of the 8th Division, that Bennett received a command. Appointed GOC of the 8th Division on 24 September 1940, the real or imagined slights of this year, together with Bennett's already unstable temperament, contributed to difficult relations within the division and between Bennett and his superior, Blamey. His relations with his British counterparts in Malaya, especially the GOC Malaya, Lieutenant-General A. E. Percival, were to be equally negative.

Bennett had only two of his brigades in Malaya, the third being distributed in small groups throughout the islands to Australia's north. When Japanese troops landed in Malaya on 8 December 1940 they were met initially by British and Indian troops who proved no match for them. Given responsibility for the defence of north-west Johore early in January 1941, Bennett demonstrated that he had little real understanding of the Japanese or of the conditions with which he was faced. Despite some local successes, such as at Gemas on 14 January, which were the responsibility of local commanders, Bennett's dispositions were unsound (he refused to allow prearranged lines of withdrawal, which meant that the wounded had to be abandoned to the Japanese), and the Australians performed no better than the British and Indians in northern Malaya.

Lieutenant-General Gordon Bennett (right) with British General Sir Archibald Wavell, C-in-C, ABDA Command, in Singapore, 8 January 1942. (AWM P1182/16/16)

The remaining forces withdrew into Singapore itself at the end of January, and the Japanese attack which began on 8 February quickly carried all before it. On 15 February, with surrender imminent, Bennett handed over command of his division and slipped out of Singapore by sampan, arriving in Australia on 2 March. Bennett's abandonment of his troops created considerable controversy within the army, although under wartime conditions much of this was withheld, at least initially, from public gaze. Promoted and given command of III Corps in Western Australia, his actions in leaving Singapore ensured that he would never be given an active service command again. He retired in what can only be described as a fit of pique in May 1944, and wrote an elaborate defence of his actions in *Why Singapore Fell* (1944).

At the end of the war his conduct was subject to scrutiny, first through a military investigation which criticised his actions, and then through a formal commission of inquiry under Justice Sir George Ligertwood, which likewise failed to vindicate him. His assertion that he alone possessed the key to defeating the Japanese was neither borne out by his own performance in Malaya nor helped by the fact that he had earlier dispatched a number of officers to Australia to communicate tactical lessons to Army Headquarters, whilst his overweening arrogance and ambition to lead the Australian Army in war clouded his judgement.

Although he found many supporters among former members of the 8th Division, and although he continued to justify himself in postwar writings and other forums, Bennett was entirely discredited by his performance in the Malayan campaign. His unrelenting prejudice against regular officers won him few friends, even among citizen soldiers, while his propensity to blame others for his own failures isolated him still further. A career that had promised much at the end of the Great War was destroyed in the opening rounds of the war in the Pacific by prejudice, arrogance and an inability to adapt to circumstances of the present war.
(See also MALAYAN CAMPAIGN 1941–42.)

A. B. Lodge, *The Fall of General Gordon Bennett* (Allen & Unwin, Sydney, 1986).

BERLIN AIRLIFT was carried out by the Western powers in response to the Soviet land and water blockade of Berlin that was imposed on 25 June 1948 (becoming a total blockade on 10 July) in an attempt to force the whole of Berlin into the Soviet camp. Ten RAAF crews (41 personnel) from No. 86 Wing, Richmond (q.v.), flew RAF Dakotas (q.v.) from Lübeck, northern Germany, to Berlin during the period 15 September 1948 to 29 August 1949, on a total of 2062 sorties, carrying 6964 passengers and 15 623 364 pounds of freight. One RAAF pilot, Flight-Lieutenant M. Quinn, was killed on an instrument approach to Berlin while on exchange with the RAF.

BERRYMAN, Lieutenant-General Frank Horton (11 April 1894–28 May 1981). An early graduate of RMC Duntroon, whose commissioning was accelerated in order to meet the expanding needs of the AIF, Berryman was appointed lieutenant in June 1915 and went overseas as adjutant to the 4th Artillery Brigade. He served in France in a number of capacities, commanding a battery in late 1917. He was wounded in the last days of the war and then served on the headquarters of the 7th Infantry Brigade, returning to Australia in August 1919.

He was fortunate to see considerable service overseas in the 1920s, in Britain at the Artillery College, Woolwich, between 1920 and 1923, and at the Staff College, Camberley, in 1927–28, followed by a period as the administrative staff officer and then the Army Representative in the High Commissioner's Office, London. He returned to Australia in 1932 as brigade-major of the 14th Infantry Brigade, essentially the position and rank he had held at the end of the Great War. In the second half of the 1930s he was in the operations branch at Army Headquarters, while the outbreak of war in 1939 found him as senior operations staff officer of the 3rd Division of the CMF.

Seconded to the 2nd AIF in April 1940, Berryman was appointed GSO1 of the 6th Division, and planned the operations of the division's first campaign in Libya. He was given command of the 7th Division artillery in late January 1941, and fought in the Syrian campaign where he commanded a composite force, 'Berryforce', in the Merdjayoun sector during the advance into Lebanon. Promoted to brigadier, he became the senior staff officer (Brigadier General Staff) at Headquarters I Australian Corps in August 1941. In this capacity he served briefly in Java in 1942 in the Netherlands East Indies campaign, as part of the hopeless attempt to stop the Japanese offensive. He returned to Australia, and in April was again promoted and made MGGS at First Australian Army. From then until the end of the war he occupied the senior staff posts in the army during the fighting in New Guinea, often wearing several hats simultaneously. He was deputy chief of staff to General T. A. Blamey (q.v.) at Advanced Land Headquarters (q.v.) from September 1942 until January 1944, and from August 1943 also acted as MGGS to New Guinea Force (q.v.), for the second time. Given command of the II Corps (redesignated I Corps in

April 1944) in November in succession to Lieutenant-General Iven Mackay (q.v.), he oversaw operations in the Finisterre Ranges, part of the gruelling campaign in the Ramu and Markham Valleys. He was frequently forward visiting the troops and observing the unfolding operations, which finally cleared the Huon Peninsula and pushed the enemy north to Madang. In July 1944 he became chief of staff at Advanced Land Headquarters, an appointment he was to hold until the end of the war.

By late in the Pacific War, then, Berryman was Blamey's principal planner and staff officer, and hence one of the most important officers in the Australian Army in its struggle against the Japanese. He was widely regarded as the finest staff officer in the army, even by some who did not much care for him otherwise. Lieutenant-General John Lavarack (q.v.), whom he had served at I Australian Corps in Syria and Java, wrote of him that he was 'the best combination of fighting leader, staff officer and administrator' in the army. The citation which accompanied his award of the CBE for his work with the 6th Division recorded that it 'has been outstanding and contributed very largely to the success of these actions. To his skill, systematic planning, organisation and preparation of orders he added quick comprehension, wide outlook, thoroughness and tireless energy.' Lieutenant-General H. C. H. Robertson (q.v.) acknowledged him as 'the soundest planner … in the AIF'. When the Japanese had been forced out of the Huon Peninsula, Blamey wrote praising his 'skilful planning, able supervision and vigorous leadership'. He could be cold and sarcastic with his juniors, earning the nickname 'Berry the Bastard' in the process, but to others, and especially with some of his Staff Corps contemporaries, he was a warm and engaging character, possessing a hard streak but one tempered by humour and a sense of the practical. He was a founder of the War Widows' Guild, and instrumental in helping to set up the first home for widows after 1945.

The postwar period was something of an anticlimax. Lieutenant-General Vernon Sturdee (q.v.) returned as CGS with Lieutenant-General S. F. Rowell (q.v.) as his deputy, and Berryman was given Eastern Command, which he held from March 1946 until his retirement in April 1954. He was an obvious contender for the post of CGS in succession to Sturdee, and there was some suggestion, entirely erroneous, that had Labor won the 1949 election he would have received preferment over Rowell. When Rowell was due to retire in 1954, Berryman made a half-hearted attempt to lobby the Minister for Defence Production and senior Liberal Party figure, Sir Eric Harrison, on his own behalf, but his age was against him. He was Director-General of the Royal Tour in 1954, and Chief Executive Officer of the Royal Agricultural Society of New South Wales from 1954 to 1961.

BESSELL-BROWNE, Brigadier-General Alfred Joseph (3 September 1877–3 August 1947). Soldier and businessman, Bessell-Browne was born in Auckland and migrated to Australia in the 1890s. He served in the Boer War in the 1st Western Australian Mounted Infantry, seeing action in Transvaal, Orange Free State and Cape Colony. He received the DSO and was mentioned in despatches. In the years before the First World War he served as a lieutenant in the Australian Field Artillery and was promoted captain in 1908. The next year he resigned from the public service and went into business. When the First World War began Bessell-Browne was appointed as a major in the AIF commanding 8th Battery, Australian Field Artillery. He left for Egypt in November 1914 and was at the Gallipoli landing, but his unit did not see action until May. His battery was located on 400 Plateau for most of the campaign. He was promoted to command the 2nd, then 3rd AFA Brigades. In 1916 his brigade was posted to the Western Front where it provided cover for the attack on Pozières. In January 1917 he was promoted colonel and temporary brigadier-general and was given command of the 5th Divisional Artillery. He saw service at Polygon Wood, Villers-Bretonneux, and took part in the last Australian actions of the war at the Hindenburg Line and the River Selle where his guns supported an American attack. He was awarded the American DSM and was appointed CB for his services in France and Flanders, and was mentioned in despatches nine times. During the Second World War he commanded the WA Volunteer Defence Corps and retired as a brigadier-general in 1942.

BEST, Colonel Kathleen Annie Louise (28 August 1910–15 November 1957). With a background in nursing in the Sydney area before the war, Best was appointed as matron to the 2/5th Australian General Hospital in May 1940, having earlier volunteered for the Australian Army Nursing Service (q.v.) Reserve. She served in Egypt and Greece, was evacuated to Crete and thence to Palestine, and was awarded the RRC. She returned to Australia in March 1942 and in July was appointed Controller of the Voluntary Aid Detachments, which in September became the Australian Army Medical Women's Service (q.v.). In February 1943, by which time she had attained the rank of lieutenant-colonel, she was made Assistant

Adjutant-General (Women's Services) at Land Headquarters in Melbourne, a post she held until placed on the Reserve of Officers in September 1944. She became Assistant Director of the Re-establishment Division in the Department of Post-War Reconstruction, responsible for the reintegration of servicewomen into peacetime Australia, and followed this with a similar position in the Medical Rehabilitation Scheme. In February 1951 she was appointed Director of the newly raised Women's Australian Army Corps, a post which she held until her early death in 1957. A memorial gateway commemorating her service was constructed at the entrance to the WRAAC School at George's Heights in Sydney.

BIBLIOGRAPHIES Five main bibliographies list sources of material on Australian military history. They are the *Australian Military Bibliography* by C. E. Dornbusch (Home Farm Press, New York, 1963), *A Select Bibliography of Australian Military History* by Jean Fielding and Robert O'Neill (ANU Press, Canberra, 1978), *Bibliography of Australia* by John Ferguson (National Library of Australia, Canberra, 1986), *A Bibliography of Armed Forces and Society in Australia* by Hugh Smith and Sue Broome (3rd edition, ADFA, Canberra, 1987) and the MIHILIST (q.v.) database.

BIRDWOOD, Field Marshal William (13 September 1865–17 May 1951) was born at Kirkee, India, where his father was serving in the Indian Civil Service. Birdwood was educated at Clifton College, Bristol (as were Haig and C. E. W. Bean [q.v.]) and at RMC Sandhurst. Before the First World War he saw service on the North-West Frontier of India and in South Africa where he served on Lord Kitchener's staff. He was promoted to major-general in 1911 and was serving as Secretary to the Army Department in India when war broke out.

In November 1914, Kitchener, who was Secretary of State for War, appointed Birdwood to command the forces raised by Australia and New Zealand for service in Europe. However as the idea of an attack on Turkey at the Dardanelles developed it became clear that these forces, then training in Egypt, would be used in some capacity against the Turks (see GALLIPOLI). In February 1915 Birdwood was sent by Kitchener to report on the progress of the naval attack on the Dardanelles. Birdwood was always sceptical of the ability of the navy to force the straits unaided and it was in large measure his reports that convinced Kitchener that a combined operation would be required. When Sir Ian Hamilton was appointed by Kitchener to overall command of the Mediterranean Expeditionary Force Birdwood, somewhat to his disappointment, returned to command of the Anzac contingent.

The plan for the Anzac landing was made by Birdwood and his staff. There has been much debate about the circumstances in which the troops came to be landed to the north of their intended beachhead. Was Birdwood to be blamed for making an elementary blunder or to be praised for avoiding strong Turkish defences near the original landing point? This whole debate misses the point that wherever the Anzacs were landed, they were not sufficiently strong to accomplish their goal of blocking Turkish troops from the north and advancing across the peninsula to the Narrows. So the failure of the plan cannot be laid at Birdwood's door, although he never questioned if the force given him was appropriate for the designated task. Birdwood was also responsible for planning the great August offensive. The plan was imaginative. The Turkish positions would be outflanked to the north and the main ridge captured. Once again, however, it was beyond the powers of the available troops and their confused commanders, and relied on split-second timing in country that rendered such timing impossible.

During the campaign Birdwood's most valuable role was in maintaining the morale of his force. As a commander he was not a great intellect; nor did he have an instinctive feel for strategy and tactics. However, he was happy to be seen among his men and was a frequent visitor to the front line. His popularity with the Anzacs, although perhaps not as great as he imagined, was well above the average for a First World War commander. Of all the corps commanders at Gallipoli, Birdwood was the only one who opposed evacuation. When the decision was made, he and his chief staff officer, Brudenell White (q.v.), organised the operation with such thoroughness that the troops left the peninsula unhampered by the Turks.

Birdwood's experiences on the Western Front, where the Anzac Corps was moved in March 1916, started badly. His troops were engaged in the Battle of the Somme from July 1916 but control of operations passed to Haig and Gough, the latter being commander of the Reserve (later Fifth) Army in which the Australians operated. The bloody encounters at Fromelles and Pozières were neither planned nor welcomed by Birdwood but once more he entered no dissenting note at the high casualties, and meagre gains. The following year, 1917, opened just as badly for Birdwood and the Australians. Still under Gough, they were called upon to take part in the British spring offensive with diversionary activities against the Hindenburg Line at Bullecourt. These battles were costly fiascos. Birdwood was uneasy about them but never saw himself as being able to dissent from higher authority.

From this point matters improved. The Anzac troops were split into two corps, Lieutenant-General A. J. Godley (q.v.) taking charge of II Anzac Corps, Birdwood of I Anzac Corps. In this capacity he participated in the successful battles of General Sir Herbert Plumer's Second Army in the Third Ypres campaign. On these occasions, however, his initiative was much less than at Anzac. The critical factor in the battle was now the artillery plan which was decided upon at the higher Army level. Nevertheless as a trainer of troops and in his caution Birdwood began to emerge as a competent corps commander. He also began to be more assertive. After the success at Broodseinde on 4 October he insisted that his corps should be withdrawn for rest. This was not granted but after his unsuccessful Battle of Poelcappelle on the 9th he became more strident in his demands and from then on his corps took only a subsidiary role in the ghastly mud phase at Passchendaele.

In 1918 Birdwood seemed destined to play a big role. After the German March offensive the Australian troops were sent to various parts of the front to help stabilise it. By 4 May the Australian divisions were around the old Somme battlefield and talk of a counter-strike from that area was in the air. At that moment Birdwood was promoted to command the Fifth Army and left the Australian Corps to Lieutenant-General Monash (q.v.). Birdwood's army played a minor role in the German defeat, most of the time following-up a retreating enemy.

Birdwood had been given temporary administrative as well as operational command of the AIF in May 1915 after the death of Major-General W. T. Bridges (q.v.), and he was formally appointed to this post in September 1916. Generally he was a success in this role. He appointed Australians to command positions within the corps wherever possible and saw that any matters concerning the corps were brought to the attention of the Australian government. Surprisingly Birdwood retained his administrative role after he left the AIF and became Fifth Army Commander. Only the termination of the war prevented this anomaly from becoming an issue of contention between Birdwood and the Australian government. Birdwood, however, retained the affection of the troops throughout the war, and in 1920 toured Australia to great acclaim. He became C-in-C of the Indian Army in 1925 and retired in 1930. It was an open secret that he wished to become Governor-General of Australia but in 1931 the Scullin government insisted on appointing an Australian to the post and Birdwood's moment passed.

BISMARCK SEA, BATTLE OF, which took place between 2 and 3 March 1943, came about as a result of a Japanese attempt to reinforce and supply their troops fighting in the New Guinea campaign. A convoy of light transports escorted by eight destroyers and carrying 6000–7000 men set out for Lae from Rabaul on 28 February. On 1 March the convoy was sighted by a US reconnaissance aircraft and the next day it was attacked by American B-17 bombers. During the night an RAAF Catalina (q.v.) tracked the convoy which the next morning faced a coordinated attack by American B-17 and B-25 bombers and RAAF Beaufighters (q.v.) of No. 30 Squadron. By the end of the day all the transports had been sunk along with three of the destroyer escorts (the remaining escorts were all damaged). Of the troops 2890 were drowned.

'BLACK LINE' see **ABORIGINAL ARMED RESISTANCE TO WHITE INVASION**

BLACKBURN, Brigadier Arthur Seaforth (25 November 1892–24 November 1960) was educated at the University of Adelaide and after completing articles was admitted to legal practice in 1913. He enlisted in the 10th Battalion of the AIF in October 1914 and was at the landing at Gallipoli; the official historian, C. E. W. Bean, later contended that Blackburn and a colleague penetrated further inland than any other Australian soldiers that day, at least among those who survived the day's action. He was commissioned in August. The following year, at Pozières, he won the VC after leading a strong raiding party which captured 400 yards of trench and destroyed an enemy strong point. Blackburn led bombing parties against concentrations of German troops on four occasions, in which many of those around him were killed. He was invalided back to Australia in March 1917, and took no further part in the war although he was active in the political campaign favouring conscription (q.v.) in 1917. He held the seat of Sturt in the South Australian House of Assembly for the Nationalist Party between 1918 and 1921, but did not contest the election in the latter year. He was active in the RSSILA from its foundation in South Australia. Placed on the Reserve of Officers in 1920, Blackburn resumed citizen soldiering with the 43rd Infantry Battalion in 1925. He was posted to light horse regiments from 1928, and in July 1939 was promoted to lieutenant-colonel and given command of the 18th Light Horse (Machine Gun) Regiment. One year later he took command of the 2/3rd Machine Gun Battalion AIF, which he led through the campaign in Syria; he took the surrender of Damascus in June 1941. Returning to Australia in early 1942, he and

his unit were landed in Java to assist the Dutch in a futile attempt to stave off the Japanese advance through the Netherlands East Indies. Promoted to temporary brigadier, Blackburn was placed in command of a scratch composite force, designated 'Blackforce'. Powerless to stop the Japanese advance, Blackburn and his command were overwhelmed in three weeks and the survivors taken prisoner. Blackburn himself was liberated in Mukden, Manchuria, in 1945. He was awarded the CBE for his efforts in Java. He returned to business interests after the war, and was State president of the South Australian branch of the RSL in 1946–49. In June 1949 he became honorary colonel of the Adelaide University Regiment.

BLACKHAWK, SIKORSKY (Transport helicopter). Rotor diameter 16.36 m; length 15.26 m; armament 2 × door-mounted 7.62 mm machine-guns; load 11 fully equipped troops; speed 290 km/h; range 600 km; power 2 × General Electric 1543 shaft h.p. engines.

The Blackhawk entered service with No. 9 Squadron RAAF in 1989 as a battlefield transport helicopter replacing the Iroquois (q.v.). They were soon transferred to the army and are now operated by the 5th Aviation Regiment in Townsville and the School of Army Aviation at Oakey. The Blackhawk is the army version of the Seahawk (q.v.) used by the RAN and was assembled in Australia by Hawker De Havilland. In early 1995 serious problems with the maintenance cycle and a shortage of spare parts led to the grounding of most of the army's Blackhawk fleet, occasioning questions in parliament and opening again the question of whether the army or the RAAF should have control of rotary wing aviation.

BLADIN, Air Vice-Marshal Francis Masson (26 August 1898–2 February 1978). Narrowly missing service in the First World War because his parents refused him permission to enlist, Bladin entered RMC Duntroon and graduated in the class of 1920. Opportunities in the postwar army were few for regular officers, and after an attachment with the British Army in 1921–22 he transferred to the infant RAAF, being appointed flying officer in January 1923. He was thus a member of one of the two streams of officers who helped to found the RAAF, younger postwar appointees and older, former Australian Flying Corps or RAF members with wartime flying experience.

Bladin's interwar service prepared him well for wartime responsibilities. In 1929 he attended the RAF Staff College at Andover, followed by unit commands in Australia throughout the 1930s.

Appointed Director of Operations and Intelligence in March 1940, he was made responsible for air operations in northern Australia in March 1942 when promoted to AOC North-Western Area with headquarters near Darwin. Throughout 1942–43, when much of the combat action in this area was conducted in the air, Bladin built up offensive air power in his command and provided protection against aerial interference for the Australian and American troops concentrated further east in northern Australia and Papua New Guinea. In August 1943 he was sent to England and served as Senior Air Staff Officer to No. 38 Group RAF. In both these appointments he continued to fly operationally himself. The end of the war found him back in Australia as Deputy CAS.

His postwar career was, inevitably, much less eventful, although from January 1946 to June 1947 he was chief of staff at the headquarters of the British Commonwealth Occupation Force (q.v.) in Japan. Returning again to Australia he was Air Member for Personnel on the Air Board (q.v.) before retiring in 1953. He was then active in RSL affairs until his death.

Bladin was fortunate in a number of his appointments, offering as they did opportunities for active flying with senior rank and responsibility. Like Air Chief Marshal Frederick Scherger (q.v.), Air Marshal Colin Hannah (q.v.) and a number of other younger officers of their generation, Bladin's career showed the professional benefits that regular officers were able to bring to a relatively new service like the RAAF in its first major wartime testing.

BLAMEY, Field Marshal Thomas Albert (24 January 1884–27 May 1951). Starting life as a schoolteacher, Blamey applied for a commission in the Commonwealth Cadet Forces in 1906, coming third in the nationwide examinations and receiving appointment as a lieutenant in the Administrative and Instructional Staff. He took to soldiering with alacrity, and in 1910 transferred to the AMF with the rank of captain.

Two years later he was sent to the Staff College, Quetta, British India (now in Pakistan), after passing the competitive entry examinations. He graduated at the end of 1913 with a B pass, and was sent to England for attachment to the British Army; his wife and family went back to Australia. Like many young dominion officers attending the imperial staff colleges, he had started with a considerable handicap in that he lacked both wide regimental experience and general background and 'polish' of the kind possessed by his British contemporaries. He more than made up for this through determination to succeed; his biographer suggests that at

Quetta he also discovered 'a broader, freer' social life, one that perhaps laid the ground for a pattern of personal behaviour which would bring him fierce criticism in later life.

After a brief stint in the War Office during the opening months of the war, Blamey arrived in Egypt on 10 December 1914 to serve with the AIF as GSO3 (Intelligence) on the headquarters of the 1st Division. He landed on the first day at Gallipoli in company with Major-General W. T. Bridges and Colonel Brudenell White (qq.v.), and that afternoon was ordered forward to lead the 4th Battalion to assist J. W. McCay's (q.v.) hard-pressed 2nd Brigade. In May he led a three-man patrol at night to locate an enemy battery, and was engaged in a short, sharp firefight in which they killed six Turks before returning to their own lines without loss. He admired White and studied his methods and behaviour closely; of Bridges he later wrote that he had 'never admired and disliked a man so much'. In July he was returned to Egypt in the temporary rank of lieutenant-colonel to form the staff of the 2nd Division, then being raised. He was back on Gallipoli in September as the Assistant Adjutant and Quartermaster-General of that formation, and remained there until the evacuation in December 1915.

With the expansion of the AIF and the move to the Western Front came promotion. Blamey went to France with the 2nd Division, but in July became GSO1 of the 1st Division in succession to White. Closely involved in planning the division's attacks on the Somme, he nonetheless wanted a unit command, and at the end of 1916 was briefly given the 2nd Battalion and then temporary command of the 1st Brigade. The divisional commander, General H. B. Walker (q.v.), insisted on his return as GSO1, however, and Blamey served in this capacity until May 1918, when he was promoted to temporary brigadier-general and selected as chief of staff of the Australian Corps, once again in succession to White.

The six months as chief of staff to General John Monash (q.v.) were among the most instructive in Blamey's career. He drove himself and his subordinates hard in planning the attacks at Hamel, the offensive of 8 August and the assaults on the Hindenburg Line, and he and Monash complemented each other well. The latter wrote of Blamey later that 'our temperaments adapted themselves to each other in a manner which was ideal ... Blamey was a man of inexhaustible industry, and accepted every task with placid readiness ... He worked late and early, and set a high standard for the remainder of the large Corps Staff of which he was the head ... I was able to lean on him in times of trouble, stress and difficulty, to a degree which was an inexpressible comfort to me.' Blamey himself clearly reciprocated both the loyalty and regard of his chief, something which he did not, even at this stage in his career, necessarily extend to all his seniors, notably some British ones. AIF commander General William Birdwood (q.v.) wrote that Blamey was 'an exceedingly able little man, though by no means a pleasing personality', which probably says as much about Birdwood as it does about Blamey.

Blamey returned to Australia in late 1919 after eight years abroad, and was appointed Director of Military Operations at Army Headquarters. Within a few months he became Deputy CGS, while in August 1922 he was sent to London again to be the Colonel, General Staff, in the High Commissioner's office and Australian representative on the Imperial General Staff. While in London in late 1923 he was appointed 2nd CGS (a post akin to Deputy CGS), and he assumed the responsibilities of this office when he returned from London in March 1925. On 1 September he left the regular army and was sworn in as chief commissioner of the Victorian Police Force.

His decision to switch careers was prompted by institutional factors, not by a diminished enthusiasm for soldiering. Although his rise had been steady, even spectacular, there were many other officers senior to him with equally distinguished war records, and many of these had commanded troops in battle, which he had not. The political and economic climate of the 1920s did not engender much confidence in soldiering as a profession in Australia, and Blamey was only 41 years of age, and with a young family. He accepted a starting salary of £1500, and transferred to the militia.

Blamey's first task was to restore relations within the force after the police strike of 1923, tensions from which still simmered. Before he had begun, however, he became embroiled in a controversy over a raid on a brothel in which a man was apprehended with Blamey's police badge in his possession. The affair eventually fizzled out, unsatisfactorily, but would continue to dog him throughout his career. Blamey himself was blameless, but his handling of the matter indicated a blind spot in his perception of the divisions between the public and private behaviour of public men.

Blamey feuded openly with the press, successive governments, and the police union, which he broke. His personal conduct was not above reproach — he flouted Victoria's ridiculously restrictive liquor laws with open disregard — and he made enemies within and without the police force with his autocratic 'military' style. The ALP in particular reviled him as a representative of 'reaction'; there are

suggestions that for a time he led the clandestine White Army (see SECRET ARMIES). His period as chief commissioner was brought to an end in July 1936 when the Country Party administration of Albert Dunstan, reliant on the Labor Party for its survival, sacked him after a press campaign arising from his having issued an untrue statement in the course of protecting the reputation of a senior police officer.

His public and private lives were in pieces. His eldest son was killed in an air accident in 1932 while serving in the RAAF; and his wife, an invalid, died in 1935. Sacked from the police, in 1937 he relinquished command of the 3rd Division of the militia, which he had held for six years, and went on to the Unattached List as a major-general. By the beginning of the war, however, he had retrieved his position. In April 1939 he married again. In September 1938 he had been appointed chairman of the Commonwealth's Manpower Committee and Controller-General of Recruiting, largely at the behest of Frederick Shedden (q.v.), and was given the task of raising the strength of the militia and laying the basis for Australia's coming war effort. On 28 September he was selected to command the newly approved 6th Division, and soon after was promoted to lieutenant-general. When the 7th Division was raised at the beginning of 1940, Blamey was given the resultant corps command. His selection, over a number of more fancied regular and militia generals, was ultimately based on the government's perception that he would handle the difficult political-military problems that would arise from serving with the British better than anyone else. The charter he was issued in April 1940 set this out clearly together with his responsibilities to his own government.

It is probably fair to say that Blamey was better prepared for his later responsibilities as a senior commander than he was for operational control in the field. He had had little contact with recent developments in tactics or technology — but then neither had most other Australian officers. He conducted the evacuation of the ANZAC Corps from southern Greece with skill, but some alleged that the rigours of the campaign broke him physically and psychologically, at least temporarily, a charge disputed vigorously by others who were with him. He made some mistakes in exercising his command early on, especially in relation to the planning of the Greek campaign, which he believed did not have 'a dog's chance' of success. His failure to inform his government fully and directly of his doubts allowed the British authorities to overcome Menzies' concerns over the use of the 6th Division in the campaign, which turned into a disaster. On the other hand, he fought tenaciously against his British superiors, successively Sir Archibald Wavell (q.v.) and Sir Claude Auchinleck, when they attempted to disperse the AIF's component parts to suit their own needs. He insisted that the weary 9th Division be relieved in Tobruk against Auchinleck's inclination to leave it there, which made him highly unpopular in British circles both in Cairo and London.

The outbreak of the Pacific War saw Blamey and the majority of his command returned to Australia. Blamey was appointed C-in-C, Australian Military Forces, in March 1942, a post he was to hold for the rest of the war. In the face of the Japanese threat he presided over a force of some 12 divisions, mostly militia in composition and seriously deficient in both training and equipment. On an earlier brief visit to Australia in November 1941 he had commented critically on the 'Indian garrison' atmosphere he had observed in Australia, and the lack of commitment to the war effort. There were divisions within the army, not only between the militia and the regulars but amongst those who had served in the Mediterranean and who were for and against the C-in-C himself.

The appointment of the American General Douglas MacArthur (q.v.) as Supreme Commander of the South-West Pacific Area (q.v.), and Blamey's appointment in turn as commander of Allied Land Forces brought further problems. MacArthur became Prime Minister John Curtin's (q.v.) principal military adviser, which Blamey resented. MacArthur himself did not respect his own chain of command, but tended to exercise control of the ground forces directly; the pressure he applied to Blamey over the perceived performance of the Australians along the Kokoda Track (q.v.) in September 1942 led to Lieutenant-General S.F. Rowell's (q.v.) sacking from New Guinea Force (q.v.) even as the Japanese advance was checked. The following month Blamey sacked the 7th Division's commander, Major-General A. S. Allen (q.v.), because MacArthur believed his advance was insufficiently rapid. The Allied clearing of the Japanese beachheads at Buna–Gona–Sanananda retrieved Blamey's standing, not least because the American formations performed badly in their first major actions. Nonetheless, MacArthur resolved to keep command of US forces out of the hands of his Australian land commander, a position maintained for the rest of the war.

The offensives of 1943 saw the Australians mount operations against Salamaua, Lae, Finschhafen and Sattelberg. Responsibility for the planning and execution of these tasks was Blamey's. Blamey quickly grasped the requirements of war in New Guinea: the need for joint-force operations, the difficulties of climate and terrain, the requirement for thorough and appropriate training and for high levels of fitness, the importance of logisitics and intelligence. The operations in 1943 demonstrate clearly how far the Australian Army had come since 1939 in terms of capability and effectiveness under Blamey's overall command.

As the Pacific War entered its final stages, the disparity in aims between the Australians and Americans became more pronounced. MacArthur intended to marginalise the Australians as he began his drive to reconquer the Philippines, in which he succeeded. Australian forces were diverted to operations in Bougainville, Borneo and around Wewak, where they sometimes encountered fierce Japanese resistance but did not help to shorten the war. As operations they were not entirely without merit, however, and in the absence of strategic advice from the government Blamey went ahead and planned within the limits that MacArthur had imposed on him.

His stewardship of the Australian Army attracted criticism also. Undoubtedly he abrogated too many functions to himself. Administrative control of the army, operational responsibilities in New Guinea, and nominal command of Allied Land Forces were too much even for a man with Blamey's extensive grasp of detail. In his defence, Blamey maintained that he needed to retain his authority in all these areas to safeguard Australia's interests against the Americans, and that there was no other Australian officer who could do so. Certainly Gavin Long (q.v.), the official historian, considered that no other Australian general could be 'the politician Blamey is. Blamey, I think, carefully reconnoitres the political battlefield, and plans in that field as meticulously as on the battleground'. But he played favourites with some of his appointments, maintaining Savige in a corps command when it was clear that he had outstripped his abilities, and relegating able men like Rowell and Laverack, and less able ones like Bennett, to the sidelines for as many personal as professional reasons. On the other hand, Berryman, his chief of staff, was one of the most able officers in the army, while Herring and Mackay proved highly competent commanders of New Guinea Force. He was careful with Australian lives, and even more protective of Australian interests.

Blamey had made it clear in 1942 that he would retire at the end of the war. Following the surrender of the Japanese in Tokyo Bay on 2 September 1945, he tendered his resignation, which was declined. In early November, however, he was informed peremptorily that his services were no longer required. There were many in the government who did not care for him, and with Curtin gone there was no-one senior enough to safeguard his interests. (He said once of his relationship with Curtin that there was 'no need to worry about rear armour'.) He was replaced by Lieutenant-General Sir Vernon Sturdee (q.v.), soon to become the post-war CGS, but the manner of his retirement left a sour taste with many in the army.

He tended to his business affairs after the war, and became involved again in shadowy anti-communist, para-military politics. After Menzies became prime minister, he was promoted to field marshal in June 1950. Presented with the baton in hospital during his final illness, he died on 27 May 1951. His funeral attracted a crowd of 250000 in Melbourne.

The military historian David Horner has argued that Blamey's career was marked by 'year upon year of wise decisions, stubborn determination to further the interests of Australia, and a deep concern for the well-being of his soldiers'. He suffered from a number of deficiencies of character: he could be tactless, harsh, and unfair. His critics made too much of his social behaviour, but he never took the point that a public figure's private life is not his own. As Horner has concluded, however, it would be difficult to think of another Australian general with the prestige, personality or grasp of politics necessary to lead the army in its greatest trials. Without doubt, he faced far greater challenges than had ever beset his old chief, Monash, and it is only fitting that he should remain the highest-ranking soldier in Australia's history.

John Hetherington, *Blamey: Controversial Soldier* (Australian War Memorial, Canberra, 1973).

BLAND, Henry Armand (28 December 1909–). Bland entered government service through the New South Wales Public Service in 1927, qualifying as a solicitor in 1935 and serving as an alderman on Ryde Council between 1937 and 1939. He was acting Agent-General for New South Wales in London in 1940–41, and moved into federal government service thereafter. Between 1952 and 1967 he was Secretary of the Department of Labour and National Service, and thus presided administratively over many features of the national service schemes that operated between 1951 and 1959 and from 1964 to 1972. He was knighted in 1965. In 1967 Prime Minister Holt appointed him Secretary of the Department of Defence in succession to Sir Edwin Hicks, despite the fact that he had no defence background or connection with the services. He undertook a number of reforms of the joint service committee structure, joint intelligence functions, defence science, and the rational-

isation of defence facilities. Bland was also instrumental in founding the Joint Services Staff College in Canberra to bring service officers together in preparation for senior staff and policy jobs within Defence. Underlying all these reforms was Bland's desire to assert the pre-eminence of the Department of Defence as the central policy formulating and coordinating entity within the defence group of ministries. Much of this work was to be extended and consolidated in the 1970s by Sir Arthur Tange (q.v.). Various appointments to government committees and boards followed Bland's retirement from Defence, including a short period as Chairman of the Australian Broadcasting Commission in 1976.

'BLUE ORCHIDS' is a phrase from the Second World War used to describe the men of the RAAF who wore a distinctive dark-blue uniform that was considered to be better cut and more glamorous than those worn by the other forces (the army in particular); 'blue' denoted the uniform, 'orchids' the notion of glamour or good looks. Its first recorded use in an Australian context was in 1940, in a unit newspaper in the Middle East, but it was also used in 1941 to describe RAF aircrew training in South Africa. It was almost always used in the plural, and there were variations: members of the Empire Air Training Scheme (q.v.) were dubbed 'Menzies' Blue Orchids'. It could also be used derisively, especially by members of the 2nd AIF in referring to their air force opposite numbers, and was coupled with other expressions of contempt for men who were thought to be avoiding the dangers faced by the foot soldier. One such expression, originating in the Middle East, gave the definition of the acronym RAF as 'Rare As Fuck', a reference to the perceived lack of close air support for the ground forces in that theatre.

BOASE, Lieutenant-General Allan Joseph (19 February 1894–1 January 1964). Graduating from RMC Duntroon in the first class in August 1914, Boase went to the 9th Battalion AIF and took part in the Gallipoli landing. Wounded in the early fighting, he was evacuated and did not return to his unit until September. Proceeding to France the following year, he served with the 12th Battalion and on the staff of the 5th Division and of various brigades for the remainder of the war.

Between the wars he held the usual variety of regimental and staff positions, attending the Staff College, Camberley, in 1924–25. For two years in 1937–39 he was an exchange officer with the British Army in India, and upon his return was appointed Commandant of the Command and

Staff School in Sydney. He sailed with the 7th Division for the Middle East in October 1940 as Assistant Adjutant and Quartermaster-General, and once there was made responsible for administration of the AIF Base Area as Brigadier-in-charge of Administration.

Boase had no active command experience during the Mediterranean campaign, although he took command of the 16th Brigade in August 1941. He was given command of the two brigades that temporarily reinforced the garrison in Ceylon in early 1942, but this in no way made up for his frustration at missing out on an active command. The official historian, Gavin Long (q.v.), thought him 'a little bitter about his treatment and inclined ... to brood over the contrast between his misfortune and the (he considers) undeserved good fortune of some others'. When the 16th and 17th Brigades returned to Australia, they were sent to New Guinea, but Boase, until then their commander, was given a staff position on Headquarters First Australian Army in southern Queensland. Active service continued to elude him until late in the war. He finally commanded the 11th Division from September 1943, and led that formation in the fighting in the Ramu Valley and Finisterres in the New Guinea campaign from April 1944 until appointed to command Western Command, by then a backwater, in April 1945.

In the postwar period he served in London between 1946–48 as Australian Army Representative in the High Commission, and from 1948–49 as Defence Representative. Promoted to major-general in October 1948, his final posting was as GOC Southern Command until retirement in February 1951.

BOER WAR, SECOND (or South African War), 12 October 1899–31 May 1902, broke out over Boer resistance in the Transvaal Republic to British support of the demands of the Uitlanders (foreigners) who had flocked to the Witwatersrand goldfields from 1886. Five years before, in 1881, the First Boer War had been fought between British and Boer forces, as the latter sought to regain the independence they had surrendered to the British in return for protection against the Zulus. British forces suffered a decisive defeat at Majuba Hill on 27 February 1881, and rather than continue a military campaign for a cause in which he did not personally believe, the British Prime Minister, William Gladstone, recognised a conditional form of independence for the Boer republics in the Pretoria Convention later that year. The influx of large numbers of foreigners and of huge amounts of capital, together with the rapid industrialisation of the goldfields, threatened to submerge the Boer values and characteristics of the Transvaal, and provoked a fierce political resistance to perceived British inroads. This resistance turned to outright military confrontation in 1899, after four years of a virtual armed truce following the abortive Jameson raid of 1895, which convinced many Boers that the British government was irredeemably hostile to the existence of the Boer republics. Much to the surprise of the British government the President of the Transvaal, Paul Kruger, rejected British terms for a settlement of the Uitlander grievances (which were, in truth, merely the surface of the much deeper question of the control of the whole of South Africa), and in October 1899 the Boers launched military strikes against the British in Natal and the Bechuanaland extension of Cape Colony.

In the first phase of the war, which lasted until the end of January 1900, the mounted infantry forces of Boer commandos thrust into British territories and inflicted a series of crushing defeats on the poorly led and cumbersome British forces, culminating in the multiple British disasters of 'Black Week' in December 1899. These set-backs galvanised the British war effort: a large force was raised in England, and the two leading generals of the time, Lord Kitchener and Lord Roberts, were sent to South Africa. By August British forces had apparently turned the tide against the Boers: the two republics were annexed as British colonies, Kruger had fled, and the Boer forces seemed completely broken. They did not surrender, however, but turned to guerrilla warfare, using small bands of mounted infantry to attack British positions and then retreat. Not until after May 1901 did the British finally and painfully grind down Boer resistance. To do this they developed their own mounted columns and, much more controversially, began a widespread, even indiscriminate, campaign of harassment against the Boer civilian population. Twenty thousand civilians, mainly women and children, died as a result of the appalling conditions in the 'concentration camps' (the first use of the term) that Kitchener established.

Australian colonial authorities reacted cautiously at first to the outbreak of war, and rejected British suggestions that they should offer small contingents to serve in South Africa. It was left to the colonial commandants, meeting in late September and early October 1899 in response to overtures from London before the formal declaration of war, to propose the raising of a combined contingent of 2000

Map 5 Areas of conflict during the Boer War, South Africa, 1899–1902.

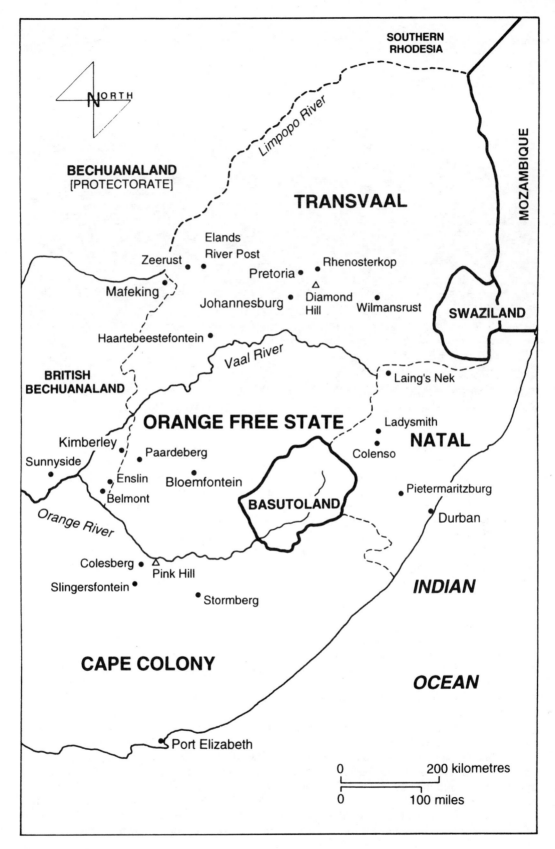

Australian horsemen in South Africa during the Boer War, c. 1900. (AWM P0175/462/058)

men. This proposal, however, was in turn rejected by the British government, which wanted colonial forces that could be broken up to reinforce British units. The action of the commandants has been criticised by later historians as an unwarranted interference in the domestic affairs of the Australian colonies, and it has been argued that the initial Australian response was 'manufactured' by imperial agents rather than springing from any genuine Australian sentiment. In acting as they did, however, the commandants were fulfilling their role in the greater scheme of imperial defence and military cooperation, and they could have argued, with some justification, that theirs was a much wider understanding than the parochial views that prevailed in so many of the debates over defence matters in the colonies. Anxious to have some Australian presence in South Africa, the British government agreed to pay the full costs of a contingent of the New South Wales Lancers, then completing training in England (at their own expense) and preparing to return to Australia. The agreement of the New South Wales government to this offer encouraged the other Australian colonies to give similar undertakings.

The first Australian contingent arrived in South Africa in December 1900 and for the first few weeks was engaged in lines-of-communications duties. However, the stunning defeats of Black Week, combined with the first Australian action — at Sunnyside in the northern Cape on 1 January 1900, in which 200 men of the Queensland Mounted Infantry performed well — did much to boost what until then had been slow recruitment for service in South Africa, and press reports began to laud the fighting qualities of Australian mounted infantry. Five infantry units were converted to mounted infantry, and won further credit for the growing Australian reputation in actions at Slingersfontein and Pink Hill. In the first engagement, on 9 February 1900, a troop of Western Australia Mounted Infantry (WAMI), accompanied by a squadron of the Inniskilling Dragoons, encountered a Boer commando of 300 to 400 men. The Boers attempted to turn the British flank by concentrating the bulk of their force against the flank that was held by the 20 members of the WAMI. Despite constant attacks by the Boers, and the loss of two dead and five wounded, the Australians held their position and carried out an orderly retreat as dusk fell. At Pink Hill, three days later, Major G. A. Eddy of the Victorian Rifles commanded a combined force of 75 Victorians, 25 South Australians, and 100 British troops drawn equally from the

Wiltshire and Inniskilling Regiments. The Boers attacked both flanks of the British position, and eventually the British were forced to withdraw, though only after offering prolonged resistance that resulted in the loss of six Australians, including Eddy, with a further 23 wounded. By the scale of the second phase of the Boer War this was a major defeat, but while there was some criticism of Eddy's decision to hold on to an untenable position, there was far more praise for the stoicism shown by Australian troops. Defeat assumed its own form of victory.

The second contingent arrived in South Africa in February 1900, and about 500 Australian troops, from the Queensland Mounted Infantry and the New South Wales Lancers, took part in the huge British cavalry sweep to relieve Kimberley that began on 13 February. Following the relief of Kimberley, Roberts moved his combined infantry and mounted force to defeat a 6000-strong Boer army moving towards Bloemfontein. The New South Wales Mounted Rifles (NSWMR) formed part of a British force of 2000 mounted infantry that tried to cut off the Boers but instead ran into severe difficulties. In the battle at Paardeberg (18 February 1900) which followed — the last major battle of the war — the NSWMR were involved in skirmishes on the edges of the Boer force, but once the enemy had been pinned down, there was little call for mounted infantry, and Australian forces played virtually no part in the battle. Although Paardeberg, which ended the second phase of the war, was a major defeat for the Boers, the war was not yet over. It then entered a third phase, one which called for a different military response from the colonies and which also provoked a different response from the public. Already by mid-1900, as men from the first contingent began to return to Australia, there was a noticeable decline in public interest. The first groups to return had been welcomed as heroes, and while far too much had been made of their relatively minor role in various minor engagements, there was at least some basis for the claims of publicists. Those claims became harder to sustain after Paardeberg, which seemed to much of the public, as indeed it seemed to Lord Roberts himself, to mark the end of effective Boer resistance.

The first two contingents from Australia had been raised from the part-time militia and volunteers, a restriction that resulted in relatively slow rates of enlistment. The third and subsequent contingents, which were raised as a result of a direct appeal from the British authorities, were recruited from good horsemen who were also good shots but who did not need to have any previous military training. The potential pool was therefore far

greater than for the first two contingents, and notwithstanding the steadily declining public interest in the war (together with a minority opinion that was increasingly hostile to the tactics being employed by Kitchener against an enemy which Roberts had advised the British government was now decisively defeated), vacancies were heavily oversubscribed, almost 2000 accredited bushmen applying for the 500 positions in New South Wales alone. The original intention was that the Bushmen's Contingent would be financed entirely from private sources, thereby demonstrating the depth of public support, but the response was uneven at best. A small number of wealthy individuals in New South Wales contributed almost the entire sum themselves, and public appeals for donations in South Australia were successful, whereas in Victoria, Queensland and Western Australia the governments were forced to cover the costs in varying degrees, wholly so in Queensland and pound for pound in Western Australia. Despite the lack of public interest in contributing financially to the cost of the Bushmen's Contingent, large crowds turned out to see the troops off in March, but significantly the size of the crowds was smaller than it had been for either of the first contingents in Sydney and Melbourne; only in Adelaide was it thought to be larger than on the first two occasions.

Even while the Bushmen's Contingent was being raised, there were further calls from the British government for more of the same. The Colonial Secretary, Joseph Chamberlain, asked for an additional 2000 bushmen, who would be paid by the British government which would also meet all other expenses associated with the force. The fourth contingent, which attracted an even greater number of applications than had the original bushmen's recruiting drive, was named the Imperial Bushmen, and left Australia in April 1900. The fifth and sixth contingents were raised as drafts to reinforce earlier contingents, the fifth being in the field by mid-1901, while half of the eight battalions of Australian Commonwealth Horse (q.v.) — the first post-Federation Australian units — which made up the sixth contingent, had not arrived in South Africa by the time the war was formally over, and of the other four battalions, only two saw any action.

The third and subsequent contingents were largely deployed in sweeping the countryside and in enforcing Kitchener's policy of denying support to the Boer guerrilla forces. There was much hard riding across the long distances of the veldt where it was scorching during the day and freezing at night, occasional skirmishes with small Boer parties, and extensive burning of farm houses and confiscation of Boer horses, cattle and wagons. The 3NSWMR, for example, covered 1814 miles in 153 days in the latter half of 1901. The romantic image that the raising of the bushmen contingents had conjured up sat uneasily with the realities of this stage of the war. For all the lack of glamour, there were dangers: skirmishes inevitably brought casualties — the 3NSWMR lost five killed and suffered 19 wounded in five months in 1901. There were no major engagements, but Australians were involved in actions at Elands River (4–16 August 1900), Rhenosterkop (29 November 1900), and Haartebeestefontein (21 March 1901) where they acquitted themselves well and reinforced their reputation for courage and boldness.

One incident stood out in stark contrast. On 12 June 1901, 350 men of the 5th Victorian Mounted Rifles camped at Wilmansrust in eastern Transvaal after failing to make contact with Boers known to be in the general area. Despite the knowledge that the enemy was in the vicinity, the commanding officer (a British officer) failed to secure the perimeter of the camp adequately, and at 7.30 p.m., while the men were resting, the Boers attacked at point blank range. The Boer attack lasted only five minutes: 18 Australians were killed, 42 wounded, and about 50 fled to avoid being captured. Much more useful to the Boers were the Australian horses: 100 were seized, and more than 100 killed. It was a humiliating performance by the Australian forces, which received considerable, and unfavourable, publicity in Australia, especially in Victoria. A similar lack of sympathy for individuals found wanting was shown in the case of Lieutenants Morant and Handcock (q.v.), Australians serving in the irregular Bushveldt Carbineers, who were sentenced to death and executed by firing squad on 27 February 1902. At the time, the Australian press was virtually unanimous in its condemnation of the men and their actions, and the suggestion that 'the episode … did more than anything else to foster Australian disenchantment with the war' cannot be sustained. When Morant and Handcock did achieve the qualified status of quasi-heroes, it was several generations later, when the Boer War was used to illustrate the dilemmas of military service in a much more controversial conflict, Vietnam.

There was, however, considerable opposition at the time to the war in general. The new Australian Labor Party strongly opposed both the war and Australian involvement, as did much of the Catholic church, whose largely Irish clergy naturally sympathised with fellow victims of British imperialism. Much of the opposition emerged at the very beginning of the war and centred on the question of whether or not the Australian colonies should become involved by sending contingents. Once that decision had been made, and Australians were engaged in fighting, the public debate largely fell away, although committed advocates on either side continued to voice their opinions. The overwhelming success of the recruiting drives for the third and subsequent contingents suggests that there was still strong support from sections of the community but, as in Britain, there developed outspoken pro-Boer advocates within Australia, even before 1900 when Kitchener embarked on his scorched earth policy and the war departed from the popular image of heroic battles. The words of the leader of the Liberal Party in Britain, Sir Henry Campbell-Bannerman — 'When is a war not a war? When it is carried out by methods of barbarism in South Africa' — undoubtedly struck a responsive chord in some Australians, but they were very much in the minority, and the passionate debates that raged in Britain had little counterpart in Australia.

The Boer War ended five months after the achievement of federation by the Australian colonies. A total of 16 175 men left Australia in the six contingents, and an indeterminate number of other Australians who had gone to work in the goldfields served in local units. Deaths in action or from battle wounds amounted to 251, with a further 267 dying from disease, and 43 reported missing. Five Australians won the VC, and lesser decorations were awarded on a liberal, even extravagant, scale. The bare statistics, and even an outline narrative, do not convey the significance of these events in Australian military history. The Boer War marked the birth of an Australian military reputation, one renowned for dash and courage and, in the guise of the bushmen and draft contingents, relying more on natural skills developed in the Australian bush than those instilled on the parade ground. Paradoxically, while the hardest fighting was undertaken by the first two contingents, which were largely drawn from the part-time militias of the colonies, it was the individualism of the bushmen which captured the popular imagination. Yet while they were praised for their horse-riding abilities, the Australian contingents, especially the third to sixth, came in for their share of criticism. Despite an emphasis during recruiting on shooting skills, many recruits proved to be barely adequate in this regard. Their horsemanship was clearly demonstrated, but they were thought to be inefficient horse masters. At times their discipline was fragile, and their officers were regarded as poorly trained. Those observers, British and Australian, who saw in the raw material of the Australian volunteers the makings of first-class soldiers if proper training, discipline and leadership were brought to bear were undoubtedly correct, but the popular view was the one that prevailed, namely that Australians were natural soldiers who neither needed nor responded well to traditional training methods. The Boer War thus left a bitter-sweet and lasting legacy for the new Federation: the tradition of the citizen soldier in action.

L. M. Field, *The Forgotten War: Australian Involvement in the South African Conflict of 1899–1902* (Melbourne University Press, Melbourne, 1979; reprinted 1994).

BOFORS GUN Range 2500 yards; shell weight 2 pounds; size of shell 40 mm; rate of fire 150+ rounds per minute.

The Bofors gun was of Swedish origin and was used by Australian naval forces in the Second World War as an anti-aircraft gun. It was either single- or twin-mounted. During the war 290 guns were made in Australia, and between 1941 and 1945 they were fitted to almost every type of warship. In the 1990s some were still in use on *Fremantle* Class patrol boats (q.v.) and various support ships.

In the army light anti-aircraft regiments were equipped with Bofors guns in the Middle East, and in the Pacific composite anti-aircraft regiments included 18 of these weapons. They were also deployed in coastal fortifications from 1943.

BOMBARD, HMAS see **ATTACK CLASS PATROL BOATS**

BOOMERANG, CAC (Single-seat fighter). Wingspan 36 feet; length 25 feet 6 inches; armament 2 × 20 mm cannon, 4 × 0.303-inch machine-guns; maximum speed 305 m.p.h.; range 930 miles; power Pratt & Whitney Twin Wasp (q.v.) 1200 h.p. engine.

The Boomerang was a fighter design rushed through in 1942 and built by the Commonwealth Aircraft Corporation (q.v.) to meet the Japanese sweep southwards at the beginning of the Pacific War. To ease production the Boomerang used many parts of the Wirraway (q.v.) trainer also being manufactured and entered service over Lae and Salamaua during the New Guinea campaign in 1943. Boomerangs were used by No. 4, 5, 83, 84 and 85 Squadrons, but in combat against Japanese aircraft were found to be slow and to perform badly at altitude. Fortunately supplies of Spitfire and Kittyhawk (qq.v.) fighters had begun to arrive in 1942, enabling the Boomerang to be relegated to the army cooperation role where they attacked ground forces and dropped smoke bombs to mark targets. They were used in this role during the Bougainville and Borneo campaigns. The

Boomerang is significant as the first combat aircraft designed and built in Australia.

BOOMERANG, HMS see **AUXILIARY SQUADRON, AUSTRALIA STATION**

BOONAROO see **SUPPLY SHIPS**

BORNEO CAMPAIGN The second series of mopping-up campaigns fought by Australian forces in the latter stages of the Pacific War were all in the partly Dutch and partly British island of Borneo. Tarakan Island and Balikpapan were part of the Netherlands East Indies, Brunei Bay was in the small British-protected state of Brunei, while Labuan Island was a British Crown Colony.

The later periods of wars are frequently untidy: the war in the Pacific was no exception. This was partly a result of the search for a new strategy to follow the end of the war in Europe. Throughout 1942 and 1943 the 'beat Hitler first' policy (q.v.), adopted by President Roosevelt and Prime Minister Churchill, had given a consistent thrust to grand strategy, and could always be appealed to in settling arguments about the apportionment of scarce means between different theatres. Formally, the pre-eminence of the European theatre's requirements over those of the Pacific was seldom questioned. But, de facto, even within the Combined Chiefs of Staff, senior Americans like Admiral Ernest J. King, who was a self-proclaimed Pacific firster, had ensured through their advocacy that the legitimate claims of the Pacific were given far greater recognition than the prevailing strategy might seem to have demanded. In the second half of 1944, however, there was some optimism that the war in Europe could be over by Christmas, and much thought was given to how the Allies' by now highly mobilised military strength could be re-directed to the task of beating Japan.

For Australia, despite the fact that an emerging strand in the Curtin government's war aims was to see British influence restored and the 'Union Jack in the Pacific', there was the immediately practical realisation that the wartime arrangements with the United States had served Australia well, and the command of the South-West Pacific Area (q.v.) by General Douglas MacArthur (q.v.), whatever partial surrender of Australian sovereignty it may have involved, was nevertheless a spectacular and continuing success.

It has never been firmly established whether MacArthur ever intended Australian troops and airmen to take an active and effective part in the re-conquest of the Philippines. But as the Australian government's expected that at least part of its force

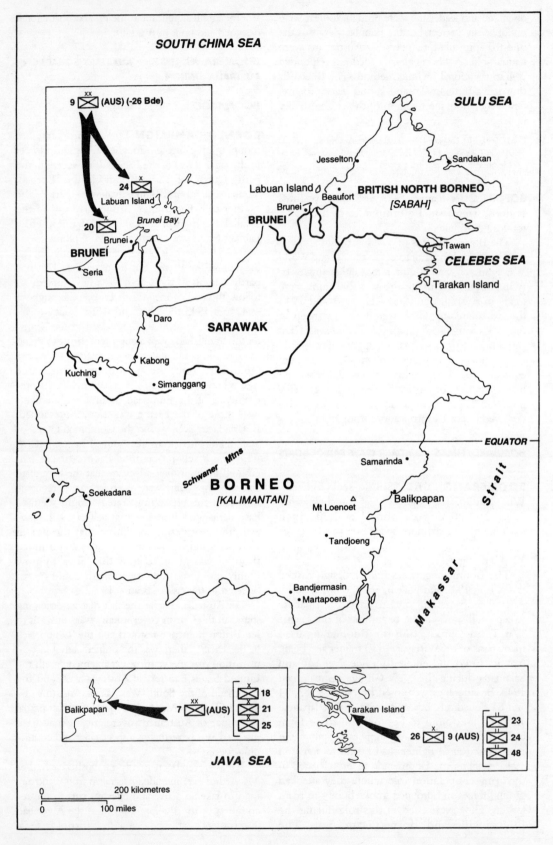

SOUTH CHINA SEA

9 XX (AUS) (-26 Bde)

24 X
Labuan Island

20 X
Brunei Bay
Brunei
BRUNEI
Seria

SULU SEA

Jesselton
Sandakan

Labuan Island
Brunei
Beaufort
BRITISH NORTH BORNEO
[SABAH]

BRUNEI

Tawan
CELEBES SEA

Tarakan Island

Daro
SARAWAK

Kabong

Kuching
Simanggang

EQUATOR

Samarinda

Schwaner Mtns
B O R N E O
[KALIMANTAN]

Soekadana

Mt Loenoet
Balikpapan

Strait

Tandjoeng

Bandjermasin
Martapoera

Makassar

Balikpapan

7 XX (AUS)

18 X
21 X
25 X

Tarakan Island

26 X 9 (AUS)

23 X
24 X
48 X

JAVA SEA

0 ___ 200 kilometres
0 ___ 100 miles

would play a visible and effective role against Japan, a worthy participation had to be found. Finally, despite the Australian government's unease, a lack of enthusiasm on the part of the US Joint Chiefs of Staff, and the misgivings of the Australian C-in-C, General Thomas Blamey (q.v.), the Borneo operations were decided upon and MacArthur issued orders for their initiation.

Thus, these operations, code-named OBOE (q.v.), were 'MacArthur's brainchild'. His overall intention was to secure bases in Borneo and, from there, move into Java to restore Dutch rule. In no sense were they later seen to have the advantages that MacArthur claimed for them at the time. The seizure of significant Borneo oil bases would not further denude Japan of oil: it was already cut off from that source by Allied operations further north. Britain had made it clear that Brunei would not be useful as a naval base during the war. The Tarakan airfields could not be got ready in time to support subsequent operations at Brunei and Balikpapan. And, finally, Blamey believed that the Balikpapan operation was strategically unsound and represented this to the Australian government, which in the end ignored this advice and bowed to MacArthur's wish. The three operations went ahead accordingly, and while they were of doubtful value strategically, tactically, they were skilfully conducted.

Tarakan

The 26th Brigade Group, commanded by Brigadier D. A. Whitehead (q.v.), landed at Red, Yellow and Green beaches at Tarakan on the morning of 1 May 1945. MacArthur had instructed Lieutenant-General Sir Leslie Morshead (q.v.) commanding I Australian Corps on 21 March 1945 that, 'using the 26th Brigade Group he was to seize and hold Tarakan Island, and destroy the enemy's forces there'. The Netherlands East Indies government was to be re-established, the oil-producing and oil-processing installations conserved and the airfields put into operation as soon as possible. The Tarakan operation was code-named OBOE 1.

The landing from the ships of Amphibious Group Six (Rear-Admiral Royal, US Navy), was very successful. The landing itself had been rehearsed at Morotai, where a combined planning headquarters had worked together from the end of March. Whitehead's force was much larger than normal: at 11 804 troops, its strength approached a light division, and included 40 units and sub-

units. A Netherlands East Indies Civil Affairs Unit accompanied the landing.

Tarakan Island is shaped like an inverted pear; 24 kilometres long and 18 kilometres wide. Inland, it rose steeply from a swampy coastal plain into a tangle of hills and steep gullies that were covered with dense rainforest and secondary growth. Only one beach, at Lingkas, the port of Tarakan town, was suitable for an assault landing. It was commanded by strong defences. The enemy force of about 2100 men included the Japanese 455th Battalion (860), 2nd Naval Garrison Force (900), and a company of the Kure Special Naval Landing Force. They manned five gun-positions on the southeastern tip of the island, 15 anti-aircraft guns and a number of light and medium guns about the airfield. Lingkas Beach itself was protected by parallel rows of posts, rails and pipes driven into the mud, and a wide anti-tank ditch.

Preliminary operations included a clandestine landing by amphibious scouts who brought out useful information; the pre-positioning of an artillery battery and commandos on to Sadau Island to suppress enemy fire; and the insertion of groups of engineers to create gaps through the beach defences with explosives prior to the Allied assault. Despite difficulties, the engineers completed their task and the landing, on a two battalion front with 2/23rd Battalion on to Green Beach, and 2/48 Battalion on to Red Beach, went according to plan. There was massive supporting fire for the landing: a preliminary bombardment by two cruisers and six destroyers; an aircraft attack which concluded just before the landing craft beached; craft with rockets and mortars accompanied the assault waves; and tanks came ashore immediately after the landing. As an observer described it: 'The beach appeared to be an inferno and was continually aflame from the crimson flashes of bursting bombs and shells'.

There was no opposition on the beaches, but because of the narrowness of the frontage, there was congestion on the shore. Then sharp fire came from Lingkas Hill. The two leading battalions, 2/23rd Battalion on the right and 2/48th Battalion on the left, pushed inland towards Tarakan town itself, overcoming opposition as they went. By last light, the 26th Brigade Group held an area 2.5 kilometres wide and 2 kilometres deep. On the left the objectives had been reached, but one nicknamed 'Metho' held out on the right. Whitehead ordered the 2/24th Battalion to push through to the airfield on the following day. Morshead's prudence in including so many engineers in the force — two pioneer battalions and three field companies — paid off as demands for removing mines and

Map 6 Borneo Campaign, 1945, showing the areas of the three Australian landings.

booby traps, road and track repairs, and efforts to get the tanks forward in support of the infantry all increased.

The characteristically stubborn Japanese defence needed every form of support including flame-throwers to overcome it. An attempt to capture the airfield by the night of 2/3 May was delayed when the Japanese fired two enormous buried charges. It was taken on 5 May 1945. By the following day, Whitehead, who had redeployed and regrouped his units with considerable skill, had established that continuing resistance was likely to come from the Mount Api–Fukukaku area, Sesanip, Otway, the Amal River area and Cape Juata. The enemy having now lost the airfield, water-purifying plant and hospitals, 'was displaying a decided disinclination to hold ground', and mostly was employing mines, booby traps, suicide raids and isolated parties fighting to the death in tunnels. In one action near a series of knolls nicknamed 'Freda', Lieutenant T. C. Derrick VC DCM was mortally wounded. Derrick had become a legend in the Australian Army and his death the following day was widely felt. Attempts to induce the enemy to surrender by offering food and treatment for the wounded were just as unsuccessful here as they had been in other operations. From then until operations ceased with the Japanese surrender, the latter fought on in increasingly smaller, but equally determined groups.

The last Japanese stronghold on the island was in the Hill 102–Hill 105 area, and Whitehead decided to roll it up from the north in a deliberate attack. The Japanese on Hill 102 were attacked on 1 June by 18 Liberator bombers, each with nine 500-pound bombs, and 24 Lightning fighters armed with napalm. The Australians then attacked with flame-throwers. The attack was successful and smaller attacks were then carried out on other positions. An attempted Japanese counter-attack on 10 June 1945 disintegrated for want of coordination and ammunition. On 20 June, the 2/24th Battalion took the last main objective, Hill 90, after 2100 artillery rounds and 600 mortar bombs had been used in the attack. After this, Whitehead divided the island into six areas of responsibility to flush out the survivors. It was later established that organised defence had ceased from 14 June 1945; but small actions continued to be fought with stubborn groups, many of whom had been attempting to escape to the Borneo mainland on improvised rafts.

Tarakan emphasised lessons that had been learned earlier: the effectiveness of well-directed artillery fire, the need for large numbers of engineers in tropical warfare, the use of tanks to reduce defences, and most of all the importance of good infantry in cooperation with other arms. White-head (like Major-General E. J. Milford, with the 7th Division at Balikpapan later) had been critical of the composition of the force, particularly the disproportionately large number of RAAF personnel used to support a relatively small number of aircraft. He wrote later that the air force component had had little or no experience of slimming to assault scales and landed with a ridiculously large amount of impedimenta, to the embarrassment of all, in particular the engineers who had to move it off the beach.

The main reason for the campaign, to establish a usable fighter airfield, was not fulfilled. It was found to be so difficult to repair that fighter support for later operations at Brunei Bay and Balik-papan was provided from Tawi Tawi (which had been occupied by American troops without a fight), and by carrier aircraft. When fighting ceased on Tarakan, the Australian force had lost 14 officers and 211 other ranks killed, and 40 officers and 629 other ranks wounded. About 1540 Japanese dead had been counted, 252 had surrendered before the cease-fire and 300 after it. It had been a short, tough campaign. Australian losses in the brigade on Tarakan had been almost as high as those suffered by the entire 6th Division in the whole of the First Libyan Campaign in 1941.

Labuan Island

The second Australian operation into Brunei Bay–Labuan Island was code-named OBOE 6 and was mounted by the 9th Australian Division (less the 26th Brigade) while the Tarakan operation was still in progress. The 9th Division's tasks were 'to secure the Brunei Bay area of north Borneo, to permit establishment of an advanced fleet base, and to protect oil and rubber resources therein'.

Brunei Bay is the best harbour on the north-west coast of Borneo. It is protected by a group of islands of which Labuan is the largest. On the mainland 40 kilometres east of Labuan was the town of Weston from which a light railway ran north-east through Beaufort to Jesselton, 90 kilometres away. At the time it was the headquarters of 37th Japanese Army. The mountainous interior of north Borneo is inhabited largely by indigenous people known as Dyaks, who distrusted the Japanese and had assisted a number of British and Australian agents of the Services Reconnaissance Department (SRD) (see ALLIED INTELLIGENCE BUREAU) who regularly reported Japanese dispositions and movements.

Japanese garrisons in Borneo that made up part of 37th Army included the 56th Independent Mixed Brigade in north Borneo, 71st Independent

Men of the 2/32nd Battalion leap from a Buffalo amphibious vehicle during the landing at Weston, near Labuan, Borneo, 17 June 1945. (AWM 109614)

Mixed Brigade in the south, and possibly 25th Independent Regiment at Jesselton. By late May 1945 Allied intelligence believed 650 Japanese were on Labuan Island, 1550 in the Brunei Peninsula, and 6600 around Jesselton. Most of these were from 56th Independent Mixed Brigade. As well, there were two or three Japanese cruisers and five destroyers in the area, and 12 enemy aircraft in Borneo. Since there were also as many as 340 Japanese aircraft in the south-east Asian region, air attack and light surface naval attack were always possible.

Because of complications in moving troops and equipment from Australia to Morotai to mount the operation, Lieutenant-General Morshead commanding I Australian Corps recommended that either the operation be postponed or reduced in scope; it was postponed from 1 to 10 June. Although based on a division, the force was much larger. In fact, Major-General George Wootten's (q.v.) command numbered more than 29 000 men. The supporting naval force was commanded by Vice-Admiral Dan Barbey, by now well known to the Australians. It included Sixth Amphibious Force, a cruiser covering force, and nearly 200 other vessels. Air support was provided by First Tactical Air Force (q.v.), RAAF.

The assault force landed at Brown Beach (Labuan) at 9.15 a.m. on 10 June 1945; landings then followed at Green and White beaches on Brunei Peninsula and Muara Island. Wootten's fireplan included creeping concentrations of naval gunfire which moved inland ahead of the infantry as they advanced. Aerial bombardment used a grid square system which, in the absence of more precise targets, suited the circumstances.

Initial objectives for the force were: 24th Brigade, commanded by Brigadier S. H. W. C. Porter (q.v.), to capture Labuan Island; and 20th Brigade, commanded by Brigadier Victor Windeyer (q.v.), to capture the Brunei–Muara Island area. After Labuan Island and Brunei town had been seized, these forces were redeployed to other objectives. Both brigade landings were successful. Many of the enemy later admitted to being disheartened by the preliminary aerial and naval bombardment. In the succeeding weeks they were given greater reason for dismay by the skill, resolution and enterprise of the troops and by the excellence of the planning that had preceded the whole operation.

After its landings near Brooketon and Muara Island, 20th Brigade moved quickly to seize Brunei town. Muara Island was unoccupied and little opposition was experienced on the main peninsula. Brunei airstrip was occupied on 12 June. A company crossed the river by boat to cut the road to Limbang. By the fourth day, 13 June, the town was occupied. On Labuan Island, consolidation had been equally rapid. The two battalions, using tanks, had cleared the island by 14 June, except for a Japanese garrison in a tangled area of jungle and swamp nicknamed 'the pocket', where a force of several hundred Japanese made a last stand. Before a final attack was made, it was extensively probed by patrols, then heavily neutralised by artillery, mortars and 8-inch shells from the cruiser HMAS *Shropshire* (q.v.), and attacked from the air. The final assault from two directions was well-supported by tanks including flame-throwing 'Frogs'. Almost 200 Japanese dead were counted; 389 had been killed on the island as a whole.

Major-General Wootten decided that, as the enemy in most areas were withdrawing, he would accelerate the second phase of operations. The following operations were mounted in quick succession by landing craft and boat: the 2/32nd Battalion secured Weston after a minor action; the 2/43rd Battalion landed at Mempakul and advanced north along the Klias River; and the 2/13th Battalion group landed and secured the Miri–Lutong area. On 26 June 1945 the 2/43rd and the 2/32nd Battalions launched a combined attack on Beaufort which was secured on 29 June. In this action Private L. T. Starcevich of the 2/43rd Battalion was awarded the VC. The two battalions lost 7 killed and 38 wounded; 93 Japanese were killed and two prisoners were taken. The Weston–Beaufort railway line was now open, and carriages pulled by a jeep began to run. Papar was occupied on 12 July 1945. In the south, the 2/13th Battalion patrolled south and west while the 2/17th Battalion, which since landing had advanced 120 kilometres on foot to Tutong, Seria and Kuala Belait, secured the area between Brunei town and Lutong. The Limbang–Lawas–Malaman area was secured by the 2/15th Battalion. It took the Australian engineers and oil experts three months to put out the oil fires in the Seria field which had been sabotaged by the Japanese.

It needs to be remembered that a parallel guerrilla war was being waged between the indigenous Dyak people (led in many cases by SRD agents), and the Japanese. The Australians' ability to help the Dyaks, although proffered where possible, had to be limited. The military objective was the coastal strip with its ports, oilfields and rubber plantations, not the entire area; successful efforts were made, however, to keep the Japanese remnants at a distance and forward patrol-bases were established on the Balai, Ridan and Marudi areas in the south through which some Japanese were escaping towards the interior mountains, along the Pandaruan and Limbang Rivers south of Brunei town, and along the Beaufort to Tenom railway line.

By the second half of July, much of the Australian effort was civic action. There were about 69 000 civilians within the area of effective control. Many were now able to return to productive work. Many civilians had been admitted to army hospitals, others were helped at aid posts. Rebuilding was being done, which included the restoration of town water supplies. Schools throughout Brunei State were functioning; rubber was being collected and sent away; and the restoration of oilfields was continuing. The Australian troops got on splendidly with the local people with whom a strong trust was developing. Martial law eventually gave way to civil administration under the British Borneo Civil Affairs Unit, which by the time of the Japanese surrender had effectively taken over civil administration.

In the OBOE 6 operation, 114 Australians were killed or died of wounds, and 221 were wounded but survived. Some 1400 Japanese were killed, and 130 were taken prisoner. The SRD groups believed that their guerrillas had killed more than 1800 Japanese throughout north Borneo.

Balikpapan

Balikpapan, the last of the three Borneo campaigns, was also the one most questioned by General Blamey. Perhaps recognising the degree of controversy it had aroused, MacArthur gave the operation massive support, unrivalled in Australian military history. It was also the last large-scale Allied operation on land of the Second World War.

The Australian force was built around the 7th Division, AIF. In fact, Balikpapan was to be the first time in the war that this division fought as a complete formation. Planning for the operation had begun on 26 April 1945, and its commander, Major-General Milford, had decided to assault on a two-brigade front between the Klandasan Besar River and Klandasan. Milford saw many advantages in this approach. He could make maximum use of his overwhelming fire support; reduce the duration of the operation against an enemy who thought nothing of sacrificing troops in a drawn-out attritional struggle; and take advantage of the fact that the Japanese were notably slow to concentrate by seizing the vital ground ahead of their reaction. This approach would also enable him to overcome the problem of bridging and engineer works during a longer coastwise advance on a narrow front.

He also appreciated that the enemy would not expect him to hazard a landing in its most strongly held area. The enemy was not alone in this thought. Air Vice-Marshal William Bostock (q.v.), who was to control air support, wrote to the Prime Minister that Balikpapan's carefully prepared defences might cause AIF casualties comparable to those at Gallipoli. The difference, which he failed to discern, was that the 7th Division was a highly experienced formation.

Intelligence concerning the Japanese order of battle and dispositions was reasonably accurate. In the town itself there were about 3900 Japanese in mobile units, 1100 Japanese workers, 2400 Indonesians and 1000 Taiwanese labourers. All were armed and could fight. There were 18 coast-defence guns, 26 dual-purpose guns (which could also be used in an anti-tank role), and 78 medium and light anti-aircraft guns. The beach defences consisted of rows of stout poles driven into the mud, and an anti-tank ditch between Stalkudo and Sepinggang. As well, a line of bunkers fringed the beach and the heights above Klandasan were honeycombed with bunkers and trench systems. A deception plan was mounted to convince the enemy that the landing would be at Manggar.

Milford's landing force, including RAAF and US air elements, numbered more than 33 000. It included 247 separate headquarters. The senior naval commander Vice-Admiral Barbey objected to the plan and would have preferred landing at Sepinggang, away from the coast defences. Milford, whose speciality was artillery, was able to overcome these objections by his well-reasoned approach. As he wrote later: 'why land up the coast and have to fight miles through jungle, which suits the enemy, when you can go straight in under heavy supporting fire, which the enemy can't stand, in comparatively open and favourable country'.

The pre-landing bombardment and target preparation were begun 20 days before the landing and were the longest in any amphibious operation of the war. A large number of mines had been laid by the Japanese, and out of four mine-sweepers deployed, three were sunk. American Underwater Demolition Teams, did a magnificent job clearing 1000 metres of the intended assault beach, as well as 750 metres of neighbouring beaches as a deception measure.

The landing was made at 9 a.m. on 1 July 1945 on three beaches: Red, Green and Yellow, with two brigades forward on a frontage of 2000 metres. The initial object was to secure enough ground for a beach maintenance area; then 18th Brigade on the left was to seize the high ground dominating the town and free the harbour for use, and 21st Brigade on the right was to advance along the coast and take the airstrips. The 25th Brigade, in reserve, was

to land on 2 July, with the probable role of thrusting along Milford Highway towards Batuchampar.

Neither brigade was landed on its correct beach but this did not disrupt the assault. With the help of tanks and flame-throwers, Japanese positions were overcome in turn, and by last light on 1 July, the bridgehead was more than 2000 metres in depth. On the following day, the 2/14th Battalion had advanced and seized Sepinggang airfield on the right. The 2/27th Battalion had taken the high ground north-west of Stalkudo. The 2/16th Battalion took Record, Owen and Oxley (nicknames of key positions), without loss. In the 18th Brigade sector, many tunnels in which Japanese were sheltering were blown in. Throughout the whole operation, engineers helped to destroy 110 tunnels and pillboxes, and more than 8000 mines and booby traps were disarmed. By 3 July, the bridgehead was 8000 metres wide, and on the right had moved towards the Manggar strip. Then, while the 18th Brigade continued to secure the town and harbour, the 25th Brigade, landing on 2 July, advanced inland and reached Nobody, Nurse and Nail on the following day. Japanese 75 mm and 120 mm naval guns, sited to cover the airstrip at Manggar, were attacked by naval bombardment, Liberator bombers and 6-pounder anti-tank guns, but not before they had destroyed two tanks and caused other casualties. They were eventually attacked and overrun. As the Australian Official History comments: 'Seldom in the war against the Japanese had Australians such opportunities for clearing the way with the fire of guns as at Balikpapan, and Milford and his subordinates made the most of the opportunities.'

By 9 July both Sepinggang and Manggar airfields were in Australian hands and the object of capturing the town and port area was rapidly being accomplished. The Japanese were using every stratagem to delay the inevitable: raiding parties, remote-controlled mines (using 500 kilogram bombs), and sniping with machine-guns. As well, the country towards Batuchampar was covered in thick scrub, and the Japanese were dug in under logs and the feet of large trees. Both Frog tanks and gun tanks were used to destroy these positions. Because it was proving difficult to land sufficient stores over the open beach, Milford wanted to open the port. However, because of the possibility of guns on the western side, Rear-Admiral Noble demanded that this first be cleared. Thus, as a matter of necessity, Milford landed a battalion of the 18th Brigade on the western shore of Balikpapan Bay to secure the area. By 25 July 1945, the Japanese had been forced out of their stronghold in Batuchampar and their honeycombed defences above Manggar, and the Australians were carrying out deep patrols beyond Balikpapan. Clashes continued to occur until the Japanese surrendered.

The Balikpapan action, the largest amphibious assault by Australians, was concluded reasonably quickly. About 1783 Japanese dead were counted, and 63 were captured. A total of 229 Australians were killed or died of wounds and 634 were wounded. The contrast, particularly in terms of air and naval support, firepower and movement of resources between I Australian Corps fighting its three actions on Borneo, and First Australian Army in New Guinea, could hardly have been more marked. Whereas the former landed divisions and brigades and had the continuing support of lavish machinery, vessels and weaponry, the best the latter could do in its operations was to redeploy a company here and there by barge or small ship. The First Australian Army was also frequently without adequate air support for long periods of time. For these reasons alone, the New Guinea operations have appeared more snail-like and pedestrian, although the troops there were acting with the same vigour.

JOHN COATES

BOSTOCK, Air Vice-Marshal William Dowling (5 February 1892–28 April 1968). Having served as a ship's wireless operator from the age of 19, Bostock joined the AIF as a signaller at the outbreak of war, and was posted to Egypt. He served on Gallipoli until evacuated with dysentery in August 1915. After recuperating he was posted as a sergeant to the Anzac Mounted Division signal squadron and took part in the campaign that cleared the Turks from Sinai. Discharged from the AIF in February 1917, Bostock immediately received a commission in the Royal Flying Corps Special Reserve, and underwent flying training in England. He flew with No. 48 Squadron in France until invalided in March 1918; while convalescing in England he transferred to the newly created RAF, from which he retired in October 1919 to return to Australia.

Like a number of former RFC officers, Bostock was appointed a flying officer in the newly formed RAAF in September 1921, and in his case was posted to No. 1 Flying Training School at Point Cook. His interwar career was one of continuous achievement and promotion. He attended the RAF Staff College at Andover in 1926–28, and in December 1929 became Director of Training at RAAF Headquarters. He commanded No. 3 Squadron between 1931–33, was sent on exchange back to Britain for two years in 1936, and by the eve of the war in Europe had been appointed Deputy Chief of the Air Staff under Air Chief Marshal Sir Charles Burnett (q.v.) and promoted to the rank of air commodore.

An able and industrious man, Bostock was intended by Burnett to succeed him upon the expiry of the latter's term of office. However, Burnett fell out badly with his minister, A. S. Drakeford, over command and administration of the wartime RAAF, and Prime Minister John Curtin (q.v.) decided that command of the air force would be divided between the headquarters of the commander, Allied Air Forces and RAAF headquarters in Melbourne. Bostock, now an air vice-marshal, became chief of staff to the American Lieutenant-General George H. Brett, who exercised operational control over the RAAF, in May 1942. Air Commodore George Jones (q.v.) was promoted and appointed as CAS, responsible for the air force's administration.

Brett was replaced by Major-General George C. Kenny in August 1942, and Kenny promptly reorganised his command into two subordinate formations: the US Fifth Air Force, and RAAF Command. In September, he appointed Bostock as AOC, RAAF Command, a post which he held for the rest of the war. The divided command of the RAAF which resulted from these arrangements proved impossible to reconcile. Jones and Drakeford argued for overall authority to reside with the CAS; Bostock, encouraged by Kenny and General Douglas MacArthur (q.v.), pressed for administrative responsibility for the squadrons under his command to be vested in him. Curtin was unable to mediate the differences successfully, and the high command of the RAAF was characterised by disagreements and disputes, often petty in nature, for the rest of the Pacific War.

RAAF Command was charged with the defence of Australia other than the north-east and for operations against Japanese forces in the Netherlands East Indies. As the allied forces went over to the offensive, Bostock's squadrons were also responsible for protection of the sea lanes to Papua New Guinea. At times he commanded not only Australian units, but Dutch and American ones as well, and was praised by Kenny for his 'exceptional tactical ability'. In March 1945 he was given responsibility for supporting the invasion of Borneo by I Australian Corps. With his headquarters forward at Morotai, Bostock coordinated the operations of the RAAF's First Tactical Air Force (q.v.) and strong elements of the US Fifth and Thirteenth Air Forces in support of the landings at Tarakan, Balikpapan and Labuan, participating in the latter two personally.

After representing the RAAF at the surrender ceremony in Tokyo Bay at Kenny's invitation,

Bostock was retired from the RAAF in April 1946, ahead of time. He turned to grazing in Victoria, and maintained his interest in air force affairs through regular articles in the Melbourne papers. He was elected as the Liberal member for Indi in the 1949 federal election, and held the seat until 1958. While in parliament he was an active member of the Joint Parliamentary Committee on Foreign Affairs, and an occasional critic of his own government on defence matters.

The acute personal dislike between Jones and Bostock made any resolution of the highly unsatisfactory command structure of the RAAF almost impossible, and both parties deserve some censure. Equally, it is clear that the Americans, and especially Kenny and MacArthur, preferred to maintain the dual command as it kept the RAAF firmly subordinate throughout the war. An able officer, regarded by Kenny as a first-rate combat leader and field commander, Bostock contributed in roughly equal proportions to the RAAF's effectiveness against the Japanese, and to its relative lack of influence in Allied decision-making.

BOSTON, DOUGLAS (3-crew light bomber). Wingspan 61 feet 4 inches; length 47 feet 6 inches; armament 8 × 0.303-inch machine-guns, 2000 pounds bombs; maximum speed 304 m.p.h.; range 1020 miles; power 2 × Wright 1600 h.p. engines.

Bostons were used by the RAAF in the South-West Pacific Area (q.v.) between 1942 and 1944. Thirty-one Dutch Bostons were put into service by the RAAF after the fall of the Netherlands East Indies, while a further 38 aircraft were delivered from the United States. Bostons were operated by No. 22 Squadron, who used this effective light bomber against Japanese targets during the Papuan and New Guinea campaigns. They also played an important role in the Battle of the Bismarck Sea. They were later used against the Japanese in the Netherlands East Indies and the Philippines until a Japanese raid on the Australian base on Morotai in November 1944 damaged most of the squadron's aircraft, leading to a re-equipment with Beaufighters (q.v.). The only VC awarded to an RAAF airman in the Pacific War was awarded posthumously to a Boston pilot, Flight Lieutenant William Ellis Newton, in March 1943.

BOWEN, HMAS see **BATHURST CLASS MINESWEEPERS (CORVETTES)**

BOXER REBELLION was an attempt by armed Chinese secret societies to resist foreign influence in China. A number of European nations, the United States and Japan formed the China Field Force to suppress the uprising. In June 1900 the British government sought and received permission from the Australian colonies to send the Auxiliary Squadron (q.v.) ship *Wallaroo* and two Australian Squadron ships to China. The colonies also offered additional military assistance, and Britain accepted 200 men from the Victorian Navy, 260 from the New South Wales Navy and the South Australian Naval Forces ship (q.v.) *Protector* with its 100-strong crew. Arriving in China in September, they were too late to see any significant fighting. The New South Wales contingent, based in Peking (Beijing) and the Victorian contingent, based at Tienstin (Tianjin), did participate in some minor offensives and punitive expeditions, but spent most of their time carrying out routine guard and police duties. They left China in March 1901, having suffered only six fatalities (none of them in combat). *Protector*, under the command of Captain W. R. Creswell (q.v.), was used mostly for survey, transport and courier duties in the Gulf of Chihli (Bohai) before leaving China in November 1900.

Bob Nicholls, *Bluejackets and Boxers: Australia's Naval Expedition to the Boxer Uprising* (Allen & Unwin, Sydney, 1986).

BRACEGIRDLE, Rear-Admiral Leighton Seymour (31 May 1881–23 March 1970). After joining the New South Wales Naval Brigade (militia) as a cadet in 1898, Bracegirdle saw service as a midshipman with the New South Wales contingent to the China Field Force during the Boxer Rebellion (q.v.) in 1900–01. He then joined the South Africa Irregular Horse and saw action in the final months of the Boer War. Returning to Australia, he remained active in the naval militia until 1911 when he joined the newly created RAN as a lieutenant. In August 1914 he enlisted in the Australian Naval and Military Expeditionary Force, and served as a staff officer during Australia's capture of German New Guinea. The short campaign completed, Bracegirdle was appointed to command the 1st RAN Bridging Train in February 1915. The Bridging Train, which served on Gallipoli from August 1915 until the evacuation, was occupied at Suvla Bay building and maintaining piers, supervising the landing of troops and stores and providing the beach water supply. Bracegirdle was wounded in late September, although he declined evacuation, and was hospitalised after the evacuation of Gallipoli with malaria and jaundice. He was twice mentioned in despatches for his work in the Dardanelles. For the rest of the war the Bridging Train served in Egypt, Sinai and Palestine, engaged in construction work, the control of traffic on the Suez Canal and the conveyance of stores. Bracegirdle was awarded the DSO in June 1916 and again mentioned in despatches for his work during the Sinai campaign. The Bridging Train was disbanded in early 1917, and Bracegirdle returned to Australia. He was promoted to commander, serving as district naval officer in Adelaide in 1918-21 and Sydney 1921–23. He was president of the Commonwealth Coal Board 1919–20 and in 1923 he became Director of Naval Reserves. In 1931 he was appointed military and official secretary to the Governor-General, Sir Isaac Isaacs, and he remained secretary to successive Governors-General until 1947 when he was appointed KCVO. He retired from the navy in 1945 with the rank of rear-admiral.

T. R. Frame and G. J. Swinden, *First in, Last out: The Navy at Gallipoli* (Kangaroo Press, Sydney, 1990).

BRAND, Major-General Charles Henry (4 September 1873–31 July 1961). A school teacher by profession, Brand was commissioned in the Queensland Volunteer Infantry in 1898, and enlisted in the 3rd Queensland Contingent for the Boer War in 1900. Commissioned in South Africa after service in Rhodesia, Transvaal and the Orange River, he returned to South Africa in mid-1901 as a member of the 7th Commonwealth Horse, seeing no action on the latter occasion. He joined the permanent forces in 1905 as a member of the Administrative and Instructional Staff, with which he served until the outbreak of war in 1914. Major-General W. T. Bridges (q.v.) selected him as brigade major of the 3rd 'All States' Brigade, with whom he went ashore at Gallipoli on the first morning, 25 April. He was active in the first days during which the beachhead was established, was wounded during the Turkish counter-attack in mid-May, and was promoted to lieutenant-colonel in July in command of the 8th Battalion. His battalion garrisoned Steele's Post, one of the most highly exposed positions on Anzac, from July until the evacuation in December, with a brief respite for leave in November. He went to France with the rest of the infantry in 1916, and in June was given command of the 6th Brigade and then of the 4th, which he led for the rest of the war. The brigade went through Pozières, Messines, Polygon Wood, Passchendaele and, most notoriously, Bullecourt, where it suffered more than 2000 casualties, easily its heaviest loss in a single action during the entire war. Brand opposed the half-baked plan for the latter action, was himself wounded at Messines, and took a leading part in the devolved battles during the Australians' advance from July 1918 in which his brigade captured Morcourt and Méricourt, after a critical period in the line at Gommecourt during the German spring offensive. He was

part of the advisory mission to the American Expeditionary Force, led by Sinclair-MacLagan (q.v.), in September–October, after which, as an original Anzac, he returned to Australia under a leave scheme overseen by Prime Minister W. M. Hughes (q.v.). After the war he resumed regular soldiering, becoming 2nd CGS in 1926 followed by a period as Quartermaster-General of the AMF. He retired in 1933. He was elected to the Senate from Victoria in the election of 1934, and remained in politics until 1947, taking a particular interest in ex-servicemen's issues.

BREN LIGHT MACHINE-GUN (LMG) was developed during the 1930s by the Czechoslovakian Small Arms Factory at Brno and the British Small Arms Factory at Enfield from the Czech Zb series of LMGs. The name BREN comes from the linking of the names Brno and Enfield. To meet the British requirements considerable redesign of the Zb series LMG was required, which included a change from firing rimless 7.92 mm calibre ammunition to firing rimmed 0.303-inch calibre ammunition. The use of the 0.303-inch calibre rimmed ammunition was responsible for the distinctive curved magazine of the BREN.

In 1937 Australia made the decision to produce the BREN LMG to replace the Lewis LMG, but it was not until 1941 that the first Australian-made BREN was completed. In the meantime Australia Army units in the Middle East and Asia were issued with British-made BRENS or used Lewis LMGs. The BREN Mark 1 0.303-inch calibre served in the Australian Army from 1940 to 1960 when it was replaced by the GPMG M60 (q.v.). During that time it saw active service with Australian soldiers in the Second World War, Korea, and the Malayan Emergency.

The BREN is a gas-operated, air-cooled, magazine-fed LMG with a reputation for great accuracy and reliability. It can be carried and fired by one man but requires a second man to assist with the carriage of ammunition, tools and spare barrel. The BREN is normally fired from its bipod and has a distinctive appearance with its top mounted magazine. Modifications were made to the BREN design during the Second World War to simplify manufacture; these included simplifying the body, replacing telescoping, adjustable bipod legs with non-adjustable legs, and dispensing with butt grips. Australian-made BRENs retained the drum-operated backsight. The BREN Mark 1 weighed around 25 pounds with a filled magazine. The box magazine capacity was 30 rounds, but it was normally only filled with 28 rounds.

The effective range of the BREN from the bipod was up to 600 yards and from the tripod up to 1600 yards. The tripod was used in the ground role to allow the BREN to fire on fixed lines and could also be used in an anti-aircraft role by fitting the ack-ack leg. The cyclic rate of fire (i.e. the rate of fire that theoretically could be achieved given a continuous supply of ammunition) was around 500 rounds per minute, with a practical rate of fire of up to 120 rounds per minute fired in bursts of three to five rounds.

British issue 7.62 mm calibre L4 series BREN LMGs were used by Australian soldiers on operations in Malaysia and Borneo during the mid 1960s, and in 1971 Australia purchased a limited number of British L4A4 BREN LMGs to supplement the GPMG M60, especially in infantry units for patrolling tasks. The L4A4 BREN weighs around 23 pounds 8 ounces with a full magazine. It has a straight rather than curved magazine with a capacity for 30 rounds and can also accept the magazine from the L1A1 SLR.

IAN KURING

BRENNAN, Louis (28 January 1852–17 January 1932). Born at Castlebar, County Mayo, Ireland, Brennan migrated to Melbourne with his family in 1861. Here he was apprenticed to an engineer. In 1874, at the age of 22, he invented the Brennan torpedo (see TORPEDOES), a shore-based weapon propelled by a steam-powered winch connected to the torpedo by fine steel wires (which also enabled the warhead to be steered towards the target and even retrieved if the torpedo missed). Brennan patented his design in 1877 and, aided by a £700 grant from the Victorian government, successfully trialled a working model in Hobson's Bay in 1879. Brennan went to England in 1880 at the request of the British government for further torpedo trials and never returned to Australia. The British government awarded him a lump sum of £5000 and £1000 a year while the torpedo was being tested, and, when, after five years' work, it was accepted by the War Office, Brennan was given a further £110 000, an enormous amount of money, which indicates the importance placed on his invention. A factory was set up to manufacture the torpedo in Gillingham, Kent, with Brennan as superintendent, and the weapons were installed in the defences of all the major British ports as well as strategic ports in the British Empire. The British government considered the torpedo so important that it refused an order from the Victorian government on the grounds that no torpedoes could be spared. Brennan also designed a monorail system and a helicopter, which crashed on trials in 1925. During the First World War he worked in the Ministry of Munitions, and from 1919 to 1926 he worked at the Royal Aircaft Establishment at Farnborough. He became a foundation member of the National Academy of Ireland in 1922. Bad health forced Brennan to travel to warmer climates in winter and in January 1932 he was killed in a car accident in Switzerland.

JOHN CONNOR

BRIDGEFORD, Lieutenant-General William (28 July 1894–21 September 1971). Commissioned in June 1915 as a member of the third class to graduate from RMC Duntroon, Bridgeford served with the AIF in Egypt and France as a machine-gun officer. Awarded the MC after the disastrous battle of Fromelles in 1916, he spent time on the staff of the 8th Brigade and on the 5th Division headquarters before returning to his unit. He was gassed in 1918, and ended the war on the staff.

Bridgeford's career between the wars demonstrated all the depressingly familiar characteristics of regular service at that time. A reduction in wartime rank and a succession of staff and instructional posts with militia units was followed by an appointment at RMC Duntroon and then two years at the Staff College at Quetta, British India (now in Pakistan), between 1926 and 1927. He attended the Imperial Defence College in London in 1938, which suggested that he had already been marked for high rank, possibly even as CGS. At the outbreak of war he returned to Australia and duty with the 2nd AIF, first in England, where he raised the 25th Brigade, and then in the Middle East, where he took part in the Greek campaign.

Returning to Australia he was appointed Deputy Adjutant and Quartermaster-General of First Australian Army in April 1942, and served in New Guinea in the same capacity with Headquarters New Guinea Force (q.v.). He finally gained a command, that of the 3rd Division, in July 1944, and led it during the final campaign on Bougainville. The official historian, Gavin Long (q.v.), believed that Bridgeford was denied a command for most of the war because he had earned the enmity of the C-in-C, Blamey (q.v.), through his criticism of the latter during the Greek campaign. As a result his relative seniority and recognised merit at the beginning of the war did not earn him the positions or rapid advancement which he undoubtedly deserved.

In 1946 he was appointed Quartermaster-General and Third Military Member of the Military Board (q.v.) at Army Headquarters in Melbourne. In July–August 1950 he led an Australian military mission to Malaya to advise the Australian government on options for supporting the British in their fight against communist insurgents, but the outbreak of the Korean War meant that assistance here was diverted to the United Nations Forces under General Douglas MacArthur (q.v.).

In November 1951 Bridgeford was sent to Tokyo to succeed Lieutenant-General Sir Horace Robertson (q.v.) as C-in-C of the British Commonwealth Occupation Force (q.v.) and non-operational commander of Commonwealth forces in Korea (see KOREAN WAR). In May 1952 he was embroiled in a major dispute between the Canadian, British, American and Australian governments over the use of Commonwealth troops on the island of Koje-do, where large numbers of Chinese and North Korean POWs were held, and which had been the scene of POW riots and violence on the part of US and South Korean guards. Bridgeford had approved the transfer of British and Canadian companies to Koje-do, and the Canadian government, which did not wish to be involved in any unpleasantness, tried to have him removed as a consequence. He received the full backing of the British and Australian authorities, but the incident demonstrated the often fragile relations that existed between the Commonwealth governments involved.

Bridgeford returned to Australia in February 1953 and retired shortly thereafter. In the same year he was appointed chief executive officer to run the 1956 Melbourne Olympic Games. He was active in RSL affairs until his death.

BRIDGES, Major-General William Throsby (18 February 1861–18 May 1915). Born in Scotland, the son of a RN officer, Bridges was educated in England and Ontario, Canada. In 1877 he entered the recently established Royal Military College of Canada in Kingston, but failed to complete the course and in 1879 followed his family to New South Wales, where he joined the roads and bridges department of the colonial civil service. His attempt to volunteer for the Sudan contingent in 1885 was made too late, but that year he was commissioned in the permanent artillery. He entered the School of Gunnery at Middle Head in 1886, becoming Chief Instructor in 1893 (after several years on courses in England), a position he held for the next nine years. In the period 1893–96, at the direction of Major-General (later Sir) Edward Hutton (q.v.), who commanded the New South Wales military forces, he acted as secretary to three committees and conferences on defence issues. From December 1899 until May 1900 he served on special secondment with the British Army in the Boer War until evacuated to England through illness. Returning to Australia in September 1900 he resumed his position at the School of Gunnery and

also served on several committees concerned with the military forces of the new Commonwealth.

In 1902 he became Assistant Quartermaster-General under Hutton (now GOC Australian forces) with responsibility for military intelligence, defence planning and organisation. Three years later he was appointed Chief of Intelligence on the Military Board of Administration (q.v.) and a member of the five-man Council of Defence (q.v.). Increasingly Bridges became an advocate of the 'imperial' approach to defence questions as opposed to those who sought a policy more directly centred on local Australian issues, a stance that brought him into conflict with such strong advocates of an Australian navy as Captain (later Sir) William Creswell (q.v.). In January 1909 he became first CGS, but after only five months was sent to London to attend the Imperial Conference (q.v.) and then to become Australia's representative on the Imperial General Staff. A year later he was chosen to establish a military college in Australia, which opened at Duntroon in the Australian Capital Territory on 27 June 1911. Bridges remained as commandant of RMC Duntroon until May 1914, when he was appointed Inspector-General (q.v.) of the Australian Army.

When war was declared in August 1914, he was ordered to raise a force for service in Europe and was given command of it. Training of the AIF, a name that Bridges himself chose, followed a strict regimen in Egypt, leading to some disciplinary problems, but Bridges insisted on high standards, not least because he was determined that Australian troops would fight as an entity rather than be subsumed within British formations. Bridges landed with the 1st Division at Gallipoli on 25 April 1915, was quickly convinced that the position was untenable and urged withdrawal, but was overruled. For the next three weeks he visited the firing lines daily, constantly exposing himself to danger, until he was hit by a sniper on 15 May, dying three days later *en route* to Egypt. His body was returned to Australia (the only war fatality to be so treated at the time), and after a service at St Paul's cathedral in Melbourne was buried on the slopes of Mt Pleasant overlooking Duntroon, in a grave whose design was supervised by the architect of Canberra, Walter Burley Griffin.

The first Australian officer to reach the rank of general, Bridges was essentially an administrator and policy planner rather than a field commander. To some extent a protégé of Hutton, himself committed to imperial rather than local objectives (whether in Canada or Australia), Bridges combined an acceptance of British priorities with a fierce determination to impose and uphold first-rate military standards in Australia, nowhere more so than at Duntroon, which has never entirely shaken off the somewhat austere image he established for it. Personally reserved and detached, he nevertheless won widespread respect, even affection, for his conduct on Gallipoli, and the emphasis that he had placed on the training of officers and men, over the whole of his military career and especially in Egypt, stood the Australian Army in good stead for the rest of the war.

C. C. Coulthard-Clark, *A Heritage of Spirit: A Biography of Major-General Sir William Throsby Bridges, K.C.B., C.M.G.* (Melbourne University Press, Melbourne, 1979).

'BRISBANE, BATTLE OF', was a riot on the night of 26 November 1942. Accounts differ, but it probably started when Australian troops came to the aid of an American soldier who was being harassed by US military police. It involved several thousand Australians; one Australian was killed, at least nine Australians and 11 Americans were seriously injured, and many more suffered minor injuries. Isolated disturbances continued the next day. Although often depicted as an anti-American riot, the main targets seem to have been US military policemen.

John Hammond Moore, *Over-Sexed, Over-Paid and Over Here: Americans in Australia 1941–1945* (University of Queensland Press, Brisbane, 1981).

BRISBANE (I), HMAS see **TOWN CLASS LIGHT CRUISERS**

BRISBANE (II), HMAS see *CHARLES F. ADAMS CLASS DESTROYERS*

'BRISBANE LINE', a term first used by General Douglas MacArthur (q.v.) at a press conference in March 1943, referred to an alleged plan to concentrate defence in south-eastern Australia and to abandon the rest of the country in the event of a Japanese invasion. Minister for Labour and National Service E. J. Ward, who had claimed in October 1942 that such a plan was developed by the previous Menzies government, repeated his allegation after MacArthur's press conference, much to the embarrassment of Prime Minister John Curtin (q.v.). Curtin appointed a Royal Commission which reported in July 1943 that no such plan had ever been official policy. However, although Ward's and MacArthur's wilder allegations about the pursuit of a defeatist policy were unfounded, the War Cabinet (see AUSTRALIAN WAR CABINET) had decided on 13 December 1941 to give top priority to the vital Sydney–Newcastle–Port Kembla–Lithgow area and second priority to the defence of Darwin and Port Moresby. Lieutenant-

General Iven Mackay (q.v.), GOC-in-C, Home Forces, had to take account of this directive in his planning. He also had to consider the absence from Australia of four AIF divisions, the probable defeat in the Malayan and the Netherlands East Indies campaigns, and uncertainty about the US Fleet's ability to achieve supremacy in the Pacific. Accordingly, he produced a memorandum on 4 February 1942 recommending that troops should be concentrated between Brisbane and Melbourne. It was never Mackay's intention to simply abandon the rest of Australia, and his planning involved the use of demolition, a scorched-earth policy and guerrilla tactics if a withdrawal was necessary. Furthermore, the government never accepted Mackay's recommendations. By late February, planners were able to take into account the imminent arrival of two AIF divisions and one US division, and by early March the government had agreed to a Pacific strategy in which Australia would be a base for offensives against the Japanese.

BRITISH ARMY IN AUSTRALIA The service in the Australian colonies of 25 British infantry regiments and a number of smaller 'veteran' artillery and engineer units has remained largely neglected by students of Australian and British military history. British military historiography has long focused on the army in war, to the exclusion of its experience as an imperial garrison, while the British Army's lack of connection with an Australian military tradition has left its history the province of a few, often antiquarian, historians. This neglect, which has only recently begun to be redressed, has tended to conceal both the army's role in colonial society and the impact of colonial service on the regiments.

The early Australian colonies projected a distinctly martial atmosphere. Their governors were naval officers up to Bligh's time and thereafter, until the 1850s, were mainly military men. The colonial capitals, and especially Sydney and Hobart, maintained large garrisons housed in imposing barracks (q.v.), while the colonial economy benefited from the Commissariat's purchasing power. Marines (not 'Royal' Marines, a title awarded in 1802) protected the infant settlements at Sydney Cove and Norfolk Island, but were relieved in 1790 by a unit specifically recruited for colonial service. The deficiencies of the New South Wales Corps (q.v.), and especially the coup mounted by its officers in 1808, persuaded the War Office that the colony of New South Wales and Van Diemen's Land required a more reliable garrison, and in January 1810 the 73rd Regiment of Foot (usually shortened to 73rd Foot) became the first line regiment to serve in Australia.

Demands on the regular regiments of the garrison changed over the succeeding 60 years, but troops were ostensibly always primarily concerned with protecting the colonies from external attack, though no actual threat was ever manifest. Troops were used to found strategic settlements to establish the British claim to the continent: in 1825 (in King George's Sound in what became Western Australia), in 1826 (at Western Port Bay in what became Victoria), and in 1824 (at Fort Dundas on Melville Island and later at Fort Wellington in what became the Northern Territory). All three settlements were abandoned, in 1831, 1828 and 1829 respectively. Troops constructed and occasionally manned batteries protecting Sydney harbour, but not until the 1850s did defence against external threats become of greater importance. Rather, the troops' secondary purpose in practice became paramount, in that they acted as a guarantor of order, deterring or suppressing civil unrest and acting, as the colonial ecclesiastic John Dunmore Lang expressed it, as 'a fulcrum for the authorities to work on'.

The number of regiments posted to Australia reflected the growing extent of the Australian colonies and the duties imposed upon its military garrison. Under Governors Macquarie and Brisbane one regiment attempted to provide small garrisons in New South Wales and Van Diemen's Land. From the mid-1820s until the early 1830s three or four regiments served, and by the late 1830s and 1840s there were five. Despite the end of transportation to New South Wales in 1840 troop strengths remained high, with four or five regiments maintained until the late 1840s (partly a consequence of the need to dispatch and relieve troops in the first Anglo-Maori war). Although the late 1850s saw four regiments in Australia (possibly a reaction to the Eureka Stockade [q.v.]), the force declined by the early 1860s, and from 1862 to 1866 the British Army was represented by only a small force of garrison artillery. For most of the period regiments served in Australia for an average of just under seven years, but the 99th Foot spent fifteen and the 11th thirteen years posted to Australia.

Three times the authorities sought to employ 'veteran' units formed of old soldiers. From 1810 to 1823 former members of the New South Wales Corps comprised an Invalid Company, Royal New South Wales Veteran Companies served from 1825 to 1832, and in Western Australia from 1850 'Pensioner Guards' were used to guard convicts and later became settlers. None of the expedients produced efficient units, and soldiers were regarded as poor prospects as settlers because military discipline eroded the initiative and resource necessary to farm successfully.

Garrison duty involved the distribution of a regiment between anything up to a dozen military stations; in the 1820s scattered as far apart as King George's Sound or northern Australia and Sydney. By the 1830s regiments tended to be stationed either in Van Diemen's Land or based on Sydney, but in each case were located in many small detachments, from 1832 often at iron-gang stockades. Officers deplored the resultant 'contamination' by convicts — though convicts and soldiers were largely drawn from the same parts of British and Irish society, and found themselves in red coats or in irons for similar reasons — and the detachments were frequently rotated between headquarters and outlying stations in order to reassert discipline. The effects on a regiment's discipline and efficiency were often severe, especially in the late 1830s in New South Wales. Commanding officers repeatedly complained of the hazards of Australian service — Colonel Breton of the 50th described Australia as 'the worst country in the world for a soldier'— and in 1839 men of the 80th Foot mounted a short-lived protest over the curtailment of privileges.

Whatever the difficulties they had faced in Australia, however, on moving to India at the end of their Australian service (the standard 'tour' throughout the period) the troops regretted leaving, anticipating in India the ravages of disease and war. Many officers retired to or in Australia, and soldiers purchased discharge or deserted in order to remain. One soldier told his officer that he would 'rather be a private in Sydney than a general in India'. Soldiers often claimed that convicts, eligible for a ticket of leave and the opportunities it offered, enjoyed a relatively easier lot. This belief underlay the notorious Sudds and Thompson case of 1826, when two men of the 57th Foot, Joseph Sudds and William Thompson, stole cloth hoping to be discharged after a brief prison sentence. Governor Darling, alarmed by recent similar attempts, made an example of the pair and sentenced them to be imprisoned in irons, in which Sudds died, providing a cause on which Darling's colonial opponents seized. Although prosperous free colonists appreciated soldiers in the abstract, they were widely reviled by the colonies' poor, and especially by emancipists: 'anything in a red coat is hated', recorded a subaltern in Sydney in 1838.

Internal rather than external potential threats, then, preoccupied the army in Australia. Units were detailed as convict guards at places of secondary punishment, at Macquarie Harbour, Norfolk Island, Port Macquarie, Moreton Bay and, from 1830, at Port Arthur. Duty at remote stations of secondary punishment was hard, and the tension of constant vigilance against individual or collective rebellion led to the troops gaining a reputation, often deserved, for brutality. Although the authorities remained mindful of the Castle Hill uprising of 1804, actual convict rebellion was rare — occurring only once, in the 'Ribbon Gang' outbreak near Bathurst in 1829 — and troops were actually used to combat bushranging outbreaks and to suppress Aboriginal resistance (q.v.) to the extension of White settlement.

Troops were used to quell outbreaks of bushranging, particularly in Van Diemen's Land from 1814 to 1820, and in New South Wales from the mid-1820s. Small military patrols saw hard service patrolling the bush of the settled areas, engaging in small but often savage fights. Bushranging imposed such severe demands on the infantry of the garrison that in 1825 a military Mounted Police was formed, at first of 12 men detached from the garrison, but by 1830 numbering over 100 troopers. Governor Darling described the Mounted Police as 'much dreaded by … Runaways and Bushrangers'. The force, distributed across the 19 counties of New South Wales and later extended into Australia Felix, became a visible and relatively efficient arm of authority, and its members were correspondingly unpopular — notably through colonial folk songs such as 'Bold Jack Donahoe'.

Regular troops and later Mounted Police operated intermittently against Aboriginal resistance in most colonies. While troops killed many fewer Aborigines than did settlers, disease or the cultural despair associated with the occupation of their land, as the colonies' only organised armed force troops were called upon to operate as the ultimate arbiter of British settlement. In New South Wales, they engaged in significant operations on the Hawkesbury against the Daruk in the 1790s, on the Hunter and against the Wiradjuri around Bathurst in the mid-1820s, and against the Kamilaroi in north central New South Wales in the late 1830s. The operations against the Wiradjuri involved the declaration of martial law and a series of extensive drives against recalcitrant Aborigines, while the 'expedition' against the Kamilaroi, comprising a party of 25 Mounted Police, culminated in the massacre at 'Slaughterhouse' or 'Waterloo' Creek in January 1838. Troops in Van Diemen's Land conducted an intermittent war in the 1820s, culminating in 1830 with the notoriously unsuccessful 'black line', in which troops and auxiliaries drove south from Launceston towards the Tasman Peninsula. At Moreton Bay (Queensland), Captain Patrick Logan's brutality to Aborigines led to him being speared, while in the Swan River Colony Aboriginal resistance led to the 'battle' of Pinjarra in 1834.

With the end of transportation, by the 1840s in New South Wales and the 1850s in Van Diemen's Land, the original need for troops passed, particularly as colonial civil police forces were formed (the Mounted Police was disbanded in 1860). Pressure within Britain to reduce imperial garrisons mounted, and troop strengths diminished markedly. The use of troops to suppress the Eureka (q.v.) uprising emphasised the military's role in acting as a bulwark of order, ironically after transportation had ended, but within several years the military infrastructure so prominent in convict colonies earlier in the century had been rapidly dismantled. In the army's final decade in Australia the main role of the 'imperial troops' (as they were termed, in distinction to colonial military forces [q.v.]) was to serve the guns protecting Sydney and Melbourne, their costs largely borne by colonial legislatures. Finally, in 1870, the colonies were required to assume responsibility for their own defence, and the last regiment to serve in Australia was withdrawn. Ironically, soldiers were never more popular in Australia than when they were ordered home.

With the departure of the last regiment the British Army's influence on Australian military institutions operated through the secondment of regular officers as colonial commandants, and through British NCOs occupying positions within the colonial forces' permanent staff. The absence of any institutional continuity enabled Australian military tradition to develop in emulation rather than at the direction of the British, and few direct links were maintained between colonial and imperial forces before those formed during the conflicts of the twentieth century. Although the Australian Army was and remains recognisably modelled on the British, and though much terminology, tradition and thinking is common, the British Army's service in colonial Australia has been remembered by neither army, and the relationship remains largely unexplored.

The primary sources available for the study of the British Army in Australia are not extensive, and those that exist are largely untapped. Except for the official correspondence and returns in the War Office papers copied under the Australian Joint Copying Project and colonial newspapers, very little primary material exists in Australia. Virtually no official documents and very few private papers (and those largely from officers) exist in Australia, with the exception of colonial records in State archives and the National Library, such as an excellent and under-exploited letter book of the mounted police.

The historiography of the army in Australia was long dominated by the regimental tradition that permeates the army of the nineteenth century, though few regimental histories dealt adequately with their subjects' Australian service. Maurice Austin's *The Army in Australia 1840-50: Prelude to the Golden Years* (Australian Government Publishing Service, Canberra, 1979) deals in detail with the 1840s — arguably the least significant decade — while his research papers, held in the Australian War Memorial, provide a valuable resource, particularly on the New South Wales Corps. Peter Stanley's *The Remote Garrison: The British Army in Australia 1788–1870* (Kangaroo Press, Sydney, 1986), provides the only general account of the army in Australia. Although based on primary research it was aimed at family historians, and consequently lacks scholarly documentation. Stanley's chapter, ' "Soldiers and Fellow-Countrymen" in Colonial Australia', in M. McKernan and M. Browne (eds), *Australia: Two Centuries of War and Peace* (Australian War Memorial/Allen & Unwin, Canberra/Sydney, 1988) amplifies and documents the relationship between the British Army and colonial society. Regiments in Australia have been variously but largely poorly treated, with Clem Sargent's unpublished study of the 48th in Australia a notable exception. While F. H. Broomhall, *The Veterans* (Hesperian Press, Perth, 1989), documents the service of Pensioner Guards in Western Australia, the experience of former British soldiers in Australia is still undocumented. Much work remains to be undertaken and published, on the service of particular regiments in Australia, on the experience of troops in Australia, and on their relationships with the constituent groups of colonial society. While Aboriginal resistance to European settlement has recently become a fecund area of research, the involvement of soldiers in that conflict has yet to be investigated. Good overviews of the subject are available in Richard Broome's chapter, 'The struggle for Australia: Aboriginal-European warfare', in *Australia Two Centuries of War and Peace*, and in Jeffrey Grey's *A Military History of Australia* (Cambridge University Press, Melbourne, 1991). Roger Milliss's *Waterloo Creek: The Australia Day Massacre of 1838, George Gipps and the British Conquest of New South Wales* (McPhee Gribble, Melbourne, 1992), a study of the mounted police's massacre of the Kamilaroi in northern New South Wales in January 1838, provides an authoritative exception to that neglect.

PETER STANLEY

BRITISH COMMONWEALTH FORCES, KOREA

(BCFK) was the overall command organisation responsible for the administration and non-operational control of all British Commonwealth forces engaged in the Korean War as part of the United Nations Command (UNC). It was set up in December 1950 to regularise the relationship between the British Commonwealth Occupation Force (q.v.) in Japan and the various individual national components in Korea contributed by Britain, New Zealand, India, Australia and Canada (and although No. 2 Squadron South African Air Force served in Korea, South Africa was not involved). The C-in-C of BCOF was also nominated as C-in-C, BCFK, and his existing organisation in Japan was responsible for much of the logistic support of Commonwealth units. In time BCFK had a sizeable establishment, and many of the units under its command were composite Commonwealth ones. When the state of occupation in Japan officially ended with the signing of the peace treaty in September 1951 and BCOF was disbanded, the Commonwealth presence in Japan was maintained through BCFK, and the organisation remained in existence until disestablished in late 1954 following the cease-fire in Korea.

BRITISH COMMONWEALTH OCCUPATION FORCE (BCOF)

Discussions on Commonwealth participation in a force for the occupation of Japan began even as that country surrendered in August 1945, although the early stages were marked by some discord between Australia and Britain over the fact of participation within a Commonwealth structure. Reflecting a growing independence in policy and action, the Australian government initially asserted its right to dispatch an independent force to be answerable only to the Supreme Commander for the Allied Powers (SCAP), General Douglas MacArthur (q.v.). The result almost certainly would have been the overwhelming of a tiny force within a vast American organisation, but the threat of separate action on Australia's part seems to have been sufficient to bring about British recognition of Australian claims, with the result that much of the responsibility for command and administration of the force was allocated to Australia.

The Commonwealth elements of the Allied army of occupation began to arrive in Japan in February 1946, about six months after the first American units. They were allotted the largely rural and severely devastated prefecture of Hiroshima, and their presence in Japan was governed by the MacArthur–Northcott agreement (q.v.), reached in December 1945, which limited the role to be played by Commonwealth forces to a largely token presence; the important functions of military government and the democratisation of Japan were to be kept firmly in American hands. The first GOC, Lieutenant-General John Northcott (q.v.), was offered the post of governor of New South Wales in February 1946, and was replaced in April by another Australian, Lieutenant-General H. C. H.

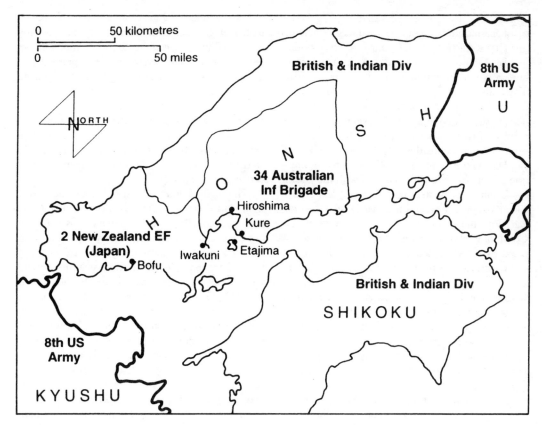

Map 7 Outline of the British Commonwealth Occupation Force area in Japan, 1946–47.

Robertson (q.v.), who commanded the force until November 1951.

At its height in 1946 BCOF consisted of the 34th Australian Infantry Brigade Group, the 9th New Zealand Brigade Group, the British and Indian Division (Brindiv) which consisted of a British and an Indian brigade, a separate air force of two wings, and shore-based naval personnel, a total of about 36 000 men and women. The command and many of the headquarters personnel were supplied by Australia, and the force was overseen by the Joint Chiefs of Staff in Australia (q.v.), essentially the Australian Chiefs of Staff Committee augmented by British, New Zealand and British Indian representation, which met in Melbourne.

Initially BCOF was kept busy on the other two tasks of the occupation army, demilitarisation and demobilisation. Large numbers of Japanese soldiers were repatriated to the home islands and demobilised there, while huge stocks of war materiel, including chemical agents and tons of ordnance, had to be rendered safe and then disposed of, usually destroyed. There was some involvement in enforcing regulations against black marketeering (made difficult by the involvement of soldiers in illegal currency transactions and the disposal of foodstuffs to the Japanese), and intelligence kept tabs on Japanese veterans' groups who for a while were thought to pose a potential source of disaffection with the conditions of the occupation. By 1947, however, it was clear that the major tasks of the occupation force had been largely completed.

For this reason among others reductions were made in the strength of BCOF even as it reached its high point, taking over neighbouring prefectures from the Americans, who were themselves bent on demobilisation of the great army that had won the Pacific War. The British had never been greatly in favour of a lengthy commitment, and citing pressures elsewhere in the Empire they began to withdraw their forces in April 1947. With independence and partition in the offing the Indians followed suit the same year. In October 1948 it was the turn of the New Zealanders to rationalise their overseas commitments and withdraw their contribution to Japan, despite pressure from the Australians to maintain a token presence. From the end of 1948 BCOF was very largely an Australian exercise, although it became subject to further reductions as the Australian authorities themselves sought to deploy limited military resources back to Australia to build up the postwar regular army.

After the election of the Menzies (q.v.) government and introduction of a national service scheme (see CONSCRIPTION), MacArthur was served notice in May 1950 that the remaining Australian contribution would be withdrawn within six months. The outbreak of the Korean War in June reversed that decision, and the remaining army and air force units in Japan soon found themselves committed to combat in Korea. Thereafter, although the occupation continued until the signing of the Treaty of San Francisco in September 1951 and the resumption of normal diplomatic relations with the Japanese, occupation matters took less and less of the Allied authorities' time as they prosecuted a major war. With the end of the state of occupation, the Commonwealth organisation in Japan was redesignated British Commonwealth Forces Korea (q.v.), and continued to supply and administer the Commonwealth forces fighting in Korea itself. These were progressively scaled down and withdrawn in the two years after the cease-fire of July 1953.

The record of BCOF was mixed, generally through no fault of those who participated in it. The Americans had clearly decided to remake Japan in a more democratic image, and to do so without interference from other parties, especially the Russians. The British, frustrated by MacArthur in their attempts to regain the level of economic penetration of Japanese markets that they had enjoyed before the war, soon lost interest, and with the withdrawal of most of the other Commonwealth components the diplomatic utility of BCOF became limited. In Australia itself the general public took little interest in the force, other than in the admittedly high rates of venereal disease during the first year of the deployment and the issue of fraternisation with Japanese women. Personnel who served in Japan received no formal recognition by way of a medal or service clasp, which contributed to a feeling of disaffection and lack of appreciation on the part of many. This latter circumstance was remedied in 1994 through the award of the Australian Service Medal to those veterans who served between 1945 and 1947.

BRITISH FREE CORPS, AUSTRALIANS IN see **AUSTRALIANS IN THE SERVICE OF OTHER NATIONS**

BRITISH NAVY IN AUSTRALIA see **AUSTRALIA STATION**

BRITISH OFFICIAL HISTORIES (SECOND WORLD WAR): Attitudes Towards Australia The British official history of the Second World War was an enormous publishing undertaking, directed from the Historical Section of the Cabinet Office and divided into three series: military, civil, and medical. The military series, under the general editorship of J. R. M. Butler, began publication in 1952 and was only completed in 1992, running to 36 volumes, some in several parts. The civil series was edited by the distinguished Australian historian long resident in Britain, Sir Keith Hancock; the first volume appeared in 1951, the last in 1964. The civil series consisted of 34 volumes, and drew on some distinguished university historians as well as civil service writers. The medical series, the least known of the three, was edited by Sir Arthur McNalty; 22 volumes were published between 1952 and 1969.

The medical and civil histories were exclusively British in their focus, and the dominions intrude only occasionally in the narratives. This is clearly not the case in some of the military series, in which the participation of dominion ground, air and naval forces is treated at length in those volumes dealing with, for example, operations in the Mediterranean in 1939–42, or the Far East in 1941–42, or the strategic air offensive, or the war at sea. In the main, the British official histories confine themselves to factual description of events and generally eschew controversy, notoriously so in some cases.

The controversy and disagreements occurred largely behind the scenes. There were disagreements between the British official historians and the official historians of all of the dominions at some point: with the Canadians over Dieppe and Hong Kong, and with the New Zealanders over Crete. The most significant sources of dispute with Australia over its official history were those concerning the Greek campaign and the fall of Singapore. The Cabinet Office was very unhappy over criticism of British planning and command in Malaya, and considered making official representations to the Australian government in 1955 to have the text of Lionel Wigmore's (q.v.) volume changed. 'It is not a balanced account', wrote one official, 'and gives the Australians more than their fair share of credit.' Wiser counsel prevailed: 'I am satisfied that there is nothing in them to which we can take exception', wrote another, 'bearing in mind that these proofs have not been sent to us for official scrutiny.' British concerns were motivated principally by Colonial Office fears that criticism of British policy for the defence of Malaya in 1941–42 might in some way jeopardise the moves towards self-government and independence then under way in the colony, and the British official historian of the war in the Far East, Major-General S. Woodburn Kirby, laboured under similar difficulties.

The points of disagreement over Greece revolved much more around personalities. Field Marshal Sir Henry Maitland Wilson, who had overall command in that campaign, took exception to Gavin Long's (q.v.) criticisms of his relationship with General Sir Thomas Blamey (q.v.) and the New Zealander, Bernard Freyberg. General Sir Richard O'Connor, commander of Western Desert Force during the First Libyan campaign, on the other hand, thought the Australian history of that campaign 'far nicer to me than I deserved', and commented in his turn that 'their complaint that senior [British] officers did not understand them … was a fair criticism'. For his part, Gavin Long strongly refuted the old British complaints about alleged Australian indiscipline: 'These complaints were often serious and nearly always inaccurate', he wrote to the Cabinet Office in 1950. The First World War official historian C. E. W. Bean (q.v.), he suggested, had also tackled them in his volumes dealing with 1918. In general, such disagreements were argued out in the correspondence between the various official historians, and do not in general find explicit statement in the final volumes as published.

BROOME, HMAS see *BATHURST* CLASS MINE-SWEEPERS (CORVETTES)

BROWNELL, Air Commodore Raymond James (17 May 1894–12 April 1974). A prewar member of the CMF, Brownell enlisted for service at the outbreak of the Great War and served in Egypt, Gallipoli and France. Allotted to the artillery, he missed the early phases of the Gallipoli campaign because the army needed his civilian administrative abilities as an accountant in Alexandria, but he found his way to the front in July 1915 and remained there until the evacuation. Proceeding to France in 1916, he won the MM for courage under fire at the battle of Pozières. Selected for the RFC, he was commissioned after flying training in March 1917, and flew operationally in France and Italy, winning the MC. He returned to Australia in April 1918 and, resigning his commission in the postwar RAF, returned briefly to civilian life. In September 1921 he was appointed a flying officer in the infant Australian air force. Brownell enjoyed a varied career in the interwar period and held a succession of prominent posts. He commanded No. 1 Squadron between 1926 and 1928 and was then Director of Personnel Services at RAAF Headquarters from 1928 to 1934. He was on exchange in England in 1934–36, and returned to Australia to command the first RAAF establishment to be based in Western Australia, RAAF Base Pearce (q.v.), which he took up early in 1938. With circumstances growing more threatening in the Pacific following the advent of war in Europe, Brownell was sent to Singapore in August 1940 to

establish RAAF Headquarters at Sembawang. He returned to Australia before the Japanese attacked Malaya and was given command of No. 1 Training Group. He spent most of the war in Australia, commanding Western Area again from 1943 to 1945, and finished the war on Morotai where he was AOC No. 11 Group. He retired in March 1947.

R. J. Brownell, *From Khaki to Blue* (Military Historical Society of Australia, Canberra, 1978).

BRUCHE, Major-General Julius Henry (6 March 1873–28 April 1961). The son of German immigrants, Bruche originally qualified for the Victorian Bar as a barrister and solicitor in 1898. A keen amateur soldier, he had been commissioned into the 1st Battalion, the Victorian Rifles, in 1891, and in 1898 he gave up the law for a commission in the Permanent Military Forces. He served in the Boer War briefly with the British Army before appointment as quartermaster with the Australian Regiment. He then saw service as adjutant of the Victorian Mounted Rifles before returning to Australia in December 1900. A further brief stint in South Africa with the 2nd Battalion, Australian Commonwealth Horse (q.v.), followed at the war's end in early 1902.

For the next 10 years Bruche served on the Administrative and Instructional Staff, eventually being appointed as Deputy Assistant Adjutant-General, first in Tasmania and then in Queensland between 1911 and 1914. He enjoyed a year's exchange with the British Army in 1910, but appears to have become dissatisfied with soldiering, applying for the position of chief commissioner of police in Victoria in 1912, a position for which he had already contended unsuccessfully in 1907. He was to fail in applications again in 1920 and 1922.

At the outbreak of war in 1914 he was appointed commandant of Western Australia, a backwater posting almost certainly explained by his German background. He was appointed to the AIF in June 1916, and served with AIF Headquarters (see ADMINISTRATIVE HEADQUARTERS, AIF), London until appointed to the Headquarters 5th Division as Assistant Adjutant and Quartermaster-General, in which position he demonstrated 'great professional ability'. General John Monash (q.v.) thought highly of his administrative skills, and believed him badly treated by reason of his birth; he appointed Bruche director of non-military employment on the Repatriation and Demobilisation staff in London to assist in the return of the AIF to Australia.

His war record seems to have wiped out the concerns about his parentage, and for the rest of his career Bruche held a succession of increasingly senior positions, culminating in appointment as CGS in 1931. Unfortunately he presided over the army in the worst depths of the Depression, and was entirely unable to persuade the governments of the day to allocate more resources to the army or to clarify the strategic tasks that the army might be called on to perform in time of war. His close questioning of British policy over the Singapore base, especially after the visit to Australia of Sir Maurice Hankey (q.v.) in 1934, earned him few friends in government although it reflected the prevailing views within the army's upper ranks. He retired in 1935.

BUCCANEER, HMAS see *ATTACK CLASS PATROL BOATS*

BUFFALO, BREWSTER (Single-seater fighter). Wingspan 35 feet; length 26 feet 4 inches; armament 4 × 0.5-inch machine-guns; maximum speed 255 m.p.h.; range 650 miles; power 1 × Wright Cyclone 1200 h.p. engine.

The Buffalo was originally designed as a carrier-based fighter for the US Navy and entered service in 1940. Buffaloes purchased by the British government were not considered good enough for use in Europe and were supplied to No. 21 and 453 Squadrons RAAF in Malaya in 1941 and used against the Japanese during the Malayan campaign. Despite their slow rate of climb and a fuel system that required the pilot to hand-pump fuel to the engine if flying higher than 6000 feet, RAAF Buffalo pilots managed to shoot down at least 20 Japanese aircraft. In the end, however, they were outnumbered and outclassed by the Japanese machines, and the few surviving Buffaloes were withdrawn before the fall of Singapore. Seventeen Buffaloes of the Netherlands East Indies Air Force, which had been withdrawn to Australia in 1942, equipped No. 25 Squadron and were for a critical period the sole air defence for Perth. These Buffaloes were transferred to the US Fifth Air Force by 1944.

BULLDOG TRACK ran from Bulldog in Papua through mountainous and inhospitable terrain to Wau in New Guinea. The track, blazed early in 1942, was used to evacuate civilians from the Bulolo Valley and then to maintain Kanga Force (q.v.), based at Wau. It proved inadequate as a supply line, but in January 1943 construction of a jeep track between Bulldog and Wau began in order to meet the need for a north–south overland supply route across New Guinea. Built by Royal Australian Engineers (q.v.) units and indigenous labourers, the Wau–Bulldog road was a massive and difficult undertaking which was not completed until August

1943. By that time fighting in the New Guinea campaign had moved on and air transport had improved, so the road was used for only a few months before being closed in mid-1944.

BULLWINKEL, Lieutenant Vivian (18 December 1915–). Born in Kapunda, South Australia, Bullwinkel trained as a nurse and midwife at Broken Hill and, on the outbreak of the Second World War, enlisted as a Staff Nurse in the Australian Army Nursing Service. She served with the 2/13th Australian General Hospital during the Malayan campaign and was among the final group of 65 nurses evacuated from Singapore on 12 February 1942, three days before the island fell. Their ship, the *Vyner Brooke*, which was also carrying troops and civilians, was attacked by Japanese aircraft and sunk off Banka Island in the Straits of Sumatra, where a party of survivors, including Bullwinkel, came ashore and surrendered to Japanese soldiers. The Japanese immediately massacred this group, Bullwinkel and a British soldier being the only accidental survivors. Bullwinkel then gave herself up to Japanese sailors who sent her to an internment camp. She was imprisoned in a series of camps in the Netherlands East Indies until 16 September 1945. On her return to Australia, Bullwinkel nursed at Heidelberg Repatriation Hospital in Melbourne and Fairfield Hospital in Sydney, retiring as Matron. She married Colonel F. W. Statham in 1977 and occasionally makes public appearances, such as launching the first of the new *Anzac* Class frigates (q.v.) in 1994. Like 'Weary' Dunlop (q.v.), Bullwinkel has come to symbolise Australian determination and strength in the face of Japanese wartime brutality. The perpetrators of the Banka Island massacre remain unknown and escaped any penalty for their crime, though in 1992 a memorial was unveiled on Banka Island to remember their victims.

BUNBURY (I), HMAS see *BATHURST CLASS MINESWEEPERS (CORVETTES)*

BUNBURY (II), HMAS see *FREMANTLE CLASS PATROL BOATS*

BUNDABERG, HMAS see *BATHURST CLASS MINESWEEPERS (CORVETTES)*

BURDEKIN, HMAS see *RIVER CLASS FRIGATES*

BURGESS, Francis Patrick 'Pat' (17 March 1925–23 January 1989). An Australian journalist and war correspondent, born in Sydney and educated at Christian Brothers, Waverley, and at the

University of Sydney, Burgess began his career as a crime reporter for the *Daily Telegraph*, but by the 1960s had moved into foreign affairs. He covered the end of Dutch rule in West New Guinea, and won a Walkley Award for a story on local elections in the Highlands in pre-independence Papua New Guinea. In 1965 he went on the first of four tours as a war correspondent to Vietnam, following 6 Platoon B Company 1st Battalion Royal Australian Regiment, which had been trained by Sergeant Kevin 'Dasher' Wheatley VC. His reporting, in print, on radio and television, concentrated on the experiences of Australian soldiers, but his sharp observations and analyses enabled him to see the broader picture in an increasingly critical light, and his reports were used by the federal Opposition to attack the Australian involvement in Vietnam. In 1980, he won a second Walkley Award for a television program on the whole of Vietnam, filming from north to south along Highway One.

Pat Burgess, *Warco: Australian Reporters at War* (Heinemann, Melbourne, 1986).

BURMA–THAILAND RAILWAY was constructed by the Imperial Japanese Army (IJA) using Allied POW labour. Fears that their position in Burma was increasingly vulnerable to allied attack led the IJA to decide in early 1943 to speed up completion of the 420 kilometres railway linking Thanbyuzayat in Burma with Bampong in Thailand. They resolved to work simultaneously from both ends of the project, aiming to complete the railway by August 1943, thus enabling the IJA to launch an attack on India. A labour force of 51 000 British, Dutch and American POWs, 9500 Australian POWs and 270 000 conscripted Asian civilians laboured on the railway in appalling physical conditions, made worse by the Japanese demands for speed at all costs and by the brutal treatment of the prisoners, especially at the hands of the mainly Korean guards. Daily individual work quotas were steadily increased, until each man, regardless of his physical condition, was required to shift more than two cubic metres of rock and soil a day. Escape was punished by execution. Food was in very short supply, medicine even more so. The result was widespread malnutrition and exposure to tropical diseases. In such conditions the medical staff, most notably Lieutenant-Colonel Edward 'Weary' Dunlop (q.v.), played a vital part in saving life, not least by trying to protect the sick from having to work under extreme conditions of hardship. Even so, 2646 Australians died on the railway, together with 10000 other Allied prisoners and 70000 Asian labourers. The railway was completed on 16 October 1943, and within two months the Australian survivors had returned to Changi (q.v.) in Singapore. The Burma–Thailand Railway has come to epitomise the Australian POW experience of brutal captivity at the hands of the Japanese.

BURNETT, Air Chief Marshal Charles Stuart (3 April 1882-9 April 1945). Born in the United States of Scottish parents, Burnett joined the Imperial Yeomanry under age in 1899 in order to serve in the Boer War. Commissioned in October 1901 in the Highland Light Infantry, he fought in Nigeria in 1904–06 before leaving the army to pursue business interests. He rejoined the army at the outbreak of war in 1914, and quickly transferred to the RFC as a pilot. He served with No. 17 Squadron in the Middle East, and commanded in turn No. 12 Squadron in France and No. 5 Wing in Palestine. In 1919 he accepted a permanent commission as a wing commander in the RAF. Between the wars he commanded operations in Iraq and held a succession of staff and command positions in Britain, culminating in his appointment as C-in-C of Training Command in 1936, a job he held for three critical years during which time the RAF expanded to meet the needs of the coming war. In 1939 he was made Inspector-General.

Prime Minister R. G. Menzies' (q.v.) decision to appoint British officers to head the Australian services after the outbreak of war in 1939 saw Burnett selected as CAS to the RAAF in December, and he arrived the following February. He knew little of Australian conditions, although he had served alongside Australian flyers in Palestine in the previous war, and his appointment was resented by many. During his time in Australia he presided over the expansion of Australia's role in the Empire Air Training Scheme (q.v.), itself a controversial effort. He tried to build up the numbers and types of aircraft available to the RAAF in Australia with less success, although it must be said that attempts by successive prime ministers to achieve this end were equally frustrated. He intervened decisively to support the formation of the Women's Australian Auxiliary Air Force (q.v.) in 1941, possibly influenced by his daughter Sybil, who had served in the British women's service and was to do so again in the WAAAF. He did not enjoy good relations with the Labor Minister for Air, A. S. Drakeford, and was replaced as CAS in May 1942 after making proposals to abolish the Air Board (q.v.) and reform the administration of the RAAF. The air force now entered an unhappy period of command relations at the top. He retired from the RAF, but was recalled as Commandant, Central Command, Air Training Corps in Britain in 1943.

Burnett's tenure as CAS in Australia was marked by a number of successes, although the context in which he worked occasioned considerable disagreement about the correct deployment of the RAAF. Although the service increased significantly in numbers under his command, over 40 per cent of these personnel were destined for service with the EATS in Europe. Given the problems which had beset command of the RAAF before the war, the decision to appoint him was probably correct, and the disputes and difficulties that followed his replacement were not at all of his making.

BURNIE, HMAS see *BATHURST* CLASS MINE-SWEEPERS (CORVETTES)

BURRELL, Vice-Admiral Henry Mackay (13 August 1904–15 February 1988). An early graduate of HMAS *Creswell* (q.v.), the RAN College at Jervis Bay, Burrell's career highlights the gradual shift in the RAN away from total orientation towards the RN to a position more nearly approximating that of an independent service. Burrell entered the RAN as a 13-year-old cadet midshipman in 1918; he became a midshipman in 1922, and was posted to HMAS *Sydney* (I) (q.v.). As was the habit of the time, he went to Britain for extensive further training, and served on the British ships HMS *Caledon* and HMS *Malaya*. He specialised in navigation, doing the navigation course in 1930, and passing the staff course in 1938. The outbreak of war in 1939 found him as Staff Officer (Operations) at Navy Office in Melbourne.

The need for closer consultations with the US Navy in the Pacific led to staff conversations between British and American naval officers in October 1940, and Menzies nominated Burrell, with his 'full knowledge of Australian naval plans and resources', to represent the RAN. He was subsequently appointed the first Australian naval attaché to Washington at the end of that year. The decision to base a naval officer in the US capital itself arose from the staff talks, which had revealed the 'large gaps' in the US Navy's knowledge of the south-west Pacific, and Burrell's work laid the basis for close cooperation between the two navies during the Pacific War. In September 1941 Burrell took command of the 'N' Class destroyer (q.v.) HMAS *Norman*, which he captained until 1943, following which he became Director of Plans at Navy Office. The end of the war found him as captain of the Tribal Class destroyer (q.v.) HMAS *Bataan*.

His postwar career involved successively more senior and varied appointments, as captain of the 10th Destroyer Flotilla, captain of HMAS *Australia*

(II) in 1948–49 and of HMAS *Vengeance* (qq.v.) in 1953–54, as well as another spell in London as the Assistant Defence Representative from 1951 to 1952. Appointed Deputy CNS in 1954, he became Second Naval member in 1956 and Flag Officer commanding the Australian fleet in 1958, and was finally appointed CNS in 1959.

Burrell's period as head of the navy coincided with some important changes in naval affairs, and he presided over considerable activity in the acquisition of ships and equipment. The planning and acquisition of *Ton* Class minesweepers (q.v.), Wessex (q.v.) anti-submarine warfare helicopters, the *Oberon* Class submarines (q.v.) and the survey ship (q.v.) HMAS *Moresby* were all initiated during his tenure of office. The two most significant decisions, however, involved the reversal of the government's intention to disband the Fleet Air Arm (q.v.), built up at great expense over the previous decade, and the decision to acquire the *Charles F. Adams* Class destroyers (q.v.) from the United States in preference to the County Class destroyers from Britain. This marked a major reversal in policy to that time, and was the cause of considerable pressure from both the RN and the British shipbuilding industry, which was resisted. Burrell worked well with the Minister for the Navy, John Gorton (q.v.), a far more interventionist minister than the navy had been accustomed to. He retired in February 1962.

Henry Burrell, *Mermaids Do Exist* (Macmillan, Melbourne, 1986).

BURSTON, Major-General Samuel Roy (21 March 1888–21 August 1960). A graduate of Melbourne University (1910) in medicine, Burston was appointed as a regimental medical officer to the Army Medical Corps in 1912, having served as a bugler in the Victorian and Australian Military Forces between 1900–05. Transferring to the AIF in March 1915, he served in Egypt and Gallipoli with the 7th Field Ambulance until evacuated in November with enteric fever. He went to France in November the following year with the 11th Field Ambulance, and won a DSO for his actions with an advanced dressing station during the battle of Messines. In November 1918 he became a colonel in charge of the 3rd Australian General Hospital, and in April 1919 was appointed Assistant Director of Medical Services for the AIF depots in Britain during the repatriation of the force.

During the interwar period he resumed civilian practice and maintained his involvement in the CMF. A distinguished physician with a considerable military record, he was appointed Assistant Director of Medical Services on the 6th Division Headquarters when that division was raised for war in 1939. He held increasingly senior positions on General

T.A. Blamey's (q.v.) staff in the Middle East, and was often seen at the front. His report on the deteriorating health of the 9th Division in besieged Tobruk (q.v.) was an important factor leading to that formation's relief.

Back in Australia with Blamey he was made Director-General of Medical Services at Allied Land Headquarters (q.v.), and was centrally involved in the army's efforts to forestall malaria (see SCIENCE AND TECHNOLOGY) among the troops fighting in the New Guinea campaign. This led to the setting up of the Land Headquarters Medical Research Unit at Cairns in March 1943, which did much valuable work in the field of tropical medicine. He enjoyed Blamey's total confidence, and that of the senior ranks of the AIF generally. He remained Director-General of Medical Services for the army until his retirement in January 1948.

BUTLER, Arthur Graham (25 May 1872–27 February 1949) was born at Kilcoy, Queensland, the son of a station manager. He was educated at Ipswich Grammar School and then at St John's College, Cambridge. There he completed a BA in 1894, a degree in surgery in 1897 and a degree in medicine in 1899. He returned to Australia in 1902 to work as a general practitioner. Butler had always been interested in military matters and he enlisted in a militia regiment as a part-time medical officer. On the outbreak of the First World War he enlisted as a captain and was appointed Regimental Medical Officer in the 9th Battalion. He sailed for Egypt in October 1914.

Butler landed with the leading waves at Gallipoli and was in charge of a forward aid post just beyond 400 Plateau. His energy and bravery earned him the DSO. In February 1916 Butler (now a major) was appointed Deputy Assistant Director of Medical Services to I Anzac Corps. He served with distinction in France, rising to the command of the 3rd Field Ambulance in 1917. During the year he served at Bullecourt and at the Third Battle of Ypres and was twice mentioned in despatches. In November 1917 he was posted to London to help collate the medical records of the 1st AIF. He was unhappy leaving the front and returned in July 1918 to become commander of the 3rd Australian Field Hospital at Abbeville, an appointment he retained until it closed in June 1919.

His experience with medical records proved to be the decisive event in Butler's life. After he returned to Australia he agreed, with much reluctance, to write the official history of Australian medical services in the Great War. He began work on the histories in November 1922 at Victoria Barracks in Melbourne, and in 1926 he was relocated to Canberra. Butler did not find the experience of official historian a happy one. He had sacrificed a secure and lucrative income to take a job for which in some ways he was temperamentally unsuited. He was painstaking in his approach to the point of paralysis, became easily bogged down in detail, proved incapable of meeting any deadline and became the despair of C. E. W. Bean (q.v.), who was overseeing the production of all the volumes of the official history. Butler was also not aided by the miserly attitude of various governments under which he worked who refused to provide him with any permanent assistance in sifting an enormous quantity of records.

As a result of all this the medical history which was meant to be completed in two years took 20, and the extent of *The Australian Army Medical Services in the War of 1914–1918* expanded from one volume to three. *Gallipoli, Palestine and New Guinea,* published in 1930, *The Western Front,* published in 1940, and *Problems and Services,* published in 1943, all sold poorly, many being given away as presentation volumes. The aim of the project, to instruct future military medical practitioners, was certainly not realised, the volumes appearing at a time when advances in medicine had made much of their clinical information obsolete.

Nevertheless, in his own way, Butler had made a contribution of another kind. His volumes are masterpieces of scholarship and sources of the utmost importance not just to medical historians but to social and military historians as well. On matters such as shell-shock, venereal disease (q.v.) and self-inflicted wounds, he was much more direct than Bean felt able to be in the military volumes of the official history. His statistical tables in volume three are extremely valuable sources concerning the number of wounded, types of wounds and effectiveness of treatment received. He also made pioneering attempts to trace the effect of war on demobilised men. Butler died in 1949, partially blind and impoverished by his twenty years' labour. The three volumes of the medical history remain his monument.

BUTLIN, Sydney James Christopher Lyon (20 October 1910–14 December 1977). One of Australia's most distinguished economic historians, Butlin was educated at Sydney University on a scholarship after his father's early death impoverished the family. He went to Trinity College, Cambridge, in 1933 and commenced an academic career at Sydney in 1935. During the war he was director of the economic division of the Department of War Organisation and Industry, where

Gavin Long (q.v.) commissioned him to write two volumes entitled *War Economy* for the official history. The first appeared in 1955, the second, co-authored with C. B. Schedvin, in 1977. The volumes are marked by a fluent style, meticulous attention to detail and the skilful handling of vast amounts of often complex data. After the war he returned to Sydney University, serving for many years as professor of economics and dean of his faculty, before accepting a personal chair at the Australian National University in 1971. His other major publications concentrated on the monetary system in the Australian colonies during nineteenth century, and on the banking sector.

S. J. Butlin, *War Economy 1939–1942* (Australian War Memorial, Canberra, 1955); S. J. Butlin and C. B. Schedvin, *War Economy 1942–1945* (Australian War Memorial, Canberra, 1977).

BUTTERWORTH, RAAF BASE, was established in 1955 in Province Wellesley, north Malaya, on the mainland opposite the island of Penang. The Australian government committed an infantry battalion, various naval units, a fighter wing of two squadrons and a bomber squadron to the newly formed Far East Strategic Reserve (q.v.). The base at Butterworth, located on the site of a Second World War airfield, was to be upgraded and expanded by an Australian airfield construction squadron. The construction of a 6300 foot runway, with accompanying hard stands, was completed by No. 2 Airfield Construction Squadron on 26 February 1958. From July 1958 Butterworth was home to No. 2 Squadron RAAF, equipped with Canberra (q.v.) twin-jet bombers, which, although primarily committed to the strategic reserve, were also used in air strikes against communist insurgent targets in Malaya, as were the Sabres (q.v.) of No. 3 and No. 77 Squadrons, which were successively based at Butterworth. The base was formally handed over to Malaysia on 31 March 1971, but RAAF squadrons continued to be based at Butterworth under the Five Power Defence Agreement until 1988, when the last of the Mirages was withdrawn. F/A-18 (q.v.) aircraft now rotate periodically through Butterworth, and the base is used by RAAF Orion (q.v.) aircraft conducting surveillance operations in the Indian Ocean and the South China Sea. What for 30 years had been a large Australian personnel presence in peninsular Malaysia has been reduced to a very small number.

C

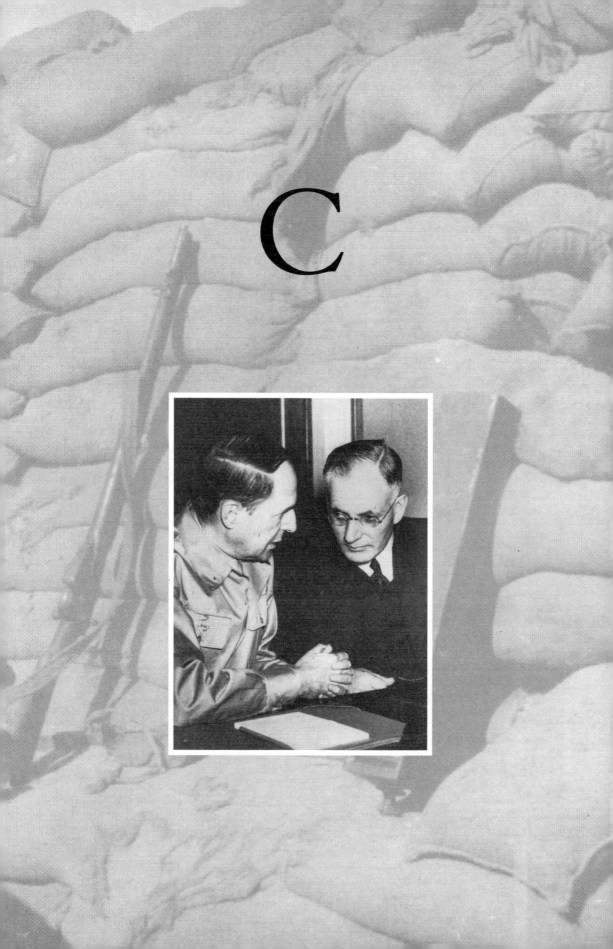

CA-15, CAC (Single-seat fighter prototype). Wingspan 36 feet; length 36 feet 2.5 inches; armament 6 × 0.5-inch machine-guns, 10 rockets or 1000 pounds bombs; maximum speed 448 m.p.h.; range 2540 miles; power 1 × Rolls Royce Griffon 1540 h.p. engine.

The CA-15 was a prototype fighter designed and built by the Commonwealth Aircraft Corporation (q.v.). Conceived in 1942 as a replacement to the Boomerang (q.v.), the CA-15 was a long-range, high-performance fighter incorporating the lessons learnt in the Pacific War and the latest innovations in piston-engine fighter design. Production was delayed in 1944 when shipments of Wright engines from the USA became unavailable. The CA-15 did not fly until March 1946, six months after the end of the Second World War, but it flew as well and as fast as any other piston engine fighter in the world. The RAAF operated the CA-15 with No. 1 Aircraft Performance Unit from 1946 to 1950. The CA-15 never entered production because it had been made obsolete by jet fighters such as the Meteor (q.v.), which had been in operational service since 1944. The sole CA-15 was broken up in 1950 but it had shown how, in a few years, aircraft production in Australia had gone from nothing to the creation of a world-class design.

CADETS see **AUSTRALIAN CADETS CORPS**

CAIRNS, HMAS see *BATHURST* **CLASS MINE-SWEEPERS (CORVETTES)**

CALDWELL, Group Captain Clive Robertson (28 July 1910–5 August 1994). Having learnt to fly before the war, Caldwell lied about his age (the age limit for trainee pilots being 28) and joined the RAAF at the outbreak of war in 1939. He sought a discharge when he discovered that his intake were destined to become instructors, and rejoined in April 1940 as a member of the first intake of the Empire Air Training Scheme (q.v.). He was commissioned as a pilot officer in January 1941.

Posted to the Desert Air Force, he served in the Syrian campaign with No. 250 Squadron, RAF, flying American P-40 Tomahawk aircraft. Initially assigned to escort and ground-strafing duties, Caldwell scored his first victory on 26 June 1941 when he shot down a German Messerschmitt Bf-109.

Squadron Leader Clive Caldwell (left), photographed by Damien Parer while commanding officer of No. 112 Squadron RAF. With him are pilots of the squadron and one of the squadron's sharkmouth-painted Kittyhawk fighters. North Africa, 1942. (AWM 011944)

Four days later he accounted for two Stukas and an Me-110. On one occasion, in December, he shot down five German aircraft in a single sortie, the rest of the squadron claiming six more. In January 1942 he was given command of No. 112 Squadron, flying Kittyhawks (q.v.), and by the time he had completed his tour in May that year he had amassed a total of 20 1/2 enemy aircraft shot down. He was awarded the DFC and bar, as well as the Polish Cross of Valour in recognition of his cooperation with Free Polish pilots in the desert.

Returning to Australia he was posted as an instructor, but returned to combat flying in January 1943 as a wing commander with No. 1 Fighter Wing, flying Spitfires (q.v.) out of Darwin. In eight months' operational flying he was credited with shooting down eight Japanese planes. After a further spell of instructing, he returned to flying duties in May 1944 with No. 80 Fighter Wing, part of the 1st Tactical Air Force (q.v.). The tasks here were in support of ground operations, and Caldwell was one of those who tendered his resignation in April 1945 as part of the 'Morotai mutiny'. Disenchanted with the air force by the war's end, he declined the offer of a permanent commission in the postwar RAAF and returned to civilian life and his prewar business interests. Dubbed 'Killer', a nickname he disliked, his principal advice to young fighter pilots was that, '[a] pilot who cannot shoot straight might as well remain on the ground, for he is useless in a fighter squadron'.

CANBERRA, GAF (2-seat tactical bomber). Wingspan 64 feet; length 65 feet 6 inches; armament 8000 pounds bombs; maximum speed 541 m.p.h.; range 3800 miles; power 2 × Rolls Royce Avon turbojets.

The Canberra was the RAAF's first jet bomber and was manufactured under licence in Australia by the Government Aircraft Factory with jet engines built by the Commonwealth Aircraft Corporation (q.v.). They entered service with No. 2 Squadron in 1953 and later also equipped No. 1 and 6 Squadrons. No. 2 Squadron's Canberra flew attacks against communist terrorists during the Malayan Emergency and served in the Vietnam War from 1967 to 1971 attached to the US 35th Tactical Fighter Wing. While in Vietnam, No. 2 Squadron flew 11 000 missions and dropped 76 389 bombs for the loss of two aircraft. The Canberra was to be replaced by the F-111 (q.v.) in 1968 but, because of delays in its delivery, Canberras flew with No. 1 and 6 Squadrons until 1970 when they were replaced with Phantoms (q.v.) leased from the USAF. After their return from Vietnam No. 2 Squadron used Canberras in a pho-

tographic reconnaissance role until 1982, when they were finally withdrawn from service.

CANBERRA (I), HMAS (*Kent* Class heavy cruiser). Launched 1927; displacement 9850 tons; length 630 feet; beam 68 feet 4 inches; armament 8 × 8-inch guns, 4 × 4-inch guns, 4 × 3-pounder guns, 4 × 2-pounder guns, 12 machine-guns, 8 × 21-inch torpedo tubes, 1 × Supermarine Seagull V (q.v.) seaplane; speed 31.5 knots.

Built at Clydebank in Scotland, HMAS *Canberra* was sister ship to HMAS *Australia* (II) (q.v.) and joined the RAN in 1928. As an economy measure at the height of the Depression in 1930–31, *Canberra*, *Australia*, *Albatross* and *Anzac* (q.v.) were the only ships kept in commission by the RAN. From the outbreak of the Second World War in September 1939 to February 1940, *Canberra* patrolled the Western Australian coast for German raiders. Then in June 1940, following the fall of France and entry of Italy into the war, the Australian government offered use of the *Canberra* to the British Admiralty and she served on the South Atlantic Station and later in the Indian Ocean. It was here on 4 March 1941 that *Canberra* with HMS *Leander* intercepted the German raider supply ship *Coburg* and the tanker *Kitty Brovig*. Both German ships scuttled themselves after *Canberra* shelled *Coburg* and *Canberra*'s seaplane bombed *Kitty Brovig*. In January 1942, following Japan's entry into the war, *Canberra* escorted troop reinforcements to Singapore, and in May took part in the Battle of the Coral Sea (q.v.) as part of Rear-Admiral John Crace's (q.v.) Task Force 44. On 7 August 1942 *Canberra*, with 48 other warships including *Australia* and *Hobart*, three US aircraft carriers and a battleship, provided support as US marines landed at Guadalcanal in the Solomon Islands. It was off Guadalcanal just before 2.00 a.m. on 9 August that a Japanese fleet of seven cruisers and a destroyer took the Allied fleet by surprise in what became known as the Battle of Savo Island. *Canberra* was hit 24 times in less than two minutes, and 84 men were killed, including Captain Frank Getting (q.v.). The order to abandon ship was given and *Canberra* was sunk the next day by torpedo by a US destroyer. The British government offered the *Shropshire* (q.v.) to the RAN as a replacement for the *Canberra*, while the US Navy commemorated the ship, and the memory of those lost upon it, by renaming the *Pittsburgh*, then under construction, the USS *Canberra*.

CANBERRA (II), HMAS see *OLIVER HAZARD PERRY CLASS GUIDED MISSILE FRIGATES*

see **AUSTRALIA–
NEW ZEALAND AGREEMENT**

CANNAN, Major-General James Harold (29
August 1882–23 May 1976). Cannan was a well-
educated young man who worked in various businesses, with an emphasis on the insurance industry,
in Queensland and New Zealand and served in the
militia before being commissioned into the 1st
Queensland (Moreton) Regiment in 1903. He was
promoted to lieutenant-colonel in command of
the 8th (Oxley) Infantry Battalion in May 1914.
Volunteering for service with the AIF, Cannan took
the 15th Battalion, part of Colonel John Monash's
4th Brigade, to Gallipoli. He took a leading part in
the fighting during the August offensives, in which
his brother was killed while serving under him, was
invalided from the peninsula in October, and
rejoined his unit in Egypt early in 1916. Proceeding to France, he participated in the battles of Pozières and Mouquet Farm, and in August was given
command of the 11th Brigade, which he led
through the campaigns of 1917. For his service in
these two years of hard soldiering he was awarded
the CB and CMG. His leadership of his brigade
during the defensive battles in the first half of 1918,
in particular at the first battle of Amiens, and during the offensive fight at Hamel in July, earned him
the DSO. In addition, during the course of the war
he was mentioned in despatches six times. After the
war, he was president of the Queensland branch of
the RSSILA between 1920 and 1921, and pursued
his business interests in the insurance industry. He
held brigade commands in the CMF between
1920 and 1925, at which time he was placed on
the Unattached List. With the Second World War,
he became Inspector-General of Administration in
the Department of Defence Coordination in May
1940, was appointed to command the 2nd Division
in July 1940, and in October that year was made
Quartermaster-General and Third Member of the
Military Board (q.v.), in part at least a testament to
his considerable business and administrative experience. He served in this capacity throughout the
war, often visiting forward areas in New Guinea
and the South-West Pacific Area (SWPA) and dealing with the myriad demands of both the
expanded Australian Army and its American allies.
Although General Blamey (q.v.) recommended
him twice for a knighthood, the Labor government
declined to act on the recommendation. He retired
from the army in 1946, but served as Director for
SWPA of the United Nations Relief and Rehabilitation Administration until 1947, and took a strong
interest in the placement of European displaced
persons as migrants to Australia. Between 1950 and
1951 he was chairman of the Queensland division
of the Australian Red Cross.

CANTEENS FUNDS were set up after both world
wars to distribute the profits from canteens that had
operated during wartime in military camps and on
troop-ships. The AIF Canteens Fund was established by the federal government in 1920 to relieve
distress among returned soldiers and their dependants. The £750 000 of canteen funds was
exhausted by 1932 after relieving 50 000 cases. The
fund's trustees, however, had also been made
responsible for a substantial bequest, left by wealthy
pastoralist Sir Samuel McCaughey, for the agricultural and technical education of deceased and incapacitated soldiers' children, and distribution of these
monies continued. The £5 million of canteens
funds available after the Second World War were
administered by the Services Canteens Trust Fund,
established in 1948. Its funds were divided equally
between the relief of ex-Second World War service
people and the education of their children. It was
abolished in 1987.

CANUNGRA LAND WARFARE CENTRE, formerly
the Jungle Training Centre (JTC) is located in
south-east Queensland. In November 1942 the
army established the Land Headquarters Training
Centre (Jungle Warfare), consisting of a Reinforcement Training Centre, an Independent Company
Training Centre (formerly at Wilson's Promontory
in Victoria) and a Tactical School, to meet the needs
of combat in the demanding environment of Papua
New Guinea. Its first commandant was Colonel A.
B. ('Bandy') MacDonald, prewar commanding officer of the Darwin Mobile Force (q.v.). By May
1943, when it was at the height of its activity, there
were 2000 reinforcements organised into eight
training companies from which 500 soldiers a
week graduated for service in New Guinea,
together with a Commando Training Battalion
supplying reinforcements for the independent
companies and an officer-training program which
turned out 60 platoon commanders every six
weeks. Training was realistic and physically
demanding, and instructors were drawn from men
with recent combat experience in either the Middle East, South-West Pacific, or both. The centre
was closed in 1946.

It was reopened in 1954 to meet the army's
training needs for service in south-east Asia. The site
was expanded and the centre divided into three sections: one to train officers and NCOs in jungle tactics, a second to train units in operations under
jungle conditions, and a third to test doctrine and
produce training manuals. From 1955 to 1957 the

commandant was Colonel F. P. Serong (q.v.), later first commanding officer of the Australian Army Training Team, Vietnam (q.v.). The calibre of instructors was again high, the chief instructor being Lieutenant-Colonel George Warfe, who had won an MC in New Guinea and a DSO in Borneo, and who had also served in the Malayan Emergency. In 1960 the School of Tactics and Administration was relocated to Canungra from Seymour in Victoria, thus broadening the centre's functions as all officers now attended promotion courses there.

During the Vietnam War Canungra became the major training venue for the army as it prepared units for active service. Companies were put through increasingly demanding training exercises culminating in a full battalion exercise in which the units were assessed for readiness for active service. 'The training at Canungra was of the greatest value', wrote one battalion intelligence officer. The height of the war saw Canungra at its peak expansion, able to deal with up to 10 000 students on courses annually. With the wind-down after the withdrawal from Vietnam and the gradual reorientation away from jungle warfare towards the defence of Australia, JTC was renamed the Land Warfare Centre in June 1975. It now also incorporates the School of Military Intelligence and the Warrant Officer and Senior Non-Commissioned Officer Wing, and hosts students on military cooperation programs from various regional and allied armies.

CARIBOU, DE HAVILLAND CANADA (2-seat Short Take Off and Landing [STOL] transport). Wingspan 29.15 m; length 22.13 m; load 3000 kg cargo, 32 troops or 26 paratroops; maximum speed 350 km/h; range 2100 km; power 2 × Pratt & Whitney Wasp 1450 h.p. engines.

Designed in Canada with the ability to take off and land from short, rough airstrips, the Caribou entered service with the RAAF in 1964. Caribous were used during the Vietnam War from 1964 to 1972 by RAAF Transport Flight Vietnam (later renamed No. 35 Squadron), with three aircraft lost. Caribous were also flown by No. 38 Squadron and were used for humanitarian aid in East Timor in 1975 and for peace-keeping (q.v.) operations on the Indian–Pakistani border between 1975 and 1978. Caribous are still flown by the RAAF.

CARRIERS, NATIVE see '**FUZZY-WUZZY ANGELS**'

CARTWHEEL see **NEW GUINEA CAMPAIGN**

CASTLEMAINE HMAS see *BATHURST* **CLASS MINESWEEPERS (CORVETTES)**

CATALINA, CONSOLIDATED (8–9-crew reconnaissance flying boat [PBY-5]). Wingspan 104 feet; length 63 feet 10 inches; armament 4 × 0.303-inch machine-guns, 4 × 650-pound depth charges or 2

A 'Black Cat' Catalina of No. 11 Squadron undergoing maintenance at the RAAF Flying Boat Repair Depot, Lake Boga, Victoria, 31 August 1944. (AWM P0448/212/039)

× Mark 13 torpedoes; maximum speed 180 m.p.h.; range 2545 miles; power 2 × Pratt & Whitney Twin Wasp 1200 h.p. engines.

Catalinas first entered service in 1936 and were even manufactured under licence in the USSR from 1938. The first 18 Catalinas for the RAAF were ordered from the United States in 1940. Under the US Neutrality Acts (1935–37) then in force, they could not be shipped to Australia but had to be flown to Australia under civilian registration by Qantas pilots. Catalinas replaced Ansons and Hudsons (qq.v.) as maritime patrol craft, but during the desperate days of the early part of the Pacific War they were even used as bombers and eight of the original order were lost to enemy action. Australian Catalinas shadowed the Japanese fleets before the Battles of the Coral Sea and the Bismarck Sea. Black-painted Catalinas, known as 'Black Cats', were used for minelaying operations that reached as far distant as the South China Sea, and Catalinas were also used for air–sea rescue. Catalinas remained in RAAF service until 1950. Qantas flew them during the Second World War (see QANTAS AT WAR) and ex-RAAF machines were purchased by several airlines after the war and used as passenger aircraft.

CATERING CORPS see **AUSTRALIAN ARMY CATERING CORPS**

CAVALRY (COMMANDO) SQUADRONS see **INDEPENDENT COMPANIES**

CENOTAPH, SYDNEY, which stands in front of the General Post Office in Martin Place, was unveiled in 1927. The design, by Sir Bertram Mackennal, consists of a cenotaph, or empty tomb, guarded by life-size bronze statues of a soldier and a sailor in the uniforms of the First World War. Despite the opening of the larger Anzac Memorial in Hyde Park in 1934, the cenotaph has remained the focal point for wreath laying and Anzac Day ceremonies.

CENSORSHIP Censorship is a seemingly inevitable concomitant of modern war, and in the twentieth century its expansion has been one of the visible signs of the considerable increase in the powers of central government, which has itself been a consequence of modern industrial warfare. In Australia, there was no mechanism for the censoring of news and information at government behest before 1914, although the government took steps in early August of that year to establish a system of censorship that would operate from the outbreak of war, which by then was imminent. The system was devised by Major-General Bridges (q.v.), and was implemented from the evening of 3 August 1914 with a view to controlling 'all cable and wireless communication throughout the Commonwealth'. At the same time, newspapers were affected by the imposition of censorship of cables at the London end, on which all were still reliant for coverage of overseas news. The office of the Deputy Chief Censor was established in Melbourne, the post being held initially, and briefly, by Colonels McCay and Monash (qq.v.) in succession. The Australian office worked closely with the Chief Censor in the War Office in London, and considerable efforts were made during the war to coordinate the various censoring policies throughout the Empire. It employed 124 people in 1914, rising to 187 by the war's end, based on the military districts (q.v.). The Deputy Chief Censor and his staff had responsibility for censorship of the press and the post, while interpreters were also employed for translation of material in various European languages that had attracted the attention of the censors. As well as the control of information, the censors fulfilled an economic and political intelligence function, alerting Commonwealth authorities to attempts to continue trading with the enemy through shipments of raw materials dispatched through neutral channels, or through attempts to remit payments to German companies. Newspaper proprietors initially complained about the slowness and inexperience of many of those charged with supervising the control of the press but, in general, for much of the war appear to have accepted the broad principles on which the censorship was based. The growth in information technology that accompanied the war was reflected in the censors becoming responsible, from 1916, for screening news and motion picture footage before its release to the cinemas. As the conscription issue became more heated, and the politics of recruiting more divided throughout the course of 1916, Prime Minister W. M. Hughes (q.v.) increasingly used the system of censorship as a political weapon against those opposed to his policies. Trade-union offices and the labour press in particular were raided and materials confiscated and destroyed; while in the lead-up to the vote on the referendum in October the government acceded to a request from the War Office in mid-August not to publish details of the casualties suffered by the AIF in the fighting on the Somme. Censorship was also used in a heavy-handed manner against members of the Industrial Workers of the World, a number of whom were gaoled on a pretext in response to their campaigns against both the conduct of the war and the government. During the course of the First World War the original purpose of the censorship system, to safeguard information of possible interest to the enemy and to control the flow of information to the public in order not to endanger morale, was subordinated to a considerable extent to Hughes's more narrowly political purposes.

During the Second World War censorship, and its corollary, propaganda, were largely the responsibility of the Ministry of Information, created on 4 September 1939. As in the previous conflict, its staff and resources expanded considerably throughout the course of the war. Its responsibilities included the censorship of all media, but it played a strongly proactive role as well, producing newsreel footage and other films for the cinema, publishing pamphlets and other forms of short-term information, and providing editorial material for the news services. The department's censorship functions were streamlined during 1942–43, but it continued to influence the nature of what was reported on Australia's war effort through its dissemination of news overseas and its authorisation and control of Australian war correspondents (q.v.). Once again, although military necessity was usually invoked as the basis of censorship, political considerations also came into play. In April 1942 Prime Minister Curtin (q.v.) set up the Press Censorship Advisory Committee to smooth relations with the newspaper editors and proprietors. While Curtin chaired it, the system appears to have worked tolerably well, but Curtin's replacement by the most junior minister in the government, Arthur Calwell, in September 1943 led to renewed dissatisfaction. Calwell had been critical of the newspapers in the parliament, and the editors tried unsuccessfully to persuade him to adopt a more liberal regime of censorship. When the Sydney *Daily Telegraph* published comment on the coal strikes in November that year, Calwell issued them with an 'order-to-submit', a mechanism that had not been invoked since 1940. On 17 April 1944 Calwell ordered that that day's issues of the major newspapers be suppressed for challenging the censorship system openly following a further round of disagreements with the Sydney newspapers, and especially the *Sydney Morning Herald*. (The censor was careful to ban only the last edition of each, so as not to endanger the newspapers' advertising revenue.) The matter went to the High Court and, following the intervention of the Chief Justice, Sir John Latham, the parties agreed to negotiate a settlement of their differences out of court. The result was the 'code of censorship principles' which was added to the Press and Broadcasting Censorship Order and which while it appeared to give victory to the newspaper editors in fact hardly altered the powers of the censors at all. At heart, the real argument came down to the Curtin government's concern to present

Australia to its allies in the best possible light, and disagreement between it and the newspapers over the best means of doing so. Although the Chief Censor, E. G. Bonney, appears to have had a largely adversarial relationship with the newspapers, in general both the press and the Department of Information cooperated closely with each other and benefited equally from the arrangement. As before, the need to control information in the interests of the war effort gave way over time to other considerations, in this case the government's desire to influence perceptions of Australia's war effort overseas.

In the field, soldiers' mail was censored locally by their officers, although officers themselves were trusted to self-censor. General MacArthur exercised his own censorship at General Headquarters South-West Pacific Area, which caused some friction with Australian journalists and editors. The Australian Army Public Relations Directorate had a separate censorship role, but in practice subordinated this to the requirements of MacArthur's headquarters, while the services themselves had only an advisory relationship to the Censorship Division and the Chief Censor. The Ministry of Information was abolished in 1950, to be replaced by the Australian News and Information Bureau and its successors. Formal censorship on defence issues during peacetime is exercised through the system of 'D notices'. Certain issues or stories may attract a D notice on grounds of their extreme sensitivity, and the press is expected to comply, although there are no penalties for failing to do so. In practice, D notices are issued sparingly by the Australian government.

John Hilvert, *Blue Pencil Warriors: Censorship and Propaganda in World War II* (University of Queensland Press, Brisbane, 1984).

CENTAUR (2/3rd Australian Hospital Ship), a motor passenger ship, was converted in January 1943 for use in the New Guinea area. It was on its second voyage as a hospital ship when, on 14 May 1943, it was sunk by a torpedo from a Japanese submarine off Brisbane (the wreck being discovered off Moreton Island in 1995). The ship was brightly lit and properly marked in accordance with the principles of the Geneva Convention, and visibility was good at the time of the attack. Of the 332 people on the ship (none of whom were patients) there were only 64 survivors who clung to debris for about 35 hours before being rescued. Australian government protests were rejected by the Japanese government. Japanese records do not acknowledge responsibility for sinking the *Centaur*, and the War Crimes Tribunal was unable to establish which submarine was responsible.

CENTRAL ARMY RECORDS OFFICE (CARO) was established on 30 July 1948 to handle the personal records of all members of the AMF, with the exception of the CMF. It grew out of the need to maintain personal records for rapidly increasing numbers of soldiers, beginning in October 1914 when the 20 000 men of the 1st Division, AIF, embarked for overseas. The Base Records Office was established in Victoria Barracks, Melbourne, on 20 October 1914, with responsibility for maintaining duplicate copies of all personal records relating to individuals serving in the AIF, the originals being held by Australian Base Records Sections 3rd Echelon GHQ, in Egypt, France and London. At the end of the war the Base Records Office collected all copies of individual records, which were subsequently located in the basement of the Australian War Memorial (q.v.), Canberra, where the Base Records Office moved in 1941. Between the wars there was no central records organisation, records being maintained on the basis of military districts (q.v.), while militia records were held by militia units.

When war broke out in September 1939, the 2nd Echelon Organisation was formed, eventually being located adjacent to Land Headquarters, Melbourne, with subsidiary offices in each military district as well as offices in overseas theatres and a Base Records Office in Canberra which was responsible for maintaining detailed records of service, including the registration of graves, for each individual serving in the AMF. When demobilisation ended on 16 February 1947, the 2nd Echelon was reorganised, emerging as CARO in July 1948, its main role, as before, being the maintenance of personal records. In 1972, after the ending of the National Service Scheme (see CONSCRIPTION) and the withdrawal of Australian troops from Vietnam, CARO was charged with implementing personnel management policies on promotion, posting, discharge and re-engagement, as well as continuing to maintain records. Its name was changed to Soldier Career Management Agency (SCMA) in December 1989, which reflected its important role in personnel management within the Army. Besides maintaining records for army use, CARO has also been a rich source of information for family historians, whose access to personal files has however been restricted in accordance with army procedures. Personal records from the First World War, which were transferred to CARO from Canberra in 1958, were transferred back to Canberra in 1993, to the Mitchell Repository of the Australian Archives where they are readily accessible as they are no longer regarded as active personal files.

CENTRAL BUREAU see ULTRA

CERBERUS, HMAS (Flinders Naval Depot), at Crib Point, Victoria, opened in 1920 as a training depot. Among the subjects taught in *Cerberus*'s early years were gunnery, communications, electrical engineering, seamanship and physical training. In 1925 it took over initial training of recruits from HMAS *Tingira* (q.v.), and in 1930 the RAN College transferred to *Cerberus* before returning to HMAS *Creswell* (q.v.) in 1957. In 1986 *Cerberus* contained the Recruit, Gunnery, Communications, Weapons, Electrical Engineering, Marine Engineering, Seamanship Training, Supply, Physical Training, Management, Medical Training, Dental Training, Training Technology, and Nuclear, Biological and Chemical Defence Schools.

CERBERUS, HMVS see VICTORIAN NAVY SHIPS

CESSNOCK (I), HMAS see BATHURST CLASS MINESWEEPERS (CORVETTES)

CESSNOCK (II), HMAS see FREMANTLE CLASS PATROL BOATS

CHAMBERS, Cyril (28 February 1897–2 October 1975). A dentist by profession, Chambers was involved in Labor politics throughout the interwar years. He received a reserve commission in the Dental Corps in March 1940, and after call-up to full-time duty served mostly in Australia with a brief period of overseas service in Papua in early 1942. He was elected to the House of Representatives in the federal election of 1943, and became minister for the Army in the postwar Labor government in November 1946. He was a solid, though not particularly imaginative, Army minister, characterised by one general of the day as believing that all senior officers were bent on self-aggrandisement at the expense of their men and that all virtue resided in the ranks. A public campaign which raised concern about the moral climate of the occupation of Japan saw him make an official inspection of the British Commonwealth Occupation Force (q.v.) area in late 1946, which helped to dispel public, or at least newspaper, concern about the circumstances confronting Australian troops there. He was a strong Catholic (which did not prevent him from marrying three times, on the second occasion to a divorcee) and member of the Labor right, opposed to communist influence in the labor movement and a foe of H.V. Evatt. He was expelled from the Labor Party as a result in 1957, and although readmitted the following year, failed to retain his seat.

CHANGI All Australian soldiers who became prisoners of war following the surrender of Singapore on 15 February 1942 were housed in a huge camp at Changi, at the eastern end of the island. Before the war the Changi area had been the site of several British Army barracks, one of which, Selerang Barracks, became the Australian POW area. Beginning on 17 February, the 14 972 Australians were marched to Changi, where they were told that their own officers would be held responsible for the maintenance of discipline. Within a month the routine of the camp had been set in place. Under orders from the Japanese for the prisoners to become largely self-sufficient in food, 50 acres of vegetable gardens were planted (subsequently expanded to 120 acres), all other spare ground within the camp area was soon under cultivation and a poultry farm had been established. The perimeter was fenced in by wire by 17 March and the Australian area sealed off from the British and other compounds, wood-cutting parties were set up to fuel the kitchen fires, and stringent sanitary measures were taken to combat the possibility of diarrhoea and dysentery. Food supplies from the Japanese were meagre, but they could be supplemented by whatever could be stolen by those (8500 by May) who were sent to the docks and warehouses on work details. Individuals also carried on a thriving trade with sympathetic locals, for although the perimeter fence was patrolled, it was relatively easy to slip out at night. To ward off the inevitable boredom of POW life, a concert party was established within two days of the Australians' arrival at Changi; for the rest of the war it staged both original shows and productions of classic theatre pieces. Much more short-lived was the 'Changi University', which began on 1 March 1942 with three departments — Agriculture, Business Training and General Education — to which were added Languages, Law, Engineering, Medicine and Science. Some of the courses had an obvious immediate and practical application, others laid the foundation for postwar careers or made up for the lack of formal education in the years of the Depression: all contributed to the maintenance of morale. As large work parties left the camp from May 1942 on, it became impossible to sustain the level of instruction, and the 'University' collapsed before the end of the year, but less ambitious schemes continued for the rest of the war.

Although food rapidly became the main preoccupation of the prisoners, the first several months in Changi were not typical. The ravages of malnutrition had not yet set in, there were many organised sporting and cultural activities, and by and large the Japanese left the POWs alone. By May, however, the demands for labour on the Burma–Thailand railway (q.v.) were such that the Japanese looked to Changi to supply their needs. On 14 May 1942, 'A' Force of 3000 men left Changi to work on the railway, followed by further contingents over the next year (see PRISONERS OF WAR). All officers of the rank of colonel and above were removed to Formosa in August, and Lieutenant-Colonel F. G. Galleghan (q.v.) was left in charge of AIF POWs at Changi. The new Japanese commandant then tried to compel all soldiers to sign a statement promising not to attempt to escape. As a demonstration of their determination to enforce the 'no escape' policy, the Japanese executed four soldiers, including two Australians, who had escaped several months earlier, and forced the senior Allied officers to witness the procedure, which was particularly gruesome because the Sikh firing party did not aim properly and it took many shots before the prisoners were killed. Following this display, the Japanese herded the British and Australian prisoners, 15 400 in all, into the Selerang Barracks Square, a small area of eight acres where they remained with minimum water and food for three days until, on 5 September, they agreed that the document in question could be issued as an order by the Allied commanders rather than as a request by themselves. Honour satisfied, the men signed and moved back to their former quarters. Galleghan was determined to maintain discipline and cohesion among the Australian POWs as the only means of guaranteeing the survival of the greatest possible number. His methods sharply divided the camp: many praised him for standing up to the Japanese and refusing to compromise, others (then and now) regarded him as an egocentric bully. Japanese demands on those who remained at Changi increased in 1943, when 800 prisoners were detailed to undertake heavy work repairing the Changi airfield. In March 1944, when new Japanese units arrived to man the airfield, they were housed in the Selerang Barracks and all prisoners were moved to Changi Gaol and its immediate environs. Overcrowding reached new heights as eventually 11 700 Allied prisoners were moved into the area, 5000 of them in the gaol building designed for 650 civil prisoners. Galleghan's authority over the Australian other ranks was revoked by the Japanese in July 1944, and restrictions on camp life increased, culminating in the concert party being closed down and the theatre demolished in March 1945.

When the war ended, Changi Gaol reverted to its status as a civil prison, which it remains to this day, while the name 'Changi' is now more familiar to Australians as the civil airport in Singapore. In 1988 one of the small outdoor chapels that had been built in 1944 in the grounds of the prison complex was re-erected near the Chapel of St Paul at RMC Duntroon as the National Memorial to Prisoners of War. In popular memory Changi has become synonomous with the brutality of the Japanese treatment of POWs but, in fact, for all the privations endured by its inmates, those who remained in it for the duration of the war were, in relative terms, considerably better off than those who made up the work gangs elsewhere in south-east Asia.

CHAPLAINS see **ROYAL AUSTRALIAN ARMY CHAPLAINS DEPARTMENT**

CHARLES F. ADAMS **CLASS GUIDED MISSILE DESTROYERS** (*Perth, Brisbane, Hobart*). Laid down 1962 (*Brisbane* 1965), launched 1963 (*Brisbane* 1966); displacement 3370 tons; length 437 feet; beam 47 feet; armament 2 × 5-inch guns, 2 Ikara (q.v.) anti-submarine missile launchers, Tartar anti-aircraft missile system; speed 35 knots.

These ships, which constituted the First Destroyer Squadron, took the RAN into the missile age. Between 1967 and 1970 both *Perth* and *Hobart* served on three tours of duty in the Vietnam War, the former engaging in naval gunfire support actions. HMAS *Brisbane* undertook one tour of duty and operated with the US 7th Fleet. All ships underwent extensive refits in the 1970s and 1980s and the *Brisbane* participated in the Gulf War in 1990–91.

CHARLESWORTH, Air Vice-Marshal Alan Moorhouse (17 September 1903–21 September 1978) graduated from RMC Duntroon in 1923 and served with the 2nd Light Horse in Queensland before secondment to the infant RAAF in 1925. He was posted to No. 1 Squadron after flying training, and in 1928 became a permanent member of the air force. He spent time in Britain with the RAF between 1928 and 1930, and in 1932 made a round-Australia flight for aerial survey purposes, for which he was awarded the AFC. He held a number of command and staff posts during the Second World War, including that of AOC North-Western Area, based in Darwin, between 1944 and 1945. In February 1947 he became Commandant of the School of Air/Land Warfare at Williamtown in New South Wales. Between 1949 and 1951 he was chief of staff at Headquarters, British Commonwealth Occupation Force in Japan. Charlesworth was successively AOC South-

ern (Training) Command from 1951 to 1953 and headed the RAAF Overseas Headquarters in London between 1954 and 1955, retiring in the latter year. In retirement he was Director of Recruiting between 1958 and 1959.

CHAUVEL, General Henry 'Harry' George (16 April 1865–4 March 1945). Born in Tabulam, New South Wales, the second son of a grazier and cattle-breeder, Chauvel was educated at Sydney Grammar School and Toowoomba Grammar. His ambition to join the British Army was thwarted by his father's straitened financial circumstances, and his military career began in 1886 when he was commissioned into the Upper Clarence Light Horse, a volunteer unit raised by his father. When the family moved to Queensland in 1888 he resigned his commission, and was commissioned into the Queensland Mounted Infantry in 1890, six years later transferring to the Queensland Permanent Military Forces with the rank of captain. After a year's courses and attachments to regular infantry units in the United Kingdom, he commanded the 1st and 2nd Contingents of the Queensland Mounted Infantry in South Africa, 1899–1900, where he distinguished himself in several actions and was mentioned in despatches. He returned to Australia to take command of the 7th Australian Commonwealth Horse, but it was a disappointing experience because the regiment arrived in South Africa three weeks after the peace settlement had been signed. For his services in the war he was appointed CMG. During the next decade in Australia Chauvel held a series of staff appointments, mainly in Queensland, developing a reputation as a trainer of soldiers. He was closely associated with the GOC, Australian Military Forces, Major-General Edward Hutton (q.v.), and became known as a strong supporter of the militia. Appointed Adjutant-General and Second Member of the Military Board in 1911, he played a major part in implementing the compulsory military training scheme (see CONSCRIPTION). In early 1914, he came to the favourable attention of General Sir Ian Hamilton (q.v.), Inspector-General of Overseas Forces, who in the course of an inspection of the AMF over several weeks was impressed above all by the light horse.

In mid-1914 he was posted to London, to be the Australian representative on the Imperial General Staff, but the outbreak of war in August changed this arrangement. He was appointed to command the 1st Light Horse Brigade, which was earmarked for service in Europe, but after Chauvel convinced the High Commissioner and, through him, the War Office, that the planned camps on

Salisbury Plain would not be ready to receive the AIF, the decision was taken to disembark the force in Egypt. The light horse continued to train in Egypt, and did not accompany the original force that landed at Gallipoli on 25 April 1915. Chauvel and the other brigade commanders of the light horse insisted, when they were called upon to provide reinforcements, that the light horse should serve as complete units, albeit as dismounted infantry, rather than be broken up piecemeal among other units. Chauvel landed at Gallipoli on 12 May and took command of the sector around Pope's, Quinn's and Courtney's Posts, which were subject to heavy Turkish counter-attacks. After briefly commanding the New Zealand and Australian Division, he was appointed GOC 1st Australian Division in December 1915, taking the division through the successful evacuation from Gallipoli and its part in the subsequent expansion of the AIF in Egypt. He was promoted to major-general in December 1915, and made CB the following month.

Rejecting offers to command one of the infantry divisions due to move to France, Chauvel chose to stay in Egypt as GOC of the Anzac Mounted Division. Retraining and regrouping were still underway when Turkish forces attacked British outposts guarding the northern approach to the Suez Canal on 23 April 1916. Chauvel's force stopped the Turkish advance at Romani (4–5 August), but lacked the strength to destroy the retreating Turks. Nevertheless the battle marked the beginning of a series of victories which ultimately drove the Turks from Egypt and Sinai. At Romani 4000 Turks were taken prisoner, while more than 1200 were killed. British forces in Egypt were reorganised in December 1916, those along the Canal and in Sinai becoming the Eastern Frontier Force (usually called the Eastern Force), while the advanced forces in Sinai were renamed the Desert Column, commanded by Lieutenant-General Sir Philip Chetwode with Chauvel as his second in command. Successes against the Turks at Magdhaba (23 December) and Rafa (9 January 1917) opened the way for the final expulsion of the Turks from Sinai. After Rafa, Chauvel was made KCMG, having failed to be awarded any honour for his pre-eminent role in the victory at Romani. Two attempts to take Gaza were unsuccessful, the first (26–27 March) having to be aborted when Chauvel's forces were ordered to withdraw in the face of the expectation of Turkish reinforcements, while the second (17–19 April), in which Chauvel's forces played only a minor part, was decisively repulsed by the Turks. As a result of these setbacks, the commander of the Eastern Force was replaced by

Chetwode, and Chauvel became GOC of the Desert Column, thus becoming the first Australian to command a corps.

Shortly afterwards General Sir Edmund Allenby became C-in-C, Egyptian Expeditionary Force, and reorganised the army into three corps. The Desert Mounted Corps, of three divisions, was commanded by Chauvel, who was promoted to lieutenant-general in August 1917, the first Australian officer to attain that rank. When Allenby launched a major offensive against the Turks on 31 October, Chauvel's role was to attack Beersheba (q.v.) from the east, seize its vital water supplies, and then threaten the Turkish forces as they withdrew in the face of the advancing Eastern Force. A charge of the 4th and 12th Light Horse Regiments, the men using their bayonets as swords, overwhelmed the Turkish defences with light losses (53 killed, 144 wounded) incurred by the attacking force. It was a major blow to Turkish morale, and even though the toll of the advance to Beersheba had so exhausted his troops that Chauvel was unable to deliver the planned decisive blow, the Desert Mounted Corps continued to play a major part in Allenby's drive to Jerusalem, which was taken by British infantry forces on 9 December. For his role in these victories Chauvel was made KCB in the 1918 New Year's Honours list.

In the spring of 1918 there was a major reorganisation of British forces in the Middle East and on the Western Front, and as a result Chauvel's Desert Mounted Corps gained a fourth division (the 5th Cavalry Division). Allenby launched another offensive in April into the Jordan Valley. Chauvel's force of only two divisions plus the Camel Brigade was unsuccessful in seizing and holding its objectives in the face of determined Turkish resistance and was forced to withdraw, but the persistence of Chauvel's attempts had the important effect of convincing the Turks that further British attacks would be made in the same area and by the same troops. Five months later, on 19 September, Allenby launched his second offensive. Chauvel secretly moved three cavalry divisions to a position just north of Jaffa, and was thus placed, in the battle of Megiddo (19–25 September), to destroy all those Turkish forces which were unable to cross the Jordan. Allenby attributed 'the completeness of our victory … to the Desert Mounted Corps under General Chauvel'. Giving the Turks no respite, Chauvel's Anzac Mounted Division pursued the retreating enemy forces, which Chauvel destroyed around Haifa and Lake Tiberias before pushing on to Damascus which he took on 1 October. After a short rest to restore his exhausted troops, who in the advance had suffered badly from malaria and typhus, he captured Aleppo in northern Syria on 25 October; five days later the Turks signed an armistice and the war in the Middle East ended.

Chauvel returned to Australia in late 1919 and was appointed Inspector-General, a position he held continuously until 1930, simultaneously being CGS from 1923. This was a period of financial stringency and military contraction. Although the government had accepted the recommendation of Chauvel and a number of other senior generals in 1920 that Australia should plan for a field force of two cavalry and six infantry divisions, with half the manpower to be provided through the compulsory training scheme, it imposed severe financial restrictions in 1922, after the new scheme had been in operation for only a year, arguing that the Washington Naval Conference had brought about a marked improvement in the international situation. The compulsory scheme was suspended the same year, on financial grounds. Chauvel insisted that the divisional structure be maintained, if only in skeleton form, but continuing reductions in army expenditure, and the government's policy of restricting the army's role to one of local defence against minor raids, made it a dispiriting period in the army's short history. Faced with these circumstances Chauvel sought to maintain the standards of Australian staff officers (whose work had played an essential part in his successes in the Middle East) by arranging regular postings to British staff colleges and the Imperial Defence College (q.v.). In 1924 he was able to persuade the government to increase the training requirements under the compulsory scheme, but was not consulted five years later when the Scullin Labor government replaced the compulsory scheme with a much smaller volunteer force. Nevertheless Chauvel worked tirelessly in support of the new scheme and could claim some credit for the success it enjoyed in its early period. He was promoted to general (again the first Australian to reach that rank) in November 1929, and retired in April 1930. For the next decade he was active in returned servicemen's circles. In 1940, at the age of 75, he was appointed Inspector-General of the Volunteer Defence Corps (q.v.), a position he held, despite failing health from 1942, until his death in 1945.

Chauvel was undoubtedly one of the leading commanders that Australia has produced. For all the romance associated with the Desert Mounted Corps, with its reputation for courage and dash, Chauvel's successes were based as much as anything on a close attention to staff work. Always conscious of the smallness of his force, he was loath to risk heavy casualties that were disproportionate to the objectives, even if on several occasions this brought him criticism for failing to act decisively. In his period as Inspector-General and CGS he fought to preseve the essentials of the army — a trained officer corps — and it could well be argued that when the need arose for such men from 1939, it had been Chauvel who had ensured that they would be available to meet the developing crisis.

A. J. Hill, *Chauvel of the Light Horse: A Biography of General Sir Harry Chauvel, G.C.M.G., K.C.B.* (Melbourne University Press, Melbourne, 1978).

CHEMICAL WARFARE see **AGENT ORANGE; GAS, FIRST WORLD WAR; GAS, SECOND WORLD WAR**

CHIEFS OF STAFF The chiefs of staff are the professional heads of the three services, and occupy the most senior single-service positions in the defence structure. The first Chief of Naval Staff (CNS), Rear-Admiral Sir William Cresswell (q.v.), was appointed at the founding of the RAN in 1911, and the appointment of the first Chief of the Air Staff (CAS), Wing Commander Richard Williams (q.v.), occurred with the founding of a separate air service in 1921. The first Chief of the General Staff (CGS), Colonel William Throsby Bridges (q.v.), held his appointment from January 1909 following a reorganisation of both the Australian and imperial armies and the introduction of a general staff system. With the exception of Admiral Sir George Hyde (q.v.), whose appointment covered the period 1931–37, the CNSs from Cresswell until after the Second World War were all officers of the RN, further accentuating the dependence of the Australian service on its imperial partner. The great majority of CGSs were Australians, a difference explained by the much larger and more independent role that the AIF played in the First World War, which produced a great many highly experienced and relatively senior army officers who were then equipped to lead the army in the interwar period. Between the wars the air force was led by just two officers, Williams and Air Commodore S. J. Goble (q.v.), both Australians. On only two occasions, at the outbreak of the Second World War and between 1952–54, following Air Marshal Sir George Jones's (q.v.) very long tenure of office, has the RAAF drawn its professional head from the British service.

The chiefs of staff presided over their respective service boards of administration and were responsible for their service to the government of the day, to whom they provided advice concerning that service. Until the early 1960s the CGS was also styled the Australian representative on the Imperial General Staff, and during the interwar years liaised closely with his British opposite number. The CNS, usually a British officer in any case, kept in close touch with British policy and strategic thinking throughout this period, and can scarcely be considered to have acted with much independence. Although the three chiefs of staff could and did act corporately on occasions, there was no formal mechanism for this, and on various occasions they came into sharp conflict, especially over roles and missions and budget allocations. On at least two occasions between the world wars, the army and navy attempted to dismember the air force and divide its assets and capabilities between themselves.

The Chiefs of Staff Committee was established by a cabinet decision in September 1939 at the same time as the War Cabinet (see AUSTRALIAN WAR CABINET). Its function was to provide advice on operational matters and strategic appreciations, and its members customarily attended the War Cabinet, although not by right. The chiefs of staff were also responsible for the operations of the services and for issuing instructions based on War Cabinet decisions. In practice, however, and especially after the beginning of the Pacific War, they enjoyed less real authority than was suggested by their place in the higher defence machinery: they exercised no responsibility for operations and increasingly found themselves confined to the administration of their individual services, a situation made explicit in the case of the army by General Sir Thomas Blamey's (q.v.) appointment in 1942 as C-in-C, AMF, and for the air force by the creation of RAAF Command under Air Vice-Marshal William Bostock (q.v.). They were much less powerful than their British counterparts, a fact made plain at the end of the war when the Chief of the Imperial General Staff, Field Marshal Sir Alan Brooke, visited Australia and commented that although 'most able', the Secretary of the Department of Defence, Frederick Shedden (q.v.), 'has been allowed to assume far too much power' relative to the chiefs of staff.

In the postwar period the Chiefs of Staff Committee was increasingly displaced as the principal source of defence advice to government by the Defence Committee, which included the three service chiefs in its membership. The Chiefs of Staff Committee was now largely confined to consideration of more narrowly military matters, but there was some dissatisfaction with its workings nonetheless, and in 1957 the Morshead Committee (q.v.) recommended that a distinct position of Chairman, Chiefs of Staff Committee, be created and that this officer was to be the principal source of military advice to the minister for Defence. He was drawn from among the service chiefs themselves, and from March 1965 when the third incumbent, Sir Frederick Scherger (q.v.), was promoted to Air Chief Marshal, held four-star rank and was thus senior to his colleagues. In practice, however, the system still had serious limitations. The chairman's was a non-statutory position, which meant that he could not command the armed forces in any sense; during the Vietnam War, the Commander, Australian Force Vietnam, was answerable to the Chiefs of Staff Committee although he received guidelines from the chairman.

The 1973–76 reorganisation associated with the Secretary of the Department of Defence, Sir Arthur Tange (q.v.), established a joint central staff within the department and the Chairman, Chiefs of Staff Committee, was replaced by the Chief of the Defence Force Staff, who now commanded the forces, although he continued to do so through the individual service chiefs. In October 1984 the title was amended to Chief of the Defence Force (CDF), and his military staff was reorganised as Headquarters, Australian Defence Force. From July 1985 the CDF commanded the forces through three environmental headquarters, Maritime, Land and Air, each commanded by a two-star officer. The chiefs of staff were thus removed from the chain of command for operations, much as they had been during the Second World War. They remain responsible for the raising, training, equipping, and maintenance of the services, and through the Chiefs of Staff Committee still maintain a strong indirect influence on the conduct of operations, as the deployment to the Gulf War of 1990–91 demonstrated.

CHILDERS, HMVS see **VICTORIAN NAVY SHIPS**

CHILTON, Brigadier Frederick Oliver (23 July 1905–). Chilton was commissioned into the Sydney University Regiment in 1926 while studying law, becoming a major by the time he was 31 and serving in both regimental and part-time staff positions. At the outbreak of war in 1939 he was seconded to the AIF and served as a junior major in the 2/2nd Battalion; by the time the battalion went into action, at Bardia in January 1941, Chilton had been promoted and was the unit's commanding officer. For his leadership in the 6th Division's first battle he was awarded the DSO. With Lieutenant-Colonel Ivan Dougherty (q.v.), commanding the 2/4th Battalion, he was one of the youngest battalion commanders in the 2nd AIF.

Chilton led the battalion through the first Libyan campaign and then in Greece, escaping through Turkey and thence to Palestine after the Allied defeat. Having passed through the staff college at Haifa, he was next appointed as GSO2 on the headquarters of I Australian Corps, and returned with the AIF to Australia in May 1942. He went to New Guinea Force (q.v.) the following month, and during the Battle of Milne Bay was Major-General Cyril Clowes's (q.v.) senior staff officer. By the war's mid-point he had thus commanded a battalion against the Germans and Italians and had held senior staff positions in critical battles against the Japanese.

After three months as chief instructor at the Land Headquarters Tactical School, Chilton was promoted in March 1943 and given command of the 18th Brigade, an appointment he held until the end of the war. The brigade took part in the Ramu Valley battles of the New Guinea campaign, including the vicious fighting for Shaggy Ridge in early 1944, for which Chilton received a bar to his DSO. He then led the brigade in the Balikpapan campaign in Borneo in 1945.

After the war he was appointed to a senior position in the expanded Department of Defence, becoming Deputy Secretary in 1950, a position which he held for eight years. Leaving the Defence department in 1958, he became chairman of the Repatriation Commission. He was awarded the CBE in 1963 and was knighted in 1969. He retired from the Repatriation Commission in 1970.

Chilton is widely regarded as one of the outstanding leaders of the 2nd AIF. The official historian, Gavin Long (q.v.), thought him 'meticulously honest, cool and determined above the average, a well-educated officer and a conscientious leader'.

CHINA CONTINGENT see BOXER REBELLION

CHINOOK, BOEING VERTOL (2-seat medium-heavy transport helicopter). Rotor diameter 18.29 m; length 15.54 m; load 9700 kg, 33 troops or 24 stretchers; maximum speed 235 km/h; range 185 km with full payload; power 2 × Lycoming 3750 shaft h.p. turboshaft engines.

The twin-rotor Chinook transport helicopter was used in the RAAF by No. 12 Squadron at Amberley (q.v.) between 1974 and 1989. They were never used in action but were used often during emergencies (see CIVIL POWER, AID TO). The Chinook can carry heavy loads such as vehicles or other helicopters slung from a large hook beneath the fuselage. Four upgraded Chinooks entered service with the Australian Army Aviation Corps (q.v.) in 1995.

'CHOCKO' (also 'choco', 'choc') is an abbreviation of the term 'chocolate soldier', meaning a soldier who will not fight, which comes from a character in George Bernard Shaw's play *Arms and the Man* (1898) who carries chocolates in his pack instead of ammunition. In the Australian context it was originally applied during the First World War to soldiers of the 8th Infantry Brigade of the AIF, who

arrived in Egypt after the end of the Gallipoli campaign in December 1915. In the Second World War it referred to militiamen or conscripts who could not serve outside Australia before 1943 (see CONSCRIPTION). Such men were also, less commonly, called 'koalas' because under Australian law it was forbidden to shoot at these animals or to export them. In both wars these terms, although originally used derisively, came to be embraced by many of the men who were so labelled.

CITIZEN MILITARY FORCES (CMF) In the twentieth century the CMF, or militia as it has been alternatively known, has made up the main part of the Australian Army, at least in terms of numbers. Until after the Second World War, it was intended to provide the main land force for use in the defence of Australia, although in fact at the beginning of both world wars a separate expeditionary force, the AIF, had to be raised for overseas service. In March 1901 the new Commonwealth government took over control of defence matters, and inherited the various colonial military forces (q.v.) which, at 31 December 1901, were slightly under their establishment of 1665 officers and 28 385 other ranks, by a total of 185 and 2512 respectively. Of these, the Permanent Forces (staff, field and garrison artillery, and engineers, medical and service corps units) numbered just 115 officers and 1323 other ranks; the militia and volunteers numbered 1365 officers and 24 550 other ranks. For the first four years of the Commonwealth the CMF was administered by the GOC, Australian Military Forces, Major-General E. T. H. Hutton (q.v.), assisted by a small headquarters staff. With the passage of the *Defence Act 1903* (q.v.), former colonial units were redesignated as Commonwealth ones, in some cases being created from the merger of existing battalions. Thus the 1st Regiment of the New South Wales Infantry became the 1st Australian Infantry Regiment, the 1st Battalion of the Militia Infantry Brigade (Victoria) became the 5th Australian Infantry Regiment, the 1st and 2nd Battalions of the Western Australian Infantry came together to form the 11th Australian Infantry Regiment (Perth Regiment), the 3rd Battalion of the Tasmanian Infantry Regiment became the Tasmanian Rangers (Volunteers), and so on around the country. Light horse (q.v.) and mounted infantry units underwent the same process. The 16th Australian Light Horse Regiment (Militia) was formed from the South Australian Mounted Rifles, the 13th Light Horse was formed from the 1st Battalion of the Queensland Mounted Infantry and the 4th Infantry Regiment, while the 7th and 12th Light Horse were raised from elements of the Australian Horse which had been activated for service in the Boer War after Federation.

This reorganisation was undertaken by Hutton, against sometimes concerted opposition on the part of older colonial units like the New South Wales Lancers, which resented losing their identities. Hutton himself was an enthusiast for mounted infantry after his experience in South Africa, and his expansion of the mounted arm caused some resentment among existing citizen soldiers, as in order to accommodate new mounted units on the order of battle (and within the budget), existing infantry units, especially in country towns, would have to be disbanded; men who wanted to continue with citizen soldiering would then be required to purchase a horse if they wanted to enlist in the new regiment. The relationship between militia and regular which was to be a feature of the Australian Army for nearly half a century was set from the beginning: the Royal Australian Regiment of Artillery provided the garrison for defended localities such as Sydney, Newcastle and Perth, while the remainder of the permanent force provided instruction and administration for the citizen forces. Hutton's plan, modified and put into effect from 1905, organised the army into a field force and garrison troops. The field force consisted of five light horse brigades, two brigades of infantry and four mixed brigades, furnished from the militia. Enlistment was voluntary and service was for a three-year engagement. The citizen forces themselves maintained a distinction between the militia, who were paid, and the Volunteers, who were not, and turnover in the latter was always high, few Volunteers serving for more than three years. (The Volunteers disappeared at the end of 1910, partly in recognition of the fact that under the provisions of the Defence Act they were now subject to the same discipline as the militia, and might as well be paid like them.) New units of light horse, infantry, artillery and engineers were raised between 1906 and 1912, and there was some reorganisation and renumbering of existing ones, and in the latter year all fit, eligible adult males between the ages of 18 and 26 served in a compulsory militia, a scheme first announced in December 1907, enacted in 1909 but only proclaimed in 1911 (see CONSCRIPTION).

The scheme, reported on by Lord Kitchener during his tour of inspection in late 1909, called for a force of 28 light horse regiments, 56 field artillery batteries and 92 infantry battalions (although it never reached this strength). The obligation for adult service lasted eight years, commencing on 1 July each year, and involved periods of home training and continuous training at annual camps. Men who had enlisted under the old voluntary scheme were paid 8 shillings a day while on service, whereas those serving under the new scheme were entitled to only 3 shillings a day in the first year and 4 shillings a day thereafter. Light horsemen received £4 a year for upkeep of the horse (which they had to provide themselves). Officers received 15 shillings a day as lieutenants, 22 shillings and sixpence as captains rising to 45 shillings for colonels. Organisation was territorial, and divided into divisional, brigade, battalion and training areas. The 92 battalion areas formed 23 brigades, with three brigades to a division. Each area furnished a battalion, a proportion of other troops, and for administrative purposes each area was subdivided into two or three training districts. Training was supervised by instructors from the Administrative and Instructional Staff of the Permanent Force, which numbered 210 officers and 779 warrant and NCOs, and 214 area officers, who held temporary commissions and also had responsibility for the training of senior cadets. The reorganised citizen forces had only been established for a couple of years when the First World War broke out, and in any case the Defence Act precluded overseas service on other than a voluntary basis. Major-General W.T. Bridges (q.v.) had intended that half the initial contingent of the AIF should be drawn from the citizen forces, but in the haste with which it was raised and dispatched this was not achieved, although a significant minority of the early enlistments were either current or former militia soldiers. Of the 631 officers of the 1st Division, 402 came from the militia and another 58 had received commissions under the new compulsory scheme. The militia thus influenced the shape of the AIF, even though no militia unit served overseas. (A partial exception to this was provided by 500 members of a militia battalion in Queensland, who volunteered for service with the Australian Naval and Military Expeditionary Force to German New Guinea in 1914. Accepted initially, the force commander, Colonel William Holmes [q.v.], eventually left them behind because they were poorly equipped and largely untrained.) The forgotten army of the First World War was the citizen one which stayed at home for the defence of Australia, and which at the war's end in 1918 numbered over 105 000 men. Compulsory training, however, was largely suspended for the duration in 1915. It was revived with the war's end, but did not long survive the pressures of the peace.

A committee comprising most of the country's distinguished senior generals from the AIF was convened by the Minister for Defence in January 1920, and its report recommended a postwar force structure of two cavalry and four infantry divisions, additional troops for local defence (the equivalent of the old garrison troops) amounting to a fifth division, personnel for coastal defence establishments, and appropriate allocations of corps and army troops for a total force of 130 000 men in peacetime and 270 000 in the event of war. The unit numbers and identities of the AIF battalions were transferred to the citizen force when the AIF was officially disbanded in 1921, but within a year the government had reduced the peacetime establishment of the army from 128 000 in 1921 to 37 000 in 1922, allegedly in response to moves towards disarmament associated with the Washington conference on naval arms limitation. The divisional structure was maintained, but in cadre form only, and in his report to parliament in 1922 the Inspector-General (q.v.), Lieutenant-General Sir Harry Chauvel (q.v.), noted that the army now retained a capability to field a force for the defence of the nation only if sufficient lead time existed to train the rank and file. Compulsory training in the 1920s was an attenuated affair. It was abandoned entirely in rural Australia, and the age of eligibility was narrowed to those aged from 18 to 19 years. The reductions led to stunted career opportunities for Staff Corps officers, which was to create a gulf epitomised by the 'two armies' policy of the Second World War and led to internecine feuding between Staff Corps and militia officers throughout the 1930s and 1940s. But militia soldiers suffered too. Militia units found it hard to attract and keep recruits, turnover was high, rates of pay poor, training opportunities few and modern equipment increasingly scarce. The official historian, Gavin Long (q.v.), noted that service in the militia between the wars 'conferred little prestige; indeed, an Australian who made the militia a hobby was likely to be regarded by his acquaintances as a peculiar fellow with an eccentric taste for uniforms and the exercise of petty authority'. The strength and composition of units fluctuated wildly, and this remained a problem even as war approached again in the 1930s. Major-General Sir Jack Stevens (q.v.), later to command the 6th Division in New Guinea and in the 1930s commanding officer of the 57/60th Battalion, wrote of the period 1935–39 that 'at our lowest ebb we went into one annual camp at Seymour only 120 strong, making training impossible'. In the reductions of 1921 five infantry battalions, two light horse regiments and two field artillery batteries were disbanded; in July 1930, after the Scullin government abolished compulsory training altogether, another two light horse regiments and nine infantry battalions disappeared. So serious was the manning problem in the early 1930s that a lower peacetime establishment was

introduced. But although the citizen soldiers monopolised the command of the forces on which Australia was supposed to rely in the event of war, the senior administrative posts — membership of the Military Board (q.v.) and command of the six district bases, for example — were held by regulars. There was no CMF member of the Military Board until February 1948.

With war imminent, on 2 September 1939 the CMF was called up for war service in successive drafts of 10 000 men, to serve for 16 days at a time. Before the first draft had completed its obligation, the government decided to call up the whole of the militia in two drafts of 40 000 men, for a month's training, to be followed by a further period of three months' training from January 1940. In October 1939 compulsory training was reintroduced, for home service only. The government decided that half of the new AIF should be drawn from volunteers from the militia, but the actual proportion of such enlistments in the 6th Division was less than a quarter. There were several reasons for this. The daily rate of pay for an unmarried private was lower in the AIF than the CMF, and the government actually appeared to talk down militia volunteering in the early months of the war: on 13 October the Minister for Defence, G. A. Street (q.v.), commented in a radio broadcast that 'the withdrawal of militiamen from industry had created a serious problem for employers'. Many militiamen held back because they hoped to serve with their existing units in the expectation that these might be sent, while some commanding officers declined to release their best junior officers, and did not enlist themselves because they would be forced to take a reduction in rank. Others, like Lieutenant-Colonel Ivan Dougherty (q.v.), cheerfully accepted a reduction (in his case to major and second-in-command of the 2/2nd Battalion) in order to get away to the war. With many of the youngest and best officers, both militia and regular, sailing for Palestine, and with the equipment needs of the AIF postponing the planned modernisation of the militia, after the flurry of activity in the opening months of the war the CMF settled back into a degree of lassitude which was to be broken in late 1941 by the impending war in the Pacific. Lieutenant-General Iven Mackay (q.v.) returned as C-in-C, Home Forces, to revitalise measures for the defence of Australia, and after the Japanese attacks on Pearl Harbor and Malaya a number of middle-ranking and senior officers from the AIF were returned to reinvigorate the training and leadership of CMF formations. Older CMF officers, many of whom had rendered good service in difficult circumstances, were retired, sometimes peremptorily, and their places taken by men with experience of the modern battlefield. When Brigadier H. C. H. Robertson (q.v.) was posted, on promotion, to take command of the 1st Cavalry Division, then in the process of mechanisation, he succeeded Major-General J. D. Richardson, a militia officer, whose first notice of his supersession was the arrival of his replacement. Militia units were deployed to Papua in combat against the Japanese and, after mid-1943 and the extension of the area to which compulsorily enlisted soldiers might be deployed, to New Guinea, and in the course of the war three militia formations served overseas, the 3rd, 5th and 11th Divisions. The extension of the area to which militiamen might be deployed had been a source of friction both between the CMF and AIF (who referred to the former as 'koalas', i.e., not to be exported or shot, or 'chockos' [q.v.]), and within the higher defence machinery. There was a strong feeling within some sections of the government and high command that Australia's position in Allied councils was weakened in circumstances where American conscripts could be sent to die in defence of Australia while Australian conscripts could not. The CMF produced some of the ablest commanders in the Australian Army, Dougherty, Chilton and Eather (qq.v.) in particular, and in the course of the war a quarter of a million men served in the CMF, not including the 200 000 who voluntarily transferred to the AIF.

The strategic lessons of the Second World War were reflected in plans for the shape of the postwar army, drawn up in 1946–47, and from these emerged the modern regular army and a reversed relationship between the regulars and the militia. Henceforth the CMF existed to support the regulars, not the other way round. The postwar history of the CMF can be divided into a number of phases. Re-established on a voluntary basis in 1948, by July 1949 it had reached a strength of 16 202, well below the target establishment of 50 000. Pay rates were now the same for both regular and citizen soldiers, and CMF obligation involved attendance at a 14-day camp and 24 days' home training a year (only 12 days of which were obligatory). With the change in government in December 1949 came further changes in defence policy, and that which had the most impact on the CMF was the decision to reintroduce national service for home defence. Beginning in 1951, 18-year-old males were called-up for 98 days' training in the regular army, to be followed by 14 days' continuous training and a further 12 days home training in each of the succeeding three years. National servicemen were eligible for full-time call-up in the CMF in the event of war. The strength of the

CMF increased rapidly, reaching 87 291 in 1956. But the scheme was in many respects a failure. Voluntary enlistment dropped substantially, from 18 700 in July 1951 to just 6200 by August 1957. The period of initial training was insufficient to turn national servicemen into effective soldiers, while the variance in training standards within CMF units had a deleterious effect on unit efficiency. The scheme was reduced in July 1957 and suspended in 1960, and the CMF reverted to voluntary enlistments once again. It underwent radical reorganisation in 1960 with the introduction of the Pentropic division (q.v.) into the army. The old unit titles were abolished and a new 'Royal' regiment created in each state (Royal New South Wales Regiment, Royal Queensland Regiment, etc.). Of the two Pentropic divisions created, one was entirely CMF in composition, the other partly so; by 1963 there were nine of the new 'battle groups' in the CMF, each consisting of a new Pentropic battalion and supporting arms. There were now two CMF battalions in Queensland, Victoria and New South Wales, and one each in South Australia, Western Australia and Tasmania. The destruction of old units with strong local loyalties and traditions, which in some cases went back before Federation, was widely resented both by former CMF members and by local communities, and initially enlistments declined, from 37 921 in 1960 to 21 958 the following year. More complex organisation and training requirements also placed greater pressure on resources because of the geographic dispersal of sub-units, especially in rural areas. Out of this period, however, came the 'One Army' concept, which emphasised the importance of the CMF in the army organisation in time of war, and this has remained a plank of reserve force policy, even though some older CMF officers believed it little more than a sop to CMF opinion.

With the abolition of the Pentropic structure in 1964 the role of the CMF was broadened to include a capability for call-out in a defence emergency, that is, a state short of war. Members of the CMF could now also volunteer for full-time duty, and this permitted the use of a few CMF personnel (medical officers, for example) with the army in Vietnam. In addition, the CMF could now be sent overseas. New battalions were created in Queensland, Victoria and New South Wales, and although the State regiment system was retained, battalions were given numbers intended to revive local association with former militia units. Thus in Queensland the 1st and 2nd Battalions of the Royal Queensland Regiment, based in Brisbane and Townsville, were joined by the 51st Battalion, based in Cairns and reviving the regimental number of the old Far North Queensland Regiment. In Sydney the 3rd Battalion of the Royal New South Wales Regiment in Sydney became 17RNSWR, the old North Sydney Regiment, while a new 3rd Battalion was raised and based in southern New South Wales, where the old Werriwa Regiment had recruited in two world wars. The reintroduction of national service in November 1964 again altered the composition of the CMF. Young men could opt to serve in the CMF instead of the regular army, but they had to volunteer before the ballot for their age group was held (see CONSCRIPTION). They would then not be called up for regular service and the possibility of being sent to Vietnam, but were required to serve six years. If their birthday was not one called up, they had still to complete their CMF obligations, and if they failed to do so could be called up for full-time duty. Although CMF strength increased, such that by 1968 about half of the 35 000 members of the CMF were 'optees', the scheme had a deleterious effect. Many of these men became ineffective when it transpired that their birthday had not been called, and in some circles the CMF came to be regarded as a haven for draft-dodgers, which lowered morale among genuine CMF men. There was a feeling too that CMF units should have been called up for overseas service, and some resentment that this had not occurred. Enlistments for men other than those exercising the option under the national service scheme declined in the late 1960s.

When the ALP resumed office in December 1972, it abolished national service. In March 1973, acting on a general perception that the CMF suffered from problems of morale, organisation, and, fundamentally, lack of a meaningful role, the Minister for Defence, Lance Barnard (q.v.), instituted a committee of review under the chairmanship of the distinguished academic defence commentator Professor T. B. ('Tom') Millar (see MILLAR COMMITTEE). Among many other recommendations, Millar suggested the renaming of the CMF as the Army Reserve (q.v.), and this was adopted in 1980. Its function remains to augment the regular force in time of war or defence emergency 'so that together they provide the basis for the expansion of the whole army on mobilisation'. Defence policy now emphasises the 'Total Force' concept ('One Army' by another name). The Chief of the Army Reserve is a major-general and there is an Army Reserve Advisory Council within the Army Office committee structure. There has been some integration of reserve and regular formations, and the three regional force surveillance units in northern Australia are entirely reserve, and draw as well upon members of the local Aboriginal population (see

ABORIGINES AND TORRES STRAIT ISLANDERS IN THE ARMED FORCES). As part of the reorganisation and reduction of the regular forces in the early 1990s attention has once again focused on locating certain functions within the Army Reserve, while the Wrigley Report (1990) called for a greatly expanded part-time force, to be called the militia. In the mid-1990s there was no sign that this recommendation would be adopted.

CIVIL–MILITARY RELATIONS involve Assistance to the Civil Community (q.v.) and Aid to the Civil Power (q.v.). Democratic theory insists that the military serve society, and should be at all times under the control of the civilian authorities. This leads to some complexity, however, as civil–military relations involve politicians in parliament, the government, the military, public servants, the media and finally the general public. The attitude of Australian political parties towards the military has changed over the years. Labor introduced compulsory military training for young men in 1909 (see CONSCRIPTION), but during the First World War the party split over the great conscription referendums, which created a trauma that coloured its thinking for the next five decades. Until 1938 it consistently opposed defence expenditure. The Second World War changed this attitude, but the Vietnam War aroused old feelings, and only when it had ended Australian participation in that war could the Whitlam ALP government rationalise higher defence planning. Non-Labor parties have adopted a public stance of caring more for the defence of Australia, and of being on better terms with returned service organisations. They have also had more members with war service or previous military careers. Yet they too have economised on defence, and from the 1980s onwards the military background of members of parliament from all parties has been very similar.

Disagreements between government and military leaders have not been uncommon. Major-General E. T. H. (q.v.), GOC 1902–04, was foiled by parliamentary concerns with economy. Between 1933 and 1939 a dispute arose over reliance on Britain and 'imperial defence', which Australian Army leaders publicly doubted, with the result that the career of Sir John Lavarack (q.v.), CGS in 1935, was temporarily clouded. Similarly, after the Second World War, General Sir Thomas Blamey (q.v.) was summarily replaced, while during the Vietnam War there was tension between the army and the government.

Military policy is controlled by the government through a civilian inner cabinet of the prime minister, minister for Defence, treasurer and minister for

Foreign Affairs. Apart from a short period when a separate Department of the Navy existed, until 1939 there was a single Department of Defence (q.v.). The creation of separate service departments of the army, the navy and the air force (qq.v.) during the Second World War complicated the situation. By the early 1950s it was clear that the Department of Defence was unable to coordinate activities. The Menzies government tried a number of expedients, such as giving more power to the permanent civil heads of the defence and service departments, tightening treasury control through the Defence Division of Treasury (q.v.), and strengthening the authority of the minister for Defence. Menzies would not, however, accept the main recommendation of the 1957 Morshead Committee (q.v.) that all sections be fused into a single Department of Defence under one minister. The situation was not helped by the chiefs of staff (q.v.) remaining in Melbourne until 1959, while the government had ruled from Canberra since 1927.

The whole structure was changed under the Whitlam Labor government. The 1973 report by Sir Arthur Tange (q.v.) led to the *Defence Force Reorganisation Act 1975*. The separate service departments were merged into one department, in which the minister presided over a 'diarchy'. The Chief of the Defence Force Staff (CDFS) represented the military heads and had direct access to the minister, but had to conform to the government's defence policy, especially in bidding for equipment, which was linked to the Cabinet's strategic assessment and the five-year defence program, itself created by mixed teams of service personnel and civilians. The amalgamation was not without difficulties, for the education, experience, interests and working methods of politicians, civilian bureaucrats and the military are very different, and the ramifications of 'administration' and 'command' overlap. The Utz Report, October 1982, noted that weaknesses and misunderstandings remained, with long debates on demarcation issues. In 1984 the CDFS became simply the Chief of the Defence Force (CDF) and gained extra staff, while the departmental structure was simplified. The Cross Report, October 1984, wanted to promote joint command and political coordination, but the government was more concerned with economies, and announced yet another review. The Dibb Report (q.v.), March 1986, again noted conflict between the department and the ADF, mainly over strategic assessments. Defence decision-making is clearly immensely political, being a process of bargaining and compromise. Ironically, the earlier structure with three separate service departments had in some ways produced more open debate. Moreover, in defending Australia the government needs support from the wider community, such as financial experts, economists, scientists, systems, intelligence and strategic analysts, and procurement specialists — not to mention the business community — which has led to the creation of various other structures from time to time both within and outside the ADF. The Strategic and Defence Studies Centre at the Australian National University in Canberra was set up in 1968 to provide feedback between academics and the defence community.

Public attitudes towards the defence forces have varied throughout the twentieth century. There was strong patriotic support at Federation, with the Boer War continuing, newspaper interest in camps and staff rides, and multi-party support for compulsory military training. The dominant ideology was support for the British Empire and acceptance of the 'adventure' or 'glory' of war. The First World War ended this romanticism and polarised Australian society. Nationalists and ex-service organisations such as the RSL (q.v.) came to represent the more patriotic section of the public, opposed by pacifists, humanitarians and the labour movement (see DISARMAMENT). This did not help the Australian military in the interwar years, when the Depression and drastic government economies led to a decline in interest and morale. The Second World War revived military standing in the community, and in 1947 the Australian Regular Army (q.v.) was established, and received increased attention during the Korean War. But soldiers and their officers were not granted high status or public esteem, although, ironically, at the same time the record of Australians in war was being romanticised. Bitter controversy followed over participation in the Vietnam War, with mass demonstrations and occasional hostility to officers in uniform. Nevertheless, Australians fought well and avoided public atrocities, and the status of service personnel gradually increased, perhaps because of the glamorous presentation of them on film and television (see FILM AND TELEVISION, WAR IN AUSTRALIAN; POPULAR CULTURE, WAR IN AUSTRALIAN), or improved public relations. The Australian armed forces have now become an accepted part of the Australian community. They are, however, subject to political control, expected to be loyal servants of the government and to carry out its policies — in the long-standing tradition of parliamentary democracies.

ERIC ANDREWS

CIVIL COMMUNITY, ASSISTANCE TO (ACC) In all countries defence forces are used to help civil authorities, especially during times of natural disaster, such as (in Australia) the great Maitland flood in the Hunter Valley in 1955, and the bushfires around Hobart in 1967. Several arms of the defence force are usually involved: in Hobart the army sent fire-fighting teams and equipment and provided relief accommodation; the RAAF used Hercules (q.v.) transports; and the RAN sent the River Class frigate (q.v.) HMAS *Derwent* and two *Oberon* Class submarines (q.v.) to provide emergency radio links. There was, however, much duplication and confu-sion, and to coordinate future efforts the Whitlam government created the Natural Disasters Organ-isation (NDO) within the Defence Department in 1974. The intention was to use the defence forces to support the States' own civil defence organisa-tions. The crisis that immediately followed, how-ever, when Cyclone Tracy devastated Darwin that December, involved not a State but a Territory, and the timing — just after midnight on Christmas morning — caused problems. The head of the NDO, Major-General Alan Stretton, therefore assumed responsibility himself. The relief opera-tion, the largest in Australian history, became marred by controversy. The RAAF and the RAN transported supplies, equipment and personnel and the army provided a field force of 650 soldiers, but because Stretton wished to use civilians as much as possible to boost Darwin's morale, he refused to allow the defence forces to take over the city. This provoked army hostility and caused confusion as to whether he was acting in a civilian or military capacity. He offended several ministers, the Secre-tary of the Department of Defence Sir Arthur Tange (q.v.), and two consecutive chiefs of staff, Lieutenant-General Sir Francis Hassett (q.v.) and especially Lieutenant-General Sir Arthur MacDon-ald (q.v.), with whom Stretton had stand-up ar-guments. As a result, the States' Civil Defence Organisations were renamed State Emergency Ser-vices (SES) and were expanded and coordinated.

In 1978 new army regulations divided ACC into five categories based on the seriousness of the emer-gency, with greater initiative being given to local commanders in more pressing situations. Defence force personnel were to act in support of the SES, yet controlled wherever possible under the normal army groupings. Army command produced PLAN CONSUL, whereby 100 men and equipment (taken from headquarters in rotation) were placed on stand-by. Despite this, similar problems occurred when an earthquake struck Newcastle, the biggest non-capital city in Australia, on 28 December 1989 — once again during the Australian Christmas hol-iday period. With electricity and telephone com-munications disrupted, the threat of looting and fear of aftershocks, Major B. Jordan, Army Reserve, took charge as the senior engineer officer on the spot.

Although he was supported on a visit by the Deputy Prime Minister Lionel Bowen, State politi-cians, local police and the RAAF Base at Williamtown (q.v.), his initiative roused concern at Headquarters 2nd Military District (see MILITARY DISTRICTS), anxious that he follow the procedures in the ACC pamphlet. All authorities, civil and mil-itary, seem to have underestimated the urgency of the situation. PLAN CONSUL was replaced by PLAN EVERGREEN, but the details remained similar and, as the title implies, still focused on slowly developing threats like bushfires. The ADF response to the 1994 Sydney bushfires was cred-itable: 400 men at a time with engineer equipment; 38 helicopters flying 800 hours; and controlled from Land Command (see COMMANDS, ARMY), but with local commanders protecting their own areas. The amount of initiative a local commander can exercise in an emergency remains a problem, but the ADF is clearly a much-needed bulwark of the Australian community during natural disasters, and gains great public support in the execution of that role.

ERIC ANDREWS

CIVIL CONSTRUCTIONAL CORPS was established on 14 April 1942 to supply labour for works undertaken by the Allied Works Council. These works included docks, aerodromes, roads, gun emplacements, stores, barracks, fuel storage installa-tions, hospitals, workshops, pipelines, wharves, mills and factories. The corps was under the control of the Director-General of Allied Works, and consisted both of volunteers and of men directed to serve in it by the Director-General and his agents. All men between the ages of 18 and 60 could be made to serve in the corps except for those employed in protected industries or undertakings, those serving in the defence forces of Australia or its allies, and diplomatic representatives or staff in Australia. Of the 53 500 men who served in the corps all over Australia to June 1943, about 8 500 were volun-teers, about 28 000 had already been working for the Allied Works Council when they were enrolled, and about 16 600 had been called up for service. Corps members' pay was based on civilian award rates, but they did not enjoy the rights of ordinary civilian workers, as they could not refuse work and their conduct on the job was regulated. The corps was abolished on 1 July 1946.

CIVIL POWER, AID TO THE (ACP) Governments use military forces, not only against external ene-mies, but also to maintain order in their own coun-try. This creates problems for democracies, which traditionally allow their citizens considerable free-dom. This tension was seen in Australia during the

First World War, when the army controlled censorship (q.v.), watching Australians of German origin (see ALIENS, WARTIME TREATMENT OF) and Australian opponents of Prime Minister W. M. Hughes (q.v.). Its most famous action occurred in 1942 during the Second World War, when it arrested members of the Australia First movement. This role was ended, however, in 1949, when the Australian Security and Intelligence Organisation was established. The military may be also used to support the police during strikes. Colonial military forces (q.v.) were used in this way during the nineteenth century, but in 1901 military forces came under the federal government, which was reluctant to agree to the States' appeals for help. The other use of the military — to perform the work of strikers — was unpopular and usually impractical, and was avoided until the New South Wales miners' strike in 1949. It was only made possible then by the peculiar situation at the time, with the fear of communism, strong public support for action and a divided union movement. Similarly Malcolm Fraser's (q.v.) Liberal–National Country Party government in 1981 and Bob Hawke's ALP government in 1989, both provoked by long-running airline strikes, used RAAF Hercules (q.v.) transports to lift stranded airline passengers. Soldiers have not been used against riots in Australia — apart from a parade of strength by troops and armoured cars against unemployed Italian migrants in Bonegilla camp in July 1952.

The 1964 ACP pamphlet stressed the primacy of civilian authorities, that troops should only operate when called out by the Governor-General, and that minimum force should be employed. Fear of terrorism overcame this caution in February 1978, when a bomb exploded outside the Sydney Hilton Hotel, then housing delegates to a Commonwealth Heads of Government Regional Meeting. The government called out troops and helicopters in a major military operation. This roused hostility among journalists (provided with misinformation), lawyers and others. Two inquiries were held. Sir Victor Windeyer (q.v.), a retired High Court judge, considered the powers and rights of the troops, suggesting only modifications which concentrated on their conduct. Call-out procedures were revised in September 1978. Mr Justice Hope's review of protective security in May 1979 discussed the legal issues in the relationship between the military and civil authorities when ACP was invoked. A new pamphlet on ACP in 1983 encompassed anti-terrorist work and became more selective in the use of force, which was to be carefully graded and controlled at all times. Commanders were to act under the direction of the minister of Defence, and troops

remained personally responsible at law for their actions. Liaison with the civil authorities and the need for good public relations were stressed. Nevertheless, three Ananda Marga suspects wanted for the Hilton bombing were arrested in a spectacular ambush, and in November 1983 men from the army instruction team attached to the Australian Secret Intelligence Service were involved in a violent raid on the Melbourne Sheraton Hotel. Public criticism, however, has meant that the incident has not been repeated. The Special Air Service Regiment (q.v.) has been given the additional job of counter-terrorist work, and has been doubled in size and modelled on its British counterpart. Its use is also carefully regulated in a series of three steps, which allow civilian control, while the regiment remains answerable for its actions at law and responsible to army command and the CGS (see CHIEFS OF STAFF). It has not been used often, and never with the full force available. Australian politicians and military have approached ACP cautiously, and Australia has no recorded experience of martial law. This is not surprising: State police forces have ample power to deal with demonstrations, commonly carry arms, and have used them in the past, as the shooting of Norman Brown by the New South Wales police during a Rothbury riot in 1929 revealed. The Australian Federal Police have also recently adopted a more aggressive role. Military leaders fear they will lose public support if service personnel are used too readily or brutally in actions against their own compatriots. They are therefore happy to assume that in Australia the maintenance of law and order is usually a police matter.

ERIC ANDREWS

CLAYMORE MINE is a weapon used to prevent the movement of infantry and non-armoured vehicles such as jeeps or trucks. The American-designed M18A1 mine currently in service with the Australian Army is rectangular in shape, 20 cm long, 3.5 cm thick and 8 cm high, is made of fibreglass and weighs 1.5 kg. Inside the mine are 700 steel balls, a layer of explosive, and two detonators which can be fired remotely or fitted with trip wires. On detonation the steel fragments are propelled in a fan-shaped pattern 2 metres high and 50 metres wide. These fragments will inflict lethal casualties at 50 metres and severe injuries at 100 metres.

CLOWES, Major-General Cyril Albert (11 March 1892–19 May 1968). Clowes and his brother, Norman, were in the first graduating class from RMC Duntroon in August 1914. Posted to the 1st Field Artillery Brigade, he served on Gallipoli from the landing, being wounded on the first day while directing naval gunfire support. In June 1916, while serving in France as a divisional trench mortar officer with the 2nd Division, he was awarded the MC for assisting raiding parties under enemy fire. Twice mentioned in despatches during the war, he earned the DSO for work with the artillery at Villers-Bretonneux in August 1918.

Back in Australia after the war he spent five years as an instructor at RMC Duntroon between 1920 and 1925, before filling a round of staff and training positions in Brisbane, Sydney and Darwin. He attended the gunnery staff course at Larkhill in Britain in 1936–38, and was appointed to command the 6th Military District (Tasmania) in August 1939.

Highly regarded by his contemporaries, he was seconded to the AIF in early 1940 and appointed artillery commander of I Australian Corps. He distinguished himself in command at Pinios Gorge during the Greek campaign and was one of the senior officers of the AIF brought back to Australia ahead of the main force in January 1942 to help prepare Australia's defences against the Japanese. Clowes's finest hour came in August that year. Sent to command 'C' or Milne Force just four days before the Japanese attempted a landing at Milne Bay in Papua, Clowes skilfully orchestrated ground and air forces in shattering the invasion convoys and the Japanese landing parties. He had a number of advantages tactically, but faced numerous problems with weather, terrain, the lack of a workable signals net, poor maps, and interference from higher headquarters. Together with the developing fight along the Kokoda Track (q.v.), the battle at Milne Bay was an early and locally decisive setback for the advancing Japanese (see NEW GUINEA CAMPAIGN).

The rest of the war was an anticlimax. Clowes received few accolades at the time for his victory, and much criticism, generally uninformed. He seems to have suffered in General T. A. Blamey's (q.v.) estimation because of his close friendship with General S. F. Rowell (q.v.). Returning from leave in December 1942, he was debilitated by malaria. He commanded Milne Force, now redesignated the 11th Division, until September the following year, but his division had now been bypassed by the allied advance. For the rest of the war he was GOC, Victorian Lines of Communication Area. He served briefly as Adjutant-General in 1946 before appointment to command Southern Command between 1946 and 1949. He retired in June 1949.

Clowes's career in the Second World War failed to live up to the promise established during the First, and he appears to have fallen victim to the feuding and rivalry within the senior ranks of the AIF and to command pressures exerted by a nervous Supreme Commander, MacArthur, in his headquarters in Brisbane. His achievement at Milne Bay was considerable, however, and was an important step in the process by which the Australian Army demonstrated its ascendancy over the Japanese.

CMF see **CITIZEN MILITARY FORCES**

COASTWATCHERS were originally unpaid civilian coastal residents who were to report on shipping movements and suspicious events in time of war. Formed following a 1922 Inter-Services Committee's decision, the organisation was placed under the control of the Naval Intelligence Division (see INTELLIGENCE). Until the 1930s the scheme could only cover the more densely populated coastal areas where people could pass messages by telegraph, but the invention of the pedal radio allowed it to be extended to the more remote coasts of northern Australia and by 1939 there were over 700 coastwatchers in Australia, Papua, New Guinea and the Solomon Islands. On the outbreak of war, Director of Naval Intelligence R. B. M. Long (q.v.) appointed Lieutenant-Commander Eric Feldt to recruit more coastwatchers from among the planters, missionaries and administrators on the islands to Australia's north and north-east. By mid-1941 Feldt had completed this task and was supervising more than 100 coastwatching stations in a 4 000 kilometre crescent from the western border of Papua New Guinea to Vila in the New Hebrides (Vanuatu). Intelligence officers stationed at Fremantle, Darwin, Thursday Island, Port Moresby, Rabaul, Tulagi and Vila were to control coastwatching activities in their areas. They reported to Feldt's headquarters in Townsville and to Naval Intelligence headquarters in Melbourne. The Japanese invasion of the islands of Papua New Guinea and the Solomons, however, forced a reorganisation of the coastwatching service, as some of its personnel were killed, some were forced to flee and some began operating behind enemy lines. As a result, coastwatchers were given ranks or ratings in the hope that their combatant status would protect them if captured. The coastwatching service in the South-West Pacific Area (q.v.) became part of the Allied Intelligence Bureau (q.v.), while in the South Pacific Area it remained part of Australian Naval Intelligence, but overall coordination was maintained by putting Feldt in charge of coastwatchers in both areas. Feldt's headquarters moved to Brisbane early in 1943, then after suffering a heart attack Feldt was replaced as Supervising Intelligence Officer by Lieutenant-Commander J. C. B. McManus in August 1943. Intelligence-gathering behind enemy lines had by now assumed great importance as coastwatchers kept Allied forces informed of Japanese naval and aircraft movements around the Pacific Islands and of activity at Japanese bases. Relying heavily on indigenous Islanders as spies, guards, couriers and labourers, the coastwatchers established hidden bases from which they transmitted their intelligence to coastwatchers' headquarters. They were also involved in mapping, in guiding Allied attacks and in rescuing or evacuating many civilians and servicemen (including future US President, then naval lieutenant, John Kennedy whose life was saved by an Australian coastwatcher). By late 1944 their intelligence-gathering was no longer needed, so some coastwatchers abandoned their instructions not to fight and led very successful guerrilla campaigns against Japanese troops. Although they received little publicity during the war because much of their work had to be kept secret, the coastwatchers' important role in many key events of the Islands campaigns, including the Battle of the Coral Sea and the struggle for Guadalcanal, won high praise from the commanders who relied on them for intelligence.

Eric Feldt, *The Coast Watchers* (Penguin Books, Melbourne, 1991; first published 1946).

COBBY, Air Commodore Arthur Henry (26 August 1894–11 November 1955). Although he saw active service in France for less than a year, Cobby became the leading ace in the Australian Flying Corps (q.v.) during the First World War; his Second World War career was to end, however, amid controversy, in removal from his command.

Cobby was commissioned into the 46th Infantry Regiment (Brighton Rifles) in 1912, and joined the AIF in 1916. Posted to the Central Flying School at Point Cook, he went overseas in January 1917 with No. 4 Squadron AFC, and arrived in France in December that year after a period of further training in England. Between February and September 1918 he shot down 29 aircraft and 13 balloons, and was awarded the DFC in June, two bars in July, and the DSO in August for leading two large raids on German airfields at Lomme and Haubourdin. Gallant to the point of recklessness, it was of Cobby, among others, that one British historian wrote that, 'The most outrageous airmen of the war, using the term as a compliment, were the Australians. They had more than brains and courage … they had flair.'

Returning to Australia at the war's end, Cobby accepted a commission in the Australian Air Force in March 1921 as a flying officer. He left the RAAF as a wing commander in 1935 to become a member of the Civil Aviation Board and controller

of operations, demonstrating clearly the close link between service and civilian flying in the interwar period. He rejoined the air force in 1939, and was Director of Recruiting and commandant of the RAAF Staff School, before being appointed to command No. 10 Operational Group, based at Noemfoor in the Netherlands East Indies, in August 1944. This became 1st Tactical Air Force (q.v.) in October, and it was here at the headquarters on Morotai in April 1945 that Cobby's career came under a cloud.

On 19 April eight prominent fighter pilots tendered highly critical, identically worded letters of resignation which complained of the sidelining of the 1st TAF and the resultant waste of aircrew lives in operations of little real military value. The allocation of roles came from the highest echelons of Allied (i.e. American) command in the South-West Pacific Area, and although Cobby acknowledged the secondary nature of his squadrons' tasks, he regarded those tasks as nonetheless being both necessary and legitimate. The 'Morotai mutiny' prompted an official inquiry under Mr Justice Barry, which found that 'a widespread condition of discontent and dissatisfaction' existed in his command, and that Cobby, as AOC, had 'failed to maintain proper control over his command'. He was relieved on 10 May.

After the war he returned to the Department of Civil Aviation, serving as Regional Director for New South Wales and then as Director of Flying Operations a year before his death. His gallant service flying career, capped in 1943 when he was awarded the GM for rescuing injured flying crew from a crashed Catalina flying boat, became another victim of the weaknesses and feuding within the RAAF's senior leadership during the Pacific War.

A. H. Cobby, *High Adventure* (Robertson & Mullens, Melbourne, 1942).

COCKATOO ISLAND lies in the mouth of the Parramatta River in Sydney, and from 1839 was used as a penal settlement. Fitzroy Dock, built entirely by prisoners, was completed in 1857, and until the penal settlement was abolished in 1870 supervised convict labour was employed on all vessels in dry dock. It was visited regularly by RN ships until the much larger Sutherland Dock opened in 1890. Development of a shipbuilding yard and engineering workshops was completed in 1908, and in 1913 the Commonwealth government took over Cockatoo Island as a naval dockyard. During the First World War over 500 vessels were docked at the island, while four River Class torpedo boat destroyers (q.v.) and the Town Class cruiser (q.v.)

HMAS *Brisbane* were built there. In 1923 Cockatoo Island came under the control of the Australian Commonwealth Shipping Board, but in 1933 it was leased to a private company, Cockatoo Docks & Engineering Co. (later called Vickers Cockatoo Dockyard Pty Ltd). Naval shipbuilding continued under the company's control, including the construction of the Tribal Class destroyers (q.v.), and during the Second World War there were 750 dockings, 350 of them naval ships. Wartime repairs and refits of both Australian and Allied naval vessels were carried out at the shipyard, as were conversions of liners to troop-ships. After the war Cockatoo Island pioneered all-welded ship construction in Australia when building two *Daring* Class destroyers (q.v.), and in the 1960s the island became Australia's main submarine refitting yard. The RAN's decision to split the fleet between the Pacific and Indian Oceans left too many dockyards on the east coast and forced Cockatoo Island's closure in 1992.

R. G. Parker, *Cockatoo Island: A History* (Thomas Nelson, Melbourne, 1977).

CODE-BREAKING or cryptography is an aspect of signals intelligence that has assumed considerable importance in twentieth century warfare. Australian involvement in code-breaking began on the outbreak of war in 1914, when RAN intelligence activated a wireless-interception cell under the command of Frederick Wheatley, an instructor from the Royal Australian Naval College. Using a code-book seized from a German merchant ship, Wheatley was able to crack the German code used for supply communications, Zeppelins, small ships and merchant ships. The RAN passed the code on to the Admiralty, and over the next few months it was to prove very valuable, because although it was a merchant code it was used by warships and U-boats when arranging re-supply. The RAN continued to support the RN's decryption operations throughout the war.

After the war Australia did not maintain its own cryptographic organisation, and it was not until the late 1930s, when a small organisation under the Director of Naval Intelligence began working with the British Far East Combined Bureau, that Australian involvement in code-breaking resumed. After the start of the Second World War the Australian Army set up a small cipher section in January 1940 and in 1941 recruited several academics from the University of Sydney to work on breaking Japanese codes. A RAN cryptographic unit was formed in July 1940 under Paymaster Commander Eric Nave, an Australian with 15 years' experience of signals intelligence work in the RN. In mid-1941 the team from Sydney University moved to Melbourne to join Nave's unit, which was formally established as the Special Intelligence Organisation (SIO) in November 1941.

In February 1942 the US Navy's 'Cast' code-breaking unit was forced to leave the Philippines and re-established itself as the Fleet Radio Unit, Melbourne. Nave's SIO worked in the same building as the US Navy unit, and liaised with them. Another cryptographic organisation was set up in Melbourne in April 1942 on the initiative of South-West Pacific Area (q.v.) commander General Douglas MacArthur (q.v.). This was a combined Allied organisation known as the Central Bureau, and it involved Australian Army and RAAF personnel as well as a small section of the RAN's SIO (including Nave). The rest of the SIO staff remained under Australian control and concentrated on diplomatic intelligence. The US Navy's Fleet Radio Unit also continued to operate separately. The Australian contribution to wartime code-breaking seems to have been of minor importance, but like other examples of inter-Allied intelligence activity it probably provided Australians with valuable experience and helped pave the way for postwar intelligence co-operation. Today cryptographic analysis is the responsibility of the Defence Signals Directorate (see INTELLIGENCE).

COLAC HMAS see *BATHURST CLASS MINESWEEPERS (CORVETTES)*

COLE, Air Vice-Marshal Adrian Lindley Trevor (19 June 1895–14 February 1966). Cole enlisted in the AIF in January 1916 and was posted to No. 1 Squadron, Australian Flying Corps, with the intention of becoming a pilot. He was commissioned in June that year and began flying training in August. He had resigned his commission in the 55th (Collingwood) Infantry Regiment, obtained in August 1914. He flew operationally in Palestine in 1917, and was awarded the MC. On two separate occasions he was forced to land in enemy territory: on the first, he was rescued by a fellow pilot, Captain Richard Williams (q.v.), while on the second, having landed himself to rescue downed aircrew, he and they were forced to walk back to their own lines. In May 1918 he was sent to No. 2 Squadron, operating on the Western Front. He shot down a number of enemy aircraft in operations between July and October, and was awarded the DFC for his part in a large-scale aerial raid on Lille.

Returning to civilian life briefly after the war, Cole accepted a commission in the Australian Air Force in March 1921. He attended the RAF Staff College at Andover, England, between 1923 and

1925, and this was followed by a succession of staff and command appointments: command of the Flying Training School at Point Cook between 1926 and 1929, commanding officer of RAAF Station Richmond (q.v.) between 1936 and 1937, and a period as Air Member for Supply on the Air Board (q.v.) in 1933–36. He was selected for the Imperial Defence College (q.v.), London, the only Australian to attend the 1938 course. At the beginning of the Second World War he was commanding the RAAF Base at Laverton. In 1941 Cole went to the Western Desert Air Force, where he commanded No. 235 Wing briefly before going to Headquarters, No. 11 Group in Britain. He coordinated air support for the raid on Dieppe in August 1942 from the destroyer HMS *Calpe*, and was seriously wounded when the bridge was strafed by German fighters. He was awarded the DSO for his service, and appointed AOC Northern Ireland.

Returning to Australia in May 1943 he took over command of North-Western Area, with headquarters in Darwin. Here he was responsible for the air defence of the region and, later, for offensive operations in support of Allied troops in the New Guinea campaign. Appointment as the Air Member for Personnel in October 1944 was followed in January 1945 by a posting as RAAF Liaison Officer with Mountbatten's South East Asia Command headquarters in Kandy, Ceylon. In this capacity he took part in the Japanese surrender in Singapore on 12 September 1945. He retired in April 1946.

COLLINS, Robert Henry Muirhead (20 September 1852–19 April 1927). Born in England, Collins served in the RN from 1866 to 1872, including a posting to the Australia Station in 1876. He retired from the RN in 1877 with the rank of lieutenant, and was appointed lieutenant in the Permanent Victorian Naval Forces. He assumed command of the gunboat *Albert* in 1884, and was promoted to commander the same year. He became Secretary for Defence in 1888, retiring from the navy (where he had been on the Unattached List since 1888) in 1896 with the rank of commodore. As Secretary for Defence Collins argued for the strengthening of local naval forces at the expense of military forces; for the lessening of colonial dependence on British naval power, to the extent of calling for the termination of the 1887 Anglo-Australian Naval Agreement (q.v.); and, as a committed federalist, for the concentration of powers over defence exclusively in the hands of a federal government. With the creation of the Commonwealth in 1901, Collins became secretary to the Department of Defence, a position he held until 1906. He was appointed CMG for his administrative services during the Boer War. As Secretary he sought to bring a sense of order into the amalgamation of the colonial forces, and to impose financial responsibility and stringency in the process. He also resisted efforts by the imperially-minded GOC, Australian Military Forces, Major-General Edward Hutton (q.v.), to allow for the control of Australian forces to be transferred to British command in time of war. In early 1906 he was seconded to London as official representative of the Commonwealth, responsible for the purchase of large amounts of defence stores and for keeping the Australian government informed of developments in British defence matters. With the appointment of a High Commissioner in 1910, Collins became his official secretary. He served on several commissions and inquiries on maritime matters, and retired in 1917. He received a knighthood in 1919 and, apart from a brief visit to Australia the same year, lived in England for the rest of his life.

COLLINS, Vice-Admiral John Augustine (7 January 1899–3 September 1989). One of the original intake to RAN College, Jervis Bay, in 1913, Collins became a midshipman in January 1917 and saw service with the Grand Fleet aboard HMS *Canada*. He specialised in gunnery and between the wars served on HMAS *Australia* (II) (q.v.) and commanded HMAS *Sydney* (II) (q.v.) during the Abyssinian crisis. He served for two years at the Admiralty in London in the Plans Division and was assistant CNS and Director of Naval Intelligence at the beginning of the Second World War.

Between November 1939 and May 1941 he commanded HMAS *Sydney* in the Mediterranean, winning plaudits for sinking the Italian destroyer *Espero* and then saving 160 of the survivors in July 1940. Later the same month *Sydney* sank the Italian cruiser *Bartolomeo Colleoni* (q.v.) and damaged the *Giovanni Delle Bande Nere* in the same action, an action which earnt both the ship and its captain great acclaim. In June 1941 he was appointed chief of staff to the C-in-C, China, and the following January became Commodore China Force, taking part in the Malayan campaign (q.v.) and evacuating civilians and naval personnel from Java during the Netherlands East Indies campaign. In April 1943 he went to England to take over HMS *Shropshire* (q.v.), loaned to the RAN to replace HMAS *Canberra* (q.v.) lost against the Japanese the previous year. While in command he took part in the operations at Bougainville, Cape Gloucester, the Admiralties, and Hollandia.

As commodore commanding the Australian Squadron, with HMAS *Australia* as his flagship, he was involved in operations at Morotai and in the

Philippines. He was badly wounded at Leyte Gulf in October 1944 and did not resume command of the Australian Squadron until July 1945. He was the RAN's representative at the Japanese surrender in Tokyo Bay in September 1945.

The Australian government had intended to appoint him CNS, a move which found favour with General Douglas MacArthur and other senior American officers, but his wounds and convalescence prevented this and Admiral Sir Louis Hamilton, RN (q.v.), was appointed instead. British authorities had already been arguing that Collins lacked the experience for such a senior post, and these arguments were advanced once more in 1946–47, when the Australian government again planned to appoint Collins, this time in succession to Hamilton. Hamilton seems to have erred in equating the post of CNS with that of First Sea Lord, which clearly carried much greater responsibilities, and to have misread the political realities attendant on the Labor government's desire to appoint an Australian. Collins attended the Imperial Defence College (q.v.) in 1947 and succeeded to the post of CNS in February 1948.

He served in that position until March 1955, and was an important figure in the postwar modernisation of the RAN. In his period the navy acquired the Fleet Air Arm (q.v.) and two aircraft-carriers, one of which saw active service in the Korean War. He sustained a shipbuilding program in the face of reduced defence expenditure, and oversaw increases in recruiting — including from UK sources — and training. He also undertook administrative reforms, especially in the area of naval stores and supply, and the creation of a pension scheme for officers. As CNS he also played a part in the strengthening of regional security arrangements in the early Cold War period, and in February 1951 led a delegation in Hawaii to discuss naval cooperation with the Americans, led by Admiral Arthur W. Radford, the US C-in-C, Pacific. The Radford–Collins agreement (q.v.) that eventually resulted was more limited than the Australian government had hoped for, as it represented no widening of joint defence planning under the newly signed ANZUS treaty (q.v.); it merely recognised the responsibilities of each side for naval matters in their respective areas, which in the RAN's case was the ANZAM (Australia–New Zealand–Malaya) area. On the other hand, along with ANZUS, it suggested a growing American predominance in naval affairs and pointed ahead to the supplanting of the RN as the RAN's principal partner. It also constituted the first direct defence negotiations between the two parties since the Second World War.

Although he sought a senior fleet command in the RN as the end of his appointment approached, none was found and he subsequently served as Australia's High Commissioner to New Zealand between 1956 and 1962. G. Hermon Gill (q.v.), the official naval historian, thought Collins 'a good mixer, ambitious and able'. The new generation of Type 471 submarines are named the Collins Class (q.v.) in his honour.

J. A. Collins, *As Luck Would Have It* (Angus & Robertson, Sydney, 1965).

COLLINS CLASS SUBMARINES (*Collins, Farncomb, Waller, Dechaineux, Sheean* and *Rankin*). Length 77 m; beam 8 m; displacement (submerged) 3350 tonnes; armament six weapon tubes for either Mark 48 torpedoes or Harpoon missiles; speed (submerged) 20 knots.

Entering service in the 1990s, the *Collins* Class submarines are designed to provide the RAN with a long-range submarine capability. Based on a Swedish design, components of the *Collins* Class are built in various parts of Australia and assembled in Adelaide by the Australian Submarine Corporation. The submarines have many technological and design improvements over the *Oberon* Class (q.v.) they are superseding, including the replacement of exposed sleeping areas with six-bunk cabins. This increased privacy will allow for the introduction of female submariners. The *Collins* Class submarines have been named after six important Australian sailors: Vice-Admiral John Collins (q.v), Rear-Admiral Harold Farncomb (q.v.), Captain Hec Waller, Captain Emile Dechaineux (q.v.), Ordinary Seaman 'Teddy' Sheean and Lieutenant-Commander Robert Rankin.

COLLISIONS, NAVAL Although naval ships are often technically better equipped than merchant ships to avoid collisions, the risk of collision is increased by naval manoeuvres under real or simulated combat conditions which involve moving at great speed, close to other vessels, and often at night. On 22 June 1916, HMAS *Australia* (I) (q.v.) collided twice with HMS *New Zealand* within the space of a few minutes in the North Sea. The repair of the damaged armour plates meant that *Australia* missed the Battle of Jutland, the only major sea battle of the First World War. HMAS *Australia* later collided with HMS *Repulse* in the North Sea on 12 December 1917, again suffering only minor damage. In fact most collisions involving RAN vessels have caused no loss of life, and some vessels have escaped almost unscathed, but there have been two dramatic exceptions: the *Voyager* incident (see *VOYAGER* [II], HMAS) of 1964 and the collision between HMAS *Melbourne* (q.v.) and USS *Frank E. Evans*, which killed 74 American sailors in the South China Sea in June 1969. These two incidents severely damaged public confidence in the RAN. At the time of the *Voyager* disaster it was usual, though not mandatory, for a board of inquiry to be convened following collisions. This was a fact-finding body made up of naval officers, not a judicial body, but it would be followed by a court martial if it established a prima facie case of misconduct against an officer or sailor. In the *Voyager* case, however, the level of concern and distrust of naval disciplinary procedures expressed in newspapers and in parliament was such that a Royal Commission was called to investigate the incident. To counteract the loss of public confidence the Naval Board (q.v.) appointed a Co-ordinator of Naval Safety at the end of 1964 and, seeking to avoid Royal Commissions in future, advocated the creation of a Naval Court of Inquiry, though legislation providing for such a court was not passed until 1982. In the case of the *Melbourne–Evans* collision the *Melbourne*'s commanding officer, Captain J. Philip Stevenson (q.v.), was cleared of all charges in a court martial following a United States Navy–RAN Joint Board of Inquiry which had sought to play down the American destroyer captain's responsibility for the collision.

COLONIAL DEFENCE COMMITTEE (CDC) was established in London in 1885 at a time when a combination of military setbacks in South Africa and Afghanistan raised fears both about Britain's military capabilities and about the designs of foreign powers, notably Russia in Central Asia. The defeat of General Gordon at Khartoum in the Sudan in 1885 led to offers of military support from the colonies (see SUDAN CONTINGENT), which in turn raised again the question of how the defence of the Empire ought to be coordinated. For the first 10 years of its existence, the CDC under its permanent secretary, Captain George Sydenham Clarke, Royal Engineers (later Lord Sydenham), confined its attentions to purely colonial questions, that is, to matters affecting the colonies rather than to broader considerations of imperial defence. Its central concern in its first decade was the defence of ports and other vital localities, and it did not at first involve itself in questions impinging upon the United Kingdom. Even with that limited scope, however, the CDC was remarkably active. By the time of the first Colonial Conference in 1887, it had produced 26 memoranda, mainly on specific matters of local colonial defence, and had made recommendations on a further 53 colonial defence questions. It had

also collected enormous amounts of information about the defence capabilities and plans of the colonies, the first time that such details had been systematically assembled in London. Each colony was required to update its defence plans annually, and to forward them to the CDC where they could be subjected to expert scrutiny.

By 1895 the CDC was poised to move beyond its restrictive focus on local colonial defence issues. The Colonial Office asked it to prepare a memorandum that could be used to 'educate colonial public opinion in the right principles of defence'. Memorandum #57M, dated 3 March 1896, laid down as a fundamental principle 'the maintenance of sea supremacy ... as the basis of the system of imperial defence', and upheld the RN's 'claim ... [of] absolute power of disposing of their forces in the manner they consider most certain to secure success'. At the very time when several of the Australian colonies were developing their own naval forces and insisting on the right to place limits on where they could be deployed, such a claim for absolute centralised control could not help but raise questions in the colonies about competing priorities and interests. Memorandum #57M went even further in staking a claim, if only a moral one, to the defence resources of the colonies: should the situation arise in which Britain was at war in circumstances that posed no direct threat to the colonies, 'the offer of assistance from them would be prized, not only for its real value, but also as evidence of that real solidarity on which the greatness of the British Empire must ultimately rest'. It was precisely the increasing possibility of such a conflict between Britain and Germany that led to the creation of the Committee of Imperial Defence (q.v.) in 1902, of which the CDC ultimately became a subcommittee.

COLONIAL MILITARY FORCES were maintained, though at first not continuously, in what is now called Australia from December 1788 (eleven months after white settlement began) to March 1901, when they were transferred to Commonwealth government control.

The forces were not armies, if such entities are defined as military bodies of all arms able to fight anywhere as independent units. Nor were they composed of professional, full-time soldiers. Rather, colonial military forces were made up largely of citizen soldiers: ordinary men who changed from civilian clothes into military uniforms once or twice a week to practise drill and marksmanship, and who were neither housed in barracks nor subject to full military discipline. If war came, the forces were not generally obliged to

do more than help secure their colony, or later their continent, from invasion; Britain's navy and regular army, perhaps aided by expeditionary forces specially raised in the colonies, would engage the enemy wherever the seat of the war might be.

The forces' part-time composition and auxiliary military role derived from the colonies' inheritance of the British citizen soldier tradition. By the late eighteenth century, Britain alone of the great powers relied for home defence largely on citizen soldiers, organised both as a militia, which the government established, paid to train and sometimes compelled men to serve in, and in volunteer corps, which men themselves raised and, for almost no pay, trained in as they saw fit. This tradition of citizen soldiering, with its militia and volunteer strands, was reproduced in all Britain's settler colonies. It took firm root in new lands whose provincial status, small populations, acquisitive culture and great distance from any serious military threat made the raising of regular armies redundant, uneconomic or even abhorrent.

The first colonial military units in Australia were tiny militias raised to help stand against possible French attack, and also to quell convict rebellion and buttress gubernatorial authority against challenge by the volatile New South Wales Corps (q.v.). On 31 December 1788 Norfolk Island's commandant, Lieutenant Philip Gidley King, ordered his free male settlers, then numbering six, to practise musketry on Saturdays. In January 1792 he briefly embodied (i.e. called out) 44 free male settlers as a militia when some of the New South Wales Corps fell out with local civilians and refused to obey orders.

The first military unit raised on the Australian mainland appeared early in September 1800 when Governor Hunter asked 100 free male settlers in Sydney and Parramatta to form Loyal Associations, as English volunteer units, whose main purpose was to oppose riot or revolution, were then called, and called on them to train in case the colony's Irish convicts rebelled. Several years later King, now Governor of New South Wales, supplemented the associations by recruiting six ex-convicts as a mounted bodyguard, and so created the first full-time military unit to be raised in Australia. The New South Wales Corps objected to the existence of the bodyguard and the loyal associations; nevertheless, all turned out together to help suppress a convict rebellion near Parramatta in March 1804. When the Corps overthrew Governor Bligh in 1808, Bligh had too little time to call out the asso-

ciations to help him. They faded away by the end of the decade, made redundant by the arrival in 1810 of the first of the British regular regiments that would garrison the Australian colonies until 1870 (see BRITISH ARMY IN AUSTRALIA), and by victory at Waterloo in 1815, which extinguished any external threat to the colonies and almost extinguished citizen soldiering around the British Empire for 40 years. The governor's bodyguard, rarely numbering more than a dozen men, survived until 1860.

Apart from the governor's bodyguard, the regulars garrisoned the Australian colonies, almost unaided by colonial forces, until 1854. Colonists and governors, occasionally supported or even prompted by the imperial government in London, sometimes proposed sanctioning citizen-organised volunteer corps or, more often, raising a militia. Support for these proposals grew during the 1840s and early 1850s, when Britons and North Americans, moved by new ideals of duty, community involvement and self-help, renewed their interest in citizen soldiering, and their governments re-formed their ailing militias as voluntarily recruited forces. Before the gold-rushes, however, there were too few free male settlers in Australia to form a militia, and never sufficient threat of invasion to justify the cost. In any case, the occasional presence of companies of retired veterans and pensioners from the regular army offered some assurance that the redcoats would not have to fight unaided if war came, while from 1824 a semi-military mounted police force relieved the garrison of the squalid task of beating back the Aborigines (see ABORIGINAL ARMED RESISTANCE TO WHITE INVASION).

Early in the 1850s war seemed imminent, at first with France and then with Russia. If militias were too expensive for colonial governments to establish, should not respectable colonists, some argued, at least be allowed to form volunteer corps to prepare themselves for any danger? When in May 1854 news of war with Russia came to the colonies, some Victorian men went further than mere argument and started rifle clubs, hoping their government would permit their transformation into volunteer corps. In Western Australia and Tasmania the Russian threat was deemed too distant to require a local response, but the other colonial governments disagreed. In August New South Wales permitted volunteer corps; South Australia followed in September, and Victoria in December. The South Australian government was unwilling to trust entirely to its citizens' personal sense of patriotism, and legislated in December that if too few men joined the volunteers it would raise a militia and ballot men into it. The simultaneous promise to pay South Australia's volunteers probably proved a more powerful inducement to drill.

Privates Brock, Hopkins and McCerry (left to right) of the Castlemaine Company of the Victorian Rifles, c. 1890. (AWM P0744/16/13)

Most volunteers of 1854 were prosperous English immigrants: clerks, lawyers, merchants and skilled tradespeople. Most raised rifle corps: foot-soldiers who dressed in loose clothing and fought as skirmishers rather than ordinary infantry. Some trained as artillerymen to attend the guns that guarded the capital cities from attack by hostile ships; a very few drilled as light cavalry. All saw themselves as auxiliaries who in a war would support the regular garrison.

Most volunteers ceased to drill after peace returned in 1856, but this proved to be a temporary hiatus in citizen soldiering in Australia. In 1859 Napoleon III seemed ready to invade England and the United States was on the verge of civil war. Thousands of Britons and North Americans began forming volunteer corps. At the end of the year Australian men began to copy them, claiming, less credibly, that they too were threatened with war. By the early 1860s almost every suburb and town in Australia supported a volunteer unit, usually a rifle corps, filled by a truer cross-section of colonial male society than had come forward in 1854. Men like barrister George Higinbotham and poet Adam Lindsay Gordon drilled beside blacksmiths, carters and labourers. By the end of 1861 around 7000 White Australian men — one in 50 of military age — were volunteers. They were not the only locals under arms. Progressive schoolteachers had begun to imitate their counterparts in England and form cadet corps among their boys (see AUSTRALIAN CADET CORPS). Most towns had rifle clubs, which, until well into the twentieth century, were not only recreational associations but also semi-military ones for men who disliked drill but still wished to practise accurate rifle-shooting, then the essence of military art, for when war came.

True to the British citizen soldier tradition, the volunteers of the 1850s and early 1860s were free men who devoted their leisure time to preparing themselves as best they could to be useful soldiers should war ever reach their homes. They could not be forced to campaign outside their colony. They were not subjected to military discipline or a set program of drill. They could choose their own officers and dress themselves as they saw fit. Yet the sudden emergence on private initiative of hundreds of units of armed men across Australia offered not the smallest challenge to governments or to law and order. Indeed, the movement indicated growing peacefulness and acceptance of legitimate authority within White colonial society. The volunteers never became an armed mob. Nor, conversely, did they become instruments of government control over the people, whether White or Black.

Britain and France did not go to war, but, while volunteer numbers declined, many units remained together. Britain and its colonies began to realise their vulnerability before the growing military and naval strength of other great powers. Conscription or a great expansion of the regular army by voluntary recruitment were not deemed to be the answers though; the first would have limited personal liberty, the second, some feared, might have ended it by armed insurrection. Citizen soldiering, therefore, remained as the British compromise between defencelessness and military despotism. Britain's colonies in Australia thus found themselves possessing enduring military forces.

But the compromise required reform, for unregulated volunteers were unlikely to make good soldiers in war. Over the next 40 years Australian colonial governments, generally following British and Canadian initiatives, strove to nudge their citizen soldiers toward as much efficiency as might be demanded of men who voluntarily gave up their leisure time to drill. In the long process of reform, private initiative and the financial and moral support of local communities were gradually replaced by submission to the directives of professional soldiers and payment from government coffers. Sometimes reform was supported, even ardently sought, by citizen soldiers themselves; occasionally it was sponsored by citizen officers who were also members of parliament. Generally, though, the slow choking of the spontaneity and free atmosphere of volunteering's early days was unpopular, occasionally resisted, and almost completely avoided by those men who remained in the rifle clubs.

Reforms to the colonial military forces during the 1860s generally followed those imposed on Britain's volunteers. Professional soldiers, called permanent staff, were appointed and paid by governments to help instruct and administer the local citizen soldiery. Volunteers who attended a certain number of drills and fired a certain amount of ammunition at a target every year were classified efficient, and their corps awarded a small sum, called a capitation grant, to help them buy uniforms and equipment. Corps were gradually obliged to conform to a standard size, and grouped into regiments, battalions and batteries. Volunteers were encouraged to assemble in annual camps, almost invariably at Easter, for three- or four-day camps of intense military training. Men chosen to be officers and NCOs were made to pass military exams before their promotions were allowed. Commandants, at first retired regular officers, were appointed to lead each colony's military force in war and to advise governments on its administration and training in peace.

One local reform, though, clearly was not derived from Britain: some colonial governments offered free land to volunteers who were efficient for five consecutive years (see LAND GRANTS, MILITIA). The offers were withdrawn when it was learned that men were selling their land for profit rather than taking it up themselves.

South Australia introduced a reform derived from the new Militia Acts of Britain and Canada. The militiamen of these countries, in return for being paid not merely a capitation grant but also the equivalent of civilian wages for each hour they spent in uniform, were expected to drill for a fixed number of hours or whole days every year, and to do so under military discipline, for a certain number of years before passing, at least nominally, into a militia reserve. While militia service was now voluntary, the three governments retained the power of compulsory recruitment, and threatened to ballot men into the forces if enlistments fell below a set quota. Thus their militias, at least in theory, were as stable and professional a military body as a regular army whose members happened to be on leave. Since 1854 South Australia had paid its volunteers to train and kept on its statute books the possibility of a militia ballot. In 1866 it pushed its citizen soldiers toward the voluntary militia model by prescribing minimum hours of drill, fixed terms of service, and nominal service in a reserve. Only popular resistance prevented the government from introducing military discipline as well.

Colonial military reform accelerated from the 1870s, prompted at first by the withdrawal of the last regular troops from Australia in 1870. If an invader eluded the RN and landed on Australian territory, the citizen forces would now have to fight alone for the weeks or months it would take for the regular army to arrive to help them. Thus the first great reform begun that decade was the expansion of the permanent forces to include artillery and engineer units to ensure that the guns and fortifications of the capital cities were tended by full-time professional soldiers. The expanded permanent forces remained too small and unbalanced to count as regular armies — brief experiments in raising infantry and mounted corps failed — and never numbered more than 7 per cent of colonial military establishments. Service in them entailed subjection to military discipline and the tedious routine of the drill square and barracks, all for lower pay than most colonial workers earned and without the compensations that Britain's regulars enjoyed of smart uniforms and official respect. Yet the permanent forces nevertheless adopted toward the citizen soldiery the contempt generally held by regular soldiers for amateurs. The regulars

clearly formed, at least in their officers' minds, a colonial military élite.

The second great reform begun in the 1870s was a general adoption of parts, and sometimes the whole, of the voluntary militia model. The measure was driven by fear of Russian raids in 1877–78 and 1885 and by the advice given in 1877–81 by visiting defence experts Colonel Sir William Jervois and Lieutenant-Colonel Peter Scratchley (qq.v.). New South Wales's government moved first. In 1878 it began to pay its volunteers to drill, though it ignored its commandant's calls to raise a voluntary militia. Victoria's more adventurous government waited until the retirement of its hidebound commandant of 20 years and in 1883–84 prescribed minimum hours of drill and fixed terms of service for its citizen soldiers, who were now paid to train and renamed militiamen, though they were not subject to military discipline. In 1885 Queensland and Tasmania passed virtual copies of Canada's Militia Act, nominally enrolling every fit man in their colonies into a defence force, threatening to conduct a militia ballot if too few came forward voluntarily to drill, and imposing on those who did partial military discipline (see DEATH PENALTY), set training and terms of service, all in return for pay. The following year, South Australia consolidated its volunteer and militia acts into a copy of the Canadian model, but still avoided imposing military discipline. Western Australia's government tried but failed to legislate for a militia during 1893–94, and like New South Wales began instead simply to pay its volunteers to train. Accompanying these reforms was a gradual adoption of the tactics and uniforms of the regular army, reflecting the new view that Australian citizen soldiers could no longer expect merely to skirmish for the regulars during an invasion, but must now be ready to take the regulars' place.

Some members of the reformed colonial military forces thought the money they were now paid was not worth their new obligations. Yet the paid volunteers of New South Wales and Western Australia and the voluntary militiamen of the other colonies belonged to forces that differed little from the old unpaid volunteers. To enact a reform was not always to effect it. Literate, confident men who could vote governments in or out could not be induced to enlist by the threat of a militia ballot, nor could they be forced to stay with their units for a fixed term or accept even modified military discipline. And while drill continued to take place more often after a long day's work rather than away at annual camps, they could not be made to train hard or often.

Not only did easy volunteer ways survive in the reformed forces; reform did not extinguish unpaid volunteering. In 1885 the British Empire's Christian hero General Gordon was killed at Khartoum, New South Wales packed off the Sudan Contingent (q.v.) raised from volunteers to help avenge him, and the Russians seemed ready to attack Australia. Filled with excitement and fear, thousands of men in every colony pressed once more to be allowed, as their fathers had been, to form traditional volunteer corps. The authorities buckled. Troops of cavalry sprang up across rural Queensland, New South Wales and Victoria; in Melbourne, dock workers formed the Harbour Trust Artillery Company, while Sydney Scots formed an infantry corps. The authorities rationalised their surrender. Unpaid volunteering, they said, if less efficient than paid citizen soldiering, at least was cheap, and encouraged free expression of the martial spirit which some feared might wilt in the sunny and prosperous colonies. The new unpaid volunteers were trained as second-line troops behind the paid volunteers and voluntary militiamen. Ten years later in South Australia a government finally dared legislate no longer to permit unpaid volunteering; but this was part of an unusual, unpopular and unsuccessful attempt by the colony's radical premier, Charles Cameron Kingston, and his eccentric commandant, Joseph Gordon, to impose full military discipline short of corporal punishment on citizen soldiers and, for the first time in Australia, make part-time military training compulsory for every fit man whether war was imminent or not.

The revival of unpaid volunteering in 1885 also yielded the beginnings of what most now consider to be an Australian military style. The new horse-soldiers of that year trained not as cavalry but as mounted riflemen, who fought with rifles rather than swords and no longer dressed in red and blue with hard helmets but in dull khaki with soft slouch hats. Somehow Australians forgot that the new mounted tactics were beginning to be used by citizen horse-soldiers throughout English-speaking countries, and that the dull new uniform was copied from India and spread at the insistence of visiting regular officers. During the 1890s a myth, a now-forgotten subspecies of the bush legend, began to grow about 'buff-coloured boys' whose untutored ability to ride and shoot was revealing a natural talent of Australian colonists for war. A formal attempt to establish an Australian military identity was launched in 1896–97 when, in response to the raising of Irish and English volunteer corps in Sydney, Australianists founded two volunteer corps from native-born men which they called the Australian Rifles and the 1st Australian Horse.

If an Australian military style seemed to be emerging, albeit slowly, during the 1890s, plans to form a single Australian citizen army that decade were frustrated. Major-General Bevan Edwards (q.v.) visited the colonies in 1889 and echoed local wisdom that they should federate their military forces. Conferences of colonial commandants in October 1894 and January 1896, and a proposal drafted by New South Wales Commandant Edward Hutton (q.v.) early in 1895, promoted schemes for combining part of each colony's military forces into a federal voluntary militia. The scheme was sunk by the smaller colonies' suspicions of New South Wales and Victoria, whose premiers and military advisers would outnumber their own on the proposed council which was to control the federal militia.

Although a federal militia was not to be, during the 1890s the colonies did join together to fortify and garrison Thursday Island and Albany, two points that they supposed a hostile power might want to seize, and in July 1899 they amalgamated their permanent forces to form the Royal Australian Artillery Regiment. They also continued to pursue piecemeal military reform. Khaki became common, though it remained unpopular with city-based citizen soldiers, who felt that scarlet represented 'actual soldiering'. Plans for rapid wartime mobilisation were drafted. Specialist units of doctors, supply troops and machine-gunners began to be formed; when a tiny corps of nurses was formed in New South Wales in 1898, Australian women began to join Australian men in imbibing the martial spirit.

By then some of the forces had seen active service, though not against a foreign foe. Several thousand citizen soldiers were mobilised in eastern Australia to maintain order during the great maritime and shearing strikes of 1890–91 (see CIVIL POWER, AID TO THE). In Melbourne, Colonel Tom Price (q.v.) of the Victorian Mounted Rifles warned his men that if the civil authorities ordered them to use force, they must 'fire low and lay out the disturbers of law and order'. But no shots were fired, the strikes were broken peacefully, the forces acted more as policeman than as soldiers, and governments and commandants fulsomely thanked them for their discipline and efforts. A few Australians were shocked by what they thought was Price's eagerness to shoot down his fellow men. Theirs was a minority view. The citizen soldiery's brief excursion into civil control had been so unusual and conducted so peacefully that no breach was to emerge between the colonial military forces and the societies that supported them.

In 1885 the forces had numbered almost 21 000 men, nearly three in every 100 of military age; at the close of 1900 the figure exceeded 58 000, nearly seven in every 100, if rifle club members are included. Such poor percentages indicate the marginal role which citizen soldiering played in the life of the Australian colonies. Nevertheless, since 1860 men of all classes, occupations, regions and religions had continued to drill. Each had his own reason for doing so. Conservatives like grazier James Macarthur Onslow wished to defend British institutions in Australia from outside attack. Radicals like Edward O'Sullivan wanted to prevent a foreign power interrupting the process of social reform. Irish Australians such as crown prosecutor Hubert Murray wished to demonstrate their patriotism to a sceptical English–Australian establishment. Ambitious outsiders like Jewish engineer John Monash (q.v.) hoped for the social advancement which a commission could bring. Countless others simply enjoyed the now-forgotten pleasures of government-subsidised rifle-shooting and strutting past the town hall in gorgeous scarlet or rugged khaki on a fine Saturday afternoon. Above all, though, many Australian men loved a stoush; and some at least saw one coming and, by training as citizen soldiers, wished to be ready for it.

The outbreak of the Boer War in October 1899 gave them their opportunity. Colonial governments responded by raising contingents and dispatching them to South Africa; nearly one in 10 citizen soldiers joined them, though soon London was asking not for men with conventional military training but for the most skilful shots and riders across the Empire. While these rugged worthies responded to the call, a new volunteer wave began at home. Thousands of men rushed to join existing units in the colonial forces or to form new ones. Mounted corps were even more popular than in 1885. But was a traditional upsurge in volunteering the best way to channel the martial spirit of Australians? Had South Australians been wrong to oppose Kingston and Gordon's plan for compulsory military training? Some Australians began to argue that every man should learn drill, or at least every boy. In February 1900 Brisbane's mayor convened a rowdy meeting which resolved that after Federation they would press the national government to legislate for compulsory training. Other Australians claimed that the war was proving that drill, uniforms and discipline were superfluous, and that all that was needed to build a formidable military force was the cultivation of existing interest in riding and shooting. This claim was particularly strong in Victoria, where 18 000 men joined rifle clubs during 1900–01 and hoped to be incorporated into the colony's military forces as a guerrilla army.

It was clear that after Federation in 1901 the new Commonwealth government would retain citizen military forces as the basis of Australia's military

system. But would citizen soldiering be voluntary or compulsory? If voluntary, would there be more militiamen than unpaid volunteers? Or would conventional categories give way to a new force of men encouraged to fight as guerrillas? These questions were unresolved when the colonial military forces passed to Commonwealth administration (see CITIZEN MILITARY FORCES) on 1 March 1901.

CRAIG WILCOX

COLONIAL NAVAL DEFENCE ACT (*An Act to Make Better Provision for the Naval Defence of the Colonies*) *1865*, gave colonial legislatures the power to raise naval forces for local defence, and to man such ships either by drawing on personnel of the RN or by providing crews from the local population, in either case at no expense to the British government. Any ships acquired under the act could, with the assent of the particular colony, be transferred to Admiralty control in time of emergency, at which point personnel enlisted in the colonial force would fall under the disciplinary procedures of the RN. While some critics in Britain argued that by allowing the development of local naval forces the act encouraged a weakening of the ties between Britain and the colonies, the Secretary of State for the Colonies, Edward Cardwell, insisted that the imperial relationship would be immeasurably enhanced for being based on an acceptance of mutual benefit: the defence of the colonies would be strengthened, and at no cost to the British government, whose naval expenditure could be directed to other areas. In the event, the consequences of the act were somewhat different from what its supporters had intended. Apart from Victoria, no other colony made any significant move towards developing local naval forces: they were either insufficiently wealthy to be able to afford the substantial cost, or they did not feel the sense of isolation from the imperial centre that made Victoria so concerned about its defence. Far from unleashing a spate of colonial naval activity, therefore, the act had almost no immediate effect.

In the case of Victoria, the outcome was precisely the opposite of what had been intended. When the Victorian Treasurer, accompanied by the newly appointed captain of the volunteer naval forces of the colony, went to London in 1866, he managed to persuade the Colonial Secretary of the new Conservative government, Lord Carnarvon, to provide an ironclad warship to Victoria at a cost of £100 000, a concession which Carnarvon thought eminently appropriate as a tangible means of encouraging colonial defence efforts, even though the act had specified that such forces as were raised were not to involve any expense to the British government. Apart from this one venture, very little flowed from the act, although by the end of the century the contributory arrangement of the Anglo-Australian Naval Agreement of 1887 (q.v.) had fallen into disfavour. When the question of separate colonial forces emerged again, the Colonial Naval Defence Act of 1865 provided the legal basis for their establishment, and thus underpinned the creation of the RAN.

COLONIAL NAVIES From 1788 onwards the naval defence of the Australian colonies was the responsibility of the RN, but the colonial war scares (q.v.) of the 1850s led New South Wales and Victoria to acquire warships of their own. Colonial Office and Admiralty concern about these acquisitions was only increased when the *Victoria* was sent to the New Zealand Wars (q.v.) in 1860–61, as it seemed unlikely that colonial vessels operating outside colonial waters would be recognised by foreign powers. In 1865, following Victoria's continuing requests for a more powerful warship and debate in Britain about the high cost of maintaining warships in the colonies, the British government passed the *Colonial Naval Defence Act 1865* (q.v.). The act authorised the acquisition, operation and main-

tenance of warships by self-governing colonies, thus meeting colonial demands while relieving Britain of some of the financial burden of naval defence. Colonial authorities would control these ships directly, but in time of war they could be commissioned into the RN.

The Victorian government's lobbying efforts were rewarded in 1867, when it received the *Nelson* on permanent loan from Britain, thus becoming the first colony to acquire a ship after the passing of the act. Operation of Victoria's naval ships was the responsibility of the Permanent Naval Forces established in 1870, supported by the Volunteer Naval Reserve (commonly known as the Naval Brigade), which had been in existence since 1859. By 1885 there were 243 officers and men in the permanent force and 350 in the reserve. In New South Wales a Naval Brigade was formed in 1863, and by 1864 it consisted of five companies and 200 men. However, since the *Spitfire* had been transferred to Queensland in 1859, the brigade had no ships until it acquired the *Wolverene* in 1882. Nevertheless, the brigade continued to grow, reaching a strength of 614 by Federation. A New South Wales Torpedo Corps was also formed in 1878 to operate the torpedo boats *Acheron* and *Avernus*.

The crew of Her Majesty's Queensland Ship *Gayundah*, 1898. (AWM P0444/214/129)

Queensland's Naval Brigade, stationed in major ports on the Queensland coast, was established in 1885, and the South Australian government also set up a Naval Brigade to operate its single ship, the *Protector*. Tasmania had only a small Torpedo Corps, while Western Australia, which acquired no warships of its own, had the tiny Fremantle Naval Artillery as its only naval defence unit.

Of all the colonies, Victoria's naval force, at its peak in the 1880s, was the most powerful. With the exception of *Victoria*'s role in the New Zealand Wars, however, none of the colonies got the chance to test their ships in battle. Naval personnel from New South Wales and Victoria took part in the suppression of the Boxer Rebellion (q.v.), but the South Australian ship *Protector*, which was also sent to China in 1900, saw no action there. For the most part, colonial naval ships were used only in exercises. Under the Anglo-Australian Naval Agreement (q.v.) of 1887, the colonial navies were supplemented by the Auxiliary Squadron, Australia Station (q.v.), but although this squadron was designated specifically for colonial defence it was under British control and did not exercise with colonial warships. At Federation the men, ships and other assets of the colonial navies were combined into the Commonwealth Naval Forces, which in turn became the Royal Australian Navy (q.v.) in 1911.

(See also NEW SOUTH WALES NAVAL FORCES SHIPS; QUEENSLAND MARINE DEFENCE FORCE SHIPS; SOUTH AUSTRALIAN NAVAL FORCES SHIPS; TASMANIAN TORPEDO CORPS SHIP; VICTORIAN NAVY SHIPS.)

COLONIAL WAR SCARES The Australian colonies were prey to various alarms and the apprehension of war throughout the course of the nineteenth century, usually a function of distance, isolation and uncertain communications. The earliest efforts of the governors of the new colony at Sydney were directed towards the fortification of the growing settlement, and for the rest of the century the 'bricks and mortar' school of colonial defence oversaw the construction of often elaborate fortifications (q.v.) all around the Australian coastline. Although most war scares were exactly that, and lacked any substance, the fears of colonial Australians were not always groundless. In June 1802 two French ships entered Port Jackson; at this stage the long war between France and Britain was in recess under the provisions of the Peace of Amiens, but one of the French officers reported to the military governor of Mauritius, De Caen, that the tiny colony was vulnerable to attack and that an attempt should be made. Nothing came of this until 1810, when Napoleon instructed De Caen to mount an expedition against Sydney, but the military resources by then were lacking, and in any case Mauritius itself was captured by a British expedition later the same year. During the War of 1812 between Britain and the United States, a convict ship bound for New South Wales was captured at sea by an American privateer, but no attempt was made to attack the colony directly although several of its prominent citizens feared that this would eventuate. The inadequacy of the harbour defences at Port Jackson was demonstrated as late as 1839 when two American warships quietly dropped anchor in Sydney Harbour one night, to the consternation of Sydney when it awoke next morning. The discovery of gold and the greatly increased wealth of the colonies, especially Victoria, which followed increased the level of apprehension, which was never far below the surface. In 1854 a minor incident involving the firing of a few ship's signal rockets by the steamer *Great Britain* in Port Phillip Bay led to a call out of the garrison and anxious preparations to repel a Russian squadron which, it was reported, was busy sinking every ship in the port. It was the Americans, again, who posed the most realistic threat to Melbourne when in early 1865 the Confederate commerce raider CSS *Shenandoah* anchored in Port Phillip Bay for over three weeks. There was little likelihood that the Confederate Navy would attack a British colony, but a chance that it might bring the Civil War to the South Pacific: after leaving Melbourne, the *Shenandoah* resumed commerce raiding in the North Pacific. In the 1870s and 1880s the presumed enemy was the Russians, and war scares swept the colonies in 1878, during the victorious Russian advance towards Constantinople in the Russo-Turkish War of 1877–78, and again in 1885 during a period of heightened tension between Russia and Britain over Afghanistan.

War scares were thus of two kinds: those, especially in the early years of the colonies, that appeared to threaten attack or invasion of the colonies themselves, and those that presumed the outbreak of a wider war involving Britain and another power which might, or might not, project power into the South Pacific. War scares usually provided a useful boost to recruiting for the colonial militias, but such increases in strength as followed generally proved short-lived, and within a year or two of the alarm being sounded numbers would decline again to previous levels.

COLOUR PATCHES A form of distinguishing insignia uniquely Australian, colour patches were first adopted in the AIF to differentiate units, although Australian members of Rimington's mounted column during the Boer War appear to have tied red ribbon to their left shoulder straps in order to identify themselves. In the First World War all units overseas wore a distinguishing patch. In an infantry division, the shape of the patch indicated the division, the base or rearmost colour (depending on the shape of the patch) the brigade and the top or foremost colour the battalion. Patches were worn 1 inch from the top of the jacket sleeve. There were nonetheless some variations. The 4th Brigade, as part of the first contingent raised, was permitted to retain its rectangular patch even though it served in the 4th Division, whose patch was round. Men who had served at Gallipoli wore a small metal 'A', for Anzac, on their patch. These became increasingly rare as the war went on. The 14th and 15th Brigades were formed from the 1st and 2nd Brigades in Egypt in early 1916, and thus wore the 1st Division's colour patch on its side; their battalions were sometimes referred to by men of the original brigades as 'pup battalions'. When CMF battalions were issued with colours in 1921, the regimental colour contained the AIF unit colour patch instead of the traditional regimental crest.

The colour patch system was adopted again at the beginning of the Second World War. Units of the 2nd AIF were to wear the corresponding patch of the original AIF unit, on a grey background in order to distinguish the 2nd AIF from CMF units that wore the same patch and, some said, in memory of the original force. The system broke down completely, however, for a variety of reasons. In early 1940 brigades were reduced from four battalions to three, and the 'surplus' battalions of the 6th and 7th Divisions regrouped into new brigades, completely destroying the basis of the system. There was additional confusion from 1942 when the 6th and 7th Divisions returned from the Middle East, because militia units which volunteered for service outside Australia and thus 'went AIF' were also entitled to the grey background: the result was different units with identical colour patches. The 25th Brigade, which was formed in England from units and reinforcements diverted from the Middle East, chose its own patch, which was identical to that of the 24th Brigade. After the siege of Tobruk, Lieutenant-General Leslie Morshead (q.v.) redesigned the 9th Division's patch into a 'T' shape to commemorate the formation's service there. When the 8th Division was divided between Malaya and the islands to Australia's north, some soldiers in that formation placed a black line down the centre of their patches, calling themselves the 'Broken Eighth'. There was also the problem of units that had not existed during the First World War, such as the armoured divisions and brigades, for whom colour patches had to be created. To complicate matters still further, soldiers in the 2nd AIF who were veterans of the 1st AIF were entitled to wear a miniature colour patch from their former unit above the patch of their current one, and this extended to men with service in other units of the 2nd AIF. As a consequence of all this, a simple and original system of unit identification ceased to work effectively.

Colour patches were discontinued with the formation of the Australian Regular Army, units and corps being assigned distinguishing badges instead, most of which were derived from their British opposite numbers. In 1987 colour patches were reintroduced for all units that could demonstrate some form of lineal descent from the AIF. They are now worn on the side of the pugaree (i.e. outer hatband) on the slouch hat.

COLOURS Unit colours have their origins on the pre-modern battlefield, where they were used in the confusion of the action, accentuated by the dense smoke generated by black powder munitions, to signify the position of the regimental commander and his headquarters, and to rally men during or after the fighting. They carry the battle honours of a unit and, like guns in the artillery, are the most sacred and prized possession of a regiment. They have not actually been carried into battle in British-pattern armies since the late nineteenth century, although the last documented occasion is disputed.

A regiment possesses two colours, Sovereign's (King or Queen's) and Regimental. They measure 3 feet 9 inches by 3 feet, not including the fringe which is 2 inches deep. The pike on which they are carried is 8 feet 7½ inches long, including the royal crest which surmounts it. The Queen's Colour of every unit was originally the Union Flag, bearing a crimson circle with the name of the unit inscribed in gold. In 1969 these began to be replaced with a colour based on the Australian national flag, the first of the new designs being presented to RMC Duntroon (q.v.) by Queen Elizabeth II personally. Regimental Colours are dark blue for Royal Regiments, dark green for all others. They bear a regimental badge or crest on a crimson background, encompassed in a wreath of leaves of a national plant, in the case of Australian regiments the wattle, and surmounted by a crown. The number of the battalion is embroidered in gold in the top-left hand corner. On the ground to either side of the crest are the battle honours (q.v.) awarded to the regiment.

Some colonial units were awarded colours before Federation, but these were often not standardised, either against each other or consistent with imperial practice. The infantry and pioneer battalions and light horse regiments of the AIF were presented with silk colours in 1920; these were laid up or passed to their successor units in the CMF as appropriate after the demobilisation of the AIF was completed in 1921. Until 1921 very few CMF units were presented with colours (the first being given to the 1st Battalion of the Australian Infantry Regiment in May 1906). With the reorganisation of the army in 1921 permission was given for Australian units to carry colours, and the regimental colours carried the unit colour patch rather than a regimental crest. All of these had green fields because only one CMF unit before 1960 carried the 'Royal' title. Cavalry and armoured units carry guidons, while the Royal Regiment of Australian Artillery (q.v.) parades the King's Banner, held in the custody of the 1st Field Regiment.

The RAN carries two Queen's Colours, one held by the flagship of HMA Fleet, the other at HMAS *Cerberus* (q.v.). Permission to use colours in the navy was granted by George V in 1924, and the first King's Colours were presented to the RAN in March 1925. New colours were presented in 1957 and, based on the national flag, in 1968. The RAAF carries a Queen's Colour for the service as a whole. Individual squadrons carry standards with a sky-blue field and the squadron crest embroidered on it, while other RAAF establishments have banners. Colours are now carried only on ceremonial occasions such as unit birthdays, beating of the retreat ceremonies or the exercise of the freedom of a city or town.

COLVIN, Admiral Ragnar Musgrave (7 May 1882–22 February 1954) joined the RN as a midshipman in 1896 and was commissioned in 1902. He specialised as a gunnery officer, and during the First World War was the executive officer on HM Ships *Hibernia* and *Revenge*, serving at the battle of Jutland. He finished the war as a captain at the Admiralty in the Directorate of Plans. After the war he served at sea in the Mediterranean and as naval attaché in Tokyo. In 1927 Colvin was appointed Director of the Naval Tactical School at Portsmouth and, promoted to rear-admiral in 1929, became chief of staff to the C-in-C, Atlantic Fleet. In 1934 he was appointed president of the Royal Naval College, Greenwich, and commander of the Royal Naval War College, and was made KBE in 1937. Colvin's was a model RN career, and in October 1937 he was sent to Australia to become CNS. Even had he been aware of the growing doubts in Whitehall concerning the efficacy of the 'Fleet to Singapore' strategy, he would by that late date have been unable to effect any

substantial change in Australian defence policy and, in any case, by the time he arrived in Melbourne the movement for Australian rearmament had already begun. During his term of office the RAN acquired three modern cruisers and began construction of Tribal Class destroyers (q.v.), but his health failed in the course of 1940 and in March 1941 he resigned his post. Back in London he was naval adviser to the Australian High Commission between 1942 and 1944.

COMBINED FIELD INTELLIGENCE SERVICE see **ALLIED INTELLIGENCE BUREAU**

COMMAND AND CONTROL Command is the function of the commander, whatever his level, and is the authority exercised over formations, units, or individuals subordinated to him. In an operational environment, control is a less complete form of authority exercised by a commander through a subordinate. At the higher levels of military organisation, command may be defined as the military direction of operations, while control constitutes their political, financial and administrative management. The two terms are not, and are not intended to be, synonymous. Current American usage refines these terms somewhat so that control encompasses the planning, direction, coordination and control of forces and operations in the accomplishment of the mission.

COMMANDO SQUADRONS see **INDEPENDENT COMPANIES**

COMMANDS, ARMY From Federation the army was organised throughout Australia on the basis of commands and military districts that were defined by geographical areas, encompassing one or more individual states of the Commonwealth. Thus Eastern Command was based on New South Wales, Southern Command on Victoria, Northern Command on Queensland, and so on. The main limitation of this system was that units could be employed on operations only by divorcing them from the existing command and administrative structure.

In 1972 the government accepted the recommendations of the committee of review chaired by General Francis Hassett (q.v.) and approved the creation of three functional commands for the army: Field Force Command, Logistics Command and Training Command. These are each commanded by a major-general. Field Force Command became Land Command in 1987. Its role is to raise, train and maintain forces for operational deployment, and it incorporates regular, reserve and ready reserve units and formations. Its headquarters are in Sydney. Logistics Command provides logistic support to the army and takes in transport, supply, ordnance, electrical and mechanical engineering, and quality assurance support; its headquarters are in Melbourne. Training Command is responsible for individual training before attachment to a unit, and the development and dissemination of doctrine, though unit training is the responsibility of Land Command. Training Command oversees all the army's schools and training establishments with the exception of RMC Duntroon (q.v.) and the Australian Defence Force Academy (q.v.). Its headquarters are in Sydney.

COMMANDS, RAAF Like the army the RAAF was organised on an area or geographical basis until after the Second World War. The RAF, on the other hand, had been organised on a functional basis for many years, and when Air Marshal Sir Donald Hardman (q.v.) became CAS in 1952 he resolved to reorganise the Australian service along more practical and efficient lines. In May 1953 he proposed the establishment of three functional commands under an air force headquarters: Home Command was to command all operational units; Training Command was responsible for recruitment and individual training; Maintenance Command was responsible for supply and technical services. The new organisation was approved by the Minister for Air and implemented during Hardman's remaining term of office.

In 1959 the functional command system was rationalised, and reduced from three commands to two. Home Command was redesignated Operational Command, while Training and Maintenance Commands were merged into Support Command. Operational Command is now called Air Command, and after a 1986 reorganisation has responsibility for five Force Element Groups: Tactical Fighter Group, Strike Reconnaissance Group, Maritime Patrol Group, Air Lift Group and Operational Support Group. Its headquarters are at Glenbrook in New South Wales. Support Command was disbanded in 1990, its place being taken by Logistics Command and Training Command.

COMMANDS, RAN The navy has two functional commands, Maritime and Naval Support, and a system of Area Commands similar to the army's military districts (q.v.). Maritime, formerly Fleet, Command, based in Sydney, has overall operational command of all seagoing vessels and closely related shore establishments. It is commanded by a rear-admiral. Naval Support Command, also based in Sydney, is responsible for all matters of logistic sup-

port. The six area commands are established in the States to provide administrative support to naval activities within their boundaries, with Naval Support Command providing this function in New South Wales and the Australian Capital Territory. In Queensland, Western Australia and the Northern Territory they also have limited operational responsibility for locally based patrol boats.

COMMEMORATION OF WAR see **ANZAC DAY; AUSTRALIAN WAR MEMORIAL; WAR MEMORIALS**

COMMONWEALTH AIRCRAFT CORPORATION (CAC) The CAC was formed on 17 October 1936 by a consortium of BHP, Broken Hill Associated Smelters, General Motors-Holden, North American Aviation and the Commonwealth government. The Corporation's factory was at Fishermens Bend in Melbourne and its managing director was Lawrence Wackett (q.v.). The CAC was one of the pioneers of Australian aircraft production (q.v.) and built the first all-metal aircraft in Australia, the Wirraway, which was test-flown in March 1939. The corporation came into its own in the Second World War during which the CAC produced over 1000 aircraft including 717 Wirraways, 200 Wackett trainers, and 250 locally designed Boomerang fighters. In the immediate post war period the corporation went into rapid decline. From 1946 to 1951 it produced only a few Wirraways and Mustangs, and the CA-15 fighter prototype did not go into production. Its fortunes then improved with a contract to manufacture the Sabre jet fighter. In the 1960s the CAC built Mirage jet fighters and Macchi jet trainers. In the early 1970s the CAC built Kiowa helicopters for the Australian Army Aviation Corps (q.v.). CAC was taken over by Hawker de Havilland in May 1985 and was renamed Hawker de Havilland Victoria.

(See also entries on individual aircraft.)

COMMONWEALTH WAR BOOK, a detailed statement of action to be taken by government departments on the outbreak of war, was begun within the Defence Department (q.v.) in 1938, though discussion papers of a similar nature had been prepared in 1927 and 1928. Modelled on the British War Book, it set out the measures required at each stage of the process of moving from peace to war, but did not deal with the conduct of the war itself. By the time Australia entered the Second World War the war book still lacked chapters on manpower and supply and had other significant gaps in coverage. Its economic sections were based on outdated ideas of the type of war Australia would be involved in, and one senior Commonwealth official commented later, 'When war came I locked my copy of the War Book away in a cupboard. It was just bloody irrelevant.' Similar volumes were prepared by State governments and by each fighting service. A Commonwealth War Book was also compiled in 1956.

COMMONWEALTH WAR GRAVES COMMISSION see WAR GRAVES

COMPULSORY MILITARY TRAINING see CONSCRIPTION

CONDAMINE, HMAS see BAY CLASS FRIGATES

CONFRONTATION Confrontation, or *Konfrontasi* in Indonesian, was a small undeclared war fought from 1962 to 1966 through which President Sukarno of Indonesia sought to destabilise and ultimately destroy the new Federation of Malaysia which emerged during the course of 1963. Sukarno, whose rule was becoming increasingly erratic as his own health declined, as the Indonesian economy worsened, and the conflict between the army and the Communist Party intensified, argued that the creation of Malaysia was a means of maintaining British colonial rule in south-east Asia behind a guise of independence for its former colonial possessions. His recent success in forcing the Dutch to relinquish their last remaining colonial territory in West Irian, a campaign largely of bluff backed by an untested military threat, undoubtedly emboldened him in his 'confrontation' of Malaysia, a term first used by the Indonesian foreign minister, Dr Subandrio, in January 1963.

The trigger for Confrontation was supplied by the Brunei rebellion of December 1962, in which a small party of armed insurgents, members and supporters of Azahari's People's Party, who were known as the TNKU or North Borneo National Army, attempted to seize power in the small, oil-rich independent enclave of Brunei which lay between the new Malaysian territories of Sarawak and Sabah. The rebellion was quickly and easily put down by British forces flown in from Singapore, and although no Indonesian personnel were involved directly it was clear that Azahari, who had fought in the anti-Dutch struggle for independence after 1945, had received support and encouragement from Jakarta. In early 1963 military activity increased along the Indonesian side of the border in Borneo, and small parties of armed men began infiltrating Malaysian territory on propaganda and sabotage missions. Although they masqueraded as members of the TNKU, this force was generally not an effective military organisation, and

the border-crossers increasingly comprised parties of Indonesian 'volunteers', although at this early stage regular Indonesian military personnel were not normally involved.

The first casualties were inflicted in a raid on a police post at Tebedu in April 1963. Further armed incursions of increasing strength culminated in a major assault on a post at Long Jawi on 28 September — the day on which Malaysia officially came into being — held by Border Scouts led by Gurkha soldiers, a number of whom were killed. At the end of the year there was another major raid against a Malaysian army post at Kalabakan, which again inflicted casualties. Although spectacular at times, these efforts did not bring about a general revolt against the Malaysian authorities. During the course of 1964 the Indonesians began to increase the tempo of operations by introducing regular army units into cross-border raids, and in the second half of the year they launched attacks of various strengths against peninsular Malaya itself, increasing the risk of a general war between the two countries.

Australian forces were based in Malaysia as part of the British Commonwealth Far East Strategic Reserve, specifically within the 28th Commonwealth Infantry Brigade Group (q.v.) which had fought against the communist terrorists during the Malayan Emergency in the 1950s. The Australian government was extremely wary of involving its troops in conflict with Indonesia, not least because of concerns over the possible extension of any such fighting to the long and generally indefensible border that Indonesia shared with Papua New Guinea. Thus despite several requests from both the British and Malaysian governments in 1963–64 for the deployment of Australian soldiers to Borneo to help meet the increasing military threat there, the Australian government decided that its troops could only be used for the defence of peninsular Malaya against external attack. This in fact occurred on two occasions in September–October 1964, when the Indonesians launched paratroop and seaborne raids against Labis and Pontian, and members of 3RAR were used to help mop up the invaders. The two assaults, easily dealt with in themselves, nonetheless posed so serious a risk of major escalation in the fighting that in January 1965 the Australian government relented and agreed to the deployment of its battalion for service in Borneo.

The fighting in the difficult terrain and debilitating climate of Borneo was characterised by several features. Extensive use was made of company bases, sited along the border to protect centres of population from enemy incursions, often close to

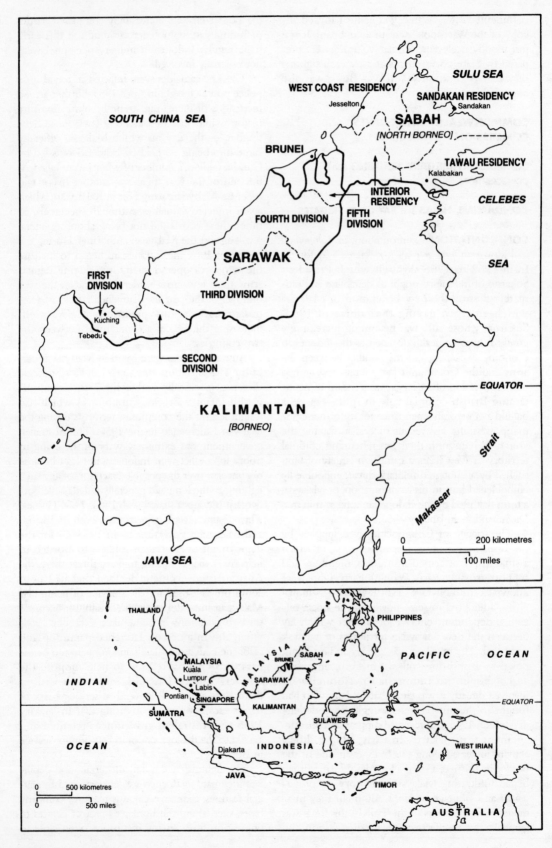

known incursion routes. After considerable deliberation, the British government had given approval for the war to be taken to the enemy, and cross-border operations, codenamed CLARET, were mounted by the security forces to procure intelligence and, more importantly, to force the Indonesians back behind their side of the border and off balance. Uncertain about where or when the British, Australian and New Zealand troops might strike next, the Indonesians devoted more of their resources to protecting their own positions and correspondingly less to offensive operations, although these still continued. Two other significant features of the border war were the extensive use of helicopters for troop movement and resupply, and the role of intelligence, including signals intelligence, in helping to divine enemy movements and intentions.

The first Australian battalion, 3RAR, arrived in Borneo in March 1965, and served in Sarawak until the end of July. In this time the Australians mounted extensive operations on both sides of the border, had four major contacts with Indonesian units, and suffered casualties in two mine incidents. In 1966 their replacement in the 28th Brigade, 4RAR, served in Sarawak between April and August. They had a somewhat less active tour, although they too operated on the Indonesian side of the border and were involved in clashes with parties of Indonesian regulars. In addition to the infantry battalions, two squadrons of the Special Air Service (q.v.), several artillery batteries and parties of the Royal Australian Engineers, and ships of the RAN were involved in various forms of activity in Borneo and surrounding waters.

Twenty-three Australians were killed during Confrontation, only seven of them on operations, and eight were wounded. Because of the extreme sensitivity of the cross-border operations, which were never admitted to at the time, very little coverage of Confrontation occurred in the Australian press. The Indonesians and Malaysians signed a peace treaty in Bangkok in August 1966, bringing the conflict to a close.

Peter Dennis and Jeffrey Grey, *Emergency and Confrontation: Australian Military Operations in Malaya and Borneo, 1950–1966* (Allen & Unwin, Sydney, 1996).

CONSCIENTIOUS OBJECTORS

CONSCIENTIOUS OBJECTORS With the passing of the Defence Act (q.v.) in 1903 Australia became the first country to provide in national legislation for the granting of total exemption from military service on the grounds of conscientious belief. The

---

Map 8 Confrontation 1964–66. (Top) The division of the Malaysian territories Sarawak and Sabah into administrative areas. (Bottom) Malaysia and Indonesia.

---

act said that exemption could be granted to men who could prove that 'the doctrines of their religion forbid [them] to bear arms or perform military service'. When the Defence Act was amended in 1909 to provide for compulsory military training (see CONSCRIPTION) it granted exemption from combat duties to those with a religious objection to bearing arms; however, such people could still be made to perform non-combatant duties (such as clerical work or stretcher-bearer duties) the nature of which would be determined by military authorities. The Defence Act of 1910 still did not extend the exemption to cover non-combatant military service, but it did alter the basis of exemption by removing the reference to religious doctrine. Now objectors had to show only that their 'conscientious beliefs' prohibited them from bearing arms. The 1910 Defence Act also spelt out that exemption would be determined by the courts, so that decisions would be made by civilian magistrates and could be appealed to higher courts.

Because conscription was not introduced during the First World War conscientious objection did not become an issue again until the introduction of the Defence Act of 1939. This act clearly stated that conscientious beliefs did not have to be of a religious character to qualify someone for exemption, but did not exempt conscientious objectors from non-combatant military duties. It also provided for appeal to the Supreme Courts or the High Court on questions of both fact and law. From July 1941 the government allowed exemption from all military service, combatant or non-combatant, but this change in policy was not formalised until February 1942 when new National Security Regulations (q.v.) were gazetted. Further amendment of the regulations in 1942 meant that men exempted from military call-up could be directed to civil work under civilian control by the minister for Labour and National Service. In fact they were largely left undirected because of their small number and the administrative problems involved in making provision for those who objected even to civilian work related to the war effort. During the Second World War courts determined the status of 2726 applicants for exemption from conscription on grounds of conscientious belief, of whom 636 were rejected, 1076 were directed to non-combatant duties, 973 were granted conscientious objector status on the condition that they were liable to perform civilian work and 41 were granted unconditional status.

The *National Service Act 1951* repeated the 1939 Defence Act's broad definition of conscientious beliefs. It created two categories of conscientious objectors: those whose beliefs did not allow them to perform any kind of military service and those who objected only to combatant duties. These provisions operated during the Vietnam War, when the definition of conscientious objection was clarified in various court decisions. It was held that conscientious belief must be deep-seated and compelling, so much so that it could be expected to have a broader impact on the applicant's life. It must be arrived at by a process of reasoning in which conclusions were drawn from basic moral premises, though emotional responses could also play a part in such a belief. Courts tended to reject applicants who showed signs of having been instructed rather than developing their beliefs on their own. Men were allowed to make successive applications, and if this happened magistrates were not required to compare past and present beliefs. Most magistrates were not prepared to grant conscientious objector status to those who objected specifically to the Vietnam War rather than to war in general, and when status was granted in such cases the Crown appealed and invariably won its case in higher courts. The Vietnam War also raised the question of how to deal with 'draft resisters' who refused to cooperate with the National Service scheme, and procedures were introduced which allowed a small number of resisters to be examined compulsorily by courts to determine whether or not they qualified as conscientious objectors. Of 1012 men who had their applications for complete exemption from military service considered between 1965 and June 1971, 733 were granted total exemption, 142 were granted exemption from combat duties and 137 were refused exemption.

During the Vietnam War the Labor Opposition had tried unsuccessfully to get the National Service Act amended so that opposition to a particular war rather than to all wars was sufficient grounds for the granting of conscientious objector status. When Labor came to power in 1983 a private member's bill to provide for conscientious objection to particular wars was introduced and referred to a senate committee which in 1985 came out in support of the change. Legislation passed in 1992 incorporated this change; it also made determination of conscientious objector status a matter for special tribunals headed by a lawyer rather than courts, and it defined conscientious objection more clearly. Conscientious objectors should hold fundamental convictions of what is morally right and wrong, though these convictions need not be religiously based, and their beliefs must be so compelling that they feel duty-bound to follow them.

Hugh Smith, 'Conscience, law and the State: Australia's approach to conscientious objection since 1901', *Australian Journal of Politics and History*, vol. 35, no. 1 (1989), pp. 13–29.

CONSCRIPTION (or compulsory military service, or national service) has been a controversial issue in Australian history. Although the creation of an efficient defence force in place of the uncoordinated efforts of the individual colonies had been one of the driving forces of the federation movement (see TENTERFIELD ORATION), the intention of the 1903 Defence Act (q.v.) did not extend beyond requiring men between the ages of 18 and 55 to serve in the military forces of the new Commonwealth, and then only for home defence. The Japanese victory in the Russo-Japanese War of 1904–5 and the gradual withdrawal of main RN forces from the Pacific to face the greater threat from Germany gave rise to fears that Australia was vulnerable to attack. While much of the Labor Party was opposed even to the limited requirements for service under the Defence Bill, one member, William Morris Hughes (q.v.), was a passionate advocate of compulsion — in peacetime as well as in war. Hughes believed that a strong military force, available for home defence, was essential if Australia's racial purity, especially in the face of Japan, was to be maintained. Just as supporters of compulsory military service in the United Kingdom argued that without a powerful home defence capability Britain was open to invasion or large-scale raids, Hughes increasingly gained support for his views within the Labor Party, so that in 1908 the federal conference of the party voted by a majority of 24 to 7 to endorse compulsory military training in peacetime. Prime Minister Alfred Deakin had also come to support compulsion, and when he introduced a bill for compulsory peacetime training, there was little opposition. Parliament's views were confirmed, at least in part, by Field Marshal Lord Kitchener, who visited Australia in 1909 at Deakin's invitation. Kitchener recommended the raising of an army of 80 000, enlisted through compulsory service and trained by graduates of the military college, whose establishment also formed part of his proposal. The Labor government that had been elected in 1910 accepted Kitchener's advice: RMC Duntroon (q.v.) opened in 1911, and universal compulsory military training came into operation on 1 January 1911, with three levels of training. Boys who turned 12 or 13 that year (or 12 in subsequent years) joined the junior cadets; 14 to 18-year-olds joined the senior cadets; and 18 to 26-year-olds were enrolled in the CMF.

Medical exemptions were granted to a small number of males over the age of 14 each year: in 1911 4 per cent were permanently exempted because of physical handicap, and a further 3 per cent were given temporary exemption because of illness at the time of examination. Permanent exemptions were granted to those who lived in sparsely populated areas, while those who lived more than 5 miles from designated training centres (which were towns and surrounding districts that could produce a minimum of 60 eligible boys) were given temporary exemptions. These two categories accounted for 10 per cent and almost 25 per cent respectively, and added to their number were those exempted because they were schoolteachers engaged in training junior cadets, members of the regular military forces, or theological students, so that of the 155 000 youths who registered in 1911 for the senior cadets and the CMF, only 92 463 were liable for training. No provision was made for the exemption of conscientious objectors, although the regulations did suggest that religious objectors might be accommodated in non-combatant roles within the cadets or the CMF.

Junior cadets, who did not wear uniforms, trained at school (which required the cooperation of the State education departments) for a total of 90 hours a year, which involved a minimum of 15 minutes' physical education a day, marching drill, and a choice of two of the following activities: first aid, swimming, miniature rifle shooting and running exercises in organised games. Junior cadet units were inspected annually by a military officer. Senior cadets were obliged to train for 64 hours a year (reduced in December 1911 from the original requirement of 96 hours), made up of four whole-day (four hour) drills, 12 half-day (two hour) drills and 24 night (one hour) drills, although voluntary camps and additional training often added substantially to this minimum requirement. Training consisted of physical drill, company exercises, field training and musketry. For those between 18 and 20, the CMF obligation entailed 16 whole-day drills or their equivalent, half of them spent in an annual camp, with the rest spread over a variety of drills (no less than six hours for a whole-day drill, three hours for a half-day drill, and one and a half hours for a night drill). Once a young man had turned 20, his CMF obligation consisted merely of an annual registration or muster parade. Most 18 to 20-year-olds served in the infantry in the CMF; entry into the Light Horse was restricted to those who could provide their own horse, while for trainees in the engineer or artillery corps, the annual obligation was for 25 days a year, 17 of which had to be spent in camp. A small number of senior cadets was selected by the RAN (which had first choice of the annual intake): 3000 in 1911, and about 1000 in following years. For naval cadets, the annual obligation was for 25 days, 17 of which had to be spent in continuous training on board ship. Senior cadets were paid a daily rate of 3

shillings in the first year of training, and 4 shillings a day thereafter. Men who were enrolled in the militia in 1911 were permitted to complete their three-year term of engagement, but from 1911, apart from officers and NCOs in the old militia who were allowed to continue serving, entry to the CMF was restricted to those who moved up each year from the senior cadets. Australia was divided into 219 training areas, and the Regular instructional staff was expanded to over 400 to provide warrant officers or NCOs to assist the officer (eventually a Regular, but originally usually a temporary or part-time officer) appointed to take charge of registration and training.

The opposition to the concept of compulsory military training did not die away once the scheme was established, and non-compliance, which was punishable by a fine of not less than £5 and not more than £100, became a problem from the very beginning, although the extent of non-compliance and prosecution has been exaggerated. In the first year of the scheme about 10 000 of the eligible youths did not register, and many others did not complete the required number of drills. The fines imposed were considered extremely harsh, and in the second year of the scheme a new scale was introduced, £5 being the maximum, with provision for fining youths after they had missed only one parade and for detention (for non-payment of fines) in places other than normal gaols. In the four years 1911–15, there were 34 000 prosecutions and 7000 detentions, out of a total enlistment of 636 000; that is, 1 in 20 youths was prosecuted, although many of the prosecutions were for multiple offences, thus reducing the number of individuals involved. Three-quarters of those who appeared in court for non-compliance pleaded guilty, while most of the rest argued that their particular circumstances made regular attendance at drill difficult; very few sought to defend their position on the grounds of principle. The compulsory service scheme continued throughout the Great War, though on a much reduced basis. In the postwar period it barely survived, the junior cadets being abolished in June 1922, with registration for the senior cadets being restricted to 16 and 17-year-olds, and training in the CMF restricted to 18 and 19-year-olds. The Labor government that came to power in 1929 suspended the scheme by regulation rather than by act of parliament.

Labor's adoption of a policy of abolition in 1922 had been prompted in part by the bitter divisions within its ranks that had been caused by the two referendums over conscription during the First World War. In October 1915, when voluntary recruitment for the AIF had begun to fall dramatically, W. M. Hughes became leader of the Labor Party and prime minister. Although only several months before he had seemingly ruled out conscription, he had also said that it was not possible to foresee what the future course of the war might demand. While in Britain in early 1916 he had been very critical of the British conduct of the war, and shortly after his return in July British forces sustained enormous losses on the Somme, with the AIF suffering especially heavy casualties at Pozières and Fromelles. Hughes was warned by the British that unless 16500 men enlisted in each of the following three months, in place of the approximately 6000 who had joined up in each of the previous months, the 3rd Australian Division then training in England would have to be broken up in order to maintain the other divisions in the field. Facing considerable opposition from within his own party, Hughes decided to go to the people in a referendum, asking the electorate to approve the government's extension of its existing powers under the Defence Act to require men to serve overseas. The word 'conscription' was not mentioned in the question. Hughes attempted to manipulate the result by withholding the votes of men who had not answered a pre-referendum proclamation calling them up for home service; and he received virtually unanimous support from the metropolitan press, the Protestant churches, and the State governments, only the Labor government of Queensland refusing to back his cause. He was opposed by the rank and file of the Labor Party and by the trade unions, and by much of the Catholic community, led by the Irish-born Archbishop of Melbourne, Dr Daniel Mannix, whose opposition to the war had been crystallised by the British treatment of the leaders of the abortive Easter uprising in Dublin. The referendum was held on 28 October 1916, and resulted in narrow rejection of conscription with 1160033 votes against, and 1087557 in favour. New South Wales, South Australia and Queensland voted against conscription; Victoria, Tasmania and Western Australia voted for it. Within the AIF, a narrow majority supported it: 72399 to 58894, the 'yes' vote among the soldiers being attributed to those not actually at the front. The political impact was immediate: the Labor caucus passed a vote of no confidence in Hughes who, with one-third of the caucus, left to form a breakaway Labor government which merged with the Liberals in January 1917 to form the Nationalist Party government. Voluntary recruitment continued to lag badly behind estimates of manpower needs throughout 1917, and by the end of the year the situation was said to be desperate, given the losses the AIF had sustained in the Third Battle of Ypres, following heavier losses earlier in the year. Pledging to resign if defeated a second time, Hughes initiated another referendum to permit a ballot of single men aged 18–44 in order to produce a monthly reinforcement rate of 7000 recruits. Again, the question that was put to the electorate on 20 December 1917 did not mention the word 'conscription'. After a bitter and heated campaign, the vote went against Hughes: 1181747 said 'no', 1015159 said 'yes'. Victoria switched from its earlier support for conscription and joined the States opposed to Hughes's proposal. With the loss of the second referendum, conscription disappeared from the political agenda, and the AIF remained the only all-volunteer force in the First World War apart from South Africa.

Compulsory military service for duty within Australia was revived by the United Australia Party government in 1939, shortly after the outbreak of the Second World War. Prime Minister R. G. Menzies (q.v.) announced on 20 October that unmarried men who would turn 21 in the year ending 30 June would be called up for three months' training with the militia, whose strength had to be maintained in the face of predicted transfers to the 2nd AIF and the withdrawal of men who were in reserved occupations. Compulsion in wartime for service within Australia was explicitly allowed under the Defence Act, but the ALP, scarred by the conscription controversy of the First World War, declared its total opposition to any form of military conscription. On several occasions the government assured the country that there would be no conscription for overseas service, but insisted that the definition of 'Australia' for this purpose included New Guinea and the adjacent islands. The liability for conscripted men to fight on Australian territory beyond the continental limits remained a theoretical one until the outbreak of war with Japan. The distinction between the AIF and the militia began to break down and, in order to bring the CMF quickly to full war establishment, in early 1942 the new Labor government suspended the right of CMF members to enlist in the AIF and compulsorily transferred surplus AIF recruits in training depots to CMF units. Throughout 1942 the Opposition urged the government to create a single army whose members had a common obligation to serve in the defence of Australia, a defence that could no longer be based on repelling invasion but which had to be secured by offensive action beyond the continental shores. Prime Minister John Curtin (q.v.) rejected the pressures as being purely motivated by partisan political motives, but he himself was already moving towards accepting the implications of a 'one-army' policy. At the

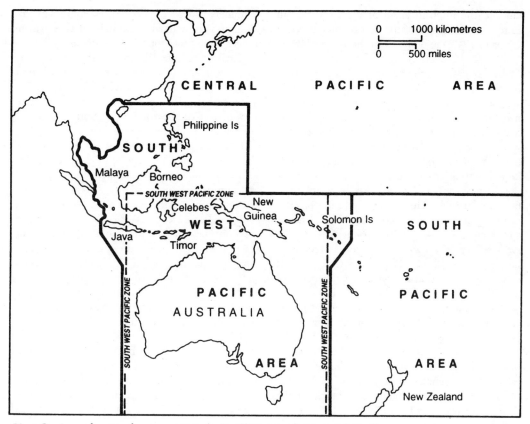

Map 9 Area of service for conscripts in the South-West Pacific Zone, 1943–45.

Special Federal Conference of the Labor Party in November 1942, Curtin argued that Allied unity with the United States, and the need to redress anomalies that had already arisen in the application of the Repatriation Act (AIF men qualified for certain benefits even though they had not actually served overseas, while militia soldiers who had fought in New Guinea did not receive these same benefits), made it necessary to extend the definition of territories under the Defence Act to include those areas of the south-west Pacific in which it was likely that Australian troops would be involved. Despite opposition from the Victorian and Queensland branches, which rejected conscription for overseas service, the reassembled Special Federal Conference in January 1943 accepted Curtin's proposals by 24 votes to 12. The Defence (Citizen Military Forces) Bill was introduced into parliament on 29 January. The Opposition attempted unsuccessfully to remove the geographical limits on the employment of conscripts outside continental Australia, and on 19 February the bill became law. Under the act members of the CMF could be required to serve in the South-West Pacific Zone, which was defined as the area bounded on the west

by the 110th meridian of longitude east, on the east by the 159th meridian longitude east, and on the north by the equator (see Map 9). The legislation was to remain in force for six months after the ending of hostilities. Despite the power to compel members of the militia to serve overseas, Curtin did not move to create one army. Far from being the means by which the Australian war effort could be widened, it is clear that Curtin conceived of Australia's role as a strictly limited one, in which Australian forces would occupy territories and bases vacated by American forces as they moved forward on their offensive drive against the Japanese. While the imposition of an overseas liability on the militia removed one ground of criticism of Australia's war effort, there is no evidence of direct American pressure for the change in policy. Nor was it a military necessity, because there were more than enough voluntary transfers from the militia to the AIF to maintain the latter in whatever role the government decided for it.

A third compulsory military service scheme was introduced in 1951 by the Liberal government. In part this initiative was a response to the deterioration in the international scene, in particular to the intensification of the Cold War in Europe, the triumph of the communist forces in China in 1949, the outbreak of the Korean War in 1950, and the declaration of the Emergency in Malaya in 1948, prompted by the Malayan Communist Party's launching of an armed insurrection against the civil power. Australia's military forces had run down substantially since the end of the Second World War, and the National Service Act was designed to produce large numbers of partially trained men who, on the outbreak of war, could be brought rapidly to an acceptable level of operational efficiency without the need for a prolonged period of training. The second declared objective of the scheme was to improve the physical fitness and discipline of the nation's young men, which was seen as part of a wider preventive and remedial health strategy. All men of 18 years were required to register for National Service, and to undertake an active training obligation of 176 days, which was completed in one continuous period in the RAN and the RAAF, whereas in the army it was broken into a period of 98 days in the Regular Army followed by 78 days in CMF units, where national servicemen mixed with volunteer CMF recruits. After completion of the prescribed period of training, trainees were required to remain on the Reserve for five years from the date of their initial call-up. The choice of service was open to trainees, but there was a significant difference between the army and the other two services. National servicemen who chose to complete their obligation in the army could be required to serve only within Australia; RAN and RAAF trainees had to volunteer to serve outside Australia if required as a precondition for their enlistment in either of those services. For the first six years, about 33 000 young men were trained annually under the scheme, all but 5000 choosing to enter the army. There was widespread public acceptance of National Service, and little evasion among 18-year-olds. In 1953 the total active service obligation was reduced to 154 days for the RAN and RAAF, and to 140 days for the army.

Much more sweeping changes were made to the scheme in 1957, when as a result of a defence review it became apparent that the objectives of National Service training had little relevance to Australia's defence needs. The commitment of ground forces to Malaya in 1955, primarily as part of the Far East Strategic Reserve but also to assist in the suppression of the communist insurgency in Malaya, indicated that in place of the threat of global war that had in part prompted the original scheme, the most likely requirement was for much smaller but more highly trained military forces.

Numbers had become much less important than skill, and the partially-trained amateur was no longer a substitute for the professional soldier. The original scheme had been a heavy drain on the manpower of the Regular Army, as several thousand of its members were involved in National Service training, which consumed a sizeable proportion of the total army budget, and was money that could be spent far better on the Regular Army, especially on equipment. In April 1957 the government abolished the universal obligation, and introduced a selective system of training, reducing the annual intake to 12 000 and ending the option of serving in the RAN or the RAAF. This freed about 2 000 Regular Army personnel to serve in the newly created mobile brigade group. These proposals attracted considerable criticism from all sides of politics. The drain on the manpower of the Regular Army still continued under the revamped scheme, albeit in a reduced form, but the defence benefits were still questionable at best. The shift away from the universal basis of the original scheme raised questions of social equity, and the failure of the government to impose an overseas obligation on national servicemen in the army meant that even if they had achieved a level of proficiency they could not be deployed outside Australia, while the constant influx of national servicemen into the CMF had a serious impact on voluntary enlistments. In other words, by 1957 the National Service scheme had outlived its usefulness in defence terms. The government, however, retained some faith in its social objectives, and was reluctant to confront the RSL (see RETURNED AND SERVICES LEAGUE), which was a staunch supporter of universal National Service. The needs of the Regular Army, especially for equipment, were so pressing, however, that on 26 November 1959 the government announced the end of National Service, using much the same arguments as it had mustered three years earlier in reducing the scheme. Almost 3000 Regular soldiers would be released from running the National Service training programs, and the annual cost of £9 million, which was one-seventh of the total army budget, could now be spent on repairing the serious deficiencies in equipment that had accumulated during the 1950s. The CMF was to be more fully integrated with the Regular Army in the new Pentropic division (q.v.), and this new role, it was claimed, made it impracticable to have both volunteers and national servicemen in CMF units. The strength of the CMF was to be increased by 12 000, a figure thought to be achievable once CMF units no longer suffered from the dilution effect of large numbers of national servicemen.

There was little opposition to these changes, except from the RSL, for the military arguments in favour of National Service had long since become unconvincing. Until the end of the scheme, public support remained high, with polls regularly showing that about 85 per cent of those questioned favoured some form of National Service.

A fourth compulsory military service scheme was introduced in 1965, when events in south-east Asia again raised fears about the security of Australia. The increasing instability of the government of South Vietnam, Indonesia's policy of 'confrontation' (q.v.) with the new Malaysian Federation, and the possibility of direct conflict between Australia and Indonesia over the common border in New Guinea led the Australian government to reactivate a variation of the previous scheme, which was announced on 10 November 1964, with the National Service bill becoming law on 24 November. On 3 February 1965, the government announced that Australian troops in Malaya would be sent to Borneo to assist in repelling Indonesian incursions across the border. Less than three months later, on 29 April, the government stated that an Australian battalion would be sent to Vietnam, and a week later, on 6 May, it introduced amendments to the National Service Act to make conscripts liable for overseas service. There was a pressing need in the army for additional manpower. Employment was high in Australia and recruiting had sunk to a very low level, both in terms of absolute numbers applying and, more importantly, in terms of those finally accepted. If the army was to reach its target of 28 000 by the end of 1967 it had to attract an average of 3670 annually for each of the three years from 1964–65 to 1966–67, allowing for normal wastage. However, recruiting had averaged only 2400 annually for the period 1961–64; these were the figures given at the time, but figures produced several years later indicated that the recruiting pattern was not as bleak as originally painted. What was alarming, from the army's point of view, was the high rate of rejection or withdrawal from the recruitment process. Each year over 70 per cent of those who applied to join the army did not finally do so, with an average of 8 and 12 per cent respectively rejected on medical or educational grounds. Selective national service enabled the army to draw more fully on the educated and fit elements among 20-year-olds than it could otherwise have done in a time of full employment, although equally it could have tried to compete more vigorously in the labour market, as indeed it did in the mid-1960s with significant improvements in pay and conditions. (The resultant increase in recruiting rates then allowed the army to decrease significantly its expenditure on recruiting.)

Exemption from registration for National Service was given to Aborigines (see ABORIGINES AND TORRES STRAIT ISLANDERS IN THE ARMED FORCES), members of the armed services, and non-naturalised migrants, the third category attracting considerable criticism in public polls, one of which, in April 1966, showed that 83 per cent of the public wanted them included in the registration net. In August 1966 the government bowed to this public pressure and the legislation was changed to make resident aliens liable to be conscripted from 1 January 1967. Thereafter, ministers of religion, theological students and the disabled were the only categories exempted from registration. Conscientious objectors could be granted exemption from military service if they could convince a magistrate that their beliefs precluded them from undertaking military service, or they could be granted a partial exemption and required to serve in a non-combatant capacity. Conscientious belief, however, had to be to military service in general; objection to a particular war was not grounds for being granted an exemption. Besides exemption, there were a number of circumstances in which an individual could be granted a deferment, temporary or indefinite. Men who could show that military service would cause them or their families undue hardship could be granted a temporary exemption, which (after 1968) could be renewed indefinitely in 12-month blocks. Temporary deferments could also be obtained by apprentices and university students, to enable them to finish their studies. Those who were married before they were required to register were given indefinite deferments, as were those who had volunteered to join the CMF before they were required to register, or who had been accepted to serve in it by the time their ballot had been held; they were required to serve six years in the CMF. Not until 1971 was a loophole relating to the CMF closed. For the first six years of the scheme men could join the CMF before they registered, and if their birthdate was not chosen, could then resign from the CMF and have no further obligation.

Selection for military service was by ballot, in which one marble for each day of a six-month period was placed in a Tattersall's lottery barrel. A certain number of marbles were drawn according to calculations made by the Department of Labour and National Service of how many would be required to produce the targeted numbers for the two call-ups in each half of the year. The ballot, variously dubbed 'Russian roulette', the 'birthday ballot' or (later) the 'lottery of death', was carried out privately by government, public service and army officials, and the dates that were drawn were not published, a source of growing criticism and suspicion that the system was not as neutral as it was claimed to be. There were inequalities built into the system, but they were virtually unavoidable. For example, those who registered in the first half of 1966 had a far greater chance of ultimately being called up than those in subsequent ballots: the pool was smaller because those who turned 20 in the first half of 1966 did not represent the full impact of the postwar rise in the birth rate, and also because in the first ballot, and less so in the second and third ballots, there was no carry-over of deferments to increase the overall numbers. In the first registration, there was a 53 per cent chance of an individual's marble being chosen; by the fifth registration, that had declined to 25 per cent. Over the life of the scheme, 1965–72, 804 286 men registered for National Service, of whom 63 735 were called up to serve in the army. National servicemen were fully integrated into Regular units, where they comprised no more than 50 per cent of any unit, and a National Service officer training school was established at Scheyville (q.v.), New South Wales. A total of 19 450 conscripts served in the Vietnam War, compared with 21 132 Regular soldiers (some of whom served on more than one tour). Two hundred national servicemen were killed in Vietnam, and a further 1279 became nonfatal casualties; the equivalent figures for the Regular soldiers who served in Vietnam were 242 and 1553 respectively.

The two-year term of service was reduced to 18 months in August 1971, following the Liberal government's announcement of the withdrawal of Australian troops from Vietnam, and the scheme itself was abolished on 5 December 1972 as one of the first administrative decisions of the new Labor government. By then opposition to the scheme had emerged on several fronts. First, the highly selective nature of the call-up procedure attracted vehement criticism from those who supported a universal system of national service, whether or not it involved an overseas liability. Second, opposition to National Service became inextricably bound up with opposition to the Vietnam War, although once Australian forces were withdrawn from Vietnam protest marches and demonstrations largely disappeared, suggesting that much of the opposition was to Australian involvement using conscripts rather than to the war itself. The granting of exemptions to conscientious objectors was subject to considerable variation, and as the war in Vietnam entered a new phase following the Tet offensive of February 1968, the fact that opposition to a particular war was not grounds for an exemption fuelled public opposition

to the conscription process. Evasion of registration became widespread, and a number of men who refused to register or to report for military duty were jailed in highly publicised incidents. Student activists were at the forefront of the protest movement, but they enjoyed widespread support in the broader community. Equally, those men who were conscripted overwhelmingly fulfilled their obligations, and served creditably in Australia and in Vietnam alongside their Regular counterparts. The ALP opposed both conscription and the Vietnam War, but could not command sufficient electoral support until 1972 to put an end to a scheme that, whatever else it achieved in military terms, had become bitterly divisive in the Australian community.

CONVOYS In wartime, where merchant ships sail in groups accompanied by warships, convoys are used for protection against enemy warships and submarines. The first contingent of the AIF, with the New Zealand Expeditionary Force, sailed for Egypt from Albany, Western Australia, on 1 November 1914 in a convoy consisting of 38 transports escorted by four warships, the Australian cruisers HMA Ships *Sydney* (I) (q.v.) and *Melbourne*, the British cruiser HMS *Minotaur*, and the Japanese cruiser *Ibuki*. During the voyage across the Indian Ocean, HMAS *Sydney* was required to leave the convoy to engage the German cruiser *Emden* (q.v.) at the Cocos (Keeling) Islands. The destruction of the *Emden*, and the confining of the German cruiser *Königsberg* to a river estuary in East Africa, meant that no German ships remained at large in the Indian Ocean to attack Australian troop-ships. The second convoy, which arrived in Egypt in February 1915, had a naval escort consisting of one 'E' Class submarine (q.v.), the *AE2*, towed by one of the troop-ships, while the third — and last — convoy, which arrived in March, had no escort at all. For the remainder of the First World War troop-ships travelled from Australia alone, without the need of convoys or escorting warships.

During the Second World War the convoy system was revived, and on 20 January 1940, a convoy consisting of 11 troop-ships escorted by HMA Ships *Australia* (II), *Canberra* and *Sydney* (II) (qq.v.) left Port Phillip Bay. Following the sinking of five ships in three days by the German raider *Komet* in the South Pacific Ocean in December 1940, a convoy system was introduced for ships travelling between Australia and New Zealand. The destination of the convoy carrying the 6th and 7th Divisions from the Middle East in 1942 became a subject of intense dispute between the British Prime Minister Winston Churchill and the Australian Prime Minister John Curtin (q.v.). Initially the divisions were to reinforce the Netherlands East Indies, and the first ships landed their troops in Java, where they were captured within a month. On 15 February — the same day Singapore surrendered — Curtin, on the advice of the CGS, Lieutenant-General Vernon Sturdee (q.v.), cabled Churchill requesting that the remainder of Australian troops destined for the Netherlands East Indies be diverted to Australia. Churchill, who was being advised by General Sir Archibald Wavell (q.v.), C-in-C, ABDA Command (q.v.), also decided the troops should be diverted, but to Burma, then under Japanese attack, and in fact ordered some ships to set a course for Rangoon. Curtin, however, would not be swayed and the ships were turned back. Eventually all the Australian troops reached Australia, although some spent time on garrison duty in Ceylon. Following the sinking of the SS *Iron Chieftain* by a Japanese submarine off the Australian coast on 3 June 1942 a convoy system was introduced in Australian waters, but with a lack of warships for convoy escorts, aircraft often had to be used for protection. This required large numbers of aircraft, and during 1943 the RAAF had more squadrons serving in Australia on anti-submarine patrols than at the front in New Guinea.

CONYERS, Matron Evelyn Augusta (1 March 1870–6 September 1944). Born in New Zealand, Conyers migrated to Victoria in the early 1890s, and trained as a nurse at the Children's Hospital and at Melbourne Hospital. She joined the Army Nursing Service (see AUSTRALIAN ARMY NURSING SERVICE) as a sister in 1903, and in October 1914 was appointed senior sister of the 1st Australian General Hospital which departed for Egypt that same month. By August 1915 she had been appointed acting matron of the 3rd Australian Auxiliary Hospital at Heliopolis. Following the reform of the nursing service, provoked in large part by Matron Jane Bell (q.v.), Conyers was appointed Matron-in-Chief of the Australian Army Nursing Service in December 1915, with responsibility for all Australian nurses other than those in India and Salonika (who came under British authority). She moved to AIF Headquarters in Horseferry Road, London, in May 1916, and her personality and administrative skills helped to minimise existing differences between the nursing service and medical officers, and smoothed relations between the Australian and British nursing organisations, which in some theatres were very poor. She returned to Australia in December 1918 and her AIF appoint-

Evelyn Conyers (centre) with nurses returning to Australia on the troopship *Orvieto* following the end of the First World War, 1919. (AWM H06813)

ment was terminated in March 1920. She then resumed her militia career. In recognition of her war service she was awarded the RRC (1st Class) and bar, was made OBE and CBE, and was mentioned in despatches. Her professional contribution was recognised by the award of the Florence Nightingale Medal and Diploma by the International Committee of the Red Cross in 1921.

COOTAMUNDRA, HMAS see *BATHURST* CLASS MINESWEEPERS (CORVETTES)

CORAL SEA, BATTLE OF THE One of a series of crucial engagements between Allied and Japanese forces in 1942 that stopped the Japanese advance southwards, the battle of the Coral Sea was fought between 5 and 8 May. The Japanese planned to consolidate their advance by clearing Papua New Guinea of Allied forces and capturing Port Moresby, while simultaneously attacking American positions in the Solomon Islands. With this achieved, the C-in-C of the Imperial Combined Fleet, Admiral Yamamoto, would then be free to deal with the remaining American naval presence in the Pacific by bringing it to battle at Midway, planning for which was already advanced. A striking force of three carriers under the command of Admiral Inoue entered the Coral Sea on 6 May, providing cover for an invasion fleet that was to mount an attack on Port Moresby. Opposing them was the US Navy's Task Force 17, commanded by Rear-Admiral Frank Fletcher, which consisted of two carriers, three heavy cruisers, and a support group Task Force 44, including the Australian cruisers HMA Ships *Australia* (II) (q.v.) and *Hobart*, under the command of Rear-Admiral John Crace (q.v.), an Australian-born officer of the RN then commanding the Australian Squadron.

Neither side was certain of the other's position as carrier and land-based aircraft searched for signs of their opponents. Crace and Task Force 44 were ordered to intercept the invasion force bound for Port Moresby, while the carriers went in search of their Japanese opposite numbers. On the morning of 7 May American aircraft located and sank the Japanese light carrier *Shoho*. Land-based aircraft from Rabaul launched several attacks on Crace's ships, without managing any hits, while in the early evening aircraft from the carrier *Yorktown* intercepted a force of Japanese bombers, shooting down nine of them. On the following day each side launched simultaneous aerial strikes against the other's carriers. American attacks damaged the *Shokaku* to the extent that it was unable to recover aircraft and was ordered out of the action. The first Japanese attacks inflicted hits on both American

carriers, the *Lexington* having to be abandoned at the end of the day when fires caused by an internal explosion could not be controlled. The Japanese, however, had suffered significant losses in aircraft and this, together with uncertainty about remaining American strength and Inoue's cautious nature, led the Japanese commander to postpone the assault on Port Moresby and recall the invasion force. The result was an Allied tactical victory, one which weakened the Japanese force at Midway the following month, as a result of which battle they lost the initiative in the Pacific and, ultimately, the war. It was also a strategic victory, in that the Japanese were now forced to try to capture Port Moresby by means of a ground campaign (see NEW GUINEA CAMPAIGN) which they were ill equipped to mount, while their wider plan to cut the sea lines of communications between Australia and the United States could not proceed.

Coming so soon after the spectacular Japanese victories in Malaya, Singapore, the Netherlands East Indies and the Philippines, the Allied victory in the Coral Sea provided an important psychological boost to the Australian population. While it is not true that the battle saved Australia from direct invasion, as at that stage the Japanese had not yet decided to attempt such a task, it is true that the Coral Sea battle was one of the decisive victories of 1942 which ultimately led to the total defeat of Japan. After the war, celebrations in Coral Sea week were used to demonstrate and symbolise the strength and importance of the postwar Australian–American alliance. While the more exaggerated statements about the battle common to such public occasions should be taken with a grain of salt, this should not be allowed to obscure the important role that the battle played in Allied victory.

Chris Coulthard-Clark, *Action Stations Coral Sea: The Australian Commander's Story* (Allen & Unwin, Sydney, 1991).

CORONET Code-name for General Douglas MacArthur's (q.v.) planned assault on the Japanese home island of Honshu, scheduled for early 1946. In his initial planning in April 1945 he informed General Sir Thomas Blamey (q.v.) that he intended to use the three AIF divisions but that they would have to use American weapons and supplies to ease logistic complications. In a burst of typically MacArthurian hyperbole, he stated that if ships of the RAN were still under his command, he intended to command the operation from one of them. The plan was modified in July: MacArthur now intended to use a British Commonwealth corps of three divisions — British, Canadian and Australian — as the assault reserve, but again they would have to use American equipment and operate as a corps within one of the three American armies (First, Eighth and Tenth) earmarked for the invasion. It is not clear whether MacArthur would in fact have agreed to the inclusion of non-American forces in the last great operation of the Pacific War, but the dropping of the two atomic bombs and the Japanese surrender in August made the question academic.

CORPS This term has two distinct organisational meanings. In the first, a corps is a grouping of two or more divisions, usually regarded as a lieutenant-general's command. In the First World War the formation of the Australian Corps in November 1917 brought together the five divisions of the AIF in France which had previously served, with the New Zealand Division, in I and II Anzac Corps. In the Second World War the 6th Division fought as part of the British XIII Corps in the first Libyan campaign in January 1941; when the 7th Division arrived in the Middle East the two were grouped as I Australian Corps. By the end of the Pacific War there were two corps overseas as part of First Australian Army: I Corps, which contained the 6th, 7th and 9th Divisons of the 2nd AIF, and II Corps, which contained the 3rd, 5th, and 11th Divisions of the AMF. There is no corps maintained in the modern peacetime regular army.

In the second meaning, the old distinction between 'arms corps' and 'service corps' has been replaced with a more sophisticated categorisation of the 22 corps that make up the Australian Army. The manoeuvre corps, those which have a direct involvement in battle, are the Royal Australian Armoured Corps and the Royal Australian Infantry Corps (qq.v.). The combat support corps enable the manoeuvre corps to operate effectively in the field: they are the Royal Regiment of Australian Artillery, the Royal Australian Engineers, the Royal Australian Survey Corps, the Royal Australian Corps of Signals, the Australian Army Aviation Corps and the Australian Intelligence Corps (qq.v.). The combat service support corps are not directly involved in combat but provide a wide variety of services and functions to the army as a whole; this group consists of the Royal Australian Army Chaplains Department, the Royal Australian Corps of Transport, the Royal Australian Army Medical Corps, the Royal Australian Army Dental Corps, the Royal Australian Army Nursing Corps, the Royal Australian Army Ordnance Corps, the Royal Australian Electrical and Mechanical Engineers, the Royal Australian Army Education Corps, the Australian Army Catering Corps, the Royal Australian Army Pay Corps, the Australian Army Legal Corps, the Royal Australian Corps of Military Police, the Australian Army Psychology Corps and the Australian Army Band Corps (qq.v.).

These corps vary enormously in size and variety of function. The Women's Royal Australian Army Corps (q.v.) was abolished in 1990 when women were integrated into the rest of the army and made eligible for service in combat-related roles. The Royal Australian Army Service Corps (q.v.) was abolished in 1973 and its varied functions transferred to other corps, or embodied in new specialist corps. The Australian Army Veterinary Corps (q.v.) was likewise abolished when the army ceased to rely on horses and other forms of animal power. The Australian Cadet Corps (q.v.), established in 1951, was abolished in 1975. It is not to be confused with the Corps of Staff Cadets at RMC Duntroon or the Corps of Officer Cadets at the Australian Defence Force Academy (qq.v.), neither of which are corps of the army despite their titles. All officers of the manoeuvre corps and combat support corps are assigned to the Australian Staff Corps (q.v.), which takes precedence in the Army List (q.v.).

CORPS OF AUSTRALIAN ELECTRICAL AND MECHANICAL ENGINEERS see **ROYAL CORPS OF AUSTRALIAN ELECTRICAL AND MECHANICAL ENGINEERS**

CORVETTES see **BATHURST CLASS MINESWEEPERS (CORVETTES)**

COTTON, Frank see **SCIENCE AND TECHNOLOGY**

COUNCIL OF DEFENCE was established on 12 January 1905 by regulations under both the *Naval Forces Regulations* and the *Military Regulations*. This duplication of regulations reflected the fact that the main role of the council was to ensure cooperation between the navy and the army. Its responsibilities were to consider matters referred to it by the minister for Defence in the areas of general military and naval defence policy, measures necessary to defend Australia in time of war, and defence expenditure. As constituted in 1905, the council's regular members were the minister for Defence, the treasurer, the Inspector-General of the Military Forces, the Director of Naval Forces and the Chief of Intelligence. In 1918 Defence Minister Senator George Pearce (q.v.) decided that the scope and powers of the council were too limited for the task of coordinating naval, military and civil resources for the defence of Australia. Accordingly, the council was abolished and re-established on 23 April 1918. The reconstituted council was to enquire into specific areas relating to the preparation for and prosecution of war. It could appoint subcommittees to consider these areas; the full council would only meet to appoint and direct committees and to consider larger questions connecting policy and the forces. The council now consisted of the prime minister, the ministers for Navy and Defence, and two members each representing the defence departments.

In 1923 it was decided to designate council meetings either General (chaired by the prime minister with wide ministerial representation) or Ordinary (chaired by the minister for Defence). At the same time the functions of the council were outlined in greater detail. It was to ensure that Australian defence planning was consistent with imperial policy, to act as the liaison body with the Committee of Imperial Defence, to oversee the efficient and economical application of defence policy, to coordinate the activities of the armed forces and civil organisations in preparation for mobilisation, and to advise upon and supervise defence expenditure. In 1935 the division of council meetings into General and Ordinary was abandoned, and the council's functions were simplified to considering and advising on questions of defence policy or organisation referred to it by the prime minister or minister for Defence. Up to 1939 meetings of the council were held very irregularly, with large gaps from 1905 to 1911, 1915 to 1918, and 1929 to 1935, when no meetings were held.

On the outbreak of war in 1939 the council was replaced by the War Cabinet (q.v.) and was not reconstituted until 28 February 1946. After the war meetings of the council became very intermittent, and for much of the time it seems it was not actively functioning. The council was reconstituted in February 1976 with the function of considering matters referred to it by the minister for Defence relating to the control and administration of the ADF and its arms. In 1985 its members were the minister for Defence, ministers assisting the minister for Defence, the secretary of the Department of Defence, the Chief of the Defence Force, and the Chiefs of Staff.

Robert Hyslop, 'The Council of Defence 1905–1939', *Canberra Historical Journal*, new series no. 27 (March 1991), pp. 40-7.

COUNTESS OF HOPETOUN, HMVS see
VICTORIAN NAVY SHIPS

COURTS MARTIAL see **LAW, MILITARY**

COWRA, HMAS see *BATHURST* **CLASS**
MINESWEEPERS (CORVETTES)

COWRA BREAKOUT occurred on 5 August 1944, when Japanese POWs, armed only with improvised clubs and kitchen knives, stormed the perimeter fences of the detention camp at Cowra, New South Wales. Many escaped, but they were recaptured over the following days. Of the 1104 Japanese in the camp, 234 were killed and 108 wounded; four prison guards also died. Factors that possibly led to the breakout were the prisoners' feelings of shame at being captured, their uncertainty about the future, and their opposition to a decision by Australian authorities to separate prisoners who were officers and NCOs from private soldiers. After the war the Japanese graves were relocated to a new cemetery, a Japanese garden was established, and the citizens of Cowra became some of the leading advocates of greater Australian–Japanese understanding.

Charlotte Carr-Gregg, *Japanese Prisoners of War in Revolt: The Outbreaks at Featherston and Cowra during World War II* (University of Queensland Press, Brisbane, 1978).

COX, General H. Vaughan (12 July 1860–8 October 1923) was educated at Sandhurst and commissioned into the King's Own Scottish Borderers in 1880, but transferred to the Indian Army in 1883. He served in the Afghan War of 1879–80, in Burma in 1885–89, on the North-West Frontier in 1897, in China during the Boxer Rebellion of 1901–02, and on the Tirah Expedition of 1907–08. A highly experienced colonial soldier, he was mentioned in despatches twice and made Companion of the Star of India in 1911. He commanded the 69th Punjabis between 1902 and 1907, was a member of the Military Committee for the great Durbar of 1911, and commanded the 4th Indian Infantry Brigade at Rawalpindi from 1912 to 1914. At the outbreak of the Great War he was given the Gurkha Brigade, which he led in the wintery campaign of 1914–15 in France and Flanders, after which all Indian Army units were removed from the Western Front. Cox then commanded the 29th Indian Brigade through the Gallipoli campaign, and after the evacuation was promoted and appointed by Birdwood (q.v.), a fellow Indian Army officer, to command the newly raised 4th Australian Division, an appointment that caused some unfavourable comment in Australian government circles, although not through any questioning of Cox's credentials. He was mentioned in despatches a further three times between 1914 and 1917, made KCMG in 1915, and KCB at the war's end. He was replaced in command of the 4th Division in January 1917 by Holmes (q.v.), in keeping with the Australian government's policy that commands of Australian formations should go, where possible, to Australians. Bean described Cox as 'that capable veteran' officer, who had 'splendidly commanded' the division through a difficult year. Appointed Secretary of the Military Department of the India Office between 1917 and 1921, Cox was a member of the Esher Committee from 1919 to 1920, and retired from the Indian Army in 1921.

COX, Major-General Charles Frederick (2 May 1863–20 November 1944). Cox became a clerk in the New South Wales State railways in 1881, and enlisted in the New South Wales Lancers in 1891, being commissioned in 1894. He travelled to England twice with detachments of his unit, in 1897 and again in 1899; on the latter occasion he and most of his men volunteered to return via South Africa and saw active service following the outbreak of the Boer War. He served in the column commanded by Major Edmund Allenby, and returned to Australia at the end of 1900. He went back to South Africa with the 3rd New South Wales Mounted Rifles the following April, and by June had been promoted locally to lieutenant-colonel and was serving with Rimington's column, where he remained until April 1902. He was mentioned in despatches, awarded the CB, and earned the nickname 'Fighting Charlie', which followed him in his service during the First World War. He commanded the 1st Australian Light Horse between 1906 and 1911, and in 1914 raised the 6th Light Horse and took them overseas; they fought on Gallipoli as part of the 2nd Light Horse Brigade. Cox was wounded in May, and succeeded Chauvel (q.v.) in command of the 1st Light Horse Brigade in November. Illness forced him to miss the campaigning in the first half of 1916 in Sinai, but he led his brigade at Magdhaba in December and was instrumental in the defeat of the Turks there which finally cleared them from Sinai. He campaigned throughout 1917–18 in Palestine and Syria, was awarded the CMG and DSO, and had numerous mentions in despatches. No great intellect, he led from the front and enjoyed considerable success as a brigade commander, although his superiors were well aware that this was his optimum level of command. After the war he was elected as a Nationalist senator for New South Wales in the election of 1919. He spoke occasionally on defence issues and matters concerning the railways, but made little mark in politics, although he held his seat until 1938. He continued a militia career immediately after the war, commanding the 1st Cavalry Division between 1921 and 1923, but was placed on the Retired List in the latter year. He was made honorary colonel of the 1st/21st Light Horse (his old unit, the New South Wales Lancers) in 1929, and worked tirelessly on behalf of returned servicemen until his health failed.

COXEN, Major-General Walter Adams (22 June 1870–15 December 1949). A clerk and draughtsman by trade, Coxen was commissioned in the Queensland militia garrison artillery in 1893, while unemployed, and in June 1895 transferred to the permanent artillery. He undertook the long course at the gunnery school at Larkhill in England in 1897, returning in early 1898 to command the Queensland garrison on Thursday Island. He did not serve in South Africa (in common with many artillery officers), and in July 1902 became chief instructor at the School of Gunnery in Sydney. He trained in England again in 1907–10, and from January 1911 was the Director of Artillery at Army Headquarters in Melbourne. He joined the AIF in May 1915, to command the 36th Heavy Artillery Group, which formed the Siege Brigade of the AIF, a unit entirely made up of regular officers and other ranks. It served in France with the British XVII Corps in early 1916 before the AIF reached the Western Front (q.v.), and Coxen commanded it in the battles of Serre, Ovillers and Pozières, being awarded the DSO for his service in the campaigns of that year. In January 1917 he was promoted to command of the 1st Division artillery, serving at Bullecourt and during the 3rd Battle of Ypres, for which he was made CMG. With the creation of the Australian Corps in November 1917, Coxen became the senior gunner in the AIF, retaining the position until after the Armistice and planning some of the largest artillery deployments of the war. He returned to Australia in August 1919 and became Chief of Ordnance and Fourth Member of the Military Board (q.v.), while in 1924 he was made Chief of Artillery. He held a number of senior appointments on the Military Board until his appointment as CGS April 1930. He held the position for only 18 months, being retired early due to a change in government policy in October 1931. He thus headed the army at the beginning of the Depression, but held the senior office in the service too briefly to influence policy during a period that would see the army suffer severe reductions for the rest of the decade. In retirement he headed the council responsible for Victoria's centenary celebrations in 1934.

CRACE, Vice-Admiral John Gregory (6 February 1887–11 May 1968). Born at Gunghalin, New South Wales (now in the Australian Capital Territory), Crace pursued a career in the RN, entering the training ship HMS *Britannia* in 1902. He rose steadily through the ranks of the RN in the interwar years and at the outbreak of the Second World War was appointed to command the Australian Squadron with the rank of rear-admiral. As most

Australian ships had been committed overseas, Crace's command was a small one, and in 1941 he sought to resign his post out of frustration. In February 1942 he was given command of Anzac Force, a squadron of Allied ships operating in New Guinea waters in support of an American carrier task force. In May 1942 Crace's squadron (renamed Task Force 44) was detached from the US Carrier Group to intercept Japanese troop-ships heading for New Guinea. The result was disappointing: his squadron was bombed, it saw no transports and being in the west missed the Battle of the Coral Sea, which was taking place in the east. Crace returned to England in June 1942, was knighted in 1947, and lived there until his death.

CRESWELL, HMAS After the establishment of the RAN in 1911 it was decided to set up a naval college for the education of officers in the new service. Jervis Bay was chosen as the location because of its excellent deep-water port for fleet exercises, because it was on federal land and because it was cheap. Construction of facilities began immediately but the isolation of the site imposed many delays and it was not completed until 1914. Until then the activities of the college had been carried out at Osborne House, North Geelong. The college was commissioned as HMAS *Franklin* in 1915 (it was apparently named after one of the transport ships associated with the construction of the buildings). After the First World War the reduced naval intake led to the closure of the college and the transference of its functions to Flinders Naval Depot in Victoria. Between 1930 and 1957 the buildings at Jervis Bay were leased for use as a tourist resort. In the mid-1950s, with the navy expanding again, Cabinet decided to resume the old site and in 1958 the college returned to Jervis Bay, now called HMAS *Creswell* after Vice-Admiral Sir William Creswell (q.v.). Additions were made to the college in the 1960s in the form of a science block and new accommodation blocks. Initially cadets at the college entered at 13 and undertook four years of academic and professional training. Subjects taught encompassed science, humanities, languages and seamanship. In 1956 this scheme was altered. The entrance age was raised to 16, and two years' matriculation study was required plus one year of post-secondary and professional training. Provision was also made for entry at 18. These cadets were required mainly to undertake the last year of the education and training schedule. In 1968 there were further changes. The first year of science and engineering degrees was taught at the college, which was now associated with the University of New South Wales. Students in their second year and above attended the main campus in Sydney, as did all Arts, Commerce and Surveying students. While these changes were taking place the internal courses run by the college were upgraded to a two-year Diploma of Applied Science. This came to an end with the opening of the Australian Defence Force Academy (q.v.) in 1986. *Creswell* now concentrates solely on professional naval training.

CRESWELL, Vice-Admiral William Rooke (20 July 1852–20 April 1933) was born in Gibraltar, the son of the colony's deputy postmaster-general. He entered the navy in 1865 and served in various postings until 1878 when he resigned in disgust over the slowness of promotion. In 1879 he migrated to Australia and spent several years in the Northern Territory exploring and cattle droving. A turning point in his career came in 1885 when on holiday in Adelaide. The Commandant of the South Australian Defence Forces, Commander John Walcot, was an old friend. He offered Creswell an appointment as lieutenant-commander of HMCS *Protector,* the only warship in the colony. Five years later Creswell succeeded Walcot as commandant. He had to wait 10 years to see action. In 1900 Creswell (now commandant of the Queensland Naval Forces) took the *Protector* to China to aid the British Government in crushing the Boxer Rebellion. He remained in China waters until January 1901 largely carrying out survey and despatch work and gaining favourable comments from the C-in-C.

Creswell had been an articulate advocate of an Australian naval force since the 1880s and 1890s. The creation of the Commonwealth in 1901 gave added impetus to this movement, and Creswell as Officer Commanding was ideally placed to take up the cause. In 1904 he strongly advocated the scrapping of the coastal defence craft and their replacement by three modern destroyers to be placed under Australian, not Admiralty, command. Half-heartedness on the part of a succession of Australian Administrations and opposition from the Admiralty insured that Creswell's schemes came to nothing.

A shift in attitude on the part of the Admiralty occurred in 1906 with the election in Britain of a Liberal government. The new First Lord of the Admiralty, Lord Tweedmouth, and his First Sea Lord, Sir John (Jackie) Fisher, declared themselves no longer opposed to dominion navies as long as the ships acquired were compatible with types in the RN. A further shift occurred in 1909. In that year a naval crisis developed over the rapidly growing strength of the German navy. At the Imperial Conference (q.v.) later in the year (which Creswell attended as naval adviser to the Australian delega-

Creswell (front row, left), while a lieutenant-commander, with a group including the Earl of Kintore, Governor of South Australia (front row, centre), Adelaide, c. 1890. (AWM P0444/214/185)

tion) the value of the 'fleet unit' proposal was developed. Each dominion would have a force consisting of an armoured cruiser, three lighter cruisers, six destroyers and three submarines. In the event of war the unit would deal with hostile forces in local waters and then merge with the RN as part of an imperial force operating outside home waters.

Creswell at first opposed the idea, but the danger of being sidelined by the Deakin administration, which had enthusiastically taken it up, soon brought him around. Indeed Creswell soon found himself a key player in the development of the fleet unit idea. In 1911 Admiral Sir Reginald Henderson produced a report into Australian naval defence. As part of the reforms he recommended the creation of a new Naval Board. Creswell was made the board's First Naval Member and promoted to rear-admiral. He became, in effect, the first C-in-C of the newly named Royal Australian Navy.

The next few years were the busiest of Creswell's life. Recruits had to be raised for the new force, dockyard facilities established or expanded, a naval college founded and training commenced. It is a credit to Creswell's industry that all was ready when the principal units of the fleet, led by the bat-

tlecruiser HMAS *Australia* (I) (q.v.), steamed into Sydney Harbour in October 1913. Less than a year later Australia was at war and it was again largely due to Creswell that the fleet was ready and would prove itself an efficient fighting unit. The war was in fact to prove an anticlimax for Creswell. After 1914 most Australian ships had been dispersed to various British squadrons overseas. Creswell remained in Australia as an administrator with virtually no fighting ships to administer. He did, however, play an active role in convoy organisation, ship construction and smoothing the sometimes difficult relations between the Admiralty and the Australian government.

For Creswell the war confirmed the need for the continued development of the RAN, especially in light of Japan's expansion into the central Pacific in the postwar period. Creswell's work, however, was largely over. He relinquished office in 1919, was appointed KBE and was placed on the Retired List. He was refused a pension by the Australian government, which, however, promoted him to vice-admiral in 1922. Creswell played a seminal role in the emergence of the RAN. For 30 years he argued its cause and helped educate the public and politicians in matters of naval defence. It is appropriate that the Australian Naval College at Jervis Bay is named after him.

CRETE When the Greek forces and their British Commonwealth allies were withdrawing down the length of Greece, General Sir Archibald Wavell reported to Prime Minister Churchill in London that the situation was fast deteriorating and that their troops would probably have to be evacuated. He then added that he assumed Crete, an island south-east of the Greek mainland, would be held. Churchill replied that it would indeed. The British Official History makes clear that until that time neither the chiefs of staff in London nor the British C-in-C in the Mediterranean had been working on any clear policy for the defence of Crete against the Germans.

Moreover, this impasse could not have occurred at a worse time. Greece was falling. The Axis partners, Germany and Italy, were bolstering their position in North Africa, where Rommel had forced the weak British line back, and had shut the 9th Australian Division into an enclave at Tobruk. British forces also had to suppress a revolt in Iraq, where Britain's influence was vital in ensuring that the oil pipeline from Persia, which traversed Iraq to Haifa, remained open. The revolt was suppressed, but German influence in the region had not been completely contained. This eventually led to a British decision to commit troops to Syria. Churchill had been contemplating this probability as early as November 1940, and had written to the Foreign Secretary: 'We shall most certainly have to obtain control of Syria by one means or another in the next few months. The best way would be by a Weygand or a De Gaullist movement, but this cannot be counted on, and until we have dealt with the Italians in Libya we have no troops to spare for a northern venture.'

The Italians had been dealt with, but this very success impelled Hitler to come to the aid of his Axis partner. In part, he had done this by his successful invasion of Greece and the Balkans. He had not only reinforced Mussolini's force in North Africa, he had also, with some reluctance, taken over control of that campaign from him, and, in order to control the supply route from Sicily to Tripoli across the narrows, he committed air groups to Sicily. Now, as an extension of his campaign in Greece which, among other things, reduced the possibility of a British bombing threat to the important Rumanian oilfields, control of the island and waters around Crete promised him an even tighter strategic grip. For the British, sea and air control of Crete was an almost exact corollary of the German position. Thus, when the British Commonwealth force was evacuated from the Greek mainland, many of the troops found themselves committed to an even more desperate defence in Crete.

This was the British dilemma. While the campaign on the Greek mainland was continuing, that country's claim for support was insistent. But the bottom of the Middle East barrel had been scraped to provide units to send there and, as a consequence, the provision of forces and equipment to defend Crete had been relegated to the lowest priority. Now the situation had been reversed, but there was not sufficient time to put the defence of the island on to a sound footing.

When General Maitland Wilson who had commanded the force in mainland Greece arrived in Crete at the beginning of May, he estimated that in addition to the forces already there a further three brigade groups each of four battalions plus a motor battalion would be needed. Already available were the weak British 14th Brigade, a Royal Marine Mobile Naval Base Defence Organisation (MNBDO) of about battalion size, an anti-aircraft force of 60 guns, and, most disturbing of all, a weak air element of six Hurricanes and 17 other assorted aircraft, with the important proviso that this air element could not be increased.

This then was the situation when General Freyberg, who had commanded 2nd New Zealand Division in mainland Greece, arrived in Crete and was directed to defend the island by Wavell. The immediate problem was how to feed, clothe and shelter some 50 000 men who had been evacuated from the mainland. Many had to bivouac among olive trees where, lacking tents, blankets and greatcoats, they had to sleep on the ground. Personal kit and shelter was gradually produced from Egypt. But the items most needed for the defence of the island — heavy weapons, motor vehicles, tanks, aircraft and reliable communications — could not be obtained.

When re-formed into their units the troops evacuated from the mainland included seven New Zealand battalions, and four Australian battalions together with elements of a fifth battalion and a machine-gun battalion. There was also part of the British 1st Armoured Brigade, minus almost all its tanks, and 10 000 Cretan recruits of the Greek Army, who were very inexperienced. A further force of four British battalions was formed from a number of their units who had escaped from Greece. These troops now had to be deployed to defend the island, and it became clear from intelligence reports (which, based principally on ULTRA [q.v.] decrypts, proved to be remarkably accurate) that the Germans would invade with an airborne and seaborne force.

Crete is approximately 250 kilometres long and 65 kilometres in width at its widest point. It has a backbone of barren mountains rising to over 2000 metres. At the time, its development and infrastructure were backward. Of chief interest to an invader were three airfields at Heraklion, Retimo and Maleme, all on the north coast. There were small ports at Suda and Heraklion capable of taking two ships at a time. At Canea and Retimo, ships had to discharge into lighters. There were no railways. The road connecting towns on the north coast was poor, as was the road and track connecting the Canea–Suda area to Sfakia on the south coast. Similarly primitive were the telegraph, telephone and transport systems. Worse for the defenders, no guns were saved from Greece and no transport. Some guns were provided from Egypt, without sights, and shells without fuses were also received. The men improvised, but in assessments made after the campaign even sixty 25-pounder guns, which would have enabled the three field regiments to be fully equipped, and were available in Egypt, could have transformed the situation, so narrow was the margin between victory and defeat. As it was, the guns on Crete did not have gun-tractors and had to be man-handled or ferried about. There was also a shortage of anti-aircraft guns which, in the absence of a force of defending fighters (those few remaining were flown out immediately before the invasion), proved critical.

On 3 May Freyberg issued an instruction to 'Creforce', as he termed it, allotting forces to the following areas: Heraklion sector (Brigadier Chappel), the 14th British Brigade, 2/4th Australian Battalion, 7th Medium Regiment (armed with rifles) and two Greek battalions; Retimo sector (Brigadier Vasey), four Australian battalions (2/1st, 2/7th, 2/11th and 2/1st Machine Gun) and two Greek battalions; Suda Bay sector (Major-General Weston), the MNBDO, 1st Battalion, The Welch Regiment, 2/8th Australian Battalion and one Greek battalion; and Maleme sector, the New Zealand division comprising the 4th Brigade (Brigadier Kippenberger), the 5th Brigade (Brigadier Hargest) and three Greek battalions. The 4th New Zealand Brigade (less one Battalion) and 1 Welch were in force reserve but administered by their sector commanders. The sector commanders were instructed to dispose one-third of their troops on or around the potential landing grounds and two-thirds 'outside the area which will be attacked in the first instance'. Freyberg's instruction warned that the attack would probably take the form of intensive bombing and machine-gunning of the airfields and

their vicinity, a landing by paratroops to seize and clear the airfields, and finally the landing of troop-carrying aircraft. In addition there were likely to be seaborne attacks on beaches close to the airfields and the harbour at Suda Bay. There is some evidence that Freyberg was constrained in his deployment of troops by a need to protect the ULTRA source of his best intelligence of likely German intentions. Whether, and to what extent, this actually prejudiced deployments, for example to the west of Maleme, remains an open question.

The German plan for the airborne invasion of Crete was put forward by the *Luftwaffe* in mid-April, and agreed to by Hitler. Operational command was given to General Lohr, the commander of the Fourth Air Fleet. His force included General Student's XI Air Corps (glider and parachute troops) and General von Richthofen's VIII Air Corps (a purely air formation). Student's force consisted of an Assault Regiment (three battalions of parachute and one of glider-borne troops), and the 7th Air Division, which was made up of three parachute regiments and divisional troops. Attached from the Twelfth Army (from Greece) were three rifle regiments from the 5th and 6th Mountain Divisions, a panzer battalion, a motor-cycle battalion and anti-aircraft detachments. In all, more than 22 000 men, with the rest of the 6th Mountain Division, were at call. In effect, the defenders were more numerous than the attackers, but they were less well equipped. The greatest disparity was in the air. When the invasion came the defending aircraft had been withdrawn and supporting aircraft from North Africa needed long-range tanks and had only short endurance over the island. In contrast, the German aircraft assembled for Operation Merkur were formidable. They included 228 bombers, 205 dive bombers, 114 twin-engined and 119 single-engined fighters and 50 reconnaissance aircraft. For landing troops there were between 600 and 750 Junkers transport aircraft and from 70 to 80 towed gliders, together able to carry 5000 to 6000 men and their equipment in one lift. Furthermore, in a short time, German engineers and construction workers had created more than a dozen airfields within range of Crete, the one on Milos island being built in three days.

The German plan was for 'softening up' air attacks to begin on 14 May, and the actual air assault to commence on the 20th. The Assault Regiment was given the task of securing Maleme, the 7th Air Division was to capture Canea–Suda, Retimo and Heraklion. Roughly one parachute rifle regiment (of three battalions) was allotted to each of the air division's objectives. The attacks on Maleme and Canea were timed

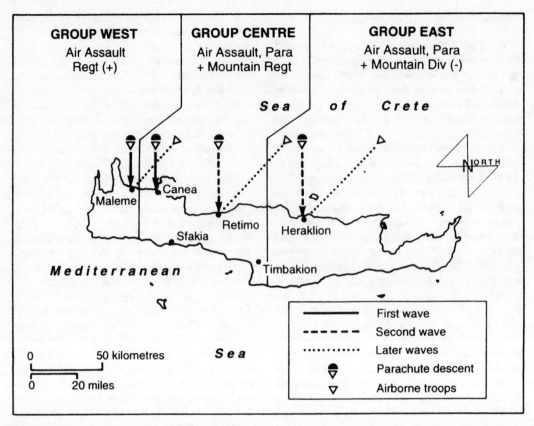

| GROUP WEST | GROUP CENTRE | GROUP EAST |
| Air Assault Regt (+) | Air Assault, Para + Mountain Regt | Air Assault, Para + Mountain Div (-) |

Sea of Crete

NORTH

Maleme · Canea

Retimo · Heraklion

Sfakia

Mediterranean

Timbakion

	First wave
	Second wave
	Later waves
	Parachute descent
	Airborne troops

0   50 kilometres

0   20 miles

Sea

Map 10a German assault on Crete, May 1941.

for the early morning of the 20th, and on Retimo and Heraklion for the afternoon. It was intended that all three airfields would be taken by the end of the first day.

When the assault began on the morning of 20 May, the German commanders quickly realised how grossly they had underestimated the number of defenders in each drop zone. For example, at Maleme the Assault Regiment dropped on either side of the airfield, intending to make a converging attack on it. Instead, because they were unaware of the New Zealand troops hidden in cover, many of them were quickly mopped up. However a part of the German force that landed by glider in the dry bed of the Tavronitis River west of Maleme managed to consolidate and this nucleus, which survived and beat off the defenders, became the key to the whole German operation. Other units that dropped in an area called Prison Valley, south-west of Canea, also managed to stand firm. German command of the air was devastating. Any movement by the defenders was pinned by fire and the bombing and machine-gunning cut telephone lines and put Freyberg out of contact with some sectors from the start.

Because of a threat to Suda, Freyberg moved Brigadier Vasey and two Australian battalions (2/7th and 2/8th) to reinforce, leaving Lieutenant-Colonel Campbell with two battalions to defend Retimo. Unfortunately, Campbell had no cipher equipment and poor radios, which meant that contact with him had to be in plain language and communication with his force was handicapped from the start. After three days outside contact with his force was lost, and it remained isolated until the end of the battle.

The difficulty for the defence was that Freyberg, having prudently disposed his force to cover the main centres with their airfields and harbours in the north at Maleme, Canea–Suda, Retimo and Heraklion, was unable through lack of armour and transport to form a mobile reserve. Thus, once the attacks came and German air superiority was simultaneously brought to bear, there was little he could do to re-dispose parts of his force to reinforce threatened areas. Each coastal enclave therefore had to fight it out as best it could. There was more flexibility in the western sector where the line of defences from Maleme to Suda was more or less continuous. But Retimo and Heraklion were entities on their own and, although they were vig-orously and successfully defended from the start, it was not long before both were separated from the western sector, and from each other, by German blocking forces.

Events turned in favour of the Germans on the second day. As has been established since, German morale had been severely shaken by the ferocity of the defence. Student had to decide where to land the reserve parachute troops that were still uncom-mitted. He correctly assessed that the key was Maleme airfield where, if it could be captured and opened, he could begin to land the troops of the mountain division. He also realised that the almost continuous line of defences from Maleme to Canea–Suda would be thin and have suffered sig-nificant casualties. He therefore concentrated attention on Maleme to capture the airfield, and on the Prison Valley and reservoir area south-west of Canea.

These operations were successful. On the sec-ond day, 21 May, a German transport aircraft, bravely landed under fire on Maleme airfield, unloaded its cargo and took off again: a significant event. Other aircraft landed on the beach west of Maleme, and a fairly rapid build-up of German forces began. The New Zealand 22nd Battalion, the original defender of the airfield, had been reduced to less than half its strength and been forced to the east. A deliberate counter-attack to recapture the airfield was organised slowly and was not mounted until the early hours of 22 May, by which time the German hold was too firm to be prised loose.

Student now committed his forces from Maleme and south-west of Canea to a converging drive to the east to roll up the defending force. Freyberg formed a large defended perimeter south-west of Canea and, when that had to be abandoned, a fur-ther line west of Suda. By 26 May, after his troops had been fighting for a week under continuous air attack and against an enemy who was being con-tinuously reinforced, Freyberg cabled Wavell that:

> I regret to have to report that in my opinion the limit of endurance has been reached by the troops under my command here at Suda Bay. No matter what decision is taken by the Commanders-in-Chief from a military point of view our position here is hopeless. A small ill-equipped and immobile force such as ours cannot stand up against the concentrated bombing that we have been faced with during the last seven days . . . Suda Bay may be under fire within twenty-four hours. Further, casualties have been heavy, and we have lost the majority of our immobile guns.

Even then, an unrealistic reply came back from Cairo, which misunderstood Freyberg's situation and that of his troops, suggesting that the

Suda–Maleme force should retire on Retimo and hold the eastern part of the island. Freyberg had to point out that the force could only continue to survive at all if food was landed at Sfakia on the south coast at once and that the chance of saving some of the force lay in withdrawing to that area. The Greek commander on the island also said that the position of his forces was difficult and they had begun to disintegrate at several points.

The withdrawal to Sfakia was begun, covered initially by the New Zealand 5th Brigade, then by 'Layforce', a force of commandos, and finally by Vasey's 19th Brigade, which inflicted a sharp rebuff to the German troops following up just north of Sfakia. At Heraklion, the force defending the air-field and port was evacuated by sea on the evening of 29 May, although many of the troops successfully taken off were subsequently lost when the ships carrying them to Alexandria were sunk or dam-aged by German air attack. Freyberg's efforts to get word to Lieutenant-Colonel Campbell at Retimo that Creforce was to be evacuated from Sfakia were all unsuccessful. Having put up a tremendous fight to defend Retimo airfield with his mixed force of two Australian battalions, about 3000 Greeks mostly ill-armed and untrained, but including a fine improvised unit of 800 Cretan police, Camp-bell was forced to capitulate on the evening of 30 May. His force was almost out of ammunition and down to its last day's ration. Campbell surrendered his own battalion — 2/1st Battalion — and most of the force. The CO of the other battalion, 2/11th Battalion (Lieutenant-Colonel Sandover), advised his men to scatter and try to escape. Subsequently, 13 officers and 39 other ranks of his battalion reached Egypt.

Out of a British Commonwealth force on Crete of over 33 000, about 16 500 were evacuated. Of the Australians, almost 800 were killed or wounded and over 3000, including the greater part of three battalions (2/1st, 2/7th and 2/11th), became pris-oners. The effort to hold Crete had cost the British Commonwealth force about 15 900 men (dead, wounded and missing). German casualties, how-ever, were also high: more than 7000 killed, wounded and missing, including 3000 killed from the 7th Air Division alone. About 220 German air-craft were destroyed, many by ground-fire. For gal-lantry on Crete, the German mountain troops were awarded 5019 Iron Crosses of various grades, and 971 War Service Crosses.

Could Crete have been held? It was clear in ret-rospect that from November 1940 onwards, valu-able time was lost because of higher command failure to define a policy, then appoint a comman-der to the island and leave him for long enough to

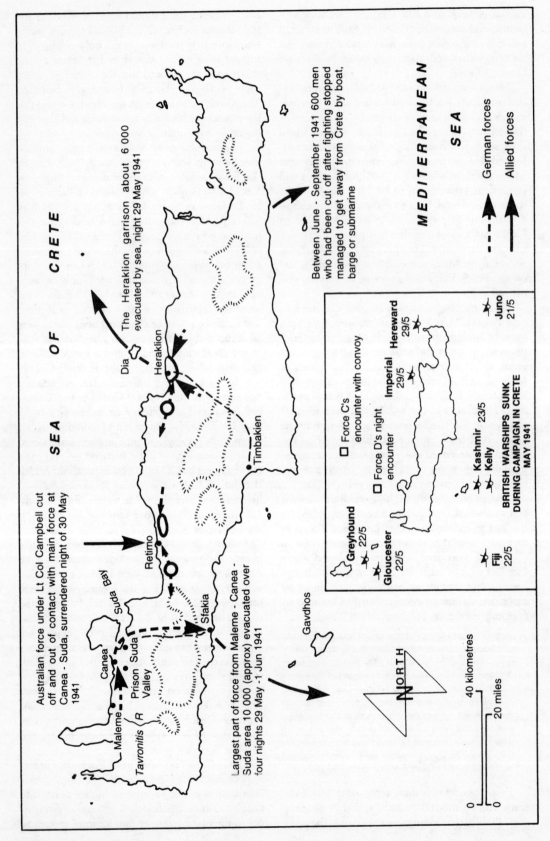

SEA OF CRETE

MEDITERRANEAN SEA

Australian force under Lt Col Campbell cut off and out of contact with main force at Canea - Suda, surrendered night of 30 May 1941

The Heraklion garrison about 6 000 evacuated by sea, night 29 May 1941

Between June - September 1941 600 men who had been cut off after fighting stopped managed to get away from Crete by boat, barge or submarine

Dia

Heraklion

Timbakien

German forces

Allied forces

□ Force C's encounter with convoy

□ Force D's night encounter

Imperial 29/5

Hereward 29/5

Juno 21/5

Kashmir 23/5

Kelly

Greyhound 22/5

Gloucester 22/5

Fiji 22/5

BRITISH WARSHIPS SUNK DURING CAMPAIGN IN CRETE MAY 1941

Retimo

Suda Bay

Canea

Maleme

Tavronitis R

Prison Suda Valley

Sfakia

Gavdhos

Largest part of force from Maleme - Canea - Suda area 10 000 (approx) evacuated over four nights 29 May -1 Jun 1941

NORTH

0   20 miles

0   40 kilometres

organise a coherent defence. From November until the invasion, six officers in succession held command. In 1941 an Inter-Services Committee was appointed to report on the campaign. Although the evidence before it was not comprehensive, some judgments remain valid. 'Six months of comparative peace,' it reported, 'were marked by inertia for which ambiguity as to the role of the garrison was in large measure responsible ... There was a marked tendency at one time to regard Crete as a base for offensive operations against the Dodecanese without any apparent regard as to the advisability of being able to operate from a secure base.'

In addition to the skilful and dogged fighting of the troops, many of whom had just undergone an arduous campaign on the Greek mainland, there were two positive outcomes in an otherwise unsuccessful battle. First, were the magnificent efforts of the navy to protect the troops from German seaborne landings — which it did on two occasions by destroying or scattering invasion caiques loaded with troops; by running the gauntlet of German air attacks to keep the garrison supplied; and, finally, by evacuating a large part of the garrison, 16 000, over four successive nights. In the battle for Crete, the Mediterranean fleet lost three cruisers and six destroyers sunk, as well as many other ships including two battleships and an aircraft-carrier damaged. About 2000 sailors were lost. Second, the severe losses among the German airborne force convinced its high command not to attempt the venture again. A forthcoming invasion of Cyprus, which had already been planned along similar lines, was cancelled as a direct result of their experience in Crete.

JOHN COATES

CRUISER MARK I TANK see **SENTINEL TANK**

CULGOA, HMAS see **BAY CLASS FRIGATES**

CURLEW, HMAS see **TON CLASS MINESWEEPERS**

CURTIN, John Joseph Ambrose (8 January 1885–5 July 1945). With a background in trade-union and electoral branch politics, and labour movement journalism, Curtin was elected to the federal parliament for the seat of Fremantle in 1928, which he held until December 1931. He had been an activist against conscription (q.v.) in Victoria during the referendum campaign of 1916, and

Map 10b Escape of Allied forces from Crete, 1941. *Inset* British warships sunk.

was gaoled briefly by the Hughes (q.v.) government. In 1924 he was a delegate to the annual conference of the International Labour Organisation in Geneva.

He won the seat of Fremantle back in the election of September 1934, and to the surprise of many was elected to the leadership of the parliamentary party in October the following year. His leadership before 1939 was characterised by attempts, largely successful, to mend the divisions in the party that had arisen during the Depression. The worsening international situation in the late 1930s meant that defence and foreign policy could not be ignored, as Labor had largely managed to do since 1918. Curtin supported increased defence spending, especially on the air force, but the legacies of the Great War precluded ALP support for the possibility of overseas service in September 1938, or for the reintroduction of compulsory military training in September 1939.

Curtin consistently rejected Prime Minister R. G. Menzies' (q.v.) offers to form a national government of all parties, proposing instead the creation of the Advisory War Council (q.v.), which he maintained after he assumed the prime ministership in October 1941 following the fall of the Fadden government. On becoming prime minister Curtin also took on the Defence Co-ordination portfolio. He presided over a cabinet and a government that were still divided between left and right wings, and which were very short on any form of military experience.

The outbreak of the Pacific War galvanised the Australian war effort and provided Curtin with the stimulus he needed to bring the ALP and the nation behind him. His famous statement that 'Australia looks to America, free of any pangs as to our traditional links or kinship with the United Kingdom' angered British Prime Minister Churchill, with whom his relations were always strained, but was no more than a statement of the obvious. Backed by the chiefs of staff, the caucus and the War Cabinet, Curtin pressed successfully for the return of the 6th and 7th Divisions from the Middle East in February 1942, and requested the appointment of an American C-in-C for the region. General Douglas MacArthur's (q.v.) arrival in March from the disastrous campaign in the Philippines met this requirement, and the two men were to establish a close working relationship for the remainder of the war, one which at times eclipsed the position of the government's principal military adviser, General Sir Thomas Blamey (q.v.).

The war led Curtin and Labor to decisions that contradicted long-standing Labor positions, such as the direction of labour in industry and the extension of service overseas for conscripts in the army.

It is testimony to Curtin's knowledge of the ALP and powers of leadership that he managed to take his government with him in directions that he had found disagreeable in peacetime. The ALP triumph in the 1943 election, in which they won 49 seats to the Opposition's 24 and gained over 60 per cent of the armed forces' vote, was a measure of the popular support which Curtin's leadership engendered.

Curtin's greatest contribution to Australia's war was during the crisis period of 1942–43. Thereafter, as his health declined, and as the divergence between Australian and American postwar aims became clearer and strategic policy less certain, Curtin's mastery of the war effort became less sure. Planning for postwar reconstruction (q.v.), a full-employment economy and improved social welfare assumed greater prominence. While the government remained active internationally in the series of great conferences held in 1944–45 to decide the shape of the postwar world, the effort produced few results. His health in serious decline from November 1944, Curtin died in office the following July.

Curtin is generally regarded as the greatest of Australia's prime ministers, and while the judgment is probably a fair one it obscures the difficulties that he faced and only partly overcame. The ALP gained office in October 1941 because Curtin was its leader and because there was no viable alternative after the implosion of the United Australia Party under Menzies. In the first 18 months of the Pacific War his government was generally united and purposeful and Curtin's contribution to the prosecution of the war against Japan was a significant one; it is this period of the war that is generally remembered when Curtin's premiership is invoked. Perhaps his greatest contribution was to show that the Labor Party was capable of effective national leadership in a time of crisis. Above all, he was respected for his unassuming, dignified and thoroughly Australian manner.

CUTLACK, Frederic Morley (30 September 1886–27 November 1967). Cutlack migrated from England as a child, and attended the University of Adelaide. He became a journalist in 1904 and moved to London and the staff of the *Daily Chronicle* in 1911, returning to Australia as a correspondent on the maiden voyage of HMAS *Australia* (I) in 1913. He returned to England to study law, and at the outbreak of war in 1914 enlisted in King Edward's Horse, an imperial mounted regiment which drew many of its officers and men from the colonies and dominions. He was commissioned and served in France in 1915–16 before attachment to the headquarters of the 3rd Australian Division in April 1917. C. E. W. Bean (q.v.) recruited him as assistant official war correspondent in January 1918. He worked closely with Bean and wrote reports from the front line, risking death or injury on numerous occasions. He was seriously injured in an accident in July, and spent the rest of the year convalescing. During this time he published an account of the activities of the Australian Corps, and in 1919 after his demobilisation was called to the English Bar. Back in Australia he resumed his journalistic career, and in 1920 was commissioned by Bean to write the volume of the official history dealing with the Australian Flying Corps; this appeared in 1923 and went through a number of editions. For the rest of his career he alternated journalism with brief periods of legal practice and public service. He was on the staff of the Prime Minister, S. M. Bruce, at the Imperial Conference in 1923 and formed part of J. G. Latham's mission to north-east Asia in 1934. As well as writing a volume of the official history he edited a volume of the *War Letters of General Monash*, published in 1935, and wrote a book seeking to vindicate 'Breaker' Morant and P. J. Handcock (q.v.), which appeared in 1962. He returned to live permanently in England in the 1950s.

John Curtin (right), meeting General Douglas MacArthur for the first time, at an Advisory War Council meeting, Canberra, 26 March 1942. (AWM 042774)

D

D-DAY, 6 June 1944, marked the Allied invasion of occupied Europe in Operation OVERLORD. A combined naval–army force, in one of the largest fleets ever assembled and backed by overwhelming air superiority, attacked German positions in Normandy, France. In overall terms the Australian contribution was small but significant. Six RAAF squadrons, No. 453 Squadron flying Spitfire (q.v.) fighters, No. 456 Squadron flying Mosquito (q.v.) fighter-bombers, No. 466 Squadron flying Halifax (q.v.) heavy bombers and No. 460, 463 and 467 Squadrons flying Lancaster (q.v.) heavy bombers, were heavily involved and there was a total of 1816 Australian aircrew engaged in operations out of Britain in the week before the landings. A major part of the air campaign was directed against rail systems in France, and of the 53 raids carried out in the three months before D-Day, 25 involved Australian squadrons. On D-Day itself, Australian fighter pilots flew more than 50 sorties over the beachheads. At sea about 500 Australians of the RAN and Royal Australian Naval Volunteer Reserve were serving with the RN in many positions including the command of a destroyer, a minesweeper, and a flotilla of tank-landing craft. The Australian army participation was on a much smaller scale and consisted of 13 officers, including Major Henry (Jo) Gullett (q.v.), on attachment to the British Army who took part in the landings to gain experience for the Australian amphibious operations in Borneo. The Australian war correspondent Chester Wilmot (q.v.) landed with British airborne troops on D-Day. From 25 June No. 453 Squadron was flying from an airfield within the beachhead. By 1944, however, notwithstanding its earlier involvement in the land war against the Axis powers in North Africa and its longer commitment to the air war over Europe, the focus of Australia's war effort was firmly on the campaigns in the Pacific.

DAKOTA, DOUGLAS (2-seat transport). Wingspan 95 feet; length 64 feet 6 inches; load 28 troops, 18 stretchers or 7500 pounds cargo; maximum speed 230 m.p.h.; range 1510 miles; power 2 × Pratt & Whitney 1200 h.p. Twin Wasp engines.

The C-47 Dakota is the military name for the DC-3 passenger aircraft of which over 10000 were built during the Second World War. On the outbreak of war in September 1939 the RAAF impressed four Australian National Airways DC-3s and flew them with No. 8 Squadron on anti-submarine patrols before returning them to the

Spitfires of No. 453 Squadron at an airstrip on the Normandy beachhead, France, as part of D-Day operations, 1944. (AWM 100821)

airline in June 1940. The RAAF began using Dakotas delivered from the United States in 1943 and eventually received 124 aircraft. These were flown as transports by No. 33, 34, 35, 36 and 37 Squadrons during the New Guinea campaign, where because of the supply drops they made to the front-line troops they became known as 'biscuit bombers'. Dakotas were subsequently used by the RAAF in the Malayan Emergency and the Korean War. RAAF crews flew RAF Dakotas during the Berlin Airlift (q.v.) and aircraft equipped with skis and jet-assisted takeoff (JATO) were used in the Antarctic. Dakotas were replaced by Hercules and Caribou (qq.v.) aircraft in the late 1950s and early 1960s. The RAN used Dakotas from 1949 to 1977. Two of these aircraft are preserved by the RAN Museum and another Dakota is jointly owned by the RAAF Museum and the Australian War Memorial. The Dakota was probably the most valuable transport aircraft ever built and is still flown by the RAAF's Aircraft Research and Development Unit.

DALY, Lieutenant-General Thomas Joseph (19 March 1913–   ). The son of a pre-1914 militia officer who served throughout the Great War with the 9th Light Horse, Daly graduated from RMC Duntroon in December 1933 into an army reduced by Depression-inspired budget cuts. He served in staff and training appointments with the 4th and 3rd Light Horse Regiments of the CMF, before being posted on exchange to the British Army in India in 1938–39 where he served with the 16th/5th Lancers.

He was transferred to the 2nd AIF as adjutant to the 2/10th Battalion, and served with the unit in North Africa before becoming brigade major to Brigadier George Wootten's (q.v.) 18th Brigade. After several further staff appointments and attendance at the Middle East Staff School at Haifa, he returned to Australia and became the senior operations staff officer to the 5th Division, a CMF formation. Near the end of the war he was appointed commanding officer of the 2/10th Battalion, and led them with great distinction in the fighting at Balikpapan. In the words of the official historian, Gavin Long (q.v.), he emerged 'with full marks — the outstanding C.O. of the campaign'. He was also the only Staff Corps officer commanding a battalion in 1945.

After the war Daly served in staff appointments and as an instructor at the Staff College, Camberley, before becoming a student at the Joint Services Staff College in Britain in 1948–49. Returning to Australia in June 1952, first to RMC and then as Director of Infantry, he was sent to Korea to take command of the 28th Commonwealth Infantry Brigade (q.v.), the first Australian officer to hold the position. The brigade was involved in vicious positional warfare while truce talks went on at Panmunjom, and Daly's abilities were fully tested operating in a multinational environment dominated by the United States. He was succeeded by Brigadier John Wilton (q.v.) in March 1953.

Attendance at the Imperial Defence College (q.v.) in 1956 was followed by promotion to major-general and a further succession of senior staff appointments, culminating in his appointment as CGS in May 1966 in succession to Wilton, who was then promoted to Chairman of the Chiefs of Staff Committee. Daly was knighted in 1967.

Daly's period as CGS was a highly active one. The Australian Army was progressively more fully committed to the war in Vietnam, and earned unfavourable attention because of the operation of the national service scheme. To minimise the risk to army personnel, in 1969 Daly instructed officers and other ranks not to wear uniform on public transport in order to avoid unpleasant incidents developing with the minority of anti-war protestors who might be inclined to violence. As CGS he also became involved in the public brawl between the Minister for Defence, Malcolm Fraser (q.v.), and the Prime Minister, John Gorton (q.v.), over the army's policy towards civic action in Phuoc Tuy province. Daly believed that Fraser had shown gross public disloyalty to the service for which he had ministerial responsibility. A man of absolute integrity and independence of mind, his willingness to stand up to Fraser as minister, it has been suggested, cost him the post of Chairman, Chiefs of Staff Committee, following Wilton's retirement.

His time as head of the army was marked by a number of positive achievements. He saw through the creation of the position of Vice-Chief of the General Staff and of functional commands to replace the old territorial districts as part of the reorganisation of the army which came out of the Vietnam experience. He also argued successfully for the acquisition by the RAAF of heavy troop-lifting helicopters to improve the air mobility of infantry units, again a consequence of the experience gained in Vietnam. He visited the soldiers in Vietnam regularly, and was concerned that the lessons and benefits gained in the war should not be lost as the army returned to a peacetime footing. He retired in May 1971. In retirement he was Colonel-Commandant of the Royal Australian Regiment (q.v.) and the Pacific Islands Regiment (q.v.), and chairman of the Council of the Australian War Memorial from 1974–82.

DARING CLASS DESTROYERS (*Vampire, Vendetta, Voyager, Duchess*). Laid down 1948–52, launched 1951–56, commissioned 1952–59, displacement 2800 tons; length 388 feet; beam 43 feet; armament 6 × 4.5-inch guns, 6 × 40 mm anti-aircraft guns, 1 × Limbo (q.v.), 5 × 21-inch torpedo tubes (later removed); speed 30.5 knots.

These ships were designed during the Second World War but did not enter service until the late 1950s when they took over the naval gunfire support role formerly carried out by the Tribal Class (q.v.). *Vampire, Vendetta* and *Voyager* had routine careers serving with the Strategic Reserve in Singapore from 1958 on. After HMAS *Voyager* collided with HMAS *Melbourne* (q.v.) and sank in February 1964 (see *VOYAGER* [II], HMAS), the RN ship *Duchess* was lent to Australia as a replacement. HMAS *Vendetta* engaged in gunfire support with the US 7th Fleet in Vietnam in 1969–70. In the early 1970s *Vendetta* and *Vampire* underwent extensive modernisation at Williamstown dockyard (q.v.) which equipped them with air-warning radar, while *Duchess*, purchased outright in 1972, was converted to a training ship. HMAS *Duchess* was paid off in 1977, HMAS *Vendetta* in 1979, while in 1986 HMAS *Vampire* was transferred to the Australian National Maritime Museum.

DARWIN After several unsuccessful attempts at White settlement in northern Australia, Port Darwin was established in February 1869 by an expedition led by the South Australian Surveyor–General, George Woodroofe Goyder. In 1870 it was decided to build an overland telegraph link to Darwin which would join up with the submarine cable from Java, thus linking the Australian colonies telegraphically with Europe; this was completed in 1872. The other main form of communications infrastructure with a defence application, the railway linking Darwin with the southern capital Adelaide, has been under discussion for more than 100 years, and seems no closer to realisation. The defence of Darwin, as terminus of the underseas cable and therefore an important communications link in time of war, first came under consideration in 1887 when an RN officer drew attention to the need to garrison the port or provide 'one or two corvettes in the neighbourhood', suggestions which were rebuffed by the authorities of the day. The South Australian government had constructed forts at Semaphore (1880) and Largs Bay (1885), on the outer reaches of Adelaide, but the Admiralty did not consider Darwin to be of strategic significance, and little more was done. When the Northern Territory was transferred to the new Commonwealth government in January 1911, the defence needs of

northern Australia were advanced as one justification, though not the main one, and more time was spent in deciding the degree of financial compensation payable to South Australia after the transfer. A Naval Reserve Sub-district was established there in the same year and the local Rifle Club raised a guard to protect the cable terminus. The military presence in the Northern Territory during the First World War was tiny, although about 250 men volunteered for service with the AIF. The Cable Guard was disbanded at the end of hostilities, to be followed by the Naval Reserve group in 1922. Between the wars the imperatives of development were of greater concern than those of defence, and it was not until 1932 that any further military force was deployed to Darwin. Construction of fuel storage tanks for the navy began, and in September 1932 five officers and 42 other ranks from the Royal Australian Engineers and Royal Australian Artillery arrived to supervise the construction of coastal batteries for the defence of the oil installations. Four 6-inch guns, salvaged from a scrapped cruiser, were installed in two batteries, and in 1936 the defences were completed with an anti-aircraft battery. By 1936 the garrison consisted of four officers and 84 other ranks, organised as a headquarters, and the 9th Heavy Battery of the Royal Australian Garrison Artillery, the whole commanded from Brisbane as part of the 1st Military District. After the 1937 Imperial Conference, the Darwin Mobile Force (q.v.) was raised to bolster the defences of the area, and additional construction was undertaken. In March 1939 following the recommendation of the Squires (q.v.) Report, the Northern Territory became a military district in its own right, 7th Military District (7MD), and in July 1939 No. 12 General Purpose Squadron RAAF equipped with Avro Ansons (q.v.) was deployed there. The outbreak of war brought an intensification of effort in constructing defences in the area, and the Darwin Overland Maintenance Force was established in September 1940 to oversee the convoying of supplies by road. The first army farms (q.v.), later to feature as a major source of food supplies to forces based in northern Australia, were established at Adelaide River the same year, and moves were made to guarantee and extend the water supply to the town, by now boasting a greatly increased population.

The outbreak of the Pacific War found the garrison of Darwin considerably augmented, and the town was used in the first weeks as a base for deploying forces to the north to aid in the defence of the Netherlands East Indies territories of Ambon (q.v.) and Timor. On 19 February 1942 Darwin was the target of a heavy Japanese air attack in which at least 243 personnel were killed for the loss of between 5 and 10 Japanese aircraft. Considerable damage was done to shipping in the harbour, while an American squadron of P-40 aircraft was destroyed on the ground. The scenes of confusion and panic which followed, and the exodus of both civilians and military personnel southwards in apprehension of a Japanese invasion, became cause for scandal and ultimately an official inquiry conducted by Mr Justice Lowe of the Victorian Supreme Court. Darwin was bombed on 64 occasions by the Japanese between March 1942 and November 1943, but the first raid was the heaviest and most destructive (see DARWIN, BOMBING OF). The forces in the Darwin area were greatly strengthened by the movement north of four of the returning AIF brigades and detachments of American troops, and the air force presence was greatly increased as Darwin became a centre for mounting aerial attacks against Japanese positions in the islands and seas to the north. Three brigades remained in Darwin until July 1944 when they were reduced to one, and while these forces had a defensive role, the Allied Intelligence Bureau (q.v.) mounted special operations out of Darwin as did the navy, which based the 24th Minesweeping Flotilla there from early 1942. In the course of 1945, as the war drew to a close and the fighting

moved well north of Australia, the military resources based in northern Australia were gradually reduced.

In the postwar period and so long as Forward Defence (q.v.) was the basic strategic policy underpinning Australian defence, Darwin and the Northern Territory once again became relatively unimportant. The naval communications station, HMAS *Coonawarra*, built before the war, played a significant role in the joint intelligence agreement known as UKUSA, but the defence of northern Australia and the shift of defence installations and assets northwards only re-entered defence thinking in the 1980s.

Peter Donovan, *Defending the Northern Gateway: The Northern Territory and the Defence of Australia: An Historical Overview* (Strategic and Defence Studies Centre, Canberra, 1989).

DARWIN, BOMBING OF, began on 19 February 1942 when a force of 188 Japanese aircraft was launched from four aircraft-carriers in the Timor Sea. At 9.37 a.m. the RAAF operations room in Darwin received a message that a large number of aircraft had flown over Bathurst Island to Darwin's north, but the station commander, probably thinking the planes could be American and concerned to avoid damaging morale with a false alarm,

Australian soldiers standing in a bomb crater in front of the damaged 119th Australian General Hospital after the first Japanese bombing raid on Darwin, 19 February 1942. (AWM P01791.003)

sounded the alarm almost at the same time as the attack began just before 10 a.m. First high-level bombers pattern-bombed the harbour and town, then dive-bombers and fighters bombed and machine-gunned the harbour, RAAF and civil aerodromes and the army hospital. The only fighter aircraft available to defend the town were 10 US Kittyhawks (q.v.) just back from an aborted trip to Java, and all but one of these were shot down before they could engage the attackers. The raiders left around 10.30 a.m., but just before noon another group of 54 bombers from Ambon (q.v.) arrived and spent 20 minutes pattern-bombing the RAAF base. The raids, which probably cost the Japanese between 5 and 10 aircraft, left Darwin devastated. Eight ships were sunk in the crowded harbour and two more near Bathurst Island, at least 243 people were killed, the wharf was severely damaged, and the post office, administrator's office, police station and police barracks were destroyed. The raids were aimed at eliminating the threat of Allied counter-attack against the Japanese invasion of Timor, but most residents believed at the time that Darwin was being 'softened up' for invasion and by mid-afternoon many were heading south in an exodus that became known as 'The Adelaide River Stakes'. Later allegations of mass panic were exaggerated, but probably at least half of the civilians living in Darwin at the time of the bombing fled (the town's population had already been halved by the evacuation of most of its women and children in the months since Japan entered the war). Breakdowns in discipline resulted in many air force men joining the exodus and in soldiers, including military police, looting the town. The government, concerned about the effect on morale of such an attack coming only days after the fall of Singapore (see MALAYAN CAMPAIGN), announced that only 17 people had been killed by the bombing. A hurried Commission of Inquiry led by Mr Justice Charles Lowe was very critical of RAAF performance during the raids, and almost all senior RAAF officers in Darwin were replaced. From 23 February the military commandant took over administrative control of the Northern Territory north of Birdum. The Japanese bombed Darwin 64 times in all, and raids continued until November 1943. But the attack on 19 February was the most destructive and has assumed a special place in popular consciousness as a symbol of Australia's supposed vulnerability.

(See also JAPANESE ATTACKS ON AUSTRALIA.)

Alan Powell, *The Shadow's Edge: Australia's Northern War* (Melbourne University Press, Melbourne, 1988); Douglas Lockwood, *Australia's Pearl Harbour: Darwin, 1942* (Ian Drakeford, Publishing, Melbourne, 1988).

DARWIN, HMAS see *OLIVER HAZARD PERRY CLASS GUIDED MISSILE FRIGATES*

DARWIN MOBILE FORCE (DMF) Cabinet approved the raising of the Darwin Mobile Force in October 1938 not, as is sometimes assumed, as a consequence of the Squires Report (q.v.) since that had not in fact yet been presented, but arising from the deliberations of the 1937 Imperial Conference and from a study made by the CGS, Major-General J. D. Lavarack, in September 1936. The recommendations of the Chiefs of Staff subcommittee at the conference prompted the government to announce in July 1938 its intention to station 750 personnel from the three services at Darwin as part of a move to improve the defences of northern Australia in anticipation of Japanese aggressive moves in the Pacific. The DMF was the result, and had the distinction of being the first field force unit raised in the Permanent Military Forces. Although it possessed a rifle company, machine-gun and mortar sections, and a troop of 18-pounder guns, the men were enlisted as gunners and the unit classed as artillery because the provisions of the Defence Act precluded the raising of permanent units for other than administrative, instructional or technical tasks.

The force was recruited in November 1938 and trained for three months at Liverpool camp. It wore a highly original uniform of patrol jacket, shorts, long socks and boots and slouch hat and, for wet weather, a cape suspended from shoulder straps. It also attracted a large number of capable and talented men. During the Second World War more than 100 were commissioned from the ranks and as many again became senior NCOs. Its original officers included F. G. Hassett (q.v.) and R. L. Hughes, both of whom became generals, four subsequent brigadiers including Maurice Austin, a noted military historian of the colonial period, and D. R. Jackson, and two full colonels including the first commanding officer, A. B. MacDonald, who went on to command the Jungle Warfare Centre at Canungra (q.v.) from 1943.

The DMF ultimately numbered 12 officers and 245 other ranks, and moved to Darwin in March 1939. As a result of the Squires Report and as a further part of the moves to improve the defences of Australia, the Northern Territory became a separate military district, 7th Military District (7MD), in the same month, and a Staff Corps (q.v.) officer, Colonel H. C. H. Robertson (q.v.), was appointed its first commander. From then until the outbreak of war in September Robertson made strenuous efforts to improve Darwin's defences. The DMF was initially housed in the old Vestey's meatworks outside the town, and engaged in a continuous round of training, field exercises and construction tasks. They also assisted in the training of the CMF garrison units, although these were generally ineffective. They moved into Larrakeyah Barracks in Darwin itself in August 1940, by which time detachments of officers and men were being siphoned off to join newly raised units in the 2nd AIF. On 20 August 1940 the unit was disbanded. The infantry became the Darwin Infantry Battalion, and from 1 November 1941 was renamed the 19th Battalion.

DE BAVAY, Auguste see **SCIENCE AND TECHNOLOGY**

DE MOLE, Lancelot see **SCIENCE AND TECHNOLOGY**

DEAKIN, Vera (25 December 1891–9 August 1978). The youngest daughter of Australian prime minister Alfred Deakin, Vera Deakin attended Melbourne University and the Melbourne Conservatorium of Music before travelling to Europe to study singing and the cello in Berlin and Budapest. Returning to Australia shortly before the outbreak of the First World War, Deakin contacted Norman Brookes, the famous tennis player and a commissioner of the Australian Red Cross (q.v.) in Cairo, in 1915, offering her services. Her offer was accepted, and on 21 October 1915, the day after her arrival in Cairo, Deakin set up the Australian Red Cross Wounded and Missing Inquiry Bureau. The aim of the bureau was to provide information for relatives of Australian soldiers who had been listed by the army as wounded or missing. Deakin herself heartily described the bureau as 'a buffer between the often over-anxious relatives and friends and the tired but cheery man who would sooner fight all day than write a letter'. Deakin remained as head of the bureau until the end of the war, by which time the bureau had made over 10 000 enquiries and replies, and received the OBE for her work. Soon after the Armistice, Deakin met Captain Thomas White for the first time, although they had already written letters to each other as White was a member of the Australian Flying Corps captured by the Turks in Mesopotamia (an experience he later related in his book *Guest of the Unspeakable*), and Deakin had corresponded with him as a POW as part of her duties with the bureau. Within six weeks of their meeting they were engaged and later married in Australia. White entered politics in 1920, and later became a minister and Australian High Commissioner in London. During the Second World War, Deakin-White was Director of the Wounded and Missing Inquiry Bureau of the Victorian Division of the Red Cross, and was President of the Victorian Society for Crippled Children and Adults from 1961 to 1966. She remained active in the Red Cross until the 1970s.

DEATH PENALTY for Australian soldiers was enshrined in Section 98 of the *Defence Act 1903*, and could be awarded for four offences — mutiny, desertion to the enemy, treachery leading to the fall of a garrison, and treasonous correspondence — but the sentence had to be confirmed by the Governor-General before it could be carried out. The main source of opposition to the imposition of the death penalty on Australian soldiers was not the execution of Morant and Handcock (q.v.) in the Boer War, which at the time evoked little public sympathy. Section 98 of the 1903 Defence Act was based on a number of pieces of colonial legislation (New South Wales 1871, Queensland 1884, Tasmania 1885), which in turn were modelled on Canadian legislation of 1868, all of them reflecting colonial concerns in the wake of the withdrawal of British garrison troops that local forces should remain under local control. Broader public discussions of the possible abolition of the death penalty, and questions about military justice, may also have influenced the drafting of the Defence Act.

Growing concerns over the discipline of Australian troops in the First World War brought Section 98 under critical scrutiny. In March 1916, shortly after the arrival of Australian troops in France, Sir William Birdwood (q.v.), GOC, Australian Imperial Force, argued that Australian soldiers should be brought under the British Army Act, which provided the death penalty for 17 offences. British officers, and increasingly senior Australian officers, insisted that discipline could only be maintained in the AIF if the same sanctions applied as in the British Army, in which about 346 soldiers were executed in the course of the war. Pressure on the Australian authorities increased in 1917 with claims that the desertion rate in the AIF was reaching unprecedented levels, and in July 1917 Birdwood again urged the introduction of the death penalty by abolishing the provision that required the Governor-General to confirm any death sentence. Birdwood's recommendation was unacceptable to the Australian government, then facing a second referendum on conscription after the failure of the first referendum in October 1916 to endorse conscription for overseas service, and it declined to act. The following year the government proposed to amend the Defence Act by allowing for the death penalty in the case of a soldier convicted of murder, but the amendment was withdrawn in November 1918. It was mooted again in 1920 but not pursued formally.

A provision similar to Section 98 of the Defence Act was included in the *Naval Defence Act 1911*, which prohibited the death penalty being carried out on any member of the Commonwealth Naval Forces until it had been confirmed by the Governor-General.

Richard Glenister, Desertion without execution: Decisions that saved Australian Imperial Force deserters from the firing squad in World War I (BA Honours thesis, La Trobe University, 1984).

DECHAINEUX, Captain Emile Frank Verlaine (3 October 1902–21 October 1944) graduated from RANC Jervis Bay in 1919, and served aboard HMA Ships *Australia* and *Anzac* before proceeding to Britain for further training. He returned to Australia in 1924 and served a further period at sea before specialising as a torpedo officer and a naval air observer. In 1935–36 he was squadron torpedo officer aboard HMAS *Canberra*. He attended the RN staff college in 1937, and thereafter worked at the Admiralty in the Tactical and Minesweeping Divisions. He commanded a British destroyer during the evacuation from Dunkirk, and was awarded the DSC in 1941 for service in the North Sea in command of a destroyer flotilla leader. Dechaineux returned to Australia at the end of 1941 and became Director of Operations at Navy Office. He returned to sea duty in command of HMAS *Warramunga* in November 1942, and commanded a Task Group in actions at Arawe, Cape Gloucester, Saidor and the Admiralty Islands. He was selected for early promotion to captain in December 1943. In command of the heavy cruiser HMAS *Australia* from March 1944, Dechaineux participated in further actions at Hollandia, Biak and Morotai before sailing for the Philippines. While supporting American landings at Leyte Gulf, the *Australia* took a direct hit near the bridge from a Japanese kamikaze on 21 October 1944. Dechaineux was mortally wounded; the ship's total losses from the hit were 30 officers and seamen killed. He was awarded the American Legion of Merit posthumously. A *Collins* Class submarine (q.v.) is named in his honour.

DECHAINEUX, HMAS see *COLLINS* **CLASS SUBMARINES**

DECORATIONS Honours and awards are conferred on members of the armed forces in recognition of gallant or distinguished conduct or service. Service decorations, such as campaign medals, are awarded for participation in a designated area of operations within a defined period. An honour usually confers membership of an order of chivalry which carries a title, such as a knighthood. Some of these, such as the Order of the British Empire, have a military as well as a civil division. Most of those in the imperial honours system were open to civilians as well, and were usually conferred on senior officers, often as recognition of having held a particular office. For many years, for example, all the chiefs of staff were knighted at some point during or immediately after their term of office. Australian honours were introduced by the Whitlam government in the form of the Order of Australia, which has a general and a military division and which, for a short period, contained a class of knighthood, which was abolished in 1986. Its four classes are Companion (AC), Officer (AO), Member (AM), and recipient of the Medal of the order (OAM).

Under the imperial honours system Australian service personnel were eligible for the range of awards for gallantry and distinguished service within the sovereign's gift. These ranged from the highest award for gallantry, the Victoria Cross (q.v.), through a range of awards such as the Distinguished Conduct Medal and Distinguished Service Order, the Military Medal and Military Cross, to the Meritorious Service Medal and the Commendation for Brave Conduct. Only the Victoria Cross and the Mention in Despatches (q.v.) could be awarded posthumously, and although eligibility extended to all ranks (and from 1868 to soldiers of colonial contingents), the award of the Victoria Cross required the testimony of an officer present at the scene for a recommendation to be forwarded. Within this system there were also a number of awards such as the Distinguished Flying Cross, the Distinguished Flying Medal and the Distinguished Service Cross that were service specific. Gallantry decorations were usually awarded by rank: as a generalisation, crosses for officers and warrant officers, medals for other ranks. The award of gallantry decorations was always covered by a formula which rationed the number of awards of each class of decoration within a six-month period in proportion to the size of the force deployed. At times Australian servicemen felt that this led to unfairness, since Australian combat units serving as part of larger British formations often did a greater proportion of the fighting (and incurred more of the casualties) than many British units on rear-area duties, but their smaller numbers overall led to fewer awards per head. At the other end of the spectrum, the scales of awards in South Africa during the Boer War were such as to bring the whole system into disrepute: there were 23 awards of the CB and no fewer than 62 DSOs, which led to at least one celebrated libel suit back in Australia. It was also possible for men recommended for a higher award (say, the Victoria Cross) to have it downgraded (to, for example, the Distinguished Service Order) or, in rarer cases, upgraded by higher authority. During the Vietnam War the commanding officers of battalions were awarded the DSO, irrespective of the distinction or otherwise of their service, while the Headquarters, Australian Force Vietnam, was the second most highly decorated unit after the Training Team. In general, thought John Monash (q.v.), commander of the Australian Corps in 1918, 'most men receive rather more or rather less than they deserve'.

As part of the new system of national awards, the Whitlam government also instituted new Australian bravery awards: the Cross of Valour, the Star of Courage, the Bravery Medal, and the Commendation for Brave Conduct. In addition, in January 1986 the Hawke government introduced the Australian Star of Gallantry, the Australian Gallantry Medal and the Australian Commendation for Gallantry, and all these together with the Victoria Cross, which has been retained in the Australian honours system, are the only awards for gallantry to which members of the ADF may now aspire. For distinguished service on operations, earlier awards such as the Distinguished Service Order have been replaced by the Distinguished Service Cross, Distinguished Service Medal and the Commendation for Distinguished Service. Non-operational awards for service now comprise the Conspicuous Service Cross and Conspicuous Service Medal, first awarded in 1991. Also in 1991, and taking a lead from long-standing American practice, the government introduced the Unit Citation for Gallantry and the Meritorious Unit Citation; the latter were awarded to HMA Ships *Sydney* and *Brisbane* for their involvement in the Gulf War (q.v.). In September 1988 the Australian Active Service Medal (AASM) and the Australian Service Medal (ASM) were instituted, to fulfil the function of earlier campaign medals such as the British General Service Medal (GSM) in covering minor campaigns, which many Australians involved in the postwar conflicts were awarded, often with more than one campaign clasp. The ASM is awarded specifically for service in peace-keeping operations (see PEACE-KEEPING FORCES, AUSTRALIAN INVOLVEMENT IN). To date the AASM has been awarded with two clasps, 'Kuwait' and 'Somalia', the ASM with 15. To take the place of earlier long service and good conduct medals, and the Efficiency Decoration and Reserve Force Decoration awarded to the CMF, the Defence Force Service Medal, Reserve Forces Decoration and Reserve Forces Medal were introduced in April 1982; these also replaced the National Medal, introduced in 1975 to fulfil the same purpose and now restricted to civilian uniformed services such as the police and fire services. Two other awards denoting operational service are

the Returned from Active Service badge, issued over and above campaign or war medals, and the Infantry Combat Badge, given to members of the Royal Australian Infantry Corps (q.v.) who have served an aggregate of 90 days service in a designated infantry posting in a warlike situation. Jealously prized, there was some concern expressed about its award to members of 1RAR for their tour of duty in Somalia in 1993.

Foreign awards have always raised difficulties within the imperial and Australian honours systems. Foreign decorations awarded under a quota system between allied nations in the two world wars could be worn, in the technical parlance 'accepted', but foreign awards given to Australian servicemen in other circumstances could not be. Early advisers in Vietnam were instructed to 'receive but not accept' honours from the Vietnamese government alongside whose soldiers they were fighting, which was absurd and insulting. When Warrant Officer Kevin Conway was killed in action in July 1964, the first adviser to be killed, the Vietnamese awarded him the Cross of Gallantry and Knighthood of the National Order. Because the Australian government was 'not at war' officially, he received nothing from his own service. In February 1989 new guidelines were issued permitting the wearing of foreign awards received for gallantry or service. The whole issue of the extension of awards to groups excluded under earlier regulations and practices was examined by a Committee of Inquiry into Defence and Defence Related Awards, which reported in 1994.

DEDMAN, John Johnstone (2 June 1896–22 November 1973). Born in Scotland, Dedman was commissioned into the 10th Battalion, the Border Regiment, in March 1915, and served in Gallipoli, Egypt and France. He was wounded in 1917 and transferred to the Indian Army, serving then in Mesopotamia and Iraq and being promoted to captain. He resigned in 1922 and migrated to Australia.

Dedman joined the ALP in 1928 and was elected to federal parliament in 1940. When Curtin (q.v.) became prime minister in October 1941 he had Dedman elevated to cabinet as minister for War Organisation of Industry, and from December 1941 he sat in the War Cabinet (see AUSTRALIAN WAR CABINET) as well. He was the only member of the Curtin cabinet with military experience. In February 1945 he succeeded J. B. Chifley as minister for Post-War Reconstruction (q.v.).

From 1 November 1946 he was also minister for Defence. This portfolio interested him much less, and his critics claimed that Dedman was more concerned to rebuild Australia than he was to defend it. Postwar reconstruction saw him charged with implementing Labor's commitment to full employment, the retraining of demobilised servicemen, and the creation of a balanced economy. In the defence area, he helped to set up the Anglo-Australian Joint Project on missile research (see LONG RANGE WEAPONS ESTABLISHMENT, WOOMERA) and approved the introduction of aircraft-carriers and a naval aviation capability into the RAN (see FLEET AIR ARM). The defence budget in the immediate postwar years was kept under tight restraint, and some of the aims of the Chifley government's five-year defence program, introduced in 1947, were not met because of other pressures in the economy, principally in manpower. Dedman lost his seat in the 1949 election. In retirement he wrote a number of accounts of the wartime administration of government.

DEFENCE, DEPARTMENT OF One of the original departments of the Commonwealth of Australia, the Department of Defence was established on 1 January 1901, and two months later it assumed control of defence and military matters from the States. For the first four years, the land forces of the Commonwealth were administered by a GOC responsible to the minister for Defence, with a Naval Officer Commanding similarly responsible for naval forces. In 1905 the Council for Defence was established, and a Military and a Naval Board were created to administer land and naval forces. With the vast expansion of Australia's military forces from August 1914, and with the ships of the RAN placed under Admiralty control at the beginning of the war, the responsibilities of the Military and Naval Boards within a single Department of Defence became so divergent that on 12 July 1915 a separate Department of the Navy was established with control over all naval matters formerly administered by Navy Office within the Department of Defence. This arrangement ceased on 21 December 1921 when the two departments were amalgamated into a single Department of Defence. As well as control of general defence matters and of the three services, the department also controlled civil aviation through administering the Air Navigation Act.

The department underwent a major reorganisation in 1938–39, losing a number of its previous functions to new departments: the Department of Civil Aviation (established on 24 November 1938), the Department of Supply and Development (26 April 1939), the Department of Information (4 September 1939), and, most significantly, the Departments of Air, Army and Navy (13 November 1939). The new Department of Defence Co-ordination, created on 13 November 1939, now had to balance the needs and demands of the three services, each

backed by its own department. The deteriorating situation facing Australia at the beginning of 1942 led Prime Minister Curtin to announce on 14 April 1942 that the department would resume its former title of Department of Defence, to reflect better its central role in policy formulation and joint-service coordination. Curtin was also Minister of Defence until his death in July 1945, thereby centralising in himself both the link between the government and the Supreme Commander of Allied Forces in the South-West Pacific Area, General MacArthur (q.v.), and between Australia and the British War Cabinet and the Pacific Council in Washington. After his death, the two positions were separated, J. B. Chifley and J. A. Beasley becoming Prime Minister and Minister of Defence respectively.

In February 1945, looking ahead to the needs of policy formulation in the postwar world, Curtin took steps to establish more effective joint-service and interdepartmental machinery within the department. Despite a number of subsidiary changes over the next decade, there was growing concern in government and defence circles that the defence policy formulation arrangements were ill-suited to the growing complexities of the international environment. In November 1957 the cabinet appointed a committee chaired by Lieutenant-General Sir Leslie Morshead (q.v.) to review the organisation of the six defence-related departments (Defence, Army, Navy, Air, Supply, and Defence Production). The committee recommended that Supply and Defence Production be amalgamated, and that the three service departments be integrated under the Department of Defence. The first proposal was relatively uncontroversial, and was readily accepted by the government; the second raised much more fundamental questions of service autonomy, questions that were bound to provoke a vigorous response from the individual services which had enjoyed their separate departments since 1939. Although the committee had directed much of its argument towards meeting the government's wish to enhance the power of the minister of Defence in relation to the service ministers, the government was not prepared to accept the integration of the single service departments, opting instead to establish the paramountcy of the minister of Defence by administrative order and creating the position of Chairman, Chiefs of Staff Committee (CCS), as the Government's principal military adviser.

Throughout the next decade the committee structure for the development of defence policy and its attendant service implications underwent a number of changes, but there remained the problem, according to the Secretary of the Department of Defence, Sir Henry Bland (1968–70), of the tendency of joint-service committees to reinforce and entrench service parochialism, and in the 'calibre, orientation and lack of continuity of the service members'. Accordingly Bland strengthened the role of the central Defence policy organisation at the expense of the individual services, and made the posting of service personnel to the central Defence organisation less influenced by service considerations and more driven by the requirements of Defence. The introduction of systems analysis for planning, programming and budgeting within Defence further enhanced the status and power of the department in relation to the service departments. Nowhere was this more evident than in the field of intelligence, previously, by and large, the preserve of the service departments. Following the recommendations of a committee chaired by the CCS, General Sir John Wilton (q.v.), the processing of strategic intelligence was centralised through the creation of the Joint Intelligence Organisation in 1970.

The most far-reaching changes to the Department of Defence came about as a result of the Tange Report of November 1973. As Secretary of the Department of Defence, Sir Arthur Tange (q.v.) was charged by the Minister of Defence in the new Labor government, Lance Barnard (q.v.), with rationalising the Defence organisation, centralising those functions dealing with management, supply, expenditure and personnel, and bringing the three services firmly under central control while retaining their separate identities. Tange recommended that the three service departments be integrated within a single Department of Defence under a single minister, who would be assisted by a junior minister within the department. The three service portfolios would be abolished, as would the service boards. The latter had been of relatively little importance for some years, but the service ministers had, at least in theory, been powerful advocates for their own services. Tange was determined to change that and to develop a defence, rather than a service, orientation to security questions. The integrated Defence Force was to be commanded by the Chief of the Defence Force Staff (CDFS) who, together with the Secretary of the Department of Defence, was to exercise administrative control over the Defence Force. The Tange Report was accepted by the government, and despite considerable criticism over the diminution of the separate service positions and the perceived enhancement of the civilian elements within Defence at the the expense of service personnel and interests, there was little sustained opposition and the *Defence Force Reorganisation Act 1975* was passed in September

1975. The following February the individual service boards were abolished, and the service chiefs became the commanders of their respective services, answerable to the CDFS.

The Defence Review Committee was established in 1981 to assess the effectiveness of the higher defence machinery in the light of the changes brought about as a result of the Tange Report. In 1983 the new Labor government accepted the committee's recommendation for even closer coordination between the Department of Defence and the Defence Force, not least through increasing the powers and strengthening the position of the CDFS, whose title was changed in October 1986 to Chief of the Defence Force.

For the first 60 years of its existence, the Department of Defence was based in Melbourne, but in the 1960s it moved to a complex of buildings in the Canberra suburb of Russell, which has become a commonly used name for the department.

DEFENCE (VICTORIAN), DEPARTMENT OF, was formed as part of a general reorganisation of Victoria's military forces in 1883–84 under the first Minister of Defence, Sir Frederick Sargood (q.v.). The first Secretary of the department, Major-General M. F. Downes, was appointed in April 1885. This was an astute move on the part of Sargood, who had been trying to assert his authority over the military commandant, Colonel Disney, as Disney was obliged to defer to Downes as a senior officer. Sargood's position was further strengthened by a recommendation from the British colonial secretary that the minister should have clear control of the department on all subjects except matters of discipline. After Federation the colonial department was absorbed into the federal one, and the head of Victoria's Defence Department, R. H. M. Collins (q.v.), became the first Secretary of the new Department of Defence.

DEFENCE, HIGHER ORGANISATION OF This term refers to the administrative, policy formulation and decision-making structure which runs the defence group of departments and the armed forces. It encompasses the relationship between the minister for Defence, junior ministers in the defence area, the chiefs of staff, and the major commands, headquarters and staff divisions of both the service headquarters and the Department of Defence. It is an organisation which has undergone considerable change, reform and reorganisation (the three terms not necessarily being synonymous), invariably in the direction of greater complexity as the conduct of defence matters has become itself more complex and more bureaucratic. The most significant changes to the higher defence machinery this century have been the relative decline in the powers of the individual service chiefs of staff (q.v.); the growth in the authority of the Chief of the Defence Force and the division of that authority with the civilian secretary of the Department of Defence (q.v.) in an essentially bipolar organisation; the development of joint functions within the armed forces, and of functional commands; the abolition of individual ministers for each of the armed forces; and the growth of centralised control represented by the enhancement of the role and authority of the minister for Defence. This tendency in the Australian higher defence machinery is one shared by most Western armed forces, especially in the postwar period.

DEFENCE ACT (*An Act to provide for the Naval and Military Defence and Protection of the Commonwealth and of the several States*) *1903,* came into force on 1 March 1904, three years after the creation of the Commonwealth. It laid down the structure of the army and the conditions under which it could be deployed, and while it has been amended many times since, it remains the basis of the military forces in Australia. The passing of the act was preceded by a protracted debate in political and military circles, a debate in which Major-General E.T.H. Hutton (q.v.), GOC, Australian Military Forces, argued powerfully for the inclusion of a provision for compulsory overseas military service in time of war, which Hutton saw as an essential requirement in his plan for an imperial approach to defence. Hutton lost, and the act severely circumscribed both the size and the possible deployment of Australia's military force. The pre-eminent role of the CMF was entrenched, for the act prohibited the raising of any permanent military force except for 'Administrative and Instructional Staffs, including Army Service, Medical and Ordnance Staffs, Garrison Artillery, Fortress Engineers, and Submarine Mining Engineers'. Section 35 laid down that the Defence Force was to be raised and maintained by voluntary enlistment only, while section 48 prohibited the use of the military forces overseas. (Following the creation of the RAN in 1911, naval personnel were liable for service overseas in keeping with the 1909 agreement that in time of war the Australian Squadron would be placed under the control of the Admiralty for use anywhere that it was required, as indeed happened in 1914.) Section 49 specified that members of the military forces could not be compelled to serve beyond the Commonwealth or its Territories. This prohibition on compulsory service overseas applied also to members of the Permanent Military Forces, and was not removed until 1951. Until then, as for example in the case of members of the British Commonwealth Occupation Force (q.v.) in Japan required for service in the Korean War, men could only be sent overseas if they specifically volunteered for such duty. Section 59 made all men between the ages of 18 and 60 (except for those exempt because of religious conscientious objection) liable to serve in the militia in time of war (which was amended in 1964 to include a 'time of defence emergency', and further amended in 1965 to enable men called up to be posted to the army, navy or air force rather than only the militia as originally laid down). Section 98 limited the death penalty to four offences (mutiny, desertion to the enemy, treachery resulting in a garrison or ship being handed over to the enemy, and treasonous correspondence with the enemy), compared with 17 in the British Army, and required that before a death sentence imposed by a court martial could be carried out, it had to be confirmed by the Governor-General (see DEATH PENALTY).

The act was amended in 1909 to enable the government to establish the peacetime compulsory training scheme (see CONSCRIPTION), but the prohibition of overseas service meant that the expeditionary force that was raised in 1914, and which became the AIF, had to be raised on the basis of voluntary enlistment. As the casualties mounted in 1916 and the need for reinforcements became more pressing, the restrictions of the act were seen by Prime Minister W. M. Hughes (q.v.) as fundamentally damaging to Australia's war effort. In an attempt to influence the first referendum on conscription, Hughes used the powers available under the Defence Act to call-up all men between 21 and 35 for compulsory home service; this merely confirmed the views of his opponents that while he was bent on full-scale conscription for overseas service he would not use his powers to impose conscription by gazetting it under the War Precautions Act for fear of risking parliamentary defeat. Pressure from the British High Command, supported by most senior Australian commanders, for the death penalty to be carried out on Australians found guilty of mutiny or desertion could not bypass the requirement that the Governor-General, on advice from his Australian ministers, had to confirm such sentences, with the result that of all British forces, only Australia did not execute any of its own combatants.

The restrictions of the act were further demonstrated in 1938 when the CGS called for the creation of permanent field units that could supply trained men to militia units once war had broken out and which could also be used to guard vital points. Because the act prohibited the raising of permanent infantry units, the infantry of the Darwin Mobile Force (q.v.) had to be enlisted and serve as artillerymen. In June 1939 the act was amended to make the CMF liable for service in Papua New Guinea, but when Prime Minister Curtin acceded to pressure from General MacArthur to resolve the anomalous distinction between the AIF and the CMF, he did so not by amending the Defence Act as such but by introducing the Defence (Citizen Military Forces) Bill, which set aside the Defence Act's prohibition on compulsory military service but only for the duration of the war and a period of six months beyond the conclusion of peace. Thus the principle of voluntary service overseas was preserved in the essential legislation.

It was removed, however, in 1964 when section 49 was repealed and was replaced by 50C, which made all members of the military forces ('except for persons compulsorily called up for service'), liable for service overseas. That exception was removed by a further amendment in 1965, with the result that conscripts called up under the National Service scheme were able to be sent to Vietnam. The provisions regarding conscientious objection, which had previously been restrictive in that an applicant had to have a conscientious objection to war in general, were changed in 1992 to allow for conscientious objection to a particular war.

(See also CIVIL POWER, AID TO THE.)

DEFENCE COMMITTEE was established by ministerial direction in May 1926 and more formally on 8 March 1929 by regulation. It consisted originally of the CNS, the CGS and the CAS, with the senior member to serve as chairman, and with an officer from the Department of Defence to act as secretary. The role of the committee was to advise the minister for Defence on all matters affecting defence policy, including the relationship of civil aviation to defence requirements, and 'all matters of policy or principle affecting the organization and distribution of such Air Forces as may be established'. Further regulations were promulgated in 1946 and 1960, and the functions of the committee were expanded to include advising the minister on inter-service matters and the 'co-ordination of military strategic, economic, financial and external affairs aspects of defence policy'. The committee is now chaired by the Secretary of the Department of Defence, and its members are the Chief of the Defence Force and the three service chiefs, together with the secretaries of the departments of Prime Minister and Cabinet and Foreign Affairs and Trade, and the Treasury.

DEFENCE COOPERATION Cooperation on defence issues is a common feature of relations between allied or friendly countries which share significant security interests, and is not necessarily, or in Australia's case generally, related to meeting a specific threat. The basic objective is 'to promote Australia's security interests in the South Pacific and Southeast Asia', and to do so through developing bilateral relations in the region and helping to ensure political stability and economic growth. Australia's Defence Cooperation Program (or DCP, a more politically acceptable version of the older term, military aid) has the twin virtues of meeting the objectives outlined above while at the same time providing useful opportunities for members of the ADF to exercise skills which might not otherwise receive the same level of use, and providing Australian industry, and especially defence industries, with opportunities to promote the sale of equipment manufactured in Australia. In addition to the sale or supply of equipment, DCP assistance takes one or more of five basic forms: joint projects, designed to meet the need for a particular capability in the recipient country's armed forces; joint operations, which undertake particular activities, often with a developmental aspect, such as mapping or channel-clearing; advisory/consultancy assistance, provided by uniformed or civilian defence personnel; training, undertaken by ADF personnel and which may occur either in Australia or the recipient country, and which also includes educational assistance to members of these armed forces, usually undertaken in Australia; combined exercises, involving the ADF and other regional armed forces. Very little DCP assistance involves the sale of armaments or equipment, most going on civil aid. Between 1978 and 1983 the budget outlay on DCP was about $176 million, and between fiscal years 1977–78 and 1993–94 the proportion of the defence budget expended in this area has averaged 0.86 per cent, with actual expenditure ranging between a low of $24.6 million in 1978–79 and a peak of $93 million in 1990–91.

Assistance programs are developed individually in consultation with the recipient government, and monitored regularly. In recent years the allocation of military aid to the micro-states of the south-west Pacific has greatly increased (through, for example, the Pacific Patrol Boat Program), while that to ASEAN nations has declined considerably and Papua New Guinea's share has also fallen. Critics charge that DCP assistance sometimes imposes technological solutions to problems which are inappropriate in the context of small Third World economies (the running and maintenance costs associated with the patrol boat program being a case in point). Other criticisms range from the allegedly destabilising effect of weapons transfers within the region to concerns about the internal uses of weapons and training provided by Australia.

DEFENCE COORDINATION, DEPARTMENT OF, was established on 13 November 1939 along with the separate departments of Air, Army and Navy (qq.v.) whose activities it was to coordinate. Its responsibilities included defence policy; administrative, financial and works coordination and review; administration of the National Security Act (q.v.) and Regulations and maintenance of the Commonwealth War Book (q.v.); and civil defence issues. With the prime minister holding the portfolio, the department was the central component of the machinery for the higher direction of the war, providing the secretariat for the War Cabinet (see AUSTRALIAN WAR CABINET), the Defence Committee and the Chiefs of Staff (qq.v.) Committee. Its responsibility for civil defence was taken over by the Department of Home Security in June 1941, while the Treasury Defence Division (q.v.) became responsible for financial coordination and review in December 1941. The department was abolished on 14 April 1942 and its responsibilities divided between the Department of Labour and National Service and the Department of Defence (q.v.).

DEFENCE FORCE REMUNERATION TRIBUNAL (DFRT) was established in 1985 as an independent body to determine pay and allowances within the Defence Force. The case for the Defence Force is argued by the Defence Force Advocate, a position established within the Defence Force, while the case for the employer — the government — has since 1987 been argued by the Department of Industrial Relations, thus removing a conflict of interest that had existed within the Department of Defence when it originally had been responsible for arguing both sides of any pay case. The tribunal has three members, and is required to review ADF pay at least every two years, as well as conducting ongoing reviews of service allowances and the impact of national wage-case increases on service conditions. Its determinations are binding on the government and the ADF.

DEFENCE FORCE RETIREMENT AND DEATH BENEFITS (DFRDB) AUTHORITY arose out of dissatisfaction with the Defence Forces Retirement Benefits Scheme, which was introduced in 1948 as the first superannuation scheme specifically for service personnel. The old scheme was complicated and far from comprehensive and, following the report of a joint parliamentary committee (the Jess Committee) in 1972, a new scheme was established in 1973 under the DFRDB Authority, chaired by the Commissioner of Superannuation, with a representative nominated by each of the three service chiefs and one nominated by the minister of Defence. The new scheme was based on contributions from all serving members at a flat rate of 5.5 per cent of an individual's annual rate of pay, and the level of retirement pay depended on the number of years of service, 20 years' service producing 35 per cent of final pay. Members retiring after 1 October 1972 were entitled to commute their retirement pay to a lump sum if this was done within one year of retirement, there was more generous treatment of invalidity pay, and in certain circumstances reversionary rights were to be extended to de facto wives and the children of de facto relationships. The scheme was reviewed in 1990 in light of the changes in government policy towards superannuation schemes in general and its intention to move greater numbers in the community towards self-funded retirement income in place of the aged pension.

DEFENCE INTELLIGENCE ORGANISATION (DIO) see **INTELLIGENCE, MILITARY**

DEFENCE POST-HOSTILITIES PLANNING COMMITTEE The shape of Australia's postwar defence arrangements was under consideration even before the Second World War had ended. As part of a move to improve the joint service machinery within the Department of Defence and to provide advice on future cooperation with the British Commonwealth and the new international organisation which was to become the United Nations, in February 1945 Australian Prime Minister John Curtin (q.v.) approved the creation of four new committees within the Defence Department, answerable to the Defence Committee (q.v.). The Defence Post-hostilities Planning Committee had three sub-committees: Armistice and Peace Terms, World Organisation, and Postwar Forces. Among its earliest tasks was consideration of the deployment of Australian forces outside Australian territory after the Japanese surrender, since this had obvious implications for demobilisation (q.v.) of the forces.

DEFENCE PRODUCTION, DEPARTMENT OF, established on 11 May 1951, took away responsibility from the Department of Supply (q.v.) for the manufacture and supply of munitions (including armaments, ammunition, war chemicals, machine tools and radar), aircraft production and defence production planning. Abolished on 23

April 1958, its powers reverted to the Department of Supply.

DEFENCE SCIENCE AND TECHNOLOGY ORGANISATION (DSTO),
part of the Department of Defence with headquarters in Canberra, was established in 1974 to take over the scientific and technical research activities previously conducted by the armed services and other government agencies such as the Department of Supply and the Department of Manufacturing Industry. Originally it consisted of three divisions — the research laboratories and establishments of the Australian Defence Scientific Service, Military Studies and Operational Analysis, and the Policy and Programme Planning Division — but its structure has been modified to suit changing trends. It is Australia's second-largest research organisation.

(See also SCIENCE AND TECHNOLOGY.)

DEFENCE SIGNALS DIRECTORATE (DSD) see INTELLIGENCE, MILITARY

DEFENSIVELY EQUIPPED MERCHANT SHIPS see DEMS

DELORAINE, HMAS see *BATHURST* CLASS MINE-SWEEPERS (CORVETTES)

DEMOBILISATION of a military force means breaking up its component formations and units and discharging its personnel from military control and responsibility. A major part of demobilisation is the return of troops to their country of origin, a process usually referred to as repatriation. In Australia, however, 'repatriation' (q.v.) came to mean assisting soldiers to fit back into civilian society, and this confusion of terminology caused problems when a Demobilisation and Repatriation Section of Administrative Headquarters, AIF, in London began its work in April 1918. The Minister for Repatriation, Senator E. D. Millen, was concerned that the word Repatriation in the section's title signalled an encroachment by the Defence Department on his portfolio, so the section's staff were told to wait for instructions from the Australian government before formulating any plans. Finally, in November 1918, they were told to take their directions from Prime Minister W. M. Hughes (q.v.) who was then in London. (After Hughes left for the Paris Peace Conference Senator George Pearce [q.v.] went to London to take over this role.) In the meantime, the Australian government had bought more time for itself by changing the AIF's period of service from the duration of the war 'and six months thereafter' to a period to be determined by proclamation by the Governor-General. On 21 November General Monash (q.v.) was appointed Director-General of the newly formed Department of Repatriation and Demobilisation which was part of the AIF but was largely independent of AIF Administrative Headquarters. At this time there were about 185 000 men to be repatriated: 95 000 in France and Belgium, 60 000 in the United Kingdom and 30 000 in Egypt and other minor theatres of war. The Australian government favoured the return of troops in existing formations, but Monash and Hughes agreed to organise repatriation on a 'first to come, first to go' basis. Accordingly, Monash developed a scheme which saw divisions classifying their members into quotas of 1000 (a normal trainload or shipload) on the basis of length of service. Each quota, though drawn from all services, would be organised as a battalion under the command of officers known to the men. As quotas were withdrawn, the longest-serving leaving first, remaining units were amalgamated once they fell below 40 per cent of establishment and eventually they were reduced to 'cadres' of specialists who would hand over equipment, camp sites and buildings to British and French authorities. The depots in Britain having been cleared by repatriating first men discharged from hospital, quotas were brought to the United Kingdom where they took two weeks' leave then awaited their turn for embarkation. Most of the paperwork relating to their demobilisation was completed in Britain. While waiting in France or Britain they could attend classes as part of the AIF Education Scheme which also provided for the 'Non-Military Employment' of those who wished to gain skills by attending educational institutions or by working in industry or agriculture. The fact that the AIF was the only major force in Britain not to have anti-demobilisation riots can be attributed to Monash's demobilisation scheme, which combined fairness with the maintenance of discipline, as well as to the AIF administration's efforts to explain the system to the troops and to keep them informed about the reasons for any delays in the process. It took 176 voyages, leaving between December 1918 and December 1919, to return all the troops to Australia. The 40 000 convalescents went separately under medical control, and 'family ships' were arranged at intervals throughout the process to carry men with wives and young children. Only the troops in the Middle East were shipped as units. Most of the voyages back to Australia were uneventful, but there were some disturbances on troop-ships both at sea and ashore, especially when men found themselves quarantined on reaching Australia because of the influenza epidemic. Finally soldiers were discharged either in a capital city or in their place of enlistment and were

A protest march held on Moratai on 10 December 1945, three months after the end of the Second World War, to complain about delays in repatriation to Australia. About 4500 men took part in the protest which broke up peacefully after representatives met with officers and discussed grievances. (AWM 124202)

given the portion of their pay that had been compulsorily deferred until their discharge.

During the Second World War the Reconstruction Demobilisation Committee was established in March 1943 under the Minister for Post-War Reconstruction (q.v.) to coordinate the planning and administration of demobilisation. In June 1944 the War Cabinet (see AUSTRALIAN WAR CABINET) approved the committee's recommendation that priority of discharge be determined by the allocation of points according to length of service, age at enlistment and marital status. Accelerated discharge was possible on medical or compassionate grounds and for those whose employment in civilian occupations was urgently needed or who were accepted for full-time training. In September 1945 Lieutenant-General Savige (q.v.) was appointed Coordinator of Demobilisation and Dispersal, and general demobilisation began on 1 October. Returning service personnel to Australia was not such a major undertaking as in the First World War since in August 1945 there were 310600 servicemen in Australia, 224000 outside Australia in the South-West Pacific Area (q.v.) and only 20100 serving outside this area. Almost all the 43600 servicewomen were within Australia. Men with low priority for demobilisation, as well as those who

requested deferment of discharge, were kept on to occupy captured territory, disarm the Japanese, sweep for mines, transport others back to Australia, administer demobilisation and perform routine duties. As personnel were demobilised, units were declared redundant and their low-priority men transferred to interim forces which were to merge into the postwar armed forces. Repatriation proceeded smoothly; most men were back in Australia by the end of January 1946, and 80 per cent of them had been discharged by mid-1946. Demobilisation was carried out in four stages, with the number to be discharged in each stage set by cabinet; by the time it was completed in February 1947 the services had been reduced to an interim strength of 60133. Army and air force personnel were discharged through dispersal centres set up in each State capital, while navy discharge was carried out at existing depots because of the small numbers involved. Before being discharged service members were interviewed by a rehabilitation officer, given a medical examination and provided with information about repatriation benefits available to them. The Manpower Directorate was responsible for placing them in employment, for civil registration and for issuing them with identity cards and civilian ration entitlements.

DEMS is an acronym for Defensively Equipped Merchant Ships. During the Second World War hundreds of Australian and Allied ships were fitted in Australia with a variety of guns, mainly to assist them in fighting off attacks from surfaced submarines or aircraft. Guns fitted included some form of anti-aircraft weapon and a 4-inch naval gun. During the war over 1000 RAN Reserve and General Services personnel were drafted to man the guns.

DENTAL CORPS see **ROYAL AUSTRALIAN ARMY DENTAL CORPS**

DERWENT, **HMAS** see **RIVER CLASS ANTI-SUBMARINE FRIGATES**

DEXTER, David St Alban (8 January 1917–15 March 1992). Dexter enlisted in the AIF in 1940 and served throughout the war until demobilisation in 1946. He then pursued a career in the diplomatic service, and was Australian delegate to the United Nations General Assembly between 1947 and 1948. He held various posts in the Department of External Affairs between 1952 and 1960, culminating in a posting as Counsellor in the High Commission in Delhi, before moving into university administration with the Australian Universities Commission between 1960 and 1967. He retired as Registrar (Property and Plans) at the Australian National University in 1978. In 1961 he published volume six of the Army series of the official history of the Second World War, *The New Guinea Offensives*, a work of massive detail which was informed, to its benefit, by his own service in that theatre.

DIAMANTINA, **HMAS** see **RIVER CLASS FRIGATES**

DIBB REPORT In February 1985 Kim Beazley (q.v.), Minister for Defence in the Hawke government, commissioned Paul Dibb, an external consultant and former member of the Department of Defence, to analyse the current content, assumptions and priorities of Australia's defence planning, and provide a series of recommendations for future developments. Published in March 1986, the report argued for a greater focus on 'the area of direct military interest', encompassing parts of south-east Asia and the south Pacific, a concentration on the direct defence of Australia (which implied the final abandonment of any notions of forward defence [q.v.]), and the protection of Australian interests, including economic interests, in the waters surrounding Australia. Australia's military posture was to be based on a strategy of 'denial', and the approaches to Australia

were to be protected by a layered defence of over-the-horizon radar (see JINDALEE), long-range patrol aircraft, limited-range maritime strike capacity provided by the air force and navy and, lastly, should the other layers be penetrated, units of the army's Operational Deployment Force. Force structure and equipment acquisition programs were to be predicated on these roles. Dibb's report operated on the basic assumption that 'Australia is one of the most secure countries on earth'. Although he allowed for the possibility of regional conflict, he argued that longer-term intentions and capabilities of other countries in the region were 'imponderable' and thus not a basis for defence planning. Greeted in some quarters as 'the high point in Australia's most recent attempt to install a more independent and self-reliant defence posture', the report was criticised for its essentially defensive strategic character, its alleged repudiation of certain Australian military capabilities, and an overly optimistic interpretation of warning time and readiness. Many of the assumptions of the report underpinned the 1987 White Paper *The Defence of Australia*, but the Hawke government modified certain of Dibb's key recommendations, especially those concerning the emphasis to be placed on the American alliance and a more active military role for Australia in the Pacific.

Paul Dibb, *Review of Australia's Defence Capabilities* (Australian Government Publishing Service, Canberra, 1986).

'DIGGER', a term applied to miners on the Australian goldfields since the 1850s, has been used of Australian and New Zealand soldiers since at least 1916. It became popular among troops constantly 'digging in' during trench warfare in France, and was applied particularly to infantrymen and privates. 'Digger' and 'dig' were used by soldiers as friendly terms of address equivalent to 'cobber' and 'mate', and after the war this usage spread to the civilian population. The term has tended to be defined in highly value-laden ways from *Aussie* magazine's 1918 definition 'a white man who runs straight' to 'a man for whom freedom, comradeship, a wide tolerance, and a strong sense of the innate worth of man, count for more than all the kingdoms of the world and the glory of them' in A. G. Butler's (q.v.) 1945 exploration of 'diggerism', *The Digger: A Study in Democracy*. Prime Minister W. M. Hughes (q.v.) took political advantage of his nickname 'the Little Digger', and though in the years between the two world wars conservatives and radicals struggled for possession of the word, its power lay in its very ambiguity. Underlying the mundane meaning are what Butler called 'the ghosts of memory that surround the word', associations which allow it to signify simultaneously seemingly contradictory ideas such as egalitarian 'mateship' and militaristic elitism.

(See also ANZAC; MILITARY SLANG.)

DIRECTORATE OF AIR FORCE INTELLIGENCE
see INTELLIGENCE, MILITARY

DIRECTORATE OF MILITARY INTELLIGENCE see
INTELLIGENCE, MILITARY

DIRECTORATE OF NAVAL INTELLIGENCE see
INTELLIGENCE, MILITARY

DIRECTORATE OF RESEARCH AND CIVIL AFFAIRS (DORCA) was created on 6 October
1943. Reporting directly to Australian commander General T. A. Blamey (q.v.), DORCA was headed by Colonel Alfred Conlon (1908–61) and contained many distinguished academics. It was an advisory body only, undertaking research on matters of interest to the army and keeping Blamey and other officers informed on relevant current events. The ambitious Conlon sought to extend DORCA's influence as far as possible, leading External Affairs Minister H.V. Evatt to warn Conlon 'to keep out of foreign affairs or he'd break his wrist'. Conlon's attempts to gain a dominant position for DORCA in civil affairs operations in British Borneo, Papua and New Guinea provoked animosity from other authorities there. DORCA staff prepared plans for Papua and New Guinea embodying their reformist outlook, but their only lasting result was the establishment of the School of Civil Affairs (which became the Australian School of Pacific Administration) in 1945 and the creation of a climate for reform in relation to Papua New Guinea. DORCA was disbanded in January 1946.

Brian Jinks, 'Alfred Conlon, The Directorate of Research and New Guinea', *Journal of Australian Studies*, no. 12 (1983), pp. 21–33.

DISARMAMENT refers to the complete abolition
by international agreement of nation-states' weaponry and armed forces (other than the minimum required to maintain internal law and order), as well as partial measures aimed at reaching this goal. Less ambitious limitations of weaponry, armed forces or military budgets which are not part of a process of complete disarmament are also often labelled as disarmament measures, though they are more correctly referred to as arms control. Australian involvement in international disarmament debates began after the First World War. The end of the perceived German threat to Australia, public disgust at the horror of the recent war, and a desire

to direct economic resources away from the armed forces and towards more productive areas all combined to make Australian governments generally sympathetic to disarmament in the 1920s. They were particularly supportive of naval arms limitation, seeing an end to US–Japanese naval competition as essential to Pacific security. Accordingly, news that the British Empire, the United States, Japan, France and Italy had agreed at the Washington conference of 1921–22 to limit their strength in capital ships was received with elation in Australia. This was despite the fact that the Five Power Treaty required the scrapping of HMAS *Australia* (I) (q.v.), the first and only cutback of Australia's armed forces to result from a disarmament agreement. There was also bipartisan support in the Australian parliament for the 1925 Geneva Gas Protocol's outlawing of the use of biological and chemical weapons. Australia signed the protocol in 1930, with the proviso that gas could be used against an enemy who used it first. However, Australian governments between the wars still relied on Britain to guarantee Australia's security, and only supported disarmament initiatives which complemented this guarantee. They generally saw disarmament as an issue for the great powers to deal with and, despite lobbying by organisations such as the World Disarmament Movement (1928–33), were reactive rather than proactive on disarmament issues.

After the use of atomic bombs against Japan in 1945, nuclear disarmament was the main disarmament issue of concern to both governments and the public. At the height of the Cold War, from about 1949 to 1965, a tiny but energetic peace movement made up of an odd alliance of communists, Christians and liberal idealists advocated multilateral, general and complete disarmament. The conservative Menzies government, committed to alliances with Britain and the United States which relied on nuclear deterrence, paid the peace movement little heed, and its support for the 1963 Partial Test Ban Treaty was in line with the American and British position. Subsequent Liberal–Country Party coalition governments were not enthusiastic about the 1968 Nuclear Non-Proliferation Treaty, which was not signed by Australia until 1970 and only ratified in 1973 by the Whitlam Labor government. The Whitlam government also took previous governments' protests about French atmospheric testing in the Pacific further by pursuing a successful case against France in the International Court of Justice. The Fraser coalition government which replaced Whitlam was more sympathetic to nuclear arms control than earlier conservative governments, and under Fraser Australia joined the United Nations' disarmament treaty negotiating body, the Committee (later Conference) on Disarmament.

Renewed Cold War tensions from around 1978 also produced an Australian peace movement which was numerically stronger but more fragmented than ever before. Principally concerned with nuclear disarmament, which it felt Australia could promote by ending uranium-mining and withdrawing from ANZUS (q.v.), the movement never attracted the majority support which sustained New Zealand's anti-nuclear policies in the 1980s. But its influence within the Australian Labor Party led the Hawke Labor government to adopt a proactive disarmament policy. To the Fraser government's policies of supporting strategic nuclear arms control, a comprehensive nuclear test ban and nuclear non-proliferation, it added initiatives such as the appointment of a disarmament ambassador, the strengthening of the Department of Foreign Affairs and Trade's capacity to deal with disarmament issues, and the negotiation of the 1985 South Pacific Nuclear Free Zone Treaty (rejected by the United States, the United Kingdom and France, the treaty is directed solely against nuclear proliferation, testing and waste-dumping in the region). Despite being the first Australian government to commit itself to 'complete nuclear disarmament' and the replacement of nuclear deterrence by 'common security', the Hawke government's disarmament policy was constrained by its commitment to the US alliance and its support of US nuclear policy. With the end of the Cold War in the late 1980s and the resulting easing of nuclear fears, other disarmament issues have come to the fore. The Australian government has been active in UN chemical weapons negotiations, and has initiated informal arrangements aimed at curbing the proliferation of chemical weapons and missile technology. However, since Australian disarmament policy is aimed at controlling arms which Australia does not itself possess, and since it is a relatively small, militarily and economically weak, and geographically isolated country, Australia's ability to influence international disarmament debate is limited.

Trevor Findlay, 'Disarmament and arms control' in F. A. Mediansky (ed.), *Australia in a Changing World: New Foreign Policy Directions* (Maxwell Macmillan, Sydney, 1992).

DIVISION A division is a military formation combining the necessary arms and services required for sustained independent operations, and is usually commanded by a major-general. In armies of the British pattern it usually comprises three infantry brigades with three artillery regiments and supporting arms and services; two armoured brigades (or one armoured and one motorised brigade) plus divisional troops in the case of an armoured division; three light horse brigades plus divisional troops in the case of a mounted division. Its establishment has changed over time: in the First World War a full-strength infantry division numbered approximately 18000 men, which had declined to about 15000 in the Second World War. More significant in some ways than the strength of the division was the 'divisional slice', which is the number of men required to deploy and maintain a division in the field. In the First World War it was 30000 for the AIF; in the Second World War it had risen to 52000, which indicated the greater technical sophistication required to support operations but which, like the corresponding figure from the previous war, was low by comparison with most of Australia's allies.

In the First World War Australia fielded five infantry divisions (1st to 5th, formed in November 1917 into the Australian Corps) and the best part of two mounted divisions (Australian Mounted and Anzac Mounted, which contained a New Zealand Mounted Rifles Brigade, both in the Middle East as part of the Desert Mounted Corps). The 6th infantry division was formed in the United Kingdom in March 1917 but because of manpower shortages was disbanded in September. The divisional structure was maintained in the peacetime army between the wars on the recommendation of the conference of senior officers which met in 1920, but the repeated cuts to the defence budget ensured that this was a skeleton organisation at best. During the Second World War Australia raised five divisions in the 2nd AIF (four infantry, 6th to 9th, plus the 1st Armoured Division), seven divisions from the militia (1st to 5th plus 11th and 12th), not all of which saw service outside Australia, together with another armoured and a motor division (2nd Motor and 3rd Armoured), all of which proved beyond Australia's manpower resources to maintain. In the postwar period the army maintained the divisional structure at three divisions, one regular and two CMF or reserve. By 1965, when the army prepared for the commitment to Vietnam, a division numbered 988 officers and 13989 other ranks in a far more complex organisation than its First World War forebear. The regular 1st Division today comprises three brigades (1st, 3rd and 6th), but these have only two battalions each and most are under strength. Australia has not actually deployed a division in the field on operations since the Second World War.

DOCKYARDS see **COCKATOO ISLAND; GARDEN ISLAND; WILLIAMSTOWN DOCKYARD**

DOUGHERTY, Major-General Ivan Noel (6 April 1907–   ). Trained as a schoolteacher, Dougherty completed an economics degree at Sydney University, and was commissioned into the Sydney University Regiment in July 1927. He served in the CMF throughout the 1930s, and by 1938 had received command of the 33rd Battalion. At the outbreak of war he enlisted in the 6th Division, taking a drop in rank to major in order to do so, and initially receiving appointment as second-in-command of the 2/2nd Battalion under George Wootten (q.v.). In August 1940 he became commanding officer of the 2/4th Battalion, part of the newly constituted 19th Brigade under H. C. H. Robertson (q.v.). The latter had wished to make his own selection, and at first treated Dougherty coldly. By dint of his considerable ability, and his willingness to stand up to his sometimes overbearing superior, Dougherty soon won Robertson over, who thereafter became one of his greatest supporters. When the brigadier was absent, Dougherty usually acted as brigade commander despite the fact that he was the youngest and most junior of the three unit commanders.

He commanded through the first campaign in Libya and then in Greece (qq.v.). Returning to Australia in March 1942 he was one of a number of able young lieutenant-colonels with recent experience of battle who were promoted to command brigades, in his case the 23rd Brigade, which he took to the Northern Territory. In October 1942 he was given the 21st Brigade, and he led this through the rest of the war, taking part in the fighting for Gona, in the Ramu and Markham Valleys in the New Guinea campaign, and finally in Borneo in 1945. A warm and humane man, he consistently demonstrated a concern for his soldiers and a refusal to risk their lives unnecessarily. Protesting the shortages of mortar ammunition necessary for the coming landings at Balikpapan to his divisional commander, Milford, he wrote: 'I … consider it criminal that Australians of any division should be asked, at this stage of the war, to go into what might well be a difficult operation with less ammunition than that which is considered necessary by their commanders.'

Returning to the Education Department after the war, he continued in the CMF, commanding the 8th Brigade in 1948–52 and the 2nd Division in 1952–54. He was CMF Member of the Military Board (q.v.) between 1954 and 1957. In 1955 he left the Department of Education and became Director of the Civil Defence and State Emergency Services of New South Wales. In 1960 the Minister for the Army, J. O. Cramer, sought to appoint Dougherty as CGS in succession to Lieutenant-General Sir Ragnar Garrett (q.v.), in an attempt to maintain the position of the CMF against the increasing dominance of army affairs by the regulars. Both Prime Minister R. G. Menzies and the Minister for Defence, Athol Townley (q.v.), opposed the idea and it did not proceed. A great believer in the 'one army' concept, Dougherty would certainly have been an able CGS, but the precedent established would probably have been a poor one. From 1958 to 1966 he was Deputy Chancellor of Sydney University and was knighted in 1968.

G. H. Fearnside and Ken Clift, *Dougherty: A Great Man Among Men* (Alpha Books, Sydney, 1979).

DOWLING, Vice-Admiral Roy Russell (28 May 1901–15 April 1969). Dowling entered the Naval College at Jervis Bay in 1915 and became a midshipman in 1919. He served on HM Ships *Ramillies* and *Venturous*, specialised in gunnery and became gunnery officer on HMS *Excellent* in 1925. He served on various Australian and British ships throughout the 1930s, and commanded HMAS *Swan* in 1937–38.

At the beginning of the war he was executive officer on HMS *Naiad*, and served with that ship in the Mediterranean until it was sunk in March 1942. He returned to Australia and was appointed Director of Plans in Navy Office. In November 1944 he was given command of HMAS *Hobart*, which he held until 1946, and took part in operations at Cebu, Tarakan, Wewak, Labuan and Balikpapan. He was awarded the DSO in November 1945 for 'courage, skill and initiative' in these actions.

After the war he held a variety of staff and seagoing posts, including captain of the aircraft-carrier HMAS *Sydney* in 1948–50 and Second Naval Member 1950–52, before attending the Imperial Defence College (q.v.) in London in 1953. He commanded the Australian fleet 1953–54, and was appointed CNS in 1955.

Dowling led the navy in a period of change, some of which he found uncomfortable. The process of weaning the RAN off the RN and reorienting it towards the United States, in such matters as ship acquisition and naval cooperation under ANZUS (q.v.), began during Dowling's tenure of office although it was completed under his successors. He sought to maintain a 'very close relationship in everything that really matters' with the British, and was partly successful in this by coordinating his own advice to the Australian government with that being given by the First Sea Lord, Mountbatten, in London. In March 1959 he was appointed Chairman, Chiefs of Staff Committee, and was involved in the planning for possible

SEATO (q.v.) action during the second Laotian crisis in 1960–61. He retired from the navy in May 1961.

DOWNES, Major-General Rupert Major (10 February 1885–5 March 1945). The son of Major-General M. F. Downes, a British officer who had served in the Australian colonies towards the end of his career, the younger Downes read medicine at the University of Melbourne between 1907 and 1912, and joined the Victorian Horse Artillery and the Melbourne University Rifles before accepting a commission in the Australian Army Medical Corps in 1908. He took command of the 2nd Light Horse Field Ambulance on the outbreak of war in 1914, the youngest lieutenant-colonel in the AIF. In March 1916 he became the senior medical officer of the Anzac Mounted Division, and implemented a number of original ideas aimed at improving the care of the wounded in the theatre of operations. In August 1917 he became Deputy Director of Medical Services for the Desert Mounted Corps. He pioneered early anti-malarial treatment during the operations in the Jordan Valley in 1918, was appointed CMG and mentioned in despatches six times. A. G. Butler (q.v.) invited him to write the section of volume one of the official medical history which dealt with the Sinai and Palestine theatres, and he complemented this with private medical practice during the interwar years. He continued his army association as well, and was involved in a number of significant measures aimed at reform of the Army Medical Service, and in 1934 was appointed Director-General of the Medical Services. This was a regular appointment which necessitated his relinquishing private practice. He placed particular emphasis on training of medical officers, medical surveys of areas such as New Guinea, and the development of the Voluntary Aid Detachments. He also advocated the building of major hospitals for the treatment of ex-service personnel, ultimately expressed in the system of repatriation hospitals such as those at Heidelberg, Concord and Keswick. In November 1940 he was appointed Director of Medical Services to the AIF, but Blamey preferred Brigadier Samuel Burston, a colleague of Downes, for the position, and the appointment was overturned, with Downes becoming Inspector-General of Medical Services. With the return of the AIF to Australia in early 1942, Burston was given the senior medical post in the army and Downes was appointed to the subordinate position of Director of Medical Services in the Second Army. Nearing retirement, he was nominated to write the official medical history of the war, in succession to Butler, and it was while engaged on preliminary work for this task that he was killed in an air crash off the Queensland coast which also killed Major-General George Vasey (q.v.) in March 1945.

DUBBO (I), HMAS see *BATHURST* **CLASS MINESWEEPERS (CORVETTES)**

DUBBO (II), HMAS see *FREMANTLE* **CLASS PATROL BOATS**

DUCHESS, HMAS see *DARING* **CLASS DESTROYERS**

DUMARESQ, Rear-Admiral John Saumarez (26 October 1873–22 July 1922). Born in Sydney, Dumaresq lived in England from the age of two. After joining the RN in 1886 he developed an interest in the science of naval warfare, inventing a range-finder and several fire-control devices. During the First World War he took part in the Battle of Jutland before transferring at his own request to the RAN to command of HMAS *Sydney* (I) (q.v.). Under his command, the *Sydney* survived a zeppelin attack, and during its subsequent refitting Dumaresq commanded HMS *Repulse* in the Battle of Heligoland. Dumaresq's campaign to get aircraft-launching platforms fitted to light cruisers paid off when one was installed as part of the *Sydney's* refitting, and the *Sydney's* planes quickly proved very useful in driving off enemy aircraft. In March 1919 Dumaresq became the first Australian-born officer to command the RAN and was promoted to rear-admiral in June 1921. His time in command of the Australian Fleet was marked by conflict with the government over severe cutbacks in naval funding, but also by considerable success in improving the RAN's efficiency and performance. Dumaresq reverted to the RN in April 1922.

DUNERA see **ALIENS, WARTIME TREATMENT OF**

DUNLOP, Lieutenant-Colonel (Ernest) Edward 'Weary' (12 July 1907–2 July 1993). With a background in the cadets and the militia while at school and university, Dunlop was commissioned into the Australian Army Medical Corps as a CMF officer in July 1935, and was seconded to the AIF in November 1939 while working in London at St Bartholomew's Hospital. He was posted as a medical officer to the Australian Overseas Base in Jerusalem, and subsequently served on the staff of the Australian Corps Headquarters and AIF Headquarters in Gaza and Alexandria. He became Medical Liaison Officer on the Headquarters of British Troops Greece, and survived evacuation from both Greece and Crete (qq.v.). A somewhat turbulent subordinate whose main wish was to return to a surgical position, he refused command of the 2/2nd Casualty Clearing Station, opting to remain as a major in a surgical post within the unit. Transported on the *Orcades* for return to Australia, Dunlop's unit and a number of others were disembarked in Java as part of a desperate and futile attempt to halt the Japanese offensive in the Netherlands East Indies. Temporarily in command of No. 1 Allied General Hospital in Java, Dunlop became a prisoner of war in April. As a medical officer in the various camps and along the Burma–Thailand Railway (q.v.) he laboured tirelessly to save wounded, sick, malnourished men and interceded personally many times with the Japanese to prevent the further brutalisation of prisoners by their captors.

Although he was not the only officer, nor the only medical officer, to act in this selfless manner, his name became synonymous in postwar Australia with the sacrifice and barbarism of the experience of captivity under the Japanese. Long before modern Australia discovered him in the 1980s, however (largely through publication of his quietly understated wartime diaries which have gone through at least six editions since first publication in 1986), Dunlop had become an iconic figure for the thousands of men who survived captivity, both for his heroic labours during the war and for his advocacy of POW health issues to government after 1945. Following demobilisation he resumed medical practice, became a leader in numerous community organisations and pioneered philanthropic contact between Australia and south-east Asia. During the Vietnam War he led a civilian surgical team working with injured civilians in South Vietnam. At the end of the Second World War he was mentioned in despatches, but it was not until later that he began to receive numerous honours for his work, including a knighthood in 1969. Towards the end of his life the Dunlop legend sometimes appeared to overshadow the man, a fact with which he was not always entirely comfortable. He advocated forgiveness of the Japanese and the importance of international cooperation and understanding, and at his death, which was greeted by an enormous popular outpouring of emotion, was probably the single best-known ex-serviceman in Australia. The reverence with which he came to be regarded, both before and after his death, is an interesting comment on the suggestion that the old-fashioned martial and personal virtues which he embodied have no place in contemporary Australian life.

E. E. Dunlop, *The War Diaries of Weary Dunlop: Java and the Burma–Thailand Railway 1942–1945* (Penguin Viking, Sydney, 1989).

DUNSTAN, Lieutenant-General Donald Beaumont (18 February 1923–   ). Dunstan was a wartime graduate of RMC Duntroon in June 1942. He held a number of junior staff positions with infantry brigades during the war in the Pacific and served as a platoon commander on Bougainville (see NEW GUINEA) in 1945. He went to Japan with the British Commonwealth Occupation Force (q.v.) in 1946 as adjutant of the 66th Battalion, and attended the Australian Staff College in 1948. He filled a succession of staff jobs thereafter, interspersed with some regimental service with 1RAR and a year as military assistant to the CGS, Lieutenant-General Sir Henry Wells. He was an instructor at RMC Duntroon, the Australian Staff College and the Staff College, Camberley, between 1955 and 1961, and commanded 1RAR in 1964–65.

After commanding the 1st Recruit Training Battalion during the build-up of the Australian Army during the Vietnam War between 1965 and 1968, Dunstan was deputy commander of the 1st Australian Task Force in Vietnam from 1968 to 1969 during the Tet offensive, and then of the 10th Task Force in Australia, before being sent to the Imperial Defence College (q.v.) in London in 1970. From March 1971 he was Commander of Australian Forces Vietnam, and in this position oversaw the gradual running down of the Australian commitment in Phuoc Tuy province. This was a sometimes difficult command because of the lack of direction and consultation with his superiors in Australia, and Dunstan found that he had to formulate his own policy on issues such as casualties and operations outside the province in the absence of any clear indication of Canberra's views. Appointed GOC of the Field Force in February 1974, he became Deputy CGS in January 1977 and succeeded as CGS in April that same year. He was created KBE in 1980, retired from the army in 1982 and served subsequently as governor of South Australia until 1990.

DUNTROON see **ROYAL MILITARY COLLEGE, DUNTROON**

E

'E' CLASS SUBMARINES (*AE1, AE2*). Laid down 1911–12, launched 1913; displacement 725 tons (810 submerged); length 181 feet; beam 22 feet 6 inches; armament 4 × 18-inch torpedo tubes; speed 15 knots (10 submerged).

These two submarines were the first to serve with the RAN. Identical to the British 'E' Class, they arrived in Sydney just a few weeks before the outbreak of war and were manned by composite Australian and British crews. Early in the war they served in New Guinea waters, aiding in the capture of German New Guinea (q.v.). *AE1* was lost in these operations, presumed sunk by an uncharted reef.

AE2, commanded by Lieutenant-Commander Stoker, proceeded to the Mediterranean in 1915 to assist in the Gallipoli operations. On 25 April 1915, after negotiating Turkish minefields and avoiding destroyers and fire from land-based guns, she became the first submarine to penetrate the Dardanelles. On the way she torpedoed a Turkish cruiser and scared away a Turkish battleship which had been shelling the anchorages off the Gallipoli beaches to prevent their reinforcement. After reaching the Sea of Marmara she spent five days carrying out her orders to disrupt Turkish shipping, but failed to sink any more vessels. On 30 April, with all her torpedoes expended, *AE2* was attacked by Turkish gunboats. Stoker was forced to scuttle the ship, and the crew was captured, remaining in Turkish captivity until 1919.

EATHER, Major-General Kenneth William (6 June 1901–9 May 1993). Eather was commissioned into the 53rd Battalion of the CMF in May 1923, and had a model militia career in the interwar period. Unlike his regular contemporaries, he received regular promotion in the 1920s, and by July 1935 was a lieutenant-colonel commanding the 56th Battalion.

At the outbreak of war in 1939 he volunteered immediately for the 2nd AIF, receiving the number NX3 and raising and commanding the 2/1st Battalion, 'The City of Sydney's Own'. He led the battalion in the Australians' early battles at Bardia and Tobruk, but was hospitalised and thus missed the Greek campaign, in which many of his unit were taken prisoner. In December 1941 he was promoted to brigadier along with a group of able young unit commanders like Frederick Chilton and Ivan Dougherty (qq.v.), and given the 25th Infantry Brigade. In the fierce fighting on the

Brigadier K. W. Eather (centre) discussing an attack on Japanese positions with Major-General A. S. Allen, GOC 7th Division (left), Papua, September 1942. (AWM 026750)

Kokoda Track in September Eather's brigade halted the Japanese at Imita Ridge and began the advance back up the track which culminated in the battles for the Japanese beachheads. He commanded the brigade in New Guinea until July 1945, when he was again promoted and given command of the 11th Division. He was only 44 years of age, but the War Cabinet had decided that a younger officer should gain experience of divisional command 'in view of the desirability of his experience . . . being conserved to the AMF during the postwar period'. At the end of the war he led the Australian contingent at the Victory parade in London, and retired from the army in 1947.

Nicknamed 'Phar Lap' because of the speed and determination which characterised his leadership, Eather was one of the group of highly able CMF brigadiers who took a leading role in the fighting in New Guinea. Gavin Long (q.v.), the official historian, wrote of him that he was 'quiet but very definite and commanding, reputed to read all the paperwork very carefully, seldom hammers officers down, but when he does, hammers them in like a tack'. Until his death he was a familiar sight at Sydney Anzac Day marches, frequently leading the parade.

EATS see **EMPIRE AIR TRAINING SCHEME**

ECHUCA, HMAS see *BATHURST* **CLASS MINESWEEPERS (CORVETTES)**

EDINBURGH, RAAF BASE, near the Adelaide suburb of Salisbury, is the main RAAF base in South Australia. Named for the Duke of Edinburgh, the base was opened in 1954 and was used by British personnel during atomic tests in Australia (q.v.) and operations of the Long Range Weapons Establishment (q.v.) at Woomera. Edinburgh is currently home to No. 92 Wing, consisting of No. 10 and 11 Squadrons using Orion (q.v.) maritime reconnaissance aircraft and support squadrons.

EDUCATION see **MILITARY EDUCATION**

EDUCATION CORPS see **ROYAL AUSTRALIAN ARMY EDUCATIONAL CORPS**

EDWARDS, Major-General James Bevan (5 November 1834–8 July 1922), a graduate of the Royal Military Academy at Woolwich, served as an officer of the Royal Engineers in the Crimean War, the Indian Mutiny, with 'Chinese' Gordon in China and in command of engineers during the Sudan expedition of 1885 before, on promotion to major-general, being sent to Hong Kong as commander of British troops in China. In 1889, at the request of Australian colonial governments, he spent four months inspecting the colonial military forces (q.v.) to recommend how they might best be organised. In October he delivered his verdict. Though lukewarm about involving New Zealand, Edwards was adamant that the other colonies synchronise their defence efforts. Their permanent forces should be amalgamated and stationed in the capital cities, on Thursday Island and in Albany. Unpaid volunteers should support the permanent troops. Western Australia and Tasmania must follow the other colonies and pay some of their citizen soldiers to train. Conspicuous red and blue uniforms should be replaced by dull khaki. And, most importantly, the four most populous colonies should form their paid citizen soldiers into brigades, which in war could serve together under a single general as a federal militia able to strike wherever an invader lodged. To allow their brigades to unite, the colonies should include in their defence legislation a clause making citizen soldiers liable to serve outside their borders during war or grave emergency. They might also wish to compel their young men to join a rifle club, for in war the militia would draw on these for reserves. Edwards privately advocated a further reform: that service outside one's colony entail service outside the Australian continent so that in a major war Australians could seize enemy bases in the South Pacific. Edwards's strategic vision was not discussed publicly, but his advice for military reorganisation was readily accepted. Henry Parkes used the issue of common defence in his Tenterfield oration (q.v.) in October 1889. However Edwards's plan had not been realised by 1893, when he retired from the army, nor by 1895, when he was elected as a Conservative MP in the British parliament. The economic depression of the 1890s and the colonial governments' decision in 1896 that civil federation would precede military federation stalled reform. But when in 1903–04 Major-General Edward Hutton (q.v.) amalgamated the colonial military forces into an Australian army, he largely followed Edwards's plan. Conscription (q.v.), lacking popular support before 1900, was not introduced until 1911.

CRAIG WILCOX

EGYPT, AUSTRALIANS IN Egypt figured largely in the experiences of two generations of Australian soldiers who sailed for active service in 1914 and 1939, and Australian soldiers made a considerable, and frequently negative, short-term impact on Egypt. The first contingents of the AIF were diverted to Egypt from their intended destination in Britain when the accommodation on Salisbury Plain (q.v.) was deemed inadequate in the approaching winter for the large numbers of newly enlisted soldiers. The Turkish entry into the war in October 1914 also prompted concerns for the security of the Suez Canal zone, and the Australians and New Zealanders *en route* for Europe were the nearest available and easily dispensable source of troops. The main Australian camp was at Mena, which was the site of the Egyptian Army's manoeuvre grounds and was linked by tram with the capital, Cairo. Conditions, especially at first, were austere, and leave was taken liberally, whether approved or not, and spent in Cairo. At this stage the men of the AIF were soldiers mostly in name alone, and the behaviour of a minority left much to be desired, although at the same time it should not be exaggerated, as it frequently was by the British authorities in Egypt. The most notorious example of indiscipline was the famous battle of the Wazza (q.v.), which broke out on Good Friday 1915. Drunken Australian and some New Zealand soldiers rioted in the brothel district, assaulting Egyptians, breaking up shops and setting fire to buildings, while fighting running battles with the military police. A second, lesser outbreak in the same area occurred in the last week of July. The origins of these disturbances lay in mutual antagonisms between soldiers and the overseers of the brothels, and had been simmering for several months, on the first occasion in April exacerbated by bad liquor and the imminent departure for active service. At a wider level, however, they were but the most obvious sign of a fundamental clash of cultures. Few Australians had travelled overseas before the war, and Egypt was exotic and very foreign. The 'Gyppos' (q.v.) were widely despised by the Australian troops, and basic racism also played a role in the relationship. Discipline again became a problem towards the end of 1915, when thousands of reinforcements were held in Egypt pending the evacuation from Gallipoli, but the departure of the majority of the AIF for France at the beginning of 1916 and the movement of the front against the Turks across Sinai and into southern Palestine in 1916–17 largely removed the problem. The light horse were used as enforcers of empire in 1919 to help put down the Egyptian revolt, which they did with efficiency and considerable brutality, culminating in a series of destructive raids on Egyptian villages.

In the Second World War Australians were again based in Egypt, but this time the British and Australian authorities were alive to the dangers of having large numbers of under-employed soldiers within close proximity of the delights of Cairo, and early on moved the force into training grounds in southern Palestine, a decision possibly prompted by uneasy memories on the part of the Egyptian authorities as well. But indiscipline remained a problem, and leave to Cairo and Alexandria was often restricted. An attempt to restage the battle of the Wazza in 1940 failed for various reasons, although a 'respectably sized fire was started'. Very few troops were involved, and the military police were much more effective than they had been in 1915. During the Suez crisis of 1956, Egyptian rioters destroyed the light horse memorial at Port Said (a replica may be seen on Anzac Parade in Canberra, just down from the Australian War Memorial), although whether on this occasion they distinguished between Australians and the British seems doubtful. A popular saying in Egypt, 'like an Australian mule', referred to the animals disposed of by the AIF at the end of the war: they required twice the fodder to complete half the work.

Suzanne Brugger, *Australians and Egypt, 1914–1919* (Melbourne University Press, Melbourne, 1980).

18-POUNDER GUN Range 6500 yards (Marks I and II), 10 900 yards (Mark IV); weight of shell 18 pounds; rate of fire 20 rounds per minute; total weight 2900–3500 pounds (approx.).

The 18-pounder gun was the standard British field artillery piece during the First World War. It could fire shrapnel, high explosive and smoke shell. The main task of the gun was to cut barbed-wire entanglements which protected enemy trenches, demolish small field fortifications and fire the protective barrage of shells which crept in advance of the infantry in attack in all battles on the Western Front (q.v.) from September 1916. Australia had over 100 of these guns before the First World War and many of these accompanied the troops to Gallipoli in 1915. After the war an Australian version with rubber tyres and a modified carriage was developed for use with motor traction. The guns remained in service throughout the Second World War. However, in the Middle East most of these guns had been replaced in the AIF by the 25-pounder by the end of 1941. Field regiments of the 8th Division were armed with 18-pounders in Malaya, and there was some use of them in operations in New Guinea.

S. N. Gower, *Guns of the Regiment* (Australian War Memorial, Canberra, 1981).

ELIZA see **TASMANIAN TORPEDO CORPS SHIP**

ELLINGTON, Marshal of the RAF Edward Leonard (30 December 1887–13 June 1967) visited Australia in June–July 1938 at the invitation of the Australian government. A series of accidents involving Hawker Demon fighter bombers in late 1937 suggested that flying standards in the RAAF were inadequate, and the government sought outside advice. A former CAS (1933–37), and then Inspector-General of the RAF (1937–40), Ellington concluded that the RAAF had failed to develop plans to reinforce Singapore. This was a direct, if not entirely fair, criticism of the CAS, Air Vice-Marshal Richard Williams (q.v.), who rightly responded that he had worked within the strategic guidelines for the deployment of the RAAF laid down by the government, guidelines that saw the role of the RAAF in terms of home defence. Much more damaging was Ellington's finding that the Air Board (q.v.) had been negligent in overseeing training standards and flying procedures, which again rebounded on Williams, as he had given himself responsibility for operational training in 1934.

On 16 January 1939 Prime Minister J. A. Lyons announced that Williams was to be sent on an exchange posting to the RAF for two years, and that he would be succeeded as CAS by his fellow member of the Air Board (and hated rival), Air Commodore S. J. Goble (q.v.). Ellington's report, and its underlying methodology, was heavily criticised, and much was made of the fact that he had spent only 10 days at RAAF Headquarters and only four with operational squadrons. Yet the broader question of the command of the RAAF was an open one. Williams had been CAS since 1922, and 10 years before Ellington's visit an earlier report by Sir John Salmond (q.v.) had raised doubts about Williams's command ability. By effectively banishing Williams from the RAAF, the Ellington report undoubtedly opened the way to a period of intense interpersonal feuding within the most senior ranks of the RAAF, which severely hampered its wartime performance.

ELLIOTT, Brigadier Cyril Maurice Lloyd (12 January 1899–8 January 1983). Graduating from RMC Duntroon in 1920, Elliott proceeded to Singapore for training with the British Army for a year. Upon his return he served in administrative posts with CMF battalions until 1932, when he was sent to the Staff College, Quetta for two years. This was followed by further posts with CMF brigades until the outbreak of the Second World War, which led to his move to Army Headquarters as a staff officer.

He went overseas with the headquarters of I Australian Corps in April 1940, and served on that headquarters throughout the Mediterranean campaigns. He was one of the small party of officers chosen to accompany Lieutenant-General Blamey (q.v.) out of Greece when the latter was ordered to leave to avoid capture. Returning to Australia at the beginning of 1942, he served as GSO1 or principal operational staff officer of the 7th Division during the early fighting in Papua, before returning to Land Headquarters as Deputy Quartermaster-General (Plans) until March 1943. He held senior staff posts as a brigadier on Land Headquarters and at Headquarters, Second Australian Army in 1943–44, and was then appointed chairman of the Base Planning Committee. This body prepared a 200 page report on the possible basing and supplying of a British force of 675 000 men in Australia for participation in the war against Japan once hostilities had ceased in Europe. This would have been a major undertaking, placing enormous strain on Australian resources, and was not finally proceeded with. After the war he served with the Joint Chiefs of Staff Australia (q.v.), the body based on the Australian Chiefs of Staff Committee in Melbourne and charged with responsibility for overseeing the British Commonwealth forces taking part in the occupation of Japan. From November 1947 he was Deputy Adjutant-General, and he finished his military career as Brigadier in charge of Administration in Southern Command from July 1949. He retired from the army in 1954, and was Director of Housing and Catering for the 1956 Olympic Games held in Melbourne.

ELLIOTT, Brigadier-General Harold Edward 'Pompey' (19 June 1878–23 March 1931) was born in West Charlton, Victoria, the son of a farmer. He first demonstrated interest in military affairs at the University of Melbourne where he joined the university regiment. In 1900 he enlisted in the 4th Victorian Contingent for service in the Boer War. During the conflict he was mentioned in despatches, received the DCM and was especially congratulated by Lord Kitchener. After the war he returned to Australia and completed his legal studies, while maintaining his interest in the military. In 1904 he became a second lieutenant in the militia. By 1913 he was lieutenant-colonel commanding the 58th Battalion.

On the outbreak of the First World War Elliott was given command of the 7th Battalion in the 2nd Brigade. He was soon to become one of the great characters of the 1st AIF. He had a massive frame, an explosive temper and a habit of plain speaking and writing that did not always endear him to his superiors. He was nicknamed 'Pompey' by his men; a name that endured despite his dislike of it. His career at Gallipoli was sporadic. He was wounded early on the first day and did not return until the stalemate in June. His brigade was in reserve in

Brigadier-General H. E. Elliott standing at the door of a German headquarters captured by the 15th Brigade at Harbonnieres, France, 8 August 1918. (AWM E02855)

August but saw heavy fighting when it relieved 1st Brigade at Lone Pine on 8 August. At the end of the month he was evacuated sick. He returned in November but sprained his ankle and therefore to his great irritation was evacuated before his men.

After Gallipoli Elliott was given command of 1st Brigade but with the raising of the 5th Division was placed in charge of the 15th (Victorian) Brigade with the rank of brigadier-general. The brigade did not have a happy introduction to warfare on the Western Front. It was in the thick of fighting at the poorly conceived and executed diversionary attack at Fromelles where it lost 1450 casualties in 24 hours. Elliott greeted the survivors with tears streaming down his face.

Elliott's brigade played an important role in following-up the great German retreat to the Hindenburg Line in March 1917. His next major action, however (and possibly his finest), took place in September of that year at Polygon Wood. The day before that operation was to take place (25 September), the Germans had launched a spoiling attack on the British division on Elliott's right. By rapid action Elliott redeployed his brigade and helped stabilise the situation. On the day of battle he then carried out an intricate manoeuvre whereby his troops sidestepped into the British zone and captured the objective originally assigned to the British. It was typical of Elliott that his brilliant actions were overshadowed by controversy. He castigated the actions of the British troops in a harshly worded report to General W. Birdwood (q.v.), written largely in ignorance of the situation that had confronted them. Birdwood ordered all copies of the report destroyed.

In March 1918 Elliott's brigade returned to the front and Elliott to controversy. The 5th Australian Division was one of the formations sent in to stabilise the front after the German March offensive. Confronted with British troops falling back from this shattering blow Elliott, with typical intemperance, ordered any stragglers shot who refused to be rallied. This was immediately repudiated by Major-General Talbot Hobbs (q.v.), his divisional commander. A lesser soldier might have been sacked for such action, but Elliott was soon to prove his worth. Ordered to retake the village of Villers-Bretonneux, he organised a night attack which proved brilliantly successful. Elliott's brigade played a significant role in the advance to victory from 8 August until the end of the war. He fought at Amiens, at Péronne, where he managed to fall in the River Somme while urging his troops to quicken their advance, and later at the Hindenburg Line.

Elliott's equanimity, however, had been shattered by the fact that he had been overlooked for higher command. He wrote unwisely forceful letters on the subject to the higher command which can only have reinforced their decision not to promote him. After the war he entered politics as a National Party senator and used this platform to air grievances against Birdwood, Major-General Brudenell White (q.v.), the British High Command and all who had conspired to thwart his promotion. His life ended in tragedy when he committed suicide in March 1931.

EMDEN, SMS, a German cruiser, wreaked havoc with Allied shipping in 1914 until it was destroyed on 9 November 1914 by HMAS *Sydney* (I) (q.v.).

The wreck of the *Emden*, following her battle with HMAS *Sydney*, Cocos (Keeling) Islands, Indian Ocean. (AWM P1236/20/03)

The most modern ship of the German Cruiser Squadron based at Tsingtao (now Qingdao) in China, the *Emden* had been launched on 26 May 1908, had a displacement of 3600 tons and was armed with with ten 10.5 cm guns. On the outbreak of the First World War, the *Emden* sailed first to the Marianas Islands in the Pacific Ocean to rendezvous with other German warships and then sailed into the Indian Ocean. The *Emden* sank or captured 25 merchant ships and shelled the port of Madras in India on 22 September. She sank the Russian cruiser *Zhemchug* and the French destroyer *Mousquet* at Penang in Malaya on 28 October 1914. When the *Emden* attacked the wireless and telegraph station at Cocos (Keeling) Island, she was intercepted by HMAS *Sydney,* which had been escorting the first Australian and New Zealand troop convoy. The *Sydney's* superior size, speed and firepower gave it the advantage, and about 95 minutes after the action began the *Emden's* captain ran his shattered vessel aground on a reef. Only four of *Sydney's* crew died in the battle compared with 134 from the *Emden.* News of this victory, the RAN's first, was received with jubilation in Australia, where it was seen as confirmation of Australia's fighting spirit, even though most of *Sydney's* crew were British.

EMPIRE AIR TRAINING SCHEME (EATS) was established in December 1939 in response to the outbreak of war with Germany, whose air force significantly outnumbered that of the United Kingdom and her dominion allies. Discussions several years earlier on the desirability of coordinating air training across the Empire had come to nothing, but the stark realities of the situation in September 1939 cast the matter in a different and more urgent light. Already, in the months prior to the outbreak of war, the British government had approached the governments of Canada, Australia and New Zealand for assistance in training aircrew, but these approaches were piecemeal and the scale of assistance requested quite inadequate. Following suggestions from both the Australian and Canadian High Commissioners in London (Stanley Bruce and Vincent Massey respectively), British requirements were spelled out in a realistic if startling manner in a cablegram sent to each of the three dominions' governments on 26 September 1939. Britain lacked the space to establish the necessary number of flying schools, and was in any case extremely vulnerable to German air attack. Moreover, it could produce from its own population base only four-ninths of the target 50 000 aircrew a year. The remaining five-ninths, the British government suggested, should be contributed by the three dominions in proportion to their population. A total of 50 flying schools should be established by the dominions, again on a basis proportional to population, to provide elementary training, but all advanced training would be undertaken in Canada.

The Australian War Cabinet (q.v.) accepted these proposals in principle, although the Air Board concluded that the requirements for instructors and aircraft meant that the plan could only be regarded as a 'long-range one'. When concrete British proposals were unveiled at a conference in Ottawa, Canada, at the beginning of October, however, the fine detail seemed to depart significantly from the original plan. The leader of the British delegation, Lord Riverdale, based his requirements on the formation and reinforcement of 100 squadrons, which would be contributed by Canada, Australia and New Zealand on a 40:40:20 basis which, as the leader of the Australian delegation, the Minister for Air and Civil Aviation, J. V. Fairbairn (q.v.), pointed out, departed radically from the original calculation based on population; that would have divided the dominions' relative contributions on a 57:35:8 ratio. There were similar objections to the proposed division of the costs. Fairbairn suggested that a more equitable distribution of costs would be achieved if a substantial proportion of Australian aircrew were trained, fully or partially, in Australia, thus saving scarce dollar resources that would otherwise have been spent on training in Canada. The Canadians also objected strongly to the cost-sharing formula proposed by the British delegation, and it was only after several weeks of hard, even bitter, bargaining that agreement was reached. Australia undertook over the three-year period of the agreement to provide a total of 28 000 aircrew, which represented 36 per cent of the whole number that the scheme was designed to produce. Elementary and advanced flying schools would train 1120 men every four weeks, of whom 194 would go to Canada for further training. The British government agreed to pay half the cost of engines that would have to be imported for the Tiger Moths to be used for training, and to bear a proportion of other costs associated with EATS operations in Australia. The costs of RAAF personnel training in Canada were the responsibility of the Australian government, and once aircrew embarked for the United Kingdom or an operational area, the United Kingdom became liable for all their costs.

The final agreement, signed on 17 December 1939, has been described by the official history of the RAAF in the Second World War as 'a document that was unique in military history', pooling as it did the resources of four countries in a common cause. In retrospect, the disputes over the financial basis of the scheme were less significant than the manner in which the agreement was implemented. Article XV of the agreement set down that Australian, Canadian and New Zealand aircrew who passed through EATS to join the RAF would serve in distinct Australian, Canadian and New Zealand squadrons. Only where Australians constituted a majority was a squadron designated an RAAF squadron; many Australians served in RAF and other squadrons. Their respective experiences in the Great War had underlined to each of the dominions the necessity of safeguarding their national interests through protecting the identity of their forces and by not allowing their manpower to be absorbed indiscriminately into the larger British military machine. The original wording of Article XV was vague on how this distinctiveness would be achieved, and differences over what the final outcome should be soon emerged. The British authorities did not intend to allow the growing proportion of EATS personnel within the RAF to lead either to a direct dominions' voice in the formulation of strategic air policy, or to a situation whereby largely British ground staff in squadrons might be commanded by officers from the dominions. To avoid this situation the British Air Ministry argued that using a ratio of 11 support staff to every aircrew member, and taking into account the manpower being used in each dominion to train EATS personnel, it was possible to determine how many operational squadrons each dominion would have maintained under wartime conditions: Canada 25, Australia 18, and New Zealand 6. These were designated dominion squadrons under the provisions of Article XV; any EATS-produced personnel in excess of the numbers required to man these squadrons could be posted to RAF squadrons. Thus the Air Ministry was able to dilute the RAAF presence in the RAF, to the point that by April 1945, 1488 Australian aircrew were serving in Article XV RAAF squadrons (Nos 450–467), while 10 532 had been posted to RAF squadrons.

The provisions for safeguarding Australian interests within Article XV squadrons seemed adequate at the time the agreement was signed, but they were never fully implemented. Although it had been laid down that British officers would command RAAF squadrons only until suitably qualified RAAF officers became available, the Air Ministry kept a tight control over such appointments, and very few Australian officers were placed in charge of RAAF squadrons, and then on several occasions only after protests from the Australian authorities when an Australian candidate was passed over in favour of a more junior British offi-

cer. Similarly, the Air Ministry ensured that the provision for a senior RAAF officer to have access to the senior levels of the RAF did not extend to any right of influence over strategic policy. The Air Ministry also refused to consult with the RAAF on the deployment of RAAF squadrons, except in a theatre that was designated as 'new', as opposed to the extension of the boundaries of an already existing theatre. By these sorts of methods, the Australian authorities were prevented from having any but the most minimal control over the disposition of the aircrew that, in Prime Minister Menzies' own words, Australia had so willingly 'surrendered' to the United Kingdom.

The agreement was renewed for an additional two years in March 1943, even though there was evidence in the Air Ministry that there was already a surplus of aircrew in every category. In February 1944, without any warning to the Australian authorities, the Air Ministry declined to accept the March–April EATS contingent. In June 1944 drafts from Australia to Canada of partially trained men were stopped, and in October it was decided that all Australian aircrew training in Canada would be returned to Australia. The EATS was wound back in stages, and its cancellation was not officially announced until 31 March 1945. By late 1944 the RAF had a total of 53 240 surplus aircrew, almost three times the projected requirement of 18 840. Even then, the Air Ministry was unwilling to release Article XV squadrons, which were clearly surplus to the war effort in Europe, for operations in the Pacific, and it was not until April 1945, a year after Prime Minister Curtin raised the question, that two RAAF bomber squadrons were included in the British bomber force earmarked for Okinawa in the final assault against Japan. A scheme that had been regarded as the most productive way of harnessing the resources of the Empire for the war effort had become, by the time of its demise, a wasteful and inefficient use of manpower.

John McCarthy, *A Last Call of Empire: Australian Aircrew, Britain and the Empire Air Training Scheme* (Australian War Memorial, Canberra, 1988).

ENCOUNTER, HMAS (*Challenger* Class light cruiser). Laid down 1901, launched 1902; displacement 5880 tons; length 376 feet; beam 56 feet; armament 11 × 6-inch guns, 9 × 12-pounder guns, 6 × 3-pounder guns, 2 × 18-inch torpedo tubes; speed 21 knots.

The *Encounter* was built for the RN and was deployed on the Australia Station (q.v.). In 1912 she was lent to the RAN. HMAS *Encounter*'s war service was spent in New Guinea waters (where she captured the German steamer *Zambesi* in 1914), the Pacific and South-east Asia, and later as a convoy escort in Australian waters. She was transferred to the RAN in 1919, was renamed *Penguin* in 1923 and served as a depot and receiving ship until she was paid off in 1929. Her hull was sunk by gunfire off Sydney Heads in 1932.

ENCOUNTER, HMAS, located in Port Adelaide, South Australia, is a Naval and Reserve Training Establishment. It provides naval representation and coordination of naval tasks in South Australia, and is the headquarters of the Naval Officer in Charge, South Australia. Commissioned under the name HMAS *Torrens* on 1 August 1940, it was renamed on 1 March 1965 so that a new destroyer escort could be named *Torrens*. The name *Torrens* came from the river on which Adelaide stands and had been the name of a destroyer, while the name *Encounter* comes from Encounter Bay, South Australia, named to commemorate the meeting on 8 April 1802 between Matthew Flinders and the French explorer Nicholas Baudin.

ENGINEERS see **BRITISH ARMY IN AUSTRALIA; ROYAL AUSTRALIAN ENGINEERS**

EUREKA STOCKADE was a crude fortification constructed by miners on the Ballarat goldfields who were discontented about the mining licence fee and attendant police harassment. Victorian Governor Sir Charles Hotham, determined to be severe with the miners, reinforced the troops and police in Ballarat and instructed the local Goldfields Commissioner to preserve order at all costs. The arrival of the reinforcements and a provocative licence hunt by police led to skirmishes which increased tension on the goldfields. On 30 November 1854 the most militant miners built the stockade and, under a flag depicting a white Southern Cross on a blue background, they pledged to defend their liberties and each other. The next day about 1000 miners began drilling, gathering weapons and supplies and making pikes, but their number quickly dwindled. Nevertheless, local officials, fearing attack by the miners and convinced that they must crush what they saw as sedition, ordered an assault on the stockade early on the morning of 3 December. About 150 miners were attacked by a government force of 296, made up of troops of the 12th and 40th regiments as well as police. The miners opened fire when troops were about 150 metres away, then the soldiers returned

fire and charged. Within 15 minutes the battle had been won by the government forces, who proceeded to bayonet the wounded and to shoot or arrest bystanders. Five soldiers and 20–30 miners died in the clash. Though militarily insignificant, Eureka has gained a place in Australian folklore largely because it was one of the few instances of organised violence between Whites in Australia. The stockade itself and the Eureka flag have been seen by Australian nationalists and by sections of the labour movement as symbols of Australian independence and resistance to tyranny.

John Molony, *Eureka* (Viking, Melbourne, 1984).

EXERCISES, COMBINED Exercises are conducted by the ADF with the armed forces of other nations for the purpose of improving inter-operability, encouraging familiarity between service officers at various levels, exchanging information on technical and operational matters, and providing training support and assistance in the development of other forces, especially within the Asia and Pacific regions. As part of the emphasis on greater involvement in those regions in recent years, the level of joint and combined exercises with the defence forces of the ASEAN nations has increased substantially. They include the STARFISH series of annual maritime exercises with the nations of the Five Power Defence Arrangements (q.v.) in the South China Sea; the Integrated Air Defence System (IADS) series held several times a year, again under the auspices of FPDA; the HARINGAROO exercises with the Malaysian Army; the AUSINI series conducted with the Indonesian Navy about four times a year; the *Penguin* patrol boat exercises held with the armed forces of Brunei; and regular passage exercises (PASSEXs) held between the RAN and regional navies. There are also numerous exercises conducted with the US armed forces, such as the RIMPAC naval exercises which began in 1971 and which include the Canadians and Japanese, and with South Korea, Papua New Guinea, the island states of the Pacific, and with British Forces Hong Kong. During the lifetime of SEATO there were also regular exercises sponsored by that organisation and involving the armed forces of the signatory nations. Following New Zealand's decision in 1985 to suspend the provisions of ANZUS (q.v.) over nuclear ship visits to New Zealand ports, the New Zealand armed forces were excluded from tripartite exercises with the other two partners in the alliance.

F

F-111, GENERAL DYNAMICS (2-seater strike bomber). Wingspan, extended 70 feet, swept 34 feet; length 73 feet 6 inches; armament 1 × 20 mm cannon and a variety of other weapons including Sidewinder missiles, Harpoon anti-shipping missiles, bomb load up to 5000 pounds; power 2 × Pratt & Whitney TF30-P-3 turbofans; speed Mach 2.4 (2550 km/h); range 5950 km.

The F-111 is the most controversial plane in Australian military aviation. It was ordered off the drawing board from General Dynamics in October 1963 as a replacement for the ageing Canberras (q.v.). They were scheduled for delivery in 1967 but problems with the 'variable geometry' or swing wings continually delayed their acceptance until 1973. The cost of the 24 aircraft was $324 million. They are flown by No. 1 and 6 Squadrons based at Amberley (q.v.) in Queensland. An additional four aircraft were purchased from the United States in 1981 to replace the four lost to crashes. In 1992 it was announced that a further 15 were to be acquired in order to rotate craft through the fleet and thus prolong their life until 2020. The first of these arrived at Amberley on 8 October 1993. Four F-111Cs have been converted to RF-111C standards for use in the photo-reconnaissance role. As part of the upgrade program the Pave Tack laser designator/tracking system has been introduced to enhance the platform's capabilities.

F/A-18 HORNET, McDONNELL DOUGLAS (Single-seat fighter [F/A-18A]). Wingspan 11.43 m; length 17.07 m; armament 1 × 20 mm six-barrel cannon, 7700 kg of Sidewinder or Sparrow air-to-air missiles, Harpoon air-to-sea missiles, Maverick air-to-ground missiles or laser-guided bombs; maximum speed 1910 km/h or Mach 1.8; range 3700 km; power 2 × General Electric F404 engines with 16 000 pounds thrust.

The F/A-18 is currently the RAAF's fighter and fighter-bomber aircraft and is expected to remain in service until 2010. It first entered service with the US Marine Corps in 1982 and was chosen to replace the RAAF's Mirages (q.v.) in 1981. Assembled in Australia with components produced by several companies including Aerospace Technologies of Australia (formerly the Government Aircraft Factory) and Hawker de Havilland Victoria (formerly Commonwealth Aircraft Corporation [q.v.]), the first aircraft entered service in 1985. F/A-18s are flown by No. 3 and 77 Squadrons and No. 2 Operational Conversion Unit at Williamtown (q.v.) and No. 75 Squadron at Tindal (q.v.). The F/A-18 can be refuelled in mid-air by the RAAF's 707s, which greatly increases their range. Fifty-seven F/A-18A aircraft were ordered as well as 18 two-seater trainers.

FAIRBAIRN, James Valentine (28 July 1887–13 August 1940). Born in England and educated at Geelong Grammár School, Fairbairn was commissioned into the RFC in July 1916. He was shot down and captured by the Germans in February 1917, permanently injuring his right arm in the process. He returned to Australia in 1919, pursuing a career in the pastoral industry and becoming increasingly active in public life. He was elected as a member of the United Australia Party to the Victorian Legislative Council in 1933, and the following year moved into federal parliament. Already recognised through practical experience as an authority on civil aviation, Fairbairn was appointed Minister for Civil Aviation in the new Menzies government on 26 April 1939. He led the Australian delegation to Ottawa in late 1939, which resulted in the establishment of the Empire Air Training Scheme (q.v.), and while in Ottawa was sworn in as the minister for Air, thus becoming responsible for all matters involving the RAAF. He was killed, along with two other cabinet ministers and the CGS, when the aircraft in which he was a passenger crashed on its approach to Canberra airport. RAAF Base Fairbairn, which shares the runway with Canberra airport, is named after him.

FAIRHALL, Allen (24 November 1909–   ). An electrical engineer by profession, Fairhall developed business interests in broadcasting in the Newcastle region in the 1930s and 1940s. He entered local politics in 1941, and held the federal seat of Paterson for the Liberal Party between 1949 and 1969. He was minister for the Interior and Works between 1956 and 1958, chairman of the Public Works Committee between 1959 and 1961 and minister for Supply from 1961 and 1966. He became minister for Defence under Harold Holt in 1966 and held the portfolio under John Gorton (q.v.) until the federal election of 1969, at which he retired. He presided over the defence portfolio at the height of the Vietnam War, but appears to have made little impact on its affairs. He was appointed KBE in 1970.

FANTOME, HMAS (*Espiegle* Class sloop). Laid down 1900, launched 1901; displacement 1070 tons; length 210 feet; beam 33 feet; armament 2 × 4-inch guns, 4 × 12-pounder guns (later 2 × 4-inch guns, 1 × 3-pounder gun, 2 machine-guns); speed 13.5 knots.

The *Fantome* originally appeared in Australian waters as a RN survey ship in 1907. She was commissioned into the RAN in November 1914. Her war service was spent in the Bay of Bengal, the Persian Gulf and Malayan waters on patrol duties.

She returned to Sydney in 1917 and in 1918 carried out punitive expeditions against the indigenous people of the New Hebrides (Vanuatu). After the war *Fantome* returned to RN survey duties and was paid off in 1924.

FAR EAST LIAISON OFFICE (FELO) see **ALLIED INTELLIGENCE BUREAU**

FAR EAST STRATEGIC RESERVE (FESR) was established in 1955 to provide forces specifically dedicated to the defence of Malaya and Singapore, and the waters around them. It was first mooted in 1953, when the imminent end of the Korean War made it possible to consider the redeployment of forces committed there, but it was not until 1955 that the FESR was created. It consisted of an infantry brigade (mainly, in the beginning, of British and Australian troops), bomber and fighter squadrons (of which Australia was to contribute one squadron each of bombers and fighters), and a carrier group, to which Australian ships — destroyers or frigates — would be attached, with the occasional visit of an Australian aircraft carrier. With the FESR's primary role designated as the defence of Malaya against external communist aggression, Australia's decision to participate effectively marked the end of its focus on the Middle East as its main theatre of potential operations. The secondary role of the FESR was, 'Without prejudice to its primary role, to assist in the maintenance of the security of the Federation of Malaya by participating in operations against the Communist Terrorists', and it was in this capacity that Australian forces, especially ground forces from late 1955 to 1960, undertook operations in Malaya. The FESR also provided forces for SEATO (q.v.), but these only took part in training exercises in Thailand.

FARNCOMB, HMAS see *COLLINS* **CLASS SUBMARINES**

FARNCOMB, Rear-Admiral Harold Bruce (28 February 1899–12 February 1971). One of the original entry to the Naval College at Jervis Bay in 1913, Farncomb was the first graduate to reach flag rank and was widely regarded as the most able naval officer of his generation.

Promoted midshipman on 1 January 1917, he went to Britain for further training and served on HMS *Royal Sovereign* with the Grand Fleet for the remainder of the war. After service in a succession of British and Australian ships, he went to London in 1930 to undertake the course at the Imperial Defence College (q.v.). He was junior in rank for the posting, but the Australian government wanted

to save the costs of sending a more senior officer. He completed the Senior Officers' Course and Tactical Course in early 1939, and the outbreak of war found him in command of HMAS *Perth*.

During the war Farncomb commanded HMAS *Canberra* (q.v.), and in this post was also chief of staff to Rear-Admiral J. G. Crace (q.v.), the commander of the Australian Squadron (q.v.). From December 1941 to March 1944 he captained HMAS *Australia* (II) (q.v.), and participated in the Battle of the Coral Sea (q.v.), the landing at Guadalcanal, the Battle of the Eastern Solomons, and the landings at Arawe and Cape Gloucester. He was awarded the DSO for 'skill, resolution and coolness' during the Solomons campaign, and was three times mentioned in despatches. When Vice-Admiral John Collins (q.v.) was appointed to command the Australian Squadron in early 1944, Farncomb went to Britain on courses and was then appointed to command the aircraft carrier HMS *Attacker*, which took part in the landings in the south of France and in operations in the Aegean. When Collins was wounded on board *Australia* in December 1944, Farncomb was brought back to take his place until he recovered, and in this capacity took part in naval actions off Luzon in the Philippines. He also took part in the landings at Wewak in May 1945, and at Brunei and Balik-papan.

At the end of the war he was appointed Commodore Superintendent of Training, and in 1946 relieved Collins as Commander of the Australian Squadron. He commanded the Australian fleet until October 1949, and then headed the Australian Joint Service Staff in Washington DC from 1949 to 1951. He retired from the navy in February 1951, ahead of time.

Farncomb commanded at sea for almost the whole of the Second World War, a most unusual pattern of postings which probably more than anything else contributed to the drinking problem which was to mar his later career. Concerns about his fitness probably explain why he did not reach the position of CNS, which was held then by Collins for an unusual seven-year period. The official naval historian, G. Hermon Gill (q.v.), thought him 'cold, and harder than Collins', while Commander R. B. M. Long (q.v.), Director of Naval Intelligence, thought him the abler of the two, 'and this was saying a great deal'. After retirement from the navy Farncomb qualified as a barrister and practised at the New South Wales Bar, becoming a Judge's Associate with the Supreme Court.

FEDERATION AND DEFENCE see **DEFENCE ACT; TENTERFIELD ORATION**

FILM AND TELEVISION, WAR IN AUSTRALIAN

Almost as soon as cinematography originated in the late nineteenth century, Australians were depicted in short films documenting their involvement in the Boer War. Some of these were documentaries in the strictest sense using what is now known as actuality footage, others, seeking to capture the enthusiastic audience of the new medium, 'recreated' African circumstances in the studio. During the first decade of federation a number of Australian films (including the Asian invasion scare film *Australia Calls* [1913]) dealt with real or imagined military conflict. But it was the outbreak of war in Europe in 1914 that produced the first significant batch of Australian war films, broadly divisible into two categories: the anti-German propaganda films such as *The Martyrdom of Nurse Cavell* (1916) and *If the Huns Came to Melbourne* (1916); and action films such as *Within Our Gates, or Deeds that Won Gallipoli* (1915) and *How We Beat the Emden* (1915), all of which mix the genres of fiction and documentary. Not surprisingly many had the cooperation of the military authorities, and all were openly propagandistic. The Anzac legend (q.v.) made its film debut barely three months after the Gallipoli landing in the popular *Hero of the Dardanelles*, which promised to portray 'THE "IMPERISHABLE GLORY" WON BY THE GALLANT AUSTRALIANS AT GALLIPOLI'. The film contains a typical mix of fictional and front-line footage, the latter supplied by the AIF's first official photographer, Frank Hurley (q.v.).

Between the wars, films such as *Ginger Mick* (1920), *Fellers* (1930), *Diggers* (1931) and *Diggers in Blighty* (1933) took a more light-hearted approach to their material, while still upholding the digger (q.v.) ethos. By the late 1930s real footage of the wars in Spain and China was being censored both for its supposed political sensitivity and because the public was considered unable to cope with the true horrors of these wars.

The Second World War, however, made heroic war films popular again, and gave the leading local director Charles Chauvel the chance to make the film he had been planning for a decade, *Forty Thousand Horsemen* (1940). Chauvel was the nephew of Australian Light Horse commander General Sir Harry Chauvel (q.v.), presumably one incentive for making the popular and highly patriotic film, but his other non-military films also display a well-developed patriotic sense, often demonstrating a strong love of country and heritage and a cast replete with a stock array of digger and larrikin male leads. Flag waving aside, *Forty Thousand Horsemen* is far more than a well-made adventure film. Its climax features an exciting recreation of the charge at Beersheba (q.v.), which should be seen as among the world's best battle sequences for authenticity and complex film editing. With its spectacle and emphasis on mateship among a small group of male leads it was to influence films like *Gallipoli* (1981) and *The Light Horsemen* (1987).

The role of one of the stars, Chips Rafferty, as the laconic, irreverent, humorous larrikin, a personification of aspects of the Anzac legend, was to be copied in almost every later war film. More importantly *Horsemen* signals the evolution of an already developed trend in the public depiction of Australian military culture, in its focus on certain events or engagements that are deemed to have heroic import and which are fixed for future reference. It is not unusual for nations to dwell on battles, heroes or leaders, but the Australian trend appears to be fixated on events that seem to epitomise digger heroics and give them mythical status. In 1940 it may have been rational to look back to the Sinai campaign for inspiration, but Beersheba and Gallipoli offered no models for military engagement in Asia. Chauvel went on to make *The Rats of Tobruk* (1944), a more war weary and less popular film in which the stereotypes seem less fit for the context. The only other major war film of the Second World War, *The Power and the Glory* (1941), is notable mainly for featuring air force heroes and scenes of aerial combat involving RAAF Wirraways (q.v.).

Feature directors such as Chauvel and Ken G. Hall also produced semi-dramatised and documentary items for the newsreels. Much of the actuality footage used in the newsreels was shot by formerly civilian professional cameramen who were designated as accredited war correspondents for the Department of Information's Film Division. The most well known was Damien Parer (q.v.). Together with Hall, Parer perfected a strongly personalised style with an increasing emphasis on revealing the human face of the soldiers with ever more adventurous, and at times foolhardy, footage of front-line action. With his own voice providing commentary, supplemented at times by his own appearance urging support, Parer's Cinesound reports were outstanding accounts of the Pacific War. This was rewarded when *Cinesound Review: Kokoda Front Line* (1942), filmed and narrated by Parer, received Australia's first Academy Award. As a measure of the changing allegiances between the Allies, Parer's other assignments for Cinesound, *Assault on Salamaua* (1943) and the compilation feature *Sons of Anzacs* (1944), sought to redress what Hall had seen as General MacArthur's chauvinistic use of front-line footage. In Hall's eyes the Americans seemed to edit non-US forces out of Pacific operations. That feeling of being a poor-relation returns in later joint operations — most notably during the Vietnam period's developing anti-imperialism.

Two postwar films that dealt with war experience were *A Son is Born* (1946), a melodramatic tale of urban family life which culminates in reasonably authentic scenes of the hero fighting in New Guinea, and *Always Another Dawn* (1948), which features naval action. Numerous Australian service personnel characters also appeared in British films made during and after the war. In some cases they were fictional characters, while other films, such as *The Dam Busters* (1954), depicted historical figures. These films reflected the national mix of the British-Allied services themselves. Two of these British-produced films, however, reflect the more typical characteristics of Australian masculinity. *The Overlanders* (1946), set during the early phases of the Second World War, develops Chips Rafferty's laconic 'she'll be right mate' attitude from Chauvel's films while adding a strongly nationalistic element, quietly urging patriotism in the face of the Asian peril. At once reflecting the strong fear of Asia inherited from the gold-rushes, the film anticipates the even stronger resurgence of this fear under Robert Menzies' (q.v.) anti-communist politics throughout the 1950s and 1960s. Although appearing more overtly racist, *A Town Like Alice* (1953) provides a contrast through the gentle strength of its central Australian larrikin (played by Peter Finch — a role later reprised in the 1981 television mini-series by Bryan Brown, another actor who frequently portrays the archetypal Australian male).

After these films it took many years for Australian film-makers to return to the subject of war. The Australian film industry was in deep recession until the early 1970s, and was being overwhelmed by Hollywood. Apart from the relatively few British and European films still depicting and analysing the Second World War (and even the First World War) what there was of Korea was largely American product. Hardly any film industry was bothering to depict the Asian anti-communist engagements before Vietnam, and Australia and Australians are scarcely seen in those except as the odd mercenary, landowner or tourist.

A number of documentaries about the Vietnam War were made while that war was in progress, and of course the armed services made training films through the Commonwealth Film Unit and in cooperation with the commercial industry, but the main effect of Vietnam on Australian film and television would occur later. As early as 1966 the training films showed potential soldiers their dual tasks of soldiery together with the low-level diplomatic mission — demonstrating the values of western culture to the susceptible Vietnamese peasantry —

which becomes designated as WHAM (the Americanism 'Winning the hearts and minds'), though the Australians saw that job in a different way from their allies. Films such as *Australian Task Force Vietnam* (1966–67), *Diggers in Vietnam* (1966–67), *Action in Vietnam* (1966) and *The Third Generation* (1966) are notable for their appeal to the continuity of the Anzac tradition as their titles suggest, while depicting a style of warfare that was radically different from any previous commitment of troops. The other major feature of the Vietnam War was the effect of television. It is a commonplace of the Vietnam War to see it as won or lost through its political effect as television spectacle. The argument is no longer debated in the United States; the majority view accepts that the nightly broadcasting of the war on American television hastened the American withdrawal. The same cannot be said of Australia, or at least not to the same degree. The Australian public was not as densely bombarded by media images of Vietnam as was its larger ally; certainly in the first phases little visual media coverage was forthcoming; although as the war evolved it became a television war in Australia too, despite the fact that the images that dominated the Australian media were largely those of American involvement. Certainly the numerous current affairs shows burgeoning at the time, such as 'This Day Tonight', covered the war regularly, though not as common memory would have it exclusively from an anti-war stance. Anti-war satire from 'The Mavis Bramston Show' and other comedy programs may have aided the gradual shift in Australian attitudes, but their effect seems equivocal as they targeted politicians, particularly of the right, and Americans, rather than moral issues.

One serious depiction of the effects of Vietnam occurred in the early 1970s television mini-series 'You Can't See Round Corners', in which the hero evades the draft. Here too there was an interesting change. The novel on which the series was based was set in the Second World War; the modernised setting hardly alters the rationale of the hero in resisting call-up — a personal focus unallied to moral or political dissent. The popular depiction of the Australian attitude to the war remained naive. The espionage adventure series 'Spyforce' was also offering images from an earlier war. Whether this was a conscious sublimation or evasion of the current Asian circumstance or not, that series signals an early indication of the habits of the 1980s in both cinema and television, namely the presentation of Australians at war in those engagements in which equivocation in moral and political terms was unheard of. Once the moratorium marches had ceased dominating the television documentaries and current affairs programs from 1970 to 1973, Vietnam disappeared from the nation's screens almost throughout the seventies.

The film revival of the 1970s coincided with a new nationalism among Australian intellectuals and a feeling, provoked in large part by the involvement in Vietnam, that Australia's dependence on imperial powers was leading it to ruin. The first fiction film to deal explicitly with Vietnam, *The Odd Angry Shot* (1979), reflected this mood. Focusing on a single group of Australian soldiers from the Special Air Services (q.v.), rather than on broader political issues, its characters were, in the words of the promotional poster, 'Aussies being Aussies' who coped with the war through bitter, aggressively sexist humour. In a number of set-piece speeches, the film's central characters outline the politics of the soldier's war. While not directly anti-American, these speeches attack the class system both within the Australian Army and the wider community of Australia. Only the working class or psychologically distressed fight wars. At the same time the final blame for the war is located elsewhere — Australia is seen as fighting someone else's war.

This typical anti-imperial stance is taken up with more fervour in later films, initially in two films dealing with much earlier wars: Bruce Beresford's *Breaker Morant* (1980), and Peter Weir's *Gallipoli* (1981). *Breaker Morant's* version of events leading up to the Morant and Handcock (q.v.) executions during the Boer War has its protagonists (Edward Woodward as Morant and Bryan Brown as Handcock, representing the flamboyant and laconic aspects of Australian 'national character' respectively) sacrificed to protect British interests. While *Breaker Morant* is polemical, *Gallipoli* has been aptly described as a 'mystery play', a repackaging of myth through the story of two young men's journey to Gallipoli in 1915. Written by prominent playwright David Williamson with historical advice from historian Bill Gammage, *Gallipoli's* debt to the First World War official historian C. E. W. Bean (q.v.) is openly acknowledged. Almost all the elements of the classic Anzac legend (and above all the celebration of mateship) are there, but with one important difference. In Weir's film, innocent Australia, represented by golden-haired country boy Archy (Mark Lee), is pointlessly butchered in the service of the cold-hearted British.

Breaker Morant and *Gallipoli* are just two of the films which spring from a renewed interest in myths of identity; films that focused in particular on legendary moments in Australian history. Why that focus fell on the Edwardian–Federation period is still being discussed. One of the more obvious explanations may be that the 'blooding' of the

nation at Gallipoli and its subsequent status as iconic event overlapped with the reassessment of imperial allegiance which has become a major part of Australian culture since the end of the Vietnam War. To this may be added the observation that the imminent demise of the surviving veterans from the First World War has made their story more fascinating as we capture their memories in the brief time left. Certainly the emphasis on 1915–18 dominated other art forms in the late 1970s and early 1980s when a number of major novels, such as Roger MacDonald's *1915* and Malouf's *Fly Away Peter*, were published and new histories were in train. This interest was reflected in a number of television mini-series including '1915' (1982), from the novel by MacDonald, and 'Anzacs' (1985). These two mini-series added important elements to the public reception of the myths of Anzac, focusing not just on the typical events in the Middle East but moving into the Australian experiences on the Western Front, which had been virtually ignored in popular understanding. Equally they played the home front against the war, and emphasised, particularly in '1915', the horrendous wounds and traumas associated with war.

Similar issues though less graphically depicted were being played out over several years in the long-running television drama 'The Sullivans' (1976–82), with home-front versus front-line elements developing over many episodes. Interestingly this series concentrated on the Second World War and charted both the developing status of women within Australian society during that war and in its aftermath, and the continuing representation of the Japanese as a semi-human enemy. Again Australia had resorted to myth and popular representations derived from an earlier time to comment indirectly on issues current in the 1970s and early 1980s.

By 1987 film makers were ready to approach Vietnam. The Sydney 'Welcome Home March' and a continuous inpouring of American cultural products both ameliorating and critical of what was becoming known as the 'Vietnam experience' seemed to empower Australia to look at what it had more or less ignored since 1973. The *Odd Angry Shot* had sunk with little trace (partly as a result of the overwhelming attention paid to Francis Ford Coppola's *Apocalypse Now* of the same year) and socially engaged films like Martha Ansara's *Changing the Needle* (1981), which dealt with the effects on the families of veterans suffering from post-traumatic stress disorder, had a relatively and lamentably narrow audience. In 1987, however, two mini-series, 'Vietnam' and 'Sword of Honour', broadened the scope of previous fictional film coverage, in both cases linking the lives of several families at war and on the home front with the events of the war on the large scale of allied and world politics. Both allegorise the familial events as marking national development. As well as the relationships between generations, notions of nationalism and the status of women are developed as parallel plots. The plight of the veteran community on its return is, at least in 'Vietnam', of central concern, the series not willing to evade that issue by taking a simplistic politically correct stance against the war.

Both also attempted something that was essentially missing from almost all American depictions of the Vietnam War — a humanised depiction of the enemy, or rather the broad view of the Vietnamese as people caught up in a war forced on them by imperial forces. Neither polemic, nor overly apologetic about the war, both series sought to achieve some kind of balanced view. As television products of the 1980s, they tend to be rather sentimental at times, and the characterisation of events and personnel is wooden at worst, and often stereotypical when it comes to types of soldiers or types of anti-war activist, but rarely dull. The best of the two, 'Vietnam', occupied 10 hours of screen time and was the highest rating mini-series (at the time of its showing) on both occasions when it has been shown.

Mainstream Australian cinema has yet to move on from nationalistic questioning of imperial ties to look at the effects of Australian involvement in imperial adventures on the Boers, Turks or Vietnamese, although the documentary industry has attempted this on several occasions. The prevailing nationalism now turning, if slightly, towards kinds of multiculturalism also means that conflicts within Australian society are often glossed over in war films.

In the wake of the resurgence of interest in Vietnam in the late 1980s there has been a considerable rise of interest in our military past, and many television documentaries dealing with war and its associated social effects have been produced. Indeed the 1990s are studded with the celebration of one event or another from the world wars, beginning with the 75th anniversary of the landing at Gallipoli. Since then television and print media have been full of specials celebrating some major event, battle, or date from those wars. As the generations of Second World War veterans are rapidly growing older, urgency for recognition is heightened. Cinema has not followed suit, but television is voracious for documentaries: series such as 'Warriors, Friends and Foes' (1988) and 'When the War Came to Australia' (1990), which sought to reassess earlier wars from a benign viewpoint were countered by critical analyses of supposedly continuing Asian involvement from the (often) conspiracy-theory riddled journalists of the John Pilger camp. Documentaries on the aftermath of Vietnam War and Vietnam's intervention in the Cambodian crises, 'Cambodia Year 10' (1990), 'Cambodia: Fields of Hope' and 'Cambodia: The Betrayal', spawned many more suitably outraged specials and some debate. More socially concerned were the documentaries devoted to the refugee problems, the boat people and the effects of the war on the Vietnamese. Bretherton's 'As the Mirror Burns' and D'Arcy's 'Children of Krouser They' are just two of this heart-tugging brigade. Tran Huu Hanh's filmed autobiography 'Broken Journey: Mending Dreams' provides a model for documentaries that depicted the life of Vietnamese refugees in their new homes.

The general public interest in things military continued throughout the early 1990s spurred in part perhaps by the CNN coverage of the Gulf War, but also by the increasing nostalgia of the generations of veterans and their families looking for reinforced recognition of their service and sustained and renewed national identity at a time when the announced government policy was a political, economic and social move into Asia. Surprisingly then the number of Australian war films or documentaries which looked at Asian matters seemed bent on redress of grievances rather than reconciliation, which had at least been the focus of the 'Vietnam' mini-series. The reprise of 'A Town Like Alice' in 1981 as a mini-series starring Bryan Brown added little to the imagery of conventional Japanese cruelty, whereas by 1990 *Blood Oath*, also starring Bryan Brown, presented a harrowing picture of the atrocities committed on Ambon (q.v.). The film's re-enactment of them and of the trial that followed the war seemed determined on challenging Japan itself to account for its behaviour. Indeed the film caused some minor problems in Japan.

There has been little fictional treatment of the armed services in peacetime, except for the television series 'Patrol Boat' (1979), one telefeature depicting the lives of service wives, 'Army Wives' (1986), and an exposé-type program showing the training of Australian Army women, 'Ladies in Lines', which caused a small scandal on its release in 1993. The documentary splurge of the 1990s often focused on the home front with documentaries devoted to the Women's Land Army, and the *Women's Weekly* magazine's handling of war, 'The Weekly's War', or other social issues. The nostalgia which these and other pieces demonstrate found a major expression in the celebrations and subsequent telefeatures surrounding the anniversaries of the Burma–Thailand railroad. This reached something of a peak with the virtual apotheosis in film (and elsewhere in print) of Edward 'Weary' Dunlop (q.v.) leading up to and following his death. There is little sign that this nostalgic recreation of events and icons of Anzac is in sharp decline as the celebration of the Second World War is still under way and likely to result in more significant and popular reassessment of Australians at war.

Perhaps the most significant continuing absence from Australian film is the topic of warfare between Aboriginal and European Australians, with only isolated incidents appearing in films such as *Caloola, or The Adventures of a Jackeroo* (1911), *Heritage* (1935), *The Chant of Jimmy Blacksmith* (1978) and the mini-series 'Women of the Sun' (1982), though the participation of Aborigines in the services has been dealt with in the television series 'Blackout' in 1994, and several Aboriginal characters have appeared in war films, most notably in *Sword of Honour* (1987).

EWAN MORRIS AND JEFF DOYLE

FINN, Major-General Henry (6 December 1852–24 June 1924). Born in England, Finn enlisted in the British Army as a trooper in the 9th Lancers in 1871. He was awarded the DCM, and mentioned in despatches, for services during the Second Afghan War of 1870-71. Commissioned into the 21st Hussars in 1881, he served as adjutant with his regiment in India from 1887, on the staff from 1890, and saw further service in Burma. He went to Egypt with his unit in 1898, taking part in the campaign which culminated in the battle of Omdurman, after which he was promoted to brevet lieutenant-colonel and mentioned in despatches again. After returning briefly to the United Kingdom, he was offered the post of Commandant of the Queensland Defence Force, which he accepted, taking up his duties in April 1900. With the advent of Federation, he first presided over the Defence Pay Committee for the emergent Commonwealth, before accepting the position of Commandant of New South Wales, with the local rank of brigadier, in January 1902. In this capacity, he worked with Major-General E.T.H. Hutton (q.v.), the new GOC, Australian Military Forces, who thought Finn a 'valuable officer'. Their relationship was not always happy, a statement which applied to many of Hutton's dealings with both military and civilian figures in the early Commonwealth period. When Hutton left Australia after a difficult period in Commonwealth civil–military relations, Finn took temporary command of the AMF until, in December 1904, he was appointed Inspector-General (q.v.), now the senior post in the army. He proved a conscientious,

though somewhat unappreciated, senior officer. With minimal staff he wrote a range of reports on the readiness and efficiency of the AMF in 1905, and chaired the committee charged with drafting new administrative arrangements for Commonwealth defence. He did not get on at all with the Minister for Defence, Thomas Playford, and in 1906 he retired on grounds of ill health and returned to England. He returned to Australia in a number of civil posts, as private secretary to the Governor of New South Wales, Sir Gerald Strickland, between 1913 and 1917, and to the Lieutenant-Governor, Sir William Cullen, between 1923 and 1924. He died in Sydney and was buried in South Head Cemetery.

Finn played an important role as a professional soldier, and one, moreover, who had risen from the ranks, in regulating the early affairs of the nascent Australian Army. He was popular with the other ranks and respected by all, save perhaps his direct political masters, who were not, in general, of the highest calibre in the period during which he held the principal military office of the Commonwealth.

FIREFLY, FAIREY (2-seat carrier-borne fighter-bomber [FR4]). Wingspan 41 feet 2 inches; length 27 feet 1 inch; armament 4 × 20 mm cannon, 2 × 1000-pound bombs or 16 × 60-pound rockets; maximum speed 386 m.p.h.; range 580 miles; power 1 × Rolls Royce Griffon 2250 h.p. engine.

The Firefly, with the Sea Fury (q.v.), was the first combat aircraft flown by the Fleet Air Arm (q.v.) and served between 1949 and 1966. During the Korean War No. 817 Squadron flew Fireflies from the aircraft-carrier HMAS *Sydney* on bombing and strafing raids from October 1951 to January 1952. Fireflies were used for anti-submarine warfare until they were replaced by the Fairey Gannet (see GANNET, FAIREY) in 1955 and were then used for training and target-towing until they were withdrawn from service.

FIRST TACTICAL AIR FORCE (1st TAF) Formerly designated No. 10 Operational Group RAAF, the First Tactical Air Force was activated in October 1944, and was intended originally to remain in New Guinea in support of Australian ground forces as the main operations moved further north, until the commander of RAAF Command, Air Vice-Marshal Bostock (q.v.), protested the role assigned. The formation moved to Morotai in November and commenced operations against Japanese positions on Halmahera and the Celebes, followed by raids as far north as Mindanao in the Philippines. At this stage 1st TAF had eight squadrons under command organised into two fighter and one attack wings, with a further three under orders to join it. Like the army, however, the RAAF was being left behind in the South-West Pacific Area (q.v.) to mop up the remains of by-passed Japanese forces while the Americans moved north to the Philippines and on to the Japanese home islands, and discontent at this subsidiary role grew in First Tactical Air Force as elsewhere. Increasingly concerned that Australian aircrew were being killed on operations which contributed little or nothing to ending the war, in April 1945 a group of eight senior officers of 1st TAF sought to resign their commissions after representations to their commander, Air Commodore Cobby (q.v.), had failed to produce results. The 'Morotai mutiny' prompted the removal of Cobby and his replace-ment by Air Commodore Frederick Scherger (q.v.) and an inquiry headed by J. V. Barry, QC. This latter confirmed the dissatisfaction within the formation and assigned the blame to Cobby and his senior staff. General Kenney, commanding Far East Air Force, interviewed the eight dissenters and concluded that they had acted in good faith; when the RAAF considered taking disciplinary action against them, Kenney threatened to appear for the defence. The 'mutiny', like many other problems in the wartime RAAF, was prompted in the final analysis by poor leadership at the senior levels. 1st TAF took part in the Oboe operations (q.v.) against the Netherlands East Indies and in direct support of Australian landings in Borneo. At its peak by the end of June 1945 1st TAF numbered 21 893 all ranks, although this had declined to about 17 000 by the end of the Pacific War. The events of 1944–45 demonstrated both the prob-lems of working under MacArthur's command and the inherent weakness in the RAAF's senior ranks.

FISHER, Andrew (29 August 1862–22 October 1928). A Scottish immigrant, Fisher was a found-ing father of the ALP and prominent in the fed-eral Labor administrations before and during the Great War.

Fisher first became prime minister in Novem-ber 1908. His administration, which held office only until May 1909, resolved a number of defence issues which had been under consideration by suc-cessive federal governments, including compulsory military training (see CONSCRIPTION) for young men under 20 years of age, and the placing of the Australian Navy under Admiralty control in wartime. Fisher himself, however, was not passion-ate on defence matters, and other senior figures in the party, especially J. C. Watson and W. M. Hughes (qq.v.), were more outspoken on these issues.

Labor won the election in April 1910, and Fisher returned to the prime ministership. This second period in office, between 1910 and 1913, saw con-siderable legislative activity, mostly directed towards social and economic issues. At the 1911 Imperial Conference (q.v.) he attempted to reassert former Prime Minister Alfred Deakin's arguments on con-sultation with the dominions on empire foreign policy, but got nowhere. In 1912 he refused to send troops into Brisbane to break the general strike, despite the direct request of the State Premier.

Campaigning for a federal election in July 1914 just before war broke out, Fisher made his famous statement that Australia would stand beside Britain to its 'last man and last shilling' (q.v.). This phrase was to haunt Labor during the conscription crises in 1916–17, although Fisher was not the first to use it and it received much less prominent attention when originally uttered. Labor won the election, and Fisher presided over the early dispatch of Aus-tralian forces overseas, but he saw little advantage to Australia from participation in the war, and unlike Hughes did not glory in martial association. The strain of leadership in wartime, coming on top of that generated by leading the Labor Party in its early years, prompted him to resign in October 1915. He was appointed High Commissioner to London, where he was a member of the Dar-danelles Commission, visited Australian troops in France, and formed a good working relationship with the AIF's commander, General Sir William Birdwood (q.v.). He was increasingly marginalised by Hughes during and after the conscription cam-paigns.

FIVE POWER DEFENCE ARRANGEMENTS (FPDA), an informal defence agreement between Australia, New Zealand, Malaysia, Singapore and the United Kingdom, replaced the Anglo-Malayan Defence Agreement (AMDA). AMDA, negotiated in 1957 after Malayan independence, guaranteed continuing British assistance for the defence of Malaya against external aggression in exchange for the right to maintain UK forces and bases in Malaya. An RAAF squadron had been based in Malaya since 1950, and since 1955 Australia had contributed to the British Commonwealth Far East Strategic Reserve (see MALAYAN EMER-GENCY), so in 1959 Australia became associated through an exchange of letters with those provi-sions of AMDA which related to the reserve. Aus-tralia was not, however, obliged to defend Malaya (or Malaysia from 1962), and the Strategic Reserve was intended for use within the purposes of SEATO (q.v.), of which Malaysia was not a mem-ber. The lack of a mutual security agreement

between Australia and Malaysia caused some concern to the ALP during Confrontation (q.v.).

In 1967 the British government announced that, as part of a general withdrawal from east of Suez, half of its forces in Malaysia and Singapore would be withdrawn by 1971 and the rest within five years after that. However, Australian Prime Minister John Gorton's (q.v.) 1969 announcement that Australia would retain its forces in Malaysia and Singapore, followed by the British government's agreement to continue basing a small force in the two countries, set the stage for the negotiation of a new defence arrangement. AMDA ceased to apply from 31 October 1971, the FPDA came into effect the next day, and for perhaps the first time Australia and Britain became equal parties to a defence agreement. Unlike AMDA, the FPDA is not a formal treaty but a loose framework involving consultative, administrative and planning measures, embodied in exchanges of letters between the five governments. The principles of the arrangements are spelled out in a communiqué issued after a five-power ministerial meeting in London in April 1971. The ministers declared that in the event of an external attack or threat to Malaysia or Singapore their governments would immediately consult together to decide on a response.

There is no joint command structure as part of the FPDA, but political consultation takes place through the Joint Consultative Council and the Integrated Air Defence System (IADS). Perhaps the FPDA's most important feature is that it is supervised by an Air Defence Council. When the FPDA first came into operation the three external powers (Australia, New Zealand and the United Kingdom) formed their forces in Malaysia and Singapore into a single ANZUK force. Australia's contribution to this force was two squadrons of Mirage (q.v.) fighter aircraft, one infantry battalion, one destroyer or frigate, and a submarine deployed on rotation with Britain. The force was disbanded on 1 January 1975 after first the Australian and then the British government decided to withdraw their ground forces. Australia's two squadrons of Mirages remained at Butterworth (q.v.) air base in Malaysia, however, and Australia has continued to provide the IADS commander. In 1988 the Mirages were withdrawn from Butterworth, leaving only a detachment of Orion (q.v.) reconnaissance aircraft, support personnel and an infantry company, supplemented by rotational deployments of F/A-18s

(q.v.) from Tindal (q.v.) air base. Nevertheless, despite the Dibb Report's (q.v.) assessment that the FPDA reflected 'the concerns of a previous era' and could now be justified only by political, not military, considerations, the Labor government in the late 1980s reasserted its commitment to the FPDA as an important part of Australia's regional defence policy.

Chin Kin Wah, 'The Five Power Defence Arrangements: Twenty years after', *The Pacific Review*, vol. 4, no. 3 (1991), pp. 193–203.

FLEET AIR ARM Australian naval aviation has taken three forms in the history of the RAN: small aircraft largely for reconnaissance purposes and carried on cruisers; the projection of naval airpower from aircraft-carriers; and the operation of helicopters in a variety of roles from frigates. The first lasted from 1917, when HMAS *Brisbane* first carried a Sopwith Baby seaplane on its aft deck, until 1944, when HMAS *Australia* (II) (q.v.) landed its amphibian, a Supermarine Seagull Mark V (q.v.). The early experiments with naval aviation in the First World War, advocated strongly by Captain J. S. Dumaresq (q.v.) among others, led nowhere initially, with the aircraft carried on the cruisers *Sydney* (I), *Brisbane*, *Melbourne* and *Australia* (I) being returned to Britain at the end of the war.

Interest in naval airpower received a setback with the formation of the RAAF in 1921, but in 1928 a seaplane-carrier, HMAS *Albatross*, equipped to carry nine Supermarine Seagull III aircraft, was launched. *Albatross* was paid off in 1933, a victim of cost-cutting in the Depression. In the Second World War Seagull V aircraft flew off Australian cruisers, and some were shot down in aerial combat, but in general their role remained strictly limited. The impact of carrier warfare in the Pacific was felt in Australia, and in 1947 the RAN began acquiring a naval air arm of two light aircraft-carriers, two naval air stations and three air groups flying Fairey Fireflies and Hawker Sea Furies (see AIRCRAFT, ROYAL AUSTRALIAN AIR FORCE). The carriers were British-built. The first, HMAS *Sydney* (III) (q.v.) (formerly HMS *Terrible*), arrived in Australia in December 1948 and received the first carrier air group, No. 20 Group. *Sydney* relieved HMS *Glory*, due for a refit, as part of the US 7th Fleet and served in the Korean War between September 1951 and January 1952, providing the first operational testing of the new Australian naval aviation capability. The RN loaned the carrier *Vengeance* (q.v.) to the RAN in 1952 until the arrival of HMAS *Melbourne* (q.v.) (formerly HMS *Majestic*) in 1955. *Melbourne* never saw operational service, and

---

Andrew Fisher (left) with William Morris Hughes, Australian Prime Minister (centre), visiting an AIF camp in England, 1916. (AWM H16101)

Supermarine Seagull IIIs of No. 101 Fleet Co-operation Flight being hoisted aboard the seaplane carrier HMAS *Albatross*. (AWM P01817.006)

was involved in two collisions (q.v.) at sea, with HMAS *Voyager* (q.v.) in 1964 and with the USS *Frank E. Evans* in 1969.

The creation of a naval aviation capability placed enormous strain on the small RAN establishment in a period of generally straitened resources, and the RN lent large numbers of British officers and ratings and provided considerable training support in the process, without which it must be doubted that the RAN could have succeeded. At its height, in the late 1950s, the Fleet Air Arm comprised 100 aircraft in five squadrons with a light fleet-carrier as flagship of the fleet. Both carriers, however, were Second World War designs, and as new generations of larger, heavier, high-performance jet aircraft came on stream the limitations of the Australian carriers became more obvious. Plans to upgrade *Sydney* were abandoned for lack of funds in 1954, and it ceased to operate as a carrier in May 1958. Recommissioned in May 1962, it functioned as a troop transport, the 'Vung Tau ferry' during the Vietnam War. HMAS *Melbourne* was to be retired in 1963, but had its role modified to anti-submarine warfare (q.v.) and embarked Westland Wessex ASW helicopters in

place of its De Havilland Sea Venoms. Modernised in 1967, *Melbourne* then deployed Douglas Skyhawks and Grumman Trackers until serious technical problems led to its retirement in 1981.

Fleet Air Arm pilots served in Vietnam as part of the RAN Helicopter Flight Vietnam (q.v.) with a US Army Assault Helicopter Company, and with RAAF units flying out of Vung Tau. The Australian government was given the option to purchase the modern British carrier ('through deck cruiser' in contemporary parlance) HMS *Invincible* in 1981. The Falklands War delayed this, and in the event the Fraser government decided not to replace *Melbourne*, a decision confirmed by the incoming Hawke government in 1983 largely on ground of cost, and because the RAAF argued that its new generation of strike aircraft, represented by the F/A-18 Hornet, could provide air cover at sea. This brought to an end the aircraft-carrier and fixed-wing aviation phases of Australian naval aviation. Since 1985 naval aviation has concentrated on helicopters. Sikorsky Seahawks are flown from the new generation of frigates and served during the Gulf War. The RAN also deploys Westland Sea King, Westland Wessex,

Bell Iroquois, Bell Kiowa and Aérospatiale Squirrel helicopters. The Fleet Air Arm's training and maintenance base remains at HMAS *Albatross* at Nowra in New South Wales.

(See also entries on individual aircraft.)

FLINDERS, HMAS see **SURVEY SHIPS**

FLINDERS NAVAL DEPOT see *CERBERUS*, **HMAS**

FLOWER CLASS MINESWEEPING SLOOPS (*Marguerite, Geranium, Mallow*). Laid down *Mallow* 1914, *Marguerite* and *Geranium* 1915, all launched 1915; displacement 1250 tons; length 268 feet; beam 33 feet 6 inches (*Mallow* slightly smaller than figures given for other two ships); armament 1 × 4.7-inch gun, 2 × 3-pounder guns (*Mallow* 2 × 4-inch guns, 3-pounder gun); speed 16.5 knots.

These minesweeping sloops were built for the RN during the First World War and acquired by the RAN in 1919. HMAS *Mallow* and HMAS *Marguerite* were used as training ships for Naval Reservists. HMAS *Geranium* served as a minesweeper until 1921 when she was fitted out as a survey ship (q.v.). All were sunk for gunnery practice in 1935.

FOOD see **RATIONS**

FORBES, Alexander James 'Jim' de Burgh (16 December 1923–   ). The son of a soldier (a graduate of the first class at Duntroon), Forbes was commissioned from RMC Duntroon into the artillery in one of the wartime short courses in 1942, and served with the AIF in New Guinea and Bougainville, where he won the MC. After the war he did tours with the occupation forces in Germany and Japan, and was on the Australian Army Staff in London in 1946. He subsequently undertook degrees at the universities of Adelaide and Oxford, and between 1954 and 1956 lectured in political science at the University of Adelaide. A member of the South Australian State council of the Liberal and Country League, he entered federal parliament as a Liberal for the seat of Barker in 1956, having contested Kingston unsuccessfully the previous year. He was appointed minister for the Navy, in succession to John Gorton (q.v.), between 1963 and 1964, and minister for the Army between 1963 and 1966. He thus presided over the early military commitment to Vietnam, but left the portfolio before it became a politically dangerous one to hold. Because of his own and his family's military background, he enjoyed good relations with many senior officers, and thought that 'there was a general feeling that I … understood their problems, difficulties and aspirations'. He was minister

for Health under three prime ministers between 1966 and 1971, and for Immigration between 1971 and 1972. He retired from the parliament in 1975, and served as a member and subsequently as chairman of the Council of the National Library. He was federal president of the Liberal Party from 1982 to 1985.

FORDE, Francis Michael (18 July 1890–28 January 1983). Born in Mitchell, Queensland, Forde entered State politics as Labor MLA for Rockhampton, and moved into the federal parliament in 1922 as member for Capricornia. In 1935 he lost the ballot for leadership of the Labor Party to John Curtin by one vote. When Curtin became prime minister in October 1941, Forde became deputy prime minister and minister for the Army, despite having no knowledge of military affairs or having demonstrated any interest in them. His greatest characteristic was loyalty, which he gave to Curtin unreservedly while wielding little influence in his own right over military policy. In October 1942, after a two day visit to Port Moresby, Milne Bay and the Kokoda Track, he wrote a report that was highly critical of the administration of the army in New Guinea. The report in turn was criticised by General Blamey, who thereafter largely ignored Forde. Following Curtin's death on 6 July 1945, Forde was prime minister for six days until Chifley succeeded him, Forde remaining as minister for the Army. In November 1945, he abruptly and ungraciously terminated Blamey's appointment, and refused to endorse Blamey's recommendations for honours for senior officers. He left the government in 1947 to become High Commissioner to Canada, and served again briefly in the Queensland State parliament in 1955–57.

FORREST, John (22 August 1847–3 September 1918). Born near Bunbury, Western Australia, Forrest was a successful explorer and government surveyor before being elected to the Legislative Assembly and becoming premier of Western Australia in 1891. He was appointed KCMG the same year, and served as premier until 1901 when, with the creation of the Commonwealth, he moved into federal parliament as member for Swan. He was appointed GCMG and became minister for Defence almost immediately, the first incumbent dying in office after only nine days. Forrest had no experience in defence matters and he was poorly prepared to shepherd through parliament a Defence Bill which he had played no part in formulating. Forrest himself accepted the imperial argument that as long as the RN held command of the sea, Australia was effectively safe from invasion.

It followed for him that Australia had no need of either large and expensive military forces or costly local naval forces, but that it should contribute to the maintenance of the RN while maintaining small, militia-based forces for local defence. The Defence Bill of 1901 had been drafted in the main by the colonial military commanders meeting as the Federal Military Committee, and it reflected their view that Australia's military forces should, in time of war, be made available for the imperial cause. This was rejected by the parliament, which passed so many amendments that the colonial commandants dissociated themselves from the bill, and the government eventually withdrew it in 1902. The Defence Act of 1903 (q.v.) curbed the power of the commandants, restricted the size of Australia's permanent forces, and prohibited the use of Australian forces overseas except on a voluntary basis. Although Forrest stepped down as minister in 1903 he had exercised some influence over the drafting of the bill, which reflected to some extent his dual imperial and national loyalties. Forrest was created Baron Forrest of Bunbury in 1918, and died at sea off Sierra Leone on a troop-ship bound for London.

FORTIFICATIONS can be divided into permanent fortifications, which are built in locations thought to require ongoing defence, and field fortifications, which are built to gain temporary defensive advantage in conditions of actual or imminent armed conflict. Australian soldiers fighting overseas have constructed field fortifications at various times. During the First World War Australians built and used parts of the massive trench systems on the Western Front, which were originally built as temporary defences but became permanent as the opposing armies became stalemated. Within Australia itself a crude fortification erected by protesting miners at Ballarat in 1854 achieved lasting fame as the Eureka Stockade (q.v.), but this was probably the only occasion on which fortifications were used in battle in Australia. Neither Aborigines nor Whites made much use of fortification in their struggle for possession of the continent (see ABORIGINAL ARMED RESISTANCE TO WHITE INVASION). Unlike the Maori of New Zealand, Aborigines did not have an established tradition of fortification, nor did they develop one in the course of their war against the invaders. They preferred to use natural rather than human-constructed defensive positions, and do not seem to have employed any but the most elementary of defensive barriers, such as felled trees. Many Whites in frontier areas fortified their houses and other buildings to some extent, and the wooden forts built at the short-lived trading posts at Fort

Dundas and Fort Wellington near present-day Darwin were presumably intended for defence against Aborigines as well as against foreign powers. There was, however, no Australian equivalent of the extensive chains of forts with which the United States government asserted its authority during the westward expansion in North America.

All of Australia's permanent fortifications were built to defend against attacks from the sea, though the earliest defences in Sydney were built with the threat of convict insurrection also in mind. Australia's coastal defences varied greatly in design, but all consisted essentially of an artillery battery surrounded on the seaward side by strong walls to defend the guns and the gunners from the fire of attacking ships. Magazines, stores and barracks were often incorporated within the fort itself, and some forts had elementary defences on the landward side as well. The first fortifications in Australia were built as early as 1788 at Bennelong and Dawes Points on the eastern and western sides of Sydney Cove. The inner defences of Sydney were constructed over the following decades, but for a long time a battery at George's Head remained the harbour's most forward defensive position. Hobart was also defended by gun batteries from the early days of White settlement, but it was not until the colonial war scares (q.v.) of the 1850s that colonial governments began to develop more comprehensive plans for Australia's coastal defence. Even then, there was little concrete progress until the early 1870s, when construction began on Sydney's outer defences, intended to stop enemy ships from entering Port Jackson at the heads.

It was the reports of Sir William Jervois and Peter Scratchley (qq.v.) in the late 1870s that were responsible for the most energetic period of fortification-building in Australian history. Asked to survey the defences of the Australian colonies, Jervois and Scratchley produced recommendations based on the assumption that the RN would shield Australia from full-scale invasion but that the colonies might be vulnerable to raids or their capitals held to ransom by small numbers of ships. Jervois and Scratchley advocated defending major ports and cities by employing a combination of fortified gun batteries, submarine minefields, and the ships of the colonial navies (q.v.). These recommendations were, by and large, accepted by colonial governments, and Scratchley stayed on to supervise the resulting program of construction. The fortifications which were built in the late 1870s and 1880s had to have guns which were capable of penetrating armoured warships while being protected from the fire of attacking vessels, and technological advances made this possible. Rifling, longer barrels,

Table 1 Australian fortifications, 1901

Queensland

Thursday Island
Green Hill Fort	3 × 6-inch breech-loading guns, 1 × 0.45-inch Maxim machine-gun. Completed 1892.
Milman Hill Fort	1 × 4.7-inch quick-firing gun. Under construction in 1901.

Townsville
Magazine Island Fort	2 × 4.7-inch quick-firing guns, 1 × 0.45-inch Nordenfelt machine-gun. Completed 1892.
Kissing Point Battery	2 × 6-inch breech-loading guns, 2 × 0.45-inch Nordenfelt machine-guns. Completed 1891.

Brisbane
Fort Lytton (13 miles below Brisbane, 2 miles above Moreton Bay)	2 × 6-inch breech-loading guns, 2 × 64-pounder rifled muzzle-loading guns, 2 × 6 pounder quick-firing Hotchkiss guns, 1 × 1-inch Nordenfelt gun, 2 × 0.45-inch Nordenfelt machine-guns. Completed 1881.

New South Wales

Newcastle
Fort Scratchley	1 × 8-inch breech-loading gun, 3 × 6-inch breech-loading guns, 3 × 1.5-inch quick-firing Nordenfelt guns, 1 × 80-pounder rifled muzzle-loading gun. Completed 1882.
Shepherd's Hill Fort	1 × 8-inch breech-loading gun. Completed 1891.

Sydney
Dawes Battery	5 × 42-pounder smooth-bore guns, 4 × 12-pounder howitzers. Completed 1859 (original fort built 1788).

Middle Head Section
Inner Middle Head	2 × 6-inch breech-loading guns. Completed 1887?
Outer Middle Head	2 × 10-inch rifled muzzle-loading guns, 4 × 80-pounder rifled muzzle-loading guns. Completed 1877?
Obelisk Bay	2 × 1-inch Nordenfelt machine-guns.
George's Head	1 × 6-inch quick-firing cannon, 1 × 6-inch breech-loading gun, 4 × 1.5-inch quick-firing Nordenfelt guns, 2 × 80-pounder rifled muzzle-loading guns. Completed 1886 (original fort built 1801).
George's Heights	2 × 6-inch breech-loading guns, 2 × 80-pounder rifled muzzle-loading guns. Completed 1873.

South Head Section
Steel Point	3 × 5-inch breech-loading guns. Completed by 1890.
Green Point	3 × 1.5-inch quick-firing Nordenfelt guns. Completed by 1890.
Minefield Battery	3 × 6-pounder quick-firing Hotchkiss guns, 1 × 80-pounder rifled muzzle-loading gun.
Outer South Head	2 × 6-inch breech-loading guns, 3 × 6-inch quick-firing cannon, 1 × 10-inch rifled muzzle-loading gun, 4 × 80-pounder rifled muzzle-loading guns, 1 × 9.2-inch breech-loading gun. Completed 1893?
Signal Hill Fort	1 × 9.2-inch breech-loading gun. Completed by 1890.
Ben Buckler Fort (Bondi)	1 × 9.2-inch breech-loading gun.
Shark Point Fort (Coogee)	1 × 9.2-inch breech-loading gun.

Botany
Bare Island Fort	1 × 6-inch breech-loading gun, 1 × 10-inch rifled muzzle-loading gun, 1 × 9-inch rifled muzzle-loading gun, 2 × 80-pounder rifled muzzle-loading guns. Completed 1885.
Henry Head	2 × 6-inch breech-loading guns. Completed 1880s.

Wollongong
Signal Hill Fort	1 × 6-inch breech-loading gun. Completed 1880s?
Smith's Hill Fort	2 × 80-pounder rifled muzzle-loading guns, 1 × 1.5-inch quick-firing Nordenfelt gun. Completed 1880s?

Victoria

Port Phillip Heads
Eagle's Nest Battery	1 × 10-inch breech-loading gun. Completed by 1889.

cont. next page

Table 1 Australian fortifications, 1901 *cont.*

Victoria *cont.*

Port Phillip Heads *cont.*

Fort Nepean	2 × 9.2-inch breech-loading guns, 3 × 6-inch breech-loading guns, 1 × 4.7-inch quick-firing gun, 1 × Nordenfelt machine-gun. Completed 1884.
Fort Franklin	1 × 10-inch breech-loading gun, 2 × 5-inch breech-loading guns, 1 × 4.7-inch quick-firing gun. Completed by 1889.
South Channel Fort	2 × 8-inch breech-loading guns, 2 × 4.7-inch quick-firing guns, 2 × 6-pounder quick-firing Nordenfelt guns, 4 × 1-inch Nordenfelt machine-guns. Completed late 1880s.
Swan Island Fort	8 × 5-inch breech-loading guns, 2 × 6-pounder quick-firing guns, 2 × Nordenfelt machine-guns. Completed 1884.
Fort Queenscliff	2 × 9.2-inch breech-loading guns, 3 × 6-inch breech-loading guns, 2 × 6-inch breech-loading guns, 4 × 14-pounder quick-firing guns, 4 × Nordenfelt machine-guns. Completed 1883.
Crow's Nest Battery	1 × 8-inch breech-loading gun, 1 × 6-pounder quick-firing gun. Completed 1880s.

Warrnambool, Port Fairy, Portland

Warrnambool	1 × 5-inch breech-loading gun, 2 × 80-pounder rifled muzzle-loading guns.
Port Fairy	2 × 80-pounder rifled muzzle-loading guns.
Portland	1 × 5-inch breech-loading gun, 2 × 80-pounder rifled muzzle-loading guns.

Melbourne

Fort Gellibrand	1 × 5-inch breech-loading gun, 1 × 9-inch rifled muzzle-loading gun, 3 × 80-pounder rifled muzzle-loading guns, 1 × 6-pounder quick-firing gun.

South Australia

Port Adelaide

Fort Glanville	2 × 64-pounder rifled muzzle-loading guns, 2 × 10-inch rifled muzzle-loading guns. Completed 1880.
Fort Largs	2 × 6-inch breech-loading guns, 2 × 9-inch rifled muzzle-loading guns. Completed 1886.

Western Australia

Albany

Princess Royal Battery	2 × 6-inch breech-loading guns. Completed 1893.
Plantagenet Battery	1 × 6-inch breech-loading gun. Completed 1893.

Tasmania

Hobart

Alexandra Battery and Redoubt	2 × 6-inch breech-loading guns, 1 × 5-inch breech-loading gun, 2 × 7-inch rifled muzzle-loading guns, 1 × 70-pounder rifled muzzle-loading gun, 1 × 6-pounder quick-firing Nordenfelt gun, 1 × 0.45-inch Nordenfelt machine-gun. Completed 1885.
Queen's Battery	2 × 70-pounder rifled muzzle-loading guns, 3 × 64-pounder rifled muzzle-loading guns, 1 × 32-pounder smooth-bore gun, 1 × 8-inch smooth-bore gun. Completed pre-1858.
Bluff Battery	2 × 8-inch rifled muzzle-loading guns, 2 × 80-pounder rifled muzzle-loading guns, 1 × 6-pounder quick-firing Nordenfelt gun, 1 × Nordenfelt machine-gun. Completed 1884.

more effective use of the propellent charge, and armour-piercing projectiles all combined to make the penetration of armoured vessels possible. Protection of mounted guns from enemy fire was made easier by the invention of 'hydro-pneumatic' (h.p.) or 'disappearing' guns. When the h.p. gun was fired, it recoiled downwards below the parapet, where it could be reloaded under cover. A hydraulic ram absorbed the energy of recoil, which could then be used to return the gun to the firing position. Thus, the gun was only exposed briefly to the enemy. Soon forts had been constructed to

defend Sydney, Newcastle, Melbourne, Hobart, Adelaide and Brisbane, and in 1885 Scratchley declared himself satisfied with the defences of the Australian colonies. Gaps remained, however, and in the 1890s the colonies jointly funded the construction of fortifications at Thursday Island in the Torres Straits and at Albany on King George's Sound in the southern part of Western Australia.

In all the time they were in use, Australia's coastal fortifications were never tested in battle, though a gun at Fort Nepean was used on two occasions, at the beginning of the First and Second World Wars,

Fortifications at Middle Head, Sydney, 1891. (AWM P0991/121/031)

to challenge German ships leaving Port Phillip Heads. The shot fired at noon on 5 August 1914 was the first shot fired by the British Empire in the First World War. Fears of Japanese invasion during the Second World War briefly brought about a new flurry of fort building, with the coast between Newcastle and Wollongong being particularly well fortified. After the war, though, these new forts and the earlier ones fell into disuse; vulnerability to air attack had made such fortifications well and truly obsolete. Fort Queenscliff went on to a new life as the location of the Army's Command and Staff College (see STAFF COLLEGES), but most other forts were neglected and many of their guns were sold for scrap. Today, however, a number of the old forts around the country are maintained by the National Parks and Wildlife Service as buildings of historical interest, and are open to visitors.

The list of Australian fortifications in Table 1 comes from the 1901 report of the Military Committee of Inquiry into Australia's defences. As such, it is a list of fortifications in use in 1901, not a complete list of all forts constructed throughout Australia's history. The armament, likewise, is that in

use in 1901, and includes only mounted guns. Where possible, the year in which the fort was completed is also given, but even when an exact date is given this should be considered approximate, as it is difficult to say exactly at what point a fort was complete.

FORWARD DEFENCE is the policy of countering perceived threats to Australian security as far from the shores of Australia as possible. In practice this has involved deploying Australian armed forces overseas, basing and exercising Australian forces in South-east Asia in peacetime, forming alliances with great powers and smaller regional powers, and encouraging great power involvement in South-east Asia in particular. Though in a sense Australian defence policy has been based on forward defence since Federation, the term is usually associated with the period after the Second World War when Australian defence was no longer integrated into that of the British Empire. The policy was motivated by the belief that Australia did not have the population or resources to defend itself alone and by a perception of communism as a monolithic threat. Should the gap between Australia and 'the southward flow of communism' narrow, warned Defence Minister Sir Philip McBride (q.v.) in 1954, 'the nature and scale of attack on Australia would become intensified as distance shortened'. Therefore, he argued, Australia should cooperate with like-minded governments to keep that gap as wide as possible. In the late 1960s the withdrawal of British forces from south-east Asia and the US government's policy of requiring its Asian allies to conduct their own defence (the Guam doctrine) forced a rethinking of the forward defence policy. The level of forces required for Australia to conduct a forward defence in South-east Asia on its own would be unlikely to gain widespread acceptance in Australia, but no government has yet adopted a coherent alternative policy.

FOSTER, Brigadier-General Hubert John (4 October 1855–21 March 1919). Graduating from the Royal Military Academy, Woolwich, in 1875, Foster was commissioned into the Royal Engineers and saw service in Cyprus, Egypt and Ireland. Between 1898 and 1901 he was Quartermaster-General of the Canadian militia, and was closely involved in the preparation of Canadian contingents for the South African War. Between 1903 and 1906 he was military attaché to Washington and Mexico City. His involvement with Australia began in the latter year when he was offered the newly created post of Director of Military Science at the University of Sydney, for an initial period of three years and with an annual salary of £800. The course he implemented lasted three years and resulted in the award of a Diploma of Military History and Science. Foster intended that in the demands it made on students it should be 'suited to the University spirit. It was to give a general knowledge of the principles and practice of war, such as could only be drawn from a study of past campaigns.' The course was aimed at those seeking a commission and others interested in military studies as part of a general education. By 1908 over 30 Arts students were enrolled in Foster's subjects. The course was regarded very favourably by senior officers of the Australian forces, and in 1909 Foster's appointment was extended for a further three years. In October 1912 he was placed on the Retired List in the British Army. The outbreak of war, however, rapidly diverted students to other pursuits, and in 1915 the course attracted no enrolments. In January 1916 Foster was appointed CGS, holding the office until September 1917 when Major-General J. G. Legge (q.v.) returned from active service in France. From October 1917 until the end of the war he was Director of Military Art

at RMC Duntroon. He retired due to ill health, and died soon after. The University of Sydney was unable to attract a suitable officer to replace Foster as director, and the course lapsed. Foster made an important contribution in laying the early foundations of military studies in Australia, and he served as CGS during the most exacting period of the war, but his work is hardly known.

4.5-INCH HOWITZER Range 6600 yards; weight of shell 35 pounds; total weight 3000 pounds (approx.).

The 4.5-inch howitzer was the standard British light howitzer during the First World War. It could fire high explosive and smoke shell. It was first used by the Australian Army on the Western Front in the Somme battles of 1916. Typically it was used to cut wire and demolish minor trench defences. It continued in use in the Australian Army during the Second World War in the Middle East, Malaya and New Guinea. Because of its light weight and high-angle fire it was most useful in difficult jungle terrain, especially at Buna, and was not withdrawn from service until 1945.

FRASER, John Malcolm (21 May 1930–   ). Fraser entered parliament in 1956, but did not receive preferment as a minister until his appointment to the Army portfolio a decade later. As Army minister from January 1966, Fraser was at the centre of a difficult period in which the services' commitment to the Vietnam War was increasing, along with popular opposition to that commitment and to the deployment of national servicemen there on active service (see CONSCRIPTION).

He held the Army ministry until 1968, and although he had clear ideas on defence issues, the three service ministries were the most junior posts in the government, and he thus lacked any base of power or authority from which to have many of his ideas accepted. He nonetheless was judged to have handled a potentially testing job well. He was a prominent exponent of government policy on Vietnam, and handled the minor crises in parliament posed by anti-conscription activism, the Gunner O'Neill (q.v.) case, and an unfortunate speech critical of government policy on the war by the secretary of his department, Bruce White (q.v.), with skill and judgment. He was promoted to the portfolio of Education and Science in February 1968.

He returned to defence issues in November 1969, taking up the position of minister for Defence which he held in sometimes controversial circumstances until March 1971. He appointed the determined and highly able Sir Arthur Tange (q.v.)

Malcolm Fraser (centre), while Minister for the Army, is shown a carbine by Captain Vin Murphy, Australian Army Training Team Vietnam, during a visit to Vietnam in 1966. At the right is Bruce White, Secretary of the Department of the Army (AWM FOR/66/538/VN)

as secretary of the department, and the two men between them began the process of galvanising the defence sector, a process which came to fruition with the Tange Report in 1973 and its implementation under Prime Minister Whitlam.

Fraser could claim a number of achievements during his time in defence. He successfully renegotiated the contract for purchase of the F-111 (q.v.) from the Americans on new terms highly advantageous to Australia. In March 1970 he presented the first serious and systematic defence review for some years to the parliament, over the objections of William McMahon, the Minister for Foreign Affairs, who felt that it trespassed into areas of concern to his own department. Fraser also took steps to review the state into which service pay and

conditions had declined by appointing a Committee of Inquiry under the chairmanship of Mr Justice John Kerr of the Commonwealth Industrial Court, although again, the full fruits of this were to be realised only under Whitlam. He also initiated planning for what was to become the Australian Defence Force Academy (q.v.), a project to which he and Tange were committed over the objections of many in the services and elsewhere.

Fraser clashed with senior service officers over a number of issues, but the most important and best publicised was his attempted intervention with the army, and specifically the CGS, Lieutenant-General Sir Thomas Daly (q.v.), over proposals in early 1971 to wind back the Civic Action program in Phuoc Tuy province, something in which Fraser had taken considerable interest since his days as minister for the Army. The fundamental issue was Fraser's unhappiness with the army's responsiveness to him and his department as the central body responsible for defence matters, an issue which had been building for at least a year. The issue was badly handled, involving unattributable briefings of journalists and consultations between the Prime Minister, John Gorton (q.v.), and the CGS, at which the minister was not present. The press reporting fuelled the disagreement by sensationalising the story, usually inaccurately. Charging disloyalty to himself as a minister on the part of his prime minister, Fraser resigned on 8 March 1971, setting in train the final destabilisation of Gorton's prime ministership.

In administrative, as opposed to party-political, terms Fraser's difficulties were a product of the too gradual shift in the organisation of the defence group of ministries, which would not be resolved satisfactorily until the adoption of the Tange Report brought Australia into line with practice in other Western countries through abolition of single-service ministries and rationalisation of military–political responsibilities and authority through a single Defence minister and Department of Defence.

During his prime ministership between 1975 and 1983, Fraser presided over the issue by his Defence minister, Jim Killen (q.v.), of a Defence White Paper in 1976 which talked of the need for increased defence self-reliance in the post-Vietnam era. Fraser's term as prime minister was marked by a renewed emphasis on the Soviet threat in Asia and the Pacific. However, acquisition of major equipment was deferred in this period, threatening the services with block obsolescence by the early 1980s. This left the Hawke government which came to office in 1983 with a major problem in matching its defence budget to the largest capital equipment acquisition program since the Second World War.

Philip Ayres, *Malcolm Fraser: A Biography* (William Heinemann Australia, Melbourne, 1987).

FREMANTLE CLASS PATROL BOATS Displacement 220 tonnes; length 42 m; beam 7 m; armament 1 × 40 mm bofors gun (q.v.), 2 × 0.5 machine-guns, 1 × 81 mm mortar; speed 30 knots.

The *Fremantle* Class patrol boats were laid down between 1978 and 1983 and commissioned between 1981 and 1984. In all 15 were built. Their duties are general patrol around the Australian coast, Bass Strait oil rig surveillance and assisting survey ships. Three of the ships starred in the second series of the ABC's television drama 'Patrol Boat' (see FILM AND TELEVISION, WAR IN AUSTRALIAN).

FREMANTLE (I), HMAS see *BATHURST* CLASS MINESWEEPERS (CORVETTES)

FRENCH, Major-General George Arthur (19 June 1841–7 July 1921). Educated at Sandhurst and Woolwich, French was commissioned into the Royal Artillery in 1860. Between 1862 and 1876 he served in a variety of capacities in Canada, and remained behind on attachment to the militia when the British Army withdrew the last of its units in 1870. He was the first commissioner of the newly formed North-West Mounted Police (forerunner of the Royal Canadian Mounted Police). In September 1883 he was appointed commandant of the Queensland Defence Force, with the rank of colonel. His report on the state of the colony's defences revealed numerous deficiencies, and in 1885 the force was reorganised in accordance with recommendations that he had drafted, and along lines similar to those pertaining in Canada. In 1891 he employed troops to break up the bitter shearers' strike, on one occasion personally leading a line of soldiers, with bayonets fixed, against a strikers' picket. He returned to England that year, having served three terms in Queensland. After further service in Britain and India, French was made commandant of the New South Wales forces in March 1896. He created a number of new units during his tenure of command, and volunteering flourished. He resented the fact that the government would not release him for service in the South African War, but was rewarded for his work in 1901 with appointment as president of the Federal Military Committee charged with drafting the Defence Act (q.v.). He returned to England in February 1902, and retired from the army. He was made KCMG. Like a number of the imperial officers who served in command of the colonial and dominion forces at the end of the nineteenth century, French distrusted colonial politicians, but his ability as an administrator served the governments of Canada and Australia well.

FRENCH FOREIGN LEGION, AUSTRALIANS IN see AUSTRALIANS IN THE SERVICE OF OTHER NATIONS

'FURPHY' is slang for a rumour, false report or absurd story. It was in use among Australian troops at Gallipoli in 1915, later passing into general Australian usage. The term comes from the name of the company J. Furphy & Sons Pty Ltd, manufacturers of water-carts on which the name 'Furphy' appeared. Because the drivers of these carts moved freely over a wide area as they transported water to troops at the front they were useful sources of rumour, gossip and information (see MILITARY SLANG).

'FUZZY-WUZZY ANGELS' were the estimated 50 000 indigenous inhabitants of Papua and New Guinea who worked for the Australian Army during the Second World War as supply-carriers, stretcher-bearers, scouts, construction workers and general labourers. The term comes from a 1942 poem by Australian sapper Bert Beros. The conscription of indigenous labour began on 15 June 1942 with an order from the GOC, New Guinea Force (q.v.). This order also established a pay scale for such work of between 5 and 15 shillings a month plus rations and other necessities. Recruitment was the responsibility of the Australian New Guinea Administrative Unit (ANGAU) (q.v.) field officers, who often used 'harsh methods because no other methods would achieve the results', as one of them told Gavin Long (q.v.) in 1943. At first the rations, clothing, shelter and medical care provided to these labourers was grossly inadequate, and yet they had to perform gruelling work, often without sufficient rest. Not surprisingly, sickness rates in some places were 25 per cent or more, and the overall death rate in 1943 was about 2 per cent, compared to the prewar average for indentured labourers of 1.5 per cent. In 1944 a new ration scale was introduced and other conditions also improved, with the result that sickness rates

dropped to 4 per cent and the death rate to 0.8 per cent. In the period 1942–45 only 46 indigenous labourers were killed by enemy action, but 1962 died of other causes. By the time labourers were released from their contracts on 15 October 1945 the absence of so many able-bodied male workers from the labour-intensive village economy had caused hunger, ill health, increased infant mortality and a reduced birthrate. Contrary to the myth of the 'loyal native', many Papuans and New Guineans had little choice but to cooperate with whoever was in control, whether it was the Australians or the Japanese, who also conscripted indigenous labour. Many Australian overseers continued the beatings which had been a routine method of labour control before the war; 'deserters' were severely punished and some indigenous labourers were executed for 'collaborating' with the Japanese. However, although prewar residents of Papua and New Guinea working in ANGAU and elsewhere urged troops to keep their distance from the 'natives' and to uphold the illusion of White superiority, many Australian soldiers whose lives depended on the work of indigenous carriers and stretcher-bearers were more informal, developing a genuine affection for these men that was sometimes reciprocated. The 'fuzzy-wuzzy angels' became White Australia's first Black heroes, and probably made Australians more sympathetic towards the needs of Papua and New Guinea after the war.

G

'G FOR GEORGE' was an Avro Lancaster (q.v.) Mark I bomber, serial number W4783, delivered to No. 460 Squadron, RAAF, in October 1942. By the time it was withdrawn from operational service in April 1944 it had flown 90 operations against targets in Germany, Italy and France, and was the last of its type to be retired. It was flown to Australia to promote the Third Victory Loan, arriving in Brisbane on 8 November 1944. It was declared surplus in July 1945 and 10 years later was installed in the Australian War Memorial's Aeroplane Hall. In 1977 its full wartime colours, including unofficial nose art 'awarding' the aircraft the DFM, the CGM and the DSO, were restored.

GALLEGHAN, Brigadier Frederick Gallagher (11 January 1897–20 April 1971). Galleghan enlisted in the AIF in January 1916, and served in France with the 34th Battalion. He was promoted to sergeant and was badly wounded at Messines, but failed to achieve the commission which appears to have been his ambition, possibly because he was of 'mixed race'. On return to Australia he obtained an appointment as lieutenant in the 2nd Battalion of the CMF, and thereafter threw himself into militia soldiering with enthusiasm. By 1932 he had been promoted to lieutenant-colonel and was in command of a CMF battalion.

At the outbreak of the Second World War Galleghan was commanding the 17th Battalion and was desperate to take a unit to war. He was the senior CMF lieutenant-colonel, but at first found it impossible to gain a command in the 2nd AIF. Following the intervention of William Morris Hughes (q.v.), Galleghan was given the 2/30th Battalion of the 8th Division in October 1940. He led this unit through the brief Malayan campaign and the battle for Singapore where along with the rest of the division and the British and Indian forces there it surrendered to the Japanese. Galleghan endured four years of captivity at Changi (q.v.) as the senior Australian officer following the removal of all other senior officers by the Japanese. After the war he was appointed head of the Australian Military Mission in Berlin, which involved him as the Australian representative on the International Refugee Organisation in Geneva. He returned to Australia in 1949, and in 1959 was appointed honorary colonel of the Australian Cadet Corps (q.v.).

Galleghan, known as 'Black Jack', became a significant figure among ex-prisoners of the Japanese, many of whom lionised him. Unlike the commander of the 8th Division, Gordon Bennett (q.v.), Galleghan had endured Japanese captivity with his men, while his battalion had fought important delaying actions against the Japanese at Gemencheh and Ayer Hitam during the fighting in Johore. He thus represented an unambiguously positive figure for former POWs, many of whom were tormented by their experiences and the fact of their surrender. In the process, he became somewhat larger than life, his faults minimised and his achievements overstated. He was a competent battalion commander with a strong sense of duty, an overbearing, sometimes bullying, manner, and average military gifts. His work in Berlin and Geneva in helping to resettle Displaced Persons and processing many for migration to Australia was distinguished, recognised by his election in 1949 as chairman of the International Refugee Organisation.

Stan Arneil, *Black Jack: The Life and Times of Brigadier Sir Frederick Galleghan* (Macmillan, Melbourne, 1983).

GALLIPOLI The commitment of troops from Australia and New Zealand to the Gallipoli Peninsula arose out of the deliberations of the War Council in London and the desire of the First Lord of the Admiralty, Winston Churchill, to seek a more offensive role for the RN. The War Council was a committee of the British Cabinet, formed to advise the Prime Minister (Herbert Asquith) on matters concerning the War. It consisted of Asquith himself, Field Marshal Lord Kitchener (Secretary of State for War), Churchill (First Lord of the Admiralty), David Lloyd George (Chancellor of the Exchequer), Sir Edward Grey (Foreign Secretary), Arthur Balfour (former Conservative prime minister) and a floating group of other ministers and service advisers.

In December 1914 various members of this body had expressed concern about the lack of progress of military operations against the Germans on the Western Front. The French counter-attack after the Marne had stalled; Britain's original six divisions had suffered heavy casualties around Ypres; trench lines had now been dug from the channel to the Swiss border; and no Allied military figure seemed to have a clear plan of how the war was to be won. Into this strategic vacuum stepped the politicians. Lloyd George suggested that Germany could best be defeated by attacks on her weaker allies — Austria-Hungary and Turkey. Maurice Hankey (q.v.), who was a Royal Marines Colonel and Secretary of the War Council, also suggested an attack upon Turkey in combination with such Balkan states as could be rallied to the Allied cause.

The most prolific source of ideas was Churchill. Since the outbreak of war he had been dissatisfied with what he determined to be the passive role played by the RN. Even before the Western Front had solidified he had put forward schemes to capture an island off the German coast, thus forcing the German Fleet to sally forth to retake it. The consequent great naval battle would see the Germans crushed, and the British Fleet enter the Baltic where in combination with Russian forces it would dominate the north German coast and force the German High Command to divert troops from the Western Front. That scheme had met with stiff resistance from Churchill's naval advisers. It would risk the Grand Fleet in shallow waters dominated by mines and submarines; the Germans might not come out and get sunk; and Russia had no troops to spare nor the expertise to carry out combined operations against the German north coast. Churchill then suggested using the navy to blockade Holland into the war, or using the firepower of the battleships to assist the British and French armies with operations along the Belgian coast. He also suggested using some of the older battleships to attack Turkey on the Gallipoli Peninsula.

The first two schemes soon collapsed for much the same reason as the German island scheme — they risked the Grand Fleet in mine and submarine dominated waters. The Gallipoli scheme, however, received some impetus from an outside source. On 2 January the Russians asked the British to mount a diversionary operation against the Turks who were pressing them hard in the Caucasus. Churchill seized on this message to ask the Commander of the British squadron in the Aegean, Admiral Carden, if the Dardanelles could be forced and Constantinople dominated by ships alone.

Churchill later stated in his memoirs that he detected a remarkable convergence of opinion at this time in favour of operations against Turkey. Lloyd George, Hankey, the Russians, Kitchener (who had passed on the Russian request to Churchill) and himself, he wrote, all favoured such action. In fact there was no such major convergence. The scheme put forward by Lloyd George was mainly concerned with operations against Austria–Hungary. Hankey's plan involved large-scale military operations carried out by a coalition of Balkan States with minor Allied assistance. The Russian request soon vanished in the mists, as the Tsar's armies pushed back the Turks in the Caucasus. Only Churchill had suggested a purely naval operation centred around forcing the Dardanelles and overawing Constantinople.

Nevertheless, for reasons that are not hard to identify, it was the latter scheme that recommended itself to the War Council. It promised the defeat of a significant enemy for very little cost. It would not require many British troops, the vast majority of which were committed to the Western Front as reinforcements for the shattered remains of the British Expeditionary Force. The only troops required would be occupation forces needed to garrison the Gallipoli Peninsula and Constantinople — following the success of the fleet. It would not risk the Grand Fleet, as Churchill had assured the War Council that only Britain's old, surplus battleships would be used. The operation was also said to have the advantages of opening a supply route to Russia and galvanising the Balkan States into action on the Allied side. This last action could see a mighty army of Allied states advancing up the Danube and attacking the Central Powers from the rear.

On 3 January 1915 the War Council heard from Churchill that Admiral Carden was also in favour of the plan. In fact Carden had placed many reservations on the practicality of the plan. But these reservations — that a great many ships would be needed, that the minefields in the straits would prove difficult to sweep, that the whole affair could be a protracted one — were not passed on to the War Council by Churchill. So the War Council sanctioned the naval attack on the Dardanelles. Operations commenced on 19 February. By mid-March it had failed decisively. The fleet proved incapable of overcoming the combination of minefields and mobile howitzers that protected the forts at the narrows. On 18 March a concerted attack on the defences by 18 battleships resulted in one-third of them being sunk or disabled.

While naval operations were in progress, a debate had sprung up in the War Council concerning the role of troops in the campaign. This debate was instigated by Churchill who, immediately the naval attack commenced, started calling for troops to 'reap the fruits' of the fleet's operations. It will be recalled that Churchill had sold the Gallipoli idea to the War Council largely on the fact that no troops would be needed. What had changed his mind? This can never be known with any certainty. But it is possible that Churchill, who had some knowledge of military matters, had attempted to convert the government to a major operation in stages — first the naval attack which would appeal because of its limited liability. Then, when the fleet was engaged, call for troops — some of which were always going to be needed even if the fleet succeeded — but which might also be required to help the fleet through the straits and to convert the whole operation into a real alternative to the Western Front.

If this was Churchill's motivation he can certainly be accused of involving the government by stealth in a much larger undertaking than they originally contemplated. And it would be an operation to some extent under his own control and for which he would get the credit if it succeeded. To

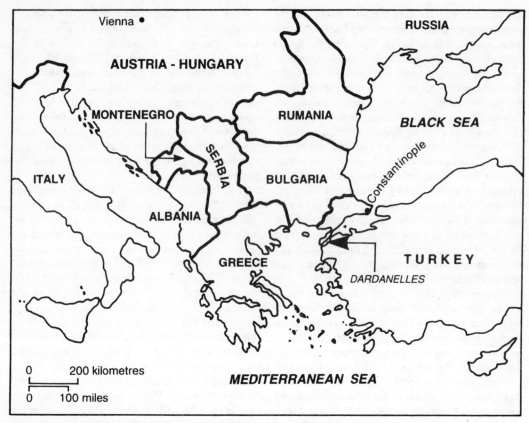

Map 11a   Gallipoli, theatre of operations, 1915.

this extent his policy in initiating the Gallipoli campaign can be said to be self-serving. However, there is little doubt that his motives were more complex than this. He had been one of the few members of the War Council to visit the Western Front, and to have witnessed firsthand the carnage that resulted from operations there. He had also been mightily unimpressed by the plans of the generals to break the deadlock. So Gallipoli seemed to him a way of waging war without the frightful toll of the Western Front. Seapower and a few troops and Britain's (alleged) traditional way of warfare could be restored. Gradually the War Council became convinced that troops would have to be landed on the Gallipoli Peninsula to eliminate the forts and mobile howitzers. This would enable the minefields to be swept and the fleet to proceed to Constantinople whereupon the Turks would no doubt surrender.

What troops were to be used for these purposes and who would command them? Churchill insisted that the 29th Division (the only British Regular Army division not committed to the Western Front) be an integral part of the force. Kitchener at first refused to release this division. Eventually, under pressure from other members of the War

Council, he agreed that it should go. In addition Churchill promised the Royal Naval Division (a miscellaneous collection of surplus officers and men from the RN). The French (wary of British postwar aims in the Middle East) also offered a division. These forces were still deemed insufficient however. The only other troops on offer were the Australian and New Zealand troops in Egypt. These men (approximately two divisions), under the command of General Birdwood, were training in Egypt (because of lack of suitable facilities in Britain) before being sent to the Western Front.

In all the forces to be arrayed against the Turks amounted to approximately 70 000 men. The fact that at an earlier War Council meeting Kitchener had considered 150 000 men insufficient to defeat Turkey had apparently been forgotten. Seventy thousand were available and 70 000 were therefore deemed to be enough. As Birdwood was considered too junior to command even a force of this size, Sir Ian Hamilton, a retired general then in command of home forces in Britain, was despatched on the rather inappropriately named HMS *Foresight* to take charge.

Hamilton's main problem was how to dispose his forces on the narrow confines of the Gallipoli

Peninsula, the southern section of which was just over 20 miles long and 5 miles across at its widest point. In selecting areas to land Hamilton was very circumscribed. Much of the coastline consisted of rugged cliffs with few beaches or paths through the cliffs which could provide exit routes for large bodies of troops. He finally decided to make his main landing with the 29th Division on five beaches at the toe of the peninsula around Cape Helles. Once a landing had been established they would be assisted by the gunfire of the fleet to advance on the Narrows forts. To the north, the Anzac forces would make a secondary landing. A covering force would first seize the rugged Sari Bair Ridge. Then the remaining troops (protected by the troops on the heights from Turkish intervention to the north) would dash across the peninsula to Maidos. This would prevent reinforcements from being sent to the Turks at Helles, cut off any troops being driven back by the British, and assist the attack on the Narrows. Meanwhile the Royal Naval Division would attempt to deceive the Turks by a feint landing at Bulair (an obvious place to land at the narrow neck of the peninsula but because of that heavily defended). Meanwhile the French would land on the Asiatic coast at Kum Kale to destroy some heavy batteries which might interfere with the British landings at Helles.

This was an imaginative plan — possibly the best that could have been devised in the circumstances. However little thought was given as to whether 70 000 men were sufficient to defeat the Turkish Army or whether such forces as the Anzacs had provided enough men to occupy a long, precipitous ridge and advance across the peninsula. Hamilton, who had been Kitchener's Chief of Staff in the South African War, never questioned his chief about the sufficiency of his force. Nor did anyone else in authority. It was merely assumed that the force at hand would defeat the incompetent Turks.

Because the troops for the assault had to be tactically loaded on the transports in the order in which they were to disembark, the landings could not take place until the end of April. Much has been made of this. It is claimed that the Turks greatly strengthened their defences in the five week interval. In fact the Turks were remarkably supine in this period — a few extra troops were placed on the peninsula, and some extra defences constructed. However, rough weather in March and April would have almost certainly delayed the landings anyway. Troops could not have been landed or reinforced and supplied in a gale.

Meanwhile Birdwood was given the task of developing the Australian plan in detail. He had at his disposal one Australian division (Bridges [q.v.]) and the New Zealand and Australian Division (Godley [q.v.]), about 25 000 men in all. Birdwood decided that to achieve surprise the forces must be assembled at night and landed at dawn. The 3rd Brigade (Sinclair-MacLagan [q.v.]) would land first as the covering force, push inland and establish positions on Gun Ridge. The 2nd Brigade (McCay [q.v.]) would then land and seize the higher points on the Sari Bair Ridge (including the highest, Hill 971). The 1st Brigade (MacLaurin [q.v.]) would land immediately after the 2nd and act as a divisional reserve. Finally the New Zealand and Australian Division would be landed and a force assembled for the advance across the peninsula.

The Anzac forces began leaving Egypt in early April to join the 3rd Brigade on the Island of Lemnos, where it had been since 4 March. On the 12th they arrived and carried out some rudimentary practice landings. But there was little time for much training. The operation was to commence on the 23rd. In the event bad weather delayed it until the 25th. On the eve of battle the 3rd Brigade was loaded on battleships and destroyers — the remaining brigades on conventional transports. From these ships the troops would first be loaded in ships' boats and then towed by steamboats until they were close to the coast. They would then be rowed ashore and landed on a 3000 yard front, the left of which was located just south of Ari Burnu Point.

At 3.30 a.m. on the morning of 25 April the battleships anchored about 3500 yards off the coast of the Gallipoli Peninsula. It was pitch dark; not even the outline of the coast was visible. The order was given to land and the steamboats set off. At 4.00 a.m. the tows were cast off and the naval ratings began to row towards the beaches. As the troops approached it became obvious that the tows had bunched together and that the landing was about to take place about a mile to the north of the designated beaches, in a small bay to the south of Ari Burnu later known as Anzac Cove.

The reason for this mistake has given rise to endless speculation. It was suggested that the tides had carried the tows to the north, then it was found that in this area there were no tides of any significance. The battleship commanders were blamed for anchoring in the wrong spot. Naval charts show this was not the case; their navigation was correct. It has even been suggested that Birdwood unilaterally changed the plan at the last moment to avoid underwater obstructions protecting the original landing beaches. The evidence for this is weak. The best explanation seems to be that the naval ratings guiding the tows lost direction in the dark and veered to the left. The result was that the whole

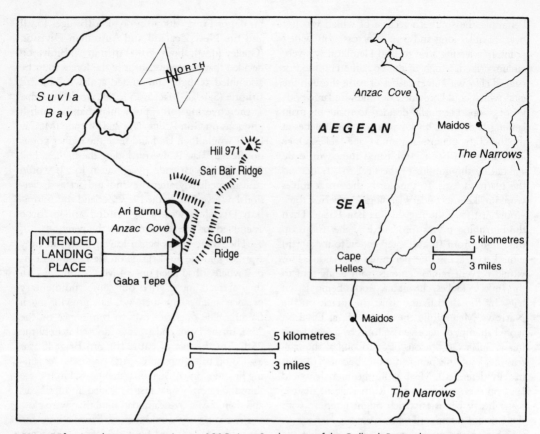

Map 11b Initial operations at Anzac, 1915. *Inset* Southern tip of the Gallipoli Peninsula.

force was landed on a narrow front with units bunched and intermixed from the start. The troops were also confronted by steeply rising ground instead of the more open country to the south.

In the event the covering forces failed to achieve their original purpose — the capture of Gun Ridge. In the first instance this was not because of strong Turkish opposition. Confronting the Anzacs were just two companies of enemy troops armed only with rifles. (Because of their commanding position, however, these men managed to inflict many casualties before they were driven back.) What prevented a coherent drive inland was the country rather than the enemy. The area the Anzacs found themselves in was unfamiliar. The maps available were so inaccurate as to be useless. So tangled and tortuous were the ravines and spurs in front of them that the groups that began to advance inland were soon split into small sections which lost contact with each other almost immediately. A few parties did manage to advance to the second ridge and a smaller number still reached their objective. But they were isolated and scattered and could provide no nucleus for any follow-up force.

Back on the beaches chaos reigned. The 2nd and 1st Brigades and later the formations of the

New Zealand and Australian divisions had all landed around Ari Burnu. Most of the efforts of the officers were expended, not in organising a force for a drive inland as intended, but in trying to sort out the men into their original units. When some semblance of order was established, groups of men were sent forward to where the firing line was thought to be.

From mid-morning, however, the whole force was confronted by Turks in much increased numbers. On the morning of the landing the local Turkish commander, Mustapha Kemal (later, as Kemal Ataturk, to become President of Turkey), had been intending to exercise his force near Hill 971. He now redistributed his units for counter-attack. At 10.30 a.m. the right of the small Anzac perimeter was driven in and most of 400 Plateau lost. In the afternoon and evening the left of the position was assaulted and the line driven back from Baby 700 and the Nek.

By evening the situation seemed serious. Bridges and his brigadiers had only a hazy idea of the exact position of the troops inland. Counter-attacks by the enemy were obviously under way. Many men, separated from their units and unclear of the situation, began to drift back towards the beach. Under

Men of the 12th Battalion in Leane's Trench, Gallipoli, during the August offensive, 1915. The man with the periscope may be watching the progress of the attack on Lone Pine as he is wearing a white patch on the back of his shirt, which was used for identification during the Lone Pine attack. (AWM P1436/10)

these conditions Bridges recommended evacuation. Birdwood, who had landed in the afternoon, agreed. Hamilton, offshore on the battleship *Queen Elizabeth*, under naval advice that an evacuation would be impossible to organise, overruled his commanders and ordered the troops to dig in. This the men were already doing. They had driven off the Turks and were holding a position from Walker's Ridge in the north starting between the first and second ridges to around Shell Green in the south. All told the first day had secured the Anzacs 2.25 square miles of Turkish Territory at a cost of around 2000 killed and wounded.

The tenacity of the Anzac troops, fighting in unfamiliar and difficult country, split up into a confusion of units, subjected to counter-attack and overlooked almost everywhere by the enemy, was certainly worthy of praise. However this should not disguise the fact that the whole northern plan had failed. Before an advance could be organised the Turks had arrived in force and were occupying the high ground. Nevertheless, it may be doubted if Birdwood and Hamilton's plan ever had a chance

of success. It was absurdly ambitious to expect a force of two divisions to capture a succession of difficult positions along a ridge 4 or 5 miles long, advance 4 miles across the peninsula, capture more difficult positions and beat off counter-attacks after this had been done. Walking over the ground after the war C. E. W. Bean (q.v.) found the objectives almost impossible to attain in reasonable time, even with the help of a Turkish guide.

While the Anzac landing was stalled the 29th Division was faring no better at Helles. They had suffered heavy casualties on the 25th and while the Turks had proved incapable of driving them into the sea, the regulars had proved just as incapable as the Anzacs of mounting a major advance. For the moment, however, Hamilton decided to concentrate his efforts in the south. A mixed force of Australian and New Zealand troops was sent to Helles to spearhead a new attack on 8 May. The result was a fiasco. Under the incompetent control of General Hunter-Weston (GOC 29 Division) the troops were thrown in without proper orders and with little notice. Many had no idea as to where the Turkish front line was situated. After the war it became clear that the main advance did not even reach the enemy trenches but had been stopped by a line of skirmishers. The Anzac brigades suffered 1800 casualties for no result.

In the days after the landing little progress was made at Anzac. Some ground was regained near 400 Plateau and in general the perimeter trenches were strengthened. There was never any chance, however, of a significant advance being made. In fact the initiative now lay with the Turks. Before they could exercise this initiative, the Anzac force suffered a considerable loss. On 15 May Bridges was wounded in the leg by a Turkish sniper. He was immediately placed on a hospital ship and evacuated to Egypt. Before the ship arrived, Bridges had died. His body was brought to Australia and buried on a hill overlooking RMC Duntroon.

With the Anzac advance successfully stalled, the Turks began building up forces around the perimeter to throw the invaders into the sea. By 18 May they had accumulated four and a half divisions, about 42 000 men. Next day they attacked. The result was the greatest disaster for Turkish arms in the course of the whole campaign. The Anzacs had been given time to consolidate their position. They had brought machine-guns to the front line and had coordinated fire support with the fleet. By these means the Turks were beaten back with great slaughter. At the end of the day 10 000 of their troops lay dead and wounded on the the battlefield. So bad was the stench that a truce was arranged to bury the dead. Never again would the Turks mount a general counter-attack against the Anzacs.

For the next 10 weeks the campaign in the north was characterised by small-scale operations designed to capture a trench here and there or generally improve the line. It was not these small-scale operations, however, that proved most trying for the troops. As the heat increased in June and July so did the rate of sickness. By the end of July, the force was losing through illness about the same number every fortnight as were lost in the initial assault. The major problem was dysentery. Sanitary conditions were poor; flies gathered in millions around latrines and rotting corpses. The medical services were overwhelmed by the number of cases and because the entire peninsula could be brought under enemy fire, had difficulty in setting up substantial hospital facilities. Nor was the food provided appropriate. The main fare was salty stew, hard biscuit and jam. Under these conditions the physical condition of the troops (superb at the landing) began gradually to deteriorate. This would have some effect on the next series of operations which were being planned.

Hamilton had decided as early as May that the growth of trench lines in front of the 29th Division and the French made large-scale operations at Helles prohibitive. The problem was essentially one of artillery. As had been discovered on the Western Front, entrenched defenders could only be blasted from their positions by a deluge of heavy artillery shells. At Gallipoli, there was never enough artillery to mount a considerable bombardment. The guns of the fleet, which were supposed to make up for this deficiency, could not fire shells with the plunging trajectory necessary to destroy trenches. Nor were their armour piercing shells the optimum type to fire against earthen defences. So Hamilton's attention turned towards Anzac. Here, in the centre and the south there were trenches aplenty. But to the north the flank was open — no defences barred the way to Hill 971. If columns could outflank the Turkish defences by stealth, there would be no need to bombard their trenches. The shortage of artillery might be overcome. The next attack therefore would be launched from Anzac.

Reluctantly the War Council (now known as the Dardanelles Committee after the reconstitution of the Asquith Government in May) agreed. As a result of much debate it was decided to reinforce Hamilton by no less than five divisions (three from the Kitchener armies and two Territorial divisions). They would arrive in time to commence the attack in early August.

The main attack would be made from Anzac and was planned by Birdwood. His principal aim was to seize the high ground between Hill 971 and Chunuk Bair. Once this ridge was secure a drive across to the straits would be made and the Narrows forts captured. For the operation Birdwood was reinforced by the 13th Division and a brigade from 10th Division (both New Army formations) and one Indian brigade.

At the same time Hamilton made preparations for the continued supply of the northern forces through the winter. Anzac Cove was very small and most supplies had to be landed at one pier. With rough weather in prospect and additional troops to supply, Hamilton proposed a second landing of two divisions on the flatter ground around Suvla Bay. This area could then be used to accommodate stores to supply the Anzacs to the south. To the confusion of almost all historians since, who have not appreciated that the primary aim of the operation was to establish a base at Suvla Bay, Hamilton specified that when it had secured Suvla Bay as a base, this force should, if circumstances permitted, aid Birdwood's attack in any way possible.

The main attack was launched from Anzac on the night of 6 August. The plan was for two columns of troops to seize the foothills that commanded the routes to the summit of the ridge. Then two more columns would pass through them and capture the three major heights (from north to south Hill 971, Hill Q and Chunuk Bair). Once these summits had been taken a force would move along the crest line down towards the Old Anzac perimeter. At the same time troops from Old Anzac would attack towards the main ridge. In this way the formidable positions of Baby 700, Battleship Hill and the Nek would come under simultaneous assault from two directions. When these attacks joined hands all of the main ridge would be in Allied hands and a drive towards the Narrows could be organised.

A diversion at Lone Pine by the 1st Brigade began on the 6th. It was designed to draw Turkish reserves away from the main battle. In every way it was a débâcle. The attack was costly (over 2000 men killed) and by the time the Turkish reserves arrived it was clear that they were only facing a diversion. The Turkish troops attracted to Lone Pine were eventually used to help thwart the main advance. As for the main attack, the covering force succeeded in clearing the foothills, but the incredibly rugged nature of the country, the lack of good maps and the sheer confusion caused by a night attack ensured that the operation was well behind schedule when the main columns began their approach march. The left column, which was to capture Hill 971 and which had the furthest to go, never had a chance. The leading troops (4th Australian Brigade, commanded by Monash [q.v.]) were soon lost in the unmapped gullies and by dawn were nowhere near their objective. The right column (29th Indian Brigade and a New Zealand brigade) eventually got within a few yards of Chunuk Bair but was too weak to advance further. The first attempt on the ridge had failed.

After resting the exhausted troops on the 7th, a new assault on the summit was organised for the 8th. Some success was obtained with a New Zealand force actually occupying Chunuk Bair. On the 9th a further attempt saw Hill Q fall to a contingent of Gurkhas. Two summits of the ridge were now in Birdwood's hands. Neither was to remain so for long. The Gurkhas were shelled off the ridge by the navy who in attempting to support the troops dropped a salvo of shells short. Then the Turks counter-attacked in strength and drove the New Zealanders from Chunuk Bair. The main ridge was now completely in Turkish hands and was to remain so.

One further action of the August attack has burnt itself into Australian consciousness. This was the attack on the Nek. In the early morning of 7 August the 10th and 8th Light Horse, who acted as infantry, assembled for the attack. A bombardment of the Turkish positions commenced. Then seven minutes before the attack was to go in the bombardment stopped. The operation was ordered by the local command to go ahead anyway, despite the lack of protecting shellfire and the absence of the New Zealand infantry who were supposed to assault simultaneously down the ridge. Three lines of Light Horsemen went over the top only to be shot down immediately by the unhampered Turks. At the end of the day 375 of the 600 attackers had become casualties. Of these 234 were dead. Their sacrifice was useless. The New Zealanders, as we have seen, were not in any position to attack down the ridge to support them.

On the August operation as a whole there has been much debate. Some authorities have claimed that the August attack came within a narrow margin of success. While it is true that for short intervals troops did hold two of the key summits of the Sari Bair Range, these troops were very few in number. In no sense can it be said that these forces 'occupied the ridge'. After the machine-guns of the New Zealanders on Chunuk Bair had been knocked out, they had nothing more than rifles with which to defend themselves. There was of course no artillery support. In these circumstances it was only a matter of time before these positions were lost to Turkish counter-attacks. To hold the positions in strength almost the entire force at Anzac would have been needed on the crest of the ridge, and given the nature of the country this would have taken many more days to organise than Birdwood had at his disposal. Turkish reinforcements could always reach the high ground faster from their side of the range than the Anzacs could from theirs.

The point also needs to be considered that it would have taken a very large force to occupy the entire range from Hill 971 to Battleship Hill. Even if this had been accomplished, supplying such a force with food and ammunition across the rugged and twisted ravines might have proved impossible. The small number of men above the Apex on Chunuk Bair had received little food, less water, a negligible amount of rifle ammunition and no bombs. In any case, capturing the ridge and supplying the troops would not have marked the end of the battle. Birdwood's original plan saw this only as the first stage. The second stage involved pushing reinforcements across the peninsula under the cover of the force on Sari Bair. In August no reinforcements were available for that task, but it is doubtful if they could have been supplied with water had Hamilton been able to offer Birdwood an extra division. Finally, pushing forward from Sari Bair would not have been a simple matter.

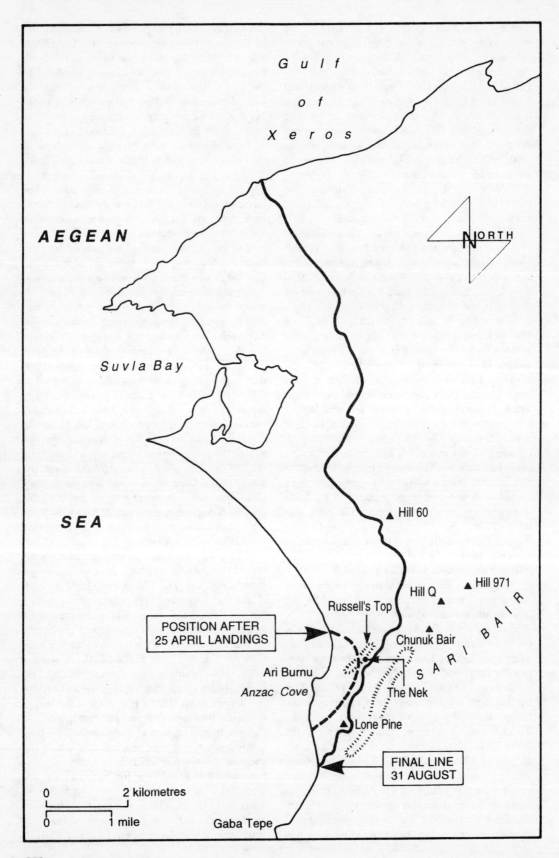

Gulf
of
Xeros

AEGEAN

Suvla Bay

SEA

▲ Hill 60

▲ Hill 971

Hill Q
▲

Russell's Top

POSITION AFTER
25 APRIL LANDINGS

Chunuk Bair

S A R I   B A I R

Ari Burnu
Anzac Cove

The Nek

Lone Pine

FINAL LINE
31 AUGUST

0   2 kilometres

0   1 mile

Gaba Tepe

North

Although Sari Bair Ridge was the highest on that sector of the peninsula, there were still many difficult ridges to traverse before the Narrows were reached. And the Anzac troops had proved that an enemy in control of the high ground did not necessarily mean that the troops lower down the slopes were automatically doomed.

The other debate concerns the role of the British forces landed at Suvla Bay. The troops landed on the night of 7 August. They suffered heavy casualties in the initial phase, especially among junior officers, and this factor, along with lack of water and clear instructions, delayed their advance on the hills overlooking the bay. By the time an organised advance was attempted the Turks were in control of the heights. Any further British advance was blocked. The failure of this force has often been said to have denied the Anzacs the support they needed to succeed. This scenario misunderstands both operations. As has been shown, the Anzac attack had not even come close to success. In addition, it was the main task of the Suvla force to establish a base for the winter. Then they were to secure the overlooking ridge. Only when they had accomplished these tasks were they to turn their attention to Anzac. In fact there was never any chance — even if these objectives had been taken with expedition — that troops in sufficient strength would have been available to negotiate the tortuous country between Suvla and Hill 971 and make a material contribution to Birdwood's operations.

The attempt to capture the Sari Bair Ridge in early August was really the last hurrah of the Gallipoli expedition. There was a great deal of confused fighting around Hill 60 at the end of August — badly handled and causing high casualties for a feature of dubious tactical value. But there were to be no more large attacks.

Back in London the Dardanelles Committee debated the possibility of evacuation. In October they replaced Hamilton with General Sir Charles Monro who recommended that the whole force be removed. Hesitation in the face of this resolute advice led to Kitchener being sent out to confirm that Monro was correct. After further hesitation he did. Once more the politicians baulked — this time because of a naval opinion that after all they could force the straits unaided. This was rejected for the nonsense it was and it was finally decided in December to withdraw the troops. The wisdom of this decision had already been confirmed by a great storm which struck Anzac in November. Gales, sleet and finally snow caused over 3000 cases of frostbite and trench foot, washing away some of the

trenches and causing a great deal of discomfort and misery.

The evacuation, which was largely planned by Birdwood's Chief of Staff, Brigadier-General C. B. B. White (q.v.), has been hailed as the best-executed operation of the campaign. It was certainly a masterpiece of planning. The troops were removed over a period of several weeks, various ruses being employed to convince the Turks that the line was still solidly held. Eventually, on 19 and 20 December, the evacuation of Anzac and Suvla was complete. There was hardly a casualty. Whether the success of the operation was due to the planning or whether the Turks, remembering their experience of 19 May, were reluctant to attack in the open an enemy that was in any case departing, cannot be stated for certain. Whatever the reason the Gallipoli experiment was at an end. The Gallipoli operation cost 26 111 Australian casualties of whom 8141 were killed. In addition New Zealand lost 7571 men of whom 2431 were killed. (Britain lost a total of 120 000 casualties at Gallipoli and the French 27 000. The Turkish total is very speculative — probably about 220 000 with a much higher percentage killed.)

To what end did these men die? It has to be said that Gallipoli had no influence on the course of the war as a whole. More depressing still, even if the expedition had succeeded in its aims, it is doubtful if the war would have been shortened by a single day. The problem was that the premises upon which the operation was based were very dubious. The planners never doubted that if the fleet got through to Constantinople the Turks would surrender. There seems to be no evidence for this. The Turkish government was certainly making every sign of defending its capital. What would have happened had the fleet arrived and the Turks fought on? Could the fleet have been supplied? Would the British have reduced Constantinople to rubble?

Even had the city fallen little need have followed. The Turkish government might have decamped into the interior of Anatolia and fought on. In this scenario would the Balkan States have been so impressed by the fall of Constantinople that they would have rallied to the Allied side? It seems unlikely. It is far more likely that they would have been impressed by the series of mighty defeats inflicted on the Russian armies to their near north by the Germans in this period.

Even had the Balkan States joined the Allies, their accession would have been a dubious advantage. Their peasant armies were under-equipped (and the Allies had no surplus equipment with which to supply them), ill-trained, poorly led and fractious to a degree that must have placed any long-term cooperation between them in doubt. Nor were communications along the Danube sufficient to sustain large armies for considerable periods of time. In any case the Austrians had the line of the Alps to sustain them in the unlikely event of these peasant levies driving them back. Finally, none of this had any implications for the German Army on the Western Front. This army was the main engine of the war for the central powers and the Western Front was the main theatre. The Germans would have to be defeated in north-western Europe before the war could end.

As far as the Anzac troops were concerned, the only thing to be said for the Gallipoli operation was that it kept them away from the Western Front for at least a year. Had they arrived in France in early 1915 their casualty list (as operations around Pozières would subsequently prove) would certainly have been much higher than it was at Gallipoli.

GANNET, FAIREY (3-crew carrier-borne anti-submarine aircraft [AS4]). Wingspan 54 feet 4 inches; length 43 feet; armament 2 torpedoes or 2 × 1000-pound mines, 16 × 60-pound rockets; maximum speed 299 m.p.h.; range 662 miles; power 1 × Armstrong Siddeley Double Mamba 3035 effective h.p. coupled turbo prop engine.

The Gannet was the first aircraft equipped with a double airscrew unit, which gave it the qualities of a twin-engine aircraft in a single-engine configuration. When on patrol one propeller and half the Double Mamba engine could be shut down for increased economy, while in combat both engines and propellers would be used. The double-propeller design also offered advantages over conventional twin-engine machines when used on aircraft-carriers. The Gannet was used by No. 816 and 817 Squadrons Fleet Air Arm aboard the aircraft-carrier HMAS *Melbourne* (q.v.) for anti-submarine warfare (q.v.) from 1955 to 1967 when it was replaced by the Grumman Tracker (q.v.).

GARDEN ISLAND in Sydney Harbour, associated with the navy since White settlement began, became an imperial naval depot after land was dedicated for that purpose in January 1865 and June 1866. The dockyard and other facilities were completed by 1896, then the Commonwealth was given use of the island from 1 July 1913. During the First World War the dockyard was used by 852 ships, but afterwards it declined and only three ships were built there in the next two decades. In 1929 the Privy Council found in favour of New South Wales in a dispute between that State and the Commonwealth over ownership of the island, but in 1939 the Commonwealth resumed the land under wartime regulations before buying it toward the end of the war. The Captain Cook Graving Dock, which joined the island to the mainland and is large enough to accommodate four destroyers at once, was begun in 1940 and opened on 24 March 1945. Today the dockyard is part of Australian

A Gannet, both propellers turning, about to land on HMAS *Melbourne*. (AWM 301040)

Defence Industries, a registered public company created by the government in May 1989 to replace the Office of Defence Production.

Tom Frame, *The Garden Island* (Kangaroo Press, Sydney, 1990).

GAS, FIRST WORLD WAR The use of chemical weapons in warfare had been outlawed by the Hague Convention of 1899. Despite being a signatory to this convention the Germans introduced chemicals into warfare in 1914 when they used an irritant gas against the French in October of that year. They followed this by using the same tear gas on the Eastern Front in February 1915. The first use of poisonous gas occurred in April 1915, when 150 tonnes of chlorine were released in the form of a poisonous cloud against the French and Canadian troops in the Ypres salient. The British retaliated by employing chlorine against the Germans at the Battle of Loos in September 1915.

Until 1916 gas was always employed in the form of a cloud. The gas was transported to the front in liquid form in cylinders and then discharged under pressure to convert it into gas. In this form it was a treacherous weapon, as the British found at Loos where the wind blew the cloud back on their own troops. To protect the troops against cloud gas both sides developed respirators fitted with chemicals that filtered out the dangerous substances in the gas. The early types were primitive, could only be worn for short periods, and made fighting very difficult.

The next development came in 1916 when artillery shells were filled with gas instead of high explosive or shrapnel. This enabled gas to be delivered to specific locations on the battlefield, to be delivered well behind the front line and to be used in such tasks as disrupting enemy artillery fire when it was found that gunners had the greatest difficulty in fighting effectively in respirators.

As the war progressed different and more deadly types of gases were employed. Both sides developed phosgene in 1916 which was more lethal than chlorine. Bromine, arsenic and cyanide were also commonly used substances. In 1917 the Germans began to use a new substance, dichlor-ethyl-sulphide, commonly known as mustard gas. It was difficult to detect, caused severe blistering on exposed skin and was persistent, contaminating the ground for days. (The Allies did not develop mustard gas until September 1918.)

The AIF encountered gas in every major action from its arrival in France until the end of the war. Gas was first unleashed against the 2nd Division near Messines in June 1916 but there were no serious casualties. During the Somme campaign the Australians occasionally suffered from gas shelling, notably at Pozières. Nevertheless casualties were low

— 156 of which 10 were fatal. In the first six months of 1917 gas casualties rose ten-fold. At Bullecourt they were deluged on several occasions with phosgene. Then at Messines the 3rd Division were caught in a random German gas bombardment of a wood as they moved up to attack. In all 1374 casualties were suffered of which 310 were fatal.

Mustard gas had first been used against the British during the preliminary bombardment phase of the Third Battle of Ypres. So when the Australians came into the line in September precautionary measures in the form of unproved respirators and protective clothing were already in place. Nevertheless, a much increased volume of gas (not only mustard but arsenic-based substances) was used by the Germans in this campaign and Australian gas casualties rose accordingly — 3702 casualties, 700 deaths. The culmination of gas warfare came in 1918. The Germans used it freely against the Australians in the north at Ypres and Messines and later on the Somme at Villers-Bretonneux. When the allies moved to the offensive the Australians suffered heavily when they tried to penetrate the German defences at the Hindenburg Line in September and October. Casualties in this year were 15 757 of whom 1915 were killed.

To disseminate information about gas-warfare, officers were appointed to each division and eventually each brigade. The medical officers were of outstanding ability according to the Official History; they helped the troops to employ efficient preventive measures and played a major role in reducing casualties. The AIF employed gas in all of its offensives. As the balance of resources tilted in favour of the Allies on the Western Front, the Germans received four gas shells for every one they fired. Particularly fierce gas bombardments were carried out by Australian batteries at Hamel, Amiens and the Hindenburg Line.

The aftermath of gassing for a soldier could be protracted and painful. The number of cases specifically attributed to gas was never high: 729 in 1916, and 662 in 1939. Gas, however, no doubt played a part in many of the diseases recorded as 'pulmonary' (3109 in 1926, and 9348 in 1938). By 1939 this category accounted for the largest number of cases treated, exceeding even general wounds.

GAS, SECOND WORLD WAR After the widespread use of gas as an adjunct to military operations in the First World War it was generally expected that this experience would be repeated in any future war. In Europe Hitler was thought to have various gases for use against civilian populations and this led to the universal distribution of gas masks in countries such as France and Britain. In

Australia attention focused on the Japanese. Operationally the army prepared a two-pronged approach to gas warfare. During 1941 and 1942 five chemical warfare companies were raised. Their duties were to enact preventive measures should the Japanese use gas, and to prepare for offensive chemical measures should this become necessary. These measures largely consisted of preparing 4.2-inch mortars for mustard-gas shells, which were available from Britain if required. The most controversial aspect of gas warfare in Australia arose from a perceived need to test the effect of mustard gas on army volunteers. The tests were conducted by a British scientist from the British Porton chemical warfare establishment, Major F. Gorrill, and were designed to train personnel in treating casualties and in preventive measures. These tests were conducted at Townsville along similar lines to a long series of tests at Porton during the interwar years. In all about 134 volunteers took part in the tests, which were fully supported by the Australian government. Gorrill's tests were designed to illuminate the value of auto-gas clothing, the efficacy of protective ointments and the ability of human skin to recover from mustard-gas burns. The results were unexpected. Under the high temperature and high humidity conditions prevailing at Townsville mustard gas was found to be at least four times more toxic than under the experimental conditions at Porton. The result was that many of the volunteers suffered severe burns, nausea and vomiting. A number had to be admitted to hospital for extended treatment. After the tests a protest was made by Colonel A. J. L. Wilson, the Assistant Secretary of the Department of Defence, to his superior Frederick Shedden (q.v.). Wilson was, however, overruled, the information produced by the tests being thought more significant than the effects on the volunteers. The findings identified the increased toxicity of mustard gas in the tropics and found British protective clothing not only to be useless but also toxic, and it was determined that troops in protective clothing could barely function under battle conditions. The irony was that the Japanese never developed mustard-gas capability and therefore the only victims of gas warfare in the Pacific were the Australian volunteers.

Bridget Goodwin, The top secret chemical warfare field trials carried out in North Queensland by Australia in association with Britain and the United States during World War II (PhD dissertation, James Cook University, 1994).

GASCOYNE, HMAS see **RIVER CLASS FRIGATES**

GAWLER (I), HMAS see **BATHURST CLASS MINE-SWEEPERS (CORVETTES)**

GAWLER (II), HMAS see **FREMANTLE CLASS PATROL BOATS**

GAYUNDAH, HMQS see **QUEENSLAND MARINE DEFENCE FORCE SHIPS**

GEAKE, Squadron Leader William Henry Gregory (23 February 1880–14 March 1944). Having migrated to Australia from England as a young man, Geake enlisted in the AIF as a private in January 1916 and was sent to the 13th Battalion. In the middle of the year he was posted to RMC Duntroon to undertake the short commissioning course run through the Officer Training School for AIF officer candidates. Gifted with an inventive mind and practical skill, evidenced by the more than 120 patents for which he applied over the course of his adult life, while at RMC Duntroon Geake assisted Mr Alfred Salenger in the design of a bomb-thrower which was taken up by the Ministry of Munitions in London. Geake went to Britain at the end of that year, and in March 1917 took up a post as a lieutenant on AIF Headquarters in Horseferry Road with responsibility for a new inventions research section. He was awarded the Albert Medal (precursor of the GC) for saving a number of men after an explosion in one of the section's facilities in September 1917. He was injured severely during a weapons demonstration the same month, and spent several months convalescing. He was awarded the MBE for his contribution to the war effort. Between the wars he worked as an industrial research engineer. In the Second World War he lied about his age and joined the RAAF, and was involved again in assessing inventions and experimental work at RAAF Headquarters in Melbourne.

GEELONG, HMAS see **FREMANTLE CLASS PATROL BOATS**

GELLIBRAND, Major-General John (5 December 1872–3 June 1945). Educated in England and Germany, Gellibrand entered the Royal Military Academy, Sandhurst, in 1892 and passed out a year later at the top of his class. Commissioned as a second lieutenant into the South Lancashire Regiment, he served with its 1st Battalion in the Boer

Major-General J. Gellibrand breakfasting with officers in a shell hole near Pozières, France, 1 August 1916. From left to right around table: Major E. C. P. Plant, Lieutenant D. N. Rentoul, Gellibrand, Lieutenant-Colonel R. Smith, Lieutenant Stanley Savige, Captain R. H. Norman, Lieutenant W. R. Gilchrist, Lieutenant J. Roydhouse. (AWM EZ0075)

War, taking part in the relief of Ladysmith. In May 1901 he transferred as a captain to the 3rd Battalion, the Manchester Regiment, and served on St Helena in 1902–03 in charge of the detachment guarding Boer POWs. When the 3rd Battalion was disbanded in 1906, Gellibrand was selected to attend the Staff College, Camberley, graduating in 1907. He was posted on the staff to Ceylon, but reductions in the army led to his being placed on half-pay; he resigned his commission and returned to Tasmania, his birthplace, in June 1912.

At the outbreak of war in 1914, the commander of the AIF, Major-General W. T. Bridges (q.v.), appointed Gellibrand as Deputy Adjutant and Quartermaster-General on the Headquarters of the 1st Australian Division. He was one of a small handful of Australians with formal staff college training, making him a valuable asset in the hastily raised and woefully under-trained AIF. The official historian, C. E. W. Bean (q.v.), thought it 'a constant wonder' that the British Army, itself very short of trained staff officers, should have let Gellibrand go in the first place.

Gellibrand distinguished himself on Gallipoli, organising the beach parties and supervising the landing of ammunition and reinforcements and their dispatch to the tenuously held Australian positions on the ridges. He was wounded twice within the first three weeks, was evacuated to Egypt the second time but was back on Gallipoli by the end of May, and was mentioned in despatches and awarded the DSO. In August he became Deputy Assistant Adjutant and Quartermaster-General to the 2nd Division, and in December was promoted to lieutenant-colonel and given command of the 12th Battalion.

His period of unit command was short-lived, because in the reorganisation and expansion of the AIF undertaken in Egypt in early 1916 he was promoted again and given command of the 6th Infantry Brigade, which he led on the Western Front. He was again wounded in May, but took part in the fighting at Pozières and Mouquet Farm in 1916, at Bapaume in early 1917, and then at Bullecourt in May that year.

Bullecourt was a disaster, partly at least of the Australians' own making. Despite the fact that his own brigade had performed well, Gellibrand asked to be relieved of his command. Historian Arthur Bazley has coyly suggested that this arose because of 'misunderstandings between himself [Gellibrand] and the staff of the 2nd Division', while Bean makes no reference to it at all. Gellibrand's own papers make it clear that he was unhappy with the command of the 2nd Division and with the conduct of the 5th Brigade. He was clearly unhappy with the preponderant influence of British officers in the division, many of whom he regarded as incompetent. The commander of the AIF, General Sir William Birdwood (q.v.), tried to mollify him, and then sent him to England, where he overhauled the AIF depots and reinforcement training system with characteristic vigour. He returned to France in November 1917 and was given command of the 12th Brigade. When General John Monash (q.v.) was appointed to command of the Australian Corps in May 1918, Gellibrand succeeded him as GOC of the 3rd Division, which he led for the rest of the war. For his service in France he was awarded a bar to his DSO, the CB and later the KCB, and was mentioned in despatches several times.

After the war he served briefly as public service commissioner in Tasmania, and then became chief commissioner of police in Victoria from 1920 to 1922. Neither experience was a happy one, the respective State governments on each occasion shelving his recommendations for reform of the organisations he had been appointed to head. In 1925 he was elected to federal parliament as a Nationalist member for the seat of Denison, but he was defeated in the elections of 1928 and 1929. He was centrally involved in the foundation of Legacy (q.v.).

Gellibrand was one of the finest operational commanders produced by the AIF. Deeply respected by officers and men alike, his command of the 3rd Division in the final months of the war was exemplary. His willingness to speak his mind did not endear him to Birdwood, and his habit of dressing like the other ranks bemused many British officers although it commended him to Bean, who treats him very sympathetically in the official history.

GERALDTON, HMAS see **FREMANTLE CLASS PATROL BOATS**

GERANIUM, HMAS see **FLOWER CLASS MINE-SWEEPING SLOOPS**

GERMAN NEW GUINEA On the outbreak of the First World War the Australian government received a request from Britain to destroy the German wireless network in the western Pacific and to take possession of German New Guinea, a German colony since 1884. A motley force of Royal Australian Naval Reserve men, infantry and machine-gun sections was raised under the command of Colonel William Holmes (q.v.). It was dispatched from Syd-

Map 12 German New Guinea, 1914. *Inset* Blanche Bay, site of the first Australian landings.

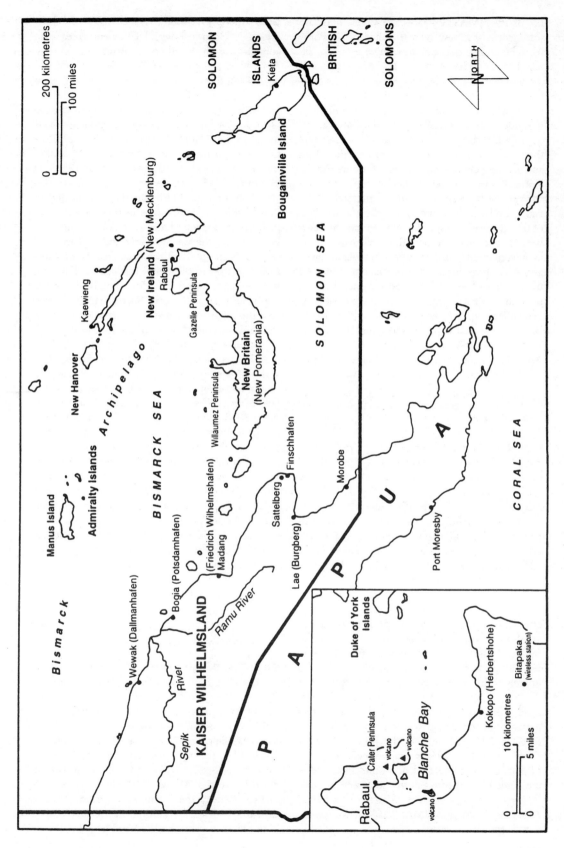

ney on the HMAS *Berrima* on 19 August 1914 and was escorted to New Britain (called Neue Pommern by the Germans) by the cruiser HMAS *Sydney* (I) and the battle-cruiser HMAS *Australia* (I) (qq.v.). The first landings took place on 11 September at Blanche Bay. After some initial German resistance which caused 10 casualties (6 dead, 4 wounded) the wireless station at Bitapaka was destroyed. Rabaul capitulated two days later and the surrender of all German troops on the island and on New Guinea was accepted from the Acting German Governor Herr Haber.

A military government for the new territories was set up by Holmes and by October departments of the Treasury, Works, Lands and Surveys, and Law were functioning. The period of military government continued until 1921 when Australia received a mandate from the League of Nations to govern the country. After Holmes had returned to Australia the military governors were: Colonel S. A. Pethebridge (q.v.) (January 1915–October 1917); Brigadier-General G. J. Johnson (April 1918–May 1919); Brigadier-General T. Griffiths (May 1919–April 1921); and General E. A. Wisdom (April 1921–June 1932).

S. S. Mackenzie, *The Australians at Rabaul,* vol. 10 of the *Official History of Australia in the War of 1914–18* (Angus & Robertson, Sydney, 1927).

GETTING, Captain Frank Edmund (30 July 1899–9 August 1942) was one of the original intake to RANC Jervis Bay in 1913. He served on HMS *Glorious* as a midshipman in 1917–18, and thereafter specialised in the new submarine branch, completing the commanding officer's course in Britain in 1926 and then serving on HMAS *Oxley.* Getting completed the course at the Royal Naval Staff College in 1933, and then attended the Imperial Defence College (q.v.) the following year. He was given command of HMS *Kanimbla* at the beginning of the war. An RN armed merchant cruiser, the ship was largely manned by Australian naval reservists, some of whom were under the misapprehension that they were not liable for overseas service in wartime. 'I have had to correct this impression in quite certain terms', he reported to the Naval Board, and the official historian, G. Hermon Gill (q.v.), commented drily that in this period of his captaincy, *Kanimbla* 'was not a happy ship'. In 1941–42 he held the post of Deputy CNS, and in June 1942 was appointed captain of HMAS *Canberra.* He died of wounds received at the Battle of Savo Island in the Solomons, a major Japanese tactical victory in which a number of Allied ships, including HMAS *Canberra*, were sunk, with heavy loss of life.

GILL, (George) Hermon (8 March 1895–27 February 1973). Born in England, Gill went to sea as an apprentice with the Aberdeen White Star Line in April 1910, and served during the First World War on ships of that line taken up from trade for use as troop transports. He migrated to Australia in 1922 and worked for the Australian Commonwealth Government Line for seven years. Appointed lieutenant in the RAN Volunteer Reserve in 1927, he turned to journalism and joined the literary staff of the Melbourne *Argus & Australasian* in 1933. He was mobilised for war service in September 1939 and in March the following year was posted to Navy Office in Melbourne in the Naval Intelligence Division, with duties in censorship and publicity. During the war he edited the *HMAS* series of annuals and in 1943, with the creation of a historical section in the Naval Intelligence Division, became naval historian with responsibility for the Naval Historical Records Section. Demobilised after the war, he remained in the naval reserve with the rank of commander until 1953, writing the official history of the RAN in the Second World War. He returned to journalism, and for 20 years was editor of the South Melbourne *Record.* His official history appeared in two volumes, *Royal Australian Navy 1939–1942* (1957) and *Royal Australian Navy 1942–1945* (1968). His background in journalism lent readability to his history, while he was more able than most of the official writers to place his subject within the global context in which it operated. On occasions he exaggerates the importance of the RAN in Australia's war effort, and although his argument that the Pacific War was decided by sea power is broadly correct, the decisive exercise of that sea power belonged to others. His history, while rarely descending to special pleading, is the most clearly advocatory of the Australian histories, explained in part by the need to justify prewar naval policy and the relatively minor role played by the RAN.

G. Hermon Gill, *Royal Australian Navy*, 2 vols (Australian War Memorial, Canberra, 1957, 1968).

GILLISON, Douglas Napier (12 February 1899–1 November 1965). Born in England, Gillison was the son of the Reverend Andrew Gillison, the first Presbyterian chaplain appointed to the AIF in 1914, who was killed serving with the 13th Battalion on Gallipoli on 22 August 1915. The younger Gillison became a journalist, and was a sub-editor on the *Age* between 1937 and 1941, and on the *Argus* between 1941 and 1942. He served as general president of the Australian Journalists Association in 1936-37. He enlisted in the RAAF as a pilot officer in February 1942, and served in the air force for the remainder of the war, including stints as a public relations officer in New Guinea in 1942–43 and at Overseas Headquarters RAAF in London in 1944–45. In April 1945 he returned to civilian life. In 1945–46 Gillison was press secretary to the Minister for Air, A. S. Drakeford, and later became public relations officer for the Department of Civil Aviation from 1953 to 1959. In the latter year he was appointed director of the Australian News and Information Bureau in London, retiring from that post in 1964. Regarded as Australia's foremost aviation correspondent, Gillison was appointed by the official historian, Gavin Long (q.v.), to write the first volume of the air series of the official history, published in 1962 as *Royal Australian Air Force 1939–1942*. A long and enormously detailed book, it dealt mainly with the organisation and prosecution of the air war against the Japanese from December 1941 to the end of the fighting in Papua in early 1943. Its treatment of the genesis and workings of the Empire Air Training Scheme (q.v.) is felt by some historians to be relatively uncritical of a highly contentious episode.

GLADSTONE, HMAS see *FREMANTLE* **CLASS PATROL BOATS**

GLASGOW, Major-General (Thomas) William (6 June 1876–4 July 1955). Glasgow enlisted in the Wide Bay Regiment in the Queensland colonial forces as an enthusiastic teenage soldier in 1893, and represented the colony at Queen Victoria's diamond jubilee in London. He was commissioned in the Queensland Mounted Infantry in 1899 and served in South Africa with the first Queensland contingent, taking part in the relief of Kimberley and the occupation of Bloemfontein. He was mentioned in despatches and awarded the DSO. He continued militia soldiering, organising the 13th Light Horse Regiment in the Gympie district in 1903. By the outbreak of war in 1914, already a major in rank, he volunteered for the AIF and was appointed second-in-command of the 2nd Light Horse Regiment. He landed at Gallipoli on 15 May with the light horsemen sent to reinforce the depleted units of the 1st Division, and took command of Pope's Hill from Lieutenant-Colonel F. M. Rowell, cousin of S. F. Rowell (q.v.), later CGS. In the August offensive he led an assault by 200 of his men in which all but 46 were killed or wounded; Glasgow withdrew carrying a wounded man. The following day he was given command of his regiment, which he exercised until after the evacuation in December.

With the reorganisation of the AIF in Egypt early in 1916, Glasgow was appointed to command the newly raised 13th Infantry Brigade, which he led in the battles of Pozières, Messines, Passchendaele and Dernancourt. For his leadership in the fighting of 1916–17 he was awarded the CMG and the CB. Together with H. E. Elliott's (q.v.) 15th Brigade, Glasgow's brigade took the town of Villers-Bretonneux in a successful counter-attack on 25 April 1918, an action later described by Monash as the turning point of the war (which was an exaggeration), and which was conducted in a manner that defied the orders of their British superiors. Ordered to attack at 8.00 p.m., Glasgow demurred, in forceful terms ('If it was God Almighty who gave the order, we couldn't do it in daylight'), and the attack was carried out two hours later. At the end of June he was appointed to command of the 1st Australian Division, which he led through the rest of the AIF's campaign on the Western Front. He returned to Australia in mid-1919, was awarded the KCB, and demobilised in August that year. He was mentioned in despatches nine times, and also received a number of foreign decorations.

C. E. W. Bean (q.v.) described him as 'the most forcible of the three brigadiers of the 4th Division', while Monash thought he succeeded 'not so much by exceptional mental gifts, or by tactical skill of any very high order, as by his personal driving force and determination'. John Gellibrand (q.v.), his contemporary in command of the 3rd Division, thought he possessed 'cold pluck and staying power … free from personal ambition … his methods were simple and direct, shrewd, sound commonsense, an easy man to work with'. He advocated the extension of the death penalty for Australian soldiers in 1917 in exceptional cases. In 1919 he was elected to the senate as a Nationalist candidate in Queensland, and served as a minister in the Bruce–Page government between 1926 and 1929. Between 1927 and 1929 he was minister for Defence. He was deputy leader of the Opposition in the senate during James Scullin's short-lived Labor government, but lost his seat in the 1931 election. In December 1939 R. G. Menzies appointed him High Commissioner to Canada, where he played an important part in overseeing the implementation of the Empire Air Training Scheme (q.v.), and took care of Australian servicemen's interests. He returned to Australia and his pastoral concerns in 1945.

GLOSSOP, Vice-Admiral John Collings Taswell (23 October 1871–23 December 1934). Born at Twickenham, England, Glossop served with the RN in Australasia, the Pacific and England from 1887 to 1913. Glossop's appointment to the command of HMAS *Sydney* (I) (q.v.) came after he was reported by the Australian naval representative in London to be anxious to command an RAN ship and sympathetic to 'the Australian Navy movement'. In 1914 he commanded the *Sydney* against the *Emden* (q.v.) in the RAN's first sea battle, but he became embroiled in a battle of a different kind in 1919 after he, as captain-in-charge of naval establishments, Sydney, presided over the court martial of mutineers from HMAS *Australia* (I) (q.v.). Controversy over the severity of the sentences and attacks on Glossop in federal parliament may have contributed to his failure to be appointed Australian naval representative in London despite being recommended for the post by the Navy Board (q.v.). Glossop went back to the RN in October 1920.

GOBLE, Air Vice-Marshal Stanley James (21 August 1891–24 July 1948). Born at Croydon, Victoria, Goble tried to enlist in the AIF at the outbreak of war in 1914 but twice failed the medical examination. He paid his own passage to England, and in July 1915 joined the Royal Naval Air Service (RNAS) as a trainee airman with the rank of temporary flight lieutenant. His wartime career was distinguished: he was awarded the French Croix de Guerre and the DFC in October 1916 and the DSO in February 1917, he was appointed MBE in 1917 and OBE in 1918, and was twice mentioned in despatches. With the merger of the RNAS and the RFC into the RAF in April 1918, he ended the war as a major in the RAF.

He returned to Australia in 1919 and with the temporary rank of wing commander served as the RAN's representative on the joint army–navy board to advise on the establishment of an Australian air force, subsequently being nominated by the navy for the position of Chief of Air Staff. When that position went instead to Wing Commander Richard Williams (q.v.), there developed a rivalry that was to poison the senior levels of the RAAF for the next 20 years. The two men were simply unable to work together. Goble attended a training course in England in 1921, and returned in 1922 to become acting CAS in Williams' absence, during which time he pressed for the creation of a separate fleet air arm, much to the displeasure of Williams who had the project suspended on his return. In 1924 Goble won the prestigious Britannia Trophy and was appointed CBE for the first around-Australia flight. With Williams back in Australia, Goble was posted to attend the Imperial Defence College (q.v.) and the RAF Staff College in 1926 and to act as Air Liaison Officer at the Australian High Commission from May 1926 to September 1927. He was pro-

moted to group captain in 1928, and made temporary air commodore in 1932. He was again acting CAS in 1932–33 while Williams was in the United Kingdom, and was then seconded to the RAF in 1935–37 to a position in the Air Ministry and subsequently to command of a bomber group. Promoted to temporary air vice-marshal in 1937, he returned to Australia in 1938, and was able to use the Ellington (q.v.) report on deficiencies in the RAAF to help discredit Williams, who was removed as CAS. Goble assumed the position in an acting capacity, his own position as Second Air Member for Personnel being filled by an RAF officer, Air Commodore J. C. Russell. Relations between Goble and Russell quickly broke down, and Goble had little sympathy for the government's enthusiasm for the Empire Air Training Scheme (q.v.), preferring instead to concentrate on local defence and on the provision of a self-contained RAAF expeditionary force. When he tried in December 1939 to resign both his commission and his position as CAS, the government refused the first request. Instead he was replaced as CAS by an RAF officer, Air Chief Marshal Sir Charles Burnett (q.v.), and, in an ironical twist, was posted to Canada as Australian Liaison Officer to the EATS, where he remained until 1945. He was retired by the Air Board in April 1946.

Goble's undoubted ability and courage as a wartime pilot were not translated into the capacities necessary at the senior administrative levels of the infant RAAF. His relations with Williams were a serious impediment to the unity of the new air force, and as long as Goble was thought to be unduly disposed towards the naval position on air matters, he was unable to develop any strong position on air power. When he did become CAS following Williams' demise, he had little influence on government defence policy.

GODLEY, General Alexander John (4 February 1867–6 March 1957). Born in England, the eldest son of a regular soldier, Godley was commissioned in the Royal Dublin Fusiliers in 1886 and served with the mounted infantry in the Boer War. In 1910 he was promoted to major-general and given command of the New Zealand Territorial Force. He was created KCB in 1913 and KCMG the following year. On the outbreak of war Godley was appointed to the command of the New Zealand Expeditionary Force. In Egypt he became GOC, NZ and A Division, a composite force consisting of two New Zealand brigades and one Australian brigade (the 4th under Monash). At the landing at Gallipoli Godley had little scope to exercise much influence on the battle, that burden falling in the first instance to the junior officers and the men. On the peninsula Godley was indefatigable in touring the trenches but his stern unbending attitude did little to commend him to the troops.

Godley has been much criticised for losing grip on the northern part of the battle in the August attack and allowing the operation to lose impetus. This is probably unfair. No commander in this period could exercise much control of his troops once they had gone over the top and this was especially true in the twisted ravines to the north of the Gallipoli perimeter. The fact is that the plan, although imaginative, was too ambitious and was almost bound to fail whatever the quality of the command.

In June 1916 Godley took the newly formed II Anzac Corps to France. His first major command took place at Messines in June 1917 when II Anzac consisted of the New Zealand Division and 3rd Australian Division with 4th Australian Division in reserve. Later he commanded II Anzac at Broodseinde and Poelcappelle in the Third Battle of Ypres. In 1918 he took charge of XXII Corps which had no Australian troops (except a few light horsemen as corps cavalry). He remained administrative head of the New Zealand Expeditionary Force until the end of the war.

It is difficult to assess Godley's role as corps commander on the Western Front. He was one of many who held this level of command at a time when the major plans and the framework in which operations would take place were being decided by the army commanders. Godley was fortunate to fight his first two battles under General Sir Herbert Plumer of Second Army, who was noted for his meticulous planning. At Messines Godley's corps performed well and it was not due to his command that heavy casualties were suffered in the concluding phase of the battle. (General Sir Douglas Haig, against Plumer's wishes, had extended the objective to be captured thus leading to much confusion in the artillery support given to the troops.) At Broodseinde on 4 October 1917 Godley's force fitted well into the structure of the Second Army and achieved the greatest advance made that day.

Godley's final battle with Australian troops was his least successful. (He commanded British divisions in II Anzac Corps at the equally unsuccessful Battle of Poelcappelle on 9 October.) At Passchendaele on 22 October 1917, Plumer insisted on attacking in adverse conditions. The result was heavy casualties and no ground gained. This was hardly Godley's fault but there is no record of his protesting the decision. Once more he acted as a cog in a much larger machine. After the war Godley was appointed C-in-C, British Army of the Rhine, 1922–24, and Governor and C-in-C, Gibraltar, 1928–33. When he was created GCB in 1928 he chose a New Zealand infantry soldier as one of the bearers on his coat of arms.

Alexander Godley, *Life of an Irish Soldier* (John Murray, London, 1939).

GORDON, Major-General Joseph (José) Maria (Jácobo Rafael Ramón Francisco Gabriel del Corazón de Jesús y Prendergast)

(18 March 1856–6 September 1929) was born in Spain, but little is known about his early years other than that he left about 1867 and moved to Scotland. He was educated at Woolwich and was commissioned into the Royal Artillery in February 1876. He was stationed in Ireland, but resigned his commission on grounds of ill health in 1879. In search of a healthier climate, he migrated to New Zealand and became a drill instructor in the armed constabulary. He tried unsuccessfully for a commission in Victoria, was employed for a time as a constable in South Australia, and in January 1882 was commissioned into the permanent artillery of that colony. By 1892 he had been promoted to lieutenant-colonel, and in the following year became commandant of the colony's military forces. He served in South Africa from January 1900 as chief staff officer for Overseas Colonial Forces. Following Federation Gordon was transferred to Victoria, where he commanded the new Commonwealth Military Forces in that state until 1905, followed by a similar command in New South Wales between 1905 and 1912. He was passed over for senior appointments several times, but in May 1912 became CGS without promotion to major-general in order not to extend his time until retirement. He was on his way to England and retirement when the First World War broke out; he offered his services to Australia, but these were rebuffed, probably because of his age although Australia was not over-endowed with knowledgable regular officers at that stage. Gordon commanded several reserve formations in England between 1914 and 1915, and served with the Army of Occupation in Germany in 1919. He was finally retired from the Australian Military Forces in 1921. He published a memoir in that year. In his service in Australia Gordon was to some extent unlucky. South Australia was a militarily under-developed colony with few opportunities for further advancement, and he became commandant at a time when London was preferring imperial officers on active duty for colonial appointments rather than retired or half-pay officers, who by definition had less current military knowledge. But Gordon was an able and intelligent officer, and he contributed to the early foundation

of military aviation and the setting up of the Lithgow Small Arms Factory.

GORDON, HMVS see **VICTORIAN NAVY SHIPS**

GORTON, John Grey (9 September 1911-   ). Gorton's wartime service in the RAAF as a flight lieutenant ended in 1944 when he was badly wounded, and following the war he entered local politics, becoming President of Kerang Shire Council in 1949 after first being elected to the council in 1947. Gorton was elected to the senate as a Liberal in 1949, and was made minister for the Navy in December 1958, a post he held until the end of 1963. He had a strong interest in defence matters, and was an excellent minister. He took his responsibilities seriously, personally chairing meetings of the Naval Board and talking to responsible officers within the department on a range of issues, and proved a successful advocate for the navy in parliament and cabinet. He enjoyed good personal relations with both the Secretary, T. J. Hawkins, and the CNS, Vice-Admiral Sir Henry Burrell (q.v.), and during his tenure of office the RAN began to procure capital equipment from sources other than Britain.

Gorton held a succession of ministries thereafter, including Education and Science between 1966 and 1968. To the surprise of many, he emerged as leader of the Liberal Party and prime minister after the sudden death by drowning of Harold Holt at the end of 1967. He was prime minister during the most intense period of Australian commitment to the war in Vietnam, but although it seems that he understood that the policies which he had inherited were no longer appropriate, he had little more idea than anyone else on the conservative side of politics of what to put in their place. The war indirectly brought about his downfall in March 1971 when the Minister for Defence, Malcolm Fraser (q.v.), resigned over Gorton's willingness to interfere, as Fraser saw it, in his portfolio by talking directly to the CGS, Lieutenant-General Sir Thomas Daly (q.v.), on matters relating to civic action programs in Phuoc Tuy province. Gorton cast the deciding vote against himself in a vote of no-confidence within the party room, and was succeeded by the lacklustre William McMahon. He held the Defence portfolio for six months in 1971 under McMahon, and retired from his seat of Higgins, to which he had switched from the senate on becoming prime minister, in 1975. In personal style Gorton was somewhat iconoclastic: this trait served him well as Navy minister, much less so as prime minister.

GOVERNMENT AIRCRAFT FACTORY (GAF) see **INDUSTRY**

GRANT, Admiral (Edmund) Percy Fenwick George (?1867–8 September 1952). As a British naval officer, Grant served in Egypt in 1882 and thereafter held various commands afloat, until appointment in 1915 as chief of staff to the second-in-command of the Grand Fleet, Admiral Sir Cecil Burney. In this capacity he was present at the Battle of Jutland in 1916. He was appointed First Naval Member and CNS in Australia in 1919, a post he held until 1921, and was made KCVO in 1920. He was then C-in-C of the Australia Station (q.v.) from 1921 to 1922. In this capacity he was one of three British commanders-in-chief called on to advise the British government on the efficacies of building a naval base at Singapore. He also acted as defence adviser to the Prime Minister, William Morris Hughes (q.v.) at the Imperial Conference in London in 1921. On completion of his tour as CNS he received the official thanks of the Australian government 'for services rendered'. He retired from the RN in 1928, although he returned to service ashore for the duration of the Second World War.

GRANT, Brigadier-General William (30 September 1870–25 May 1939). Educated at the University of Melbourne, Grant was commissioned in the Queensland Mounted Infantry in 1901. He demonstrated an aptitude for the army despite the relatively advanced age at which he took up soldiering, and by 1911 was a lieutenant-colonel in command of the 14th Light Horse Regiment. In March 1915 he was given command of the 11th Light Horse Regiment with the AIF, but the unit was disbanded in Egypt in August and Grant was allotted to the 9th Regiment; when its commanding officer was killed in the heavy fighting for Hill 60 at Gallipoli, Grant took his place. The 11th Light Horse was re-formed in the reorganisation of the AIF in early 1916, again with Grant as its CO. He took part in the Sinai campaign, being awarded the DSO in December 1916, fought at the second Battle of Gaza, and was then promoted and given the 4th Light Horse Brigade just before the third Battle of Gaza in October 1917. It was Grant's brigade which was allotted the task of taking the Turkish defences at Beersheba (q.v.) and seizing the wells. Chauvel selected the 4th Brigade for the task — 'Put Grant straight at it' — but the decision to remain mounted and charge the Turks like cavalry appears to have been Grant's own. He received a bar to the DSO for the battle. He led his brigade throughout the campaign of 1918, was mentioned in despatches four times and awarded the CMG. He returned to his grazing interests after the war, but continued to soldier in the militia until placed on the Retired List in 1928.

GRANT, Douglas (?1885–4 December 1951). From the Bellenden Ker Ranges, Queensland, Grant, an Aborigine, was adopted as an infant by a White couple after his parents were killed, possibly in a police massacre. During the First World War, he enlisted in the AIF in January 1916, but was discharged from the 34th Battalion as it was about to embark for overseas because of regulations preventing Aboriginal people from leaving the country without government permission. He re-enlisted later in 1916, joined the 13th Battalion in France, and was captured by the Germans in April 1917. While being held as a POW in Germany, Grant was studied by German doctors and anthropologists (see ANTHROPOMETRY). He returned to Australia in July 1919 and involved himself in ex-servicemen's affairs after the war but, frustrated by racism, developed an alcohol problem later in life.

GREECE The Greek campaign, in which forces from the United Kingdom, Australia and New Zealand supported the Greeks, was an almost complete misjudgment of ends and means. Apart from its nobility in assisting a small State to resist Fascist and Nazi aggression, it was probably the worst-judged piece of Churchillian strategy of the war. The relatively meagre support given to the Greeks not only curtailed a successful Western Desert offensive under General O'Connor, which could conceivably have secured the port of Tripoli ahead of the arrival of Rommel's German force in North Africa, but was also too small and poorly balanced to provide effective assistance to the Greeks who had become drained of resources during their successful operations against the Italians in Albania, and now had to face the vastly superior Germans. Indeed, the British official history of the campaign states frankly that: 'The British campaign on the mainland of Greece was from start to finish a withdrawal'.

It began on 6 April 1941 when German forces attacked Greece and Yugoslavia simultaneously. It ended on the night 30 April/1 May when the last organised group of Commonwealth troops was evacuated. (Although Crete is also part of metropolitan Greece, the short 10 day campaign on that island is treated separately.) The genesis of the campaign lay in the long-standing British guarantee to

An Australian soldier in Greece with a BREN gun mounted for use against German aircraft, April 1941. Mount Olympus is in the background. (AWM P1166/08)

support Greece if it was attacked without provocation. This had been invoked when Italy invaded Greece on 28 October 1940. The Greeks promptly defeated the initial thrust into their territory through Albania, while the British sent assistance in the form of an air contingent, a weak brigade group to Crete, and a force of anti-aircraft guns.

When it became clear that Germany would support its Italian ally in Greece, Churchill sought to form a Balkan front that would include Greece, Turkey and Yugoslavia. To further this he sent a mission to the area in February 1941. It included Anthony Eden (Foreign Secretary) and the Chief of Imperial General Staff (General Dill). Churchill spoke airily of 'a Balkan front ... comprising about seventy Allied divisions'. In the event, the front did not eventuate. Turkey remained neutral and Yugoslavian resistance to the German assault was overcome in a few days. The dominion governments, Australia and New Zealand, who ultimately provided most of the troops, were not apprised of the venture until planning was well advanced.

When Prime Minister Menzies and Lieutenant-General Blamey (commanding Australian troops in the Middle East) were finally informed, they were each told separately of the proposal, and in such a way that each concurred in the venture. Despite the fact that communication between Menzies and Blamey was less than satisfactory, it appears that Churchill and General Wavell, C-in-C, Middle East Command, deliberately obscured the truth in order to ensure Australian support. As the evidence mounted that it was likely to be a rash and unsuccessful initiative, Menzies sought to have it re-evaluated on 4 March. His uneasiness was met by assurance from the Eden–Dill Mission (joined by Wavell) that the troops would have a 'reasonable fighting chance', and that they were completely confident that the Benghazi front, where German units were now appearing, could be held without interfering with the Greek enterprise. In the circumstances, the belief that Britain knew best was probably natural, given Australia's lack of independent strategic experience and remoteness from the theatre of operations.

It was planned that the force for Greece would consist of the 6th and 7th Divisions (AIF), the 2nd New Zealand Division, a Polish brigade, one armoured brigade and probably a second brigade; a total of 100 000 men. However German success in North Africa, and later the requirement to send a further force to Syria, meant that the 7th Division, the Polish Brigade and the second armoured brigade were never sent. As a consequence, the planned force — already small and ill-balanced for the nature of the commitment — was even further

denuded of combat power, particularly in tanks and aircraft, and never made up its deficiencies before the fighting started.

The original Greek plan had included the premise that, in order to assist Yugoslavia, which it was hoped would became an ally, the key port of Salonika in north-eastern Greece, vital to the Yugoslavs, would have to be held. General Papagos, the Greek C-in-C, therefore intended to hold initially along the Metaxas or Doiran–Nestos Line, that is, virtually along the Bulgarian frontier. As a successful German assault through Yugoslavia would outflank this line and thus open the possibility of a fast-moving armoured drive towards the Plain of Thessaly in south-central Greece, General Sir Henry Maitland Wilson, who had been appointed by Churchill to command the British Commonwealth force (designated 'Lustreforce' by the British and 'W Force' by Papagos), wanted the main defence to be further back, to take advantage of the mountains. A satisfactory agreement was never reached, which highlighted the difference in approach: the Greeks to protect their territory, Churchill's to promote a Balkan Front.

Papagos temporised. Some Greek troops were allotted to Wilson, who was given overall command along the Vermion–Mount Olympus Line (Map 13a). But the essential truth was that with the bulk of the Greek Army engaged against the Italians, the addition of a small British Commonwealth force was insufficient to form a significant reinforcement against the numbers and types of divisions the Germans were about to commit against Greece. Given that the Germans were operating on interior lines (as Map 13a shows they could enter Greece via both the Bulgarian and Yugoslavian frontiers then rapidly move to bolster the Italians on the Albanian front), the nature and extent of their assault was limited only by the terrain and the capacity of the Greek road and rail communications to allow passage of the force.

These facts were perceived by the Australian War Cabinet in late February when in concurring, as it thought, with the commitment of two divisions of the AIF to Greece, it did so on the basis that its consent 'must be regarded as conditional on plans having been completed beforehand to ensure that evacuation, if necessitated, will be successfully undertaken and that shipping and other essential services will be available for this purpose if required'. In other words, the evacuation was being considered even before the advance into the country had begun.

The German assault on 6 April confirmed this pessimistic view. In the north-eastern part of Greece that was to become the British Commonwealth area of operations, there were four Greek divisions on the frontier, and a further two as part of W Force under Wilson in the Katerina–Veria Pass area. The single road and rail link up-country from Athens and the entry port of Piraeus slowed the build up of the force. Only two of Mackay's brigades of 6th Australian Division had arrived when the German attack came. Against this force, Germany had a potential 27 divisions, although only 10 were eventually used in Greece. They were remarkably well balanced. As well as five experienced infantry divisions, there were three armoured divisions and importantly, in view of the terrain, two divisions specially organised and trained for mountain warfare. The combined arms cooperation achieved by the Germans was matched by the Allies only in the last year of the war. In Greece, the German mountain formations very skilfully cleared paths for the faster moving armoured forces to exploit. In the air, the British force had 80 aircraft to meet a German force of 800. Only at sea were the Allies stronger. The convoys of troops from north Africa to Greece were scarcely interfered with. And in a brilliant naval action off Cape Matapan, ships of the British and Australian navies scored such a notable success against an Italian fleet that sea control of the Eastern Mediterranean was not contested by enemy surface forces until the end of the campaign.

The nature of the operations in Greece and the tactics of the land battle were determined by a number of interlocking factors.

- Although General Papagos was appointed to overall command of the Allied force, effective coordination was never realised. Language difficulties, poor liaison, the inadequate equipment of the Greeks and the general unsuitability of Wilson's force for the terrain meant that at best the Allies could offer little more than stubborn resistance to the German assault. Wilson's command structure was clumsy and, while he insisted on retaining tactical command at the beginning, he devolved command of the withdrawal to Blamey at a time when the latter was still concentrating his troops.

- The Greek plan to hold forward on the frontier, while understandable for national morale, meant that, as in France in May 1940, the German armoured spearhead quickly broke through the crust. What could have been a well-developed defensive line from the beginning, was vitiated by a complex series of redeployments when Papagos, who had directed Wilson to deploy in the Vermion–Olympus area, then attempted to shorten the defensive line by withdrawing to the Aliakmon Line, and

at the same time switch two Greek divisions from right to left across Wilson's front. But in fairness to the Greek forces, who fought well with rudimentary equipment and almost no transport, when the main breakthrough came it did not occur in the Greek area but against Wilson's force in the Tempe Gorge south east of Mount Olympus.

- The primitive road and rail system, together with a devastating German air attack on Piraeus Harbour on the first day, meant that the build-up of a strong defensive forward position was not achieved in time. The Allies lost the initiative at the beginning and never regained it. Despite this, and the terrain and the weather (the Australians were digging-in in snow at the Vevi and Veria Passes), the subsequent fighting withdrawal over 480 kilometres, virtually the length of Greece, was skilfully executed. There was always the possibility that the narrow and predictable withdrawal corridor would be cut behind the force by air attack, or by parachute troops — which were used at the Isthmus of Corinth and later more decisively in Crete. To the credit of commanders at all levels, and the troops themselves, the force did not lose its overall cohesion although out-numbered and out-gunned. Moreover, all the logistic needs of the force had to be brought in, with difficulty. The Greek people offered goodwill in abundance, but the cupboard was bare; and there was also a shortage of labour.

- Most disturbing of all was the knowledge that, if Yugoslavia was quickly overcome (which occurred), a German advance from Monastir across the border towards Veve and Florina would enable the German armour to get through to the Upper Aliakmon valley and thus towards the River Pinios and the Plain of Thessaly. If this was achieved, a wedge would be driven between 'Lustreforce' and the Greek forces, and the major Greek supply route to its force facing the Italians would be cut. This was an essential part of the German plan that precipitated a further, final withdrawal to the Thermopylae–Brallos area, and recognition that Wilson's force would have to be evacuated. When the Greek Army of Epirus, on Wilson's left flank, capitulated, it merely confirmed the fact that complete German domination of the peninsula was only a matter of time.

Yet, as in almost every Allied campaign in the early part of the war, the worst mistakes of the politicians and strategists were moderated by the bravery, fighting qualities and sheer dogged determination of the troops. Greece was no exception.

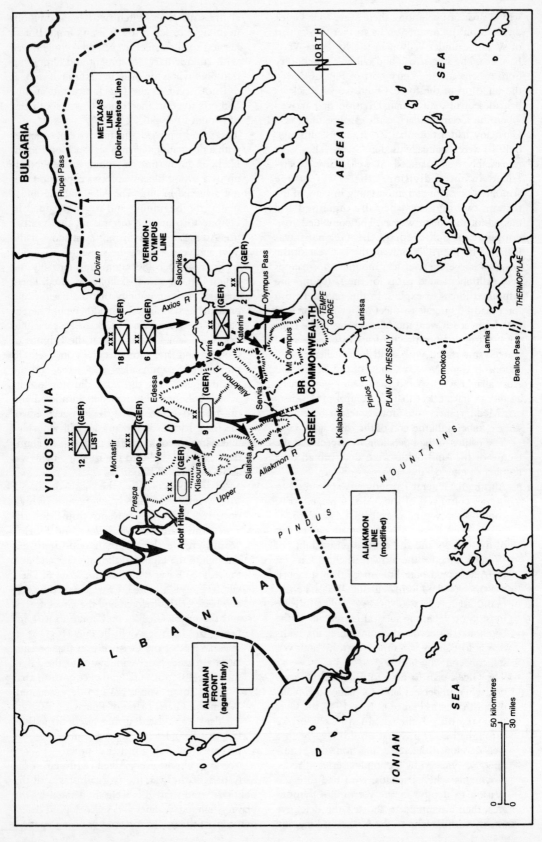

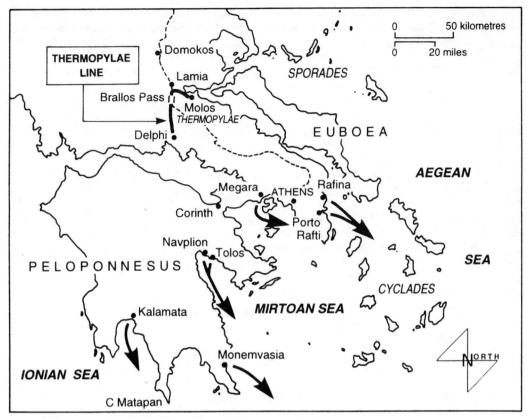

Map 13b Evacuation of Allied forces from Greece, 1941.

General Freyberg's 2nd New Zealand Division was deployed initially in generally open country to the north of Mount Olympus. At one stage two brigades were on a 25 000 metre front. It was clear that the division could not stay where it was and Papagos soon agreed to its redeployment. But valuable time had been lost and they could not prepare a strong position. Thus, when later a German thrust between Mount Olympus and the sea threatened a breakthrough in the Tempe Gorge, the major part of Brigadier Allen's 16th Australian Brigade had to be moved there quickly to restore the position.

The British Commonwealth force was able to do little more than plug leaks in the dam wall. Major-General Mackay with one brigade of his 6th Australian Division was sent almost immediately on arrival to the Veve Pass to prevent a likely thrust from the Monastir area. His force included the 19th Brigade (Brigadier Vasey [q.v.]), part of the British 1st Armoured Brigade with worn-out tanks, as well as medium and anti-tank artillery.

The Germans used two divisions against this force and the sustained pressure compelled it to

Map 13a Operations in northern Greece, 1941.

withdraw, with most of the infantry forced to walk 80 kilometres through the mountains. In turn, the new line through Servia–Mount Olympus was not held for long and General Papagos, as mentioned, elected to withdraw to the Aliakmon Line. From this time, despite some vigorous and successful local actions by the Australian and New Zealand troops, who had been re-designated the Anzac Corps, and highly effective shooting by the artillery, at times point-blank against German tanks, the force could do little more than disrupt German plans and slow them down. When German forces broke through the Greek defences in the Klisoura and Siatista passes, Wilson decided that the Aliakmon Line would also have to be abandoned and he obtained Papagos's consent to withdraw to a new series of positions in the Thermopylae–Brallos area.

Remarkably, the German aircraft, though far more numerous, were not able to interdict effectively the main withdrawal route at its many bottlenecks between Larisa and Lamia, even though the withdrawing column of vehicles and guns was frequently more than 16 kilometres long. Meanwhile Mackay's third brigade, 17th Brigade (Brigadier Savige [q.v.]), had at last been ferried forward. But, as if to underline the by now piecemeal nature of the total operation, it came no further forward than Kalabaka, where it was immediately deployed in a blocking force role.

By 20 April, after almost two weeks of constant withdrawal, the Thermopylae-Brallos Pass area was occupied and Brigadier Vasey gave the ringing order to his brigade: 'Here we bloody well are, and here we bloody well stay.' But agreement had already been reached with the Greek government that the force should be evacuated, and planning for its withdrawal from Greece was already under way. In the parlous circumstances that then existed, it was perhaps surprising that Blamey, who was to become Wavell's Deputy C-in-C in the Middle East, and Major-General Mackay were both ordered out of Greece in the middle of the withdrawal, when their physical presence was important to the troops. In contrast, it was as well that Freyberg refused to leave. He wrote later: 'I cabled back to G.H.Q., Athens, and told them I was being attacked by tanks, fighting a battle on a two brigade front, and asked who was to command the New Zealand troops if I left. I was given the answer of "Movement Control". I naturally went on with the battle. After that I never received an order as to my disposal.'

The final withdrawal to the evacuation beaches was skilfully carried out, despite the fact that a German parachute force seized the narrow five kilometre isthmus of Corinth that connected the Peloponnese peninsula to the main part of Greece. The evacuation in Admiral Cunningham's ships began on 24 April and over five successive nights more than seven brigades (over 50 000 troops) were taken off. Many individuals and small groups who had been cut off in earlier fighting found their way via Turkey and the Greek Islands to Crete or North Africa, frequently at great risk to those who befriended and helped them.

In retrospect, abiding controversy continues to surround this campaign. It did not succeed in its primary object to assist the effective defence of Greece. Australian and New Zealand losses in this campaign, and the one that followed in Crete, were high. Both the 6th Australian and 2nd New Zealand Divisions were out of action for some time, and virtually had to be rebuilt. Moreover, there is little evidence for the popular claim that the Allied operation in Greece helped to postpone Hitler's invasion of Russia, thus ensuring the defeat of his armies in the Russian winter. When the invasion of Greece began, Hitler had not fixed a date for BARBAROSSA, the invasion of Russia, and it was only confirmed after the Greek campaign had ended. By the middle of May most of the Balkan force was already redeploying and by the end of the month most of the armoured units had refitted. In any case, any disruption to Hitler's plans can only be regarded as serendipitous fallout for the Allies: and not a result of Churchill's strategic insight. Perhaps Blamey can be given the last word:

> The outstanding lesson of the Greek campaign [wrote General Blamey in his despatch] is that no reasons whatever should outweigh military considerations when it is proposed to embark on a campaign, otherwise failure and defeat are courted. The main principles that must be satisfied are that the objects to be secured should be fully understood, the means to achieve the objects should be adequate and the plan should be such as will ensure success. All three essentials were lacking in the campaign in Greece, with the resultant inevitable failure. As far as my limited knowledge goes, the main reason for the dispatch of the force appears to have been a political one, viz., to support the Greeks to vindicate our agreed obligations.

JOHN COATES

GRIMSBY CLASS SLOOPS (*Yarra, Parramatta, Swan* and *Warrego*). Laid down 1934–39, launched 1935–40; displacement 1060 tons; length 266 feet; beam 36 feet; speed 16.5 knots; armament 3 × 4-inch anti-aircraft guns, 4 × 3-pounder guns, 1 machine-gun, 2 depth-charge throwers (*Yarra* and *Swan*); 3 × 4-inch anti-aircraft guns, machine-guns, 2 depth-charge throwers (*Parramatta* and *Warrego*).

These ships were built at Cockatoo Island (q.v.) as general coastal protection vessels. HMAS *Parramatta* saw service in the Mediterranean during the Second World War and was involved in supplying Tobruk (q.v.) in 1941. She was sunk by the German submarine *U559* on 27 November 1941. HMAS *Yarra* also saw service in the Mediterranean in 1941–42 and then sailed to fight the Japanese. She was sunk by Japanese surface craft while escorting a convoy off Java on 4 March 1942. HMAS *Swan* and HMAS *Warrego* served in New Guinea, Darwin, the Philippines and other areas of the South-West Pacific Area (q.v.). HMAS *Warrego* was present at the Lingayen Gulf and Balikpapan landings. In the postwar period both ships were converted to training and survey duties. They were sold for scrap in 1964–65.

GROUPS, AIR FORCE A group is an administrative and tactical unit comprising two or more squadrons, and may apply to combat support and service support units as well as combat ones. The RAAF has rarely fielded large enough forces on operations to justify the formation of groups. In the Second World War in the south-west Pacific, RAAF Command had three operational groups, Nos 9 (subsequently designated Northern Area), 10 (later 1st Tactical Air Force [q.v.]), and 11 (with a largely administrative function although it commanded some combat squadrons), deployed on operations against the Japanese. It failed to press for the formation of an Australian group within RAF Bomber Command in England, as the Canadians did with No. 6 (Canadian) Group and as it was entitled to do under the terms of the Empire Air Training Scheme (q.v.), thus ensuring that RAAF squadrons in England were scattered among RAF groups and denying Australian officers the opportunities for senior command experience. The RAAF today disposes of five operational groups: Strike and Reconnaissance Group, Tactical Fighter Group, Maritime Patrol Group, Air Lift Group and Operational Support Group, totalling 13 squadrons.

GUINEA GOLD see **SERVICE NEWSPAPERS**

GULF WAR Australian forces were deployed in the Gulf War of 1990–91, designated by the Americans Operations DESERT SHIELD, DESERT STORM, and DESERT FAREWELL, under the auspices of the United Nations and in support of American policy following Iraq's invasion of neighbouring Kuwait. Australia provided naval vessels to the Multi-National Naval Force, which acted as a maritime interception force in the Gulf to enforce sanctions against Iraq ordered by the United Nations. Designated RAN Task Group 627.4, it comprised two of the recently acquired *Oliver Hazard Perry* Class frigates (q.v.) and the underway replenishment ship HMAS *Success*. Four warships, HMA Ships *Sydney* (IV), *Adelaide*, *Brisbane* and *Darwin* served tours in Gulf waters. Because of the threatened use of chemical weapons, the crews underwent intensive training in defensive measures necessary for chemical and biological warfare, while *Success*, which had no air-defence capability, was assigned a detachment from the army's 16th Air Defence Regiment, equipped with RBS70 missile-launchers. An RAN Clearance Diving Team (q.v.), CDT3, was also dispatched for explosive ordnance demolition tasks. During the operational period of the deployment, the Australian warships formed part of the anti-aircraft screen of the carrier battle groups of the US Navy. In addition to naval units, assorted Australian personnel took part on attachment to various British and American air and ground formations, and a small group of RAAF photo-interpreters was based in Saudi Arabia together with a

An Australian Squirrel helicopter from HMAS *Darwin* (right) provides cover while a British Lynx helicopter lowers a boarding party onto the Iraqi ship *Tadmur* to enforce United Nations sanctions against Iraq, Gulf of Oman, September 1990. (AWM P1575.002)

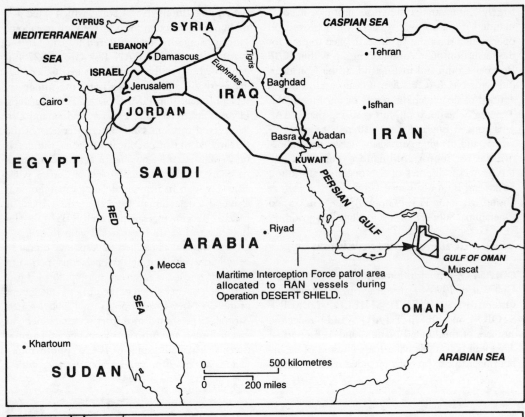

Maritime Interception Force patrol area allocated to RAN vessels during Operation DESERT SHIELD.

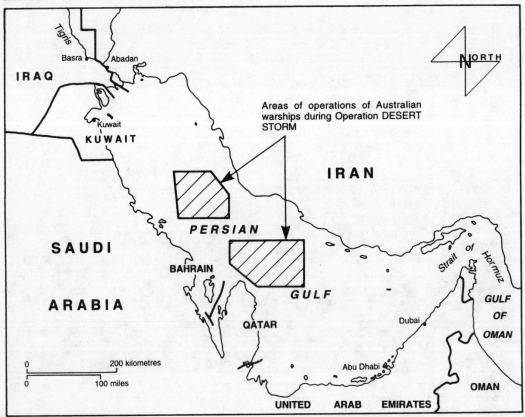

Areas of operations of Australian warships during Operation DESERT STORM

detachment from the Defence Intelligence Organisation. Four medical teams were also dispatched at American request, while in January 1991 the replenishment tanker HMAS *Westralia* left Fremantle to relieve *Success*. Although the ships and their crews were in danger from mines, which the Iraqis sowed liberally, and from possible air attack, the Australians' war was relatively uneventful, and there were no casualties. In the aftermath of the rapid victory over the Iraqi forces, 75 Australian personnel were sent to northern Iraq to assist in humanitarian aid to the Kurds in the exclusion zone declared there, while ships of the RAN remained on station, at American request, to maintain sanctions against the Iraqi government. The deployment to the Gulf caused some dissent in Australia, most of it of a minor nature, although some unfortunate film of sailors fooling about in mock Arab dress caused offence to the local Arab community. The Gulf commitment demonstrated that the ADF could make a minor contribution to a distant commitment at short notice, and it gave the Australian command and control systems very valuable testing. Government nervousness at times over public reaction to the deployment, which involved no direct threat to Australia (although it clearly posed a potentially very serious threat to Australian interests), underlined the difficulties in the post-Vietnam era of involving Australian service personnel in distant obligations.

David Horner, *The Gulf Commitment: The Australian Defence Force's First War* (Melbourne University Press, Melbourne, 1992).

GULL, HMAS see *TON CLASS MINESWEEPERS*

GULL FORCE, composed of the 2/21st battalion and ancillary forces, was sent to Ambon (q.v.) on 17 December 1941 to assist in the defence of the Netherlands East Indies. Under the command of Lieutenant-Colonel W. J. R. Scott (q.v.) who replaced Lieutenant-Colonel L. N. Roach in controversial circumstances on 17 January 1942, it was defeated within three days of the Japanese attack on 30–31 January. Those units defending the airfield at Laha were executed shortly after capture; 36 Australians escaped to Australia. The remainder were interned on Ambon and Hainan Island (to which 263 were shipped in October 1942) until the end of the war in the Pacific. Malnutrition and maltreatment resulted in the deaths of 74 per cent of the force, one of the highest death rates of any Australian group of POWs (q.v.).

Map 14 Positions of Australian warships in Operations DESERT SHIELD (top) and DESERT STORM (bottom) during the Gulf War.

Joan Beaumont, *Gull Force: Survival and Leadership in Captivity 1941–1945* (Allen & Unwin, Sydney, 1988); Courtney T. Harrison, *Ambon: Island of Mist* (T.W. & C.T. Harrison, North Geelong, 1988).

JOAN BEAUMONT

GULLETT, Henry Baynton Somer 'Jo' (16 December 1914–). Son of the cabinet minister and official historian Sir Henry Gullett (q.v.), 'Jo' Gullett was educated at the Sorbonne and Oxford Universities before becoming a journalist on the Melbourne *Herald* in 1937. He enlisted in the 2nd AIF in 1939 and was allotted to the 2/6th Battalion. As a sergeant he took part in the battalion's attack on Post 11 at Bardia (see LIBYAN CAMPAIGN, FIRST), heavily defended by its Italian garrison. He was seriously wounded, and many of his platoon became casualties.

Commissioned into the battalion in January 1941 as a result of the action, he fought in the Greek campaign before returning to Australia. He fought in New Guinea as a captain, and was awarded the MC in the Wau area in July 1943 for leading a rearguard action after an unsuccessful attack on Japanese positions. From March 1944 until the end of the war in Europe he was attached to the British Army, and participated in the landing at D-Day (q.v.). He finished the war as a major and was placed on the Reserve of Officers in September 1945.

In 1946 he won the federal seat of Henty, and held it until 1955. He was Chief Government Whip for the Liberal Party from 1950 to 1955, a member of the Standing Committee on Public Works in 1946–47, and took an interest in defence issues. After retiring from parliament he farmed in the Tharwa area, and was Australian ambassador to Greece from 1965 to 1968. His memoir, *Not As A Duty Only* (1976), dealt with his wartime service and provides one of the most eloquent testimonies to the Australian experience of the Second World War.

Henry Gullett, *Not As A Duty Only* (Melbourne University Press, Melbourne, 1976).

GULLETT, Henry Somer (26 March 1878–13 August 1940). With a background in journalism on the *Sydney Morning Herald* and Sydney *Daily Telegraph*, Gullett was selected in 1915 as official Australian correspondent with the British and French armies on the Western Front. In 1917 he was appointed by C. E. W. Bean (q.v.) to head the subsection of the Australian War Records Section in Egypt, and he was commissioned into the AIF in August that year. Twelve months later he was appointed Australian official correspondent to the AIF in the Middle East, and covered General Sir

Edmund Allenby's final offensive against the Turks in Palestine.

He attended the Paris peace conference as a press liaison officer to the Prime Minister, William Morris Hughes (q.v.). He returned to Australia and was briefly the first director of the Australian War Museum (later Memorial) (q.v.), but in 1920 Hughes appointed him head of the Australian Immigration Bureau. He resigned in 1922, returned to journalism in Melbourne for a time, and in 1925 won the seat of Henty for the Nationalists.

He held a number of portfolios when the conservative parties were in government in the 1920s and 1930s, and in April 1939 was appointed minister for External Affairs in the first Menzies (q.v.) Cabinet. From 15 September 1939 he was a member of the War Cabinet. He was killed in an air crash along with the CGS and two other ministers near Canberra, a disaster which robbed Menzies of able colleagues and destabilised his government.

Gullett was one of Bean's principal collaborators on the official history, writing one volume, *The Australian Imperial Force in Sinai and Palestine 1914–1918,* which was published in 1923 and went through eleven editions by 1941, and with Bean editing the photographic volume of the series, published in the same year. As a correspondent in the Middle East he had shown a willingness to stand up to senior officers over fair recognition of Australian achievements, and this attitude is displayed in his account of that campaign.

GUNNER O'NEILL INCIDENT After going absent without leave on 16 January 1966, Gunner Peter O'Neill, a regular soldier serving in South Vietnam with 105 Field Battery, Royal Australian Artillery, refused to attend a field punishment parade and was placed under close arrest on the orders of his commanding officer, Major Peter Tedder. He was subsequently handcuffed to a star picket in a weapon pit for a period of seven days, an anti-quated but still permissable form of field punishment. The matter was reported in the press on 8 March and caused a storm of indignation, leading to close questioning of the Minister for the Army, Malcolm Fraser (q.v.), in the parliament and pressure for an inquiry, which was resisted. Tedder was court martialled as a result, but was acquitted. O'Neill was court martialled for failure to attend the disciplinary parade and was sentenced to six months' detention; on his release from Holsworthy he was discharged from the army. The public clamour died down as Fraser gradually released information in the House of Representatives on O'Neill's previous disciplinary record, while admitting to short-comings in the exercise of the service disciplinary code. O'Neill later claimed that his treatment resulted from allegations he had made concerning the disposal of medical supplies on the black market, but these were never proven. The case provided an early example of the potential sensitivity of reporting from Vietnam, and provided a test of Fraser's abilities as a minister. The incident was briefly notorious again in mid-1990 when a Sydney newspaper ran a story alleging that treatment of the matter was being suppressed in the official history of the Vietnam War; the official historian sued and the matter was subsequently settled out of court.

GUNS see **ARTILLERY**

'GYPPO' or 'gyppie' were derogatory terms used by British soldiers of Egyptian soldiers since the 1890s and routinely used by Australian troops in Egypt when referring to the local people. Captain R. Hugh Knyvett in his 1918 book *'Over There' With the Australians* wrote of 'the animated lump of muck that the "Gyppo" is' (see EGYPT, AUSTRALIANS IN). 'Gyppie tummy' was a colloquial expression descriptive of a common bowel disorder.

H

HALIFAX, HANDLEY PAGE (7-crew heavy bomber [Mark III]). Wingspan 104 feet 2 inches; length 71 feet 7 inches; armament 9 × 0.303-inch machine-guns, 13 000 pounds bombs; maximum speed 282 m.p.h.; range 1030 miles; power 4 × Bristol Hercules 1615 h.p. engines.

The Halifax entered service with the RAF in 1940 and with the Lancaster (q.v.) was the most common aircraft used in Bomber Command during the Second World War. Halifaxes were used by two Australian squadrons whose crews had trained through the Empire Air Training Scheme (q.v.). No. 462 Squadron began flying Halifaxes in north Africa in 1942. In 1943 No. 466 Squadron in the United Kingdom replaced its Wellington (q.v.) bombers with the Halifax. From 1944 to the end of the war No. 462 Squadron also flew from Britain. Halifaxes were used for bombing raids on railways, factories and German cities.

HAMILTON, Admiral Louis Henry Keppel (31 December 1890–22 June 1957). From a naval family, Hamilton entered the RN in 1903 after education at Osborne. He served in the Great War and was present at Jutland, having already won a DSO for service in the Cameroons in 1915. In the Second World War he served with the Home Fleet between 1939 and 1943, winning a bar to the DSO at Narvik, and commanding both destroyers and a cruiser squadron. Between 1943 and 1945 he served in the Mediterranean.

Hamilton was appointed CNS to the RAN in June 1945, although the intention had been to break the tradition of appointing the professional head of the Australian service from the ranks of the RN and to appoint Commodore John Collins (q.v.) instead. Collins, however, had been wounded at Leyte Gulf the previous year and was on sick leave and a shore posting until July 1945; furthermore it was felt in some quarters, especially British ones, that he still lacked sufficient experience to meet the demands of the job. During his term as CNS, Hamilton began the process of acquiring a fleet air capability (see FLEET AIR ARM) for the RAN. He lacked a sense of Australian political realities as they affected his position, and found dealing with Australian ministers a frustrating experience. In 1948 he refused any extension of his term, on grounds that he had been out of England since 1939, and retired. He was the last RN officer to head the RAN.

HAMILTON, General Ian Standish Monteith (16 January 1853–12 October 1947) was commissioned in 1872, saw action in many of the small frontier wars of the late nineteenth century, and served in India almost continuously until appointed Commandant of the Musketry School at Hythe, in England, in 1898. During the Boer War he served in various capacities, as a column and divisional commander, and as Kitchener's chief of staff. He was appointed KCB in 1900. A man of markedly superior intellect, and with a military record matched by few of his contemporaries, Hamilton held a succession of major posts in the following decade: military secretary at the War Office from 1902, head of a military mission to the Japanese armies in the field during the Russo-Japanese War of 1904–05, GOC of Southern Command between 1905 and 1909, Adjutant-General in 1909, and C-in-C of the Mediterranean Command in 1910, to which was added responsibility as Inspector-General of Overseas Forces. In this latter capacity he visited Australia in 1914 to report on the state of the Australian Military Forces, a visit requested by the Australian government during the 1911 Imperial Conference (q.v.). He returned to Britain in 1914. On the outbreak of the First World War he was given command of the Central Force, which was responsible for repelling a German invasion of Britain. In March 1915 Kitchener selected him to command the Mediterranean Expeditionary Force. Later in the month Hamilton was instructed to prepare a plan to land Anzac and British forces on the Gallipoli Peninsula to capture the forts at the Narrows and thus clear the way for the fleet. Hamilton never questioned whether his force was sufficient for the job but went about his task with a light-hearted equanimity that could either be considered a sign of balance and calm consideration or gross irresponsibility to the troops under his command. The plan for landing was imaginative but impossible to achieve with the force available. Once battle was joined Hamilton became a rather remote figure, hardly ever venturing on to the peninsula. His despatches to the British government were highly literate and spiced with good humour. This was hardly the spirit to describe an operation that was bogged down and going nowhere; the troops surely deserved some plain speaking rather than poetry. The government eventually arrived at this conclusion. After the failure of the August attack Hamilton was relieved of his command and never offered another one. His account of the campaign, entitled *Gallipoli Diary* (1920), was not in fact a diary but was written up after the war from notes. It is nonetheless a revealing self-portrait. He was appointed Lieutenant of the Tower of London in 1918, was for many years colonel of his family regiment, the Gordon Highlanders, and was active on behalf of the major organisation of British veterans, the British Legion.

HAMPDEN, HANDLEY PAGE (4-crew medium bomber [Mark I]). Wingspan 69 feet 2 inches; length 53 feet 7 inches; armament 6 × 0.303-inch machine-guns, 4000 pounds bombs; maximum speed 265 m.p.h.; range 1990 miles; power 2 × Bristol Pegasus 965 h.p. engines.

The Hampden was flown by No. 455 Squadron, the first Australian squadron to join Bomber Command, in 1941. The crews were trained through the Empire Air Training Scheme (q.v.). The Hampden had entered service with the RAF in 1938 and was one of the main types used by Bomber Command in the early part of the Second World War. No. 455 Squadron flew bombing raids over German cities such as Cologne, Frankfurt and Berlin until early 1942 when the Hampdens, which had been suffering losses because of their lack of defensive armament, were transferred from Bomber Command to Coastal Command and used for the safer role of maritime patrol. In September 1942 No. 455 Squadron delivered Hampdens to Murmansk in the USSR (Russia) and trained Soviet crews in their use. In late 1943 the Hampdens were replaced with Beaufighters (q.v.).

HANCOCK, Air Marshal Valston Eldridge (31 May 1907–   ). An important postwar head of the RAAF, Hancock began his career by entering RMC Duntroon in 1925 intending to make his way in the army. Only when he discovered, several months before graduating as the senior cadet of his class in December 1928, that no vacancy existed in his preferred corps (Engineers) did he decide to try for the RAAF. Like other Duntroon graduates in the air service, he found that a disproportionate number of administrative duties came his way. Completing pilot training in 1930, he filled staff posts at Air Force Headquarters in Melbourne for most of the next 10 years, interrupted by attendance at the RAF Staff College at Andover in 1937. In August 1940 he was given command of the wartime Bombing and Gunnery School at Evans Head in New South Wales, becoming a group captain from April 1941, but from January 1942 he returned to filling appointments on planning staffs in Melbourne, and at Allied Air Forces Headquarters in Brisbane; other administrative postings with Western Area and at Air Force Headquarters followed in 1943–44. Only in the last year of the war did he receive the opportunity of operational experience, being given command of No. 100 Squadron, a Beaufort (q.v.) unit in New Guinea, in January 1945. After less than three months in this post, he took command of 71 Wing in late March and led it for the last six months of the war. With the return of peacetime conditions, Hancock's career resumed its prewar course with staff postings at headquarters. During this period he began pressing his superiors to address the RAAF's needs in officer education. When establishment of a cadet college at Point Cook, Victoria, was approved in 1947, he was appointed its founding commandant as an air commodore. In shaping the new institution, he drew on his experience of Duntroon to replicate the best features of the army college while attempting to avoid what he considered its shortcomings. Leaving Point Cook at the end of 1949, he went to London to attend the Imperial Defence College (q.v.). On his return early in 1951 he was given acting rank of air vice-marshal and became Deputy CAS. This was the first of a succession of senior posts which would take him to the top RAAF position within 10 years, being followed by appointments as Air Member for Personnel on the Air Board in 1953; Head of the Australian Joint Service Staff at Overseas Headquarters, London, in 1955; on loan with the RAF as officer commanding Far East Air Force at Singapore and then 224 Group in Malaya from 1957; and, in July 1959, AOC Home Command (which from September that year became known as Operational Command). In May 1961 he was promoted to air marshal and succeeded Sir Frederick Scherger (q.v.) as CAS, the second occupant of the post with an ex-Duntroon background. His four-year tenure coincided with a program already under way to re-equip the air force totally with modern aircraft, and with the government's decision to send Australian forces (including air elements) to the Vietnam War. Hancock was opposed to the latter commitment, fearing the financial impact this would have on the RAAF's modernisation objectives. In particular, he was anxious to acquire for the service an effective strategic bomber to replace the obsolescent Canberra. In 1963 he personally led a procurement team on an overseas visit which ultimately led to Australia's controversial decision to buy the radical new American F-111C swing-wing (q.v.) bomber. Hancock retired in May 1965. When Scherger was due to leave the post of Chairman of the Chiefs of Staff Committee a year later, he was proposed as successor by the Minister for Air, Peter Howson. Regarded as 'reserved, thorough and hardworking', he was, however, not favoured by Prime Minister R. G. Menzies and the appointment went to General Sir John Wilton (q.v.). Hancock returned to his home State of Western Australia, becoming involved in the local branches of the Australian Defence Association and the Royal Commonwealth Society.

V. E. Hancock, *Challenge* (Access Press, Perth, 1990).

CHRIS COULTHARD-CLARK

HANKEY, Maurice Pascal Alers (1 April 1877–26 January 1963) was Secretary of the Committee of Imperial Defence 1912–38, Secretary of the British War Cabinet 1916–18, and Secretary to the Cabinet 1919–38. The creator of the cabinet secretariat and long-standing secretary of the Committee of Imperial Defence, Hankey played an important role for several decades in the formulation of British defence policy. His position in the cabinet machinery and in a number of subsidiary committees gave him almost unrivalled command of defence issues. In October 1934 Hankey visited Australia, ostensibly to take part in the centenary celebrations in Victoria, but also to advise local politicians of developments in British defence planning. The issue of greatest concern to Australian leaders was British intentions regarding the naval base at Singapore (see SINGAPORE STRATEGY). Within Australian defence circles there was a fierce debate between those, mainly in the RAN, who argued for the defence of Australia to be based on a wholehearted embrace of naval power and for acceptance of British assurances about the reinforcement of their naval presence at Singapore, and those in the army and RAAF who doubted both Britain's willingness and ability to develop the full capacity of the Singapore base, and who wanted a much greater emphasis placed on the defence of Australia against invasion. Hankey had hoped to be able to reassure Australian leaders of Britain's intention to complete the Singapore base and to commit to it the naval forces necessary to protect imperial interests in the Far East. However, just as opinion in Australia was divided over these defence issues, so too in Britain there was widespread disagreement in political and defence circles over the priorities that should be accorded to the various calls on the defence budget. The Chancellor of the Exchequer (and future prime minister), Neville Chamberlain, vigorously opposed Hankey's endorsement of the Singapore strategy, and the best Hankey could take with him to Australia was a commitment to continue work on the base so that in a major emergency it would be capable of handling a large fleet. This fell far short of the guarantee that Australian leaders had expected. It enabled Hankey to urge a continued reliance on the Singapore naval strategy, but it did not silence those critics in Australia who continued to doubt that British promises would provide security for Australia against Japan.

HANNAH, Air Marshal Colin Thomas (22 December 1914–22 May 1978). After brief service as a CMF gunner in the 8th Field Artillery Brigade in his home State of Western Australia, Hannah became an air cadet in January 1935 at the RAAF Flying Training School, Point Cook. Commissioned as a pilot officer in July 1936, he was posted to No. 22 Squadron, a Citizen Air Force unit, and thence to No. 23 Squadron, which moved to the new RAAF Base Pearce (q.v.) in Western Australia in March 1938.

Hannah was on a course in England when the Second World War began. Returning to Australia in March 1940, he worked in a number of technical postings before promotion as Deputy Director of Armament in 1941. He assumed command of No. 6 Squadron, flying Beauforts (q.v.) at Milne Bay in November 1943, and commanded No. 71 Wing in early 1944 until evacuated for medical reasons. In September that year he became Senior Air Staff Officer on the Headquarters of Western Area in Perth, and commanded the Area from June 1945 to May 1946.

His postwar career involved a succession of staff and command posts, with the former predominating. He attended the RAF Staff College in 1947, then commanded No. 82 Wing, flying Lincoln (q.v.) bombers in 1950–51. A number of staff posts followed in the Department of Air before attendance at the Imperial Defence College (q.v.) in London in 1955. In January 1956 Hannah was posted as Senior Air Staff Officer with the RAAF element to Headquarters Far East Air Force in Singapore, where he dealt with operational matters relating to the growing insurgency mounted by the communists during the Malayan Emergency. Appointed Deputy CAS in December 1961, he succeeded Air Marshal Sir Alister Murdoch (q.v.) as CAS in January 1970.

Hannah's tenure of office was relatively short, although he presided over the final stages of the RAAF commitment in support of the 1st Australian Task Force in Vietnam. His most significant contribution to the air force had come earlier. He had been one of those responsible for making the RAAF an 'all through jet' force, with reliance on propeller aircraft minimised, and had been a strong advocate of the RAAF acquiring more and better helicopters. He retired in March 1972 to become Governor of Queensland. He created controversy with his criticisms of the Whitlam government in late 1975, and as a result the British government declined to support an application for a second term in its advice to the Queen. He retired again in March 1977.

HARBOUR DEFENCE LAUNCHES These were small, cheap, easily constructed craft which carried a light artillery piece (2- or 3-pounder) or machine-guns. They were used in Australia between 1942 and 1945 for the defence of Australian ports and waterways. They could also be used as hydrographic vessels and some saw service in this capacity during US assaults on the Admiralty Islands, Hollandia, Morotai and the Philippines. Australia employed locally built craft and obtained other types from Britain and the United States. A typical harbour defence launch was 72 feet long, had a beam of 15.5 feet, a speed of 11–12 knots, and was made of wood. In all Australia had 28 of these ships, the last being laid up in 1982.

HARDMAN, Air Chief Marshal (James) Donald Innes (21 February 1899–2 March 1982). Educated at Oxford, Hardman enlisted to serve in the Great War in 1916 and joined the RAF on its formation in 1918. Reaching the rank of air vice-marshal by the end of the Second World War, he held a senior staff post with Air Command Southeast Asia in 1946–47 before returning to England and the position of Assistant CAS (Operations). He was Commandant of the RAF Staff College between 1949 and 1951 and commanded RAF Home Command before appointment as CAS of the RAAF in January 1952, when he was made KCB.

The decision to appoint a British officer to head the Australian service arose out of the wartime disputes between Air Commodore George Jones and Air Vice-Marshal W. D. Bostock (qq.v.) and the damage this had done to the organisation and development of the RAAF. Jones had remained as CAS after the end of the war, and by 1952 had been in the post for 10 years. The Australian government placed a great deal of pressure on the British, who reluctantly acceded to their request and, in contrast to the appointments made in the interwar period, on this occasion sent one of their best serving officers to Australia.

Hardman's selection was resented by many inside and outside the RAAF, a fact of which he was well aware. His principal contribution during his tenure of office was the reorganisation of the RAAF into functional commands, a system which had existed in the British service for years. Before returning to Britain and further senior appointment in the RAF in 1954, he gave a widely published interview which castigated the operational effectiveness of the other two services, highlighting the rivalry over budget allocations which continued to bedevil relations within the defence sector. He retired from the RAF in 1958 and was appointed GBE.

HARMAN, HMAS, located 12 kilometres southeast of Canberra, is part of Naval Communication Station Canberra, which links Australian command authorities with ships at sea and with allied commands throughout the world. HMAS *Harman* also provides accommodation and support services for naval personnel in the Canberra area. Originally a receiving station, it was completed along with a transmitting station at Belconnen in December 1939. Its name is a contraction of the surnames of the Director and Assistant Director of Signals at the time. Commissioned on 1 July 1943, it became a control centre after a new receiving station was completed at nearby Bonshaw in 1958.

HARRINGTON, Vice-Admiral (Wilfred) Hastings (17 May 1906–17 December 1965). On graduation from RANC Jervis Bay in May 1924, Harrington went to the RN for further training and served on board HMS *Malaya* with the Mediterranean Fleet. He did further sea time on the Australian ships *Australia*, *Canberra* and *Albatross* before returning to Britain, and served a further three years on the China Station. In September 1939 he was commanding HMAS *Yarra*, and served in the Mediterranean and the Netherlands East Indies. He spent most of the war at sea, serving with HMAS *Australia* between March 1942 and July 1944 in the South-West Pacific Area, before receiving command of HMAS *Quiberon*. He was twice mentioned in despatches and received the DSO in 1943. After the war he spent a period ashore in Navy Office, Melbourne, and attended the Imperial Defence College (q.v.) in 1952. In 1955 he was given command of HMAS *Sydney*, and in 1957 was promoted to rear-admiral. The following year he became Second Naval Member and in December 1959 assumed command of Her Majesty's Australian Fleet. In February 1962 he was appointed CNS. Easily recognisable in public because of the prominent cheekbone whiskers which he cultivated, he retired from the RAN in February 1965, and died suddenly before taking up a position as Australian Commissioner-General to the Montreal Exposition.

HARRISON, Eric John (7 September 1892–26 September 1974) enlisted in the AIF in 1916 and reached the rank of sergeant by the war's end. A keen sportsman and member of the New South Wales champion rowing crew in 1914, he was a member of the AIF rowing team that won the King's Cup against crews from other Allied forces at Henley in 1919. He was elected to parliament for the federal seat of Wentworth in 1931 as a member of the United Australia Party, and in 1934 entered the ministry as minister for the Interior. He was Postmaster-General and minister for Repatriation in 1939–40, and between 1942 and 1943 rejoined the army and served as a liaison officer with US forces in Australia. He was a founding member of the Liberal Party, becoming deputy leader of the party in 1944. When Menzies led the Liberals into government in 1949, Harrison became minister for Post-War Reconstruction and for Defence until the 1951 election. He was minister for Defence Production between 1951 and 1956 and also had responsibility for the army and navy portfolios between 1955 and 1956. He retired from his seat in the latter year, and was High Commissioner in London between 1956 and 1964. He was made KCVO in 1954 and KCMG in 1962.

HASLUCK, Paul Meernaa Caedwalla (1 April 1905–10 January 1993). Writer, journalist, official historian, foreign minister and Governor-General, Hasluck's career in public life was a long and distinguished one. A Western Australian by birth, he worked on the *West Australian* newspaper, and wrote a Master of Arts thesis on Black–White relations in nineteenth century Western Australia, before secondment to the Department of External Affairs in early 1941. He took part as a member of the Australian delegation in a number of significant wartime international conferences, such as those at Bretton Woods in 1943 and the San Francisco conference in 1945 which set up the United Nations; he resigned in 1947 while head of Australia's mission to the UN. His relations with H.V. Evatt were less than cordial, and accounted in large part for his decision to leave the foreign service.

Selected by the official historian Gavin Long (q.v.) to write the civil volumes of the official history of the Second World War, Hasluck returned briefly to academic life before seeking Liberal preselection for the federal seat of Curtin in the 1949 election, which he won. In 1951 Prime Minister R. G. Menzies (q.v.) appointed him to the portfolio of External Territories, which he held for the next 12 years. He was briefly minister for Defence before taking over External Affairs, which he held at a critical period of Australia's gradually increasing involvement in the Vietnam War. He was appointed Governor-General in early 1969 and knighted, and retired in 1974. He was made KG in 1979.

His two volumes of official history, *The Government and the People,* appeared in 1952 and 1970, the long gap between them explained by his many other pressing public duties. His work has been criticised for being rather more about the government than the people, and certainly it largely predates, and takes little account of, trends towards social history from a 'bottom-up' perspective. Hasluck's work displays his strong belief in the virtues of the parliamentary system, and despite his status as a Liberal MP and minister his work was

judged generally 'fair and accurate'; his treatment of John Curtin and other senior figures in the Labor government was scrupulous. With 'a flair for the ironic phrase and flowing prose', the volumes form a major scholarly achievement which nonetheless, in the view of one commentator, 'made few concessions to superficial popular interest'.

Paul Hasluck, *The Government and the People*, 2 vols (Australian War Memorial, Canberra, 1952, 1970).

HASSETT, General Francis George (11 April 1918–   ). A graduate of RMC Duntroon in 1938, Hassett was posted to the first regular infantry unit raised in the Australian Army, the Darwin Mobile Force (q.v.), where he commanded the mortar platoon. At the outbreak of war he became adjutant of the 2/3rd Battalion, and served in the First Libyan campaign in early 1941, during which he was wounded. He was brigade major of the 18th Brigade and, while still only 23, was promoted to lieutenant-colonel and a senior staff job on the Headquarters, II Australian Corps. He finished the war with the 3rd Division on Bougainville.

After the war he instructed at the Staff College, and in April 1951 was given command of 1RAR. Within months he was posted to Korea to command 3RAR there, and led them through the fierce and successful fighting at Maryang San (see KOREAN WAR), part of Operation COMMANDO. His service in Korea marked him as one of the best battalion commanders the postwar regular army has produced.

Returning from Korea in September 1952, he became chief instructor at the School of Infantry and then Director of Military Art at RMC Duntroon. Appointment as Military Secretary followed in 1958, and in 1960 Hassett was sent to Malaya to become the first Australian commander of the 28th Commonwealth Infantry Brigade since the end of the Korean War. During this period, from 1960 to 1962, the brigade formed part of the Far East Strategic Reserve (q.v.) and trained for operations in a SEATO role. He attended the Imperial Defence College (q.v.) in 1963, spent two years as Deputy CGS, and served as GOC Northern Command from 1968 to 1970.

His most important overall contributions to the Army were still to come. In 1970–71 he headed the Army Reorganisation Planning Staff, charged with modernisation of the army's structures, many of which had been in place since Federation. The principal areas of overhaul were the system of geographic commands, a rationalisation of logistic services, and the structure of Army Headquarters. Appointed Vice-CGS in 1971 to oversee the implementation of reform, Hassett introduced functional commands and improved the army's ability to mobilise rapidly in an emergency. He became CGS in November 1973, the first professional head of the army in the post-Vietnam era. In 1975 he was selected as Chairman, Chiefs of Staff Committee, an office which in February 1976 was redesignated Chief of the Defence Force Staff. He retired in April 1977, having been appointed AC in 1975 and KBE in 1976.

HAWK, HMAS see *TON* **CLASS MINESWEEPERS**

HAWKESBURY, HMAS see **RIVER CLASS FRIGATES**

HERCULES, LOCKHEED (5-crew transport [C-130H]). Wingspan 40.41 m; length 29.79 m; load 92 troops, 64 paratroops, 74 stretchers or 19 500 kg cargo; maximum speed 602 km/h, range 7876 km; power 4 × Allison 4508 effective shaft h.p. turboprop engines.

The Hercules has been the RAAF's heavy transport aircraft since 1958. During the Vietnam War Hercules aircraft ferried supplies and reinforcements to the Australian troops and flew home many of the wounded. They have also been widely used to provide supplies to areas hit by natural disasters (see CIVIL COMMUNITY, ASSISTANCE TO) and as passenger aircraft during airlines strikes (see CIVIL POWER, AID TO). Hercules are flown by No. 36 and 37 Squadrons from Richmond (q.v.). In 1966 the initial order of 12 C-130A aircraft was augmented by 12 improved C-130E aircraft. The C-130A aircraft were replaced by 12 C-130H aircraft in 1978 and one C-130A is held by the RAAF Museum at Point Cook.

HERINGTON, John (3 June 1916–22 January 1967) was born in England and educated at Cambridge, graduating in 1937. He migrated to Australia in 1938, and enlisted in the RAAF in 1940. Herington completed pilot training in Canada as a member of the Empire Air Training Scheme (q.v.), and served with Coastal Command on Catalinas and in operational intelligence, rising to the rank of squadron leader. In 1945 he accepted the invitation of the official historian, Gavin Long (q.v.), to write two volumes in the air series of the Official History. These appeared as *Air War Against Germany and Italy 1939–43* (1954) and *Air Power over Europe 1944–45* (1963). Having remained in the air force reserve postwar, in 1948 he rejoined the RAAF as a squadron leader and worked in the Directorate of Intelligence. In 1954 Herington left the air force and was employed by the Department of Supply as the chief security officer. This was followed by appointment as chief departmental representative at Australia House in London in 1964, where he died three years later at the age of 51. His official histo-

ries are solid and detailed. Intended originally to be encompassed in a single volume, the story of Australian involvement in the air war in Europe, principally though not only as part of Bomber Command, eventually took two volumes, at Herington's urging. His account of the EATS, and of its weaknesses, is generally better than that provided by Douglas Gillison (q.v.) in the first volume of the same series.

John Herington, *Air War Against Germany and Italy 1939–43* (Australian War Memorial, Canberra, 1954); *Air Power over Europe 1944–45* (Australian War Memorial, Canberra, 1963).

HERRING, Lieutenant-General Edmund Francis 'Ned' (2 September 1892–5 January 1982). A Rhodes scholar for Victoria in 1912, Herring served in the Officers' Training Corps at Oxford University. Commissioned into the Royal Field Artillery, he spent the war in France and Macedonia, where he won the DSO and the MC and contracted malaria. He took a law degree from Oxford after the war, and returned to Australia at the end of 1920.

He was a citizen soldier (see CITIZEN MILITARY FORCES) between the wars, accepting a commission as a legal staff officer with the 2nd Cavalry Division in October 1922. He soon moved into command positions, however, taking over an artillery battery in August 1923. An enthusiastic and capable part-time soldier, he was also politically active in the clandestine world of the paramilitary White Army, being a regional commander for part of the Mornington Peninsula, and in more conventional conservative party politics. His legal career flourished, and in February 1936 he was appointed King's Counsel. By the beginning of the war he was a colonel, and Commander, Royal Artillery, of the 3rd Division of the CMF.

General Sir Thomas Blamey took him away with the 2nd AIF as CRA of the 6th Division, and in this capacity he served in the early campaigns in the Mediterranean. He succeeded Lieutenant-General Iven Mackay in command of the 6th Division in August 1941, and commanded Northern Territory Force in the early months of the Pacific War back in Australia. When Blamey sacked Lieutenant-General S. F. Rowell (q.v.) as GOC, New Guinea Force, and I Australian Corps in September 1942, he immediately recommended that Herring be appointed to succeed him. Although Rowell had led the forces which stalled the Japanese advance towards Port Moresby, Herring was responsible for pushing them back to the coast and in the fighting for the beachheads at Buna, Gona and Sanananda. In particular, he enjoyed excellent relations with the Americans and

ensured a high level of smooth cooperation between the Allied forces at a critical time in the ground fighting. In recognition of his services in the Papuan campaign he was appointed KBE in 1943.

He commanded the corps until October 1943, conducting the successive offensives against Salamaua, Lae and Finschhafen. He was appointed Chief Justice of Victoria in January 1944, a position he held until retirement in 1972. In June 1950 he briefly returned to active duty as Director-General of Recruiting pending the return from Japan of Lieutenant-General Sir Horace Robertson (q.v.).

Herring was not immune from the personal jealousies which characterised the senior officers of the Australian Army in the Second World War, although he was highly regarded by many, including many Americans. One of the keys to his success as an operational commander was the excellent personal relationship he enjoyed with Blamey, which went back to their association in the 1920s. A highly able artillery commander, he was one militia officer who generally avoided the staff corps/militia divide, making no secret, for example, of his great admiration for George Vasey (q.v.).

Stuart Sayers, *Ned Herring* (Hyland House, Melbourne, 1980).

HETHERINGTON, John (?1907–18 September 1974). A journalist, like so many involved in the writing of Australia's official and popular histories of the two world wars, Hetherington worked for a number of Melbourne papers before serving as a war correspondent in Greece for British and Australian news organisations. After the war, he was editor of the Adelaide *News* between 1945 and 1949, and from 1952 to 1967 worked in different capacities for the Melbourne *Argus* and *Age*. A man of wide-ranging interests, he wrote several books during the war including an account of the campaign on Crete, entitled *Airborne Invasion* (1943), and *The Australian Soldier* (1943), and books on the artist Norman Lindsay, the soprano Nellie Melba, Australian painters, and several volumes on Australia's architectural heritage. He is probably best known as the biographer of Field Marshal Sir Thomas Blamey. An early and abbreviated volume was published in 1954, followed by a more substantial but still unsatisfactory book published in 1973. Hetherington was able to discuss Blamey with many of the latter's contemporaries, as his papers in the Australian War Memorial make clear, and benefited accordingly, but clearly felt unable to deal with his subject critically. As a result,

Major-General E. F. Herring (right) meeting Emir Abdullah of Jordan at Jericho, Palestine, February 1942. (AWM 023585)

the final work has a defensive tone, not helped by the lack of scholarly apparatus such as footnotes, and leaves the historiography of Australian military history without a satisfactory study of Australia's most senior soldier. It is probably unfair to blame Hetherington for this entirely, as he wrote at a time when many of those involved were still alive, and military history had not at that stage achieved respectability in historical circles.

John Hetherington, *Blamey: Controversial Soldier* (Australian War Memorial, Canberra, 1973).

HEWITT, Air Vice-Marshal Joseph Eric (13 April 1901–). A graduate of the Royal Australian Naval College, Jervis Bay (q.v.), Hewitt served in the RAN between 1915 and 1928 before transferring to the RAAF. In the interwar years he attended the RAF Staff College at Andover, commanded an RAF bomber squadron while on exchange in Britain, and held several senior staff and administrative posts.

Appointed Deputy CAS in 1941, he was sent the following year to US General George Kenney's headquarters as Director of Allied Air Intelligence. He made a markedly good impression on his American superior and contemporaries, and worked hard to foster good relations with US Army Air Force authorities. Virtually alone of Australian air force officers, Hewitt obtained Kenney's ear, and was privy to the highest levels of USAAF planning in the theatre.

In February 1943 he was posted to command No. 9 Operational Group. Although many of the Australian officers who worked for him regarded him highly as both an administrator and an operational commander, he was sacked in November in controversial circumstances. While the official reason for his removal was 'morale and discipline' problems within his command, Hewitt himself believed that he was the victim of a smear campaign orchestrated by a dissatisfied former staff officer. He returned to his former post as Director of Air Intelligence. From 1945 until his retirement in 1956 he was successively Air Member for Personnel and for Supply and Equipment. After his service career he went into business.

Hewitt's personality contributed to the controversy surrounding his removal from command of No. 9 Group. While middle-ranking RAAF officers who had served in both the Mediterranean and Pacific theatres rated him highly, and Kenney thought his sacking 'bad news', others regarded him as abrasive, outspoken, even 'cocky'. The official historian, Gavin Long, thought him 'a live wire, very emphatic, and . . . one whose snap judgments and decisions would sometimes be contradictory'.

The fact remains, however, that his removal made a bad impression on senior American officers, and further emphasised the poor quality of leadership and decision-making at the top of the RAAF.

J. E. Hewitt, *Adversity in Success: Extracts from Air Vice Marshal Hewitt's Diaries 1939–1948* (Langate Publishing, Melbourne, 1980).

HICKS, Edwin William (9 June 1910–14 May 1984). Hicks entered public service in 1928, and after war service with the RAAF was a highly successful Secretary of the Department of Air between 1951 and 1956. He succeeded the formidable Sir Frederick Shedden (q.v.) as Secretary of Defence in the latter year, and remained there until retirement in 1968. His principal contributions were made in the areas of higher administration and policy, and he was responsible for a number of attempts to modernise a department which had changed very little under his predecessor's long reign. He was a member of the Morshead Committee (q.v.) in 1957, and enjoyed good relations with the chiefs of staff. Hicks was noted for the personal touch, dropping in to talk over issues with his subordinates in their offices, and the first staff Christmas party was held in the department after he took over, with a bar set up in the Secretary's office. After retirement he was appointed High Commissioner to New Zealand between 1968 and 1971, and was a member of the Interim Council at RMC Duntroon.

HIXSON, Captain Francis (8 January 1833–2 March 1909) was born at Swanage, England, to a family with a long seafaring tradition. Hixson was a member of various RN survey expeditions to Australasia between 1848 and 1862. He commanded the New South Wales Naval Brigade throughout its existence, from 1863–1902, to which was added the New South Wales Naval Artillery Volunteer Company in 1886. Under his command the brigade was able to maintain reasonable numbers, but the level of training and state of equipment gradually declined. When a New South Wales naval contingent was sent to help put down the Boxer Rebellion in 1900, Hixson was considered too old and lacking in naval warfare experience to command it. Instead, before returning to Australia, he was placed in charge of the contingent during its voyage to Hong Kong. In addition to these duties, Hixson also functioned as President of the Marine Board of New South Wales between 1872 and 1900, and was chairman of the Sailors' Home in Sydney. Three of his sons, lieutenants in the Naval Brigade, volunteered to go to China, while a fourth son was a member of the Queensland navy.

HOAD, Major-General John Charles (25 January 1856–6 October 1911). Virtually nothing is known of Hoad's childhood and background until his appointment as a schoolteacher in Victoria in 1878. Like many regular soldiers of the period, he first joined the militia, in his case as a lieutenant in the 1st Battalion, Victorian Rifles in December 1884, before transferring to the permanent staff in 1886. He served under the command of Colonel Tom Price (q.v.) before embarking for England in October 1889 to undertake a series of courses in musketry, signals and military engineering. Returning to Australia in early 1891, by 1895 he had been promoted to the rank of lieutenant-colonel and the position of Assistant Adjutant-General of the Victorian colonial forces. He went to the South African War with the first of the Australian colonial volunteers in late 1899, and received command of the 1st Australian Regiment (formed by combining contingents of 'outer states' soldiers), which soon joined the Kimberley Relief Force and subsequently formed part of Hutton's (q.v.) 1st Mounted Infantry Brigade. He was invalided back to Australia in August the following year, and was made CMG and mentioned in despatches. Between 1902 and 1906 he was aide-de-camp to the Governor-General. His association with Hutton paid dividends during the latter's period as C-in-C of the Australian Military Forces, and Hoad was his principal staff officer; for a few months in 1903-04 he also temporarily commanded the 6th Military District, based on Tasmania. In 1904 he was sent by the Australian government as an observer of the Russo-Japanese War, attached to the Japanese forces in Manchuria. With the reorganisation of the Australian Army following Hutton's departure, Hoad successively held the posts of Deputy Adjutant-General and Inspector-General (q.v.) before representing the Australian government in discussions in London in 1908 aimed at the creation of an imperial general staff. 'Politically powerful' in the view of one observer on the Committee of Imperial Defence, Hoad disliked Bridges (q.v.), and their positions on army reorganisation and the implementation of Hoad's recommendations concerning the Australian section of the Imperial General Staff were based as much on personal antipathy as anything else. Bridges was appointed the first Chief of the Australian General Staff in January 1909, with Hoad as his deputy, and when Bridges was sent to London in the middle of that year Hoad succeeded him to the post. His health broke down early in 1911, and he died of heart failure before the year's end.

HOBART (I), HMAS see **LEANDER CLASS LIGHT CRUISERS**

HOBART (II), HMAS see **CHARLES F. ADAMS CLASS DESTROYERS**

HOBBS, Lieutenant-General (Joseph John) Talbot (24 August 1864–21 April 1938). Hobbs was born in London in 1864 and migrated to Perth in 1887. He set up practice as an architect and in the next few years designed a number of important colonial buildings, private and public. His other great interest was the army. He joined the Volunteer Field Artillery in Perth in 1887 and by 1903 was commanding the 1st Western Australian Field Battery. Hobbs was an assiduous student of his craft. He attended gunnery courses in England in 1902 and 1906 and enrolled in the University of Sydney's military science course in 1909.

On the outbreak of war Hobbs was selected for the command of the artillery of the 1st Australian Division. He landed at Gallipoli but with the meagre artillery resources available, was able to do little to affect the course of any of the battles. In October he temporarily commanded the 1st Division but was evacuated in early November with dysentery.

In 1916 Hobbs accompanied the 1st Division to France. He took part in the bloody and indecisive battles around Pozières that were part of the Somme campaign. This was a frustrating period for gunners, for their equipment was not yet accurate enough to ensure close support of the infantry; accordingly even the best gunners could only provide unreliable support to their troops. At the end of the year Hobbs was given command of the 5th

Division and in January 1917 promoted to major-general. He quickly proved a popular commander who was frequently in the line with his troops. His first battle as divisional commander was, unfortunately for him, the mishandled affair at Bullecourt, where his protests and representations against the operation were ignored by higher commanders. Later in 1917 his division was involved in the successful phase of the Third Battle of Ypres. His men entered the line in late September to take part in the Battle of Polygon Wood. The day before the battle, the 25 September, the Germans launched a strong counter-attack on the troops to Hobbs's right. Only rapid action by the divisional commanders and the redoubtable H. E. Elliott (q.v.), who was one of his brigadiers, restored the situation and allowed the attack on the 26th to proceed to a successful conclusion.

After Polygon Wood Hobbs's division was pulled out of the line to rest. The German offensive in March 1918 saw its next employment. In April the tactically important village of Villers-Bretonneux, thought to be the key to the major Allied rail junction at Amiens, had fallen to the Germans. Hobbs was ordered to recapture it. Two brigades of the 5th Division accomplished that feat on 27 April in a brilliantly executed night attack. This action effectively halted the German advance in this area and was the beginning of the close Australian association with Villers-Bretonneux. Later Hobbs's 5th Division played a major part in the advance to victory. After the initial success at Amiens on 8 August,

Colonel Talbot Hobbs (right) at Gallipoli with a camouflaged 18-pounder gun, 1915. (AWM G01055)

HOGUE

the 5th Division helped to capture Péronne on 2
September and break the Hindenburg Line at Bullecourt on the 29th. When the Australian Corps
were withdrawn to rest after 5 October Hobbs
became temporary commander. He succeeded
Monash in command on 28 November 1918 with
the rank of lieutenant-general, and was made
KCMG in January 1919. He retained the command until May 1919.

As an architect Hobbs had an interest in the
design of commemorative memorials to the Australians in France and Belgium. Four of the five
designs for divisional memorials were his, and he
was responsible for the placement of the Australian
National Memorial at Villers-Bretonneux. On his
return to Australia he was appointed to a committee advising the government on army matters. In
1921 he was appointed to command the 5th Division and the 13th Mixed Brigade, which appointments he held until his retirement in 1927. He
died at sea in 1938 on his way to attend the
unveiling of the war memorial at Villers-Bretonneux. Short in stature, brave, and blessed with a
measure of common sense sometimes lacking in
First World War generals, Hobbs was regarded with
affection by his men and respect by his superiors.
He was one of the comparatively few soldiers
whose reputation was enhanced by the events of
1914–1918.

HOGUE, Major Oliver ('Trooper Bluegum') (29
April 1880–3 March 1919) was the son of a prominent Sydney newspaper editor and politician. An
active and athletic youth, he explored much of Australia on bicycle and worked as a commercial trav-
eller before joining the *Sydney Morning Herald* in
1907. Considering himself a bushman, he became
the principal celebrant of the light horsemen as the
ideal type of bushman hero. He enlisted in the 6th
Light Horse in September 1914, and while serving
on Gallipoli wrote articles for the *Sydney Morning
Herald* under the name 'Trooper Bluegum'; these
were collected in the book *Trooper Bluegum at the
Dardanelles* (1916). Another 'Trooper Bluegum'
book, *Love Letters of an Anzac* (1916), consisted of
sentimental love letters from a light horseman to his
Australian sweetheart, while a third book, *The
Cameliers* (1919), intersperses a fictional love story
with accounts of the Imperial Camel Corps in
action. Hogue had transferred to the Imperial
Camel Corps in November 1916, and served with
them through the campaigns in Sinai and Palestine.
Promoted to the rank of major in July 1918, he
commanded a squadron of the 14th Light Horse
through the last phases of the war in the Middle
East. Hogue also published poems in the light horse
newspaper *Kia Ora Coo-ee* and other publications.
His wartime writings blur the boundaries between
fiction and reportage and attempt to boost the pro-

Captain Oliver Hogue (centre) with other members of 4th Battalion, Anzac Section, Imperial Camel Corps,
Abassia, Egypt, 1916. (AWM H02695)

file and reputation of the light horse. Hogue drew on many of the clichés of Australian war writing, especially those of bushmen as natural soldiers and of war as a sport at which Australians excelled. He died in London of influenza contracted during the postwar pandemic.

HOLMES, Major-General William (12 September 1862–2 July 1917). From a military family, Holmes joined the New South Wales Public Service and worked at the Sydney Mint before moving to the Department of Public Works. In 1895 he became Secretary and Chief Clerk of the Metropolitan Board of Water Supply and Sewerage, a position he held until his death. He became a citizen soldier at the age of 10, joining the 1st Infantry Regiment of the New South Wales colonial forces as a bugler, was commissioned in 1886, and in 1903 was appointed to command of his original unit. He served in the South African War with the first contingent from New South Wales, and joined the Australian (Mounted Infantry) Regiment. He saw action at Colesberg, Pretoria and Diamond Hill, was awarded the DSO, mentioned in despatches and promoted to brevet lieutenant-colonel. He was wounded at Diamond Hill and returned to Australia in August 1900. From 1912 he commanded the 6th Infantry Brigade, and on the outbreak of war was selected to command the Australian Naval and Military Expeditionary Force. The force conquered German New Guinea (q.v.) quickly, and Holmes remained to administer the captured colony until relieved in January 1915. He took command of the 5th Infantry Brigade of the AIF in March 1915, and saw service at Gallipoli from August until the evacuation. He accompanied the brigade to France and led it through the Somme campaign in 1916, before being promoted to major-general and given the 4th Division. He was mortally wounded while conducting a group of visiting Australian politicians to the Messines battlefield on 2 July 1917. He had been awarded the CMG and was mentioned in despatches four times for his war service. Holmes was regarded as the epitome of the successful citizen soldier, and Bean among others cited his example when arguing the merits of citizen versus regular officers for senior commands in war.

HOLSWORTHY Part of the Liverpool Military Area (along with Moorebank and Ingleburn) in Sydney's south-west, Holsworthy has seen military activity since the early nineteenth century. A permanent military training area was gazetted in 1910, and during the First World War an internment camp for enemy aliens was established there. Con-

ditions in the camp were primitive for a population which by the middle of 1918 had reached 6000, and a number died, including two men shot while trying to escape. In both world wars the camp was used as a training and concentration area for troops departing for overseas service. Today Holsworthy provides facilities for live firing and field training, and is home to the army's 1st Brigade, which consists of 3RAR and 5/7RAR. The outward spread of Sydney has imposed limitations on the types of training now possible, particularly on the use of large-calibre weapons.

HONNER, Lieutenant-Colonel Ralph (17 August 1904–14 May 1994) was educated at Perth Modern School and became a schoolteacher. He was admitted as a barrister in 1935 after studies at the University of Western Australia. Commissioned into the 11th/16th Battalion of the CMF in June 1936, he volunteered for the AIF in October 1939 and served as a company commander with the 2/16th Battalion. Honner took part in the First Libyan campaign against the Italians, and came to official notice for his spirited leadership of C Company of the battalion during the fighting at Derna. Sent to Greece, he won the MC for action on the Thermopylae Line, and was praised by his commanding officer as 'the best company commander I have known in this or the last war'. Evacuated to Crete, Honner led a group of survivors from his battalion through the rugged back country until they made rendezvous with a RN submarine which evacuated them to Alexandria. Returning to Australia in May 1942, he was ordered to Papua to take command of the 39th Battalion, a militia unit of mostly very young soldiers which had already had three commanding officers in quick succession. Driven back gradually by superior numbers of Japanese, the 39th fought a two-day delaying battle at the village of Isurava on the Kokoda Track until relieved by the 2/14th Battalion. Heavily depleted by disease and enemy action, the battalion was rested until December when it was again sent in against strong Japanese resistance around Gona and Sanananda, for which Honner was awarded the DSO. After being sent back to the Atherton Tableland in Queensland for refitting, the 39th Battalion together with the other units of the 30th Brigade was disbanded and its men distributed as reinforcements for other formations. As commanding officer he knew ahead of his men what was to be done to them, but could say nothing. Two of his company commanders recalled later that in the days before the announcement he 'literally took to drink', something he normally never did, so upset was he over the fate of his battalion. Honner was posted to the 2/14th Battalion as commanding officer and led it through the fighting in the Ramu and Markham valleys of New Guinea in 1943–44, until seriously wounded in the advance towards Dumpu. He was evacuated and did not return to active service. In August 1944 he was appointed GSO1 in the Directorate of Military Training in Melbourne, and in January 1945 was placed on the Reserve of Officers. After the war Honner held posts with the War Pensions Assessment Appeal Tribunal, was president of the United Nations Association (NSW Division), and State president of the Liberal Party of New South Wales between 1961 and 1963, the first Catholic to hold that position. From 1968 to 1972 he was Australian Ambassador to Ireland.

HOPKINS, Major-General Ronald Nicholas Lamond (24 May 1897–24 November 1990). A graduate of the last class of RMC Duntroon to see service in the Great War, Hopkins went to the light horse in Palestine, serving with the 6th Light Horse and on the staff of the 3rd Light Horse Brigade during the repatriation of the force from Egypt in 1919. He returned to a succession of staff and instructional jobs with the militia which, while professionally stultifying, enabled him to maintain his horses. A period as company officer at RMC Duntroon in 1926 was followed by a two-year period at the Staff College, Quetta, British India (now in Pakistan). Between 1937 and 1939 he was sent to Britain for instruction in armoured warfare, the first Australian Army officer to specialise in the area. On his return in May 1939 he was given the appointment of GSO2 (Mechanisation and Armoured Fighting Vehicles) at Army Headquarters, but little serious progress was made before the outbreak of war. He led the divisional cavalry of the 7th Division between 1940 and 1941, and then returned to staff duties concerned with armoured vehicles and the mechanisation of the army, experience which saw him appointed GSO1 of the newly created 1st Australian Armoured Division, under Major-General John Northcott (q.v.), in May 1941. Thereafter he served on Land Headquarters and as BGS to New Guinea Force (q.v.) during the critical fighting from September 1942 which saw the Japanese pushed out of Papua (see NEW GUINEA CAMPAIGN). From February 1943 to September 1944 he was on the staff of the 7th Amphibious Force, a US Navy command, but this was his last direct involvement in the war. From September 1944 to April 1946 he commanded the Australian Staff School at Cabarlah in Queensland, and its successor, the Australian Staff College. From there he was chosen for command of the 34th

Australian Infantry Brigade, the major Australian component of the British Commonwealth Occupation Force (q.v.) destined for duty in Japan. His selection was probably a mixture of availability, a demonstrated ability to work with the Americans, and previous service with BCOF's first commander, Northcott. He was to command in Japan until December 1948, and made a considerable success of a difficult command, difficulties occasioned first by the trying material circumstances in which his soldiers operated initially, and then by the gradual realisation that they fulfilled little obvious purpose in the occupation force. After returning from Japan, Hopkins commanded the 4th Military District before appointment as Deputy CGS in May 1950, and was Commandant of RMC Duntroon from 1951 to 1954. He was Chief Executive Officer of the first Adelaide Festival of the Arts in 1960, and was later an Honorary Fellow of St Mark's College at the University of Adelaide. In retirement he wrote a serious study of the development of the Australian armoured corps, which rose above the usual parochial preoccupations of regimental histories to analyse the effectiveness and purpose of Australian use of armoured forces from the Second World War to Vietnam.

R. N. L. Hopkins, *Australian Armour: A History of the Royal Australian Armoured Corps 1927–1972* (Australian Government Publishing Service, Canberra, 1972).

HORSEFERRY ROAD see **ADMINISTRATIVE HEADQUARTERS, AIF**

HORSES see **ANIMALS; WALERS**

HOWSE, Major-General Neville Reginald (26 October 1863–19 September 1930). Born and educated in England, Howse migrated to New South Wales where he eventually set up his medical practice in Orange. Commissioned as lieutenant in the New South Wales Medical Corps in January 1900, he sailed to South Africa with the 2nd Contingent in the Boer War, and in an action at Vredefort in the Orange Free State rescued a wounded man under heavy crossfire, for which he was awarded the VC in June 1901. He was promoted to captain in October 1900, captured by the Boers, released as a noncombatant and returned to Australia. He went back to South Africa as an honorary major in the Australian Army Medical Corps (see ROYAL AUSTRALIAN ARMY MEDICAL CORPS) just before the end of the war. He remained in the AAMC Reserve and in August 1914 sailed with the Australian Naval and Military Expeditionary Force to German New Guinea (q.v.) as principal medical officer, in which position he was responsible, by providing appropriate drugs and treatment, for minimising serious illness. Returning to Australia, Howse embarked with the first contingent of the AIF as staff officer to the director of medical services. At Gallipoli he personally organised the evacuation of the wounded from the beaches, and instituted a series of measures designed to improve health and fitness. In November 1915 he became director of medical services of the AIF, an appointment that enabled him to maintain the independence of the AAMC (even against the attempts of Australian senior officers to bring it under their control) and to insist, as he did when the bulk of Australian forces moved to Europe, that he have direct access to the GOC of the AIF. With his headquarters in London, Howse controlled medical services in Egypt, Palestine, France and Belgium. He introduced surgical teams in the field, and reorganised the field ambulances to increase their efficiency. In January 1917 he was appointed KCB and promoted to major-general. He returned to Australia in October 1918 to brief the Minister for Defence on crippled returned soldiers, and went back to London in February 1919 to help direct the repatriation program (q.v.). He was mentioned in despatches in 1919, and appointed KCMG and a Knight of the Order of St John of Jerusalem. In 1920, after a brief return to private practice, Howse was appointed Director-General of Medical Services as a regular major-general, but since that appointment prevented him speaking on wider issues of public health, he resigned in November 1922 and was subsequently elected as a Nationalist MP for the seat of Calare. From January 1925 until April 1927 he was minister for Defence and Health and minister in charge of repatriation, but ill health forced him to give up the Defence and Health portfolios, although he continued to administer the repatriation program. He returned to increased ministerial responsibilities in February 1928, but lost his seat in the 1929 election. The following year he went to England for medical treatment, and died of cancer in London.

Howse was responsible for the establishment of the AAMC on a truly professional basis. Shocked by what he regarded as the 'criminal negligence' of the British authorities in providing for the medical relief of casualties at Gallipoli, Howse strove to minimise the effects of war on soldiers. His concern for the health of the AIF carried over in his efforts on their behalf in the repatriation program, and he attempted to widen the provision of health services to the Australian public.

HUDSON, LOCKHEED (5-crew medium bomber [Mark II]). Wingspan 65 feet 6 inches; length 44 feet; armament 5 × 0.303-inch machine-guns, 750 pounds bombs; maximum speed 246 m.p.h.; range 2160 miles; power 2 × Pratt & Whitney Wasp 1050 h.p. engines.

The Hudson was a military version of the Lockheed Model 14-passenger aircraft designed in 1938 for the British government as a maritime patrol and reconnaissance bomber replacement for the Avro Anson (q.v.). The Australian government placed its first order of 100 Hudsons to replace its Ansons in late 1938 and they were the main bomber type used by the RAAF until 1943. No. 1 and 8 Squadrons, equipped with Hudsons, were stationed in Malaya and Singapore in 1941, and Hudsons flew the first attack by Australian airmen against the Japanese during the Malayan campaign (q.v.). Their slow speed made the Hudson relatively easy for Japanese fighters to shoot down, but because they were the only aircraft available they equipped 12 RAAF bomber squadrons. Hudsons were used to support Sparrow Force in Timor and during the Papuan and New Guinea campaigns. They were replaced by Beauforts (q.v.) from 1943. No. 459 Squadron used Hudsons successfully in the Mediterranean between 1942 and 1944 as maritime patrol aircraft, a role for which the Hudson was more suited. In August 1940 a Hudson used as a passenger aircraft crashed on the approaches to Canberra airport, killing J.V. Fairbairn (q.v.), Minister for Air, Brigadier G. A. Street (q.v.), Minister for the Army, H.S. Gullett (q.v.), Vice President of the Executive Council, and General Sir Brudenell White (q.v.), CGS. It has been speculated that Fairbairn, a First World War combat pilot, was piloting the aircraft when it crashed.

HUGHES, William Morris (25 September 1862–28 October 1952). Born in London of Welsh parentage, and educated largely in Wales, Hughes migrated to Queensland in 1884 with an assisted passage, and lived for a time in acute poverty before setting himself up in business in Sydney. An important figure in the organisation of the early Labor Party in Sydney, he won a federal seat in 1901 and in the short-lived Labor government of 1904–05 he held the External Affairs portfolio. In the decade before the outbreak of war Hughes became a leading figure in the parliamentary and broader Labor movement. Long interested in defence issues, he was an enthusiastic and very public advocate of the Australian National Defence League and a strong supporter of the compulsory military training scheme introduced in 1909 (see CONSCRIPTION).

Hughes was attorney-general in a Labor government at the beginning of the war; he suggested abandoning the federal election then under way and the formation of an all-party government, which was unconstitutional. In the event, Labor regained office and Hughes quickly became a leading advocate of a vigorous prosecution of the war. When Andrew Fisher (q.v.) resigned as prime minister in October 1915, Hughes was the party's unanimous choice to succeed him.

The Labor movement was a coalition of often disparate interests, and Hughes's combative style of leadership and advocacy of the war effort placed it under serious strain even before the conscription referendum of 1916. As attorney-general he had devised a referendum campaign to give the Commonwealth control over commerce and industrial relations, something which found little favour among many Labor leaders in the States, but which was ultimately granted in 1916 by a decision of the High Court. His three-month visit to Britain in March–July 1916, including time spent with the AIF, convinced him of the need for conscription for overseas service as the only satisfactory means of maintaining the strength of the AIF in the face of the enormous casualties sustained on the Western Front. Hostility within the Labor Party and the Senate towards the proposal closed off the legislative option, and Hughes decided on the expedient of a referendum which, if carried, would give him the authority needed to enable conscription legislation.

Views on conscription had polarised in his absence, with strong opposition building in both the industrial and political wings of his party. Hughes campaigned vigorously for the 'Yes' case, but his own abrasive personality and the suppression of the Easter Rising in Dublin by the British diverted attention much of the time from the substantive issues. The ill-judged decision to invoke the call-up of single men for service in Australia under the Defence Act confirmed the suspicions of many, and the referendum held on 28 October was lost by a small majority of votes and in three States.

The impact of the first conscription campaign was felt most in the Labor Party which, after expelling Hughes, split. Many of those who followed Hughes represented the older generation, socially but not necessarily politically radical and nationalist rather than internationalist in outlook. Hughes soon merged with the Opposition in order to form a Nationalist government. The Senate forced him to the polls in May 1917; against predictions, his new party won a smashing victory with majorities in both houses. Hughes himself was returned with a large majority in a different seat, Bendigo, in Victoria. The election result represented two things: on the one hand, the electorate had voted in favour of Hughes's prosecution of the war effort by all means short of conscription; on the other, Hughes was now isolated from 'his political,

social and geographical roots'. And the election result explicitly did not provide a mandate for the introduction of conscription.

The disasters of 1917, however, predisposed Hughes to raise the conscription issue once again, in November 1917. The second campaign, set to culminate on 20 December, was even more hysterical and spiteful than the first. The Irish issue again played a prominent part, as did the spectre of Bolshevism and the activities of the Industrial Workers of the World, many of whose tiny membership were arrested. Hughes himself polarised opinion even further, especially through his impassioned public denouncements of Archbishop Mannix of Melbourne. The referendum was lost by a larger majority than before. Hughes resigned, but was recommissioned by the Governor-General with the same cabinet as before. The domestic political atmosphere for the rest of the war was poisonous.

Hughes was in England when the Armistice was signed, and helped to ensure that the dominions would be represented by their own leadership at the peace conference. At the conference itself in Paris the following year, he took a strong line on reparations, the annexation of German New Guinea, and the racial equality clause of the proposed covenant of the League of Nations; he clashed famously with the American President, Woodrow Wilson, on a number of occasions. It is open to argument whether he succeeded in getting his way on matters of importance to Australia through his skilful and unorthodox diplomacy, or because issues in the Pacific were of little interest to the Great Powers.

He returned to great popular welcome, but was forced from office after the election of 1922 when the Country Party, holding the balance of power, refused to work with him. In the interwar years he played a variety of roles in and out of office, often seemingly more intent on a spoiling role. In the 1930s he was an often lone voice warning of the need to prepare for war against growing Japanese aggression in the Pacific. He held a number of portfolios in the United Australia Party ministries at the beginning of the Second World War, and sat on the Advisory War Council. He later joined the Liberal Party and in the 1949 election was returned for the seat of Bradfield; at his death he had been a member of parliament continuously for more than half a century.

Hughes is a highly controversial figure. Regarded by many in the Labor Party as a 'rat' for creating the first of the great splits in that party, he was nonetheless an energetic and vigorous wartime leader. Lord Bruce, by no means an admirer, said of him that 'he had two qualities which are very rare and very important in a politician: he had imagination and he had courage'. Lionised by many in 1919 as 'the little digger', his political career went into eclipse increasingly thereafter. His diplomatic achievements were significant in helping to assert a separate place for Australia within the councils of the Empire at that time. His legacy, on the whole, however, was mixed.

L. F. Fitzhardinge, *William Morris Hughes: A Political Biography*, 2 vols (Angus & Robertson, Sydney, 1964, 1979).

HUMOUR has played an important part in Australians' images of their military heroes. C. E. W. Bean (q.v.), reporting on the actions of the AIF at Pozières in 1916, reassured his readers that Australian troops were 'just the men that Australians at home know them to be; into the place with a joke, a dry cynical Australian joke as often as not'. Indeed, such was the wartime appetite for stories of laughing diggers that official war artist (q.v.) Will Dyson complained that, while it was 'proper and wise' for soldiers to adopt a humorous attitude towards their hardships, for people at home 'to be so preoccupied with discovering the humours of the soldier's lot is scarcely seemly'. Naturally, people on the home front dwelt on the light side of military life because they did not want to consider the possibility of their loved ones suffering; but believing that Australian soldiers brought a uniquely humorous approach to war also reinforced Australians' image of themselves as a people with a dry, sardonic, self-deprecating and pretension-busting humour.

The question of whether or not Australian military humour can be taken as evidence of a distinctively Australian comic tradition is a difficult one. Certainly there is some continuity with the humour of bush workers, who shared with soldiers a harsh, routine-dominated existence and an overwhelmingly male social environment. Soldiers' jokes about English officers have something in common with the mocking of upper-class Englishmen in 'new chum' jokes, while soldiers and bush workers shared another favourite target in the much maligned cook. The popular army joke in which an officer's query, 'Who called the cook a bastard?' receives the reply, 'Who called the bastard a cook?', has also been anthologised as a shearers' yarn. Despite this continuity, though, Australian military humour does not seem particularly distinctive. The stereotypical characteristics of Austra-

Australian soldiers changed the name of this Port Moresby hotel from the 'Starview' to the 'Starve-U' Hotel, Papua, July 1942. (AWM 025636)

lian humour are there in abundance, but also present is the sort of humour that is common to the armed forces of other countries.

Humour is an excellent coping mechanism for people in stressful situations, and has been used as such by Australian service personnel both in peacetime and at war. Humour works in several ways to make hardship easier to bear. For a start, it relieves tension. Simple silliness is often most effective in this regard: through incongruity, the making of seemingly illogical connections and comparisons, soldiers made light of their own anxieties. Thus, during the landing at Gallipoli one boat was convulsed with laughter after a man recalled an old soldier's earlier comment that bullets sounded like birds flying overhead, while on other boats men joked that 'They want to cut that shooting out, somebody might get killed.' Under bombardment in the trenches men shouted 'Open the door and pay the rent; that's the landlord.'

Humour can also act as a safety valve, a form of protest and complaint that stops short of actually trying to change things. Bad food, pompous officers, uncomfortable conditions, the never-ending war against rats, lice, mosquitoes and other pests; all were routinely the subjects of jokes and humorous songs. This song from the Vietnam War is typical of many such humorous complaints:

> Driving through the mud in a jeep that should be junk
> Over roads we go, half of us are drunk.
> Wheels on dirt roads bounce, making arses sore,
> Lord I'd sooner go to hell than finish out this war.
>
> Jingle bells, mortar shells, VC in the grass,
> We'll get no merry Christmas cheer until this year has passed.
> Jingle bells, mortar shells, VC in the grass,
> Take your merry Christmas cheer and shove it up your arse.

A soldier at Tobruk used the great Australian adjective to express his feelings with equal directness:

> This bloody town's a bloody cuss
> No bloody trams no bloody bus
> And no one cares for bloody us
> Oh bloody bloody bloody.

While the trials and discomforts of military life were dwelt on and even exaggerated, however, the real dangers of warfare were played down, as in the following song from the First World War:

> We're all waiting for a shell
> (Send us a nine-two) —
> Please don't keep us waiting long,
> For we want to go to Blighty,
> Where the Nurses change our nighties,
> When the right shell comes along.

Such attempts to laugh away fear and horror reveal the limits of the protest expressed through humour. Most of those who engaged in humorous complaint accepted their situation, and their protests were rarely translated into revolt even over the conditions they complained about, let alone more radical action against the war or the military hierarchy.

One group who were definitely unwilling to accept their situation, however, were prisoners of the Japanese during the Second World War. Unable to mount any effective protest against their captors, they turned instead to humour to maintain their sanity and reassure themselves that they had not lost their fighting spirit. On the Burma–Thailand Railway, Australians would sing 'Go home you mugs, go home' or 'They'll be dropping thousand pounders when they come', taking great delight in the Japanese guards' enjoyment of these subversive songs. One of the most famous prisoner-of-war stories, told here by George Aspinall, shows the pleasure Australians took in 'getting one over' their gaolers:

> Now we had one particular Japanese lieutenant who had just enough English to get himself into bother. He decided that he was going to address us about scrounging cigarettes and food. First of all he got a digger's slouch hat, put a tin of condensed milk on the ground and covered it with the hat. Then he mounted a box and started to talk to us: 'Now you Australian soldier you think we know fuck nothing that's going on here. You are wrong, we know fuck all!' He got down and said, 'You have your hat there and under it you are stealing things.' He went to pick up the hat and show the tin of condensed milk; but the tin was gone. Somebody had picked it up before he got there!

Here the Japanese officer's comical misuse of English simultaneously reinforces his foreignness and his impotence in the face of Australian ingenuity: he really *does* know fuck all. The story ends with an Australian victory, rather than with the punishments that might be expected to have followed such an incident. Symbolically, the tables have been turned, and the Australians have ended up where they feel they should be, on top. In such stories POWs drew on Australian anti-authoritarian traditions to help overcome their humiliation at being captured by people they had been taught to regard as inferior.

Another important role of humour in military life is in socialisation into the military community. Humour can be used to draw social boundaries; the significance of the 'in joke' lies in its ability to alienate those who are outside the group and therefore lack the knowledge necessary to get the joke, while reinforcing the feeling of those inside the group that they share a common language, experiences and beliefs. The use of in-group slang is one way of reinforcing group identity. Thus, when First World War soldiers were asked how they spent their time in the trenches they liked to reply, 'Oh, we just sit around chatting.' Only those who knew that 'chats' were lice would get the joke. Employing a particular style of humour is also an effective way of developing group solidarity. Those who were unable to participate in the bawdy, aggressive, competitive style of humour characteristic of all-male groups in Australia were likely to be ostracised to some degree within the armed forces. More recently, though, since the armed forces have had to address problems of sexual harassment and bastardisation, this style of humour has become less acceptable, at least at an official level. Calling female naval personnel SWODs (Sailors Without Dicks) is a type of joke, but not one which the women who are so labelled can be expected to find funny, as it is deliberately intended to humiliate them. Bastardisation rituals at RMC Duntroon, once a crucial part of socialisation into the Corps of Staff Cadets, have been defended on the grounds that they are funny, not cruel. As one former cadet put it, 'I was a fourthie [Fourth Class cadet] for a year and it was the best year of my life, certainly the funniest! Humour to remove the pressures of College is the key — not bastardisation.' However, following embarrassing publicity about bastardisation rituals in 1969, military officials have made it clear that such 'humour' is unacceptable.

Humour in the Australian armed forces has taken a wide variety of forms, including jokes, stories, songs, mimicry, teasing and practical jokes. It has ranged from off the cuff quips to stories and songs that have passed into military folklore. It could be crude, but it could also be remarkably inventive: in some of their puns, for example, Australian soldiers displayed a capacity for sophisticated and creative wordplay. The people responsible for the sign which renamed a French street 'Roo de Kanga', or for calling the 1st AIF's 48th Battalion 'the Joan of Arc Battalion' (commanded by Raymond Leane, the 48th also contained a number of his relatives; hence, 'Made-of-all-Leanes') were comic wordsmiths of the highest order. Most military humour was verbal or physical in nature, but some of it took the form of cartoons, stories, mock-advertisements and so on, which appeared in service newspapers (q.v.) and other publications. This written material, combined with the recollections of ex-service personnel, and some humorous material recorded in private letters, is the source of our knowledge about the nature of military humour. Unfortunately, however, unless such pub-

lications were truly ephemeral and intended only for distribution within the military unit, the writers and cartoonists would have engaged in self-censorship, leaving out material which might offend people back in Australia.

Although military humour is very diverse, there are a few general characteristics that are common to much of it. Many jokes highlighted the Australian soldier's stereotypically laconic and insouciant nature. One such was the story of the Australians on the Western Front who found a man buried to the neck in mud. Their efforts to help him were to no avail, until at last he drawled, 'Wait a minute mates and I'll take me feet out of the stirrups.' Even more common in Australian military humour are expressions of anti-authoritarianism and impatience with military formality. As early as March 1915 a story was in circulation which had a hapless sentry at a camp in Egypt, on challenging a stranger with 'Halt! Who goes there?' and receiving the reply 'What the —— has that got to do with you?', resignedly saying 'Pass, Australian'. Stories of Australians mocking stuffy British officers (usually equipped with that symbol of upper-class ostentation, the monocle) are legion. There is the story of the two lost Australians who, happening on a British general, ask where they are. 'Do you know who I am?' responds the general angrily, whereupon one Australian turns to the other and says, 'Cripes, Bill, here's a bloke who's worse off than we are. We don't know where we are, but this poor blighter doesn't know *who he is*.' One notable exception to the general contempt for officers expressed in military humour is the treatment of General Sir William Birdwood (q.v.), who generally seems to appear as a sympathetic figure. Rather than being the butt of the joke, Birdwood plays the role of straight man, as in the story which has him, at Gallipoli, seeing an Australian soldier using a wheelbarrow. 'Excellent work,' says Birdwood, 'did you make it yourself?' 'No, Mr Birdwood,' the Australian replies, 'I bloody well didn't, but I'd like to find the bastard who did!' Other generals, though, came in for much harsher treatment. During the Second World War campaigns in North Africa, Australians sang scathingly of 'these crazy bastards Churchill sends out here':

Now the generals that they sent us
Had not a bloody clue.
They ought to round the bastards up
And put them in a zoo.

They said, 'Keep your eye on Rommel,
Don't let the bastard pass.'
But he'd sneak around behind them,
Then he'd slam it up their arse.

Hostility was not just reserved for those fighting on the same side as the Australians; enemies came in for their share of ridicule too. An Australian soldier recorded this verse about the German Emperor in 1916:

Here's to the Kaiser, the son of a bitch,
May his balls drop off with the seven years itch,
May his arse be pounded with a lump of leather,
Till his arsehole can whistle 'Britannia' forever.

Other jokes were directed against the enemy 'race' as a whole, as part of the process whereby enemies are dehumanised in order to make them easier to kill. An example of this kind appeared in a 1984 issue of an underground satirical magazine produced by cadets at Duntroon, but it presumably originated during the Vietnam War:

Q: What has 700 balls and fucks gooks?
A: A claymore.

There can be little doubt that much Australian military humour concerned women and sex, but it is in the nature of such bawdy humour that it tends not to get written down. Probably the sex in Australian military humour was like that in British armed forces songs described by Martin Page, which feature sex 'of a grotesque and fantastical kind ... it is virtually impossible to sing them and keep in one's mind what real sex is like'. It seems likely, too, that much of it concerned what Australians imagined to be the sexual practices of the people in whose countries they were fighting (the song 'You can't fuck Farida if you don't pay Farouk' is one example of this) or expressed concern about the morals of women back in Australia (POWs in Singapore, for example, referred to dried fish as 'modern girls': two-faced and no guts).

Australian service personnel have not only mocked others, but have also turned their humour against themselves. Many humorous songs have an element of self-deprecation, although this may simply be false modesty. The famous 'We are the Ragtime Army' song of the First World War had its counterpart in Korea, where Australians sang:

We're a pack of bastards,
Bastards are we,
We come from Australia,
The arsehole of the world
And all the universe.
We're a pack of bastards,
Bastards are we,
We'd rather fuck than fight
For Syngman Rhee.

The regular army, navy and air force have their own traditions of humour which are largely inaccessible to outsiders but which are passed on to new recruits as part of their socialisation into the military environment. The humour of the citizen soldiers of the First and Second World Wars, on the other hand, was recorded after the wars in newspapers such as *Smith's Weekly* and in numerous collections of digger songs and stories. Many people, also, would have heard humorous stories told by friends and relatives who were in the wars. Such storytelling was part of a process whereby memories were reconstructed to emphasise the humour rather than the horror of war, and a bowdlerised version of military humour at that. In George Johnston's semi-autobiographical novel *My Brother Jack* (1964), the character David Meredith recalls returned servicemen who would gather at his parents' house to exchange 'badinage ... in bad soldier-French' and 'interminably repeated and curiously esoteric jokes about saps and support-trenches and furphies and bedpans and whizzbangs and entrenching tools'. In such private get-togethers, as well as in more formal reunions and 'smoke nights', returned men (and occasional returned women, such as the character of David Meredith's mother) told such esoteric jokes to reinforce their common bond and their feeling of difference from those who had not been to war, but they also told them to try to forget a large part of what war had really been like. As Bill Langham told the historian Alistair Thomson, 'You're taboo if you start talking about war if you go to a smoke night. You pick out all your funny incidents. It's like, you want to forget it, see. You want to forget the bad parts, which we all do.'

Graham Seal, 'Two traditions: The folklore of the digger and the invention of Anzac', *Australian Folklore*, no. 5 (September 1990), pp. 37–60.

HUON, HMAS see **RIVER CLASS TORPEDO BOAT DESTROYERS**

HURLEY, (James) Francis (15 October 1885–16 January 1962) joined the AIF in August 1917 as official photographer, with the honorary rank of captain. Hurley had made his name as a photographer on several expeditions to the Antarctic: with Douglas Mawson's expedition from December 1911 to March 1913, and with Sir Ernest Shackleton's expedition from October 1914 to November 1916. In France and Belgium (see WESTERN FRONT) Hurley tried to convey the magnitude of the military struggle through stark images of the front line, including shots of exploding shells which he took at great personal risk. More an artist than a mechanical recorder, Hurley clashed heatedly with the official historian, C. E. W. Bean (q.v.), over his intention to combine several photographs into one.

Hurley embellished his images by adding detail where necessary; 'judicious manipulation' he called it, but to Bean, 'combination printing' was little short of forgery. If the finished product had a certain staged appearance, it was nevertheless immensely powerful. One of the most famous photographs, of an Australian infantry attack at Zonnebeke, measured 6.5 by 4.5 metres, and used 12 different negatives. Criticism of Hurley's methods and what he saw as censorship of his work led him to resign. He moved to the Middle East and spent some time recording the exploits of the light horse, most notably in the battle for Jericho, an experience which led him to photograph the cavalry charge sequence in Charles Chauvel's film *Forty Thousand Horsemen* (1940). In the Second World War Hurley volunteered three times for overseas service before being accepted, and was appointed head of the Department of Information's Cinematographic and Photographic Unit in the Middle East, a position he held until 1943 when he became director of the British Army's Features and Propaganda Films Unit, where he remained for three years. Apart from several documentary films (for example *Siege of Tobruk*) in the early stages of the war, his work attracted relatively little attention, and his style, with its reliance on careful craftsmanship and attention to detail, lacked the immediacy of the work of a younger generation of war photographers such as Damien Parer (q.v.).

HUTTON, Lieutenant-General Edward Thomas Henry 'Curley' (6 December 1848–4 August 1923), born in England and educated at Eton, entered the British Army as an ensign. Campaigning in Africa from 1879 to 1885, he served in the new mounted infantry arm, and on returning to England raised and commanded mounted infantry units. It was then customary that British regular officers commanded the military forces of Britain's settler colonies (see COLONIAL MILITARY FORCES), and Hutton was to hold three such appointments from 1893 to 1904. He was guided in this work by a vision of imperial military federation, believing that the Empire's citizen military forces, which outnumbered the small British Army, should be organised and trained to form on war's outbreak a vast militia, well trained and disciplined and strong in mounted troops who used rifles rather than swords. Thus he sought to form the colonial soldiers under his command into forces of all arms, to mount many of them on horses, to train and discipline them toughly, and to persuade their governments to commit them to serving outside their colonies in war.

Despite informal support for his efforts from the imperial government, Hutton's frantic energy, intemperate outbursts and inability to understand local resistance to his vision ensured his failure. Hutton's command in New South Wales (1893–96) was his least controversial. He led only a tiny force, and financial depression restricted what he could make of it. Still, he made training more rigorous, formed regiments into brigades, raised new corps, and imposed a khaki uniform. He led an unsuccessful campaign to inspire Australian colonial governments to act on Major-General Bevan Edwards's (q.v.) advice and form their paid citizen soldiers into a federal militia obliged to serve anywhere in the south-west Pacific. During his appointment in Canada (1898–1900) Hutton began to turn its disconnected and dispirited military units into a modern army. He raised new corps, converted the cavalry from swordsmen to riflemen, and retired aged and incompetent officers. His new broom soon alienated both soldiers and politicians. Worse, he went behind the government's back and drafted secret plans to send a force to South Africa to fight in the Boer War. Early in 1900 the Canadian government forced him to resign. Hutton escaped to the Boer War, led a mounted brigade of soldiers from around the Empire, and restored his reputation. He now claimed that the war had proved him right: colonials obviously wished to participate in imperial wars, and clearly the horse's mobility was the key to success on the modern battlefield.

He enthusiastically took up his final colonial appointment late in 1901 as GOC in Australia. Hutton's task was to amalgamate as cheaply and easily as possible the military forces of the six Australian colonies into an army, and to train and discipline it for modern war; his hope, and that of the imperial government, was that he would organise the army for operations overseas within an imperial force. But the Australian government affirmed its power to decide itself at war's outbreak whether it would raise a force for service overseas.

Hutton was also largely frustrated elsewhere. He amalgamated the six colonial forces much as Bevan Edwards had proposed, putting paid citizen soldiers into federal brigades, most of which were to ride horses, and unpaid citizen soldiers and permanent soldiers into garrison forces. But the amalgamation had to be accompanied by financial retrenchment, and the disbandments and lower pay this sometimes caused, together with dislike of new drill and tougher discipline, provoked a wave of protest among citizen soldiers. Hutton's hounding from the army of several citizen officers, some for incompetence, others for breaches of discipline, made matters worse. Meanwhile, with the end of the Boer War, martial enthusiasm declined and recruiting fell away. By July 1904 Hutton had created an Australian army, but it existed more on paper than in reality. Hutton's term of office ended later that year with a furious quarrel and his resignation. He returned, a failure, to England. He left behind him, however, a tiny clique of devotees. Its most talented member was William Throsby Bridges (q.v.). Hutton was appointed KCB in 1912.

CRAIG WILCOX

HYDE, Admiral George Francis (19 July 1877–28 July 1937). Born at Southsea, Southampton, in England, Hyde's burning ambition to enter the RN direct was thwarted by his family's financial circumstances; instead he had to enter the merchant marine, gain a commission in the Royal Naval Reserve (RNR), and be commissioned in the RN through a supplementary list. He duly served in the merchant marine from 1894 to 1898, and in the RNR as a midshipman from 1896, being promoted to sub-lieutenant in 1901 and lieutenant the following year. Although in the RNR, he was able to undertake service continuously in the RN from 1899, finally being gazetted lieutenant in the RN in 1905. After several sea commands he was seconded to Australia in 1910 to command the Commonwealth Naval Force destroyer flotilla. He transferred to the RAN in 1912 with the rank of commodore, and the following year joined the HMAS *Australia* and sailed on her maiden voyage to Australia. He served on *Australia* for the first year of the war, pursuing the German Pacific Fleet, and then returned to the United Kingdom to command HMS *Adventure* in the Coast of Ireland Command from July 1915 to December 1917. He was promoted to captain and mentioned in despatches. Following two shore-based postings he returned to Australia in August 1918 to become Director of the War Staff in Navy Office, a position he held for a year. From 1923 to 1924 he was Second Naval Member of the Naval Board, and commanded the Australian Squadron from 1926 to 1928. He was promoted to rear-admiral in 1928, and from 1930 to 1931 he commanded the 3rd Battle Squadron of the British Home Fleet. He became First Naval Member of the (Australian) Naval Board (i.e. CNS) on 20 October 1931, and remained in the position until his death in 1937. Hyde's tenure in that office was marked by the competing demands of financial stringency and a deteriorating international situation. He believed absolutely in the closest possible cooperation between the RAN and the RN, and accepted without demur the strategic policies articulated by the Admiralty. As naval expenditure doubled in the period 1931–37, Hyde was responsible for guiding the naval rearmament program: the new cruiser HMAS *Sydney* was acquired, and five destroyers replaced the obsolescent 'S' Class, so that when war broke out in 1939 the RAN was as well prepared for war as the restrictions under which it had laboured during the 1930s would allow. Hyde was promoted to vice-admiral in 1932, and admiral in 1936, and was appointed CBE in 1926, CVO in 1927, and KCB in 1934. He was the first officer of the RAN to reach the rank of admiral, and the first of its seagoing officers to become First Naval Member. His beginnings in the British merchant marine and his own decision to transfer to the RAN made his achievements all the more remarkable.

HYDROGRAPHIC SURVEYING in Australia was conducted largely by British ships until 1860, when the Admiralty began conducting surveys in colonial vessels before returning to the use of RN ships from 1880 to 1926. The RAN Surveying Service (later Hydrographic Service) was formed in 1921, and in 1946 the Commonwealth government officially gave responsibility for surveying and charting Australian waters to the RAN. Before the Second World War the British Admiralty had maintained complete chart coverage of Australian waters, but with charts increasingly being produced in Australia, an agreement was reached in 1963 that each country would reproduce the other's charts. The introduction of echo-sounding equipment in the late 1920s and electronic position-fixing equipment in the late 1950s made the production of much more accurate charts possible. By 1983, however, 64 per cent of Australia's continental shelf had still not been surveyed since the invention of sonar and echo sounder.

(See also SURVEY SHIPS.)

I

IKARA Dissatisfaction with US and British anti-submarine missiles in the late 1950s led to the Australian government ordering the development of its own system in 1960. Shipboard trials commenced in 1963 and the system entered service in 1965. The system developed was named the Ikara, a pilotless miniature fixed-wing missile. The Ikara is launched from a surface vessel (frigate or destroyer). It is propelled toward its target by rocket motors and guided by radar or radio to the near vicinity of the enemy craft. It is then dropped by parachute on receipt of a command, enters the water and homes in on the target. It has a range of 18 kilometres and has been sold to the RN and the navies of Brazil and New Zealand.

IMPERIAL CONFERENCES were a consultative device intended to coordinate British and dominion policy, especially in the area of defence. They grew out of the Colonial Conferences, the first of which was held in April 1887 to coincide with Queen Victoria's Golden Jubilee. While founded in an atmosphere in which notions of imperial federation were actively discussed, the conferences were in fact marked by a strong concern on the part of colonial, and later dominion, governments that their prerogatives not be compromised by London. The 1887 conference was notable for producing the Naval Defence Agreement with the Australasian colonies, which established the principle of colonial contribution (though not of unlimited contribution) towards the costs of imperial defence. A further conference was convened in Canada in 1894, and unusually for gatherings before 1914 did not deal with defence or external policy issues (although the record of its proceedings fill some 400 printed pages). The Queen's Diamond Jubilee in 1897 was the occasion of a further gathering, which decided, in response to Joseph Chamberlain's ruminations on the attractions of a formal council of the Empire, that the existing relationships were 'generally satisfactory'. Thus even before dominion status had been conceded to most, colonial statesmen were showing themselves disinclined to accept a greater degree of centralised direction from London; and this was to be the pattern of Anglo-dominion relations until the Second World War rendered the whole issue academic. The 1902 conference (the last to carry the 'colonial' designation) convened for the coronation of Edward VII and was held in the context of the South African War, in which British military abilities had been sorely tested. Indeed, Chamberlain and others were at pains to reassure the delegates concerning Britain's reliability as a great power, while at the same time, and perhaps in slight contradiction, appealing for greater colonial contributions to help ease the burden of imperial defence. The New Zealand proposal for a joint colonial and imperial reserve, however, found little favour with other delegates and was not proceeded with.

From 1907 the nature and purpose of the conferences changed, along with the name. As the British historian, Nicholas Mansergh, has observed: 'Gone was the indeterminate character of the 1887 gathering and in its place was a Conference with a defined composition and an established status.' Delegates now represented dominion, not colonial, governments and conferences were to be scheduled every four years to discuss 'questions of common interest … as between His Majesty's government and His governments of the self-governing dominions beyond the seas'. This shift in status led to administrative changes in the way in which the British government dealt with the dominions; in December 1907 the Colonial Office was reorganised into three departments the first of which, the dominions department, forerunner of the Dominions Office created in 1925, was responsible for relations with the overseas dominions.

The 1907 conference dealt with two of the major issues with a capacity to bedevil relations, trade and defence. On the first, the conference reaffirmed an earlier position on trade preferences opposed to the creation of an imperial system. On defence, however, the growing naval race with Germany and the emergence of Japan as a major power pushed dominion governments closer to an accord with London. Reforms in the British Army pioneered by R. B. Haldane led to the creation of a General Staff, and in 1907 the dominions agreed to the exchange of staff officers, which enabled the foundation of an Imperial General Staff. Agreement was also reached on the three principles of imperial defence as they were now conceived: that the RN was the first line of defence; that the dominions should maintain forces sufficient to safeguard their own defence; and that in time of great emergency all should subscribe to mutual support. This last point was resolved only in 1911, when the dominions agreed, at a meeting of the Committee of Imperial Defence (CID) held during the main conference, to coordinate the training, organisation and equipment of their forces. All were still adamant, however, that this did not imply a commitment of forces in advance, a position which the British government fully accepted. A supplementary conference was held in 1909 against the background of the Dreadnought crisis, from which emerged agreement, after many years of Admiralty objection, to the principle of individual dominion navies, and which laid the basis for the Royal Australian Navy (q.v.), founded in 1911. As well as agreement on coordination of ground forces, the 1911 conference produced a further agreement governing the coordination of naval forces and their command and control in wartime. By the time the next Imperial Conference was due to be held, the empire was at war. The British Prime Minister, David Lloyd George, convened an imperial conference, often designated the Imperial War Conference, in March 1917, although it had little to do with the conduct of the war. For the first time, India was represented. The major outcome was the creation of the Imperial War Cabinet, which then sat with a floating membership for two sessions in 1917 and 1918. It was in fact 'a cabinet of governments rather than of ministers', as the Canadian Prime Minister, Sir Robert Borden, pointed out, and its influence on the conduct of the British, much less the Allied, war effort was minimal. It did, however, result in individual dominion representation at the Paris Peace Conference in 1919.

Postwar conferences were held in 1921, 1923, 1926, 1930 and 1937. The 1921 conference was concerned with the status of a Free Ireland within the Empire, and the broader constitutional arrangements within which the dominions, of which Ireland for a time became one, might operate. At the 1923 conference the basis was laid for the Singapore strategy (q.v.), which was so to bedevil considerations of imperial defence for the rest of the interwar period. Those in 1926 and 1930 took place within the context of the Balfour Report and the promulgation of the Statute of Westminster, which altered the relationship between the dominions and the British government forever. Only in 1937 did considerations of defence and external policy again come to the fore, driven this time by the growing crises in Europe and Asia occasioned by German and Japanese aggression. The beginning of the Second World War brought further changes to the consultative arrangements of the Empire and Commonwealth. Churchill flatly refused to recreate the Imperial War Cabinet that Lloyd George had inspired, despite strong urging on the part of the Australian Prime Minister, Robert Menzies. In 1944 the British convened the first Prime Ministers' Conference, followed at regular intervals by others convened in 1946, 1948 and 1949 (to consider the issue of Indian membership of the Commonwealth under republican constitutional arrangements), and so on. In time these gave way to the modern Commonwealth Heads of Government Meetings.

Although matters of Commonwealth defence cooperation continued to feature on the agenda throughout the 1950s, the broadening of Commonwealth membership and the changing nature of the Commonwealth gradually precluded any consideration of defence coordination and cooperation between member States, at least within the broad forum of the whole. At the first Colonial Conference in 1887, the Prime Minister, Lord Salisbury, remarked that the business of the gathering 'may seem prosaic, and may not issue in any great results at the moment', but that it was 'the beginning of a state of things which is to have great results in the future'. While colonial, imperial and commonwealth conferences have never satisfied the expectations of those who saw them as a vehicle for the central coordination of the Empire, or later Commonwealth, there can be little doubt that, frustrating as their deliberations frequently were, they nonetheless paved the way for that level of cooperation in defence, foreign policy and trade which was ultimately achieved, to the benefit of the whole, not least in the two world wars.

Nicholas Mansergh, *The Commonwealth Experience* (Weidenfeld & Nicolson, London, 1969).

IMPERIAL DEFENCE The doctrine of imperial defence emerged gradually from the middle of the nineteenth century, as Britain sought to make the self-governing colonies of settlement (subsequently the dominions) accept some share of the burden of providing for the defence of the Empire on a global scale. The doctrine was codified at the Imperial Conferences (q.v.) held in the first decade of the twentieth century, and came to rest on three basic principles: that the RN provided the first line of defence, and that accordingly every effort should be made to establish and maintain supremacy at sea; that the colonies/dominions should sustain such forces from their own resources as were necessary to ensure their own defence; and that in time of general war or major international crisis all parts of the Empire would cooperate in the defence of the whole. There was always a tension between the demands of increasingly assertive dominion governments, developing mature interests of their own, and the imperial centre in London which, especially after the losses suffered in each of the world wars, became increasingly reliant upon the resources of the Empire and Commonwealth to maintain Britain's status as a great power. After 1945 the doctrine was revived in modified form as Commonwealth defence cooperation, but in general this did not survive the 1950s, although examples of combined military action, especially in the Far East during the 1950s and mid-1960s, could still be found.

IMPERIAL DEFENCE COLLEGE (IDC) Established in 1927, the Imperial Defence College catered for the higher education of senior officers of the three British services, the civil service, and a limited number of dominion officers. In the interwar period Australia was allotted an annual quota of two places on each year-length course, which in most years was not filled. The curriculum covered the general principles of defence and organisation for war, issues of imperial defence, strategy and international relations, together with consideration of concrete problems referred to it by the chiefs of staff, usually closely related to issues of imperial defence. The size of the course varied between a minimum of 20 in 1929 to a maximum of 32 in 1938, and many of the students concerned went on to the most senior appointments in their services during the Second World War, as was indeed the intention. Reopened after the Second World War, the aims of the IDC varied little from its prewar form, except that current problems were no longer referred for consideration. An attempt was made to appoint an Australian senior officer to the directing staff, but this was not successful and with the exception of a single Canadian officer, Lieutenant-General G. G. Simmonds, the staff was always British. From 1946 a system of overseas tours became part of the syllabus, involving students in visits to various countries of defence interest, and from 1948 visits to industry were introduced as well. By 1960 the size of the annual course had increased to 63, the proportion of overseas students had doubled, and the rank required for attendance had increased to colonel or brigadier equivalent. In 1971 the college was renamed the Royal College of Defence Studies (RCDS), better reflecting its nature and Britain's changed role in the world. Between its opening in 1927 and its fiftieth anniversary in 1977, 192 Australians attended the college, and many of these went on to senior positions in the services and Defence Department, including the long-serving wartime secretary of the department, Sir Frederick Shedden (q.v.). In later years some aspects of the RCDS course, with its strong emphasis on NATO and European defence problems, became of less relevance to Australian officers, but the course retains a role in exposing Australian students to a wider range of ideas and experience than it is possible to provide within their own service establishments.

T. I. G. Grey, *The Imperial Defence College and the Royal College of Defence Studies 1927–1977* (Her Majesty's Stationery Office, Edinburgh, 1977).

IMPERIAL GIFT AIRCRAFT were surplus First World War military aircraft offered by the British government to the Australian and other dominion governments in June 1919 to enable them to form their own air forces. Australia, Canada, South Africa and New Zealand all accepted the British offer and all except New Zealand received over 100 aircraft. Australia acquired 128 DH9, DH9A, SE5A and Avro 504 aircraft (see AIRCRAFT, RAAF) as well as 191 engines, 259 transport vehicles, 14 hangars and other equipment worth about £1 million which were shipped to Melbourne in 1921. It has often been written that when the RAAF was formed it had more aircraft than men, but most Imperial Gift aircraft remained packed in their crates until 1925 when the first RAAF squadrons were formed. The Imperial Gift aircraft served until 1930 when they were written off for disposal due to old age.

IMPERIAL WAR GRAVES COMMISSION see **WAR GRAVES**

INDEPENDENT COMPANIES Following the decision of the British Army in the latter part of 1940 to form commando units to carry out raids and guerrilla operations in German-occupied Europe, a British training team known as Military Mission 104 was sent to Australia to set up similar units in the AIF. The Australian units became known as Independent Companies, because of their ability to fight independently of other units. Military Mission 104, commanded by Colonel J. C. Mawhood, who had served with the AIF during the First World War, arrived in December 1940, and established No. 7 Infantry Training School at Wilson's Promontory, Victoria, in February 1941. The school's first commander was Major William Scott (q.v.) and the 2/1st Independent Company was formed after eight weeks' training in June 1941. An Independent Company was commanded by a major, with a captain as second-in-command and another captain as medical officer. There were three platoons, each of 67 men, each platoon commanded by a captain, and each platoon was divided into three sections each commanded by a lieutenant. In addition to the infantry platoons, each company also had its own engineer and signals sections commanded by lieutenants. An Independent Company therefore consisted of 273 other ranks and 17 officers, which was a higher ratio of officers to men than in a normal infantry company. When the threat of war with Japan became greater in July 1941, the 2/1st Independent Company was scattered in outposts (where many would be taken prisoner) in the New Hebrides (Vanuatu), the Solomon Islands, New Ireland and Manus Island, while the 2/3rd went to New Caledonia. When war broke out, the 2/2nd was sent to Timor (q.v.), where, following the Japanese occupation of the island, they retreated to

Men of the 2/3rd Independent Company in action against Japanese forces at Orodubi, New Guinea, 29 July 1943. Corporal R. R. S. Good dresses the arm wound of Private H. W. Robins. (AWM 127978)

the interior and carried out guerrilla operations for almost a year. After the Japanese landed at Lae and Salamaua in New Guinea in March 1942, they were harried by the 2/5th Independent Company, which formed part of Kanga Force (q.v.). Independent Companies were used throughout the New Guinea campaigns for patrols ahead of the main forces. In October 1942, the 2/7th Australian Cavalry Regiment, which had been the 7th Division's armoured unit, was made the administrative headquarters for Independent Companies and they were renamed Cavalry (Commando) Squadrons (this was later simplified to Commando Squadrons). This change was resented by men of the units, for whom, in the words of one soldier, the word 'commando' conjured up images of 'a blatant, dirty, unshaven, loud-mouthed fellow covered with knives and knuckle-dusters'. When the cavalry regiments of the 6th and 9th Divisions were disbanded in January 1944, they too were

re-formed as four Commando Squadrons. Commando Squadrons were serving in Bougainville, New Guinea and Borneo when the war ended.

INDONESIAN CONFRONTATION see **CONFRONTATION**

INDUSTRY Manufacture in Australia of munitions and other defence equipment developed in a haphazard way from the mid-1830s. Not until the 1880s was serious interest shown in laying the foundations for continuous production to meet local defence needs. In 1888, after exploring its options for establishing rifle ammunition and artillery factories, the Victorian government reached agreement with the Colonial Ammunition Company (CAC), a New Zealand-based firm, to set up an ammunition factory at Footscray, on the western outskirts of Melbourne. Once production started in 1890, this

Munition workers heat the barrel of a 3.7-inch anti-aircraft gun before it is straightened, Bendigo, Victoria, 21 April 1943. (AWM 138692)

facility supplied the military forces of Victoria and other colonies with their full requirement of cartridge rounds.

Following Federation in 1901, moves began to increase the range of military items which could be made in Australia, beginning with those that were essentially the same as made for civil use, such as camp equipment, saddlery and horsed vehicles. Following the 1907 Imperial Conference (q.v.) in London, however, a concerted policy was pursued to achieve self-containment or at least reduced reliance on British arms manufacturers. Plans were developed to establish government factories for the production of rifles at Lithgow, New South Wales, cordite at Maribyrnong, Victoria, saddlery and leather accoutrements at Clifton Hill, Victoria, and uniforms and other clothing items such as head-gear at South Melbourne. These facilities were all in production by the time the First World War began in August 1914; construction of a woollen cloth factory at North Geelong was by then underway too.

The foundations of the RAN had also been laid in the immediate prewar years, under a scheme which included provision for developing a local naval shipbuilding capability. Among vessels initially acquired for the Australian fleet was the River Class destroyer (q.v.) HMAS *Warrego*, which was assembled at Cockatoo Island dockyard (q.v.), Sydney, during 1910–11 from imported parts. From 1913 construction was begun on the Town Class light cruiser (q.v.) HMAS *Brisbane* and another three destroyers which were built entirely in Sydney, and other vessels followed.

During the war years the government factories underwent considerable expansion and production was extended to civil industry as well. More than 20 firms across Australia became involved in 1915–16 in a scheme to begin the local manufacture of 18-pounder shells, until advice was received that production shortfalls in Britain had been overcome. Although plans for the manufacture of machine-guns were abandoned, a superior new type of hand grenade known as the 'Welch-Berry' was successfully developed and 15 000 of these bombs were produced and shipped to Britain; the War Office, however, declined to adopt the type, ostensibly on the grounds of equipment standardisation.

The First World War also gave rise to a scheme for centralising the production of all Australia's arms requirements in a single facility to be located on federal territory at Tuggeranong, south of the national capital site at Canberra. Considerable effort was expended on the arsenal project, but by late 1917 it was beginning to be recognised that war experience in Britain had shifted thinking more towards dispersing production to major population centres where civil industry was already established and available for utilisation, and where labour and supporting services were more readily available. By 1919 the idea of one great arsenal was formally dropped. Before the scheme was abandoned, the opportunity was presented from late 1918 to purchase machine tools and equipment from British wartime factories then being dismantled. Seizing this chance, a plant reportedly worth £1.5 million was obtained at one-tenth this price and shipped to Australia for re-erection in the new arsenal facility when constructed.

Sharp cutbacks in production and the need for economy, combined with the availability of stocks of surplus war stores (some of which was presented by Britain as an 'imperial gift'), changed the basis of Australia's self-containment policy in the postwar period. Although the government acquired control of the Colonial Ammunition Company's plant at Footscray in January 1921 under a lease arrangement, in 1923 the woollen factory was sold to private enterprise and the harness factory closed down. A wartime facility for fermenting acetate of lime (to produce acetone needed in cordite manufacture) also ceased production in 1925.

Organisationally, in 1920 the factories had come under the control of a board of administration within the Defence Department. The next year all the factories except clothing were transferred to a new Munitions Supply Board (q.v.). The policy pursued by the MSB chairman, Arthur Leighton, was to reduce all the facilities to nucleus operations to save on maintenance costs and devote available funds to establishing new capabilities. Under this approach, in 1922 the cordite factory was converted to producing TNT and other explosives, and from that year a new ordnance factory was progressively established at Maribyrnong, using plant equipment purchased in Britain at the end of the war. In 1925 a new gun ammunition factory was constructed at Footscray, adjoining the works leased from the CAC; at the end of 1926 the Common-

wealth exercised an option to purchase the CAC's interest and merged the two facilities.

Leighton's policy had enabled growth in Australia's defence industrial capacity despite severe economic conditions, but by 1928–29 declining funding levels made it impossible for this to continue. Following the election of the Scullin Labor government in 1930, Leighton obtained approval for the factories to accept orders for business in limited competition with private enterprise, to supplement their small outputs of defence goods. Cockatoo Island dockyard, denied this avenue under a 1926 High Court ruling on its participation in a powerhouse project, was leased by the Commonwealth in 1933 to a private company for a 21-year period. The activities of the government factories were resented by commercial firms following the onset of the Depression, and in 1935 another challenge was made, this time in the High Court against the clothing factory's acceptance of orders outside the defence forces. It was unsuccessful.

In the process of rearmament begun from 1933, the fortunes of the factories began to revive. New defence orders led to a lessening of reliance on commercial work and increased funding for a program of capital development. By 1939–40 the value of production from the factories was 10 times the amount it had been five years earlier. From 1938 measures were also taken in hand to begin the organisation of civil industry for defence purposes in the event of war occurring.

In the same period, steps were finally taken for the commencement of aircraft production in Australia. Although some local manufacture had taken place previously, activity in this field had been confined to components, repairs or assembly of imported parts. In 1935, however, a consortium of Australian industrial firms, acting under the chief general manager of BHP Co. Ltd, Essington Lewis (q.v.), responded to a government suggestion by forming the Commonwealth Aircraft Corporation (q.v.) with works at Fishermens Bend, Melbourne. The new enterprise secured a government contract for 100 advanced trainer aircraft required by the RAAF, the type offered being an American design, the NA-33, known locally as the Wirraway (q.v.). Responding in alarm to this departure from traditional arrangements regarding supply of defence equipment, the British government in 1939 reached agreement to establish a separate factory at Fishermens Bend for the construction of British-designed Beaufort bombers (q.v.). The De Havilland company also started producing Tiger Moth basic trainer aircraft (q.v.) at Bankstown, Sydney, so that by the time the Second World War began in September 1939 the basis of a viable industry had been laid.

In the months before the war, the MSB was absorbed within a new Department of Supply and Development. By the middle of the following year, a Department of Munitions (q.v.) had been set up with Essington Lewis as director-general. This new body controlled all government factories connected with munitions and aircraft production, but left the clothing factory with Supply and Development. In June 1941 aircraft production was also divided off into a separate department, but with Lewis — described as 'virtually an industrial dictator' — as its director-general too.

Wartime expansion of munitions production in Australia was both rapid and large scale. By 1945 the Commonwealth had built 47 factories in addition to 244 annexes attached to privately-owned factories or state instrumentalities such as railway workshops. Many of the items produced in these improvised facilities had never before been made in Australia and were entirely new to local industry. Moreover, many of the factories and annexes had been established in rural areas where previously there had been little industrial activity of any sort.

With the Pacific a major theatre of operations, unlike during 1914–18, there was a much increased requirement for naval support. Shipbuilding had been practically a moribund industry in Australia in 1939, with Cockatoo Island the only large construction and engineering works in existence. While continuing under the management of its civilian lessees, this yard played a pivotal role by assisting in the transfer of skills to new centres at Adelaide and Whyalla in South Australia, Newcastle in New South Wales, and Brisbane and Maryborough in Queensland, all of which subsequently contributed to programs for the building of *Bathurst* Class corvettes and River Class frigates (qq.v.).

Adding to the naval engineering resources available in Australia was the decision in 1940 to construct a large graving dock at Garden Island (q.v.), Sydney. Opened in March 1945, this was able to accommodate ships up to and including aircraft carriers. A second but smaller such facility was undertaken at Brisbane in 1942, and in that year, too, the Commonwealth took control of the former State government-owned dockyard at Williamstown (q.v.), Melbourne, and turned it into a naval facility.

In anticipation of eventual victory, the level of defence munitions output was scaled down from 1943 and diverted to food production instead. Throughout 1944–45 factories and annexes were closed and transferred to other departments or uses.

At those facilities which remained, production was slowed and increasingly turned to non-defence orders. The government moved quickly to recoup part of its expenditure in this field by selling off or leasing many of the more than 4300 buildings across Australia that were now vacant.

The departments of Munitions and Aircraft Production were abolished in 1946, being reabsorbed by Supply and Development (which became simply Supply from March 1950). By the time the process of rationalisation was complete, Australia still possessed an industrial base for its defence effort that was much larger than before the war. Now, in addition to the original four munitions factories (small arms, ammunition, explosives and ordnance) and the clothing factory, there were: the former Beaufort division at Fishermens Bend, now called the Government Aircraft Factory (GAF), and various engine and repair workshops; additional ordnance works at Bendigo and Echuca, Victoria; a marine engine works at Port Melbourne; and a second explosives factory at Mulwala, New South Wales. As well, part of an explosives factory at Penfield, South Australia had been retained in reserve, along with four synthetic ammonia plants. The dockyards at Garden Island and Williamstown remained to serve the RAN's needs.

A period of heightened strategic uncertainty following the outbreak of the Korean War in 1950 brought renewed attention to Australia's defence preparedness. The munitions and aircraft production activities of Supply were again formed into a separate Department of Defence Production in May 1951, and the various facilities enjoyed a resurgence in work orders. Several facilities that had been closed after the Second World War were also brought back into use, these being the explosives factory at Albion, Melbourne, and the munitions filling plant at St Marys, Sydney.

By the late 1950s the defence facilities faced renewed contraction, with Defence Production once more reincorporated into the Supply organisation in April 1958. The next year the ordnance factory at Echuca which was making ball-bearings was sold as a going concern to a group of foreign manufacturers, and two of the synthetic ammonia plants were sold and a third closed. The absence of defence orders caused reductions in production and workforce levels, and a return to pursuing commercial work. By the mid-1960s the pendulum had swung again, in conjunction with Australia's participation in the conflict in South Vietnam. This improvement proved purely short-term, however, and by the early 1970s conditions had resumed a depressed level.

From this slump there was no significant upswing for more than a decade. The situation is demonstrated by the plight of the munitions group of factories, which saw their workload decline by over 70 per cent between 1969–70 and 1976–77. Predictably, in this period the factories found themselves subjected to increasing scrutiny over matters of productivity, efficiency, and management. Although criticism on many of these aspects were justified, often there was little regard given to the fundamental diseconomies and structural problems associated with the special position and role of these establishments.

Among the difficulties endured by the facilities from the mid-1970s was that of shifting ministerial responsibility within the government administrative structure. After the Department of Supply was abolished in 1974, control of the factories was transferred between four different departments over the next six years. Notwithstanding the uncertainty, discontinuity and disruption associated with such changes, attempts were made to adopt a more commercial approach to the conduct of business within the factories, by adopting many of the management and planning practices common in private enterprise. Modern technology, particularly involving computers, was also embraced, in an attempt to ensure the facilities were not left behind in comparison with civil industry and to improve performance.

In reality, though, many of the factories had severe problems of obsolesence which needed huge injections of capital expenditure to cure. Funding for replacement plants had, however, declined from over $5 million a year at the end of the 1960s to only $1.5 million for the years 1974/75–1976/77. The factories continued to survive solely through heavy taxpayer subsidisation.

In the face of such conditions, attention during the 1970s and early 1980s repeatedly focused on the future of the defence facilities, not only individually but also collectively. Consideration was given to various schemes of rationalisation and redundancy, or for ensuring that — to the maximum extent possible — defence equipment orders were placed with the factories rather than overseas suppliers, even though the latter may be more economical. Some closures inevitably occurred, the marine engine works ceasing operation in July 1979. In May 1981 a newly created Department of Defence Support assumed control of all the defence facilities munitions, aircraft and clothing factories and, for the first time, naval dockyards as well. This new arrangement did little to improve the overall position, with the department's staff of 15 000 making it one of the largest in the public service, and only 40 per cent of its expenditure of almost $600 million in 1983–84 being recovered through sales to customers. In December 1984 Defence Support was abolished, its function transferred to the Department of Defence under the title of the Office of Defence Production (ODP).

As the financial position of the factories continued to deteriorate, action was finally taken to achieve radical restructuring. In December 1985 the decision was taken to dispose of the Aircraft Engineering Workshop maintained at Pooraka, South Australia, by sale to private interests. Closure of the Albion explosives factory was announced at the same time; from September 1986, the functions of this latter facility were to be transferred to Mulwala over a six-year period.

In March 1986, attention turned to cutting the running costs of ODP's largest establishments — Williamstown and Garden Island dockyards, and GAF — through staff reductions. At the same time, the intention was made public to create a government-owned company to take over the operations of the aircraft factories and continue these on a commercial basis. The company, called Aerospace Technologies of Australia (ASTA), assumed control of the assets of GAF in October 1987.

In April 1987 it was announced that Williamstown would be put up for sale to private interests. Purchased by the Australian Marine Engineering Corporation (later Transfield Shipbuilding), the yard was closed down while restructuring took place but survived to successfully undertake construction of two American-designed *Oliver Hazard Perry* Class guided missile frigates (q.v.) and win a contract for 10 *Anzac* Class frigates (q.v.).

In August 1988 the Minister for Defence, Kim Beazley (q.v.), announced the corporatisation of the remainder of ODP as another government-owned company to be called Australian Defence Industries (ADI). In May 1989 this new body took over the property assets of ODP along with nearly 7000 of its employees. ADI immediately became Australia's largest single supplier of defence equipment and services, with an annual turnover of $400 million. Despite its size and position in the market, ADI has continued to restructure to maximise its performance in a field where it now has a number of large commercial competitors. In 1993 the government advertised ASTA for sale to private interests. Consideration has also been given to the sale of ADI, although the outcome of this debate over privatisation was not known in 1995.

E. Scott, *Australia during the War* (Angus & Robertson, Sydney, 1936); D. P. Mellor, *The Role of Science and Industry* (Australian War Memorial, Canberra, 1958).

CHRIS COULTHARD-CLARK

INFANTRY BATTALIONS, AUSTRALIAN ARMY

The battalion is the basic unit of organisation in the infantry arm. It consists of between 800 and 1000 men organised into sections (commanded by a corporal), platoons (commanded by a lieutenant) and companies (commanded by a captain or a major), with organic (i.e. integral to the battalion's establishment) specialist support, and the whole is commanded by a lieutenant-colonel. Originally comprised of foot-soldiers, the application of technology has produced variations (mounted, mechanised, parachute etc.) on the original theme.

Australian infantry battalions have usually been organised on the British pattern. Between Federation and 1914 this was the eight-company organisation, each company made up of three officers and 117 men and consisting of four sections each of two squads as the 'fire unit'. Some early units of the AIF were raised in 1914 on the eight-company pattern, but changed to the new four-company organisation before sailing for Egypt. The new organisation called for four companies of six officers and 221 men consisting of four platoons each of four sections; this is characterised as a 'square' organisation. The older assumptions which considered the company to be the basic tactical unit gave way in the face of the demands of the First World War battlefield to acceptance of the primacy of the battalion, but at least initially in the First World War infantry doctrine changed slowly. Australia raised 60 battalions of infantry for overseas service (together with 14 regiments of light horse [q.v.], who were in fact mounted infantry). The initial contingents of 16 battalions (1st-4th Brigades) were used as the basis for the expansion and reorganisation of the infantry in Egypt after the Gallipoli campaign, and in recognition of the different demands of the fighting in France (see WESTERN FRONT) new specialist functions were added to the traditional infantry ones.

During the First World War an infantry battalion consisted of 35 officers and 970 other ranks (this was the ideal, known as the battalion establishment; as the war went on battalions were very rarely at full strength). Each battalion had a transport establishment (known as first-line transport) of 25 vehicles (wagons) and 55 horses and/or mules for carrying food and ammunition forward to the unit. A light trench mortar battery was added in 1916. It numbered four officers and 46 other ranks and provided lightweight mobile fire support from eight 3-inch Stokes mortars (q.v.). A machine-gun section of one officer and 32 other ranks and fielding two guns had been vital to the infantry battalion establishment at the beginning of the war,

although on Gallipoli these were often unofficially grouped into companies to maximise fire support in the assault. In early 1916 they were withdrawn from the battalions and brought together into machine-gun companies of six officers and 200 other ranks with 16 Maxim or Vickers guns (q.v.) and attached to infantry brigades. (In 1918 these were further grouped into machine-gun battalions [q.v.] attached to the divisions.) The loss of machine-guns from the battalions was compensated through the issue of Lewis guns (q.v.), initially 30 to a battalion, although as the war went on the Australians tended to 'acquire' additional guns by whatever means to boost the firepower of under-strength companies. In response to the worsening reinforcement situation in the last year of the war, the Australian authorities began to disband some battalions and redistributed the men among the remaining units.

By the Second World War the infantry battalion had evolved again, partly in recognition of the changes in the nature of combat wrought by advancing technology. A battalion now comprised 21 officers and 752 men organised into three rifle companies of four platoons of three sections each, a support company of three machine-gun and one 3-inch mortar platoon, and a headquarters wing with signals, transport, intelligence, quartermaster and administrative staff. The early battalions of the 6th Division were raised in this manner, but as in 1914 the basic infantry organisation was changed soon after the outbreak of war, and the subsequent AIF divisions were raised on this basis from the start while the 6th Division's units were converted. There were now 35 officers in a battalion (not counting the Salvation Army representative), and the battalion comprised four rifle companies of three platoons of three sections each and a headquarters company with six platoons (signals, mortar, carrier, pioneer, anti-aircraft, and transport and administrative). Machine-guns were controlled at the divisional level while an anti-tank company was formed for each brigade of the 6th Division during the First Libyan campaign when the division's anti-tank regiment was diverted to Britain. Both the strength and internal organisation of infantry battalions changed constantly throughout the war, making further generalisation difficult. Following British practice, brigades now consisted of three rather than four battalions, and as in the previous war there were specialist machine-gun and pioneer (q.v.) battalions, and anti-tank regiments (although these had little application in the Pacific) together with independent companies and special units which did some infantry work. In the 2nd

AIF there were 36 battalions, but with the militia units mobilised for the defence of Australia and operations in New Guinea together with the Papuan Infantry Battalions and New Guinea Volunteer Rifles (see PACIFIC ISLANDS REGIMENT) the army had raised more than 100 battalions of infantry by the middle years of the war. The establishment of the regular army after the war saw that portion of the infantry organised into the Royal Australian Regiment (q.v.), of which initially there were three battalions (1RAR, 2RAR, 3RAR), joined by a fourth raised in 1952, which acted as a depot battalion for the regiment until it was disbanded at the end of the decade.

Within the Australian establishment there was also a battalion of the Pacific Islands Regiment (q.v.), based in Papua New Guinea. There was no major organisational change to these, however, until the implementation of the Pentropic (q.v.) divisional experiment in 1960. The impact on infantry battalions was significant: a division now comprised five battalions, each one and a half times larger than the standard British-pattern unit but able to deploy twice the firepower, since each battalion now contained 80 sections rather than the former 36. Battalion headquarters was augmented with additional officers (since the old brigade structure, and with it brigade headquarters, had been abolished), and a battalion now comprised a headquarters company, support company (anti-tank, mortar, signals and assault pioneer platoons), and five rifle companies, each of four platoons. Separate pioneer battalions had been abolished, and machine-guns were now distributed to each platoon.

As a result of the reorganisation and its extension to the CMF the old militia infantry battalions, many of which could trace their lineage to Federation, were abolished, and new State regiments established in their place. This caused widespread discontent within the CMF and led to the loss of many experienced part-time soldiers through resignation. Pentropic itself proved unsatisfactory and was unpopular with many, though by no means with all. Many officers were simply too conservative to accept such a sweeping organisational change, and one which was difficult to master initially, but there were also criticisms of the size and unwieldy nature of the new units and sub-units. In any case, while no other army used the system it was a luxury which the Australian Army could not afford, as was made clear whenever the battalion serving in Malaysia with the Far East Strategic Reserve (q.v.) had to be replaced. A Pentropic battalion had to be stripped down to standard British pattern and the returning battalion built up. In

1964 the army reverted to a standard Tropical Warfare organisation for the infantry battalions. As Australia's commitments overseas increased, it was not the size of the units which mattered so much as the number of them, as the CGS, Lieutenant-General Sir John Wilton (q.v.), noted. Reducing the Pentropic battalions enabled the army to create new infantry battalions virtually overnight to meet the deployments in Malaysia, New Guinea and South Vietnam. A battalion now comprised 37 officers and 755 other ranks organised into four rifle companies, with three platoons of three sections, a support company and an administration company, and with 29 vehicles organic to establishment. The support company had assault pioneer, signals and mortar platoons and an anti-tank platoon equipped with the new Carl Gustav 84 mm anti-tank weapon. This was the organisation which the army took to the Vietnam War. Between 1965 and 1967 the RAR expanded from four to nine battalions, assisted by intakes of national servicemen (see CONSCRIPTION), and the Pacific Islands Regiment doubled in size to two battalions. With the end of the Vietnam War, Confrontation and national service, this size could not be sustained, and the RAR was reduced to six battalions through amalgamations (the four original battalions became 1RAR, 2/4RAR, and 3RAR, and were joined by 5/7RAR, 6RAR and 8/9RAR). The regular force of six battalions was arrived at as part of the deliberations of the Hassett (q.v.) committee in 1972, on the assumption that this was the minimum size necessary to act as a base for expansion in war and the minimum necessary to justify the retention of the 1st Division on the army's order of battle. It was to be backed up by a reserve force of 15 battalions in the redesignated Army Reserve (q.v.). At present 1RAR and 2/4RAR, based in Townsville, form the Operational Deployment Force, while 3RAR is a parachute battalion and 5/7RAR maintains the army's mechanised infantry capability. In 1991 6RAR and 8/9RAR were designated battalions of the Ready Reserve (see ARMY RESERVE). In late 1994 the army decided to split 2/4RAR into separate units again, to help meet perceived needs for the rotation of battalions in the event of further overseas commitments such as that to Somalia (q.v.).

INSPECTOR-GENERAL The position of Inspector-General of the military forces was established under section 8 of the *Defence Act 1903*, and between the departure of the first GOC, AMF, Major-General E. T. H. Hutton (q.v.), in 1904 and the appointment of Colonel W. T. Bridges (q.v.) as the first CGS in 1909, was the most senior officer in the Australian Army and a member of the Military Board (q.v.), which was responsible for the administrative control of the army. The first incumbent was a British officer, Major-General H. Finn (q.v.), but in September 1906 the post was filled by an Australian, Major-General J. C. Hoad (q.v.). The Inspector-General's function was essentially to review and report on the efficiency of the military forces and their administration by the Military Board. In December 1919 Lieutenant-General Sir Harry Chauvel (q.v.) became Inspector-General, a post he was to hold until his retirement in 1930. When General Sir Brudenell White (q.v.) retired as CGS in 1923 Chauvel accepted that appointment as well, and for the next seven years acted in a dual, and slightly contradictory, capacity. Chauvel's annual reports to the minister for Defence, published as parliamentary papers, provide an excellent guide to the parlous state of the army in the decade after the First World War, and Chauvel used them to warn successive governments about the consequences of long-term neglect of the military — to little apparent effect, it must be said. The post of Inspector-General was abolished in 1930, to be revived in 1938 with the appointment of Lieutenant-General E. K. Squires (q.v.). It lapsed again on the outbreak of war. A new Inspector-General Division within the Department of Defence, responsible to both the Secretary and the Chief of the Defence Force, was created in 1987, responsible for administrative and financial audit and the review of defence programs.

INTELLIGENCE, MILITARY, is the systematic collection, processing and dissemination of information about the actual or potential enemy and area of operations. It includes estimates of the enemy's capabilities and intentions and the conduct of activities to collect the required information by the use of surveillance and reconnaissance, including visual observation, aerial photography and more recently satellite and other forms of imagery; the capture and exploitation of enemy personnel, equipment and documents; and the interception and decryption of enemy communications. From their earliest deployments overseas in the nineteenth century Australian forces included reconnaissance officers whose duties included the acquisition and interpretation of information about the enemy. A more structured approach to intelligence began in the army and the RAN soon after Federation. The Australian Intelligence Corps (q.v.)

was formed in 1907 under the directorship of Lieutenant-Colonel James Whiteside McCay (q.v.). In 1912, within a year of the RAN's formation, Commander Walter Thring was appointed assistant on intelligence matters to the First Naval Member, Rear-Admiral William Creswell (q.v.). The capture of a code-book from a German ship in Port Phillip Bay in August 1914 enabled the RAN wireless room in Melbourne (whose small staff included future official historian Captain A. W. Jose [q.v.]) to decipher German naval messages. This information was passed to the British Admiralty and assisted in the location and destruction of the German Cruiser Squadron off the Falkland Islands on 8 December 1914. Major Edmund Piesse (q.v.) became AIF Director of Military Intelligence in 1915, though during the First World War the Intelligence Corps worked totally under British Army supervision.

Between the wars the position of Director of Naval Intelligence (DNI) was created by the RAN and a network of coastwatchers (q.v.) was formed throughout the Pacific in 1922. However the intelligence organisations of the army and the RAN (the RAAF as yet had no intelligence unit) were wound down and some of their functions were subsumed for a time by the government censorship board and the police forces. On the outbreak of the Second World War the DNI's office was expanded, with Commander Rupert Long (q.v.) appointed as Director. A Director of Military Intelligence was appointed in 1940 and AIF signals intelligence units were formed and saw service in the Middle East. The RAAF's involvement in signals intelligence began in August 1941 when secret listening posts, manned by RAAF personnel in civilian clothing, were established in the Netherlands East Indies to intercept Japanese messages.

Australia's efforts in the field of intelligence in the Pacific War were integrated with those of the Allies and made some contributions of great importance. The coastwatching system gave important information on Japanese air, shipping and troop movements. Special Operations Australia (SOA) (q.v.), also known as the Inter Allied Services Department, was formed under the intelligence umbrella in 1942, although its operations were more in the nature of special operations than intelligence. A branch of the British Secret Intelligence Service was formed in Australia known as Secret Intelligence Australia (SIA), while the Far East Liaison Office (FELO) was formed to spread propaganda to raise the morale of the people under occupation and to lower the morale of Japanese soldiers, though there is little evidence that either goal was achieved. In July 1942 the Allied Intelligence Bureau (AIB) (q.v.) was formed to coordinate these various organisations, all of which became sections of the AIB.

Australian personnel also formed part of inter-Allied intelligence units attached to General Douglas MacArthur's (q.v.) Headquarters. The Allied Translator and Interpreter Service (ATIS) (q.v.) translated captured Japanese documents and interrogated Japanese POWs, while the Central Bureau had the extremely important task of decoding Japanese signals (often intercepted by army and RAAF wireless units) to provide intelligence reports known as ULTRA (q.v.). During the Second World War the services (principally the army) also provided considerable resources to security and counter-intelligence activities on the home front and in areas captured from the enemy.

In the postwar era strategic intelligence staffs were maintained in the Directorate of Military Intelligence, the Directorate of Naval Intelligence and the Directorate of Air Force Intelligence. These directorates provided operational intelligence for their respective services. Their strategic intelligence role was incorporated into the formation of the Joint Intelligence Organisation in 1970.

Service intelligence personnel served as unit intelligence officers in Korea, Malaya and Vietnam, and as integrated members of allied intelligence staffs and units in Korea and the Far East Land Forces (FARELF) and in the SEATO (q.v.) military organisation. The commitment in Malaysia and Singapore continued until the withdrawal of Australian forces in the 1970s. Australian intelligence personnel made significant contributions to the collection and interpretation of information during Confrontation.

During the Vietnam War there was a significant increase in the army's intelligence effort. Initially Intelligence Corps officers provided the intelligence officers of 1RAR and 173 (US) Brigade and along with officers of other corps served in intelligence-related positions as part of the Australian Army Training Team Vietnam (q.v.). Intelligence specialists were called upon to establish and man intelligence staffs in HQ Australian Force Vietnam, the 1st Australian Task Force, the Logistic Support Group, the Detachment of the Divisional Intelligence Unit and 547 Signals Troop. A commensurate professionalisation of the training of Intelligence Corps personnel occurred at this time in order to provide and prepare personnel for service in Vietnam.

Intelligence specialists of all services have continued to provide a contribution to Australia's overseas commitments. During the Gulf War, army and RAAF officers served as integrated officers on the allied headquarters and specialist staffs in operations DESERT SHIELD and DESERT STORM.

Chris Coulthard-Clark, *The Citizen General Staff: The Australian Intelligence Corps 1907–1914* (Military History Society of Australia, Canberra, 1976); Wayne Gobert, 'The evolution of service strategic intelligence 1901–1941', *Australian Defence Force Journal*, no. 92 (January/February 1992), pp. 56–64.

CRAIG WOOD

INTER-ALLIED SERVICES DEPARTMENT (ISD)
see **SPECIAL OPERATIONS AUSTRALIA**

INTERNMENT see **ALIENS, WARTIME TREATMENT OF**

IPSWICH, HMAS see *FREMANTLE* CLASS PATROL BOATS

IROQUOIS, BELL
(2-seat medium transport helicopter). Rotor diameter 14.63 m; length 12.77 m; load 11 fully equipped troops or six stretchers; maximum speed 200 km/h; range 418 km; power 1 × Lycoming 1400-shaft h.p.

More Iroquois have been produced than any other helicopter in the world, and it has been used by all three Australian services. The Iroquois entered service with the RAAF in 1964 and was used by No. 5 Squadron in Malaysia from 1964 to 1966. No. 9 Squadron used the Iroquois during the Vietnam War from 1966 to 1971, losing six helicopters and seven men during this time. Iroquois were mostly used for troop deployment and medical evacuation (q.v.) from Vung Tau. In 1969 some Iroquois were converted to gunships (called 'bushrangers') armed with machine-guns and rockets and operated in support of the 1st Australian Task Force. RAAF Iroquois were used in the Sinai during 1976–79 and 1982–85 on peace-keeping (q.v.) duties. In 1989 the RAAF's Iroquois were transferred to the Australian Army Aviation Corps (q.v.) and now equip 171 Squadron, 5 Aviation Regiment. The RAN began using Iroquois for search and rescue, training and transport in 1964. These Iroquois were replaced by the Squirrel (q.v.) in 1984. Four ex-RAAF Iroquois were given to the Papua New Guinea Defence Force (q.v.) in 1989 and they have been used in Bougainville in controversial circumstances. The Iroquois' nickname 'Huey' comes from the letters UH in the helicopter's model numbers.

IRVING, Colonel Sybil Howy
(25 February 1897–28 March 1973) came from a military family, with both her father and brother reaching general officer rank. She was secretary of the Girl Guides Association from 1920 to 1940, and assistant secre-

tary of the Victorian Division of the Australian Red Cross from 1940 to 1941. With the formation of the Australian Women's Army Service (q.v.) in 1941, Irving was appointed Controller on the recommendation of the Adjutant-General, Major-General Victor Stantke. She was promoted to lieutenant-colonel in January 1942 and colonel in February 1943, the first woman to attain that rank in the army's history. She was awarded the MBE for her wartime services. Irving retired from the army with the dissolution of the AWAS in 1947, and became general secretary of the Victorian Division of the Red Cross, a position she held until 1959. When the women's service was re-formed as the Women's Royal Australian Army Corps (q.v.) in 1950, Irving was appointed honorary colonel of the corps.

ITALIAN RESERVISTS, FIRST WORLD WAR see **AUSTRALIANS IN THE SERVICE OF OTHER NATIONS**

J

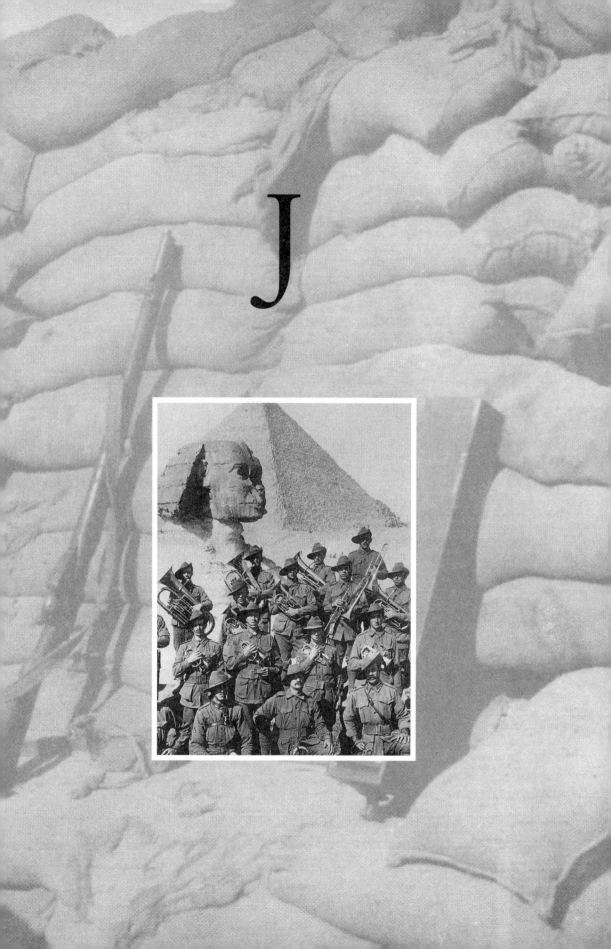

'J' CLASS SUBMARINES (*J1, J2, J3, J4, J5, J7*). Displacement 1210 tons (surfaced); length 275.5 feet; beam 23 feet; speed 19.5 knots surfaced, 9.5 knots submerged; armament 6 × 18-inch torpedo tubes, 1 or 2 × 4-inch guns.

These were British submarines laid down from 1915 to 1917. In 1919 six of them were transferred to the RAN. On reaching Australia they were found to be in poor condition and had to undergo extensive refits. Five of them were subsequently based in Geelong. Their only major cruise in the RAN was in 1921 when they visited Tasmania. They were decommissioned in the 1920s.

JACKA, Captain Albert (10 January 1893–17 January 1932) was born at Winchelsea, Victoria, and was employed as a forestry worker before the war. He enlisted in September 1914 and served with the 14th Battalion. His actions on 19 May 1915, when he shot five Turks and bayoneted two more while recapturing a section of trench, won him the first VC awarded to the 1st AIF. He immediately became a national hero in Australia, and his reported words to the first officer he saw after his VC-winning feat ('I managed to get the buggers, Sir') passed into legend. He went on to receive the MC for his role in retaking part of the line near Pozières in August 1916, an action in which he was seriously wounded and which C. E. W. Bean (q.v.) called 'the most dramatic and effective act of individual audacity in the history of the A.I.F'. As the 14th Battalion's intelligence officer he won a bar to his MC by capturing two Germans by hand after his gun misfired. By March 1917 he had risen to the rank of captain, but he may have been held back from further promotion by frequent disputes with his superior officers. As an officer he earned the respect of his men through his actions. Although best known for his bravery he also possessed tactical ability, as he showed when he led the 14th Battalion against German pillboxes at Polygon Wood in September 1917. In May 1918 he saw his last action of the war at Villers-Bretonneux, where he was badly gassed. On his return to Melbourne in 1919 he was greeted by a large crowd and hailed by the *Herald* as 'the symbol of the spirit of the Anzacs'. The cult built up around Jacka had been encouraged by the use of his name and image on recruiting posters and by both sides in the conscription (q.v.) debate. Bean also lauded Jacka, who seemed a perfect example of his thesis that the rigours of bush life produced superior soldiers. Throughout the 1920s Jacka remained in the public eye through his participation in Anzac Day marches and his election as Mayor of St Kilda in 1930. He worked tirelessly on behalf of the unemployed, including many ex-soldiers, during the Depression, and his early death resulted from a combination of overwork and the debilitating effects of war service. At his funeral he was described as 'Australia's greatest front-line soldier', and thousands turned out to witness the funeral procession. As a man known for his aggressive fighting ability, Jacka and his legend stood in contrast to the story of Simpson and the donkey (q.v.). It is perhaps indicative of changing public attitudes that while the care-giver Simpson's legend was resurrected in the 1960s, Jacka faded from sight after the Second World War.

Ian Grant, *Jacka, VC: Australia's Finest Fighting Soldier* (Macmillan, Melbourne, 1989).

JANDAMARRA see **ABORIGINAL ARMED RESISTANCE TO WHITE INVASION**

JAPAN, OCCUPATION OF see **BRITISH COMMONWEALTH OCCUPATION FORCE**

JAPANESE ATTACKS ON AUSTRALIA The Japanese advance southwards between December 1941 and February 1942 was rapid and unexpected. By the end of January 1942 Japanese forces had taken Rabaul in the Australian mandated territory of New Guinea, while in February Singapore capitulated and the 8th Division went into captivity. On 19 February 1942 a force of 90 Japanese aircraft, followed soon after by a second wave of 54, bombed Darwin, inflicting heavy damage and killing 243 people (see DARWIN, BOMBING OF). Thereafter, northern Australia was the target for Japanese air attack until September 1943, bombs being dropped on Townsville, Katherine, Wyndham, Derby, Broome and Port Hedland as well as Darwin, which remained the principal target and received 64 enemy air raids. The other form of attack on mainland Australia was by submarine. The attack by three Type A midget submarines on Sydney Harbour on the night of 31 May 1942 is well known, and resulted in some damage and the loss of 19 sailors aboard HMAS *Kuttabul* (q.v.). Japanese submarine activity along the Australian coast continued until June 1943, at which point the remaining boats were withdrawn to defend Japanese island positions further north. The submarine campaign sank 19 ships (including the hospital ship *Centaur* [q.v.]) amounting to 80 874 tonnes, and claimed 503 lives. They were not a precursor to invasion but intended rather to isolate Australia and hinder its war effort. Although Japanese submarines were technically good (their torpedoes were excellent), and the

Map 15 (Top) Japanese attacks on Australian territory 1942–43. (Bottom left and right) Details of submarine attacks.

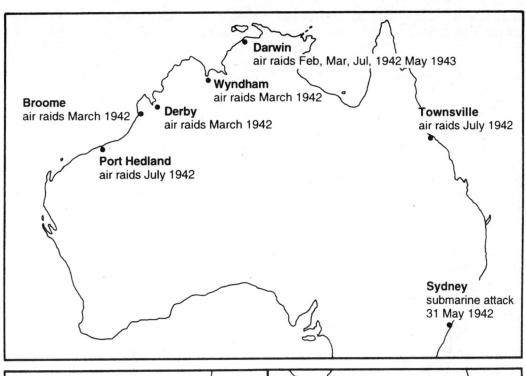

Darwin
air raids Feb, Mar, Jul, 1942 May 1943

Wyndham
air raids March 1942

Broome
air raids March 1942

Derby
air raids March 1942

Townsville
air raids July 1942

Port Hedland
air raids July 1942

Sydney
submarine attack
31 May 1942

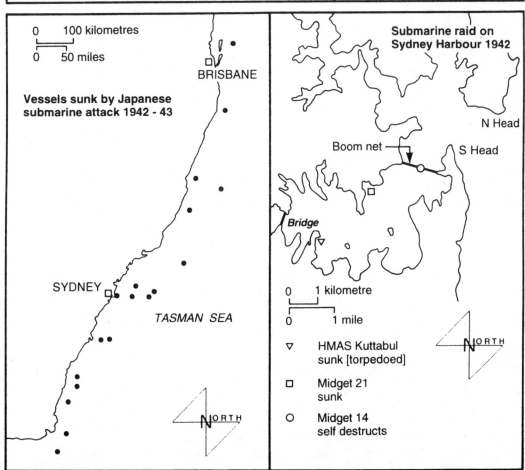

0   100 kilometres

0   50 miles

**Vessels sunk by Japanese
submarine attack 1942 - 43**

BRISBANE

SYDNEY

TASMAN SEA

NORTH

**Submarine raid on
Sydney Harbour 1942**

N Head

Boom net

S Head

Bridge

0   1 kilometre

0   1 mile

NORTH

▽   HMAS Kuttabul
     sunk [torpedoed]

□   Midget 21
     sunk

○   Midget 14
     self destructs

crews brave and skilful, Japanese submarine doctrine did not stress aggressive anti-shipping activities (unlike the German and US navies, for example), which made the campaign in Australian coastal waters less serious than might have been the case.

David Jenkins, *Battle Surface: Japan's Submarine War Against Australia 1942–44* (Random House Australia, Sydney, 1992).

JAPANESE THREAT In the 100 years up to the outbreak of the Second World War, White Australians felt chronically insecure about their country's position as an outpost of the British Empire located far away from Britain in a part of the world which was densely populated with non-White peoples. This insecurity was reflected in fears about the military threat posed to Australia by Britain's imperialist rivals as well as the social threat of Asian immigration, and these two threats became one with the rise to power of Japan. Australians first became aware of Japan as a possible threat after Japan's victory in the Sino-Japanese war of 1894–95, but it was the Japanese defeat of Russia in 1904–05 which led to a widespread perception of Japan as Australia's principal military threat. Though Japan had become an ally of Great Britain in 1902, Australians came to realise that the existence of a potentially hostile Pacific power with a strong navy meant they could no longer rely on their continent's isolation as a defence. Defence spending was increased, and compulsory military training was introduced with the full support of the Labor Party, which had abandoned its anti-militarist stance in the face of the perceived Japanese threat. By the time of the renewal of the Anglo-Japanese alliance in 1911 there was almost universal apprehension about Japan expressed in newspaper editorials, for while the renewal itself was cause for relief there was great concern about what would happen when the treaty expired in 1921.

In the First World War the Japanese government honoured the Anglo-Japanese alliance by entering the war on the side of Britain. The Japanese navy helped to patrol the Australian coastline and to escort troop convoys, but the Australian government was less than enthusiastically grateful. Censorship restricted the expression of anti-Japanese opinions, but Prime Minister W. M. Hughes (q.v.) privately indicated his anxiety that Japan still had designs on Australia. Hughes's opponents in the bitter conscription (q.v.) debates played on such fears by claiming that if large numbers of men were conscripted and sent overseas Australia would be left open to invasion. Despite his concerns, however, Hughes reluctantly supported the secret Anglo-Japanese agreement of February 1917 which divided Germany's Pacific Island colonies, those north of the equator going to Japan and those to the south going to the British Empire. At the Paris Peace Conference in 1919 Hughes defended Australian imperial ambitions against those of Japan by gaining a 'C' class mandate over New Guinea (which would allow Australia to prohibit Japanese settlement there) and by blocking a Japanese-sponsored declaration of racial equality.

In 1919 E. L. Piesse (q.v.) was appointed as director of the newly created Pacific Branch in the Prime Minister's Department, providing the Australian government for the first time with relatively sober and informed assessments of Japanese intentions. Apart from the brief period of Piesse's appointment, both official and popular opinions about Japan in the years between the Russo-Japanese War and the Second World War developed on the basis of a profound ignorance about the Japanese government, military and society in general. For the most part, Australian governments relied on the information about Japan which the British government chose to send them, using their own racist prejudices to fill in the gaps.

After the First World War there was less agreement than there had been previously about the nature and extent of the Japanese threat. The trauma of the conscription debate had made the Labor Party return to its anti-militarist roots, and while they remained deeply racist, its members now spoke of Japan as a bogey used by the capitalists and militarists to support their own interests. Hughes's government, on the other hand, was worried about the increasing rivalry in the Pacific between Japan and the United States, both of which had been dramatically increasing their naval expenditure. It was, therefore, somewhat relieved when the Washington Conference of 1921–22 limited the tonnage of naval capital ships (ships above 10 000 tons) to a ratio of UK 5: USA 5: Japan 3. In fact, however, this treaty assured Japan of naval superiority in the western Pacific because it was the only Pacific power which did not have naval obligations in other parts of the world. Following the Washington Conference, the British Foreign Office blithely assured the 1923 Imperial Conference (q.v.) that there was 'not a cloud in the sky' of British–Japanese relations, and, for the most part, Australian politicians shared this sense of security in the 1920s.

Within the armed forces, however, there was considerable debate between the army and the navy about the threat from Japan. The army, unhappy with its role of merely defending Australia against small raiding parties, emphasised the threat of Japanese invasion, while the navy defended its privileged position within an Australian defence based on the Singapore strategy (q.v.) by playing down this threat. The army claimed that Australia could not rely on the British fleet and the Singapore base to deter a Japanese invasion, so a stronger army should be built up to hold off the invaders until British help arrived. The navy stressed that the British fleet at Singapore would prevent any invasion, that Australia was too far away for Japan to mount an effective invasion, and that the most Australia could expect was small raids. Nevertheless, both services agreed that Japan was Australia's main threat, and in 1930 the Defence Committee noted that 'Japan is obviously the only country at present which can seriously threaten Empire or Australian interests in the Far East.'

After the Japanese invasion of Manchuria in 1931 put an end to any hopes that the Japanese government might not pursue a policy of aggression, the Australian government felt constrained to display a conciliatory attitude towards Japan in public. It refused to condemn the Japanese military's advance into China and as late as 1940 it was lobbying Britain to accede to a Japanese request to close the Burma Road to China. War with Japan had to be avoided if at all possible because Australian defences were weak, it was not clear that British ships could arrive in Singapore in time to prevent an invasion, and it seemed unlikely that the United States would come to Australia's defence. A friendly relationship would also ensure the continuation of Australian trade with Japan, which had been booming since the war and which was very much in Australia's favour, though the Australian trade diversion policy which imposed massive tariffs on non-British imports from 1936 provoked a Japanese response which drastically reduced Australian exports to Japan. Despite its official stance, the government shared public concern about the alleged Japanese 'southward advance' policy, and External Affairs minister George Pearce (q.v.) admitted privately in 1935 that his government hoped Japan's westward expansion would prevent it from moving south towards Australia. The Australian delegation to the 1937 Imperial Conference sought a clarification of the British Empire's 'strategical objects ... in a war with Japan, or Japan and another first-class Power' and asked if a Japanese invasion of Australia was possible. Growing fear of Japan was also reflected in the increase in defence expenditure from about £3 million in 1932–33 to about £17 million in 1938–39.

Despite several decades of anti-Japanese hysteria in Australia, the actual entry of Japan into the Second World War in December 1941 with the attack on Pearl Harbor and the sinking of the *Prince of Wales* and *Repulse* came as a shock to many Australians. The rapid advance of Japanese troops down the Malayan Peninsula, culminating in the fall of Singapore, was even more shocking (see MALAYAN CAMPAIGN). Few had believed that the Japanese would be capable of such daring actions, and the US Consul in Adelaide reported a city close to panic, with 'staid businessmen who only the day before were complacent about the menace of the "yellow dwarf" [being] reduced almost to wringing their hands'. The Australian military, however, does not seem to have underestimated the Japanese armed forces before December 1941 as much as both Australian popular opinion and British military analysts did.

It is now known that the Japanese military never planned to invade Australia, but there was no way of knowing this in 1942 when fear of invasion was almost universal. The popular stereotype of Japanese soldiers as short-sighted and second-rate had been challenged to such an extent that Australian military propagandists now had to stress that the Japanese military was not invincible. The attitudes towards the Japanese which developed during the Second World War were a complex mixture of the old stereotypes and new ones, of bitter hatred and grudging respect. Probably the favourite characterisation of Japanese soldiers was one which had been used as early as 1933 by the then Director of Military Operations and Intelligence, Lieutenant-Colonel V. A. H. Sturdee (q.v.): they were 'fanatics who like dying in battle'. Linked to this view was the idea that the Japanese were incapable of thinking for themselves, and could only operate *en masse*. Thus Australians were able to think of all Japanese people as a single unit, and there was little attempt to distinguish between Japanese soldiers and civilians, between the Japanese people and the Japanese government, or between Japanese culture and the ideology of Japanese imperialism. The enemies in Europe were 'the Nazis', but the enemy in Asia was simply 'the Jap'. Unlike the war in Europe, the Pacific War was widely seen as a 'race war', as Prime Minister John Curtin (q.v.) made clear when he sought parliamentary approval of the government's declaration of war on Japan by proclaiming the nation's determination to maintain the 'principle of a White Australia'.

These attitudes (and parallel ones on the Japanese side) were used to justify the extreme brutality of the Pacific War. The frequent comparison of the Japanese to animals could lead to exterminationist rhetoric, as when General Sir Thomas Blamey (q.v.) told troops in 1943 that 'we have to exterminate these vermin if we and our families are to live'. Yet even White Australian racism had its limits. When the Department of Information ran its 'hate' campaign in March 1942, with advertisements proclaiming 'We've always despised them — NOW WE MUST SMASH THEM!' and radio broadcasts promising to clear 'such loathsome creeping creatures from the face of the earth', a Gallup Poll found that 54 per cent of those surveyed opposed the campaign. But though a bare majority of Australians seemed prepared to accord some respect to Japanese humanity, they still did not trust the Japanese and when the war ended the question became whether or not the former enemy could be reformed. In the United States there was a very rapid shift in attitudes towards Japan after the war, as propaganda images of the Japanese as brutal apes gave way to images of them as friendly monkeys. Australian attitudes seem to have taken longer to change, perhaps because Australia's weaker military position left Australians still feeling vulnerable. Australians also found it hard to forgive Japanese treatment of Australian POWs (q.v.), which seemed to support the belief that the Japanese 'race' was uniquely cruel.

In the immediate aftermath of the war the Australian government shared the popular reluctance to trust the Japanese, and pushed strongly for a thoroughly demilitarised Japan which could no longer pose a threat to the region. However, as the Cold War began to take shape the government became more ambivalent in its attitude towards Japanese rearmament. As long as they stuck to their original position they would be out of step with their powerful American ally whose government was, from 1947, supporting a rearmed Japan as an anti-communist stronghold. The strongly anti-communist Menzies government, which came to power in 1949, accepted the US position, but with reservations, arguing unsuccessfully that all weapons which could only be used in long-range, offensive operations should be denied to Japan. With the signing of the San Francisco peace treaty in 1951 the Australian government decided simply to accept the US policy, having been reassured by the signing of the ANZUS Treaty (q.v.) that same year.

Trying to explain Australian attitudes towards Japan in a 1952 issue of the American journal *Foreign Affairs*, Prime Minister Robert Menzies pointed to the perception that the Japanese had been 'uncivilized' enemies and 'brutal and inhuman' gaolers who had come close to invading Australia. As a result, he said, 'the instinctive reaction of Australia to any proposal for a Japanese peace settlement is, "Keep them down! Don't let them rearm! Don't trust them!"' But over time this attitude has softened, and by the late 1960s less than 10 per cent of Australians were naming Japan in polls as a threat to Australia's security. Apart from the fact that memories of the Second World War are fading, possible reasons for this change include the fact that

the Japanese government stood solidly on the side of the United States throughout the Cold War, and the importance of the Australia–Japan trade relationship. In 1970–71 Japan became Australia's principal trading partner, while Australia was Japan's second largest trading partner. From the 1960s Australian governments have welcomed Japanese moves to become more involved politically in the Asia–Pacific region, but have discouraged any Japanese military involvement. A 1983 survey of 200 members of Australia's foreign policy élite found little support for any significant Japanese defence build-up, and their positive view of Japan's role in the region was based on the perception that such a build-up is unlikely to occur. Popular fears of Japan, some still coloured by wartime memories, are most often reflected now in concern about Japanese investment in Australia.

JAYWICK/RIMAU raids against shipping in Japanese-occupied Singapore were carried out by members of Special Operations Australia (q.v.) in 1943 and 1944. The two operations were devised and led by a British officer, Major Ivan Lyon, who had escaped from Singapore in 1942 and had joined up with a Special Operations Executive unit in India. His plan, for canoeists to enter Singapore harbour at night and attach limpet mines to the Japanese shipping at anchor, was approved by General Sir Archibald Wavell (q.v.), C-in-C, India. Lyon argued that the attack would need to be launched from Australia rather than India, as the route would be safer, and so he was sent to Australia in August 1942. Lyon took some time in gaining approval in Australia for his operation, now called JAYWICK, but on 2 September 1943 set sail from Exmouth, Western Australia, with four British and 11 Australian army and naval personnel aboard the *Krait,* a wooden ship about 20 metres long which had sailed Singapore waters before the war. The *Krait* passed safely north to Subor Island, within 11 kilometres of Singapore. Here, on the night of 26 September, six men in two-man rubber and canvas canoes called Folboats were sent to paddle into Singapore harbour. The teams placed timed limpet mines on several merchant ships without being discovered and escaped to a nearby island where they rested. The mines exploded as planned and sank seven ships of a combined tonnage of 35 000 tons. The canoeists then paddled 80 kilometres for their rendezvous with the *Krait* on 2 October, arriving safely back at Exmouth on 19 October.

After the success of JAYWICK, Lyon perhaps unwisely decided to attack the same target again in 1944, this time using newly constructed Motor Submersible Canoes (MSC), which enabled the one-man crew to approach the target underwater. This operation, named RIMAU, the Malay word for 'tiger', began on 11 September 1944 when Lyon, with five British and 17 Australian personnel plus a British observer from South East Asia Command, left Fremantle aboard the submarine HMS *Porpoise*. The submarine, which had been converted for the mission, carried 15 MSCs and 11 Folboats. The plan was to sail to the waters around Singapore and capture a boat to which the team and the canoes would be transferred. A boat called the *Mustika* was captured off the Borneo coast on 29 September. The *Mustika*, however, was found not to have an engine, and the operatives were forced to spend a day learning from the crew how to use the sails before HMS *Porpoise* departed, taking with them the crew of the *Mustika* who were summarily detained until the end of the war. The operation went awry when the *Mustika* was observed by water police about 20 kilometres south of Singapore. The RIMAU operatives opened fire and killed all but one of the Malay policemen, who escaped to raise the alarm. Lyon then decided to cancel the operation, scuttled the *Mustika* and the secret MSCs, and attempted to escape to their rendezvous point using the canoes. Japanese troops methodically searched the islands south of Singapore and made contact around 16 October when Lyon and three comrades were killed while providing covering fire to enable the rest of the team to escape. A British submarine, HMS *Tantalus,* missed the first rendezvous on 8 November, and did not return until 21 November, but no member of RIMAU been at the rendezvous since 4 November. Eventually all the operatives were either killed or captured by the Japanese. After being imprisoned in Singapore, the surviving prisoners were beheaded by the Japanese on 7 July 1945. Three weeks later, on 31 July, a third raid on Singapore Harbour, this time by two British midget submarines, sank the Japanese cruiser *Takao*. Although it has been written that the success of JAYWICK gave a much needed boost to Allied morale, this was not so, as the raids did not become public knowledge in Australia until August 1945. Since then JAY-WICK and RIMAU have become the best known exploits of Australian Special Forces and have been retold several times in books and films. However, without denying the bravery of the men involved, the operations had little effect on the Japanese war effort and far more enemy shipping was destroyed by the mines dropped by RAAF aircraft and by US submarines. The *Krait* is now exhibited at the Australian National Maritime Museum in Sydney.

JEPARIT see **SUPPLY SHIPS**

JERVOIS, Major-General William Francis Drummond (10 September 1821–17 August 1897). Jervois graduated from the Royal Military Academy, Woolwich, and was commissioned into the Royal Engineers in 1839. He held several important positions in Britain and the colonies before being appointed Director of Works for Fortifications in 1862. Jervois visited the United States and Canada several times to study their respective fortifications, he advised the Canadian government on the construction of forts, and supervised similar projects in Britain and a number of imperial outposts. Knighted in 1874, he was appointed governor of the Straits Settlements the following year. In 1877 he was asked to survey and report on the defences of Australia and New Zealand. While in Melbourne with Colonel P. H. Scratchley (q.v.), he was told that he had been appointed governor of South Australia, which was designed to remove him from the Straits Settlements where he had antagonised much of the local population. Jervois's reports on New South Wales, Victoria and Queensland were submitted in the period June–August 1877; those on South Australia and Tasmania followed in December 1877 and February 1878 respectively.

Jervois concluded that while the defence of the Australian colonies ultimately depended on the supremacy of the RN, which made a large invasion impossible, the colonies were susceptible in local terms to raids, bombardment and demands for ransom. He therefore recommended that fortifications be built to protect the approaches to the capital cities and to such important coaling stations as Newcastle, and that gun emplacements located within coastal forts be supported by infantry and mobile field artillery which would guard against flanking movements by any forces that could be landed from enemy warships. The danger of bombardment of Sydney from an enemy ship could be countered by the provision of an ironclad (warship), while Melbourne could be protected by a combination of fortifications at either side of the entrance to Port Phillip Bay, minefields and the guns of the *Cerberus* (see VICTORIAN NAVY SHIPS). Local naval power he regarded as even more important for South Australia, because of Adelaide's relatively exposed position, but again, as with Queensland and Tasmania, he also recommended the raising of local infantry and field-artillery forces to support the garrison guns.

The importance of Jervois's reports lay in the fact that while he accepted that British naval power was the ultimate guarantee of the security of the colonies, he also recognised that there were legitimate local concerns that could be met through the provision of local defence forces. When schemes for imperial defence were articulated in the 1880s they owed much in spirit, if not in detail, to the combination of imperial and local responsibilities set out by Jervois.

JESS, Lieutenant-General Carl Herman (16 February 1884–16 June 1948). Son of a German immigrant, Jess became a school teacher, although reputedly largely self-taught himself, and served in the Victorian Volunteer Cadets, reaching the rank of colour sergeant. In 1902 he enlisted in the 5th Battalion of the Victorian Infantry, a militia unit, and again rose to the rank of sergeant before becoming a regular in the Permanent Military Forces in June 1906. He qualified and was commissioned in July 1909, and held staff positions in Victoria and New South Wales. In 1912 he enrolled in the diploma course in military science at the University of Sydney. Monash (q.v.) took him as staff captain on the headquarters of the 4th Brigade of the AIF in September 1914, which occasioned some nasty invective about their joint German backgrounds but which did not prevent Jess from serving through the Dardanelles campaign, from May as brigade major of the 2nd Brigade.

The expansion of the AIF in Egypt early the following year saw him promoted again and given command of the 7th Battalion, which he led until March 1917, being awarded the DSO for the Battle of Pozières, during which he was gassed but refused to leave his command. He was sent to England in early 1917 as the first dominion officer on the instructing staff of the Senior Officers' School at Aldershot, but stayed only six months before returning to the AIF in staff positions with I Anzac Corps and Monash's 3rd Division. In October 1918 he received command of the 10th Infantry Brigade, making him one of the youngest brigadier-generals in the Empire armies (a claim also advanced for Henry Gordon Bennett [q.v.]).

His earlier connections with Monash saw him retained in England throughout 1919 in various administrative commands and as Monash's successor as Director-General of Repatriation and Demobilisation. Like most regular officers, he reverted to peacetime rank at the end of the war, but stayed in England to complete the staff college course with the brilliant and often very senior first postwar classes at that institution. He received an A pass, then as now a difficult achievement. His service during the remainder of the 1920s was anticlimactic. He returned to the instructional staff

briefly, and by 1925 was Commandant of the 6th Military District (q.v.) in Tasmania. He commanded the 4th Division of the militia from 1932 to 1933, and by July 1935 had reached the rank of major-general and the appointment of Adjutant-General at Army Headquarters. He brawled in private with the CGS, Major-General J. D. Lavarack (q.v.), leading the government to revive the post of Inspector-General (q.v.), filled by a British officer, Lieutenant-General E. K. Squires (q.v.), in 1938. Jess was saved from retirement by the outbreak of the Second World War, and became chairman of the Manpower Committee, a post which he had held for a brief time as Adjutant-General, and had relinquished to Blamey (q.v.) in November 1938, whom he now succeeded. He remained chairman of the Manpower Committee until March 1944, and deputy chairman of the Manpower Priorities Board to which it reported. He organised the Australian Women's Land Army (q.v.) in 1942, and in 1943 also became Director of Women's National Services in the Department of Labour and National Service. In the last year of the war he compiled a study of the AMF between 1929 and 1939, and retired from the army on grounds of invalidity in April 1946.

Jess's rise from the ranks was a striking achievement, explained to some extent perhaps by S. F. Rowell's (q.v.) observation that he had 'hardly an interest outside the service'. His career between the wars was disappointing in some respects, but his managerial and bureaucratic abilities were employed to the full during the Second World War.

JINDALEE OVER THE HORIZON RADAR (OHR)

Jindalee is a long-range surveillance and early-warning capability developed by the Defence Science and Technology Organisation (q.v.) with assistance from the United States. It works by bouncing shortwave radar impulses off the ionosphere from transmitter stations, which are caught in turn by receiver stations which then read the echo from ships and aircraft hundreds or thousands of kilometres distant. The OHR station at Alice Springs began operations in 1989, and additional stations are to be housed at Charters Towers in Queensland and Merredin in Western Australia by 1996, for a total projected cost of $500 million. Because beams from Jindalee descend on their targets, they are almost impossible to avoid, but the system does have some shortcomings: its minimum range is 400 kilometres, it relies for success on stability in the ionosphere, and it does not replace, although it clearly complements, existing Air Borne Warning and Control capabilities which the RAAF is keen to acquire.

JINDIVIK, GAF (Pilotless target aircraft [Mark 3B]). Wingspan 7.92 m; length 7.11 m; maximum speed 900 km/h; range 1000 km; power 1 × Rolls Royce Viper 2500-pound thrust engine.

The Jindivik pilotless aircraft was originally ordered in 1948 by the British government to assist in guided missile development. It was designed and built by the Government Aircraft Factory in Australia and first flew in 1952. The main purpose of the Jindivik is to act as a high-speed target for air- and ground-launched missiles. Jindiviks have been used by the Australian, American, British and Swedish defence forces and are still being ordered.

JOHNSTON, Lieutenant-Colonel George (19 March 1764–5 January 1823). Born in Scotland, Johnston joined the 45th Company of Marines in 1776, serving in North America and the East Indies before sailing for New South Wales with the marine detachment in the First Fleet. When the marines returned to Britain in 1790, Johnston was selected by Governor Phillip to raise a company in the colony which would join the incoming New South Wales Corps (q.v.). He went on to hold a number of responsible positions in the colony and earned popular admiration by suppressing the convict rebellion at Castle Hill in 1804, but he quarrelled with Governors King and Bligh over what he saw as their intrusion into military administration. In 1800 he was arrested for breaches of liquor regulations and was sent for trial to England, where the trial was cancelled due to lack of evidence. In 1808, however, he risked more serious charges when on 26 January he arrested Governor Bligh and assumed the lieutenant-governorship. Claiming a popular mandate for his actions, Johnston declared martial law. The causes of the rebellion are complex and unclear, but it seems to have been largely the result of a power struggle. Bligh's reforms in such areas as land grants and the liquor trade threatened the wealth of an influential clique of landowners and officers whose most prominent member was John Macarthur. Johnston was relieved as lieutenant-governor in July and, with Macarthur, travelled to London where he hoped that an investigation could be organised and his conduct vindicated. Instead he was court-martialled in June 1811, found guilty and dismissed from the army. Johnston returned to New South Wales, arriving in May 1813. He went on to become a substantial landowner, a successful farmer and a well-respected colonist.

John Ritchie (ed.), *A Charge of Mutiny: The Court Martial of Lieutenant Colonel George Johnston for Deposing Governor William Bligh in the Rebellion of 26 January 1808* (National Library of Australia, Canberra, 1988).

JOINT CHIEFS OF STAFF IN AUSTRALIA (JCOSA) Formed at the end of 1945 as an advisory body for the oversight of the British Commonwealth Occupation Force (q.v.) in Japan, this was essentially the Australian Chiefs of Staff Committee (q.v.) augmented by senior representatives of the other contributing nations, Britain, India and New Zealand, supported by its own planning staff. The committee met in Melbourne, convening for the first time on 4–5 December 1945. The control and administration of BCOF were provided by the Australian chiefs of staff, and the C-in-C, BCOF, had the right of direct access to JCOSA (outside the American chain of command in Japan) on administrative matters affecting his force. JCOSA in fact did not work altogether well, since the British were suspicious of the dominant position which the Secretary of the Department of Defence, Sir Frederick Shedden (q.v.), exercised over the Australian chiefs of staff, and on occasions they proved uncooperative. Basically, the British wished to influence the running of the Commonwealth part in the occupation to a degree which the Australians, who had been allotted the executive role in commanding the force, would not accept, and this was never resolved. The committee was dissolved on 31 December 1947 after the withdrawal of British and Indian contingents from Japan.

JOINT DEFENCE FACILITIES is the term used to describe US installations based in Australia in which the Australian government has an equal or at least significant role, providing some staff and resources towards the running of the bases. It does not, however, encompass the full range of American military, strategic and other research installations based in Australia, on which few precise details have been released. The three most important facilities are the naval communications station at North West Cape in Western Australia (named for the late prime minister, Harold Holt); the joint defence space research facility at Pine Gap, near Alice Springs; and the joint defence space communications station, Nurrungar, near Woomera in South Australia. North West Cape was opened in September 1967 and forms a critical link in the US global communications system, mainly through maintaining communications with America's nuclear submarine fleet. Pine Gap, opened in 1969, comes under the operational control of the Central Intelligence Agency and in the past has played an important role in monitoring the former Soviet nuclear missile testing and launching capabilities. Nurrungar, which began working in 1971, is a ground station for the US satellite early warning system. All three, and no doubt other facilities as well, thus played a crucial part in American Cold War strategy and nuclear war capabilities. As such they were naturally the subject of debate, sometimes intense, especially on the left wing of Australian politics. The existence of these facilities on Australian soil undoubtedly made Australia a potential nuclear target, to the extent that all three would have been high on Soviet targeting priorities in the event of global war, but contrary to some claims such targeting priority probably did not extend any further to encompass purely Australian targets. In the context of the decisions about hosting these bases, made in the 1950s and 1960s at the height of Cold War confrontation, such risks were accepted. The extension of 'joint' status helped to answer some of the criticisms made about the compromising of national sovereignty implicit in hosting foreign bases, although this was seen by many critics as a sop. With the end of the Cold War, the collapse of the Soviet Union and, more importantly, advances in satellite communications, the importance of the bases to American global capabilities is receding, and it seems likely that they will be phased out in the medium term.

Des Ball, *A Suitable Piece of Real Estate: American Installations in Australia* (Hale & Iremonger, Sydney, 1980).

JOINT INTELLIGENCE ORGANISATION (JIO)
see **INTELLIGENCE, MILITARY**

JOKES see **HUMOUR**

JONES, Air Marshal George (22 November 1896–24 August 1992). Born in Rushworth, Victoria, Jones left school at 14 to become a motor mechanic. He served as a private in the 9th Light Horse at Gallipoli, and as a corporal in the Imperial Camel Corps before joining No. 1 Squadron, Australian Flying Corps (AFC), as an air mechanic, second class. By the end of the war he was a captain in No. 4 Squadron, AFC, and had been awarded the DFC. After a short period as a civilian, he rejoined the RAAF in 1921, and by 1939 had become Assistant CAS and was heavily involved in the administration of the Empire Air Training Scheme (q.v.). In 1942 he was a controversial choice to succeed the British officer, Air Chief Marshal Sir Charles Burnett (q.v.), as CAS. The chain of command for RAAF units deployed in the South-West Pacific Area under General Douglas MacArthur became a matter of heated dispute between Jones and Air Vice-Marshal W. D. Bostock (q.v.), and the controversy was not satisfactorily resolved by the time the focus of US strategic interest had passed beyond the areas in which Australian forces were committed. Jones supervised the demobilisation of the RAAF at the end of the

war, and remained as CAS until 1952, being responsible for the deployment of RAAF forces in the Berlin airlift (q.v.), the Korean War and the Malayan Emergency. He was appointed KBE in 1953, and for five years after retiring from the RAAF was Director of Coordination of the Commonwealth Aircraft Corporation (q.v.).

JOSE, Arthur Wilberforce (4 September 1863–22 January 1934). A journalist, teacher and publisher's reader, Jose had known difficult times after the loss of his father's fortune in the 1880s and had as a consequence worked for a time as a sleeper-cutter for the Victorian Railways. Educated at Clifton College, England, and employed as an assistant master at All Saints College, Bathurst, between 1885–87, he knew the Bean (see BEAN C. E. W.) family well before the First World War.

His writing career began in the 1890s with the publication of *The Growth of the Empire* (1897), followed by his *History of Australasia* (1899), which went through 15 editions by 1929. He saw brief service in South Africa as a war correspondent in 1899–1900. In 1904 he returned to Sydney as correspondent for *The Times,* for which he wrote regularly on cricket, Australian politics, and the unity of the Empire. A founder of the Australian National Defence League (q.v.) in 1905, he held a commission in the Australian Intelligence Corps (q.v.) from 1909. From 1915 he was attached to the intelligence branch of the RAN in order to compile a history of naval operations in the war, and to analyse intelligence on China, Japan and South-east Asia.

In 1920 he was appointed by Bean to write the official volume on the RAN. The volume was delayed repeatedly through Jose's many other activities, and because of the determination of the Naval Board to censor it, especially on intelligence matters. There was also a concern to limit Jose's criticisms of the unpreparedness of the Admiralty in the early months of the war and of its handling of the Australian fleet in 1914. Bean undertook the final revision, which appeared in 1928. Jose's history is reliable and scholarly in tone, although the implications of some of his criticisms for interwar naval strategy appear to have escaped most reviewers at the time.

A. W. Jose, *The Royal Australian Navy* (Angus & Robertson, Sydney, 1928); Stephen Ellis, 'The Censorship of the Official Naval History of Australia in the Great War', *Historical Studies*, vol. 20, no. 80, April (1983), pp. 367–82.

JUGOSLAV BATTALION, FIRST WORLD WAR see AUSTRALIANS IN THE SERVICE OF OTHER NATIONS

JUNGLE TRAINING CENTRE see CANUNGRA LAND WARFARE CENTRE

K

KANGA FORCE Formed in April 1942 as a means of reinforcing the New Guinea Volunteer Rifles (NGVR) (see PACIFIC ISLANDS REGIMENT) operating around Japanese positions at Lae and Salamaua, it consisted of the NGVR, the 2/5th Independent Company and one platoon of the 2/1st Independent Company, under the command of Major N. L. Fleay, a veteran of the fighting in the Middle East. Its main purpose at this stage was reconnaissance, although in addition it staged a number of surprise raids against Japanese positions in the Wau–Bulolo area before being pulled back in the face of the Japanese advance in August. In October it was reinforced by the 2/7th Independent Company under Major T. F. B. MacAdie and continued to operate against Japanese positions around Wau, raiding and destroying enemy facilities preparatory to the main Australian advance. In January 1943 the force joined Brigadier M. J. Moten's 17th Brigade, which then fought a stiff action against Japanese forces advancing on Wau from Mubo. On 23 April 1943 Major-General S. G. Savige (q.v.) assumed command of the Wau–Bulolo area with his 3rd Division, and Kanga Force ceased to exist.

KANIMBLA, HMAS see **ARMED MERCHANT CRUISERS**

KAPOOKA At present home to the 1st Recruit Training Battalion (1RTB) of the Australian Regular Army (q.v.), the site of the camp located near Wagga Wagga in southern New South Wales was first acquired for military purposes in 1940. During the Second World War it was an Army Engineer training camp and subsequently a transit camp until closed in 1946. It then became a migrant camp administered by the Department of Immigration until 1951, when the facility was resumed for military use and 1RTB was established. During the Korean War it was joined by the 2nd Recruit Training Battalion, which was disbanded in 1953. The present permanent camp was constructed in 1965–66. During the Vietnam War it handled national servicemen (see CONSCRIPTION) as well as regulars, and from 1985, when the Women's Royal Australian Army Corps (q.v.) School closed, it has processed female recruits as well. In recent years Kapooka has handled up to 3000 recruits annually.

KARRAKATTA, HMS see **AUXILIARY SQUADRON, AUSTRALIA STATION**

KATOOMBA, HMS see **AUXILIARY SQUADRON, AUSTRALIA STATION**

KAUPER, Henry see **SCIENCE AND TECHNOLOGY**

KEOGH, Colonel Eustace Graham (24 April 1899–9 November 1981). Enlisting under age in the AIF in May 1916, Keogh served in the First World War as a driver in the 1st Australian Wireless Squadron in Mesopotamia. He received a commission in the postwar AMF in 1924, and served in the interwar period in a variety of postings. He won the AMF Gold Medal Essay prize for 1931, and again in 1937.

At the outbreak of the war in Europe in 1939 he was seconded to the AIF, and held a variety of staff and regimental appointments within the 7th Division and I Australian Corps. He served in the Middle East, Australia and New Guinea, and finished the war in the Directorate of Military Training as a GSO1 with responsibility for reinforcement training and the training of the women's services.

Demobilised in mid-1946, he soon reappeared in the directorate as a civilian employee, and from 1948 became the editor of the *Australian Army Journal* (q.v.) which began publishing in June–July of that year. An active writer as well as editor, he wrote a number of texts adopted for military instructional purposes, including studies of the Second World War campaigns in the Middle East, Malaya and the south-west Pacific, and the Shenandoah Valley campaign of 1861–62. He won the AMF Gold Medal Essay competition for the third time in 1948, and presided over the *AAJ* until his retirement in April 1964. 'Intellectual, disciplined and methodical', he founded and sustained the first successful professional periodical in the history of the Australian armed forces, and his emphasis on the importance of history in the professional development of army officers influenced a generation. In retirement he acted as historical adviser to Crawford Productions Pty Ltd in the making of the highly successful and long-running television series *The Sullivans*.

KERR, John (24 September 1914–24 March 1991) was a research officer in the A (Personnel) Branch at Land Headquarters from October 1942 to October 1944. From April to September 1945, as a lieutenant-colonel, he served in the Directorate of Research and Civil Affairs where he was an adviser to General Blamey on civil affairs in Papua New Guinea. After the war he returned to his legal career, and served as Chief Justice of New South Wales from 1972 to 1974. He was appointed Governor-General of Australia in 1974, and the following year, in circumstances of considerable controversy, he dismissed the Labor Prime Minister, E. G. Whitlam. One of his last public duties was in December 1977, when as Governor-General and C-in-C, he reviewed the Passing Out Parade at

RMC Duntroon, the last incumbent of the office to do so wearing full morning dress.

KIA ORA COO-EE see **SERVICE NEWSPAPERS**

KILLEN, (Denis) James (23 November 1925–). Killen saw service as an air gunner with the RAAF between 1943 and 1945, and worked in a variety of jobs before winning the federal seat of Moreton for the Liberal Party in 1955. He was minister for the Navy between 1969 and 1971, and following the return of the Liberal Party to office was minister for Defence between 1975-82 in the government of Malcolm Fraser (q.v.). Killen oversaw the 1976 Defence White Paper which first articulated the demise of 'forward defence' (q.v.) as a strategic basis for Australian policy, and argued that future operations involving the ADF might occur in Australia's own region. Although the paper spoke of 'increased self-reliance' as a future basis for development of the services, the defence budget did not increase significantly. Decisions regarding major equipment acquisitions for the air force and navy were largely deferred during Killen's period as minister (although he was not necessarily to blame for this), and the period has been well characterised as 'one of uncertainty'. Killen was Vice-President of the Executive Council and Leader of the House in 1982–83 before leaving politics. He was appointed KCMG in 1982.

Jim Killen, *Killen: Inside Australian Politics* (Methuen Haynes, Sydney, 1985).

KIOWA, BELL 206 (2–5 seat light observation helicopter). Rotor diameter 10.7 m; length 9.93 m; maximum speed 220 km/h; range 550 km; power 1 × Allison 420 shaft h.p. engine.

The Kiowa was ordered in 1971 as a replacement for the Bell Sioux (see SIOUX, BELL). In 1971, during the Vietnam War, eight Kiowas leased from the US Army were used by 161 Flight, while 56 helicopters were built under licence by the Commonwealth Aircraft Corporation (q.v.) between 1972 and 1977. Kiowas are used by the RAN and by No. 161 and 162 Squadrons of the Australian Army Aviation Corps (q.v.).

KIRKPATRICK, Private John Simpson see **SIMPSON AND HIS DONKEY**

KITTYHAWK, CURTISS (Single-seat fighter-bomber [P-40E]). Wingspan 37 feet 3.5 inches; length 31 feet 9 inches; armament 6 × 0.5-inch machine-guns, 1000 pounds bombs; maximum speed 364 m.p.h.; range 810 miles; power 1 × Allison 1325 h.p. engine.

The most widely used RAAF aircraft during the Second World War, the Kittyhawk played a vital part in the defence of Australia in 1942. Shipped from the United States, Kittyhawks were used in the defence of Port Moresby and the defeat of the Japanese at Milne Bay (see NEW GUINEA CAMPAIGN). More suited to flying at low than at high altitude, they were used as ground-attack aircraft rather than fighters and supported ground forces throughout the South-West Pacific Area (q.v.), including the Admiralty Islands, New Guinea, New Britain and the Netherlands East Indies (q.v.); they were flown by No. 75, 76, 77, 78, 80, 82, 84 and 86 Squadrons. In 1945 Kittyhawks of the 1st Tactical Air Force (q.v.) based at Morotai provided fighter cover during the Australian landings in Borneo. Kittyhawks were operated in the Mediterranean by No. 3 and 450 Squadrons from 1941 to 1945 and were flown by Group Captain Clive Caldwell and Squadron Leader 'Bluey' Truscott (qq.v.).

KNIGHT, Corporal Albert see **ABORIGINES AND TORRES STRAIT ISLANDERS IN THE ARMED FORCES**

KOKODA The first Japanese attempt to capture Port Moresby was defeated in the Battle of the Coral Sea (q.v.) in May 1942, and their defeat at the Battle of Midway in June, in which they lost four aircraft-carriers, meant that an ambitious amphibious assault was now beyond them. This left as the only option an advance from positions on the northern coast at Buna and Gona across the Owen Stanley Ranges, a difficult undertaking for which, as it transpired, the Japanese were not well equipped. Their planned offensive was assisted, however, by intelligence failures at the headquarters of General Douglas MacArthur (q.v.), which played down the likelihood of an early renewal of Japanese activity.

The garrison at Port Moresby was increased to two militia brigades in May, and the commander of New Guinea Force, Major-General B. M. Morris (q.v.), responded to orders to push troops up the Kokoda Track towards Buna, but the 39th Battalion met strong Japanese forces which had landed there on 21 July, and within a week the battalion had been forced out of Kokoda itself. Reinforcements from the 7th Division and another AIF brigade under Major-General Cyril Clowes (q.v.), destined for Milne Bay, would not arrive until mid-August, but fortunately enemy attention was distracted by the Marines' landing at Guadalcanal, and the planned offensive towards Port Moresby was postponed, to be coordinated now with an attack at Milne Bay. In the meantime, Lieutenant-General S. F. Rowell (q.v.) was sent up to take command of New Guinea Force.

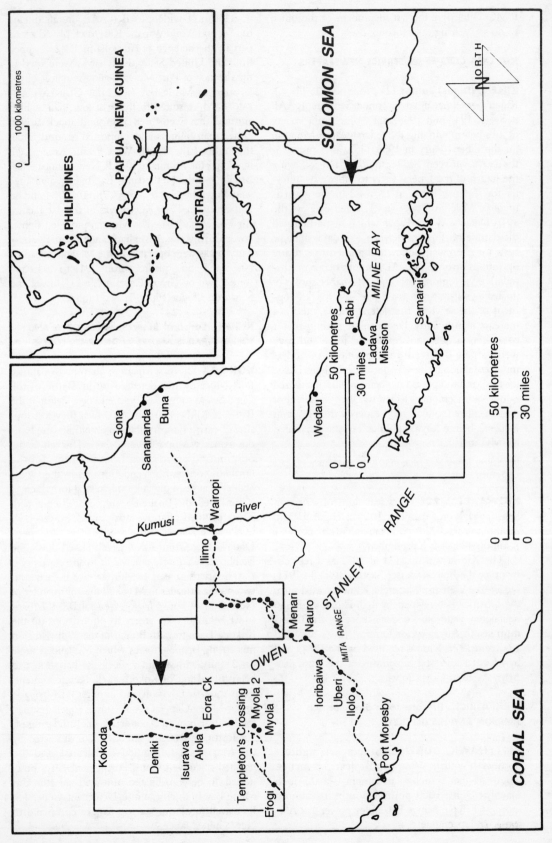

On 26 August a two-pronged offensive began against Milne Bay and Isurava. By 6 September Clowes had defeated the landing at Milne Bay in difficult circumstances, but the threat along the Kokoda Track now assumed greater importance. The 39th Battalion led by Lieutenant-Colonel Ralph Honner (q.v.), a veteran of the campaigns in the Middle East, fought a series of desperate delaying actions along the track while waiting for reinforcement from the units which were arriving in Port Moresby. The Australians were forced out of Isurava after four days, and for the first two weeks of September fought back along the track to positions at Efogi, where they again fought a bloody series of engagements before withdrawing to Ioribaiwa. From here, on 16 September Brigadier K. W. Eather (q.v.), commanding the 25th Brigade, withdrew to strong positions at Imita Ridge after a planned counter-attack went wrong. This withdrawal, militarily sound as it was, created panic in the remote high command in Australia, who throughout the fighting had in any case demonstrated that they did not possess a sound grasp of what was actually happening, and MacArthur ordered Blamey (q.v.) forward to investigate the situation. As a result Rowell and Brigadier A. W. Potts (q.v.), who had held the position at Efogi with the 21st Brigade against strong enemy attacks, were relieved of their commands, unjustly.

The Japanese, however, were at the end of their resources. They had failed at Milne Bay and their operations on Guadalcanal were going badly, while the force on the Kokoda Track was exhausted at the end of a long and difficult supply line. On 25 September the Japanese high command ordered a withdrawal, and their forces established defensive positions at Templeton's Crossing and Eora Creek, which were overcome by hard and costly fighting in October. The command in Australia seemed no more aware of the conditions facing Australian troops now than they had been in August, and Major-General A. S. Allen (q.v.), commanding the 7th Division, was repeatedly ordered to increase the speed of his advance. Tiring of the pressure from his rear and facing supply difficulties which his superiors clearly did not comprehend, Allen responded to one message urging action with the signal, 'If you think you can do any better, come up here and bloody try', but was persuaded not to send it. He was relieved on 27 October, just as his forces were gaining the initiative, which they were not to lose again. His replacement was Major-General G. A. Vasey (q.v.). On 2 November Kokoda was retaken, and the Australian commanders now prepared for

Map 16 The Kokoda Track, 1942. *Inset* Milne Bay.

the advance to the coast and the savage fighting for the beachheads at Gona and Buna.

Kokoda was an important battle because it defeated an enemy offensive which, had it been successful, would have altered the strategic situation in the South-West Pacific Area. The Japanese lost because they underestimated the task they had set themselves, because they were distracted by the operations on Guadalcanal, and because the Australians along the track bought time by their stubborn defence. The command decisions on the Australian side, which emanated from headquarters in Brisbane, were unsatisfactory at best, while the heavy casualties suffered by the 39th Battalion and the other units first thrown into the fighting showed just how inadequate had been the preparations for the defence of Australia.

Peter Brune, *Those Ragged Bloody Heroes: From the Kokoda Trail to Gona Beach 1942* (Allen & Unwin, Sydney, 1991).

KONFRONTASI see CONFRONTATION

KOREAN WAR The Korean War began in the early hours of 25 June 1950 when the (North) Korean People's Army invaded the Republic of (South) Korea across the 38th parallel. The attack was preceded by a growing civil struggle in which 100 000 Koreans were killed on both sides. The Korean War was a consequence of growing Cold War tensions between East and West and the desire for reunification of the country, artificially divided by the Russians and Americans at the end of the Second World War. Australia was involved in Korean affairs before 1950 as part of United Nations initiatives, and played a part in the UN Temporary Commission on Korea (UNTCOK) and the UN Commission on Korea (UNCOK), which had monitored the elections of May 1948 in the southern republic. In the week before the North Korean attack two Australian officers, Lieutenant-Colonel F. S. B. Peach and Squadron Leader R. J. Rankin, had been provided as members of a UN monitoring team, and their report on conditions along the 38th parallel immediately before hostilities led the UN Security Council to declare the war an act of North Korean aggression and invite UN member states to send forces to restore the situation. Australia was one of the first nations to commit units (from all three services) to the fighting, and these played small but occasionally significant roles as part of the United Nations Command led, until April 1951, by General Douglas MacArthur (q.v.). In addition, Australians remained involved in UN commissions such as the UN Commission on the Unification and Rehabilitation of Korea (UNCURK), which acted at times to moderate

Clad in winter clothing, Sergeant Chaperlin, of 3RAR, fires a Vickers machine-gun at Chinese positions at Chipyong-ni, Korea, February 1951. (AWM P1479/07)

the authoritarian domestic policies of the southern president, Syngman Rhee. In the United Nations itself, Australia's role was generally one of supporting American policy, although the divergences between the Americans and British over recognition of China and the threatened use of atomic weapons placed Australian representatives in a difficult position.

The first Australian unit committed to the fighting was No. 77 Squadron, equipped with P-51 Mustangs (see MUSTANG, NORTH AMERICAN AND CAC). The last of the air force units still based in Japan as part of the British Commonwealth Occupation Force (q.v.), the squadron had been preparing to return to Australia when the war began, and the Australian government hesitated at first to deploy any of the forces based in Japan. MacArthur gained the use of the Australian planes only by placing public pressure on the Australians, but within a few days permission was given for the squadron to fly in support of the sorely pressed American and South Korean ground forces, and the first missions were flown on 2 July. Air power was critical to the survival of the ground forces in the first months of the war, and the Australian

squadron was kept busy against enemy ground targets and North Korean fighters. The appearance of modern, Soviet-supplied MiG-15 jet aircraft in November 1950 reduced the value of the Australian squadron, however, because its piston-engine aircraft were now seriously outclassed, and it was re-equipped with the British Meteor Mark 8 (q.v.), an inferior plane which failed to match the MiG in performance. As a result, the Australian squadron was reassigned to less glamorous fighter-sweep and ground-attack roles for the rest of the war. The RAAF also fielded base and maintenance squadrons, and in addition to the fighter squadron deployed No. 36 Transport Squadron, flying Dakotas (q.v.), and No. 30 Communications Flight, the whole organised as No. 91 Composite Wing.

Ships of the RAN were involved in Korean waters from the early days of the war. On 29 June 1950 the Australian government committed the frigate HMAS *Shoalhaven* and the destroyer HMAS

Map 17 Korean War 1950–53.
Inset 1st Commonwealth Division positions on the Jamestown Line 1951–53.

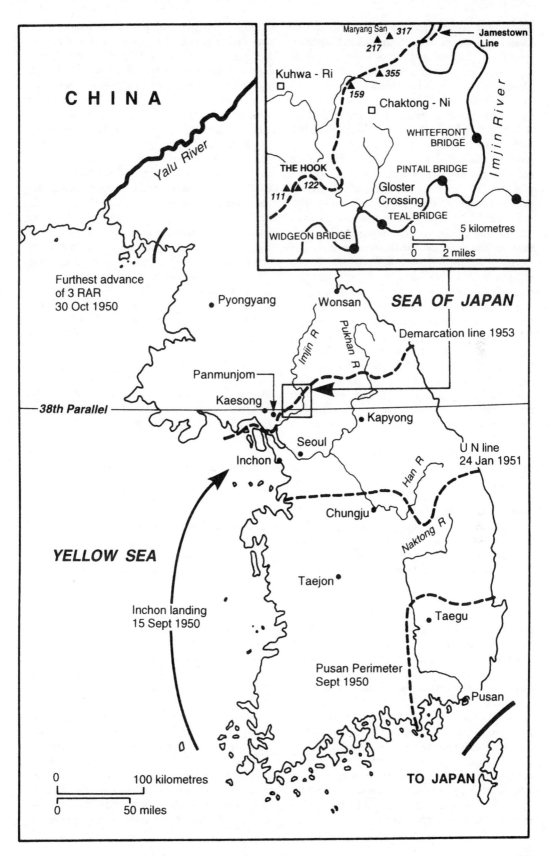

CHINA

Yalu River

Furthest advance
of 3 RAR
30 Oct 1950

Maryang San ▲ 317
217 ▲

Kuhwa - Ri □

▲ 355

159

Chaktong - Ni □

WHITEFRONT
BRIDGE

THE HOOK

PINTAIL BRIDGE

▲ 122
111

Gloster
Crossing

TEAL BRIDGE

WIDGEON BRIDGE

Imjin River

Jamestown
Line

0   5 kilometres

0   2 miles

Pyongyang

Wonsan

SEA OF JAPAN

Demarcation line 1953

Panmunjom

Kaesong

Imjin R

Pukhan R

38th Parallel

Kapyong

Seoul

Inchon

Han R

U N line
24 Jan 1951

Chungju

YELLOW SEA

Naktong R

Inchon landing
15 Sept 1950

Taejon

Taegu

Pusan Perimeter
Sept 1950

Pusan

0   100 kilometres

0   50 miles

TO JAPAN

Bataan, both then in Japan as part of the occupation force. The ships joined the British Commonwealth naval force which formed part of Task Group 96.8, the West Coast Support Group. In late August *Shoalhaven* was relieved by HMAS *Warramunga,* and for the remainder of the war the RAN maintained two destroyers or frigates on station in Korean waters. In October 1951 the aircraft carrier HMAS *Sydney* (III) with the 20th Carrier Air Group replaced the British carrier HMS *Glory,* which sailed to Sydney for refit, returning to Korean waters in February the following year. This was the first (and only) active service deployment of Australia's newly developed carrier aviation capability.

The major Australian commitment was provided by the army, and like the other two services the initial deployment was based on forces already in Japan. By June 1950 the only ground-force unit remaining in Japan was the 3rd Battalion of the recently formed Royal Australian Regiment (q.v.), seriously under strength with an establishment of 500–550 men and in fact preparing to return to Australia. On 26 July the government announced that Australia would commit ground units, after first intending otherwise but changing its mind when the British reversed an earlier decision not to commit their own ground units either. Although the decision was thus made within a context of coordinating responses within the Commonwealth, the main pressure within the Australian cabinet came from the Minister for External Affairs, Percy Spender, who believed that a strong show of support for United States policy would assist in the negotiation of a security pact, which ultimately found expression in ANZUS (q.v.). The battalion in Japan was in need of reinforcement and retraining before being sent into battle, and the C-in-C in Japan, Lieutenant-General Sir Horace Robertson (q.v.), insisted that 3RAR be brought up to strength before it was dispatched. As a result, the Australian unit did not arrive in the southern port city of Pusan until late September, where it joined two British battalions as part of the 27th Commonwealth Infantry Brigade. For the rest of the war Australian units served within Commonwealth formations; when the original British battalions were withdrawn in April 1951, the brigade was redesignated the 28th Commonwealth Brigade (q.v.). From April 1952 3RAR was joined by a second Australian battalion, successively 1RAR and 2RAR on one-year tours, and from that time command of the brigade reverted to Australia. In July 1951 the 28th Brigade was combined with the 25th Canadian and 29th British Brigades to form the 1st Commonwealth Division.

The Australians took part in the long advance into North Korea which began with the Inchon landings in September 1950, and in the equally long and desperate retreat which was occasioned by Chinese intervention in October–November. During the Chinese Fifth Phase offensive in April 1951 3RAR and the 2nd Battalion, Princess Patricia's Canadian Light Infantry, fought a hard defensive battle at Kapyong which stopped an enemy division. Further hard fighting followed in October during Operation COMMANDO, during which UN forces extended their positions along the Jamestown Line and in which 3RAR took the strongly defended Chinese positions on Maryang San. Although truce talks had commenced, first at Kaesong then Panmunjom, a bitter positional war continued and this static phase, which lasted almost two years, saw Australians involved in major actions at the Hook and Hill 355, as well as conducting a war of trench raids and patrols reminiscent of the Western Front. Total casualties for all services were 1584, with 339 killed and 29 taken prisoner. The main consequence of the war for Australia was the signing of the ANZUS Treaty in September 1951 and the successful negotiation of a $250 million loan under unusually favourable conditions. The war made little direct impact in Australia itself, although the North Korean attack and the prolonged fighting contributed to the heightened Cold War atmosphere which came to characterise domestic politics in the first half of the 1950s.

Robert O'Neill, *Australia in the Korean War 1950–1953,* 2 vols (Australian War Memorial, Canberra, 1981, 1985).

KRAIT see **JAYWICK/RIMAU**

KUTTABUL, HMAS (1) A Sydney Harbour ferry built in 1922, requisitioned by the navy in 1940 and commissioned on 26 February 1941 as a depot ship and accommodation vessel. It was moored at Garden Island (q.v.), where on the night of 31 May 1942 it was sunk by a torpedo fired by a Japanese midget submarine which was trying to hit the USS *Chicago.* The 21 sailors who died as a result were the only casualties of the midget submarine attack on Sydney Harbour (see JAPANESE ATTACKS ON AUSTRALIA).

(2) The navy's administrative support centre for the Sydney area, located at Garden Island and commissioned on 1 January 1943. The establishment was later moved a short distance to a new building in Wylde Street, Potts Point, which was completed in October 1966. In addition to its administrative functions, HMAS *Kuttabul* provides accommodation for hundreds of Sydney-based naval personnel.

L

L1A1 SELF-LOADING RIFLE (SLR) In 1954 the United Kingdom, Australia and Canada decided to adopt a version of the Fabrique Nationale (FN) FAL automatic rifle to replace their bolt action 0.303-inch calibre Lee Enfield rifles. During the mid-1950s extensive trials were carried out with FN trial rifles known as the X8 model in the United Kingdom, Europe, Canada and Australia, and on operations in Malaya. From these trials the L1A1 evolved as a self-loading rifle rather than an automatic capability rifle. The first Australian production L1A1 SLR came off the production line at the Small Arms Factory in Lithgow, New South Wales, in 1959. The L1A1 SLR was the standard service rifle in the ADF from 1959 until its replacement by the Steyr AUG (F88) assault rifle (q.v.) in the early 1990s. During its service the SLR has been used by Australian soldiers on operations in South Vietnam, Malaysia and Borneo. It has an excellent reputation for reliability, robustness and hitting power.

The SLR is a gas-operated, magazine-fed, semi-automatic rifle. It fires 7.62 mm calibre ammunition and has a 20-round capacity magazine. An SLR with a filled magazine fitted weighs 11 pounds 3 ounces (5.07 kg). The effective range for an individual firer is 300 yards and 600 yards for section fire against area targets. The rate of fire is up to 20 aimed shots per minute. The SLR can be fitted with a grenade projector, a blank firing attachment, and the L1A2 knife bayonet with an 8-inch blade. The Small Arms Factory produced two other rifles related to the SLR — the shortened L1A1-F1 SLR for the Papua New Guinea Defence Force and the L2A1 heavy-barrel automatic rifle for the ADF. The L2A1 has an automatic capability, fires from a bipod and has a 30-round magazine.

IAN KURING

LACHLAN, HMAS see **RIVER CLASS FRIGATES**

LADAVA, HMAS see **ATTACK CLASS PATROL BOATS**

LAE, HMAS see **ATTACK CLASS PATROL BOATS**

LANCASTER, AVRO (7-crew heavy bomber [Mark I]). Wingspan 102 feet; length 69 feet 6 inches; armament 10 × 0.303-inch machine-guns, 14 000 pounds bombs; maximum speed 287 m.p.h.; range 2530 miles; power 4 × Rolls Royce Merlin 1145 h.p. engines.

No. 460, 463 and 467 Squadrons RAAF flew Lancasters during the Second World War from 1942 to 1945. The Lancaster enabled Bomber Command to carry out its bombing campaign against German cities and factories. The first Lan-casters, which flew in 1941, could carry 14 000 pounds of bombs, in comparison to the mere 4000 pounds carried by both the twin-engined Wellington (q.v.), which it replaced, and the four-engined B-17E, then the largest American bomber in service. The Australians who crewed Lancasters were trained under the Empire Air Training Scheme (q.v.). A development of the Lancaster called the Lincoln (q.v.) was produced in Australia after the Second World War. The Lancaster 'G for George' (q.v.) of No. 460 Squadron is held by the Australian War Memorial, while 'S for Sugar', which served with a British squadron and No. 467 Squadron, is preserved at the RAF Museum at Hendon.

LAND ARMY see **AUSTRALIAN WOMEN'S LAND ARMY**

LAND GRANT SCHEME for colonial volunteers originated in Victoria in 1865. Before it was cancelled in 1869, the Victorian scheme rewarded volunteers who had served for five years with land warrants of £50 each. The New South Wales government introduced its scheme in 1867 as an aid both to recruiting and to land settlement. It provided for grants of 50 acres of land to volunteers who had completed five years' continuous 'efficient' service. Certificates of efficiency were issued by commanding officers. Volunteers who were already in the forces could count part of their service since 1860 towards qualifying for a grant. The scheme was abolished in 1878. In Queensland, a land grant scheme operated between 1868 and 1876. It entitled volunteers who had completed five years' efficient service to grants of 10 acres of suburban land or 50 acres of rural land. The last scheme to be introduced was that of Western Australia, which began in 1873. It granted rural lots of not more than 50 acres to volunteers after five years' of efficient service, with fee simple to be granted if improvements to the value of £25 had been made on the land after three years (otherwise the land reverted to the Crown). These schemes seem to have had some positive effect in boosting the volunteer forces' numbers, but, in New South Wales at least, they provoked public opposition when it was discovered that many volunteers were selling their land to large landholders (see COLONIAL MILITARY FORCES).

LAND SETTLEMENT see **SOLDIER SETTLEMENT**

LANDAU, Samuel (19 January 1915–4 January 1983) was educated at the University of Melbourne, and was one of the first graduate entrants to join the Department of Defence in 1936. He travelled overseas as a defence adviser with Menzies in 1941 and with Chifley in 1946, and by 1950 was assistant secretary of the department. As an official he was involved in the negotiations over ANZUS (q.v.) in 1951 and at the Manila conference in 1954 which led to the SEATO (q.v.) pact. Landau was appointed first assistant secretary in 1957 and attended the Imperial Defence College (q.v.) in 1958. He was appointed Secretary of the Department of the Navy in 1963, a position he held for 10 years. In an unusual step, the leader of the Opposition, E. G. Whitlam, criticised Landau personally in parliament over the *Frank E. Evans* collision (see COLLISIONS, NAVAL). In 1974–75 Landau was Minister (Politico-Military Affairs) and Head of the Defence Staff in the Australian Embassy in Washington. He retired in 1976.

LARK FORCE, consisting of the 2/22nd Battalion AIF and supporting units, garrisoned Rabaul (q.v.), the administrative centre of the Mandated Territory of New Guinea, in April 1941. Its tasks were to protect the airfields at Lakunai and Vunakanau, the seaplane base in Simpson Harbour, and to form part of 'an advanced observation line', to give early warning of Japanese movements. On 8 October 1941 Colonel J. J. Scanlan arrived from Australia to command Lark Force, which also included a coastal defence battery of two 6-inch guns and searchlights, a detachment of New Guinea Volunteer Rifles, an Anti-Tank Battery, and a detachment of the 2/10th Field Ambulance with six nurses. Two anti-aircraft guns defended the airfields, where No. 24 Squadron, with Hudson and Wirraway (qq.v.) aircraft, was based.

The Japanese South Seas Force, commanded by Major-General Horii, attacked Rabaul on 23 January 1942. A landing force in brigade strength overwhelmed Lark Force and captured Rabaul in a single day. Scanlan authorised an 'every man for himself' withdrawal for which no forward plans had been made. About 400 escaped successfully and 800 became prisoners, of whom about 160 were massacred at Tol Plantation.

JOHN COATES

'LAST MAN AND LAST SHILLING' was a phrase used most famously by Labor leader Andrew Fisher (q.v.) while campaigning in a general election on the eve of the First World War. After Liberal Prime Minister Joseph Cook pledged all Australia's resources to the Empire on 31 July 1914, Fisher responded that same night by promising that, in the event of war, 'Australians will stand beside our own to help and defend her [Britain] to our last man and our last shilling'. The phrase received little attention at the time; it had been used in a 1907 Empire Day speech by a member of the New

South Wales Legislative Assembly and may well have been a cliché of imperialist rhetoric. It was, however, taken up by anti-conscriptionists in 1916–17 as an indication that the government planned to bleed the working class white in the service of the Empire. Speaking in December 1917, for example, Archbishop Daniel Mannix accused 'the wealthy classes' of being 'very glad to send the last man, but they have no notion of giving the last shilling, nor even the first'. Since then 'the last man and the last shilling' has become the most famous Australian phrase of the First World War, and has come to symbolise the imperial fervour of 1914.

LAUNCESTON, HMAS see *FREMANTLE* CLASS PATROL BOATS

LAVARACK, Lieutenant-General John Dudley (19 December 1885–4 December 1957). Lavarack joined the regular army through direct appointment as a lieutenant in the artillery in August 1905. The method of selection was by application and competitive examination, and Lavarack scored highly in the latter. He filled a variety of regimental postings within his corps before being selected to attend the Staff College, Camberley, in 1913–14.

The outbreak of war terminated the two-year course prematurely, and for six months Lavarack served as a junior staff officer at the War Office. He was sent to France with the British 22nd Division in September 1915, and from there served in Salonika with the British XVI Corps. The time in Salonika he considered one of the most interesting parts of his war service, but in July 1916 he was recalled to France for service with the AIF. As one of the very few Australian officers with staff college training, he served in a variety of capacities on the staffs of the 2nd and 5th Division Artillery and the Headquarters of the 1st, 3rd and 4th Divisions. He returned to Australia in late 1919.

In the small interwar regular army Lavarack developed a reputation for intellectual ability and unorthodox views. He believed, contrary to government policy, that the CMF 'was the medium by which the Regular officer was trained, rather than the reverse', and he in fact only held one peacetime posting with militia troops, on the staff of the 2nd Division between 1924 and 1925. Thereafter he served as Director of Military Training before attending the second course at the Imperial Defence College (q.v.) in London in 1928. The other Australian on the course that year was a young public servant, F. G. Shedden (q.v.). He then spent three and a half years as Director of Military Operations and Intelligence, was Commandant of RMC Duntroon (q.v.), and in April 1935 succeeded Major-General Sir Julius Bruche (q.v.) as CGS.

Lavarack's period as CGS was stormy and controversial and, it must be said, largely ineffectual in terms of preparing the army for impending war. In January 1933 he had published an article in the *Army Quarterly* (q.v.) on imperial defence and the Far East which was critical of the Singapore strategy (q.v.) and prevailing policy. As CGS he enjoyed generally poor relations with the Minister for Defence, Sir Archdale Parkhill (q.v.). These were both exacerbated and highlighted by Parkhill's controversial demotion of Lieutenant-Colonel H. D. Wynter (q.v.) in 1936–37, and relations with his successor, H. V. C. Thorby, were not always much better. Although Lavarack was able to carry through some necessary reforms, such as the establishment of a permanent School of Signals, the addition of a Directorate of Staff Duties to Army Headquarters and the opening of a Command and Staff School, his efforts were hampered by the numerical depletion of the Staff Corps and by an inability to increase greatly the army's share of the defence budget allocation.

In May 1938 the government revived the position of Inspector-General of the AMF, and appointed a British officer, Lieutenant-General E. K. Squires (q.v.), for a period of two years. The move was interpreted as a criticism of Lavarack's tenure, and Squires's first (and only) report was severely critical of some important aspects of army organisation and administration, many of which were far from being Lavarack's fault. In May 1939 Lavarack departed for Britain to confer with the Chief of the Imperial General Staff on the latest developments in training, tactics and administration; the visit was cut short by the impending outbreak of war, and he returned to Australia in late August 1939.

In October Squires was appointed CGS, and Major-General Sir Thomas Blamey (q.v.) was given command of the 2nd AIF. At first Lavarack was given Southern Command, based on Victoria. When the decision was made to raise another division, the 7th, in April 1940, Lavarack was appointed to its command. His division was held back from the disastrous Greek campaign and one of its brigades was sent to garrison Tobruk with the 9th Division under Major-General Leslie Morshead (q.v.). Lavarack was appointed to succeed the British General Neame as GOC Cyrenaica Command in the face of a renewed offensive by enemy forces, this time including German formations under the command of Rommel. He organised the defence of Tobruk (q.v.) against the Axis attack of 13-14 April, but was returned to command of his division in

mid-April at Blamey's behest, the latter arguing, against the evidence, that Lavarack was unsuited to high command.

In June–July he led the 7th Division through the campaign against Vichy French forces in Syria. Although much of the higher-level operational planning was carried out by the British, Lavarack was able to influence the campaign's conduct on a number of occasions and to the advantage of the soldiers involved. In mid-June he was appointed to command I Australian Corps, but continued to conduct the operations in Syria until the armistice on 12 July. He continued in command of I Corps until April 1942, and was involved in this capacity briefly in the disastrous Allied attempts to defend the Netherlands East Indies against the Japanese. He argued strongly and successfully with the government in Australia that commitment of the Australian Corps to operations in defence of Java would result merely in their loss. His support of Wavell's proposal that they be sent instead to Burma suggested, however, a lack of appreciation of the grave situation then facing Australia, which he himself admitted, and was to be used against him by his enemies (of whom he had many) on his return.

Lavarack did not receive another fighting command during the war. From 1942 to 1944 he commanded the First Australian Army in Queensland, which did not leave Australia until after Lavarack relinquished it in February 1944. An appointment as head of the Australian Military Mission in Washington DC between 1944 and 1946 followed. He retired from the army in September 1946, and served as governor of Queensland until 1956.

Lavarack's career contained more than its share of disappointments, explained more by his personality than by his ability, which was considerable. His temper often got him into trouble, and he made it plain too how little he was prepared to suffer fools, even when they were his political or military superiors. Blamey's dislike of him led to his being 'shelved' for much of the war, but his performance in the Mediterranean demonstrated that regular officers were as capable of command as citizen officers if given the chance.

LAW, MILITARY, is often associated solely with disciplinary aspects of military forces. However, military law encompasses all aspects of law that directly affect military forces. The most important issues in Australian military history concern constitutional law, disciplinary law and international law, including law of armed conflict. Military law also covers the ancillary legislation which pertains to the administration of Australian military forces, as well as overseas forces in Australia. This legislation includes the

Crimes Act 1914, Defence (Special Undertakings) Act 1952, Defence (Visiting Forces) Act 1963, and *Defence Force Re-organization Act 1975,* and a large number of regulations made under various acts which govern such things as military practice areas; remuneration; conditions of service, entry and discharge; and military courts of inquiry.

Since Federation legal concerns regarding command and control of Australian military forces have been a continuing theme. While the desire to ensure inter-operability with allies has been a prime concern, the need to ensure that Australian forces are available to defend Australia and are not under the control of a foreign power has been equally important. The Australian soldiers who volunteered to serve in the Sudan in 1885 and South Africa during the Boer War illustrate this. These men and women (nurses) were volunteers who engaged to serve in defence of the British Empire and were accordingly subsumed into the legal regime that governed all British forces. The New South Wales contingent who went to the Sudan could not be covered by New South Wales legislation because it did not have extraterritorial effect, that is, no New South Wales laws had any force beyond the borders of the colony. Therefore the British Army Act applied.

During the Boer War, some 16 000 Australian soldiers and 60 Australian nurses served under the command and control of Lord Kitchener. The well-publicised court martial of Lieutenants Harry Morant, Peter Handcock and George Witton demonstrates how the Australian government had lost control of its military manpower. An earlier trial which resulted in the conviction and sentence of death (later commuted to life imprisonment) of three Australian soldiers had been carried out without any notification of the Australian government. As a result, Prime Minister Edmund Barton took steps to ensure his government would be informed of future courts martial involving Australians. Despite this action, Lieutenants Morant and Handcock were convicted and executed in Pretoria on 27 February 1902, following court-martial proceedings. At no stage was the Australian government consulted about the case and few details were provided by the British authorities. This gave greater public prominence to the incident although there was little sympathy at the time for the two men. Accordingly, control of Australian soldiers overseas was an important issue which affected the Defence Bill that the government was trying to pass, although earlier colonial legislation was ultimately more influential (see DEATH PENALTY).

A consequence of Federation was the transfer of responsibility for defence matters to the new Commonwealth government. This necessitated the incorporation of the various colonial military forces into a new Australian military and naval force. This was done by a proclamation of the Governor-General on 1 March 1901. There was an obvious need for Commonwealth legislation which would provide cover and guidance for these forces. Until this was done the forces in each State would be governed by six different sets of colonial legislation. The newly installed Defence Minister, Sir John Forrest (q.v.), introduced a Defence Bill which had a controversial drafting history because of the provisions relating to the command of Australian forces by British officers, conscription for overseas services, the role of the Governor-General, application of British disciplinary law and service overseas. These matters continued to plague the Defence Bill, which was submitted to the parliament, and ultimately led to its rejection. Forrest withdrew the bill in 1903.

The new bill was passed in 1903 and proclaimed in March 1904. The act provided the basis for the raising and maintenance of the Australian military forces and serves as the ultimate source of military law, operating both in peace and war. The Governor-General was authorised to make regulations for the discipline and good government of the Defence Force. The act prohibited overseas service, unless the individuals volunteered, was silent on the role of the Governor-General as C-in-C, stated that Australian soldiers could be commanded by British officers, if the individuals volunteered, and provided for compulsory service, if required for the defence of Australia. The Defence Act also contained a bar on a large standing army other than permanent members who were restricted to administrative and instructional staff. The act clearly served national requirements for the security of Australia rather than an imperial need for a dominion expeditionary force.

One compromise in the act which did merit criticism was the disciplinary code that was to be applied to the military forces. It has been described as one of the most eccentric legislative arrangements ever brought before a Commonwealth parliament. A disciplinary code for a military force should provide for the maintenance and good order of its members by establishing a mechanism whereby offences are described and breaches of the regulations can be punished following a trial. The 1903 Defence Act detailed few offences but left the power to establish which code should apply to regulations to the government of the day. The initial effect was to provide two codes: one for the navy, which incorporated the British Naval Discipline Act and the King's Regulations and Admiralty

Instructions; and one for the army incorporating the provisions of the British Army Act and its Rules of Procedure. In peacetime the harshness of the British code was ameliorated by Australian regulations; on active service the complete British codes applied. This scheme was later mirrored by the 1923 Air Force Act and its development.

While this nineteenth-century based English legislation was amended from time to time it caused unnecessary administrative difficulties and complex command and control arrangements. By the end of the period of its operation in 1985, this disciplinary scheme meant that discipline law for the Australian Defence Force was contained in three United Kingdom acts, two of which had ceased to operate in the United Kingdom; four sets of United Kingdom rules or regulations, all of which had ceased to operate in the United Kingdom; three Australian acts; and nine sets of regulations under the Australian acts. This maze of legislation was replaced by the *Defence Force Discipline Act 1982*, implemented in 1985. For the first time the Defence Force Discipline Act brought together the disciplinary law applicable to the three arms of the Defence Force.

As a result of the Imperial Conference on Defence in 1909, the naval agreement with the British government was terminated and an independent Australian Navy was to be created. Following this conference, the Minister for Defence, Senator George Pearce, introduced, and parliament passed, the Naval Defence Act in 1910. This repealed many provisions of the Defence Act and made fresh provisions that applied only to the RAN. In 1921, Pearce introduced a similar bill designed to govern the new air force which had been created, independent of legislation, the same year. The Australian Air Force came into being on 31 March 1921 as part of the Naval and Military Forces of the Commonwealth, under section 30 of the Defence Act. It was not until July that year that the prefix 'Royal' was authorised, with the new name, Royal Australian Air Force Force, coming into effect in August 1921.

Pearce's Air Defence Bill of 1921 met similar objections to those raised to the initial Defence Bill, and for similar reasons. After heated debate the bill lapsed in the House of Representatives and the RAAF was governed by regulations made under the Defence Act. Another attempt to pass an Air Defence Bill was made in 1923 but after continued opposition it too was withdrawn and a skeleton bill which applied the Defence Act and its regulations was passed as an interim measure. The *Air Force Act 1923* was passed for the establishment and organisation of the Royal Australian Air Force as a distinct part of the Defence Force of the Commonwealth. This interim measure was to continue until 1939 when the search for an acceptable Air Defence Bill was abandoned. By this stage all the other air forces in the British Commonwealth had adopted the legal code laid down by the Imperial Air Force Act. It was thus decided in Australia, by amendment of the *Air Force Act 1923*, that the Imperial Air Force Act, for the time being in force, would apply to the RAAF. This ensured that future amendments of the Imperial Act did not affect the RAAF. This situation was finally remedied by the passage of the Defence Force Discipline Act 1982.

Despite several attempts by the government to achieve uniformity within the three Australian services, to update the disciplinary codes applicable to these services and to break away from British law, no modern, uniform, Australian code was established before 1985. This contrasted to the situation in the United Kingdom, Canada and the United States, all of whom updated their military disciplinary codes after the Second World War. The RAN was the only Australian service that modernised its disciplinary code when the government adopted the 1957 United Kingdom Naval Discipline Act. Despite attempts to introduce a new indigenous Australian disciplinary code the RAN quickly accepted and applied the new British legislation, which meant that it had to introduce a new disciplinary code and replace all the existing regulations and modify the British law to the RAN. The Whitlam Labor government pursued a new discipline bill in 1972, but on the government's fall in 1975 interest in the project diminished until 1982, when the Minister of Defence James Killen (q.v.) submitted the Defence Force Discipline Bill which was passed in April 1982 and took effect on 3 July 1985.

Apart from discipline of the Defence Force a number of other legal issues has arisen with respect to the use of the three services. Apart from the issue of conscription (q.v.) there is the question of the use of the armed forces for non-defence purposes. Instances where this has occurred have included the use of soldiers to break strikes in the northern New South Wales coalfields in 1949, and on the waterfront at Bowen in 1953, the use of an F-111 aircraft to photograph building of Tasmanian dams in 1983, and the use of RAAF transport aircraft in 1989 to break an airline pilots' strike. Opinion indicates that the use of permanent forces on such occasions is a legitimate exercise of the Commonwealth's constitutional power. The Defence Act provides that Reserve forces, however, can only be called out in time of war, a defence emergency or for the defence of Australia. This latter category is a recent 1987 amendment that imposes significant limitations on the period of time during which Reservists can be called out.

While international law has been an important factor affecting Australian military actions, many of the strategic decisions relating to the use of Australian forces during armed conflict have been made by other nations. Up until the end of the Second World War, Britain took many decisions as to the legality of military actions in which Australian forces were engaged. An example of this was the strategic bombing campaign against Germany in the Second World War. Here was a case where Australian service personnel played a large part but the strategic control was in the hands of the major Allied powers. Similar situations existed in the Korean War, where the United Nations authorised the United States to lead UN forces, and in Vietnam where the United States again took the lead. While the Australian government contributed forces, the actual planning and conduct of operations was left to the United States. In contrast, Australian commanders took responsibility for tactical actions taken by units and individuals under their command. These responsibilities primarily related to the observance of the law of armed conflict.

Many of the older laws were inherited from treaties to which the United Kingdom was a party, and include the Hague Conventions of 1907, the 1925 Gas Protocol and the early versions of the Geneva Conventions. After the Second World War Australia took a more independent position, and this is reflected in its membership of the United Nations and ratification of the 1949 Geneva Conventions which covered the protection of wounded, sick and shipwrecked service personnel, POWs and civilians. These four conventions were ratified and thus became Australian domestic law with the passage of the Geneva Conventions Act of 1957. In 1991, new legal obligations, which were contained in the 1977 Protocols Additional to the Geneva Conventions of 1949, were made Australian law by amendments to the Geneva Conventions Act. Australia has also become party to many other international agreements that impose obligations on the members of the Defence Force during armed conflict. These obligations were evident in the use of military legal officers to advise operational planners and commanders who were participating in the 1991 Gulf War.

Australia participated in the enforcement of international law when it played a large part in the prosecution of Japanese military personnel at the conclusion of the Second World War. The Far East War Crimes Tribunal sat in Tokyo and other Australian courts martial tried Japanese military personnel at a number of different locations in south-east Asia. These tribunals sat in judgement of Japanese military personnel and officials who were accused of war crimes and crimes against humanity. In all, Australian courts tried 924 Japanese of whom 644 were convicted and 148 sentenced to death and executed. Other Japanese were tried by similar United States and British courts. The Australian courts martial were conducted pursuant to the 1945 War Crimes Act. The trials attracted wide interest in Australia heightened by the fact that Australian military personnel sat as judges with Australian legal officers acting as prosecutors.

International obligations stemming from the United Nations Charter have also affected Australia's participation in peace-keeping operations. These reflect Australia's responsibility as a party to the charter, a good international citizen and member of the United Nations. The United Nations has never required its members to contribute military forces as a compulsory requirement and all forces who have participated in United Nations operations have done so on a voluntary basis. To assist with legal matters arising on such operations Australian Army legal officers have participated as members of Australian deployments on UN missions in Namibia, Somalia and Rwanda.

ENRICO CASAGRANDE

LEANDER CLASS LIGHT CRUISERS (MODIFIED)

(*Hobart, Perth* and *Sydney* [II]). Laid down 1933, launched 1934; length 550 feet; beam 57 feet; displacement 7000 tons (approx.); armament 8 × 6-inch guns, 8 × 4-inch guns, 4 × 3-pounder anti-aircraft guns, 8 × 21-inch torpedo tubes; speed 32.5 knots.

Originally ordered for the RN in 1933, these ships were taken over by Australia in 1935 and renamed HMA Ships *Hobart, Perth* and *Sydney*. All served in the Second World War. HMAS *Sydney* served in Mediterranean waters where in July 1940 it helped sink the Italian cruiser *Bartolomeo Colleoni* (q.v.) and damaged the *Giovanni Delle Bande Nere*. *Sydney* was sunk in Western Australian waters in November 1941 with the loss of all 645 men on board while engaging the German raider *Kormoran*. HMAS *Perth* also served in the Mediterranean where it fought in the Battle of Matapan in March 1941. It was damaged by German air attack during the evacuation of Crete (q.v.) in May 1941. In 1942 *Perth* joined ABDA Command (q.v.) and on 1 March, with USS *Houston,* engaged a large Japanese fleet at the Battle of Sunda Straits. Both *Houston* and *Perth* were sunk. The *Perth* lost 357 lives when it went down, and of the 320 men who survived and became prisoners of the Japanese, 106 died in captivity. After brief service in the Mediterranean HMAS *Hobart* was transferred to Australian waters in January 1942. In May *Hobart* took part in the Battle of the Coral Sea and in August it supported the American landings at Guadalcanal and fought in the Battle of Savo Island. HMAS *Hobart* was struck by a torpedo from a Japanese submarine in the Coral Sea in July 1943 but was repaired in time to support the Australian landings at Wewak in the New Guinea campaign and in Borneo in 1945. HMAS *Hobart* was at the Japanese surrender in Tokyo Bay on 2 September 1945. The ship was decommissioned in 1947 and sold for scrap to Japan in 1962.

LEEUWIN, HMAS, began life in 1926 as a drill hall and headquarters for the District Naval Officer at Croke Lane, Fremantle. Commissioned on 1 August 1940, it moved to nearby Preston Point on 1 July 1942 following the appointment of the first Naval Officer-in-Charge Fremantle. After the war it was mostly used for training reserves and national servicemen until it was selected as the site for junior recruit training in 1959. Junior recruits, aged $15\frac{1}{2}$–$16\frac{1}{2}$ years on entry, had a year of academic and naval training at *Leeuwin* before being sent elsewhere for specialist training. The first trainees entered on 18 July 1960, and at its peak in the early 1970s there were just over 800 recruits under training. By 1984, the last year of the scheme, 12 074 junior recruits had graduated from *Leeuwin*. The Fremantle Port Division of the Naval Reserve, based at HMAS *Leeuwin* since December 1949, continued to use the site after it was decommissioned on 11 November 1986.

LEGACY was a Melbourne lunch club for returned servicemen founded in September 1923 by Sir Stanley Savige (q.v.) and modelled on the Remembrance Club formed in Hobart by Sir John Gellibrand (q.v.) earlier that year. Other Legacy clubs were soon established around the country and they followed the Melbourne club's lead when in 1926 Legacy adopted the care of deceased ex-servicemen's dependants as its primary function. The creation of the Legacy Co-ordinating Council in 1938 gave the national movement greater coherence, and the increased profile and financial support it gained during the Second World War allowed it greatly to expand its welfare work. In 1946 it was decided to restrict Legacy membership to men who had served with the armed forces overseas in wartime. Legacy benefits were to be available to all dependants of deceased ex-servicemen who had served overseas in wartime (regardless of whether or not their deaths resulted from war service), but this rule was applied flexibly. By 1963 the 28 Legacy

clubs and 137 smaller groups had 5500 members ('legatees') who spent £700 000 supporting nearly 46 000 widows and 32 000 children. As well as providing income support and assistance in areas such as housing and education with money raised mainly through public appeals, legatees have acted as advisers to the families of deceased ex-servicemen and have organised leisure activities for the children of such families ('junior legatees') in Junior Legacy Clubs. Today, as the average age of legatees increases and as aged widows without dependent children make up an ever greater proportion of Legacy's beneficiaries, Legacy's future appears uncertain.

Mark Lyons, *Legacy: The First Fifty Years* (Lothian, Melbourne, 1978).

LEGAL CORPS see **AUSTRALIAN ARMY LEGAL CORPS**

LEGGE, Lieutenant-General (James) Gordon (15 August 1863–18 September 1947). Legge was first commissioned into the New South Wales forces during the Russian scare of 1885 (see COLONIAL WAR SCARES), while employed as a schoolteacher. He resigned the following year, but was reappointed in October 1887. Admitted to the Bar in 1891, he practised law before joining the permanent staff of the New South Wales Military Forces in 1894 and departing for a short period of service with the British Army in India. He served in the Boer War with the first contingent from his home colony, and remained there until late 1902 in a succession of regimental and staff posts.

In the years before the outbreak of war in 1914, Legge held a succession of influential positions in the tiny AMF, culminating in his appointment as CGS, with the rank of colonel, in May 1914. Along the way he helped draft the Commonwealth's first military regulations under the 1903 Defence Act (q.v.), worked with W. T. Bridges (q.v.) on the proposals for a compulsory military training scheme (see CONSCRIPTION) in 1907, and served on the Military Board as Quartermaster-General between 1909 and 1912 and as Director of Operations between 1910 and 1911. From 1912 to 1914 he was the Australian representative on the Imperial General Staff in London.

He was appointed to succeed Bridges on the latter's death at Gallipoli in the face of considerable opposition within the senior ranks of the AIF. He commanded the 1st Division for about a month before being returned to Egypt in late July to raise the 2nd Division there. At the same time he forfeited administrative command of the AIF to Lieutenant-General Sir William Birdwood (q.v.). He commanded the 2nd Division only briefly before being evacuated sick, having already been tor-

Major-General J. G. Legge photographed by C. E. W. Bean at Gallipoli, October 1915. (AWM G01131)

pedoed off Lemnos on his way back to Gallipoli in September.

Legge commanded in action again in France the following year. His division's attack at Pozières on 28–29 July was a failure which cost 3500 casualties. The British C-in-C, General Sir Douglas Haig, thought him 'not much good' and his division 'ignorant'. A second, much-delayed attack on 4–5 August succeeded after very heavy fighting, and the division was engaged again at Mouquet Farm in late August and at Flers in November. In January 1917, however, he was sent to England on grounds of ill health although he protested the decision. Denied a further operational command, he returned to Australia in April 1917 and was appointed Inspector-General, AMF, resuming the post of CGS in October, in which he saw out the war.

In June 1920 he became commandant of RMC Duntroon (q.v.) in a period of crisis in the college's history. He disagreed strongly with the minister over changes to the administration of RMC, and was retrenched on economic grounds in August 1922. Forced to retire before the prescribed age, he was denied a military pension and turned to farming. His elder son became a general officer in the Second World War; his younger son was killed in France in the First World War.

Legge, whose career had seemed so full of promise in 1914, came to grief in the face of the demands of the Western Front, both military and political. All conceded his ability and intellect, but there was considerable division over his capacity as

an operational commander. He had undoubtedly made enemies within the tiny Australian regular establishment which had existed before 1914, and he did not get on with Birdwood, who commanded the AIF overall. A good trainer of troops and an able administrator, the circumstances of his forced retirement embittered him; his treatment by C. E. W. Bean in the official history is almost certainly unfair in that Legge's period of command of the 2nd Division is treated harshly.

C. D. Coulthard-Clark, *No Australian Need Apply: The Troubled Career of Lieutenant General Gordon Legge* (Allen & Unwin, Sydney, 1988).

LEIGHTON, Arthur Edgar (17 June 1873–6 November 1961). Born in Surrey, England, Leighton became an industrial chemist and in 1903 was appointed to the explosives factory at Aravunku in Southern India. In 1909 he was appointed designer and manager of the planned cordite factory at Maribyrnong in Victoria (see INDUSTRY) and brought the plant into production. In 1915, while on a visit to the United Kingdom, Leighton was appointed by the British government as a technical adviser to the Ministry of Munitions. While there, he organised a scheme by which 6000 Australian chemists, technicians and tradesmen travelled to the United Kingdom to work in British munitions factories. He returned to Australia in 1919, having been appointed as general manager of a proposed — but never built — arsenal at Tuggeranong, south of Canberra. In 1921, with the creation of the Munitions Supply Board (q.v.), he became the first Controller-General of Munitions and held this position until 1938. During this period, Leighton was responsible for the development of a body of technicians, facilities and an industrial capacity not seen in Australia before. Australian raw materials that could substitute for imports were found, the production of artillery and machine-guns was begun and a system of quality control using precision measuring equipment was introduced. In the words of R. G. Casey, Leighton was 'the father of munitions production' in Australia, and the ability of Australia to produce defence materials for its own needs during the Second World War was a result of his work.

LEWIS, Essington (13 January 1881–2 October 1961). Lewis was blocked in his desire to enlist in the First World War with his three brothers by the influence that his superior at Broken Hill Proprietary (BHP), G. D. Delprat, exercised with the Federal Munitions Committee and the Department of Defence. In 1921 he succeeded Delprat as BHP general manager at a salary of £4000 per annum. After reforming the company's steel operations in 1922–23 he was designated managing director and became the first executive officer in the history of BHP to achieve a seat on the board. In the 1930s he pioneered modern production techniques in the Australian steel industry, making BHP one of the world's leading companies in the field.

Around 1934 during an overseas trip he became concerned at the growth of Japanese military power and thereafter advocated preparations for military production within the company. In 1935, at his urging, BHP formed a consortium to build aircraft, the Commonwealth Aircraft Corporation (q.v.), and began production of Wirraways (q.v.) at Fishermens Bend the following year. The company also began the manufacture of artillery shell casings and built a shipyard at Whyalla, South Australia.

In May 1940 Prime Minister R. G. Menzies (q.v.) made Lewis Director of Munitions. In this position he controlled the production of all ordnance, explosives, munitions, small arms, aircraft and vehicles and all machinery and tools used in such production. He was a member of the Defence Committee and had access to the War Cabinet (q.v.) on the same basis as the chiefs of staff of the armed forces. When Labor came to office Prime Minister John Curtin (q.v.) extended his authority by making him Director-General of the Department of Aircraft Production. Although critics would later charge that Lewis had presided over excess production, his meticulous planning ensured that the armed forces were not faced with short-

ages of essential equipment or munitions. The end of the war in sight and his main tasks completed, Lewis resigned his government appointments in May 1945 and returned to BHP.

During the war Lewis became, in the historian Geoffrey Blainey's words, an 'industrial dictator'. He wielded enormous power within the war effort, almost entirely to the benefit of Australia's ability to prosecute the war. Many of the methods pioneered in wartime industry were the precursors of complex industrial processes developed in Australian manufacturing industry during the 1940s and 1950s, and his influence continued to be felt in areas such as the Long Range Weapons Establishment (q.v.) at Salisbury and Woomera, South Australia. A bleak, eccentric personality committed to hard work and thrifty virtues, he made an important contribution to Australia's fight against the Japanese.

Geoffrey Blainey, *The Steel Master* (Sun Books, Melbourne, 1981).

LEWIS LIGHT MACHINE-GUN (LMG) was a 0.303-inch calibre, gas-operated, magazine-fed, air-cooled, automatic-fire weapon. It was in Australian military service from 1916 to 1945 and during that time saw active service in the two world wars in a variety of roles including as an infantry LMG, in the anti-aircraft role on land and sea, and as armament on some military aircraft and vehicles. The Lewis was obsolete as an LMG at the beginning of the Second World War but was pressed into service until it was replaced by other more modern machine-guns, such as the BREN (q.v.) in the infantry LMG role.

The Lewis was designed in America by Samuel MacLean and his design was modified during 1910–11 by Colonel Isaac Lewis of the US Army. The Lewis LMG officially went into service with the British Army in October 1915 and was manufactured by the Birmingham Small Arms Company in England. In its infantry role, the Lewis was distinctive with its horizontal drum magazine, large cylindrical barrel shroud (which housed a unique finned aluminium radiator cooling system), and bipod.

The Lewis was the first LMG issued to British and Commonwealth infantry at platoon level and its automatic firepower permitted the development of fire and movement tactics at section and platoon level. The Lewis could be carried and operated by one man, but required at least a second man to carry spare magazines and to assist with the operation of the gun. The Lewis could be fitted with a 47- or 97-round magazine, both of which had a reputation for causing stoppages as they were easily damaged. The 47-round magazine was normal for infantry use.

The weight of an empty Lewis was 26 pounds (11.8 kg) and the weight of a loaded Lewis (with 47-round magazine) was 30 pounds 20 ounces. The rate of fire was 550 rounds per minute cyclic (i.e. the rate of fire that theoretically could be achieved given a continuous supply of ammunition) and a practical rate of fire was around 120 rounds per minute. The effective range of the Lewis was around 600 yards and it was sighted to 1900 yards.

IAN KURING

LIBERATOR, CONSOLIDATED (10–11-crew heavy bomber [B-24J]). Wingspan 110 feet; length 67 feet 2 inches; armament 10 × 0.5-inch machine-guns, 12 800 pounds bombs; maximum speed 290 m.p.h.; range 2100 miles; power 4 × Pratt & Whitney Twin Wasp 1200 h.p. engines.

Liberators were flown by the US Air Force from Java during the Netherlands East Indies campaign (q.v.) in 1942 but did not enter service with the RAAF until 1944. They equipped five squadrons that had previously flown the failed Vengeance (q.v.), finally giving the RAAF a heavy bombing force in the South-West Pacific Area

(q.v.). They were used for long-range sweeps against shipping, raids on isolated Japanese garrisons in the Netherlands East Indies and bombing in support of the Australian landings on Borneo (see NEW GUINEA CAMPAIGN). Liberators of No. 200 Flight dropped members of Special Operations Australia (q.v.) and storepedoes (q.v.) behind enemy lines.

LIBYAN CAMPAIGN, FIRST (December 1940– February 1941), was the first major ground campaign involving Australian troops in the Second World War. The Australian 6th Division, which with the British 7th Armoured Division comprised Lieutenant-General Sir Richard O'Connor's XIII Corps in the Western Desert, played a key role in the defeat of numerically greatly superior Italian forces, providing Australia and the Empire with a successful campaign after a succession of disasters the previous year and setting a standard of performance in battle against which other Australian formations were to judge themselves in later campaigns.

The 6th Division had arrived in Egypt piecemeal in the early months of 1940, and spent that

Italians captured at Bardia receiving medical attention at a 2/1st Australian Field Ambulance casualty clearing station, 4 January 1941. (AWM 005246)

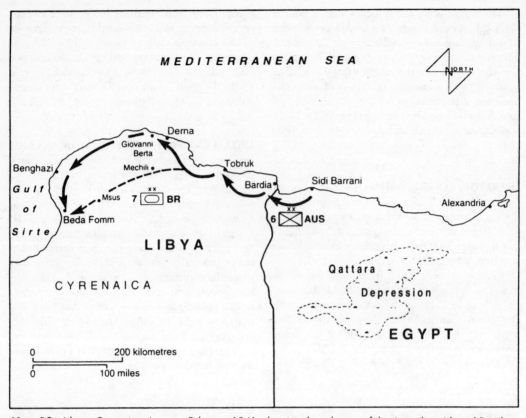

Map 18 Libyan Campaign, January–February 1941, showing the advance of the Australian 6th and British 7th Armoured Divisions.

year consolidating its training and organisation and making good the numerous deficiencies in weapons and equipment from which it suffered at virtually all levels. Commanded by Major-General Iven Mackay (q.v.), the 6th Division was detailed for the second phase of Operation COMPASS when the 4th Indian Division, which had taken part in the fighting which had pushed Graziani's invading Italian force back across the Egypt–Libya border at Sidi Barrani, was transferred to the Abyssinian front on 11 December. The 16th Brigade under Brigadier A. S. 'Tubby' Allen (q.v.) moved forward on the 12th, but the attack on the fortified Italian town of Bardia was postponed until 3 January owing to delays in the forward movement of the remaining Australian brigades, the 17th and 19th, caused by shortages of motor transport in the Australian formation.

The attack on Bardia was necessarily devoid of surprise, and the 16th and 17th Brigades were involved in some stiff fighting against strong enemy positions before cutting the defences in two on the evening of 4 January. The garrison surrendered next day, with 38 000 Italian soldiers and large stocks of guns and equipment taken. Most of the 500 casualties suffered by the attackers were Australian.

Even while the positions at Bardia were consolidated, the 19th Brigade, which had played little part in the earlier assault, was advancing on the fortified port city of Tobruk (q.v.) in concert with the British 7th Armoured Brigade. Severing the western approaches, the armour and infantry laid seige to the town's defences by 9 January, but the attack was not launched until 21 January in order to provide time to move ammunition forward and thoroughly reconnoitre the Italian defences. The 19th Brigade under Brigadier H. C. H. Robertson (q.v.) played the leading role in the assault, which was launched in darkness, and by late morning had secured a bridgehead about 4.5 miles deep inside the enemy positions. With the 16th Brigade in support, these battalions continued to press their advantage, aiming for the Italian headquarters at Forts Pilastrino and Solaro. Early the following morning, the Italian 61st Division surrendered together with a large group of senior officers, and by mid-afternoon resistance had largely ceased. A further 25 000 prisoners were taken, together with additional plentiful stocks of materiel.

The campaign now involved the pursuit of the Italian forces who were attempting to make good their escape into Cyrenaica. The usual shortages of motor transport meant that only one infantry brigade could advance at a time, and Robertson's 19th Brigade was sent forward by Mackay to invest and capture Derna, in cooperation once again with the dwindling number of tanks of the 7th Armoured Division. Derna fell on 28 January after the 4th Armoured Brigade had taken Mechili to the south and following stout Italian resistance. Plans for a more protracted investment of Benghazi were shelved when intelligence revealed the haste with which the Italians were evacuating Libya, and forward elements of 7th Armoured were pushed rapidly south-west to Msus and then Beda Fomm to intercept them. Meanwhile, units of the 6th Division pursued the Italians south from Giovanni Berta, entered by the 17th Brigade on 2 February. The Australians entered Benghazi on the evening of 6 February but the Italians, outflanked and cut off to the south, surrendered next morning.

Never more than two divisions strong, XIII Corps destroyed the Italian Tenth Army of some 10 divisions in a campaign which lasted about six weeks, capturing 130 000 prisoners and immense amounts of equipment in an advance of about 500 miles. Total casualties were approximately 500 killed (of which the Australian share was 256), 1373 wounded and 55 missing. Contrary to popular belief, the Italian army fought hard and well in retreat, making good use of such opportunities as presented themselves. They suffered from equipment and leadership deficiencies while operating at the end of a long and vulnerable supply line and this, together with British tactical air superiority and clear-sighted generalship on O'Connor's part, largely explains the outcome of the campaign. Under pressure from General Sir Archibald Wavell (q.v.) in Cairo and from London, O'Connor was forced to transfer forces back to Egypt in preparation for the ill-starred campaign in Greece; the enemy high command believed at the time that a further British offensive would have conquered all remaining Italian territory, thus ending the North African campaign.

LIGHT HORSE Mounted units made their appearance among the colonial militias and volunteers as early as 1854, and before 1870 they comprised a motley collection variously identified as cavalry, yeomanry, and mounted rifles. By 1885 all six colonies had some mounted units in their military establishments, units that were often, though not exclusively, based in rural areas with the headquar-

Light horsemen at a well on Mount Zion, Jerusalem, 22 January 1918. (AWM B01518)

ters centered on a country town. The Australian mounted arm may be said to have come into its own during the Boer War, when large numbers of Australians served in mounted rifle and mounted infantry units fielded by the various colonies, or in the battalions of the Australian Commonwealth Horse, the first units of the new Commonwealth's military forces. Many British officers of the day held that the settler dominions were peculiarly suited to the provision of mounted infantry, and a sizeable proportion of the mobile arm of the British Army in South Africa was provided by men from Australia, New Zealand, Canada, and the South African colonies.

Following Federation, the Commonwealth reorganised the mounted forces into a number of light horse brigades on a State basis: the 1st and 2nd Brigades in New South Wales, the 3rd and 4th in Victoria, the 5th in Queensland, and two unnumbered formations in South and Western Australia, making a total of 18 light horse regiments. Before the First World War light horse regiments were organised into half squadrons with two troops to each recruiting centre (this changed in the reorganisation of 1921 with most centres reduced to supporting a single troop, in part a reflection of the decline in both the numbers and quality of horseflesh in a gradually industrialising society). The light horse underwent considerable reorganisation and standardisation during Major-General E. T. H. Hutton's (q.v.) term as GOC Australian Military Forces. On 1 July 1902 Hutton issued the 'Mounted Service Manual for Australian Light Horse and Mounted Infantry', which prescribed new standards for training and efficiency. Hutton abolished most of the existing cavalry units, got rid of officers who regarded their service as merely an adjunct to their social position in local society, and divided mounted units into two types, light horse and mounted infantry. The distinction has been lost on more recent generations, exemplified by the line given to a Turkish soldier in the 1987 film, *The Lighthorsemen*, in which he explains for the audience's benefit that the troops opposite are 'Australian light horse ... mounted infantry'. The distinction in fact formed an important part of a now long-lost doctrinal debate within the British Army during the late Victorian era. Light horsemen were horsemen trained to fight on foot but capable of providing a limited number of mounted capabilities relating to reconnaissance and screening of larger formations, while mounted infantry were merely infantrymen who had been provided with a means of temporarily increased mobility. The former could be an operational level asset, the other merely a tactical one. Hutton made himself unpop-

ular by making mounted training more rigorous, and by increasing the number of light horse units through the disbandment of rural infantry battalions and their replacement by mounted units: rural men wishing to soldier part-time in the militia were now put to the greatly added expense of providing their own horses and horse furniture.

Between 1905, after Hutton's departure, and 1914, the light horse establishment was increased from 18 regiments to 23, and many of the units formed in the post-Federation period drew traditions and titles from earlier colonial units. A great deal of redesignation also went on during this time. The brigades were renumbered, so that the 5th Brigade in Queensland became the 1st Brigade, the 1st Brigade in New South Wales became the 2nd Brigade, and so on, and a total of eight brigades was formed including two mixed brigades in Western Australia and Tasmania. By 1913 all light horse units had adopted territorial titles and many had been renumbered. The 13th Light Horse Regiment, formed from the Queensland Mounted Infantry after Federation, became the 2nd Light Horse (QMI), and ultimately the Moreton Light Horse; the 11th Light Horse, which drew on the former Victorian Mounted Rifles, was divided in 1912 to form the 19th and 20th Regiments, the latter being designated the Corangamite Light Horse the following year; the famous 10th Light Horse grew out of the 18th Regiment, itself formed from the Western Australian Mounted Infantry and reformed as the 25th Light Horse in 1912; and so on. By the beginning of the war in 1914, the Australian light horse regiments mustered some 9000 men in all ranks, with each regiment assigned an establishment of 29 officers and 552 other ranks and fielding some 579 horses. A regiment was commanded by a lieutenant-colonel, and each squadron by a major. Troops in each squadron were numbered from one to four, unlike an infantry battalion where the platoons are numbered consecutively across the unit.

The finest hour of the Australian mounted arm undoubtedly occurred during the First World War in the Middle East, a theatre which gave maximum scope to the combination of mobility and shock that such a force was able to combine. The mounted force, of which the light horse was a significant part, was led ultimately by a commander of genius, Lieutenant-General Sir Harry Chauvel, a pre-Federation colonial soldier whose father had founded the Upper Clarence Light Horse in northern New South Wales in 1885. Four regiments of light horse were raised initially, but within weeks of the beginning of the war a second brigade was formed, and by the time the second contingent left Australia a third had been established. The 1st Brigade sailed with the 1st Australian Division for Egypt, joined by the 2nd Brigade in January 1915 and by the 3rd some weeks later. The pressing need for reinforcements following the landing at Gallipoli saw the light horse units deployed in the dismounted role, which brought some disadvantages as a light horse regiment was smaller than an infantry battalion, and was reduced even further when horse handlers were taken off the strength to care for the mounts left back in Egypt. The men were not equipped on the infantry scale either, and many spent several weeks in the trenches before receiving adequate supplies of personal equipment. They suffered heavy losses during the Gallipoli campaign, exemplified by the slaughter of the 3rd Brigade at the Nek during the August offensive. They were relieved in November–December, greatly reduced in strength, and were reformed and retrained in Egypt early the following year.

The 13th Regiment and part of the 4th went to France with the infantry as divisional mounted troops (and ultimately were re-formed as the 1st Anzac Mounted Regiment), but the bulk of the light horse (12 regiments) stayed in Egypt as the strike force (with the New Zealand Mounted Rifles and some Yeomanry and Indian cavalry units) of the army protecting Egypt and the Canal Zone from the Turks, who mounted an offensive in mid-1916 before being driven back over the remainder of the year to the border with Palestine. Attempts were made to move the light horse to France and use them as infantry reinforcements as the manpower crisis deepened in 1916–17, but these were resisted successfully and the Australian Light Horse, under Chauvel's command and organised into the Desert Mounted Corps, took a leading role in the long, arduous but ultimately decisive defeat of the Turkish armies in Palestine and Syria which led to the Turkish surrender in late October 1918. While being repatriated from Egypt, light horsemen performed as enforcers of empire during the Egyptian revolt in 1918–1919, earning a reputation for brutality against the native population explained perhaps as much by lengthy war service as by contemporary racism.

In the reorganisation of the postwar army recommended by the conference of senior officers (which included Chauvel) in 1920, the mounted arm was to comprise two cavalry divisions and three separate light horse regiments earmarked for local defence tasks. The reductions of the interwar period particularly affected the light horse: four regiments were removed in 1921 (from the original 30 in 1919) when the divisional cavalry regi-

ments were dispensed with; the amendments to the universal training legislation restricted compulsory military training to major centres of population, with obvious implications for rural units; and in 1929 two more regiments were removed from the order of battle, while units generally were placed on restricted establishments. Regiments at this time generally consisted of two squadrons of three troops each, together with a machine-gun squadron and a headquarters wing, although this changed again with the addition of a headquarters squadron in 1929, which consisted of a machine-gun troop and a signals troop. In 1937 a third 'sabre' squadron was added and in 1938 squadrons had their fourth troop restored to the establishment. More seriously, mounted units suffered from a decline in the quality of horse stock and the number of young men familiar with horses. In 1937 four regiments were converted to machine-gun regiments, equipped with Vickers guns and issued a bewildering variety of mostly inadequate vehicles. As with attempts to motorise and mechanise the cavalry in the British Army, this was a slow process which was only completed once the war in the Pacific had begun. The mystique of the light horse legend, however, was as powerful a conservative factor in the Australian Army as the cavalry influence in Britain, and the demise of the horse was resisted stubbornly in many quarters. When the 6th Division was raised at the end of 1939, a divisional cavalry squadron was formed, equipped with carriers; it had been suggested that a mounted squadron be attached instead, but fortunately this was not proceeded with. On their return from the Middle East the divisional cavalry of the AIF were reorganised as cavalry (commando) regiments and, under various nomenclature, fought in New Guinea and the Pacific islands. In Australia, the two cavalry divisions were mechanised as armoured divisions, fated never to see service as formations outside Australia.

The CMF light horse began the war with 21 regiments; three additional machine-gun regiments were raised, together with an additional armoured regiment. Of the original 21 regiments, five were converted to motor regiments, two to reconnaissance battalions, the four prewar machine-gun battalions remained as they were and 10 units retained their light horse designation, although they gradually lost their horses. Although units did not necessarily see service outside Australia, many of their members did: to take but one example, the 21st Light Horse Regiment, based on the Riverina, did not leave Australia, but it provided cadres for units in New Guinea and one of its number, Corporal Reg Rattey, won the VC. There was little role for the light horse after 1945, the Royal Australian

Armoured Corps (q.v.) subsuming many of its functions and traditions. By 1966, seven CMF units of the RAAC maintained titles and associations from the light horse, but the majority of units are now commemorated only by the guidons laid up in local churches.

R. J. G. Hall, *The Australian Light Horse* (Dominion Press, Melbourne, 1968).

LIMBO ANTI-SUBMARINE MORTAR This weapon was developed by the RN in the 1950s. It is a 12-inch mortar which can throw a projectile weighing 200 kg from 1000 to 2000 metres. The fuses can be set to explode at a predetermined depth. Limbo was used for anti-submarine warfare (q.v.) by 'Q' Class anti-submarine frigates (q.v.) from the mid-1950s. Considerable technical modifications introduced by the RAN reduced the cost, weight and noise of the system, and simplified its use so that it could be operated by a crew of three instead of seven.

LINCOLN, GAF (7-crew heavy bomber [Mark 30]). Wingspan 120 feet; length 78 feet 4 inches; armament 2 × 20 mm cannon, 4 × 0.5-inch machine-guns, 14 000 pounds bombs; maximum speed 305 m.p.h.; range 1470 miles; power 4 × Rolls Royce Merlin 1750 h.p. engines.

The Lincoln was a development of the famous Lancaster (q.v.) bomber which was manufactured under licence in Australia by the Government Aircraft Factory and entered service with the RAAF in 1948. No. 1 Squadron flew Lincolns on bombing raids during the Malayan Emergency (q.v.) between 1950 and 1958. Fourteen Australian Lincolns were specially upgraded for use by the RAAF and RAF on exercises in 1949–50 simulating the bombing of Moscow from bases in the United Kingdom. They were also used to chase the radioactive dust clouds caused by atomic weapons tests (q.v.). Other Lincolns were given extended fuselages and used for maritime reconnaissance by No. 10 Squadron. Lincoln bombers were replaced by the Canberra (q.v.), while the maritime patrol version was replaced by the Neptune (q.v.) in 1961.

LITERATURE, WAR IN AUSTRALIAN Until recently, Australia's definition of itself as a nation has drawn disproportionately on its involvement in war. The literature generated by overseas conflicts in which Australians have participated — from the Maori Wars of the 1860s till the withdrawal from Vietnam in 1972 — has been significantly coloured by an obligation which authors have sporadically but strongly felt, to seek or create expressions of the national spirit. Australia's historical

circumstances have meant that literature has been alive to a contradiction. When Australians went to the Sudan in 1885, to the Boer War from 1899 to 1902 or to the Great War, their contributions were apt to be spoken of as exemplifying both loyalty to the British Empire and to the emerging Australian nation. William Dalley justified his zealous offer of New South Wales troops for the war in the Sudan by saying that this testified 'to the readiness of the Australian colonies to give instant and practical help to the Empire'. This prompted James Service, Premier of Victoria, enviously to observe that the contingent 'precipitated Australia, in one short week, from a geographical expression to a nation'. For 'Banjo' Paterson (q.v.), whose first published poem appeared in the *Bulletin* in 1885 to voice opposition to the dispatch of Australian troops, 'fair Australia, freest of the free,/ Is up in arms against the freeman's fight'. 'England's degenerate generals', rather than the Mahdi (leader of the Sudanese resistance) to whom Paterson's poem is addressed, are the villains whom he perceives.

Almost from its inception, Australian war literature featured works written in opposition to particular wars. Henry Lawson fulminated against the Boer War in 'Who'll Wear the Beaten Colours?' Poets such as Andrew Taylor attacked Australia's involvement in the Vietnam War. Thoroughgoing pacifist sentiments were much less common. Many writers whose subject was Australia at war were agitated by a presumed (not always actual) military dependence which to them exemplified a broader and shaming cultural reliance, whether on Britain or, later, the United States. Many of these authors would be drawn to the discovery of how uniquely Australian values found their richest expression in combat. The sources of that last assumption in particular need to be examined. Australians have rejoiced in their freedom from history, and at the same time wished to be enlisted in it. Having no hereditary enemies and no tradition of border disputes; having subdued their Aboriginal peoples relatively soon (although with more resistance than was once acknowledged); never having suffered invasion, nor begun a war and happily lacking the experience of civil war, Australians have been spared the kinds of brutal historical knowledge familiar to authors in Europe and the United States. But as they would be assured soon after the fact, Australians took their distinctive place among the nations through their part in the Great War.

One of the primary myths of that conflict — in Australia and elsewhere — was that the ante-bellum era had been an age of innocence which was lost forever because of the war. Perhaps the myth had a peculiar force in Australia. 'Is war very big?/

As big as New South Wales?' These are the ingenuous questions that conclude Les Murray's poem, 'The Trainee, 1914', a work in itself part of the 1970s' literary revival and revision of Australia's role in the Great War. Numbers of authors besides Murray — Thomas Keneally, Roger McDonald, David Malouf, Geoff Page among novelists, Chris Wallace-Crabbe in his long poem 'The Shapes of Gallipoli' — examined the presumed golden age which the war destroyed. They were aware of how an old Australia could be celebrated in retrospect because of the advent of the Great War, while coincidentally the nation rejoiced in its new incarnation through blood sacrifice.

Drawing on pastoral traditions of the resilient bushman as national archetype, writers such as Paterson in his *Happy Dispatches* (collected in 1934) from the Boer War, and then the official historian of Australia in the Great War, C. E. W. Bean (q.v.), argued that this figure of nineteenth-century rural legend effortlessly refitted himself as the heroic Anzac soldier. By his literary exertions, Bean was concerned not only to write a history of the 1st AIF, but also to forge the myth of its martial origins which he felt the new nation to need. Thus he contended that the essential qualities of the Australian (male) which had been generated by work in the country's inhospitable outback, were tested and triumphantly asserted in the battles of the Great War. The Anzac legend draws both on the popularising of an idealised rural type and on nineteenth-century German Romanticism, in which nations rejoice in their formation through the storm and strife of war. Australia therefore borrowed much of its nationalist ideology from Germany, its principal foe on the Western Front during the Great War, although Social Darwinist influences were not confined to Germany.

The literature of Australian involvement in colonial wars has suffered scholarly neglect, as has the Korean War and the commitment of troops to Malaya in the 1950s. Poetry and prose which treat of colonial conflicts far from Australia are sparse, but of historical, as well as of premonitory importance. Paterson's 'El Mahdi to the Australian Troops' sounded a sceptical note about following Britain's call to war without question. In fact, the New South Wales government had foisted its contingent upon reluctant imperial authorities. Australians — in Paterson's idealising phrase, 'freest of the free' — were presumably people untouched by what Lawson called 'old world errors, and wrongs, and lies'. This was all very well, but splendid isolation could not be sustained when later writers and propagandists required the testing of the proto-national mettle in conflict abroad.

Paterson's complaint concerning the Sudan expedition foreshadowed stronger objections to the exertion of British will at the time of the Boer War. The young radical, W. M. Hughes (q.v.), inveighed against 'a wanton deed of blood and rapine'. Randolph Bedford mocked the bard of Empire, or as he called him, 'the Empire's bagman', who had 'yarded God, and branded him "JB over E" on the flank'. His *Bulletin* readers could relish 'The Rhyme of Rudyard K.' Most famously, 'Breaker' Morant (q.v.), an English-born, Australian-enlisted soldier (like so many in the 1st AIF some years later), wrote his last verse testament, 'Butchered to Make a Dutchman's Holiday', on the eve of his execution for killing Boer prisoners. Morant's alleged accomplice, George Witton, called his memoir of the affair *Scapegoats of the Empire* (1907). The bellicose Australian nationalists of the 1970s, left without a war, found Morant's censure of imperial arrogance a fit subject for book and film, as they would also enjoy retelling the supposed bungles of British generals in the Great War, for example in such films as Peter Weir's *Gallipoli* (1981).

Australia's first military incursion into Asia coincided with the South African War. A naval contingent was dispatched to China at the time of the Boxer Rebellion. Predictably, newspaper and magazine versifiers joined the affray, if at a distance. With a relish for apocalyptic battle that animated so much literature of imagined wars written in the generation before 1914 (and which hastened the Great War by making it seem inevitable), R. Stewart lectured readers of the *Bulletin* on the Oriental threat. His poem, 'The Sword of Genghis Khan', crystallises the exultant dread of Asian invasion which is also expressed in vivid turn-of-the-century cartoons. 'The Mongol has awakened', Stewart declared, 'the ancient Dragon flaming biddeth forth the Yellow man'. Still a distant prospect when Stewart wrote, that 'Yellow man' (Japanese rather than Chinese) would veritably become the stuff of Australian nightmares before and during the Second World War.

The most influential Australian writer about the Great War was C. E. W. Bean. Having discovered 'the real Australia' in western New South Wales during his travels as a journalist, he witnessed, and celebrated, the national apotheosis at Gallipoli and in Flanders. His endorsement of Australian heroism was enthusiastically seconded by such Anzac memoirists as 'Trooper Bluegum' (Oliver Hogue [q.v.]), whose literary exploits were sardonically analysed by Robin Gerster in *Big-Noting: The Heroic Theme in Australian War Literature* (1987). Many of these writers tried to translate the circumstances of conflict in the Great War into pre-industrial kinds of battle, better to realise opportunities for old-style heroics.

In fact, the preferred weapon of Anzac literary self-celebration, the bayonet, inflicted few casualties compared with shot and shell, dispatched from a lethal distance. The terms of warfare now meant that endurance and resignation, rather than reckless personal aggression, became the better part of heroism. Frank Dalby Davison's account of the last successful Australian cavalry charge in military history, *The Wells of Beersheba* (1933), was a happy literary exception.

As was the case with their American, British, French and German counterparts, the notable Australian novels of the Great War appeared a decade and more after its conclusion. These were Leonard Mann's *Flesh in Armour* (1932) and Frederic Manning's *The Middle Parts of Fortune* (1929). The former deals sensitively with the experience of neurasthenia, or 'shell-shock', which was a ubiquitous, but long-suppressed feature of the torment of trench warfare. The latter ranks with the novels of Hemingway and Remarque in its evocation of the muddle of battle, the euphoria of leave that punctuated long stretches of boredom and fear, the accommodations of ordinary men to what would be a common fate for so many.

Poetry of the Great War ranges in tone from the bitter solemnity of Manning's 'Grotesque', which makes the analogy of the battlefront with the 'damned circles that Dante trod', to demotic jesting in C. J. Dennis's 'The Push'. There is no epic; Bean's *Official History* would most nearly fill that office, in prose. Harley Matthews (who would be interned in the Second World War as a member of the Australia First Movement) wrote with an irony that he could not have expected, but would come sourly to savour, of 'True Patriot'. From the homefront, Christopher Brennan's disenchantment with his German-born wife spiced his portentous 'A Chant of Doom'; Mary Gilmore implored due remembrance for the dead in 'Gallipoli', while Lawson could manage no more than the doggerel of 'England Yet'. The nearest to poetic resistance to the war in Australia can be found in 'To God: From the Weary Nations' (1916), by 'Furnley Maurice' (Frank Wilmot).

More than half a century after the war had ended, Australian poets and novelists returned to it. They were intent on examining, and on occasions burnishing, a myth of national innocence forgone for the sake of nationhood. Their varied works also examined the peculiar ways in which Australian servicemen's experiences of overseas countries in time of war strengthened their parochialism. This type of negative self-definition is an insufficiently remarked component of Australian nationalism. Few of the 'six bob a day tourists' who went to the Great War,

or their counterparts in the Second World War, came back with a cosmopolitan understanding of life overseas that they wished to communicate to their fellows. Instead, their racial prejudices, their insularity, were more often reinforced.

To a degree, the novels of the Great War by McDonald, *1915* (1976), Malouf, *Fly Away Peter* (1982) and Page, *Benton's Conviction* (1985) may have chosen the divided, angry and anguished climate of that time as their setting as a means of dealing, indirectly, with Australia's part in the Vietnam War, where similar social schisms greeted Australian involvement. That conjecture reminds us that much war literature is belated and retrospective. In literary terms, the shock of battle is always an aftershock, its most typical issue the carefully crafted (and sometimes merely conventional) prose reminiscence. Thus Mann's, Manning's and Davison's fictions appeared between the two world wars, when the Official History which Bean edited and partly wrote also began to be published. These works take on an uncanny aspect of prediction as well as remembrance, in the same fashion that the military figures in war memorials seemed to stand to for the war to come, rather than merely looking back to 'the war to end war'.

Although Australian military casualties were fewer in the Second World War than in the Great War, in other respects its social consequences were more shocking. Poets died: Pat Gallaghan, Arnold Gardner, Tip Keleher, D. B. Kerr and James Picot among those whom Ian Mudie included in his anthology, *Poets at War* (1944). Australians knew of a defeat — the fall of Singapore — which could not readily be transmuted, as Gallipoli had been, into consoling legend. Following that cataclysm, 30 000 Australian troops and nurses passed into Japanese captivity, of which the literary result was one of the most distinctive, and distinguished, bodies of national war writing. Invasion threatened Australia also. Its consequences had been airily imagined in late nineteenth-century popular literature. In the 1940s they were not lightly entertained. The prospect of what might have been, had the Japanese taken northern Australia, is the subject of John Hooker's novel, *The Bush Soldiers* (1984). He was examining by exaggeration the crucial shocks that the Second World War delivered to Australia: the revelation of its military vulnerability, the proximity of Asia and the recognition of how — in peace and war — these linked circumstances must be confronted if the national identity was to be sustained.

Novelists of the Second World War seem anxious about the minatory presence of the generation of their fathers, many of them members of the 1st AIF. Their characters are versed in what Eric Lambert, in *The Twenty Thousand Thieves* (1951), calls 'the legend of the terrible laughing men in the slouch hats'. But the conditions of war in the Pacific theatre especially, meant that qualities of endurance and comradeship are emphasised, rather than daring deeds of arms. This is the tone of T. A. G. Hungerford's *The Ridge and the River* (1952) and 'David Forrest's' (David Denholm's), *The Last Blue Sea* (1959), as well as Peter Ryan's brilliant memoir, *Fear Drive My Feet* (1959). Novels of the war in New Guinea present the jungle as the main enemy, although the Japanese receive derogatory attention. They are depicted as atavistic, barbarian, 'apes with pants on' (in Norman Bartlett's *Island Victory*, 1955), out of 'the dark ages' for the protagonist of *The Last Blue Sea*.

Yet the soldiers who fought the Japanese in New Guinea could not match the animus of those who were imprisoned by them. Only the Second World War produced a sizeable prisoner-of-war literature. Australia has none of the 'missing-in-action' chronicles which American writing of the Vietnam War has to show. Instead the reader of its novels of wartime captivity is confronted with stories of prisoners' shame (or experiences unassimilable within the Anzac tradition) and of betrayal (in which the martial prowess of Australia is sacrificed by the weakness and incompetence of its allies). Male mateship is confirmed, but also called into question. The travail of imprisonment — as registered in such works as Rohan Rivett's *Behind Bamboo* (1946), Russell Braddon's *The Naked Island* (1952), Ray Parkin's *Into the Smother* (1963, second volume of a trilogy, the first of which was one of the few accounts of conflict at sea in Australian war literature) and Norman Carter's *G-String Jesters* (1966) — can occasionally be comic, at other times the stuff of adventure story. If related as tragedy, this is more to do with the individual casualties of the prisoner-of-war experience than the honour of the nation. The tone of these books modulates from violent anti-Japanese feelings in the first two, to Parkin's benign strategy of coping by introspection.

In the tradition of the revenant figures who are of such importance in Australian literary culture, the POW has a notable, if ambiguous place. All returning personnel are potentially disruptive, with their tales of strange places to disclose to unwilling listeners who have adjusted to their long absence. In Patrick White's *The Twyborn Affair* (1979), Eddie Twyborn elects to be silent about his experiences in the Great War, as does White's Stan Parker in *The Tree of Man* (1956). Many former POWs also dammed up what was to them a shaming tale. When their stories burst forth, the purpose was sometimes indictment of the Japanese, as for Brad-

don and Rivett. In John Romeril's play, *The Floating World* (1974), the protagonist, Les Harding, suffers a mental breakdown when his memories of imprisonment overpower him. The POW often has a perception of the enemy which may be both more resentful, and sentimentally dependent, than that of those not taken into captivity, thence into the humiliating, semi-security of the enemy's embrace. The former POW endures a second, interminable and metaphorical captivity, from which memoirs may offer partial release. The abiding popularity of prisoner-of-war fiction of the Second World War (although the derring-do of escape stories belongs rather to Europe than Asia), indicates a broader pattern in the nation's war literature. Significant elements of Australian literary culture have long been the prisoner of past wars.

The revival of the Great War as a literary subject in the late 1970s was not conducted through the channels of popular fiction. These had never ceased to be conduits for Second World War reminiscence, whether as individual memoir, unit history, or fiction (in so far as these kinds can readily be distinguished). One of the most famous of Australian novels, *My Brother Jack* (1964), by the former war correspondent George Johnston, opens with the narrator's recollections of the casualties of the Great War who gathered in his parents' home. The quintessential Anzac, Jack Meredith, misses the military career for which he had apparently been so well fitted. Instead it is his brother David who has a good war. Johnson's novel was published in the mid-1960s, in a decade which — far from deserving its radical reputation — is notable in Australian literature for introspection, for the outpouring of works of autobiography and of social diagnosis — by Donald Horne, Graham McInnes, Hal Porter, James McAuley, Robert Hughes and Bernard Smith, among many others. *My Brother Jack* is a seminal retrospect on Australian self-valuations and delusions, after one war, during the next, and between the two. The shadow of war falls across all the novel.

Of contemporary novelists, David Malouf has written perhaps the finest account of the consequences of the Second World War, focused on the experiences of two men who were prisoners of the Japanese in Changi. This is *The Great World* (1990). Thomas Keneally, writing under his own name, or as 'William Coyle', has been the Australian author to dwell most extensively and incisively on war. His fiction has ranged in time from the Hundred Years' War, in *Blood Red, Sister Rose* (1974), to the American Civil War in *Confederates* (1979), then the war in Eritrea in *Towards Asmara* (1989). He is sensible of how such civil wars most terribly rend the social fabric. While Keneally regards the Great War as the crucial divide in the history of this century, only in *Gossip From the Forest* (1975) — his account of the armistice negotiations of 1918 — has it been a setting for his fiction. It is the Second World War which he has most fully treated: the domestic and political homefront in *The Fear* (1965) and *The Cut-Rate Kingdom* (1980) respectively; the conflict between Germans and partisans in what was Yugoslavia in *Season in Purgatory* (1976) and the Pacific War in his two pseudonymous novels, *Act of Grace* (1988) and *Chief of Staff* (1991).

In popular fiction, the Second World War lives on lustily. Novels such as John Yeomans's *Much Curious Pleasure* (1987) and Joe White's *Pringle's War* (1988) loyally preserve the stereotypical figures of the Anzac fighting man, as he was once to be found, pristine, in fiction. The publishing industry built on Australia's part in the Second World War seems indomitable. Each year dozens of new books — memoirs, diaries, novels, stories of forgotten campaigns, rehabilitations of discredited generals — appear. John Millett's collections of poems, *Tail Arse Charlie* (1982) is altogether different. Based on long meditation over his experiences in the air war over Europe (and as such an uncommon contribution to the Australian literature of war), this book unblocked Millett as a poet. Much more typical was the career of a poet such as David Campbell, which gained an early impetus from war service. Campbell, Vincent Buckley, John Manifold, Kenneth Mackenzie and Geoffrey Dutton are among authors who first found assured voices during this war. McAuley and Harold Stewart put their leisure time in the Victoria Barracks in Melbourne to use by concocting the Ern Malley hoax poems. More sombrely, Manifold's 'The Tomb of Lt John Learmonth, AIF' is one of the most eloquent of wartime elegies. Only war correspondent Kenneth Slessor's (q.v.) 'Beach Burial', occasioned by his sight of bodies washed ashore on the coast of North Africa in 1943, is more renowned among Australian poems of the Second World War. And rather than marking a beginning, this benediction for all the dead in war was nearly the end of Slessor's self-truncated poetic career. Only the satirical 'An Inscription for Dog River', where he demolished the grandiose pretensions of the Australian General Blamey, remained to be published. It is a poem unusual also for the acerbic, combative note that it strikes in the midst of so much oddly quietist poetry produced by Australians during the Second World War.

Enemies had begun to disappear from Australian war poetry, if not from its fiction, some time before Slessor wrote. In Australian war literature, the earliest enemies are the recalcitrant Boer, the

awakening Mongol, the Hun 'at the gate'. Subsequently, and as part of the seismic shift in perception that the Great War occasioned, the real enemy began to be discerned 'to the rear', in the form of incompetent generals, politicians and profiteers. In 'True Patriot', Matthews inveighed against the 'old man in a ship's saloon' who directs the battle. In Martin Boyd's novel, *Where Blackbirds Sing* (1962), Dominic Langton — who becomes an objector to the Great War but not a pacifist — declares that 'the Germans are only my artificial enemy ... They are just like people you see in the street'.

The presence of any kind of adversary, or agent of the harms which men suffer in battle, gradually and curiously ebbs out of Australian war poetry. One might never know who causes 'The Death' in Leon Gellert's Great War poem of that name. Although it is 'the hour of battle' in Vance Palmer's 'The Camp', there is no sign of an enemy. In David McNicoll's 'Ski Patrol' (a poem of the Second World War), death is 'the sport', but the poem's wintry landscape is bereft of enemies. It is as if the strangers against whom one must fight are being willed out of existence. Self-communion, rather than antagonism, increasingly occupies Australian war poets, at least until the Vietnam period, when polemical verse such as that collected in *We Took Their Orders and Are Dead* (1971) assails the United States as the true national enemy.

When Australian advisers were committed to Vietnam in 1962, the country's longest military involvement began. Responsive to the peculiarly taxing conditions of combat, in which Australian soldiers died for an increasingly unpopular cause, allies were deeply distrusted and the task of distinguishing friend from foe was acute (and not always attempted), novelists of the Vietnam War produced disappointingly conventional narratives. In *When the Buffalo Fight* (1980), 'David Alexander' (Lex McAulay) stridently defended what he saw as an imperilled military tradition. His novel is 'inspired by the Australian soldier, by his matter-of-fact day-to-day courage, his sense of humour and his sense of comradeship'. When so fondly recalled, the Anzac legend looks fragile. Novels such as Rhys Pollard's *The Cream Machine* (1972) and William Nagle's *The Odd Angry Shot* (1975; later filmed) were accounts of first things: meeting the platoon, basic training, departure for Vietnam (for many their first overseas trip to a place whose whereabouts was a mystery); and then war: the wounding and loss of friends, killing in battle. An Australian army officer, John Rowe, wrote a novel, *Count Your Dead* (1968) which — by its iconoclastic view of national military self-esteem — helped to end his career in the army. Rowe became a professional writer, specialising in thrillers that imagined the approach to the end of the world. So too did John Carroll, author of *Token Soldiers* (1983). Nagle has written film-scripts, including that for the Vietnam War movie, *Fire Base Gloria*. Among the more recent fiction concerning the war and its protracted domestic consequences are such accomplished novels as Robert Allen's *Saigon, South of Beyond* (1990) and Georgia Savage's *Ceremony at Long Nho* (1994).

The Vietnam War is the hidden subject of many of the novels which treat of ingenuous men and women in other Asian countries. This long and burgeoning list includes Christopher Koch's *The Year of Living Dangerously* (1978), Robert Drewe's *A Cry in the Jungle Bar* (1979), Blanche D'Alpuget's *Turtle Beach* (1981) and Ian Moffitt's *The Retreat of Radiance* (1982). Often the protagonists of these novels seek Asian mentor figures. Their quests seem to be ways of tacitly making amends for Australian interference in, and ignorance of, Vietnam during the war. Recently, some Australian writers have dared to hope for a mutual healing process. As her Vietnamese guide Sanh says to Savage's heroine Fiona: 'You are part of our history now and we shan't let you go'.

For established poets, the war led to a reassessment of Australian society. Anti-war verse redefined the (poetic) enemy once more, this time in the guise of American culture, language and imperialist ideology. These are forms of hegemony less concrete than the military alliance that, although resented, could not easily be repudiated. No such simple gratification of personal anger was entertained by David Campbell or Bruce Dawe, for whom the war led to a reorientation of their careers. Dawe's 'Homecoming' is the finest threnody in Australian literature of the Vietnam War, while his neglected 'The Fate of Armies' imagines men's escape, as if in dream, from battle.

Australian writers, in many mediums and at different levels of accomplishment, show no signs of wishing to escape from those key periods in its history when the nation has been embroiled in war. The revisionist body of poetry and fiction of the 1970s that scrutinised the Great War now seems a comprehensive remythologising of a lost Australia, in reaction to then present discontents. At the same time, conventional narratives of the Second World War and the Vietnam War were being produced in substantial numbers, upholding, if at times desperately, an older mythology of Anzac military prowess. The steady stream of novels concerned with Asia (a few of them set in Vietnam) indicate how Australian involvement in the Vietnam War eventually prompted an anguished and general reappraisal of Australia's place in the region, the rel-

evance of its old allegiances, the quality of its culture. Australian literature of war is now all and inevitably retrospective, but its dynamic linkage to the definition of Australian national identity — however expedient, morally compromised, sentimental or triumphalist this has been from time to time — is no less important, or troubling, than at any stage in the country's history.

John Laird, 'War literature' in William H. Wilde, Joy Hooton and Barry Andrews, *The Oxford Companion to Australian Literature*, 2nd edition (Oxford University Press, Melbourne, 1994); *Australian Literary Studies*, vol. 12. no. 2, October 1985, special issue 'Australian Literature and War'.

PETER PIERCE

LITHGOW SMALL ARMS FACTORY see **INDUSTRY; SCIENCE AND TECHNOLOGY**

LITTLE, Captain Robert Alexander (19 July 1895–27 May 1918). Having failed to gain a place at the No. 1 Flying School at Point Cook, Victoria, Little sailed to England in July 1915 and qualified as a pilot at his own expense. In January 1916 he entered the Royal Naval Air Service (RNAS) as a sub-lieutenant, and in October went to the Western Front with No. 8 Squadron, RNAS, flying Sopwith Pups. He shot down his first enemy aircraft on 1 November, and never looked back; by March 1917 he had accounted for nine enemy planes. By August 1917, when he went on leave to England, his tally stood at 37, with many more damaged, and he had been awarded the DSO and bar, the DSC and bar, the Croix de Guerre, and was mentioned in despatches. He returned to France in March 1918, and when the RNAS and RFC were amalgamated to produce the RAF, became Captain Little of No. 203 Squadron. He was killed at the age of 22 on the night of 27 May 1918 while flying solo against enemy bombers. His final score of enemy aircraft stood at 47, making him the most successful Australian ace of the war, and the fourteenth ranking ace from all sides, but his exploits are hardly known in Australia, probably because he flew with a British service.

'LITTLE BOY AT MANLY' On 4 March 1885, the day after the New South Wales contingent left Sydney for the Sudan (q.v.), nine-year-old Ernest Laurence wrote to the Acting Premier, W. B. Dalley: 'yesterday I begged to be allowed to do something, even if it was only to send what little money I had saved up to the Patriotic Fund for the poor widows and orphans, and at last father has not only consented, but gave me something to add to it. Now, dear Mr Dalley, I have written to you with it all, and want you to send it on with my best wishes from a little boy at Manly.' Laurence enclosed £25, and four days later his letter and his name were published in the *Sydney Morning Herald*, despite his father's insistence that his son be identified only as 'Ernest, a little boy at Manly'. Thereafter cartoonists and writers mercilessly mocked the sentiments that Dalley was accused of exploiting, and the 'Little Boy at Manly' became a symbol of a certain Australian innocence, but one that was ridiculed by some as unrepresentative of Australian aspirations, although other illustrators used the image in a more uncritical manner. Not until 1915, the historian K. S. Inglis suggests, did there emerge a 'new Australian myth-figure, drawn from real life, national, classless and virile. Soon he was named the Digger' (q.v.), a lineal descendant of the 'Little Boy at Manly.'

K. S. Inglis, *The Rehearsal: Australians at War in the Sudan, 1885* (Rigby, Sydney, 1985).

LLOYD, Major-General Charles Edward Maurice 'Gaffer' (2 February 1899–31 May 1956) graduated from RMC Duntroon at the end of 1918, too late to see active service. As an artillery officer he held the usual round of junior staff and regimental postings in the 1920s, and attended the Staff College, Camberley, between 1932 and 1933. Although a regular officer, he found time to complete a law degree through the University of Sydney in 1925, and possessed a breadth of vision and widespread interest in matters outside the army which was unusual in a junior officer of his generation. He worked in the Directorate of Artillery at Army Headquarters from 1938 to 1939, and was seconded to the 2nd AIF as soon as war broke out. After a number of increasingly important administrative staff jobs with the 6th Division and I Australian Corps in the Middle East in 1940, he became the senior operational staff officer on the 9th Division headquarters in December 1940, and went through the siege of Tobruk (q.v.). Seconded to a senior staff position on Wavell's headquarters in the short-lived ABDA Command (q.v.), he returned to Australia in April 1942 and in July was made Director of Staff Duties at Land Headquarters (q.v.). In February 1943 he was appointed Adjutant-General, a post he held until 1946. He had been promoted from major to major-general in less than two and a half years, making him the youngest general officer in the Australian Army. He transferred to the Reserve of Officers in February 1946, essentially turning his back on the army, and became a senior newspaper executive with the *Argus & Australasian*. In civil life he was also chief of the United Nations Refugee Organisation in Australia and New Zealand from 1948 to 1951, and

Chief of Mission of the United Nations Korean Reconstruction Agency between 1951 and 1953. Lloyd's departure at the early age of 47 was a loss that the postwar army could ill-afford. The official historian, Gavin Long, described Lloyd as 'one of the ablest of the particularly promising group' of young regular officers to graduate from RMC at the end of the First World War. The correspondent Chester Wilmot (q.v.), who observed him with Lieutenant-General Leslie Morshead (q.v.) on the 9th Division, thought him 'one of the ablest staff officers and most colourful characters in the AIF. . . Big and bluff, Lloyd has a manner that is a strange mixture of bluntness and friendliness … He is no respecter of persons and is essentially a realist who sees a job to be done and goes about it in the most direct way'.

LOGISTICS is the science of planning and carrying out the movement and maintenance of forces, and is as old as warfare, although only recently has it been accorded the formal recognition which is its due. For much of their history the Australian forces were dependent on their Great Power allies for logistic support in the field (on the British in France during the First World War, on the Americans in the Korean War and to a lesser extent in the Vietnam War, for example), and often paid directly for services which they could not provide themselves. All three services now have functional commands responsible for the logistic support of that portion of the ADF (Naval Support Command, [Army] Logistics Command, [RAAF] Support Command). Australia operates under a Logistic Support Agreement with the United States, signed in November 1989, which provides access to and priority within the American logistics system for the ADF in both peace and war. The agreement covers the provision of 'defence articles', which includes weapons, munitions and the implements of war and the ability to manufacture, service and repair them, and 'defence services', which includes services, training and technical assistance and covers computer software and technical data. The system enables the ADF to reduce the need for stockpiling large quantities of munitions and other high-volume consumable items, which would otherwise make a considerable impact on already tightly stretched resources, but has also been criticised for precisely this reason, and for the reliance on the United States which it builds into Australian defence policy. The potential dangers of the system are demonstrated by the Swedish-produced Carl Gustav anti-tank weapon with which the army was equipped in the 1960s. The Swedes tried to discontinue supplies of ammunition for the weapon because it was being used by Australian troops against bunker systems in Vietnam, and supplies were only maintained through a face-saving formula.

LONE PINE MEMORIAL, situated in the Lone Pine Cemetery at Anzac, is the main Australian memorial on Gallipoli, and one of four memorials to men of the New Zealand Expeditionary Force. It was designed by Sir John Burnet, the principal architect of the Gallipoli cemeteries. The memorial is designed as a thick tapering pylon 14.3 metres high on a square base. It is built from limestone mined at Ilgardere in Turkey. The Memorial commemorates the 3268 Australians and 456 New Zealanders who have no known grave and the 960 Australians and 252 New Zealanders buried at sea after evacuation through wounds or disease. The names of Australians are listed on a long wall of panels in front of the pylon and to either side, while those of the New Zealanders are inscribed on stone panels mounted on the south and north sides of the pylon. Names are arranged by unit and rank. The memorial stands over the centre of the Turkish trenches and tunnels which were the scene of heavy fighting during the August offensive. Most cemeteries on Gallipoli contain relatively few marked graves, and the majority of Australians killed on Gallipoli are commemorated on the Lone Pine Memorial.

LONG, Commander Rupert Basil Michel (19 September 1899–8 January 1960) was born in Melbourne and joined the RAN in 1913 through RANC Jervis Bay. In the last year of the First World War he saw sea duty on HMA Ships *Australia* (I) (q.v.) and *Huon*. Appointed Assistant Director of Naval Intelligence between 1936 and 1939, he became Director of Naval Intelligence (DNI) at the outbreak of war, a post he was to hold until 1945. Long's achievements as wartime DNI included initiating the expansion of the coast-watchers (q.v.) and the formation of the Allied Intelligence Bureau (q.v.). Though extremely able and hard-working, and reported as never losing either his sense of humour or basic courtesy, he remained an acting commander throughout the war, and was only promoted to commander on retirement. Gavin Long (q.v.), who was no relation, attributed his failure to gain promotion to his 'incisive, critical mind, which appears not to be an asset in a service in which high marks are awarded for orthodoxy'. G. Hermon Gill (q.v.), the naval historian, also thought that Long was 'by far the ablest officer in the RAN', but that he had been passed over and anticipated that he 'would leave the navy as soon as the war ended', which is what he did.

LONG, Gavin Merrick (31 May 1901–10 October 1968). After a brief period as a schoolteacher, Long turned to journalism, finding employment initially with the newly established *Daily Guardian* in Sydney in 1926 and, later that same year, with the Melbourne *Argus*, where he stayed until 1930. He returned to Sydney and worked at the *Sydney Morning Herald* as a sub-editor and critic and, from 1936, as chief cable sub-editor. His writings on foreign policy and defence, which had always been areas of personal interest, now increased considerably. In 1938 he accompanied the Governor-General, Lord Gowrie, on an official visit to the Netherlands East Indies; the trip produced a number of articles on the defences of the Netherlands East Indies and on the Singapore base, which he correctly predicted was vulnerable to attack from the landward side. Late in 1938 he went to England for what was intended to be a two-year appointment in the *Sydney Morning Herald's* London bureau.

At the beginning of the war Long was among the first correspondents accredited to accompany the British forces to France, and he covered the 'phoney war' and the fighting leading to the fall of France. Later in 1940 he went to the Middle East as a correspondent for Australian newspapers, and covered the Greek campaign before being recalled to Sydney in the middle of 1941. He continued to act as a defence correspondent until January 1943, when he was appointed official historian by the War Cabinet, at C. E. W. Bean's (q.v.) earnest behest.

For the remainder of the war Long divided his time between visits to the fighting fronts in the Pacific and the organisation of the official history in Australia. While overseas he used the opportunity presented to quiz officers and men not only about events as they were occurring in the present, but to help fill in his knowledge of earlier campaigns, before he had been appointed: he was acutely aware that, unlike Bean in the previous war, he had not been present at many of the Australians' significant actions, either in the Mediterranean or in the early fighting against the Japanese.

The official history ultimately ran to 22 volumes divided into five series: Army, Navy, Air, Civil and Medical. Long himself wrote three volumes in the Army series, *To Benghazi* (1952), *Greece, Crete and Syria* (1953), and *The Final Campaigns* (1963), and selected a diverse team of fourteen writers for the rest. Progress was slower than expected, which helped to underline the unrealistic initial forecasts that the whole enterprise could be completed within a few years. The final volume was not published until 1977, nearly a decade after Long's death.

Long was especially concerned that censorship, a necessary feature of wartime, should not extend to the writing of the history, and he extracted from the government in 1945 a charter which limited the exercise of censorship 'to the prevention of disclosure of technical secrets of the three Services ... which it is necessary to preserve in the postwar period'. The most obvious of these was the existence of ULTRA signals intelligence, the existence of which was not made public until 1974. It is highly doubtful that Long and his writers even knew of its existence; it is a tribute to their skill as historians that the general thrust of their volumes does not noticeably suffer from the omission.

The official histories written in the Commonwealth and the United States after the Second World War were pursued on a basis of informal cooperation, with regular contact between the historical sections of the various governments and some sharing of important documents. Long himself commented on a number of drafts from the British and Indian official histories, and entered into several memorable exchanges with British military figures who were concerned to put the best possible light on the failures in Greece, Crete or Singapore. He also resisted strong domestic political pressure from one or two senior figures within his own government, who attempted to influence the treatment of the controversial commander of the 8th Division in Malaya, Gordon Bennett (q.v.).

Long's approach was inspired in large part by that of Bean: 'I originated little in the technique of writing the war history', he wrote. 'I merely followed Bean's principles, modifying these to meet the special requirements of the World War II history.' Such modifications included a very substantial expansion of the space accorded the domestic front, with two additional volumes devoted to the war economy. Long's original intention to include a volume dealing with strategic policy and the higher direction of the war effort did not go ahead, however, for reasons which are now unclear but which may well have included opposition at the political level to too close an analysis of policies and decisions with which many postwar political figures had been closely associated. Long continued Bean's practice of focusing on the deeds and experiences of individuals, especially of ordinary soldiers. Like Bean's history, Long's is replete with the footnoted annotations detailing personal circumstances which are one of the hallmarks of the 'democratic' history which Bean pioneered. There is, however, no detailed discussion of logistics, training, administration, doctrine or the other military technicalities which determined the effectiveness of modern military organisations. The result, as John

Hetherington noted, was that Long's history was 'wonderfully readable', a quality not to be dismissed lightly, but that it was also less useful as a work of reference. This is especially frustrating when the published volumes are read against Long's detailed diaries and notebooks, which reveal a very deep knowledge of the Australian Army and its wartime workings, much of which did not find its way through to the printed page.

In early 1963 Long decided that the official history no longer required the services of a full-time series editor, and resigned his position. He accepted a research fellowship at the Australian National University, and helped to establish the *Australian Dictionary of Biography* (q.v.). He completed a single-volume study of Australia's war, *The Six Year's War* (1973), which was published posthumously, and a penetrating analysis of General Douglas MacArthur's generalship, *MacArthur as Military Commander* (1969). He was on the Commonwealth Archives Committee, various boards of the Australian War Memorial, and the Battlefields Memorials Committee. His health, never particularly robust after the war, declined and he died at the relatively young age of 67.

Long has generally failed to emerge from under the shadow of C. E. W. Bean, and his history is less well known and, in consequence, less authoritative than that for the First World War. This is unfortunate, because in some respects Long was the better historian. Well aware of the importance of his task as a national historian, he wrote at one stage that 'the smaller partners in alliances can be sure that if they do not write their own histories their individual experiences will be written very small in the pages of recorded history'.

LONG RANGE WEAPONS ESTABLISHMENT, WOOMERA A joint Anglo-Australian project established in 1946, the Long Range Weapons Establishment was in part a reaction to the experience of German V weapons used against London in 1945, and was conceived within a context of increasing Cold War tensions. It was also established on the assumption that a joint Commonwealth defence policy was possible after 1945. A testing range was established at Woomera, in South Australia, which, at 1250 miles in length, became the largest land testing range in the Western world. A research and development centre was established in the Adelaide suburb of Salisbury. Britain provided all the weapons for testing together with much of the equipment and many of the scientific personnel, while Australia provided the facilities and all support, together with some personnel. Across the life of the project, from 1947 to 1980, Australia's share of the costs was $2262 million, with Britain bearing a further $982 million as its contribution. A range of air-to-air and surface-to-air weapons was developed, including Seawolf, Rapier, Sea Dart, Bloodhound and Blue Steel (a cruise missile), together with the Australian designed and developed weapons Malkara (an anti-tank weapon) and Ikara (an anti-submarine weapon) and the target aircraft, Jindivik. The project began to founder, however, with the cancellation in 1960 of the prototype ballistic missile, Blue Streak, designed to carry the British independent nuclear deterrent but rendered obsolete by the development of the Polaris system in the United States. By 1964 there were considerable doubts in Britain over the value of expensive testing in Australia of weapons that were not necessarily adopted by the British services, and concern over the competition with larger and often more efficient American programs. As a consequence British interest in the joint project declined. Woomera was not used solely for weapons testing, however, and some of its most successful work in the 1950s was conducted with large research rockets, such as Black Knight, which led to the development of satellite launch systems including the Australian WRESAT system. The

direct benefits of the project to the Australian armed forces were insubstantial, as they did not choose to take up many of the British-designed weapons developed at Woomera, and had little use for the extensive range facilities themselves; while the opportunity to use the facilities and technical expertise in launch and recovery systems for civil and commercial purposes was allowed to lapse through lack of imagination on the part of the government. Woomera's principal function in the 1990s is as a support centre for the ground station at Nurrungar, a joint facility operated with the United States (see JOINT DEFENCE FACILITIES).

Peter Morton, *Fire Across the Desert: Woomera and the Anglo-Australian Joint Project 1946–1980* (Australian Government Publishing Service, Canberra, 1989).

LONSDALE, HMVS see **VICTORIAN NAVY SHIPS**

LOOTING see **'SCROUNGING'**

LUSTREFORCE was the code-name for the force used in Operation LUSTRE, the British deployment to Greece (q.v.) in April 1941. The force was originally intended to comprise the 2nd New Zealand Division, 6th and 7th Australian Divisions,

The commander of Lustreforce, Lieutenant-General Sir Henry Maitland Wilson (centre), with Lieutenant-General Sir Thomas Blamey, commander of the 6th Division (left), and Major-General Bernard Freyberg, commander of the New Zealand 2nd Division (right), Greece, 1941. (AWM 128425)

the British 1st Armoured Brigade and an Independent Polish Brigade Group. Command of the force was invested in the British Lieutenant-General Sir Henry Maitland Wilson, despite the fact that Australia was to contribute the largest single proportion of the field force. When questioned on Wilson's appointment, General Sir Archibald Wavell, GOC Middle East, argued that the Australians comprised only one-third of the ration strength (42 000 out of 126 000) and 40 per cent of the fighting force; while the New Zealand divisional commander, Freyberg, thought that Wilson was appointed because 'he could handle the King of Greece better'. The force actually sent was smaller than envisaged (the 7th Division was not included), but the incident seemed to suggest again that the British were not prepared to appoint dominion generals to combined commands.

LYNCH, Arthur see **AUSTRALIANS IN THE SERVICE OF OTHER NATIONS**

LYON, Lieutenant-Colonel Ivan see **JAYWICK/RIMAU**

M

M SPECIAL UNIT see **ALLIED INTELLIGENCE BUREAU**

M60 GENERAL PURPOSE MACHINE-GUN (M60 GPMG) is a gas-operated, belt-fed, automatic-fire, air-cooled weapon with a quick-change barrel system. It was designed and manufactured in the United States, incorporating design features from the operating system of the German FG42 assault rifle and the belt-feed system of the German MG42 machine-gun.

The M60 went into production for the US Army in 1960 and was selected for the Australian Army in 1961 to replace the BREN LMG and the Vickers MMG. As a GPMG, the M60 was designed to be light enough to be carried and fired by one man in the light role, and to be fired from a tripod in the medium role and provide sustained fire by employing the quick-change barrel system. Although the M60 could be carried and fired by one man it required a crew of at least two men to maintain it in action with ammunition, spare barrel and tools.

In Australian military service the M60 has been employed by the infantry in the light and medium roles and as the defensive armament for helicopters. It was used on active service in South Vietnam and was gradually phased out of service during the late 1980s and replaced by the FN MAG GPMG and the F89 Minimi LSW. The M60 had a mixed reputation, and was generally regarded as not being a good example of American design and engineering. However, a good, reliable, well-maintained M60 was valued for its firepower.

The M60 fired 100-round belts of disintegrating metal link, 7.62 mm ammunition. Each belt weighed 7 pounds. An unloaded M60 weighed 23 pounds (10.48 kg) and its spare barrel (complete with bipod and gas system) weighed 8 pounds 4 ounces (3.75 kg). The M60 had a cyclic rate of fire (i.e. the rate of fire that theoretically could be achieved given a continuous supply of ammunition) of 550 rounds per minute and was fired in short bursts of 3 to 5 rounds, long bursts of 5 to 10 rounds, or sustained bursts of up to 20 rounds. The maximum range was 3725 metres. In the direct-fire role the effective range was 500 metres from the bipod and 100 metres from the tripod. In the indirect-fire role, firing from the tripod with the C2 sight, the effective range was 2000 metres with a capability for harassing fire out to 3000 metres.

In Australian service the M60 has used two different tripod mounts for its employment in the medium role. The original tripod was the American M122 tripod (weight 8.5 kg), which was replaced in the late 1970s by the British L5A1 tripod (weight 14.5 kg). The C2 sight unit weighed 4.5 kg.

IAN KURING

MACANDIE, George Lionel (26 June 1877–30 April 1968). Macandie began his working life as a clerk with the Queensland colonial navy. After Federation he transferred to the Commonwealth Department of Defence and helped set up the Navy Office in Melbourne. Between 1903 and the outbreak of war he worked as Vice-Admiral Sir William Creswell's (q.v.) principal administrative assistant. From 1914 until his retirement in 1946 he was Secretary to the Australian Naval Board.

In March 1919 he became the Secretary of the first Department of the Navy. When this was merged again with the Department of Defence (q.v.) in 1921 he continued to preside over the Navy Office, and from June 1920 spent a year on attachment to the Admiralty in London to widen his experience of naval administration and policy. During this early postwar period there were a number of attempts to unseat him by senior naval officers arising from the duality of his position: on the one hand, his was a senior public service position, but many senior naval officers regarded him as the senior administrative officer of the Naval Board and therefore under navy, that is, service control. This impression was perhaps exacerbated by the fact that he held honorary rank as fleet paymaster (commander equivalent), which was granted him in 1916.

He was central to the administration and finances of the RAN in the interwar period, and played an important role in the re-equipping of the navy in the 1930s with modern cruisers. In 1939 a separate Department of the Navy was again created, but by this time Macandie was 62, and a younger man, A. R. Nankervis, was appointed secretary. Macandie nonetheless stayed on beyond formal retirement age as Secretary of the Naval Board until after the end of hostilities. In 1950 he published *The Genesis of the Royal Australian Navy*, which retains its value as a detailed guide to early naval policy and administration.

MacARTHUR, General of the Army Douglas (26 January 1880–5 April 1964). Son of a famous father, MacArthur was commissioned from the United States Military Academy, West Point, in 1903 and served in various parts of America's growing empire before spending the years from 1913 to 1917 on the War Department General Staff. He became chief of staff of the 42nd 'Rain-

General Douglas MacArthur inspects the recently-shot body of a Japanese sniper, Labuan, Borneo, 10 June 1945. Lieutenant-General Sir Leslie Morshead, commander of I Corps, is on MacArthur's left. (AWM 109147)

bow' Division, and served with it in France during the last stages of the First World War as a brigade and divisional commander; he was wounded twice and decorated. MacArthur served as Superintendent of West Point and was appointed chief of staff of the army in 1930, a post he held until 1935. On relinquishing his position he went to the Philippines as military adviser to the government, and reorganised its defences; in 1936 he was made field marshal in the Philippine Army, and he retired from the US Army the following year.

Recalled to active duty in July 1941 he was given command of US Army forces in the Far East, based on the Philippines. The early Japanese advance quickly overwhelmed the Philippines, and MacArthur was ordered to leave by the President, Franklin D. Roosevelt. He arrived in Australia in March, intending to prepare and lead a force to relieve those he had left behind in defensive positions at Bataan and Corregidor, but the last two remaining American–Philippino forces surrendered, and in any case there was little in Australia at the beginning of 1942 from which to fashion a relieving armada. He was appointed to command the newly created South-West Pacific Area (q.v.) in April 1942, and this point marked the renaissance of his military career.

His instructions from the Combined Chiefs of Staff were to hold the 'key military bases of Australia as bases for future offensive action', and to stem the tide of the Japanese advance to Australia's north and east and interdict Japanese communications in the theatre in preparation for resuming the offensive. MacArthur's early decisions as C-in-C, SWPA, were to set the tone for his exercise of command throughout the rest of the war, especially so far as the Australians were concerned. Blamey (q.v.) was appointed Commander, Allied Land Forces, in recognition of the fact that the overwhelming preponderance of army units in the theatre at that stage (and for some time to come) were Australian, but all other principal commands went to Americans. MacArthur's senior staff was entirely American in composition, and with few exceptions was made up of men who had served with him in the Philippines, the so-called 'Bataan gang'. The US Army chief of staff, General George C. Marshall, remonstrated with MacArthur and urged that he appoint some Australian and Dutch officers to senior posts on his headquarters, but MacArthur refused, claiming that the Australians did not possess sufficient qualified senior officers for their own needs, and by implication dismissing the Dutch as soldiers altogether. Neither position was fair, but within Australia at least MacArthur's own standing was such that no one was disposed to argue with him.

Because MacArthur chose to direct the conduct of operations himself, he was able to negate Blamey's notionally allied appointment by keeping American ground forces under separate command. As a consequence, the higher direction of the war was conducted very much on MacArthur's own terms. His relationship with the Australian Prime Minister, John Curtin, was in many respects a close one, and the two worked in concert not least because it was in each other's interest to see that the SWPA was accorded a higher priority for reinforcements, equipment and supplies. Although General Sir Thomas Blamey was the Australian government's principal military adviser, he was frequently excluded from discussions and decisions arrived at between MacArthur and Curtin; the only other Australian generally present at all their deliberations was the Secretary of the Department of Defence and the War Cabinet, F. G. Shedden (q.v.).

MacArthur's style and massive egotism was the cause of much ill feeling both at senior levels and among the soldiers in the field. His communiqués rarely mentioned subordinate commanders, and credit for successful operations was usually shared among 'Allied' troops, when it might more properly have been ascribed to Australian units and formations. When Lieutenant-General Robert L. Eichelberger arrived to take over command of the US I Corps, MacArthur instructed him to introduce himself to his Australian opposite numbers and then have nothing more to do with them; Eichelberger in fact found Australian senior officers a competent and experienced group, and ignored his superior's instruction. There is some evidence that towards the end of the war Curtin came to regret the authority over Australia's war effort which his government had ceded to MacArthur in 1942, but by then MacArthur's interests and attention lay elsewhere than in Australia. In any case, the almost complete lack of military experience within the Curtin cabinet meant that the government and the Prime Minister would have been heavily reliant for military advice upon someone, and MacArthur represented the conduit through which American troops and American supplies would be forthcoming to assist in the safeguarding of Australia's territory and interests, at least in the short term. MacArthur carefully coordinated his appeals to Washington with Curtin's own communications, to the undoubted benefit of both sides at least until 1944. Thereafter, however, with his view increasingly fixed on the return to the Philippines — which he seems to have considered at one level as the consummation of a sacred trust — MacArthur had little time for or interest in Australian goals and aspirations. He excluded Australian ground forces from the campaign to reconquer the Philippines, despite having indicated at one stage that there was a major role there for them, which led to the diversion of Australian divisions to minor campaigns in Borneo, Bougainville and northern New Guinea.

The plans for the invasion of the Japanese home islands, which MacArthur was to command, called for a British Commonwealth corps to which Australia was to contribute at least one division, but in the arrangements for the surrender ceremony in Tokyo Bay in September 1945, the Supreme Commander for the Allied Powers (as he now became) had to be prevailed on to admit Australian and New Zealand representatives to the signing ceremony. During the occupation of Japan between 1945 and 1950 he again had Australian troops under his command, but the provisions of the MacArthur–Northcott agreement (q.v.) which governed their involvement meant that they had little of consequence to do. At the outbreak of the Korean War in June 1950 MacArthur successfully appealed for the use of Australian air, sea and land contingents then stationed in Japan, but had little or nothing to do with them directly. He was relieved of all his commands in the Far East in April 1951 by President Harry S. Truman, after he persistently defied the cardinal principle of civilian control of the military. He retired into outspoken bitterness, discarded and, by the time of his death, largely overlooked.

MacArthur deserves his reputation as one of the most brilliant of America's generals, and as one of its most controversial. Although flawed in important respects, he possessed the ability to captivate even those who approached him somewhat sceptically. The Australian official historian, Gavin Long, wrote of him after an audience in 1944 that he was 'a man entirely at unity with himself, fulfilled and confident. Sensitive. Never in doubt as to his own imposing stature. Convinced that he is born to command, and enjoying this conviction sensuously … This is a man of mind and feeling rather than a man of iron.' Long's final conclusion, published after his own death, was that 'here, patently, was a great commander'.

Gavin Long, *MacArthur as Military Commander* (Angus & Robertson, Sydney, 1969); D. Clayton James, *The Years of MacArthur*, 3 vols (Houghton Mifflin, Boston, 1970, 1975, 1985).

MacARTHUR–NORTHCOTT AGREEMENT Negotiated between General of the Army Douglas MacArthur (q.v.), appointed Supreme Commander for the Allied Powers (SCAP), and Lieutenant-General John Northcott (q.v.), C-in-C of the British Commonwealth Occupation Force (q.v.), this agreement governed the basis on which

BCOF operated within the Allied structure for the occupation of Japan. Discussions began on 12 December 1945, draft proposals were referred to the Joint Chiefs of Staff in Australia (q.v.) five days later, and a memorandum of understanding was signed by Northcott and Major-General R. J. Marshall, chief of staff to SCAP, on 18 December. It assigned BCOF responsibility for Hiroshima prefecture under the direction of the commanding general, Eighth United States Army (at that stage Lieutenant-General Robert L. Eichelberger); the Commonwealth air component came under the command of the US 5th Air Force. On administrative matters Northcott was to have direct access to JCOSA, but questions of operations or policy were to be dealt with by the Australian government directly with Washington, any issues requiring action then being referred to MacArthur. In matters of occupation policy and dealings with the Japanese, American policy would be followed. After further discussion between the participating Commonwealth governments, these principles were accepted and the agreement was announced publicly on 31 January 1946. In the view of Northcott's successor, Lieutenant-General Sir Horace Robertson (q.v.), the agreement was 'a very hard bargain', since it relegated BCOF to a minor place in the occupation and limited its role in Japan in a way which some felt impinged on the sovereignty of the Commonwealth governments concerned. It was, however, in Robertson's view again, probably the best deal that Northcott, or anyone, could have negotiated given the dominance which MacArthur exercised over occupation matters. It was probably fortunate for the Japanese as well, since MacArthur's approach to rebuilding Japan was much less retributive than that of Australia, in particular, especially in the early period immediately after the war.

MacARTHUR-ONSLOW, Major-General Denzil (5 March 1904–30 November 1984). Scion of an old established family with a long tradition of military service, MacArthur-Onslow was commissioned into the Australian Field Artillery in 1924 at the age of 20. In the interwar years he held a variety of regimental appointments in militia artillery units and a junior staff post on the headquarters of the 1st Cavalry Division. He was sent to England to undertake a course in armoured warfare and on return served with the newly raised 2nd Armoured Car Regiment. He learnt to fly in his youth and made his first parachute jump in the early 1930s. As a businessman he advocated the manufacture of parachutes in Australia and in 1935 began manufacturing them for the RAF.

When the Second World War began he was one of the first to volunteer for the 2nd AIF (his army number was NX135). He raised and commanded a squadron of the 6th Divisional Cavalry Regiment in the first desert campaign and in Syria. He was awarded the DSO for actions at Bardia where he led his squadron in a frontal attack over the escarpment against Italian positions (see LIBYAN CAMPAIGN). On another occasion he was observed attacking an Italian tank single-handed with a Boyes anti-tank rifle, 'crawling all over the tank … trying to find a place to blow a hole in it'.

After commanding his regiment from June 1941 he returned to Australia in early 1942 and was sent to the 1st Australian Armoured Division, then being formed and equipped, as a brigade commander. He remained in command of armoured brigades for the rest of the war, and was responsible for a great deal of the experimentation and adaptation of armour to tropical conditions. He also pioneered many of the armoured tactics used against the Japanese. He contested federal elections in 1943 and 1946 on behalf of the Liberal Democratic Party, and came within a few hundred votes on each occasion. After the war he returned to his business interests, but remained a keen part-time soldier, being promoted to major-general in 1955. Between 1958 and 1960 he was the CMF Member of the Military Board (q.v.). He was knighted in 1964.

MacArthur-Onslow was widely regarded as one of the very best of the young CMF officers produced by the war. He thought deeply about the problems of armoured warfare, and many of his ideas on the operations of armour in the defence of mainland Australia would still find favour today. A dashing, brave and enthusiastic officer, his slightly daredevil behaviour concealed a keen intellect and a puritanical streak where personal behaviour was concerned.

McBRIDE, Philip Albert Martin (18 June 1892–14 July 1982). McBride held the federal seat of Grey in South Australia for the United Australia Party between 1931 and 1937, when he retired from the House of Representatives to enter the senate. He was a member of the War Cabinet between 1939 and 1940, was briefly minister for the Army for two months in the latter year, and became minister for Munitions in 1941 until leaving office when Labor came to power. He was defeated for his senate seat in 1943, but was returned to the House of Representatives for the Liberal Party in the seat of Wakefield in the 1946 elections. Menzies, to whom he was close both personally and politically, made him minister for

the Interior in 1949, and he held the Defence portfolio from 1950 and 1958, when he was again defeated in the elections. Between 1960 and 1965 he was Federal President of the Liberal Party, and at one stage in his career was spoken of as a possible successor to Menzies.

McCARTHY, Dudley (24 July 1911–3 October 1987). From 1933 to 1935 McCarthy was a patrol officer in the Morobe and Sepik areas of New Guinea. He returned to Australia and earned his living as a freelance writer and high school history teacher. When war broke out he enlisted as a private soldier in the 2/17th Battalion, and went with them to the Mediterranean. He was commissioned in July 1940, and served as the battalion intelligence officer through the siege of Tobruk (q.v.). He returned to Australia on the headquarters staff of the 6th Division, and from there went to Headquarters Northern Territory Force before becoming brigade major of the 23rd Brigade in March 1943. This was followed by staff jobs at Land Headquarters in 1943–44, Headquarters II Australian Corps, and Headquarters New Guinea Force in 1944–45 in the rank of temporary major.

After the war he served on the Universities Commission, and in May 1952 was appointed to the Department of Territories. He was assistant secretary of the department between 1958 and 1963, which carried with it responsibilities as Australian senior commissioner on the South Pacific Commission from 1959 and 1962 and Australian special representative to the UN Trusteeship Council in 1961-62. He then moved to the Department of External Affairs, and between 1963 and 1966 was Australian minister to the United Nations. His final diplomatic appointments were as ambassador to Mexico from 1967 and 1972 and to Spain between 1972 and 1976. He was later chairman of the Australian Film Board of Review.

The publication of an article in the Sydney *Bulletin*, describing the Australian retreat to Tobruk in March–April 1941 in which he was a participant, led to the invitation from Gavin Long (q.v.) in January 1946 to write a volume in the army series of the *Official History of Australia in the War of 1939–45*. Volume 5 in the series, entitled *South-West Pacific Area — First Year: Kokoda to Wau* (1959), dealt with the desperate fighting in the Papuan campaign (see NEW GUINEA CAMPAIGN) during which the Japanese had advanced across the Owen Stanley Range to threaten Port Moresby, and culminated in an account of the bitter combat which eradicated the Japanese beach-heads on the northern coast. It benefited greatly from the author's personal knowledge of the country, and from his understanding of the way the army worked at that time. It is characterised as well by detailed treatment of the Japanese side, and by its author's ready acknowledgement of the skill and prowess of the enemy.

McCarthy contrived to combine a diplomatic career with his writing, and published in addition a novel, *The Fate of O'Loughlin* (1970), and, in retirement, a biography of the First World War official historian C. E. W. Bean (q.v.), *Gallipoli to the Somme: The Story of C. E. W. Bean* (1983). The latter is disappointing, confining itself to a narrative account of Bean's war experiences and stopping well short of any analysis of the official history or even an account of its writing. The author denied having written a 'warts and all' biography on the grounds that 'Bean had no warts'.

McCAULEY, Air Marshal John Patrick Joseph (18 March 1899–3 February 1989) graduated from RMC Duntroon in 1919 and served in the army until 1923, including a year on exchange in Britain. In 1924 he transferred to the RAAF, training as a pilot. He went back to Britain in 1926, attending various courses including the RN College, Greenwich, graduating with an 'A' pass. He returned to Australia in 1928 and a position at Air Force Headquarters. Needing something more to keep his mind active, McCauley became a part-time student at the University of Melbourne in the following year, completing a Bachelor of Commerce (with interruptions) in 1936. He attended the RAF Staff College, Andover, in 1933, and followed this by graduating from the demanding course at the Central Flying School at Wittering. He then spent 1934 on attachment to the Air Ministry before again returning to Australia, as Director of Training, a position he held until 1938. With the outbreak of war in 1939, McCauley was initially allocated to training duties, organising No. 1 Engineering School at Melbourne Showgrounds; it functioned right through the war, graduating more than 15 000 airmen through its courses. Like other regular air force officers, McCauley found himself kept back in Australia to provide the necessary training and administrative experience for a rapidly expanding wartime force, and did not get overseas until June 1941, when he was given command of RAAF Base Sembawang, on Singapore Island. He commanded four squadrons of obsolete aircraft (Buffaloes and Hudsons [qq.v.]) largely crewed by recently arrived airmen with little experience of tropical flying. In December 1941 McCauley moved his squadrons up to their war stations in peninsular Malaya, and it was Hudsons of No. 1 Squadron that first located large concentrations of Japanese transports and escorts off the coast of

southern Indochina on 6 December. Outclassed and outnumbered, by January McCauley gathered what was left of his aircraft and flight crews and prepared to withdraw to Sumatra. He was appointed to command a RAF station near Palembang, and RAAF aircraft under his command again gave first warning of the Japanese invasion force as it headed towards Sumatra on 13 February. With some Hurricanes at his disposal, the air forces were able to put up a creditable performance against their opposite numbers, but isolated action by aircraft alone was insufficient to achieve more than temporary delays, and McCauley again withdrew, this time to Java. He and his aircrews were evacuated to Australia at the end of February. After a period as senior air staff officer with Headquarters North-Western Area, providing experienced guidance to the newly arrived 49th Fighter Group of the US Army Air Force, McCauley was appointed Assistant CAS, and later Deputy CAS, in May 1942. In November 1944 he again found himself on exchange duty with the RAF, as Air Commodore Operations with the 2nd Tactical Air Force operating from Belgium. He remained there until the end of the European war. He returned as Deputy CAS between 1946 and 1947, and then went to Japan as chief of staff to Robertson (q.v.) at British Commonwealth Occupation Force Headquarters until 1949. Appointed CAS in 1954, McCauley retired in 1957, and was made KBE in 1955. As one commentator noted at the time of his appointment to head the RAAF, 'seldom has a better-equipped officer led a branch of the Australian services'. In retirement he worked for various charitable and philanthropic organisations, and between 1964 and 1974 served as federal president of the Air Force Association.

McCAY, Lieutenant-General James Whiteside (21 December 1864–1 October 1930). McCay (or M'Cay) migrated from Ireland as an infant. A man of considerable scholastic achievement, he was educated at Scotch College and the University of Melbourne, worked as a teacher in several private schools, and in 1885 bought Castlemaine Grammar School. He took out a Masters of Law in 1897 and practised as a solicitor. A solid middle-class citizen, he was commissioned into the 4th Battalion of the Victorian Rifles in 1886, and reached the rank of lieutenant-colonel in 1900. He also found time to become active politically, and was a member of the Victorian parliament between 1895 and 1899. He won the federal seat of Corinella as a Protectionist in March 1901. McCay held the Defence portfolio between August 1904 and July 1905 (a relatively brief period in keeping with the instability and short-lived nature of several of the early Commonwealth ministries), and presided over a number of important reforms. He made Hutton's (q.v.) recommendations acceptable to a majority of the parliament, ensuring that they would be enacted, and established the Council of Defence (q.v.) and the Military and Naval Boards of Administration (qq.v.), in part reactions against Hutton's autocratic style and penchant for ignoring civilian control. His electorate was abolished in 1906 and he was trounced in the electoral contest for the seat of Corio; he stood for the senate in 1910, and again was defeated soundly, results explained in part by his unappealing persona and sarcastic manner. He returned to the law, and to citizen soldiering. In December 1907 he was appointed to command the Australian Intelligence Corps (q.v.), and promoted to colonel. For all his personal unpopularity, he worked well and closely with Monash but his relations with the Military Board were bad, and got worse, and his appointment was terminated in March 1913. In August 1914 McCay was appointed Deputy Chief Censor in Melbourne, but this lasted less than a fortnight and in the middle of the month he was given command of the 2nd Infantry Brigade of the new AIF. He trained his men hard in Egypt, which paid off in the first days on Gallipoli although the brigade lost half its strength in dead and wounded before being relieved on 29 April 1915. A few days later the brigade was sent south to Helles and suffered further heavy casualties in a futile assault on Krithia. McCay had narrowly avoided being wounded on three occasions during the landing at Gallipoli; at Krithia his leg was broken by a bullet and he was evacuated to Egypt. He returned to the peninsula before full recovery, and in July was sent back to Australia after his leg gave way again. He was appointed Inspector-General of the AIF in Australia in the temporary rank of major-general, a job which can hardly have satisfied him.

He enjoyed a change of luck, however, with appointment to command the 5th Division in Egypt, through the government's insistence that Australians should command the new formations being raised there, and without having to face a medical board, which might have failed him. His division was the last to sail for France, but the first to see a significant battle, at Fromelles, in July 1916, in which it incurred more than 5500 casualties in less than 24 hours. The whole operation from the Australian viewpoint was badly planned and badly managed, and while much of the blame resided with the British corps commander, McCay must accept some measure as well. Never popular with his men, he was widely blamed by them for the disaster, as he had been after Krithia. In January 1917 he was relieved of command, probably due to ill

health and bad relations with his subordinates. He was appointed to command the Australian base depots in England, although he spent the rest of the war scheming to get himself either the overall administrative or combat command of the AIF in France, without success. He spent the rest of the war training reinforcements, recognised by the award of the KCMG and the KBE at the war's end. He returned to Australia and was demobilised in August 1919, and retired from the army in 1926 in the honorary rank of lieutenant-general. McCay was a knowledgable and intelligent soldier whose career in senior command during the war was blighted, and finally ended, by a mixture of bad luck and personality deficiencies.

MACCHI MB-326H, CAC (2-seat basic and advanced trainer). Wingspan 10.57 m; length 10.67 m; speed 800 km/h; range 1500 km; power 1 × Rolls Royce Viper 2500-pound thrust turbojet.

The Macchi was ordered in 1964 as an 'all through' jet trainer that would replace both the Winjeel (q.v.) basic trainer and the Vampire (q.v.) advanced trainer. The Macchi, as it was normally called, was designed by Aermacchi in Italy and built under licence by the Commonwealth Aircraft Corporation (q.v.). But only two 'all-through' jet courses were conducted and most pilots did their initial training on propeller aircraft before graduating to the Macchi. Because of their reliability and sharp handling, the RAAF's Roulettes aerobatic team flew Macchis from 1970 to 1989. They were replaced in the applied training role by the PC-9 (q.v.) in 1989 and are now flown by No. 25 and 76 Squadrons as fighter lead-in trainers for F/A-18 (q.v.) pilots. The RAN flew Macchis from 1970 to 1983.

McCLEMANS, Chief Officer Sheilah Mary (3 May 1909–10 June 1988) was educated at Perth Modern School and the University of Western Australia where she gained degrees in Arts and Law. Before the war she was the first woman to appear as a barrister before a supreme court anywhere in Australia, in her case the Supreme Court of Western Australia. She joined the WRANS (q.v.) in January 1943 as a writer, was selected to attend the first WRANS officer training course in February 1943, and was appointed to the staff of the Director of Naval Reserves and Mobilisation (then responsible for the WRANS) with the rank of Third Officer. McClemans was appointed the first Director WRANS in August 1944, and was promoted to the rank of Chief Officer in January 1945; she remained Director until the disbandment of her service in 1947, although the role was more advisory than executive and she had little real or direct control over the women's service. Highly capable and extraordinarily hard-working, her role as Director WRANS was personally frustrating because of the lack of support for women within the navy from the senior ranks of the service, and her attempts to maintain a postwar WRANS organisation were unsuccessful, at least in the short term. She was awarded the OBE in 1951 for her wartime services. Back in civilian life she married in 1949, and from 1969 was Executive Director of the Law Society of Western Australia.

MacDONALD, General Arthur Leslie (30 January 1919–28 January 1995). MacDonald graduated from RMC Duntroon in 1939, in time to take part in the Second World War. He held a variety of command and staff positions with the 61st Battalion of the militia, the 2/15th and 2/9th Battalions of the 2nd AIF, and on the headquarters of the 14th Infantry Brigade, and served in Australia, the Middle East and New Guinea. In 1944–45 he went to England for a period as an instructor at the Staff College, Camberley. His postwar career in many ways exemplified the opportunities that presented themselves to able regular officers, and encompassed most of the staff and command positions of importance together with several periods of active service overseas. MacDonald commanded 3RAR in Korea in 1953–54 during the last months of combat and the uneasy early months following the signing of the cease-fire, before returning to Army Headquarters as Director of Military Operations and Plans. He was senior Australian planner at SEATO (q.v.) Headquarters in Bangkok between 1957 and 1959, and then Commandant of the Jungle Training Centre at Canungra (q.v.) from 1959 to 1960, a time when the army was busy incorporating the lessons of jungle warfare acquired in Malaya during the Emergency into its doctrine. Appointments as Director of Staff Duties between 1960 and 1961, deputy commander of the 1st Division during the Pentropic (q.v.) experiment from 1963 to 1964, and in command of Papua New Guinea Command in 1965–66 during the nervous period of Confrontation (q.v.) with Indonesia, culminated in appointment as Deputy CGS in 1966, just as the army began its major deployment of forces to South Vietnam. Between 1968 and 1969, a period of optimum military activity for the 1st Australian Task Force, MacDonald was Commander, Australian Forces Vietnam. He became CGS in 1975, and was Chief of the Defence Force Staff from 1977 until his retirement in 1979. In retirement he was Colonel Commandant of the Royal Australian Regiment from 1981 to 1985, and a member of the Defence Review Committee in 1981–82. Known as 'A. L.', he possessed a fierce temper and an often irascible nature, which, combined with considerable intelligence and a capacity for hard work, made him a sometimes difficult superior.

MACHINE GUN BATTALIONS In 1914 machine-guns were attached to infantry battalions (q.v.) of the AIF on the British establishment, two (and subsequently four) to a battalion and manned by the machine gun section of one officer and 32 other ranks. On Gallipoli, to compensate for the lack of artillery support and the diminished firepower of units reduced by wounds and disease, the machine gun sections were grouped unofficially and came under the command of the brigade staff, an arrangement which lasted until the end of the campaign. When the AIF was reorganised at the beginning of 1916 the opportunity was taken to form the machine gun sections in companies, one per brigade and re-equipped with the Vickers medium machine-gun (q.v.), an effective, water-cooled weapon first introduced into the British Army in 1912 to replace the Maxim. Each company consisted of a headquarters and four sections of four guns with an establishment of 10 officers and 142 other ranks, and often with infantry attached semi-permanently as ammunition-carriers. (One of the factors which limited the tactical handling of medium machine-guns was both the weight of the gun and the weight and bulk of ammunition.) In February–March 1917 an additional company for each division was raised in England, and this brought the divisional machine gun establishment to 64 guns manned by 50 officers and 870 other ranks. The logical final step was taken in March 1918, with the four companies being brought together as machine gun battalions and controlled by the divisional commander. As a consequence, infantry were no longer attached as carriers. A barrage fired by massed machine-guns could be a highly effective means of assisting the infantry forward. In the attack by the 6th Brigade at Ville-sur-Ancre in May 1918, 33 machine-guns took part in the barrage, 'sprinkling with their fire selected back areas during the first three hours of the fight', noted Bean. Each gun fired 70 rounds per minute in the first 10 minutes of the attack, and 100 rounds per minute for the next hour and 50 minutes. Over the three-hour period the guns fired a total of 340 000 rounds, 11 750 per gun. To compensate the infantry companies for the centralising of machine-guns in this manner, the Lewis gun (q.v.) was issued from 1916. At first, these replaced the medium machine-guns on a similar establishment, one section of four guns per battalion. In the course of the war this increased to eight, then 12, and finally to 20 or more in the final year of the war. The organisation of machine-guns in the light horse (q.v.) was broadly similar. Beginning the war with a section of two guns, an officer, 24 men and two pack-horses, in June 1916 the sections were grouped into brigade machine gun squadrons, and their place at unit level taken by Lewis guns. A light horse machine gun squadron consisted of eight officers and 225 other ranks, organised into a head-quarters and six sections of two guns. In addition, each squadron disposed of 305 horses and nine mules. The 1st and 2nd Squadrons were part of the Anzac Mounted Division, the 3rd, 4th and 5th Squadrons of the Australian Mounted Division. (The 5th Light Horse Brigade was formed from the Imperial Camel Corps and its machine gun squadron was actually organised from New Zealand machine-gunners of that formation.) Machine gun battalions, like the Machine Gun Corps in the British Army, disappeared with the Armistice. In 1937 four regiments of light horse on the CMF order of battle were converted to machine gun regiments and motorised. In large part these were units which found it difficult to maintain large numbers of horses because of the motorisation of civilian transport. They were equipped with Vickers guns and a variety of trucks and vehicles. Three addi-tional machine gun regiments were raised from light horse units during the war as part of the con-version of the cavalry divisions to motor divisions. In the 2nd AIF four new machine gun battalions were raised for service with the four divisions. Because the 2/1st Machine Gun Battalion was in the convoy with the 18th Brigade which was diverted to England in 1940, the 6th Division entered its first campaign, in Libya in January the following year, supported by the 1st Battalion of the Royal Northumberland Fusiliers, a British regular army machine gun battalion which also supported the 9th Division in Tobruk. Additional battalions saw service in the south-west Pacific, the 7th in Papua and the 6th in New Guinea, but all were dis-banded at the war's end. In the postwar army the machine-gun was again devolved back to unit level.

MACKAY, Lieutenant-General Iven Giffard (7 April 1882–7 September 1966). Active in the cadets while at school at Newington College, Syd-ney, Mackay was commissioned in July 1913 into the 26th Infantry, a unit which recruited many young professional men from the inner-city Syd-ney suburbs of Surry Hills and Redfern. A keen militia soldier (see CITIZEN MILITARY FORCES), he volunteered for the AIF at the war's beginning, and was appointed adjutant of the 4th Battalion, which

recruited largely from east Sydney. A riding injury prevented his embarkation with his battalion, and he sailed for Egypt with reinforcements for the 13th Battalion in February 1915; he did not join his unit on Gallipoli until early May, and thus missed the landing.

Serving as a company commander and battalion intelligence officer, Mackay took part in the fierce fighting during the August offensive, and was twice wounded at Lone Pine. He was evacuated to Eng-land and saw no further service in the Dardanelles. Rejoining his battalion in France, he was promoted to lieutenant-colonel and commanded the unit in the battles of Pozières and Mouquet Farm in July–August 1916 (see WESTERN FRONT), at Flers in November the same year, and throughout 1917. During the second Battle of Bullecourt in May 1917 he temporarily commanded the 1st Infantry Brigade. With the grouping of battalion machine-gun companies into battalions in early 1918 (see MACHINE GUN BATTALIONS), Mackay was chosen to command the 1st Machine-Gun Battalion, formed from the companies of the 1st Division. In June he was promoted to brigadier-general and took over the 1st Brigade, which he led in the Allied advance until the AIF was pulled out of the line in October 1918.

Mackay lectured in physics at Sydney Univer-sity from 1920 to 1922, and then held a number of administrative posts. From 1933 to 1940 he was headmaster of Cranbrook School, and served as the Commonwealth Film Appeal Censor. He contin-ued with militia soldiering, and held a number of brigade commands in the 1920s and 1930s culmi-nating in command of the 2nd Division from 1937 until his appointment to the AIF in early 1940.

With the decision to raise a second division for the AIF and General Blamey's consequent elevation to command I Australian Corps, Mackay was chosen to replace him as GOC of the 6th Division. He led it in its first campaign in Libya and through the rigours of Greece, impressing his subordinates and superiors alike. The British commander in Libya, General Sir Richard O'Connor (q.v.), wrote of him that 'behind a rather diffident and shy manner [he] possessed an extremely strong and resolute character. He was full of commonsense and … carried the full confidence of myself and his division'. On first appearances Mackay had impressed his command as a stern disciplinarian, reserved, cool and rather pedantic. They nicknamed him 'Mr Chips' in recog-nition of his peacetime profession. His courage and imperturbability in the face of enemy air attack in Greece soon changed that appraisal.

In August 1941 he was recalled to Australia to become C-in-C of the Home Forces, responsible for trying to right 20 years of defence mismanagement in the face of the impending Japanese threat. In this capacity in February 1942 he wrote the appreciation which gave rise notoriously (and wrongly) to the 'Brisbane Line' affair. Mackay had decided that in the event of invasion he would concentrate his forces within the Brisbane–Melbourne area, and would not reinforce strategically peripheral areas like north Queensland or Western Australia. There was no suggestion that the forces already there would be withdrawn, nor that the areas would simply be abandoned. Nor was any line drawn. It was superseded within weeks by the knowledge that two divisions of the AIF were returning, and thereafter became, in the words of one historian, 'a plaything of American propagandists and Australian politicians'. The issue was made much of in parliament in 1943 by the Labor member for East Sydney, E. J. Ward, whose campaign to use it against R. G. Menzies (q.v.) and the conservatives was disavowed by his own leaders. It is an issue which is still given a sensationalised run in the Australian press in slow news periods. (See also 'BRISBANE LINE'.)

Following the reorganisation of early 1942 Mackay was given command of the Second Army based in Sydney. Between January and May 1943 he commanded New Guinea Force (q.v.) in the operations culminating in the capture of Wau (see NEW GUINEA CAMPAIGN). In August General Blamey again sent him to New Guinea, where he oversaw the opening stages of the battles for Sattelberg and the Ramu Valley. Disagreements with the Americans over provision of shipping for reinforcements led to delays in taking Sattelberg, and some senior Australian officers, Lieutenant-General Sir Edmund Herring and Major-General George Vasey (qq.v.) in particular, felt that Mackay was not forceful enough and disinclined to bother Blamey with his difficulties. At 61 and after several hard campaigns, Mackay probably was getting 'a bit old [and] a bit slower' in the face of the rigours of campaigning in New Guinea, and may have welcomed the recommendation of the government that he become Australia's first High Commissioner to India. He took up the post in February 1944, and retired in 1948. In retirement he accepted the post of chairman of the recruiting committee in New South Wales during the government's heightened recruiting campaign between 1950 and 1952, and was active in ex-servicemen's associations.

Ivan Chapman, *Iven G. Mackay: Citizen and Soldier* (Melway Publishing, Melbourne, 1975).

MACKENZIE, Seaforth Simpson (9 August 1883–20 October 1955). A New Zealander by birth, Mackenzie practised as a barrister and solicitor before moving to Melbourne and spending a period as editor of a literary magazine. In February 1914 he joined the Attorney-General's department as a clerk. He was commissioned into the Australian Naval and Military Expeditionary Force in March 1915 and sent to the former German New Guinea (q.v.) as deputy judge advocate general and principal legal adviser to the Administrator, the former Secretary of the Defence Department, Colonel Samuel Pethebridge (q.v.). He favoured more lenient treatment of the remaining German settlers and planters, with which his superiors did not agree, but did sufficiently well in his appointment to be named acting Administrator upon Pethebridge's departure in November 1917. He left New Guinea in January 1921 to return to the Attorney-General's Department, and in June the following year was made principal registrar of the High Court of Australia. C. E. W. Bean (q.v.) selected him to write the volume of the official history of the First World War dealing with the conquest and administration of Germany's New Guinea territory, and *The Australians at Rabaul* appeared as volume 10 in the series in 1927, although Bean complained about the slowness with which he went about his task. His delay may have been due to financial preoccupations, because his property investments in New Guinea soured and by 1932 he owed over £26 000 on three plantations. He was tried for forging and uttering in 1936, convicted, and jailed for four years. He was released in October 1940.

McKENZIE, William (20 December 1869–26 July 1947). His family migrated to Queensland from Scotland when McKenzie was 15, and farmed near Bundaberg. In 1887 he attended a Salvation Army meeting, heard the call, and was commissioned into its ranks two years later. McKenzie applied for a chaplain's position as soon as the war broke out in 1914, and was selected, with initial attachment to the 4th Battalion. Unlike many of the chaplains who attended to the AIF's pastoral and spiritual needs, he mixed widely with the men, took part in their sports (he was a ferocious boxer) and devoted himself in Egypt to bettering their material conditions and providing for their off-duty hours. He earned the nickname 'Fighting Mac' for his willingness, even eagerness, to be at the front with his charges. McKenzie was one of the first chaplains ashore at Gallipoli, and carried stretchers, tendered the disabled and buried the dead for months. In France he was present at all the AIF's big battles — Pozières, Bullecourt, Mouquet Farm and the rest — throughout the course of 1916–17. A physically imposing man, the strain told and his health was undermined, leading to his repatriation at the end of 1917. He was awarded an MC in June 1916, and received a succession of hero's welcomes on his return to Australia. In the late 1920s he undertook Salvationist work in China, and on returning to Australia was promoted within the organisation. McKenzie was awarded the OBE in 1935, and retired from active involvement in 1939. Dubbed by some 'the most famous man in the AIF', he was a regular participant in Sydney Anzac Day marches in the 1930s and 1940s.

Michael McKernan, *Padre: Australian Chaplains in Gallipoli and France* (Allen & Unwin, Sydney, 1986).

MacLAURIN, Brigadier-General Henry Normand (31 October 1878–27 April 1915) was a barrister and part-time soldier, and a member of a distinguished Sydney family. He had been promoted to lieutenant-colonel in July 1913, and at the outbreak of the war was commanding the 26th Infantry Regiment. He joined the AIF in mid-August 1914, and was given command of the 1st Infantry Brigade with the rank of colonel. The official historian, C. E. W. Bean, described him as 'a man of lofty ideals, direct, determined, with a certain inherited Scots dourness … an educated man of action of the finest type that the Australian Universities produce'. At the landing at Gallipoli MacLaurin's brigade comprised Bridges's sole reserve during the chaotic fighting of the first few days. Bean notes that he showed himself a brave and energetic commander, but like most Australian officers at this stage of the war, he still had a lot to learn about the conditions of modern war. While observing the fighting from a position south of Steele's Post, behind the ridge which was to be named after him, he was shot from behind by a Turkish sniper and killed. His brigade major had been sniped from the same position only 10 minutes previously. He was buried at Beach Cemetery.

McNAMARA, Air Chief Marshal Neville Patrick (17 April 1923–   ). A wartime entrant into the RAAF, McNamara enlisted as an aircrew trainee in October 1941 and graduated as a pilot. He flew fighters with No. 75 Squadron in the South-West Pacific Area during the Second World War and, after a period as a flight instructor with the Central Flying School, flew again operationally with No. 77 Squadron in Korea in 1953.

His postwar career was marked by a succession of increasingly responsible and senior appointments, both staff and command. In addition to commanding No. 25 Squadron and No. 2 Operational Conversion Unit between 1957 and 1961, from 1966 to 1967 he was the officer commanding the RAAF Contingent Thailand, based at Ubon as part of a SEATO commitment. On the staff side he was Director of Personnel (Officers) from 1964 to 1966 and before that Senior Air Staff Officer on the RAAF Staff in London between 1961 and 1964. His highest active service command was as commander of RAAF Forces, Vietnam, and deputy commander of the Australian Force Vietnam in 1971–72, towards the end of the Australian commitment (see VIETNAM WAR). Following Vietnam service he was Air Attaché in the Australian Embassy in Washington DC, becoming deputy CAS in 1975.

In 1979 McNamara became CAS. The late 1970s to early 1980s was a period of considerable flux in Australian strategic thinking, and during his tenure of office the RAAF developed the first significant operational concept for the air defence of Australia, largely written by the then Chief of Operations, Air Vice-Marshal S. D. Evans. The antilodgement strategy was the beginning of a process of formulation of Australian air-power doctrine which culminated in the *Air Power Manual* of 1990. McNamara also oversaw the selection of the American F/A-18 Hornet as the next generation of front-line fighter aircraft for the RAAF. He was appointed KBE in 1981, and in 1982 he became Chief of the Defence Force Staff, only the second RAAF officer to hold this position. He retired in April 1984.

McNAMARA, Air Vice-Marshal Francis 'Frank' Hubert (4 April 1894–2 November 1961) trained as a teacher and had joined the senior cadets while still at school himself. In 1913 he was commissioned in the 46th Infantry Battalion (Brighton Rifles), and as a militia officer was mobilised in 1914 for duty within Australia. He was an instructor at the AIF Training Depot at Broadmeadows in Victoria, and in August 1915 was selected for pilot training at Point Cook. McNamara was sent to No. 1 Squadron as adjutant; the unit went to the Middle East but he went to England and flew with No. 42 Squadron RFC before rejoining the Australian Flying Corps in the second half of 1916. It was while serving again with No. 1 Squadron in the Middle East in March 1917 that he became the first Australian airman to win the VC while attempting to rescue a colleague whose aircraft had been downed by Turkish ground-fire. McNamara attempted to take off with the second pilot hanging from the wing struts of his single-seater aircraft, but the weight proved too great and the aircraft crashed. They managed to get the downed aircraft started, and despite wounds and loss of blood McNamara piloted them to safety, a distance of 122 kilometres. He was promoted to captain but was invalided home to Australia and demobilised in early 1918. He transferred to the new RAAF when it was formed in March 1921. For the next few years he held a succession of staff and instructional posts, and attended the Imperial Defence College (q.v.) in 1937. He was on the staff at Australia House when the war broke out in 1939, and in 1942 was appointed AOC, RAAF, in Britain with the rank of air vice-marshal. Between 1942 and 1945 he served on attachment to the RAF in Aden. He retired from the air force in 1946, and was a member of the National Coal Board in London between 1947 and 1959. He died in Britain.

McNICOLL, Vice-Admiral Allan Wedel Ramsay

(3 April 1908–11 October 1987) graduated from RANC Jervis Bay in 1925, and served in Britain on HMS *Repulse*. He specialised as a torpedo officer, undertaking the long course in Britain in 1933–34. He was in Britain at the outbreak of war in 1939 and served on HMS *Fiji* and HMS *Medway* with the Mediterranean Fleet. He was awarded the GM and a C-in-C's Commendation for gallantry in July 1941, for removing the inertia pistols and thus disarming eight Italian torpedoes in a captured submarine. In the course of 1942–43 he was squadron torpedo officer on HMS *King George V*, before spending a year at the Admiralty. McNicoll returned to Australia at the end of 1944, and joined HMAS *Hobart* shortly before the war's end. He then spent a period at Navy Headquarters, as Director of Plans between 1948 and 1949, before further sea postings including command of HMAS *Warramunga*. In 1952 he was chairman of the planning committee for the British nuclear testing program in the Monte Bello islands. He was captain of HMAS *Australia* between 1952 and 1954, attended the Imperial Defence College (q.v.) in 1955, and was head of the Australian Joint Service Staff in 1957–58. Appointed Flag Officer Commanding Her Majesty's Australian Fleet between 1962 and 1964, he became CNS in 1965. During his period as CNS the RAN adopted the Australian White Ensign (q.v.) in place of the Royal White Ensign, dominated by the Cross of St George. He also oversaw an extensive modernisation of the fleet, although he had to cope with the turmoil occasioned by the investigations into the *Voyager* (q.v.) disaster. McNicoll retired in 1968 and was Ambassador to Turkey between 1968 and 1973. He had published his first volume of poetry, *Sea Voices*, in 1932, and in retirement edited a selection from the odes of Horace, published in 1979. His brother Ronald was a senior Royal Australian Engineers officer in the Australian Army, who subsequently wrote three volumes of the history of his corps, while another brother,

David, became a distinguished journalist, war correspondent, and poet.

MACQUARIE, HMAS see RIVER CLASS FRIGATES

MADANG, HMAS see ATTACK CLASS PATROL BOATS

MAGIC was the code name for intelligence intercepts of American origin based on the reading of enemy codes and ciphers. The term came from references to the code-breakers in the American Secret Intelligence Service as 'magicians' by the Chief Signal Officer, Brigadier-General Joseph O. Mauborgne. The term referred only to decryptions of Japanese diplomatic traffic which was encrypted on the PURPLE machine. There is some evidence that Australian ministers were made aware of both MAGIC and ULTRA (q.v.) sources before the beginning of the war against Japan. During the Pacific War there was a considerable Australian signals intelligence effort directed against Japanese forces.

Edward J. Drea, *MacArthur's ULTRA: Codebreaking and the War Against Japan, 1942–1945* (University Press of Kansas, Lawrence, 1992).

MALARIA see SCIENCE AND TECHNOLOGY

MALAYAN CAMPAIGN, which grew out of a muddled British policy to protect its naval base at Singapore, is among the shortest, most spectacular campaigns of military history. For the Japanese attackers it was a triumph of rational planning, considered preparation and brilliant execution. In 70 days, a Japanese army of three divisions advanced the length of Malaya against a larger body, and forced the capitulation of a defended island that was thought by many to be impregnable. For the then defenders of the British Empire, disunited under a lacklustre command, it was Britain's greatest humiliation since Cornwallis surrendered at Yorktown in 1781. For Australia, the entire experience, coming as it did less than a year after the disastrous operations in Greece and Crete, represented the country's final disillusionment with Churchillian strategy.

The Singapore base

In retrospect, it is clear that too much was expected of Singapore as a base. What had started as a British Admiralty initiative to defend a key nodal point came to be, particularly in Australian and New Zealand eyes, the centrepiece of a whole Singapore strategy (q.v.). Almost from the start, the proposal to build the base was plagued by imprecise aims and vacillating policy. A site was selected early in 1923, and work begun, only to be terminated; then progress was slowed as officials argued whether guns or aeroplanes should be its principal defence. The same process was repeated later for different reasons with other issues.

Virtually from the start, the Australian government had greater confidence in the Singapore strategy than its military advisers. The RAN advocated a more secure base at Sydney. The Senior Officers' Conference convened in 1920 by Defence Minister George Pearce (q.v.) recommended a balanced policy based on a significant contribution to a Far Eastern Fleet while also maintaining a strong army based predominantly on a citizen force that could mobilise fairly quickly to prevent an aggressor from obtaining a decision on shore.

But other happenings served to accelerate a blue-water strategy based on Singapore. Britain terminated the Anglo–Japanese Naval Treaty in 1921 against the wishes of Australia and New Zealand, which had both seen it as guaranteeing stability in the Pacific (see JAPANESE THREAT). The Washington Conference of the same year, as well as showing dismay among the British dominions over not being consulted during the Chanak crisis of 1922 (in which the dominions refused to back a unilateral British ultimatum to Turkey over the control of the Dardanelles), created a level of uncertainty that led some members of the Empire to call for collective security measures. At the Imperial Conference of 1923, Australia's Prime Minister Stanley Bruce was won over to the Singapore concept by specious British Admiralty argument that has since led one Australian historian to comment:

> In one of the most unfortunate statements of prime ministerial judgment ever, Bruce observed at the Conference that while he 'was not quite as clear as I should like to be as to how the protection of Singapore is to be assured, I am clear on this point, that apparently it can be done'.

As against Bruce's statement, the Australian Inspector-General, Lieutenant-General Sir Harry Chauvel (q.v.), was later to write presciently, in his *Annual Report* of 1928:

> The traditional policy of the most dangerous potential enemy in Eastern waters is to commence hostilities without warning and to attack the foundations of their opponent's sea power from the start … While the ration of capital ships stands at 5:5:3 it would appear to be impossible to contemplate a state of affairs which would permit the peacetime concentration of naval forces at Singapore capable of meeting the Japanese fleet. It is therefore open to serious question, bearing in mind Japan's traditional policy, whether it might not be quite feasible for Japan to attack and destroy the Singapore Base within the six

weeks envisaged as the minimum within which the Royal Navy could make its power effectively felt in the Pacific.

Contrary to later criticisms, there had always been some consideration of the possibility of a landward attack on the base. But until the late 1930s it had always been occluded by other factors, most of all the long sea passage which any invading force would have to make from Japan, which would have permitted its detection and interception. However, as Britain's straitened economic circumstances and wide strategic commitments caused the interval it would take a fleet to reach and operate from Singapore to extend from 42 days, to 70 days, to 90 days, and finally to six months, the likelihood that the putative fleet could arrive in time to meet and defeat its Japanese opponent was receding. Conversely, it was becoming equally apparent that Japan could land a force in Thailand or Malaya to take the Singapore base from the rear.

From 1931 onwards, when Japan occupied Manchuria and began forays into China, possibilities of this nature began to be recognised and, at the 1937 Imperial Conference, were voiced formally by Australia and New Zealand. Reassessments followed, and led eventually to an acknowledgement that to defend Singapore it would be necessary to defend the whole of Malaya. How to do so became the question which from then on occupied officials both in Singapore and London until the Japanese attack. The assessments made in Singapore were vexed by inter-service rivalry and by the fact that Malaya's tin and rubber earned dollars that would have to be forgone if the country was to be put on an imminent war footing. In London, particularly after Dunkirk and the fall of France, the Far Eastern problem ran a poor third behind the requirements of the Middle East and, from June 1941, of the need to keep the Soviet Union in the war.

To bring some coherence to its planning in the region, Britain, almost at the eleventh hour, appointed a C-in-C Far East, Air Chief Marshal Sir Robert Brooke-Popham, on 17 October 1940, whose powers to effect coordination were severely constrained, but whose appointment reflected the fact that the future defence of Malaya, in the realistic absence of a main fleet, was to be based principally on air power. The problem then became how to find the front-line aeroplanes that could intercept and destroy a Japanese invasion fleet, or destroy any landings it made. The planners believed that between 336 and 566 were needed. Since nothing like this number could possibly be sent, a provision was made that more troops would be provided in the meantime, the garrison to be reduced progressively as the aircraft arrived. More troops did arrive, including the 8th Australian Division (initially less two brigades), but most of the units from India were under-trained and under-equipped, and had been 'milked' of a high percentage of their officers and NCOs in order to expand the Indian Army. Thus when Lieutenant-General A. E. Percival became GOC Malaya in May 1941, he inherited a complex series of problems which, brilliant staff officer though he was, he was unable to overcome in the relatively short and turbulent time before the Japanese landed. Japanese planning and preparation for the Malayan campaign were, in contrast, both shorter and infinitely more effective.

Japanese preparatory directives for a southward drive went out from Imperial General Headquarters in October 1940. The three divisions subsequently employed in Malaya, the 5th Division, Imperial Guards and 18th Division, were all ordered to train for landing operations in tropical areas. Much of this had to be simulated. The Taiwan Army Research Department, hurriedly established in January 1941, carried out experimental landing operations off the Kagoshima coast, and in great secrecy in South China. Through hard work at a frantic tempo it conceived a practical method of landing over open beaches, which was then communicated to the Japanese Army. A sixty-page pamphlet, *Read This Alone—And The War Can Be Won*, was as good a primer for troops about to embark on complex operations as can be found in any language. In contrast, life in British Malaya and Singapore moved at its leisurely prewar social pace right up to the first air attacks on the country.

Japanese and British plans

The invasion was to be carried out by 25th Japanese Army under General Tomoyuki Yamashita, who held a final conference for the attack on Malaya and Singapore at Saigon on 15 November 1941. As part of a comprehensive timetable, he was given 100 days to take Singapore. His plan was for the 5th Division to land two regiments (brigade equivalents) in Thailand at Singora, and one regiment at Patani, move swiftly southwards to occupy Perlis and Kedah in Malaya and then advance down the west coast using 'a combination of expensive British roads and cheap Japanese bicycles — every person not riding in a vehicle will ride a bicycle'. The 56th Regiment of the 18th Division was to land at Kota Bharu, seize the airfields there and at Gong Kedah and Machang, and then advance southwards along the east coast to Kuantan. It was to be followed up by a second regiment. The task of the Imperial Guards Division was first to capture important points in Thailand. It was then to move

down the Kra Isthmus by rail, and to concentrate in the rear of the 5th Division in the Taiping area. A fourth division, the 56th, was to stand by in Japan, to be called forward if needed. It was not used by Yamashita. Air support for the 25th Army was provided by the 3rd Air Division operating from Phu Quoc island west of the Mekong Delta. Close naval cover for the convoy was provided by the 7th Cruiser Squadron. Yamashita's force sailed from Hainan Island on 4 December 1941. His army numbered 60 000 troops, 400 guns and mortars and 80 tanks. On the east coast his supply echelons were horse-drawn. In air support he could rely on 459 army and 158 navy aircraft.

Against this relatively modest but well-integrated Japanese force, Brooke-Popham's command, while imposing on paper, suffered from the worst aspects of an ad hoc strategy. The British plan, as outlined earlier, was to base the defence on aircraft which were to operate from forward airfields protected by the army. If Japanese forces threatened to land in Thailand, Plan 'Matador' would pre-empt the landing by seizing the landing beach at Singora. However, since the force would not be strong enough to also seize Patani, a separate group, 'Krohcol', would protect 'Matador's' right flank by seizing a position called 'The Ledge' across the Thai border. Brooke-Popham's ground commander, Lieutenant-General Percival, was constrained in his dispositions by the need to protect airfields in northern Malaya, particularly at Alor Star, Sungei Patani and Kota Bharu on opposite sides of the central spine of mountains. He had also to protect airfields at Gong Kedah, Kuantan and Kluang on the east coast, at several places in Johore, at Butterworth, Ipoh and Port Swettenham in the west, as well as on Singapore Island. In short, this faulty strategy forced Percival to adopt a wide area defence without the corresponding ability to deploy mobile forces to counter penetrations. Percival had a total of 31 battalions but no tanks. He had several major formations: III Indian Corps, commanded by Lieutenant-General Heath, consisted of two divisions, the 9th and 11th Indian, and was responsible for the defence of northern Malaya, with the 28th Brigade in reserve; his second major formation, the 8th Australian Division, commanded by Major-General Gordon Bennett (q.v.), consisted of two brigades, the 22nd and 27th, stationed in Johore and Malacca. Singapore Fortress, commanded by Major-General Simmons, included troops equivalent to a further division. As a command reserve, Percival had the 12th Indian Brigade located at Port Dickson.

In the air, where it should have been strong Malaya was weak. Instead of the needed 336–566 operational aircraft, when Japan attacked there were 158 aircraft with 88 in reserve, including the RAAF's No. 1 and 8 Squadrons flying Lockheed Hudsons (q.v.) and No. 21 and 453 Squadrons flying Brewster Buffaloes (q.v.). None of these aircraft could match the Japanese Navy's Mitsubishi A6M2 'Zero' in aerial combat. Nor was there an air-warning system for Malaya, although one existed for Singapore Island. Since the chiefs of staff had accepted the various assessments, why did the results fall so far short of the requirement? As mentioned, Malaya had low priority. But, even more importantly, the British government hoped Japan would not attack, and strove to prevent it by appeasing the Japanese. The forces thus left over for Malaya were meant by Prime Minister Churchill to be a deterrent, not a coherent defence. They were more than token, but they were far short of realistic. The results were quickly apparent.

The invasion

Unless it could have been rapidly reinforced, particularly by strong sea and air forces, the fate of Malaya — and also of Singapore — was decided within four days of the initial Japanese landing at Kota Bharu at 1.45 a.m. on 8 December 1941. The landings there, and at Singora and Patani, were all successful. Brooke-Popham did not order 'Matador' and, through a breakdown in personal communications between him and Percival, 'Krohcol' was launched too late, and was too slow. Using a series of well-trained advance guards built around infantry battalions and including tanks and engineers, the Japanese kept up a high rate of advance on both approaches through Thailand into northern Malaya. Their advance on the east coast was slower, but still effectively tied up one of the two British Indian divisions in northern Malaya.

There is a fallacy that the Japanese forces used in Malaya were greatly experienced in jungle warfare. They were not. Their advantage in experience lay in the fact that most of their formations had been battle-tested in China and Manchuria. True, they had comprehensively studied tropical warfare, but that is not necessarily the same thing. In any case, they used jungle to screen movements only when they needed to. When they struck opposition they did two things. On the main avenue they tried hard to break through directly using tanks and infantry, often successfully. The British Indian linear style of defence assisted them. Simultaneously and as a drill they launched two infantry envelopments, one shallow, one deep, to get behind the enemy and set up roadblocks. On almost every occasion this forced their opposition to react by withdrawing prematurely. Since they had forces on both sides of the central spine of Malaya, and also on more than one route, this meant that a withdrawal by one group of defenders forced the others to conform or risk being cut off. Moreover, almost immediately they gained virtually unimpeded control of the sea and air.

On 10 December 1941, *Prince of Wales* and *Repulse*, Britain's only capital ships in the area, were sunk. By 12–13 December, not only had the Japanese broken through 11 Indian Division's partially prepared defensive position at Jitra, they had forced 'Krohcol' west of Kroh, destroyed half the operational aircraft in northern Malaya and forced the abandonment of the key airfield. The defenders had lost the initiative, which they never regained. From here until they closed up to the Johore Strait, the Japanese, either by landing troops behind the defence, or threatening to do so, controlled their own rate of advance and effectively dictated the course of the campaign.

The defence was not helped by frequent changes of command at almost every level. At the top, Brooke-Popham was replaced by Lieutenant-General Sir Henry Pownall on 23 December 1941. Pownall in turn was replaced almost immediately by General Sir Archibald Wavell (q.v.) when the latter became commander of the ABDA Area (q.v.). Wavell decided to make a clean break in northern Malaya and withdrew all the way to Johore. He there — over Percival's head — entrusted the defence to Bennett whom, although he had not been tested, Wavell believed showed greater fighting spirit than either Percival or Lieutenant-General Heath of III Corps. In the event, Wavell's confidence in Bennett was misplaced. Bennett, who in this campaign became the most controversial senior commander in Australian history by abandoning his troops, fought an operation flawed from the start by disposing a practically untrained Indian brigade with only one battery of artillery along 38 kilometres of river with the further direction that two companies of each forward battalion had to be on the north bank. Importantly, this brigade, thinly spread, was astride the coast road, protecting the coastal approach. Yamashita exploited this weakness by attacking it in front by the Imperial Guards, and flanking it from the sea by infiltrating a battalion on to the defenders' rearward communications. As a consequence, the defence of northern Johore collapsed. Despite some brilliant tactical efforts, such as 2/30th Australian Battalion's ambush at Gemas, a fighting stand at Bakri by two Australian battalions

---

Map 19 (Top) The Japanese advance during the Malayan campaign, 1941–42. (Bottom) Japanese operations leading to the fall of Singapore, February 1942.

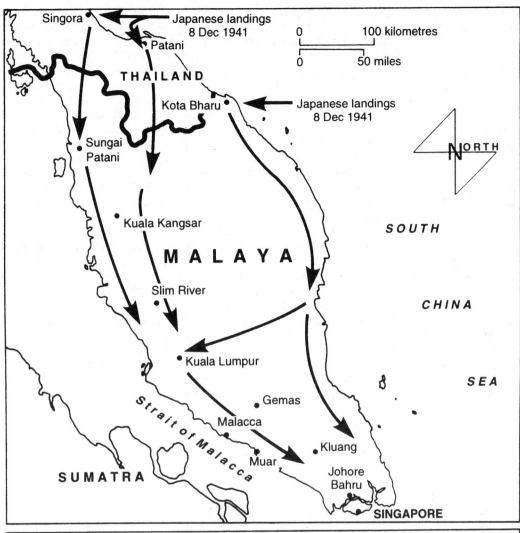

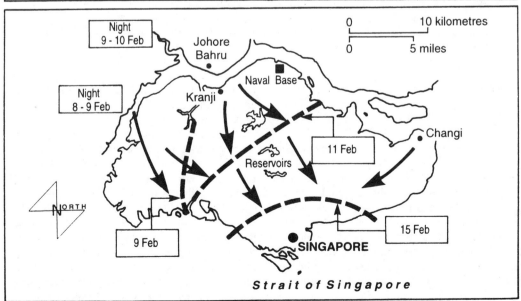

and an artillery battery, the 22nd Australian Brigade's ambush north of Jemaluang, and the action at Muar where Lieutenant-Colonel Charles Anderson (q.v.) won the VC, Japanese progress was impeded only temporarily. By 31 January 1942, the last troops had withdrawn to Singapore Island and the connecting Causeway had been cratered.

The battle for Singapore Island

Even in the preceding critical weeks and months, Singapore Island had not been developed to resist a landward attack across the Johore Strait. And, in addition the British Empire forces had lost heavily during their withdrawal down the length of Malaya. Percival, whose name is most often associated with the ultimate surrender, has in that sense been treated unfairly, because he was never in overall command. However, his plan for the ground defence of Singapore was maladroit and deserves brief discussion.

Wavell had suggested that the most likely area of Japanese attack would be the north-west, where the Johore Strait was narrowest and a series of river mouths on the north bank provided cover for launching amphibious craft. He suggested that the fresh British 18th Division be put there and the 8th Australian Division be put in the next likely area, the north-east. Instead, Percival put the Australians, most of whose units had been reduced to half-strength, in the north-west, with both brigades forward in the 'shop-window'. Thus once the battle started it was already committed. As the British official historian put it:

> By trying to defend the whole coast when it was obvious the Japanese would concentrate on one carefully selected point, Percival was weak everywhere, no formation had any reserve for immediate counter-attack, the command reserve was too small to be of any value, no proper plans had been made to prepare the two switch lines [defensive positions in depth at Jurong and Serangoon] for defence and the vital naval base — Woodlands–Kranji — area was defended by two different formations … Once they had effected a landing the Japanese were thus given the opportunity of driving deep into the vital central area of the island before a reserve could be collected to check them.

In the air, although Singapore had been reinforced by more than 50 Hurricane fighters, the four airfields on the island were under constant attack, and the defenders outnumbered by more than four to one. Yamashita's reconnaissance of the Singapore defences by aircraft (and by a tethered balloon at Johore Bahru) was excellent and was virtually unopposed. Percival, in contrast, was blind. Even a few stay-behind parties with radio sets could have told him where the Japanese were concentrating, but he did not use them.

Preceded by a heavy bombardment which effectively cut the defenders' communication lines, and by a preliminary operation which sought to divert attention by capturing Puau Ubin Island in the north-east, the Japanese put 13 battalions across the Johore Strait into the 22nd Australian Brigade's sector at 10.30 p.m. on the evening of 8 February 1942. Yamashita gave the 5th and 18th Divisions Tengah airfield as their first objective and a north–south line through Bukit Panjang as their final objective. His third division, the Imperial Guards, was to attack a night later in the Causeway–Kranji sector, seize Mandai, then move eastwards along the Mandai road to Nee Soon and finally southwards to prevent the British garrison from falling back to Changi.

Many of the Australian-defended posts took a heavy toll of the Japanese, and the artillery sank some barges, but during the night of 8–9 February they were forced back towards Ama Keng and Tengah. The flaws in Percival's design for battle immediately became obvious, and the lack of a real ability to influence effectively any stage of the ensuing action was apparent from here on. His hesitancy to commit reserves from unthreatened parts of the island was now compounded by two other happenings. Bennett prematurely released to his subordinates a forecasted order for withdrawal from Percival, which resulted in the loss of the Jurong Line, when it could still have been held. Moreover, when he realised what had occurred he did not order Brigadier Taylor's 22nd Brigade to immediately counter-attack and regain it. When he himself was later so ordered, and attempted to do so, the Japanese had been reinforced with tanks and the moment had passed. Again Bennett's failure to get forward and see what was happening in the 27th Australian Brigade's sector in the Kranji area resulted in a mistake in the defence against the Imperial Guards who, although shaken by the demolition of oil storage tanks which set fire to some of their assault craft crossing the strait, quickly rallied and took the area.

As key points, including all but one airfield and the water reservoirs, were lost, it was a matter of time before the island succumbed. When Wavell left on 11 February 1942, he ordered the remaining aircraft to withdraw to the Netherlands East Indies. And although Churchill from London had earlier decreed that the fight would go on 'among the ruins of Singapore city', he recoiled from this and, through Wavell, Percival was advised that he should surrender when satisfied that resistance could no longer be effective. After meeting with Yamashita at the Ford motor factory north of Bukit Timah village, Percival ordered his troops to cease hostilities at 8.30 p.m. on 15 February 1942.

In a classified report until then closely held by the British government and not released until 1992, it was alleged to Wavell that indiscipline among Australian troops contributed directly to the loss of Singapore. The report has little foundation. There *was* indiscipline among troops in the last days of Singapore, particularly among poorly trained, or untrained, reinforcements from Britain, India and Australia, but while detrimental to public order and morale, especially civilian morale, it was irrelevant to considerations of the loss of a 'fortress', the defence of which had been actively, if inconsistently, pursued for almost two decades. The Australian troops of the 8th Division fought well and were among the few who earned the respect of the Japanese, and while they constituted only 14 per cent of the British Empire's ground forces, they sustained 73 per cent of its battle deaths.

The reasons for the loss of Malaya and Singapore belong to British rather than Australian history. Successive Australian governments had acquiesced in a defence policy that sought to buttress the imperial position with minimal expenditure, and to that extent Australian policy-makers must share some of the blame for the Singapore débâcle. The greater responsibility, however, lay with Britain, which for so long had enjoyed an empire 'on the cheap' in Asia. In Malaya and Singapore, as in Hong Kong and Burma, its bluff was called by Japan, whose forces it had consistently under-estimated. In neither planning, preparation nor execution did the British command in Malaya match the Japanese. The latter had advanced more than 1000 kilometres in 70 days, at an average rate of 15 kilometres a day from Singora to Singapore's southern coast, and destroyed several formations in the process. By any standards it was a spectacular achievement. Overall the casualties among the defenders were: British 38 496, Australian 18 490, Indian 67 340, local volunteer troops 14 382 — a total of 138 708 of whom more than 130 000 became prisoners. With few exceptions the latter endured a captivity at least as humiliating and deprived as any similar group in modern history. More than one-third died.

JOHN COATES

MALAYAN EMERGENCY was declared on 18 June 1948 in response to the murder of three estate managers in Perak, northern Malaya, by guerrillas of the Malayan Communist Party (MCP), which had determined on the violent overthrow of the British colonial government. The military wing of the MCP had its origins in the Malayan Peoples'

Anti-Japanese Army (MPAJA) that had been established during the Second World War and which, despite only limited forays against the Japanese, had maintained a reputation for active resistance. The MPAJA, with strong roots in the Chinese community in Malaya, was disbanded on British orders at the end of the war, but its structure remained largely intact and it retained many of its weapons despite having undertaken to surrender them. Although never more than a few thousand in number, the MCP's guerrilla force (whose name was changed in 1949 to the 'Malayan Races Liberation Army' [MRLA]) posed a serious threat to the colonial government. Its military strength was magnified many times over by the fact that it could draw on the active support of significant sections of the Chinese community through the 'Min Yuen' (masses' organisations), while the rest maintained a 'wait and see' policy towards both the guerrillas and the government. The colonial government had done little before the war to bring the Chinese community within the fold of government services, especially the large numbers of squatters who eked out an existence on the fringes of the jungle, working plots of land for which they held no formal title. Postwar plans for constitutional change had been a fiasco: the Malayan Union proposals of 1946, which offered the Chinese a realistic schedule for gaining citizenship, had enraged the Malays who saw themselves becoming a minority in their own country. Within three months the British had withdrawn the plan, bringing charges of betrayal from the Chinese while doing nothing to repair the suspicion and distrust that the Union proposals had engendered among Malays. The political circumstances that created a favourable climate for anti-government agitation were matched by harsh postwar economic and social conditions. The British military administration had proved incapable, in the immediate postwar period, of providing minimum food rations or basic law and order in the face of thriving corruption and gang warfare. Malaya was ripe for revolution, at least in the Chinese community.

The initial response of the government was slow and measured. The military threat posed by the MCP was largely discounted, and there was virtually no recognition of the fact that the MCP's program was nothing less than the overthrow of the colonial government. The government's efforts to counter the military activities of the MCP were poorly coordinated, and it was not until March 1950 that a Director of Operations was appointed. Seven weeks after arriving in Malaya, Lieutenant-General Sir Harold Briggs (rtd) presented his report; it became the basis for all government

action against the MCP thereafter. Briggs emphasised the need for the closest possible coordination of all facets of government policy, based on the realisation that only a recognition of the underlying political, social and economic grievances of the Chinese community would enable the government to choke off sustenance for the MCP guerrillas. In order to repair the neglect of many largely Chinese areas, to enable the government to provide services and security to win over those who were wavering over which side to support, and to prevent MCP sympathisers in the Chinese community from giving direct assistance to the guerrillas, Briggs recommended the movement of fringe dwellers into 'New Villages', where they could enjoy the benefits of government control (elementary schools and health facilities) and where they could be physically isolated from contact with the guerrillas. In military terms Briggs argued for a south–north policy of systematically clearing Malaya of communist influence by eliminating their bases of support, thereby eventually forcing the MCP to attack the security forces on the latter's own ground.

Briggs's plan was not expected to produce immediate results, and for several months it seemed that the situation was deteriorating markedly, the guerrillas striking a major blow on 22 October 1951 with the assassination in an ambush of the High Commissioner, Sir Henry Gurney. Drastic measures were needed, and they were taken in the appointment of General Sir Gerald Templer as High Commissioner and Director of Operations. Templer galvanised the anti-guerrilla operations into focused action. The coordination of government policy that Briggs had called for was finally achieved with a tight decision-making structure being created, and with a new vigour injected into the battle on all fronts by the sheer force of Templer's personality and dedication. By 1954, when Templer left Malaya, the pattern for government operations against the communist terrorists (or CTs, as the guerrillas were now officially known) had been set: large-scale prolonged operations in designated areas, aimed at destroying the CTs' base of support in the local community and driving them into jungle areas where even a bare existence was difficult and where constant patrolling by the security forces would break up the concentrations of CT strength and gradually wear down and eliminate small groups. By 1955, when Australian ground forces arrived in Malaya, the outcome of the Emergency was no longer in doubt. The essential military and political struggle had been won,

but there remained a long and wearying 'clearing up' stage to complete.

Australian involvement in the Emergency began in June 1950 when, in response to a request from the British government, six Lincolns of No. 1 (Bomber) Squadron, RAAF, and a flight of Dakotas from No. 38 (Transport) Squadron, RAAF, arrived in Singapore as part of the Far East Air Force (FEAF). The Dakotas were deployed on cargo runs, in troop movement and in paratroop and leaflet dropping within Malaya, the latter an important part of the security forces' psychological warfare strategy against the CTs. They were also, because of their commitment to the FEAF, deployed on cargo flights in east and south Asia — to Japan, Korea, Hong Kong, the Philippines, French Indochina, Borneo and Ceylon (Sri Lanka). By the time No. 38 Squadron returned to Australia in December 1952, it had dropped 1 669 798 pounds of supplies in Malaya, airlifted 326 sick and wounded troops to treatment centres, and carried 17 000 passengers throughout its theatre of operations.

The Lincolns, operating from bases in Singapore and from Kuala Lumpur, remained the backbone of the air war against the CTs. There was at the time, and has been since, considerable controversy over the use of air power in Malaya. It was difficult to establish firm links between air activity and CT kills, and much of the justification for bombing raids in the first two years of the campaign lay in the fact that the ground forces were incapable at that stage of undertaking major offensive action against the CTs. Air power enabled the government to harass CT forces, to attack their camps when they had been identified, and even to claim that bombing tracts of jungle kept the CTs 'on the move', thereby disrupting their contacts with the Min Yuen. There was some truth in these claims, but once the army and police forces, relying on carefully built up intelligence reports, had become sufficiently strong to launch long-term offensives against CT dominated areas, the need for air action became progressively less, and when it was used, it was directed against specific targets. The greatest number of air strikes was flown in 1951, and thereafter the rate declined until by 1953 it was less than half that of the year before (114 strikes in 1953 compared with 364 in 1952). Increasingly the Lincolns were deployed as part of a combined air–land assault against CT targets, such as in Operation TERMITE east of Ipoh, Perak state, in July 1954. In that operation five Lincolns from No. 1 Squadron, together with six from No. 148 Squadron (RAF), made simultaneous attacks on two CT camps, followed by paratroop drops and a ground attack and further bombing runs 10 days later. This

Map 20 Malaya during the Emergency, 1948–60.

Ground crew preparing a Canberra of No. 2 Squadron for a bombing mission during the Malayan Emergency, Butterworth, Malaya, c. 1956. A sun shade has been placed over the canopy to protect the cockpit, and the aircrew who will later sit there, from the hot tropical sun. (AWM MAL1060)

resulted, by November 1954, in 13 CTs being killed, one surrender, and 181 camps found and destroyed. This was one of the major successes in the Emergency, and the Lincolns undoubtedly played a part in that success, but there always remained the question of whether the results justified the huge cost of the air campaign. Opinion in the RAAF, then and since, is that at the very least, the commitment of RAAF aircraft to the Emergency provided unparalleled training experience.

Australian ground forces arrived in Malaya in October 1955 to form part of the Australian contribution to the recently established Far East Strategic Reserve (q.v.), which had been created in April 1955. The primary role of the FESR was to deter external communist aggression against South-east Asia, especially in Malaya and Singapore and in the seas around them. The secondary role of the forces committed to the Strategic Reserve was to assist in the suppression of the communist insurrection in Malaya, but only if this could be done without interfering in the training and availability of troops for their major role. The creation of the FESR and the introduction of further British and British Com-

monwealth forces into Malaya carried political risks: although the British government had announced its intention to move Malaya towards political independence as quickly as possible, critics, both in Malaya and outside, charged that the FESR was designed to entrench British influence after independence and that it would drag an independent Malaya into the big-power rivalries that beset the region.

When the 2nd Battalion, The Royal Australian Regiment (2RAR), arrived in Penang in October, the popular view as expressed in the Australian and indeed Malayan press was that it had come to fight the CTs. That was not in fact the case: the battalion received relatively little training in guerrilla warfare (although the Jungle Training Centre at Canungra, Queensland, had been reopened to prepare it for operations in Malaya), and had not been given any copies of the standard British manual, *The Conduct of Anti-Terrorist Operations in Malaya (ATOM)*. After training on Penang Island, 2RAR was several times warned for anti-CT operations, only to have the order cancelled at the last minute. Not until 1 January 1956 did it cross to the mainland to begin what most (mistakenly) saw as its primary role. For the next 20 months, as part of the 28 Commonwealth Brigade (headquartered first at Butterworth and then at Taiping), 2RAR was engaged on a variety of operations, mainly in Perak, which was one of the main areas of CT activity. The battalion's work consisted of extensive patrolling in and near the jungle areas, watching for contacts between CTs and the Min Yuen in the rubber plantations, and mounting a perimeter guard on New Villages. Contacts were sparse and the results meagre, and the battalion had a mixed record, losing three members in the 'pipeline ambush', on 25 June 1956, for only two 'kills'. The battalion was withdrawn from anti-CT operations in August 1957 (the month that Malaya gained political independence) for training in its primary role, and left Malaya in October.

It was replaced by 3RAR which arrived in October 1957 and immediately underwent six weeks' training at the Far East Land Forces Training Centre at Kota Tinggi in Johore, southern Malaya. Thereafter it was committed to food denial operations in Perak and Kedah, which involved extended patrols and ambush positions. Success was measured in single kills, which were all the more important when the CT eliminated was a significant figure in the local or regional organisation. That was the case in two of the battalion's actions. In the first, on 27 July 1958, a CT was wounded. When he surrendered the following day, he was found to be a leading member of the branch committee of the MCP, and agreed to lead the security forces to a number of camps in the area. As a result two key CT groups were destroyed, a number of their members being killed and rest being forced across the border into Thailand. On the second occasion, 20 November 1958, a patrol from 3RAR uncovered a CT camp and killed three of its occupants, one of whom turned out to be a member of the District Committee and one of the group who had killed the High Commissioner in 1951. The importance of these successes was that they demonstrated the growing control of the security forces while further destroying the organisation of the CTs and making support for them increasingly futile and dangerous. In April 1959 the state of Perak, previously one of the main centres of the MCP and the MRLA, was declared secure, a major achievement in the struggle to break the communist campaign. 3RAR left Malaya in October 1959, and was replaced by 1RAR.

By 1959 operations against the CTs were in their final phase. The strength of the MRLA was estimated at 698, of whom only 243 were thought to be in Malaya, the rest having moved into the safety of Thailand. Intelligence reports suggested that the CTs would avoid contact with the security forces but that they would fight if cornered. 1RAR operated in the border region, but made no contact with the enemy, and was strictly prohibited from crossing into Thailand, even when tracks suggested that penetration across the border would uncover active CT camps. 1RAR's tour ended in October 1961, more than a year after the Emergency had been declared officially over (31 July 1960). 2RAR returned for a second tour in October 1961, and was committed to anti-CT operations in Perlis and Kedah in August 1962, where it remained for only two months before being withdrawn in October 1962 for six months' training as part of the Strategic Reserve. Several companies of the battalion undertook further anti-CT operations in May 1963, while B Company took part in SEATO exercises in northeast Thailand. 2RAR returned to Australia in August 1963.

As well as air and infantry forces, Australia also provided artillery and engineering support, an airfield construction squadron that built the major runway at Butterworth (q.v.), and signals personnel. A number of RAN ships were part of the FESR, and on two occasions (HMAS *Anzac* on 29 September 1956, and HMA Ships *Queenborough* and *Quickmatch* on 22 January 1957), while off the east coast of Johore, fired on suspected CT positions.

Fifty-one Australian servicemen were killed in Malaya, although only 15 of these deaths occurred as a result of operations, and 27 were wounded, the majority of whom were in the Army. Australia's participation in the Malayan Emergency marked a transitional stage in the development of the Australian Army as a professional force, a process which began in Korea. Malaya demonstrated that reputations gained in one war, against the Japanese, could not necessarily be transferred to another in quite different circumstances, and the success that Australian ground forces eventually achieved against the communist insurgents came at a high price in terms of self-inflicted casualties. The lessons that could be transferred from Malaya to Vietnam were not those of the specifics of anti-CT operations but more general ones — the importance of training, of field discipline and skills, and of good leadership at every level. The political benefits for Australia were considerable: Malaya, an area of vital concern to the security of Australia, remained free from communist domination.

MALLOW, HMAS see **FLOWER CLASS MINESWEEPING SLOOPS**

'MAN WITH THE DONKEY' see **SIMPSON AND HIS DONKEY**

MANOORA see **ARMED MERCHANT CRUISERS**

MANUNDA (2/1st Australian Hospital Ship) made many trips during the Second World War to the Middle East, as well as to Port Moresby and other parts of the South-West Pacific Area (q.v.). At the war's end she returned many POW patients and army medical personnel to Australia. She narrowly escaped being hit by enemy mines and shells in the Middle East and at Milne Bay, but during the first Japanese air raid on Darwin (q.v.) she was lying in the harbour surrounded by military targets, and as a result was hit by bombs which killed 12 members of the crew. After the sinking of the *Centaur* (q.v.) she was refitted as a defensively armed vessel while retaining its role as a hospital ship, but was soon reconverted when it was realised that this meant an abandonment of the Geneva Convention (which precludes vessels sailing under the protection of the Red Cross from being armed or used to transport warlike stores), which would affect not only hospital ships but all other Australian medical units.

MANUS ISLAND, an Australian mandate in the Admiralty Islands north east of New Guinea, was developed as a major American naval base following its recapture from the Japanese in January 1944. After the war the Australian government, and in particular the Minister for External Affairs, Dr H.V. Evatt, tried to use Manus as a bargaining tool in order to persuade the United States to commit its forces to the ongoing defence of the south-west Pacific. In the immediate postwar period, the United States had shown considerable interest in Manus, but it was not prepared to undertake a wider defence role in the area, and by 1946 the concept of perimeter defence, in which Manus Island had been a key site, was replaced by 'containment'. The American strategic focus shifted well away from bases south of the equator, and Australia's offers to renegotiate terms for continued use of Manus Island came to nothing. The Americans removed and sold off large quantities of materiel (much of it to the Nationalist Chinese), and it was left to Australia to redevelop the base, albeit on a much smaller scale. HMAS *Seeadler* was commissioned as a naval base in 1950, and shortly afterwards was recommissioned as HMAS *Taragau*. Its size declined steadily until 1965 when it was designated as a base for the development of an independent Papua New Guinea naval force. It was decommissioned on 14 November 1974 and handed over to the Papua New Guinea government, and immediately recommissioned as the PNG Defence Force patrol boat base.

MAORI WARS see **NEW ZEALAND WARS, AUSTRALIAN INVOLVEMENT IN**

MARALINGA see **ATOMIC WEAPONS TESTS**

MARGUERITE, HMAS see **FLOWER CLASS MINE-SWEEPING SLOOPS**

MARIBYRNONG ORDNANCE FACTORY see **INDUSTRY; SCIENCE AND TECHNOLOGY**

MARINES, BRITISH see **BRITISH ARMY IN AUSTRALIA**

MASCOTS see **ANIMALS**

MATADOR PLAN see **MALAYAN CAMPAIGN**

MAUGHAN, (David Wilfrid) Barton (7 October 1912–15 July 1988). After reading politics, philosophy and economics ('PPE') at Oxford, Maughan practised as a barrister at the Sydney Bar and was a militia soldier before the war. He enlisted in the 2nd AIF in May 1940, and was commissioned into the 2/13th Battalion in April 1941. He served with the 9th Division at both the siege of Tobruk and the Battle of El Alamein (qq.v.), and returned with the 9th Division to take part in the New Guinea campaign. Between July 1943 and October 1944 Maughan was intelligence officer on 20th Brigade Headquarters, and was awarded the MC for the operations at Finschhafen. For the last two years of the war he served with the British South-East Asia Command (SEAC) as a staff officer at the headquarters of the Supreme Allied Commander, Mountbatten, and in the Liaison Section. Following demobilisation Maughan joined The Zinc Corporation in Broken Hill and later retired in Coffs Harbour. Gavin Long, the official historian, commissioned him to write the volume of the official history dealing with the 9th Division's campaigns in the Mediterranean theatre, and this appeared in 1966 as *Tobruk and El Alamein*. It is massively detailed and overly long when one remembers that it deals with the activities of a single division for only 18 months of the war. On the other hand, Maughan generally transcended the temptation to write parochially about his subject, and his is the most extensive account of the siege of Tobruk published in any of the Commonwealth official histories. Long had intended originally that the volume should be written by the journalist, Chester Wilmot (q.v.), but after the latter's death in an air crash in 1954, had to find a replacement. He was not altogether satisfied with Maughan's progress, complaining on a number of occasions that the latter appeared to neglect work on the manuscript for long periods of time, and Maughan's was indeed the last volume to be published in the Army series of the history.

Barton Maughan, *Tobruk and El Alamein* (Australian War Memorial, Canberra, 1966).

MAWHOOD, Colonel John see **INDEPENDENT COMPANIES**

MEDALS see **DECORATIONS; VICTORIA CROSS**

MEDICAL CORPS see **ROYAL AUSTRALIAN ARMY MEDICAL CORPS**

MEDICAL EVACUATION is the process by which sick and more especially wounded are transported from the front for treatment. Major developments in modern military medicine began with the First World War, and these provided the first test of the Australian Army Medical Corps' procedures. On Gallipoli these procedures did not work at all well, although much of the blame for the breakdown in the medical services belonged with the British, while the complexities imposed by an ambitious amphibious operation did not help matters. In theory, wounded would be taken out of the line by unit stretcher-bearers to the regimental aid post about 100 metres to the rear, where they received elementary attention from the Regimental Medical Officer. From there they were evacuated to the field ambulance and, if the decision was made to evacuate rather than treat locally (a decision depending on the seriousness of the case), to the casualty clearing station on the beach. At that point they would be taken by small naval craft to hospital ships for passage to the advanced base hospital on Mudros, and from there to general hospitals in the base areas in Malta and Egypt. Serious cases or those requiring discharge would proceed to hospitals in England or Australia. In practice, however, there were a series of bottlenecks and grave shortcomings in the provision of transport for the wounded which led directly to the deaths of many wounded men, and to the dispersal as far as Egypt and in many cases England of less seriously wounded cases who should have been returned to the line after a short period of convalescence. This latter in particular contributed greatly to the reinforcement problems experienced by the AIF as the campaign wore on. The number and type of hospital ships provided was woefully inadequate for the casualties actually incurred, another result of poor planning and coordination before the landing, while conditions aboard the ambulance carriers, ships converted to some extent to cater for casual-

ties and quickly dubbed 'black ships', were indescribable.

The medical arrangements on Gallipoli generally were in a more or less constant state of rearrangement throughout the campaign, and although they improved over time they were never adequate to cope satisfactorily with the growing numbers of sick and wounded. To compound the problems, there were frequent shortages of everything from medical supplies to stretchers. So serious were the deficiencies in the medical arrangements that they came to occupy much of the time and attention of the Dardanelles Commission, which met in London in 1916–17 and whose final report was properly and highly critical of them.

In France (see WESTERN FRONT) casualty evacuation proceeded along more 'normal' lines. Unit stretcher-bearers bore the wounded to aid posts where ambulances, horse-drawn or motorised, took them to casualty clearing stations where, increasingly, treatment was centralised and, except during periods of very heavy casualties, most urgent surgery was carried out. Situated a few miles from the front, they greatly improved the prospects of wounded soldiers. From there as necessary they were transported by ambulance convoy or even ambulance train to the various hospitals in the lines of communications zone or, by hospital ship, to England and even Australia in cases of invalidity. Urgent surgery aside, however, as the medical historian A. G. Butler (q.v.) noted, the casualty clearing stations (as their name implied) and to an extent even the general hospitals in France were more concerned with the evacuation of wounded than with their longer-term treatment. In England, where major treatment was concentrated, the hospital and command depot systems worked on a 'six months policy' of returning a man either to Australia or the front within that time.

In Sinai and Palestine the evacuation problem was compounded by the often considerable distances traversed between the site of wounding and the final destination for treatment. Camel-borne cacolets were used in Sinai, a form of travel exquisite in its agony for wounded men because of the nature of the animal's movement, as were sand sledges. Light horse field ambulances were attached to light horse brigades, with whom they tended to function as an integral part of the brigade establishment. The bearer division of a light horse field ambulance was responsible for conveying casualties from the regimental aid post to an advanced dressing station a mile or so from the firing line. By camel or horse-drawn ambulance or even by train, the wounded would then be taken back to the casualty clearing station or stationary hospital in the lines of communication area for onward transport to general hospitals in Egypt.

In the Second World War this chain of evacuation system was applied once more, although again there were local variations imposed by terrain, climate and the tactical situation. Thus during the first Libyan campaign the field ambulances moved forward with the advancing troops and relayed casualties back to the 2/2nd Casualty Clearing Station in the rear; at the beginning of the campaign during the battle of Bardia this was located at Mersa Matruh, but by the time the advance to Benghazi had been completed in February 1941 it had moved forward to Tobruk. During the siege of Tobruk the smallness of the area necessitated much greater concentrations of medical facilities within the defensive perimeter, and shorter distances between the regimental aid posts and the hospitals which were situated around the port. In Greece medical evacuation was made more difficult by the retreat of the Allied forces in the face of constant German pressure and the chaos which attends defeats. During the first battle of El Alamein in July 1942 ambulances came under sometimes heavy fire from German shelling, and evacuation had at times to be suspended. Always in the desert environment there was the problem of locating wounded men passed by in the advance. In the fighting in Malaya additional hazards were posed by Japanese disregard of the Red Cross symbol; ambulances and trucks bearing wounded were fired on regularly by Japanese troops, while many non-ambulatory wounded were massacred when their aid stations or hospitals were overrun by the enemy advance. In New Guinea wounded were often transported to the rear by native bearers, the famous 'fuzzy-wuzzy angels' (q.v.) of contemporary popular song, and increasingly by air ambulances — light Auster aircraft equipped to carry stretcher cases.

In the Korean War the responsibility for casualty evacuation was initially held by the regimental medical officer of 3RAR until the arrival of the 60th Indian Field Ambulance as part of the 28th Commonwealth Infantry Brigade (q.v.), although a section of an Australian field ambulance was attached to the battalion to assist in evacuation. In Korea itself use was made of the American medical system, and casualties were transported to a Mobile Army Surgical Hospital (the famous MASH), the equivalent of a casualty clearing station. Casualties then went south to Pusan for movement by air or sea to the British Commonwealth Occupation Force (q.v.) General Hospital in Japan, although it was not unknown for Australians to end up in the American hospital system also. The rugged terrain, poor road system and, in winter, extreme tempera-

tures all placed stress on the evacuation system — Australian 3-ton ambulances, for example, were unheated, unlike American ones which had central heating. From early 1951 helicopters were used for serious cases, or in circumstances where the terrain precluded rapid evacuation by other forms of transport, although their lift capacity was limited to two stretcher cases at a time.

After the static period of the war began, evacuation policy changed and treatment was administered as far forward as possible. Cases likely to take three weeks or more to recover were taken to Japan, often by air in instances of surgical need, while those who were likely to recover within three weeks were held at the reserve field ambulance in Korea and less severe cases stopped at the advanced dressing station. Advances in aeromedical evacuation, the increased use of blood plasma and the greater availability of antibiotics meant that many men now survived wounding and evacuation who would have died of their wounds in earlier conflicts. In Malaya during the Emergency and in Borneo during Confrontation problems were again posed by terrain, since it could take up to six hours to carry a stretcher case 2 kilometres through primary jungle, and again helicopters provided a solution, although on occasions this was achieved only at the price of a tactical penalty in identifying the Security Force's position. In Borneo in particular, the weather could also impose limitations on the availability of helicopters. Soldiers requiring advanced medical treatment ended up in the British military hospital system. In Vietnam casualties were routinely evacuated by the helicopter operation 'Dust Off', and in the early period of Australian involvement helicopters were often supplied by the Americans. Aeromedical flights were mounted by the RAAF from Vietnam to an Australian hospital at Butterworth (q.v.) in Malaysia. Although always reliant to some extent on American resources, increasingly from 1967 helicopter evacuation of casualties was carried out by No. 9 Squadron RAAF. Aeromedical procedures in Vietnam ensured that 97.5 per cent of Australian casualties evacuated survived their wounds.

MEDICAL PROBLEMS see **AGENT ORANGE; GAS, FIRST WORLD WAR; GAS, SECOND WORLD WAR; VENEREAL DISEASE**

MELBOURNE (I), HMAS see **TOWN CLASS LIGHT CRUISERS**

MELBOURNE (II), HMAS (Modified *Majestic* Class Light Fleet Aircraft-Carrier). Displacement 16 000 tons; length 701 feet; beam 80 feet; armament 25 (later 12) × 40 mm guns; speed 24 knots.

HMAS *Melbourne* sailing in close formation with the destroyers *Vendetta* and *Voyager, c.* 1959. On 10 February 1964 *Melbourne* hit and sank *Voyager* off Jervis Bay, New South Wales. (AWM 301014)

HMAS *Melbourne* was laid down in the United Kingdom as an RN vessel in 1943 but was incomplete when construction halted with the end of the Second World War in 1945. Construction resumed in 1949, incorporating design improvements such as mirror landing aids and an angled deck. She was finally completed in 1955 and transferred to the RAN in 1956 where she became the flagship. From 1957 to 1963 *Melbourne* was on regular deployment with SEATO forces or with the Far East Strategic Reserve (q.v.). In 1964 the *Melbourne* rammed the destroyer HMAS *Voyager* (q.v.) in controversial circumstances off Jervis Bay. In 1967, after the Indonesian 'Confrontation' (q.v.), it was decided to refit the ship in order to extend her operational life. She was recommissioned in 1969 and sailed for exercises in the South China Sea. During these exercises she struck the USS *Frank E. Evans*, a destroyer, and sank her with the loss of 74 American lives (see COLLISIONS, NAVAL). *Melbourne's* subsequent career was less eventful. She regularly took part in exercises with the US fleet, and assisted in Darwin after Cyclone Tracy (see CIVIL COMMUNITY, ASSISTANCE TO). HMAS *Melbourne* was originally equipped with De Havilland Sea Venom fighters and Fairey Gannet anti-submarine aircraft (qq.v.). In 1963 Westland Wessex (q.v.) anti-submarine helicopters were added. After her recommissioning in 1969 she flew Douglas Skyhawk fighter-bombers and Grumman Tracker anti-sub-

marine aircraft (qq.v.). Westland Sea King (q.v.) helicopters replaced the Wessexes in 1975. She was decommissioned in 1982 in preparation for the arrival of her replacement HMS *Invincible*. But the British government decided to retain their carrier and Australia has had no aircraft-carrier from that time (see FLEET AIR ARM).

MELLOR, David Paver (19 March 1903–9 January 1980). Mellor was a research scientist with a distinguished career before the Second World War that included appointments to the Commonwealth Solar Observatory from 1928 and, unusually for an Australian at that time, in the United States, at the prestigious California Institute of Technology in 1938. After the war he pursued his career at the University of New South Wales (as it became), ultimately serving as Dean of the Science Faculty between 1968 and 1969. He retired in 1970. In the course of his professional life he won numerous awards and served as both president and secretary of the Royal Society of New South Wales in the 1940s. Gavin Long (q.v.), the official historian of the Second World War, selected him to write the volume *The Role of Science and Industry*, which was part of the Civil series and which appeared in 1958. The book broke new ground in an Australian official history in dealing with equipment development and acquisition and the involvement of industrial and scientific concerns in the war effort, especially against the Japanese. It may be criticised, however, for echoing too closely at times the views of certain senior figures closely involved in these processes, especially those of the Deputy Master-General of the Ordnance, Brigadier John O'Brien (q.v.).

MENE, Corporal Charles see **ABORIGINES AND TORRES STRAIT ISLANDERS IN THE ARMED FORCES**

MENIN GATE MEMORIAL, so named because the road led to the town of Menin, was constructed on the site of a gateway in the eastern walls of the old Flemish town of Ypres, Belgium, where hundreds of thousands of Allied troops passed on their way to the front in the Ypres salient, the site from April 1915 to November 1918 of some of the fiercest fighting of the First World War. The memorial was conceived as a monument to the 350 000 men of the British Empire who fought in the campaign. Inside the arch, on tablets of Portland stone, are inscribed the names of 56 000 men, including 6178 Australians, who served in the Ypres campaign and who have no known grave. The opening of the Menin Gate Memorial on 24 June 1927 so moved the Australian artist Will Longstaff that he painted *Menin Gate at Midnight* (q.v.). Two medieval stone lions which had stood at the gateway of the Menin road and were damaged during the war were presented by the city of Ypres to the Australian War Memorial (q.v.) in 1936. Since the 1930s, with the brief interval of the German occupation in the Second World War, the city of Ypres has conducted a ceremony at the memorial at dusk each evening to commemorate those who died in the Ypres campaign.

MENIN GATE AT MIDNIGHT, also known as *The Ghosts of Menin Gate,* is perhaps the best known example of Australian war art (q.v.). The painting, by Will Longstaff, shows the Menin Gate Memorial (q.v.) at Ypres, Belgium, under a night sky surrounded by ghostly outlines of soldiers. Longstaff painted the work shortly after attending the unveiling of the Memorial on 24 July 1927 and claimed it was based on a vision he had while on a midnight walk in Ypres the night of the opening. In a society grieving the loss of so many men in the First World War, the painting, with its spiritualist overtones, was immediately popular. It was purchased by Lord Woolavington in 1928 for 2000 guineas and presented to the Australian government. After exhibition in the United Kingdom, including a private viewing by King George V at Buckingham Palace, the painting toured Australian cities and joined the collection of the Australian

War Memorial (q.v.), where it is on permanent display. Longstaff followed the success of this painting with other works featuring war memorials and the ghostly images of British, Canadian and New Zealand soldiers. Prints of *Menin Gate at Midnight* became a common feature of Australian homes. Four hundred reproductions, made in the United Kingdom, signed by Longstaff and costing 10 guineas each, were sold in Australia in 1928. A cheaper, unsigned copy was later made by the Australian War Memorial and sold by door-to-door salesmen.

JOHN CONNOR

MENTION IN DESPATCHES (MID) The award of a mention in despatches is made where the conferring of a gallantry or other decoration is deemed inappropriate or is not otherwise available, and is signified by an oak leaf worn on the ribbon of the relevant service or campaign medal. It may also be awarded for continuous good service over a period of time, and is sometimes known disparagingly as the 'staff officer's decoration' in consequence. Implemented formally with royal approval in British Army Orders in May 1919, mentions in despatches have existed in British armies since at least the Revolutionary and Napoleonic Wars when general officers commanding wrote despatches or reports to their superiors in London, but until the twentieth century brought with them no formal symbol to mark the fact. The initial form of recognition was a certificate, but this did not apparently satisfy the requirement for a tangible symbol of the 'mention', and in January 1920 the now-familiar oak leaf was instituted. Only one is worn, no matter how many times a recipient received a 'mention' in the campaign concerned (and in the First World War, for example, some Australians were mentioned three or four times during the course of the war). For service in the First World War it was worn on the ribbon of the Victory Medal. MIDs were relatively rare to members of the AIF, some 5798 in total being awarded (Gallipoli, 559; Palestine, 1150; the Western Front, 3935). An MID awarded for service during the Second World War is worn on the ribbon of the War Medal 1939–45, while that earned during the Korean War is displayed on the ribbon of the Korea Medal. Mentions awarded on campaign between the wars, or to Australians during the Malayan Emergency or Confrontation, are recorded on the ribbon of the General Service Medal. Those for Vietnam are displayed on the Vietnam Medal. In the latter conflict 606 were awarded, 421 to the army, 137 to the RAAF and 48 to the RAN. Where no campaign medal is authorised, the oak leaf emblem is worn on the jacket following the last medal on the medal bar. Together with the Victoria Cross, the MID is the only other award that may be granted posthumously; this has led on occasion to acts of singular gallantry being recognised by the relatively humble award of the mention in despatches because no other scale of decoration has been available.

MENZIES, Robert Gordon (20 December 1894–15 May 1978) was born in the small Victorian town of Jeparit and educated at the University of Melbourne. While at university he served in the Melbourne University Rifles and was commissioned. (Contrary to popular belief, he did not resign at the outbreak of war; he did not in fact join the unit until 1915, serving until 1919.) He was admitted to legal practice in May 1918, and quickly built a substantial reputation through his appearance in the Engineers' Case in 1920. He entered politics in 1928, as a Nationalist member of the Victorian Legislative Assembly, and within a few months had entered the ministry. Menzies took silk in 1929 as the youngest King's Counsel in the country, and in 1932 became attorney-general and solicitor-general for Victoria. In 1934 he moved to the federal parliament at the invitation of the Prime Minister, Joseph Lyons, holding the seat of Kooyong for the United Australia Party. Menzies became attorney-general, deputy leader, and Lyons's designated successor, and following Lyons's sudden death in April 1939 he became prime minister. As the first of Australia's wartime prime ministers he held office during the 'phoney war', which was followed by the disasters in western Europe in 1940 and the string of reverses in the Mediterranean the following year. Menzies lost office in August 1941 as a result of vicious personal intrigues within his own party, which was in the process of collapse as a political force. He returned to the leadership of the UAP in 1943 and in 1944–45 began to fashion a new conservative party which became the Liberal Party. He fought and lost the 1946 election, but won office in December 1949, and was to remain prime minister for 16 years. The Menzies era, as it has been dubbed, was characterised by economic expansion, general prosperity, growth in the public sector, and especially in the universities (which perhaps form Menzies' most enduring legacy), and a strengthening of alliance ties with the United States. He was made a Companion of Honour in 1951, and knighted in 1963.

Menzies' wartime prime ministership was more important than some historians have suggested, and the notion that he did little or nothing to further Australia's war effort, leaving the country woefully under-prepared to meet the Japanese threat in 1942, cannot be sustained. As his biographer, A. W. Martin, has noted, 'with the best will in the world, it was not easy during the "Phoney War" to stir a sense of urgency about mobilisation in a country far from the scene of real hostilities'. Two decades of neglect of the armed forces could not be made up overnight. Defence expenditure increased dramatically, troops were called up for local defence, and four divisions were raised for overseas service. (This last point is important as had the three divisions of the 2nd AIF which fought in the Mediterranean theatre not acquired the hardening and experience of modern warfare provided by that service, it is difficult to see how else it might have been acquired before they faced the Japanese. The officers of the 2nd AIF who returned to Australia in late 1941 and early 1942 were also able to spread that experience among militia formations, which otherwise lacked it entirely.) Menzies was sensible of the threat posed by Japan; he had referred previously to the fact that Britain's Far East 'was Australia's near north', and it was partly concern over Japanese intentions which took him to Britain in early 1941 for consultations with Churchill and his advisers. The British decision to use Australians in Greece and Crete ended in disaster, but Menzies cannot be blamed for this as it is clear that he was, at best, misled over the chances of success in the campaign and over the views of the senior Australian commander, Major-General Thomas Blamey (q.v.). The first slogan of the Australian home front during the war was, it is true, 'Business as usual', but it is generally overlooked that the slogan which most people identify with Australia during the war, 'All in!', was his as well, not Curtin's. Finally, for all the later suggestions that he was overly deferential to British leadership, the fact is that he stood up to Churchill, in London and by cable, far more resolutely than John Curtin (q.v.) was to deal with General Douglas MacArthur (q.v.) and the Americans. On the whole, the tendency among historians has been to read Menzies' first period as prime minister, between 1939 and 1941, in terms of their beliefs concerning his later period, between 1949 and 1966, and this distorts the record.

In his second period as prime minister Menzies presided over Australian involvement in the Korean War, the Malayan Emergency, Confrontation with Indonesia and the early stages of Australia's commitment to Vietnam. The decision to commit forces to Korea was essentially made by the External Affairs minister, Percy Spender (q.v.), in his absence, but Australian assistance in Malaya, beginning in 1950, was largely at Menzies' initiative and reversed the ALP policy of non-engagement pursued under his predecessor, Chifley. He and his cabinet wavered over Confrontation, concerned that President Sukarno's ambitions in Borneo should be thwarted, but uncertain that direct Australian involvement on the ground might not make matters worse and tempt the Indonesians to launch similar operations against Australian territory in Papua New Guinea. On Vietnam Menzies was convinced that American policy was correct and that the critics of greater Australian involvement were wrong. In this he was to some extent guided by his reading of the Malayan Emergency, over which many of the same critics had indeed been wrong. The ANZUS Treaty (q.v.) was largely the work of Spender, but could not have been negotiated without Menzies' active support within cabinet. Australian support for the Anglo-French-Israeli conspiracy during the Suez Crisis in 1956 was a sorry episode, but provides the one concrete example in his public life of Menzies allowing sentiment to overrule national interest. Defence budgets in the 1950s were generally low, with his governments devoting themselves to the development of national infrastructure and the economy. They were able to get away with this for so long not least because of the careful manner in which Menzies tended the American alliance. In that sense, the commitment to Vietnam was the price paid for years of defence 'on the cheap' through the American security relationship.

Perhaps the most sustained, and some of the nastiest, criticism to attach itself to Menzies concerns conscription. During his second period in office he introduced national service schemes between 1950 and 1959 and again in 1964 (which lasted until 1972), the second with a provision for overseas service. Menzies had been a pro-conscriptionist during the First World War, but had not himself served with the AIF, a fact which was thrown back at him by his opponents, particularly in the 1960s. In fact, what prevented Menzies from enlisting was not want of patriotism, or courage, but familial duty. He had two older brothers in uniform, Frank and Les, and a family conference decided that two was enough and that he should stay home to provide for his parents and sister should anything happen to the others. This was, in fact, a not uncommon arrangement, but it was obviously a difficult one for a man who regarded himself as a British Australian and loyal citizen of the empire, and some testimony suggests that it remained a personal regret throughout his life. It says much about Menzies' sense of dignity and internal strength that he never bothered to explain this private arrangement in the face of detraction. It is also worth noting that the allegations concerning his lack of war service did not in any obvious or material manner impede his public career.

Menzies was Australia's longest serving prime minister, a record unlikely to be exceeded. He was prime minister in two important periods of Australian history: the beginning of the Second World War, and the long period of armed engagement in South-east Asia. His wartime leadership laid the foundations for Curtin's successful harnessing of Australia's effort against the Japanese, while his reading of Cold War realities in the 1940s and 1950s served Australian interests well until American efforts in Indochina began to unravel in the following decade. By the time that became obvious, however, he was no longer in public life.

A. W. Martin, *Robert Menzies: A Life* (Melbourne University Press, Melbourne, 1993).

MERMAID, HMAS see **SURVEY SHIPS**

METEOR, GLOSTER (Single-seat ground-attack fighter [Mark 8]). Wingspan 37 feet 2 inches; length 44 feet 7 inches; armament 4 × 20 mm cannon, bombs 2 × 500-pound or 8 rockets; maximum speed 590 m.p.h.; range 980 miles; power 2 × Rolls Royce Derwent turbojets.

The first jet aircraft to fly with the RAAF was a Meteor Mark 3 on loan from the RAF used for tropical trials in 1946–47 until it crashed at Darwin. During the Korean War, when it became necessary to replace No. 77 Squadron's Mustangs (q.v.) with jets, the Australian government ordered the Meteor Mark 8 because the more advanced Sabre (q.v.) was unavailable. The Meteor Mark 8 was a development of the original Meteor design, which was the first British jet and had seen service during the Second World War. The Meteor entered service in Korea in 1951 but was no match for the MiG-15 in air-to-air fighting. Within six months the Meteor had been withdrawn from air-to-air combat and assumed a ground-attack role from January 1952 to the end of the war. Only three MiGs were shot down for the loss of 53 Meteors. After replacement by the Australian-built Sabre, some served with the Citizen Air Force until 1960.

MIDDLE STRATEGY was one of four strategic options discussed by Prime Minister John Curtin (q.v.) during his visit to London in the first half of 1944 as the basis for British Commonwealth participation in the final stage of the Allied war against Japan. It assumed an advance by British forces from northern Australia by way of Ambon (q.v.) and Borneo to Saigon. It was in fact a compromise predicated on the assumption that the American advance northwards through the Philippines would not be as rapid as the American planners assumed, and was itself subsequently modified to begin from the northern coast of reconquered New Guinea and bypass Ambon. The alternatives were CULVERIN, based from India and designed to move down through the Indian Ocean to Malaya and Singapore and then northwards to Formosa, and the Pacific strategy, based on Australia and involving a movement northwards through the islands towards Japan itself. These discussions were held in the context of decisions on the future direction of both the British and Australian war efforts in the Pacific after the defeat of Germany.

MIDGE, HMQS see **QUEENSLAND MARINE DEFENCE FORCE SHIPS**

MIDGET SUBMARINES see **JAPANESE ATTACKS ON AUSTRALIA**

MIHILIST, a database developed by the Australian Defence Force Academy (q.v.) Library, contains over 11 000 records for books, journal articles and other sources relating to the military history of Australia, New Zealand and Papua New Guinea. MIHILIST can be used either at ADFA or via remote dialling from elsewhere, and keyword searching allows users to find sources rapidly by referring to specific topics.

MILDURA, HMAS see **BATHURST CLASS MINESWEEPERS (CORVETTES)**

MILDURA, HMS see **AUXILIARY SQUADRON, AUSTRALIA STATION**

MILFORD, Major-General Edward James (10 December 1892–10 June 1972) was educated at Wesley College in Melbourne and RMC Duntroon, graduating in July 1915. He served with the AIF as an artillery officer, and on the staff of the 3rd Brigade and the 2nd Division, and spent two years with the British Army on attachment between 1919 and 1921 before returning to a succession of staff jobs in Australia. For his war service he was awarded the DSO and was mentioned in despatches. After a lengthy period as staff officer to the High Commissioner's Office in London between July 1925 and February 1929, Milford attended the Staff College, Camberley, in 1930–31. He was Assistant Director and Director of Artillery at Army Headquarters from 1936 to 1940. Highly experienced as both a gunner and a staff officer, he was seconded to the 2nd AIF in April 1940 and commanded the 7th Division artillery, although without seeing action. He returned to Australia in January 1941 to become Master-General of the Ordnance and 4th Military Member of the Military Board; he stayed at Army Headquarters until April 1942. Returning to active command, he was given the 5th Division, a militia formation, and led it in the New Guinea campaign until January 1944. A six-month period as Major-General, General Staff on the headquarters of New Guinea Force (q.v.) in the first half of 1944 was followed by appointment to command the 7th Division, which he led in the landing at Balikpapan, Borneo, on 1 July 1945. He was made Deputy Chief of the General Staff in March 1946, and Adjutant-General on the reconstituted Military Board (q.v.), a post which he occupied until he retired in 1948.

MILITARY BANDS, in musical terminology, are bands made up of brass, woodwinds and percussion, and the musicians need not be military personnel.

The first armed forces band to arrive in Australia was that of the British Marines, who came out with the First Fleet, and they were followed by other British regiments who also brought their own bands. At this time the role of bands was simply to pass on orders musically, to provide a beat for marches and manoeuvres and to entertain the civilian population. The first band of an Australian volunteer force was raised in 1861, and the Garrison Artillery Band of Sydney (today known as the Australian Army Band, Sydney) became the first Permanent Army band when it was formed in 1874. From the 1880s until the 1960s, the army maintained mostly brass bands, and band members were also trained as stretcher-bearers. In addition to the 60 infantry battalion bands which served overseas in the First World War and the 65 which were raised during the Second World War, Australian Army bands have served in the Sudan, Japan, Korea, Malaya, Borneo and Vietnam.

The appointment of Captain R. A. Newman as the army's first Director of Music in 1951 and the establishment of the Army School of Music at Balcombe in 1953 marked the beginning of the modern development of army bands. All Regular Army bands were converted from brass to military instrumentation between 1964 and 1974, and the Australian Army Band Corps was created in 1968. Musicians in battalion bands were transferred to area bands, and the battalions created new bands (mostly pipe and drum) from within their own ranks. Bands of the Army Reserve are maintained part-time in area and regimental roles.

The first RAN bandsmen were trained in Britain in 1913. Members of the navy's Band Branch, all of whom are trained to undertake other tasks at sea in addition to their musical duties, served in both world wars and in the Korean War. The navy's first Director of Music, Commissioned

The band of the 1st Battalion at Mena Camp with the Sphinx and a pyramid in the background, Egypt, 9 January 1915. (AWM P1143/01)

Bandmaster C. G. Maclean, was appointed a few years after the end of the Second World War, and the RAN School of Music at HMAS *Cerberus* (q.v.) was established in 1951. Australian Navy bands maintain their strong links with the British Royal Marines Band Service, and an exchange program was instituted in 1955. As well as serving at sea, naval bands are stationed at some establishments on land, and there are Naval Reserve bands in the capital cities of all Australian States.

The RAAF Central Band was formed in 1952 at the RAAF base at Laverton, Victoria, under the direction of Squadron Leader L. H. Hicks. Another band, based at RAAF Base Richmond, has had several names over the years, but is now known as the Air Command Band. It began as a part-time brass band during the Second World War, and it was not until 1969 that the band became a full-time unit with military instrumentation under the direction of Pilot Officer Mike Butcher.

In 1983 the Army School of Music moved to Watsonia, near Melbourne. It was redesignated as the Defence Force School of Music in 1984 and given the responsibility of training musicians of all services. Music continues to play an important role in service life, and bands provide music for ceremonial and other public occasions. Australia has yet to produce a distinctive style of military music, and its bands remain in the British tradition. Australian composers have, however, been active in this field, most notably the internationally renowned 'Sousa of the Antipodes', Alex Lithgow (1870–1929).

MILITARY BOARD OF ADMINISTRATION, established on 12 January 1905, was charged with the administration of the AMF. The regular members of the board were to be the minister for Defence, the Deputy Adjutant-General, the Chief of Intelligence and CGS, the Chief of Ordnance and a civilian Finance Member. Consultative members from the CMF could also be brought in when required. No military member was supposed to have supreme authority, and the board was made subject to the control of the minister, thus ensuring that senior officers could not exercise command independent of the government. In July 1942 new regulations stated that in time of war the Military Board's functions would be assumed by the commander of the military forces, so from July 1942 to March 1946 the board was dormant and General Sir Thomas Blamey (q.v.) exercised its powers. The Military Board was abolished on 9 February 1976 along with the Air and Naval Boards (qq.v.).

MILITARY DISTRICTS The military district has been the basic administrative unit of the army for most of the century. Australia is divided into seven military districts, each of which equates more or less to the boundaries of a State or territory. After Federation the former colonies were designated as military districts, which further emphasised the continuity in military practice between pre- and post-Federation eras. Queensland became the 1st Military District (styled 1MD); New South Wales 2MD; Victoria, 3MD; South Australia, 4MD; Western Australia, 5MD and Tasmania, 6MD. The Northern Territory was initially unallotted, being the administrative responsibility of South Australia until transferred to the Commonwealth, at which point it was administered by 1MD as, subsequently, was Papua New Guinea following the First World War. During the interwar period the administrative designation was district base (2DB,

3DB, etc.), but little else changed. At the beginning of the Second World War there was some reorganisation into geographic commands, with 2MD becoming Eastern Command, 5MD designated Western Command, Northern Command covering Queensland, and Southern Command encompassing the three southern States. In 1939 the Northern Territory had been designated a separate military district in its own right (7MD), informally sanctioned in March 1939, while Papua New Guinea became 8MD with the extension of the provisions of the Defence Act in October that year. This system lasted only until the reorganisation of the army command structure in early 1942, after which Queensland, New South Wales, Victoria, Tasmania and South Australia were redesignated Lines of Communications Areas, Western Command remained unchanged, 7MD became Northern Territory Force and 8MD became New Guinea Force (and, ultimately, First Australian Army). The geographical commands were reestablished after the war and existed until the system of functional commands was set up in 1972. In the 1990s the seven military districts are organised on much the same basis as previously, with the exception that 7MD now covers both the Northern Territory and the Kimberley region of Western

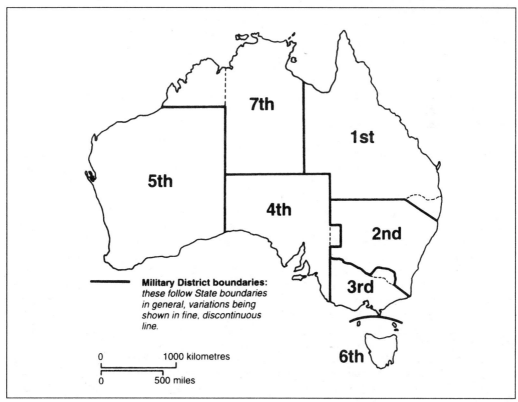

Map 21 The seven military districts of Australia in the 1990s.

Australia. Their function is management of army assets and to provide general support to units, Regular or Reserve, within their area, and their commanders may exercise some of the functions delegated by the functional commands. They are also the point of contact between the army and State government.

MILITARY EDUCATION An important distinction must be drawn between education and training, the latter generally being intended to impart specific skills or service qualifications. The military education system is largely confined to officers and the length of courses varies from six months to four years. Education in the military begins with the entry of an officer cadet or midshipman into either the Australian Defence Force Academy (q.v.), or one of the three single-service colleges: RMC Duntroon, RANC Jervis Bay and RAAF Academy Point Cook (qq.v.). Education at ADFA is conducted by the University College of the University of New South Wales and leads to the degrees of Bachelor of Science, Engineering or Arts, with the option of honours for those suitably qualified. Graduates then pass to the single-service institutions for pre-commissioning training and for education in requirements specific to their service. Officer cadets or midshipmen who enter these latter institutions directly may already have degrees from civilian universities, or possess no qualifications at all, and the course for them is longer than for those who have attended the Defence Academy.

Before the establishment of ADFA in 1986, degree education was conducted for the army by the Faculty of Military Studies at RMC Duntroon, a program established in 1967. From its opening in 1911 until the mid-1960s, the four-year course at Duntroon had no independent standing, although in 1921 the first year was recognised by the universities as equivalent to matriculation (a much rarer phenomenon then than now) and in 1920 the Universities of Sydney and Melbourne agreed to admit graduates of RMC into engineering degrees at third year level, to complete the last two years of that course. As a result of recommendations made in 1946, courses in arts and science were introduced in 1947, which were intended to bring graduates to a standard equivalent to the first two years of comparable university courses, while the earlier arrangement covering engineering was strengthened with the introduction of a department of engineering in 1948, graduates then proceeding as before to the University of Melbourne to complete the final two years there. These programs were incorporated under a Director of Civil Studies from 1951.

The air force eventually established a degree course at Point Cook which offered courses in science and engineering and which was organised under the auspices of the University of Melbourne. Before this RAAF officers had been trained at RMC Duntroon between the wars, and a number of such graduates was instrumental in the foundation of a separate RAAF Academy in 1947.

The original courses at RANC Jervis Bay consisted of four years of secondary schooling for the entrants, who enlisted at the age of thirteen until 1955 when that system was abolished, although in the 1950s several variations on entry were introduced. Future engineering officers studied subjects approved by the University of Melbourne, although from 1922 the Passing Out examination completed before graduation was generally regarded as the equivalent of matriculation. Engineering officers continued to study at the University of Melbourne until 1960, when the Naval College entered into an agreement with the University of New South Wales in Sydney. Under further revisions to the system of education, from 1968 the first year of study was spent at Jervis Bay before the midshipmen moved on to the Sydney campus to study full time in arts, science or various branches of engineering. Those not wishing to undertake degree education took the 'Cresswell Course', introduced in 1968. From 1981 it received accreditation as a Diploma of Applied Science. All these separate degree arrangements were terminated with the opening of the Defence Force Academy.

Higher military education has been provided by a range of service institutions. The army's staff college was founded in Sydney in 1938, before which time officers attended either Camberley or Quetta (as a small number have continued to do each year). RAAF officers went to the RAF Staff College at Andover, RAN officers to the RN Staff College at Greenwich. The army's Command and Staff College is now situated at Queenscliff outside Melbourne, the RAAF Staff College is located at RAAF Base Fairbairn in Canberra, having been established in 1949 at Point Cook, while the RAN Staff College is in Sydney. The courses vary in length from 6 to 12 months and are intended to provide selected officers with higher level military skills to equip them for staff officer positions (especially in the various headquarters in Canberra) and to broaden their understanding of contemporary military, political and social issues sufficient to prepare them for promotion to higher rank. As with the gradual process of upgrading the qualifications conferred by the service colleges, the staff colleges have been assessed for accreditation of their courses sufficient to lead to the award of graduate tertiary qualifications. The same has occurred at the Joint Services Staff College, located in Canberra, which fulfils the same range of functions as the staff college for more senior officers and at a higher level. From 1995 a new National College of Defence and Strategic Studies began operations in Canberra, aiming to provide the type of high level preparation for very senior rank formerly obtained through attendance at the Imperial Defence College (later called the Royal College of Defence Studies [q.v.]) in London. All these colleges also cater for small groups of foreign students, many of them under Defence Cooperation (q.v.) Programs. In addition to education delivered through service institutions, all three armed forces make provision for civil schooling, enabling selected officers to undertake degree-level studies at civilian universities, while the army maintains a junior officer education scheme intended to raise the educational level of the officer corps through a structured program of external study administered by the University of New England.

MILITARY HISTORY AND HISTORIANS Published military history in Australia was represented for a long time by the official histories (q.v.), edited by C. E. W. Bean and Gavin Long (qq.v.), and by little else. Various explanations have been advanced for this: the overwhelming nature of the two massive official series is alleged to have stifled further investigation; the tiny size of the Australian university sector until after the Second World War (in the 1930s there were six professors of history in Australian universities with perhaps a dozen teaching staff, mostly part-time or temporary); the fact that Australian history generally, with a few notable exceptions like Keith Hancock and Ernest Scott (q.v.), was not taught until after the Second World War; that the leftist tradition which allegedly characterised many aspects of Australian intellectual life was inimical to the serious study of war and the military; and, at a practical level, that the provisions of relevant archival legislation until the 1970s restricted treatment of issues less than 50 years old. Probably it was a mixture of all of these, but the effect, until fairly recently, was to result in the writing and teaching of a national history which failed, in the words of one historian from the Left, Duncan Waterson, to place 'defence and foreign policy in its proper context as one of the dynamic … issues in Australian history in the twentieth century'.

Before the 1960s the treatment of war and the military was occasional, episodic and unconnected. As a result of moves in Britain to improve the general education of officers, in November 1904 the universities of Sydney, Melbourne, Adelaide and Tasmania were asked if they would offer instruction in subjects with a military orientation as part of a putative Empire-wide scheme. Only Sydney encouraged the idea, and in December 1905 set up a Department of Military Science to oversee its diploma-level course. About 25 diplomas were awarded before the whole idea faded away in the mid-1920s. At the newly established Royal Military College, Duntroon (q.v.), the original syllabus contained 148 hours of instruction in military geography and history in the four-year course, and when revised in 1926 a total of 146 hours was set aside for the study of military history — but this was available only to staff cadets at the college, who were few in number in the interwar period and who had little or no opportunity to develop an interest in the subsequent course of their careers, and in any case much of what was taught dealt with non-Australian subject matter, the Shenandoah Valley campaign in the American Civil War being a hardy perennial in the syllabus for years. The navy and the air force, more technically oriented, did little or nothing for decades to encourage the study of military history generally or their own history specifically. From time to time officers would write papers of historical interest for British military journals (there being no Australian journal of the profession of arms until after the Second World War) or RSL publications, while the interwar years saw a steady trickle of unit histories (q.v.) and some memoirs. But works of history such as G. L. Macandie's (q.v.) on the genesis of the RAN were exceptions to the rule, while with the further exception of the *Cambridge History of the British Empire*, volume 7 of which, devoted to Australia, edited by Ernest Scott and published in 1933, dealt at some length with the First World War, general histories of Australia like Keith Hancock's *Australia* (1931) had little to say on the subject.

When Australian history began to feature on university courses after 1945, the emphasis at least initially was on the colonial period, with little or nothing to say about the British forces in the colonies before 1870, the growth of notions of imperial defence (q.v.), or the establishment of colonial military and naval forces (qq.v.). The teaching and writing of military history underwent a fundamental change from the 1960s, partly as a consequence of the growth of university history (by 1971 there were about 750 historians employed in the tertiary sector), partly through the influence of the Vietnam War and the debate over conscription (q.v.), partly through the influence and example of a small number of key figures. The Vietnam War prompted academics and students to look at the earlier conscription debates of the First World War, resulting in some productive historical writing which looked not only at issues of sectarianism and class but at the whole basis of the raising of the AIF. In this the Melbourne University historian Lloyd Robson was central. From the early 1970s he offered the first university course specifically devoted to the Australian experience of war, followed soon after by Jim Main at Flinders University in South Australia. Earlier still the Australian National University's Ken Inglis had published a number of seminal articles on C. E. W. Bean, Anzac, and the Gallipoli campaign which helped to define the study of Australia in the First World War, at least, for the next several decades. Much of this teaching and research was, and remains, firmly focused on the home front and adopts a 'war and society' approach, but the growth in research theses and published work has led to a broadening in the areas of military history now deemed suitable for scholarship.

By 1990 there were at least seven university history departments offering undergraduate courses in Australian military history, while these and other departments offered courses in other areas of the subject: Monash University's first-year subject on the Second World War, for example, was its single most heavily subscribed offering, while in the 1980s Sydney University had mounted a number of later-year units in, for example, the Hundred Years' War. On the other hand, it remained the case that operational history and the study of strategy and military policy were much less widely apparent, and military history generally tended to remain the interest of a distinct minority of university historians. Some of the isolation inherent in such a situation was broken down in the 1980s through the efforts of the Australian War Memorial (q.v.), which began to modernise itself and reach out to the scholarly community not least through its series of annual conferences and the publication of its journal. Publishers likewise developed an awareness of the appeal of military history for the book-buying public, with Melbourne University Press and Allen & Unwin in particular building substantial lists in the field.

In the armed forces there were developments also, broadly in keeping with the movement in civilian education. RMC Duntroon became a degree-conferring institution from 1967, and under the tutelage of A. J. Hill, Gerald Walsh, Robert O'Neill (q.v.) (a future official historian) and L. C. F. Turner (himself a former South African official historian), young army officers were encouraged to consider research degrees in history, and inevitably for many, given their chosen profession, in military history. From the 1970s

books by D. M. Horner, Chris Coulthard-Clark, Peter Pedersen and others began to appear, while staff in the history department at RMC, notably Hill, John Robertson and John McCarthy, began to fill in the numerous blanks on the map of Australian military history. This legacy has been built upon in the history department at the University College, Australian Defence Force Academy (q.v.), since 1986 arguably the centre for academic military history in the country. The RAAF established the Air Power Studies Centre in Canberra in 1989, part of whose mission is to facilitate the study of the history of the air force and of air power generally. The RAN has lagged badly behind the other services in this area, but in 1994 it finally appointed a historian with responsibility for fostering naval history within its Maritime Studies Project. The army also supports research through its Army History Research Grants scheme. Outside the universities and the ADF, interest in military history has grown steadily through membership of the Military Historical Society of Australia and publications by popular and amateur historians. While a lot of this involves the retelling of well-known events for new readers (often heavily based on the official histories) or satisfies a minority specialist or antiquarian interest, some of it is genuinely useful in opening up neglected areas of enquiry, and occasionally it results in first-class historical work.

MILITARY LAW see **LAW, MILITARY**

MILITARY MISSION 104 see **INDEPENDENT COMPANIES**

MILITARY SLANG A writer in *Army Magazine* (June 1944) commented on the experience of soldiers in northern Australia and in the islands to the north of Australia: 'thousands of Diggers complain humorously that they are "going troppo", which means degenerating into mild imbecility through tropical conditions. When the war ends there won't be so many to whom those conditions apply, but the man with a fishy gaze and sluggish limbs is almost certain to be for ever described as "troppo"'. The writer was correct, and the term, especially in the phrase 'to go troppo' has found a permanent place in Australian idiom. The First and Second World Wars produced many such slang terms which have enriched the lexicon of Australian English. The Vietnam War, however, produced few if any distinctively Australian terms. The history of Australian military slang is a significant index of the wider public's attitudes towards, and involvement in, Australia's wars.

Military slang, as with any kind of slang, has the function of uniting groups and defining their values. Since the Australian military is made up of many groups, each of these has its own special language, although there are overlaps in terminology between these Australian groups and worldwide military slang. Navy slang, especially, tends to be international (originally British) slang. Even so, many distinctively Australian Navy terms have been produced; they include: 'beagle' a steward; 'boy scout's leave' a brief shore-leave; 'dimple' a hole in a ship's hull caused by a torpedo; 'drain the bilge' to be extremely seasick; 'macaroon' a new rating; 'molly' a malingerer; 'pump packing' tough steak; 'redders' tomato sauce; 'squarie' girlfriend; 'squid' female trainee. Air Force slang is also greatly influenced by British traditions, but there have been many Australian terms: 'blear' (when lost) fly about in search of a landmark; 'blind stabbing' blind flying; 'emu' member of the ground staff; 'flaming arsehole' the large red circle painted on the side of a Japanese plane; 'nest' an aerodrome (to which all the little aircraft fly home); 'wags' signallers. Such language is generally known only to the members of the service or to groups within a particular service. The wider community has access to this language only in exceptional circumstances, as when in 1993 the publicity surrounding incidents on board HMAS *Swan* brought into the public arena the Navy slang term 'swab' (sailor without balls, i.e. a female sailor), and when in 1992 charges of bastardisation at the Australian Defence Force Academy similarly made the term 'woofing' (the application of a high-powered vacuum cleaner to the genital area) widely known.

Most of the RAN and RAAF terms noted above were in use during the world wars. But it is the army which carries the numbers, and it is the army which has produced the bulk of the military slang which has found its way back into the wider Australian community. The huge scale of the two world wars suddenly threw together people from vastly different backgrounds, people who had no other reason than the fact of war itself for living together in extraordinarily close and intimate circumstances. In the Introduction to his *Digger Dialects* (1919) Downing comments: 'By the conditions of their service, and by the howling desolation of the battle-zones, our men were isolated during nearly the whole of the time they spent in theatres of war, from the ways, the thoughts and the speech of the world behind them.' This opening sentence of Downing's Introduction is interesting because it demonstrates the lexical inventiveness of wartime experience (this is the earliest recorded occurrence of the term 'theatre of war'), and because it explains the reason for that inventiveness: since warfare is a

liminal experience, those involved in it need a new language to adapt to their new situation, and to construct ways of coping with it. When Tom Skeyhill in 'Soldier Songs from Anzac' (1915) wrote 'We've forgotten all our manners / And our talk is full of slang', he similarly points to the liminality of the war experience.

It is inevitable that most terms do not survive their wartime contexts, for the end of a war brings to an end the need for the existence of such terms. This is illustrated by the following terms from Downing's *Digger Dialects* (most of them confirmed by the 1924 typescript 'Glossary of Slang and Peculiar Terms in Use in the AIF' by A. W. Bazley et al., held at the Australian War Memorial): 'Anzac button' a nail used in place of a trouser button; 'Anzac soup' shell-hole water polluted by a corpse; 'Anzac stew' an urn of hot water and one bacon rind; 'Anzac wafer' a hard biscuit supplied to the AIF in place of bread; 'beer-swiper' a drunkard; 'belly-ache' a mortal wound; 'boy-with-his-boots-off' a shell which bursts before the sound of its passage through the air is heard; 'broken-doll' an inefficient staff-officer returned to his unit; 'camouflaged Aussy' an Englishman serving with the AIF; 'to go into cold storage' to be killed during the 1916 winter; 'dugout king' an officer who remains at the bottom of a dugout, while his men are exposed to danger; 'feed bag' a variety of gas helmet used early in the war; 'floating kidney' a soldier unattached to any unit, or without definite duties; 'goldies' teeth; 'lance-corporal bacon' bacon consisting of fat through which runs a thin streak of lean; 'Noah's doves' reinforcements who were at sea on their way towards a war zone at the time the Armistice was signed; 'Parapet Joe' a German machine-gunner, who attempts by continuous fire to prevent our men from looking over the parapet; 'smudged' killed by being blown to pieces by a shell. Most of the terms picked up by the soldiers from foreign languages also did not survive the war's end.

Yet while these terms have been lost, the First World War produced a number of major Australian cultural icons, especially the terms 'Anzac', 'digger' (qq.v.), and 'Aussie'. The term 'Anzac' appears in 1915 (in C. E. W. Bean's [q.v.] diary) as an acronym formed from the initial letters of *Australian and New Zealand Army Corps*, originally used as a telegraphic code name for the corps. In the same year it was used as an abbreviation for 'Anzac Cove' at Gallipoli, and then as a term for the 'Gallipoli campaign' (in *'Honk!' the voice of the benzine lancers and organ of the gear-box musicians*, France, 1915: 'The whole Italian Press praises the valour of the Australasian troops in the Dardanelles at Anzac'). In 1916 it was first used to refer to a member of the

Australian and New Zealand Army Corps who served in the Gallipoli campaign. By the end of the war the term was being used emblematically to reflect the traditional view of the virtues displayed by those who served in the Gallipoli campaign, especially as these virtues are seen as national characteristics. The term 'digger' in the military sense is a transferred use of the meaning 'a miner on the Australian goldfields'. Throughout the century it has retained the military associations established in the First World War (it was widely used during Second World War, and during the Vietnam War the Americans still knew the Australians as 'diggers'). The term has also undergone a widening of meaning — in many contexts 'digger' and its abbreviated form 'dig' are used devoid of their military connotations as a synonym for 'cobber' or 'mate'. It was the First World War which produced the term 'Aussie' for Australia (1915: 'A farewell dance for the boys going home to "Aussie" tomorrow'), for an Australian soldier (in 1918 the Sydney *Truth* writes: 'We consider the term Aussie or Ossie as evolved is a properly picturesque and delightfully descriptive designation of the boys who have gone forth from Australia'), and then more generally for 'an Australian' or 'Australian' (1916: 'One of our Aussie officers').

Many other common Australian terms had their origin in the First World War. The firm J. Furphy & Sons Pty Ltd operated a foundry at Shepparton, Victoria, and water-carts were included among their products. These water-carts, bearing the name 'Furphy', were used in the First World War. Very quickly the term 'furphy' (q.v.) came to mean a rumour or false report, an absurd story — perhaps because drivers of the carts were notorious for bringing rumours into the camps, or because the conversations which took place around the cart were sources of gossip and rumour. The term 'oil' in the sense 'information, news' (a transferred use of 'oil' as the substance essential to the running of a machine) and its compounds 'dinkum oil', 'straight oil', and 'good oil' all gained wide currency as First World War services slang. The term 'possie' for 'position of supposed advantage to the occupant; a place; a job' is now so entrenched in Australian English that few realise it had its origin in trench warfare as the term for an individual soldier's place of shelter or firing position. It is in First World War Australian military contexts that 'souvenir' in the sense 'to appropriate; to steal; to take as a souvenir' first appears. The term 'plonk' (probably a corruption of French *blanc* in *vin blanc* 'white wine') appears to have begun its Australian career during the First World War. It is in First World War Australian military contexts that many Australian idioms

are first recorded: 'his blood's worth bottling', 'give it a burl', 'hop in for one's chop', 'come a gutzer', 'rough as bags'.

During the Second World War, many earlier terms were revived. The term 'chocolate soldier' and its abbreviation 'choco' (q.v.) (the term derives from George Bernard Shaw's 'chocolate cream soldier', a soldier who will not fight, in *Arms and the Man*) appeared in the First World War for a soldier in the 8th Infantry Brigade, so called because the Brigade arrived in Egypt after the Gallipoli campaign. Downing reports that while it was originally an 'abusive term' it became 'an honourable appellation'. In the Second World War it was used of a militiaman or conscript, called up for home service and unable, before 1943, to serve outside Australia and its territories. Yet the Second World War proved to be as productive as the first in the creation of distinctively Australian terms. The phrase 'wouldn't it' (elliptical for 'wouldn't it root you') as an exclamation of dismay, exasperation, or disgust, had its origin in the Second World War, and summed up for many the frustrations of military life.

Terms which had their origin in Second World War slang, and which have found their way into the wider Australian lexicon include: 'acre' arse, anus; 'animal' as a term of contempt; 'back-up' second helping of food'; 'blot' anus; 'blue' a fight, altercation, disagreement; 'bore it up (someone)' attack with vigour, let (someone) have it; 'break it down!' desist!; 'bronze' anus, backside; 'bronzed (Aussie)'; 'come the raw prawn' attempt to deceive; misrepresent a situation; 'cruet' head (especially 'silly in the cruet'); 'dinger' anus; 'doover' thing for which the name is unknown or forgotten; 'cut-lunch commando' soldier serving with a home base unit; 'drack' dreary, unprepossessing; 'fuzzy-wuzzy' indigenous inhabitant of New Guinea; 'fuzzy-wuzzy angel' (q.v.) one who gave assistance to Australian military personnel, especially as a stretcher-bearer; 'game as a pissant' brave; foolhardy; 'giggle-suit' fatigue dress; 'go through' make a speedy departure (originally, go AWOL [absent without leave]); 'goldfish' tinned fish; 'grog on' engage in protracted drinking; 'homer' a wound of sufficient severity to ensure a recipient's repatriation; 'it's on (for young and old)' description of a battle, party, argument, et cetera, characterised by the participants' lack of inhibition or restraint; 'jungle juice' any crude alcoholic drink (originally 'Services' speech' in New Guinea); 'Lady Blamey' beer bottle with its neck removed, used as a drinking vessel (after the wife of General Sir Thomas Blamey [q.v.]); 'no-hoper' an incompetent or ineffectual person; a failure; 'nong' a fool; 'pongo' an Englishman (transferred use of the sense 'soldier, marine'); 'rooted' exhausted; 'shiny-arse' an office worker (originally one who had a desk-job at a base unit); 'shit-kicker' now an unskilled worker, a person of little consequence; originally a ranker, one of the boys; 'shoofty' a look; 'shoot through' a variant of 'go through'; 'retread' retired person (originally, a discharged soldier) who is re-engaged; 'spine-bash' to rest or loaf.

Compared with that of the two world wars, the Australian slang of the Korean and Vietnam Wars has been very thin. The term 'nog' (also 'noggie', 'Nigel the nog', 'Nigel'), which had its origin in the Korean War, was well known to the Americans as an Australian term for a Vietnamese soldier, but the wider public is probably more familiar with the American terms for the enemy: 'charlie', 'dink', 'gook', 'slant', 'slope', 'zip'. The Australians used the term 'hutchi' for a soldier's tent in the field, rather than the American 'hooch' or 'hoochie' (although the American form often appears in Australian writing on the War). This term was also established during the Korean War (a corruption of the Japanese word 'uchi', meaning house), but it is not one which has found its way into the wider Australian lexicon. There were revivals of earlier terms such as 'back up' for a second serving at a meal, 'base-bludger' for anyone not at the front line, and the widespread synonym for base-bludger, 'pogo' (an abbreviation of 'pogo-stick' rhyming slang for 'prick'). Perhaps the only distinctively Australian term widely known to both serving personnel and the general public was 'nasho' (a person who undergoes compulsory military training as introduced under the National Service Act [see CONSCRIPTION]; a conscripted soldier). Indeed, the language of the Vietnam War is really the language of the United States. It is the language of 'fragging', 'zapping' and 'wasting'. This emphasis was perhaps inevitable, given the nature and extent of the American involvement in the war, but it also points to the increasing Americanisation of both the English language and the language of war. The ultimate Americanisation of the Australian experience of Vietnam occurred when the term 'Vietnam vets' ('vet' being the American abbreviation of 'veteran') became the standard appellation for 'returned Vietnam Service personnel'. On the right-hand stele of the Vietnam Veterans' National Memorial (q.v.) in Canberra, unveiled in 1992, are 33 quotations which were chosen from various submissions. A news release of 23 July 1990 asserted: 'These quotations will represent to future generations the unique vocabulary developed during the war.' Yet, compared with the language of previous wars, the language of the chosen inscriptions is remarkably bland, and it is dominated at the colloquial level by Americanisms. The American term 'dust-off' (referring to medical evacuation missions using helicopters) occurs on three occasions; the term 'Puff the Magic Dragon' for a gunship is similarly American. If the Vietnam War produced the 'unique vocabulary' of that 1990 press release, that vocabulary was fundamentally American; or, if it was 'uniquely Australian', the Australian lexicon of the Vietnam War, unlike the Vets themselves, has yet to receive its 'homecoming'.

W. A. J. Downing, *Digger Dialects*, 1919 (republished with additional material by J. M. Arthur and W. S. Ramson, Oxford University Press, Melbourne, 1990); Sidney J. Baker, *The Australian Language*, 2nd edn (Currawong Publishing Company, Sydney, 1966); Bruce Moore, *A Lexicon of Cadet Language, Royal Military College, Duntroon, in the Period 1983 to 1985* (Australian National Dictionary Centre, Canberra, 1993).

BRUCE MOORE

MILITIA see **CITIZEN MILITARY FORCES**

MILLAR COMMITTEE Chaired by the distinguished academic defence analyst Professor T. B. (Tom) Millar, this committee of inquiry was set up by the Labor Minister for Defence, Lance Barnard, in April 1973 to review the role, organisation and functions of the CMF. Such a review had been coming for many years, and was foreshadowed in the recommendations of the Hassett (q.v.) Committee in its 1972 report, which the then Liberal minister, David Fairbairn, had declined to take up. The Millar Committee deliberated for 12 months, at the end of which its chairman moved into a Canberra hotel room in order to write up the committee's report and recommendations, which he did in near-record time. While arguing that the CMF retained an important role in the defence forces, the report recommended that it be completely reorganised and redesignated the Army Reserve (q.v.). It should be functionally organised, and seriously under-strength units (defined as those at 70 per cent or more below establishment), of which there were many, were to be amalgamated. Headquarters were to be reorganised as well and a Reserves Branch created within Army Office. At least initially this involved the abolition of the CMF divisions and their removal from the army's order of battle. This and the revelations of ineffectiveness led to bitter and impassioned condemnation of the report and its chairman by a number of retired senior CMF officers, who tried to make play of the fact that Millar was an RMC Duntroon graduate (class of 1943) by depicting the report as the revenge of the Staff Corps for past slights. This was too ridiculous to be taken seriously, and Millar's reforms were broadly welcomed by younger CMF officers. Other recommendations, however, involving additional expenditure on the Army Reserve, were shelved. The report also made recommendations concerning the Army Cadet Corps (see AUSTRALIAN CADET CORPS). Millar continued to enjoy a distinguished academic career in Australia and Britain, and died in London in 1994.

Report of the Committee of Inquiry into the Citizen Military Forces (Australian Government Publishing Service, Canberra, 1974); *Committee of Inquiry into the Citizen Military Forces: Report on the Army Cadet Corps* (Australian Government Publishing Service, Canberra, 1974).

MILLEN, Senator Edward Davis (7 November 1860–14 September 1923) migrated to Australia from England as a young man in 1880 and worked as a journalist in western New South Wales. He entered the New South Wales Legislative Assembly for the seat of Bourke in 1894 as a Free Trader, supported the move for Federation while opposing many of the details, and lost his seat in 1899. He was elected to the senate in 1901, and held his seat until his death. Millen was minister for Defence in the Cook administration of 1913–14, and coordinated the government's response to the outbreak of war in August 1914, which included the raising of the first contingent of the AIF. Following the Labor victory at the election in September he became Opposition leader in the senate, and in 1915 became a member of the parliamentary subcommittee on repatriation, an issue which was to occupy him increasingly for the rest of his career. The Labor split in 1916 brought him back into office as Vice-President of the Executive Council, with responsibility for repatriation, and in September 1917 his responsibilities as minister for Repatriation (q.v.) were formalised. He created a new government department from scratch, one, moreover, whose functions were without precedent in Australia, and staffed it largely with returned servicemen. The early administration of repatriation programs was deficient in a number of respects, and Millen was attacked bitterly both inside and outside the parliament. He was acting prime minister in July 1919 while Hughes (q.v.) was overseas, but his health began to fail under the strain of his responsibilities. He continued in public life, however, representing Australia at the first meeting of the General Assembly of the League of Nations in November 1920 and reorganising the administration at Australia House in London. Writing once that repatriation was 'an emanation of the heart', Millen more than any other single individual may be regarded as the progenitor of the system of 'parallel welfare' which has served Australia's returned service personnel and their dependants for three-quarters of a century.

MILLS COMMITTEE (Select Committee of the House of Commons on Colonial Expenditure, chaired by Arthur Mills MP) was established in 1861 to resolve a long-running dispute between the War Office and Colonial Office over the apportionment of the costs of colonial defence. The War Office argued that where imperial garrisons were maintained essentially for the purpose of colonial, that is, local, defence, the costs ought to be shared between the imperial and the colonial government uniformly throughout the Empire, the imperial government's contribution being half the total cost. This was rejected by the Colonial Office, on the grounds that there could not be a clear distinction between colonies in that all were of varying degrees of value and interest to the imperial government, and also on the grounds that a uniform contribution system would be impracticable because it would not take into account local circumstances. These questions were debated within a wider agenda — pressure for military reform made all the more urgent by the manifest inadequacies of the British Army in the Crimean War, and demands for greater economy in government, defence expenditure being an obvious target for reductions.

The report of the committee rejected the proposal for uniform colonial contributions to defence costs, and concluded that colonies (as distinct from colonial territories that were military dependencies) should bear the main responsibility for and cost of their military defence. The committee argued that British troops should be concentrated as far as possible for the defence of the United Kingdom, while naval supremacy would provide the best protection for the 'distant dependencies' of the Empire. When the report was tabled on 4 March 1862, the House of Commons unanimously resolved that self-governing colonies should be entirely responsible for their internal order and security, and 'ought to assist in their own external defence'. While the report of the Mills Committee eventually led, in 1869–71, to the withdrawal of the imperial garrisons from Canada and the Australian colonies, the Commons resolution, the final words of which went further than the committee had ventured, opened another debate, which arose out of differing views of the extent to which the self-governing colonies should contribute to their external defence, what form that contribution should take, and what measure of control such contributions should bestow on the colonial governments. These discussions between the imperial government and the colonial governments led to the Anglo-Australian Naval Agreement of 1887 (q.v.).

MINING OPERATIONS (FIRST WORLD WAR) The first Australian involvement in mining and counter-mining occurred at Gallipoli in May

1915. The Turks began tunnelling towards Quinn's Post on 9 May and the Australians were only partially vigilant in their counter-measures, enabling the Turks to explode the mine on 29 May during an unsuccessful attack on the outpost. Both sides continued tunnelling and the Australians began using former miners to undertake or at least supervise the digging. Mines were exploded underneath Turkish positions as part of the evacuation from Anzac in December 1915. The Gallipoli experience led to the raising of two battalions of picked miners under the Professor of Geology at Sydney University, T. W. Edgeworth David, and their dispatch to England in February 1916.

As soon as the armies on the Western Front (q.v.) settled into trench warfare, attempts to undermine the enemy trenches began. The Germans struck first in late 1914 when they blew up a section of the British line near Kemmel. The British countered by forming a tunnelling force, which arrived in France in February 1915. The Australians arrived at the front in May 1916 and were engaged during the remainder of the year in minor operations in the Armentières area and at Fromelles. Their main work, however, did not begin until they were dispatched to Ypres sector later in the year. By then the original battalions had been reinforced and the whole force reorganised into three tunnelling companies. In October the lst Australian Tunnelling Company relieved the Canadians in the Hill 60 area. Mining and counter-mining had been under way in this area since 1915. The Australians greatly extended the Canadian tunnels, deepening and lengthening them. This work paid off in June 1917 when the Hill 60 mines were incorporated into the Second Army plan to blow the top off the Messines Ridge as part of a major attack. The Hill 60 mine was blown (with 18 others) at 3.10 a.m. on 7 June 1917. As a result the infantry were able to move forward and secure Hill 60 which had been in German hands since early 1915. Other operations in 1917 were not as successful. The 2ATC was sent to the Belgian coast near Nieuport later in June to assist in a British attack which was to be part of the great Third Ypres campaign. But the Germans struck first. On 10 July they attacked the British positions and gained some ground. Forty-six tunnellers were cut off in this operation and taken prisoner or killed.

During the German March offensive in 1918 various sections from the three tunnelling companies served as infantry. When it became the Allies' turn to advance the tunnellers proved useful in defusing booby traps and German mines. It can be fairly said that the Australians had an influence on mining operations out of proportion to their numbers. Professor David was influential in converting mining into a science. He was the first to recognise the importance of taking geological surveys of areas to be mined, and with Belgian help he located many water tables along the British front which enabled trenches to be resited before winter rains raised the water levels and washed the trenches away. His expertise was recognised when he became adviser to General Headquarters on geological matters for the entire British Expeditionary Force.

MIRAGE, GAF (Single-seat fighter [IIIO]). Wingspan 26 feet 11.5 inches; length 49 feet 3.5 inches; armament 2 × 30 mm cannon, 1 × Matra air-to-air missile, 2 × Sidewinder air-to-air missiles, 2000 pounds bombs; maximum speed Mach 1.14 at sea level, Mach 2.05 at 30 000 feet; range 745 miles; power 1 × SNECMA Atar 13670-pound thrust turbojet.

The Mirage was a delta-wing fighter able to fly at twice the speed of sound, designed by Dassault in France. The Mirage IIIO was a modification of the French IIIE design built by the Government Aircraft Factory, with engines built by the Commonwealth Aircraft Corporation (q.v.). The Mirage was the RAAF's first supersonic aircraft and entered service in 1964, replacing the Sabre (q.v.) It was flown by No. 3, 75, 76, 77 and 79 Squadrons at Williamtown (q.v.) and Butterworth (q.v.) but was never used in combat by the RAAF. The Mirage was withdrawn from service in 1989 and replaced with the F/A-18 (q.v.). In 1990 50 ex-RAAF Mirages were sold to the Pakistan government.

MITCHELL, NORTH AMERICAN (5-seat medium bomber [Mark III B-25J]). Wingspan 67 feet 7 inches; length 52 feet 11 inches; armament 12 × 0.5-inch machine-guns, 3000 pounds bombs; maximum speed 230 m.p.h.; range 1355 miles; power 2 × Wright Cyclone 1700 h.p. engines.

After the Japanese conquest of the Netherlands East Indies (q.v.) in 1942, 30 Mitchells ordered for the Netherlands East Indies Air Force were diverted to Australia and issued to the Dutch crews of No. 18 (NEI) Squadron RAAF. On 5 June 1942 one of these Mitchells sank a Japanese submarine off the south coast of New South Wales. More Mitchells were obtained from the United States, and in 1944 these equipped No. 2 Squadron. The Mitchell was used for bombing and strafing raids against airfields, military positions and shipping in Portuguese Timor and the Netherlands East Indies and served with the RAAF until the end of the Second World War. A former US Navy Mitchell repainted in RAAF colours is in the collection of the Australian War Memorial (q.v.).

MONASH, General John (27 June 1865–8 October 1931) was born in Melbourne, the eldest child of Louis and Bertha Monash who were of Prussian–Jewish origin. He was educated at Scotch College and the University of Melbourne where he graduated in engineering in 1893 and arts and law in 1895. It was at university that Monash first became interested in military matters. In 1887 he joined the Metropolitan Brigade of the Garrison Artillery which was engaged in port defence. In this capacity Monash could bring some of his mathematical and engineering skills to the science of gunnery. For many years Monash combined the professions of arms and engineering. His progress in the garrison artillery was painfully slow. He was promoted to captain in 1895 and to the command of a battery two years later; he was still a battery commander in 1907. His career seemed to be at a dead end. Then Colonel J. W. McCay (q.v.) offered him the post of Officer Commanding Victorian Section of the Australian Intelligence Corps. Monash accepted and threw himself into his new duties, studying military mapping and attending military science schools at the University of Sydney. In June 1913 he was promoted to colonel and given command of the 13th Brigade. It was during this time that Monash's *100 Hints for Company Commanders* was written. It was later to become a regularly issued training document.

On the outbreak of war Monash acted briefly as Deputy Chief Censor before he was given command of the 4th Brigade of the AIF. He sailed for Egypt with the second contingent on 22 December 1914. The 4th Brigade was located at Heliopolis as part of the New Zealand and Australian Division. At the Gallipoli landing Monash's force was in reserve. He landed early on the morning of the 26th and his brigade was given the left-centre of the perimeter to hold — near Pope's Hill and Quinn's Post. Of the early fighting in this area little need be said. Monash deprecated an attempt to push the line forward on 2 May and was vindicated by the unfruitful result. The 4th Brigade played a role in defeating the major Turkish attack on 19 May. On 29 May a much depleted 4th Brigade was relieved and was 'rested' in Reserve Gully. Monash's next chance came with the August offensive. This was a major attempt to outflank the Turkish positions facing the Anzacs, seize the Sari Bair Ridge which overlooked them, and then press on across the straits to the Narrows. The 4th Brigade was to form the spearhead of the left assaulting column. After a covering force had seized the approaches,

this column was to pass through and establish itself on Hill 971 and Hill Q. The brigade started well enough, although it was behind schedule due to the covering force taking longer to establish itself than originally planned. Soon, however, things went awry. The guides lost their way in the dark, the maps were found to be unreliable and the tortuousness of the country had not been fully grasped. Dawn found the column well short of its objective and unsure of its position. Sensibly Monash ordered his scattered force to dig in and await events. Next day (7 August) an attempt was made to renew the advance with Monash remaining at his headquarters and delegating the command to Colonel H. Pope. The result was a fiasco: Monash remained out of touch, and Pope was confronted with heavy Turkish reinforcements. The remnants of the brigade were forced to withdraw on the 8th. Simmering hostility between some of Monash's battalions arising out of mutual perceptions of poor performance resulted in some of the dead and wounded being abandoned. The morale of the brigade plummeted. Certainly this was not Monash's finest hour. Yet it must be stressed that no action, or lack of it, could have saved the offensive. The task given the 4th Brigade was beyond the capacity of any troops. In any case the attack to the south had also collapsed and this alone would have doomed the operation to failure. Monash's role in the August offensive was not quite over. British forces had landed at Suvla Bay and the main force at Anzac had only managed to establish tenuous contact. A small outcrop between them, Hill 60 which was held by the Turks, was considered essential to make the link between the bridgeheads secure. The success of the attack on 21 August depended on the elimination by the artillery of enfilading Turkish machine-gun fire. The artillery missed the Turks and the attacking waves of the 4th Brigade were shot down in large numbers. A second attempt four days later also failed. On 13 September the remnants of the brigade were embarked for rest on Lemnos. After a three week rest Monash returned to the peninsula. There was to be no more heavy fighting. In November a decision was taken to evacuate the troops. Monash left with the 4th Brigade in December 1915.

Back in Egypt Monash spent January 1916 training his brigade, which had largely been reconstituted. The brigade unit was then moved to the line of the Suez Canal and spent two uneventful months waiting for a Turkish attack which never materialised. In June the brigade, now part of the 4th Australian Division, moved to the Western Front where it went into the line at Armentières. In the same month Monash was promoted to major-general and given command of the new 3rd Division forming at Salisbury Plain in England. The next five months were spent in intensive training. Monash was assiduous in all training matters. His orders were comprehensive and lucid, the exercises held on the plain realistic. By November when 3rd Division went into the line it was as ready for battle as any new formation could hope to be. During the next five months the division became accustomed to trench warfare and prepared itself for battle. Many raids were carried out against the Germans and cost the division 2500 casualties during this period. Viewed in retrospect, most of these raids seem to have served little purpose. Monash was, however, an enthusiastic supporter of them. He considered they raised morale and acted as a 'powerful stimulant to all ranks'. Whether his men agreed can be considered doubtful.

Monash's first battle was at Messines — the attempt by General Sir Herbert Plumer's Second Army to capture a ridge overlooking the Ypres salient. The 3rd Division orders for this battle were notable for their comprehensive nature. Monash did not hesitate to designate tasks down to platoon level or to enumerate the number of picks and shovels each battalion would require. The British C-in-C, General Sir Douglas Haig, was particularly impressed by Monash's preparations, noting that 'Monash is in my opinion a clear-headed, determined commander. Every detail has been thought of. His brigadiers were equally thorough. I was most struck with their whole arrangements'. However, once battle was joined at dawn on 7 June Monash was as much confused by the fog of war as any other commander on the Western Front. Some of his instructions were unfortunate — he ordered the artillery at one point to shorten the range, not realising that this would result in its shelling his own troops. Other orders reached the front at a time when the situation to which they related no longer applied. Nevertheless Monash and his new division performed well in their first action. They captured all their objectives, 314 prisoners and 11 guns.

The next major British operation for the year was the Third Battle of Ypres. It began on 31 July when some ground was gained and continued through the mud of August when no progress was made. The Australians entered the battle on 20 September when, as part of Plumer's Second Army, they captured a section of the Gheluvelt Plateau during the Battle of the Menin Road. This was followed by a second success on 26 September at Polygon Wood. Monash's 3rd Division entered the campaign at this point as part of II Anzac Corps. They were to prepare for a third action at Broodseinde on 4 October. Once again Monash's plan was characterised by its detail and precision. The attack was divided into a succession of advances, each shorter than the last, each to be carried out by fresh forces. Each objective was carefully chosen to give the consolidating infantry protection from German fire and was clearly delineated for the follow-up waves. The battle, which was dominated by the artillery, went largely according to plan. Monash's arrangements worked well. The 3rd Division accomplished the deepest penetration of the German defences that day and successfully beat off German counter-attacks.

Monash's division was relieved after the battle and so missed the futile Battle of Poelcappelle on 9 October when rain rendered the ground unsuitable for operations. Worse was to come. The next operation for which the 3rd Division was designated, was to take place after only a three-day interval. Rain was still falling. There was on this occasion no time for careful preparations. Monash did not respond well to this situation. It was not known with any clarity after the attack on the 9th where the front line had come to rest. Australian patrols established that it was almost in exactly the same position as before the battle. Monash seems to have received these reports but took no action on them. Consequently the artillery barrage commenced where it was thought the front line was *before* the patrol reports — some hundreds of yards in front of the attacking infantry and falling short of the enemy lines. Without artillery protection the troops had no chance. The small advance cost 1800 casualties. To the extent that Monash ignored the patrol reports or refused to reconnoitre the front position himself, he can be held partially responsible for the toll. However he had no responsibility for ordering the attack in the first place and to that extent the responsibility must be shared by Plumer and Haig. After the débâcle on 12 October (the First Battle of Passchendaele) all Australian divisions were withdrawn. They re-entered the line in a quiet sector south of Messines where they spent the winter. In this period the divisions were reorganised into the Australian Corps with General Sir William Birdwood (q.v.) as commander.

The corps was intended to fight as a formation but the Ludendorff offensive on 21 March 1918 delayed that moment for some months. From March to May Australian forces were thrown piecemeal into battle primarily to help stem the German tide. The offensive found Monash resting on the French Riviera. On 24 March he was recalled and his division sent south to the Somme. There, it played its part in stopping the German drive on Amiens. In these actions Monash exercised little influence. Action often had to be improvised at a lower level without reference to higher command. In May Monash's career took a sudden upward turn. Birdwood was promoted to command the Fifth Army. A new corps commander was needed. The Australian official correspondent C. E. W. Bean (q.v.) and journalist Keith Murdoch (q.v.) favoured General Brudenell White (q.v.); but Birdwood insisted White accompany him and recommended Monash. Despite the best efforts of Murdoch and Bean over the next few months this appointment was approved. Monash was to command the Australian Corps in the climactic period of the war.

His first chance to exercise his new office came in July. It was a small affair involving the strengthening point of the Australian line near Hamel. The result was a triumph of careful planning. Monash's instructions, with the considerable aid of Major-General E. G. Sinclair-MacLagan (q.v.), the commander of the 4th Division, which was to carry out the operation, were models of their kind — detached, lucid and communicated to all ranks via a series of pre-battle conferences. Nevertheless, careful planning does not necessarily win battles. The Australians were supported by such firepower that the dispirited and poorly protected German units holding the line opposite them never stood a chance. The line was straightened, casualties were light. The great series of victories in which the Australian Corps participated from the Battle of Amiens on 8 August 1918 to the breaching of the Hindenburg Line (29 September–5 October) are described in the entry on the WESTERN FRONT. Here it is merely necessary to describe Monash's role in them and identify the characteristics of his command.

It must be said from the outset that none of the overall plans was devised by Monash. They were planned at the Fourth Army Headquarters by General Sir Henry Rawlinson and his staff. Nevertheless, the Australian Corps played an important role in all the actions and Monash's attention to detail and training ensured that his divisions went into action well-briefed and with clearly identified objectives. At Amiens the corps instructions ran to some 80 pages of detail — little was left to chance. The most controversial aspect of the campaign was Monash's conduct of operations after Amiens and before the attack on the Hindenburg Line. Since Amiens, casualties had mounted. Most divisions were well under-strength and in need of rest. Rawlinson was quite prepared in these circumstances to slacken the offensive. Monash ignored him. With Haig-like tenacity he relentlessly drove the Australian Corps forward against an enemy that he rightly judged was losing even defensive capa-

bility. Monash might have judged the enemy correctly but in some cases he drove his own men beyond endurance. Elements in the 1st Battalion mutinied in early September — more because of exhaustion than ill-discipline.

Nevertheless most troops continued to fight, in Bean's words, 'to the extreme limit of their endurance'. In these last campaigns Monash was not at his best. No doubt exhaustion was taking its toll on the Corps Commander as well. His plan for capturing the main Hindenburg Line was deeply flawed — and only redeemed by the intervention of the Fourth Army Staff. His plan to exploit a breakthrough with the cavalry was ludicrous — a throw back to the darkest days of the Somme. In the end the sheer weight of artillery and the determination of the troops finished the job. The last German line in the west was breached, and the Australians were withdrawn. The war ended before they could return. After the Armistice Monash continued to perform valuable services. He oversaw the harmonious demobilisation (q.v.) of 160 000 troops in the minimum possible time, overcoming considerable obstacles, including the attitude of the government. In postwar Australia he occupied a number of important positions, including management of the State Electricity Commission of Victoria, but he held no further military appointment. He died in 1931 laden with honours. He was appointed KCB in 1918, and GCMG later that year, mentioned in despatches five times, decorated by the French, Belgian and American governments, and awarded honorary doctorates by Oxford, Cambridge and Melbourne universities. In 1958 Monash University was named after him.

Monash was perhaps fortunate in the timing of his various levels of command. As divisional commander he served under Plumer, a most meticulous planner and in the main a careful commander. When he succeeded to the corps the Germans had squandered their last reserves in the March offensives, British industry ensured munitions in unprecedented quantities and technical experts had at last solved the problem of providing adequate fire-support for attacking troops. Despite these favourable circumstances, his personality should not be underestimated. He had a cool head, an ability to make rapid decisions, a facility for logical exposition, a warranted obsession with detail and the determination and ruthlessness to obtain the maximum effort from his troops. His reputation as Australia's greatest commander is secure.

MOOREHEAD, Alan (22 July 1910–29 September 1983). Moorehead was born in Melbourne in 1910 and educated at Scotch College and the University of Melbourne. During the 1930s he was a reporter on the *Herald*; he left for London in 1936 and became consecutively the Gibraltar, Rome and Paris correspondent for the *Daily Express*. On the outbreak of war Moorehead was transferred to Cairo where he became a war correspondent. Later he covered the European campaign from D-Day to the German surrender. His despatches identified him as a superb recorder of war. They were incisive, moving, witty and accurate. His work on the Western Desert campaign was later turned into three books: *Mediterranean Front* (1941), *A Year of Battle* (1943) and *The End in Africa* (1944), followed by *Eclipse* (1945) which dealt with the end of the war in Europe. In the postwar years he directed his considerable talents to historical writing. His biography *Montgomery* (1946) and his *Gallipoli* (1956) were definitive in their time and still yield many rewarding insights. In later years he turned his attention to historical travel writing, enjoying much success with *The White Nile* (1960) and *The Blue Nile* (1962).

MORANT, Lieutenant Harry Harboard 'The Breaker' (?9 December 1864–27 February 1902) and **HANDCOCK, Lieutenant Peter Joseph** (17 February 1868–27 February 1902) were born in England and Peel, New South Wales, respectively. After migrating to Queensland in 1883, Morant gained a reputation as a horseman and bush balladeer; Handcock was a blacksmith and labourer. Both enlisted separately in 1899 in Australian contingents bound for the Boer War and by early 1901 were part of the Bush Veldt Carbineers (BVC), an irregular unit. Morant commanded a BVC detachment, which included Handcock, in the wild Spelonken district of the Transvaal from August to October 1901, when he and other BVC officers were arrested. Courts martial subsequently found Morant, Handcock and Lieutenant G. R. Witton guilty of killing 12 Boer prisoners. Morant accepted full responsibility for the killings, claiming they were justified by the difficulties of guerrilla warfare and by alleged orders from General Kitchener to take no prisoners. Morant and Handcock were acquitted of a further charge of killing a missionary, Heese, but this may nevertheless have been the charge which led to their sentence of death by firing squad. Other British units which shot Boer prisoners were never even charged, and though the case provoked little controversy in Australia at the time, later writers argued that Morant and Handcock had been scapegoated for British war crimes or that they had been executed to appease the German government (Heese was of German extraction). Witton, whose death sentence was

commuted to life imprisonment, argued their case in *Scapegoats of the Empire* (1907), but admitted privately in 1929 that Handcock had confessed to murdering Heese at Morant's instigation. As Australians executed by British authorities without the knowledge of the Australian government, Morant and Handcock became convenient symbols for those with an anti-British axe to grind, and the film *Breaker Morant* (1980) established them in the public mind as noble bushmen and courageous underdogs (see FILM AND TELEVISION, WAR IN AUSTRALIAN).

Margaret Carnegie and Frank Shields, *In Search of Breaker Morant: Balladist and Bushveldt Carbineer* (Graphic Books, Melbourne, 1979).

'MOROTAI MUTINY' see **FIRST TACTICAL AIR FORCE**

MORESBY (I), HMAS see **SURVEY SHIPS**

MORESBY (II), HMAS see **SURVEY SHIPS**

MORRIS, Major-General Basil Moorhouse (19 December 1888–5 April 1975). Morris had served in the school cadets in the non-compulsory scheme while a student at Melbourne Grammar School; although he spent a year at Melbourne University, he did not take a degree (largely precluding his entry into the Royal Australian Engineers, in which he initially expressed interest), and received a commission in the permanent forces in December 1910 in the Royal Australian Artillery. He served in the garrison artillery until secondment to the AIF in May 1915. In line with his specialisation, he first served in France with the 55th Siege Battery of the 36th Heavy Artillery Group until posting to the divisional (i.e. field) artillery of the 5th Division, with whom he served for the remainder of the war, including some time firing in support of American troops after the Australian infantry units were withdrawn from the line in early October 1918. Between the wars Morris endured the usual round of staff and command appointments, as an instructor in the artillery schools, commanding a brigade of coastal artillery between 1926 and 1931, and as Director of Supplies, Transport, Movements and Quartering at Army Headquarters from 1937 to 1939. Seconded to the 2nd AIF in November 1939, he was given command of the Australian Overseas Base in Palestine, where he remained until August 1940. He then received appointment as Australian Liaison Officer in India, based in Bombay. This was not as quixotic a job as its title might suggest, since India acted as a staging post for Australian transports bringing reinforcements from Australia for service in the Middle East, and experience had shown that the administration and discipline of Australian soldiers proceeded most smoothly when undertaken by Australian authorities. In May 1941 he was promoted to major-general and sent to command the newly created 8th Military District, based on the Australian territory of Papua. Having neglected its northern defences for years, the Australian government greatly reinforced the garrison at Port Moresby in January 1942 from a battalion to a brigade, but this was insufficient to prevent the capture of Rabaul (q.v.) by the Japanese or to prevent the latter's initial advances across the Owen Stanley Range (see NEW GUINEA CAMPAIGN). Morris was superseded in command of New Guinea Force (q.v.), as it became in August 1942, by Lieutenant-General S. F. Rowell (q.v.). He remained throughout the war as GOC of the Australian New Guinea Administrative Unit (q.v.), which he had set up in February 1942 to maintain the administration of the Australian territories following the dissolution of the civil administration, and he oversaw its important work in maintaining and reasserting Australian government among the indigenous people. In February 1946 Morris was reappointed as GOC of the 8th Military District to supervise the transition to peacetime conditions, testimony to the effectiveness of his wartime work in this area. He retired from the army in October that year. In retirement he contested Victorian State parliamentary seats for the Liberal Party in 1947 and 1950.

MORSHEAD, Lieutenant-General Leslie James (18 September 1889–26 September 1959). At the outbreak of war in 1914 Morshead was a schoolteacher, whose only exposure to military matters had been as a lieutenant of the cadets (see AUSTRALIAN CADET CORPS). That position nonetheless entitled him to apply for a commission in the AIF, and he was appointed a lieutenant in the 2nd Battalion in September 1914. At Gallipoli he led a platoon in the fight for Baby 700 and a company at Lone Pine, during which battle he was wounded. He was invalided to England with enteric fever in October 1915, and then back to Australia. Recovered, he was appointed in command of the 33rd Battalion, then being raised and trained as part of Major-General John Monash's (q.v.) 3rd Division. He took the unit to England for further training, and led it through the gruelling battles of 1917 at Messines and 3rd Ypres and in the retreats and advances of the following year. A distinguished wartime record was recognised in the award of the CMG, DSO and French Legion of Honour, and no fewer than six mentions in despatches. At the

end of hostilities he accompanied Monash to London and a position on the demobilisation staff.

Between the wars Morshead re-established himself in civil life, and continued soldiering in the CMF. He commanded a succession of battalions and infantry brigades in these years, but the experience can have added little to his knowledge of war. One of his Second World War brigadiers, Victor Windeyer (q.v.), later commented that Morshead was 'a bit "last war"' in his unwillingness to yield ground'; there were few opportunities for even senior CMF officers to keep fully abreast of developments in warfare in the 1930s.

At the outbreak of war he was selected to command the 18th Brigade, which was diverted to England in 1940 and thus later joined the 7th rather than the 6th Division. When Lieutenant-General H. D. Wynter (q.v.) was forced by ill health to relinquish command of the 9th Division, General Sir Thomas Blamey selected Morshead to succeed him. Still incomplete, ill-equipped and only partly trained, the division was sent to Cyrenaica in March 1941 (see LIBYAN CAMPAIGN, FIRST) to relieve the 6th Division for participation in the Greek campaign (q.v.). To the deficiencies within his formation were added serious shortcomings on the part of the British command which compromised the Australians' tactical positions from the outset. When the German attack began, the British were forced to retreat and the 9th Division withdrew on Tobruk (q.v.). Morshead became the fortress commander, and conducted a vigorous defence on a 'no surrender and no retreat' basis. Tobruk may have been a defensive battle for the Australians, but it was conducted in a thoroughly offensive spirit. Although on occasions the Germans pressed the perimeter hard, the defence prevailed, and the static nature of operations was advantageous to a division that had experienced some deficiencies in training and mobility. The 9th Division, less one battalion, was finally relieved in October, and Morshead was knighted early the following year.

When the rest of the 2nd AIF returned to Australia at the beginning of 1942, the 9th Division and its commander remained in the Mediterranean, participating in both the first and second battles of El Alamein; the Australians took particularly heavy casualties in the latter engagement. Thereafter they returned to Australia, and in October 1943 Morshead relieved Lieutenant-General E. F. Herring (q.v.) as GOC of II Australian Corps in New Guinea. His period as a corps commander was short; on 6 November Blamey elevated him to command of New Guinea Force and Second Australian Army, replacing him with Major-General Frank Berryman (q.v.). He remained there until the fall of Madang in May 1944. He then went to command of I Corps, training on the Atherton Tableland, leading it ultimately in the complicated amphibious operations against Borneo at the end of the war.

The highpoints of Morshead's career came at Tobruk and Alamein, and it is not coincidental that he performed at his best in more static operations reminiscent of those in 1918. Especially after Tobruk, he became one of the best-known of the Australian commanders. Dubbed 'Ming the Merciless' by his troops, he and his division were nicknamed 'Ali Baba Morshead and the 20 000 thieves' by Axis propaganda. A confident fighting commander with a fine grasp at the tactical level, he was never really called on to command at a higher one. Gavin Long (q.v.), the official historian, was 'much impressed with the breadth of Morshead's mind' and thought him, from the troops' point of view, 'easily the most trusted leader of general rank we have'.

Morshead retired from the army at the war's end. He was recalled by the Menzies government in November 1957 to head a high-level committee (see MORSHEAD COMMITTEE) of inquiry into the defence group of departments. The report was presented in December, with a supplementary report in the following February, and recommended wide-ranging reforms including the streamlining of resources and the amalgamation of the single-service departments. Most of the recommendations were shelved, and had to await realisation in the reforms associated with Sir Arthur Tange (q.v.) in the early 1970s.

MORSHEAD COMMITTEE Convened in September 1957 under the chairmanship of Lieutenant-General Sir Leslie Morshead (q.v.), this committee examined the organisation and functioning of the existing defence group of departments which numbered six: Defence, Army, Navy, Air, Supply, and Defence Production, each with their own minister. The retirement of the long-serving Secretary of the Department of Defence, Sir Frederick Shedden (q.v.), the previous year had been the necessary precondition for a wide-ranging review of defence organisation, which had changed little since the beginning of the Second World War. The committee reported in December 1957, and again in February 1958 in response to particular issues arising from the first report and taken up by the cabinet. The most important of the recommendations were the amalgamation of the Departments of Supply and Defence Production, on grounds of efficiency and to avoid duplication of function; the amalgamation of the three single-service ministries with the Department of Defence, to be headed by

a single minister and assisted by two associate or junior ministers; and the review of the Service Boards of Administration with a view to either their abolition or restructure. The government accepted the first, since it was sensible and uncontroversial. The second recommendation was not taken up, on the grounds that it would overtax the Minister for Defence, although his authority within the system was strengthened slightly through an administrative order declaring his overall authority in policy formulation. The review of the service boards was not undertaken either. To strengthen further the advice received by government, Prime Minister R. G. Menzies (q.v.) announced the creation of a new position of Chairman of the Chiefs of Staff Committee, to be drawn from the service chiefs but with responsibility for defence issues overall. The position, however, had no statutory authority and only a very small support staff. An important opportunity for necessary reform of the defence sector was missed, to be taken up in the early 1970s by Sir Arthur Tange (q.v.).

MORTARS A mortar is a smooth-bore, high-angle weapon used to drop a shot, usually at reasonably short range, on an entrenched enemy. Mortars have been used in warfare for hundreds of years. In Australia mortars are recorded in colonial stores lists as early as the 1850s, and some were cast in Sydney for use in the New Zealand wars (q.v.). But mortars became prominent mainly during the First World War. At the beginning of that war the British army had no mortars but they soon developed a variety of such weapons and many of these were used by Australian troops. The most common infantry support mortar was the 4-inch Stokes mortar. This consisted merely of a base plate, adjustable lugs and a tube. Bombs were dropped in the top of the tube and hit a striker at the bottom which detonated a percussion cap. It could throw a 10 pound bomb 350 metres at a maximum rate of 30 per minute and was useful against trenches and machine-gun strong points. These were first used by the Australian infantry at Gallipoli. Later on the Western Front heavier mortars, including a 6-inch and a 9.5-inch, were used. The heavier weapons were grouped into batteries of four and manned by artillerymen. In the Second World War Australians used 3-inch mortars in the Middle East and Malayan campaigns, and 4.2-inch mortars in the Papuan, New Guinea, Bougainville, New Britain and Borneo campaigns. The latter was a formidable weapon which could throw a 19-pound bomb over 3500 metres, The high-angle fire made it ideally suited to the steep valleys in New Guinea. Both Second World War-design mortars were also

used by Australian troops in the Korean War and the Malayan Emergency. During the Vietnam War the 81 mm F2 mortar, normally deployed mounted in an armoured personnel carrier or a wheeled vehicle, was used. The 81 mm mortar can throw high explosive or smoke bombs to a maximum distance of 4700 metres at a rate of 20 rounds per minute, and is still in use with the Australian Army.

MOSQUITO, DE HAVILLAND (2-seat fighter bomber [FB 40]). Wingspan 54 feet 2 inches; length 40 feet six inches; armament 2 × 20 mm cannon, 4 × 0.303-inch machine-guns, 2500 pounds bombs; maximum speed 415 m.p.h.; range 1400 miles; power 2 × Packard Merlin 1300 h.p. engines.

The Mosquito was originally designed in 1940 as an unarmed bomber so fast it could outfly German fighters. In Europe No. 456 and 464 RAAF Squadrons flew Mosquitos from 1942 to 1945. On 18 February 1944 No. 464 Squadron took part in a pinpoint bombing raid on the Gestapo gaol in Amiens that enabled captured French Resistance fighters to escape. No. 456 Squadron used Mosquitos as a fighter and shot down 41 German aircraft and 24 V-1 rockets. In 1942 De Havilland began constructing Mosquitos in Australia with the assistance of General Motors-Holden. But construction difficulties meant that the first Australian-built FB 40 Mosquito did not reach the RAAF until 1944. No. 1 and 94 Squadrons were equipped with the Australian-built Mosquitos and No. 1 Squadron saw action with the 1st Tactical Air Force (q.v.) before the end of the war. Between 1947 and 1953 specially modified Mosquitos carried out an aerial photographic survey of Australia.

MOSQUITO, HMQS see **QUEENSLAND MARINE DEFENCE FORCE SHIPS**

MOULD, Lieutenant John Stuart (?1910–9 August 1957) migrated to Australia from England at the age of two, and became an architect after education at Sydney Grammar School. He enlisted in the AIF in 1940, but was discharged after succumbing to pneumonia. He was commissioned into the Royal Australian Navy Volunteer Reserve in September 1940, because he held a Board of Trade Yachtsman's Certificate. Mould then trained in Britain and specialised in mine disposal (now called explosive ordnance demolition). He was awarded the GC in April 1942 for a series of actions that saw

him develop techniques for rendering safe German magnetic-acoustic and moored magnetic mines, which had killed a number of other mine disposal officers previously. He subsequently made an important contribution to the Allied landings on D-Day (q.v.) by training parties of divers (known as mine-sweepers) who cleared obstacles in German-controlled waterways. He also developed a diving suit that operated independently of external air supplies. After the war he returned to Australia and resumed his career as an architect.

MULLOKA is a sonar anti-submarine and torpedo detection device, developed specifically for the warm shallow waters around the Australian coast. Sonar works by sending sound waves through the water that bounce back to a recording device if they strike anything. The device can measure the time taken for the waves to return and thus calculate the distance and direction of the object. Mulloka was first fitted to the River Class destroyer escorts (q.v.) in 1984 and means 'water devil' in an Aboriginal language.

MUNITIONS, DEPARTMENT OF, was created on 11 June 1940 to help meet the wartime demand for military supplies which had become too great for the Department of Supply and Development to handle alone. It was responsible for operating and acquiring factories, securing supplies and employing and training workers for the purpose of munitions production. It was also to control profits when munitions were produced by private enterprise. 'Munitions' were defined in 1947 as 'armaments, aircraft, arms, ammunition, weapons, vehicles, machines, vessels of ships, including the materials necessary for the production of those things'. Other military supplies remained the responsibility of Supply and Development, and from July 1941 to November 1946 there was also a separate Department of Aircraft Production. The Department of Munitions was abolished on 6 April 1948 and its functions transferred to the Department of Supply and Development.

MUNITIONS SUPPLY BOARD, established on 13 August 1921, took over the functions of the Board of Factory Administration. The board's responsibilities included provision of munitions, research and design, inspection of military equipment up to the point of issue to the service, and administration of government munitions factories and research laboratories. In 1939 its functions were taken over by the Factory Board.

A 3-inch mortar operated by men of the 2/14th Battalion firing on Japanese positions, Manggar, near Balikpapan, Borneo, 5 July 1945. (AWM 111239)

MURCHISON, HMAS see **BAY CLASS FRIGATES**

MURDOCH, Air Marshal Alister Murray (9 December 1912–29 November 1984). The son of an army brigadier, Murdoch followed an older brother into RMC Duntroon in 1929. While his brother, Ian Thomas, was destined for the army (later attaining the rank of major-general), Murdoch was admitted as a candidate for the RAAF. In December 1930, midway through his four-year course, Duntroon was moved to Sydney as an economy measure, and all RAAF cadets were transferred to Point Cook to complete their training.

Graduating as a pilot in 1931, he was commissioned the next year and specialised in flying the types of seaplanes used in naval cooperation work. His skill in this field led to his selection in December 1935 to join a small RAAF detachment accompanying an expedition trying to locate American explorer Lincoln Ellsworth, with whom radio contact had been lost during a flight across Antarctica. In 1937 he went to England and completed the RAF's long navigation course. Following his return to Australia he assisted in instructing the first similar specialist course for the RAAF, which began in June 1938 and lasted for nine months.

After the outbreak of the Second World War, he was appointed to command No. 1 Air Observers School at Cootamundra, New South Wales, in April 1940. Murdoch was sent overseas the following year on attachment to RAF Coastal Command, and in August 1941 he was appointed commanding officer of No. 221 Squadron based in Iceland. The next year he took his unit to the Middle East where he undertook additional duties as Staff Officer Operations with 235 Wing, RAF, for several months. In July 1942 he was attached to the Combined Operations staff in London for duty in connection with the abortive Dieppe raid in August.

Back in Australia and promoted to group captain, in July 1943 he became Senior Air Staff Officer (SASO) at Eastern Area headquarters in Sydney. In January 1944 he became SASO at the Darwin headquarters of North-Western Area, during a period when missions mounted from there were bombing and mining as far afield as Java and the coast of southern China (see CATALINA); he played an important role in planning these operations and for a time held temporary command of the Area. In April of the next year he was appointed SASO at the RAAF's major formation in the field, 1st Tactical Air Force (q.v.), then based at Morotai under Air Commodore F. R. W. Scherger (q.v.). While there he firmly established his reputation as an outstanding staff officer, particularly during the Balikpapan landing.

While filling senior staff jobs at Air Force Headquarters in Melbourne after the end of the war,

Murdoch was identified as one of a small group of middle-ranking officers possessing the qualities specially needed in the postwar service and meriting accelerated promotion. He was sent to attend the Imperial Defence College (q.v.) in London in 1948, and after an initial staff job on his return was appointed commandant of the RAAF College and AOC at Point Cook in June 1952; while filling this post he was promoted to air commodore.

Given acting rank as air vice-marshal in December 1953, he became AOC Training Command. In this capacity during 1954 he led a team overseas on a six-month tour to investigate modern aircraft suitable to the RAAF's requirements for new jet fighters and bombers, transports and trainers. He returned with recommendations that the US built F-104 starfighter and C-130 Hercules transport (q.v.) were types that would galvanise the service. Funds for the purchase of the C-130 were eventually made available by the government, but plans to acquire the F-104 Starfighter were ultimately abandoned in 1960 in favour of the French-designed Mirage III (q.v.).

Seconded to the Department of Defence in January 1956 to fill the appointment of Deputy Secretary (Military), Murdoch (by now substantive as an air vice-marshal) became Deputy CAS in February 1958. He was again seconded to Defence in October of the following year and filled the post of Head of the Australian Joint Services Staff in London. On his return to Australia he was appointed AOC Operational Command in June 1962 and three years later succeeded Sir Valston Hancock (q.v.) as CAS with the rank of air marshal.

A genial character of great personal charm, Murdoch exhibited a common touch which extended to keen interest in a range of sporting activities including a fondness for betting on horse races. Although praised by Scherger as 'the last of the professionals' — a reference to the fact that he was the last of the Duntroon-trained officers who served as CAS in the fifteen years from 1954 to 1969 — he demonstrated little intellectual depth despite obvious mental sharpness and thoroughness. His narrow vision was particularly evidenced by his dealing with other services, as illustrated by an unnecessary clash in 1965 with the CGS, Lieutenant-General J. G. N. Wilton (q.v.), over suggestions that RAAF helicopters be sent to support the army contingent in Vietnam. After retiring as CAS in 1970, Murdoch retained his interest in defence matters. In 1975 he joined a private think-tank of nine prominent defence commentators who advocated the acquisition of nuclear weapons as vital to Australia's survival.

Alan Stephens, *Power Plus Attitude* (AGPS Press, Canberra, 1992).

CHRIS COULTHARD-CLARK

MURDOCH, Keith Arthur (12 August 1885–4 October 1952). A journalist and newspaper proprietor, Murdoch is chiefly of interest to military historians for his activities during the two world wars. When war broke out in 1914 he was working for the Sydney *Sun* and narrowly lost a ballot to C. E. W. Bean (q.v.) for the position of official Australian War Correspondent. He was commissioned in 1915 by the government to investigate mail services to the AIF at Gallipoli. While on the peninsula he fell in with British correspondent Ellis Ashmead-Bartlett, who persuaded Murdoch to carry to British Prime Minister Herbert Asquith a letter from Ashmead-Bartlett highly critical of the conduct of the campaign. The military got wind of this affair and Murdoch was arrested in Marseilles and the letter removed. However *en route* to England he composed his own document, if anything even more critical of the higher administration of the operation. On arrival in England he met David Lloyd George, Andrew Bonar Law, Edward Carson and Asquith, who arranged to have his letter printed as a cabinet document. It assisted those who were endeavouring to have the campaign wound up and might have been a factor in the recall of General Ian Hamilton (q.v.). In 1917 Murdoch became embroiled in controversy over the command of the Australian Corps. Together with Bean he campaigned for Lieutenant-General Brudenell White (q.v.) to replace Major-General Sir William Birdwood (q.v.). When Lieutenant-General John Monash (q.v.) was given the job he then intrigued to have Monash removed to the largely administrative role of GOC and have White replace him. These cabals were bluntly run off by Birdwood and Monash and only had the effect of discrediting Murdoch. During the Second World War Murdoch, now the proprietor of the Melbourne *Herald* and other papers, was appointed by Menzies as Director-General of Information in June 1940. His hamfisted approach to the task of censorship earned him the hostility of his fellow proprietors and he resigned his post in December. For the remainder of the war he wrote articles on military strategy (always critical of the higher direction of the war) and on the need for the spiritual renewal of the country. As his articles became more extreme his influence waned. The election of John Curtin (q.v.), whom he detested, and the sweeping Labor victory in 1943 virtually silenced him.

MURRAY, Lieutenant-Colonel Henry William (1 December 1880–7 January 1966). Murray served in the field artillery for six years in Tasmania while still a teenager, and at about the age of 20 moved to Western Australia where he held various jobs. He enlisted in the 16th Battalion in October 1914, and was posted to the machine-gun section with which he landed at Gallipoli. With his mate, Lance-Corporal Percy Black (killed as a major in France in 1917), he held back the advance of Turkish troops around Pope's Hill, refusing to retire although both men were wounded. Murray was evacuated at the end of May and returned to the battalion in July. He was awarded the DCM for actions in May, was wounded again during the August offensives, was commissioned in August and transferred to the 13th Battalion. He went to France with the battalion as a captain, and took part in all the actions of 1916. He was awarded the DSO for actions at Mouquet Farm, where although wounded twice during the battle he had led his company throughout. He won the VC for his gallantry and leadership at Gueudecourt in February 1917. During the first battle of Bullecourt he won a bar to the DSO (in the action in which Black was killed) and was promoted to major. Murray commanded his battalion at the end of 1917, was promoted to lieutenant-colonel in May 1918 and was given the 4th Machine Gun Battalion. He was awarded the Croix de Guerre and the CMG, and between 1917 and 1919 was mentioned in despatches four times. He became a grazier after his return to Australia. During the Second World War he commanded the 26th Battalion of the militia until 1942, and held appointments in the Volunteer Defence Corps. A modest man who believed that it was discipline rather than heroics that motivated men in battle, Murray was the most highly decorated infantry soldier from the Empire armies of the war.

MUSEUMS, MILITARY Because so much of Australia's military history has involved its armed forces fighting outside the continent, war has left almost no mark on the Australian landscape. While the ubiquitous war memorials (q.v.) remind Australians of the human cost of wars fought overseas, few physical traces remain of the Aboriginal resistance to White invasion (q.v.) or of the Japanese attacks on Australia (q.v.) during the Second World War. Only the old barracks (q.v.) and fortifications (q.v.) (some of which now house museums) survive as large-scale relics of the military past. Australian military museums, then, are particularly important as sites where the experience of war is interpreted to a society for which that experience seems increasingly remote. The popularity of Australia's premier military museum, the Australian War Memorial (q.v.) in Canberra, which had 880 000 visitors in 1994, shows that there is a continuing interest in trying to understand the wars that have had such a significant impact on Australian society. While much of its space is taken up with military technology and relics displayed out of context, the Memorial has recently attempted to take a more explicitly interpretive approach to military history (one which has caused some conflict with its role as a memorial) and to deal with war as a social rather than a narrowly military phenomenon. Its lead has not been followed by other military museums, which continue to function simply as repositories of equipment and memorabilia for the military history enthusiast. Some are private museums, but most are attached to army, air force or navy bases. The principal RAAF museum is located at Point Cook, Victoria, while the navy's main museum is at HMAS *Cerberus* (q.v.). The army has many small museums, mostly relating to the history of particular corps or military districts. Admission to most of these museums is either free or very cheap, and some have libraries or archival material which is available by arrangement to researchers. In addition to the museums that are open to the public the armed forces also have historical collections to which access is more limited.

MUSIC, MILITARY see **MILITARY BANDS**

MUSTANG, NORTH AMERICAN AND CAC (Single-seat fighter [CAC Mark 23]). Wingspan 37 feet; length 32 feet 3 inches; armament 6 × 0.5-inch machine-guns, 2000 pounds bombs or 8 rockets; maximum speed 437 m.p.h.; range 1700 miles; power 1 × Rolls Royce Merlin 1490 h.p. engine.

Though it had been flown by the British and Americans since 1941, the Mustang did not enter service with the RAAF until the last year of the Second World War. No. 3 Squadron flew Mustangs in Italy, while in the South-West Pacific Area (q.v.) the 1st Tactical Air Force (q.v.) was in the process of replacing Kittyhawks (q.v.) with Mustangs when the war ended. The Commonwealth Aircraft Corporation (q.v.) built 200 Mustangs between 1945 and 1951. No. 76, 77 and 82 Squadrons flew Mustangs in Japan as part of the British Commonwealth Occupation Force (q.v.), and No. 77 Squadron also flew them during the Korean War. Outperformed by MiG jets in Korea, No. 77 Squadron's Mustangs were replaced by Meteors (q.v.) in 1951, though Mustangs served with the Citizen Air Force until 1960. In October 1953 six Mustangs were exposed to an atomic test (q.v.) in South Australia to test the effect of radiation on aircraft.

MUTINIES see *AUSTRALIA* (I), HMAS; FIRST TACTICAL AIR FORCE

N

'N' Class Destroyer HMAS *Nestor* under attack from German aircraft, 15 June 1942. This attack led to her sinking on 16 June in the Mediterranean Sea about 200 km north-east of Tobruk. (AWM 044998)

'N' CLASS FLEET DESTROYERS (*Napier*, *Nepal*, *Nestor*, *Nizam*, *Norman*). Laid down 1939, launched 1940–41, scrapped 1955–58 (*Nestor* sunk 1942); displacement 1760 tons; length 356.5 feet; beam 36 feet; speed 36 knots; armament 6 × 4.7-inch guns, 1 × 2 pounder, 3 oerlikons, 45 depth-charges, 10 (later 5) × 21-inch torpedo tubes.

After transferring from RN to RAN control from 1940, these ships saw service in a wide variety of theatres during the Second World War, including the North Atlantic, Mediterranean, Indian Ocean, Australian, South-east Asian and Japanese waters. Their main duties were convoy, fleet protection and anti-submarine, but they were also used to attack shore installations and carry out land bombardments. HMAS *Nestor* suffered heavy damage from German bombers off Malta and on 16 June 1942, when it became clear that she could not reach port, she was sunk. The ships reverted to British control at the end of the war.

NAPIER, HMAS see **'N' CLASS DESTROYERS**

NARRUNGAR see **JOINT DEFENCE FACILITIES**

NATIONAL SECURITY ACT, passed on 9 September 1939, was equivalent to the War Precautions Act (q.v.) of the First World War. Making use of the Common-

wealth's constitutional power to legislate with respect to Australia's defence, the act allowed the Commonwealth government to make regulations to secure the public safety and defence of Australia. It specifically ruled out, however, the imposition of military or industrial conscription (q.v.) and the court-martialling of civilians. The act was amended in June 1940 to give the government the power to control the property and services of Australians, but the imposition of conscription for service beyond Australian territory was still excluded. Regulations made under this legislation covered a range of areas too numerous to list here. More Statutory Rules were issued under the act in 1942 than in any other year, and between the outbreak of war with Japan and April 1943 the index to the Manual of National Security Legislation doubled in length. The only checks on the Commonwealth's national security powers were public opinion and the political impracticability of exercising these powers without the cooperation of State governments. The High Court could strike down regulations on the grounds that they were not sufficiently related to the Commonwealth's defence power, but as in the First World War it rarely did so, allowing a substantial transfer of powers from the States to the Commonwealth. In June 1944 the Attorney-General, H. V. Evatt, set up an advisory committee on the National Security Regulations. The committee, which issued a number of reports up to June 1945, made no recommendations for major changes to the regulations but was useful in pointing out particular injustices and regulations which had become obsolete. The National Security Act expired on 31 December 1946, but the government used its defence power to extend its control of some economic matters to December 1949. In May 1948 the Commonwealth sought to achieve a permanent power over rents and prices by referendum, but the referendum was defeated and the remaining powers assumed under the National Security Act reverted to the States.

NATIONAL SERVICE see **CONSCRIPTION**

NATIVE POLICE see **ABORIGINAL ARMED RESISTANCE TO WHITE INVASION**

NAVAL ASSOCIATION OF AUSTRALIA was founded in Melbourne in November 1920 as the Ex-Navalmen's Association. Sub-sections were subsequently established in other States and the association adopted its current name in 1960. By 1991 it had 85 sub-sections throughout Australia and was the largest organisation specifically representing serving and former navy members. The association brings former and serving naval personnel together to support each other and works to protect their interests by making representations to government. It supports a strong navy by participating in naval-related activities, and publishes a quarterly magazine, *White Ensign*.

NAVAL AND MILITARY CLUB, MELBOURNE, was formed as the Pipeclay Club by officers of the Victorian Volunteer Force on 16 May 1881. With membership restricted to officers and former officers of the armed forces, it took on the functions of both a social club and a united service institution, featuring regular lectures on military topics. Benefiting from the location of the defence bureaucracy in Melbourne until 1959, the club was able to attract many of Australia's most senior military officers, including Sir John Monash (q.v.) and Sir Thomas Blamey (q.v.), as members. The club has undergone several changes of name and address during its existence: it was located at 7 Alfred Place from 1920 until 1967, when it moved to its present home at 27 Little Collins Street. Writing in *The Road to Gundagai* (1965) of his childhood visits to Alfred Place, Graham McInnes recalls military pictures, maps and trophies on the walls, loud and animated discussion of current events, and 'a lazy sense of cameraderie [*sic*], of some secret shared in common and which excluded me'. This camaraderie

was aided by the fact that many members were part of Melbourne's business and professional élite and that it was an all-male establishment until 1973, when members' wives were admitted on to the premises. By 1980 the club had 3822 members and was affiliated with other service clubs in each Australian State capital and overseas.

Warren Perry, *The Naval and Military Club, Melbourne: A History of its First Hundred Years, 1881–1981* (Naval and Military Club, Melbourne, 1981).

NAVAL BOARD OF ADMINISTRATION, established on 12 January 1905, was, subject to the control of the minister, charged with the administration of Australia's naval forces. It originally consisted of the minister for Defence, the Director of the Naval Forces and the Finance Member (representing the head of the Department of Defence). In 1911, following Admiral Sir Reginald Henderson's report on the Australian Navy, membership of the board was increased to five and each member was given particular areas to supervise (a structure modelled on that of the Board of Admiralty in the United Kingdom). The minister (for Defence, or Navy from 1915 to 1921 and 1939 to 1973) was president of the board; the First Naval Member (also CNS from 1919) supervised war preparations, intelligence, ordnance, exercises, naval works and appointments to senior positions; the Second Naval Member supervised personnel and reserves, discipline, stores, victualling and medical care; the Third Naval Member supervised construction and repair of ships, naval dockyards and bases; and the Finance and Civil Member supervised finance, contracts and legal matters. In 1920 the functions of the board were revised so that it was not only charged with 'control and administration' of the navy but was also given 'executive command' of the RAN. The Secretary of the Department of the Navy (q.v.) was also a member of the board from 1940, though from 1942 to 1954 he had no powers or duties as a member. From 1954 the Secretary was responsible for coordination of the board's business, financial administration and administration of civil personnel. There was also a business member (responsible for contracts, works, transport services, stores and other matters) on the Board from 1942 to 1954 and a member responsible for the Fleet Air Arm (q.v.) from 1947 to 1959. A Fourth Naval Member, added in 1959, was responsible for supply, machinery, spares, works and movement of personnel. The Naval Board was abolished on 9 February 1976 along with the Air and Military Boards (qq.v.).

NAVAL CONTROL SERVICE (NCS) was formed following the outbreak of the Second World War to control and safeguard merchant shipping in Australian waters. Members of the RAN Volunteer Reserve ran NCS offices in the major Australian ports where merchant ships received clearance on entering and leaving port. When Japanese submarines began attacking ships off the Australian coastline (see JAPANESE ATTACKS ON AUSTRALIA), NCS staff organised convoys (q.v.). The NCS cooperated with the British Admiralty and US authorities and also set up offices in Papua and New Guinea.

NAVAL DEFENCE AGREEMENTS (1887, 1903) see ANGLO-AUSTRALIAN NAVAL AGREEMENTS, 1887, 1903

NAVY, DEPARTMENT OF THE, was originally created on 12 July 1915, replacing the Defence Department's (q.v.) Navy Office. It was originally responsible for naval defence, bases, dockyards and works, as well as wireless telegraphy (until 1920) and construction and repair of vessels for Commonwealth departments. The department was abolished on 21 December 1921 and absorbed into the Department of Defence. A new Department of the Navy was created on 13 November 1939 to administer the navy and naval defence. On 30 November 1973 the department was once again replaced by the Navy Office within the Defence Department.

NAVY LEAGUE was formed in Britain in 1895 as a lobby organisation, but by the time its first Australian branch was established in 1914 its primary role was the training of naval cadets (see RAN CADETS). An Australian Navy League Council was formed in 1935 and in 1949 the Navy League of Australia became independent from the British body. In the late 1960s the league moved away from organising cadet training towards lobbying government and attempting to educate the public on naval defence issues.

NAVY LIST contains details of all permanent officers of the RAN, officers of the Australian Naval Reserve and officers on the Retired List. Permanent officers are listed alphabetically and in order of seniority according to branch (General, Ordnance Inspection, Health Services, Chaplains and Special Duties). The Navy List is published annually.

NELSON, HMVS see **VICTORIAN NAVY SHIPS**

NEPAL, HMAS see **'N' CLASS DESTROYERS**

NEPEAN, HMVS see **VICTORIAN NAVY SHIPS**

NEPTUNE, LOCKHEED (7-crew maritime reconnaissance and anti-submarine aircraft [SP-2H]). Wingspan 103 feet 10 inches; length 91 feet 8 inches; armament 8000 pounds bombs, torpedoes, depth charges or mines; power 2 × Wright turboprop engines plus 2 × Westinghouse turbojet engines; maximum speeds 305 m.p.h. (propellers only), 356 m.p.h. (all engines); range 2200 miles.

The Neptune design was developed from that of previous Lockheed maritime patrol aircraft such as the Hudson (q.v.) and the Ventura (q.v.) but it arrived too late to see service in the Second World War. In 1946 a specially modified Neptune set a long-distance flight record by flying non-stop for 55 hours from Perth to Columbus, Ohio, in the United States. Neptunes entered service with the RAAF with No. 11 Squadron in 1951 and No. 10 Squadron in 1962. From 1959 RAAF Neptunes were fitted with both piston and jet engines to enable the aircraft to vary its speed and range when needed. The Neptune's radar, though designed for maritime reconnaissance, could also pick up airborne objects, and during the Vietnam War No. 10 Squadron flew from Thailand to provide for airborne early warning of missile attacks for USAF B-52s flying bombing raids over North Vietnam. No. 11 Squadron replaced Neptunes with Orions (q.v.) in 1967, while No. 10 Squadron kept their Neptunes until 1977.

NESTOR, HMAS see **'N' CLASS DESTROYERS**

NETHERLANDS EAST INDIES CAMPAIGN By the second half of February 1942, the Japanese scheme of conquest in the 'southern resources zone' was ahead of schedule. Singapore had fallen in 70 days on 15 February, well in advance of the 100 days that General Yamashita had been allowed. In turn, the 'Malay Barrier' disintegrated very rapidly. Java's isolation followed as the islands on either flank, with their airfields, succumbed equally rapidly. Southern Sumatra was taken after paratroop and amphibious forces landed around the airstrips and oil refineries near Palembang. By 18 February, Bali had been occupied. On 19 February, a massive air raid on Darwin ensured that reinforcements would not be coming from that quarter. Timor fell on 23 February.

An attempt by Admiral Helfrich, General Wavell's ABDA Naval Force commander, to intercept the Japanese invasion of Java was unsuccessful. From the start the contest was uneven. The Japanese forces were strong both in surface power and accompanying air power. They were also cohesive and well trained. The Allied naval force, in contrast, was a heterogeneous collection of ships from four nations including Australia. It was short of fuel, short of spares, its charts were poor, its ships had not trained together, were not used to operating with each other, and had no effective air cover. Neither did they have communications to manoeuvre a combined force effectively.

The force failed to stop the Japanese Eastern Invasion Force in an action on 27 February 1942 which continued on into the night. The Dutch cruisers *De Ruyter* and *Java*, as well as other vessels, were sunk. The survivors, including the Australian cruiser *Perth* and the US cruiser *Houston*, put back to Batavia. On the evening of 28 February, the two ships attempted to pass through the Sunda Strait to the south and ran into the Western Invasion Force. They sank two Japanese transports and damaged others, but both were then sunk. This series of actions, known collectively as the Battle of the Java Sea, spelt the end of effective Allied naval strength around Java. The Allied resistance had delayed the Japanese invasion fleets for 24 hours but had not otherwise inflicted a significant setback to Japanese plans. The Eastern Force began landing at Kragan, 160 kilometres west of Surabaya, on 1 March 1942; the Western Force, whose ships the *Perth* and *Houston* had happened upon, landed in three places: at Merak and Bantam Bay, west of Batavia (Jakarta), and at Eretanwetan to the east.

As in Malaya, the principal defence of Java was meant to be conducted by the use of aircraft. When the Japanese attacked, the defending air force included 18 British fighters and 20 twin-engined aircraft fit for operations, a few American fighters and 10 Dutch squadrons, which were much depleted. They attacked the Japanese landings but not to any effect, and were quickly whittled down. Just before this, considerable change had occurred in the defence arrangements within Java itself. ABDA Command (q.v.) had been dissolved on 25 February 1942, and Wavell had handed command in the area to the Dutch. A number of Australian units of the 7th Division, which had arrived in the *Orcades* convoy and had been deployed under Wavell's orders to protect airfields, did not rejoin the main body of the division when it returned to Australia. Instead they became part of the defence of the island under the command of the Dutch C-in-C, General ter Poorten. They were grouped as 'Blackforce' under the command of Brigadier Blackburn VC, who had been promoted from CO of 2/3rd Machine Gun Battalion. Blackforce numbered 3000 Australians in all and, in turn, became part of a larger British Empire force commanded by Major-General Sitwell.

General ter Poorten's 25 000 Dutch troops were deployed in four area commands on Java but, through lack of transport, they were virtually immobile. Nor did they have effective communications. Blackforce, with two battalions — 2/3rd Machine Gun, and 2/2nd Pioneers — was ter Poorten's only mobile strike force. Blackburn's other Australian units included engineers, medical, transport and men of a Guard Battalion. He was also given a squadron of British tanks, part of a US Field Artillery Battalion and some signals troops, and he formed the whole force into a composite brigade. He came under the tactical command of the Dutch General Schilling at Buitenzorg in western Java.

The Japanese advance from its various landing points was quick and brutally efficient. In the east the 48th Division advanced on Surabaya which it occupied on 8 March. The 56th Regiment advanced across the island and captured Tjilatjap, which until then had been used as an evacuation port. In the west the force that landed at Eretanwetan advanced south-west, captured the airfield at Kalidjati and cut the route between Bandung and Batavia the capital. The 2nd Japanese Division, which had landed at Bantam Bay and Merak on the western tip of Java, began to advance towards Batavia and also south-east towards Buitenzorg. Schilling's plan was that as the Japanese advanced Blackforce, by moving rapidly west and then north, would take them in the rear. The Japanese, however, achieved such rapid progress towards Bandung from the east that the Dutch command became anxious and the original bold plan to use Blackforce was modified, modified again and finally abandoned. Indeed, so heavily had the fog of war descended on Java that on 3 March, three days after the Japanese landings, Blackforce received a Dutch intelligence report which stated: 'No Japanese landings on Java.' Five minutes later, five Japanese light tanks appeared at the Australian defensive position west of the bridge at Leuwiliang.

By this time it had become clear that the Dutch command had become demoralised and paralysed by the speed and comprehensiveness of the Japanese attack. On 4 March, ter Poorten decided to abandon Batavia and Buitenzorg and concentrate around Bandung. That movement was begun and Blackforce had withdrawn to Sukabumi and then to a position east of Bandung when Sitwell told Blackburn that the Dutch authorities intended to capitulate. On his own initiative, Blackburn reconnoitred the mountain country south of Bandung and cached food there with a view to carrying on a guerrilla struggle. Later, on medical advice that 'without drugs and without adequate shelter the health of my troops would suffer very severely if I remained in the mountains ... I therefore reluctantly decided that in the best interests of my troops and their lives I must capitulate', Blackburn joined the senior British and American officers in a formal surrender of their forces in Bandung on 12 March 1942.

On 28 March, when the Japanese finally occupied the whole of Sumatra, the Netherlands East Indies campaign, except for Australian and Dutch guerrillas holding out in places like Timor, effectively came to an end. In three months the Japanese had shattered the Malay Barrier and gained possession of the resources of the rich southern area for which they had gone to war. The total Australian casualties in Java up to the time of the formal surrender were estimated at 36 killed and 60 wounded. Eventually, 3000 Australian prisoners from Blackforce and other survivors were concentrated in POW camps in Java.

JOHN COATES

NEW CALEDONIA Like all French colonial territories, New Caledonia's status was called into question after the French surrender to the Germans in June 1940 and the subsequent installation of a collaborationist government at Vichy under Marshal Pétain. The Australian government was concerned that New Caledonia should not be occupied by the Japanese (as French Indochina was) and used as a forward base for attacks on Australia. Local feeling was divided: the governor and many senior officials were supporters of Charles de Gaulle and the Free French movement, but many others, including many military officers, had Pétainist sympathies. There were French naval vessels based in the south Pacific, and these might pose a threat not only to maritime trade but to troop convoys embarking for the Middle East. After much diplomatic consultation with Britain and New Zealand, Australia agreed to send the light cruiser HMAS *Adelaide* to escort a Free French official who was to take over as governor and thus preserve the territory for the Allies. After further hesitation occasioned by the possible intentions of French warships, the Australian ship arrived on 19 September bearing the new governor, to find the port of Nouméa held by pro-Vichy forces. A popular Gaullist demonstration followed by the arrival of more local troops and the presence of the Australian warship weakened the defenders' resolve, however, and their commander was arrested. Any likely resistance was negated by the Australian warship, and two pro-Vichy naval vessels retired from the area rather than face the more capable Australian ship. The success of the intervention in New Caledonia was in marked contrast to the British failure that same month at the French colony of Dakar in Africa, and in which the Australian cruiser HMAS *Australia* was involved.

NEW GUARD see **SECRET ARMIES**

NEW GUINEA CAMPAIGN Of all the campaigns of the Second World War that can reasonably claim to have been fought on a logistic shoestring, New Guinea ranks high. The country's relative isolation, inhospitable terrain, enervating and malarious climate, and underdeveloped infrastructure meant that the maintenance and re-supply of military forces presented a continuing challenge. After a series of false starts, the Allies generally overcame their logistic difficulties: the Japanese did not. Therein lay the most important single factor in the campaign.

In reality there were two campaigns. The first began when the Japanese captured Rabaul (q.v.) on 23 January 1942 and ended when General MacArthur's forces concluded the Vogelkop operation at the western extremity of the island on 31 August 1944. The second was a series of predominantly Australian operations to mop up those pockets of Japanese who had been bypassed in the earlier fighting. This latter phase was still continuing at the end of the war.

The first campaign

The initial Japanese war plan, which was inseparable from Japan's need to achieve victory in China, sought to cripple or destroy the US Pacific fleet, while simultaneously, through a quick advance against generally unprepared enemies, conquering a vast area of the Pacific and East Asia within a defensive perimeter of 18 000 kilometres from the Indo–Burmese border through the Netherlands East Indies and New Guinea, to the Gilbert Islands, then north to the Kuriles. An essential part of the plan called for the occupation or neutralisation 'as speedily as operational conditions permit' of eastern New Guinea, New Britain, Fiji, Samoa, the Aleutians, Midway and 'strategic points in the Australian area'. In short, Rabaul was sought as a sea and air base in order to provide defensive depth to the key Japanese centre at Truk in the Caroline Islands, while other points in New Guinea — Salamaua, Lae, Samarai Island and Port Moresby — would in their turn provide defensive depth for Rabaul. In pursuit of this strategy, Salamaua and Lae were occupied on 8 March 1942.

The first Japanese forward movement

Japan's early success was so stunning that the Japanese leaders were forced to revise their initial strategy and look at a number of competing options. The broad choices were: first, strengthen the existing defensive perimeter; second, clear the British from the Indian Ocean; third, invade Australia to prevent its use as a stepping stone for an advance towards Japan; fourth, cut communications between the United States and Australia by immediately seizing Port Moresby, New Caledonia, Fiji and if necessary islands further south; and, finally, an option strongly advocated by Admiral Yamamoto, the architect of the Pearl Harbor attack, a move into the central Pacific to occupy Midway and so force the US Navy's carriers into a decisive action in which they could be destroyed.

They decided on the second, fourth and fifth options; with the third — invade Australia – to be revisited later. And so, following the successful foray into the Indian Ocean, in late April a regrouped force including the 4th Fleet (Admiral Inouye), the South Seas Detachment (Major General Horii) and Carrier Division 5 (Admiral Takagi), including the fleet carriers *Shokaku* and *Zuikaku*, launched a multiple operation to capture Port Moresby, occupy Tulagi in the Solomons, attack air-installations in north-east Australia and occupy a series of islands as seaplane bases.

Forewarned through its ability to read the Imperial Japanese Navy's principal operational code (see ULTRA), the US Navy had deployed two carrier groups, 'Lexington' and 'Yorktown', to the Coral Sea. The opposing forces met on 7 May 1942 and in the subsequent Battle of the Coral Sea, the first in which the surface forces on either side did not sight each other, both sides suffered major losses. Importantly for the Allies, however, the Port Moresby invasion was abandoned, and the force turned back to Rabaul.

The Allied counter-action

Japan's next move towards Midway gave New Guinea a short respite. Anticipating that the Japanese would make a renewed attempt to capture Port Moresby, 'any time after June 10', General MacArthur (q.v.), who on 18 April 1942 had become C-in-C, South-West Pacific Area, now reinforced a number of initiatives that had been put in place by the Australian command. By a Joint Chiefs of Staff Directive of 30 March 1942, MacArthur had, *inter alia*, been ordered to 'hold Australia as a base for future offensive action against Japan', and to 'prepare to take the initiative'. His measures in pursuit of that directive included:

• building a series of airstrips at Milne Bay at the south-eastern tip of New Guinea and garrisoning the area with troops and a fighter squadron;
• further developing Port Moresby as a port and air base for both defensive and offensive operations;
• reinforcing Australian troops (Kanga Force [q.v.]), who were screening Japanese positions at Salamaua and Lae, with the intention of destroying those positions if possible; and,
• when it became evident that the Japanese, following their catastrophic naval loss at Midway,

were taking an interest in Buna on the north coast of New Guinea as a base from which to launch a land offensive against Port Moresby over the Owen Stanley Range, he sought to forestall it by securing vital sections of the route with Australian infantry, particularly in the Kokoda area.

The second Japanese forward movement

In the event the Japanese moved faster, and, although MacArthur had already launched an engineer reconnaissance of the Buna–Dobodura area to build airfields to strike at Rabaul more effectively, the Japanese began landing in force on 21 July 1942. Their original intention was to test the feasibility of the route through the mountains by reconnaissance; this quickly became a full-scale offensive.

In brief, the Japanese plan of 31 July 1942, concluded between 17th Army (General Hyakutake) and 8th Fleet (Admiral Mikawa), was for the South Seas and Yazawa Detachments to capture Port Moresby overland. The Kawaguchi Detachment and the 8th Fleet would take Samarai Island by amphibious assault, and then the same force would go on to attack Port Moresby in combination with the overland thrust from the north. The date chosen, 7 August 1942, was by coincidence the date also chosen for the US Marine landing at Guadalcanal, in the Solomon Islands. At the time, the Japanese were unaware of the Allied force at Milne Bay which on 21 August 1942 had been reinforced with the 18th Australian Brigade (Brigadier George Wootten [q.v.]) and a second Kittyhawk squadron. Hyakutake sought to deal with the Guadalcanal landing and the newly discovered Allied base at Milne Bay simultaneously, and became over extended in the process.

Despite some initial success, the Japanese landing at Milne Bay on the evening of 25 August 1942 was destroyed, and the survivors evacuated by sea on 4–5 September 1942. It was the first significant reverse on land that the Japanese had suffered in the war. At the same time, however, Major-General Horii's reinforced South Seas Detachment was advancing spectacularly along the Kokoda Track towards Port Moresby, against increasingly effective Australian opposition. By 16 September 1942 his troops had reached Ioribaiwa, virtually in sight of Port Moresby. Until then the Japanese force outnumbered the Australian, and had greater combat power. But, because of Japanese reverses at Guadalcanal, Horii was ordered on to the defensive and then told to hold a primary defensive position on the north coast, pending better times to launch a further counter-offensive to take Port Moresby. He began withdrawing on 24 September 1942.

The second allied counter action

The Japanese withdrawal from Ioribaiwa enabled the Allies to plan a comprehensive envelopment of the Japanese position in the Buna–Sanananda–Gona beachhead. MacArthur's plan of early October called for simultaneous actions along three lines of advance: 7th Australian Division along the Kokoda Track; 32nd Division (US) along the Kapa Kapa–Jaure route or the Abau–Namudi–Jaure route and on to the north coast (in the event, most of the American troops were flown over the mountains); and an advance by Australian troops north west along the coast from Milne Bay, capturing Goodenough Island in the process.

The Japanese withdrawal, which suffered heavy losses from the 7th Division on the ground as well as from bombing and strafing by the Allied air forces, was nevertheless skilful. However, while the Japanese who had retreated to the mountains were sick and short of food, the remainder were proceeding urgently to build up the defences in the bridgehead area to withstand a virtual siege. This was about to occur.

The bridgehead battles

The leading Australian and American elements closed up to the three Japanese enclaves at Buna, Gona and Sanananda in mid-November and it immediately became evident that the systems were well sited, heavily constructed and cunningly concealed. Efforts to take them quickly 'on the run' were uniformly unsuccessful. As well, the Japanese were able to bring in reinforcements by surface ship, barge and submarine to bolster their positions. These positions, which were on an 18 kilometre front, were systematically sited on the only dry ground in the area: the Australian and American troops, in contrast, were in waterlogged areas or swamp. After a gruelling and remorseless series of actions over many weeks, which cost both sides, but particularly the Japanese, very heavily in casualties and even greater losses from sickness and fever, especially malaria, Japanese resistance was overcome. Gona was captured on 9 December 1942, and Buna on 3 January 1943. At Sanananda, after an attack on 12 January 1943 by the Australian 18th Brigade supported by American troops and by tanks had seemed unsuccessful, the commander 18th Army (General Adachi) ordered the evacuation of Sanananda to begin on the following day. The Japanese, though disorganised, continued to resist in several places. By 19 January 1943, mopping up had begun; it was completed on 22 January 1943.

The bridgehead battles were among the bloodiest of the Pacific War. Contrary to MacArthur's communiqués on the subject, the Papuan operations had been neither cheaply won, nor conducted on the supposition that there was, 'no necessity of a hurry attack'. From 21 July 1942, when Major-General Horii's advanced force landed at Buna, and during the ensuing six months while operations ran their course, the Australians committed seven infantry brigades and a dismounted cavalry unit to action. Australian battle casualties were 5698: 1731 killed in action, 306 died of wounds, 128 died from other causes and 3533 wounded in action. The American ground commitment from mid-November was four infantry regiments. American ground casualties were 2848: 687 killed in action, 160 died of wounds, 17 died from other causes, 66 missing in action and 1918 wounded in action. Sickness, though in a majority of cases temporary, caused heavier losses. The Australians had 15 575 cases of infectious disease to the end of 1942 alone; the Americans 8659 during the course of operations.

The Japanese suffered greater losses. They committed between 16 000 and 17 000 troops; of these they successfully evacuated 1300 men from Milne Bay and 300 from Goodenough Island. About 1000 sick and wounded were evacuated to Rabaul while Japanese ships were still making the run there; 2000 sick and wounded got out of the area by sea or on foot during the closing stages. Thus, they lost approximately 12 000 men. Towards the end the survivors were starving, and some resorted to cannibalism.

What were the strategic results? As the US Army's official historian put it, 'after six months of bitter fighting and some 8500 casualties, including 3000 dead, the South West Pacific Area was exactly where it would have been the previous July had it been able to secure the beachhead before the Japanese got there'. However, despite the huge costs in casualties, the Japanese had been cleared from Papua, and the Allies were able to develop the Buna, Dobodura, Oro Bay area into a sea and air base from which operations into the Huon Peninsula could be prosecuted.

The final Japanese forward movement

Following the Japanese disaster at Midway, General MacArthur evolved a series of ambitious plans (TULSA) to capture Rabaul. These were not accepted by the Joint Chiefs who instead, in a directive dated 2 July 1942, laid down a series of

Men of the 14th Field Regiment hauling 25-pounder guns up the Kokoda Track to fire on the Japanese positions at Ioribaiwa. Uberi, Papua, September 1942. (AWM 026855)

tasks for him and for Vice-Admiral R. L. Ghormley in the adjacent Pacific Ocean Area. Briefly, these were: task one: seize the Santa Cruz Islands, Tulagi and adjacent points; task two: seize the remainder of the Solomon Islands, Lae, Salamaua and the north-east coast of New Guinea; and task three: seize Rabaul and other points in the New Guinea–New Ireland area. Task one under Ghormley was given a target date of 1 August 1942 but, as stated, the landings at Tulagi and at Guadalcanal actually took place on 7 August. The landings caused a strong reaction from the Japanese who, over the ensuing weeks and months, committed strong naval and ground forces to wrest back the area which they had previously occupied. These attempts failed at great loss until, finally, the Japanese evacuated their forces from Guadalcanal on 7 February 1943. Thus their withdrawals from Papua and Guadalcanal were almost coincident. But while they had elected to go over to the defensive in the Solomons, they had not abandoned their intention to take Port Moresby. They now mounted a series of operations designed to regain the initiative.

At Lae, the 51st Japanese Division was reinforced from Rabaul in early January and, in that month, 3000 troops set out to capture Wau. Their effort was stopped within sight of Wau airfield by Australian troops, mainly of the 17th Brigade, many of whom had just been flown in to the same airfield. Further Japanese efforts to reinforce ended in disaster. A convoy including eight troop transports bound for Lae was intercepted in the Battle of the Bismarck Sea by Allied aircraft in early March. All eight transports were sunk, as were many escort vessels. Only 850 Japanese troops reached Lae, 3000 having been killed. Never again did the Japanese try to run a large convoy through the Allied air blockade.

A final large-scale Japanese air offensive, Operation I-GO, was launched under Admiral Yamamoto's direction in early April 1943. It was designed to destroy the Allied build-up of forces at key points like Guadalcanal, Port Moresby and Milne Bay. It not only failed to achieve its objective, it suffered heavy losses. Not long after, Yamamoto's own aircraft was shot down and he was killed. It proved to be a prelude to the Allied counter-offensive which continued unbroken until the end of the campaign.

The Allied counter-offensive

Until August 1943, Allied strategic planning, and particularly MacArthur's own operational plans, were based on the assumption that the reduction of Rabaul by direct assault was an important goal. At the 'Quadrant' Conference of Allied leaders at

Quebec in August 1943, this goal was modified. Rabaul was 'to be neutralised rather than captured'. But, in addition, and for the first time, MacArthur was authorised to advance to the Vogelkop Peninsula in a series of 'airborne–waterborne operations'. For the South-West Pacific theatre, the Quadrant decisions gave assurance of a strategic continuity which until then had been lacking. Not only could MacArthur continue the sequence of CARTWHEEL operations which he had commenced on 30 June 1943, secure in the knowledge that his operations would be neither shut down nor curtailed in favour of Admiral Nimitz's central Pacific thrust, he could now advance to the western end of New Guinea with an implied assurance that, once there, he would be permitted to continue his operations to the Philippines.

The CARTWHEEL Operations, which had grown out of MacArthur's ELKTON plans, were a series of combined, mutually supporting steps by which his Australian and American forces were coordinated with those of Fleet Admiral W. F. Halsey in the South Pacific Area. They now aimed at the encirclement and neutralisation of Rabaul and, once begun, they received few setbacks. Unopposed landings by American forces of General Krueger's Sixth Army at Woodlark and the Kiriwina Islands coincided with Halsey's landing at New Georgia on 30 June 1943. MacArthur's landings were carried out in the ships of Vice-Admiral D. E. Barbey's newly created 7th Amphibious Force, which gave him a capability he had previously had to improvise. An American force was landed simultaneously, and virtually unopposed, at Nassau Bay in the boats of 2nd Engineer Special Brigade (US), whose small craft were to prove as invaluable for coastwise support as Barbeys' larger vessels were for amphibious landings.

Japanese interference with these operations was reduced by growing Allied naval and most importantly, air strength. In two days, 17–18 August 1943, General Kenney's Fifth Air Force destroyed more than half the 7th Air Division's 200 aircraft at Wewak. Now, key Japanese ground positions were reduced in rapid succession: Lae fell on 16 September 1943 after a combined amphibious, airborne and air landed operation, and Salamaua on 11 September. Finschhafen was captured on 2 October following an amphibious landing, but a subsequent operation to capture Sattelberg Mountain which dominated the area took gruelling weeks until 25 November. In the Ramu Valley, the 7th Australian Division cleared Shaggy Ridge on 23 January 1944; then, on 10 February, Australian forces advancing along the coast linked with an American seaborne assault at Saidor.

Not only were these operations remorselessly grinding down Japanese strength, but, once their forces lost control of the strategically important Vitiaz and Dampier Straits, the great strong point, Rabaul, quickly became redundant. This condition, partly achieved by the capture of Finschhafen, was reinforced when American forces landed in New Britain in December 1943, and was made complete by landings in the Admiralty Islands on 29 February 1944. The bypassing strategy was most spectacularly displayed by MacArthur's leap to Hollandia and Aitape, which left Japanese forces at Wewak and Hansa Bay to 'wither on the vine'. It was distinguished by two advantages: MacArthur's use of ULTRA which, through a newly acquired Allied capability to decipher the four-digit mainline Japanese Army code, gave him devastatingly accurate knowledge of the Japanese order of battle and dispositions; and the use, for the first time, of Admiral Nimitz's carriers to cover his landings, until his engineers could put air bases into service.

The Hollandia operation did not conclude the New Guinea campaign but it heralded its end. Further landings, some strongly contested, as at Biak Island, reminded the Allies that Japanese troops could not be brushed aside. And, in tactical counter-attack, as at the Driniumor River, the Japanese demonstrated that they had lost none of their former ferocity. But, in grand strategic terms, the South-West Pacific had caught up. MacArthur, from a Washington perspective, was no longer dragging his feet.

An American historian has put the New Guinea campaign in this perspective: '[It] is really the story of two Allied armies fighting two kinds of war — one of grinding attrition and one of classic maneuver … The series of breathtaking landings, often within a few weeks of one another, were the fruits of the Australians' gallant efforts in eastern New Guinea'.

It is interesting to speculate on what might have been the outcome had not the two principal allies in the South-West Pacific, the United States and Australia, complemented each other's strengths while reducing each other's weaknesses, in the way they did. Until the end of 1943 the brunt of the conflict on the ground was borne by the three experienced AIF infantry divisions. There was perhaps unpalatable but genuine truth in Blamey's statement of 1942 that the American division at Buna (the 32nd Division) was 'definitely not equal to the Australian militia'. That the American ground forces in MacArthur's command were given time and opportunity to condition and orientate themselves for jungle war, was almost solely due to the fact that for the first two years the brunt of the ground fighting was carried by the Australians. Lieutenant-General R. L. Eichelberger admitted this: MacArthur's excessive vanity and ethnocentrism prevented him from doing so. At the same time the Australian troops could have achieved very little had they not had massive American support in the areas of logistics, sea and air transport, and offensive air and naval support. The degree of cooperation could have been greater still, had the two countries' commands been less divided.

The sequel to the campaign in New Guinea, the mopping-up operations carried out by Australian formations from November 1944, when they relieved American formations at Aitape–Wewak, and on the islands of New Britain and Bougainville, until the end of the war, are treated in the next section. These operations, particularly the manner in which they were conducted, remain controversial. One Australian writer has referred to them as 'the unnecessary war'. It is useful, however, to remember that MacArthur carried out similar mopping-up operations in the Philippines, but at much greater cost. Moreover, unlike the Australian operations, they aroused little or no controversy. And, at the Japanese surrender, they too were continuing.

These controversies, however, had nothing to do with any lack of tenacity, resolution or courage on the part of the troops involved. Not long before the final American landing at Sansapor in the Vogelkop Peninsula, MacArthur on 12 July 1944 ordered Blamey to have Australian troops relieve Americans at several places, thus releasing the latter for the thrust to the Philippines. His decision was consistent with the Curtin government's aim of using Australian troops to regain Australian territory. Places and dates of relief were to be: Northern Solomons (Bougainville), Green Island and Emirau Island on 1 October 1944; Australian New Guinea (Aitape–Wewak) on 1 November 1944; and New Britain on 1 November 1944. Blamey was also warned that in the Philippine operations, 'it is contemplated employing initially two AIF Divisions [7th and 9th Divisions] as follows: One Division — November 1944; One Division — January 1945'.

In the event this did not occur and, amid conflicting information from MacArthur's staff and his own fluctuations, it seems unlikely that he ever intended major Australian formations to have a part in the re-conquest of the Philippines or subsequent operations towards Japan itself. A further controversy concerned the scale of relief.

Blamey's intention was to carry out MacArthur's orders using seven brigades: the latter insisted he use 12. Usually superiors have to resist subor-

Men of the 25th Battalion search the bodies of dead Japanese soldiers following fighting on Slater's Knoll, Bougainville, 6 May 1945. (AWM 090365)

dinates' demands for more troops; the reverse was true here. Perhaps MacArthur did not want it recorded that six and a half US divisions were relieved by an approximately equal number of Australian brigades. There is no other convincing explanation. But MacArthur's orders stood.

Bougainville

In Bougainville, the largest of the then Solomon Islands group, the United States formation, basically XIV US Corps, had pursued a 'live and let live' policy with the Japanese forces which numbered between 39 000 and 40 000 (compared with a United States estimate of 12 000). These comprised the bulk of the 17th Japanese Army (Lieutenant-General Hyakutake), disposed in three main areas: the 18th Independent Mixed Brigade around Numa Numa on the north-east coast; the 13th Regiment around the Jaba River–Gazelle Harbour in the south-west, with the 23rd Regiment further south again; and the 45th Regiment around Kieta in the east. In addition, there were sizeable Japanese Marine forces: 87th Garrison Force (4000) was in the Buka area, and the 6th Sasebo Special Naval Landing Force (2000) and 7th Kure Special Naval Landing Force (1500) were in the south. All were making

efforts to become self-sufficient with extensive gardens and food cultivation. When the Japanese became aware through American news broadcasts in October 1944 that Australian troops were relieving the existing American garrison, they instructed their forces to meet all patrols with offensive action. They estimated that the Australians would not attack before January 1945; in this they miscalculated.

The overall Australian commander for the New Guinea–Solomon Islands area was Lieutenant-General V. Sturdee (q.v.), with his headquarters at Lae. His four forces were deployed widely over 1600 kilometres (1000 miles from east to west). His directions to commanders differed in each area, but, as will become apparent, both his directive from Blamey and his own orders resulted in far more proactive operations on the ground than had been the case previously.

His largest force, II Australian Corps (Lieutenant-General S. G. Savige [q.v.]), was committed to Bougainville. It included 3rd Australian Division (Major-General William Bridgeford [q.v.]), consist-

---

Map 22a Australian operations on Bougainville, 1944–45. *Inset* Location of Bougainville Island.

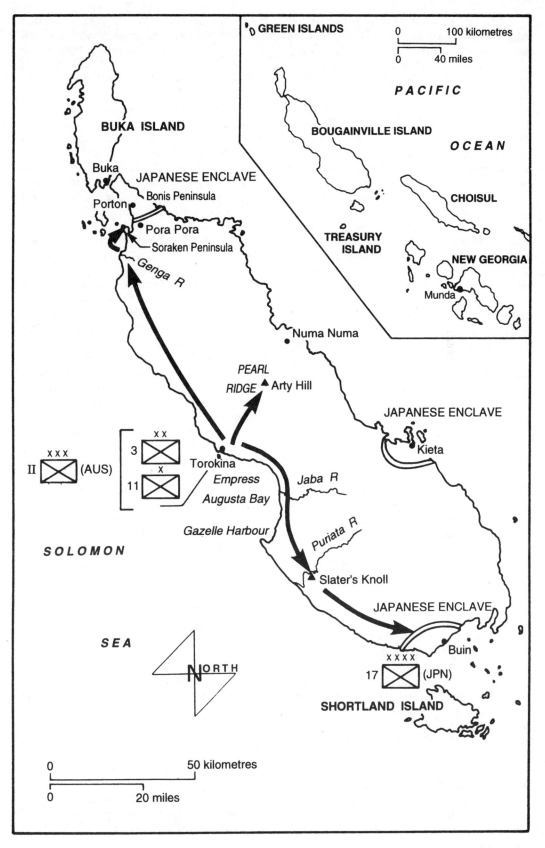

GREEN ISLANDS

0   100 kilometres

0   40 miles

PACIFIC

BOUGAINVILLE ISLAND

OCEAN

BUKA ISLAND

Buka

JAPANESE ENCLAVE

Bonis Peninsula

Porton

Pora Pora

Soraken Peninsula

CHOISUL

TREASURY ISLAND

NEW GEORGIA

Genga R

Munda

Numa Numa

PEARL RIDGE ▲ Arty Hill

JAPANESE ENCLAVE

Kieta

xx
3 ⊠

xxx
II ⊠ (AUS)

x
11 ⊠

Torokina

Empress

Augusta Bay

Jaba R

Gazelle Harbour

Puriata R

SOLOMON

Slater's Knoll

JAPANESE ENCLAVE

Buin

SEA

NORTH

xxxx
17 ⊠ (JPN)

SHORTLAND ISLAND

0   50 kilometres

0   20 miles

ing of the 7th, 15th and 29th Brigades, and two independent brigades, the 11th and 23rd. The 4th Base Sub Area (Australian) was the supporting logistic formation and the force was supported by a number of squadrons of the RAAF and RNZAF. The 23rd Brigade (Brigadier A. W. Potts [q.v.]) relieved US forces on Green Island, Emirau Island, Treasury Island, and at Munda in New Georgia on 27 September 1944. At Torokina in Empress Augusta Bay, Savige had deployed most of his force and had assumed responsibility from Major-General Griswold, US Army, on 22 November 1944.

Savige's plan, which was to destroy the Japanese forces on Bougainville, differed dramatically from his US predecessor's, and was to involve three simultaneous offensives: in the north, to force the enemy garrison into the narrow Bonis Peninsula and there destroy it; in the central sector east of Torokina, to clear the enemy from the high mountain spine near Pearl Ridge, and then by aggressive patrolling threaten important Japanese lines of communication along the east coast; and in the south, to fight the main Japanese concentration clustered round its garden areas near Buin and destroy it in the decisive battle of the campaign. Throughout its operations, the Australian force was hampered by the lack of shipping and other means of mobility and this, at least, was to render these offensives slow, and relatively ponderous.

The 7th Brigade (Brigadier Field) advanced into the central Sisivie–Arty Hill area in late November 1944. Japanese resistance stiffened until at Pearl Ridge the Japanese committed a freshly reinforced battalion and other troops with guns and 20–30 mortars. Japanese opposition was overcome by 30 December 1944, and from the vantage point so captured, the Australians could see both sides of the island. A substantial push further to the east was precluded by difficulties of re-supply. However, patrols by Australian infantry and by regular and irregular New Guinea troops led by Australian officers of the Australian New Guinea Administrative Unit (ANGAU) and Australian Intelligence Bureau (AIB) (qq.v.), continuously harried the Japanese towards Kieta and loosened their hold over the local people.

In the north the orders to 11th Brigade (Brigadier Stevenson), which was advancing along the coast, were to destroy the Japanese forces progressively. Where possible they were to get behind the Japanese rather than force them into the interior. On 29 January 1945 the Japanese launched a heavy counter-attack against a bridgehead on the north side of the Genga River. It was broken up with the aid of mountain artillery. A request for tanks to speed up the advance was denied by Savige who wanted to achieve surprise by first using them in

the southern sector. By 26 April 1945 the Soraken Peninsula had been occupied. A week later the Japanese had been forced out of Pora Pora and the intention was to advance to the Porton Plantation area where the Bonis Peninsula narrowed to 5 kilometres. An attempt to come in behind the Japanese in the plantation area by landing a company group from the sea was unsuccessful. The landing craft grounded well offshore and the Japanese defenders reacted fiercely. After four days the force was withdrawn with heavy casualties, having inflicted equally heavy casualties on the enemy. By June 1945, after six months of operations, the Japanese force in the north had been confined to Buka Island and the Bonis Peninsula.

Meanwhile, the main Australian operation was going forward in the south, where Savige's orders were 'To destroy Japanese forces in Southern Bougainville.' Savige restricted operations to a battalion at a time. Signs had been growing that the Japanese would counter-attack, and they did so in March and April 1945 when the advance reached the Puriata River. Using 2400 fresh troops and considerable artillery the 23rd Japanese Regiment launched a series of human wave attacks against the 25th Battalion at Slater's Knoll. The attacks were broken up with the help of tanks and medium artillery. On 6 April 1944, 292 dead Japanese were counted around the knoll. Although Japanese morale was shaken, Blamey, who was coming under criticism in Australia for the human cost of the operations and the manner of their conduct, cautioned subordinates on the spot: 'Take your time Hammer. There's no hurry', he told one brigade commander. Efforts to induce the Japanese to surrender by the use of leaflets and other forms of psychological warfare were uniformly unsuccessful.

When Japan surrendered in August 1945, the Australians were still short of Buin but the troops had acquitted themselves well. In all, 516 Australians had been killed in the operation and 1572 wounded. Against these figures 8500 Japanese had been killed and 9800 had died of disease; 23 571 surrendered.

The Bougainville operation, then and since, has been criticised both for its aims and methods. It should be kept in mind that MacArthur, and not Blamey, dictated the relatively large number of troops to be used. When the operation started, the end of the war was not in sight and some Allied leaders, including Churchill, believed it might continue into 1948. If Savige is to be criticised for the nature and tempo of operations, he was not restrained by Blamey, except in the mild, and indirect, reproof mentioned. In a sense, the sort of 'pacification' carried out by the Australians was remarkably similar in intent to that employed by MacArthur himself in the Philippines. Yet it was done at far less cost in human lives and with a minimum of resources. Had Savige tried to 'clean up' Bougainville faster, casualties, particularly against the Japanese, would have been far higher. In any case he could not have done so without greatly increased amphibious support, or paratroops, who were not to be had. Had he tried solely to rely on patrolling (knowing what we now know about counter-insurgency practice, to which these operations bore some similarity), he might equally have been criticised for being dilatory. His course of action was blunt, but he scarcely had the means for more imaginative operations.

New Britain

As in Bougainville, when the Australians took over from US troops in New Britain, both Americans and Japanese were observing a tacit truce. The two sides were geographically separated. Within the extensive 'no man's land', Allied patrols of indigenous troops led mainly by Australians were waging an enterprising guerrilla-style campaign.

The American enclaves were at Talasea–Cape Hoskins, Arawe and Cape Gloucester. The Japanese were mainly in the Gazelle Peninsula in the north behind Rabaul. They comprised the bulk of General Imamura's Eighth Area Army. In numbers, they were estimated by MacArthur's intelligence staff at 38 000. Actually there were 93 000. Along the south-west coast there were Japanese watching-stations as far forward as Awul. Again, as in Bougainville, the Japanese were cultivating rice, and gardening to increase their self-sufficiency. Although they were numerically very strong, their once powerful air and naval forces had been reduced to two serviceable aircraft and no ships, although they still had perhaps as many as 150 barges capable of carrying 10–15 tons or 90 men.

The 5th Australian Division (Major-General A. H. Ramsay [q.v.]), consisting of 4th, 6th and 13th Brigade Groups, began taking over from the 40th Division (US) at Cape Hoskins on 8 October 1944. From the beginning, the Australians intended to operate much further forward than the US bases. Accordingly, the 6th Brigade (Brigadier Sandover), less one battalion that was already at Cape Hoskins, landed at Jacquinot Bay on 4 November 1944. Major-General Ramsay assumed responsibility from the American commander on 27 November 1944. Ramsay's orders from Lieutenant-General Sturdee were that as information about the enemy on New Britain was conflicting and incomplete, it would be unwise to undertake major operations immediately. Accordingly, he was told 'to obtain the required information, to maintain the

offensive spirit in our troops, to harass the enemy and retain moral superiority over him'. In particular, the tasks of the division were to defend the bases; 'so far as the maintenance situation would permit with existing resources, to limit Japanese movement south-west from Gazelle Peninsula'; and to collect information on which future plans could be based. In achieving this, the cooperation quickly established between the troops of the incoming division and the AIB patrols operating forward against the Japanese, was vital. It remained of a high order throughout the subsequent operation.

Australian forces advanced on both coasts. In the north, leaving a small detachment at Cape Hoskins to protect the airfield, the 36th Battalion was moved by sea to Ea Ea in January 1945, from which it patrolled forward in company strength. By the end of April its main body, after clashing several times with strong Japanese groups, had moved to Watu Point, with elements further forward. Thus Australian troops held the western edge of the Gazelle isthmus.

On the south-east coast advanced elements of 6th Brigade had progressively driven in the Japanese picquets. On 27 February 1944, Sturdee ordered Ramsay to secure the Waitavalo–Tol area, which first involved engineer work to make a crossing of the Wulwut River, followed by a systematic series of attacks which the Japanese resisted. By April, the Australian force was firmly established across the neck of the Gazelle Peninsula. It held this line to the end, while patrolling forward of it.

In many respects this had been a classic containment campaign, aided by Imamura's extraordinary quiescence. Throughout, the Australian troops had been hampered by many shortages, particularly in regard to shipping. In addition supporting aircraft were based on the mainland and requests for air action had to be made 24 hours in advance. Until the end of April, that is for almost the entire period of offensive operations, no light intercommunication or reconnaissance aircraft were available. The shortage of shipping was so serious that the relief of a battalion on the north coast had to be carried out by a protracted overland march from the south coast across the mountain spine rather than by the easier sea route. And this occurred at a time when other Australian troops in Borneo had more sea transport and air support than they knew what to do with.

In summary, the achievements of the division and the AIB patrols were remarkable in this campaign. Against Japanese offensive forces of 53 000, mostly veteran Japanese units and 16 000 marines, the inexperienced Australian militiamen, using only a brigade at a time, had forced the Japanese out of their forward positions, bottled them up in the Gazelle Peninsula, and established an ascendancy so complete that the Japanese offered no real resistance to deep patrols in the last four months of the war. This was achieved at a cost of 53 killed in action, 21 dead of other causes and 140 wounded.

Aitape–Wewak

The third mopping-up campaign was fought in an elongated triangle of country bounded in the north by the sea, in the south by the Sepik River and in the west by a line running north and south through Aitape. Enclosed within this were the Torricelli Mountains and the fertile plateau area around Maprik.

Here in October 1944 the 18th Japanese Army (Lieutenant-General Hatazo Adachi) was deployed. It had been greatly depleted in operations against the Australian Army during the Lae, Salamaua, Huon Peninsula and Ramu Valley campaigns; then reduced again in abortive counter-attacks against the US forces at Aitape and the Driniumor River. It was estimated at 30 000 troops: the true figure was about 35 000. Adachi's headquarters were at Wewak. Other dispositions are as shown in Map 22b (top). Each Japanese division was about the strength of a regiment, and some 3000 base troops had been dispersed into small groups to forage in the countryside.

The Australian force, predominantly the 6th Division (Major-General J. E. S. Stevens [q.v.]), deployed progressively to Aitape in the latter half of October 1944. Sturdee's orders to Stevens defined the division's role as follows:

(a) to defend airfield and radar installations in the Aitape–Tadji area;

(b) to prevent movement westward of Japanese forces in the area and seize every opportunity for the destruction of those forces; and

(c) to give maximum help to AIB and ANGAU units in the area in their task of gaining intelligence, establishing patrol bases and protecting the native population.

Even before 6th Division's arrival, ANGAU long-range patrols using selected native villagers called 'sentries' — taught to use grenades and Japanese rifles — were engaged in guerrilla warfare with the Japanese and had killed numbers of them.

The American force began moving out on 10 September 1944 from the perimeter. The 2/6th Cavalry (Commando) Regiment deployed east to Babiang and then into the mountains to the south to harass and report on the Japanese; the 19th Brigade (Brigadier J. E. G. Martin) relieved the Cavalry (Commando) in the Babiang area and prepared for operations east of the Danmap River, where it

was also to prevent enemy movement west of the Driniumor River. The 17th Brigade (Brigadier M. J. Moten), while initially improving the Aitape defences, was to prepare to move into the Torricelli Mountains where large elements of the Japanese force were developing extensive garden areas and coercing the local people to support them. The 16th Brigade (Brigadier R. King) was initially to remain in reserve. Stevens believed the Japanese would defend the Danmap River area to prevent movement east along the coast and that inland they would also protect the Tong–Maprik area for gardening and foraging. As the campaign developed Stevens' planning was increasingly influenced increasingly by the following factors: the need for vigorous patrolling to gain information; the need to make good use of air support where available; and the need for an advance along the coast to be accompanied by a thrust through the Torricellis. Sturdee also elaborated his earlier instruction, by requiring Stevens to conduct minor raids and patrols by land, sea and air where possible, and to maintain an offensive spirit among the troops.

Thus, while the 17th Brigade moved into the Torricellis, 19th Brigade moved east from Babiang. Its first line was Dogreto Bay to Malin. This systematic movement east along the coast was continued with the 16th Brigade relieving the 19th in January 1945. The fighting in the coastal strip and in the Torricellis was continuous and bitter with the Japanese, mainly in small groups, resisting the Australian advance skilfully and doggedly, taking maximum advantage of the ground. The monsoon rains did not help the operations which were frequently constrained by flooding. The further they continued, the more difficult they were to support administratively. Stevens asked for paratroops, but they were not available.

As better information of Japanese strengths and dispositions were obtained, Stevens, with Sturdee's agreement, was able to revise his plans. In February, he informed his subordinates that he intended to take But, Dagua and Wewak in that order; and in the mountains to capture Maprik and advance eastward. In the mountains, the 17th Brigade, as well as driving towards the Maprik area on a wide front (11 000 yards), set out to outflank many of these villages and garden areas in the south and capture or destroy as many Japanese as possible who in these areas 'had obtained complete initiative'.

Map 22b New Guinea campaign 1944–45. (Top) Australian operations against the Japanese in the Aitape–Wewak area. (Bottom) Australian operations against Japanese positions around Rabaul.

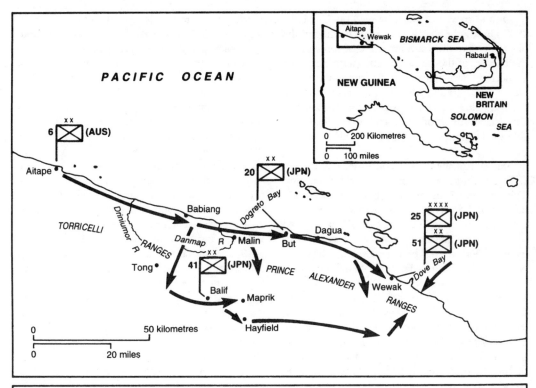

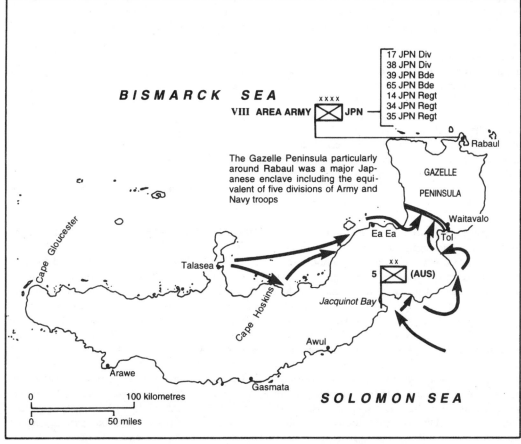

On the coast, But was occupied on 16 March 1945. During the advance to Dagua fighting in the high ground to the south proved to be some of the most severe of the campaign. The Australians used as much artillery and mortar-fire support as possible. Both Dagua and Maprik were taken in April: the latter after a series of well-conducted assaults. The establishment of an airstrip at Hayfield suitable for Dakotas (q.v.) helped ease the re-supply problem.

Sturdee had always been concerned that the 6th Division might be required in the Philippines. Therefore, in seeking ways to speed up operations, both he and Stevens believed that with an allocation of 10 extra landing craft tanks, Wewak could be taken more quickly by amphibious assault than by any other means. The extra landing craft tanks were obtained. On 10 May 1945, Wewak was taken by a landward assault suported by tanks and artillery. An amphibious force (Farida Force) was landed simultaneously at Dove Bay east of Wewak to cut the coastal road. Just as in the parallel operations on Bougainville and New Britain, the course of events showed that even modest additional logistic support paid overwhelming dividends in time and casualties saved.

By the end of May the Aitape–Wewak operation was entering its final stage. Most of what remained of the 18th Army had been driven away from the coast and broken up into ever smaller groups. However, as the Australian eastward advance continued in the mountains it encountered well-fed and well-led troops who were also well organised. Immediately before the Japanese surrender in August 1945, 18th Army had been driven into what General Adachi referred to as a 'last stand' area in the hinterland behind Wewak, which formed a semicircle 16 kilometres in diameter. Many of these Japanese in the main garden and cultivation areas were probably in better condition than the Australians.

A Japanese criticism of the Australian operations, made during interrogations after the former's surrender, was that they had been too cautious. And, that if the Australians had pressed harder, they could have separated the Japanese from the main garden area in a more emphatic fashion. The criticism, while partly valid, fails to acknowledge the relative poverty of logistic means available to the Australians, particularly all-weather re-supply means. It also fails to recognise the fact that these particular Japanese no longer had worthwhile military installations to defend: they could concentrate on growing food and looking after themselves. As it was, in 10 months the 6th Division had advanced 70 miles along the coast and driven the Japanese from 3000 square miles of terrain. It had lost 442 men killed in action and 1141 wounded. The Japanese lost about 9000 killed and 269 taken prisoner.

It has to be said, however, that Australian losses, although smaller, were made harder to bear by the conviction among many of those taking part that they were unnecessary. The operations themselves were about as physically gruelling as any that Australian troops undertook during the war. Yet as their statements and reports show, it required a supreme effort of motivation for the troops to convince themselves that their fighting and dying really contributed to winning the war. It is unlikely that this issue can ever be resolved to everyone's satisfaction. As has already been mentioned, these operations were of a pattern with MacArthur's in the Philippines, but cost far less in lives and resources. When they began, it is unlikely that anyone could have foreseen that the war would end within a year. It had been the Australian government's clearly expressed aim to clear the Japanese enemy from its territories. It was both unfortunate and galling for the troops that their efforts and sacrifice were not apparently taking them any closer to Tokyo. In this regard, however, it is useful to refer to a comment of Brigadier H. H. Hammer (q.v.). His brigade — the 15th — was then fighting on Bougainville. He wrote afterwards that its morale

> could not have been better if it had been fighting the Alamein battle or capturing Tokyo. Yet every man knew, as well as I knew, that the operations were mopping-up and that they were not vital to the winning of the war. So they ignored the Australian papers, their relatives' letters advising caution, and got on with the job in hand, fighting and dying as if it was the battle for final victory

JOHN COATES

NEW GUINEA IN THE FIRST WORLD WAR see GERMAN NEW GUINEA

NEW GUINEA FORCE Created as part of General Sir Thomas Blamey's reorganisation of the AMF in April 1942, New Guinea Force absorbed the 8th Military District (q.v.) and the troops serving in it. As with Northern Territory Force, it absorbed coastal defence, anti-aircraft and garrison units which normally came under the Lines of Communications Area headquarters. Its initial commander was Major-General B. M. Morris, a regular soldier whose experience thus far in the war had been in India and as commander of the Australian Overseas Base in the Middle East. It was assigned to General Douglas MacArthur's (q.v.) South-West Pacific Area (q.v.) in the middle of the same month. Morris was replaced by Lieutenant-General S. F. Rowell (q.v.) in late July. New Guinea Force was responsible for driving the Japanese out of south-eastern New Guinea and Papua, and at various times had command of Australian and American units, and also assumed responsibility for Milne Force, under Major-General Cyril Clowes (q.v.), during the fighting at Milne Bay in which the Japanese offensive there was defeated. It experienced a certain amount of command turbulence in August when Blamey relieved Rowell of his command and replaced him with Lieutenant-General Sir Edmund Herring (q.v.). By April 1943, with Papua cleared of the enemy, New Guinea Force had six Australian brigades in three militia divisions under the acting command of Lieutenant-General I. G. Mackay (q.v.). The latter was succeeded by Lieutenant-General Sir Leslie Morshead (q.v.) in November 1943, and the force continued to exercise administrative responsibility for Australian operations in New Guinea. Its responsibilities were assumed by the First Australian Army on 1 October 1944.

NEW GUINEA INFANTRY BATTALION see **PACIFIC ISLANDS REGIMENT**

NEW SOUTH WALES CORPS was an army unit raised in England to replace the Marines who had accompanied the First Fleet. The Marines had clashed with Governor Phillip over their refusal to supervise convicts or undertake judicial duties, so on 8 June 1789 the British government issued an order to recruit a new regiment. Major Francis Grose was appointed commanding officer, and in June 1790 the first 183 men of the corps landed in Sydney. From the departure of Governor Phillip in December 1792 to the arrival of Governor Hunter in September 1795 the corps' commanders administered the settlement, and it was at this time that some officers received large land grants. This was probably also the time when they developed virtual monopolies on trade in important goods including spirits. The officers' trading activities were an important factor in the event for which the corps is most famous, the so-called Rum Rebellion of 26 January 1808, in which soldiers led by Major George Johnston (q.v.) deposed Governor Bligh. The corps' commanders governed New South Wales from the time of the rebellion until the arrival of Governor Macquarie and the 73rd Regiment in December 1809. Its officers were then recalled to Britain, but the rank and file were given the option of remaining in the colony. In Britain, the corps was reorganised as the 102nd Regiment, was later renumbered the 100th Regiment, and was disbanded in 1818.

Largely because of its involvement in the Rum Rebellion, the corps has had an unsavoury reputation in Australian history, but the reality is more complex. A total of 1645 men served in the corps between 1790 and 1810, but on average there were only about 570 (approximately 15 per cent of the colony's civilian male population) on active service at any one time. Throughout the period most corps members were recruited in England, but there were also significant numbers of ex-Marines, ex-convicts and sons of settlers recruited in the colony. Convicts still under sentence, who were offered their freedom if they served a full term of duty, constituted at most 14 per cent of the corps. The corps made an important contribution to the colony's free population, with 422 being discharged in the colony and 447 choosing to stay in New South Wales when the regiment was recalled. Of the latter group, 265 transferred to the 73rd Regiment and 111 joined the specially created Veteran Corps. Of the 82 officers who served with the corps 37 eventually settled in the colony. Discharged soldiers were offered land grants of 25 to 30 acres, along with provisions, tools and two convict labourers, but 60 per cent of them did not take up this offer. Instead, they sought work in the towns, and since many had specific trade skills they made a valuable contribution to the colony's economic development. Even more important was the role of officers, whose trading activities helped to create a market economy in New South Wales. One former officer who got his start in business through the corps was John Macarthur, who with his wife Elizabeth founded the Australian wool industry.

Pamela Statham (ed.), *A Colonial Regiment: New Sources Relating to the New South Wales Corps, 1789–1810* (Australian National University, Canberra, 1992).

NEW SOUTH WALES NAVAL FORCES SHIPS When it was launched in 1855, New South Wales's wooden gunboat *Spitfire* became the first ship ordered to be built for defensive purposes by an Australian government, and also the first locally constructed warship. Built to allay fears of Russian invasion (see COLONIAL WAR SCARES), it was transferred to the new Queensland government in 1859, becoming a civilian pilot vessel. Russian war scares also lay behind the government's decision to form the New South Wales Torpedo Corps in January 1878, and in March 1878 *Acheron* entered service, followed shortly afterwards by *Avernus*. These second-class 'outrigger' torpedo boats, built by the local Atlas Engineering Company to a design which had only recently been adopted by the RN, were intended principally to defend Sydney Harbour. They had funnels that could be lowered in action, and they carried unusual spar torpedoes (see TORPEDOES) which were pivoted at the bow rather than extending straight over the bow as in other torpedo boats. The two ships were sold in 1902. New South Wales's largest ship was the wooden screw corvette HMCS *Wolverene*, which was launched in 1863 and served as the flagship of the RN's Australian Squadron (q.v.) from 1875. In 1882 it was given to the New South Wales Naval Brigade for use as a training ship. When not on training cruises *Wolverene* spent most of its time laid up in harbour. It was sold in 1893. There were also 14 auxiliary vessels, some armed, used at different times by the New South Wales Naval Forces. The colony's naval ships were used in exercises which usually involved challenging imaginary enemies at Sydney Heads or 'attacking' forts along the coast. (See also COLONIAL NAVIES.)

Spitfire: displacement 65 tons; length 62 feet (overall), beam 16.5 feet, draught 5.5 feet; armament 1 × 32-pounder gun.

Acheron and *Avernus*: displacement 22 tons; length 82.5 feet, beam 10.5 feet, draught 4 feet; armament spar torpedoes, 2 × 14-torpedoes.

Wolverene: displacement 2431 tons; length 225 feet (overall), beam 40 feet, draught 18 feet; speed 10 knots; armament: 16 × 8-inch guns, 1 × 7-inch gun, 4 × 40-pounder gun.

NEW ZEALAND Defence issues and military affairs have been an important element in Australian–New Zealand relations since the middle of the nineteenth century. While a number of factors predisposed the colonists on both sides of the Tasman to regard each other as natural allies, genuine closeness of purpose and cooperation has generally only occurred in wartime. Groups of Australian colonists served in the colonial militia during the New Zealand wars of the 1860s (q.v.), and the colonies were party to a succession of agreements governing colonial naval defence arrangements, but there was little close defence cooperation otherwise in the nineteenth century. In the decade before the First World War, Australia and New Zealand came to cooperate more closely, introducing compulsory military training and organising their armies on similar territorial bases. New Zealand was eager to send cadets to the new Royal Military College, Duntroon (q.v.), for training, and both sides pursued joint planning arrangements in the event of war: the mobilisation of 1914 had its origins in the discussions between the commander of the New Zealand forces, Major-General Alexander Godley (q.v.), and his Australian colleagues during his visit to Melbourne in 1912. The passage of the Australian and

New Zealand transports was coordinated in 1914 and their forces joined together by the imperial authorities into an Australian and New Zealand Army Corps. The two New Zealand brigades were joined with two Australian ones to form the New Zealand and Australian (NZ&A) Division, commanded by Godley and staffed by New Zealanders. When the AIF was reorganised after the Gallipoli campaign, Australian and New Zealand divisions were formed into II Anzac Corps, again under Godley's command, and served together until the decision to form an Australian Corps in November 1917. In Palestine, Australian and New Zealand mounted units served in the Anzac Mounted Division. In the interwar period, while common interests remained, especially a shared focus on the south Pacific and concern over Japanese intentions, direct cooperation was undercut by the Singapore strategy (q.v.) and by New Zealand's decision to maintain its naval forces as part of the RN rather than as a separate national fleet. At the outbreak of war in 1939 coordination again eluded the two sides. New Zealand announced that it would dispatch its expeditionary force as soon as it was ready; Australia had believed some delay prudent until Japanese intentions were clearer, but the New Zealand announcement now forced their hand. For most of the Second World War there was no formal connection of Australian and New Zealand forces. An Anzac Corps was formed during the Greek campaign and fought there under Lieutenant-General Sir Thomas Blamey's (q.v.) command, but subsequent attempts to re-form it were unsuccessful and were beside the point once Japan entered the war in December 1941; the Australian divisions were returned to the Pacific, while the 2nd New Zealand Division remained in the Mediterranean. Although New Zealand formed another division, the 3rd, for service in the Pacific, it fought under American command in the Solomons and was later disbanded. In January 1944, however, with issues of postwar policy assuming increasing importance, Canberra and Wellington signed an Australia–New Zealand Agreement (q.v.), known as the Anzac Pact, which asserted their right to be consulted by other powers (basically Britain and the United States) in matters of direct interest in the region. The agreement produced an angry reaction in both London and Washington, and although it seemed to promise a heightened level of coordination and planning in the two nations' external affairs, soon fell into abeyance. New Zealand committed forces under Australian command to the British Commonwealth Occupation Force (q.v.) in Japan, withdrawing them in 1948 despite requests from the Australians that they should remain. The Korean War saw New Zealand gunners serving in the 28th

Commonwealth Brigade (q.v.) in support of Australian infantry battalions, and this became the pattern of New Zealand involvement in the subsequent regional conflicts of the 1950s and 1960s. Australian and New Zealand troops took part in operations during the Malayan Emergency (q.v.) and Confrontation (q.v.) although given the low-intensity nature of the operations, rarely together, and Australian and New Zealand units served with British forces as part of the ANZUK Brigade. New Zealand contributed artillery and a company of infantry to the 1st Australian Task Force in the Vietnam War, and the Australian battalions which incorporated a New Zealand company were designated Anzac battalions. The Vietnam commitment, while it provoked dissent in New Zealand, did not attract there the level of popular disaffection and protest evident in Australia, because of differing traditional attitudes towards conscription (q.v.) and because in any case, unlike the situation in Australia, only regular New Zealand servicemen were sent to Vietnam. The two countries had joined with the United States in the ANZUS Treaty (q.v.) in 1951, and this was to remain the basis of defence cooperation between the two sides until New Zealand's abrogation of the treaty over the issue of nuclear ship visits in 1984–85. Cooperation on defence issues with Australia, in matters such as joint procurement of weapons and equipment and access to training facilities, remains important to the New Zealand armed forces because of their much smaller establishments and lack of certain complex capabilities. Although there have been differences between the two sides, it is difficult to conceive of a situation in which their defence postures would diverge radically.

NEW ZEALAND WARS, AUSTRALIAN INVOLVEMENT IN New Zealand Wars, in which British and colonial troops fought against the indigenous Maori of New Zealand's North Island, resulted from the continuing expansion of European settlers on to Maori land and the colonial government's determination to crush Maori independence. War first broke out in 1845–46, when Maori in New Zealand's far North under the leadership of Hone Heke Pokai were defeated by imperial forces which included British troops based in Australia. When a new war began in 1860 in Taranaki province British troops were again sent from Australia, and the campaign was controlled by the commander of imperial forces in Australia until the New Zealand Command was separated from Australia in 1861. The colony of Victoria also sent across its ship *Victoria* (see VICTORIAN NAVY SHIPS), which was used mainly for patrol duties and logistic support. The most significant eruption of

warfare occurred in the Waikato region, where in 1858 a chief had been chosen as the first Maori king. New Zealand Governor George Grey refused to accept the King Movement's demand for autonomy, and in July 1863 his troops invaded the Waikato, causing a majority of North Island tribes to rally to the King Movement's defence. About 2500 Australians, lured across the Tasman Sea by the promise of settlement on confiscated Maori land, joined the four newly formed Waikato Militia regiments, which were used mostly for patrolling and garrison duties. Some also became scouts and bush guerrillas in the Company of Forest Rangers. Australians were not heavily involved in major battles, and perhaps as few as 20 died as the result of action. Though armed conflict would continue elsewhere in New Zealand in the years to come, the Maoris in the Waikato region had been defeated by 1864, and the Waikato regiments were disbanded in 1867. After the war the New Zealand government failed to provide the infrastructure which Australian military settlers needed to develop their farms, so many moved to the towns or the goldfields, or returned to Australia.

NEWSPAPERS see **SERVICE NEWSPAPERS**

NIRIMBA, HMAS, located in Sydney's western suburbs, was commissioned in April 1953 and in 1956 became the centre for training naval apprentices in the fields of marine, weapons and aviation engineering. In February 1987 it was given the additional role of providing General Duties sailors with their initial training, but in 1993 *Nirimba* was decommissioned and its duties taken over by HMAS *Cerberus* (q.v.).

NIZAM, HMAS see **'N' CLASS DESTROYERS**

NORFORCE see **ARMY RESERVE**

NORMAN, HMAS see **'N' CLASS DESTROYERS**

NORRIS, Major-General (Frank) Kingsley (25 June 1893–1 May 1984). Abandoning his medical studies at the University of Melbourne, Norris enlisted as a trooper in the AIF in August 1914, and served initially with the 1st Light Horse Field Ambulance. His father, who was at that time the federal Director of Quarantine, enlisted also and was a medical staff officer with the High Commissioner's office in London and, in 1915–16, Assistant Director of Medical Services of the AIF in England. The younger Norris served in the Middle East and reached the rank of lance sergeant before being ordered home in 1916 to resume his medical studies, an outcome shared by many medical students who had enlisted early in the war.

After establishing a specialist practice in children's medicine, Norris joined the CMF in 1925 as a captain, and was posted to the 2nd Cavalry Field Hygiene Section, where he served for six years. In 1924 he was given command of the 2nd Field Ambulance, and from October 1938 till the outbreak of war he was Deputy Assistant Director of Medical Services of the 2nd Cavalry Division. Enlisting in the 2nd AIF, he commanded the 2/1st Casualty Clearing Station for six months before promotion to colonel and the position of Assistant Director of Medical Services of the newly raised 7th Division. Norris served in the Mediterranean theatre before returning to Australia at the beginning of 1942 by way of Java, avoiding capture by the Japanese in the process.

As chief medical officer of the 7th Division Norris served in the arduous Papuan campaign of 1942–43. Although 49 years of age, he crossed the Owen Stanley Range on foot, the first senior staff officer to do so. He spent three months on the Kokoda Track (q.v.), supervising medical evacuation and resupply, and at times assisting the overworked surgeons in the forward areas. In May 1943 he became Deputy Director of Medical Services of the 1st Australian Corps, and served through the gruelling campaigns for the Japanese beachheads at Buna–Gona–Sanananda and the assaults on Lae and Finschhafen. His insistence on seeing for himself in the front lines broke his health and he was evacuated to Australia in April 1944 and, after a 10-month convalescence, discharged from the army as medically unfit.

His health regained, he returned to civilian practice, but in 1948 joined the regular army as Director-General of Medical Services with the rank of major-general. He supervised medical arrangements during the Korean War and oversaw the establishment of the School of Army Health at Healesville, later named in his honour. He retired from the army in June 1955, and in retirement engaged in a wide range of professional and philanthropic activities, including a period as medical adviser to the federal Department of Civil Defence within the Ministry of the Interior. He was knighted in 1957.

F. Kingsley Norris, *No Memory for Pain: An Autobiography* (Heinemann, Melbourne, 1970).

NORTH AUSTRALIA OBSERVATION UNIT

NORTH AUSTRALIA OBSERVATION UNIT (2/1st NAOU, the 'Nackeroos') was established on 11 May 1942 with the anthropologist Major W. E. H. Stanner (1905–1981) appointed commander. With headquarters at Katherine in the Northern Territory and responsibility for a vast area of Northern Australia, the NAOU's task was to patrol the northern coastal areas (usually on horseback) looking for signs of enemy activity, to man fixed coastwatch stations and to run a signals network for Northern Australia. At its peak strength the unit consisted of nearly 550 men, including the author Xavier Herbert, and employed 59 Aboriginal workers as guides and labourers. Its operations were scaled back from July 1943, and it was disbanded in March 1945.

Richard Walker and Helen Walker, *Curtin's Cowboys: Australia's Secret Bush Commandos* (Allen & Unwin, Sydney, 1986).

NORTH RUSSIAN INTERVENTION Following the collapse of the Russian war effort and the February and October revolutions in 1917, the British dispatched a 560-strong military mission to the North Russian ports of Murmansk and Archangel. Its purpose was to train a White Russian force in that area preparatory to the creation of a new Eastern Front against the Central Powers, and to ensure that the large quantities of military supplies shipped there from 1916 onwards to equip the Tsar's armies did not fall into German hands. There were several other military missions in Russia at this time, and more were sent once the civil war began in 1918. The North Russian Expeditionary Force (NREF) included nine Australians (three officers and six sergeants) selected by AIF Headquarters in April 1918. All were experienced soldiers, with three having served on Gallipoli as well as in France. The NREF was divided into two forces, Syren and Elope, and these reached their destinations in late June. The men were split up into small groups engaged on a variety of administrative, instructional and advisory tasks, and were sometimes in as much danger from the men they led as from the enemy, who by late 1918 was no longer the Germans but the Bolshevik forces engaged in the civil war which was to last until 1920. One Australian, Captain Allan Brown, was killed on 20 July 1919 when the White Russian battalion to which he was attached mutinied and murdered its officers, and there were a number of other cases of White units killing their officers and going over to the Reds. In March 1919 the decision was made to withdraw the mission, but this could only be done safely by the provision of a covering force, no faith now being had in any locally raised units. Recruiting for the North Russian Relief Force (NRRF) began at once on a voluntary basis, and drew officers and men from every regiment of the British Army and from all the dominion forces. About 100–120 Australians enlisted, and served in the 45th and 46th (Service) Battalions of the Royal Fusiliers and the 201st Machine Gun Company. The NRRF arrived in Archangel in early June 1919 and began training for an offensive up the rail and river systems which was designed to push the Bolshevik forces back while the Allies withdrew without interference. A secondary and optimistic aim was to leave the White forces in a better military position, in the hope that they might then hold their own against the Reds. In early August the commander of the force, Major-General Edmund Ironside, launched his offensive against the 6th Red Army in the area, inflicting large numbers of casualties and taking many prisoners for negligible losses to his own force. It was during this offensive that Corporal A. P. Sullivan won the VC for saving a group of men from drowning while under enemy fire. A second VC was awarded to an Australian, Sergeant S. G. Pearse, at the beginning of September when he was killed attacking a Bolshevik blockhouse with a Lewis gun, and these were the only two VCs awarded for the campaign. Minor patrol activity continued while forward positions were evacuated and stores either removed or destroyed, and by the night of 26–27 September the Allies had withdrawn from Archangel. Murmansk was evacuated on 12 October. The involvement in North Russia was a relatively minor if unusual episode, one which had almost no impact back in Australia, and which made no difference to the outcome of the Russian civil war, other than to confirm Bolshevik mistrust of the Western powers.

Jeffrey Grey, 'A "pathetic sideshow": Australians and the Russian Intervention, 1918–19', *Journal of the Australian War Memorial*, no. 7, October (1985), pp. 12–17.

NORTH WEST CAPE see **JOINT DEFENCE FACILITIES**

NORTHCOTT, Lieutenant-General John (24 March 1890–4 August 1966). Educated at the University of Melbourne, Northcott was commissioned in the 9th Light Horse in August 1908. He transferred to the Administrative and Instructional Staff in November 1912, and volunteered for the AIF on the outbreak of war. Appointed adjutant of the 12th Battalion, he landed at Gallipoli and was severely wounded during the campaign. He held a staff position back in Australia after returning to duty, as GSO3 on the headquarters of the 5th Military District. After the war he suffered the customary reduction in rank, being appointed to junior staff jobs once again with the 5th Military District and the 13th Mixed Brigade. As a regular, however, his career improved with his selection to attend the Staff College, Camberley, between 1924 and 1925. He returned to Army Headquarters in the latter year, but was again posted to Britain in January 1933 as an exchange officer, serving with

the 44th (Home Counties) Division for two years before attending the Imperial Defence College (IDC) on the 1935 course. His overseas attachments did not end there. Graduating from the IDC, he attended the Senior Officers' School at Sheerness before serving a five month attachment to the Committee of Imperial Defence, the premier strategic policy formulation body in the Empire. After a further period in the United States and Canada, he returned to Australia in June 1937 as the principal operational staff officer on the 4th Division's headquarters.

The outbreak of the war in 1939 found him as Director of Military Operations and Intelligence, and within five weeks he was appointed Deputy CGS. With a couple of brief exceptions, his war service was to consist of critical staff positions, in which he provided valuable service. He attended the Dominions Conference in London at the beginning of the war, and was seconded to the 2nd AIF in September 1941. He did not serve overseas (other than a brief attachment to British formations in the Middle East in late 1941), however, being appointed GOC of the 1st Armoured Division, then being raised from existing cavalry formations, and destined never to serve overseas as a formation, despite the government's original intentions. Northcott's lack of direct command experience showed, and critics held that he never really 'gripped' his new command, which was still floundering when he was succeeded by Major-General H. C. H. Robertson (q.v.), newly returned from the Middle East, in April 1942. A short stint as GOC of the II Australian Corps in the middle of 1942 was followed by his promotion to the post of CGS in September, a position which he held for the remainder of the war and at which he excelled. The relationship between the CGS, who was responsible for administrative matters, and Blamey, as C-in-C of the Australian Military Forces, was a delicate one, and it was greatly to Northcott's credit that he carried off the role without the traumas and tantrums that characterised a similar command arrangement in the RAAF between Air Vice-Marshals Jones and Bostock (qq.v.). It meant, however, that he spent the whole of the Pacific War in Australia, and missed the opportunities for professional advancement given to his contemporaries who held active commands in New Guinea. This, along with his largely unrecognised contribution to the successful prosecution of the war effort, was acknowledged by Sturdee (q.v.) when the latter was asked by the Labor government to take up the post of CGS again after the war; he agreed, on the condition that Northcott be given the command of the British Commonwealth Occupation Force (q.v.), then forming for service in Japan. As such, he negotiated the MacArthur–Northcott agreement (q.v.) in December 1945, which governed the deployment of the force under overall American command; it was a hard bargain driven by MacArthur's principal subordinates but one which, in the view of Lieutenant-General H. C. H. Robertson who succeeded him in this post also, could not have been bettered by any one else. Once again, however, Northcott's lack of command experience showed, and the occupation force only really attained a sound basis under his successor, although in fairness it must be added that the problems faced by both were formidable in a devastated area of Japan. He accepted the offer of the governorship of New South Wales in 1946, as the first Australian-born incumbent, and held the post until 1957. He was appointed KCMG in 1950 and KCVO in 1954. In retirement from the army he became honorary colonel of the 1st New South Wales Lancers in 1949. He was twice Administrator of the Commonwealth.

NOVELS see **LITERATURE, WAR IN AUSTRALIAN**

NUCLEAR WEAPONS TESTING see **ATOMIC WEAPONS TESTS**

NURSES see **AUSTRALIAN ARMY NURSING SERVICE; AUSTRALIAN ARMY MEDICAL WOMEN'S SERVICE; ROYAL AUSTRALIAN ARMY NURSING CORPS**

'O' CLASS SUBMARINES (*Oxley, Otway*). Laid down 1925, launched 1926; displacement 1354 tons (surfaced); length 275 feet; beam 28 feet; speed 15.5 knots surfaced, 9 knots submerged; armament 8 × 21-inch torpedo tubes, 1 × 4-inch guns, 2 machine-guns.

These submarines were ordered from Britain in 1925 and arrived in Australia in 1929. They saw little service with the RAN and were transferred to Britain in 1931 because of economic stringencies.

OBERON CLASS SUBMARINES (*Oxley, Otway, Ovens, Onslow, Orion, Otama*). Laid down 1964–73, launched 1965–75; displacement 1610 tons (surfaced); length 295.5 feet; beam 26.5 feet; speed 16 knots surfaced, 19 knots submerged; armament 8 × 21-inch torpedo tubes.

The *Oberon* Class was ordered from Britain in 1963. In their time they were the most advanced conventionally powered submarines in the world, capable of staying at sea for months without support. The ships acted as training and assault craft and have taken part in exercises with American and Indonesian ships. They will be progressively replaced as the *Collins* Class (q.v.) submarines are brought into service.

OBOE OPERATIONS OBOE was the code-name for the second stage of the MONTCLAIR operations aimed at the reoccupation of large areas of the Netherlands East Indies, the southern Philippines and British North Borneo, and the destruction of Japanese forces there. The first phase, VICTOR, saw landings on Panay, Cebu and Negros in the Philippines, and was completed by mid-April 1945. Originally there were six OBOE operations planned: OBOE 1 against Tarakan; OBOE 2 against Balikpapan; OBOE 3 against Banjermasin; OBOE 4 against Surabaya or Batavia (Jakarta); OBOE 5 against the eastern Netherlands East Indies; and OBOE 6 against British Borneo (Sabah). In the end only the operations against Tarakan (1 May 1945), Balikpapan (1 July 1945) and British Borneo, at Labuan and Brunei Bay (10 June 1945), were carried out. Together these operations constituted the last campaigns of the Australian forces in the war against Japan, and they remain the subject of considerable controversy.

O'BRIEN, Brigadier John William Alexander (13 June 1908–29 May 1980). A civil engineer by calling, O'Brien was commissioned into a survey company of the Australian Garrison Artillery as a lieutenant in August 1928. He thrived on citizen soldiering (see CITIZEN MILITARY FORCES), and in the 1930s received rapid promotion and a range of

command and staff appointments such that by the outbreak of war he was undoubtedly one of the most highly qualified young gunners in the army.

At the outbreak of war O'Brien was immediately called up for full-time duty at Army Headquarters, serving as deputy to 'Gaffer' Lloyd (q.v.), then Assistant Director of Artillery. Lloyd and O'Brien oversaw the conversion of the artillery from horses to motorisation. In May 1940 he was posted as a major to the 2/7th Field Regiment and went overseas with the unit in November 1940. In March the following year he was promoted again and given command of the 2/5th Field Regiment, which he led ably through the Syrian campaign (q.v.). In April 1942 he was promoted to brigadier, the youngest officer of that rank in the army.

By now back in Australia, O'Brien was appointed Director of Artillery at Land Headquarters, and for the remainder of the war he occupied important technical staff positions. He was instrumental in the development of the short or 'baby' 25-pounder field artillery piece, modified for jungle conditions in New Guinea, and in January 1943 he became Deputy Master-General of the Ordnance. He remained in this position until the end of the war, responsible for the acquisition and trial of a wide range of army equipment and supplies. 'There is little honour or glory working in the MGO Branch', he later wrote, 'the main reward is one's own satisfaction of a job well done'.

After the war he remained in the army and, uniquely for a non-American, was made head of the Science and Technology Division of the Economic and Scientific Section at General Douglas MacArthur's headquarters in Japan. At first the job had an important role in helping to revive Japanese economic life, but as the occupation wore on O'Brien found he had little to do. In October 1948 he was selected as president of the military court set up to try the last chief of the Japanese Naval Staff, Admiral Soemu Toyoda, for war crimes. The trial lasted for 10 months, at the end of which Toyoda was acquitted, much to MacArthur's displeasure since the verdict struck at the basis of 'command responsibility' under which a number of other senior Japanese officers had been tried and executed. After his return from Japan in 1951 O'Brien was given another technical staff job, as senior representative at the Australian embassy in Washington with responsibility for defence supply and production. He returned to Australia in 1954 and retired from the CMF in 1963. He corresponded extensively with the official historian D. P. Mellor during the latter's writing of the volume *The Role of Science and Industry* (1958), and many of Mellor's judgements on weapons' development and

Table 1 Australian Defence Force ranks

Navy	Army	Air Force	Stars
Admiral	General	Air Chief Marshal	Four Star
Vice-Admiral	Lieutenant-General	Air Marshal	Three Star
Rear-Admiral	Major-General	Air Vice-Marshal	Two Star
Commodore	Brigadier	Air Commodore	One Star
Captain	Colonel	Group Captain	
Commander	Lieutenant-Colonel	Wing-Commander	
Lieutenant-Commander	Major	Squadron Leader	
Lieutenant	Captain	Flight Lieutenant	
Sub-Lieutenant	Lieutenant	Flying Officer	
Midshipman	2nd Lieutenant	Pilot Officer	

procurement reflect O'Brien's own views. While in Japan he also found time to write the regimental history of his wartime command, published as *Guns and Gunners: The Story of the 2/5th Field Regiment* (1950).

Gordon Rimmer, *In Time for War: Pages from the Life of the Boy Brigadier* (Mulavon Publishing, Sydney, 1991).

ODGERS, George James (29 March 1916-). Educated at the Universities of Western Australia and Melbourne, Odgers was appointed by Gavin Long to write a volume of the Air series of the Official History following his war service in the AIF. The resulting volume, *Air War Against Japan 1943–45*, appeared in 1957. Odgers had an extremely active career in defence journalism and public relations, and wrote numerous books in the area aimed at a popular readership but without sacrificing either detail or rigour. He spent time with Australian forces in Korea, Malaya during the Emergency and in Vietnam, and from his experiences produced a number of volumes that still bear reading well after their publication. Notable among these are *Across the Parallel* (1952), which examined the service of the RAAF's No. 77 Squadron in Korea; *Mission Vietnam* (1974), detailing the RAAF's service during the Vietnam War; and a series of large-format books on the past and present activities of each of the armed forces, which have gone through several editions. Between 1965 and 1975 he was Director of Public Relations first for the Department of Air and subsequently for the RAAF, while from 1975 until his retirement in 1981 he was Director of Historical Studies and Information in the Department of Defence.

OFFICER RANK STRUCTURE Officers hold commissions and can be defined both by rank and function. In the navy, flag officers are those of the rank of commodore and above, so called because they have the right to fly a special flag which denotes their rank. In the army, senior officers of the rank of brigadier and above are known as general officers. The rank of brigadier did not exist before the First World War; in 1914 a brigade was a colonel's command, and the brigade commanders were only made up to brigadier-general after they reached Egypt, in line with British practice. Senior officers in the air force (air vice-marshal, air marshal etc.) owe the grand title of 'marshal' to Winston Churchill's romantic streak when he created the rank structure for the RAF at the end of the First World War. In recent years there has been a move towards American usage, which merely defines senior officers by the number of stars denoting their rank (one-star, two-star, etc.), irrespective of their service. The most senior officer of the Australian services has been Field Marshal Sir Thomas Blamey (q.v.), promoted to the rank not long before he died in 1950. When the Officer Cadet School, Portsea (q.v.), closed in 1985, and direct entry training for army officers was concentrated at RMC Duntroon (q.v.), the rank of 2nd lieutenant disappeared from the regular army, although it is still to be found in the Army Reserve (q.v.). Australian Defence Force ranks are listed in Table 1.

OFFICIAL HISTORIES The Australian government has commissioned and underwritten four separate series of official war histories, dealing with Australian involvement in the First and Second World Wars, the Korean War, and the conflicts in Southeast Asia between 1948 and 1973. While official in the sense that the resources to produce them have been furnished by government, the Australian official historians have always proclaimed proudly that their work has been free from censorship or other interference, and while attempts to influence the judgements and conclusions have certainly been made, they have rarely been successful.

The first series, *Official History of Australia in the War of 1914–18*, was edited, and largely written by, the official correspondent with the AIF on Gallipoli and in France, the journalist C. E. W. Bean (q.v.), whose appointment as official historian dated from 1919. The work appeared in 15 volumes between 1921 and 1943, and included three volumes on the medical services written by A. G. Butler (q.v.). Bean himself wrote the two volumes on Gallipoli and four volumes on the AIF in France and Belgium between 1916 and 1918, produced the volume of photographs in collaboration with H. S. Gullett (q.v.), and had (often close) supervision of the other five in between numerous other demands on his time. Although careful to maintain his own freedom from outside direction, he appears to have intervened in the volume on the operations of the RAN by A. W. Jose (q.v.), modifying some of Jose's more acerbic judgements on British personalities (including Churchill, whom Jose regarded as no friend of the Pacific dominions) and decision-making, especially in the early months of the war. Bean's histories ran into financial difficulties until marketing was taken over by the Director of the Australian War Memorial, John Treloar (q.v.), and by the late 1930s the volumes, and especially the first six, had run to numerous printings and had sold something in the order of 20 000 sets. They remain highly influential in shaping popular views of the Australian experience of the First World War, although mostly through the work of other writers who draw heavily on them, as Bean's own great work is now scarcely read outside a research audience. He was widely credited, however, with having written 'democratic' history, and there is little doubt that the scope and nature of his history influenced not only his successors in Australia but some official historians elsewhere in the Commonwealth after the Second World War.

Bean was also instrumental in the appointment of the official historian of Australia's part in the Second World War, Gavin Long (q.v.), who took up the position of 'general editor' of the series *Australia in the War of 1939–45* in January 1943. Like Bean, Long was a journalist and had been an official correspondent with the British and Australian armies. His history, which ran ultimately to 22 volumes arranged in five discrete series (Army, Navy, Air Force, Civil, Medical), appeared between 1952 and 1977 and involved a team of 12 writers other than himself. Long wrote three volumes in the army series, and administered the project until his retirement in 1963. His administrative task was far greater than Bean's, partly because the arrangements for the history involved the Australian War Memorial, the Department of the Interior, and the army in various capacities. His plan for the series also underwent a number of changes, not least because the Second World War was a more complex experience for Australia than the previous war

had been. He originally conceived of 14 volumes: four on the army (it grew to seven), two on the navy, three on the RAAF (which became four), two on the home front (which became four, divided between politics and economics), one on war industries, one on primary industries (subsumed into the economic volumes), and a volume on general defence policy (which was abandoned). There were also four volumes of medical history.

He had difficulties with authors, and with the unreasonable expectations of his political masters who held quite unrealistic ideas about the length of time needed to assimilate records and write books. He drew up a list of 14 prospective authors in September 1944, but only seven (including himself) accepted commissions. The stipend offered was miserly, and Long had to argue, cap in hand, for supplementation to enable some of his authors to continue supporting their families while completing their volumes. Paul Hasluck's (q.v.) second volume was inordinately delayed by his election to federal parliament in 1951 and subsequent lengthy ministerial career. A. S. Walker (q.v.) resigned through ill health before completing his fourth volume and died before it appeared, while Long's original choice for the volume on Tobruk and Alamein, the journalist Chester Wilmot (q.v.), was killed in an air crash before he could start work. As Bean had done, Long wrote a synoptic volume in retirement.

His history, like Bean's, is known principally to a research audience, while its influence more generally has never matched that of his predecessor. Long stated that his aim was 'to establish a story that will carry conviction in other countries'. The small partners in a great coalition war 'must write their own stories if they wish them to be told in reasonable detail. The larger partners . . . have no inclination to do this task for them'. Where Bean's series had been a national history within an imperial context, Long's was much more assertively Australian in its emphases, and it seems a shame that his achievement is not more widely appreciated.

In the 1970s the government commissioned Dr Robert O'Neill (q.v.) as official historian for the Korean War. This appointment marked a departure in the selection of official historians and reflected the growth of academic and scholarly requirements in the writing of history generally. A graduate of RMC Duntroon, O'Neill had been a Rhodes scholar, had served in Vietnam as an infantry officer and, after leaving the army, had moved firmly into academic life as one of Australia's first university-based defence analysts. At the time of his appointment he was head of the Strategic and Defence Studies Centre at the Australian National University, and his career was to take him subsequently to head the International Institute of Strategic Studies in London and then to the Chichele Chair in the History of War at Oxford. The more limited scope of Australian involvement in Korea called for a more limited conception of an official history, and O'Neill worked alone and part-time through the 1970s, assisted by a single research assistant based in the Australian War Memorial. *Australia in the Korean War 1950–53* appeared in two volumes in 1981 and 1985, devoted respectively to strategy and diplomacy, and combat operations.

This treatment reflected the relatively greater importance of policy in the Cold War environment, the general lack of influence over the conduct of the war exercised by the smaller partners like Australia in the United Nations Command, and the fact that the three Australian services rarely operated in support of one another and could therefore be treated as discrete deployments. His first volume was and remains the single best and most detailed treatment likely to be written on the evolution of the ANZUS Treaty (q.v.), so central a part of Australian defence and foreign policy in the postwar period, while the large volume on combat operations provides a detailed account of the experiences of Australian soldiers, sailors and airmen in the first conventional conflict of the atomic era. Once again an official historian was able to declare his freedom from government interference in the writing of the history, although the Attorney-General's Department tried to influence the treatment of the Australian journalist and alleged traitor Wilfred Burchett, at that time still living, in the lengthy chapter on Australian POWs written by another author, himself a former POW of the Chinese. O'Neill's scholarly and meticulous history, like the conflict which was its subject matter, was treated with a measure of indifference when it appeared, reflected in sluggish sales.

His successor, Dr Peter Edwards, was appointed in 1983 to write the *Official History of Australia in Southeast Asian Conflicts 1948–1975*. A Rhodes scholar and established diplomatic historian, Edwards conceived of a multi-volume history which would do justice to the complexity of Australian involvement in the Malayan Emergency, Indonesian–Malaysian Confrontation and the Vietnam War (qq.v.), the last in particular involving diplomatic, operational, medical and intensely partisan domestic political issues, many of which (like Agent Orange [q.v.]) continued to resonate in the 1980s and beyond. The project has enjoyed more than its fair measure of controversy and difficulty, and certainly more than any of its predecessors. Edwards had intended originally to appoint a team

of writers, but resources proved insufficient to match staff historians to all the volumes and he was forced to call on a number of historians outside the Australian War Memorial who agreed to work on various volumes without remuneration.

Disagreements within his original team led to a certain amount of staff turnover and became highly public with the publication of the first volume, *Crises and Commitments*, in 1992. Many of the claims advanced by his critics, that the book represented a whitewash of Australian policy in the period covered and, more generally, that official history itself was an illegitimate activity, were not sustainable and were refuted by many of the reviews, both in Australia and overseas, which praised its detached and scholarly analysis of complex events. The second volume, Ian McNeill's *To Long Tan*, dealing with the early period of Australian ground force commitment in Vietnam, won the Templer Medal of the Society for Army Historical Research in Britain when it appeared in 1993. Further volumes, for a total of eight, are set to appear in the course of the 1990s.

More generally, the Australian practice of setting up an *ad hoc* official history organisation has been wasteful of both resources and through the loss of 'institutional memory', and there seems to be a case for establishing within government a small unit along the lines of the Historical Section of the Cabinet Office in Britain or the service historical branches in the United States, to administer and coordinate future works of official history, both civil and military. While the official histories themselves have generally enjoyed a limited readership, they have formed the basis for many popular works of Australian military history (q.v.). As Gavin Long himself noted, by this means 'the stories told in the official histories, often in their own words, reach a wider audience'.

OLD GUARD see **SCOTT, Lieutenant-Colonel William John Rendell; SECRET ARMIES**

OLIVER HAZARD PERRY CLASS GUIDED MISSILE FRIGATES (FFGs) (*Adelaide, Canberra, Sydney, Darwin*). Laid down 1977–81, launched 1978–82; displacement 3680 tonnes (full load); length 136 m; beam 14 m; speed 28+ knots; armament 1 × 76 mm guns, 1 Phalanx close-in weapons system, 1 × Mark 13 launcher for standard AA or Harpoon A/S missiles, 2 × Seahawk (q.v.) helicopters, 2 × Mark 32 torpedo tubes.

These frigates were ordered from the United States for fleet air defence, reconnaissance, interdiction and anti-submarine duties. Their principal weapon system is the missile-launcher which can fire a mix of surface-to-air and surface-to-surface missiles. Their main defence from missiles comes from the Phalanx anti-missile gun. They took part in deployments to the Persian Gulf during and after the 1990–91 Gulf War (q.v.).

O'NEILL, Robert John (5 November 1936–). Born in Melbourne and educated at Scotch College, O'Neill graduated from RMC Duntroon in 1958 and was commissioned into the Royal Australian Corps of Signals. He completed an engineering degree with first class honours at the University of Melbourne, and was Rhodes Scholar for Victoria in 1961. At Oxford University he gained his Doctor of Philosophy degree for a study of the relations between the German Army and the Nazi Party, which since publication has remained a standard work on the subject. On his return to Australia he transferred to the infantry, served in Vietnam with 5RAR as second in command of a company and as battalion intelligence officer, and was mentioned in despatches. He taught in the Department of History at RMC Duntroon from 1967 to 1969, left the army with the rank of major, and joined the Department of International Relations at the Australian National University. He was head of the Strategic and Defence Studies Centre, 1971–82, Australian official historian for the Korean War, on which he published two volumes (1981, 1985), and Director of the International Institute for Strategic Studies, London, from 1982 to 1987. In 1987 he was appointed Chichele Professor of the History of War at Oxford and a Fellow of All Souls. One of the most influential strategic writers and historians Australia has produced, he is a prolific author and commentator on defence and security matters.

ONSLOW, HMAS see *OBERON* CLASS SUBMARINES

ORDNANCE CORPS see ROYAL AUSTRALIAN ARMY ORDNANCE CORPS

ORION, HMAS see *OBERON* CLASS SUBMARINES

ORION, LOCKHEED (11-crew maritime reconnaissance and anti-submarine aircraft [P-3C]). Wingspan 30.37 m; length 35.61 m; armament 10 × Harpoon air-to-sea missiles, 8 torpedoes or mines; maximum speed 761 km/h; range 3835 km; power 4 × Allison 4910 effective h.p. engines.

The Orion is at present the RAAF's maritime patrol aircraft and is flown by No. 10 and 11 Squadrons at Edinburgh (q.v.), South Australia. The first 10 P-3B Orions were delivered in 1968 and were joined by 10 more advanced P-3C models in 1978. The P-3Bs were replaced by P-3Cs from 1985. The P-3C is armed with Harpoon anti-shipping missiles and Australian-made Barra (q.v.) sonobuoys which are dropped into the ocean and send back to the aircraft indications of submarines. Orions have been used in civil search-and-rescue operations (see CIVIL COMMUNITY, ASSISTANCE TO). In 1994 the American electronics company E-Systems gained the contract to replace the radar, acoustic processing systems, computers, navigation and communication equipment used on RAAF Orions with more modern systems, at an estimated cost of $700 million.

OTAMA, HMAS see *OBERON* CLASS SUBMARINES

OTWAY (I), HMAS see 'O' CLASS SUBMARINES

OTWAY (II), HMAS see *OBERON* CLASS SUBMARINES

OVENS, HMAS see *OBERON* CLASS SUBMARINES

OVERLORD see D-DAY

OWEN MACHINE CARBINE (OMC) was a blow-back-operation, magazine-fed, sub-machine-gun (SMG) capable of automatic fire and single shots. The Owen was used in the Australian Army from 1942 until the late 1960s, although it was officially replaced by the F1 SMG in 1964. During that time the Owen saw active service with Australian soldiers in the South-West Pacific Area in the Second World War, Korea, the Malayan Emergency, Confrontation in Borneo, and South Vietnam. The Owen had a reputation for being simple, robust, reliable and easy to handle. It was distinctive in appearance with a tall, overhead magazine and front and rear hand grips positioned below the weapon with a Cutts compensator built into the muzzle.

The Owen was Australian designed and produced, having evolved from a prototype 0.22-inch calibre rimfire SMG designed and built by Evelyn Owen during the late 1930s. The design was then developed by the Lysaght Works at Port Kembla during 1940–1941 with the assistance of Vincent and Gerard Wardell, who worked with Owen on the experimental models and trials. The weapon finally went into production there late in 1941 amid controversy and competition from the AUSTEN SMG. There were various versions of the Owen featuring minor modifications and

Sapper W. Whelton of the 2/6th Independent Company cleans his Owen gun after a skirmish with Japanese forces, Antirigan, New Guinea, 27 September 1943. (AWM 057655)

attempts to lighten its weight by using a skeleton steel butt instead of a solid wooden one and simplifying construction of the body. A modification carried out during the period 1948 to 1955 saw the fitting of a safety slide to trap the cocking handle and prevent the accidental discharge of a dropped weapon. Some Owen barrels were fitted with a bayonet standard to allow the fitting of a special Owen bayonet with either an 8-inch or 10-inch blade (shortened SMLE Pattern 1907 bayonets).

The Owen fired 9 mm calibre ammunition and had a maximum effective range of 100 yards for aimed fire and about 25 yards for instinctive fire. The Owen weighed 10 pounds 30 ounces empty and around 11 pounds with a filled magazine. There were minor variations on these weights depending on the version. The Owen's 33-round capacity magazine was normally filled with fewer rounds to avoid tension on the magazine spring and to avoid feeding problems. The rear face of the magazine acted as the ejector. The Owen had a cyclic rate of fire (i.e. the rate of fire that theoretically could be achieved given a continuous supply of ammunition) of around 600 rounds per minute and was usually fired in bursts of two to three rounds as required or in single shots.

IAN KURING

OXLEY (I), HMAS see **'O' CLASS SUBMARINES**

OXLEY (II), HMAS see **OBERON CLASS SUBMARINES**

P

PACIFIC ISLANDS REGIMENT (PIR) Native soldiers were enlisted into the 1st Battalion of the Papuan Infantry Battalion (1PIB) beginning in June 1940, and many of the early recruits were drawn from ex-policemen of the Royal Papuan Constabulary. No similar move was made in the mandated territory of New Guinea until 1944, and there was some unease and mistrust of the notion of providing natives with arms and military training. By the time the Japanese landed there were 300 native soldiers in 1PIB, and this increased to 700 over the next two years. They took part in the early fighting along the Kokoda Track (q.v.) in company with the 39th Battalion, in the advance to Salamaua, and in the campaign to clear the Ramu and Markham valleys. In March 1944 the 1st New Guinea Infantry Battalion (1NGIB) was raised, followed shortly by 2NGIB, the whole being incorporated into a new Pacific Islands Regiment. Two more battalions had begun enlisting men and a sixth was authorised by the time the Pacific War ended. All officers and senior NCOs were Australians, and volunteers.

In the course of the war three battalions each of 77 Australians and about 550 native soldiers saw action against the Japanese, and approximately 3500 Papuans and New Guineans passed through the ranks of the PIR before its disbandment in the course of 1946. Three soldiers of 1PIB were awarded the DCM for gallantry and a fourth went to a soldier serving with a special unit, and a further 29 men were decorated or mentioned in despatches. There were some incidents of indiscipline on New Britain during the raising of 2 NGIB and at the war's end, but these were often related to issues of pay and conditions or prompted by insensitive or deficient handling on the part of Australian officers, although they were used to 'prove' that the raising of native units had been a mistake by those in the army and the civil administration, and especially in the Australian New Guinea Administrative Unit (q.v.), who had opposed the idea all along. In the years immediately after the war consideration was given to re-establishing an army presence in Papua New Guinea, initially through the establishment of a Whites-only unit of the CMF, the Papua and New Guinea Volunteer Rifles, which was authorised in July 1949 after sustained pressure from the administrator, who nonetheless continued to oppose the raising of native units. In November 1950, after active scrutiny, the army authorised the raising of a

locally recruited regular battalion, once again to be designated the Pacific Islands Regiment, and the unit began enlisting in March 1951.

In December 1957 there were riots in Port Moresby between soldiers and Kerema men which had to be broken up by the constabulary; 153 soldiers were fined and 15 were discharged, while 117 civilian men were convicted for riotous behaviour. The organisation of the PIR was reviewed as a result of this. Henceforth officers were to serve for between four and six years with the regiment, and a few would be returned as senior officers later in their careers, thus avoiding the earlier error of posting in officers who had no experience of serving with PNG soldiers. When the Pentropic division (q.v.) was introduced into the Australian Army in 1960, PIR remained the only infantry battalion organised on the old establishment. There was a further outbreak of indiscipline in January 1961, this time over discriminatory pay scales. The disaffected were removed, but pay scales were increased and attempts, largely successful, were made to break up regional or tribal concentrations of soldiers within sub-units. By 1962 the unit numbered 660 PNG soldiers and 75 Australian officers and warrant officers. With concerns mounting over Indonesian intentions, the PIR began training for guerrilla operations, and in September 1963 a second battalion was authorised, but a third, foreshadowed the following year, was never raised, although the issue of a third battalion was to resurface from time to time until the eve of independence. At the same time, in January 1965, Papua New Guinea was reorganised as Papua New Guinea Command, thus formally ending the command link with Headquarters Northern Command in Brisbane. Beginning in 1963, NCO and junior officer ranks began to be filled by indigenous personnel, junior officers being trained at the Officer Cadet School, Portsea (q.v.), and by 1970 there were 30 PNG officers in the PIR with more on the way. Until independence in 1975, however, PIR was controlled from Australia, with no local influence or command, and there was for a time considerable ill feeling towards the PIR on the part of many PNG citizens who were moving into positions of authority in the lead-up to independence. In January 1973 the military establishment in Papua New Guinea was designated the Papua New Guinea Defence Force (q.v.) with the announcement of self-government, while defence powers were transferred formally in March 1975. The Pacific Islands Regiment became a 'Royal' regiment in 1985.

James Sinclair, *To Find a Path* (Boolarong, Bathurst, NSW, 1990, 1992).

Men of the Papuan Infantry Battalion escort carriers across the Yupna River near Singorkai, New Guinea, 21 March 1944. (AWM 071717)

PACIFIC MILITARY CONFERENCE, convened by the US Joint Chiefs of Staff, was held in Washington DC on 12–28 March 1943 to coordinate operations against the Japanese, particularly in the light of the decision taken at the Casablanca Conference (January 1943) to give the Pacific fifth priority in the Allied war plans, behind the battle of the Atlantic, aid to the Soviet Union, operations in the Mediterranean and the defence of the United Kingdom. This low priority accorded the Pacific was not intended, however, to revoke the directive issued in July 1942 to recapture Rabaul. General Douglas MacArthur's (q.v.) plan against Rabaul required an additional five divisions and almost 2000 extra aircraft, none of which could be provided as the American military build-up in Europe gathered pace. Although MacArthur's representative at the Washington conference antagonised most of those present, the outcome was satisfactory for MacArthur: the 1942 directive to retake Rabaul was replaced with a more general aim of containing the Japanese until circumstances made it possible to seize the Bismarck Archipelago. MacArthur was promised an additional two divisions and 524 combat aircraft, and the naval forces of Admiral W. F. Halsey were placed under his general control, giving him sufficient forces to prepare for a limited offensive. Even though MacArthur was denied the means at this stage to realise his grand design of cutting Japanese links with the vital oilfields of the Netherlands East Indies by establishing air and naval bases in the Philippines, the reallocation of resources within the Pacific theatre to the South-West Pacific Area was a significant victory. Ultimately, however, MacArthur's drive to the Philippines relegated Australian forces to secondary areas in which they engaged in operations subsequently described by some commentators as unnecessary.

PALESTINE CAMPAIGN The entry of Turkey into the First World War in October 1914 greatly complicated the strategic position of the British Empire by ensuring that resources would have to be diverted from the main theatre on the Western Front in order to protect the lines of communication with India and the Pacific dominions by way of the Suez Canal. The Turks made an early attempt to threaten the Canal in January 1915 but this, poorly coordinated and at the end of an over-long supply line, was easily repulsed by British forces. It was ironic that the low level of Turkish military ability which this attack seemed to suggest helped to feed the over-confidence of the planners of the Gallipoli campaign which followed a few months later.

With the evacuation of the Dardanelles in December and the transfer of many of the infantry divisions to the Western Front, the security of the Canal Zone became the primary task of the Egyptian Expeditionary Force (EEF), which included in its ranks the Australian Light Horse (q.v.) and New Zealand Mounted Rifles, combined with British and later Indian units to form two divisions, ultimately designated the Australian and the Anzac Mounted Divisions. These became an important element of the forces guarding Egypt (see Map 23a). There were only a certain number of routes across the Sinai Peninsula, dictated by terrain and the availability of wells, and in August 1916 a force of 14 000 Turks was beaten back at Romani, with losses of about 5000 killed and wounded and 4000 captured over five days of fighting. The Turks withdrew with British and empire forces in slow pursuit — slow because the advance could proceed only at the speed which it took to construct the railway and pipeline which guaranteed supplies of fodder, ammunition, food, and most critically of all, water. Water, or its absence, was the vital factor in operations, for the absence of a ready water supply limited tactical opportunities.

On this assumption, commanders were initially governed by a direction that if the first attack against Turkish positions did not succeed, they were to regard the operation as a reconnaissance and withdraw. The show of force involved could have the desired effect, however, as at Mazar in mid-September 1916, where the Turkish garrison withdrew two days after the light horse attack on them. In November the mounted units of the Desert Column were ordered to advance on El Arish, on the edge of the desert, which was occupied when the Turks abandoned it on 21 December. The retiring enemy had left blocking forces at Rafa and Magdhaba, about 30 miles apart. The Desert Column, commanded by General Sir Philip Chetwode, then moved on each in sequence. Magdhaba, to the south-east, was held by 2000 Turks in strongly entrenched positions. It was attacked on the morning of 23 December, but for much of the morning little progress was made, and withdrawal was considered, but elements of the 1st Light Horse Brigade, the Camel Corps and the 10th Light Horse Regiment (the latter making a wide approach march and charging the enemy at the gallop over the last few hundred yards) overwhelmed the central redoubts. Over 300 Turks were killed and more than 1200

Map 23a Sinai, 1916, showing the three main routes across the desert between Egypt and Palestine.

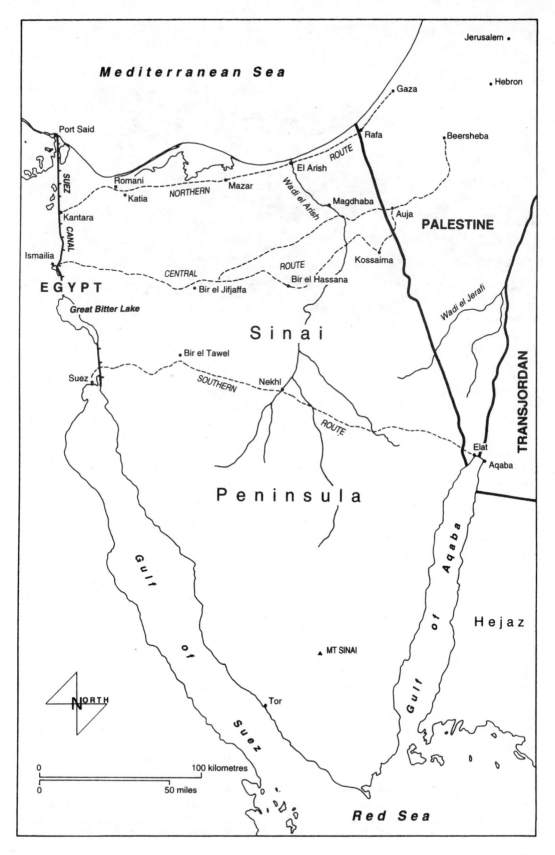

Australian light horsemen ride past the bodies of Turkish soldiers killed during the Battle of Magdhaba, Sinai, December 1916. (AWM P1034/81/11)

captured, including a lieutenant-general on a tour of inspection. It was a small action, but in its final stages a good demonstration of the continuing utility of mobile forces in open country. Rafa was then attacked by a mixed force of Australians and New Zealanders on 9 January 1917, and all 1800 defenders were killed or captured. The last small Turkish garrisons in Sinai were eliminated in February, and the EEF stood poised to take the war to the enemy in Palestine.

The key to southern Palestine was the town of Gaza, on which the Turkish defences hinged. An initial attempt to take the town on 26 March failed. Fog delayed the initial assault and the Turks put up a stout resistance. Staff work was poor, and higher headquarters lost touch with forward units and were unaware of the gains made. By evening the high ground to the east of Gaza had been taken, but Chetwode and General Sir Charles Dobell, commanding Eastern Force, ordered a withdrawal even as the Turks had begun to destroy their communications equipment preparatory to evacuating the town. The British commanders were worried about an approaching relief force from the north, and discovered too late that it had halted, and that the garrison had been on the point of surrender. British casualties were 3967 dead, wounded and missing, while the Turks lost 2437. A second attempt to take Gaza in two stages between 17 and 19 April fared no better. The German commander on the scene, the able Colonel Friedrich Freiherr Kress von Kressenstein, had used the intervening weeks to strengthen his positions, intending to fight a defensive battle. The British increased the weight of artillery in the assault, and deployed tanks and gas in the theatre for the first time, but to no avail. Turkish defensive fire was hardly affected by British counter-measures, von Kressenstein stating later that his troops were not much shaken by the enemy artillery. The battle was a fiasco for the EEF, resulting in 6444 further casualties.

Dobell and the commander of the EEF, General Sir Archibald Murray, who had performed well in the campaign in Sinai and who had carried a difficult administrative and political burden in Egypt, were relieved, the latter being replaced by a general fresh from the Western Front, Sir Edmund 'The Bull' Allenby. Allenby had won a striking victory with the Third Army at Arras, and was a highly competent professional soldier. Unlike his predecessor, he was to benefit from the renewed interest and, more importantly, material support which London now provided to the theatre. Additional divisions were transferred from Salonika, and the air force in the theatre was expanded. Allenby determined to take Gaza by attacking the further end of the Turkish defensive line, around the town of Beersheba (q.v.), and rolling up the defences from east to west, while a strong demonstration in front of Gaza itself kept Turkish attention there. The plan proved a spectacular success, with mounted units of the Australian light horse charging the

Map 23b Area of operations, Palestine, 1916–18.

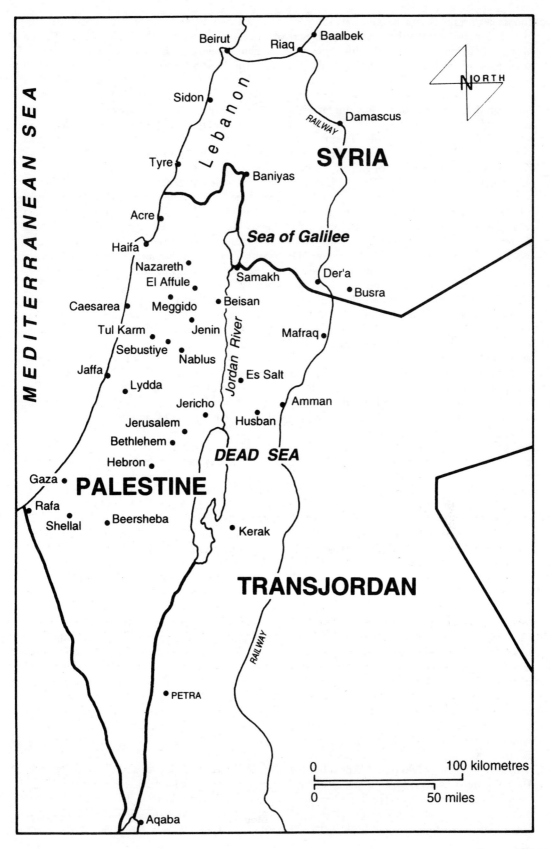

Turkish defences of Beersheba on the afternoon of 31 October and opening the way for the Desert Mounted Column to seize the high ground towards Khuweilfe. Further hard fighting followed in successive days, but on 6 November the Turks abandoned Gaza and began to retreat northwards. Although defeated, the enemy was still far from beaten, and several weeks of hard fighting followed before Jerusalem was taken in early December, the 'Christmas present to the British nation' which Lloyd George had requested of Allenby to help offset the dismal news from the Western Front that year.

While Allenby planned for operations in the coming year, the German March offensive on the Western Front led to the reduction of his forces and the withdrawal of most British units from the theatre to reinforce France. The army which was to defeat the Turks in 1918 was thus largely drawn from the Indian Army and the southern dominions. (In June 1918 London attempted to strip Allenby's force of half the light horse for infantry reinforcements in France, but his protests, and those from the Australian authorities, were enough to head this off.) An additional complicating factor in his calculations was the need to cooperate with the forces of the Arab revolt. A sizeable raid on Es Salt and Amman in late March proved abortive, for although Es Salt was taken the resistance offered by the Turks from commanding positions prevented the advance to Amman and the cutting of the railway to Damascus. A second raid, in concert with the Arabs, was made towards Es Salt at the end of April, with the aim of denying the Turks the wheat crop about to be harvested and establishing a stable link with the Arab forces. The 3rd Light Horse Brigade seized Es Salt, but its position was threatened to the rear by strong Turkish forces, and it was again forced to withdraw after several days' fighting. Although unsuccessful, the raids had the desired effect of concentrating enemy attention inland rather than on their coastal flank, which was where Allenby intended to break through at the end of the summer. Both the morale and the physical state of the Turkish forces had declined, and Allenby's forces outnumbered them heavily.

Allenby's great breakthrough battle was fought at Meggido in September. In three days of operations beginning on 19 September, the infantry divisions broke through Turkish lines and allowed the mounted formations to advance northwards to Haifa and Nazareth, while on 23 September the Anzac Mounted Division began a similar advance towards Es Salt and Amman. The Turks retreated, with large numbers surrendering in order to avoid falling into Arab hands. Nine days of operations

gained over 10 000 prisoners, and organised resistance on the part of the Turkish Seventh and Eighth Armies had largely ceased. On 26 September the Desert Mounted Corps under Chauvel (q.v.) was ordered to renew the pursuit to Damascus, which the advanced elements entered on 1 October. Mounted units and Arab forces pushed on northwards past Homs to take Aleppo on 26 October, which had been abandoned by its defenders. On 31 October the Turks signed an armistice.

The Palestine campaign ended in total victory over the Turks, whose government collapsed, but it contributed little to ending the war. The mistaken belief that its allies were the 'props' on which the German war effort relied misread the situation completely, for it was in fact the Germans who kept their Turkish, Bulgarian and Austro-Hungarian allies in the war well past a point where any of the latter believed a victory was possible. The campaign was instructive, however, for a number of reasons. It demonstrated that mobility still had a place in modern warfare, especially at the tactical level. (This was misinterpreted by many in the interwar period to mean that the horse still had a place, which was largely not the case.) Officers who served there were given a better grounding in logistics and complex administration than was available on the Western Front, and the lessons of the Middle East fighting were suggestive of subsequent developments in the Second World War. The Australian and New Zealand mounted units played a central role through the two phases of the campaign, in Sinai where their mobility enabled the British to move forces otherwise outnumbered to meet Turkish concentrations, and in Palestine, where they were a constant element of Allenby's striking force. It is important to remember, however, for all the romance of the light horse which accrued about the campaign, that Palestine had many of the same features of a modern war evident on the Western Front (massed artillery, aerial reconnaissance, tanks and entrenchments), and that victory was won by a force of combined arms (infantry, artillery and mounted) and not by any one arm alone.

PALTRIDGE, Shane Dunne (11 January 1910–21 January 1966). Paltridge served in the RAAF for two years from 1940 before transferring to the AIF as a gunner in the 2/7th Field Regiment because he had failed to qualify for aircrew. Demobilised in 1945, he became a foundation member of the Liberal and Country League (LCL) in 1946 and a member of the Western Australian state executive of the LCL in 1947. He was elected to the senate in 1951. Paltridge first entered the ministry in 1955, was deputy leader in the senate from 1959 to 1964, and leader from 1964 until his death in office two years later. Menzies made him minister for Defence in 1964, and he held office during a period when Confrontation with Indonesia was entering its most dangerous phase and as the question of an Australian commitment to Vietnam began to loom large, with a concomitant expansion in the armed forces. He was lukewarm on the question of reintroducing national service in mid-1964, sharing the view of the chiefs of staff that the voluntary system might meet the needs of an expanding army if pay and conditions were improved in a climate of full employment. He nonetheless directed that legislation enabling the creation of a national service scheme be prepared, and this was introduced in November (see CONSCRIPTION). In January–February 1965 he made an important tour of South-east Asia, which resulted in the decision to commit ground troops to Borneo, and saw additional advisers sent to Vietnam. He became ill in August 1965, but only resigned the Defence portfolio in mid-January the following year when he entered hospital.

PALUMA, HMAS see **SURVEY SHIPS**

PALUMA, HMQS see **QUEENSLAND MARINE DEFENCE FORCE SHIPS**

PAPUA NEW GUINEA DEFENCE FORCE (PNGDF) With its antecedents in the Pacific Islands Regiment (q.v.), the PNGDF came into being in January 1973 in the lead-up to independence, which followed in 1975. It is a joint force for the defence of Papua New Guinea and its territories against external attack, but with secondary functions which include nation-building and internal security tasks. At independence it numbered 3750 all ranks, not including 465 Australian personnel on loan to the new government to assist in training and technical positions in, for example, the air transport squadron. This was reduced to 3050 in 1982, largely for financial reasons. Defence accounts for 4 per cent of the annual budget and the PNGDF receives about $23 million in direct assistance from the ADF and assistance in training support from both New Zealand and the United States. Soldiers of the PNGDF have been called out in aid of the civil power (q.v.) in Port Moresby in 1983, and in operations against the OPM (Organisasi Papua Merdeka or Free Papua Movement), based in Irian Barat and fighting the Indonesian armed forces. Four hundred troops were sent to assist the Vanuatu government put down a secessionist revolt in 1980. Since 1989

the PNGDF has been fighting the secessionist Bougainville Revolutionary Army (BRA) on the islands of Bougainville and Buka. The conduct of the PNGDF on Bougainville has been criticised for its treatment of civilians. In June 1991 Colonel L. Nuia was dismissed for killing civilians and using Australian-donated helicopters to dump their bodies in the sea, and in September 1993 the PNG government offered to pay US$300 000 in compensation for Solomon Island civilians killed by PNGDF forces who had pursued BRA members across the international border. The fighting has also revealed weaknesses in command, training and force structure.

PAPUAN INFANTRY BATTALION see **PACIFIC ISLANDS REGIMENT**

PARER, Damien (1 August 1912–20 September 1944). Devoutly Catholic in belief, Parer nonetheless gave up a vocation to the priesthood in favour of photography, and in the course of the 1930s developed into one of Australia's leading young photographers and cameramen. He worked with both Max Dupain and as a cinematographer with Charles Chauvel, including on the latter's early wartime epic, *Forty Thousand Horsemen*.

Selected by the Department of Information for the AIF Film Unit, Parer was one of only three official cameramen serving with the 2nd AIF in the Middle East, taking part in both the Greek campaign and the siege of Tobruk. He returned to Australia in March 1942 and was sent to Papua. In August he accompanied the 39th Battalion during the heavy fighting along the Kokoda Track. The result was the newsreel film *Kokoda Front Line*, which won an Oscar for best documentary in 1943. He made further films with the commandos on Timor and accompanying the troops during the fighting for Salamaua, but grew tired of the restrictions imposed on him by Department of Information regulations, and accepted a job with the American company Paramount, to film American actions in the Pacific. He was killed while filming with the Marines at Peleliu. He is commemorated by a plaque on the wall of the Pentagon.

Parer's films are among the most accomplished to come out of the Second World War. He explained once to the official historian, Gavin Long (q.v.), that he wanted 'to try and do with a camera what Dyson did with his pencil in the last war'.

Niall Brennan, *Damien Parer: Cameraman* (Melbourne University Press, Melbourne, 1994).

PARKES, HMAS see *BATHURST* CLASS MINESWEEPERS (CORVETTES)

PARKHILL, (Robert) Archdale (27 August 1878–2 October 1947). A conservative machine politician, Parkhill became a close crony of William Morris Hughes (q.v.) in the creation of the National Party following the Labor Party split in 1916–17. Elected to the federal seat of Warringah in 1927, he received ministerial preferment when the conservatives returned to office in 1931. An increasingly strong supporter of Joseph Lyons, he was acting prime minister in the absence overseas of Lyons and Latham in 1934, and was one of the most senior figures in the United Australia Party.

Between 1934 and 1937 Parkhill served as minister for Defence. He favoured traditional views supporting imperial defence and the navalist strategy associated with the base at Singapore (see SINGAPORE STRATEGY), and established the Commonwealth Aircraft Corporation. He had Colonel H. D. Wynter (q.v.) removed from his post at Army Headquarters after a paper which the latter had produced was used by the Opposition as the basis for a strongly worded attack on his and the government's defence policy, and he was not regarded favourably within the army. A Norman Lindsay cartoon of January 1937 entitled 'Higher Education' depicted Parkhill looking down the barrel of a howitzer while a soldier explained, 'Yes sir, it's a gun, sir. Same as we had in France, sir'. He attended the 1937 Imperial Conference, and was defeated by P. C. Spender (q.v.) in the federal election of that year. He did not return to public office thereafter.

Conservative to the point of reaction in many of his views, Parkhill was unpopular within his own party, being beaten by R. G. Menzies (q.v.) for the deputy leadership in 1935. Arrogant, vain and foppish in dress, he was an undistinguished Defence minister in a crucial period. His removal of Wynter was petty, while his role in the foundation of the aircraft industry is disputed. His uncritical faith in the Singapore strategy, even at the 1937 conference and in the face of much evidence that the strategy was flawed from the British chiefs of staff themselves, contributed further to Australia's unpreparedness for war in 1939.

PARRAMATTA (I), HMAS see **RIVER CLASS TORPEDO BOAT DESTROYERS**

PARRAMATTA (II), HMAS see ***GRIMSBY* CLASS SLOOPS**

PARRAMATTA (III), HMAS see **RIVER CLASS ANTI-SUBMARINE FRIGATES**

PARRAMATTA (IV), HMAS see ***ANZAC* CLASS FRIGATES**

PATERSON, Major Andrew Barton 'Banjo' (17 February 1864–5 February 1941). Educated at Sydney Grammar School, Paterson took articles and was admitted as a solicitor in 1886. He began writing while still a student, and had his first verse published in the (Sydney) *Bulletin*, 'the Bushmen's Bible'. Between October 1899 and September 1900 he served in South Africa as a war correspondent (q.v.) for the *Sydney Morning Herald*, where the high calibre of his reporting attracted attention in the British press and resulted in his being employed by the news agency, Reuters, as a correspondent. From South Africa he went to China to cover the Boxer Rebellion. Paterson, who helped to establish the context for the Anzac (q.v.) legend by romanticising the hardy bushman as the embodiment of Australian national character, was already famous for his poetry by the time he left for South Africa. Travelling with the New South Wales Lancers, he was the first correspondent to ride into Bloemfontein after its surrender and the first to report on the relief of Kimberley. Twelve of Paterson's poems are based on his Boer War experiences, and though he never questioned Britain's right to be in South Africa, the poems reveal his increasing sympathy for the Boers. His journalistic reports, however, were more conservative, consisting chiefly of vivid accounts of the action which heaped praise on the courage, independence and skill of Australian troops. Having failed in his attempts to get appointed as a war correspondent during the First World War, in late 1914 he served in France as a volunteer ambulance driver. He was commissioned into the 2nd Remount Unit (see WALERS) of the AIF in October 1915, having already made several journeys with horses engaged as a vet. He commanded the Australian Remount Squadron from October 1916 until 1919. Paterson returned to journalism after the war, and in 1934 published *Happy Dispatches*, based largely on his experiences in war zones.

***PATRICIA CAM*, HMAS** see **AUXILIARY MINESWEEPERS**

PAY CORPS see **ROYAL AUSTRALIAN ARMY PAY CORPS**

PC-9, PILATUS (2-seat basic trainer). Wingspan 10.12 m; length 10.17 m; maximum speed 556 km/h; range 1642 km; power 1 × Pratt & Whitney 950-shaft h.p. engine.

The PC-9 is a Swiss-designed basic trainer built under licence in Australia by Aerospace Technologies of Australia, Dunlop Aerospace and Hawker de Havilland. The RAAF ordered the PC-9 to replace the CT-4 Airtrainer for basic training, and the Macchi (q.v.) for advanced training, and the first aircraft entered service in 1987. The PC-9 is at present the RAAF's main training aircraft and is used by No. 2 Flying Training School at Pearce (q.v.), the Central Flying School at Point Cook, and the Roulettes aerobatic team.

PEACE-KEEPING, AUSTRALIANS AND MULTI-NATIONAL

An independent but cautious member of the League of Nations, Australia in 1945 embraced its successor, the United Nations (UN), with more enthusiasm. At the United Nations Conference on International Organization, which opened in San Francisco on 25 April 1945, H. V. Evatt, the Australian Minister for External Affairs, was a leading exponent of the rights of the smaller powers. Australia supported the principles of collective security enshrined in the UN Charter, and has participated in the two most robust expressions of those principles, in Korea (1950–53) and against Iraq (1990–91).

Military observers

Leaving Korea aside, Australia was for many years cautious of committing any great resources to peace-keeping operations. Indeed, until recently UN peace-keeping in general has been a pale imitation of the enforcement envisaged in the charter. The combination of the Cold War and the great powers' veto rights on the UN Security Council forced the UN to restrict itself mostly to monitoring cease-fires once they had been negotiated (Korea was the accidental exception). Peace-keepers in general were military observers, unarmed or armed only for self-defence, their sole weapon their ability to report their observations back to New York.

In 1947, Australians were among the very first group of UN peace-keepers, military observers attached to the UN Good Offices Commission (UNGOC; later the UN Commission for Indonesia, UNCI). These observers monitored the cease-fire between the Dutch and the republican Indonesians in the Netherlands East Indies, and ultimately oversaw Indonesia's independence. In Korea, in 1950, Australian military observers attached to the UN Commission on Korea (UNCOK) were the first to report the North Korean invasion of the South.

Observer missions contribute stability, but do not of themselves promote the settlement of underlying issues. As a result, they may continue in operation over long periods, with observers rotating through them, usually (in Australia's case) on 6- or 12-month postings. In Kashmir, Australian observers participated in the UN Military Observer Group in India and Pakistan (UNMOGIP) for over 30 years before their withdrawal in 1985, while Australia has had observers on duty with the UN Truce Supervision Organization (UNTSO), monitoring cease-fire lines on Israel's borders, ever since 1956. During this time, observers have also been detached to serve with separate operations in Yemen, Syria and Lebanon.

Another long-running UN operation, in which Australia has participated since its inception in 1964, is the UN Peace-keeping Force in Cyprus (UNFICYP). In this case, Australia's contribution has been a contingent of about 20 police, originally from State forces, but more recently Federal Police. (See Table 1 for a summary of Australian participation in peace-keeping operations.)

From observation to enforcement

In other situations, peace-keepers must be armed. The first major UN peace-keeping force, the UN Emergency Force (UNEF I), was formed in 1956 in response to the Suez crisis. Australia was keen to contribute, but an Australian presence was politically unacceptable to Egypt. In the early 1960s the UN operation in the Congo, with nearly 20 000 troops, saw severe fighting. Australia contributed only a small medical team.

In recent years, Australia's peace-keeping has taken on more varied forms. There have been traditional military observers, with the UN Iran–Iraq Military Observer Group (UNIIMOG), between 1988 and 1990. A decade earlier, in 1979–80, Australia had for the first time supplied a substantial body of armed soldiers to a peace-keeping operation. The Commonwealth Monitoring Force (CMF), a non-UN operation overseeing Zimbabwe's transition to independence, operated in an environment still dangerous in the aftermath of a protracted war between government forces and guerrillas. Australia contributed a contingent of over 150.

The progress from peace observation to peace enforcement was completed in 1993 with the formation of the US-led but UN-sanctioned Unified Task Force (UNITAF) in Somalia, working alongside the UN Operation in Somalia (UNOSOM). Australia contributed a battalion group (based on 1RAR) of regular infantry for a period of five months, in an operation designed to impose sufficient order on a country fallen into anarchy to allow the distribution of aid and movement towards a political settlement. Within the area assigned to them, centring on Baidoa, the Australians were successful, though poor political direction and failures of military planning and cooperation ultimately led to a failure to establish any lasting peace in Somalia.

Table 1   Australian participation in peace-keeping operations

Name of operation	Dates of Australian involvement	Number of Australians involved	Role of Australians
UN Good Offices Commission (UNGOC); UN Commission for Indonesia (UNCI)	1947–49 1949–51	15 (maximum)	Military observers
UN Commission on Korea (UNCOK)	1949–50	2	Military observers
UN Military Observer Group in India and Pakistan (UNMOGIP)	1950–85	18 (maximum)	Military observers and air transport
UN Truce Supervision Organisation (UNTSO) [Israel and neighbours]	1956–present*	13 in 1995	Military observers
UN Command Military Armistice Commission — Korea (UNCMAC)	1956–present	1	Monitoring cease-fire between North and South Korea
UN Operation in the Congo (ONUC)	1960–64	a few	Medical team
UN Temporary Executive Authority (UNTEA) [West Irian]	1962	11	Supporting humanitarian aid
UN Yemen Observation Mission (UNYOM)	1963	2	Military observers
UN Force in Cyprus (UNFICYP)	1964–present	16–50 State and federal police	Maintenance of law and order
UN India–Pakistan Observation Mission (UNIPOM)	1965–66	4	Military observers
UN Disengagement Observer Force (UNDOF) [Syria]	1974–present	a few	Military observers detached from UNTSO
UN Emergency Force II (UNEF II) [Sinai]	1976–79	46 RAAF personnel	Monitoring a cease-fire between Israel and Egypt
UN Interim Force in Lebanon (UNIFIL)	1978	a few	Military observers detached from UNTSO
Commonwealth Monitoring Force — Zimbabwe (CMF)	1979–80	c. 150	Monitoring Rhodesian security forces, cantonment of guerrillas, and return of civilian refugees
Multinational Force and Observers (MFO) [Sinai]	1982–86 1993–present	110 in the first period; 25–30 in the second	Monitoring Israeli withdrawal from the Sinai
Commonwealth Military Training Team — Uganda (CMTTU)	1982–84	6	Training government forces
UN Iran–Iraq Military Observer Group (UNIIMOG)	1988–90	15	Military observers
UN Border Relief Operation (UNBRO) [Cambodia]	1989	2 federal police	Maintaining law and order and training Cambodian civilian guards in police work
UN Transition Assistance Group — Namibia (UNTAG)	1989–90	c. 300	Engineering support: supervision of elections

Table 1 Australian participation in peace-keeping operations *cont.*

Name of operation	Dates of Australian involvement	Number of Australians involved	Role of Australians
UN Mine Clearance Training Team (UNMCTT) [Afghanistan/Pakistan]	1989–93	13 in 1993	Mine clearance: instructing refugees and planning operations
Maritime Interception Force (MIF) [Persian Gulf, Gulf of Oman, Red Sea]	1990–93	3 RAN Ships (maximum); c. 600 personnel in 1990	Enforcing UN-imposed sanctions
Operation HABITAT [Kurdistan, northern Iraq]	1991	75	Delivering humanitarian aid
UN Special Commission (Iraq) (UNSCOM)	1991–present	5 in 1993	Inspection of Iraqi chemical, biological and nuclear weapons capabilities
UN Mission for the Referendum in Western Sahara (MINURSO)	1991–94	45	Communications
UN Advanced Mission in Cambodia (UNAMIC)	1991–92	65	Communications
UN Transitional Authority in Cambodia (UNTAC)	1992–93	c. 500	Communications, transport, assisting the election and maintaining law and order
UN Operation in Somalia (UNOSOM)	1992; 1993–95	c. 40	Movement control and other support
Unified Task Force — Somalia (UNITAF)	1992–93	c. 1100	Protecting delivery of humanitarian aid
UN Protection Force in Croatia, Bosnia–Herzegovina and Macedonia (UNPROFOR)	1992–93	a few	Military observers and liaison
UN Assistance Mission in Rwanda (UNAMIR)	1994–present	c. 300	Medical personnel (115), infantry protection, support troops
UN Operation in Mozambique (ONUMOZ)	1994–95	20 (maximum)	Police, mine clearance
South Pacific Peace-keeping Force (SPPKF) [Bougainville]	1994	50 (maximum), plus two ships	Force commander; logistic and other support
Multinational Force (MNF) [Haiti]	1994–95	30	Police monitors

* present = 1995

A diversity of contribution

The increasing complexity of peace-keeping operations ensures that all three services and many specialist areas within them have a chance to contribute. Throughout the history of Australia's peace-keeping, personnel from all three services have been posted individually to operations or have formed part of an Australian contingent. In addition, the RAAF has been particularly involved in three operations, providing a Caribou (q.v.) to carry supplies into the mountains of Kashmir for UNMOGIP from 1975 to 1979, and operating Iroquois helicopters (q.v.) in the Sinai with the UN Emergency Force II (UNEF II) from 1976 to 1979 and the non-UN Multinational Force and Observers (MFO) from 1982 to 1986. The RAN's chief involvement in peace-keeping activity has been in the imposition of sanctions against Iraq before and after the Gulf War, with frigates and support ships operating in the Gulf of Oman, Persian Gulf and Red Sea.

Lance-Corporals Dave Hurrey (left) and Paul Astbury on patrol in Thbeng Mean Chey, Cambodia, 18 February 1993. These men, of the Australian Signals Corps, were members of the United Nations Transitional Authority in Cambodia. (AWM CAMUN93/111/14)

Australia will often provide a specialist contingent to an operation, such as the 300 engineers, drawn mainly from the Royal Australian Engineers and Royal Corps of Electrical and Mechanical Engineers, who formed part of the UN Transition Assistance Group (UNTAG) in Namibia, in 1989–90. From 1991 to 1994 Australia provided signallers for the UN operation in Western Sahara, MINURSO. From 1991 to 1993, Australian signallers formed the Force Communications Unit for the UN Advanced Mission in Cambodia (UNAMIC) and the much larger UN Transitional Authority in Cambodia (UNTAC). The 500 Australian signallers in UNTAC operated all over Cambodia, often in isolation from other members of the Australian contingent.

Australian Army doctors have served in several missions, including the Congo, Western Sahara, and in Operation Habitat among the Kurds of northern Iraq in 1991. In 1994 an Australian medical contingent of over 100 medical staff and nearly 200 infantry and support troops was sent to Rwanda, as part of the United Nations Assistance Mission for Rwanda (UNAMIR). Except in Operation Habitat, the medical teams' primary role has been to treat other members of the UN force; however, medical assistance is usually also given to the local people.

The boundaries of peace-keeping are ill-defined. Training other people's armies is scarcely 'peace-keeping', yet groups such as the Australian Army Training Team Uganda (q.v.), to which Australia contributed from 1982 to 1984, can have a significant stabilising role and therefore may be considered partly a peace-keeping force. In general, it is probably best to view any impartial and multi-national force mopping up in the aftermath of war as a form of peace-keeping. An example is the UN Special Commission on Iraq (UNSCOM), charged after the Gulf War with inspecting and supervising the destruction of Iraq's facilities for manufacturing nuclear, chemical and biological weapons. Australian expertise, especially on chemical weapons, was of value to this group. From 1989 to 1993 Australians also served with the UN Mine Clearance Training Team (UNMCTT), teaching returning Afghan refugees mine awareness and mine clearance skills. In Namibia and Cambodia, Australian electoral officers helped supervise elections. The federal police have continued the tradition begun in Cyprus with postings to Mozambique and Haiti in 1994.

Commanders and casualties

Two Australians have commanded UN forces. The first was Major-General (later Lieutenant-General)

R. H. Nimmo, a Gallipoli veteran and Second World War brigadier, who was Chief Military Observer with UNMOGIP for an astonishing period from 1950 to 1966. The second was Lieutenant-General John Sanderson (q.v.), who from 1992 to 1993 commanded the military forces of UNTAC, with over 20 000 personnel under his command. In 1994, Brigadier David Ferguson became force commander for the non-UN MFO.

Only seven Australians have died on peace-keeping operations, a record which is testimony both to the excellent skills and training of Australia's peace-keepers and to good fortune. Nimmo died while still with UNMOGIP. Three policemen have died in Cyprus, two in vehicle accidents, one from a land-mine explosion. An army officer was also killed by land-mine in Lebanon, while in 1993 an army doctor was killed in a plane crash in Western Sahara and a private was accidentally killed in Somalia.

Conclusion

Australia's contribution to multinational peace-keeping has been extensive and varied, but scarcely on a par with that of many other countries. Australia has on occasion refused UN requests to provide the commanding officer for an operation, and in the mid-1980s came close to withdrawing from peace-keeping altogether, in a period when this approach fell out of favour with the ALP. Since then, especially due to Minister for Foreign Affairs and Trade Gareth Evans, the nation has become a much more committed contributor. In early 1993, over 2000 Australian peace-keepers were in action. That was an aberration, however, and is unlikely to be repeated for some time. In 1994 Australia ranked around the middle of the 75 or so providers of UN peace-keepers in terms of numbers. Yet the Australian contribution has consistently been of a high quality, and in that way Australia has made its mark.

PETER LONDEY

PEARCE, George Foster (14 January 1870–24 June 1952). A carpenter by trade, Pearce became politically active through the trade union and labour movements. He was elected to the Senate in the first federal election and was to remain a senator for 37 years, 25 of them as a minister.

Pearce entered parliament with an anti-military bent entirely typical of many early Labor politicians, who believed that war and militarism were the products of capitalism and thus inimical to the interests of the working classes. Again like not a few of his contemporaries, the Japanese victory over the Russians in 1905 converted Pearce, who in the years thereafter believed that Australia now faced the real prospect of attack and that this necessitated strong naval and military forces. He was made minister for Defence in the Fisher government of 1908, and sparked controversy over the ordering of three torpedo boat destroyers without parliamentary sanction. He was also a strong advocate of the gift of a Dreadnought to Britain following the Dreadnought crisis of 1909. In the second Fisher government of 1910 Pearce was again minister for Defence, and he built on the work of the previous Fisher government in legislating for a scheme of compulsory military service (see CONSCRIPTION). He was also responsible for the passage of the Naval Defence Act of 1910 which established the Commonwealth Naval Board and the naval college at Jervis Bay, and for the creation of the Royal Military College, Duntroon (q.v.), following the recommendations of Field Marshal Lord Kitchener.

For all his lack of formal education (he left school at 11) and of personal military experience, Pearce possessed marked administrative ability, a receptiveness to new ideas and new technology, and a sound strategic grasp. At the 1911 Imperial Conference he accepted that Australia needed to plan for an expeditionary force for overseas service in the event of a major war with Germany, and on his return initiated secret planning for mobilisation in the event of war. While in Britain he saw a demonstration flight of aircraft which led him to establish the Central Flying School at Point Cook in 1912. Politically he rode out the criticism from left-wing members of his party over the compulsory service scheme, and although his mistrust of Japanese intentions grew before 1914 he nonetheless left in place the agreements which brought the RAN under Admiralty control in the event of hostilities.

Pearce's greatest contributions as minister for Defence came during the First World War. He served in this portfolio until December 1921 in five governments led by Fisher and W. M. Hughes (q.v.), in addition to other weighty responsibilities. Until the split he was leader of the Senate; while Hughes was overseas in 1916 he was acting prime minister; along with Hughes he was the chief architect of the two referendum campaigns on conscription, and played a major role in arguing the 'yes' case; he was a key political figure in the Labor split and the creation of the National Party. His administration of the Defence portfolio came in for criticism in 1917 (criticism no doubt linked to the politics of the conscription issue), and a Royal Commission was convened. This reported shortcomings, especially in financial matters, record-keeping and the supply system, but in general found the administration of the department in good shape given the unprecedented demands placed on it. Pearce himself took steps to adopt most of the commission's recommendations.

The First World War also accentuated a growing shift in Pearce's world-view. He had insisted that the two service colleges founded before the war should be strongly Australian in character (he probably failed in this regard with Jervis Bay), and as the war progressed he became increasingly identified with an Australian perspective. Thus it was Pearce who insisted that officers in the AIF must come from the ranks after January 1915, and he intervened on a number of occasions with General Sir William Birdwood to advocate the promotion and appointment of Australians to senior positions within the AIF. He was a strong supporter of General Sir John Monash's (q.v.) claims to senior command against a nasty whispering campaign which drew attention to the latter's Jewish origins. He maintained the government's opposition to the death penalty in the face of strong pressure from Birdwood, the Governor-General Sir Ronald Munro-Ferguson, and others. At the end of the war he went to London, Hughes being preoccupied with the peace conference, to oversee Monash's procedures for the repatriation of the AIF and assist in the negotiations for shipping and the acquisition of surplus war stocks. Throughout the war he worked long hours, insisting that others do likewise; the Melbourne *Age* commented after the war began that Pearce's conduct of defence matters 'had always been marked by extremely careful preparation and a grip of detail, unusually clear and accurate'. His biographer is undoubtedly correct in his judgement that Pearce 'was the first Minister for Defence to play a positive role', not least because of his length of time in the office: before his appointment in 1908 there had been eight ministers in eight years, and the average term of office for all incumbents other than Pearce between Federation and the outbreak of war was less than 12 months.

After the Armistice Pearce turned his attention to future defence needs, instructing a board of senior officers of the AIF to draw up plans for Australia's postwar military requirements. The resultant report identified Japan as the most likely threat to Australia and advocated a field force for the defence of Australia against a major attack, but in the economic and international political climate of the time this proved too ambitious, and was not implemented. Pearce represented Australia at the Washington Disarmament Conference, and then chose himself to leave the Defence portfolio in December 1921. He was appointed KCVO in 1927, and returned to Defence in January 1932 at Lyons's direction, to 'restore confidence' in a badly run-down and demoralised area.

He succeeded in restoring the defence vote in 1934–35 to its 1920s levels, but prevaricated on the strategic basis for defence planning, as did his successors. He became minister for External Affairs in 1934, but lost his seat in the elections of October 1937. From 1939 he was a member of the Board of Business Administration within the Department of Defence, which oversaw service expenditure and encouraged a standardised approach to purchasing and procurement, and from 1940 until its abolition in 1947 he was its chairman.

Pearce was one of the most important Defence ministers Australia has had, and a politician of great ability and accomplishment. R. G. Menzies (q.v.) regarded him as one of the ablest ministers he had ever observed, and later wrote that the absence of men like Pearce from his first government had considerably weakened it. On his death *The Times* (London) headed its obituary 'The Defence of Australia'.

Peter Heydon, *Quiet Decision: A Study of George Foster Pearce* (Melbourne University Press, Melbourne, 1965).

PEARCE, RAAF BASE, lies 44 kilometres north-east of Perth and is the main RAAF base in Western Australia. The 260-hectare site at Bullsbrook was purchased in December 1934 and, despite construction delays, became operational in March 1938 when it became home to No. 23 (City of Perth) (General Purpose) Squadron. Pearce is currently used by No. 2 Flying Training School for advanced flying training using Pilatus PC-9 (q.v.) trainers. RAAF Base Pearce was named in honour of the Western Australian senator and long-serving minister for Defence, Sir George Pearce (q.v.).

PELORUS **CLASS LIGHT CRUISERS** Laid down 1897, launched 1898–99; displacement 2100–2200 tons; length 314 feet; beam 37 feet; speed 20.5 knots; armament 8 × 4-inch guns, 8 × 3-pounder guns, 3 machine-guns (after 1918 *Psyche* 2 × 4.7-inch, 2 × 3-pounder).

Psyche and *Pioneer* were two light cruisers built for the RN which saw service with the Australian Squadron (q.v.) from 1903 to 1913. In 1913 *Pioneer* was transferred to the RAN, followed by *Psyche* in 1915. HMAS *Pioneer* spent the early part of the war in Western Australian waters where she captured two German steamers. In 1915 she played a part in destroying the German cruiser *Königsberg* in East Africa, then engaged in patrol and bombardment work along the East African coast. HMAS *Psyche's* routine duties included patrol work in the Bay of Bengal, and Malayan, Indian Ocean, Chinese and Australian waters. HMAS *Pioneer* was paid off and served as an accommodation ship at Garden Island (q.v.) until 1922; she was subsequently stripped and then scuttled in 1931. HMAS *Psyche* was paid off in 1917, recommissioned and finally sold in 1918 as a tanker lighter.

PEMULWUY see **ABORIGINAL ARMED RESISTANCE TO WHITE INVASION**

PENGUIN, HMAS, named for a nineteenth-century RN ship, is situated at Balmoral in Sydney and was commissioned on 14 July 1942. It conducted basic seamanship training for many years, but since the early 1970s has provided only advanced training in diving; explosive ordnance disposal and demolitions; nuclear, biological and chemical defence and damage control; hydrographic surveying; medical aid and underwater medicine. The RAN Staff College (see STAFF COLLEGES) and an RAN hospital are located there, and it is also the base for Clearance Diving Team (q.v.) Two.

PENTROPIC DIVISION Based on American experiments with the Pentomic divisional structure, the Pentropic division whose adoption was announced in November 1959 represents the most radical attempt at reorganisation of the Australian Army this century. As its name suggests, it was a five-sided structure, with infantry battalions (q.v.) now having five companies each of five platoons (four rifle and one weapons platoon). This represented an increase in strength of roughly 50 per cent, and with more than double the number of sections, from 36 to 80, greatly increased the firepower of the unit. With supporting arms added, the battalion became a battle group, of which there were five to a division. The abolition of the brigade as the intermediate headquarters between battalion and division, and the increase in rank of the battalion commander from lieutenant-colonel to colonel, led to the augmentation of battalion headquarters with additional personnel, but the result was a complex and difficult command proposition which many officers did not like. There were problems as well at the platoon level, with platoon commanders generally being young and inexperienced in the command of their relatively large force. The reasons for introducing the new system were in part a desire to modernise the army and integrate it better with the forces of Australia's major ally, the United States, but the abolition of National Service in 1959 played a part as well. The CGS, Lieutenant-General Sir Ragnar Garrett (q.v.), wished to retain the funding which had been expended on national service training for the modernising of the regular army, and believed that a thoroughgoing reorganisation would achieve this, which indeed it did. At this time the CMF was reorganised as well along Pentropic lines, but this was a much less happy process. Eight Pentropic battalions were formed from the existing 31 CMF infantry battalions, which meant the abolition of many old units with strong traditional links into their communities. The new State regiments were granted the prefix 'Royal' as part of this process, but this was not enough to reassure the doubts of some CMF officers who professed to believe that the reorganisation was a measure of revenge on the part of the regulars for slights suffered before the Second World War. In June 1961 the US Army announced that it was abandoning the experiment, which left the Australian Army as the only one in the world with a five-sided organisation. There was an additional complication, in that battalions sent to Malaya as part of the British Commonwealth Far East Strategic Reserve had to be reorganised on standard British lines before departure, and then reorganised yet again on return. Following a report into the field force by Major-General J. S. Andersen in November 1964, the CGS, Lieutenant-General Sir John Wilton (q.v.), decided to abandon the experiment. In his view, the changing strategic circumstances facing Australia necessitated several simultaneous overseas commitments, in which case 'it's not the size of the battalions that counts: it's the number'. A reversion to a Tropical Warfare establishment (even if the brigades were dubbed 'task forces' to avoid the impression that a mistake had been made), meant that the army could straight away form three new battalions of the Royal Australian Regiment (q.v.) from men now surplus to existing battalion establishment. The Pentropic organisation had been adopted without significant evaluation and its passing was generally unlamented. It was important, however, in helping to re-equip the army with much of the more modern weaponry and equipment with which it fought the Vietnam War.

J. C. Blaxland, *Organising an Army: The Australian Experience 1957–1965* (Strategic and Defence Studies Centre, Canberra, 1989).

PERTH (I), HMAS see **LEANDER CLASS LIGHT CRUISERS**

PERTH (II), HMAS see **CHARLES F. ADAMS CLASS DESTROYERS**

PERTH (III), HMAS see **ANZAC CLASS FRIGATES**

PETER STUCKEY MITCHELL PRIZES AND SCHEMES were endowed by a wealthy New South Wales grazier and philanthropist and member of a prominent rural family, Peter Stuckey Mitchell (1856–1921). He died childless and left the bulk of his estate, some £215 000, to fund a number of schemes intended to 'encourage and help the capable, healthy and strong to develop ... their natural advantages'. Some of these schemes, such as the one for young unmarried women aimed at improving knowledge of the theory and practice of child care, but with a requirement for a 'sound and appreciative knowledge' of great swathes of the Protestant Bible, were clearly of their time. The system of prizes within the Australian Defence Force, however, continues and covers virtually all levels of training and education within the ADF, from undergraduates at the Australian Defence Force Academy (q.v.), through a variety of trades and skills courses, to a number of essay prizes for officers of the services. Mitchell stipulated that these were to be open to all members of the armed forces of the British Commonwealth. A prize for naval officers of the rank of commander and below is administered annually through the office of the Australian CNS, while an annual army essay prize is awarded for the best contribution from a student at one of the command and staff colleges of the Commonwealth armies. After Mitchell's widow died in 1954, some provisions of his will were contested, with the result that some of those parts felt to be no longer practicable, such as the scheme for women, were abandoned and that proportion of the bequest was subsequently paid annually to the Winston Churchill Memorial Trust.

PETHEBRIDGE, Brigadier-General Samuel Augustus (3 August 1862–25 January 1918) entered the Queensland Public Service as a clerk in 1876, and in 1888 was appointed secretary to the Marine Board. In 1893 he was commissioned as a sub-lieutenant in the Queensland Naval Brigade. Following Federation he transferred to the Commonwealth Public Service and became chief clerk in the Department of Defence, assisted perhaps by his connections with Creswell (q.v.), with whom he had served in the colonial militia. When the secretary of the department, Muirhead Collins (q.v.), went to London in 1906, Pethebridge became acting secretary. Following the creation of the Military and Naval Boards of Administration in 1905, he functioned as secretary to both. Like many of the first generation of federal public servants, Pethebridge was of modest educational background and conservative views; he believed absolutely in the anonymity of the official, and in keeping with this worked steadily within the Department of Defence to entrench the principle of civilian primacy over the military, which Hutton's (q.v.) period as GOC of the AMF had seemed to challenge. In 1911 he accompanied his minister, George Foster Pearce

(q.v.), to London for the Imperial Conference. He was overseas when the war broke out in 1914, and his chief clerk, Thomas Trumble (q.v.), had acted as secretary in his absence in the frenetic early weeks of the war. Through concern for continuity in a period of crisis, Pethebridge was accordingly offered command of the expedition to occupy German colonial possessions in the north Pacific. These however were occupied by the Japanese, and he suggested that his formation, designated Tropical Force, should relieve the Australian Naval and Military Expeditionary Force in former German New Guinea (q.v.) instead. In January 1915 Pethebridge succeeded Holmes (q.v.) as administrator in Rabaul. Over the next two years he did a certain amount to consolidate the communications throughout New Guinea and undertook some minor labour reforms, but was not permitted to 'Australianise' the territory as its future had not been determined. His health had begun to decline before the war, and in January 1917 he contracted malaria, which forced his departure in October. He died a few months later.

PETRE, Major Henry Aloysius (12 June 1884–24 April 1962). Trained as a lawyer, Petre gave away legal practice after qualifying as a pilot and earned his living as a flying instructor before joining the Handley Page company in 1912. He was appointed as an instructor with the aviation staff of the AMF in August 1912, and commissioned as a lieutenant. He was the first commanding officer of the Central Flying School at Point Cook, and selected the site for the school himself. In April 1915 Petre was appointed to command the Mesopotamian Half-Flight, of four pilots, sent by the Australian government at the request of the government of India. The unit subsequently became No. 30 Squadron, RFC, and continued to fly against the Turks until disbanded in December 1916. In the course of that year he was mentioned in despatches three times, and was awarded the MC and the DSO. In 1917 Petre was sent to France and flew with No. 15 Squadron, RFC, before assuming a training post with the Australian Flying Corps in Britain. He transferred to the RFC in January 1918 and remained with its successor, the RAF, in the rank of major, until demobilised in September 1919. He resumed legal practice in Britain and died there in 1962.

PETS see **ANIMALS**

PHANTOM, McDONNELL DOUGLAS (2-seat fighter-bomber [F-4E]). Wingspan 38 feet 7.5 inches; length 63 feet; armament 1 × 20 mm cannon, 4 Sparrow anti-aircraft missiles, 16 000 pounds bombs, rockets or missiles; maximum speed 1432 m.p.h. (Mach 2.12); range 1612 miles.

The Phantom was operated by the RAAF between 1970 and 1973 as a stop-gap measure because of delays in delivery of the F-111 (q.v.) bomber. Twenty-four aircraft were leased from the USAF and were used by No. 1 and 6 Squadrons at Amberley (q.v.). The Phantom replaced the Canberra (q.v.) bomber in these squadrons and provided useful experience for both aircrew and ground and technical crews in the maintenance and use of more modern supersonic aircraft. One Phantom was lost during their RAAF service. In 1972 an offer by the US government to purchase the 23 surviving Phantoms for $54 million was rejected by the Australian government and, after the arrival of the F-111, they were returned to the USAF. In 1989 the RAAF Museum received a USAF Phantom which has been painted in the colours of a RAAF Phantom.

PIESSE, Major Edmund Leolin (26 July 1880–16 May 1947) was born in Hobart and educated at the Friends' High School and the University of Tasmania. In 1909 he joined the Australian Intelligence Corps, and following the outbreak of war was transferred to the intelligence section of the Directorate of Military Operations and Plans. In 1916 he was appointed Director of Military Intelligence, with the rank of major. In the course of this employment, Piesse became concerned with Australia's Pacific policy. In contrast to many of his contemporaries, in both Australia and Britain, he was able to move beyond conventional racist fears of a Japanese threat (q.v.) by learning Japanese and studying all available information on Japan. Initially fearful of a Japanese southward advance, he urged the creation of an Australian foreign policy section because he felt British representatives were unsympathetic to Australian interests. However, after becoming head of the Pacific Branch of the Prime Minister's Department, a position he held between 1919 and 1923, and travelling around Asia, he became convinced that Japan posed no threat to Australia. He called for the removal of Australian barriers to Japanese trade and immigration but had little influence on government policy. Piesse's resignation from the Pacific branch was prompted partly by frustration that official information was not being passed on to him, but also by the hope that in the wake of the Washington disarmament conference (at which he had been an Australian delegate) detailed study of Japan was no longer necessary. The re-emergence of Japanese imperialism led him to sound a warning in *Japan and the*

Defence of Australia (1935). He then campaigned in the press and on radio against reliance on British naval protection and for the army and air force to get a greater share of the defence budget. A strong critic of the Singapore strategy, his article 'The defence of Australia', published in the *Round Table* in 1937, provided a strong critique of Australian government policy, although it failed to change it. After leaving government service he returned to legal practice, and between 1942 and 1944 was president of the Law Institute of Victoria.

'PIGEON' see **ABORIGINAL ARMED RESISTANCE TO WHITE INVASION**

PINE GAP see **JOINT DEFENCE FACILITIES**

PIONEER, HMAS see *PELORUS* **CLASS LIGHT CRUISERS**

PIONEER BATTALIONS were a wartime innovation in the British Army (although they had existed in the Indian Army before 1914), a response to the much greater emphasis on field engineering (in the form especially of trenches and defended strong points) and the enhanced needs of road and railway transport during the First World War. Authorisation for the formation of Australian pioneer battalions was issued in February 1916, and five were raised, one per division. They were organised like an ordinary infantry battalion, with a lieutenant-colonel commanding, a headquarters staff and a four-company organisation. They were trained in infantry work, and could find themselves committed to the firing line in the normal manner. Within the division the pioneer battalion was frequently assigned to the Commander, Royal Engineers, the senior engineer officer, for allocation of tasks, although it was administered separately from the engineering squadrons. It incorporated men with various trades and skills, and was in many ways the most versatile unit in the infantry division. The 5th Pioneer Battalion, assigned to the 5th Division, for example, was made up of volunteers from the division's three infantry brigades; preference was given to men with specific trades, 20 in all being represented in the completed unit, with the remainder of its strength consisting of men considered 'handy', or with mining or other 'pick and shovel' experience. Its commanding officer, Lieutenant-Colonel H. G. Carter, was an engineering graduate from the University of Sydney.

Four pioneer battalions were raised for service with the 2nd AIF and served both as infantry and in the engineering support role. In the Syrian campaign, for example, the 2/2nd Pioneer Battalion fought as infantry, as did the 2/1st Pioneers at Tobruk and the 2/3rd Pioneers at El Alamein, despite the fact that they had not received full infantry training at that stage, and were not yet fully organised or equipped on the infantry table of organisation and establishment. During the Markham Valley campaign in New Guinea, the 2/2nd Battalion advanced and built the airstrip at Nadzab before being committed to the fighting alongside an American parachute infantry regiment and an Australian infantry battalion. In the postwar army the pioneer function was devolved to the infantry battalion itself, and an assault pioneer platoon (1 officer and 31 other ranks) was formed as part of the Support Company. During the Korean War, when infantry officer replacements were in short supply, the assault pioneer platoon commander was sometimes supplied by the Royal Australian Engineers.

PIRIE, HMAS see **BATHURST CLASS MINESWEEPERS (CORVETTES)**

PLANT, Major-General Eric Clive Pegus (23 April 1890–17 May 1950). Educated at Brisbane Grammar School, Plant took a commission in the 9th Infantry Regiment in March 1908, and moved across to the permanent forces in December 1912. At the outbreak of war he was appointed ADC to Major-General Sir William Throsby Bridges (q.v.), but in February 1915 transferred to a subaltern's command with the 9th Battalion of the AIF after arriving in Egypt. As such, he went ashore at Gallipoli with the attacking force, and led his rapidly depleted platoon forward over the 400 Plateau and across a succession of ridges; realising that he was too far forward and in a dangerously exposed position, he retired his group under pressure from enemy reinforcements, having been one of those who had pressed furthest forward on the first day of the campaign. The casualties among officers in his battalion led to his assuming the post of adjutant and the rank of captain the following day. He was wounded by a premature shell burst on 3 June, but was not evacuated until September, when he was taken to England suffering from enteric fever. He rejoined his battalion in February 1916, and was transferred to the 49th Battalion as part of the expansion and reorganisation of the AIF in Egypt. Very soon after he became brigade major to the 6th Infantry Brigade under Brigadier John Gellibrand (q.v.). He served through the gruelling fighting of 1916–17, at Fleurbaix, Pozières, Mouquet Farm and in the advance to the Hindenburg Line, being awarded the DSO and a bar. He became GSO2 of the 4th Division and saw out the rest of the AIF's campaigns, at the Third

Battle of Ypres and through the 1918 fighting, until secondment to the headquarters of the Fourth Army in early October, where he remained until the end of the war. He joined the Repatriation Staff under Monash in London, a sure testimony to his administrative abilities, and left England in May 1920 to return to Australia.

Between the wars, Plant attended the Staff College, Camberley, in 1921–22, served as an instructor at RMC Duntroon, held a number of staff posts and spent a further period on exchange duty in Britain, culminating in appointment as Director of Military Training in 1937. In October 1939 he became Commandant of RMC Duntroon, and did not get away to the 2nd AIF until July 1940, when he was given command of the 24th Brigade in the newly raised 8th Division; in the reorganisation concomitant with the raising of the 9th Division he found his command reallocated to the latter formation. He missed the siege of Tobruk through temporary appointment as commander of the AIF's rear echelon, but led his brigade through the campaign in Syria and the period of garrison duty thereafter. In December 1941 he was returned to Australia with other younger officers to take command and staff jobs concerned with the defence of Australia, and was appointed GOC of Western Command. Following General Blamey's reorganisation of the Australian Army in April 1942, Plant became GOC of the Victoria Lines of Communications Area until September 1943, and then of the New South Wales Lines of Communications Area until his retirement from the army in March 1946. Plant's age, relative to officers like Vasey and Robertson (qq.v.) who were also vying for a limited number of commands, was the principal factor in his seeming relegation to rear-area jobs, rather than any perceived shortcomings in his performance in the Middle East.

PLATYPUS, HMAS, located at Neutral Bay, Sydney, and commissioned on 18 August 1967, is the home of the Australian Submarine Squadron and the RAN Submarine School. It provides the squadron's operational headquarters and communicates with submarines by very low frequency radio through North West Cape Naval Communications Station in Western Australia (see JOINT DEFENCE FACILITIES). It also provides technical and logistic support facilities, and maintains a complement of sailors to replace, at short notice, any submarine crew member who has to be landed.

POETRY see **LITERATURE, WAR IN AUSTRALIAN**

POINT COOK, RAAF BASE see **ROYAL AUSTRALIAN AIR FORCE ACADEMY**

POLICE, MILITARY see **ROYAL AUSTRALIAN CORPS OF MILITARY POLICE**

POMEROY, John see **SCIENCE AND TECHNOLOGY**

POPULAR CULTURE, WAR IN AUSTRALIAN Australian popular culture has always been dominated by cultural products imported from elsewhere (first from Britain, later from the United States), and even home-grown popular culture has tended to follow overseas trends. Since Australians in the nineteenth century were overwhelmingly of British origin, and many were actually British-born, it is not surprising that representations of war and of the armed forces in Australian popular culture reflected contemporary British attitudes.

In the early years of the Australian colonies many residents, like other Britons, looked down on soldiers, who were thought to be recruited from among the dregs of society, and such attitudes combined with convict hatred for their gaolers to produce a society which was often hostile towards the military. The wealthy landowners and the growing middle class appreciated the military's role in maintaining order, but while they mixed socially with army officers they shared popular contempt for the rank and file. When the Australian colonies began to organise their own volunteer and militia forces (see COLONIAL MILITARY FORCES) these were generally regarded with amusement, as popular songwriter Charles Thatcher showed in his satirical song of the 1860s 'The Rifle Brigade'. From the middle of the nineteenth century, however, the armed forces became increasingly popular with Britons, who feared invasion by rival European powers and approved of the army's role as guardian of Empire (the navy was already held in high regard and 'Jolly Jack Tar' was a folk hero). Australians eagerly read newspaper reports about imperial wars, and the importance of war was reinforced by the naming of streets, towns and geographical features after battles and military heroes. The spread of settlement can be followed in the patterns of names relating to British and Australian military history.

In the late nineteenth and early twentieth century patriotism and militarism were closely intertwined. Militaristic imperial patriotism, expressed in songs like 'Rule Britannia' and 'The British Grenadiers', went hand in hand with the developing sense of Australian nationalism. The original lyrics of 'Advance, Australia Fair', composed in 1878 and adopted in a highly modified version as Australia's national anthem in 1984, promise that should 'foreign foe' dare to invade Australians will 'rouse to arms like sires of yore'. The churches generally supported militaristic nationalism, with

Methodists, for example, singing 'God of battles, lend thy might/ Where e'er we fight Australia's foe'. Patriotic days such as Empire Day and Trafalgar Day promoted the celebration of British military heroism, and military contingents were a frequent feature of ceremonial occasions like the parade through Sydney on New Year's Day 1901, the day of Australia's Federation, when both imperial and Australian troops marched in pride of place just before the Governor-General. Australian idolisation of British military heroes was shown in mass expressions of grief and outrage at the death of General Charles Gordon in 1885 and in the response to the visit of Field Marshal Lord Kitchener in 1910, when Australians were said to have gone 'Kitchener mad'. As well as reading in newspapers and popular magazines about the gallant deeds of British troops, Australians could 'see' them in action at exhibitions and at 'magic lantern' slide shows, while boys could live out their military fantasies by playing with toy weapons and soldiers that had become very sophisticated by the end of the nineteenth century.

The education system also promoted the cult of British military heroism, and schools were seen as playing an important role in creating a male population which would be ready for war. They were increasingly used to give all boys a taste of military discipline with the introduction of quasi-military drill to Queensland schools in 1876, the creation of school cadet (see AUSTRALIAN CADET CORPS) units in other colonies and from 1911 the use of schools as part of the compulsory military training scheme (see CONSCRIPTION). The idea, popular with the middle and upper classes, that 'manly' sports were good training for war, was also encouraged by schools. This association between sport and war was also made in the naming of racehorses and in popular sporting plays. On the other hand, some militarists saw sport as a pernicious influence, as Randolph Bedford showed in his popular play *White Australia; or, the Empty North* (1909), a racist republican piece whose hero declares in the face of Japanese invasion, 'No more mad devotion to vicarious sport — arm yourselves and think, get guns and resolution'.

Although Australians saw themselves as inheritors of a long British military tradition, there was still concern about whether or not they would prove worthy of that tradition. The Boer War seemed to show that they would, and one song suggested that 'Australia now can hold her own / With the wide, wide world'. Australian performance in the war seemed to allow the reconciliation of Australian suspicion of invasion with traditional British fear of standing armies since, as Senator Lieutenant-Colonel J. C. Neild complained, it was widely believed that untrained Australian soldiers had proved 'a match for an indefinite number of the best trained soldiers of any nation'. The myth was born of Australians as natural soldiers, with skills and toughness acquired from generations of struggle in the bush.

This myth reached its apotheosis during and after the First World War in the Anzac legend (q.v.). The war was widely seen as simultaneously confirming Australian nationhood and Australian masculinity; indeed, the two were seen as synonymous. Since 'the national type' was identified as male, war as the supreme test of manhood was also a test of national character, and it was a test which Australians felt they had passed with flying colours. Donald Horne was brought up in the interwar years to believe that Australian soldiers were 'uniquely *men* ... the greatest men in the world ... [O]ur fathers and the many war stories in the boys' comic books told us that the Huns made a tough enemy and it took us to beat them'. It was not having fought the Huns in the trenches of France and Belgium (see WESTERN FRONT), though, that was the supreme test of manhood, but rather having faced Turkish fire on the beaches of Gallipoli (q.v.). At Gallipoli, it was said, Australian nationhood had been born, and Australian soldiers had shown their superiority as fighters. But the war also produced a reassertion of Britishness and was justified entirely in terms of Australia's imperial responsibilities. As W. W. 'Skipper' Francis wrote in 'Australia Will Be There', perhaps the most popular Australian song of the First World War:

When Old John Bull is threatened,
By Foes on land or Sea,
His Colonial Sons are ready
And at his side will be.

Thus the special qualities of Australian soldiers could be attributed to their British inheritance, while much of the imagery and rhetoric of Anzac reached back still further and portrayed the diggers (q.v.) as possessing the virtues of ancient Greek warriors. Australian soldiers were not always portrayed as noble and heroic, though; an equally popular image was of the digger as a scruffy, undisciplined larrikin, especially in the pages of *Smith's Weekly*.

The onset of war reopened the debate about the connection between war and sport, as it would again during the Second World War. As in the Boer War, many found the origins of Australian military prowess in the bush, but others found it on the playing field. Senator Gardiner claimed in 1915 that 'Australians, who had won an imperishable record in Gallipoli, owed much to their training in the fields of sport', while an Australian soldier wrote that 'Australia can hold its own against the world ... in sport ... and in fighting'. Specific appeals to sportsmen and sport spectators were made in recruiting campaigns, but many middle- and upper-class people felt that, laudable though sport might be as a method of training for war in peacetime, in time of war it was a distraction which kept fit men from enlisting. A campaign against sport, which led to both voluntary and government-imposed restrictions on many sports during the war, was resented by many working-class people who saw sport simply as harmless entertainment.

Between the First and Second World Wars the presence of war in the Australian imagination was greater than at any other period in Australian peacetime history. Children growing up at this time, particularly those whose fathers had served with the AIF, certainly noticed this presence, as Donald Horne's autobiographical work *The Education of Young Donald* (1967) and George Johnston's semi-autobiographical novel *My Brother Jack* (1964) make clear. Horne remembers the war casting 'a bright light over our house', but for Johnston it was a 'vast, dark experience' beyond his understanding which 'impregnated' his family's house. Many houses contained collections of war memorabilia: postcards from the front, medals, weapons, souvenirs of foreign lands, and photographs, both the formal posed photographs of soldiers in full uniform which were displayed proudly in living rooms and the more casual shots of troops in camp or on leave which were stored away in drawers. In parks and school grounds children played on captured German guns. Though the word 'Anzac' (q.v.) was protected by law so that it could not be used for commercial purposes, people could still make their own 'Anzac biscuits' and the word 'digger', along with images of Australian soldiers and sailors, appeared frequently in advertisements from 1915 to the Second World War.

Many returned servicemen tried to put the war behind them, but there was a section of them who emphasised their separateness from the rest of society and created a distinct digger subculture. They shared experiences which set them apart from their parents, their children, women (except those few who had been nurses) and those men of their own age who had not fought (who could easily be represented as 'shirkers' or 'traitors'). These returned men often missed the excitement, camaraderie and sense of purpose which they came to associate with the war. They became disillusioned and cynical about the Australia they returned to, wondering whether it was really worth the sacrifices they and their dead comrades had made. Nostalgia for the war could compensate for troubles in business and employment, particularly after the onset of the Depression. The digger subculture emerged in organisations such as the RSL and the secret armies (q.v.), as well as in private get-togethers like those described by George Johnston, where it was expressed in the singing of sentimental wartime songs and 'badinage exchanged in bad soldier-French'. The mouthpiece of this culture was *Smith's Weekly* (published 1919–50), 'The Diggers' Paper', a popular and populist Sydney-based newspaper. Though its readership and subject matter were not restricted to diggers, it aired their grievances, and a regular feature, the 'Unofficial History of the A.I.F.', did much to establish the larrikin image of the Australian soldier in the public mind.

Children growing up during and after the First World War were immersed from an early age in the war-fixated culture, and as in earlier times schools played an important part in this process. Parading to military music, singing militaristic patriotic songs, seeing paintings of war scenes on classroom walls, reading about British military heroes and the glory of Anzac, being taught to honour ex-pupils who had gone to war (whose names appeared on school honour boards and whose photographs were sometimes displayed on the walls), and listening to returned servicemen speak before Anzac Day were all part of schooling between the wars. Children were not taught to glorify war, but they could not escape the message that war was central to the definition of British-Australian identity. One man who was a child in the interwar years remembers growing up 'in the Methodist religion, the Anzac mystique, the culture of British imperialism, and the middle-class ethic of self-improvement. Warrior virtues occupied a key place in this weird web of ideologies'. With war also a common topic in popular children's magazines it is not surprising that many boys absorbed these profoundly masculine values, perhaps envying their fathers' warrior status, and carried the Anzac legend in their minds as they went off to fight in the next war.

The Second World War did not produce myths powerful enough to supplant those of the First in popular consciousness. The actions of Australian forces during the war were absorbed into the Anzac tradition, but Kokoda did not replace Gallipoli as the supposed birthplace of Australian nationhood despite the fact that the Second World War's most lasting myth was that of the nation fighting for the first time for its very survival. In this fight the image of the American soldier became almost as important as that of the Australian, but

Australian views of American troops showed that Australians were in two minds about their country's move towards closer ties with the United States. The 'Yanks' were the 'saviours' of Australia, but they were also seen as a threat to the supremacy of Australian manhood and the virtue of Australian womanhood (the 'over-paid, over-sexed and over here' image). Popular songs, though, painted a rosy picture of the American–Australian relationship. Instead of 'Australia Will Be There', the Second World War produced the hit song 'The Aussies and the Yanks Are Here', written by an American soldier who proclaimed 'We're all together now as we never were before, / The Aussies and the Yanks, sure we're gonna win the war'.

It was not only through the presence of US troops that American influence was felt in Australia; American cultural products were now outstripping their British and Australian equivalents in popularity. Comics were a good example, and many Australian boys grew up reading American war comics, but Australian comic artists made a valiant and sometimes successful attempt to compete. The larrikin image of the digger was still so popular that military authorities felt the need to campaign against it during the Second World War, seeing it as subversive of discipline and of respect for the military. One of the principal offenders in this respect was the popular comic strip 'Bluey and Curley', created by Alex Gurney in 1940, which was briefly dropped from an army newspaper because its humorous, anti-authoritarian portrayal of army life was thought to set a bad example. 'Wally and the Major', another strip created in 1940 by *Smith's Weekly* veteran Stan Cross, concentrated on the home-front army, using more subtle humour that deliberately avoided playing on the larrikin image. Both strips remained popular long after the war was over and their characters had re-entered civilian life. More heroic war stories could be found in comic strips about 'Wanda the War Girl', a busty redhead created by Kath O'Brien to give recognition to Australian servicewomen. Wanda, whose adventures ran from 1943 to 1951, was voted number one pin-up girl by Australian servicemen and was more popular than Superman in her time. By the 1950s Australian comic books and popular children's fiction were reinforcing the Cold War atmosphere. During the Korean War in particular they frequently featured tales of war and espionage with anti-communist and anti-Asian themes.

There were signs, however, that new images of war and of Australian soldiers were emerging. Sumner Locke-Elliott put his wartime experiences on stage in *Rusty Bugles* (1948), which depicted both the boredom and humour of life in a

Northern Territory army camp. It was phenomenally successful, but the vulgarity of the soldiers' language shocked some people. Alan Seymour's play *The One Day of the Year* (1960) stirred up even greater controversy by exploring the generation gap through the conflict between a father and son over the meaning of Anzac Day (q.v.). While the play is sympathetic to both sides, the depiction of the returned soldier father as a somewhat pathetic figure and the vigour of the son's attack on the Anzac legend were seen by many as the beginning of a process of questioning Australia's military myths. The process was continued by campaigners against the Vietnam War, but the protest movement generally looked to American popular culture for inspiration. The only Vietnam-related Australian song to become popular during the war was 'Smiley' (1969), sung by Ronnie Burns, an apolitical song of a man who goes to fight in an unnamed Asian country. After Australian troops were withdrawn from Vietnam, however, disillusionment with the war produced a new image of the returned serviceman as victim of the war. This theme was taken up by three very popular songs which tell the stories of individual soldiers: folk-singer Eric Bogle's 'The Band Played Waltzing Matilda' (1972), hard-rock band Cold Chisel's 'Khe Sanh' (1977), and radical folk-rock group Redgum's 'I Was Only 19 (A Walk in the Light Green)' (1983). The American origins of the new image, reflected in the fact that returned servicemen were now called by the American term 'veterans', is most apparent in Cold Chisel's picture of the veteran as a drug-dependent drifter rejected by society on his return from Vietnam. Bogle and Redgum, writing about the First World War and Vietnam respectively, use their depiction of the veteran as physically and mentally maimed by war to attack the Anzac legend which, in the words of 'I Was Only 19', 'didn't mention mud and blood and fear'.

Throughout the 1980s an anti-war stance continued to be expressed in popular music by one of Australia's most popular rock bands, Midnight Oil, but for most Australians war and the military had a low profile. Warlike imagery was still commonly used in sports journalism, and one Melbourne Australian rules football team was even called the Bombers, but sport was no longer spoken of as training for war. More than anything it was the increasing public relations skill of defence officials which kept the military in the public eye by means of military band concerts, displays of RAAF flying skills, open days on navy ships and commemorations of important events in Australian military history. War was still seen as an important element in the definition of Australian nationality, and, despite predictions to the contrary, Anzac Day ceremonies continued to be well attended.

PORTER, Major-General Selwyn Havelock Watson Craig 'Bill' (23 February 1905–8 October 1963). A bank officer in civil life, Porter joined the 55th Battalion (The Essendon Rifles) in December 1924 after being posted to Melbourne. He had both regimental and staff experience with the CMF, and in October 1939 joined the 2nd AIF as a major and second-in-command of the 2/5th Battalion.

He took little part in the early campaigning of the 6th Division, but took over the 2/31st Battalion in February 1941 and led it through the Syrian campaign. He was shot through the thigh by a sniper, but elected to remain with the unit for a further fortnight until ordered to the rear. Returning to Australia at the beginning of 1942, he was given command of the 11th Infantry Brigade, followed after only two months by command of the 30th Brigade in the New Guinea campaign. This was the only militia brigade then serving outside Australia, and the first to meet the Japanese in battle. When Porter took over command the units of the brigade suffered from equipment deficiencies and some poor-quality officers; Porter thought that as a militia formation they had been shabbily treated in important respects, but nonetheless removed about 20 officers and replaced them with AIF men, some of them newly commissioned NCOs. In hard fighting his brigade first held and then gradually began to push back the Japanese advance until relieved by the 21st Brigade in August.

Porter himself crossed the Owen Stanley Range during the final phase of the Papuan campaign, the fighting for the Japanese bridgeheads at Buna–Gona–Sanananda, in which the 30th Brigade was again heavily engaged. In July 1943 the brigade was disbanded and absorbed into the 16th Brigade of the AIF, and Porter was posted as chief instructor to the Land Headquarters Tactical School, where his writings and teachings on tactical doctrine became increasingly recognised throughout the army. In November 1943 he returned to active command with the 24th Brigade of the 9th Division, which he commanded until the end of the war through the Huon Gulf campaign and the Labuan landings in Borneo.

After the war Porter continued to soldier in the militia, commanding in turn the 6th Brigade and the 3rd Division. In July 1953 he became CMF member of the Military Board (q.v.), a position he held until retirement in 1954. In January 1955 he accepted the appointment of chief commissioner of the Victorian Police, following in a long tradition of retired general officers.

Porter was one of the best of the CMF brigadiers of the Second World War generation. He brought organisational skills of a high order together with courage and a fertile mind to the tasks at hand, and established a ready rapport with the soldiers under his command, many of whom he could address by name. Gavin Long (q.v.), the official historian, writing at the end of the war thought him 'a fine soldier who could command a division or a corps with distinction'.

PORTSEA, OFFICER CADET SCHOOL (OCS) opened in January 1952 in response to a perceived shortfall in the number of young officers available to the army, a shortfall that RMC Duntroon was unable to make up. It was situated on the site of the old Quarantine Station at Portsea, on Port Phillip Bay south of Melbourne. The initial course was six months long, but from January 1955 was extended to a year. A short course, it was intended to produce junior officers quickly, and graduated them as second lieutenants. Eligibility was somewhat wider than for RMC, with broader age ranges and lower educational requirements; a commission through Portsea was thus available to suitable other ranks. Portsea also educated larger numbers of foreign students, drawing them from New Zealand, Malaysia, Fiji, Singapore and Papua New Guinea. By 1977 OCS had graduated 2516 officers. Some commentators thought that OCS graduates made better platoon commanders in Vietnam than the Duntroon graduates which, if true, might be explained by the fact that their training was more narrowly focused at that level. The influx of large numbers of OCS graduates had the effect over time of diluting the proportion of Duntroon graduates in the Staff Corps (q.v.). By 1976 Portsea graduates were beginning to enter the senior ranks of the army, with the first one promoted to brigadier in April 1978. Portsea was closed at the end of 1985 and its functions transferred to the new officer course at Duntroon. The school's colours are laid up in the chapel at RMC Duntroon.

POSTWAR RECONSTRUCTION Combining several central issues in Labor Party policy, a Department of Post-War Reconstruction was established in December 1942, inheriting a number of functions from the Reconstruction Division of the Department of Labour and National Service, which had been created in February 1941. The first minister was Ben Chifley (q.v.), but the portfolio is probably most commonly associated with John Dedman (q.v.), who replaced him in February 1945. The first permanent head was the distinguished economist and public servant H. C. 'Nugget' Coombs, appointed in January 1943, and his most notable successor, from August 1949, was the equally distinguished academic L. F. 'Finn' Crisp, who presided over the department until it was abolished in March 1950. Postwar reconstruction, both as an ideal and as an administrative entity, was designed to smooth the transition from a wartime footing to a peacetime economy, both for individuals demobilised from the services and for the economy as a whole. The Labor government wished to avoid a return to the Depression conditions of the 1930s and the misery which had attended many former soldiers returned to civilian life after 1918. Both in international affairs and domestic policy, Labor argued that the war had only been worth fighting if something better emerged at the end, and postwar reconstruction was designed to ensure this within Australia itself. Thus the department, which was essentially a planning and coordinating body, was charged with overseeing the commitment to full employment and a range of practical social welfare measures such as widows' pensions, maternity allowances, unemployment and sickness benefits, the establishment of the Commonwealth Employment Service, and a range of agreements with State governments covering housing (which was critically short at the end of the war), hospitals, and subsidies to State universities. In addition, the Commonwealth created the Australian National University and inaugurated the Snowy Mountains Hydro-Electric Scheme. Attempts to control prices and exercise authority over social welfare were frustrated by the failure of a referendum on these issues in 1944. The responsibilities of the department changed over time, with functions being passed to other departments once they had become established or the requirement for them had altered. Thus, for example, the Department of War Organisation of Industry was merged with Post-War Reconstruction in June 1945; responsibility for the re-establishment of ex-servicemen was passed to the Repatriation Department (q.v.) in January 1950, while the War Service Land Settlement Division went to the Department of the Interior at the same time. Postwar reconstruction had its equivalents in both Britain and the United States, and marked the foundation of a welfare state system which was to survive largely unquestioned until the 1980s.

POTTS, Brigadier Arnold William (16 September 1896–1 January 1968) migrated with his family to Western Australia from the Isle of Man in 1904. He enlisted in the 16th Battalion of the AIF and served at Gallipoli, and was commissioned in Egypt after the evacuation. He served in France, where he won an MC and was seriously wounded at Vaire Wood in July 1918. Potts returned to Australia, recovered from his injuries, and took up a soldier-settler block at Kojonup in Western Australia. Between the wars he continued soldiering in the CMF, was appointed to the Reserve of Officers during the cut-backs in 1922, but by 1939 was serving with the 25th Light Horse (Machine Gun) Regiment. He was appointed second-in-command of the 2/16th Battalion of the 2nd AIF in May 1940, and served overseas in Palestine and Syria, commanding the battalion from August 1941. On return to Australia, Potts was promoted and given the 21st Infantry Brigade in May 1942. He commanded during the retreat along the Kokoda Track during which the Australians fought a series of desperate delaying actions against stronger Japanese forces who enjoyed a number of tactical advantages. The official historian, Dudley McCarthy, wrote that Potts shared the arduous retreat with his men, and spoke of his 'characteristic energy' and 'strength of spirit'. He was replaced in command by Brigadier I. N. Dougherty (q.v.) in September and was posted to Darwin. Potts had the misfortune to command a retreat in the early stages of the Pacific War, and although his immediate superiors, Generals Rowell and Allen (both of whom would be dismissed by Blamey), thought he had done a good job in impossible circumstances, others further from the fighting misjudged the reasons for his failing to hold the Japanese advance at an early stage.

PRICE, Colonel Thomas Caradoc Rose (21 October 1842–3 July 1911) was born in Hobart. After serving for 20 years in India as an infantry officer and police superintendent, he returned to the Antipodes in 1883 to farm at Heidelberg outside Melbourne. Two years before, Boer farmers in South Africa had beaten a British army, and Price discerned that the rural men of Victoria possessed a similar military potential. In 1885 he was appointed a permanent officer of the Victorian Military Forces (see COLONIAL MILITARY FORCES) and allowed to raise the Victorian Mounted Rifles, a volunteer corps of good shots recruited from rural rifle club members, who were to ride horses for mobility rather than the charge. Price dressed his regiment not in customary red or blue but in khaki with coloured facings, a practical military costume common in India. Three years later he briefly assumed command of a similarly conceived volunteer corps, the Victorian Rangers, which did not ride horses. In 1890 Price commanded rural volunteers who mustered in Melbourne to police the maritime strike (see CIVIL POWER, AID TO). On 30

August he instructed his men to 'fire low and lay out the disturbers of law and order' if ordered to do so by the civil authorities. No shots were fired, and the strike was broken peacefully. But some Victorians were shocked by what was misreported as Price's eagerness to shoot down fellow men. A court of inquiry was held but could not find a case against him. In 1900 Price led a Victorian contingent to the second Boer War, and, alone among Australian officers during the conflict, was given command of a force which included British regulars. In a war which vindicated Price's view of the value of good marksmanship and horsemanship, he was twice bruised by shell fire and appointed CB. He returned to Australia on the eve of Federation and in May 1901 commanded the mounted troops at the opening of the first session of the Commonwealth parliament. Price ended his military career with two appointments as State commandant, then the second-highest position in the Australian military forces: first of Victoria (March–July 1902), then of Queensland (July 1902–August 1904). After Price died he was buried with military honours.

PRIME MINISTER'S WAR CONFERENCE first met on 8 April 1942. It was established by Prime Minister John Curtin (q.v.) to give him access through the new Allied C-in-C, General Douglas MacArthur (q.v.), to the strategic decision-making processes that affected Australian interests, such access having failed to materialise by means of the Pacific War Council (q.v.). As a result of the first conference, Curtin formalised MacArthur's demand that he have direct access to the Prime Minister, and vice versa, thus replacing the chiefs of staff as his principal military advisers with the American C-in-C. The right of direct access was subsequently extended to the Australian C-in-C, General Sir Thomas Blamey (q.v.), thereby centralising military and political power over strategic matters to an unprecedented degree. The conference met whenever it was thought necessary, and normally consisted of the Prime Minister, MacArthur, the Secretary (usually F. G. Shedden [q.v.]) and other ministers as Curtin saw fit.

PRISONERS OF WAR So often marginalised in accounts of war, POWs play a major role in Australian military history. Their numbers in many conflicts were not large—the Second World War being the exception—but their experiences were often so traumatic that they have come to have a special place in Australian mythology.

In the Boer War, 104 of the approximately 16 000 Australians who served in South Africa were prisoners for any length of time. Given the guerrilla tactics adopted by the Boers, many other Australians were held for only short periods before being stripped of their arms and belongings and being allowed to make their way back to their own units. Some Australians were interned in the one large camp managed by the Boers, at Waterval near Pretoria. Poor sanitary arrangements led to outbreaks of dysentery and enteric fever, and food rations were nutritionally inadequate. However, with money sent by the Australian government through the American consul, prisoners were able to supplement their diet. A few Australians devoted themselves to escaping; the remainder were liberated by British forces in June 1900.

In the First World War about 4082 Australians were captured: 3850 by the Germans on the Western Front in France and Belgium (of which 30 per cent were captured during the First Battle of Bullecourt in April 1917); and 232 by the Turks. Prisoners' prospects varied dramatically depending on their captors and the regions in which they were interned. International legal protection for prisoners of war was limited in 1914, with the Hague Conventions of 1899 and 1907 dealing in only a rudimentary fashion with the enduring concerns of prisoners: food, work, pay, discipline, escape, religious freedom, and access to sources of support from outside the prison camp.

Nearly half of the prisoners of the Turks were light horsemen captured in Sinai and Palestine; a third were taken on Gallipoli and the remainder were members of the Australian Flying Corps and the crew of the submarine *AE2*, which made a remarkable passage through the Dardanelles in 1915 before it was disabled and captured by the Turks. After capture these prisoners endured forced marches and crowded railway journeys before reaching camps where disease, poor diet and inadequate medical facilities prevailed for most of the war. The other ranks were employed on working parties, many being engaged in railway construction in the Taurus mountains; 25 per cent of them died. Only one death was an officer, reflecting the often unremarked fact that officers, by virtue of their privileged status under international law, benefit from better conditions in captivity. Unlike the other ranks, they do not have to work, a key to survival in situations of malnutrition.

Prisoners in German hands on the Western front suffered less — 9 per cent of them died in captivity — but their treatment often fell short of the standards of the Hague Conventions. (Although technically these did not apply because not all belligerents were parties to the conventions, they provided the benchmark against which treatment of POWs was measured. Given the obvious lacunae of the Hague

Conventions, however, wartime agreements concerning prisoners of war were negotiated between the belligerents.)

Much of the hardship suffered by prisoners was a consequence of logistical difficulties — over five million prisoners were taken in this unprecedentedly massive conflict — and of the privations in Germany as the Allied blockade's stranglehold tightened in 1917–18. Towards the end of the war German civilians were, if anything, more undernourished than the Allied prisoners who had access to food parcels from external sources such as the Red Cross.

Much of the privation, however, was the product of malice and retaliation. Determined to make a mockery of the Allied claims to harmony between their multinational forces, the Germans insisted on interning prisoners of different nationalities together, with no allowance for cultural distinctions. Prisoners were also retained in the vicinity of the front line, a practice that was allegedly in reprisal for similar treatment of German prisoners by the British. While on the Western Front prisoners' labour was indisputably connected with the operations of war (despite the prohibition on such practices in international law) and they were vulnerable to shelling by their own artillery. Food supplies were far more precarious than in Germany.

It was the Second World War that ensured POWs their place in Australian national memory. The 395 deaths of prisoners in the First World War were totally eclipsed by the more than 58 000 deaths in battle. In the war of 1939–45, however, prisoners' deaths accounted for nearly 30 per cent of all Australian deaths in the conflict. The vast majority of these deaths in captivity occurred in the Pacific theatre. Only 265 Australian prisoners died in Europe, a statistic that perhaps accounts for the persistent neglect of prisoners of the Germans and Italians in the Australian historiography of the war.

This statistic also illustrates the vast disparity between the treatment Australian prisoners received from the Germans and Italians, on the one hand, and the Japanese, on the other. This was not so much a question of the European States being party to the 1929 Geneva Convention, which had been developed in response to the problems highlighted by the First World War, while the Japanese were not. The Germans' treatment of prisoners on the Eastern Front plumbed the depths of cruelty and indifference. It was only in Western Europe that, with some infringements, they observed the 1929 regulations. What gave the Eastern Front and the Pacific War their particularly brutal quality were the racial and cultural dimensions to these conflicts.

Australians from all services were interned in Europe and the Mediterranean theatre but by far

Australian POWs working on the Burma–Thailand railway, Takanun, Thailand, 1943. (AWM P0406/40/20)

the greatest number of prisoners were members of the AIF captured in north Africa (1941), Greece (2065) and Crete (3109). There were 1476 Australian airmen, mostly from Bomber Command, captured, the first prisoners being those attached to the RAF during the campaign of mid-1940.

Prisoners captured in the Western Desert of north Africa, even if they fell into the hands of the Afrika Corps, were normally handed over to Italian control. After varying periods in poorly organised transit camps in Libya, they were shipped across the Mediterranean to Italy. In the absence of any agreement between the Allies and the Axis powers about the identification of ships carrying POWs, this voyage became increasingly dangerous. The torpedoing of the Italian freighter the *Nino Bixio* by Allied submarines on 16 August 1942 resulted in the loss of 37 of the 201 Australians on board.

In Italy prisoners (from both the AIF and RAAF) were interned across the country but were progressively concentrated in a number of camps: at Sulmona (Campo 78) in central Italy, where the majority of Australian officers were accommodated; Gruppignano (Campo 57) in the north-east at the foot of the Alps; and Vercelli (Campo 106), 80 kilometres from Turin. Living conditions in these camps were generally tolerable though much depended on the character of the camp administration which could vary from positive obstruction to passive inefficiency. The Fascist commandant at

Gruppignano imposed a particularly harsh and petty disciplinary regime which resulted in several fatal shootings, including one Australian. In northern Italy the winter climate was severe and the quantity and quality of food declined as food supplies in Italy diminished. With the breakdown in supply of Red Cross parcels, officers contributed from their higher pay to funds to assist the other ranks.

The collapse of Italy in mid-1943 should have secured the Australian prisoners their freedom but in a disastrous error of judgment the British War Office sent instructions to all camp leaders that in the event of an armistice with Italy all prisoners were to remain in their camps until the arrival of Allied forces. As the prisoners by this time, been concentrated, in northern Italy (those at Sulmona having been moved to Campo 19 at Bologna in June), this meant the vast majority of prisoners fell into the hands of the German forces occupying the northern half of the country. Almost the entire camp at Gruppignano was evacuated to Germany. Thirteen Australians from Bologna escaped to Switzerland when being ferried by cattle trucks to German camps. Only at Vercelli, where there were few German troops in the area, were 400 Australians able to reach Switzerland. Some evaders stayed in Italy, living with civilians or joining the partisan bands. By agreement with the Swiss government, the Australians in Switzerland were maintained under British military organisation and control.

Prisoners taken in Greece and Crete endured at first even worse conditions than in north Africa. Whatever the conflict, prisoners are generally most vulnerable to hunger and disease when they are in transit or held in camps that are not permanent. There seems little doubt that the Germans were overwhelmed by the logistics of handling the 25 000 Commonwealth troops they captured in Greece and Crete, just as the Allies were by the over five million German prisoners taken in early 1945, who endured appalling privations in the camps on the Rhine where at least 56 000 died.

Australians captured early in the Greek campaign were packed into cattle trucks and moved to Marburg in northern Yugoslavia through a series of staging camps where sanitation provisions were primitive and diseases that are more commonly associated in the popular memory with Japanese camps — beriberi, malaria and dysentery — were widespread. In the south, prisoners from Greece and Crete were channelled through camps at Corinth and Salonika. Overcrowding, rations that were practically at starvation level, lack of medical facilities, and gratuitous brutality on the part of German guards made conditions, particularly at Salonika, harsh. Only the seriously wounded were treated with compassion.

From June 1941 onwards Allied prisoners were progressively moved to Austria and Germany by cattle trucks, where again nightmare conditions prevailed on journeys that were at least seven days long. Although Australians were scattered in multinational camps across central Europe, their main destinations were, for the other ranks, Marburg in northern Yugoslavia (Stalag XVIIID); Wolfsberg, Austria (Stalag XVIIIA); Spittal, Austria (Stalag XVIIIB, later XVIIIA/Z); Hammelberg, near Frankfurt-on-Main, Germany (Stalag XIIIC); Moosberg, near Munich (Stalag VIIA); and Lamsdorf, Silesia (Stalag VIIIB, later 344). Officers were interned separately, mostly at Biberach, Bavaria (Oflag VB), and Warburg, in north-western Germany (Oflag VIB). Non-commissioned officers also were conceded the right to have separate accommodation — and not to work — in camps such as Hohenfels, southern Bavaria (Stalag 383) and, from March 1944, Thorn, Poland (Stalag XXA, later 357), which was relocated near Bremen.

RAAF personnel were mostly accommodated in special Stalag Lufts, particularly Stalag Luft III at Sagan which from early 1942 was the central camp for air force men. Alternatively, they were accommodated in separate compounds for airmen in sta-lags such as Moosberg, Lamsdorf, Thorn, and Heydekrug on the Baltic coast (Stalag Luft VI). Naval personnel and merchant seamen, whose numbers are uncertain, were interned in the Marlag-Milag at Bremen. Just under a hundred Australians had the misfortune to be interned in concentration camps. These included airmen who had been wearing civilian clothes when captured and were sent to Buchenwald until high-level intervention transferred them to Stalag Luft III.

As the above indicates, despite the requirement of the 1929 Geneva Convention that belligerents should avoid as far as possible 'bringing together in the same camp prisoners of different races or nationalities', the Germans did not concentrate Australians as a group but treated them as part of a broader British Commonwealth category.

While conditions in these camps varied (depending on the attitude of the commandant) and were often far from ideal, they were a vast improvement on the experiences of Greece and the staging camps of Yugoslavia. The mere permanence of the camps meant that facilities which required systematic organisation, such as educational courses, entertainment, sports programs, manufacture of furniture and the production of camp newspapers, were able to be developed. Prisoners had regular access to Red Cross supplies and to representatives of the International Committee of the Red Cross, an advantage that is often undervalued by historians. The right to send and receive mail, to get relief parcels, to speak to someone objective about their conditions, to know that their families were aware of their fate: all of these helped to make captivity tolerable. Red Cross parcels, in fact, were essential for the prisoners' diet as the Germans, assuming that these supplies would be available, kept the rations they provided to a bare minimum. The ready supply of cigarettes, chocolate, soap and coffee (real not ersatz) also gave the prisoners the capacity to trade with guards and the local population.

Once in permanent camps in Germany, the other ranks were deployed on work detachments away from the main camps. Employment ranged across factory, forestry, construction, and railway work to labour on farms and in mines. Work detachments were a mixed blessing; they gave prisoners more opportunity to escape, to get better food and to trade their Red Cross supplies with civilians for services, which in some cases included prostitution. Work also countered the 'barbed wire psychosis' that afflicted prisoners unable to cope with the monotony and confinement of life in the permanent camps. On the debit side, however, men on work detachments could suffer long working hours and uncomfortable living conditions (depending on their employer), and inadequate access to medical treatment, mail and Red Cross facilities.

Contrary to the Geneva Convention of 1929 prisoners were sometimes the victims of tit-for-tat reprisals between the German and Allied governments. The most infamous of these was the manacling of Allied prisoners, including some Australians, in retaliation for the British tying the wrists of German prisoners during the raid on Dieppe in August 1942 and a commando raid on Sark in the Channel Islands in October the same year. Despite the intervention of the International Committee of the Red Cross the Germans continued the practice for more than 12 months, although with the passage of time the Germans became less rigid in applying the practice and the prisoners found ways of circumventing it.

The mass transfers of prisoners from Italy in late 1943 threw the established order in the camps in Germany into chaos, with overcrowding in many camps persisting even after further work camps were established. Australians transferred from Italy went to Gorlitz in Silesia (Stalag VIIIA), to Wolfsberg, and Lamsdorf (renumbered Stalag 344 when two new camps at Sagan [Stalag VIIIC] and Teschen [VIIIB] were created).

As 1944 progressed, Australian prisoners, now spread across Germany, became victim to the very Allied successes that they were depending on for their release: strategic bombing, the disruption of the German railway system, and the advance of the Allied armies. Rather than allow their prisoners to fall into Soviet hands the German authorities insisted on their being rapidly evacuated, often by forced marches through the winter snow. About 250 000 Allied POWs straggled across Germany in the winter of 1944–45, joining the mass of refugees and German military traffic fleeing the east. Conditions were wretched, with little or no provision for food and accommodation *en route*, and exhausted and frostbitten prisoners being harassed by their own air forces mistaking them for the enemy. The situation was little better in the reception camps they finally reached, where food was scarce now that Red Cross supplies were disrupted (but not entirely halted), and facilities were taxed to the utmost degree. The bitter irony was that no sooner had some prisoners arrived at camps in western Germany, than they were forced to take to the roads again, this time fleeing the British and American advance.

From March 1945 onwards the Allied armies progressively overtook the prison camps and columns of prisoners still marching. Owing to the speed of the Red Army's advance in the east some 50 Australians were 'liberated' by them and then faced a long journey through Poland and the Soviet Union to Odessa before repatriation. Prisoners liberated by the Western Allies were flown to the United Kingdom within a few weeks of the German capitulation. Repatriation to Australia occurred from May to August 1945.

Such mythology as there is about the prisoner-of-war camps in Europe ignores the considerable privations of captivity, in the first and last years especially, and focuses instead on escape and activities designed to outwit or bewilder the Germans and Italians. These were, however, essentially the preoccupation of officers, NCOs and airmen as these prisoners had the essential leisure in which to plan and execute escapes. Escaping from an established camp, as opposed to a staging camp or while in transit between camps, was fraught with difficulties and few prisoners managed to reach neutral territory or Britain. All prisoners from the only large-scale escape in Italy, by tunnel from Gruppignano, in October 1942, were recaptured within a few days. Of the 41 airmen who escaped from German camps only five reached Britain. The most infamous escape, that from Stalag Luft III, ended in the execution, on Hitler's orders, of 50 escapees including four Australians. Other Australians, who were persistent offenders, were transferred to the Oflag at Colditz, which was considered to be escape-proof, or punished by incarceration in concentration camps at Theresienstadt and Brno (in Czechoslovakia).

For airmen, evasion after being shot down was a further possibility, depending on where they landed. In the Western Desert the chances of evading were relatively high, while in Europe they were low in the first four years of the war. Thereafter evasion was facilitated by the emergence in Holland, Belgium and France of escape organisations. Through the efforts of the resistance a number of safe routes were set up whereby airmen could be passed from one hiding place to another until they reached neutral territory. Hence, it has been calculated, of every 12 RAAF men lost over enemy territory, eight were killed, one evaded capture and three became POWs.

The situation of Australian prisoners of the Pacific War was a dramatic contrast to that in Europe. More than 22 000 Australians were captured, the vast majority in the defence of Malaya, Singapore and the Netherlands East Indies from December 1941 to March 1942. This was the largest number of Australians taken prisoner in any campaign. Over 21 000 were from the AIF (particularly the 8th Division); 354 were from the RAN; 373 were RAAF officers (who were on occasions treated as war criminals and executed); and 71 were women from the Australian Army Nursing Service. Of these 14 792 were captured at Singapore; 2736 on Java; 1137 on Timor; 1075 on Ambon and 1049 at Rabaul.

Although there were greater differences between prison camps throughout Asia than is often conceded, Australians generally suffered from malnutrition, overwork, disease, harsh disciplinary regimes and often gratuitous brutality from their Japanese and Korean guards. Japan was not a party to the 1929 Geneva Convention and its officials' attitude towards prisoners was conditioned by a mixture of harsh pragmatism, racial misunderstanding, culturally-rooted disdain for personnel who had surrendered, and, particularly towards the end of the war, the frustration that arose from their own vulnerability and the imminence of defeat. Nearly 36 per cent of Australian prisoners — 8031 — died in captivity, nearly as many as died in action in the Pacific War.

Many deaths occurred in the early weeks after capture, a traditionally precarious time for prisoners. Massacres of Australians occurred at Tol on New Britain (160 Australians), Parit Sulong in Malaya (110); and Laha on Ambon (over 200). Twenty-one Australian nurses were executed on the beaches of Banka Island; and an unknown number of Australians elsewhere in Malaya and in Singapore, especially at the Alexandra Hospital.

In their three-and-a-half years of captivity most Australians were moved a number of times across the Asian region to meet the needs of the Japanese economy and armed forces for labour. In the first instance those captured in Malaya and Singapore were concentrated in Changi (q.v.), on the eastern tip of the island. Although Changi has become a byword for the horrors of captivity in the Far East, it was, in fact, one of the better places of internment, especially from February 1942 to May 1944 when AIF prisoners were accommodated in the Selarang Barracks. (Thereafter they were held in the very overcrowded Changi gaol with increasing restrictions on their activities and shortages of food.) In the Selarang Barracks area the prisoners were able to establish sanitation systems, relatively reasonable medical facilities, educational programs such as the 'Changi University', news bulletins, vegetable and poultry farms, even a black market — indeed, all the features of a permanent camp in Europe.

The population of Changi fluctuated dramatically during the war as it functioned as a transit camp for prisoners moving from one region to another, and as a base from which working parties moved out to labour detachments within the immediate area or further afield. The first of these working parties was assigned, only a week after the surrender of Singapore, to the wharves and warehouses at Keppel Harbour and elsewhere on Singapore island and the mainland. By May 1942, 8200 Australians had been deployed to sites such as River Valley Road, Sime Road, Mersing and Blaking Mati, on tasks that included repairing war damage, filling shell craters, unloading ships, demolishing mines and, at Bukit Timah, erecting a monument to the 'Fallen Warriors'.

In mid-1942 the Japanese began an ongoing process of moving large groups of Allied prisoners from Singapore and the Netherlands East Indies to camps elsewhere in South-east Asia and Japan. On 14 May 1942 A Force (3000 strong) left Singapore for Burma; B Force (1496) departed on 8 July. On 16 August, all senior officers above the rank of lieutenant-colonel were sent, with a force of engineers and technicians, to Formosa. The senior Australian officer at Changi, Major-General C. A. Callaghan, delegated the command of AIF troops to Lieutenant-Colonel F. G. ('Black Jack') Galleghan (q.v.). Apart from siphoning off these senior officers, the Japanese generally did not segregate officers from the other ranks, though there were exceptions to this, such as on Borneo.

Further contingents left Singapore over the next 12 months: C Force (2200 strong, including 563 Australians) at the end of November 1942 for Japan; D Force (about 5000 including 2242 Australians) from 14–18 March 1943 for Thailand; E Force (1000, of which 500 were Australian) at the end of March for Borneo; F Force (7000 including 3600 Australians) from 18–26 April for Thailand; G Force (1500, including 200 Australians) on 26 April for Japan; H Force (3000 strong including 600 Australians) in May for Thailand; J Force (900, of which 300 were Australians convalescing) on 16 May for Japan; and in June and August 1943, K and F Forces, which were small, mainly medical forces, again bound for Thailand. Concurrent with these drafts, prisoners in the Netherlands East Indies were moved north to Thailand. A force of about 900 under Colonel E. 'Weary' Dunlop (q.v.) was the first to reach Konyu on 24 January 1943.

The drafts sent to Burma and Thailand were part of a force of 61 000 Allied prisoners and countless indigenous people assigned to the construction of the Burma–Thailand railway (q.v.). Stretching 421 kilometres from Kanchanaburi (Kanburi) in Thailand to Thanbyuzayat in Burma through dense mountainous jungle, the railway was intended to eliminate the dependence of Japanese troops in Burma on the increasingly precarious supply route via Singapore and the Malacca Straits. Completed in November 1943 when the working parties from either end met at about Nieke, the railway cost the lives of more than 10 000 prisoners and an unknown (but certainly far greater) number of civilians. Of the prisoners' deaths nearly 2650 were Australian. The worst conditions were experienced by F and H Forces where 29 per cent and 30 per cent, respectively, of Australians died. In A, D and F Forces Australians survived at a higher rate than the British — though this was not the case in H Force.

The high death toll on the railway was due to a number of factors, the most significant of which were the frenetic pace at which the Japanese pushed their workforce, most notably in the 'Speedo' during the monsoon period of mid 1943; the utterly inadequate logistical system, which meant the most remote camps near the Thai–Burmese border were denied adequate supplies of food and other necessities; the impossibility of maintaining the infrastructure of the camps along the line in view of excessive Japanese demands for labour and the lack of tools, equipment and, in some cases, even cooking utensils; and, predominantly, the utter disregard of the captors for the health of their prisoners. At the height of the push to finish the railway, prisoners who had marched, in some cases, 20 days to their first camp, and who were on starvation rations and debilitated by illness, were leaving the camps in pouring rain at 10 a.m. and returning at 4 a.m. Cholera broke out at camps all along the line, and diseases of malnutrition, dysentery and flesh-devouring tropical ulcers were endemic. In the face of Japanese obstruction and denial of medical supplies, the boundless ingenuity, improvisation and dedication of medical officers such as Dunlop, Colonel J. B. Coates and Major B. A. Hunt, could only contain the death toll at levels below that which it would otherwise have reached.

The Burma–Thailand railway has dominated public memories of prisoners of the Japanese and been a potent source of mythology about Australian collective and individual behaviour in captivity. This is in some ways understandable in view of the particularly intense horror of the Speedo period and the fact that 35 per cent of Australian deaths in captivity were suffered on the railway. But at its worst the Australian death toll in the major forces on the railway was 30 per cent. Other groups of Australians, on Ambon and in Borneo, suffered far higher tolls. The railway's prominence in the national memory no doubt owes something to its effective publicists immediately after the war, such as Russell Braddon, Rohan Rivett and Wilfrid Kent Hughes. It owes something also to the sheer megalomania of the Japanese conception of the project. As possibly the most famous of the books about the railway (by a Frenchman about British prisoners), *The Bridge on the River Kwai* suggests that at the end of all the horror, there was at least a railway. Although, as the phrase has it, a life had been paid for every sleeper, it remained a remarkable engineering achievement created against the odds with primitive tools and sheer physical toil. In contrast, the suffering of Australians on Ambon, about which almost nothing was published until 1988, had a particular futility.

After the completion of the railway in late 1943 the bulk of Australian prisoners were either returned to Singapore or concentrated in camps at the Thai end of the railway. Of the latter some remained in Thailand and Burma in 1944–45 maintaining the railway, building defence positions or constructing roads across the Isthmus of Kra, in conditions that continued to be so harsh as to cause death rates of up to 25 per cent. As had been the case on some sections of the railway, Australian officers, removed from their men, were forced to work on a variety of tasks, despite their protests about their privileged status in international law.

Other contingents of Australians were taken to Saigon awaiting shipment, which did not always eventuate, to Japan. A considerable number of Australian prisoners were already stationed in Japan. In addition to C, G and J Forces, sent from Singapore in 1942–43, two drafts had been dispatched from New Britain. Conditions on the voyages to Japan, as elsewhere in Asia, were often wretched. Prisoners were crowded into small cargo vessels, with primitive sanitation and restricted water and food supplies. Voyages (the last of which sailed from Singapore in December 1944) lasted from 10 to 70 days. Unrestricted Allied submarine warfare forced Japanese ships to take increasingly evasive routes and, even then, about 1600 Australian prisoners were sunk by Allied attacks: 849 of these were from one of the drafts from New Britain, on the *Montevideo Maru* (sunk on 1 July 1942); and 543 prisoners, including the commander of A Force, Brigadier A. L. Varley, went down with the *Rokyu Maru* (sunk on 12 August 1944). US submarines rescued some survivors of the latter who provided the first authentic news of conditions in Burma and Thailand.

The experiences of prisoners in Japan were diverse. Most of the Australians who arrived in 1943 seemed to have gone to the Kobe–Osaka area on the southern coast of the main island of Honshu. As the war progressed, they were scattered over a number of camps: Zentsuji, on the island of Shikoku; Naoetsu, Yokohama, Ikuno, Oeyama, Takefu, Taisho and Hiroshima on Honshu; Nagasaki, Moji, Fukuoka, Omuta and Kanoya on Kyushu. A small number went to Tokyo and to Sakata in the north of Honshu. Six Australian nurses from New Britain were interned throughout the war at Totsuka near Yokohama.

Prisoners worked on wharves and shipyards, in factories, steel mills and zinc foundries and, most hazardous and gruelling, in coal mines. As in Europe, conditions in the camps varied, in part depending on the personality of the commandants and guards. But generally prisoners endured long working hours, harsh disciplinary regimes, bitterly cold winters and food shortages that became acute as the Allied blockade of Japan intensified. At one of the worst camps, Naoetsu, 60 of the 300 Australians died in the first 13 months. In 1945 Allied air raids became an additional hazard, in the face of which the prisoners were regularly moved. Kobe House camp was destroyed by incendiary raids on 5 June 1945, though there were no fatalities for the prisoners. A number of Australians were in the vicinity of Nagasaki when the atomic bomb was dropped on them on 9 August 1945.

In addition to Japan, Burma, Thailand and Singapore, places of internment for Australian prisoners of the Japanese included Java, Sumatra, Formosa (Taiwan), Korea, Manchuria, Ambon, Hainan and Borneo. In the eight months after the Allied defeat in the Netherlands East Indies in March 1942 many Australian prisoners — survivors of the campaign for Java, the sinking of HMAS *Perth,* and the 2/40th Battalion from Timor — were gradually concentrated in Java at Bandung, Makasura and Batavia (the Bicycle Camp). In late 1942 and early 1943 many prisoners were drafted for overseas parties, but the remainder (including about 400 Australians) remained on Java, concentrated mostly in the Batavia area at Glodok, Makasura and the Bicycle Camp. In early 1945, as conditions were becoming increasingly difficult, many of the Australians remaining on Java were shipped to Singapore to work in the River Valley and other improvised camps.

On neighbouring Sumatra about 60 Australians from all three services, who had been captured while escaping from Singapore, were interned at Pelambang in conditions typical of the camps in the region. In May 1945 they too were shipped to Singapore in vessels crowded to the point where a number of prisoners died on the voyage. Sumatra was also the place of internment for the 32 Australian nurses, including Vivien Bullwinkel (q.v.), who survived the sinking of the *Vyner Brooke* and the massacre at Banka Island in February 1942. They foiled attempts early in their captivity to make them staff a Japanese brothel, but as they rotated between Muntok, Pelambang and Lubuklinggau in the years that followed, their circumstances became ever more crushing in their misery. Eight nurses eventually succumbed to disease and malnutrition in 1945.

Formosa was the destination of the group of senior officers who left Changi in August 1942. In the next 22 months they shared camps with US and British officers of similar rank in conditions that fluctuated from bad at Karenko and Shirawaka — physical abuse, low rations, illness and 'voluntary' labour — to tolerable at Tamasata and Tiahoku. In October 1944, the Japanese, fearing invasion, moved the senior officers by air and sea to Japan, and thence to Korea and ultimately Manchuria. Treatment at Chen Cha Tung (Liayuanchow) and, from May 1945, at Mukden, was a significant improvement on Formosa. Red Cross parcels were received on a scale unheard of in South-east Asia: about one every three weeks. The cold, however, was intense, sometimes as low as minus 50 degrees Fahrenheit.

Ambon and Borneo, in contrast, were the sites of the greatest disasters of the Pacific War. The 807 members of Gull Force (q.v.) who survived the campaign of January–February 1942 and the massacre at Laha were initially treated reasonably but in October 1942 they were divided, with 263 Australians and an equivalent number of Dutch prisoners going to Hainan Island. The Australians remaining on Ambon were left, in effect, leaderless by this dilution of the force and the subsequent destruction of the camp by Allied bombing in February 1943. With a particularly mindless and vicious Japanese interpreter in control of the camp, the prisoners were starved and worked to death in the last year of the war: 77 per cent of those on Ambon died. On Hainan, where conditions teetered on the catastrophic at times but were ultimately retrieved by Japanese intervention and stronger Australian leadership, 31 per cent died.

On Borneo the members of B and E Forces were concentrated by mid-1943 at Sandakan. They quickly established an elaborate intelligence organisation linking local civilians with civilian internees on Berhala Island some kilometres off the coast. When this network was betrayed and broken in mid-1943, all officers remaining with the men were sent to Kuching in south-west Borneo. Conditions at Sandakan deteriorated thereafter until late in January 1945 when the Japanese, anticipating an Allied attack, began moving them from Sandakan to Ranau, a distance of some 256 kilometres over a steep and muddy track. Many of the already ill and starving men did not survive the journey. Those who did either died in the appalling conditions that awaited them at Ranau or were executed in the last weeks of the war. Only six of the 2500 British and Australian prisoners at Sandakan in mid-1943 survived (see SANDAKAN DEATH MARCH).

In the circumstances that confronted prisoners in Asia, escape was rarely an option. The Japanese extracted from prisoners in various camps in August–September 1942 guarantees that they would not escape, guarantees which in the case of Changi prisoners were given only after a dramatic crowding for three days of 15 400 British and Australians into Selarang Barracks Square. Such undertakings, however, were not the real deterrents to escape in the Asia-Pacific. These were the impossibility of escapees blending into the local population, the impenetrability of the jungle, the distance of places of internment from neutral territory and, above all, the Japanese policy of executing escapees. Only on Hainan and Borneo did Australians manage to escape and join local guerrilla movements. Thirty-six members of Gull Force on Ambon made their way to Australia by island-hopping before the Japanese hold on the islands of the Arafura Sea had tightened in February–March 1942.

Five years after the end of the Second World War Australian service personnel again faced capture in Korea. By this time the Geneva Conventions had been revised in 1949 to address many of the inadequacies of international humanitarian law which the sufferings of prisoners, civilians and military personnel during the Second World War had revealed. The 1949 Conventions had not been formally ratified by any of the parties to the Korean conflict but it was generally agreed that at least the principles of these agreements should apply. In practice, however, the North Koreans and Chinese violated many of the guarantees of prisoners' wellbeing. During the United Nations Command advance to the Yalu river and the subsequent Chinese offensive in 1950–51, when chaotic conditions prevailed, prisoners suffered forced marches, summary executions, torture, inadequate diet, disease and chronic shortages of medicine and medical supplies. After truce talks began in July 1951 there was a marked improvement in treatment and the threat of death through malnutrition subsided.

Of the 29 Australians who were listed as prisoners of war in Korea, only one died of starvation and ill-treatment. Those who survived, however, often endured extreme hardship and physical abuse. Given the fluidity of the front and the haphazard movements of prisoners between caves, tunnels and camps, it is difficult to generalise about Australians' experiences. Treatment depended on factors such as the date and circumstances in which they were captured, whether they chose to attempt escape or to resist the intense political indoctrination program to which their communist captors subjected them, and their rank. Whereas the other ranks were viewed as victims of a capitalist hierarchy and suitable subjects for 're-education', officers were 'reactionaries' and the repository of valuable military knowledge. Despite the prohibition in the Geneva Conventions against the use of any form of coercion to secure information from prisoners, officers were harassed and kept in solitary confinement in an effort to break their resistance.

In later years United Nations Command prisoners in the Korean War became the subject of profound controversy when it was alleged in the United States that US Army prisoners were guilty of misconduct, character defects and poor discipline, all supposedly the results of decadent modern society. Australian prisoners had no such charges levelled against them and, given their small numbers and low death rate, their experiences are comparatively unknown.

No Australians were taken prisoner in the Vietnam War. The Australian method of operating and their choice of Phuoc Tuy province as their main battleground did not normally draw them into large-scale confrontation with the enemy. Australian airmen did not fly over North Vietnam and Laos where the majority of US prisoners were taken. There were, however, obviously situations in which Australians could have been taken prisoner and the fact that they were not can be put down simply to the fortunes of war. Six Australians were declared missing in action but all evidence suggests that they were killed in action with no known grave.

The place that POWs occupy in Australian collective memory owes much to the literature that the Second World War generated. The immediate postwar memoirs were immensely popular, with Russell Braddon's 'factional' *The Naked Island* selling over a million copies and Rohan Rivett's *Behind Bamboo* and R. H. Whitecross's *Slaves of the Sons of Heaven* being reprinted five and six times respectively. The

Australian official histories of the Second World War, unlike those of other Allies (with the exception of New Zealand) devoted considerable space to POWs. An unending flow of memoirs and publications in subsequent decades has confirmed the apparently inexhaustible public appetite for the subject. Dunlop's diaries have sold around 80 000 copies since their publication in 1986. Stan Arneil's diary, *One Man's War*, also a best seller, has been an optional text for English students in New South Wales and Victorian secondary schools. Combat novels, like Lawson Glassop's *We Were the Rats* and Eric Lambert's *The Twenty Thousand Thieves,* though their initial sales were immense, appear not have the same enduring appeal.

The images of Australian prisoners presented in these publications is at times ambiguous and negative, but the dominant image of captivity has been celebratory. From Rohan Rivett on, ex-prisoners of war have consciously or unconsciously integrated their experience into the heroic tradition of Australian military writing and into the Anzac legend. The experience of captivity therefore has reinforced the enduring mythology about the character of Australian servicemen: their supposed resourcefulness in situations of hardship and stress, their adaptability, resilience, superior social organisation and capacity for comradeship or mateship. An exhibit in the Australian War Memorial's prisoner-of-war gallery, 'explaining' the higher survival rates of Australian prisoners in the Pacific War, is illustrative: 'Australians', it is claimed, 'were more cohesive as a group, offering more support to each other during hard times, strengthening each other's will to see that distant day of liberation'. Such claims have not been, and indeed cannot be, proved empirically. Undoubtedly many Australians did manifest the qualities attributed to them during captivity; but so did prisoners of other nationalities, who of necessity or intuitively acquired the same skills of survival. But national mythology is not made of such admissions.

Patsy Adam Smith, *Prisoners of War From Gallipoli to Korea* (Viking, Melbourne, 1992); Joan Beaumont, *Gull Force: Survival and Leadership in Captivity* (Allen & Unwin, Sydney, 1988); P. J. Greville, 'The Australian prisoners of war' in Robert O'Neill, *Australia in the Korean War 1950–53*, vol. 2 (Australian War Memorial and AGPS, Canberra, 1985), ch. 23; John Hetherington, *Air Power over Europe 1944–1945* (Australian War Memorial, Canberra, 1963), ch. 19; Gavan McCormack and Hank Nelson, *The Burma–Thailand Railway* (Allen & Unwin, Sydney, 1993); Hank Nelson, *Prisoners of War: Australians under Nippon* (ABC, Sydney, 1985); A. J. Sweeting in Lionel Wigmore, *The Japanese Thrust* (Australian War Memorial, Canberra, 1957), chs 23–5.

JOAN BEAUMONT

PRISONERS OF WAR IN AUSTRALIA Though many 'enemy aliens' from Australia and overseas were interned in Australia during the First World War (see ALIENS, WARTIME TREATMENT OF), few if any were true prisoners of war. During the Second World War, however, the Australian government agreed in May 1941 to accept POWs from the Middle East, with the UK government taking financial responsibility. In May 1943 the government approved plans to transfer additional Italian prisoners to Australia from India. A total of 25 720 POWs were held in Australia during the Second World War: 1651 Germans, 18 432 Italians and 5637 Japanese. They were housed in specially built camps, guarded by CMF reservists, and treated in accordance with Geneva Convention rules. From May 1943 Italian POWs were employed as farm labourers without guards but under the supervision of nearby control centres staffed by army personnel. They were paid a minimum of £1 a week, and the scheme was generally considered a success both by the Government and by farmers. There were a number of escapes by POWs, but the only serious incident was the Cowra breakout (q.v.). The prisoners were not repatriated immediately after the war because of a lack of shipping, but all except a small number of escapees had been sent home by January 1947.

***PROTECTOR*, HMCS** see **SOUTH AUSTRALIAN NAVAL FORCES SHIPS**

PROVOST CORPS see **ROYAL AUSTRALIAN CORPS OF MILITARY POLICE**

***PSYCHE*, HMAS** see ***PELORUS* CLASS LIGHT CRUISERS**

PSYCHOLOGY CORPS see **AUSTRALIAN ARMY PSYCHOLOGY CORPS**

PUCKAPUNYAL The area around Seymour in Victoria has been used for military purposes since the First World War, when a training area was located south-east of the town and later purchased under the Land Acquisition Act. An ordnance store area, known variously as Mobilisation Siding and, from 1965, Tel el Kebir Barracks, was constructed to the east of Seymour in 1921, while in 1924 a rifle range was built and continued in use until the 1950s. Puckapunyal itself became a camp at the beginning of the Second World War, when once again the existing training areas were found to be inadequate. Over 14 000 acres were purchased, supplemented by further acquisitions during the war and periodically thereafter. In 1988 the training area consisted of 39 290 hectares. Puckapunyal is the main army facility in Victoria, and provides the army with extensive range and training facilities which are under no threat of urban encirclement; this has obvious advantages for training but makes postings there less than highly prized. Over time Puckapunyal has been home to the 1st Infantry Brigade, the 3rd National Service Training Brigade during the national service (see CONSCRIPTION) scheme of the 1950s, the Royal Australian Army Service Corps (q.v.) Centre, and school cadets during school holiday camps, while during the Second World War the US 41st Infantry Division trained there before being sent north to Papua. The 1st Armoured Regiment was based there until June 1995; the Armoured Corps Centre, the School of Transport and the School of Catering remain there. Heavy use over many decades led to serious land degradation, and in the 1980s the army embarked on a successful program of rehabilitation and tree-planting.

'Q' CLASS ANTI-SUBMARINE FRIGATES Displacement 2020 tons; length 359 feet; beam 36 feet; speed 32 knots; armament 2 × 4-inch guns, 2 × 40 mm anti-aircraft guns, 2 × Limbo (q.v.) or Squid anti-submarine mortars.

The 'Q' Class destroyers (q.v.), HMA Ships *Quadrant*, *Queenborough*, *Quiberon* and *Quickmatch*, were converted to Type 15 anti-submarine frigates for use in anti-submarine warfare (q.v.) between 1950 and 1957. The ships exercised in British and Far Eastern waters with British and American units. *Quickmatch* and *Queenborough* engaged in shore bombardment during the Malayan Emergency (q.v.); *Quickmatch* was employed as an accommodation ship from 1963. *Queenborough* was used as a training ship from 1966 to 1972. HMAS *Quadrant* was scrapped in 1963, the other ships in the 1970s.

'Q' CLASS DESTROYERS Laid down 1940–41, launched 1941–42; length 359 feet; beam 36 feet; displacement 1705 tons; armament 4 × 4.7-inch guns, 1 × 2-pounder pom pom guns, 6 × 20 mm anti-aircraft guns, 4 × depth-charge throwers, 8 × 21-inch torpedo tubes; speed 34 knots.

HMA Ships *Quiberon* and *Quickmatch* were lent to Australia by Britain in 1942. They saw service in the Mediterranean, where HMAS *Quiberon* helped sink the Italian submarine *Dessie* on 28 November 1942, and in the South Atlantic, Indian Ocean, Netherlands East Indies and Australia. Their main duties were convoy escort and anti-submarine patrol. After the Second World War they served in Japanese waters as part of the British Commonwealth Occupation Force (q.v.). *Queenborough*, *Quality* and *Quadrant* were lent to the RAN in 1945, serving in postwar operations around Australia and New Guinea. After being transferred outright to the RAN in 1950, all except HMAS *Quality* were rebuilt as 'Q' class anti-submarine frigates (q.v.).

QANTAS AT WAR Qantas Empire Airways began using flying boats in July 1938 when it inaugurated the imperial 'All-up Mail Scheme'. The airline became directly involved in the war effort in 1941 when Qantas aircrews ferried 19 Catalinas from Honolulu to Sydney, opened a new air service to Dili, Timor, in January 1941, and relieved pressure on the British Overseas Airways Corporation operating the 'Horseshoe Route' (Sydney–Durban, South Africa, via Cairo) by agreeing to take over the leg from Singapore to Karachi. In January–February 1942 Qantas flying boats flew many missions evacuating civilians and some service personnel from Singapore and Java, and in May 1942 Qantas organ-

ised the successful evacuation of 78 civilians from behind Japanese lines at Mount Hagen, New Guinea. By March 1942 it had lost three of its 10 flying boats to enemy attack. Although several flying boats were taken over by the RAAF, Qantas as a whole remained an independent commercial entity, but one increasingly involved in the war effort. Its repair and maintenance facilities, which by September 1942 were employing 450 men and women, serviced RAAF and US Army Air Force aircraft until July 1944, when the front line had moved too far forward to make the use of Qantas bases practicable. The Indian Ocean Catalina service, which began in August 1943, remained in operation until July 1945, by which time it had carried 858 passengers, 90 793 pounds of priority war cargo and 207 260 pounds of mail in a total of 271 crossings. Twenty-one Qantas staff were killed during the war, with another 10 killed on active service with the RAAF.

QUADRANT, HMAS see **'Q' CLASS ANTI-SUBMARINE FRIGATES; 'Q' CLASS DESTROYERS**

QUALITY, HMAS see **'Q' CLASS DESTROYERS**

QUEENBOROUGH, HMAS see **'Q' CLASS ANTI-SUBMARINE FRIGATES; 'Q' CLASS DESTROYERS**

QUEENSLAND MARINE DEFENCE FORCE SHIPS
When the Queensland Marine Defence Force was established in 1883 the government ordered two gunboats and one torpedo boat for its new navy. The third-class steel twin-screw gunboats *Gayundah* and *Paluma* (allegedly Aboriginal words for lightning and thunder) were built in Britain and arrived in Australia in March and May 1885. Both ships sat very low in the water, had a tall single funnel amidships and had two masts (though sail was used only as an auxiliary). Both were based in Brisbane; *Gayundah* was used for training cruises along the coast, while *Paluma* was used for survey work with the RN. In 1893 the flooded Brisbane River dumped *Paluma* in the botanical gardens where it was grounded until another flood refloated it two weeks later. *Gayundah* and *Paluma* were out of commission from 1893 to 1898 and 1895 to 1899 respectively. In 1903 *Gayundah* was used for the first Australian experiments in ship-to-shore wireless telegraphy. The two ships were integrated into the RAN in 1911, but *Paluma* was little used and was sold in 1916. *Gayundah* continued to be used for patrols and training; it was paid off and recommissioned twice during the First World War before being finally paid off in 1918 and sold in 1921. The Queensland Navy also pos-

Her Majesty's Queensland Ship *Paluma*, Brisbane River, 1895–1899. (AWM 300020)

sessed the British-built steel second-class torpedo boat *Mosquito* and wooden turnabout torpedo launch *Midge*, which arrived in Brisbane in October 1884 and June 1888 respectively. Both were used in exercises and for general defence before 1913 when *Mosquito* was hulked and *Midge* was sold. Eleven auxiliary vessels, mostly armed, also served with the Queensland Marine Defence Force at different times. Queensland naval ships carried the prefix HMQS and, unlike the ships of the other Australian colonies, both of Queensland's gunboats flew the RN's white ensign (q.v.). (See also COLONIAL NAVIES.)

Gayundah and *Paluma*: displacement 360 tons; length 120 feet (overall); beam 26 feet; draught 9.5 feet; speed 10.5 knots; armament *Gayundah* originally carried 1 × 8-inch gun, 1 × 6-inch gun, 2 × 1.5-inch guns, 2 × machine-guns; 1899 — 1 × 8-inch gun, 1 × 4.7-inch gun, 2 × 12-pounders, 2 × machine-guns; 1914 — 1 × 4.7-inch gun, 2 × 12-

pounders, 1 × Maxim machine-gun, 2 × Nordenfelt machine-guns; *Paluma* originally carried 2 × 1.5-inch Nordenfelt machine-guns, 2 × machine-guns; 2 × 5-inch guns added 1899; 1 × 6-inch gun added 1901; 2 × 12-pounders added 1903; 1 × 4.7-inch gun replaced 6-inch gun 1905.

Mosquito: displacement 12 tons; length 63 feet; beam 7.5 feet; draught 3 feet (aft); speed 16 knots; armament spar torpedoes, 1 × machine-gun, later 2 × 14-inch torpedo-dropping gear.

Midge: displacement 11 tons; length 65 feet; beam 11 feet; draught N/A; speed 16.5 knots; armament 1 × 6-pounder, 2 × 14-inch torpedo-dropping gear, 2 × machine-guns.

QUIBERON, HMAS see **'Q' CLASS ANTI-SUBMARINE FRIGATES; 'Q' CLASS DESTROYERS**

QUICKMATCH, HMAS see **'Q' CLASS ANTI-SUBMARINE FRIGATES; 'Q' CLASS DESTROYERS**

R

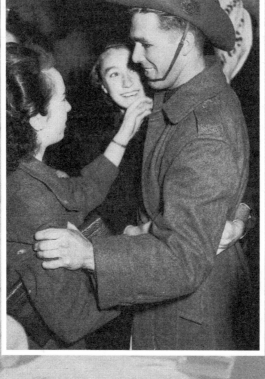

RABAUL From December 1941 to March 1942, Japanese armed forces, enjoying command of the seas, overwhelmed isolated garrisons as they steadily moved south, west and east. High on their list of priorities was Rabaul on the island of New Britain. Rabaul was the centre of administration of the Mandated Territory of New Guinea but, from a Japanese perspective, its attraction lay in the fact that it had a number of airfields and perhaps the best natural harbour in the south Pacific. Its capture would give Japan an indestructible south-eastern corner to its new defensive perimeter and provide defence in depth to the Combined Fleet's base at Truk in the Caroline Islands.

As a result of the Singapore Conference of February 1941, and discussions with the Australian War Cabinet (q.v.) during Air Chief Marshal Brooke-Popham's visit to Australia in March, the 2/22nd Battalion AIF garrisoned Rabaul in the following month. Its tasks were to protect the airfields at Lakunai and Vunakanau, the seaplane base in Simpson Harbour, and to form part of 'an advanced observation line', to give early warning of enemy movements. As events subsequently demonstrated, the defending force was inadequate for its task. Its virtue, apart from being a demonstration of Australia's overall desire to assist in its own security along the 'Malay Barrier' and help the Dutch colonial interest, was as an early warning station. This task, however, could have been performed by coastwatchers with radio sets.

The view of the Australian chiefs of staff at the time was that, although it was unlikely that Rabaul could be defended against the scale of attack liable to be brought against it, the enemy should be made to fight for it, rather than just occupy it. Lieutenant-General V. A. H. Sturdee, the Australian CGS, reluctantly agreed to the commitment of a battalion. Colonel Scanlan arrived from Australia to command the base on 8 October 1941. During December an RAAF force of four Hudson aircraft and 10 Wirraways commanded by Wing-Commander Lerew was deployed there. The anti-aircraft defence of the airfields depended on two guns. Other elements of Scanlan's force, in addition to the 2/22nd Battalion, were the coastal defence battery of two 6-inch guns and searchlights, a detachment of the New Guinea Volunteer Rifles, an anti-tank battery, and a detachment of the 2/10th Field Ambulance with six nurses.

Almost immediately, Japanese high level bombers began to reconnoitre the area. Scanlan believed he would need a brigade group to defend Rabaul and began to plan for it. In the meantime the commanding officer of the 2/22nd Battalion, Lieutenant-Colonel Carr, deployed his troops to cover the main approaches. The first Japanese air raid began on 4 January 1942 and within days the defending aircraft had been all but eliminated. Thereafter, Japanese air attacks became almost leisurely as they destroyed ships in the harbour and installations. From 21 January 1942, when Scanlan recognised that invasion was imminent, he redeployed some forces. Lerew's remaining aircraft were flown out to Lae and the airfields cratered by exploding unused bombs.

The Japanese force that attacked Rabaul on 23 January 1942, the South Seas Force (Major-General Horii), had already seized Guam in December 1941. It had expected strong resistance at Rabaul; consequently its naval covering force had been reinforced by battleships and two aircraft carriers. Prior to landing troops, Japanese planes heavily bombed and machine-gunned the Australian defences. They silenced the coastal defence battery which Scanlan then ordered destroyed. He also ordered that demolitions be carried out and the township evacuated.

Scanlan's problem in the defence of Rabaul was, in a sense, a microcosm of Percival's much larger problem of the defence of Malaya (see MALAYAN CAMPAIGN). Once the defending aircraft were evacuated and the airfields no longer needed, not only did they have to be denied to the enemy, but the forces originally there to defend them were now in the wrong place and too dispersed. Moreover, his efforts to redistribute his scattered force just before the Japanese assault meant that the new positions were less well prepared than they might have been.

Horii began landing at 2.45 a.m. He landed the 1st Battalion 144th Infantry Regiment at Praed Point and the 2nd Battalion near Nordup. Both battalions then advanced rapidly to occupy Lakunai airfield and Rabaul township. The 3rd Battalion was landed near Vulcan crater and thrust towards Vunakanau airstrip. The Australian defenders in each area put up what opposition they could but, as at Ambon and Timor, the enemy's complete command of the sea and air, as well as a landing force in brigade strength, made their position desperate from the start. By 8 a.m. all of the localities as well as Scanlan's position near Tomavatur were in danger of being overrun. Eventually, Carr was given discretion to break out in small groups. Thus, except for mopping-up operations, the Japanese had captured Rabaul in a single day.

For the Australians, the sequel was dispiriting in the extreme. No forward plans had been made for the 'every man for himself' withdrawal that Scanlan had authorised. No dumps had been established along the routes out of Rabaul, nor had men been taught to live off the land. The cessation of military control led to the survival of the fittest, or luckiest. Some moved south-east into the Gazelle Peninsula, others south-west. Some tried to make contact with friendly locals, some tried to get away by sea. The Japanese distributed messages advising men to surrender.

Scanlan's force on 23 January 1942 was 1396 strong. Casualties on that day were two officers and 26 men killed. About 120 members of the RAAF were taken off in flying boats on 23–24 January, which prompted the official historian to comment upon what could have been achieved had there been a proper plan. About 400 escaped successfully; 800 became prisoners. Some prisoners, including officers, nurses and civilians, were shipped to Japan. About 160 prisoners were massacred at the Tol plantation.

For their part, the Japanese captured Rabaul at remarkably little cost. Their losses were 16 killed and 49 injured. They proceeded to transform it into a massive sea and air base which became the main object of Allied strategy for the next two and a half years. At the end of the war, more than 90 000 Japanese soldiers and sailors surrendered there.

JOHN COATES

RADAR see **SCIENCE AND TECHNOLOGY**

RADFORD–COLLINS AGREEMENT Negotiated by the Australian CNS, Vice-Admiral Sir John Collins (q.v.), and the C-in-C, Pacific, Admiral Arthur W. Radford, in 1951, this agreement provides for the protection and control of shipping in wartime. The agreement, and the negotiations which preceded it, were a product of the early Cold War and were given a considerable fillip by the ready Australian response to the outbreak of the Korean War. The principal area of cooperation between Australia and New Zealand was the ANZAM (Australian, New Zealand and Malaya) area, and the United States judged as desirable some form of cooperation between maritime forces in the ANZAM and CINCPAC areas in the event of a general East–West conflict. The negotiations between Radford and a joint Australian, New Zealand and British delegation, led by Collins, took place between late February and early March 1951. The agreement which resulted was more limited in scope than some ministers in the Menzies government had hoped, confining itself essentially to the defence of the lines of communication in the ANZAM area. While this was of considerable importance in terms of future naval cooperation between the RAN and the US Navy, it prompted the Australian government to push for a formal peacetime security agreement with the United States, which was to result in the ANZUS Treaty (q.v.). The provisions of the Radford–Collins agreement remain secret, and it retains a current role in defence planning with the United States.

Thomas-Durrell Young, *Australian, New Zealand and United States Security Relations, 1951–1986* (Westview Press, Boulder, 1992).

RAMSAY, Major-General Alan Hollick (12 March 1895–19 September 1973) enlisted as a gunner in the AIF in 1915 and served in the 22nd and 4th Field Artillery Brigades. He was commissioned in January 1919. A school teacher in civil life, Ramsay maintained an interest in soldiering through the CMF between the wars, serving in artillery appointments and commanding the 10th Field Brigade between 1930 and 1936 and the 4th Division artillery in 1939. Seconded to the 2nd AIF at the outbreak of war, he initially took a step down in rank to command the 2/2nd Field Regiment, but was quickly promoted to colonel and appointed to command the medium artillery on the headquarters of I Australian Corps. He commanded the 9th Division artillery for Lieutenant-General Leslie Morshead (q.v.) from October 1940 to June 1943, although for a time he and his guns were separated from their parent formation. While Morshead and the division were garrisoning Tobruk (q.v.), Ramsay and his artillery units were providing fire support for the 1st South African Division in the defences around Mersa Matruh in Egypt. During the First Battle of Alamein in July 1942, in which the 9th Division played a conspicuous part, Ramsay commanded not only his own field regiments, but a British medium regiment and the guns of the 1st South African Division as well. At the Second Battle of Alamein, in October, he again commanded an augmented artillery force, this time comprising 12 field and three medium regiments, with a total of 360 guns. Morshead thought sufficiently highly of him to recommend that he command the 9th Division in the event of Morshead's becoming a casualty. He was awarded the DSO for service in the Middle East in 1942, and was mentioned in despatches three times during the course of the war. After commanding the II Corps artillery in the latter half of 1943, he was given command of the 5th Division, in succession to Major-General E. J. Milford (q.v.), in January 1944, and led it for the remainder of the war, through the fighting in the area around Madang and then in the campaign on New Britain from late 1944. Returning to civil life after the war, he was Director of Education in Victoria from 1948 to 1960, and was knighted in 1961.

RAN CLEARANCE DIVING TEAMS Diving has a number of offensive and defensive applications in naval warfare, and ships' divers were trained in the RAN at HMAS *Cerberus* (q.v.) from the early 1920s. The Clearance Diver category was established in the RAN in 1951 and given bomb and mine disposal (now termed Explosive Ordnance Disposal or EOD) as its primary function. The Clearance Diving School was located in Sydney Harbour until 1956, then moved to HMAS *Rushcutter* before settling at HMAS *Penguin* in 1968. Navy clearance divers were called on from time to time to assist in civilian tasks: clearing blockages in the underwater tunnels of the Bunnerong Power Station in Sydney; helping to overcome sealing problems in a diversionary tunnel of the Snowy Mountains Scheme; or assisting in the salvage operations after Cyclone Tracy, for example. Operationally, Clearance Diving Teams (CDTs) were deployed in the Vietnam War from February 1967, and during the Gulf War in 1991. In Vietnam CDT3 was involved in the searching of ships in Vung Tau harbour for mines placed by enemy sappers (called 'swimmers'), salvage, and explosive ordnance disposal in the Mekong Delta. It was awarded a United States Navy Meritorious Unit Citation. CDT3 was again deployed to the Gulf, working with the US Navy in ship protection tasks. Following the Iraqi surrender, the team was used to help clear the ports and beaches of Kuwait of the huge quantities of explosive ordnance placed there by the Iraqis during their occupation. In the 1990s there are three clearance diving teams in the RAN: CDTs 1 and 2 are based in Sydney, CDT4 in Perth. CDT3 was formed for overseas service in Vietnam in 1967 and re-formed for service in the Gulf in 1991.

RAN HELICOPTER FLIGHT VIETNAM (RANHFV) The only deployment of Australian naval aviation during the Vietnam War involved helicopter flight crews who served with the US 135th Assault Helicopter Company and No. 9 Squadron RAAF. The Helicopter Flight Vietnam converted to UH 1B Iroquois helicopters (q.v.) and arrived in South Vietnam in October 1967. It was integrated with the American company and flew operations in IV Corps Tactical Zone, based on the Mekong Delta. Support missions were flown frequently for the Army of the Republic of Vietnam and involved fire support, troop insertion and casualty evacuation. From November 1968 they also supported units of the Royal Thai Army. Four contingents served until withdrawal in June 1971 in what were frequently dangerous combat circumstances, since the enemy made a particular point of engaging helicopters

with ground fire. Five aircrew were killed in action, and others wounded. The placing of Australian aircrew into what was then dubbed an 'experimental military unit' was generally regarded as a success, although the Australian government's restrictions meant that during operations into Cambodia in 1970 RAN flight crew were grounded while their American colleagues flew missions. The flight's parent unit, No. 723 Squadron of the Fleet Air Arm, was awarded the battle honour (q.v.) 'Vietnam 1967–71'. The detachment serving with No. 9 Squadron flew in support of Australian ground forces in Phuoc Tuy province between February 1968 and May 1969.

RANK STRUCTURE see **OFFICER RANK STRUCTURE**

RANKIN, **HMAS** see *COLLINS* **CLASS SUBMARINES**

RATIONS Feeding the armed forces has always presented enormous problems of processing, packaging, transportation and storage, and before the Second World War the primitive state of the available technology meant that the food of the Australian armed forces remained at a fairly constant level of monotony. Furthermore, because scientific information about nutrition was not widely available, military authorities often did not understand the importance of providing troops with a balanced and varied diet. In 1885 a member of the New South Wales contingent to the Sudan reported receiving rations of 'dry biscuits and 1 lb tinned meat' while in the field, and even in camp they were given little more than tinned meat, vegetables and biscuits. Thirst was an even bigger problem than hunger in the Sudanese desert, and concerns about the safety of local water supplies meant that water had to be carried to the troops by rail and camel. In the Boer War the daily ration was four hard biscuits, a tin of bully beef (salted preserved meat) and a small quantity of tea or coffee, but troops on the march could receive even less than this. They were, however, able to supplement their diet by 'commandeering' food from Boer farmers, though such actions were officially prohibited.

By the time of the First World War the food given to Australian troops was based on the British ration scale, calculated to provide about 4000 calories a day. At Gallipoli, where troops cooked for themselves, C. E. W. Bean (q.v.) reported that daily rations were 12 oz preserved meat, 1.25 lbs biscuits, 4 oz bacon, 3 oz cheese, 8 oz onions (or onions and potatoes), 4 oz jam, 3 oz sugar, as well as tea, salt, mustard and pepper. In France (see WESTERN FRONT) the AIF had kitchens at the rear of the lines. Food was brought as far forward as possible in

Private F. R. Cowley of the 2/13th Battalion pours baked beans into a hot-box to be taken to forward troops, Lutong, Borneo, 5 July 1945. (AWM 111413)

horse-drawn wheeled boxes, carried the rest of the way by ration-carrying parties who had to avoid such dangers as shell holes, wire and enemy fire, then dumped at Company Headquarters where it would be collected by small parties from each platoon. The food delivered in this way, according to Bean, consisted of hot tea three times a day and hot meat or stew in the evening. Breakfast was tea and 'a slice of luke-warm bacon on bread or biscuit'; lunch was tea with bread, jam and cheese. When food from the cookhouse could not reach the front lines the troops were forced to cook their emergency rations (bully beef and biscuits legendary for their hardness) in 'dixies' (cooking pots) over coal or charcoal fires. While this diet was sufficient to keep the troops reasonably healthy it had a bad effect on morale. Amid the appalling conditions at Gallipoli, in particular, Bean found that 'the troops sickened of their unchanging ration. They had little appetite for the over-salted "bully", which, in the heat of midday or afternoon, slipped in its own fat across the platter or mess-tin, swamping stray flies as it went; or for the thin apricot jam on tasteless biscuit; or for the cheese, greasy from exposure to the sun and filling the dugout with an odour sicken-

ingly reminiscent of that exhaling from the corpses in No-Man's Land'. Some relief from the nauseating tedium of this diet came in the form of Australian Comforts Fund (q.v.) food parcels, food bought from army canteens, generous feeding in billets during periods of recuperation away from the front lines, and local foods which were sometimes obtained by theft.

Between the wars there were two rations scales, one for the Permanent Military Forces and one for the militia (see CITIZEN MILITARY FORCES), with the latter able to be commuted to money and used to buy food locally. When the 2nd AIF was established it was fed in the same way as the militia, with results that were often nutritionally unsatisfactory. Cooking arrangements were a unit responsibility and cooks, who were chosen by the unit commander, had very low status and were the butts of frequent jokes. Early in the war Sir Stanton Hicks, Professor of Human Physiology and Pharmacology at the University of Adelaide, was appointed to advise the army's Supply Services on nutrition and began campaigning for the application of scientific knowledge to army catering. He finally persuaded military authorities to set up the Australian Army

Catering Corps (q.v.) in 1943, and became the Chief Inspector (later Director) of Army Catering. The corps' role was to use nutritional science in reforming the feeding of soldiers, to make sure food was used correctly without waste, and to provide the army with technical advice and technicians. To these ends, soldiers were educated about nutrition, and the status of cooks was raised by improving their training, pay and promotional opportunities. On examining the army's ration scales Hicks found they contained too much fat and meat while they were deficient in fruit and vegetables, so a new ration was compiled which supplied 4500 calories and contained all essential food items. Improvements in food technology meant that canned fruit and vegetables could now be prevented from spoiling without significant loss of nutritional value or flavour, and with the adoption by the army of the mobile Wiles steam cooker food could be cooked quickly in such a way that its value was retained and the danger of burning food was eliminated. These advances, however, were in practice not always sufficient to alleviate a dull and monotonous diet on campaign in which fresh rations were still rare.

There was a much greater variety of food available to soldiers in the Second World War than in previous wars, and surveys conducted at the time showed that on the whole troops found their diet acceptable. As in the First World War, they were often able to supplement their diet with local foods and Australian Comforts Fund parcels. There were, however, major problems to be overcome which were caused by the tropical jungle conditions in which many Australians were fighting. Great advances had been made in canning and dehydration, but it was still important for morale to supply troops with fresh food, and the difficulties of transport and storage of fresh food in the tropics were only partly overcome by the establishment of army farms (q.v.). Transporting food, either fresh or processed, to soldiers in forward areas in the jungle was also a great challenge. The Air Maintenance Organisation dropped food to troops in inaccessible areas and when supplies could not get through at all they were able to fall back on operational rations. First used in 1943, these compact ration packs consisted of three ready-to-eat meals sealed in tin containers, and were used extensively in forward areas, where they met with general approval. Two such packs could sustain a soldier doing moderate work for three days. Only in exceptional cases, such as that of the Independent Companies (q.v.), were Australians completely cut off from supplies and forced to rely entirely on local foods.

After the war, major work on army food resumed in 1954 when development of combat ration packs began in Melbourne. Over the next four years emergency rations, 24-hour one-man rations, and 10-man composite rations were all developed. Other defence food science projects were also being carried out at this time, but in 1958 it was decided to end all such research except that connected with the development of rations for the armed forces. An existing research station at Scottsdale, Tasmania, was made the centre for this work and was placed under army control, though it could also undertake research for the other services. From 1971 the centre has been known as the Armed Forces Food Science Establishment (AFFSE), and in 1975 it came under the control of the Defence Science and Technology Organisation (q.v.). Its main role is to determine the energy and nutrient requirements of service personnel and to translate these requirements into ration scales and packs. The ways in which it carries out this function include experimenting with new techniques of food manufacture; performing chemical, microbiological and other kinds of scientific analysis of food; conducting troop-feeding trials under combat conditions; and producing food components when this cannot be done economically by commercial manufacturers. The Army School of Catering has been based at Puckapunyal (q.v.) since 1974.

Until the early 1970s the army provided food and board to its personnel, paying an allowance to those who lived off base, but now soldiers who decide to live on base pay a rationing and quartering allowance which allows them to eat three hot meals a day at the mess. Other soldiers choose to provide their own accommodation and food. For those being fed by the army, the Australian Defence Force Ration Scales set out the maximum quantities of various foods which each person may draw per day. The maximum energy intake per day under this scale is 17 400 kilojoules (4156 calories), which is considered enough for service people engaged in normal activities. In the field during exercises the army tries to provide at least one meal using fresh foods each day if the operational situation allows. Mobile field kitchens are used, and if necessary meals can be taken to the troops in aluminium 'TV dinner' trays. Operational ration packs, issued when the provision of fresh food is impractical, are not intended to be used for more than seven days without being supplemented by fresh food. There are two main rations that provide food for one soldier for one day: the Combat Ration One Man which consists mainly of two canned main meals and various snack foods, and the lighter Patrol Ration One Man which consists of two freeze-dried main meals and various supplementary items in lightweight packaging. There is also the Combat Ration Ten Man for group feeding.

The RAAF has its own catering staff and a catering school at Wagga Wagga, but it uses the same ration scale as the Australian Army. The RAN, on the other hand, has a completely different feeding system. The 'general messing' system introduced in 1912 involves the setting of a daily victualling allowance, the amount of money to be spent on food per person per day. Cooks on each ship and at each naval establishment devise menus to fit the total amount of money available. At first goods were issued at the actual local costs, so special allowances had to be given to ships operating in areas where food was more expensive, but from 1918 fixed prices have been laid down for all essential foodstuffs. Until 1943, when they were put under the general messing system, smaller ships used a standard ration system like that of the army. During the Second World War RAN victualling compared favourably with that of the RN, but unfavourably with that of the US Navy, whose storage and refrigeration facilities were superior to the RAN's. The naval school of cookery was established at HMAS *Cerberus* (q.v.) in 1924, and the Navy Victualling Department was using its laboratory in Melbourne to check food specifications before the Second World War. Although the victualling allowance is not based on nutritional ration scales, it is calculated to provide all naval personnel with a balanced diet. Fresh food is used wherever possible and proposed menus must be approved by medical officers. The navy has been working with AFFSE to develop a better nutritional basis for victualling.

RAWLINGS, Private William see **ABORIGINES AND TORRES STRAIT ISLANDERS IN THE ARMED FORCES**

READY RESERVE see **ARMY RESERVE**

RED CROSS see **AUSTRALIAN RED CROSS**

RED FLAG RIOTS were a series of attacks in 1918–19 by Queensland conservatives, especially ex-servicemen, on 'disloyal' elements. The most dramatic outbreak of violence occurred in Brisbane on 23 and 24 March 1919, when thousands of 'loyalists', angered by a civil liberties march by socialists carrying red flags, clashed violently with police and terrorised the local Russian community. The main outcome of the riots was an increase in the popularity among ex-servicemen of the conservative Returned Sailors' and Soldiers' Imperial League of Australia (see RETURNED AND SERVICES LEAGUE), which in April 1919 formed one of Australia's first private anti-communist armies in Brisbane (see SECRET ARMIES).

Raymond Evans, *The Red Flag Riots* (University of Queensland Press, Brisbane, 1988).

REGIMENTAL ALLIANCES Alliances between Australian army units and those of other countries, almost always Britain, began in the 1920s. Many CMF units were twinned with British regiments, often on a highly arbitrary basis such as a common numerical designation (the 18th Light Horse Regiment was allied to the 18th Royal Hussars, for example). Most alliances had no obvious basis in shared experience, or even in commonality of purpose, there being little obviously in common between, for example, the Sydney University Regiment (q.v.) and the King's Royal Rifle Corps, now the Royal Green Jackets, with whom they were allied. Exchanges of personnel were rare, and the main purpose appears to have been ceremonial: some militia battalions, for example, adopted the regimental marches of their longer-established 'twin'. In the regular army the various corps are affiliated with their British Army opposites, while the battalions of the Royal Australian Regiment (q.v.) have individual affiliations as well as a regimental one. Such arrangements are usually based on considerations of morale and *esprit de corps;* there seems little obvious connection between these sorts of traditions and the military efficiency of the Australian soldier.

RELIGION AND WAR The place of religion in Australia has been a subject long and hotly debated. Some have drawn comparison between the reasons for the European settlement of Australia with other parts of the New World, arguing that as religion played no part in that decision in Australia, its status was consequently reduced. On the other hand, war has played a very important role in the development of an Australian consciousness. Some churchmen, perhaps anticipating this, saw war as an opportunity for the churches to become more closely aligned with Australian sentiment while others have seen the bonds created for Australians by the experience of war as replacing the need for formal religion.

The decision for war and the conduct of war have traditionally been areas where religious thinking has been influential. For much of Christian history a tension has existed between pacifists who believed that the gospels forbade violence between peoples and those who believed that the State had a duty to defend its own and to expand its territories for godly reasons. The latter predominated and

Christian thinkers such as Thomas Aquinas codified the justification for war within a neat set of guidelines designed to absolve the consciences of those whose message was peace in the context of near permanent warfare in the Christian era.

There has been little pacifism in the mainstream Australian Christian churches at any time in their history although genuine pacifist ministers such as the Methodist Linden Webb were treated with regard and respect. Rather war in Australia has tended to cause Christians more political than religious problems. The resolution of these political issues affected both perceptions within and outside the church and the influence of individual churches in ways that produced longstanding and important results. A discussion of religion and war in the Australian context will necessarily deal more with politics and influence than it will with individual conscience or theological concepts.

Despite some difficulties for ministers and churches over the war in South Africa, Christians embraced the news of the outbreak of war in Europe in 1914 with unseemly enthusiasm and excitement. In this they differed little from the rest of the community but while 'war fever' reigned it might have been more sensible for Christian spokesmen to have recommended and displayed some restraint, at least until the issues had become clearer. Clergymen pronounced themselves certain that war would cleanse Australian society and produce a more noble people, a people more accustomed to sacrifice than to ease, a people more spiritual than material, a people able to recognise the importance of a cause and the need to give their all for it.

Such was the strength of these feelings that an Anglican newspaper, lamenting that Australian troops had been held in reserve and in training in Egypt while the war waged in Europe, feared for the spiritual reformation of the troops and through them of the nation. Publishing on 30 April 1915, that is, five days after the Anzac landing at Gallipoli, and within a few days of the news reaching Australia, the editor called for Australian troops to suffer 'a shattering and sledge-hammer blow' so that Australians would appreciate war's 'cleansing, its refining, or its spiritual revelation'. Blessed with hindsight, and in the knowledge of the thousands of acres of ground occupied by First World War cemeteries, it is hard to read this statement in the way that it was doubtless intended. Even so, in its callous disregard for the sufferings and deaths of some, it indicates the strength of a Christian view that from evil good could come and that the war must have been ordained by God to lead Australia to a purer and better way.

Chaplains quickly learnt another story. It was the Western Australian Catholic chaplain, John Fahey, disregarding orders and approaching the Gallipoli shoreline with the first wave of Anzacs, who was the first Australian chaplain in this war to discover what war meant in reality. 'The sailor in the stern was hit first, then another fell across me; then an oarsman dropped his oar and fell to the bottom of the boat … it was horrible … you just had to sit there and wait for your bullet.' Fahey, who had heard the confessions of all the Catholics in the 11th Battalion before they had gone ashore, was busy by the end of the day burying those who had died struggling to take the heights above him. Here was war's 'shattering, sledge-hammer blow'. So intense was the regard these chaplains developed for the men they served that when he heard the orders for the evacuation of Gallipoli nine months later, Chaplain Walter Dexter, an Anglican from Melbourne, paraded before General Birdwood and asked if he might be left behind to tend the graves of the Australians he and the others had buried to protect those graves from the expected desecration by the Turks. Denied his request, Dexter spent his last few hours on the peninsula planting wattle seeds so that there might be some reminder of Australia shading the graves of those who had given their lives for their country.

At home the clergy read accounts of these doings in the same church newspapers that were printing the sermons of their leaders which called for greater sacrifice to make sense of the sacrifice already incurred. Perhaps more than most on the home front, however, these clergymen witnessed the cost of the war to families and individuals as they delivered the casualty telegrams telling of death at the front. Asked by the government to perform this uncomfortable task, church leaders somewhat recklessly agreed. It showed, the bishops and moderators argued, the importance of the church to national life at a time of crisis. It gave the churches, they hoped, an opportunity for spiritual counselling and involvement at a time when people might be well disposed to hear words of hope and encouragement. The danger was that the churches would be seen as opportunistic, imposing an unwanted and embarrassing presence at a time of grief when families might well seek privacy.

And so it proved. Before long clergymen came to regret their leaders' easy acquiescence in the government's plans. There would have been few streets in suburban Australia by late 1916 from which at least one young man had not enlisted. The presence of a clergyman in the street, prominent in his clerical clothes, and almost invariably on foot, was enough to turn householders to panic and, should he call at the door, to terror. Clergymen reported that householders would faint away when the door was opened to them, even if their duty was no more sinister than that of regular parish visiting. Some clerics took to wearing differently shaped hats, or other easily identifiable features, when delivering casualty telegrams, to make the rest of their work endurable. They reported, too, that rarely did the householder seek words of spiritual comfort at such time, retreating instead to private grief.

Despite the mounting evidence of war's awful cost, church leaders called for greater sacrifice and joined wholeheartedly with governments in recruiting campaigns, hoping to show those who had not enlisted that only through sacrifice could they be saved. Some extraordinary arguments were heard from clergymen on these recruiting platforms and name-calling and invective replaced rational discussion in many cases. Having developed an expectation that every fit young man should enlist, clergy counted it as failure if only a percentage of the 'eligibles' in the audience came forward to the recruiting sergeant. Frustrated in their unrealistic expectations, many of these clergy were among the first to call for conscription as the only reliable means of reinforcing those at the front.

All of the major Christian churches embraced conscription even before Prime Minister William Morris Hughes (q.v.) had decided to seek the people's endorsement of this means of recruitment. Church leaders spoke of conscription as representing equality of sacrifice, as necessary to prevent a German victory which would threaten Christian values, and as honourable in view of Australia's earlier total commitment to the Empire's war effort. It is likely that Hughes decided on a referendum on the issue in part because of the strong support he had received from church leaders and synods. Only Melbourne's Catholic archbishop Daniel Mannix disrupted the unanimity of church leaders, and even his own church in 1918 sought ways of restraining Mannix's outspoken anti-government comments. While it is difficult to explain the notoriety that Mannix achieved for his hesitant and brief intervention in the first conscription referendum in 1916, it is probable that his prominence derived from the almost total absence of any other leading figure outside politics and the unions prepared to oppose the government and the anticipated significant 'yes' majority.

The defeat of conscription shocked Australian opinion-makers whose simplistic analysis identified Mannix and the 'Irish' Catholics as the leading factors in the 'no' victory. Determined to root out 'disloyalty' in an effort to show support for the troops at the front, pro-conscriptionists attacked those whom they had assumed had voted 'no'. There developed in Australia a remarkable level of sectarianism and bigotry that would come to bedevil Australian political and social life for most of the next half century. Ironically on the battlefront Australians learned to live in harmony with those from differing religious and social backgrounds because the troops genuinely believed that they had discovered a basis for a new and more cooperative Australia. Examples of sectarianism among the chaplains were few and the better chaplains quickly learned that the troops expected cooperation where in Australia even before the war there had been disharmony and suspicion. A South Australian Catholic chaplain admitted that he had never before spoken to any of the Protestant ministers working in his small home town before the war and he relished the opportunity for friendship and community with ministers of other denominations in the army. An Anglican chaplain who had discovered a new ideal on the battlefield confidently predicted that these views would prevail in Australia after the war as returning chaplains took positions of leadership in the churches.

This did not happen. In some cases absence from Australia for at least four years had denied men opportunities for promotion within the church, in other cases returning chaplains were regarded with some suspicion as if, like the troops themselves, they had been infected with unsound ideas. In yet other cases the appalling nature of warfare on the Western Front had meant that the chaplain's formal religious faith disappeared entirely or that the returning priest or minister was no longer capable of high level commitment to his calling. Thus were the churches deprived of the insights of the clergymen who could claim closest acquaintance with ordinary Australians, churchgoing and non-churchgoing alike, and instead of building on these experiences the churches moved further from the spirit that had been created and celebrated in the AIF. Far from securing a place for the churches in a reformed Australia under the impact of war, war had further marginalised Australian Christian churches, locking leaders and churchgoers into positions that the majority of Australians neither understood nor respected.

Perhaps this is best exemplified by the divisions that emerged in the annual commemoration of Anzac Day (q.v.) in the late 1920s. Intended to bring together former comrades and to remember the awful loss of life that Australia had experienced during the Great War, and to celebrate the achievements and the spirit of those who had endured,

Anzac Day achieved its aims for former comrades but became a further cause and example of division within the Christian churches and promoted a picture of the irrelevance of the churches to Australian society. Catholics were forbidden to attend the services of other Christians, and yet there was a strong sentiment that any commemoration of Australia's war dead should include a religious component. Catholic leaders argued that the celebration should avoid any religious overtones in deference to the consciences of their adherents whom they continued to insist had played their full part in the AIF. In most cases compromise dictated that the Catholics would fall out from the march of veterans before it reached the cenotaph or war memorial where prayers would be said, indicating the perpetuation of the disunity that the conscription debates had done so much to promote.

Christian response to the remaining wars of the twentieth century was largely dictated by a desire not to repeat the mistakes and errors of the Great War years. Church leaders were less prone to preach about war incessantly, again quite at odds with the preacher who in 1914 had complained 'I can't keep the war out of my sermons.' This time too they avoided 'theologising' the war as a gift of God to a reformable Australia. Casualty telegrams became the responsibility of the government and the defence forces and church leaders left political and strategic considerations to elected officials and the military. Nevertheless Australia faced enormous pressures during the Second World War, particularly on the home front during the Pacific War. Work for women, an 'invasion' of friendly forces, largely American, and the mobilisation of almost the entire society caused drastic change to manners and morals and gave churchmen much to talk about.

The Second World War destroyed, probably forever, the distinction between combatants and civilians that was at the heart of Thomas Aquinas's rules for the conduct of a just war. In the terrible bombings of British and German cities where the objective was to destroy civilian morale and thus the capacity of the nation to continue to wage the war, complex moral and theological issues arose which invited response from society's moral guardians, even when allowance is made for the defective knowledge that wartime censorship imposed. Australian Christian leaders were more likely to be debating changed regulations to allow for entertainment of the troops on Sundays or the need for moral guardians for young working women than the morality of pattern bombing or the evil of the implementation of racist ideologies. News of the use of atomic weapons against Japan attracted a subdued community response as if people recog-

nised the movement of warfare to a vastly different scale of possibilities, but only a handful of clergymen attempted to weigh the moral equation implicit in the action. It was as if church leaders had lost their voices precisely when the community looked for informed moral discussion. Where Daniel Mannix had spoken against government policy in wartime, almost regardless of the consequences, during the First World War, now almost every issue of substance passed unremarked by church leaders.

The Cold War produced even greater compliance among Christian leaders. Communism seemed the greatest possible threat to a Christian society and the place of the churches within it and Catholics, in particular, had the image of the 'churches in chains' in eastern Europe and China placed regularly before them. It seemed to them a new age of persecution and martyrs. The Catholic church promoted a 'crusade for the safety of Australia', alarming adherents with the threat of an internal or external takeover of Australian society by communists. Similar campaigns in other churches promoted 'moral rearmament' to fight off the foe and there was a certain apocalyptic feel to the 1950s in Australia. The Korean War pitted Australians against a communist enemy for the first time and alerted Australians to the strength of a movement perhaps even more dangerous than the enemies so recently defeated.

Christian response to the despatch of Australian troops to the war in Vietnam reflected the tensions and concerns of these Cold War years. Even the introduction of conscription (q.v.), by a lottery mechanism which entrenched inequality and luck, attracted little mainstream Catholic opposition, despite the sturdy arguments of 50 years earlier. Most Christian leaders, the Catholics overwhelmingly included, respected the government's sources of information, accepted the government's right to make judgments, and acquiesced in the decision for war and in the means of prosecuting the war.

The emergence of the antiwar movement caught many Christians by surprise, particularly as many of those who opposed the war in Vietnam explicitly rejected a religious basis for their position. Rather than seeing the churches as separate from the State and as standing in judgment of the State, many antiwar activists criticised the churches as the servants of the State, or at least as an instrumental force in the creation of a pro-war sentiment. In Australia, they believed, this had always been the Christian way.

Some Christians, but few of those in power and authority, joined the antiwar forces in what became one of the most powerful debates in Australian history. A priest who spoke against the war from his pulpit was removed by his superior with the agreement of the majority of his congregation and church journals were closed to those who contested the dominant view. A simple Christian pacifism merged with sophisticated historical and political analyses of the situation in Vietnam and it was a Christian influence that may have restrained protesters when violence seemed likely. A mass moratorium movement attracted widespread Christian involvement although few marched in identifiable Christian groups. The Catholic leadership condemned the marchers but the days of overwhelming adherence to the leadership had long passed in Australia, if in fact they had ever existed. No doubt few remembered, if they had ever known, that thousands of Sydney Catholics had noisily protested against their archbishop's unquestioning adherence to a pro-conscription line in 1918 but in their refusal to follow Catholic teaching during the Vietnam years these Catholics demonstrated a similar determination to strike out on their own in matters which they viewed as predominantly political.

Throughout these bitter and divisive debates in each of the wars of the twentieth century, Christian people attempted in their own lives to try to ameliorate some of the evils war caused. Chaplains gave devoted service to their troops in conditions that tested even the younger men and exhausted the older men, as chaplains often were. Welfare officers, often but not exclusively from the Salvation Army (q.v.), helped to make endurable service life and the rigours of the battlefront. With their smokes, some coffee and perhaps a radio for news broadcasts, good men toured the perimeter at Tobruk or helped out on the Kokoda Track and bore witness to a religion that promoted love and service to others above all else. But the times were against them and their leaders too often proved incapable of understanding the spirit of the Australian in times of hardship. There was a spirituality emerging in Australia, and it was assisted by good Christian people from all the churches, but its emergence owed little to a Christian leadership which failed to discern the spirit.

MICHAEL MCKERNAN

REMOUNT SERVICE see ANIMALS

REPATRIATION In Australian usage, repatriation has three meanings: (i) the process of returning men to their country of origin at the end of a war; (ii) a short-hand term for the Department of Repatriation, set up after the First World War to administer the system of veterans' and dependants' benefits; (iii) the system of veterans' and depen-

dants' benefits itself. Australian soldiers who fought in the Boer War were not provided for under any colonial or new federal legislation, and the re-establishment of returning soldiers and the care of the dependants of those who did not return was the responsibility of the patriotic funds. The *Defence Act 1903* (q.v.) made provision for death or incapacity while on active service outside Australia, and this coverage was extended in 1909 to injury or illness while on duty, but servicemen were excluded from the provisions of workers' compensation on grounds of the potential size of any such liability.

Involvement in the First World War changed all this forever. Senator E. D. Millen (q.v.), in many ways the principal architect of the repatriation system, and in 1914 minister for Defence in the Cook government, announced a limited scheme of war pensions in September 1914, one which he described as 'not unduly generous but reasonable and fair'. The proposal was not implemented by the time the Cook government left office in the elections of that year, but the Labor Party pledged that its own scheme would not be less generous, and the scheme which operated during the war was extended several times to include nurses, non-dependent wives and children, and mothers of dead soldiers. In April 1918 a Repatriation Department was set up under Senator Millen and in line with the *Australian Soldiers' Repatriation Act 1920*, with a national structure based on State Repatriation Boards and with a federal Repatriation Commission of six members designed to oversee the whole. Before the end of the year, General Sir John Monash (q.v.) was appointed administrative head of a Department of Repatriation and Demobilisation in London to expedite the return of the AIF to Australia, most of whom were home by the end of 1919 although a few, for various reasons, continued to arrive into 1920. Once the soldiers were demobilised (i.e. discharged from the army), all tasks aimed at their rehabilitation and re-establishment back into civilian life became the responsibility of Repatriation, or 'the Repat', as it was known colloquially. When the AIF was officially disbanded in March 1921, Repatriation also took over the army general hospitals, which became the basis for the Repatriation General Hospitals known to several generations of Australians.

The early history of Repatriation was mixed. Millen believed that the task of minister would kill him 'either politically or physically', and it certainly did the former. The early administration of Repatriation was disorganised, not least because efforts were made to employ former soldiers in most staff positions, and these men often lacked the necessary administrative experience. The official historian

C. E. W. Bean (q.v.) dealt with Millen especially harshly, while the populist newspaper *Smith's Weekly* regarded Repatriation as a natural target for its invective. It ran a regular page (or more) each week for the 'Sailors' and Soldiers' Parliament' which aired ex-soldiers' grievances, and dubbed the department 'The Cyanide Gang'. As well as pensions and disability benefits, Repatriation became responsible for war service homes (generally a success, although roughly handled by the Depression), the soldier settlement scheme (q.v.) (one of the great disasters in Australian rural history), various schemes for the education of the dependants and descendants of servicemen, and Australian war graves through the Imperial (later Commonwealth) War Graves Commission (see WAR GRAVES) (subsequently formalised in 1975 into an Office of Australian War Graves, now located within the successor Department of Veterans' Affairs). Despite the best efforts of Millen and others, there was legitimate cause for dissatisfaction with the repatriation system during the interwar period, not least over which conditions did, or did not, qualify for benefits, an area which was expanded gradually during the 1930s. Two areas, in particular, were accepted as war-related disabilities: pulmonary tuberculosis was linked to gassing, and the 'burnt-out digger' syndrome, which accepted that war service led to premature ageing and which was based in part on a Canadian policy, was recognised for disability purposes. In 1936 service pensions were introduced, and these were extended to Boer War veterans in 1941 (although these latter were not made eligible for free medical care without the application of a means test until 1973!). The payment of pensions peaked in 1931, with 283 322 pensioners in receipt of £7 774 806. Millen saw the purpose of repatriation as being 'a sympathetic effort to reinstate in civil life all those who are capable of such reinstatement'; the writer C. J. Dennis, in his *Digger Smith,* put it in the vernacular: 'An' all that we are askin' you / Is jist a fair square deal. / We want this land we battled for / To settle up — and somethin' more.' But for many veterans the 1930s, marked by the Depression, was a period of 'double sacrifice'; pensions were reduced and those in government employment (secured through the provisions for soldier employment preference at State and federal levels) forwent salary in line with government spending cuts. Many of those settled on the land under the soldier settlement (q.v.) schemes were forced off, through under-capitalisation, falling

commodity prices, inexperience in farming, or inability to work the land because of war-related disabilities. Even the commemoration of the dead was affected by financial stringency, with the construction of the Australian National War Memorial (q.v.) at Villers Bretonneux, approved in 1926, finally dedicated only in 1938 largely because of fiscal stringency on the part of the government.

For veterans of the Second World War the system worked relatively smoothly, although former POWs of the Japanese had a hard fight convincing authority that the privations which they had survived (and which many of their fellows had not) had marred their health irretrievably. Sir Edward Dunlop (q.v.) lobbied for years for a full survey of POWs' health to be carried out, without success.

The postwar regular forces were covered likewise by extensions to the legislation, successively the *Repatriation (Far East Strategic Reserve) Act 1957* for servicemen based in Malaya, and the *Repatriation (Special Overseas Service) Act 1962*, which extended coverage to those engaged in 'warlike operations' and was used for veterans of the Vietnam War. These were modified further by the *Veterans' Entitlements Act 1985*, which greatly streamlined the system and modernised its administration. By the time veterans of the Vietnam War sought assistance, the repatriation system was a sophisticated one of 'parallel welfare' for veterans and their dependants, one sometimes criticised for its apparent largesse (this depended very much on where one stood, as some Vietnam veterans would be quick to point out). Vietnam veterans used the provisions for war service homes, rehabilitation services, repatriation hospitals and education, while the fact that medical records were kept well and more consistently gave this generation of veterans an added advantage over those who had gone before. In two areas, Agent Orange (q.v.) and post-traumatic stress disorders, some veterans, and principally those organised around the Vietnam Veterans' Association (q.v.), expressed extreme dissatisfaction with the repatriation system, resulting in extended lobbying of a kind wholly familiar to veterans of the 1920s and 1930s, and a Royal Commission of Inquiry which failed to convince those who had called loudest for its creation. Australians who served in the Gulf War were brought under the repatriation system by entirely non-controversial amendment to the Veterans' Entitlements Act. By the late 1980s, with the demands of the Second World War generation beginning to peak, the budget allocation for the Department of Veterans' Affairs stood at $4.3 billion in fiscal 1987–88, and some moves were made to reduce the level of services provision on the part of the federal govern-

Private J. Wood, a released POW, is welcomed home in Sydney, 7 July 1945. (AWM 110607)

ment by, for example, transferring responsibility for repatriation hospitals to the States, a move opposed by many veterans. The repatriation system has generally been one of the successes of Australian social welfare policy, and looks likely to continue to provide services and support to veterans well into the twenty-first century.

Clem Lloyd and Jacqui Rees, *The Last Shilling: A History of Repatriation in Australia* (Melbourne University Press, Melbourne, 1994).

REPATRIATION, DEPARTMENT OF, was established on 28 September 1917, inheriting the functions of the Australian Soldiers' Repatriation Fund. In 1920 it took over responsibility for war pensions and hospitals from the Departments of Defence and Treasury, but in that same year the Repatriation Commission was set up and began assuming the department's administrative functions. The commission, made up of three returned servicemen, determined policy, prescribed by regulation the nature and extent of assistance offered under legislation, decided how the minister should be advised and heard appeals. Until 1947 the actual administration of the repatriation system was carried out by the Repatriation Commission or by other departments, although throughout this period the Repatriation Department was never abolished. Repatriation also remained a ministerial portfolio throughout these years, except from 1923 to 1929.

In 1947 the department took back real administrative responsibility when the Repatriation Commission's staff were transferred to the public service and the commission's Chairman became the Permanent Head of the Department. District branches were established in each State capital city, and returned service personnel were represented by State Boards consisting of three paid members. There were also local committees, linked to the district branches, which acted as subagents for the disbursement of assistance and had discretionary powers over the disbursement of supplementary assistance. Between 12 June 1974 and 22 December 1975 the department was given additional functions and renamed the Department of Repatriation and Compensation. Following the recommendations of the Toose Report on the repatriation system the government replaced the Department of Repatriation with an expanded Department of Veterans' Affairs on 5 October 1976. In addition to administration of the repatriation system the new department was given responsibility for defence and war service homes and for defence force and war graves (q.v.). Control of the Australian War Memorial (q.v.) was transferred to the department in 1984.

REPORTS AND INQUIRIES, DEFENCE see **DIBB REPORT; EDWARDS, Major-General Bevan; ELLINGTON, Marshal of the RAF Edward Leonard; HANKEY, Maurice Pascal Alers; JERVOIS, Major-General William Francis Drummond; MILLAR COMMITTEE; MILLS COMMITTEE; MORSHEAD COMMITTEE; SALMOND, Marshal of the RAF John Maitland; TANGE, Arthur**

RESERVED OCCUPATIONS A draft List of Reserved Occupations was first prepared in 1938, but it was between May and August 1939 before the list which would actually be used on the outbreak of the Second World War was put together. At first designed primarily to maintain the supply of labour to the munitions industries, the list was administered by the Manpower Committee. It set out occupations considered necessary for the maintenance of production and essential services so that men working in those industries would not be called up for service in the armed forces, unless they were called up to serve in their technical or trade capacity; it also listed ages at which reservation applied. Legally the list applied only to compulsory service, though it was extended by administrative action to cover the volunteer services (RAAF, RAN and AIF) as well. It did not act as a significant brake on enlistment, since large loopholes essentially allowed all men who wished to enlist to do so, while the commandant of each military district could overrule exemptions from compulsory service in the citizen forces when he considered it necessary.

When the Directorate of Manpower was created in January 1942 it took over responsibility for the List of Reserved Occupations, which on 9 March 1942 was replaced by a Schedule of Reserved Occupations and Industrial Priorities. The schedule differed from the list in two important ways. First, its provisions were not mandatory; it was applied flexibly and seen as only one instrument in an interrelated set of controls on manpower. Second, it was structured differently, being based on a division of industries into three grades depending on their importance to the war effort. In first-priority industries, those in which employment should be maintained or increased, there was often a complete prohibition of enlistment, at least initially, though in some cases specific reserved occupations were listed. For second-priority industries, those in which employment could decline to a limited extent, lists of reserved occupations, together with ages of exemption, were usually drawn up. Third-priority industries, those in which employment should decline substantially, generally involved only a few reserved occupations or none at all. On 7 August 1942 the employment of women in protected factories and government departments and of women with qualifications which were in short supply was reserved, but an amended list issued in April 1943 restricted the number of reserved occupations for women. By the end of 1942 the Schedule of Reserved Occupations was of only minor importance as a tool of manpower control of men, a position which would be reached in relation to women a year later. Used flexibly, the schedule had served its purpose of releasing large numbers of men and women for enlistment without harming vital wartime production.

RETURNED AND SERVICES LEAGUE (RSL) The Returned Soldiers' and Sailors' Imperial League of Australia (RSSILA, later the RSSAILA when 'airmen' were recognised as members of a separate service, but colloquially rendered from early on as simply RSL) was formed in 1916 by representatives of existing returned soldiers' organisations in Queensland, New South Wales, South Australia and Victoria who met in Melbourne in June 1916. There were several issues driving the creation of a federal body of ex-servicemen: concerns over government provision for pensions and benefits, a desire to counter the deteriorating image of returned men given by the rowdy, even violent, behaviour of a few, and a desire to further the war effort through support for the conscription proposal and the efforts of Hughes's government. The first federal congress was held in Brisbane in September the same year. The delegates had rejected forming a political party, opting instead for a non-partisan pressure group which would influence government from outside an increasingly divided party-political arena, a decision further confirmed in 1918. The league's desire to have a direct involvement in the administration of the repatriation scheme (q.v.) was blocked in 1918 with the creation of a separate government department under Senator E. D. Millen (q.v.), although in time the two bodies were to develop an habitual closeness of relations.

By 1919 the RSL had achieved a membership of 150 000, but its claim to be above politics was an increasingly hollow one, even then, and its closeness to Hughes and the Nationalist government led to a dramatic decline in its membership, such that by 1924 it represented only 9 per cent of returned men, with a membership of about 30 000. The 1920s was a period of conflict within returned servicemen's politics, and numerous rival organisations were founded to challenge the RSL's claims to represent returned men at the State and federal levels. Some, like the Returned Soldiers' and Sailors' Australian Democratic League, were politically in direct

opposition to the RSL while others, such as the Soldiers' National Political Party, were in effect ex-service political wings of major political parties. In the course of the decade, however, the RSL achieved a dominant position among returned service organisations, gaining official access at the ministerial, and often at the prime ministerial, level, and controlling official commemoration of Anzac Day, an increasingly powerful tool in shaping the popular perceptions of the war and returned men. Its federal presidents enjoyed considerable political power, especially on repatriation issues, and felt free to speak out on other issues, especially in the defence arena. The longest serving federal president was Sir Gilbert Dyett, who held office for 27 years from 1919 to 1946, not without controversy within the RSL itself. His record of involvement in the federal affairs of the league is rivalled by that of Sir William Keys, federal secretary from 1961 to 1978, and then federal president from 1978 to 1988.

Membership of the RSL declined in the late 1950s, as the Great War generation began to die off, but a number of forward-looking changes to membership eligibility have seen it remain relatively stable since the mid-1970s. These too have not been accepted unopposed within the organisation. For many years membership was open only to returned men: that is, to those who had embarked for overseas service. A man who had served throughout the Second World War but, for example, in Australia only, was ineligible for membership. This began to change from the 60th national congress in 1975, and today membership is open to all current and former members of the forces, including current reservists, a change reflected in the altered name of the organisation: the Returned and Services League.

The structure of the RSL has changed little since 1921. It consists of branches and sub-branches at the State and community levels, a national congress to which the States send delegates, and a national executive supported by a national headquarters. The RSL has long attracted controversy for its alleged influence in policy areas outside the repatriation portfolio, but it is not always clear that this is entirely deserved. Its stand on non-British immigration was generally in step with popular opinion, at least until the 1970s, and at times was ahead: the RSL lobbied for the extension of Australian citizenship to Japanese war brides in the early 1950s, and was rebuffed by the government. Its generally conservative stance on defence issues, and notably on the question of national service, has never yet influenced a federal government to move in a direction in which it did not already intend to proceed: the RSL opposed the cessation of national service in 1959, and was ignored. In the 1980s Sir William Keys attempted to modernise some of the RSL's policy statements, with some success, but as in the wider political arena was hampered by the federal structure of the organisation. Often obscured by the sensational interventions into federal issues of some State RSL officials, is the enormous amount of welfare work undertaken by the RSL on behalf of ex-service personnel and their dependants.

Peter Sekuless and Jacqueline Rees, *Lest We Forget: The History of the Returned Services League 1916–1986* (Rigby, Sydney, 1986); G. L. Kristianson, *The Politics of Patriotism: The Pressure Group Activities of the Returned Servicemen's League* (Australian National University Press, Canberra, 1966).

REVEILLE The name of the bugle or drum call used to rouse soldiers from their sleep, it is also the title of the journal of the New South Wales branch of the RSL (see RETURNED AND SERVICES LEAGUE). Originally entitled *The Reveille,* it commenced publication in August 1927 under the editorship of W. J. Stagg. It reflected the concerns of the RSL, especially over 'justice and a fair deal' for 'the returned man', his dependants and those of men who did not come back, meaning repatriation and benefits issues. In the first issue the editor declared that 'no literary geniuses are connected with this paper, and no literary gems are therefore likely to be found', but in fact in the interwar period the journal is a magnificent source of material relating to individuals who served in the AIF and Australian war service generally. From November 1928 its title was shortened to *Reveille.*

'REVOLT OF THE GENERALS' Recently returned from active service in the Middle East, and alarmed at the state of unpreparedness of Australia's defences in the face of the Japanese advance, a small group of senior officers approached the Minister for the Army, Frank Forde, in March 1942 to urge that Major-General H. C. H. Robertson (q.v.) should be made C-in-C of the Australian Military Forces, and that all officers over the age of 50 should be retired. The principal conspirators were Major-Generals George Vasey and Edmund Herring and Brigadier Clive Steele (qq.v.). Their actions were motivated by their concern at the inexperience of the Labor government, which had only been in office six months, and a fear that the wrong man might be appointed to command the Australian Army. In their view, the wrong men would be either Lieutenant-General Sir John Lavarack (q.v.), who was acting C-in-C pending the return of General Sir Thomas Blamey (q.v.), but whom many thought would tell the government what it wanted to hear rather than what it needed to hear; or Lieutenant-General Henry Gordon Bennett (q.v.), who had recently abandoned his command at the fall of Singapore to return to Australia, and who was now openly distrusted by many but who remained the senior CMF general on the Army List. Robertson was undoubtedly aware of the moves made on his behalf, but does not appear to have instigated them. Blamey's arrival in Australia on 26 March meant that the moves led to nothing, and his only response was to express surprise at their choice of candidate. Their involvement did the plotters' careers no harm, but the incident is notable for what it reveals of the personal feuding within the senior ranks of the army, and of the desperate circumstances which confronted Australian commanders in the opening months of the Pacific War.

RHODESIAN MONITORING FORCE A Commonwealth monitoring force to police the cease-fire was agreed as part of the Lancaster House agreement signed in London on 21 December 1979 between the British government, the government of Rhodesia-Zimbabwe led by Bishop Abel Muzorewa, and the two Marxist guerrilla organisations ZANLA and ZIPRA, which made up the Zimbabwe Patriotic Front and which had waged a decade-long insurgency against White minority rule. The Australian Prime Minister, Malcolm Fraser (q.v.), had played a leading role within Commonwealth councils seeking a resolution of the long-standing Rhodesian problem, and it was virtually axiomatic that Australia would contribute to the force. The monitoring force was to oversee the cease-fire and the movement of insurgent forces to designated areas in order that elections could be held. It consisted of contingents from New Zealand (76), Fiji (24), Kenya (51) and Britain (1200), which provided the bulk of the force, mostly army personnel. The Australian contingent numbered 152: 27 officers and 109 senior NCOs in the monitoring force, a medical officer, and four officers and 11 senior NCOs on the headquarters. The contingent was commanded by Colonel F. K. Cole. Its main task was the control of four of the fourteen assembly places in which Patriotic Front forces were concentrated. The Australian deployment was code-named Operation DAMON, the operation as a whole being designated AGILA by the British. The force was withdrawn over a number of days in March 1980. The cost of the Australian deployment was estimated at $2.4 million.

A Hawker Demon at Richmond RAAF base, 1940. (AWM P0484/10/07)

RICHMOND, RAAF BASE, on the western outskirts of Sydney, is currently the base for the RAAF's transport operations. Richmond is home to No. 35 and 38 Squadrons flying De Havilland Canada Caribou (q.v.), No. 36 and 37 Squadrons flying Lockheed Hercules (q.v.) and No. 33 Squadron flying Boeing 707s. An aerodrome had been built at Richmond during the First World War for a flying school run by the New South Wales Department of Education and this became the first RAAF base outside Victoria when the 71 hectare site was purchased for £9318 in 1923. No. 3 Squadron began operating from the base in 1925. During the Second World War Richmond was used for training paratroops.

RIMAU see **JAYWICK/RIMAU**

RIMPAC The abbreviation for 'rim of the Pacific', the term describes a major biennial maritime exercise between the ANZUS (q.v.) navies and those of other allied Pacific nations including Canada and, more recently, Japan. The exercise is sponsored by the C-in-C, Pacific, the senior American command in the Pacific region with headquarters in Hawaii, and began in 1971. The exercise is valuable for the RAN because it exposes Australian ships and their crews to all aspects of modern high-technology naval warfare, but has been criticised for involving deployments outside Australia's primary area of strategic interest and for rehearsing skills allegedly of use only in interventions in the Third World.

RINGAROOMA, HMS see **AUXILIARY SQUADRON, AUSTRALIA STATION**

RIOTS see **'BRISBANE, BATTLE OF'**; **RED FLAG RIOTS**; **WAZZA RIOTS**

RISING SUN BADGE was probably modelled on a trophy consisting of a collection of bayonets radiating in a semi-circle from a crown. This trophy, designed by Major J. M. Gordon (q.v.) and later presented to Major-General Sir Edward Hutton (q.v.), is now held at Army Office in Canberra. Apparently when the 1st Battalion, Australian Commonwealth Horse, was being raised for Boer War service in 1902 Hutton suggested that their badge be 'something like' the trophy. Such a badge (in three different designs) was struck and was used as the basis of a new design created in 1903 and adopted as the General Service Badge in 1911. The badge bore the title 'Australian Commonwealth Military Forces' on the scroll at the bottom until 1949 when the word 'Commonwealth' was deleted. In 1972 the inscription was shortened again to 'Australia' and other minor changes were made to the design. The inscription now reads, 'The Australian Army'. The rising sun design has also been used in regimental badges and the badge was worn on both the slouch hat and the tunic in both AIFs. It has been suggested that the association of the badge with the rising sun came from the trademark of a popular 'Rising Sun' brand of jam, but others say it represented the rising sun from the start. Rising sun designs (probably related to the image of Australia as 'a young nation' and 'a new Britannia') had appeared on colonial coins and military insignia for decades before the creation of the modern rising sun badge.

RIVER CLASS ANTI-SUBMARINE FRIGATES Displacement 2100 tons; length 370 feet; beam 41 feet; speed 30 knots; 2 × 4.5-inch guns, 1 × Seacat anti-aircraft missile, 1 × Ikara (q.v.) anti-submarine missile, 1 × Limbo (q.v.) anti-submarine mortar.

HMA Ships *Parramatta* (III), *Yarra* (III), *Stuart* (II) and *Derwent* were laid down between 1957 and 1959 and launched between 1958 and 1961. HMAS *Swan* (III) and HMAS *Torrens* (II) were laid down in 1965 and launched 1967-68. Most of the careers of these ships has been spent in routine duties in Australian waters and exercises with the RN and the US Navy. HMAS *Stuart* rescued survivors from the *Voyager* disaster (q.v.) in 1964 and in 1967 and 1968 escorted the troop-carrier HMAS *Sydney* (III) (q.v.) to Vietnam. HMAS *Yarra* also visited Vietnam in 1967.

HMAS *Swan* became embroiled in controversy in August 1992 when an allegation of sexual assault was made by a female officer against a male officer. Although the individual concerned was acquitted in a court martial, a naval inquiry censured the captain of the *Swan* and four other officers for allowing sexual harassment to occur on the ship. In August 1994 a senate committee found that this public censure had been too severe.

RIVER CLASS FRIGATES (*Barcoo, Barwon, Burdekin, Diamantina, Gascoyne, Hawkesbury, Lachlan, Macquarie*). Laid down 1942–43, launched 1943–45; displacement 1420 tons; length 301.5 feet; beam 36.5 feet; speed 19–20 knots; armament 2 × 4-inch guns, 10 × 20 mm anti-aircraft guns (later Hedgehog anti-submarine weapons and depth-charge throwers).

The River Class frigates saw service mainly in the Pacific during the Second World War. They served as convoy escorts, on anti-submarine duties and on shore bombardments. HMAS *Burdekin* accepted the surrender of Japanese forces in Dutch Borneo in 1945. HMAS *Hawkesbury* was the first RAN ship to enter Singapore at the end of the war. All ships except HMAS *Barcoo* were laid off during 1945–49. HMAS *Lachlan* was transferred to

the Royal New Zealand Navy in 1949. In the 1950s and early 1960s *Barcoo, Hawkesbury, Macquarie, Gascoyne* and *Diamantina* re-entered service for periods of several years, operating as training, patrol and survey vessels. All were sold in the 1960s and 1970s, except for HMAS *Diamantina* which served as a survey ship until 1980, when it was acquired by the Queensland Maritime Museum.

RIVER CLASS TORPEDO BOAT DESTROYERS (*Parramatta, Yarra, Huon, Torrens, Warrego, Swan*). Laid down 1909–15, launched 1910–15, sold or scrapped 1929–34; displacement 700 tons; length 250 feet; beam 24 feet; speed 26–28 knots; armament 1 × 4-inch guns, 3 × 12-pounder guns, 3 × Lewis machine-guns, 1 × machine-gun, depth charges, 3 × 18-inch torpedo tubes.

Among the first vessels to join the Australian fleet, these small ships were based on a RN design modified to meet Australian requirements. *Parramatta* and *Yarra* were built in Britain, the others at Cockatoo Island (q.v.). HMA Ships *Parramatta, Yarra* and *Warrego* began the war with the Australian Squadron under Rear-Admiral Sir George Patey. They served in New Guinean, Australian and south-east Asian waters until 1916, when they were replaced by their three sister ships. In 1917 all six were sent to the Mediterranean where they formed part of the Adriatic blockade. In August 1918 HMAS *Huon* and HMAS *Yarra* collided and were repaired in Italian ports. At the conclusion of the war all ships were serving in the Aegean and Black Seas. In December 1918 HMAS *Swan* became involved in the Russian Civil War when she accompanied the French destroyer *Bisson* on a fact-finding mission to the port of Mariapol on the Sea of Azov, an arm of the Black Sea. There a party of officers met General Denikin, the leader of the anti-Bolshevik Don Cossacks, before the *Swan* and all other ships of this class returned to Australia in early 1919. After the war the ships were involved in routine training duties. In the late 1920s a plan to use HMAS *Parramatta* and HMAS *Swan* as prison hulks had to be abandoned because of public outcry.

ROBERTSON, Lieutenant-General Horace Clement Hugh (29 October 1894–28 April 1960). A graduate of the second class at RMC Duntroon in November 1914, Robertson made his mark through service with the 10th Light Horse Regiment of the AIF on Gallipoli and in Sinai and Palestine. Awarded the DSO at Magdhaba in 1916, Robertson was promoted to major and held various staff positions during the First World War. He attended the Staff College, Camberley, in England in 1923–24, and held various posts in the interwar period, including Director of Military Art at RMC and first commandant of the 7th Military District (q.v.), Darwin, in 1939–40. He consolidated his professional attainments through distinguished performance at Staff College and the patronage of the army's Inspector-General (q.v.), Sir Harry Chauvel (q.v.), and was involved also in the strategic debate over the defence of Australia during the 1930s. After a highly creditable performance commanding the 19th Brigade during the first Libyan campaign, he fell out with the Australian C-in-C, Lieutenant-General Sir Thomas Blamey (q.v.), and this, compounded by periods of indifferent health, conspired to keep him sidelined from active commands for most of the rest of the Second World War. He commanded the 1st Armoured Division (1942–43) and III Corps (1943-44) in Australia, and commanded the 6th Division during the New Guinea campaign at the end of the war.

His lengthy period as C-in-C, British Commonwealth Occupation Force (q.v.), presented him with numerous difficulties as commander of a multi-national Commonwealth force operating within an American command and lacking a clear role in the occupation. His relations with British officials both in Tokyo and London were generally unhappy, and this situation was repeated in the early months of the Korean War, where he held non-operational command of Commonwealth forces in Korea 1950–51. Robertson retired from the army in 1954, having been appointed KBE in 1950. In narrow terms, he discharged the command of BCOF effectively and efficiently, but the broader political aims of the force remained unrealised for reasons which had little if anything to do with either the force or its commander.

Tall, red-headed and highly intelligent though given to displays of egotism and bombast, he was both widely admired and greatly disliked by contemporaries and juniors alike. Judged by many as one of the best trainers of troops in the army, his command in Japan was the high point of a varied though not uniformly distinguished career.

Jeffrey Grey, *Australian Brass: A Life of Lieutenant General Sir Horace Robertson* (Cambridge University Press, Melbourne, 1992).

ROCKHAMPTON, HMAS see **BATHURST CLASS MINESWEEPERS (CORVETTES)**

ROSENTHAL, Major-General Charles (12 February 1875–11 May 1954). The son of Scandinavian migrants, Rosenthal joined the Geelong Battery of the Victorian Militia Garrison Artillery in 1892 while articled to an architect. During the 1890s he moved between the colonies of Victoria, Western Australia and New South Wales, practising

his profession, becoming a noted musician and singer, and developing his interest in militia soldiering (see COLONIAL MILITARY FORCES and CITIZEN MILITARY FORCES). He was commissioned in 1903, and transferred to the Australian Field Artillery in 1908, commanding a howitzer battery. In 1914 he became commander of the 5th Field Artillery Brigade.

Seconded to the AIF in August 1914, Rosenthal was given command of the 3rd Field Artillery Brigade. An energetic field commander, he clashed briefly with Major-General W. T. Bridges (q.v.) during the landing at Gallipoli (q.v.) over the latter's decision not to bring the badly needed guns ashore because the perimeter was not yet secure. Rosenthal prevailed, but remained critical of Bridges thereafter. Wounded twice on Gallipoli, the second wound took him to England for hospitalisation, where he used the opportunity to study recent developments in artillery tactics in France. When he returned to Egypt at the beginning of 1916 he was given command of the 4th Division artillery, and a promotion to brigadier-general.

Rosenthal fought through the campaign on the Somme (see WESTERN FRONT), being again wounded in December. In July 1917 he was appointed to command the 9th Infantry Brigade, which he led from the front, spending a great deal of time in the forward positions. He often covered up to 20 miles a day on foot in his brigade area, was gassed at Passchendaele, and on a number of occasions took unnecessary risks for a senior officer by engaging in minor actions with enemy patrols. On the other hand, his style of command had enormous appeal for his soldiers, and his promotion in May 1918 to command the 2nd Division was a popular choice. He was again wounded soon after the battle of Hamel, but recovered in time to lead his division in the great series of battles which began on 8 August. The 2nd Division's capture of Montbrehain in the Hindenburg Line on 5 October was the last Australian action of the war.

In March 1919 he went to England and supervised the AIF Depots during the repatriation of the force. He returned to Australia in January 1920, having been appointed KCB the previous year, rebuilt his architecture practice, and resumed citizen soldiering. He commanded the 2nd Division of the militia from 1921 to 1926 and again between 1932 and 1937. He served briefly in the New South Wales State parliament between 1922–25 and 1936–37 as a Nationalist member, and for many years led the Sydney Anzac Day march, at which he was always in great demand among ex-soldiers. He was a founder and later president of the King and Empire Alliance from

1921, and may have been involved in more clandestine right-wing political activity (see SECRET ARMIES). There is some suggestion that D. H. Lawrence modelled his authoritarian leader figure, Benjamin Cooley, in the novel *Kangaroo* (1923) on Rosenthal. During the Second World War he was administrator of Norfolk Island, a post he had taken up in 1937.

Rosenthal was one of the great characters of the AIF, and one of its most able leaders. He possessed a breadth of interests, courage and energy, and wide experience with artillery which stood him in good stead for brigade and divisional command on the Western Front. Nicknamed 'Rosie', he earned his soldiers' respect as both 'a man and a fighter', and he established an authority with them which carried on well into the aftermath of the war.

ROWELL, Lieutenant-General Sydney Fairbairn (15 December 1894–12 April 1975). A member of the first class at RMC Duntroon (q.v.), Rowell, in common with so many of the early graduates, went straight to service in the First World War, in his case as a troop and squadron commander and, for a short time, as adjutant of the 3rd Light Horse Regiment at Gallipoli. Invalided sick from the peninsula in late November, he was returned to Australia in early 1916 and, under a rigid and nonsensical rule which forbade regular officers, once returned to Australia for any reason, from again embarking for overseas service, spent the rest of the war in a succession of minor and frustrating postings. As a consequence, Rowell was considerably disadvantaged in terms of command and staff experience by comparison with his contemporaries.

His career in the interwar period was mixed. He attended the Staff College, Camberley, in 1925–26, but then served in a variety of postings in Western Australia, which 'in those days could only be regarded as an outstation', as he later put it himself. This was followed in 1932 by a posting to Army Headquarters and, in 1935, by a further posting to Britain, where he was in turn exchange officer serving with the 44th (Home Counties) Division, a Territorial formation, and a student at the Imperial Defence College (q.v.). Soon after his return to Australia in 1938 he was appointed as staff officer, and hence principal assistant, to the new Inspector-General (q.v.), Lieutenant-General E. K. Squires (q.v.), and played an important part in shaping Squires's report on the state of the Australian Army.

Widely regarded as one of the army's outstanding staff officers, Rowell was selected for the post of GSO1 of the 6th Division at the beginning of the war, and quickly went from there to be Brigadier General Staff or chief operational staff officer of I Australian Corps in the Middle East. His relations with his superior, Lieutenant-General Sir Thomas Blamey (q.v.), were mixed, and after the Greek campaign became strained; 'we are not tuned to the same wave-length and are never likely to be', he wrote. After the Syrian campaign Rowell returned to Australia to become deputy CGS under Lieutenant-General Vernon Sturdee (q.v.), a further sign of the regard in which his administrative abilities were held, although it must have seemed that his operational service had been cut short once again.

This was not the case, for in April 1942 as part of the expansion and reorganisation of the army for the defence of Australia against the Japanese, Rowell was promoted to lieutenant-general and given command of I Australian Corps. With headquarters near Port Moresby he grappled with the many problems which beset New Guinea Force (q.v.) as it sought to check and then drive back the Japanese advance. To logistic shortages and the unpreparedness of many of the troops, especially of the militia, was added a lack of comprehension of the true state of affairs on the ground at General Douglas MacArthur's (q.v.) headquarters in Brisbane. Rowell blocked attempts to interfere with Major-General Cyril Clowes (q.v.) as the latter fought the battle of Milne Bay in late August, administering a serious defeat to the Japanese forces there, but the pressure on him from both MacArthur and Blamey was intense.

It was as a direct result of such pressure that Rowell relieved Brigadier A. W. Potts (q.v.) of command of the 21st Brigade on 8 September because the latter was making insufficient headway against Japanese attacks along the Kokoda Track in conditions of great difficulty. As the Japanese advance continued and Australian units withdrew again in the face of strong enemy pressure, MacArthur persuaded Prime Minister John Curtin to order Blamey north to Papua 'to energise the situation', since he believed unwarranted reports from his own staff that Rowell was a defeatist and that there was a real danger that Port Moresby would fall. The pressure on Blamey was intense, too, since there seemed a strong possibility that MacArthur would replace him as commander of Allied Land Forces if a further defeat ensued. It is clear that Rowell felt that this represented a lack of confidence in him as a commander and, given the low regard in which he held Blamey, he may have been predisposed to take offence or seek a confrontation. Blamey, on the other hand, had demonstrated a ruthlessness when his own interests were at stake which would certainly permit him to remove a senior subordinate if he thought it necessary to his own survival. The working relationship necessary if a breach was to be avoided never had a chance to develop, and on 25 September Blamey relieved Rowell of his command, ironically as the Japanese began to fall back along the Kokoda Track. In the opinion of some observers, including Rowell's own chief of staff, Rowell's temperament was at least as much to blame for his dismissal as any personal motives on Blamey's part.

For the rest of the war Rowell filled essentially 'make-work' positions on Middle East headquarters and for the last two years of the war as Director of Tactical Investigations at the War Office in London. With Blamey's retirement, however, his career was revived and he was appointed Vice-CGS, again under Sturdee, in January 1946. The latter was a shrewd and compassionate judge of people, motivated by a strong sense of fairness. Rowell's intelligence and ability were not in question, and in the difficult early postwar years of planning and building a regular army Sturdee put them to good use. In April 1950 Rowell succeeded him as the first Duntroon graduate to be CGS.

As Vice-CGS he had had the main responsibility for overseeing the use of the army in aid to the civil power (q.v.) during the 1949 coal strike. As CGS he presided over the army as it fought a major war in Korea, a national service scheme was introduced (see CONSCRIPTION), and the old relationship between the CMF and the regulars was reversed. He was appointed KBE in 1953 and retired at the end of 1954, and from 1957 to 1968 was Chairman of the Commonwealth Aircraft Corporation.

Assessments of Rowell are diverse. He was, without question, highly able, even brilliant. Reserved, dour but not humourless, proud and highly sensitive, he was not a colourful, outgoing personality, and he possessed a moralistic streak which led him to deep disapproval of Blamey's earthier nature. One contemporary observed that 'the trouble with Syd is that he expects everyone to act like a saint'. His relative inexperience as a commander contributed to the difficulties which led to his sacking in 1942, but the circumstances surrounding the incident were extraordinary ones, suggested by his return to senior positions after the war.

S. F. Rowell, *Full Circle* (Melbourne University Press, Melbourne, 1974).

ROWLAND, Air Marshal James Anthony (1 November 1922–) interrupted his studies in aeronautical engineering at Sydney University to join the RAAF in May 1942 and trained as a pilot under the Empire Air Training Scheme (q.v.). He was commissioned in 1943, and in England undertook conversion training to Halifax and Lancaster bombers. A highly skilful pilot, he was soon sent to the Pathfinder Force where he became a Master Bomber, one of the élite of the Pathfinders who controlled the bombing runs of all other planes on a particular raid. He was awarded the DFC for a raid against Dusseldorf in December 1944. In January 1945 he was taken prisoner when his aircraft was involved in a low-level collision with a Canadian bomber over Frankfurt. He parachuted safely but was captured on the ground and spent the rest of the war as a POW. Following demobilisation Rowland returned to Sydney and completed his degree, and then rejoined the RAAF. He was chief test pilot at the Research and Development Unit at Laverton in Victoria, and was awarded the AFC for his work there. He attended the RAAF Staff College in 1956, and thereafter his postings were mostly technical, and included three years as senior technical staff officer with the RAAF mission in Paris which oversaw the acquisition of the Mirage (q.v.) fighter for Australian service. After completing the Royal College of Defence Studies course in 1971 Rowland became Air Member for Technical Services with the rank of air vice marshal the following year, and in March 1975 was appointed CAS. He was created KBE in 1977, and retired in 1979, serving as Governor-General of New South Wales, 1981–89. Rowland was the first CAS appointed from those who had joined the service during the Second World War, and as an engineer the first non-pilot.

ROYAL AUSTRALIAN AIR FORCE (RAAF) Established on 31 March 1921, the Australian Air Force (the 'Royal' prefix was granted in August of that year) was the successor of the Australian Flying Corps (q.v.) and the Australian Air Corps (q.v.) and took over the aircraft and equipment, the Point Cook base and many of the personnel of that organisation. It owed its origins as an independent third service to the inability of the army and the RAN to agree on the division of funds made available by government in 1918–19 for the establishment of a separate aviation corps under their respective direct control. This led to a decision to have a single air service meeting the needs of both. To this end an Air Board (q.v.) had been formed in 1920, comparable to the bodies administering the military and naval forces, but with an Air Council above it until 1928 to ensure joint control of the air force by the two senior services.

Starting with a strength of 21 officers and 128 airmen, the force was originally planned to grow quickly to 1500 all ranks — about one quarter of whom would be citizen force. These personnel were intended to man four landplane squadrons (two equipped with fighters for the air-defence role, and two corps reconnaissance units for army cooperation work), operating from airbases near Sydney and Melbourne, and two squadrons of seaplanes and flying boats (for naval cooperation) based on Corio Bay. Five of these units had been raised at cadre strength at Point Cook when, in 1922, economy measures led to a reduction to a single mixed unit. Recruitment stopped at about 50 officers and 300 airmen and remained at this level until 1924.

In mid-1925 two squadrons (No. 1 and 3) were raised, both being composite units of fighters and bomber types used for army cooperation, and 101 Fleet Co-operation Flight was formed a year later. No. 3 Squadron was immediately transferred to a new RAAF station opened at Richmond (q.v.), outside Sydney, and was joined there by 101 Flight when it came into being. In 1928 No. 1 Squadron moved to Laverton, near Point Cook, which had housed the RAAF's main stores depot since 1926. These three units, with No. 1 Flying Training School at Point Cook (which included a seaplane training flight), were the major operational elements of the RAAF for almost a decade. The composite unit organisation was abandoned at the end of the 1920s, brought about by the retirement of the RAAF's war surplus fighters. While No. 3 became an army cooperation squadron, No. 1 now specialised as a bomber unit. Skills in the fighter role were retained in a small element (called 'Fighter Squadron' though it was little more than a flight) attached — with 'Seaplane Squadron' — to the flying training school.

Throughout the 1920s and 1930s, the focus of the RAAF's existence was on local defence and, in particular, meeting the training requirements of the naval and military forces. Apart from surveying air routes needed for the deployment of forces during some future defence emergency, the RAAF devoted considerable time and energy to maintaining its public profile at air shows and derbies. It also performed a range of non-defence tasks required by federal and State agencies, undertaking aerial photography for such purposes as furthering development of forestry, fishing, oil and mining resources; assisting exploration and scientific research: and conducting air searches for missing ships, aircraft and persons. In addition, aircraft and crews flew summer bushfire patrols, made meteorological flights for weather forecasting, and sprayed insect pests. Another significant area of involvement saw the RAAF train pilots, lend equipment and share facilities with the Civil Aviation Branch of the Defence Department, thereby playing an important role in fostering a commercial air industry.

While the service's record in this inter-war period was generally creditable, making do as best it could with limited resources and often obsolescent equipment, a number of serious deficiencies were apparent. Training in flying and navigation was often of a standard well below that necessary for service operations, and contributed to a poor accident record which at times involved the RAAF in heated public controversy. The air force was also riven by personal rivalries in its top ranks, most notably between the two senior officers, Richard Williams and S. J. Goble (qq.v.), which damaged both its standing among the other services and diminished political confidence in its leadership. Spurred on by periodic challenges mounted against the RAF in Britain, the army and RAN in Australia were prompted to propose the dismemberment of the RAAF in a series of crises which continued until 1932. Not until 1935, when Williams (the long-serving CAS) became an air vice-marshal, was the RAAF effectively able to outgrow its status as the perpetual 'Cinderella' among the services.

From the end of 1934, after the government announced a program of defence expansion, the RAAF finally began to grow rapidly. The creation of additional squadrons and the opening of new airbases at Pearce (q.v.), Western Australia, in 1937–38, and at Darwin and Amberley (q.v.), near Brisbane, in 1939, greatly enhanced the nationwide reach of air power and gave some credibility to the RAAF's role. By the time the Second World War began in September 1939, the RAAF had 12 squadrons fully or partially formed towards a projected total of 18 squadrons. Whereas personnel numbers had been less than 1000 in 1935, just four years later the RAAF had a strength of 310 officers and 3179 airmen. Aircraft numbers stood at 246, of which two-thirds were counted as operational types, although in reality they were mostly out-of-date or inferior types. Attempts to initiate the local manufacture of military aircraft had begun too late to overcome delays and shortages from Australia's traditional supplier, Britain, due to the hurried rearmament of the RAF against the German *Luftwaffe*.

On the outbreak of war, Air Vice-Marshal Goble (who had succeeded Williams) initially prepared to raise an air expeditionary force of two bomber wings each of two squadrons and a fighter wing of two squadrons for overseas service. The Australian government, however, joined with other British dominions in adopting the Empire Air Training Scheme (q.v.), aimed at providing trained aircrews to fight in and alongside the RAF; and the air expeditionary force was abandoned. Instead, a

Pilots of No. 77 Squadron RAAF relaxing at their base at Kimpo, South Korea, during the Korean War. Left to right: Squadron Leader F. Laurenson, Flight Lieutenant D. Hurst, Squadron Leader I. Hubble, Flying Officer P. Cooney, Flight Lieutenant D. Hiller. (AWM P0337/13/12)

party of Australian personnel already in England to take delivery of new Sunderland flying boats (q.v.) was expanded into a full squadron (No. 10) and attached to the RAF, initially the sole RAAF presence in the European theatre. In 1940 a squadron (No. 3) from Australia arrived in the Middle East, followed by No. l Air Ambulance Unit at the end of 1941, but the rest of Australia's air assets were put to work in preparing for the vast training commitments entailed by EATS.

Late in 1940 the first Australians trained under EATS began reaching England where, under Article XV of the agreement between governments, units which had a predominantly Australian composition were duly designated as RAAF squadrons. A total of 17 such squadrons was eventually formed, five in the Middle East and 12 in Britain itself: one of the latter subsequently served in the Middle East and two Spitfire (q.v.) squadrons were sent to Australia in 1942; eight of those raised in Britain were bomber units and took part in the strategic bombing offensive over enemy-occupied Europe. Many other Australians served in RAF formations, and in squadrons designated as part of other dominion air forces under the Article XV rule. Before EATS was ended in mid-1944, 27 387 aircrew had been trained to advanced stage in Australia before proceeding overseas; in addition, 10 351 who had received elementary training were sent to Canada for final instruction, as were 674 sent to Rhodesia (now Zimbabwe), before being shipped on to England to join the RAF. During

the Second World War the RAAF suffered 6500 combat deaths, of which 5100 occurred in the fighting in Europe.

When Japan launched its offensive in South-east Asia and the Pacific in December 1941, the RAAF was immediately engaged by virtue of four squadrons which had been sent from Australia during 1940 to join the British forces garrisoning Singapore and Malaya, plus one EATS unit (No. 453 Squadron) sent out from Britain. Despite some spirited resistance, they became swept up in the Allied rout at the hands of the superior Japanese air forces and within months had been forced to fall back to mainland Australia. RAAF aircrews and support staff subsequently worked alongside American, Dutch and British allies in defence of the Australian mainland and New Guinea, and in regaining the Netherland East Indies (now Indonesia), Borneo and the Philippines. Another 1300 combat casualties were suffered by the RAAF in the Pacific theatre.

During the course of the Pacific War, the RAAF underwent a massive expansion in Australia which saw it attain a peak of 182 000 men and women by August 1944. When Japan surrendered in August 1945, this strength had dropped to 132 000 but these personnel were still manning more than 50 air squadrons and operating more than 6000 aircraft (3200 operational types and nearly 3000 trainers). Despite this impressive size, Australia's air force had not played a commensurate role in combat operations. This was due to a perception among the Americans (who had control of the Allied command structure) that the RAAF was weak and inefficient, a view which owed much to an awareness of the severe tensions that existed between the CAS in Melbourne, Air Vice-Marshal George Jones (q.v.), and the RAAF operational commander, Air Vice-Marshal W. D. Bostock (q.v.).

After the end of the war, all three armed services underwent severe contraction. By 1947, when the strength of the RAAF had shrunk to under 11 000, the government announced plans to establish a permanent peacetime force of 13 000 members operating 144 aircraft in 16 squadrons, a quarter of which would be citizen force units. By this stage the RAAF was already contributing a wing of three fighter squadrons and some 2000 personnel to the British Commonwealth Occupation Force (q.v.) in Japan; in 1948–49 other Australian aircrew took part in the Berlin airlift (q.v.) to ensure the resupply of the former German capital throughout a Soviet land blockade.

While the five-year program was still being implemented, the outbreak of the Korean War in June 1950 caused major revision of its scope. At the same time as Australia agreed to commit a fighter squadron, and subsequently transport aircraft, to the conflict in Korea, it acceded to British requests for air units to assist in the suppression of the communist terrorists in Malaya. After initially providing Dakota transports (q.v.), a squadron of Lincoln bombers (q.v.) soon followed. As a result of the Korean and Malayan involvements, the strength of the RAAF grew to over 16 000 by August 1952 and, although hostilities on the Korean peninsula effectively ended in 1953, the personnel level of Australia's air force remained at about 15 000 for more than a decade. Adding to the pressures on the service was another overseas commitment to Malta from July 1952, when a wing of two fighter squadrons was deployed on garrison duties in the Mediterranean until January 1955. Among the important changes made in RAAF organisation during this period was the abandonment of the system of headquarters based on geographical areas and groups in favour of three functional commands — Home (later renamed Operational), Training, and Maintenance (renamed Support).

Although the terrorist emergency in Malaya was declared at an end in 1960, the RAAF presence was carried as part of the Far East Strategic Reserve (q.v.). The Lincolns were replaced by Canberra bombers (q.v.) in 1958, operating from a new base at Butterworth (q.v.) on the west coast opposite Penang Island, and a wing of two Sabre fighter (q.v.) squadrons arrived during 1958–59. The avowed opposition of Indonesia to the creation of a new federation of Malaysia in 1963 produced still further tensions in the region. Before the Indonesian policy of confrontation was abandoned in August 1966, the RAAF had sent a further detachment of Iroquois (q.v.) helicopters to join its forces operating in the Malayan area.

The advent of still more overseas commitments (a Sabre squadron to Thailand in 1962–68 and a flight of new Caribou transports [q.v.] to South Vietnam from 1964), combined with a program to almost completely re-equip the service and turn it into the most modern small air force in the region, greatly increased the size of the RAAF. From 1963 the air force added about 1000 personnel a year, reaching a peak of 22 712 in 1969 — at which level it remained for the next 20 years. Before the Australian air presence in Vietnam began winding down in 1971–72, a total of three air squadrons had been committed there (Caribous, Iroquois and Canberras) and a peak strength of about 750 personnel attained. The conclusion of Australia's involvement in Vietnam came when the southern republic was overrun by its communist neighbour in 1975, during which time the RAAF flew humanitarian relief missions for refugees.

Since the 1970s the air force has undergone a lengthy period of consolidation. Various overseas commitments have still been undertaken, mainly of a peace-keeping or humanitarian nature. Among these were support for the Red Cross in Timor (1975) and Australian forces acting for the United Nations Organisation in Kashmir (1975–79), Cambodia (1992–93) and Somalia (1993), as well as the international group of observers monitoring ceasefire arrangements between Egypt and Israel (1976–86). RAAF transports also played a supporting role for the RAN task group engaged in the Gulf War (1990–91), filling a vital link in the logistics chain supplying cargo and mail. Despite these involvements, the emphasis of RAAF planning and activity has been predominantly on the defence of Australia, most notably in the chain of new airbases — Tindal (q.v.), Curtin, Learmonth and Scherger, due to be constructed by 2000 — established across the north of the continent. Overseas, the base at Butterworth (q.v.) was handed over to Malaysian authorities in 1970 and became a joint facility: by 1988 no RAAF air units were permanently based there, Australian use being limited to periodic deployments.

Changes to the RAAF structure since the 1970s have included the abolition of the Air Board and the vesting of command of the service from 1976 in the CAS, subject to the authority of the Chief of the Defence Force (CDF). The number of field commands was also reduced to two, Air Command and Support Command, administered by five staff divisions within the Department of Defence (Air Force Office) responsible for development, materiel, personnel, technical services and supply. The head of Air Command, known as the Air Commander Australia, controls the operational assets of the RAAF which are organised in groups (tactical fighter, strike reconnaissance, air lift and maritime patrol). This officer also heads the joint service Air Headquarters of the ADF, responsible directly to CDF for specified air operations.

By the early 1990s the RAAF had undergone an extensive process of reform, involving further trimming of its organisational structure, personnel numbers and infrastructure. Support Command was renamed as Logistics Command, a Training Command created, and the number of Air Force Office divisions reduced from five to three (Personnel, Resource Management, and Materiel) as well as the Office of Deputy CAS. From a level in 1991 of some 22 000 men and women on full-time duty, plus 1200 active reservists and just over 300 aircraft in 16 combat squadrons, RAAF strength was down to 20 600 by July 1993 and is expected to reach about 17 000 by 1996. These reductions have occurred primarily in the area of supporting services, many of which have been contracted out under commercial arrangements, and there has been some rationalisation of base facilities. The object of these measures is to make the RAAF a leaner, more efficient service. The RAAF of the 1990s presents the appearance of an orderly, modern, professional and well-run organisation, but — like the other services — remains essentially untested as to its current suitability to cope with a major defence emergency.

George Odgers, *Air Force Australia* (National Book Distributors, Sydney, 1993).

CHRIS COULTHARD-CLARK

ROYAL AUSTRALIAN AIR FORCE ACADEMY opened in 1961 when the RAAF College at Point Cook became an affiliated college of the University of Melbourne. Point Cook was recognised as the birthplace of military flying in Australia, and it was appropriate that it be the site of the RAAF's main educational and training institution. The Central Flying School, Werribee, was established at Point Cook in February 1914, and Point Cook became the assembly point for Australian Flying Corps units proceeding overseas. Cadet training for the new RAAF began at Point Cook in February 1923, the course lasting 12 months. During the Second World War Point Cook became a major training base for pilots, many of the old temporary buildings were replaced, and a total of 2730 pilots

graduated from the flying school in the course of the war. In 1947, recognising that extensive training and education was required for its officers, the air force established the RAAF College at Point Cook, where the course now became a four-year program in which flying training was undertaken in the third and fourth years together with professional military studies. As the RAAF moved increasingly into areas of advanced technology, this mix of education and training was thought to be inappropriate, and the college was transformed into the RAAF Academy in 1961. In place of the previous combination of tertiary level study with intensive flying training, the academy focused on the educational needs of potential officers. The course, of four years' duration, provided graduates with a three-year science degree plus a fourth year of professional studies and some exposure to subjects in humanities. Flying training no longer formed part of the academy curriculum, and was undertaken after graduation, but in the early 1980s it was restored to the fourth year of the course. The academy closed at the end of 1985, when the provision of tertiary level studies for RAAF cadets was transferred to the new Australian Defence Force Academy (q.v.).

ROYAL AUSTRALIAN AIR FORCE ASSOCIATION was formed as the Australian Flying Corps Association in 1920. During the Second World War it was decided to broaden the membership to include officers and men who had served in any branch of the British Commonwealth's air forces at war and to adopt the organisation's current name. Its principal aims are to promote the spirit of comradeship, the development of air defence and the welfare of the association's members. To these ends it has been involved in establishing welfare facilities for airmen and in lobbying the government on issues of welfare and air defence. The RAAF Association now admits peacetime airmen and airwomen, including reservists, as full members.

ROYAL AUSTRALIAN ARMOURED CORPS (RAAC) was created as the Australian Armoured Corps on 9 July 1941 to provide personnel to use armoured fighting vehicles. The corps was granted the prefix 'Royal' in 1948. At present it has three types of units: tank regiments, reconnaissance units and armoured personnel-carrier units. RAAC training takes place at the Armoured Centre, Puckapunyal (q.v.).

(See also ARMOUR.)

R. N. L. Hopkins, *Australian Armour: A History of the Royal Australian Armoured Corps 1927–1972* (Australian War Memorial, Canberra, 1978).

ROYAL AUSTRALIAN ARMY CHAPLAINS DEPARTMENT (RAAChD)

Religious ministry within the Australian Army was undertaken well before any formal organisational structure existed to meet the spiritual needs of soldiers. Two chaplains accompanied the New South Wales contingent to the Sudan in 1885, and a further 17 are known to have ministered to the Australian contingents in South Africa during the Boer War. In 1908 regulations were promulgated establishing chaplains within the army structure and making them the responsibility of the Adjutant-General, but throughout the First World War, and for the first part of the Second, the Chaplains Department had no departmental machinery or staff structure within Army Headquarters. Four Chaplains-General were appointed at Army Headquarters in 1913 (Church of England, Roman Catholic, Methodist and Presbyterian), but the formal structure of the Chaplains Department only came into being in June 1942, answerable to the Adjutant-General and with a staff and designated administrative function. A senior Jewish chaplain was first appointed in 1942, and a Chaplain-General for the Other Protestant Denominations had been created in 1940, and in 1943 was redesignated 'United Churches', bringing the total number of Chaplains-General to five.

During the First World War chaplains were appointed to temporary commissions with army rank linked to the level at which they served: unit, formation, Principal Chaplain, and so on. In 1920 this was changed and chaplains were ranked in 'classes', a practice derived from the British Army. Many chaplains who served with the AIF were unsuited to the rigours of campaigning and army life (the two not necessarily being synonymous), and the turnover of chaplains was extremely high with many serving for a year or less. Some 414 chaplains attended to the spiritual requirements of the AIF between 1914 and 1919. Some, like 'Fighting Mac' MacKenzie (q.v.) of the Salvation Army, or Andrew Gillison, a Presbyterian chaplain killed on Gallipoli attempting to rescue wounded men, were outstanding successes, but many others were illustrative of the failure of the mainstream churches to maintain their influence in Australian society in the years before 1914 (see RELIGION AND WAR). During the Second World War 753 chaplains served in the Australian Army throughout its entire sphere of operations. Fifteen were killed in action or died of other causes and 37 were taken as POWs. These latter functions were sometimes seen as a natural corollary of the chaplain's task: when the decision was made after the action at Muar, in Malaya in Jan-

uary 1942, that the non-ambulatory wounded had to be left in order that the rest of the unit might make good its escape, the battalion medical personnel and the chaplain volunteered to stay behind to share their fate, which under the Japanese was usually immediate and final. So it proved on that occasion, with the wounded bayoneted where they lay and the doctors, orderlies and chaplains shot or bayoneted peremptorily. In the prison camps along the Burma–Thailand railway, in Changi or in countless other indescribable circumstances, chaplains both preached and practised the gospel, maintaining the spirits of dispirited men, and at times paying the price of ministry with their lives.

In the postwar conflicts the risks faced by chaplains were generally less severe than they had been previously. During the Korean War the army fielded 28 chaplains at various times, a further 31 during the years of overseas basing in Malaya/Malaysia which included the Malayan Emergency and Confrontation, and 49 in Vietnam. A total of 17 served with the Pacific Islands Regiment (q.v.) in the course of its existence. Chaplains were also involved in the 'character training' courses run by the army during the National Service scheme of the 1950s. The 'Royal' prefix was granted to the Chaplains Department in 1948. At the highest level, direction of the Chaplains Department was vested from June 1943 in the Chaplains-General in Conference, and this system continued until 1981 when it was replaced by the Religious Advisory Committee to the Services, which has responsibility for advising the ADF, and the Chief of the Defence Force in person, if necessary, on chaplaincy matters. It operates at the 'two star' level (see OFFICER RANK STRUCTURES), advises on the appointment of Principal Chaplains to each of the services, and assists in the recruitment of new chaplains. Permanent commissions for chaplains began to be offered within the army in 1971, but the army chaplaincy remains a mixture of long-serving military priests and others for whom service chaplaincy is one form of ministry undertaken among others in the course of a life in the church. From 1989 the system of 'classes' was replaced with a divisional structure.

ROYAL AUSTRALIAN ARMY DENTAL CORPS (RAADC)

was formed as the Australian Army Dental Corps on 23 April 1943. From 1915, when dentists were first appointed to provide dental treatment for the AIF, until 1943, dental personnel in the Australian Army were members of the Australian Army Medical Corps (q.v.). The AADC was granted the prefix 'Royal' in 1948.

ROYAL AUSTRALIAN ARMY EDUCATIONAL CORPS (RAAEC) grew out of the Australian Army Education Service (AAES), which was created on 29 October 1943. The role of the AAES during the Second World War was to help maintain soldiers' morale, to prevent boredom, to keep the troops in touch with events at home and abroad, and to prepare soldiers for their eventual return to civil society. To these ends, the service organised lectures, discussion groups, correspondence courses, vocational education, literacy training, recreational activities, films, publication of the *Current Affairs Bulletin* and the journal *Salt* (see SERVICE NEWSPAPERS), and establishment of libraries. The AAES became the Australian Army Educational Corps in September 1949 and was granted the title 'Royal' in 1960. Its main activities are classroom education of soldiers in literacy, numeracy and communication skills (required for promotion of non-commissioned officers); teaching communication skills to officer candidates at RMC Duntroon (q.v.) and the Australian Defence Force Academy (q.v.); providing trade-related education at the Army College of Technical and Further Education (formerly the Army Apprentices School); teaching Asian languages to defence personnel at the Defence School of Languages (formerly the RAAF School of Languages); teaching English to foreign students in Australia under the Defence Co-operation (q.v.) program; advising the army on educational matters; and handling the educational aspects of army re-establishment or demobilisation (q.v.) schemes. It also has important tasks in wartime relating to morale and rehabilitation of the wounded, and its instructors provided classroom education for soldiers serving in the Vietnam War.

ROYAL AUSTRALIAN ARMY MEDICAL CORPS (RAAMC) was created on 1 July 1902, when existing State military medical services were formed into the Australian Army Medical Corps. The corps was granted the prefix 'Royal' in 1948. In every overseas conflict in which the Australian Army has been involved RAAMC members have accompanied front-line troops, operated ambulance services, and staffed base hospitals, medical centres and first-aid posts. Currently, the corps' role is to provide medical support and advice to the army and to conduct medical research specifically in the fields of malaria (q.v.) and aviation medicine.

ROYAL AUSTRALIAN ARMY NURSING CORPS (RAANC) was formed in February 1951 from the Royal Australian Army Nursing Service, whose origins went back to the late colonial period (see COLONIAL MILITARY FORCES), and the Australian Army Medical Women's Service (q.v.), which was disbanded. In the reorganised corps, nursing duties were performed by qualified nurses who received commissions, while non-professional duties were discharged by other ranks. In 1952 CMF companies of the RAANC were formed in each command. Army nurses first served in Malaya during the Emergency from September 1955, and remained in Malaya/Malaysia until 1971. From 1966 a small detachment was deployed to Papua New Guinea to provide medical support for the families of soldiers of the Pacific Islands Regiment (q.v.). In 1967 army nurses were sent to Vietnam to join 8 Field Ambulance and, from the following year, the 1st Australian Field Hospital. Forty-three sisters served in Phuoc Tuy, the last being withdrawn in late 1971. There has been a number of changes in army nursing since Vietnam, reflecting changes in both the army as a whole and wider civilian society. In 1970 policy was changed to permit women who married to remain in the RAANC, and in following years this was widened to allow married women to join, and in 1975 for women with dependants to remain. The first male officer was appointed in 1972, although males have remained a minority in the corps. In June 1988 an instruction was issued allowing members of the corps to carry arms while on parade, and requiring RAANC members posted to field units to carry a personal weapon. In September 1988, as a result of a report supported by the Director-General of Army Health Services, the RAANC lost its other ranks, who were transferred to the Royal Australian Army Medical Corps (q.v.), and became a small, specialised all-officer corps. A number of RAANC officers served on an American medical support ship, USNS *Comfort,* during the Gulf War in 1991.

Jan Bassett, *Guns and Brooches: Australian Army Nursing from the Boer War to the Gulf War* (Oxford University Press, Melbourne, 1992).

ROYAL AUSTRALIAN ARMY ORDNANCE CORPS (RAAOC) was formed as the Australian Army Ordnance Stores Corps (AAOC) on 1 July 1902. At the same time the Australian Army Ordnance Department (AAOD), a civilian-staffed public service department, was established and given overall control of ordnance services for the Australian Army. The corps was largely employed in the inspection, repair and maintenance of equipment until 1942, when these responsibilities was transferred to the newly created Corps of Australian Electrical and Mechanical Engineers (q.v.), leaving the Ordnance Corps to concentrate on the supply of equipment. On 29 May 1942 ordnance services were placed under total military control, and assets transferred from the AAOD to the AAOC. After the Second World War ordnance services remained under military control, and in 1973 the corps assumed responsibility for specialised commodities such as engineer, medical and dental stores, thus becoming the single Australian Army Supply Service. In 1948 it was granted the prefix 'Royal'. The corps has provided support to Australian troops in all campaigns fought overseas, and was responsible for the lengthy process of de-equipping the army after both world wars. The RAAOC is at present responsible for the provisioning, receipt, storage and issue of all army items of supply. It also provides such services as ammunition repair, salvage, parachute maintenance, printing, laundry and bakery facilities.

J. D. Tilbrook, *To the Warrior his Arms: A History of the Ordnance Services in the Australian Army* (RAAOC Committee, Canberra, 1989).

ROYAL AUSTRALIAN ARMY PAY CORPS (RAAPC), first formed on 13 October 1916, was disbanded at the end of the First World War. It was re-formed in 1939, reaching a strength of 7000 before being again disbanded in 1946. In 1948 it was granted the title 'Royal', and detachments were formed as part of the CMF. The corps is responsible for finance and pay services to the army.

ROYAL AUSTRALIAN ARMY SERVICE CORPS (RAASC), created on 1 July 1903 as the Australian Army Service Corps (AASC), took over from colonial organisations involved in military supply and transport. Between the First and Second World Wars the corps was drastically reduced in size, and a 1933 decision to mechanise the AASC was implemented very slowly before the sudden mechanisation required to fight the Second World War. During the First World War 9735 AASC members served in the AIF, while about 50 000 AASC members served in the Second World War. Corps members also served in Japan, Korea, Malaya and Vietnam. The corps was granted the prefix 'Royal' in 1948. On 1 June 1973 its transport members transferred to the newly formed Royal Australian Corps of Transport (q.v.) and its supplies and clerical members to the Royal Australian Army Ordnance Corps (q.v.).

Neville Lindsay, *Equal to the Task*, vol. 1: *The Royal Australian Army Service Corps* (Historia Productions, Kenmore, Qld, 1992).

ROYAL AUSTRALIAN ARTILLERY see ROYAL REGIMENT OF AUSTRALIAN ARTILLERY

ROYAL AUSTRALIAN CORPS OF MILITARY POLICE (RACMP) replaced the Australian Army Provost Corps, which was formed as the ANZAC Provost Corps on 3 April 1916. Its roles included controlling traffic, maintaining discipline and running gaols. Its name was changed in 1918, and it was disbanded in 1920. Re-formed in the Second World War, the corps reached a strength of 4600, with provost units serving with each Australian formation. It was granted the prefix 'Royal' in 1948 and adopted its current title on 4 September 1974. The role of the RACMP is to assist with command and control, operations, and maintenance of morale and discipline. This involves such tasks as traffic control, security duties, dealing with POWs, investigation of offences and supervision of detainees.

ROYAL AUSTRALIAN CORPS OF SIGNALS (RA Sigs) was created as the Australian Corps of Signals on 14 February 1925, taking over responsibility for army communications from the Royal Australian Engineers (q.v.). There had been an Australian Corps of Signallers from 12 January 1906 until 12 July 1912, when the corps merged with the RAE and became known as Signal Engineers. The present corps was granted the prefix 'Royal' in 1948. Its members operate road and air dispatch services, and radio, microwave and satellite links in order to keep all parts of the army in contact with each other.

Theo Barker, *Signals: A History of the Royal Australian Corps of Signals 1788–1947* (Royal Australian Corps of Signals Committee, Canberra, 1987).

ROYAL AUSTRALIAN CORPS OF TRANSPORT (RACT) was formed on 1 June 1973 by amalgamating parts of the Royal Australian Engineers (q.v.) and the Royal Australian Army Service Corps (q.v.). Its function is to control and operate the army's land and water transport, other than unit transport, and to provide movements, postal and air supply support.

ROYAL AUSTRALIAN ENGINEERS (RAE) Engineer units existed in all the colonial military forces (q.v.), the first being formed in Victoria in November 1860 by Captain Peter Scratchley (q.v.). The Royal Engineers, which had played an important part in developing the infrastructure of the early colonies, was the model for them all, and the various branches of military engineering (field, fortress, telegraphic and submarine mining) were represented among the colonial corps. After Federation the small regular component was combined and granted the 'Royal' designation. The militia engineers remained simply the 'Australian Engineers' until January 1936, when both Permanent Force

Corporal Francine Rigby and Warrant Officer 2nd Class Barry James of the Royal Australian Corps of Military Police question a Cambodian shop owner about stolen goods, Phnom Penh, Cambodia, 14 August 1992. The Australians were part of the United Nations Transitional Authority in Cambodia. (AWM CAMUN92/038/20)

and CMF units were brought together in the Corps of Royal Australian Engineers. In the First World War there were 40 engineer units of various types in the AIF responsible for regular field engineering, tunnelling (especially on the Western Front [q.v.]), signals, and light railway operations. After the war the size of the corps was again reduced, in part because it lost its signalling function; a new corps was formed in 1925. In the Second World War the RAE again underwent enormous expansion, and provided technical services and support in all theatres where Australian troops were deployed, as well as throughout Australia. After 1945 it was again reduced in size, and engineers saw service in Japan during the occupation, in Korea (though on attachment to British units since no RAE unit was sent), in Malaya and Borneo and in Vietnam where the variety of tasks given to engineer units ranged from clearing Viet Cong tunnel complexes ('tunnel rats') to assisting the civilian population through the 1st Australian Civil Affairs Unit. In 1972 the organisation of the RAE was varied to permit the raising of field regi-

ments and construction regiments to provide greater flexibility than was possible with the smaller field and construction squadrons.

Ronald McNicoll, *The Royal Australian Engineers,* 3 vols (the Corps Committee, Canberra, 1977, 1979, 1982).

ROYAL AUSTRALIAN INFANTRY CORPS consists of the battalions of the Royal Australian Regiment (q.v.) and the Special Air Service Regiment (q.v.) of the Australian Regular Army, together with the commando companies and the State and university regiments (q.v.) of the Army Reserve. It was formed in 1949. In general the corps badge is only worn by infantrymen on non-regimental appointments, and even then it is generally more common for them to retain their regimental insignia. The constituent regiments each have an honorary colonel, while the colonel-in-chief of the corps is the sovereign.

ROYAL AUSTRALIAN NAVAL COLLEGE see *CRESWELL, HMAS*

ROYAL AUSTRALIAN NAVY As an island continent, there has always been an inevitable concern
in Australia for naval defence. From the arrival of
the Europeans in 1788 and the incorporation of
Australia into international strategic thinking, successive governments — imperial, national and
provincial — have struggled with the challenges of
devoting sufficient resources for naval defence and
determining the optimum deployment of ships and
support facilities. It is a struggle that will continue.

It was the Crimean War and a fear of Russian
warships in 1854 (see COLONIAL WAR SCARES) that
first prompted the Australian colonists to agitate
seriously for local naval defence which would supplement that provided by the Admiralty. The Australia Station (q.v.) had been formed in 1821 as part
of the larger East Indies Station but few warships of
any substance spent much time in Australian waters.
By the time of the gold rushes when Australia was
exporting more goods than it was importing,
colonists were concerned that commerce raiders
from Russia or America might attempt to storm
Australian ports or attack ships carrying Australian
gold in contiguous waters. The Victorian Legislative
Council responded by acquiring the steamship *Victoria* (see VICTORIAN NAVY SHIPS), which later saw
action in the New Zealand Wars, while the New
South Wales Government built the small armed
ketch *Spitfire* which was constructed in Port Jackson
and launched in 1855 (see NEW SOUTH WALES
NAVAL FORCES SHIPS). These were essentially informal initiatives because it was not until 1865 when
the imperial government passed the Colonial Naval
Defence Act (q.v.) that Britain gave all its colonies
authority to establish their own naval forces. By
1884, the Australia Station had become independent with its own dedicated squadron under the
command of a flag officer.

The issue of naval defence was the central subject of debate at the Colonial Conference held in
London in 1887. It resolved and later formally
agreed in the Australasian Naval Defence Act that
Britain should provide an Auxiliary Squadron (q.v.)
of five third-class ships and two torpedo boats to be
deployed around the Australian coast principally for
trade protection. These ships were to be separate
from the existing Australian Squadron and were
not to leave Australian waters without the express
approval of the colonial governments. The colonial
navies would continue to provide local harbour
defence although naval brigades from New South
Wales and Victoria were despatched in the South
Australian cruiser *Protector* during the Boxer Rebellion in China in 1900.

Shortly after Federation on 1 January 1901, the
Commonwealth government subsumed the colo-

nial navies which were amalgamated into the
Commonwealth Naval Forces. Its Director, Captain
William Rooke Creswell (q.v.), was a former RN
officer who had been involved in colonial naval
defence after migrating to Australia in the early
1880s. A clear thinker and persuasive advocate for
naval defence, he summed up the nation's key
defence factor in 1902:

> For a maritime state furnished without a navy, the sea,
> so far from being a safe frontier is rather a highway for
> her enemies; but, with a navy, it surpasses all other
> frontiers in strength.

But was the Australian navy to be an independent
fleet or a detached squadron from the imperial navy?
Could an independent fleet meet Australia's needs or
did it require external assistance? The Australian press,
noting the heightening wave of public enthusiasm for
naval defence, particularly after the visit of the American 'Great White Fleet' in 1908, hoped that it would
be Australian-owned and controlled, and believed
that the Australian people were prepared to meet the
necessary financial outlay. This was, not unexpectedly,
the view in London. The First Lord of the Admiralty,
Lord Selborne, stated that:

> it is desirable that the populations of the Dominions
> should become convinced of the truth of the pro-
> position that there is no possibility of the localisation
> of naval forces, and that the problem of the Bri-
> tish Empire is in no sense one of local defence. The
> sea is all one, and the British Navy, therefore must be
> all one.

Australia was virtually presented with an ultimatum. When it came to outlining a force structure
for the proposed Australian Navy, Creswell argued
from 1907 that Australia needed destroyers to protect its ports, sea lanes and seaborne cargoes.
Creswell's scheme reflected the dilemma that has
become a feature of Australian naval policy: to what
extent should Australian seapower be developed
beyond local requirements for diplomatic purposes
and as a means of controlling and influencing
events elsewhere? The Admiralty, which sought to
maximise its control of naval programs in all of the
dominions, asserted that Australia needed what it
called a 'Fleet Unit', structured for blue-water
operations, which would be led by a Dreadnought
battlecruiser and several cruisers. The Fleet Unit
also included submarines, vessels that Creswell
believed had very little relevance to Australian conditions or requirements.

A compromise was achieved in 1909. The
Admiralty set aside its demand for centralised control and Australia agreed to purchase the Fleet Unit
to which it added a number of destroyers. But as

Filling the coal bunkers of coal-powered ships was dirty and tiring work. Sailors take a break from coaling HMAS *Brisbane* at Fremantle, Western Australia, June 1917. (AWM A00096)

the force structure exceeded local needs, Australia was drawn irretrievably into imperial naval strategy. Some, such as the radical New South Wales politician Edward O'Sullivan, had feared during the debate held in 1887 that a scheme in which Australia would contribute in cash and kind to a British squadron based in Sydney could tie Australia too closely to British imperial policy and drag Australia on to the wrong side of a genuine 'struggle for independence'. But others saw the intertwining of dominion and imperial policy as a favourable outcome, as the Minister for Defence, Joseph Cook, told Parliament in 1909:

> We must remember, first of all, that Australia is part of the Empire, and that within our means we must recognise both our Imperial and local responsibilities. The Empire floats upon its fleet. A strong fleet means a strong Empire, and therefore it is our duty to add to the fleet strength of the Empire.

So while the Royal Australian Navy, which was formally established by Royal Decree on 10 July 1911, would tend to local needs — the purpose for which it was acquired — its principal function in the minds of most politicians had become reinforcing imperial relations.

The arrival of the Australian Fleet, consisting of the battlecruiser *Australia* (I) (q.v.) and several cruisers, destroyers and submarines, for the first time on

4 October 1913 — a nationally declared public holiday — was described by the contemporary press as the greatest day in Australia's history and a sure sign that the nation had reached an important level of maturity and self-sufficiency. The fleet for which there had been such intense public lobbying had finally arrived. Australia was the first dominion to create an independent navy.

The early part of the First World War was to demonstrate the value of seapower although the demonstration was probably too effective. The Australian Fleet, which came under Admiralty operational control when hostilities began, had swiftly removed any threat to Australia and its local interests within three months. By the end of 1915 there was only one major warship in Australian waters. The navy came again to be seen primarily as an important expression of Empire loyalty. But it was also an opportunity for the RAN to model itself closely on the parent service. Integration and a crude inter-operability continued to be the key themes in Australian naval thinking, as expressed by Sir Joseph Cook, minister for the Navy, on 20 December 1918:

> The RAN has been working in and with the Royal Navy. During the war it has therefore been relatively easy to work to a single standard. It is fundamental to the idea of Empire naval defence that there should be a complete standardisation of personnel and ships and

equipment and that this should be to the level of the best. Only the best is good enough for any Navy in the British Empire.

There was a reversion of naval policy after the 1914–18 war to an emphasis on Australian trade protection rather than preparing against invasion, which was considered to be unlikely, or contributing directly to imperial policy. The official historian of the navy in the Second World War, G. Hermon Gill (q.v.), remarked that 'in 1923 naval defence was visualised by Labor Party spokesmen as local defence of the coastline and approaches carried out by submarines, local defence craft and aircraft'. This concentration on local needs continued during the Depression into the 1930s when the RAN was reduced to five ships and 3200 personnel. In 1934 the Naval Board of Administration (q.v.) decided against acquiring a new battlecruiser, which would be of little use on its own but ideally suited for use in major fleet actions. With war looming and the acquisition of a battlecruiser rejected again in 1937, the Government concluded that:

the first line of security against invasion is naval defence, with the army and air force supplementing and co-operating. If the enemy attempts aggression and must be resisted, it is far preferable to fight him away from our shores than when he is seeking to land on our coasts or has actually established himself in our territory.

The defence program that began in 1937 resulted in the RAN being ready for war in 1939 with a substantial force including three modern cruisers of the modified *Leander* Class (q.v.). The other two services were neglected in favour of the navy which had the largest number of permanent members. While there was bipartisan political recognition that the surrounding seas offered Australia the best protection against invasion, a strong navy would be needed to protect trade.

The dedication of Australian warships to Europe after 1939 was in response to the collective character of imperial defence, and, more importantly in naval circles, to ensure that Australian personnel remained current with British naval practice. There was never any serious suggestion in 1939 that RAN ships deployed to the Mediterranean were there as part of an Australian maritime strategy. That they were crucial to Australian local defence became apparent as the Japanese threat became more serious during 1940.

The scale of possible Japanese attack on Australia was assessed by the British chiefs of staff in mid-1940 as cruiser raids possibly combined with a light scale of seaborne attacks against ports. To threaten the security of Australia would require the control of sea communications in the Pacific or Indian Ocean for an indefinite period. This was affirmation that the Japanese threat to Australia did require a maritime strategy but over-reliance on the Singapore strategy (q.v.) and promises of British naval power in the 1930s, and the inability of the Allied nations to achieve a joint policy in the East, hindered the development of a separate Australian approach.

Notwithstanding these factors, there was an attempt to articulate a local defence strategy. At the Advisory War Council (q.v.) meeting of 5 February 1941, Opposition Leader John Curtin (q.v.) argued that:

the danger to Australia would come in the first place from the sea and, secondly from the air, while the army would only be brought into full action if both the Navy and Air Force failed.

By the end of 1942, the Australian government and people were left in no doubt that Australia was a maritime nation and they a maritime people who relied heavily on unrestricted use of the seas and upon seaborne cargoes. The government was forced to regulate totally all incoming and outgoing shipping movements; to protect all ports, harbours and focal points for domestic and overseas shipping traffic; and build very large numbers of small *Bathurst* Class corvettes (q.v.) to escort and convoy trade and to defend its control of sovereign and contiguous waters. The RAN was on the way to becoming the fourth largest navy in the world with a peak wartime strength of 337 ships and 39 650 personnel.

The Battle of the Coral Sea in May 1942, which threatened to deprive Australia of use of its sovereign waters, the Japanese midget submarine attack on Sydney Harbour on 31 May, and the sinking of more than 30 merchant ships off the east Australian coast in less than one year, left an indelible mark on Australian consciousness (see JAPANESE ATTACKS ON AUSTRALIA). It was now clear that any power with a coastline and a navy could directly threaten Australia's sovereignty and its interests. Protecting this sovereignty and these interests led to the loss of eight major warships in 1942 alone.

However, the vastness of the waters surrounding Australia and the airspace above which could not be effectively controlled by Japan, saw the Japanese recoil from any attempted invasion of the Australian continent. They believed that overcoming the sea–air gap to the north was clearly beyond their resources although they hoped to minimise Australia's war effort by severing trade routes across the top of the continent and to America. Once local needs were met, and they had been met effectively by mid-1943, the navy was again available for deployment in response to other priorities.

The Australian naval contribution was subsequently aimed at ejecting Japan from the western Pacific and South-east Asia rather than defending the Australian continent or regaining control of its contiguous waters. Exercising control of the seas to the north of the continent and through both the Indian and Pacific Ocean basins allowed Australia to supply and reinforce its land forces in the Southwest Pacific, permitted the safe passage of vital war commodities, and allowed Australia to contribute to the Allied war effort in Europe, North Africa and the Middle East.

The Australian public were left in no doubt by 1945 as to the role played by the RAN and the importance of seapower in the Pacific War. This awareness was clearly reflected in the 1947 Strategic Basis paper:

> Australia is situated at the end of a series of islands extending from South-east Asia. Except for those islands to the north and north-west, she is surrounded by oceans. Her geographical position, therefore, is such that no hostile power without possessing command of the sea and local air superiority could successfully invade Australia.

However, the Commonwealth Defence Committee concluded that 'the defence of the Australian area is beyond the capacity of Australia and New Zealand unaided and that defence co-operation with the United Kingdom ... is essential'. The sea lanes which were vital to Australia's security could be protected 'only ... by a powerful Empire (or Allied) Fleet superior to that of any possible enemy in the Pacific'. This was an obvious case of regressive policy-making to fall back on the invasion bogey and British naval power when neither the likelihood of invasion nor need for British seapower had been properly demonstrated. Who was the possible naval enemy to be, and what assurances existed that Britain would come to Australia's aid when it had been prevented from doing so in 1942?

It was one thing for the RAN to hanker after the RN and look to it for support in the postwar period. The RN's continuing ability to respond in the way the RAN wished was quite another. It was the steady decline of the Admiralty's political muscle and the gradual exclusion of Australia from privileged information that forced the RAN to look elsewhere for patronage and assistance. The development and deployment of the US Navy was now of enormous significance to the RAN as the experience of joint operations had demonstrated during the Korean War in which the RAN partic-

ipated with rotational deployment of the aircraft carrier *Sydney* and a number of destroyers and frigates. With the creation of NATO, there were also agreed procedures which governed fleetwork, tactical signalling and combat operations. There was no longer the great wartime problem of operating with the Americans in accordance with a different set of rules.

A fundamental change in Australia's war-fighting attitude was made plain in a speech given by Prime Minister R. G. Menzies (q.v.) on 4 September 1957. He announced the beginning of a move to standardise weapons and techniques with the US. The reason was that:

> having regard to ANZUS and SEATO and to our geographical situation, Australian participation in any future war must be in close association with the forces of the USA.

It was Menzies' statement of 1957 that allowed Senator John Gorton (q.v.) as minister for the Navy to secure the approval of cabinet for the purchase of American *Charles F. Adams* Class destroyers (q.v.) in 1961, despite some resistance from within the Naval Board which was still wedded to Britain. It represented a watershed in thinking and in attitudes throughout the navy. A crucial start had been made in weaning the RAN off the RN. It did not, however, signal the start of the Americanisation of the RAN. In the same period of enormous expansion, Australia ordered *Oberon* Class submarines (q.v.) from Scotland and constructed British-design frigates at Williamstown (q.v.) in Victoria. However, the attitude which ensured that all equipment acquisitions — both capital and minor — were British had been undermined and was soon discarded.

Of all the factors that forged an intimate relationship between the RAN and the US Navy, the Vietnam War was by far the most substantial and lasting. For over five years the RAN was an integral part of an operational US Fleet and relied upon the vast American fleet train. There was cooperation and integration based on inter-operability and common logistic support. The Australian naval contribution consisted of a destroyer on six-month rotational deployments, the RAN Helicopter Flight Vietnam (q.v.) and a RAN Clearance Diving Team (q.v.). *Sydney*, converted to a troop-ship, conveyed the bulk of Australia's ground forces to South Vietnam.

At the same time, the RAN was establishing close ties with the emerging navies of the developing South-east Asian nations. The Far East Strategic Reserve (q.v.) was created on 1 April 1955 and RAN ships were deployed to Singapore on rotational deployments. Regular joint operations were also commenced under the SEATO (q.v.) umbrella. They brought the smaller regional navies and the RAN together on a regular basis within a flexible administrative and operational framework from 1961 until 1973 when they were discontinued. This coincided with the demise of the British Far East Fleet and an end to Britain's military commitments east of Suez. Half of the British forces deployed in Malaysia would be withdrawn by 1971, and the remainder by 1976.

Although the Vietnam conflict brought the RAN and the US Navy together in joint operations while also facilitating cooperation for preserving security throughout the Pacific, the postwar operational specialisation of the RAN, anti-submarine warfare (q.v.), allowed the RAN to contribute to the American global maritime strategy. In this case, the RAN was a substitute for American forces. In many respects, the postwar RAN was an ASW navy. This was principally because Australia relied heavily on overseas trade which was threatened primarily by the growing Soviet submarine fleet. This led to comprehensive local ASW research and development producing the Ikara ASW missile, the Mulloka sonar and the Barra sonobuoy (qq.v.).

It was not until the 1980s, with the demise of the RAN aircraft-carrier capability, that the navy attempted a more general approach to naval operations and dropped its primary commitment to ASW. This also prompted, or even forced, a greater commitment to equipment inter-operability with the US Navy, something that had never really been taken seriously. This commitment, coupled with Australia's great satisfaction with the performance of the *Charles F. Adams* Class destroyers, led the Commonwealth government to order three *Oliver Hazard Perry* Class Guided Missile Frigates (FFGs) (q.v.) in 1976. A fourth was later ordered and another two were built in Australia. With the commissioning of the first Australian FFG, HMAS *Adelaide* (II), in 1980, the RAN accepted the need for greater standardisation with American technology. This led to changes in ship repair and maintenance planning and practice, the need for regular exercises with the US Navy for trials and comparative assessment, and new approaches to logistic support management. These were all of substantial benefit when the RAN later combined with the US Navy in the multi-national force raised for the 1991 Gulf War.

By the end of the 1970s, the substantial global and regional involvement of Australian forces had finally been rejected as a basis for war planning or the peacetime 'force in being'. The emphasis of the government's 1979 Strategic Basis paper was developing capabilities for the defence of 'any military

convoys, our coastal shipping, focal areas proximate to Australia and our off-shore resources'. The 1987 Defence White Paper stated that 'Australia's area of direct military interest was the vast maritime area surrounding the continent'. This outlook enabled the navy to acquire the *Collins* Class submarines, the *Anzac* Class frigates, a squadron of Seahawk helicopters (qq.v.) and a class of minehunters intended to meet Australia's naval defence needs until well beyond the year 2000.

The key themes in the RAN's history are integration and inter-operability. Navies are inherently expensive and no single navy operating in isolation has the ability to counter every weapon or offensive measure that even a weaker power can acquire today. Other than those possessed by the old imperial powers or the modern superpowers, navies are of little strategic or tactical use on their own, and for them to be of any practical use in warfare they must be dependent to some degree on the navies of nations with either similar strategic concerns (i.e. cooperation) or similar technology. Thus, the RAN has attempted to achieve integration through inter-operability and the possibility of a higher level of naval defence for Australia than it could provide on its own.

T. R. FRAME

ROYAL AUSTRALIAN REGIMENT (RAR) The Australian Regiment was formed from the infantry battalions (q.v.) of the 34th Infantry Brigade on occupation duty in Japan (see BRITISH COMMONWEALTH OCCUPATION FORCE) on 23 November 1948. The existing 65th, 66th and 67th Battalions became respectively the 1st, 2nd and 3rd Battalions of the new regiment, which was granted the 'Royal' designation by George VI in March 1949. The 3rd Battalion (3RAR) remained in Japan after the other two battalions were withdrawn to Australia at the end of 1948, and was committed to the war in Korea in September 1950. It fought right through the Korean War and into the post-armistice period, while 1RAR and 2RAR served 12-month tours in 1952 and 1953. A fourth battalion, 4RAR, was raised in 1952 as a depot battalion for the regiment, but was disbanded at the end of the decade. During the Malayan Emergency battalions of the RAR served two-year tours of duty with the British Commonwealth Far East Strategic Reserve (q.v.), including operations against the communist terrorists, beginning in 1955. Following the end of the Emergency in July 1960 the Australian battalion remained in Malaya, although the introduction of the Pentropic division (q.v.) into the Australian Army complicated this as the British did not use that establishment and the battalion assigned for duty in Malaya had to be reorganised before it left Australia, and reorganised again on its return.

The Fourth Battalion was reactivated in February 1964 specifically for overseas service as part of a general expansion of the regular army in the 1960s. It served in Borneo in 1966 during Confrontation with the Indonesians; 3RAR had served a tour there the previous year. In March 1965 approval was given to expand the RAR to a strength of seven battalions, and the fifth, sixth and seventh battalions were raised from the ranks of 1RAR, 2RAR and 3RAR respectively, the six battalions then being completed with drafts of national servicemen. An eighth battalion was raised in July 1966, building on members of 1RAR recently returned from service in Vietnam, while 9RAR was raised in November 1967 as a result of the decision to expand the size of the 1st Australian Task Force in South Vietnam from two battalions to three. Unlike the other recently raised battalions, it was not built on a foundation of men drafted from existing battalions. The units of the RAR provided the backbone of the Australian forces in Vietnam, and suffered the majority of the casualties. Following the report of the Hassett (q.v.) Committee in 1972, the strength of the regiment was reduced to six battalions through amalgamation. In 1991 the two battalions of the Brisbane-based 6th Brigade were declared Ready Reserve (q.v.) battalions, while remaining part of the RAR.

The regiment carries battle honours (q.v.) for campaigns in Korea, Malaya, Borneo and Vietnam, and for specific actions at Kapyong, Maryang San, Long Tan and Binh Ba. The regiment is allied with the Royal New Zealand Infantry Regiment and with the Royal Malay Regiment, and each battalion also has a separate alliance and separate regimental marches. During the Korean War the 3rd Battalion was awarded a US Presidential Distinguished Unit Citation for action at Kapyong, and in the Vietnam War D Company of the 6th Battalion was awarded one for Long Tan. The 8th Battalion was awarded a Vietnamese honour, the Cross of Gallantry with Palm Unit Citation, for its single tour of duty in South Vietnam.

D. M. Horner (ed.), *Duty First: The Royal Australian Regiment* (Allen & Unwin, 1990).

ROYAL AUSTRALIAN SURVEY CORPS (RASvy)

came into existence on 1 July 1915, replacing the Survey Section, Royal Australian Engineers (q.v.), which had been established in 1910. Survey work was again undertaken by Survey Section RAE from 1920 until the Survey Corps was re-established in 1932. The corps undertook vital work in the Second World War, particularly in Papua and New Guinea where they produced maps of previously uncharted areas. In 1948 the Australian Survey Corps was granted the title 'Royal'. In addition to defence mapping work, the corps was involved in the national mapping program from 1955. It has also undertaken mapping in Indonesia and the South-west Pacific under the Defence Cooperation Program (see DEFENCE COOPERATION). The School of Military Survey was established at Balcombe, Victoria, in 1948.

ROYAL CORPS OF AUSTRALIAN ELECTRICAL AND MECHANICAL ENGINEERS (RAEME) was

formed as the Corps of Australian Electrical and Mechanical Engineers to take responsibility for repair of army equipment. Its formation was authorised on 16 October 1942, and members of the Australian Army Ordnance Corps' (q.v.) Mechanical Engineering Branch, as well as unit maintenance tradesmen who specialised in maintenance of mechanical equipment, were transferred to the new corps. The corps reached a peak strength of 25 000 during the Second World War, when its members served in all theatres. It was granted the title 'Royal' in 1948. Corps members operate both mobile and static workshops where army equipment is repaired and maintained.

ROYAL MILITARY COLLEGE (RMC), DUNTROON

opened on 27 June 1911, on the grounds of the former grazing property of Robert Campbell on the outskirts of Canberra, the embryonic national capital. Campbell named the property after his ancestral home of Duntrune in Argyllshire, and the college has been known popularly as Duntroon since its inception. Ever since Federation there had been calls for the establishment of a military college, the Colonial Defence Committee (q.v.) arguing in 1901 that such an institution would help the cause of imperial defence, while the GOC of the new military forces, Major-General Edward Hutton (q.v.), urged that a college should have a special branch within it to train officers in staff duties. The final impetus for the creation of the college came in the 1910 report of Field Marshal Lord Kitchener who suggested that a military college could produce the officers necessary to run the new compulsory military training scheme (see CONSCRIPTION). Kitchener's preferred model was the United States Military Academy, West Point, rather than Britain's Royal Military Academy, Sandhurst, the former based on a four-year course entailing solid academic standards, rather than Sandhurst which had a shorter course and which drew for its recruits on a small circle of select schools. The first commandant of RMC Duntroon was Colonel W. T. Bridges (q.v.), who as a cadet had failed to complete the course at RMC Canada. Bridges was determined that the college would be demanding, professionally rigorous, and open to talent, and that it should avoid the narrow elitism that was regarded as the hallmark of British and European military academies. From its beginning, however, Duntroon had a certain elitist basis, in that it was laid down that after five years of operation, its graduates alone would be eligible for commissions in the permanent forces, thereby placing Duntroon in a privileged position, and widening the gulf that already existed between the permanent and the citizens' forces.

The first class to enter the college consisted of 31 Australian and 10 New Zealand cadets. Discipline was severe and conditions bordering on the primitive, and life was made all the more demanding by the unofficial but tolerated practices of bullying and hazing which, long before they were made public in the late 1960s as 'bastardisation', became endemic at Duntroon. The college had barely settled into a routine when war broke out. Desperate for officers to command the AIF, Bridges, now Inspector-General, recommended that the senior class, which was due to graduate on 1 January 1915, be commissioned immediately. The offer was put to the cadets, who volunteered for overseas service, as was required under the Defence Act (q.v.), and on 11 and 14 August the first class, a total of 35, left for the war, the New Zealand cadets going first. They, and those who followed after them to serve in the war — only 133 Australians and 25 New Zealanders in all — did not reach the heights of promotion that might have been expected. Several factors contributed to this situation. In 1917 a confidential instruction had been issued that RMC graduates were not to be promoted beyond the rank of major. At first graduates were posted to regiments as specialist officers, but not as staff officers; not until later in the war did they move into positions such as staff captains and brigade majors. There was also the permanent versus militia rivalry, and concerns that RMC graduates who achieved command positions during the war would be difficult to employ in the postwar permanent forces. The result was that graduates were, and felt themselves to be, held back during the war, despite having passed through the very system that was designed to produce a professionally trained officer corps.

Between the wars Duntroon suffered as did the army at large. Academic standards declined and the curriculum stagnated, and it was difficult to attract recruits because of the limited career opportunities offered in the permanent forces, especially after the announcement of budget cuts in 1930. In February 1931, as part of a massive economy drive, the college was moved to Victoria Barracks, Sydney, and there was no new entry, the whole cadet strength being held at 31. Four years of straitened circumstances followed, and it was not until 1937, following an increase in defence spending, that the college moved back to Duntroon where new buildings had been erected. In 1938 the course was shortened to three years, the first of successive reductions as the demands of the coming war made the faster production of officers necessary. The course was reduced to 2.5 years in 1939 and 1940, to two years from 1941 to 1943, to one year in 1943, and was restored to three years in 1944. Two army committees in 1944 and 1946 recommended the restoration of the four-year course and the provision of a university education for cadets. Although steps were taken to raise academic standards, the implementation of a full university program was not achieved until 1967, with the establishment at Duntroon of the Faculty of Military Studies as part of the University of New South Wales, Sydney. (The name of the faculty was in some ways unfortunate: it suggested that the curriculum was 'military' in orientation, and it therefore attracted considerable suspicion from the wider academic community, whereas the name was chosen simply to differentiate the Duntroon faculty, which embraced arts, science and engineering, from the traditional faculties on the main university campus.) After 1967 the character of the college changed markedly, and not always comfortably for some. There was a much greater emphasis on academic matters, but this at the same time as the army was deployed on active service in Vietnam. Academic–military conflict was especially sharpened by the 'bastardisation' scandal of 1969, which ultimately resulted in a judicial inquiry to investigate cadet practices and the ethos of the college. The hazing of cadets, often to brutal extremes, had been an intimate part of life at Duntroon from the inception of the college, but what had been widely accepted (if bitterly resented by at least some cadets) in past generations was completely unacceptable in the university climate of the late 1960s. So great was the reaction within the University of New South Wales that for a time it seemed that the link between the college and the university might be severed. As a result of the inquiry, sweeping changes were introduced into the college, but sporadic outbreaks of harassment continued to occur, their supporters arguing that such behaviour was necessary to mould young men into the military life and test their strength of character. These arguments gradually lost credence, not least among the cadets themselves, although the very nature of military institutions leaves them open to abuses of the discipline that is necessary to the military profession.

The Faculty of Military Studies closed in 1985 and the university functions of the college were transferred to the new Australian Defence Force Academy (q.v.), Duntroon becoming a training institution. With the closure in 1985 of the Officer Cadet School, Portsea (q.v.), all pre-commissioning training is now undertaken at Duntroon, where courses contain a mixture of male and female staff cadets who have graduated after three years' university education at ADFA and those who have come direct from civilian life.

C. D. Coulthard-Clark, *Duntroon: The Royal Military College of Australia, 1911–1986* (Allen & Unwin, Sydney, 1986).

ROYAL REGIMENT OF AUSTRALIAN ARTILLERY

(RAA) was formed as the Royal Australian Artillery as a result of the amalgamation of the various colonial permanent artillery forces. This amalgamation was approved by the Colonial Office in July 1899, and after Federation the unit became a Commonwealth one. In 1911 the regiment was separated into the Royal Australian Field Artillery and Royal Australian Garrison Artillery, but in 1927 the title Royal Australian Artillery was reassumed before changing again to Royal Australian Artillery Regiment. After the Second World War it amalgamated with the Royal Australian Artillery (Militia) and again took the title Royal Australian Artillery. The present title was adopted on 19 September 1962, though the regiment retains the short title Royal Australian Artillery.

(See also ARTILLERY.)

ROYLE, Admiral Guy Charles Cecil

(?1885–4 January 1954). After service in the First World War, Royle held the post of naval attaché in the British Embassy in Tokyo between 1924 and 1927, and became a specialist in the relatively new field of naval aviation. Between 1933 and 1934 he commanded the aircraft-carrier HMS *Glorious*, followed by two years in London as Naval Secretary to the First Lord of the Admiralty. He was Vice-Admiral Commanding Aircraft Carriers from 1937 to 1939, and returned to sea duty in 1939–40 commanding the aircraft-carrier HMS *Ark Royal*. This was followed by a further period at the Admiralty, as Fifth Sea Lord and Chief of the Naval Air Service, in 1940–41.

Between 1941, when he was appointed KCB, and 1945 he held the position of CNS of the RAN. The increasing dominance of the Pacific exercised by the US Navy made this a difficult period for Royle, and he was a strong critic of the joint command arrangements in the South-West Pacific Area (q.v.). In his turn, General Douglas MacArthur made it clear that he would have preferred that Royle's

term of office had not been extended in 1943, and that a senior Australian officer of the RAN should have been brought forward to head the service, not least because of Royle's right of reporting back to the Admiralty in London, to which MacArthur objected. Royle appears to have been out of sympathy with at least some of the aspirations of the Labor government of John Curtin (q.v.), and was a strong supporter of the basing of the British Pacific Fleet on Australia late in the war. He attempted unsuccessfully to boost the size of the RAN in the war's last years in the context of extreme manpower shortages, which produced strong disagreement within the Chiefs of Staff Committee. He retired from the RN in 1946 and was appointed Yeoman Usher of the Black Rod and, from 1948, Secretary to the Lord Great Chamberlain.

ROYSTON, Brigadier-General John Robinson

(29 April 1860–25 April 1942). South African by birth and education, Royston fought in the Zulu War of 1879 and served in the Natal Mounted Rifles, in which he reached the rank of sergeant in 1884, while continuing to farm in Natal. He was a squadron sergeant major in the Natal Border Rifles when the South African War began in 1899 (see BOER WAR), was commissioned immediately and fought at the siege of Ladysmith, for which he was twice mentioned in despatches. In April 1901 he took command of the 5th and 6th Contingents of the Western Australian Mounted Infantry, and led them in action in Transvaal, Natal and the Orange Free State. He was awarded the DSO in January 1902 and was appointed CMG. His connection with Australians continued into the postwar years: in 1904 he was given command of the Border Mounted Rifles, most of whose rank and file were Australians who had stayed in South Africa after the war. During the Zulu rebellion of 1906 he raised a mounted unit, designated Royston's Horse, again mostly Australian in composition, and was again mentioned in despatches. In 1914 Royston recruited and led the Natal Light Horse, again predominantly Australian, in the early campaign against the Germans in South-west Africa. He went to England with a re-founded unit of Royston's Horse in 1915, and was persuaded by Kitchener, whom he knew already, to proceed to Egypt and an Australian mounted command. Birdwood (q.v.) gave him the 12th Light Horse in February 1916, and he was in temporary command of the 2nd Light Horse Brigade during the Battle of Romani in August. He was wounded on the first day, but refused treatment; ordered to hospital by Chauvel (q.v.) on the second day, he left within hours. His Australians nicknamed him 'Galloping Jack' for his

prodigious energy at the battle, in which he was alleged to have ridden 14 horses to a standstill. At Magdhaba in December, by now in command of the 3rd Light Horse Brigade, he came upon a party of armed Turks; he ordered them to surrender — in Bantu — and they immediately complied. Shortly before the action at Beersheba in October 1917 — which would have thrilled him — Royston was recalled to London and returned to South Africa, seriously ill. He took no further part in the war. In 1934 the 8th Light Horse invited him to visit Melbourne for the dedication of the Shrine of Remembrance, and his visit, encompassing four states, became a sort of triumphal progress. He published a memoir in 1937. A. B. Paterson (q.v.) said of him that he was 'by instinct a bandit chief and by temperament a hero'. H. S. Gullett (q.v.), the official historian, wrote of his 'remarkable qualities as a fighting leader and his personal lovableness'. A physically imposing figure and superb horseman, he was a larger than life character, enormously personally attached to his Australian commands and held in deep affection by his men.

RUM REBELLION see **JOHNSTON, Lieutenant-Colonel George; NEW SOUTH WALES CORPS**

RUSHCUTTER, **HMAS**, named for its location on the shore of Rushcutters Bay in Sydney, was originally a training site and headquarters of the District Naval Officer, Sydney. The Navy's Anti-Submarine School opened there on 13 February 1939, and HMAS *Rushcutter* was commissioned on 1 August 1940. During the Second World War it acquired several new functions, including radar work and Women's Royal Australian Naval Service (q.v.) training, but its involvement in most of these functions ceased after the war. In the postwar period it was mostly used for training reserves and national servicemen. After the Anti-Submarine School's move to HMAS *Watson* (q.v.), *Rushcutter* was temporarily paid off on 30 April 1956 before recommissioning on 1 July 1957 as a tender to *Watson*. In the 1960s it became involved in underwater research, diving, underwater medicine and mine clearance, but it closed down on 29 July 1968 after these functions were transferred to HMAS *Penguin* (q.v.).

RUSSIAN CIVIL WAR, AUSTRALIANS IN see **NORTH RUSSIAN INTERVENTION**

RYRIE, Major-General Granville de Laune (1 July 1865–2 October 1937). Ryrie joined the volunteer movement as a trooper, and was commissioned in the 1st Australian Horse in 1898. He served in the Boer War with the 6th (New South Wales) Imperial Bushmen in 1900–01, and was wounded at Wonderfontein. He returned to Australia and continued to pursue a part-time military career; by 1904 he was the commanding officer of his unit, now reorganised and designated the 3rd Light Horse Regiment. He stood for the New South Wales Parliament, successfully, in 1906, and for two federal seats, unsuccessfully, in 1910, but won the seat of North Sydney in the elections of 1911. Ryrie was given command of the 2nd Light Horse Brigade of the AIF in September 1914. At first the light horse remained in Egypt when the rest of the AIF sailed for Gallipoli, but in May they were sent forward as reinforcements in response to the heavy casualties suffered in the first weeks of fighting. Ryrie was wounded twice during the campaign, and spent much of his time in the forward positions sharing the hardships and dangers of his soldiers. After the evacuation from Gallipoli, the light horse brigades were reorganised and became the striking arm of the forces detailed for the conquest of Sinai, Palestine and Syria. The 2nd Brigade took part in the long advance of the Anzac Mounted Division, which began with the defeat of the Turkish offensive at Romani in August 1916 and culminated in the crossing of the Jordan in September 1918, participating along the way in the battles of Gaza, Beersheba and Es Salt, and the capture of Amman (see PALESTINE CAMPAIGN). He was appointed KCMG in 1919 and returned to Australia where he continued his political career, including a period as assistant minister for Defence between 1920 and 1921. He commanded the 1st Cavalry Division from 1921 to 1927, when he retired from the CMF. In that year he became High Commissioner in London, and in this capacity he represented Australia at the League of Nations. He returned to Australia and retired from public life in 1932. Ryrie was a 'soldier's soldier', blunt in manner and speech, a thoroughgoing Tory in politics and sentiment. Nicknamed 'Bull', he was a solid, courageous, uncomplicated man ideally suited to regimental command. Both Lieutenant-General H. B. Walker and Lieutenant-General Harry Chauvel (qq.v.) doubted his capacity for higher command, which explains his failure to gain promotion during the war, and as a brigade commander he was reliant on his staff to a considerable degree, especially on Gallipoli where his brigade major was the talented young regular officer, Captain W. J. Foster.

S

'S' CLASS DESTROYERS (*Stalwart, Success, Swordsman, Tasmania, Tattoo*). Laid down 1917–18, launched 1918, scrapped 1937; displacement 1075 tons; length 276 feet; beam 27 feet; speed 36 knots; armament 3 × 4-inch guns, 1 × 2-pounder pompom guns, 1 × machine-gun, 4 × Lewis machine guns, 4 × 21-inch torpedo tubes.

These ships were transferred to the RAN from the RN in 1919. Most of their service was spent on the east coast of Australia in routine duties.

SABRE, CAC (Single-seat fighter [Mark 32]). Wingspan 37 feet 1 inch; length 37 feet 6 inches; armament 2 × 30 mm cannon, 2 × Sidewinder air-to-air missiles, 1200 pounds bombs; maximum speed 700 m.p.h.; range 300 miles; power 1 × Rolls Royce Avon turbojet.

The Sabre was an American jet aircraft which first flew in 1947 and became famous during the Korean War. It made use of the Second World War German jet aircraft innovation of the swept wing, which made it superior to straight-wing British jets such as the Meteor and the Vampire (qq.v.). Sabres were built under licence in Australia by the Commonwealth Aircraft Corporation (q.v.) and were first flown by the RAAF in 1954. The RAAF's Sabres were a formidable fighter, better than the American original: they were powered by the superior Rolls Royce Avon engine (also built by CAC), with a redesigned fuselage to fit the new engine that was shorter and lighter, and they were armed with cannon rather than machine-guns. Sabres were flown by No. 3, 76, 77 and 79 Squadrons. They were stationed at Butterworth (q.v.) and at Ubon in Thailand at the time of the Vietnam War. From 1960 Sabres were armed with Sidewinder air-to-air missiles. They were gradually replaced by the Mirage (q.v.) from 1964. Sixteen ex-RAAF Sabres were given to both the Malaysian and Indonesian air forces. The Sabre retired from RAAF service in 1971 but one aircraft is still flown by the RAAF Historic Flight.

SALISBURY PLAIN An area of rolling grassland in Wiltshire, England, 20 miles north of Salisbury, which was the principal training area for many of the 1st AIF in England from 1916 to 1918. The first unit to train there was the 3rd Division under Major-General John Monash in June 1916. Units from all other divisions used the area for training at one time or another. The main camps were Perham Downs, Rollestone, Lark Hill and Tidworth. Badges of various units which trained there can still be seen carved in the chalk hills around this area.

SALMOND, Marshal of the RAF John Maitland (17 July 1881–16 April 1968). Air Marshal Sir John

Salmond visited Australia from 26 June to 20 September 1928 at the request of the CAS, RAAF, Group Captain Richard Williams (q.v.), then, as for most of the 1920s, fighting to preserve the independence of the air force against the other two services. Salmond's report was highly critical of the state of the RAAF: it was, he said, 'totally unfit to undertake war operations'. To repair the deficiencies that he identified in training, equipment and conditions of service, he recommended a nine-year expansion program of modest proportions. Even that limited scheme was abandoned by the government within a year as the Depression forced major cuts in defence spending, although when the expansion of the RAAF did begin in 1934, it was along the lines recommended by Salmond. For Williams, who as CAS inevitably came in for considerable criticism in the wake of the report's publication, it had one positive feature: Salmond had written that 'credit is due to those responsible for the great work achieved in bringing the Royal Australian Air Force to its present stage'. In the context of the poor rating of the RAAF by Salmond, that at least was a welcome endorsement of Williams's leadership.

SALT see **SERVICE NEWSPAPERS**

SALVATION ARMY, an international Protestant missionary body modelled on the military and well known for its welfare work, does not take sides in wartime and leaves its members free to decide for themselves whether or not to participate in war. Individual Australian Salvationists took part in both world wars, most commonly working as chaplains, nurses, bandsmen and in other non-combatant roles to promote the welfare of the troops. The Salvation Army's Red Shield huts and hostels, often with signs bearing a picture of a kangaroo and the words 'Hop in, it's yours!', provided refreshment and recreational facilities for troops in Australia and wherever Australians were serving overseas, even in the middle of Malayan and New Guinea jungles. Its mobile canteens were also extremely popular. Red Shield welfare officers have accompanied Australian troops to every major overseas conflict (there were 274 serving with Australian forces in the Second World War) and they still work in a welfare capacity in large Australian military camps today. Salvation Army chaplains have had a reputation for bravery and dedication ever since their first front-line chaplain, William 'Fighting Mac' McKenzie (q.v.), showed his willingness at Gallipoli to follow troops into battle, where he collected the wounded and buried the dead. The high esteem in which many soldiers held 'the Salvos' is reflected in

The 'Hop Inn Red Shield' stall run by Salvation Army Captain A. R. Hall offers tea to troops moving along the Milford Highway, Balikpapan, Borneo, 7 July 1945. (AWM 111270)

the story that when F. M. Forde (q.v.) introduced himself to a soldier as the 'Minister for the Army' he was greeted with 'Put it there! You padres are doing a good job, too!'

SAMARAI, HMAS see **ATTACK CLASS PATROL BOATS**

SANDAKAN DEATH MARCH Located on the north-east coast of British North Borneo (modern Sabah), Sandakan was the site of a prison camp occupied initially by 'B' Force of Australian POWs from Changi Prison (q.v.) in Singapore, numbering 1496 men. The camp was established in July 1942. Over time the prisoner population in Sandakan greatly increased and included large numbers of British prisoners as well. Conditions deteriorated in the second half of 1944, with the rice ration being reduced before ceasing altogether. In January 1945 the Japanese began to move batches of prisoners from Sandakan to Ranau, about 160 miles away, probably because they feared an invasion along the coast. The very low physical state of the prisoners was compounded in March when the Japanese began to withhold medical supplies from the POW medical staff: the death rate immediately increased. Following the Australian landings at Tarakan in May, the camp commander Captain Susumi Hoshijima ordered that the prisoners be organised

into groups, destroyed the facilities of the camp, and began to march them inland. Large numbers died of malnutrition, exhaustion, disease and ill-treatment. Those who survived to reach Ranau were shot summarily by their guards on 1 August 1945. Only six Australians survived out of 2500 Australian and British prisoners marched to Ranau. All of the 292 POWs left behind at Sandakan because they were too sick to travel were dead by 15 August, mostly as a result of direct Japanese action. Fifteen Japanese officers and NCOs faced trial for crimes relating to the Sandakan death march. Captain Hoshijima and General Masuo Baba, commander of Japanese forces in Borneo, were found guilty of murder and executed in Rabaul. Two other Japanese officers were also found guilty of murder and hanged, while seven officers were found guilty of the lesser charge of compelling prisoners 'to march on long forced marches under difficult conditions and when sick and underfed' causing death, and were sentenced to 10 years' prison. A warrant officer and an NCO were acquitted of murder on grounds of insufficient evidence. The Sandakan death march was one of the worst Japanese atrocities against western POWs during the Pacific War.

Athol Moffitt, *Project Kingfisher* (Angus & Robertson, Sydney, 1989).

SANDERSON, Lieutenant-General John Murray (4 November 1940–) graduated from RMC Duntroon in 1962. He served with the Royal Australian Engineers in east Malaysia during Confrontation, and commanded the 17th Construction Squadron in South Vietnam between 1970 and 1971. He filled a variety of staff and instructional positions, including a period as instructor at the Staff College, Camberley, between 1976 and 1978, Military Assistant to the CGS in 1982, and commander of the Army's 1st Brigade between 1986 and 1988. A graduate of the Royal Melbourne Institute of Technology and the US Army War College, Sanderson was charged with a number of reviews of areas of the ADF, including the Army Reserve review in 1986, the review of higher defence structures in 1989, and the Defence Force structure review in 1991. A highpoint of his career was undoubtedly appointment as military commander of the United Nations Transitional Authority in Cambodia in 1992–93, overseeing the lead-up to the first free Cambodian elections in a generation. On his return to Australia, amid considerable public interest in both the man and the task which he had recently accomplished, he became Commander, Joint Forces Australia. In July 1995 he was appointed CGS.

SARGOOD, Lieutenant-Colonel Frederick Thomas (30 May 1834–2 January 1903). Victoria's first minister for Defence, Sargood was born in London, emigrated to Melbourne with his family in 1850, and became prominent in his father's successful business. He joined the Victorian Volunteer Artillery in 1859 as a private, and formed the St Kilda Rifle Corps the same year. As a citizen soldier he rose to the rank of lieutenant-colonel, but it was as Victorian minister for Defence between 1883–86, 1890–92, and again in 1894 that he made his chief contribution to colonial military affairs. He was a member of Victoria's Legislative Council between 1874 and 1880, and again between 1882 and 1901. As minister, Sargood presided over a reorganisation of Victoria's military forces which included the creation of the Department of Defence (q.v.) and the replacement of volunteers with paid militia forces (see COLONIAL MILITARY FORCES). The navy, local fortifications and armament supplies were all built up. Rifle clubs were encouraged, a military instruction school was set up, and in 1884 the school cadet corps was formed. He enforced recognition by the Colonial Office and locally engaged British military officers of the primacy of civilian control of the colonial forces through his public disagreement with the Commandant of the Victorian

forces, Colonel Disney, in 1885. He was knighted in 1890. Having led an extraordinarily active public life in the Victorian parliament, he was elected to the first federal senate in 1901, sitting until 1903. A noted philanthropist and a man of great personal wealth, he owned the famous Melbourne mansion Rippon Lea, and was president of the Metropolitan Liedertafel, whose members sang at his funeral.

SAUNDERS, Captain Reginald Walter (7 August 1920–2 March 1990). Born in Portland, Victoria Saunders was inspired to join the army by his father's stories of service in the 1st AIF. He enlisted in the army in April 1940, and served with the 2/7th Battalion in North Africa, Crete, and New Guinea. His promotion to lieutenant in November 1944, which made him the first Aboriginal commissioned officer in the Australian Army, was considered with great caution and referred on to General Sir Thomas Blamey (q.v.) because of its 'special significance'. After the war he found that the army would not accept Aboriginal people for the British Commonwealth Occupation Force (q.v.) in Japan, so he worked in various low-paid manual jobs, and by the time he left for Korea, having rejoined the army in August 1950, his family was living in one room of a condemned house. Saunders served at the battle of Kapyong, and returned to Australia at the end of his tour in September 1950. He thrived on the excitement of active service, and frustration with his training role at Puckapunyal led him to resign from the army in 1954. Harry Gordon's 1962 biography of Saunders, *The Embarrassing Australian*, argued that his leadership skills were being wasted in blue-collar employment, but this situation was rectified in 1969 when he began working for the Office (later Department) of Aboriginal Affairs. A proud man, Saunders coped with discrimination with remarkably good grace, and always remained committed to achieving a better deal for Aboriginal people.

Harry Gordon, *The Embarrassing Australian: The Story of an Aboriginal Warrior* (Landsdowne, Melbourne, 1962).

SAVIGE, Lieutenant-General Stanley George (26 June 1890–15 May 1954) served with the junior and senior cadets between 1902–09, and was a scoutmaster in the years between 1910 and the outbreak of war. He enlisted in the AIF as a private soldier in March 1915, and was allotted to the 24th Battalion. Savige served on Gallipoli from the beginning of September until the evacuation. He was promoted to company sergeant-major on the peninsula, and at the beginning of November was commissioned second lieutenant. Service in France

Lieutenants Reg Saunders (left) and T. C. Derrick after their graduation from the Officer Cadet Training Unit, Seymour, Victoria, 25 November 1944. (AWM 083166)

followed. He was posted as intelligence officer on the headquarters of the 6th Brigade, commanded by Gellibrand (q.v.), and returned to the 24th Battalion as adjutant in February 1917. Posted back to brigade headquarters as brigade major in September, he remained there until January the following year. He was gassed on the Somme, awarded the MC for the second battle of Bullecourt in May 1917, and mentioned in despatches three times. In January 1918 he was selected along with a group of dominion officers and NCOs for special duty in Persia as part of Dunsterforce. He served there and as commander of Urmia Force until September 1918, was awarded the DSO, and returned to Australia in March 1919. A strong Baptist before the war, he wrote at the end of 1917 that although his belief in a supreme being was undoubted, 'where He is or by what means we approach is beyond me', a sentiment certainly shared by many men who had gone through the war from start to finish.

Between the wars he cultivated his business interests, and put his concerns for the rehabilitation of wounded ex-servicemen and the care of the dependants of the dead into practical form by founding Legacy (q.v.). He also pursued a career as a citizen soldier. Appointed captain in the 5th/24th Infantry Regiment in September 1920, by July 1926 he was a lieutenant-colonel in command of the 37th Battalion. He was CO of the militia descendant of his wartime battalion, the 24th, from 1928 to 1935, and then commanded the 10th Infantry Brigade as a colonel between 1935 and the outbreak of the war in 1939. Seconded to the 2nd AIF in October 1939, he was given command of the 17th Brigade in the newly raised 6th Division, and took it away to the Middle East. His performance in the first Libyan campaign (q.v.) was criticised, especially but not only by Staff Corps (q.v.) officers, for being too pedestrian and unimaginative: 'he lacks the rudiments of manoeuvre above [the level of] a company', wrote Major-General S. F. Rowell (q.v.), and others thought that he had gained his position largely through his long-standing friendship with Blamey (q.v.), which went back to the 1920s. Savige led the 17th Brigade through the disastrous campaign in Greece and the successful one in Syria, before returning to Australia. He was promoted and given command of the 3rd Division in January 1942, which he commanded until February 1944, notably during the Wau–Salamaua campaign. His period of divisional command was successful, although his dependence on his brilliant senior operations staff officer, Lieutenant-Colonel John Wilton (q.v.), a regular, was remarked on, and not only by Staff Corps officers. The official historian, Gavin Long, noted of Savige: 'Not a brilliant mind — his staff invariably beat him badly at checkers', but went on to note his 'gift of leadership, knowledge of men, great tact and much commonsense'. By 1944 he was the oldest general still on active service. He administered command of the 2nd Corps in late 1942-early 1943, and was appointed to successive commands of the 1st Corps and New Guinea Force before settling as GOC 2nd Corps in October 1944. In this capacity he commanded the fighting on Bougainville (see NEW GUINEA CAMPAIGN) until the end of the war, and took the surrender of the Japanese forces there at Torokina in September.

Savige was appointed Coordinator of Demobilisation and Dispersal within the Department of Postwar Reconstruction in October 1945, and retired from the army in June 1946. He was created KBE in 1950. In retirement he was honorary colonel of the 5th Battalion (Victorian Scottish) in the CMF. Savige was a somewhat controversial figure, at least within the army, during and after the Second World War. Widely revered by those who served under him, his professional abilities were less universally admired. He contributed to the militia–Staff Corps feud both before the war and during it, and made himself few friends among the regulars by claiming in a letter to the papers in May 1946 that many Duntroon graduates had been failures on active service. Long summed up some of the contradictions in his character when he noted in 1944 that 'depressed by the horrors of war, [he] undoubtedly finds war interesting and exciting — this one just as much as the last'.

SCHERGER, Air Chief Marshal Frederick Rudolph William (18 May 1904–16 January 1984). A leading Australian airman of the post-1945 era, he was the first officer of the RAAF promoted to four-star rank (see OFFICER RANK STRUCTURE). Born at Ararat, Victoria, of migrant grandparents, in his youth he endured intense wartime hostility over his German name. Despite this, he gained admission to RMC Duntroon in 1921 and a year after his graduation at the top of his class in 1924, he was seconded to the RAAF and subsequently transferred permanently.

Scherger embraced air force life with gusto and total commitment, revelling in flying manoeuvres in the open-cockpit biplanes. He became an outstanding pilot, mastering combat tactics in the RAAF's few Bristol Bulldog fighters and serving mainly at the training school at Point Cook, where he was chief flying instructor and frequently a test

pilot as well; he was also an early volunteer for parachute training. In 1934 he was sent to England after gaining selection to attend the RAF Staff College, and on completing the course underwent other training in the United Kingdom in navigation and instrument flying.

The outbreak of the Second World War found him at Air Force Headquarters in Melbourne as Director of Training in the rank of wing commander. In July 1940 he was appointed to command No. 2 Service Flying Training School at Wagga Wagga, New South Wales, gaining promotion in September of that year to group captain. He was made commander of the RAAF station at Darwin in October 1941 but had moved to become Senior Air Staff Officer at the Darwin headquarters of North-Western Area a month before the first Japanese air raid in February 1942. In the wake of the initial attack he reacted calmly and professionally while trying to deal with a chaotic situation at the devastated base (see DARWIN, BOMBING OF).

To the Lowe Royal Commission appointed to inquire into the events surrounding the successful Japanese strike, he gave strongly critical evidence regarding the lack of defence preparedness at Darwin. Acute annoyance at his outspokenness caused him to be virtually ostracised by the Air Board for some time, though the commissioner's findings fully endorsed his views. A variety of staff posts, each briefly held, followed during the rest of 1942 — command of the RAAF Station at Richmond, supernumerary at Air Force Headquarters, Director of Defence, Director of Training — until in July 1943 he was made commander of No. 2 Training Group at Wagga.

In November 1943 he was given the most important RAAF field appointment available, that of No. 10 (Operations) Group in New Guinea. Granted acting rank of air commodore, here he commanded Australian and American air units during the assault on Aitape in April 1944 and the landing at Noemfoor Island in June, displaying courage and leadership which earned him the DSO. He was particularly proud of the fact, as he later noted, that he had been the 'first RAAF man to lead both Australian & American Air Forces'. On 28 July he was seriously injured in a motor vehicle accident on Noemfoor, suffering a fractured pelvis which necessitated his evacuation to Australia.

Still recuperating, Scherger was filling the post of Air Member for Personnel when the so-called 'Morotai mutiny' led to the removal of Air Commodore A. H. Cobby (q.v.) in May 1945 and brought about his own return to No. 10 Group, now based at Morotai and known as First Tactical

Air Force (q.v.). By this time the formation had become a large conglomeration of units totalling some 22 000 personnel, and he set about preparing for its forthcoming use during the invasion of Borneo.

Following the end of the war, Scherger went to London to attend the Imperial Defence College (q.v.). On his return in 1947 he became Air Commodore Operations at Air Force headquarters and later that same year was made Deputy CAS, receiving the temporary rank of air vice-marshal in May 1950. In 1951–52 he served in Washington as Head of the Australian Joint Services Staff, after which he was substantively promoted to air vice-marshal and seconded to the appointment of AOC, RAF headquarters in Malaya. Here he was in command of all British air units in the country as well as RAAF squadrons involved in operations against communist terrorists. While again serving as Air Member for Personnel from 1955, he was a major force behind a review into the RAAF College to determine whether it was meeting the service's future needs. His concern to lay the foundations of a nuclear-age air force was behind the upgrading of the syllabus to university level and the renaming of the college to become the RAAF Academy (q.v.).

Promoted to air marshal in March 1957, Scherger took up the post of CAS in succession to Sir John McCauley (q.v.), who had been the first RMC graduate to head the RAAF. The principal preoccupation of the service throughout the 1960s was the program to re-equip completely with modern aircraft. Scherger overturned proposals to replace the RAAF's Sabre fighters (q.v.) with the US F-104 Starfighter, later winning approval to acquire the French Mirage III (q.v.). He also explored with the RAF the possibility of obtaining tactical atomic bombs for Australia, with the aim of keeping the RAAF at the leading edge of technology, until the government warned him off in April 1958. He nonetheless continued to adhere to the doctrine of an Australian nuclear deterrent and even after his retirement occasionally caused minor public furores by speaking out on this issue. He was appointed KBE in 1958.

While CAS he also attempted to foster RAAF interest in such 'state-of-the-art' issues as missiles, the adoption of helicopters, and the updating of radar installations in the air-defence system. Considered to be something of an iconoclast who was not afraid to seek new solutions to old problems, at the end of 1959 he challenged whether Australia's aircraft-manufacturing industry provided a genuine and worthwhile degree of self-sufficiency at an affordable cost.

Vacating the CAS position after four years, in May 1961 he succeeded Vice-Admiral Sir Roy Dowling (q.v.) as Chairman of the Chiefs of Staff Committee (CCSC). He held this post for five years and was promoted air chief marshal in 1965 before retiring in May the next year. His principal impact on defence affairs during this period was undoubtedly the role he played in bringing about direct Australian military involvement in the Vietnam conflict. Leading Australia's delegation to military staff talks with the Americans at Honolulu in March 1965, he exceeded his brief by virtually offering combat forces in pre-emption of any political decision for such a course. In the view of Prime Minister John Gorton (q.v.), Scherger was inclined to behave like 'a politician in uniform'.

As both CAS and Chairman of the Chiefs of Staff Committee, Scherger had become a prominent national identity, his personality, style and appearance all enhancing his personal appeal and authority. Behind the charisma, however, was a decisive leader of distinctively Australian character. Returning to civilian life, he became chairman of the Australian National Airlines Commission in 1966 and in 1968 of the Commonwealth Aircraft Corporation (q.v.) as well, during which time he served as a member of an Australian defence industries mission sent to the United States in 1969. He retired from these posts in 1975 and lived in Melbourne until his death.

Harry Rayner, *Scherger* (Australian War Memorial, Canberra, 1984).

CHRIS COULTHARD-CLARK

SCHEYVILLE, OFFICER TRAINING UNIT (OTU) was established in April 1965 near Windsor, New South Wales, to train selected national servicemen for short-term commissions. It was housed on the site of an internment camp for enemy aliens during the First World War, and had been a migrant hostel after 1949. The course lasted six months and was designed to produce second lieutenants capable of leading an infantry platoon in one of the battalions of the rapidly expanding Royal Australian Regiment (q.v.). Scheyville was a temporary expedient that ceased to operate after the commitment to Vietnam ended. It graduated some 1500 men, of whom 270 opted to transfer to the regular army and some of whom were still serving in the 1980s.

SCIENCE AND TECHNOLOGY Technology has always played an important part in warfare, but since the seventeenth century, and especially since the First World War, scientific research which produces both knowledge and new technology has become of paramount importance. In earlier times technological progress was manifested in various simple material forms, but in the modern age it has increasingly come to be measured in terms of advances in scientific knowledge. Indeed science-based technical innovation has been so rapid, planned and continuous in the twentieth century that scientists have become as important in war as politicians or soldiers. War stimulates and intensifies technological development where, under the pressure of circumstances, innovations are often effected in a relatively short time. The two world wars produced a series of scientific advances of increasing importance and complexity. During the First World War scientific discoveries directly influenced the development of new military technologies and weapon systems; for example, aircraft, the tank and chemical weapons. In the Second World War there was an intensification of these developments which culminated in a weapon of outstanding destructiveness — the atomic bomb. Since the Second World War science has been one of the major transforming forces of the age and the related technological advances made during that war have in a large measure been turned to peaceful use, as exemplified by the great postwar development of aviation, space exploration, nuclear energy, electronics and communications.

In the colonial period and for some years after Federation, Australia's defence depended almost entirely on British technology. However, local defence initiatives were influenced by the possibility after 1851 that the burgeoning wealth of Victoria and New South Wales might attract an aggressor (see COLONIAL WAR SCARES), the withdrawal of British land forces from Australia in 1870, and the inventive and innovative spirit so evident in agriculture and the pastoral industry. In 1855 Australia's first war vessel, the 60-ton *Spitfire*, mounted with a 32-pounder gun on a traversing carriage, was built in Sydney where, over the next 17 years, were also built two stern-wheel gunboats for use by the New Zealand government in the Maori Wars and five armed schooners for the RN for patrol work in the South Sea islands. These shipbuilding achievements did not involve new technology but two inventions of the late nineteenth century did — the Brennan torpedo (q.v.) and Alcock's range-setter and fire-control device.

Facilities for further research and development for Brennan's dirigible torpedo, patented in England in 1877, were lacking in Australia, but the British government invited Brennan to England where, after five years of testing, his torpedo was accepted by the War Office for coastal and harbour defence. In 1892, A. U. Alcock, a Melbourne contemporary of Brennan, patented an electrically operated coordinated range-setter and fire-control device for coastal artillery. This was a great improvement on the then-current Watkins position finder, where the guns had to be laid onto the target *after* its position had been found. Like Brennan, Alcock went to England to develop his invention, but failed to come to an arrangement with the War Office. Nevertheless, Alcock's invention pioneered the field of coordinated range-setting and fire-control from forts to naval vessels. Mention should also be made during this period of the pioneering work of Lawrence Hargrave in aeronautics between 1884 and 1894. While Hargrave had no appreciation of the commercial and military applications of powered flight, his investigations into curved wing surfaces and invention of the cellular box-kite won worldwide recognition and his findings were used overseas in early powered flight experiments. Hargrave's invention in 1889 of a rotary aeroplane engine anticipated the famous Gnome rotary aircraft engine of 1908.

In the colonial period some attention was paid to the manufacture of ammunition. The Victorian government helped establish the Colonial Ammunition Factory at Maribyrnong for the manufacture of small arms ammunition in 1888 and in the 1890s examined the question of the supply of explosives and gun propellants including the new substance, cordite. From 1901, when defence became a Commonwealth concern, until the outbreak of the First World War, the government took several initiatives towards the establishment of a munitions industry. In particular, two important advances were made in 1912 when the manufacture of explosives and propellants was begun at the Cordite Factory also at Maribyrnong, and the production of the Short-Magazine Lee-Enfield 0.303 rifle (q.v.), using the repetition manufacturing technique of Pratt and Whitney of the United States, began at the Small Arms Factory at Lithgow, New South Wales.

Before the First World War Australia's capacity to produce heavy ordnance was severely handicapped by the lack of an efficient steel industry, but even after the opening of the Broken Hill Proprietary Company's works at Newcastle, New South Wales, in March 1915, Australia's output of munitions steel and especially 18-pounder (q.v.) shell was infinitesimal compared with the amount produced in Britain and Canada. Apart from the lack of skilled workers, patterns and capital equipment, Australia's distance from the war meant that it was impossible to keep abreast of the constant changes and improvements in manufacturing techniques that were required. Aware of its deficiencies in the production of guns and high explosives, the government took every possible step to remedy the situation and in 1915 began to consider plans for the construction of an arsenal at Tuggeranong, near Canberra. However, it soon became evident that, apart from its fighting men, the best practical assistance that Australia could give to the Empire's defence was in the form of materials (like munitions steel, wool and foodstuffs) and skilled operatives. One of Australia's most valuable contributions was the recruitment of draughtsmen, engineers, over a hundred chemists, and hundreds of other personnel to work in munitions factories in Britain. The experiences gained by these personnel in the manufacture of explosives and in mass-production engineering techniques provided a solid foundation for Australia's subsequent highly efficient munitions industry after the war and during the Second World War.

While Australia was insufficiently industrialised during the First World War to do little more than provide her armies with food, clothing, rifles and ammunition, individual Australians showed great ingenuity in adapting established technology and inventing new devices. In addition to the well-known examples from Gallipoli — the periscope rifle sight and the delayed action self-firing rifle (q.v.) — Bottrill's partly tracked wheel (pedrail) overcame the problem of transporting guns across sandy deserts where ordinary wheels were impractical, and the low-cost, portable 'spearpoint' pump solved the watering problems for the light horse in the Sinai. One of the most notable Australian innovations solved the problem of machine-guns being jammed through the shrinkage of the ammunition belt. Corporal A. R. Muirhead of Adelaide put forward two practical solutions, the better of which — making the belts of disposable paper — was adopted. Among other inventions made at the front were William Geake's (q.v.) message rocket, the Varley smoke bomb, General Talbot Hobbs's (q.v.) steel machine-gun cupola, Lawrence Wackett's (q.v.) sight for anti-aircraft work and Worsfold's anti-aircraft indirect fire instruments. Australians also made valuable contributions to sound-ranging for artillery purposes and wireless-telephony.

At home in Australia, scientists at the University of Melbourne produced an effective canister gas mask within two months of the first gas attack in 1915, but although it probably influenced the new British design, once again distance from the theatre of war obviated its use before the British mask was mass-produced. The bacteriologist Auguste de Bavay invented a process to manufacture the solvent acetone, used in making cordite, from molasses, and New Zealand-born John Pomeroy of Melbourne perfected his explosive bullet which proved effective in the destruction of German Zeppelins.

It was an Australian, too, who suggested and designed the best means of overcoming the power of the defensive as the war bogged down in the trenches — the tank. In 1913, an Adelaide inventor, Lancelot de Mole, sent designs to the British War Office of a continuously tracked armoured vehicle, and in 1917 took a working model of his tank to England when he embarked with reinforcements for the 10th Battalion. Due to indifference and incompetence on the part of officialdom his idea was ignored. De Mole's design, however, was far superior to the crude and cumbersome tanks that went into operation on the Western Front in 1916, and after the war he was recompensed and decorated for his pioneeering concept of the armoured tank.

Aviation was another new area of military technology. In August 1915 the first Australian-built military aircraft, the Boxkite CFS-8, was test flown at Point Cook in Victoria, but production ceased after the construction of a number of wooden, fabric-covered aircraft with imported engines for training purposes In England, however, Melbourne-born Henry Kauper helped design the Sopwith Pup warplane and invented and patented the Sopwith-Kauper interrupter gear which synchronised the firing of a machine-gun through the rotating propeller.

Between the wars Australia took significant steps to establish a munitions industry and broaden her industrial base in manufacturing and heavy industry. On the advice of Arthur Leighton (q.v.), who had been the designer and manager of the Cordite Factory and technical adviser to the Ministry of Munitions in Britain in 1915–18, the arsenal project was dropped in favour of an integrated industrial complex formed with scientific support and controlled by the Munitions Supply Board (q.v.) set up in August 1921 with Leighton as Controller-General of Munitions. In 1922 a Munitions Supply Laboratory (MSL) was set up at Maribyrnong staffed by personnel from the Science Laboratory that had operated at Victoria Barracks, Melbourne, during the war and by many of the experienced young technologists who had been trained by Leighton in Britain. Among its functions were the maintenance of standards of manufacture and supply, the production of defence supplies from Australian raw materials and the inspection of defence stores and equipment. The board increased its activities in the Maribyrnong area in 1927 by the outright purchase of the Colonial Ammunition Factory (which it had leased from 1921) and by the establishment of an Ordnance Factory in 1928 to produce quick-firing field guns, thus realising the arsenal concept but in a location amply supplied with both labour and transport. At the same time, the capacity of the Lithgow Small Arms Factory was increased to produce the tools and gauges for gun production and its small arms manufacture extended from rifles to Vickers machine-guns. A factory was also set up for the production of steel high-explosive artillery shells. The exacting requirements and complexity of munitions and arms production, where interchangeable components are a desideratum, meant that strict technical and quality control had to be maintained and this was one of the chief functions of the MSL. In 1925 the MSL installed the first basic standards of imperial measurement from the National Physical Laboratory in England together with precision measuring equipment to ensure the highest standards of accuracy in mass-production techniques.

In the interwar period both the government and private enterprise took important steps to improve Australia's scientific and industrial potential. In 1926, the Commonwealth government, realising that national security and industrial development were symbiotic and required research on a scale often beyond the capacity of private enterprise, set up the Council for Scientific and Industrial Research (CSIR). Initially the Council's work was directed to problems affecting agriculture and the pastoral industry. Accordingly, in 1938, plans were drawn up for the erection of three new laboratories that greatly improved Australia's technological potential: the National Standards Laboratory (Sydney) and the Aeronautics Engine and Testing Laboratory and the Division of Industrial Chemistry (both in Melbourne). By 1939 the CSIR was the dominant force in Australian science.

In the meantime, disquiet about Japan's growing military strength and militarism led Essington Lewis (q.v.) and other prominent industrialists to investigate the possibility of setting up an aircraft and motor vehicle industry. As a result of this private initiative the Commonwealth Aircraft Corporation (q.v.) was established in 1935. After a worldwide survey for the type of aircraft most suited to Australian conditions and resources, the CAC selected the North American NA-33 advanced trainer — a low-wing monoplane with the newly developed retractable undercarriage and variable-pitched airscrew. Within three years the CAC factory at Fishermens Bend near Melbourne produced the CA-1 Wirraway (q.v.), which was test

flown in March 1939. The Wackett trainer (q.v.), designed in Australia, soon followed. As international tension rose the government, prompted by Britain, turned its attention to producing an aircraft suitable for regional defence in the south-east Asian area. The one chosen was the Beaufort torpedo bomber (q.v.). However, concern over a steady supply of British engines for the aircraft led to the substitution of the Bristol Taurus with the American Pratt and Whitney Twin Wasp engine.

The motor vehicle and other manufacturing industries received an impetus during the economic depression, 1929–32, when the government increased import restrictions in order to aid local production. In order to capitalise on the government's considerable manufacturing capacity, now largely lying idle, and preserve the skills of its workers, the government allowed its ordnance factories to accept orders from the private sector. This was particularly valuable for the automotive industry because up to that time Ford and General Motors had been concerned with little more than body-building. As the only facility equipped for making certain car parts and for testing materials was the Ordnance Factory with its supporting Supply Laboratories at Maribyrnong, the factory embarked on the mass production of motor vehicle components, most notably axles, shock absorbers and springs. This, together with the experience likewise gained by the Lithgow Small Arms Factory, the Ammunition Factory at Footscray and the Explosives Factory, had important ramifications for the Australian engineering industry as a whole, as it employed production techniques similar to those required for the production of modern sophisticated weapons of war. As economic conditions improved the automotive industry began to expand as did the capacity of the munitions industry.

One other important initiative was taken during this period. Under the aegis of the CSIR and supported by the research laboratories of the Postmaster-General's Department, a Radio Research Board was established in 1927, to encourage research on the transmission and reception of radio waves and to provide a link between university research in this field and the more industrially oriented research of CSIR. It was to be the first step in Australia's important contribution to the development of radio detection and ranging (radar) and radiophysics.

Although the outbreak of the Second World War coincided with both increased government and private efforts to build up an effective production capacity, no excessive demands were made on outputs until the Allied collapse in June 1940, which meant that Australia had to equip her own forces. Largely without overseas materials and technological support, Australia mobilised and coordinated her limited supply of scientific and technical resources to an impressive degree. The Department of Munitions set up production directorates covering explosives, gun ammunition, ordnance, armoured fighting vehicles, small craft, radio and signal equipment. Government factories and private industry began producing a wide variety of munitions, weapons, and equipment ranging from guns, tanks, aircraft, mines and torpedoes (q.v.) to optical instruments, radio sets and radar units, many of which were made for the first time in Australia (see INDUSTRY).

A vital factor in such increased production was the provision of the appropriate machine tools and gauges to ensure accurate measurement specifications and quality control. The MSL at Maribyrnong and the National Standards Laboratory in Sydney initially checked standards but as output rose and factories became more and more dispersed, the technology of gauge design and manufacture was transferred to selected private research laboratories. As part of this successful initiative, Johannsen slip gauges, accurate to one-hundred-thousandth of an inch, were produced in Australia for the first time. Another aspect of quality control that required the attention of scientists was the radiological examination of castings and welds. The MSL branch laboratory at Villawood, New South Wales, carried out this work on the AC 1, or Sentinel (q.v.), tank, the hull of which was cast in one piece from steel-armour plate made with magnesium and zircon alloy instead of nickel, which was unprocurable.

Among other important areas where research scientists contributed to new or improved technology were optics, radar, the tropic-proofing of materials, pharmaceutical chemistry and operations research. The speedy production of optical munitions during the war is one of the most notable achievements in Australian technology. Despite the fact that optical glass, suitable for making lenses and prisms for telescopes, gun sights, range finders, binoculars and the like, had never before been made in the country, by the time of Japan's entry into the war Australia had an assured supply of optical glass. Instead of the predicted four years, this was effected within 10 months of the decision to undertake local production, and at a cost of £60 000 instead of the projected £1 million.

Australia made an effective and individual contribution to the use of radar in the Pacific War. This was the result of effective and generous technology transfer from Britain and later from the United States and also because of the work of the Radio Research Board, a small group of scientists who were already familiar with pulsed radio waves, the basis of radar. Australia's first radar installation was the Shore Defence Station at Dover Heights, Sydney, in May 1940. While it embodied features of current British design an important innovation was made in the aerial system where one tower and aerial was used for both transmitting and receiving instead of two. This saved valuable materials and lowered the production cost, as did the use of the scanning beam technique, as opposed to 'floodlighting', which was used in all Australian radar units. Another Australian contribution was the building of lightweight, air-transportable sets to warn of approaching enemy aircraft. Such a set proved effective in Darwin following the initial raids of 1942, and was the forerunner of other lightweight air-warning designs (LW/AW) that could be easily dismantled, transported, and used under tropical conditions. In 1943 as the Allied offensive intensified over 200 of these sets were used by the army, RAAF and United States forces in Papua New Guinea and the Pacific islands giving the Allies a critical advantage as the Japanese did not possess radar until late in the war. The lightweight principle was adopted by other Allied countries and a seaborne version of LW/AW was used by the RAN from 1943. The development of radar in Australia was largely the work of CSIR's Radiophysics Laboratory which was established in 1939 in Sydney and which employed over 200 scientists, engineers and technicians. Production engineering was originally carried out in the workshops of the Postmaster-General's Department, but by 1942 rapid demand for radar led to its control by the new Directorate of Radio and Signal Supplies under the Ministry of Munitions.

The rapid deterioration of stores and equipment in the tropics as a result of high temperatures, and especially high relative humidity, posed scientific problems of some complexity. The enormous economic cost incurred due to wastage was, of course, nothing compared with the loss of operational efficiency occasioned by the constant need to repair unserviceable items and the increased danger to life through equipment failure at critical moments. Apart from the rotting of boots, clothing and tenting, radio and electrical equipment (especially dry-cell batteries) was invariably rendered ineffective by corrosion and the deterioration of insulating materials. Even gun sights and binoculars were degraded by fungal attacks on glass. The whole problem was investigated by the Scientific Liaison Bureau which advised better packaging and storage methods, and the treatment of fabrics with appropriate fungicides and metals and electrical equipment with anti-corrodents. These were essentially stopgap measures on tropic-proofing; a more effective solution used later involved better design and the sealing of radio circuits and electrical components during production to exclude moisture.

The war saw great advances in the pharmaceutical industry. Before the war Australia depended almost entirely on overseas supplies for vitamins, drugs and other fine chemicals; the disruption of normal supplies led, however, to a very effective cooperation between university and industrial chemists and chemical engineers to make up the deficiencies as stocks were depleted. This effort was coordinated from August 1940 by the Defence Department's Medical Equipment Control Committee under the full-time chairmanship of Sir Alan Newton. Vitamin C (the anti-scurvy vitamin) was synthesised and an alternative source of Vitamin A was extracted from sharks' livers. The Monsanto Group manufactured the sulpha drug, sulphanilamide and from it sulphaguanidine, which was effective against bacillary dysentery, a disabling condition which could at times be as great a threat to an army as the enemy. Indeed in the Pacific theatre during the period September 1943–February 1944, the ratio of battle casualties to those caused by dysentery, malaria, scrub typhus and other diseases was 1:15. The Australian Army was the first to use sulphaguanidine as a specific treatment for dysentery and it was Newton's opinion that 'If this drug had not been available there might have been a different end to the battle of Kokoda.'

Malaria posed the most serious threat to the health of Australian forces in the tropics and attracted considerable research. At first the sulpha drug sulphamerazine and atebrin were used with indifferent results, but at the end of 1943 Australian scientists made a highly significant discovery — the necessary minimum regular dose of atebrin required to ward off malaria for a considerable period. Later paludrine proved superior to atebrin. Other preventive measures led to the local production of the insecticide DDT and other repellents.

One of the most significant pharmaceutical developments of the war, and a good example of successful technology transfer, was the production of the life-saving antibiotic, penicillin. Using manufacturing technology from the United States, the Commonwealth Serum Laboratory, established in 1915, produced its first batches for the armed forces in March 1944. Among many other initiatives were the production of morphine from the opium poppy and the alkaloids atropine and hyoscine from indigenous species of the *Duboisia* genus of plants. Australia supplied the entire needs of the Allies of atropine, used in ophthalmics, and hyoscine, used to prevent seasickness among the troops landing in Normandy on D-Day.

Operations research, the application of scientific method to the analysis of operations and to the evaluation of new types of equipment, was another important application of science during the war. The Australian Army set up two small groups largely staffed by university honours graduates in mathematics and physics. The first under the leadership of Dr D. F. Martyn of CSIR began work in June 1942 and dealt largely with the efficient operational use of radar equipment and in particular with the problem of spurious echoes and 'blind' zones caused by superrefraction. The second group, composed of a small nucleus of British personnel and reinforced by Australian scientists, was mainly concerned with weapon evaluation and began work in the forward areas of New Guinea in August 1943 investigating the operational life of dry-cell radio batteries. Other problems investigated included locating survivors of crashed aircraft in tropical rainforest areas and the effect of firing small arms through jungle vegetation. The RAAF, while slower to adopt operations research, carried it out on broader and perhaps sounder lines from late 1942, and covered administrative matters as well as operations. The RAAF's main task to the north of Australia was to prevent the Japanese from using the sea lanes by preventing their ships from breaking harbour. Operations research on sea-mining operations directed against Japanese-operated ports showed that minelaying was 40 times more effective than the bombing of land targets and that fewer aircraft were lost for each ship sunk or seriously damaged (see CATALINA). Administrative matters examined by the RAAF included ways of providing effective relief for personnel after tours of duty in the tropics, the nature and incidence of occupational sickness and the causes of operational inefficiency in signal workers. The RAN relied on the research departments of the Admiralty in London for its small requirements in operational research.

As in the First World War, the war produced a flood of inventions. At the outbreak of war the army was the only service with a section dealing with inventions — those submitted to the RAN and RAAF were referred to London. A Central Inventions Board, established in 1940, proved unsatisfactory and was replaced in 1942 by the Army Inventions Directorate in Melbourne. In all the directorate examined over 21 000 submissions of which about 3700 were followed up, but of these only 175 devices were built and tested representing 0.83 per cent of total submissions. In addition to two important life-saving devices that were adopted, the Mills' cellular plywood float-raft for ships and Robertson's signal mirror for attracting aircraft, other inventions that proved useful in varying degrees were the Owen gun (q.v.), F. S. Cotton's aerodynamic anti-G flying suit, Miller's mine detector, the aeropak and storepedo (q.v.) for dropping supplies, W. L. Hallam's combined range-finder and sight, F. A. Stevenson's flash simulator and flash eliminator, R. S. Boyle's process for camouflaging wire netting, Goldner's 'triflex' zip-fastener for marquees and tarpaulins, the Wiles Mobile Steam Cooker and Pearson's service ration calculator. Among many ingenious devices not adopted because of time or because of problems of research and development were H. E. Jeffrey's robot tankette for detonating mines by remote control, P. S. Barna's bullet-proof vehicle radiator, E. T. Both's apparatus for transmitting line diagrams by telephone or radio, and R. S. Robertson's 'recoil-less' gun for aircraft. Towards the end of the war when the emphasis was on increasing food production the directorate assisted with several innovations in agriculture.

The contributions of scientists and industrialists to the war effort in such areas as optics, drugs and radar, led to a greater appreciation by the government and public alike of the importance of the universities and the CSIR, not only for defence, but also for realising the industrial potential of the country. As a result governments found it easier to support pure and applied research than they had previously: a research university (the Australian National University) was set up in Canberra in 1946, the New South Wales University of Technology was established in Sydney in 1949, and, under the *Science and Industry Research Act 1949*, the CSIR was reconstituted as the Commonwealth Scientific and Industrial Research Organization (CSIRO), to carry out research in all fields except defence, medicine and atomic energy. By this time defence science had come to be recognised as a distinct branch of applied science and from 1949 began to evolve as a highly integrated activity.

With a few notable exceptions the war had brought little new in the design of armaments but this was to change in 1946 when the government entered into an agreement with the United Kingdom to set up the Long Range Weapons Establishment (q.v.) at Salisbury, South Australia, with a firing range at Woomera, north-west of Adelaide. In 1949 the Australian Defence Scientific Service

(ADSS) was set up under the technical direction of Dr W. A. S. Butement, who remained Chief Scientist until 1966. Initially it consisted of the Long Range Weapons Establishment and the MSL at Maribyrnong, later to become the Defence Standards Laboratories and then the Materials Research Laboratories (MRL), all under the control of the Department of Supply and Development, and from March 1950 the Department of Supply (q.v.). Shortly afterwards CSIR's Division of Aeronautics, renamed the Aeronautical Research Laboratories (ARL), was transferred to ADSS.

As Australia's population and resources prevented it from undertaking a large and wide-ranging scientific effort in defence, it was decided to concentrate on guided weapons research. In 1951 the ADSS establishments were augmented by the addition of a High Speed Aeronautical Laboratory, a Propulsion Research Laboratory to work on motors for guided weapons, and an Electronics Research Laboratory to work on the problems of guiding and tracking weapons in flight, all at Salisbury. The ADSS was further reorganised in 1955 by the amalgamation of all the laboratories in the Salisbury area into one organisation, the Weapons Research Establishment (WRE). Policy for defence science became the responsibility of the Defence Research and Development Policy Committee of the Department of Defence (q.v.). In addition to the Chief Scientist, the committee consisted of eminent scientists, the deputy chiefs of staff and chief technical officers of the armed services and the Controller-General of Munitions. A further strengthening of ADSS was the foundation of the Institute of Defence Science consisting of members from ADSS, the services, universities, CSIRO and private industry to interest scientists outside ADSS in problems of defence science and to bring service officers and scientists into closer contact.

In addition to ADSS each of the armed services employed scientific advisers to interpret postwar developments in armaments and related equipment in the light of Australia's special requirements and where appropriate to work in conjunction with defence scientists and industry. Among early postwar problems investigated were thermal and other effects in the southern oceans on sonar equipment, and 'flame out' in early jet engines for which the ARL devised a unique oxygen-injection system to alleviate the problem. Involvement in the Korean War led the army design establishment and industry to produce for the first time a range of specially designed military vehicles. The army also had close links with United Kingdom defence authorities and with them helped develop a light mortar and improve the Owen gun. The adoption of the L1A1 self-loading rifle (q.v.) in 1959 also required wide-ranging research into alternative materials for both weapon and ammunition. Among the successes of the WRE in the early postwar period were the Jindivik (q.v.) pilotless target drone and the Malkara anti-tank weapon. Service scientific advisers with groups at WRE and ARL carried out operations research especially on the evaluation of weapons at Woomera. During the Vietnam War scientists were employed in the Field Operations Research Section in Vietnam.

At the time of the Vietnam War, overall policy and financial control over defence science and technology was exercised by the Defence Committee (q.v.) of the Department of Defence, while executive responsibility for the management of the various research and development facilities as well as munitions production was vested in the Department of Supply. In May 1968, however, the Department of Defence set up a small Defence Science establishment consisting of a full-time scientific staff under the Chief Defence Scientist. Comprising three branches — Programmes and Operations, International Programmes and Projects, and Policy and Equipment Requirements — its function was to advise on research and development initiatives, test and evaluation programs and scientific matters in general. It was also required to ensure that the level of scientific and technical expertise was maintained to meet present and foreseeable defence needs.

In 1974, following a major review of defence activities, government defence research was placed under the control of the Defence Science and Technology Organisation (q.v.), within the Department of Defence and under the direction of the Chief Defence Scientist. Originally it consisted of three divisions: Military Studies and Operational Analysis, Policy Programmes and Planning, and the ADSS's Services Laboratories and Trials Division which controlled WRE, MRL and ARL as well as the Engineering Development Establishment in Victoria, the RAN Research Laboratory and the RAN Trials and Assessing Unit in New South Wales, the Armed Forces Food Science Establishment in Tasmania and the Tropical Trials Establishment and the Joint Tropical Research Unit in Queensland. By 1977–78 DSTO employed about 5300 people. The Department of Supply, however, remained responsible for the production of munitions which was carried out in the government factories at Footscray (ammunition, fuses, primers and cartridge cases), Deer Park (high explosives and propellants) and Maribyrnong (propellants, explosives, rocket motors, paints and allied products) in

Victoria, and Mulwala (propellants) and St Marys (filled and explosive munitions) in New South Wales. The Small Arms Factory at Lithgow produced rifles, machine-guns, mortars and small calibre shells and fuse components, while ordnance factories at Bendigo and Maribyrnong made heavy guns, mountings, projectiles, fire control equipment and rocket motor components. In addition to these factories private industry produced a wide variety of ancillary material.

In 1987–88 DSTO was reorganised into five major laboratories and seven smaller establishments, including a Defence Information Services Branch in Canberra, and its charter was amended to enable it to establish appropriate links with sectors of industry not involved directly in defence work. As the result of the government's policy of rationalisation and downsizing of defence support services, a further reorganisation was carried out in 1994 when the then four laboratories, Materials, Aeronautical, Surveillance and Electronics, were reduced to two — the Aeronautical and Maritime Research Laboratory and the Electronics Surveillance Laboratory. DSTO has begun contracting out as much engineering work to industry as possible. Two scientific divisions (Explosives Ordnance and Guided Weapons) will form a new Weapons Systems Division in 1994–95. When structural changes are completed in 1996, DSTO will employ about 2600 people, including between 1200 and 1300 scientists and engineers, and will have a budget of just over 2 per cent of defence expenditure.

Australian defence science and technology has benefited greatly from Australia's involvement with other countries. From the time of the Joint Project Agreement with the United Kingdom in 1946 Australia has gained valuable experience through cooperative ventures and through the exchange of information with other nations. From the early 1950s Australia has been an active member of the Commonwealth Defence Science Organisation, the aim of which is to promote the best use of the defence science resources of its members. Australia is also a member of the Commonwealth Advisory Aeronautical Research Council which provides for consultation and coordination between members in a range of aeronautical technologies for both civil and military aviation. In 1962 Australia was a founder member of the European Launcher Development Organization and in 1965 joined the United Kingdom, the United States and Canada on the subcommittee of the Non-Atomic Military Research and Development Organization. Australia had benefited considerably under the Technical Co-operation Programme of the latter, as well as being able to make a substantial contribution in such areas as anti-submarine warfare (q.v.). From its very beginning Australia, because of its strategic location, has been involved with the US National Aeronautical and Space Administration in providing sites and support for satellite tracking activities. In 1967, using an American rocket, Australia became the third country in the world to launch its own satellite (WRESAT). In addition to the Technical Co-operation Programme, Australia has separate bilateral arrangements with the United States, United Kingdom, Canada, New Zealand, France and Sweden, as well as less formal arrangements with a number of other European countries, and an Agreement for Cooperation in Defence Science with Singapore.

Australia very much relies on imported technology for its defence capabilities, but certain environmental factors and Australia's size and strategic geography are sufficiently different from those of other countries to demand modified technology. Research and development carried out by DSTO in recent years has enabled the development of the Jindalee radar network (q.v.), the Barra sonobuoy (q.v.), towed acoustic arrays for surveillance, airborne lasers for hydrographic survey and the Nulka expendable decoy. DSTO has also developed specialised welding technology, designed acoustic coatings to optimise the performance of the *Collins* Class submarines (q.v.), and developed new methods of airframe fatigue testing and repair methods for aircraft and guided missile frigates. DSTO also provided valuable support for Australian defence elements in the Gulf War, including the evaluation of threat to naval vessels.

While DSTO is the main source of Australia's defence science and technology, the organisation also draws on the CSIRO, the universities and industry. The Department of Defence shares a Memorandum of Understanding with the CSIRO and draws upon its work in areas such as environmental data collection; the universities undertake a significant amount of defence research, and the Department of Defence is a member of several of the government's Cooperative Research Centres, set up to foster interaction between public sector research, the universities and industry. Each service maintains a small group of experts for testing and evaluation work and some research is carried out by the services on occupational health.

Throughout the history of Australia's defence, science and technology has played an important role in war and peace. The demands of war stimulated invention and innovation in a wide variety of fields and pushed Australia towards a firmer industrial base. In peacetime in the 1990s Australia relies for its defence on advanced science and technology and to achieve this concentrates on four main objectives: to be able to exploit future technological developments for defence needs; to ensure that Australia is always a well-informed buyer of defence equipment; to develop new capabilities as the need arises; and to support existing capabilities by increasing operational performance in a cost-effective manner.

GERALD WALSH

SCOTT, Ernest (21 June 1867–6 December 1939). Following employment as a pupil-teacher and a journalist, Scott migrated to Melbourne from England in 1892, and worked for the Theosophist movement, of which his wife, daughter of Annie Bessant, the radical freethinker, was a firm adherent. Scott worked for the Melbourne *Herald* in the early 1890s, and developed what began as an amateur interest in history into the beginnings of a serious professional calling. Between 1910 and 1914 he published three books dealing with the European discovery of Australia. He left the *Herald* in 1895 and became a Hansard reporter, initially for the Victorian parliament and, between 1901 and 1913, for the Commonwealth parliament. In 1913 he was invited to apply for the chair of history at the University of Melbourne, to which he was appointed on the strength of his published output despite his lack of formal tertiary qualifications. Over a long academic career he proved to be an inspiring teacher, an adept if somewhat autocratic administrator (in keeping with the age's taste for 'God professors'), and a scholar with a pronounced capacity for work, demonstrated by his continuing publishing record: over the next 23 years he published eight books, dozens of articles and shorter pieces, while making a significant contribution to both Bean's (q.v.) official history and to the *Cambridge History of the British Empire*. While at Melbourne he nurtured a number of young historians who went on to celebrated academic careers in their own right. He also taught some of the first university-level courses in Australian history, 20 years before Manning Clark. A man with a wide range of interests, he was sociable, personally generous, passionate about classical music and both fond of and knowledgable about wine, theatre and dogs.

Scott's contribution to Bean's history, published as volume 11, *Australia During the War*, in 1936, was notable for being the only volume written by a university historian. Scott undertook the task in 1928 after Bean's first choice for the volume had proven ill-matched to the undertaking. More than 350 000 words and over 900 pages in length in its published form, Scott was forced to abandon a planned fourth section to the book, which was to deal with popular and cultural responses to the war, when his initial estimate of a 240 000 word manuscript proved too modest. His work is immensely detailed — probably too much so for modern tastes — and has been criticised for its favourable treatment of the Governor-General, Sir Ronald Munro-Ferguson (later Lord Novar), with whom he enjoyed good personal relations going back to the war years. Other critics suggest that although Scott's brief was to discuss the impact of the war on Australia and Australians, the level of his discussion rarely descends below that of the great public figures: 'the people' are seldom in evidence. Scott himself was a conservative in his later life (a long-time member of the Melbourne Club), and he wrote conservative history, but he also wrote history which was, on the whole, judicious and fair-minded even where modern readers may disagree with some of his conclusions or assumptions. He retired from the university in 1936. There is a chair in history named in his honour at Melbourne University, and the university supervises an Ernest Scott prize awarded every two years for published work in the field of Australian or New Zealand history.

Stuart Macintyre, *A History for a Nation: Ernest Scott and the Making of Australian History* (Melbourne University Press, Melbourne, 1994).

SCOTT, Lieutenant-Colonel William John Rendell 'Jack' (21 June 1888–19 November 1956).

Scott was one of the most enigmatic officers to serve in the Australian army. Educated at Sydney Grammar School, he had a distinguished career at Gallipoli and in France during the First World War. He was awarded the DSO for his role at Flers on 14 November 1916 and was twice mentioned in despatches in 1917. At the end of the war he was responsible for arranging the shipping for the AIF's repatriation.

On his return to Sydney, Scott set up the insurance broking firm, Scott & Board, but his real interests were in returned servicemen's organisations, Empire loyalist associations and the militia. He was one of the central figures in the conservative fight against Bolshevism in the interwar years. Probably the model for Jack Callcott in the novel *Kangaroo* by D. H. Lawrence, Scott was treasurer of the loyalist King and Empire Alliance in 1921–23 and a key figure in later clandestine paramilitary preparations to meet a presumed radical threat. In 1931 he was instrumental in creating the secret army (q.v.), the Old Guard.

In the early 1930s he emerged as a public apologist for the Japanese, writing articles in the *Sydney Morning Herald*, visiting Japan in 1934 and playing an active role in the Japan–Australia Society. His pro-Japanese stance was modified — at least publicly — when, from April 1935 to late 1940, he was chief of a civilian intelligence sub-group working under the joint aegis of the New South Wales Police and Military Intelligence, Eastern Command. Allegations of his being untrustworthy on the grounds of his closeness to the Japanese, though probably false, were eventually used by his detractors within intelligence to deny Scott access to secret files.

Joining the General Staff in Melbourne in June 1940, he commanded the guerrilla warfare training centre for independent companies at Wilson's Promontory from February to May 1941. As GSO 1, General Staff Branch, Army Headquarters, from 9 May 1941, he was well placed to take over command of Gull Force (q.v.), who were assisting Dutch forces in the defence of the island of Ambon, when their commander Lieutenant-Colonel L. N. Roach was removed for insisting that the defence of Ambon was impossible. Assuming command only two weeks before the Japanese attack, Scott never established a rapport with Gull Force. Although obviously a man with considerable charisma and leadership skills in the interwar years, during captivity on Hainan Island he resorted to authoritarian disciplinary measures. Australians committing sometimes minor offences, were handed over to the Japanese for harsh punishment which included beatings and electric shocks. By early 1945 Scott's alienation from the men he commanded was so complete that the ranks took discipline into their own hands and his officers considered removing him from command by declaring him insane.

The 2/21st Battalion refused to allow Scott to lead them at Anzac Day parades after the war, preferring their old commander, Roach.

Joan Beaumont, *Gull Force: Survival and Leadership in Captivity 1941–1945* (Allen & Unwin, Sydney, 1988); Andrew Moore, *The Secret Army and the Premier* (New South Wales University Press, Sydney, 1989).

JOAN BEAUMONT

'SCRAP-IRON FLOTILLA' refers to HMAS *Stuart* and the four 'V' and 'W' Class destroyers (q.v.) transferred to Australia from the RN in 1933 and described during the Second World War by German propaganda minister Joseph Goebbels as 'another consignment of scrap-iron from Australia'.

SCRATCHLEY, Major-General Peter Henry (24 August 1835–2 December 1885). Born in Paris, the son of a medical officer in the Royal Artillery, Scratchley graduated from the Royal Military Academy, Woolwich, in 1854 and was commissioned in the Royal Engineers. He served in the Crimean War from 1854 to 1856, and in the Indian Mutiny

between 1857 and 1859, before being sent to Melbourne in 1860 in command of a detachment of engineers to construct defence works of his own design. Financial restrictions hampered his work and very little was built, but Scratchley remained in Melbourne, becoming involved in local defence matters and supporting the establishment in Victoria of a volunteer engineers corps. He returned to the United Kingdom in 1863 but retained his interest in Australian defence questions, not only in Melbourne but in South Australia, on whose defence problems he wrote a report in 1865. While in England he supervised the construction of coastal defence gun batteries at Portsmouth Harbour, and shortly afterwards was appointed Assistant Inspector and then Inspector of Works for the Manufacturing Department at the War Office, a position he held for 12 years. In 1874, newly promoted lieutenant-colonel, he was appointed to assist Sir William Jervois (q.v.) in reviewing the defences of the Australian colonies. Jervois became governor of South Australia in 1878, and the following year Scratchley was appointed Commissioner of Defences of all the Australian colonies and New Zealand. Scratchley believed in the protection that British seapower could afford the colonies, and he rejected costly schemes to raise large local forces, preferring instead to trust in strong, well-placed fortifications near ports and vital coastal installations, supported by submarine mines to obstruct shipping channels and approaches, and the offensive use of torpedoes against enemy ships. After he retired from active service in 1882 with the rank of honorary major-general, Scratchley was retained by the Colonial Office as defence adviser for the Australian colonies.

He was made KCMG in 1885 and appointed Special Commissioner for the British Protectorate of New Guinea, which was regarded as being important to the security of Australia. His administration showed signs of an unfashionable sympathy towards the indigenous population, but he died at sea of malaria before he could make any lasting impact. The defences of Melbourne, Sydney and Newcastle were his legacy to the defence of Australia.

'SCROUNGING' is a term which has been used by Australian service personnel at least since the First World War. W. H. Downing's *Digger Dialects* (1919) defines the verb 'scrounge' as 'misappropriate', but the word's full range of meanings is probably most neatly summed up in John Laffin's book *Digger* (1959): 'the acquisition of anything to which the scrounger feels he is morally entitled'. Thus it covers everything from gathering souvenirs to making use of scrap and abandoned material to outright theft.

During the First World War Australian soldiers were dedicated collectors of souvenirs (in the conventional sense of mementos; the term 'souvenir' itself gained currency as a euphemism for theft during the war). Many members of the 1st AIF were well aware of the historical significance of the war in which they were engaged; for most, also, it was their first and only chance to travel, so they were determined to bring back with them items that would help them to remember the experience and that would prove they had been there. Australian soldiers do not seem to have been such keen souvenir-hunters during the Second World War, partly because the experience was less novel and partly because conditions were less conducive to such activities.

Scrounging in the sense of making use of discarded material has been noted as a characteristic of Australian service personnel since the Boer War. The Second World War saw the construction of a working radio transmitter, 'Winnie the War Winner' (q.v.), from salvaged material, while the use by Australians in the Western Desert of equipment abandoned by enemies and allies alike earned them the nickname 'Ali Baba Morshead and his twenty thousand thieves'. The idea that Australian soldiers are consummate scroungers in this sense is profoundly flattering to Australians' image of themselves as a resourceful people able to perform miracles with the proverbial piece of No. 9 fencing wire. Air Chief Marshal Frederick Scherger (q.v.) was probably closer to the mark, though, when he pointed not to Australians' supposed bush heritage but to services starved of resources as an explanation for the tradition of scrounging.

Finally, theft has also been engaged in by Australians in every war since the Boer War, in which it was referred to as 'commandeering'. 'What a multitude of sins that word commandeering covers!' wrote one Australian soldier in 1900. 'What we call at home thieving, looting, burglary, and horse-stealing, is all called commandeering here, and is very much in fashion.' Such practices, though common in wartime, never received official sanction. Australian POWs who stole from their Japanese captors during the Second World War were able to invest their thieving with greater moral legitimacy than most, and tales of prisoners outwitting the Japanese by stealing things from under their noses are common in POW literature.

SEA FURY, HAWKER (Carrier-borne fighter-bomber [FB11]). Wingspan 38 feet 5 inches; length 34 feet 8 inches; armament 4 × 20 mm cannon, bomb load 8 × 60-pound rockets or 2 × 1000-pound bombs; maximum speed 460 m.p.h.; range

680 miles; power 1 × Bristol Centaurus 2480 h.p. engine.

The Sea Fury, with the Fairey Firefly (q.v.), entered service in 1949 and was the first combat aircraft flown by the Fleet Air Arm (q.v.). Fast and highly manoeuvrable, the Sea Fury belonged to the last generation of piston-engine fighters that was made obsolete by jet aircraft. Sea Furies from the aircraft carrier HMAS *Sydney* flew 1623 sorties for the loss of eight aircraft during the Korean War from October 1951 to January 1952. Sea Furies in Korea flew armed reconnaissance and naval gun-fire-spotting mission, ground-attack and combat air patrols. Australian Sea Furies made no contacts with enemy aircraft over Korea, but RN Sea Furies actually shot down several MiG jets. Sea Furies were replaced by the Sea Venom (q.v.) from 1956 and one is held by the Naval Aviation Museum at HMAS *Albatross*.

SEA KING, WESTLAND (4-crew anti-submarine, search-and-rescue helicopter [HAS Mark 50]). Rotor diameter 18.90 m; length 17.02 m; armament 4 × torpedoes; maximum speed 230 km/h; range 1230 km; power 2 × Rolls Royce Gnome 1535 shaft h.p. engines.

The Sea King replaced the Wessex (q.v.) as the RAN's anti-submarine helicopter in 1975. They flew from the aircraft-carrier HMAS *Melbourne* (q.v.) until its retirement in 1982. Sea Kings were then tri-alled for use as anti-submarine aircraft on the newly commisioned *Oliver Hazard Perry* Class Guided Missile Frigates (q.v.), but this was unsuccessful and the role was taken by the Seahawk (q.v.). They are now based at HMAS *Albatross,* Nowra, and are also used for search and rescue. Sea Kings are equipped with a dipping sonar for submarine detection.

SEA VENOM, DE HAVILLAND (Carrier-borne 2-seat all-weather fighter). Wingspan 42 feet 10 inches; length 36 feet 7 inches; armament 4 × 20 mm cannon, 8 × 60-pound rockets; maximum speed 560 m.p.h.; range 1000 miles, power 1 × De Havilland Ghost Mark 104 turbojet.

The Sea Venom was ordered from Britain in 1951 as a replacement for the Sea Fury (q.v.). They were delivered to HMAS *Melbourne* (q.v.) in the United Kingdom during 1955 and arrived in Australia in May 1956. Four Fleet Air Arm (q.v.) squadrons received the new aircraft, three operating from the *Melbourne* and one based at Nowra in New South Wales. The Confrontation (q.v.) with Indonesia extended their life in the mid-1960s. In 1967 they were replaced with the Skyhawk (q.v.). Eleven of the 39 delivered crashed while in service, with the loss of nine lives.

SEAGULL V (WALRUS), SUPERMARINE (3-crew reconnaissance seaplane). Wingspan 45 feet 10 inches; length 37 feet 7 inches; armament 2 × 0.303-inch machine-guns; maximum speed 135 m.p.h.; range 600 miles; power 1 × Bristol Pegasus 775 h.p. engine.

The Seagull V was specifically designed to be flown from RAN cruisers. Twenty-four were delivered between 1935 and 1937. The RN then decided to order the same aircraft but called it the Walrus. Thirty-seven additional aircraft from the British order were delivered to Australia between 1939 and 1944 and they retained the Walrus name and British serial numbers. During the Second World War HMA Ships *Australia* (q.v.) and *Canberra* (q.v.) and the *Leander* Class cruisers (q.v.) *Hobart*, *Perth* and *Sydney* were all equipped with one Seagull V which was launched from a catapult. The aircraft flown from HMAS *Perth* was shot down by a German aircraft on 28 April 1941 during the evacuation from Crete. The armed merchant cruisers (q.v.) *Manoora* and *Westralia* also carried the aircraft. They were used mainly in a fleet reconnaissance role, spotting the fall of shot in shore bombardments and in air-sea rescue work. A Walrus was used during the 1947–48 Antarctic expedition, after which they were withdrawn from service.

SEAHAWK, SIKORSKY (3-crew anti-submarine helicopter). Rotor diameter 16.36 m; length 15.26 m; armament 2 × 7.62 mm machine-guns; 2 × torpedoes; maximum speed 257 km/h; range 460 km; power 2 × General Electric 1900 shaft h.p. engines.

The Seahawk is the naval version of the Black-hawk (q.v.) chosen by the RAN to be its current anti-submarine helicopter. The first Seahawk entered service with the RAN in 1988. Eight of the sixteen helicopters ordered were assembled in Australia by Aerospace Technologies of Australia (formerly the Government Aircraft Factory). Sea-hawks are flown from the *Oliver Hazard Perry* Class Guided Missile Frigates (q.v.) and in 1990 two were used during the Gulf War.

SEATO see **SOUTH-EAST ASIA TREATY ORGANIZATION**

SECRET ARMIES such as the Old and New Guards in New South Wales, the League of National Security or White Army in Victoria, and the Citizens' League in South Australia were at their peak during the Depression, when they had well over 100 000 members across Australia. With military-style chains of command in place, they stood ready to avert an expected communist revolution and to take power themselves if they thought it necessary.

They all shared an anti-communist, anti-labour, élitist and empire loyalist outlook common among the business and professional classes from which most of their members came, and their *bête noire* was populist New South Wales Labor Premier J. T. Lang. After Lang was dismissed from office by Governor Sir Philip Game in 1932, secret army activity declined sharply. There were a large number of men with military experience involved in these organisations, including such senior officers as Sir Thomas Blamey (q.v.) (almost certainly the commander of the White Army) and Sir Edmund Herring (q.v.). It has been suggested that Sir Charles Rosenthal (q.v.) was the basis for the character of Benjamin Cooley, leader of the Sydney-based secret army depicted in D. H. Lawrence's 1923 novel *Kangaroo*.

Michael Cathcart, *Defending the National Tuckshop: Australia's Secret Army Intrigue of 1931* (McPhee Gribble, Melbourne, 1988).

SECRET INTELLIGENCE AUSTRALIA (SIA) see **ALLIED INTELLIGENCE BUREAU**

SECRET INTELLIGENCE ORGANISATION (SIO) see **CODE-BREAKING**

SELF-FIRING ('POP-OFF') RIFLE This was a delayed-action weapon, devised by Lance Corporal William Scurry and Private Alfred Lawrence, both of the 7th Battalion, for use during the evacuation of Gallipoli in December 1915. The principle was based on the action of the sand in an hour-glass. A weight attached to the trigger would fall and fire the rifle. Overbalancing of the weight occurred when water had trickled from one tin, with a small hole in it, into another, placed beneath it, the whole mechanism joined by string or wire. The size of the hole functioned as a primitive form of timing device. The rifles were placed to fire along fixed lines, 12 to a battalion front, on the night of 19–20 December. The effect was to simulate an ordinary level of night-time activity, fooling the Turks into believing that the Australians were still in possession of their trenches. Scurry was awarded a DCM and mentioned in despatches for his efforts.

SENTINEL TANK Weight 27 tons; speed 30 m.p.h.; armour 50 mm hull, 75 mm turret; armament 2-pounder gun, 2 × 0.303 Vickers machine-guns (one co-axially mounted in the turret, one hull mounted); crew 5; power 3 Cadillac petrol engines.

The Australian-built Sentinel tank (Cruiser Tank [Mark I]). (AWM 101156)

The Sentinel Cruiser tank was designed in Australia in 1941 as a response to the need for tanks for the AIF in the Middle East. To power the tank three Cadillac automobile petrol engines were combined into a single unit. The first model was tested at Puckapunyal (q.v.) in February 1942; 66 were later delivered to the army. In the event none saw service. The Middle East forces had been supplied with Grant tanks, so the Sentinels were used as training vehicles within Australia. Later in the war the Mark III Sentinel was equipped with a 25-pounder gun and a Mark IV with a 17-pounder gun. The Sentinel was an advanced design not equalled in the British and American forces until the Sherman.

SERONG, Brigadier Francis Phillip 'Ted' (11 November 1915–) entered RMC Duntroon in February 1935 and graduated in 1937. He served on the staff of the 6th Division in New Guinea between 1942 and 1945. He held a number of staff posts at Army Headquarters after the war, and in 1955, with the rank of colonel, was given command of the newly reopened jungle warfare training centre at Canungra (q.v.) in Queensland. Between 1960 and 1962 Serong acted as an adviser to the Burmese military, training them in counter-insurgency techniques. In 1962 he was recalled to become the first commander of the Australian Army Training Team, Vietnam (q.v.), where he remained until 1965. He was then seconded to the American forces, and remained in Vietnam in a number of capacities: as commander of the South Vietnamese Police Field Force between 1965 and 1967, and as an adviser on security and paramilitary operations to both the US and South Vietnamese governments between 1967 and the fall of Saigon. He retired from the Australian Army in 1968. A controversial figure inside and outside the army, he wrote widely on problems of counter-insurgency during the 1970s.

SERVICE CORPS see **ROYAL AUSTRALIAN ARMY SERVICE CORPS**

SERVICE NEWSPAPERS The many newspapers produced by Australian soldiers during the First World War varied greatly in style, content, presentation and quality as a result of the varying conditions under which they were produced. Newspapers were published on AIF troop-ships as early as August 1914, and continued to be produced on almost every troop-ship going to and from Australia until 1919. Run off on spirit or jelly copiers, most were typescript, but some were hand-written. Some could not be produced at sea, so material was collected and printed at a port of call. A few were produced on shipboard printing presses. Editors relied on the troops to provide material for the papers, and some ran for only one or two issues due to lack of contributions. Others took off and became enormously successful — 16000 copies of the *Osterley*'s paper were sold during a voyage from Australia to Egypt in 1917. Designed to combat the boredom of troop-ship life, these newspapers concentrated on humour and gossip. They poked fun at shipboard routine, the discomfort of life at sea, and the foibles of fellow soldiers, but also contained patriotic and nostalgic material. Many ships produced souvenir editions of their papers to be sent home to loved ones, and these editions concentrated on accounts of the voyage as well as serious, often patriotic, poetry and prose rather than gossip and in-jokes.

Some troop-ship papers continued to be produced as unit newspapers after the voyage was over, and even when there was not a direct continuity of publication, troop-ship papers provided the model for the style and tone of papers produced on active service. 'Trench' newspapers, as the latter were called, can be roughly divided into three categories. The first category includes those papers produced under difficult circumstances virtually in the front line. They generally consisted of one or two hand-written pages, duplicated by carbon or stencil machine. These short-lived publications were subject to little or no official control, and were usually issued irregularly at no cost. Publications in the second category were produced behind the front line, where they were subject to some official control. They were usually typed and reproduced on jelly or spirit duplicators, while a few were printed. Such papers could be four or more pages long, were produced semi-regularly and were sold to readers. In the third category were the best-known papers such as *Aussie*, *Kia Ora Coo-ee* and *Digger*, generally produced with official support by people with some journalistic experience. They appeared regularly and frequently, were printed, were of substantial length, contained photographs, and were sold to members of the AIF in general rather than to particular units. They could achieve huge circulation: 10000 copies of *Aussie*'s first issue were distributed, and by its third issue it had achieved a print run of 100000 copies.

Some of the more sophisticated publications included news, letters and verse from Australia, as well as articles written by newspaper staff or by official war correspondents. However, what distinguished all of these papers, of whatever degree of sophistication, was their reliance on contributions from the troops themselves. These contributions took such forms as short stories, verse (mostly in the style of bush balladists like 'Banjo' Paterson [q.v.] or urban larrikin poets like C. J. Dennis), mock advertisements and cartoons. They show clearly the emergence of a distinctive digger (q.v.) subculture, with its own language and traditions. Much of their content came directly from the diggers' oral culture: gossip, furphies (q.v.) and humorous complaints about army life and discipline. Such complaints were staples of troop newspapers, with favourite targets such as army cooks and pompous officers coming in for repeated lampooning. This airing of grievances in an acceptably lighthearted form acted as a safety valve, and the form of complaint, stressing the farcical rather than the horrific aspects of war, revealed an overall acceptance by soldiers of their lot. There was also a strongly sentimental strand running through much of this literature, expressed in the idealisation of Australia and longing for loved ones as well as in the celebration of mateship.

These newspapers projected an unambiguously Australian image in their content and language. *Aussie* announced that it 'aims at being a dinkum Aussie — and a dinkum Aussie uses the language of the Aussie'. As with the troop-ship journals, the more sophisticated troop newspapers were sent home as souvenirs, and this made editors and contributors more self-consciously concerned to record their experiences for posterity while also imposing restrictions on what they could write. The *Kia Ora Coo-ee*'s management committee, admitting that 'one or two contributions not quite suitable for Home readers of all classes' had been published, promised 'a general improvement in tone' which was reflected in its second series by an increase in educational features and a decrease in ironic complaints. After the war the tradition of troop newspapers was continued in civilian papers such as *Smith's Weekly* and a revamped *Aussie* which catered to a readership of returned servicemen, as well as in the journals of returned services organisations. Picking up on the sentimental strand in the digger subculture, these publications emphasised nostalgia by reprinting selections from wartime digger papers as well as jokes, stories and verse from returned soldiers.

During the Second World War troop newspapers were once again produced and were sometimes commissioned by officers as 'furphy-flushers', designed to counter the spread of rumours by printing officially sanctioned news. One such commissioned newspaper was the *Tobruk Truth*, which reached a circulation of 800 roneoed copies a day

and was distributed to all units and detachments at Tobruk. It consisted almost entirely of news from BBC radio broadcasts with occasional local news, but it did eventually start publishing cartoons. More in line with the First World War trench newspaper tradition were unit newspapers such as *Mud and Blood*, begun on a troop transport and continued at Tobruk, where it concentrated as much on local comment, verse and cartoons as on news. Much more ambitious in scale was the *AIF News*, established in March 1940 at the instigation of then Brigadier A. S. Allen (q.v.). Initially produced as a cyclostyled typescript, it soon began to be printed and sold weekly to AIF members throughout the Middle East. It reached a circulation of 40 000, and more than paid for itself. Produced by servicemen with backgrounds in journalism, drawing on material supplied by war correspondents and Australian newspapers, it concentrated on local and Australian news but also included topical pictures and verse by AIF poets. There was some conflict between the staff of *AIF News* and Army Headquarters over the paper's content, and Kenneth Slessor (q.v.) commented that the army 'doesn't seem to know whether it wants a paper or a parish weekly'. When the 9th Division returned from the Middle East they were accompanied by the *AIF News* staff, who from May 1943 to June 1944 produced the army newspaper *Table Tops* on the Atherton Tableland. Meanwhile *Army News* had begun publication in the Northern Territory in October 1941, and *Guinea Gold*, which reached a circulation of 57 000, had been published daily in New Guinea since November 1942. Both of these newspapers continued until 1946.

General Sir Thomas Blamey (q.v.) encouraged the publication of Army newspapers as a means of keeping troops informed, but he strictly forbade the inclusion of editorial comment of any kind. This restriction did not apply to the other important army periodical of the Second World War, the army education journal *Salt* (published September 1941 to April 1946). Designed to fit into a soldier's pocket, *Salt* expanded from 32 to 64 pages and its circulation grew from 55 000 to 185 000 copies during its lifetime. It was issued free weekly or fortnightly to soldiers, with the twin aims of providing troops with information and of allowing them to express themselves in words or pictures. As well as publishing material on current affairs and educational matters it encouraged and published letters, stories, jokes, verse and drawings from its readers. Though it observed strict neutrality in party-political matters it was still accused of bias by some politicians, and it was also attacked by some military leaders who felt it was subversive of discipline and decency. From June 1943 it was subject to censorship by the Director-General of Public Relations.

Troop newspapers produced in later wars, such as Korea and Vietnam, generally followed the style of their predecessors.

The tradition of producing service newspapers does not seem to have been as strong in the RAN and RAAF as in the army. *Air Force News*, a self-supporting newspaper, was produced by the RAAF Public Relations Directorate during the Second World War. After the entry of Japan into the war it ceased publication for security reasons, but it was later replaced by a fortnightly magazine, *Wings*, published from April 1943 to December 1945. During the Second World War individual ships within the RAN also produced newspapers which, like unit newspapers in the army, were written by ordinary sailors and thrived on local gossip.

Graham Seal, '"Written in the trenches": Trench newspapers of the first world war', *Journal of the Australian War Memorial*, no. 16 (1990), pp. 30–8.

SERVICES RECONNAISSANCE DEPARTMENT
(SRD) see ALLIED INTELLIGENCE BUREAU; SPECIAL OPERATIONS AUSTRALIA

SHEDDEN, Frederick Geoffrey (8 August 1893–8 July 1971). Born in Kyneton, Victoria, Shedden began a long career in the Department of Defence in 1910 as a junior clerk. He studied as an accountant and in March 1917 was appointed as a lieutenant in the Army Pay Corps of the AIF, serving in London and France until the end of 1919. He returned to Defence and completed an economics degree at the University of Melbourne, and then undertook further study at the University of London. In 1928 he attended the Imperial Defence College (IDC), where Lieutenant-Colonel E. K. Squires and Lieutenant-Colonel J. D. Lavarack (qq.v.) were among his fellow students. He remained in London for several years, attached to the War Office in order to study financial administration, and to the High Commissioner, S. M. Bruce. While in the United Kingdom, Shedden developed close and enduring contacts with a number of powerful figures in the defence establishment, notably the Commandant of the IDC, Admiral Sir Herbert Richmond, and the Secretary of the Committee of Imperial Defence (CID), Sir Maurice Hankey (q.v.). Both were committed to the ideals of imperial defence and, through a two-year attachment to the Cabinet Office and to the CID, Shedden developed a similar approach, both to the formulation and the execution of defence policy. In later years, when he had reached the pinnacle of his power and influence in the Department of Defence, he was sometimes, and not always flatteringly, called Australia's 'pocket Hankey' in the light of his undisguised emulation of his great mentor.

On his return to Melbourne, Shedden rose steadily through the ranks of the Department of Defence, becoming secretary in 1937, a position he held until his retirement in 1956. Whatever doubts, if any, he had about the Australian government's reliance on the Singapore strategy (q.v.), he firmly supported the government's acceptance of British assurances, even though as early as his year at the IDC he had known that army officers such as Lavarack had grave misgivings. He played a key role in the expansion of Australia's military forces from 1938, and in 1939, with war looming, he organised the compilation of the *War Book*, a detailed masterplan for the mobilisation of Australian resources in the event of war. With the establishment of the War Cabinet in 1939, Shedden was appointed its secretary, and became responsible for the efficient conduct of cabinet business, but his influence quickly spread to areas outside his nominal control and he became a highly influential adviser to Prime Minister Menzies and his Labor successor, John Curtin. Shedden's hitherto unwavering support of British imperial defence policy and strategy was severely shaken by the débâcle of the Greek campaign in 1941 and even more so by the fall of Singapore in February 1942. With the entry of the United States into the war, Shedden became the vital personal link between Curtin and General Douglas MacArthur, and opened a liaison office in Brisbane when MacArthur moved his headquarters there in order to maintain regular and confidential communication between the two leaders. He accompanied three prime ministers, Menzies, Curtin and Chifley, on overseas trips in 1941, 1944 and 1946, respectively, preparing detailed briefing papers for them. He was appointed CMG in 1941, and KCMG in 1943, the latter on Curtin's recommendation.

He served as Chairman of the Defence Committee from 1948 to 1956, overseeing Australian involvement in the Korean War and Malayan Emergency, and the growing complexity of Australia's postwar defence relationships. By 1956, however, his unrivalled position of power and influence, and his intensely personal control of defence matters, had aroused growing hostility within the government bureaucracy, and his 20-year tenure of office came to be seen not necessarily as a source of strength but as an impediment to fresh thinking and greater flexibility. After publicly criticising the effectiveness of defence spending before a parliamentary committee, Shedden was asked to step aside in 1956 and devote his energies to writing a history of Australian defence policy, which he proceeeded to do with typical dedication but with a lack of discrimination, so that the final huge manuscript was never published.

Shedden was one of the outstanding Defence bureaucrats in Australian history. He presided over the Department of Defence at a critical period and he established over many years exceptionally high standards of administration. Exactly what his influence was on the formulation of policy has yet to be assessed with any clarity.

SHEEAN, HMAS see *COLLINS* **CLASS SUBMARINES**

SHEPHERD, Malcolm Lindsay (27 October 1873–25 June 1960) entered the New South Wales Public Service in 1890, and moved to the Commonwealth Public Service in 1901 as a clerk in the Postmaster-General's Department. In 1904 he became private secretary to Prime Minister Alfred Deakin, and thereafter served successive prime ministers until 1911. When Andrew Fisher established a Department of the Prime Minister, Shepherd became its first secretary. He enjoyed a close personal and professional relationship with William Morris Hughes (q.v.), who appointed him official secretary to the High Commission in London; for a number of months in 1921 he became acting High Commissioner. In 1927 he became Secretary of the Department of Defence. He retired in 1937 and was succeeded by Frederick Shedden (q.v.). With limited formal education, Shepherd's rise was a notable achievement, although his perceived closeness to successive Labor leaders, Fisher and Hughes, caused comment and resentment in what was still a very conservative federal Public Service.

SHEPPARTON (I), HMAS see **BATHURST CLASS MINESWEEPERS (CORVETTES)**

SHEPPARTON (II), HMAS see **SURVEY SHIPS**

SHOALHAVEN, HMAS see **BAY CLASS FRIGATES**

SHORT-MAGAZINE LEE ENFIELD (SMLE) NO. 1 MARK III RIFLE was a manual-operation, bolt-action, magazine-fed weapon. The SMLE was the standard issue service rifle in the Australian Army from 1913 to 1959, when it was replaced by the L1A1 SLR (q.v.). During that time it saw active

service with Australian soldiers in the First and Second World Wars and Korea. The SMLE was designed at the Royal Small Arms Factory at Enfield in England during the early 1900s and the No. 1 Mark III went into production there in 1907. In 1912 the SMLE No. 1 Mark III went into production at the Small Arms Factory at Lithgow with the first rifles being issued to the army in 1913. Production modifications to the No. 1 Mark III in England in 1916 resulted in the No. 1 Mark III* which was not produced at Lithgow until the late 1930s. The main modifications were the deletion of the original long-range volley sights (on the left side of the weapon) and the magazine cut-off. During the Second World War, the Small Arms Factory produced Sniper Rifle versions of the SMLE fitted with telescopic sights and experimented with shortened versions of the SMLE for jungle warfare. Early in the Second World War, to cover the light machine-gun deficiency, Australia and New Zealand began emergency conversion of some SMLE rifles to automatic fire with the Charlton conversion. They were not, however, put into service. The SMLE could be fitted with the 17-inch (423 mm) Pattern 1907 Sword Bayonet, wire cutters and a grenade launcher. It had a fine reputation as a robust, reliable, accurate and easy to operate combat rifle.

The SMLE fired 0.303-inch calibre ammunition, was fed from a ten-round capacity detachable magazine and could be loaded using five-round charger clips. The weight of the SMLE was 8 pounds 15 ounces (4.1 kg) empty and 9 pounds 14 ounces with a full magazine, with the bayonet fitted it was 10 pounds 9 ounces. The effective range of a SMLE for individual fire was 300 yards and 600 yards for section fire. The sight range was 200 to 2000 yards and the volley dial fire sight of the earlier No. 1 Mark III had a sight range of 1600 and 2800 yards. A well-trained rifleman was capable of firing up to 15 well-aimed shots per minute.

IAN KURING

SHRINE OF REMEMBRANCE in the Melbourne Domain was opened in 1934 to commemorate Victorian men and women who served in the First World War. The 1922 competition to design the memorial was won by two architects and returned soldiers, Phillip Burgoyne Hudson and James Hastie Wardrop. The large and imposing design is based on one of the seven wonders of the classical world, the Mausoleum of Halicarnassos (a royal tomb built in what is now Turkey about 350 BC), and is decorated with sculptures by English sculptor Paul Montford. The focus of the interior is the

Stone of Remembrance sunk into the shrine floor which bears the words 'Greater Love Hath No Man'. A unique feature of the Shrine is a precisely-placed hole in the wall that allows a beam of sunlight to pass over the word 'love' on the Stone of Remembrance at 11.00 a.m. on 11 November (Remembrance Day) every year. The introduction of daylight saving in 1971 would have made the beam of light arrive one hour late, so small mirrors were installed to bend the light so the effect is still retained.

SHROPSHIRE, HMAS (*London* Class Heavy Cruiser). Laid down 1926, paid off 1947; displacement 9830 tons; length 633 feet; beam 66 feet; speed 32 knots; armament 8 × 8-inch guns, 4 × 4-inch guns (later 8), 4 × 3-pounder guns, 4 × 2-pounder guns, 8 × 21-inch torpedo tubes; 1 Seagull V (q.v.) seaplane.

After the sinking of HMAS *Canberra* (q.v.) in 1942 at Savo Island, the British Government announced the transfer of HMS *Shropshire* to the RAN. The cruiser took up her duties in Australian waters in September 1943. In December 1943, HMAS *Shropshire* as a member of Task Force 74 covered the invasion of New Britain. After that she was involved in numerous fire support missions. In 1944 HMAS *Shropshire* was present at the Battle of Surigao Strait, emerging from that conflict unscathed. HMAS *Shropshire*'s last act of the war was to support the Australian landing at Balikpapan in 1945. She was present at the Japanese surrender in Tokyo Bay in September 1945. HMAS *Shropshire* was decommissioned in 1947 and sold for scrap in 1954.

SIGNALS CORPS see **ROYAL AUSTRALIAN CORPS OF SIGNALS**

SIMPSON AND HIS DONKEY John Simpson Kirkpatrick (6 July 1892–19 May 1915), better known as 'Simpson' or 'the man with the donkey', was born at Shields in England and joined the merchant navy at 17. He deserted at Newcastle, New South Wales, in 1910, and spent the next few years as an itinerant labourer. On 25 August 1914 he joined the AIF, hoping to be returned to England, but was instead assigned to the 3rd Field Ambulance, Australian Army Medical Corps, and sent to Egypt. He was among the covering force which landed on Gallipoli at dawn on 25 April 1915. At Gallipoli he used a donkey (named 'Abdul', 'Murphy' or 'Duffy') to carry wounded soldiers to the dressing station and gained a reputation for being undaunted by enemy fire. On 19 May he was killed, and though he was mentioned in orders of the day and despatches, he received no bravery award. The myth-making began almost immediately after his death, and he soon became one of the best-known images of Anzac (q.v.). He was portrayed as an Australian Everyman, 'the Australian spirit personified' as the Melbourne *Argus* called him in 1933. His popular appeal must be understood in the context of the cult of the wounded soldier which flourished during the First World War. Promoters of this cult in the newspapers and elsewhere idealised the 'typically Australian' courage of wounded soldiers who fought on regardless. They applauded the rescue of wounded comrades as the highest form of mateship; and used the figure of the wounded soldier to shame 'stay-at-homes' into joining up or putting more energy into the war effort. Simpson's selfless devotion to saving the wounded was easily incorporated into this cult, and even those who opposed the war could not argue against the worth of such a saintly figure. Furthermore, Simpson's role as a caregiver, a gentle saviour with 'the hands of a woman', probably had special appeal to women, who so often seemed excluded from the rampantly masculine Anzac legend. The 1965 biography of Simpson by the Reverend Irving Benson, entitled *The Man with the Donkey: The Good Samaritan of Gallipoli*, only helped to promote the myth. Much of it was based on letters from Simpson to his mother and sister (now held at the Australian War Memorial), but Benson left out those letters which reveal Simpson's often fiery temper and radical labourite politics. There are statues commemorating Simpson at the Australian War Memorial and at the Shrine of Remembrance, Melbourne.

Peter Cochrane, *Simpson and the Donkey: The Making of a Legend* (Melbourne University Press, Melbourne, 1992).

SINAI CAMPAIGN see **PALESTINE CAMPAIGN**

SINCLAIR-MacLAGAN, Major-General Ewen George (24 December 1868–24 November 1948). Educated at the United Services College, Westward Ho! in Devon as a contemporary of Rudyard Kipling, MacLagan was commissioned into the Border Regiment in 1889 after a brief stint in the militia. He served in India against the Waziris and in the Boer War, where he was wounded and awarded the DSO while still a captain. He was posted to Australia in 1901 to assist in the reorganisation of the AMF under Major-General E. T. H. Hutton (q.v.), returning to England in 1904.

MacLagan returned to Australia in 1910, brought back as Director of Drill at the new RMC Duntroon by its first commandant, Colonel (later

Major-General) W. T. Bridges (q.v.). One of the cadets in the first entry, S. F. Rowell (q.v.), wrote of him many years later that he was 'the ideal type of regimental soldier ... a leader easy to follow if difficult to emulate successfully'. When Bridges raised the 1st Australian Division in August 1914, he chose MacLagan for command of the 3rd Brigade, which drew its units from the 'outer' States. Easily the most experienced infantry officer among the brigade (or divisional) commanders, his brigade was the one chosen to lead the assault on Gallipoli, and he went ashore with the second wave of the 9th Battalion. His tactical acumen and powers of command helped to impose order on an overly fluid battle in its critical early hours, and probably saved the landing force from disaster. In Monash's later view this was 'the most difficult ... phase of the landing'.

With one brief interlude after the landing, he remained on Gallipoli until evacuated in August. He returned to the command of his brigade in Egypt in January 1916, and took it to France, leading it through the Somme battles (see WESTERN FRONT). From January 1917 he was in command of the AIF Depots in Britain, and Director of Training. On the death in action of Major-General William Holmes (q.v.) in July, MacLagan was promoted and given the 4th Division, which he commanded for the rest of the war. The 4th Division had suffered heavily in the battles of the first half of the year, and was to remain seriously under strength thereafter. MacLagan led it through the fighting at the Third Battle of Ypres and in the defensive battles at Dernancourt and Villers-Bretonneux the following year. The 4th Division led the assault at Hamel, and took part in the storming of the Hindenburg Line in September. In that same month MacLagan was responsible for a training and advisory mission of AIF officers and NCOs attached to II US Corps.

MacLagan was one of the very few British officers to remain with the AIF throughout the war. Monash rated him highly, noting that 'he never failed in performance, and invariably contrived to do what he had urged could not be done. One could not afford to take him at his own modest estimate of himself'. After the war he returned to the British service, and commanded the 51st Highland Division between 1919 and 1923. He retired from the British Army in 1925, and was Colonel of the Border Regiment from 1923 to 1938.

SINGAPORE, FALL OF see MALAYAN CAMPAIGN

SINGAPORE STRATEGY Australian defence thinking between the world wars was dominated by the Singapore strategy, in which the security of British interests in Asia, including Australia and New Zealand, rested upon the construction of a massive naval base in Singapore. It was intended that in time of war a major British naval presence would be stationed in Singapore in such strength as to deter the only serious threat in the region, an expansionist Japan. The Anglo-Japanese Naval Treaty was terminated in 1921, against the wishes of the Australian government whose concerns weighed less in British thinking than the perceived need to reach an agreement with the United States. Two years later, at the Washington Naval Conference of 1923, the three powers accepted a ratio of 5:5:3 in capital ships and aircraft carriers between the United Kingdom, the United States and Japan, an insufficient margin should Britain be faced with war in more than one theatre. That possibility seemed remote in the immediate aftermath of the First World War, and the fact that construction of the base proceeded slowly did not give rise to undue concern. Nor did the fact that the effectiveness of the base, even when completed, rested on British promises to send a fleet to the east within six weeks (extended to 90 days by 1939) discourage a willingness on the part of Australian leaders to accept the basic premises of the Singapore strategy.

Throughout the nineteenth century and into the twentieth, there was general agreement that an imperial approach to defence offered the greatest measure of security, that the colonies and subsequently the dominions were best protected by accepting a blue-water strategy, which asserted that if the RN controlled the seas, outlying parts of the Empire, as well as the metropolitan centre, would be secure against major invasion. Local forces and defences could deal with minor attacks and raids. This strategic view was reasserted at the Imperial Conference (q.v.) of 1923, when the central role of the RN was reaffirmed. The Australian Prime Minister, Stanley Bruce, admitted at the conference that the precise reasoning behind the Singapore strategy escaped him, but he was nevertheless content to accept British assurances of its soundness. His Australian naval advisers preferred the base to be located in Sydney, but this was clearly special pleading based on a narrow view of what the Singapore base was intended to achieve and it was not difficult to understand that the east coast of Australia was much less suitable than Singapore, whose position would enable a British fleet to dominate both the western Pacific and the Indian Oceans. The deterrent value of the base, however, could only be realised if the base was built and if there was a fleet to operate from it. The first requirement was hampered by cuts in British defence spending throughout the 1920s and 1930s, as was the second — the construction and availability of a major fleet. More than anything, though, the Singapore strategy ultimately foundered on changing strategic circumstances that no one had foreseen, not least because they were unthinkable. From the early 1930s British planners had warned the British government that the United Kingdom could not fight a war simultaneously in Europe, the Middle East and the Far East. Britain did not have, and would never have, sufficient resources (including but not limited to military forces) to face war in three widely separated theatres. In the 1920s expectations were that this situation would, indeed could, never arise; in the 1930s, as the unthinkable edged into the realm of possibility, British efforts to repair their accumulated defence deficiencies were accompanied by increasingly urgent efforts to find diplomatic solutions to the dilemmas posed by the growing complexity of the strategic picture.

On one level the Singapore strategy rested on an element of bluff, that the naval base, both as a symbol and a tangible indication of British power, would deter Japan from attacking British interests in the Far East. Australian political leaders willingly accepted the promises, made in good faith if reasserted in a growing climate of unreality, that the base would be impregnable, that the fleet would be dispatched in time to prevent a Japanese attack, and that if necessary, as Churchill assured Menzies in 1941, Britain would sacrifice her position in the Mediterranean to reinforce Singapore and thereby protect Australia. When the time came, and Japan launched her attacks upon Hawaii, Indochina and Malaya, Britain was in no position to honour her long standing commitments, nor to do as Churchill had promised. The strategic situation had deteriorated even from the bleak position from which Churchill spoke in 1941.

In January 1942, as the position of Singapore became ever more precarious, Prime Minister Curtin wrote to Churchill that 'the evacuation of Singapore would be regarded here and elsewhere as an inexcusable betrayal', and went on to insist that Australia had been assured that the fortress would be made impregnable and that it could hold out until the arrival of the fleet. It is, perhaps, understandable that in the fearful circumstances of stunning Japanese successes harsh words should be spoken, but the charge that Britain had 'betrayed' Australia cannot be sustained. Australian politicians had willingly embraced the imperial approach to defence, they had directed defence expenditure heavily towards the navy at the expense of the army and air force, and they had paid no heed to dissenting voices in the army that argued for a greater concentration on local defence capabilities, indeed they had prohibited further discussion of such proposals. Australian defence cuts had been just as severe as those in Britain, and the illusion of security that the Singapore strategy held out had been welcomed in Australia which, unlike New Zealand, had refused to contribute towards the cost of the naval base. Neither Churchill nor successive British governments were guilty of betrayal; the fault lay as much within Australia which, like Britain, had succumbed to the allure of false assumptions. The precise circumstances of the Malayan campaign aside, a degree of collective guilt over unpreparedness and shortsightedness might have been a more appropriate reaction than outraged cries of betrayal. The controversy over the Singapore strategy was revived in an equally tendentious manner by the Australian Prime Minister Paul Keating in 1992.

SIOUX, BELL (2-seat light helicopter). Rotor diameter 37 feet 1 inch; length 32 feet 7 inches; maximum speed 105 m.p.h.; range 273 miles; power 1 × Lycoming 240 shaft h.p. engine.

The Sioux was used as a reconnaissance helicopter by the RAAF between 1960 and 1965. Sioux were then transferred to the army and was used by 161 Reconnaissance Flight in the Vietnam War for observation and other roles including medical evacuation (q.v.). The Sioux was the first helicopter used by the army and remained in service until it was replaced by the Kiowa between 1973 and 1977. A Sioux is preserved by the Australian War Memorial.

'SIX BOB A DAY TOURISTS', a phrase used to describe members of the 1st AIF, referred to the pay rate of private soldiers (6 shillings a day), which was above that of the average worker and made them the best-paid troops of the First World War. It was often used disparagingly by people who suggested that AIF soldiers would see little action and were after nothing more than a well-paid holiday; some leftist opponents of the war were more scathing, calling them 'six bob a day murderers'. However, the term was also adopted by the troops themselves, sometimes ironically but also in recognition of the genuinely touristic aspects of their experiences in famous and exotic places.

SKI TROOPS were a short-lived innovation in the Syrian campaign. In October 1941 Lieutenant-General T. A. Blamey (q.v.) persuaded Major-

Men of I Corps Ski School, Lebanon, January 1942. From left: Major James Riddell, Sergeant Due, Captain C. Parsons, Captain R. Mooney, Sergeant L. S. Salmon, Sergeant J. Abbottsmith. (AWM 011402)

General A. S. Allen (q.v.) that ski troops should be trained to operate in the snow-covered mountains of Lebanon, and in December I Australian Corps Ski School commenced in the Lebanon Mountains. It was intended that a ski company would be formed in each Australian division to patrol in areas which were inaccessible to conventional troops, but in February 1942, before the first course was even finished, the men were ordered to rejoin their units.

SKYHAWK, McDONNELL DOUGLAS (Single-seat carrier-borne fighter-bomber [A-4G]). Wingspan 27 feet 6 inches; length 40 feet 3 inches; armament 2 × 20 mm cannon, 2 × Sidewinder air-to-air missiles, 7000 pounds bombs, rockets or missiles; maximum speed 658 m.p.h.; range 600 miles; power 1 × Pratt & Whitney 9000-pound thrust turbojet.

The Skyhawk was the last fixed-wing combat aircraft to fly with the Fleet Air Arm (q.v.) and entered service in 1967 at a time of tension with Indonesia (see CONFRONTATION), replacing the Sea Venom (q.v.). The A-4G variant used by the RAN could be fitted with either bombs and rockets for strike missions or air-to-air missiles to protect ships from air attack. The Skyhawk was flown by No. 805 Squadron from the aircraft-carrier HMAS *Melbourne* (q.v.). Another 10 aircraft were delivered in 1971, but over the following decade 10 Skyhawks crashed,

some as a result of HMAS *Melbourne*'s catapult failing to launch them at the correct air speed. When the *Melbourne* was decommissioned in 1982 they were withdrawn from carrier service and in 1984 they were sold to New Zealand. Since 1990 No. 2 Squadron RNZAF, equipped with ex-RAN Skyhawks, has been stationed at Nowra for exercises with the RAN.

SLANG see **MILITARY SLANG**

SLESSOR, Kenneth Adolph (27 March 1901–30 June 1971). Poet, journalist, war correspondent, Slessor was a significant figure in Australian literature and one of the country's finest poets, although his body of published work is not large. Born in Orange, he began a career in journalism in Sydney in 1920, worked on the Melbourne *Herald* from 1925 and, for 15 years from 1927, with the notorious scandal-sheet *Smith's Weekly*, the self-appointed 'digger's friend'. In April 1940 Slessor was appointed official war correspondent from a pool of 50 applicants, the final selection being made by C. E. W. Bean (q.v.) and the Minister for Information and former AIF correspondent Sir Henry Gullett (q.v.). He accompanied the 2nd AIF through the Mediterranean campaigns in Greece, Crete and Syria, returning to Australia in March 1942. He chafed at the restrictions the army felt it necessary

to impose on correspondents, and was seriously disenchanted by his dealings with the army bureaucracy. Increasingly critical of senior officers, he was
allowed to visit the war zones in New Guinea on
only three occasions during the rest of the war. In
October 1943 he gave an interview to the Sydney
Sunday Sun describing the assault at Finschhafen,
and which was critical of arrangements for the
wounded and of intelligence estimates of Japanese
strength. This provoked considerable protest within
the army, reaching the Prime Minister, John Curtin
(q.v.), and the Minister for Information, Arthur
Calwell. Moves were afoot to withdraw his accreditation as an official correspondent, but Slessor pre-
empted such action by resigning on 21 February
1944. He returned to the Sydney *Sun* as a senior
journalist, and remained active in journalism and
literary circles until his death.

Clement Semmler (ed.), *The War Diaries of Kenneth Slessor: Official Australian Correspondent 1940–44* (University of Queensland Press, Brisbane, 1985); *The War Despatches of Kenneth
Slessor* (University of Queensland Press, Brisbane, 1987).

SLOUCH HAT is a wide-brimmed felt hat that was
worn by several Australian colonial armies with the
brim turned down all around. When a shortage of
cork helmets led to its use by British Empire troops
in the Boer War it became popular with the Australians because it provided protection from sun
and rain. After it became standard Australian Army
headgear in 1903, it was worn with the left brim
clipped up and displaying the Rising Sun badge
(q.v.) so that rifles could be held at the slope without damaging the brim. It was during the First
World War that the slouch hat became famous as
the mark of an Australian soldier, although it seems
to have been worn in several different ways:
Monash's 4th Infantry Brigade, for example, wore it
on Gallipoli with the brim turned down at all
times. In the Palestine campaign, Turkish soldiers
came to identify the light horse (q.v.) by their 'big
hats' and, especially after the charge at Beersheba
(q.v.) in October 1917, would sometimes refuse to
stand their ground when they saw them approaching. William Morris Hughes (q.v.) used the slouch
hat during his visits to the AIF during the war as a
means of identifying himself, and his government,
more closely with the soldiers, and his sobriquet,
'the little Digger', was successfully emphasised during the election campaign of 1919 by his adoption
of the head gear. The slouch hat's continued use
during the Second World War and thereafter has
made it an important Australian national symbol,
recognised by the 1993 decision by the Chief of
the General Staff, Lieutenant-General John Grey, to

emphasise the slouch hat as a uniform item, especially in preference to the beret and the officer's
peaked cap, both of which are items of British
Army provenance. The new form for the wearing
of the slouch hat has the Rising Sun badge on the
upturned brim and the corps or regimental badge
on the front. Units of the Royal Australian
Armoured Corps with an entitlement to do so
wear emu plumes behind the Rising Sun.

(See also UNIFORMS.)

SMALL ARMS are man-portable firearms, firing
ammunition of or below 15 mm calibre. They
include rifles, sub-machine guns, machine-guns
and pistols. Small arms are the basic weapons of the
defence force, especially the army, and are the primary weapons of the infantry (q.v.) arm of the
army. In combat, small-arms fire is employed by
units and sub-units on offensive and defensive
operations to kill and suppress enemy soldiers and
to damage items of enemy equipment including
light vehicles, aircraft and helicopters. It should be
noted that small arms are expected to be robust,
and to operate reliably at all times in all extremes of
climate and geography.

The rifle is the basic personal weapon of the
defence force. All military personnel regardless of
rank and trade are trained at some stage of their
career to carry, operate, fire and drill with it. The
military rifle evolved from the flintlock, percussion-
cap, single-shot, muzzle-loaded, smooth-bore musket during the eighteenth and early nineteenth
centuries. Rifling gave the musket long range and
accuracy, but did not improve its slow rate of fire.
The development throughout the middle and late
nineteenth century of breech loading, metal-case
centre fire cartridges, the bolt action, magazine
feeding and improved manufacturing techniques
revolutionised the combat capabilities of the rifle.
By 1900 the standard, long military rifle was a
magazine-fed, hand-operated, bolt-action weapon
capable of accurate, long-range, rapid fire. During
the twentieth century the rifle has become shorter
in length and gas operation has increased the rate
of fire by giving it a semi-automatic and automatic
capability. The musket and rifle have always been
capable of being fitted with a bayonet (q.v.) for
close-quarter fighting. Most twentieth-century
rifles have been capable of being fitted with a
grenade launcher or projector for the projection of
grenades. The effective range for individual rifle fire
should be up to 300 metres and up to 600 metres
for section or group fire.

Assault rifle is the term used since the Second
World War to describe lightweight, reduced size,

automatic-fire capability rifles firing short, reduced calibre ammunition. Assault rifles have given each soldier an automatic fire capability and because the ammunition is smaller and lighter have enabled the carriage of greater amounts of ammunition.

A carbine is usually a shorter, lighter, reduced range, handier version of the standard long rifle, but may also be a special design. During the nineteenth century and early twentieth century carbines were popular with mounted infantry and cavalry.

Sub-machine guns (SMG) are short, handy, magazine-fed, blowback-operated, air-cooled, automatic-fire capability weapons firing pistol ammunition. The SMG was developed towards the end of the First World War for close range combat situations such as clearing trench systems and has proved useful since then for fighting in built-up areas, close country and at night. The advent of assault rifles has reduced the need for the SMG in recent times; specialist SMGs have been retained, however, for use with Special Forces for counter-terrorist work and for silenced automatic fire. The effective range for an SMG is up to 100 metres.

Machine-guns (MG) are automatic-fire weapons capable of placing accurate bursts of automatic fire out to ranges beyond the effective range of the rifle. Machine-guns may be belt- or magazine-fed and gas, recoil or blowback operation. Machine-guns may be man or crew portable and are usually fired from a bipod or tripod. They may also be mounted on vehicles, helicopters and watercraft. There are various categories of machine-gun including: light, medium, heavy, general purpose, and light support weapon. All categories of machine-gun, apart from the heavy machine-gun, fire rifle-calibre ammunition.

The light machine-gun (LMG) is a product of the early twentieth century, whose role was developed during the First World War to provide the main firepower at section or squad level for small groups of soldiers. It is a man-portable, normally magazine-fed and gas-operated, lightweight machine-gun, fitted so that it is capable of being fired from the shoulder, hip or bipod. The LMG is likely to have a quick-change, air-cooled barrel to assist in maintaining high rates of fire in short bursts. Some LMGs can be mounted on lightweight tripod mounts to provide accurate, extended range, fixed lines fire at night and in poor visibility. It should be noted that some heavy-barrel, automatic rifles (HBAR) fitted with a bipod have been given the LMG role in some armies. The effective range of an LMG fired from the bipod is 500 metres.

The medium machine-gun (MMG) is a product of the late nineteenth century, whose role was developed during the First World War. Its usefulness declined in the late 1950s with the introduction into service of general purpose machine-guns. The MMG is a much heavier, crew-served, recoil- or gas-operated, belt-fed machine-gun firing rifle-calibre ammunition, with a heavy air-cooled or water-cooled barrel, and it is fired from a substantial, heavy tripod. It is a specialist weapon and requires a trained crew of at least four men: one to carry the gun, one to carry the tripod, and others to carry ammunition, sighting equipment, a tool kit, and water (if water cooled). The MMG is capable of firing long bursts (up to 20 rounds) of sustained, direct or indirect fire, effective to ranges of between 2000 and 4000 metres.

The general purpose machine-gun (GPMG) is a product of the 1930s, the role and concept of which was developed during the Second World War and continued to be worked on into the 1960s. The GPMG is a man-portable, gas-operated, belt-fed, machine-gun with a quick-change, air-cooled barrel. The GPMG concept was to have a machine-gun that with little or no modification could carry out the LMG and MMG roles from bipod and tripod and be fitted to vehicles, aircraft and watercraft. The GPMG concept has been a compromise, as the GPMG is heavier and more awkward to handle than an LMG and is not able to provide the same type of sustained fire capability as an MMG. The effective range for a GPMG is up to 500 metres firing from the bipod and up to 2000 metres from the tripod.

The light support weapon (LSW) is a product of the 1970s with a role at the section, squad or platoon level that is still being developed in the 1990s. The LSW is a compromise between the HBAR, LMG and the GPMG and fires light rifle-calibre ammunition, the same as that used for the assault rifle. It is man-portable, air-cooled, either magazine- or belt-fed, may be fitted with a quick-change barrel and can be fired from the shoulder, hip or bipod. It is capable of high rates of fire in short bursts for short periods of time. The effective range of the LSW is up to 500 metres.

Pistols are close-range handguns and are usually carried by officers, aircrew, weapon and vehicle crews for personal protection. They are also useful for protection when working in confined spaces such as searching buildings and tunnels. The main types of pistol are the single or double action revolver and the recoil-operated, magazine-fed, self-loading pistol. The main advantage of the magazine-fed pistol over the revolver is that it is quicker changing magazines than refilling an empty cylinder on a revolver. The effective range of a pistol is up to 50 metres.

Small arms in Australian service

During the late nineteenth century the armed forces of the Australian colonies were in a state of transition regarding small arms and small arms ammunition. The British influence meant that the new smokeless 0.303-inch calibre ammunition was coming into service to replace black powder 0.45-inch and 0.303-inch calibre ammunition at the same time as bolt-action, magazine-fed Lee Metford and Lee Enfield rifles and carbines and belt-fed, water-cooled, recoil-operated Maxim machine-guns were coming into service. These small arms were replacing the single-shot, lever action Martini rifles and carbines and the mechanical, hand-operated Gatling and Nordenfelt machine-guns. The service revolvers of the time were the 0.476-inch calibre Enfield and the 0.455-inch calibre Webley, although it was also common practice for officers to purchase their own pistols.

Australian soldiers serving in South Africa during the Boer War used 0.303-inch calibre Lee Enfield long rifles; 0.303-inch Martini Enfield and Lee Enfield carbines; 0.45-inch and 0.303-inch calibre Maxim and 0.303-inch calibre Colt-Browning Model 1895 machine-guns; and the 0.455-inch Webley revolver. Long-range rifle fire out to 2000 yards was a feature of Boer War fighting. Carbines were more popular with horse-mounted infantry and cavalry units than the standard long rifles. The effective employment of machine-guns was limited by numbers and knowledge. Australian naval infantry serving in China during the Boxer Rebellion were rearmed with Lee Metford long rifles and 0.45-inch calibre Maxim machine-guns *en route* to China.

The experience of the Boer War led to the decision to produce a shorter, lighter, easier to operate Lee Enfield rifle that would be acceptable for use by infantry foot-soldiers and horse-mounted soldiers. This resulted in the 0.303-inch calibre Short Magazine Lee Enfield (SMLE) Mark 1 bolt-action rifle being introduced into British Army service from 1903. Production of the SMLE Mark III rifle for the Australian Army was commenced at the newly established Small Arms Factory at Lithgow in 1912. This rifle, with minor modifications, was to remain in service in Australia until 1959.

During the First World War, Australian soldiers went into action at Gallipoli armed with 0.303-inch calibre SMLE rifles, 0.455-inch calibre Webley revolvers and a small number of the earliest water-cooled, belt-fed, tripod-mounted, 0.303-inch calibre Vickers MMGs — a lighter version of the Maxim machine-gun that was easier to operate and manufacture. From 1916 when Australian infantry soldiers went into action on the Western Front their SMLE rifles were supplemented at section and platoon level by the issue of air-cooled, magazine-fed Lewis LMGs. They also received increased numbers of Vickers MMGs for employment with specialist machine gun battalions (q.v.). Australian light horse units serving in the Palestine campaign were issued with the less bulky air-cooled, gas operated, strip-fed Hotchkiss Mark I LMG, when it was found that the Lewis LMG was too bulky for quick reaction carriage on horses. During the First World War, Australian service personnel were issued with a variety of 0.455-inch calibre revolvers (including Webley, Smith & Wesson and Colt) and, on occasions, self-loading pistols. First World War combat experience demonstrated the value of automatic small arms fire from machine-guns and the necessity for mobile light machine-guns at platoon level.

In 1929 the Small Arms Factory at Lithgow commenced production of the Vickers MMG and in the late 1930s a decision was made to manufacture the Czechoslovakian–British developed 0.303-inch calibre BREN LMG at Lithgow. The first Australian-made BRENs came off the production line in 1941.

Australia entered the Second World War with an array of small arms weapons exactly the same as those her forces had used in the First. Australian Army units sent to the Middle East from 1940 were equipped with Australian-made SMLE rifles and Vickers MMGs, and were then issued with British-made BREN LMGs and American-made 0.45-inch calibre Thompson SMGs on arrival there. A similar situation occurred for Australian units during the Malayan campaign. Australian units during the Papuan campaign were equipped initially with SMLE rifles, Lewis LMGs, Vickers MMGs and Thompson SMGs. The obsolete Lewis LMGs were eventually replaced by the lighter, more robust and reliable BREN LMGs during 1942. The Thompson SMGs were gradually replaced by Australian-made Owen (q.v.) and Austen SMGs from late 1942 onwards. The Owen Machine Carbine produced by Lysaghts at Port Kembla quickly became a favourite with the soldiers for its reliability and automatic-fire capability, a necessary requirement in close-range combat in jungle country. The Austen SMG produced by Diecasters Pty Ltd and W. T. Carmichael & Sons, both of Melbourne, never achieved the same reputation as the Owen and went out of service immediately after the war, while the Owen continued to serve for about another 20 years. First World War-vintage 0.455-inch calibre revolvers remained in service and were supplemented and replaced by a variety of 0.38-inch calibre revolvers including the Enfield, Colt and Smith & Wesson. It is likely that some Australian personnel obtained and used 9 mm Browning and 0.45-inch Colt Model 1911 self-loading pistols during the war.

The experience of the Second World War demonstrated that mobile, automatic firepower in the form of SMGs and LMGs was a requirement in close-range combat, especially in forest, jungle and urban environments. The heavier, water-cooled MMG was still useful for providing sustained long-range automatic fire in relatively static situations when water was plentiful, but was not as suitable for mobile operations and mounting on vehicles as the lighter, air-cooled GPMGs. The bolt-action rifle, while reliable and accurate, was disadvantageous in situations where it was opposed by an enemy who were using higher rate of fire semi-automatic and automatic capability rifles, especially at close range.

Australian soldiers went to the Korean War with the same small arms weapons that were in use at the end of the Second World War. The SMLE rifle, the BREN LMG, the Owen Machine Carbine, the Vickers MMG the 0.38-inch revolvers were used throughout the Korean War. This time, the slower rates of fire from the bolt-action SMLE rifle and the magazine-fed BREN LMG put Australian soldiers at a disadvantage during mass attacks at close range by Chinese and North Korean soldiers using semi-automatic rifles and SMGs. The severe cold of the Korean winter caused operating problems for all weapons.

During the mid-1950s the decision was made that the next generation of Australian small arms would use the standard North Atlantic Treaty Organization (NATO) 7.62 mm calibre ammunition, that the next service rifle would be gas-operated with automatic or self-loading capability and that a belt-fed GPMG would be obtained to replace the LMG and MMG.

Australian soldiers in the Malayan Emergency were equipped initially with the lightweight, bolt-action 0.303-inch calibre Lee Enfield No. 5 Jungle Carbine, until they were replaced by the Belgian-manufactured 7.62 mm calibre X8E1 and X8E2 self-loading trials rifles. These were the trials versions of the L1A1 Self Loading Rifle (SLR) that was to come into Australian service in 1959 as the

replacement for the SMLE. Other small arms used by the Australians in Malaya included the 0.303-inch calibre BREN LMG, the 9 mm Owen Machine Carbine, the 0.30-inch M1 Carbine, the 12-gauge Browning A5 Shotgun and 0.38-inch calibre revolvers.

The 7.62 mm L1A1 SLR produced at the Small Arms Factory in Lithgow came into service from 1959 and a heavy-barrel, bipod-fitted, automatic-fire version known as the L2A1 came into service from 1962. Both of these weapons were based on the Fabrique Nationale (FN) FAL automatic rifle developed in Belgium. The American belt-fed 7.62 mm calibre GPMG M60 came into service from 1961 to replace the Vickers MMG and the BREN LMG. During the 1960s, Belgian 9 mm Browning Model 1935 self-loading pistols gradually replaced the 0.38-inch calibre revolvers. In 1964 the Australian-designed 9 mm calibre F1 SMG came into service to replace the Owen Machine Carbine, a process that was not completed until the late 1960s.

During the 1960s, Australian soldiers serving in Malaysia during Confrontation were equipped with Australian-issue 7.62 mm L1A1 SLRs and 9 mm Owen Machine Carbines, British-issue 7.62 mm L4 BREN LMGs (converted from 0.303-inch) and 7.62 mm L7 GPMGs (British version of the Belgian FN MAG 58 GPMG), and American 5.56 mm Colt AR-15 Assault Rifles.

In 1965 Australian soldiers were sent to fight in the Vietnam War equipped with the 7.62 mm L1A1 SLR, the 7.62 mm GPMG M60, the 9 mm Owen Machine Carbine and the 9 mm Browning Model 1935 Self Loading Pistol. Operational experience soon showed that the 9 mm ammunition for the Owen and its replacement the F1 SMG lacked the necessary killing power at ranges beyond 50 metres and this led to the replacement of the SMG in infantry units with the American 5.56 mm Colt M-16 Assault Rifle. Throughout their service in Vietnam, Australian soldiers had a high regard for the hitting power of the 7.62 mm ammunition fired by the L1A1 SLR and the GPMG M60. During 1970–71 a relatively small number of 7.62 mm L4A4 BREN LMGs were purchased from England to supplement the GPMG M60 in infantry units, especially for patrolling tasks.

During the 1970s the Special Air Services Regiment (q.v.) was given a counter-terrorist role and during the late 1970s this led to them adopting various versions of the German 9 mm Heckler & Koch MP5 SMG. The MP5 is more accurate than a normal SMG because it fires from the closed-bolt rather than open-bolt position. Australian Special

Forces also use the British 9 mm Silenced Sterling SMG. In the early 1980s a small quantity of British 7.62 mm L7A2 GPMGs were purchased to replace the GPMG M60 in the sustained-fire machine-gun role, as the M60s were wearing out. During the mid-1980s an additional quantity of Belgian FN MAG 58 GPMGs were purchased to replace the GPMG M60 in the GPMG role and sustained-fire machine-gun role.

In the mid-1980s a new generation of small arms firing the new standard NATO 5.56 mm calibre SS109 ammunition was selected for the Australian Defence Force. The futuristic looking, bullpup configuration, largely plastic construction, Austrian Steyr AUG Assault Rifle (q.v.) was selected to replace the LlAl SLR, the M16Al Assault Rifle, and the Fl SMG. The Belgian FN Minimi LSW was selected to replace the GPMG M60, the L2Al Assault Rifle, the L4A4 BREN LMG and the MAG 58 GPMG in the light role. The Steyr AUG and the Minimi are being manufactured in Australia and came into service in 1989 and 1990 respectively, and have already seen operational service during peace-keeping (q.v.) operations in Somalia, Cambodia and Rwanda.

IAN KURING

SMALL SHIPS SQUADRON In the Second World War the army developed a maritime capability to support its amphibious operations and to assist in resupply in the islands and territories to the north. By the war's end the army operated a very large fleet of landing craft, barges, launches, workboats and tugs, almost of all of which were sold off in 1946–47. In 1959 four Medium Landing Ships (LSMs) were acquired from the United States, and these served with 32 Small Ship Squadron of the Royal Australian Engineers (q.v.) during Confrontation (q.v.), and in Vietnam between June 1966 and August 1971 (see VIETNAM WAR) in logistical support roles. The vessels themselves were named for prominent Australian generals of earlier generations: *Clive Steele, Harry Chauvel, Vernon Sturdee* and *Brudenell White*. In addition, the army deployed a small coastal cargo vessel, the *John Monash*, on resupply missions. In 1972 the LSMs and the *John Monash* were sold and plans to acquire a large army amphibious ship were cancelled, in line with a decision that all ocean-going vessels would be operated by the RAN. The army's small craft capability henceforth consisted of a number of Landing Craft Mechanised (LCM), which began to come into service from 1967, and a variety of workboats and harbour craft. In 1980 the army's amphibious capabilities were renewed and

enhanced with the commissioning of HMAS *Tobruk*, a Heavy Landing Ship (HLS).

SMITH, Admiral Victor Alfred Trumper (9 May 1913–   ). Graduating from RAN College Jervis Bay (see CRESWELL, HMAS) as a midshipman in May 1931, Smith served first on HMAS *Canberra* (q.v.). He specialised in the relatively new field of naval aviation, taking the observer's course in 1937 and serving on the aircraft-carriers HM Ships *Glorious* and *Ark Royal*, and taking part in the February 1942 attack on the *Scharnhorst*. He returned to Australia in March 1942, and initially served as a liaison officer with the US Navy before being posted to HMAS *Canberra* until that ship was lost off Savo Island.

Sent back to Britain to serve with HMS *Shropshire*, then being transferred to the RAN, he requested a posting with the Fleet Air Arm and was sent to the escort carrier HMS *Tracker*, taking part in convoy duties in the Atlantic and on the Murmansk run. This was followed by a staff position with the Flag Officer British Assault Area (Normandy), planning for the D-Day (q.v.) invasion and, in November 1944, by a logistics staff job with the British Pacific Fleet back in Australia.

Smith had served for most of the war with the RN, but he was an enormously experienced naval aviator and this stood him in good stead in the early postwar RAN. He drafted the plan governing the creation of an Australian Fleet Air Arm, which included a further period in 1945–46 at the Admiralty. Approval for the creation of the Fleet Air Arm and the acquisition of two carriers was given in 1947, and Smith became a member of the Naval Aviation Planning Staff that year. With the arrival of HMAS *Sydney* in January 1950 and the activation of the 20th Carrier Air Group, he was appointed executive officer of *Sydney*, a position he held for two years. In this capacity he took part in the deployment to Korea in September 1951, the first occasion on which the RAN Fleet Air Arm was used on active service (see KOREAN WAR).

Thereafter he held a succession of staff and command postings, culminating in command of the flagship HMAS *Melbourne*, between 1961 and 1962. In February 1966 he became Flag Officer Commanding HMA Fleet, and in April 1968 was made CNS, the first naval aviator to head the RAN. He was appointed KBE the following year. Smith led the navy during its diverse commitment to the Vietnam War, but the most potentially serious incident to arise was the collision while on exercise between the Australian carrier *Melbourne* and the US Navy destroyer *Frank E. Evans*, resulting in significant loss of life on the American ship (see COLLISIONS, NAVAL). In November 1970 he became Chairman of the Chiefs of Staff Committee, a position he held for five years. This period brought a number of changes in Australia's strategic environment and defence outlook, with the withdrawal from the Vietnam commitment, the gradual scaling down and final dissolution of SEATO (q.v.), the establishment of the Five Power Defence Arrangements (q.v.), the election of a Labor government for the first time in 23 years and the defence group reorganisation spearheaded by Sir Arthur Tange (q.v.). Smith retired from the RAN in November 1975.

P. D. Jones (ed.), *A Few Memories of Sir Victor Smith* (Australian Naval Institute Press, Canberra, 1992).

SMYTH, Major-General Nevill Maskelyne (14 August 1868–21 July 1941). Commissioned from Sandhurst in 1888, Smyth joined the 2nd Dragoon Guards in India and served in a variety of appointments in India, on attachment to the Royal Engineers, and in Egypt and the Sudan during the operations against the Mahdi in 1898. At the Battle of Omdurman in that campaign, he rescued two journalists from enemy attack and was awarded the VC. He spent a short period thereafter as acting governor of the Blue Nile District of the Sudan, while still in the rank of captain. He returned to his regiment in South Africa in 1902, and again saw battle at close quarters, on one occasion narrowly escaping capture. By 1912 he was a lieutenant-colonel in command of the 6th Dragoon Guards,

and in the following year he transferred to the Egyptian Army. With his experience of colonial soldiering and contacts with Field Marshal Lord Kitchener in both India and Egypt, he was detailed to service with the Australians in the Dardanelles in 1915 in a bid to provide a leavening of much needed experience to that force. He commanded the 1st Australian Brigade at Lone Pine and took his brigade to France early the following year. At the end of 1916, after serving through the Somme campaign, he was given command of the 2nd Division, which he led through the bloody and inconclusive fighting of 1917. In line with the moves to nationalise the Australian Corps during the course of 1918, Smyth was transferred back to the British Army in May 1918. He commanded both the 58th and 59th Divisions before the Armistice, went on to a Territorial Army command after the war, and retired in 1924. During the course of the war he was, among other awards, mentioned in despatches 11 times. He migrated to Australia in 1925, and took up grazing while dabbling in Nationalist Party politics. He was highly regarded by his contemporaries, and by the senior officers, at least, of the AIF, to whom his example of thoroughness and professionalism was an important influence in their own development as soldiers.

SNIPE, HMAS see **TON CLASS MINESWEEPERS**

SOLDIER SETTLEMENT Even before the end of the First World War, there was increasing concern among public officials about what was to be done

Table 1 Soldier settlement schemes post-1918

State	Total settlers allotted farms	Failures among original settlers (%)
New South Wales	9 302	29
Victoria	11 140	17
Queensland	6 031	40
South Australia	4 082	33
Western Australia	5 030	30
Tasmania	1 976	61
Australia	37 561	29

Table 2 Soldier settlement schemes post-1945

State	Total settlers allotted farms	Failures among original settlers (%)
New South Wales	3 057	11 (approx.)
Victoria	5 926	4
Queensland	470	10 (approx.)
South Australia	1 022	9.8
Western Australia	1 010	10 (approx.)
Tasmania	551	28
Australia	12 036	10–11 (approx.)

with returned servicemen. As early as 1915, reports of civil disorder caused by ex-servicemen were multiplying, and conservatives were haunted by the spectre of discontented, unemployed Anzacs turning to socialism to escape from their predicament. The Nationalist government of W. M. Hughes (q.v.) moved quickly to seek solutions to this potential crisis, and before long it had seized on the agrarian myth as the answer to its prayers. This myth, which had a long history in British and Australian political thought, promoted the virtues of establishing self-reliant 'yeoman farmers' on small plots of land. Such 'closer settlement', as well as being seen as the best way of developing rural areas, was considered to be morally beneficial. It would produce sturdy, property-owning individualists who would be resistant to the degeneracy and weak-minded collectivism associated by many conservatives with the urban working class.

The idea of soldier settlement was first raised at the War Committee in November 1915, and after several conferences attended by representatives of State and Commonwealth governments, a plan for such settlement was developed. The administration of the scheme was to be in the hands of the States, but the Commonwealth would supply the States with loan funds to a maximum of £500 per soldier settler (this maximum was gradually increased until it reached £1000 in 1924). The settlers would initially be charged only 3.5 per cent interest, and this would increase by 0.5 per cent a year until it reached the normal rate; they would also be paid sustenance money during the establishment period. In addition, the States were to provide inexperienced settlers with agricultural training. Thousands of returned servicemen believed the rosy picture of farming life painted by the government, and took up this offer. Government statistics on soldier settlement should be treated with caution, but the figures given in a 1929 report on the scheme give some indication of its scale (see Table 1).

Once established on their farms, soldier settlers found that they, their wives and their children had to work exhausting 14-hour days, without holidays. They could not afford to employ labour, nor to buy labour-saving machinery, and the government was reluctant to provide credit for the employment of labour, which did not fit with the self-reliant yeoman farmer model. Governments became increasingly concerned about the costs of soldier settlement, and talk of 'efficiency' and the need to be self-supporting replaced the earlier rhetoric about society's debt to the Anzacs. Now the debt ran the other way — by 1943, accumulated losses from the soldier settlement scheme for

the whole of Australia were estimated at £45 million. By then, also, less than 50 per cent of the original settlers remained on the land. The situation of many ex-servicemen and their families was summed up by one soldier settler, William Bellamy, who in 1934 said of his time farming in the Victorian Mallee: 'After eleven years of struggle I went out with nothing.'

Many reasons have been suggested for the failure of soldier settlement after the First World War, and both the federal government and every State government except Queensland appointed royal commissions or select committees to come up with such explanations. One criticism of the scheme is that it made the same mistake as every attempt at 'closer settlement' since the days of free selection: the farms were simply too small to be economically viable. Furthermore, since the best land had already been taken in most parts of Australia, much of the land given to soldier settlers was of poor quality, and where new land could be opened up or old land improved the cost of doing so was borne by the settlers themselves. Without capital of their own, most soldier settlers had a 100 per cent debt to the government, and having no equity in their property it was inevitable that increasing pressures would lead to increasing forfeitures. One such pressure was the combination of falling prices after 1920 with high costs for stock and equipment. There was also a lack of planning of soldier settlement in relation to markets, and a failure to conduct scientific surveys of such things as climate, soil and drainage in the areas selected for soldier settlement. Finally, the settlers themselves, many of whom had no previous agricultural experience and were often burdened further by war injuries, were inadequately trained and prepared for farming.

When the Commonwealth and State governments came to set up another soldier settlement scheme after the Second World War, the lessons which had been learnt from the failure of the previous attempt were embodied in a set of principles laid down in the *War Service Land Settlement Act 1945*. This time there was an effort to ensure that the land was of sufficient quality and that the holdings were large enough to allow the settlers to earn a reasonable income. The suitability of applicants for farm life was assessed, and those judged to be suitable but inexperienced in agricultural work were given thorough training. Soldier settlers were not to be excluded because they lacked capital, but they were expected to invest some of their own money in the holding. Government funds were available for the construction of houses, fences and other structures, and the purchase of livestock and machinery. During the first year of farming the settlers were not required to make repayments, and support was provided for their families; after that they were paid an allowance to cover operating costs. Perhaps the most important aspect of the scheme was that the level of repayment was not to be determined until commodity prices had stabilised after the decline which was expected in the immediate postwar period. In fact, however, prices never did decline after the war, but rose and remained high throughout the 1950s. This unexpected bonus, combined with good planning, made the scheme a success, and by 1960 most of the soldier settlers from the War Services Land Settlement Scheme remained on their holdings. Table 2, which covers both the War Services Land Settlement Scheme and smaller State schemes, shows the relative success of post–Second World War soldier settlement. Servicemen who had been in the Korean War and Malayan operations (see CONFRONTATION; MALAYAN EMERGENCY) were also eligible to take part in the scheme, which ended in the 1960s.

SOMALIA, AUSTRALIAN DEPLOYMENT TO

During the course of 1990–91 the sub-Saharan African nation of Somalia collapsed into anarchy and intertribal civil war; the situation became so grave that the United Nations decided that the established principle of non-intervention in the internal affairs of a member state no longer applied. From September 1992 UN agencies invoked external military involvement to protect humanitarian efforts in the face of massive famine and to begin to reconstitute the country's administrative infrastructure. Between October 1992 and January 1993 thirty ADF personnel were despatched to Somalia to constitute a Movement Control Unit as part of the first UN Operation in Somalia (UNOSOM I). Under United States auspices, Australia then contributed a 937-man battalion group based on the First Battalion, the Royal Australian Regiment (1RAR), a unit of the Operational Deployment Force, for duties in Somalia as part of UNITAF, the Unified Task Force in Somalia. The force was commanded by Colonel W. J. A. Mellor, and the deployment was codenamed Operation SOLACE, part of the wider US Operation RESTORE HOPE. Committed for a finite period between January and May 1993, the Australian force was assigned responsibility for a 17 000 square kilometre area centred on Baidoa, a provincial centre in the south-west of the country. The Australians were involved in a variety of tasks, including security for humanitarian

relief teams, the location and destruction of weapons caches and disarming of local armed gangs, and the gradual restitution of civil law and order, a task further reinforced by members of the Australian Federal Police after the withdrawal of the army units. One Australian soldier was accidentally killed while on patrol. After the battalion group withdrew, the Movement Control Unit remained behind to provide continued support to the second of the United Nations' Somalia operations, UNOSOM II, the Australian share in which was designated Operation IGUANA. The main Australian commitment to Somalia was of a type unusual in terms of Australia's normal involvement in peace-keeping operations (q.v.), and although it was highly successful in the short term it seems unlikely that the structure of the ADF would permit frequent deployments of this kind for any extended period. The extension of the Infantry Combat Badge to members of 1RAR aroused some adverse comment upon its announcement.

SOUTH AFRICAN WAR see BOER WAR

SOUTH AUSTRALIAN NAVAL FORCES SHIPS

Ordered in 1882 to meet possible threats from commerce raiders or Russian invasion, HMCS *Protector* was built in Britain and reached Adelaide in September 1884. The steel cruiser *Protector* was much larger than the gunboats which were bought at around the same time for the Queensland and Victorian navies, and it had all-round fire of three guns compared to one gun in the smaller gunboats. Its purchase reflected the fact that while Adelaide was relatively open and thus not easily defended by either forts or warships, the waters of St Vincent and Spencer Gulfs were very suitable for a large and powerful vessel to cruise on. *Protector* was kept in a state of combat readiness and was regularly deployed on station at Largs Bay. In 1886 Rear-Admiral George Tryon, the first admiral of the Australia Station (q.v.), remarked privately that 'the *Protector* was not only better kept than any ship on the Station, but was so much better kept that there was no ground for comparison'. In 1900 *Protector* sailed to China to take part in the suppression of the Boxer Rebellion (q.v.) but saw no action there. After integration into the RAN in 1911, *Protector* served as a tender to HMAS *Cerberus* and undertook various duties during the First World War. *Protector* was renamed *Cerberus* and became a tender to Flinders Naval Depot in April 1921, then was sold in 1924. Former Tasmanian torpedo boat TB No. 191 also served in South Australia from 1905 to 1911.

Protector: displacement 920 tons; length 185 feet (overall); beam 30 feet; draught 12.5 feet; speed 14 knots; armament 1 × 8-inch gun, 5 × 6-inch guns, 4 × 3-pounders, 5 × machine-guns; rearmed during First World War with 2 × 4-inch guns, 2 × 12-pounders, 4 × 3-pounders.

(See also COLONIAL NAVIES.)

SOUTH-EAST ASIA TREATY ORGANIZATION (SEATO) was a regional defence organisation formed as a result of the Geneva Conference of 1954, which had failed to resolve the outstanding issues arising from the Korean War and the French defeat in Indo-China. In September that year a conference in Manila resulted in the South-east Asia Collective Defence Treaty, signed by the United States, Australia, New Zealand, France, Britain, Pakistan, Thailand and the Philippines. The Australian Minister for External Affairs, R. G. Casey, played an important role in bringing it to fruition. The treaty designated any attack on a member state, or on protocol states (which meant South Vietnam, Laos and Cambodia) as a matter of common danger inviting a common response, and was thus an attempt to create a regional counterpart to NATO. The United States specified that the treaty could be invoked only in response to a communist attack, thus placing, for example, the Indo-Pakistan conflict outside its provisions. SEATO had no standing forces of its own, unlike NATO, and the United States refused to designate any forces for SEATO tasks ahead of need. The British Commonwealth Far East Strategic Reserve (q.v.), based in Malaya and formed from forces contributed by Britain, Australia and New Zealand, was earmarked for SEATO tasks, though never used for that purpose. SEATO was headed by a secretary-general, had a combined headquarters and contingency planning, and was the coordinating body for regular military exercises between member states. The Laos crisis of 1961–62 demonstrated the organisation's inability to act collectively, and in March 1962 the American Secretary of State, Dean Rusk, declared that member states' obligations were individual as well as collective; this was used as a rationale for Australian intervention in the Vietnam War in the mid-1960s. Pakistan withdrew in 1968 and France suspended membership in 1975. The last SEATO exercise was held in February 1976 and the organisation was wound up formally in June 1977. Its principal advantage to Australia was the opportunity it provided for dialogue and defence cooperation within the region and with the United States. Indeed between 1956 and 1972 most of the consultative process under ANZUS (q.v.) was conducted through SEATO, and neither the ANZUS Council nor the Military Representatives met in that period. Its main disadvantages were that it identified Australia closely with American foreign policy in the region (but then ANZUS already did that), while its particular regional grouping (White regional and former colonial powers with a small number of politically acceptable Asian nations) became increasingly inappropriate in the 1970s.

Leszek Buszynski, *SEATO: The Failure of an Alliance Strategy* (Singapore University Press, Singapore, 1983).

SOUTH-WEST PACIFIC AREA (SWPA), consisting of Australia and the islands to its north, was created in March 1942 under the command of General Douglas MacArthur (q.v.) when the British and American Combined Chiefs of Staff divided the world into operational areas including the SWPA, under US Army control, and the adjoining South Pacific Area, under US Navy control. The choice of MacArthur as Supreme Commander was supported by the Australian government to ensure the allocation of American resources to the defence of Australia. General Sir Thomas Blamey (q.v.) was appointed Commander Allied Land Forces in the SWPA, but had little practical control over American troops. Australian troops were involved in fighting in the SWPA in Papua, New Guinea and Borneo.

SPANISH CIVIL WAR, AUSTRALIANS IN see AUSTRALIANS IN THE SERVICE OF OTHER NATIONS

SPECIAL AIR SERVICE REGIMENT (SAS) The decision to form an SAS company was announced in April 1957, and came out of deliberations over the future shape of the regular army, based as well on observation of the British SAS in Malaya during the Emergency (q.v.). Its original establishment was 16 officers and 144 other ranks, but beyond that little detailed consideration was given initially to its function or purpose. During the flirtation with the Pentropic division (q.v.) in the early 1960s, the SAS became for a time part of the Royal Australian Regiment (q.v.), and its principal wartime task was to act in an infantry reconnaissance role at divisional level. Pressure to deploy the Australian SAS to Borneo to assist in operations against the Indonesians (see CONFRONTATION) led to the dispatch of the 1st Squadron for active service in 1965 and, in preparation for this, led to the separation of the SAS from the RAR and its reorganisation as a regiment consisting of a headquarters and two squadrons (since if one squadron was deployed overseas, a second would be necessary for duties in the defence of Australia), with an establishment

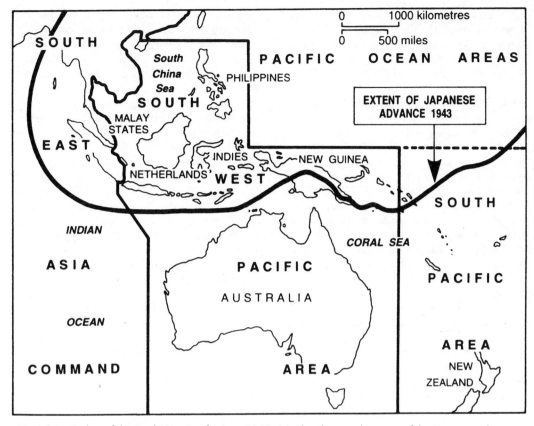

Map 24 Outline of the South-West Pacific Area 1942–44, also showing the extent of the Japanese advance in 1943.

of 15 officers and 209 other ranks. Both squadrons saw service during Confrontation, 1 Squadron between March and August 1965, 2 Squadron between February and July 1966, suffering losses of three killed in action. A third squadron was raised for service in South Vietnam at the end of 1965, and was deployed there as part of the 1st Australian Task Force in June 1966. All three squadrons served in Vietnam, completing two tours each, and acted in a reconnaissance role, providing tactical intelligence for the Task Force, and in attacks on enemy communications within Phuoc Tuy province. A further six SAS men were lost in Vietnam. After the withdrawal from Vietnam renewed emphasis was placed on the SAS's role in the defence of Australia, and on its functioning at a strategic rather than operational or tactical level as before. In the late 1970s it began to develop a major counter-terrorist capability, which remains one of its primary functions. In July 1979 the SAS became part of the Special Action Forces, controlled by its own Directorate within Army Office, along with the army's two commando squadrons and the special forces signal

squadron. Its base is at Swanbourne in Western Australia.

D. M. Horner, *SAS: Phantoms of the Jungle. A History of the Australian Special Air Service* (Allen & Unwin, Sydney, 1989).

SPECIAL LIAISON UNITS (SLU) Established in November 1944 by British intelligence as a means of further integrating intelligence activities in the South-West Pacific Area into the British ULTRA system. SLU 9, headed by a RAF squadron leader, was able to provide top-secret intelligence material directly to Australian forces and independently of the Central Bureau (see ULTRA), which had now moved forward to Leyte in the Philippines and which, although it was an Allied unit, came under MacArthur's direct control. By early 1945 there were detachments with the 1st Tactical Air Force (q.v.) and Blamey's Advanced Land Headquarters on Morotai, with Lieutenant-General Sir Vernon Sturdee's (q.v.) First Army at Lae and with the RAAF's North-West Area in Darwin. The SLUs gave Australian authorities direct access to ULTRA, and should be seen in the context of MacArthur's gradual sidelining of the Australian forces during the final drive northwards to Japan. They were probably also a means by which the British sought to reassert their influence in an area where they had had little direct involvement since the fall of Singapore in early 1942.

SPECIAL OPERATIONS AUSTRALIA (SOA) The Special Operations Executive (SOE) was a British organisation formed in June 1940 on the orders of Prime Minister Winston Churchill to conduct sabotage operations behind enemy lines and to encourage and assist occupied populations to do the same. When a senior SOE officer, Major Egerton Mott, escaped to Australia from Java in March 1942, he met senior Australian officers and successfully lobbied for the establishment of an SOE-style organisation in Australia. SOE-Australia was formed in April 1942 with the cover name of the Inter-Allied Services Department (ISD) and took over the commando training school formerly used by Independent Companies (q.v.) on Wilson's Promontory. Personnel were drawn from Australian, Dutch, British and New Zealand forces and in July 1942 ISD became part of the Allied Intelligence Bureau (q.v.). AIF personnel attached to ISD were administered by Z Special Unit (q.v.). In February 1943 ISD was disbanded and replaced by SOA which, under the cover-name of the Services Reconnaissance Department, conducted operations from that date until the end of the war. Despite the glamour attached to special operations it cannot be said that SOA missions achieved anything of significance. SOA operated in Timor, the Solomons, the Moluccas, New Guinea and Borneo. Most SOA units were, however, subordinated to local army commanders who tended to use them for tasks that were inappropriate for special forces, such as reconnaissance or advance liaison for field units. Furthermore SOA missions in Timor and Dutch New Guinea were hampered by being placed under the command of (usually unpopular) colonial officials whom the local inhabitants eagerly handed over to the Japanese. SOA was probably most successful in Borneo where it led local uprisings and gained control of some areas of occupied territory in 1945. However, by this stage, the main events of the war were overtaking such small, local operations. In the final analysis SOA operations were characterised by inefficiency, inappropriate objectives and unreliability. They did not greatly hamper the enemy and did not shorten the war by a single day.

SPENDER, Percy Claude (5 October 1897–3 May 1985). Educated at Fort Street High School and the University of Sydney, Spender entered the New South Wales Public Service in 1915. He volunteered for the AIF in 1918, but was demobilised at the war's end without leaving Australia. He worked in the Crown Solicitor's Office, was called to the New South Wales Bar in 1923, and was made King's Counsel in 1935. Maintaining an interest in citizen soldiering, and the useful connections which it might bring a young lawyer, he was commissioned as a captain in the Australian Army Legal Department in August 1923, and served as Legal Staff Officer to the 9th Infantry Brigade between 1926 and 1931. He was elected as an Independent for the federal seat of Warringah in 1937, and switched to the United Australia Party in 1938. A member of the War Cabinet from 1939, he became minister for the Army in 1940, and held office until Labor took power in October 1941. Spender then became a member of the Advisory War Council, and retained his membership until the end of the war despite the fact that his Opposition colleagues resigned *en bloc* in 1944; his failure to leave led to his being expelled from the UAP. He served in the AMF between 1942 and 1945, entirely in Australia on the staff of the Lieutenant-General-in-charge of Administration (LGA), and was honorary colonel of the Australian Army Legal Corps after the war between 1945 and 1952. He returned to government as a Liberal in 1949 and was appointed minister for External Affairs and Territories. In this capacity he undertook the crucial preliminary negotiations which led to the ANZUS Treaty (q.v.). He was also instrumental in committing Australian forces to the Korean War in Menzies' absence overseas. Perceived by some as a rival to Menzies, Spender left parliament in 1951 and became Australian ambassador to the United States, a job which he relinquished in 1958 to move to the International Court in the Hague as a justice. In 1964 he became president of the court, retiring in 1967. Highly able, he was known as 'the butcher bird' at the Sydney Bar for his style of cross-examination in court.

Percy Spender, *Politics and a Man* (Collins, Sydney, 1972).

SPITFIRE, HMCS see **NEW SOUTH WALES NAVAL FORCES SHIPS**

SPITFIRE, SUPERMARINE (Single-seat fighter [Mark LFVIII]). Wingspan 36 feet 10 inches; length 31 feet 2 inches; armament 2 × 20 mm cannon, 4 × 0.303-inch machine-guns; speed 404 m.p.h.; range 660 miles; power 1 × Rolls Royce Merlin 1700 h.p. engine.

Spitfires were operated by the RAAF during the Second World War between 1941 and 1945. No. 452 and 457 Squadrons, whose pilots had trained through the Empire Air Training Scheme (q.v.), flew Spitfires during the Battle of Britain and then returned to Australia in 1942 to defend Darwin. The RAAF operated over 600 Mark V, Mark VIII and Mark LF VIII Spitfires in the South-West Pacific Area (q.v.), eventually as part of the 1st Tactical Air Force (q.v.). Spitfires were also flown by No. 451 Squadron in the Mediterranean and No. 453 Squadron flew Spitfires over Normandy on D-Day (q.v.). Both Group Captain Clive Caldwell and Squadron Leader 'Bluey' Truscott (qq.v.) flew Spitfires. A Spitfire IIA used by No. 452 Squadron in Britain is held by the Australian War Memorial.

SPRY, Brigadier Charles Chambers Fowell (26 June 1910–27 May 1994). Spry graduated from RMC Duntroon in December 1931, and held the usual round of interwar training posts with CMF units which were the lot of most Staff Corps officers. He served on attachment with the Duke of Wellington's Regiment in India between 1935 and 1936, and at the outbreak of war was in the G or Operations Branch of Army Headquarters. He joined the 2nd AIF in October 1939, and went overseas with the 6th Division as a junior staff officer. Serving entirely on the staff during the war, he was promoted to temporary colonel in April 1942, and finished the war as a senior member of the Australian Mission to Lord Louis Mountbatten's headquarters at South East Asia Command.

After the war he moved into intelligence, and from April 1946 was Director of Military Intelligence at Army Headquarters. In this capacity he played an important part in founding the foreign and domestic security services, the Australian Secret Intelligence Service and the Australian Security Intelligence Organisation (ASIO). In July 1950 he was seconded to the Attorney-General's Department and became the head of ASIO, a post which he held for 19 years. As one historian has noted, the ethos of that organisation, at least in its early years, derived very much from his military approach and belief in the separation of the military and political realms. He played an important part in handling the defection of the Petrovs in 1954, and resented the imputation on the part of the Labor leader, Dr H.V. Evatt, that he had in some way engineered the timing of the event to obtain electoral advantage for the Prime Minister, Robert Menzies (q.v.). He was knighted in 1964 and retired in 1969 as a result of ill health.

SQUADRONS, RAAF In air force terms a squadron is a generic title for a number of different types of unit, and embraces flying squadrons, which are always designated with a number (e.g., No. 1 Squadron), base and maintenance squadrons, which

may or may not have a numerical designation, training squadrons, and reserve squadrons, the latter of which are numerically designated. Two or more squadrons combine to form a wing. There were four squadrons in the Australian Flying Corps (q.v.) during the First World War, three in France and one in Palestine. The first squadron of the RAAF — No. 1 Squadron, still operational — was formed in 1922, and the eight original flying squadrons of the RAAF had their origins in the AFC. In the Second World War the RAAF expanded enormously and fielded 82 flying squadrons, including those on operations with the RAF in Europe, and squadrons designated part of the Netherlands East Indies forces in the South-West Pacific Area. There are at present 13 operational flying squadrons on the RAAF's order of battle. No. 1 and 6 Squadrons fly F-111s (q.v.) from Amberley (q.v.) in a strike and reconnaissance role; No. 3, 75 and 77 are fighter squadrons equipped with F/A-18 Hornets (q.v.) at Williamtown and Tindal (qq.v.); No. 10 and 11 Squadrons are maritime squadrons equipped with P-3C Orion (q.v.) aircraft for anti-submarine and anti-shipping operations from Edinburgh (q.v.); while fixed-wing transport capabilities are maintained by No. 33, 34, 35, 36, 37 and 38 Squadrons, which fly C-130 Hercules (q.v.), C-7A Caribous (q.v.), Boeing 707s and other aircraft from Richmond (q.v.), Amberley, Townsville and Darwin.

SQUIRES, Lieutenant-General Ernest Ker (18 December 1882–2 March 1940). Born in India to a clerical family, Squires was commissioned into the Royal Engineers in January 1903 after graduating from the Royal Military Academy, Woolwich. He served in India from 1905 to 1910, in Aden from 1910 to 1912, and then again in India. Sailing for France with the Lahore Division of the Indian Corps in 1914, he was wounded at Givenchy in December and at Ypres in April the following year. After convalescence in England he served further in Mesopotamia during the advance to Baghdad and in the Third Afghan War in 1919. He was awarded the DSO and the MC and was mentioned in despatches six times. Between the wars he attended the Staff College, Quetta, British India (now in Pakistan), in 1920 and the Imperial Defence College (q.v.) in 1928. Promoted to major-general in 1935, he became Director of Staff Duties at the War Office in 1936. In June 1938 he accepted the post of Inspector-General (q.v.) of the AMF, revived by the Lyons government after a lapse of nearly 10 years.

Some Australians doubted the need for an outside appointment, and Squires's designated mission, to report on the army's preparedness for war, cannot have been popular with everyone. The practice of calling on a British officer to investigate the state of the Australian services was used a number of times in the interwar period (the Salmond and Ellington inquiries into the RAAF being cases in point), and while the CNS, Admiral Sir Ragnar Colvin (q.v.), opposed Squires's findings because they gave too prominent a role to the army in Australia's future defence, the CGS, Major-General J. D. Lavarack (q.v.), welcomed the appointment. The report itself was presented to the government in December 1938. Among a large number of generally practical, sound and in many cases overdue recommendations were proposals to raise a permanent force of all arms, a reform of the command arrangements within Australia, the formation of new units for the militia and an increase in the peacetime army, and a purge of the Staff Corps' senior ranks and the introduction of a more equitable promotion structure for younger regular officers. The permanent force was put on hold by the new Prime Minister, R. G. Menzies (q.v.), but a number of the other proposals were adopted in time to have an effect in the early years of the Second World War.

Squires had been acting as CGS following Lavarack's departure overseas in May 1939, and on the outbreak of war the position was confirmed. As a result, a British officer oversaw the early months of the raising of the 2nd AIF. His health had begun to fail, however, and he died in office after a brief hospitalisation.

Against some expectations, Squires's period as Inspector-General was a success and considerably to the benefit of the army as it prepared for the impending war. His report was an incisive assessment of the effects of 20 years of governmental neglect of the army. His intelligence, experience, personal charm and determination to see things for himself impressed many of his subordinates; his period of service in Australia, in the opinion of some, made an impact out of proportion to its length.

SQUIRREL, AÉROSPATIALE (2-seat trainer and air-sea rescue helicopter). Rotor diameter 10.69 m; length 10.93 m; maximum speed 230 km/h; range 720 km; power 1 × Turbomeca Arriel 641 shaft h.p. engine.

The French-built Écureuil or Squirrel entered service with the RAAF in 1983 as training helicopters with No. 5 Squadron replacing the Iroquois (q.v.). They were also used by the search-and-rescue flights at RAAF bases Williamtown, Pearce (qq.v.) and Darwin. The RAAF helicopters were handed over to the Australian Army Aviation Corps (q.v.) in 1990 and are flown by the Australian Defence Force Helicopter School at Fairbairn. The RAN flies Squirrels with No. 723 Squadron for utility and survey work and aboard the *Oliver Hazard Perry* Class Guided Missile Frigates (q.v.). During the Gulf War three Squirrels, armed with a door-mounted 7.62 mm machine-gun, were flown from RAN ships in the Persian Gulf.

STAFF COLLEGES provide staff training for officers of each of the three services. Until just before the Second World War, army officers selected for staff training were sent either to the British Army Staff College at Camberley in England (established in 1857) or to the Army Staff College at Quetta (now in Pakistan, formerly in British India, established in 1907). These were valuable postings in that they produced staff officers who could function within the British military system as a whole. Australian officers also attended the Haifa Staff College in Palestine between 1940 and 1943. The Australian Army Staff College was established in Sydney in July 1938 as a Command and Staff College, and became the Australian Staff College in 1946 when it was moved to Fort Queenscliff, Victoria. The course is one year's duration, and combines broader defence-oriented subjects with intensive operational staff training. Selected students who perform at a high level are eligible to combine the Staff College course with an external study program that leads to a Graduate Diploma in Defence Studies from Deakin University.

Since for much of its history the RAN was intimately connected with the RN, there was for many years no perceived need for a separate Australian Naval Staff College. As the focus of Australian defence ties shifted away from the United Kingdom towards the United States, it was no longer acceptable for Australian naval officers to receive all their staff training at the RN Staff College, Greenwich, as had been the case previously, and the RAN Staff College was opened at HMAS Penguin, in Balmoral, Sydney, in 1979. The course runs for six months, the difference in length between it and the army course being due to the absence from the naval curriculum of extensive operational staff studies. The RAAF Staff College opened at Point Cook, Victoria, in 1947, and transferred to RAAF Base Fairbairn, Canberra, in 1960, where the course now runs for six months. Prior to the opening of the RAAF Staff College, from 1924 RAAF officers had attended the RAF Staff College at Andover, England.

The second level of staff college in Australia is the Joint Services Staff College (JSSC), which opened in Canberra in 1972, having grown out of the Joint Services Wing which was established in 1969. The 22 week course, preceded by an orientation program for overseas students, is designed to equip officers of lieutenant-colonel or equivalent rank with a close knowledge of the workings of the Department of Defence and of strategic developments affecting Australia and its region. In 1995 a third tier was established with the opening of the Australian College of Defence and Strategic Studies (ACDSS) adjacent to the JSSC. The ACDSS course, one year in length, is designed for officers of colonel/brigadier rank or equivalent, and emphasises advanced managerial skills and strategic studies, with the strengthening of professional and personal links within the region as an underlying rationale. In all of the Australian staff colleges, the course membership (and most of the directing staff) draws on all three services as well as overseas representation and civilians from government departments and private enterprise.

Australian officers continue to attend overseas staff colleges. Traditionally these have been Camberley and Quetta for the army, Andover for the RAAF, Greenwich for the RAN, and, since 1927, the Imperial Defence College (q.v.) for more senior officers. Since the Second World War the number of institutions to which Australians have been sent has widened considerably. It now includes the Canadian Command and Staff College in Toronto; the National Defence College, Kingston, Ontario (whose closure was announced in early 1994); the US Army War College, Carlisle, Pennsylvania; the US Naval War College, Newport, Rhode Island; the US Air War College, Maxwell, Alabama; and the Royal College of Defence Science, Shrivenham; as well as many other specialist institutions for specific branches and corps of the services. Australia's growing involvement in Southeast Asia is indicated by the attendance in recent years of Australian officers at staff colleges in Indonesia and Malaysia.

STALWART (I), HMAS see **'S' CLASS DESTROYERS**

STALWART (II), HMAS (destroyer tender). Laid down 1964, launched 1966, commissioned 1968; displacement 10000 tons; length 515 feet; beam 67.5 feet; armament 4 × 40 mm anti-aircraft guns; facilities for 2 Westland Wessex (q.v.) helicopters; speed 20 knots.

HMAS *Stalwart* began service in 1968. She is fitted with workshops for repairing destroyers and frigates at sea. She has exercised with other Australian ships in Malaysian and Australian waters. Most of her life, however, has been spent in Sydney Harbour at Garden Island.

STAND-TO is a shortened form of the order 'stand to arms'. Australians in the trenches of the Western Front would 'stand-to' in the hour before dawn and dusk against the possibility of enemy attacks or trench raids. The term was adopted as the title of the journal of the ACT Branch of the Returned Soldiers', Sailors' and Airmen's Imperial League of Australia (subsequently the RSL, see RETURNED AND SERVICES LEAGUE), first published in January 1950. Its purpose was to watch over the interests of ex-service personnel, to deal with issues 'of national importance such as defence and immigration', and to publish original material about Australian experiences in the two world wars. From 1950 to 1967 it was edited by C. E. W. Bean's (q.v.) long-standing assistant on the official history (q.v.), Arthur Bazley, and is an excellent source of first-hand accounts of active service and the leading personalities of the two AIFs in particular. The title was later adopted by the RSL's national newspaper.

STAWELL, HMAS see **BATHURST CLASS MINE-SWEEPERS (CORVETTES)**

STEELE, Major-General Clive Selwyn (30 September 1892–5 August 1955). Steele graduated with a degree in civil engineering from the University of Melbourne, and was commissioned into the 60th Infantry Regiment as a second lieutenant in December 1912. On the Unattached List when war broke out, he was seconded to the AIF in the same rank in September 1915. He served with the 5th and 6th Field Squadrons of the Royal Australian Engineers (RAE) (q.v.) during the war, reaching the rank of major in late 1918.

He continued to serve as a CMF officer with the engineers during the 1920s. Placed on the Unattached List again in 1931, he commanded the 14th Battalion from July 1933 until the outbreak of war. Seconded to the 2nd AIF, he was the original Commander Royal Engineers of the 6th Division, becoming Chief Engineer on Lieutenant-General Sir Thomas Blamey's (q.v.) 1st Corps headquarters in April 1940. He returned to Australia and the post of Engineer-in-Chief, Land Headquarters, at the beginning of 1942, which he held with a brief interruption for the rest of the war. He was demobilised in 1946 and returned to his civil career. He was appointed KBE in 1953 and was honorary colonel of the RAE from January 1953 until his death. The correspondent John Hetherington described him as 'by nature a fighting man, whether in war or peace, a blunt man, terse to the point of abruptness, whose few words were always right to the point'. He played a leading role in the so-called 'revolt of the generals' (q.v.) in February–March 1942.

STEVENS, Major-General Jack Edwin Stawell (7 September 1896–20 May 1969) enlisted as a sapper in the AIF in 1914, and was commissioned in the Australian Corps Signals Company in January 1917. He went on to the Reserve of Officers in 1920 on his return to Australia, before commanding the 2nd Cavalry Division's signals in the militia between 1923 and 1926. In civil life he was a public servant, while in his militia service he made the unusual transition, for the time, from the engineers to command of the 57/60th Infantry Battalion in July 1935. These were difficult times for militia soldiering, and he later recalled going in to annual camp one year with his unit so understrength that any sensible form of unit training was rendered virtually impossible. In October 1939 he was seconded to the 2nd AIF in command of the 6th Division signals. He was given command of the 21st Brigade in April 1940, which he led until his return to Australia in April 1942. Thereafter he commanded the 4th and 12th Divisions, which served in Australia, interspersed with a period as GOC of Northern Territory Force between August 1942 and March 1943. In April 1943 he was given the 6th Division, and led that formation in the fighting in New Guinea and through periods of retraining in Australia until relieved by Major-General H. C. H. Robertson (q.v.) in July 1945. His recall was at government request, in order that he might reorganise the Postmaster General's Department (PMG) in readiness for the tasks it would face under postwar reconstruction (q.v.).

He was a Commissioner of the Public Service Board, and in postwar life held a succession of senior civil service posts, including Secretary of the Department of Supply between 1951 and 1952 and chairman of the Atomic Energy Authority between 1952 and 1956. He was CMF Member of the Military Board (q.v.) in 1950. The official historian, Gavin Long, described him as 'a highly strung little man, full of drive and fight [and] unassuming'. But he had other qualities, which were called on when he was appointed to command of the 4th Division in April 1942, at that time serving in Lieutenant-General Gordon Bennett's (q.v.) III Corps in Western Australia. Querying his departure from an AIF command, he was told by Major-General S. F. Rowell (q.v.) that 'you're the only one we

could find who would serve under Bennett without kicking up a fuss'. His son, Captain Duncan Stevens, was killed on the bridge of HMAS *Voyager* (q.v.) when his ship was sunk by HMAS *Melbourne* (q.v.) during a naval collision in 1964.

STEVENSON, Group Officer Clare Grant (18 July 1903–22 October 1988) was educated at the University of Melbourne, taking out a Diploma in Education in 1925. She worked for the YWCA before joining the Berlei Company, manufacturers of what were then known as foundation garments. She became a company executive and spent time with the firm's London and New Zealand branches, with responsibility among other things for training of sales staff. In late 1940 she resisted efforts to appoint her as head of the Women's Auxiliary Australian Air Force (WAAAF), but relented subsequently and in May 1941 was appointed Director WAAAF. The CAS had been keen to obtain a 'working girl' for the appointment rather than a socialite, as was the pattern in Britain, and there is no question that Stevenson's background and image were important in establishing the WAAAF and then in earning it the respect of the mainstream service. In this she was undoubtedly helped by a relatively less-blinkered attitude towards women's service than pertained, for example, in the RAN. But her task was by no means an easy one, and her colleague, head of the Australian Women's Army Service (q.v.), Colonel Sybil Irving, later recorded that Stevenson was the person 'who did the most pertinent pioneering work' in having women accepted in the uniformed branches of the services. Her appointment as Director WAAAF was terminated in March 1946. After the war she went on to a career in educational and welfare services for women, being a founder of the Carers Association of New South Wales. She was made MBE for her wartime services, but only in 1960, and was awarded the AM in 1988 for community service and work with veterans.

STEYR AUG (F88) ASSAULT RIFLE is a gas-operated, magazine-fed, selective-fire, optically-sighted, modular-construction, bullpup-configuration, assault rifle. It was designed and manufactured in Austria in the late 1970s by the Steyr-Mannlicher Company for the Austrian Army and was selected by the Australian Army in 1985 after extensive trials with the Colt M16A2 to replace the 7.62 mm L1A1 SLR, 5.56 mm M16A1 and the 9 mm F1 SMG. It is manufactured in Australia under licence by the Australian Defence Industries Factory at

Group Officer Clare Stevenson with other senior officers of the women's services, Melbourne, 14 November 1942. Left to right: Stevenson; First Officer Annette Oldfield, WRANS; Group Officer I. M. Lang, Matron-in-Chief RAAF Nursing Service; Matron J. Sinclair-Wood, Matron-in-Chief AANS; Lieutenant-Colonel Sybil Irving, Controller AWAS; Lieutenant-Colonel Kathleen Best, Controller AAMWS. (AWM 137133)

Lithgow, New South Wales, and production deliveries began in November 1990, although troop trials production rifles had been in service since 1988–89. Since its introduction into service Australian soldiers have used the Steyr AUG on peacekeeping operations in various locations including Somalia, Cambodia and Rwanda. The Steyr AUG is futuristic in its appearance, and extensive use has been made of high strength plastic compounds in its construction to aid maintenance and to save weight.

The Steyr AUG fires 5.56 mm SS109 ammunition with an effective range for individual fire of up to 300 metres and up to 600 metres for section fire. The magazine capacity is 30 rounds. The cyclic rate of fire (i.e. the rate of fire that theoretically could be achieved given a continuous supply of ammunition) is 680 to 850 rounds per minute, with a rapid rate of 30 aimed shots per minute and an automatic fire rate of 90 rounds per minute. The weight of a loaded Steyr AUG is 4.1 kg. The optical sight has × 1.5 magnification and is designed for ranges up to 300 metres. The rifle can also be fitted with a bayonet (q.v.). Australia is producing three models of the Steyr AUG: the standard rifle (F88) with 508 mm length barrel; the carbine (F88C)

with 407 mm length barrel; and the 'special' model with the alternative housing assembly which allows the fitting of a variety of special sights, including night vision.

IAN KURING

STIRLING, HMAS, the Naval Support Facility at Garden Island, Cockburn Sound, Western Australia, is named after Captain James Stirling who landed at Garden Island in 1827 and established the first White settlement in Western Australia. Commissioned on 28 July 1978, HMAS *Stirling* can provide maintenance support to four destroyers and three submarines, as well as assisting Australian and foreign naval vessels visiting or refitting in the Fremantle–Cockburn Sound area. *Stirling* is the home port for several navy ships, and the plan to homeport destroyers in Western Australia was implemented with the arrival of the destroyer escort HMAS *Stuart* on 20 January 1984.

STOREPEDO A watertight cylindrical container of very hard cardboard, reinforced with metal bands with a conical, hollow head attached to a parachute and used for dropping supplies to troops. As the parachute opened the head pointed down and

struck the ground first, absorbing much of the shock. It was used extensively in the Second World War and later in Malaya to deliver food, ammunition, weapons, radio equipment and medical supplies into the jungle with the minimum possibility of damage. It was invented by K. M. Frewin (1905–59), aeronautical consultant to the Army Inventions Directorate.

STRAHAN, HMAS see *BATHURST* **CLASS MINESWEEPERS (CORVETTES)**

STRATEGIC DEBATE IN THE NUCLEAR AGE, AUSTRALIAN CONTRIBUTIONS TO Possibly the first informal Australian connection to the academic debate on strategy engendered by the nuclear revolution in warfare was participation in the (International) Institute for Strategic Studies (ISS) August 1963 conference (at the RAND [Research and Development] Organisation, Santa Monica, California) on 'The Security of Southern Asia'. The institute itself was only five years old at this time and conferences were very small by later standards. Of the 29 members, five were Australian. That high proportion was undoubtedly due to the belief of the first Director of the ISS, Alastair Buchan, that Australians would be, for the foreseeable future, much preoccupied with strategic relationships in South-east Asia. Those were the days when the strategic debate over Vietnam was just beginning to become public. The French were out of combat, but the Americans were not yet fully and formally in. United States President John F. Kennedy was still alive, and a good deal of the time of the conference was spent discussing his just-leaked refusal of some military proposals for American intervention in Laos. Both official opinion in Canberra, and mainstream intellectual opinion among the small group of Australians given to discussing strategic matters, was at the time decidedly 'hawkish', eager for American intervention in Vietnam, and even prepared to contemplate the building or acquisition of nuclear weapons by Australia.

The primary origin of this frame of mind was that the four years or so from the launching of the Indonesian policy of 'confrontation' against Malaysia (which then included Singapore) by President Sukarno in 1962–63, until his violent overthrow by the Indonesian army in September 1965, was a period of profound alarm in policy-making and intelligence circles in Canberra. The concept of

the 'Peking–Djakarta' axis (China and Indonesia in revolutionary alliance) had gained considerable currency. President Sukarno appeared to be dependent for support (against the army) almost solely on the Indonesian Communist Party (PKI) which had proclaimed its brotherhood-in-arms with Mao Zedong's China, then already lurching towards the great convulsion of the Cultural Revolution, and also towards nuclear status (its first atomic test having been carried out in 1964). The forces that looked to Ho Chi Minh in Vietnam were deemed certain to be victorious within a year or so, unless the Americans intervened, and triumphant policy-makers in Hanoi were assumed likely to extend their sway over both Cambodia and Laos in short order, and move on to destabilising both Thailand and Malaysia. The Kennedy policy-makers looked much less firmly committed to the defence of South-east Asia than their predecessors in Eisenhower's time, mostly because of their alleged 'appeasement' of Indonesia over Irian Jaya. So a member of the very small intelligence community in Canberra could quite plausibly in those years make a 'worst case' analysis in which the PKI secured ascendancy in Indonesia, and stepped up the confrontation campaign against Malaysia, while Ho Chi Minh's forces took over South Vietnam, with the backing of both China and the Soviet Union, and began to turn their attention to the rest of South-east Asia.

The medium-term strategic outlook for Australia thus seemed almost as bleak to some analysts as it had done in early 1942. This was the background against which the foundation of the Strategic and Defence Studies Centre (SDSC) has to be seen. Its creator was Dr T. B. Millar, who had been a young major in the Second World War before gaining a PhD at the London School of Economics and taking up an appointment in the Department of International Relations, Australian National University (ANU). His first book, *Australia's Defence* (1965), and his later *Australia in Peace and War* (1978), were influential and much used as university texts, but his primary achievement for the forwarding of strategic understanding in Australia was undoubtedly the setting up of the SDSC, which has since provided the framework of administrative support, research facilities, funds and collegiality for most of the Australians who have subsequently worked (as civilians or defence personnel on leave) on strategic issues. Tom Millar's success was achieved against considerable resistance: many of his university colleagues as well as most of the students at the time were inclined to equate an interest in strategic matters with an insensate militarism. Without his tenacity of

Checking a storepedo before it is dropped by a Wirraway, Bougainville, 25 December 1944. (AWM 077735)

purpose and intellectual resourcefulness, the project might not have survived.

His successor as head of the SDSC was Dr Robert O'Neill (q.v.), who had been a young officer in Vietnam during the Australian involvement there, and went on to become the official historian of Australia's role in the Korean War. It was O'Neill who developed the SDSC into an alternate, independent source of strategic analysis and comment for the Australian policy debate. A major influence on his mode of strategic analysis had been Sir Basil Liddell Hart, with whom he had worked as a graduate student at Oxford. As a historian, he was able to use the past as an illumination of the present, and had no difficulty in combining historical and contemporary studies, moving easily from work on the Korean and Vietnam wars to studies of superpower policies and regional security in the Asian, Pacific and Indian Ocean areas. He had a notable level of success in expanding the international links of the SDSC, and its financial infrastructure, by securing Ford Foundation support for a visiting fellows program which recruited scholars and officials from south and South-east Asia.

O'Neill's consolidation of the enterprise in Canberra was undoubtedly a factor in his appointment in 1982 as Director of the International Institute for Strategic Studies (IISS) in London. Most of the members of the institute are American or west European, and they include many of the most important policy-makers of the countries concerned. So the advent of an Australian to the Director's role represented an 'insider's' position in the world strategic debate that no other peripheral or middle-power national has enjoyed. Moreover, O'Neill's years there coincided with a period of much controversy over central balance, defence and arms control policies: matters such as Reagan's Strategic Defense Initiative (SDI or 'Star Wars'), cruise missile deployment in Western Europe, and the Strategic Arms Reduction talks between Washington and Moscow. This dangerous and exciting patch of history (the winding down of the Cold War) engendered many passionate arguments, not least within the conferences and the study-groups of the IISS, and it was to the major credit of the Director that the institute's reputation for calm and balanced expert analysis, illuminated by his own brand of liberal realism, was never diminished. He also managed to broaden the institute's area of preoccupation from its original concerns mostly with the central nuclear debate and the European theatre of potential military operations to encompass regional conflicts round the Pacific Rim, and in Asia, Africa and Latin America, and to reinforce its independence by strengthening its financial base through gathering in impressive capital funds, and bringing its publications and other activities into the computer age. In 1987 O'Neill moved to the Chichele Chair of the History of War at Oxford University, and he remains a major figure in the central strategic debate as Vice-Chairman of the IISS, and Chairman of its Executive Committee. He also influences generations of graduate students (many of whom will go on to be policy-makers) from all over the world at Oxford. His research interests in the mid-1990s are focused on the relation between thinkers and practitioners in the field of security over the past two centuries.

The next major Australian voice in the world's strategic debate was that of Dr Desmond Ball. His road to influence might be regarded as unorthodox, as he had been a leading activist of the Vietnam protest movement while a student at ANU, but his doctoral thesis, on the alleged 'missile gap' of the Kennedy period, established him at once as a leading analyst in the realm of nuclear theory. Des Ball evinced from his graduate student days an exceptional cogency of analysis concerning nuclear-strategic relationships, as well as an exceptional talent for securing information which the governments concerned were not entirely willing to see in the hands of outsiders. His *Politics and Force Levels: The Strategic Missile Program of the Kennedy Administration* (1980) dispelled a good deal of earlier 'disinformation' about the strategic nuclear balance between the United States and the USSR, and his *A Suitable Piece of Real Estate: American Installations in Australia* (1980) greatly broadened the basis of understanding on which the presence of 'joint facilities' at Pine Gap, Nurrungar and elsewhere has been debated in Australia.

Ball's most influential work on nuclear strategy was done in the late 1970s and early 1980s, much of it undertaken at or published by the IISS. His *Can Nuclear War Be Controlled?* (1981), often described as a brilliant analysis, was a direct and effective attack on the US strategy of controlled nuclear war-fighting. His work on strategic nuclear targeting exposed the strategic bankruptcy of nuclear employment policies. In the mid-1980s, his work on command and control, intelligence, and crisis stability introduced a new dimension to the study of crises as a precursor to nuclear war.

During the period when Professor Ball was head of the SDSC, the centre was at the forefront of the conceptual transformation of Australian defence policy from 'forward defence' (q.v.) and dependence on 'great and powerful friends' to a self-reliant policy of defence of Australia. The work of Ball, Ross Babbage, J. O. Langtry and Paul Dibb on low-level contingencies, northern defence, mobilisation planning, and force structure development was important in establishing the conceptual framework of Australian strategic and defence planning for the next decade. At this time also, the SDSC moved to develop a postgraduate course in strategic studies (the first in the southern hemisphere) and initiated a major publications program of monographs and working papers. The latter has underpinned a much more comprehensive and up-to-date debate on Asia–Pacific, as well as Australian, security issues. Since the end of the Cold War, Ball's attention has been mostly focused on the strategic issues affecting South-east Asia and the Pacific Rim, especially the establishment of the Council for Security Cooperation in the Asia/Pacific (CSCAP), in which he has been one of the primary intellectual driving forces.

The present head of the SDSC, Professor Paul Dibb, has undoubtedly had more direct influence on strategic policy-making in Australia than any of his predecessors. His original work in geography (at the University of Nottingham) developed into a long-term interest in the Soviet Union, and he was among the earliest and most convincing sceptics, while the Cold War was still on, of its survival as a superpower. His knowledge of that vast area of strategic preoccupations helped bring him headship of the Office of National Assessments, and later appointments as Director of the Joint Intelligence Organisation and Deputy Secretary of Defence. His work on the Soviet system was published as *The Soviet Union: The Incomplete Superpower* (1985). However, his direct influence on the formulation of Australian defence policy stems primarily from a 1986 official study (*Review of Australian Defence Capabilities*), usually known as the *Dibb Report* (q.v.), which largely determined the shape of subsequent Defence White Papers. In Dibb's view, the essential strategic problem for Australia, since the declining years of the Cold War, has been to devise an intellectual framework that will be useful in guiding policy-makers through a prospectively fast-changing world balance. The fixed guidelines or signposts of the Cold War years have already largely become irrelevant, and should, he argues, yield to unchanging certainties such as the nature of Australia's northern geography, the geopolitics of the area, and the need to be able to discern, in the light of a detailed knowledge of regional military capabilities, which threats are credible, and what means are necessary to meet them.

Although the SDSC has been central to the study of strategic issues in Australia ever since its foundation, there have been some Australian scholars in the field who worked largely outside its aegis.

The most notable and influential of these was Professor Hedley Bull, who held a Chair of International Relations at ANU from 1967 to 1977, and was the author of an outstanding work of general analysis of international relations, *The Anarchical Society* (1977). Hedley Bull was a distinguished member of the very small band of scholars who made arms control an issue that had to be taken seriously, not only by strategists, but by the major governments of the central balance powers. Thus he helped determine the course of the Cold War.

He came to that field of interest somewhat accidentally. As an undergraduate he read philosophy at the University of Sydney in John Anderson's day as Professor (retaining the impress of that formidable scholar to the end of his days). After a B.Litt at Oxford, he was recruited as a lecturer in international relations at the London School of Economics. Strictly for financial reasons, he also took on the task of helping an old Labour stalwart of the 1930s disarmament debates, Philip Noel-Baker, put a book together. Believing Noel-Baker to be seriously wrong, and turning his mind to more contemporary problems connected with nuclear weapons, and he published an outstanding work of his own, *The Control of the Arms Race*, in 1961. It made so notable an impression on those working in the field that when Harold Wilson, as newly elected Labour Prime Minister in 1964, decided to make the arms control issue one of special interest to the United Kingdom, Hedley Bull was appointed to supply (as Director of the Arms Control and Disarmament Research Unit) the intellectual drive and understanding needed to get the project off the ground. His years at the Foreign Office, working on the problems involved, generated real intellectual excitement, and a large input into the still more important debate on the same issues in Washington, where the process which eventually produced the Strategic Arms Limitation Talks (SALT) was beginning to get underway. Eventually it produced also the SALT and START treaties, and thus the beginnings of limitation and dismantling of nuclear stockpiles. Bull's methods of recruitment and consultation were sometimes innovative enough to startle the Foreign Office: the author was a member of the Arms Control Advisory Panel which he set up to gather understanding from seismologists and nuclear physicists as well as strategists and historians. Hedley Bull must undoubtedly be accorded a place among the handful of scholars who dominated and shaped that branch of strategic theory.

When he returned to Australia as professor of international relations at ANU, Bull brought some of those he had worked with in London with him. That legacy seems to have been among the factors which account for Australia's unusual degree of activism (by comparison with other middle powers) in the field of arms control. It was particularly notable in the early years of the Hawke Labor government, from 1983. Perhaps taking a leaf from Harold Wilson's 1964 book (the domestic political need to reconcile the left of the party was quite similar in the two cases) the new government in Canberra instituted an Ambassador for Disarmament (Richard Butler was the first appointee) and provided extra resources to the relevant sector of the department, creating what eventually came to be called the Peace, Arms Control and Disarmament Division. Its efforts have borne not inconsiderable fruit at the international level. The text of the Chemical Weapons Convention, which appears likely to come into worldwide force, was largely devised by Australian officials, and Australia was also a major force in the progress of the Partial Test–Ban Treaty and the Non-Proliferation Treaty. Because of its location in the southern hemisphere and the existence of the monitoring facilities at Pine Gap and Nurrungar, and seismological facilities elsewhere, Australia is and will continue to be exceptionally well-placed to play a role in the global monitoring of arms control.

A few other Australian contributors to the world's strategic debate should be mentioned in passing. Coral Bell may perhaps have been the first to see publication on the field within hard covers: 'Atoms and Strategy' in the *Survey of International Affairs for 1954* (written while working in London for the Royal Institute for International Affairs) and her book, *Negotiating from Strength* (1962), had a major international impact. Arthur Lee Burns, who was appointed to a Chair of Political Science at ANU in 1966, published an influential monograph, *From Balance to Deterrence*, in 1956, and a more substantial work, *Of Powers and their Politics*, in 1968. He was a lively and original thinker, more at home with mathematical models than most of the Australians in the field. In more recent years the scholars associated with Peace Research Centre and the International Relations division at ANU, especially Professor Andrew Mack, have contributed substantially to the progress of arms control systems (especially via Confidence Building Measures) in the Pacific Rim area.

All these may be accounted part of the mainstream strategic debate, either global or regional. There has also been a small counter-current, sceptical or distrustful of conventional strategic assumptions, which has tended to suggest alternative modes of meeting security dilemmas, or denied their existence. Their analyses have mostly revolved round concepts of armed neutrality (David Martin), or 'defensive defence' (Graeme Cheeseman).

One point that must be raised is how far any of the theorists or analysts under consideration have influenced the actual policies of the Australian government, or any other. Doubts have often been raised as to whether they had any influence. Tom Millar, the founder of the SDSC, used to say that he did not know of any Australian academic who has had a demonstrable effect on any item of foreign or defence policy. He was, however, probably speaking ironically to provoke debate, as he often did. Theorists are frequently practitioners as well, making policy directly at some stages in their careers. Paul Dibb obviously had a large input into Australian defence policy when he was a close adviser to the minister. Hedley Bull helped create the policies of the British government at an important period for global arms control. Martin Indyk (an Australian until he adopted US citizenship) helped devise United States' policy in the Middle East as a member of the National Security Council in Washington, and later as United States Ambassador in Israel. But quite apart from such direct policy-making roles, power over opinion is an important kind of power, and institutions like the IISS and the SDSC, by bringing together the theorists and the policy-makers, have a large though almost untraceable influence in changing the climate of opinion. Thus the directorship of the IISS, as in O'Neill's time, may be regarded as more than equivalent to a direct bureaucratic or even political role.

The policy-makers, in the armed forces as well as in the bureaucracy, have very often been irritated by the theorists, regarding them as people who rock the boat, push the debate in unwanted directions, and reveal matters it is inconvenient to have exposed. Nevertheless on the whole Australia has had, in the period under consideration, a very notable set of figures among ministers (Defence or Foreign Affairs), secretaries of departments, intelligence officers and chiefs of the armed services willing to entertain the concepts put to them by 'outsiders', and most importantly support their research with government funds.

CORAL BELL

STREET, Geoffrey Austin (21 January 1894–13 August 1940) enlisted in the Australian Naval and Military Expeditionary Force from the University of Sydney, where he was reading law, in August 1914. He transferred quickly to the AIF, was commissioned in September, and sailed for Egypt with the 1st Division. He was wounded at Gallipoli and served in France in staff and regimental appointments, again being wounded in September 1918. He was awarded the MC in 1917. He married in London before the war's end, and in 1919 returned to Australia and took up grazing in the Western District of Victoria, becoming involved in local politics during the 1920s. He rejoined the army in 1931 as a major in the militia, and became CO of the 4th Light Horse in December the following year. He stood successfully as a United Australia Party candidate for the Victorian federal seat of Corangamite in the 1934 election, and quickly came to prominence. He also continued to pursue his militia career, and in 1935 was given command of the 3rd Cavalry Brigade; he was promoted substantively to brigadier in 1938.

He took a strong lead in debates on defence issues in the federal parliament, while professing that defence issues should be non-partisan, and was a critic, though a relatively mild one, of reliance upon the Singapore strategy (q.v.). He became parliamentary secretary to the Department of Defence in the Lyons government in July 1938, and was appointed minister in November that year amid disagreements within the government over H. V. C. Thorby's administration of that portfolio. This meant the end of his militia career, but saw him preside over Australia's desperate and belated attempts to rearm and modernise the armed forces as the Second World War approached.

When R. G. Menzies (q.v.) became prime minister on Lyons's death in April 1939, Street became one of the most senior members of the cabinet, a position which he retained until his death. He oversaw the final drafting of the Commonwealth War Book, which dictated the government's response at the outbreak of war, and when war came was instrumental in the allocation of command and staff appointments in the senior ranks of the forces. In keeping with the government's early, equivocal reaction to the dispatch of an expeditionary force, Street appeared to talk down voluntary enlistment in the AIF, especially from the militia, and made no formal appeal for volunteers until the raising of the 8th Division in May 1940. Along with the CGS, General Brudenell White, and his ministerial colleagues J. V. Fairbairn and Henry Gullett (qq.v.), he was killed in the Canberra air crash of August 1940, which robbed the government of those ministers most experienced in military affairs just as the conduct of the war began to go badly wrong for the Allies.

STREETER, Captain Joan (25 April 1918–14 April 1993) joined the Women's Royal Australian Naval Service (WRANS) as a writer in 1943, was commissioned as a Third Officer, and served in Cairns and Sydney as a staff officer until demobilised in 1946. She rejoined the WRANS in 1954, as a Second Officer, was promoted to First Officer in 1955 and became Director WRANS in 1958 with the rank of Chief Officer. She served in this capacity for 15 years until retirement in 1973. Promoted to Superintendent in 1968, her rank was later changed to captain as part of the gradual merging of naval women into the mainstream of the navy. She was awarded the OBE in 1964. By the time of her retirement the women's service had grown to a strength of 741 all ranks.

STUART (I), HMAS (*Scott* Class destroyer leader). Laid down 1917, launched 1918, scrapped 1947; displacement 1530 tons; length 332.5 feet; beam 32 feet; speed 36.5 knots; armament 5 × 4.7-inch guns; 1 × 3-inch guns, 7 smaller guns, 6 × 21 inch torpedo tubes.

The *Stuart* was transferred from the RN to the RAN in 1933. Until the outbreak of war the ship served in Australian waters. In 1939 she sailed for the Mediterranean and undertook patrol work with HMA Ships *Vampire*, *Vendetta*, *Voyager* and *Waterhen*. HMAS *Stuart* had an eventful period of service. She took part in shore bombardments off Libya, the battles of Calabria and Matapan in 1940 and 1941, and troop transport in the Greek and Crete campaigns. HMAS *Stuart* was ordered home in 1942 to protect convoys between Australia and New Guinea. From 1944 to 1945 she was converted into a fast store-carrier and troop transport.

STUART (II), HMAS see **RIVER CLASS ANTI-SUBMARINE FRIGATES**

STUART (III), HMAS see **ANZAC CLASS FRIGATES**

STURDEE, Lieutenant-General Vernon Ashton Hobart (16 April 1890–25 May 1966). Having served an apprenticeship to a firm of mechanical engineers, Sturdee sat the examination for entry to the Permanent Forces and was commissioned as a second lieutenant in October 1908. Seconded to the AIF in August 1914, he commanded a field company on Gallipoli and in France, where he was responsible for the Anzac light railway system in the Somme area. Promoted to lieutenant-colonel in February 1917, he commanded the 4th Pioneer Battalion for most of that year, before spending four months as Commander Royal Engineers of the 5th Australian Division. From March 1918 until the end of the war he served as a staff officer at Field Marshal Sir Douglas Haig's headquarters. His war service had thus combined extensive command in the field with a breadth of administrative duties unusual in the AIF.

Sturdee attended the Staff College, Quetta, British India (now in Pakistan), in 1922 and 1923, and was then posted as an instructor at RMC Duntroon (q.v.). Between 1929 and 1933 he was in England, first as the Australian representative at the War Office, then in attendance at the Imperial Defence College (q.v.), finally as the military representative in the High Commissioner's office. Returning to Australia at the beginning of 1933, he spent the rest of the decade at Army Headquarters, as Director of Military Operations and Intelligence and then, from March 1938, as Director of Staff Duties.

In the first months of the Second World War Sturdee presided at Eastern Command, overseeing the expansion of facilities to meet the raising of the AIF and for the provision of local defence. He was seconded to the AIF in July 1940 and given command of the new 8th Division, but his opportunity for active command was cut short by the death of the CGS, General Sir Cyril Brudenell White (q.v.), in August. Sturdee, with his wide experience and administrative ability, was a natural choice as his successor, and he was to remain as CGS until September 1942. He was for a time the only Australian member of the Chiefs of Staff Committee. It was in recognition of the dire state of national affairs in early 1942 that his British colleagues, Burnett and Royle, supported him in his recommendation to the government that the AIF be returned from the Mediterranean for the defence of Australia, Sturdee making it clear that he would resign if this was not done. He had already demonstrated a firm grasp of national interests during 1941 by backing General Sir Thomas Blamey (q.v.) over the relief of the 9th Division in Tobruk, through withholding of the third brigade of the 8th Division for the defence of northern Australia, and in demanding from the Cabinet an increase in the militia.

When Blamey returned to Australia, Sturdee was sent to Washington to head the Australian Military Mission. A vital posting at this stage of the war, with the decisions which directly affected Australian security increasingly being made by the Americans, Sturdee's move signified the confidence in him held by the C-in-C, Blamey, by the government, and by General Douglas MacArthur (q.v.).

As had been promised, in February 1944 he returned to take command of the First Australian Army, with headquarters at Lae. In this capacity he commanded operations spread from the Solomons to the border with Dutch New Guinea, and in September 1945 accepted the surrender of all Japanese forces in this area from General Imamura. Following Blamey's arbitrary dismissal in November 1945, he became for a short time acting C-in-C of the AMF, supervising the demobilisation of the formations in the islands and New Guinea and overseeing the repatriation of the Japanese.

In March 1946 he began a second term as CGS, a sign of the regard in which he was held by a Labor Party not generally enamoured of serving senior officers. His second period as the head of the Australian Army was almost as important as the first. There were numerous problems of demobilisation still to be faced. An Australian component for the occupation of Japan had to be raised and dispatched, and the command and administration of the whole force arranged from Australia, and the shape and direction of the postwar army had to be decided on and then implemented in a climate of economic restriction. Sturdee did as much as anyone could in the circumstances which faced the services immediately after the war, and a revised postwar plan was accepted by the government in 1946 which laid the basis for the establishment of the Australian Regular Army. In 1949 the army was called on to intervene in the coal strike, and Sturdee's strict policy of personal political neutrality greatly aided his dealings with the Chifley government in a very difficult political climate. He retired in April 1950, to be succeeded by his deputy, Lieutenant-General S. F. Rowell (q.v.), the first RMC Duntroon graduate to hold the post of CGS.

Only two soldiers have held the post of CGS twice, Sturdee and Brudenell White. Of Sturdee his successor as CGS wrote that he was 'a gifted officer of high professional qualifications with the benefit of the best education and experience that the army of the day could offer'. He was appointed CGS by both R. G. Menzies (q.v.), in 1940, and Ben Chifley, in 1946. Frank Forde, Minister for the Army in the Curtin government, wrote of him in 1942 that 'all had a great admiration for him'. He proved 'quite unflappable' in the circumstances of crisis in early 1942, and his clear, logical and calm advice to government on the future deployment of the AIF and measures for the defence of Australia proved fateful in helping to ensure the repulse of the Japanese advance towards Australia. He may properly be regarded as one of Australia's greatest soldiers.

SUBMARINES see *COLLINS CLASS; 'E' CLASS; 'J' CLASS; 'O' CLASS; OBERON CLASS SUBMARINES*

SUCCESS (I), HMAS see *'S' CLASS DESTROYERS*

SUCCESS (II), HMAS see **GULF WAR**

SUDAN CONTINGENT left Sydney on 3 March 1885 to join a British expeditionary force sent to the Sudan to avenge the death at Khartoum of General Charles ('Chinese') Gordon, who had been killed by forces loyal to the Madhi, a messianic leader pledged to drive the infidel British out of the country. Gordon had attracted enormous public adulation in Victorian Britain and the Empire, and there were widespread demands for his death to be avenged. A British force under Sir Garnett Wolseley was dispatched up the Nile to retake Khartoum; another force commanded by Sir Gerald Graham was to strike west from Suakin on the Red Sea. On 12 February 1885, the Acting Premier of New South Wales, William Bede Dalley, cabled the Agent-General in London offering to provide two batteries of Permanent Field Artillery and a 500-strong infantry battalion, to land at Suakin within 30 days of embarkation. The British government's initial reaction was to decline the offer, but after several days it reversed its decision, persuaded that to refuse would cause offence in Australia and further intensify criticism in Britain, where it was already under stinging attack for not having done enough to save Gordon when his plight at Khartoum had become known. Offers of assistance were also sent by Victoria, South Australia and Queensland, but were not taken up by the imperial government.

The New South Wales contingent arrived at Suakin on 29 March 1885, and within days was engaged against the enemy. Three men were wounded in an attack, but otherwise the infantry saw little action. The artillery saw even less, largely remaining at Suakin, practising drill until they had achieved a level of efficiency that probably exceeded that which they had attained in Sydney.

On 17 May the contingent embarked for the return voyage to Australia, and landed at Sydney on 23 June 1885. Three men subsequently died of disease, but otherwise the expedition was free of casualties. Its significance lay not in the scale of its commitment, its military encounters or its cost, but in the fact that a colonial society had rallied of its own accord to the imperial cause. In that sense it demonstrated the depth of imperial sentiment in colonial Australia, a sentiment personified by the 'Little Boy at Manly' (q.v.).

SUNDERLAND, SHORT (13-crew reconaissance flying boat [Mark V]). Wingspan 112 feet 10 inches; length 85 feet 4 inches; armament 2 × 0.5-inch machine-guns, 8–12 × 0.303-inch machine-guns, 2000 pounds bombs; maximum speed 213 m.p.h.;

A Sunderland (Mark V) of No. 461 Squadron landing at Pembroke Dock, Wales, following an anti-submarine patrol in the Bay of Biscay, c. 1944. (AWM P1520/01)

range 2980 miles, power 4 × Pratt & Whitney Twin Wasp 1200 h.p. engines.

The Sunderland is a military version of the Empire passenger flying boat. Nine were being collected by No. 10 Squadron to take to Australia in 1939 when the Second World War began. No. 10 Squadron stayed in the United Kingdom, later joined by another Australian Sunderland squadron, No. 461, and flew patrols over the Atlantic Ocean protecting Allied convoys against German submarines and bombers. Sunderlands from both squadrons assisted in the sinking of twelve U-boats. On 2 June 1943 a Sunderland of No. 461 Squadron was attacked by eight Junkers Ju 88 aircraft over the Bay of Biscay. With its heavy armament and solid construction, the Sunderland destroyed three Ju 88s, damaged another three and managed to limp back to England. Sunderlands were also used as transports in the South-West Pacific Area (q.v.) from 1944 by No. 40 Squadron.

SUPPLY, DEPARTMENT OF, created on 17 March 1950, took over some functions of the Department of Supply and Development including arranging the manufacture and supply of equipment for the armed forces. From 1951 to 1958 its responsibilities for munitions, aircraft production and defence production planning were assumed by the Department of Defence Production (q.v.), and its design and inspection functions were transferred to the Department of the Army (q.v.) in 1959. When the Long Range Weapons Project (q.v.) was established it too came under Supply's control. The department was abolished on 12 June 1974 and its functions were taken over by the Department of Manufacturing Industry.

SUPPLY, HMAS (Fleet replenishment tanker). Laid down 1952, launched 1954; displacement 15 000 tons; length 583 feet; beam 71 feet; speed 16 knots plus; armament 6 × 40mm Bofors guns (q.v.)

HMAS *Supply* (originally *Tide Austral*), built for the RAN in Belfast, Northern Ireland in 1952, was lent to the RN between 1955 and 1962. She then reverted to the RAN and was renamed *Supply*. Her duties included providing other vessels with fuel and water, thus enabling the fleet to remain at sea for protracted periods. In the 1960s and 1970s HMAS *Supply* took part in a number of fleet exercises. Her most controversial role was to act as support ship to a New Zealand contingent near Mururoa protesting against the 1973 series of French nuclear tests.

SUPPLY SHIPS Two Australian National Line (ANL) Cargo Vessels, the *Jeparit* (3790) tons and *Boonaroo* (3900) tons were used to ferry supplies to Australian forces in Vietnam between 1967 and 1972. The ships had to be commissioned into the navy because of the refusal of members of the Seaman's Union to man them for this purpose. The *Boonaroo* was returned to the ANL in 1967 and the *Jeparit* in 1972.

SURVEY CORPS see **ROYAL AUSTRALIAN SURVEY CORPS**

SURVEY SHIPS Countries such as Australia with long coastlines have a continuous need to map the surrounding waters by hydrographic survey (q.v.). Over the years a variety of ships has been used for this purpose. The first was HMAS *Moresby* (I), a converted RN minelayer which arrived in Australia in 1925. During the next few years *Moresby* helped carry out the first comprehensive survey of the Great Barrier Reef. The first purpose-built Australian survey ship was HMAS *Moresby* (II) (2000 tons) which became operational in 1964. *Moresby* (II) carried out extensive surveys of the Tamanian coast, the Port Hedland area of Western Australia and the Great Barrier Reef. Survey ships currently in operation are the HMAS *Flinders* (800 tons) and four survey motor launches, HMA Ships *Paluma, Mermaid, Shepparton* and *Benalla*. The latter vessels are catamarans of 310 tons built in Adelaide. They are equipped with the latest in survey equipment including a computerised processing system which enables the suveyor to see the results of a day's surveying by pressing a button.

SWAN (I), HMAS see **RIVER CLASS TORPEDO BOAT DESTROYERS**

SWAN (II), HMAS see *GRIMSBY* **CLASS SLOOPS**

SWAN (III), HMAS see **RIVER CLASS ANTI-SUBMARINE FRIGATES**

SWORDSMAN, HMAS see **'S' CLASS DESTROYERS**

SYCAMORE, BRISTOL (2–5-seat rescue and training helicopter). Rotor diameter 48 feet 7 inches; length 46 feet 2 inches; maximum speed 126 m.p.h.; range 268 miles; power 1 × Alvis Leonides 520 shaft h.p. engine.

The Sycamore was the RAN's first helicopter and entered service in 1953. They served aboard the aircraft-carriers *Vengeance, Sydney* and *Melbourne* [qq.v.] and from HMAS *Albatross* Naval Air Station at Nowra. They were used for training and rescues during floods and other natural disasters (see CIVIL

COMMUNITY, ASSISTANCE TO). The Sycamore left service in 1965 and an example is preserved at the Naval Aviation Museum at Nowra. The Sycamore was the first British military helicopter and two were used by the RAAF from 1951 to 1965 at the Long Range Weapons Establishment (q.v.) at Woomera, South Australia.

SYDNEY (I), HMAS Displacement 5400 tons; length 457 feet; beam 50 feet; armament 8 × 6-inch guns, 1 × 3-inch anti-aircraft gun, 4 × 3-pounder guns, 10 machine-guns, 2 × 21-inch torpedo tubes; speed 25 knots; range 4000 miles.

The *Sydney* was a Town Class light cruiser laid down in England and commissioned into the RAN in 1913. Originally it was designed as part of the Australian Fleet Unit (flagship *Australia*) which was to be an independent squadron of the British fleet, operating in Australian waters and joining with the RN when required. In the early months of the First World War *Sydney* operated around New Guinea and played a role in the capture of some German Pacific Islands. Then in November 1914 it joined the convoy escorting the first contingent of the AIF to Egypt. As the convoy approached the Cocos Islands in the Indian Ocean a radio message was received from the station there that a strange warship was approaching. The *Sydney* was ordered to break off from the convoy and investigate. At 9.30 a.m. on 8 November the Australian ship sighted the *Emden* (q.v.), a light cruiser from the German Far Eastern Squadron that had been wreaking havoc among Allied merchantmen in Malayan waters and the Bay of Bengal. The *Emden* opened fire first and scored some hits, but the greater firepower of the *Sydney* (it had 6-inch guns as against the Germans' 4.1-inch) soon told. Eventually the German captain, von Muller, was forced to ram his ship into a coral reef. *Sydney* immediately signalled, '*Emden* beached and done for'. After clearing the convoy safely to Egypt *Sydney* was posted to the West Indies station before joining the Grand Fleet at the main British naval base at Scapa Flow in 1916 as part of the 2nd Light Cruiser Squadron. The remainder of the war was spent in rather monotonous patrols of the North Sea. The ship was present at the surrender of the German fleet in November 1918. After service in Australian waters *Sydney* was paid off in 1928.

SYDNEY (II), HMAS see *LEANDER* CLASS LIGHT CRUISERS

SYDNEY (III), HMAS see FLEET AIR ARM

SYDNEY (IV), HMAS see *OLIVER HAZARD PERRY* CLASS GUIDED MISSILE FRIGATES

SYDNEY HARBOUR, ATTACK ON see JAPANESE ATTACKS ON AUSTRALIA

SYNNOT, Admiral Anthony Monckton (5 January 1922–). Graduating from Creswell in 1939, Synnot spent the war aboard the RN ships HMS *Barham* and *Punjabi* and the RAN ships HMAS *Quiberon*, *Stuart* and *Canberra*. He enjoyed a routine of ship and shore appointments after the war, being promoted to the rank of commander in 1954 and captaining HMAS *Vampire* between 1960 and 1961. In 1950 he was the naval member of the Bridgeford mission to Malaya which provided advice to the Australian government on assistance to the British, then engaged in the early stages of the Malayan Emergency. Synnot's separate report discussed possible naval support which, given the small role of naval forces and the dominance of the RN in the region at that time, was small. This was the first part of a longer involvement in Malayan/Malaysian affairs: between 1962 and 1965 he was the CNS of the infant Royal Malaysian Navy, heading a small Australian advisory staff. The Malaysians appear to have turned to Australian rather than British naval officers to provide technical advice and higher staff support as they built their naval service in order to distance themselves somewhat from the former colonial power and to provide relatively more independent advice on the directions which their service should take. This was a good example of the postwar defence involvement in the South-east Asian region which the Australian services did much to pioneer. After his Malaysian service Synnot commanded HMAS *Sydney* (III) in 1966 during its deployments that year to Vietnam as the 'Vung Tau ferry', and in the following year commanded the flagship, HMAS *Melbourne* (q.v.). He held further staff appointments, as Director-General of Fighting Equipment and Chief of Naval Personnel, between 1969 and 1971, before appointment as Deputy CNS in the latter year. He commanded HM Australian Fleet, the usual precursor to appointment as CNS, in 1973, and then undertook a stint as Director of the Joint Staff in 1974–76. Appointed CNS in 1976, he achieved the senior position within the ADF, Chief of the Defence Force Staff (q.v.), in 1979, with the rank of admiral, and was created KBE in 1979. He retired from the navy in 1982. In retirement he served a term as chairman of the Council of the Australian War Memorial between 1982 and 1985, and pursued interests in horse-breeding and competition carriage-driving.

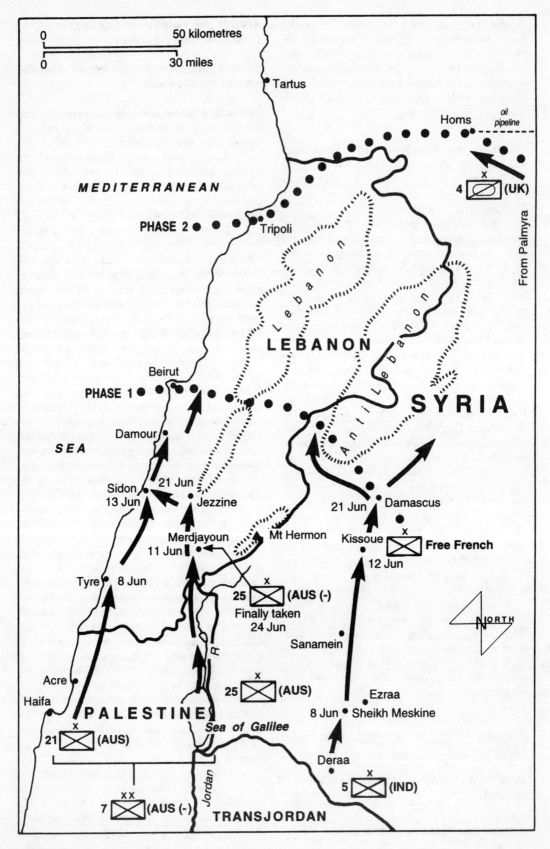

SYRIA The short, sharp, five-week campaign in Syria against the Vichy French in June–July 1941 was, from an Australian viewpoint, most notable as the first occasion on which the 7th Division was committed to action. The campaign was based on an unrealistic premise in that there was a notion, particularly in London, that the well-equipped Vichy French forces would offer only token resistance. This was not so, and from the outset the defenders not only demonstrated great bitterness towards the Free French cause and defied entreaties to succumb to the over-confident De Gaulle, they fought with courage and skill in rugged and difficult terrain. The determination of the Vichy French supported by the fact that in almost every department, with greater numbers, more armour, a more formidable air force and ample time to prepare their defences, the balance of military advantage lay decidedly with them.

The British policy that brought about operations in Syria had been enunciated on 1 July 1940, when the British government made clear that it would not allow Syria or the Lebanon to be occupied by a hostile power, or used as a base for attacks on those neighbouring countries which Britain was pledged to defend. On 3 April 1941, a *coup d'état* by Rashid Ali in Iraq not only threatened the security of British oil supplies through Haifa, it proclaimed the possibility of a deeper threat, had Hitler's Germany effectively supported the revolt. Instead, the Germans did not act decisively. Given Syria's sensitive geo-political position at the confluence of three continents, however, there existed the abiding possibility that they might.

The sequence of events that followed were Byzantine in their complexity, but the fact remained that from the perspective of London, Rashid Ali's collapse and the successful restoration of effective British control in Iraq did not remove the German threat to Syria. The British government's intention then became to secure Syria before Germany had recovered from the drain on resources that its air operations in Crete had imposed. There was also the possibility that the Turkish government, which had become alarmed by German aircraft landings at Damascus and had begun to bolster its defences on the Syrian frontier, might be induced to make common cause with the British. In any event, General Sir Archibald Wavell (q.v.), despite his own misgivings, was directed to improvise the largest possible force and be prepared to move into Syria at the earliest date.

What was not known in London at the time was that Germany had elected to turn aside from further ventures in the eastern Mediterranean in order to prepare for BARBAROSSA, the invasion of Russia. Indeed, by markedly reducing the German presence, it was striving to avoid giving the British any pretext for invasion. At the same time the High Commissioner, General Dentz, had been instructed from Vichy that he was to oppose any British attempt to enter Syria. This then was the situation when the British Commonwealth force together with Free French troops crossed the southern frontier into Syria on the night of 7–8 June 1941.

The force was commanded by the British General Henry Maitland Wilson who had recently commanded in Greece, and whose plan was to advance along three axes across a very wide front. He planned two phases: in the first phase to secure a line that included the capture of Rayak together with Damascus and Beirut, the capitals of Syria and Lebanon respectively; and in the second phase to advance to Palmyra, Homs and Tripoli. His main force was the 7th Australian Division, minus a brigade. One brigade column was to advance along the coastal route to Beirut, which he considered the main thrust. A second brigade was to advance in the centre through Metulla and Merdjayoun to Rayak along the valley of the Litani River. On the right, the 5th Indian Brigade Group and a Free French force under General Legentilhomme of six battalions, two batteries of 75 mm guns, one tank company and a cavalry detachment, were to occupy Deraa, Sheikh Meskine and Ezraa and then proceed to Damascus.

In turn, Major-General J. D. Lavarack (q.v.), the commander of 7th Australian Division, gave his brigades successive objectives: the first was a line from Merdjayoun to Sidon on the coast; the second a line joining Jezzine with Sidon to protect the coastal flank; and the third, the Rayak–Beirut road.

As it had been in Greece, so in Syria Wilson's command arrangements were tortuous. Although the Australian I Corps Headquarters was available as an operational headquarters, he chose not to use it but instead to exercise tactical command himself from Jerusalem. However, he instructed Lavarack that when the line Damascus–Rayak–Beirut had been reached, he (Lavarack) was then to take command of the operation — as a corps commander — for the second and final phase. This was to precipitate a situation similar to Greece, with the Australian corps command taking over in the middle of a desperate battle.

The complicating factors in all this were the nature of the terrain and the strength and dispositions of the enemy. In two successive conferences

Map 25 Allied operations against Vichy French forces in Syria, 1941.

before the operation began, Lavarack pointed out to Wilson that the right hand thrust through Damascus, where the terrain was easier, was likely to be much more readily forced than either the centre or coastal routes which ran through difficult terrain, which was more easily defended. Wilson chose to make three more or less equally weighted thrusts, while maintaining that the coastal advance on Beirut was the decisive one.

The Vichy French plan was also predicated on that assumption, and was sufficiently strong both in numbers and weapons to make the task of overcoming the defence a formidable one. The centre route from towering Mount Hermon along the Litani River valley was in some respects similar in ruggedness to Peshawar near the entrance to the Khyber Pass and into Afghanistan; the coastal route where the Lebanon range falls right down to the sea in a jumble of steep, rocky spurs and narrow valleys was easy to block by forces which were difficult to turn.

The forces initially available to Dentz were numerically stronger than the invaders. Dentz had two strong infantry divisions and a half division of tanks, cavalry and armoured cars — 35 000 regular troops backed up by a further 10 000 Levantine infantry of doubtful value. His defences, in the main, were well dug and well wired. In the air, Wilson's force could rely on support from 70 first line aircraft, including No. 3 Squadron RAAF which was in the process of re-equipping with Tomahawk fighters. The Vichy French force was larger. It included 100 aircraft (60 fighters), and then doubled its size as further aircraft were flown in from north Africa. Along the coast Wilson could rely on a force of cruisers and destroyers for seaborne support, including a landing ship.

For ease of reference, the operations can be considered in three phases: in the first phase (8–13 June), Wilson's expectation of a Vichy collapse did not occur and although all three thrusts made progress, the advance, particularly along the coast and in the centre, became slow, grinding, slogging matches. The second phase (14–22 June) coincided with operation BATTLEAXE in the Western Desert, which effectively curtailed needed support, particularly in the air. The 5th Indian Brigade's thrust towards Damascus was proceeding well, when two separate Vichy forces counter-attacked towards Kuneitra and from the Jebel Druse towards Ezraa, and captured both. These two initiatives posed a threat to the main supply route of 5th Indian Brigade and the Free French.

In the centre, the 25th Australian Brigade had captured Merdjayoun, but for the moment was incapable of exploiting beyond it. Instead, Lavarack chose to mask it with a small force, while transferring the bulk of the brigade to Jezzine at night over a difficult track, with the intention of bolstering the force on the coastal route. The plan misfired. The commander of the small force left at Merdjayoun determined not to be passive. But, in attempting to out-manoeuvre the enemy blocking the route to Rayak, he left Merdjayoun unguarded and the enemy promptly recaptured it. Lavarack acted swiftly and using his only reserve — two machine-gun companies and an anti-tank battery — he despatched them to hold a bridge in the rear, which effectively prevented the Vichy French counter-stroke exploiting its success to the south and into Palestine. He also ordered part of the 25th Brigade to return to Merdjayoun and sent his artillery commander, Brigadier F. H. Berryman (q.v.), to organise the defence and stabilise the situation in that area.

Wavell, who had warned his political superiors about the inadequacy of the forces available to him, but not to the point of resignation, now set about energetically to bolster the position which these multiple reverses had caused. An additional British brigade, hurried forward from the Nile delta, recaptured Kuneitra, while part of the Transjordan Frontier Force recovered Ezraa. A British battalion was sent to reinforce the depleted 25th Brigade at Jezzine which was also in difficulties.

In this situation, with various parts of the force intermingled with others, it was apparent that some positions would need to be regained and forces regrouped before the advance towards the Damascus–Rayak–Beirut line could be resumed. Accordingly, on 18 June, Lavarack was ordered to take command of all Allied troops in Syria which, in addition to the 7th Division, now included the 5th Indian Brigade, the Free French, the 16th British Brigade and the lines-of-communication troops. Wilson also ordered him to continue to regard operations against Damascus as being of secondary importance, and to collect the scattered battalions of 16th British Brigade and, using that formation as a reinforcement for the 7th Australian Division, advance as rapidly as possible up the coast and capture Beirut (that is, a return to the original plan).

Lavarack's first step was to establish a better command structure. Using Major-General J. S. Evetts, whose 6th British Division was now arriving, he appointed him to command all troops east of the Litani River, while Major-General A. S. Allen (q.v.), now commanding 7th Australian Division, would concentrate on the advance to Beirut.

He also recognised that, notwithstanding Wilson's direction about the main thrust, he should exploit success wherever it occurred. And so, when Evetts asked for the temporary return of his 16th Brigade to reinforce the 5th Indian Brigade/Free French attempt to take Damascus, Lavarack backed his own judgment that the capture of this important centre would do much to prise open the Vichy defence and erode its morale. He sought and gained Wilson's agreement. Evetts's operation was successful and after heavy and confused fighting during which the Vichy troops were forced away from Damascus towards Rayak and along the road to Homs, Damascus surrendered on 21 June. Thus, by that date the situation was that Sidon on the coast was in the hands of Brigadier J. E. S. Stevens's (q.v.) 21st Australian Brigade; Damascus had been taken; Kuneitra and Ezraa had been recaptured; Merdjayoun was resisting Berryman's attempts to recapture it; and, on the far right flank, the British 4th Cavalry Brigade advancing from Iraq was attempting to capture Palmyra but was having re-supply difficulties.

It was this last factor, the British threat from Iraq directed at Palmyra, that had earlier prompted General de Verdilhac to launch the series of Vichy French counter-attacks at the Allied lines of communication that had so successfully brought Wilson's advance to a halt. By now, however, de Verdilhac, who was acting as General Dentz's field commander, had used 13 of his 18 regular battalions, and had been forced to disperse most of the balance on defensive tasks. His reserves were almost exhausted. To counter the various Allied threats he had been forced to dispose five battalions to the Damascus area, two to the Jebel Druse, three to Merdjayoun, one and a half to Jezzine and six to the coastal sector north of Sidon, in an attempt to stop the 21st Australian Brigade.

To balance the force operating along the coast better, Lavarack reconstituted the 17th Brigade (6th Division) from the 2/3rd and 2/5th Battalions and 2/2nd Pioneers and allotted it to the 7th Division. During the third and final phase (23 June–12 July), the Allied grip along each thrust-line and in the country generally began to tighten. An important task for both the navy and air force had been to prevent the arrival of Vichy seaborne reinforcements. From the outset it was assumed that reinforcement would be attempted both by air and sea, and this assumption proved to be correct. At one point the Vichy government had hoped to use German transport aircraft to move in more troops, but the opening of Hitler's offensive against Russia on 22 June had effectively precluded this possibility.

An audacious attempt to reinforce by sea was made but the troop-ship was sunk. Not only had the Allies been able to add to their troop strength but, after BATTLEAXE was called off on 17 June, they were able to add two further fighter squadrons and three bomber squadrons. For the first time the Allied force had air superiority and could concentrate on ground targets in support of the troops. On 24 June, Merdjayoun had at last been recaptured, and although attempts to continue the advance to Rayak had been stopped by stubborn resistance, the threat to the Allies' rearward communications had been removed, and reorganisation for the next forward movement had been made possible.

On the coastal route, 21st Brigade had closed up to the Damour River while on its right 25th Brigade had advanced along the Lebanon Range. The terrain north and south of the Damour River was rocky, almost sheer in parts, and some of the most rugged in the country. The force defending Damour was equal to a reinforced brigade. It included the French Foreign Legion, Algerians, Senegalese and other troops, together with seven light tanks and ample artillery. The 7th Division's plan was for 21st Brigade to pin the defence from the front and turn it with a two-battalion attack on the inland flank. Once the defence had been broken, the 17th Brigade would pass through the 21st and drive straight up the road to Beirut. Further inland, the 25th Brigade was to conform to this movement and eventually join the main advance. The navy was to bombard the Vichy positions with four cruisers and eight destroyers, while the 7th Division was to have a bomber squadron, a fighter squadron and reconnaissance aircraft in direct support.

The initial 21st Brigade attack went in on the night 5–6 July and after heavy fighting over three days the plan worked. On the night 8–9 July, a carefully planned attack by 2/2nd Pioneer Battalion drove right through the Vichy position covering the coast road and linked up with 17th Brigade east of Damour. By midday on 9 July, all resistance in the Damour area had ceased and further inland the Vichy defence had also been overcome. In the Damascus sector Evetts's attacks were strong enough to prevent the defenders from transferring troops to the coastal battle. Faced with this situation and believing that honour had been satisfied, Dentz asked for an armistice. It was signed at Acre on 13 July. Under its terms the country would be occupied by the Allied forces, while the Vichy French forces were to be granted full honours of war, to keep their individual arms but stack all other weapons under Allied control.

In this campaign the Australian contingent numbered 18 000, the British 9000. There were 2000 Indian troops and 5000 Free French. In fact, in the wake of the Syrian operation, it is fair to say that without the three Australian infantry divisions and one from New Zealand, the British Commonwealth would not have been able to sustain a Mediterranean strategy at anything like the tempo that it did, and probably not at all. Casualties in Syria (killed and wounded) were: Australian 1600; British and Indian 1200; Free French 1100. They were nowhere near as heavy as in Greece and Crete, but coming on top of those débâcles they were still serious for an Australia that was soon to face the Japanese. To that end it was providential that the 7th Division had been thoroughly 'blooded' in Syria (and its 18th Brigade in Tobruk), and that these seasoned troops, together with some from the 6th Division, would be on hand at Milne Bay and for the Owen Stanley operations in New Guinea.

Syria had been close run, and success had come about more by good luck and hard fighting than by special prescience in the British high command. Two matters deserve final mention. As in the Greek campaign — and even though General Sir Thomas Blamey (q.v.) was now Wavell's deputy — first hand Australian involvement in planning and conducting the campaign was relegated almost to an afterthought. It was both a travesty and a waste that the Australian I Corps Headquarters should stand by idly with nothing useful to do while Wilson tried to run the campaign from the King David Hotel in Jerusalem, then, as he did in Greece, hand over the battle to the Australian command at the least propitious time. Second, during the campaign Wavell was relieved by General Sir Claude Auchinleck and, while Wavell was undoubtedly tired, the easy facility with which Churchill could hold him responsible, in part or wholly, for the loss of Benghazi 'which had undermined and overthrown all the Greek projects on which we had embarked', and also for failure both in Crete and the Western Desert, surely represents an unrealistic and unfair distribution of blame. In the light of other contemporary and subsequent events, it seems clear that responsibility for these outcomes should reasonably be shared by several others, including Churchill himself.

JOHN COATES

T

TAMWORTH, HMAS see *BATHURST* **CLASS MINESWEEPERS (CORVETTES)**

TANGE, Arthur Harold (18 August 1914–). A highly influential senior public servant, Tange served, among other appointments, as Secretary of the Department of External Affairs, 1954–65, and Secretary of the Department of Defence, 1970–79. Appointed by the new Labor government in 1972 to head a committee of inquiry into the organisation of the Department of Defence, Tange's report was accepted by the Minister for Defence, Lance Barnard (q.v.), in November 1973 and its recommendations were progressively implemented over the following three years. The Tange reorganisation was built on the reforms of the previous two decades, but went much further in centralising control of defence matters in the hands of the minister, his senior military adviser (the Chairman, Chiefs of Staff Committee, whose title was to be changed to Chief of the Defence Force Staff) and the Secretary of the Department of Defence. The Tange report identified as 'widely acknowledged deficiencies' the existence of the separate service boards, which gave rise to conflicts of interest and to a parochial approach to defence problems, and to the 'continuing adversarial relationship between the Services and the civilian Department of Defence'. Under the Tange reorganisation, the service boards were abolished, the service and military functions of the Defence Department and the services were much more closely integrated, a small number of powerful committees was established to draw together the essential parts of the defence policy formulation process, and a five year basis for planning, including financial allocation, was adopted.

There was much fierce resistance in some service circles to these changes, the main criticism being that they subjugated military expertise to civilian control, and gave too much power to the secretary of the department at the expense of the service chiefs. Tange was widely disliked, not only for his policies but for his forthright and abrupt manner: he did not suffer fools gladly and took few pains to conceal his contempt for those (particularly in uniform) whom he felt inadequate to the task of performing in a much more complex defence environment. These criticisms of the reforms have largely become muted in recent years, but there is still evidence of a lingering single service orientation rather than the coordinated defence approach that the reforms were designed to instil. Tange was a firm believer in the continuing general and professional education of the defence community, and was a strong supporter of the increased emphasis on tertiary education for officers, not least through the establishment of the Australian Defence Force Academy (q.v.).

(See also DEFENCE, DEPARTMENT OF.)

TANKS see **ARMOUR; SENTINEL TANK**

TASMANIA, HMAS see **'S' CLASS DESTROYERS**

TASMANIAN TORPEDO CORPS SHIP Tasmania's Convict Marine Service operated the armed wooden sailing schooner *Eliza* between 1835 and 1844, but its main role was simply to prevent the escape of convicts. Tasmania's only true warship was the steel second-class torpedo boat TB No. 191, ordered from Britain in 1883, which reached Hobart in May 1884. Lack of funds meant that the vessel could not be kept in good condition, and it spent much of its time laid up. Furthermore, the torpedo corps was often well below strength and was denied permission to fire practice torpedoes, apparently because of the cost of doing so. TB No. 191 was transferred to South Australian control in 1905, and was generally inactive before it was hulked in 1911.

TB No. 191: displacement 12 tons; length 63 feet; beam 7.5 feet; speed 16 knots; armament spar torpedo replaced by 2 × 14-inch torpedo-dropping gear and 1 machine-gun in 1887.

(See also COLONIAL NAVIES.)

TATTOO, HMAS see **'S' CLASS DESTROYERS**

TAURANGA, HMS see **AUXILIARY SQUADRON, AUSTRALIA STATION**

TB NO. 191 see **TASMANIAN TORPEDO CORPS SHIP**

TEAL, HMAS see *TON* **CLASS MINESWEEPERS**

TENTERFIELD ORATION was a speech by New South Wales Premier Sir Henry Parkes on 24 October 1889 in which he used Major-General J. Bevan Edwards's (q.v.) report on Australian defence (which advocated the federation of Australian colonial defence forces) to argue that political federation of the Australian colonies was necessary for the continent's defence. Though this speech is often considered to be the beginning of the campaign for federation, defence issues played a relatively minor role in the ensuing debate.

Ronald Norris, *The Emergent Commonwealth: Australian Federation, Expectations and Fulfilment 1889–1910* (Melbourne University Press, Melbourne, 1975).

TIGER MOTH, DE HAVILLAND (2-seat elementary trainer). Wingspan 29 feet 4 inches; length 23 feet

The sole ship of the Tasmanian Torpedo Corps, *Torpedo Boat No. 191*, at dock in Hobart. (AWM 300025)

1 inch; maximum speed 109 m.p.h.; range 300 miles; power 1 × De Havilland Gipsy Major 130 h.p. engine.

The Tiger Moth first entered service with the RAAF in 1939. Most of the aircraft flown by the RAAF were built by De Havilland Australia at Bankstown in Sydney. Their first use was as elementary trainers under the Empire Air Training Scheme (q.v.). By the end of the Second World War thousands of Australian pilots and navigators had trained in the aircraft. Some saw service in the New Guinea campaign as army cooperation aircraft. Tiger Moths were used by the RAN from 1948 to 1954 and remained in service with the RAAF until 1957, their longevity being a tribute to their ability to withstand the rough and tumble of pilot-training.

TIMOR In February 1941 Australia agreed with British and Dutch officials to reinforce Timor with aircraft and troops if the Japanese entered the war. It was also agreed that the force would be under Australian command. In consequence, 'Sparrow Force', commanded by Lieutenant-Colonel William Leggatt, disembarked at Koepang on 12 December

1941. The force was built around the 2/40th Battalion and also included the 2/2nd Australian Independent Company and a battery of coast artillery, totalling 70 officers and 1330 men. The only air force unit on the island was No. 2 Squadron RAAF, flying Hudson bombers (q.v.). Politically, Timor was complicated by the fact that the western part, with its centre of administration at Koepang, was governed by the Dutch, while the eastern end, with its major centre at Dili, was under Portuguese administration. A Portuguese enclave at Ocussi was also within the Dutch area. The Dutch defence in early 1942 included a force of 500 troops centred on Koepang; the Portuguese force at Dili numbered 150. Following the Japanese capture of Ambon (q.v.) on 3 February 1942, the next logical target in their eastern envelopment of the Netherlands East Indies was Timor, which, with its airstrip at Penfui, had become a key point in the ferry route between Australia and General Douglas MacArthur's (q.v.) forces in the Philippines.

When the decision was made to reinforce Timor with Australian troops and aircraft, the possibility of a large-scale attack on it was not envisaged, and so Timor, along with Rabaul (q.v.) and

Men of the 2/2nd Independent Company with a Timorese carrier during their guerrilla campaign against the Japanese, Timor, 1942. (AWM P0707/27/02)

Ambon, became another melancholy example of token defence. In addition, a complex agreement had been reached with the Portuguese that, in the event of Japanese aggression against its territory, the Portuguese Government of Timor would ask for Australian and Dutch help. Realising that by waiting for the Japanese to attack, such help would almost certainly be too late, Leggatt convinced them to allow him to station a force there.

Meanwhile, General Sir Archibald Wavell (q.v.), the Commander of ABDA Command (q.v.), recognising Timor's importance as an air link in the 'Malay Barrier', sought to have it reinforced. As a result Brigadier W. C. D. Veale arrived from Darwin with a small staff to command an expanded force. Five hundred Dutch troops were added, as was a British light anti-aircraft battery. A further Australian–American force was on its way, but was cut off by the Japanese attack of 20 February 1942.

As with their assaults on Rabaul and Ambon, the Japanese used carrier-based aircraft to reinforce their other fire support. The Japanese force from XVI Army was built around the 228th Regimental Group of the 38th Division which had previously attacked Ambon. It also included the 3rd Yokosuka

Special Naval Landing Force (a paratroop and infantry unit about 1000 strong). The Japanese landed two battalions with light tanks at the Paha River south of Koepang in the early hours of the morning. They advanced in two columns towards Koepang which was heavily bombed. On the same morning 300 paratroops landed 22 kilometres east of Koepang with the intention of cutting off the defending force from the interior.

The Australians' difficulty in planning their defence had been how to provide enough force in two separate areas and retain a mobile reserve to deal with the main thrust when it came. To add to the problem, Japanese aircraft had been attacking the island from 26 January to such effect that most of the defending aircraft had been destroyed, and those remaining had been withdrawn to Australia on 19 February. From then on, the airstrip at Penfui was a refuelling stop only, and the original reason for the defence of Timor had largely disappeared.

Before the Japanese landing, Leggatt, with Veale's concurrence, had laid out the defence with most of the Dutch troops and two companies of 2/40th Battalion in the Koepang area and nearby beaches. The two remaining companies of 2/40th Battalion with added transport, were kept as a mobile reserve. A combined defence headquarters, which had been set up at the airfield with the RAAF, was now redundant with the latter gone. A force administration base had been established at Champlong, 40 kilometres east of Koepang in the foothills. Leggatt, like Lieutenant-Colonel L. N. Roach on Ambon, concerned at how thinly his force was stretched, had despatched the 2/2nd Independent Company to near Dili, a highly malarious area, and was on the point of redeploying it to Koepang when the Japanese attacked.

The defence of Koepang, while spirited, and exacting a high toll in Japanese casualties, particularly of the paratroops, who were almost all wiped out, nevertheless could not be sustained. As in almost every area where Allied forces met the Japanese at this stage of the war, the latter were better prepared, better equipped, and better instructed in the sort of fighting that island operations called for. Their control of the sea and air gave their ground forces inestimable advantages both in reconnaissance and fire support which the defending force could not counter. There was a similar situation with communications. Australian radio sets were poor, and the Japanese preliminary bombardment and aerial bombing cut communication lines so effectively that messages had to be sent laboriously and hazardously by runner, or not at all. Very early in the fight, Veale at Champlong lost contact with Leggatt and the bulk of the troops, and with a Japanese force between them, Leggatt's options were limited. The longer term Australian plan, once forced away from Koepang, had been to withdraw towards Su, which could perhaps have been held for some time. But, not only was Leggatt cut off from Su, he was also cut off from his ammunition reserves and food supplies at Champlong.

In a series of well-executed small attacks, he was able to capture Babau, and was on the point of breaking through to Champlong when his column was attacked from the rear by the mass of the Japanese force led by tanks. The Japanese, both on the ground and from the air, began systematically to destroy his force and the large group of wounded it was carrying. Almost out of ammunition and food he surrendered at 9 a.m., 23 February 1942. His troops had fought for four days against a greatly superior force with heavier firepower. Some escaped and joined Veale's force, which moved into the Portuguese part of the island.

From this point on, the fighting on Timor became the story of the 2/2nd Independent Company's guerrilla campaign against the Japanese. Timor, with an area of 14 874 square kilometres, is a much larger island than Ambon. It also has a rugged

interior which one Australian officer described as 'one lunatic contorted, tangled mass of mountains', in which 'the mountains run in all directions and fold upon one another in crazy fashion'. It offered the possibility of guerrilla warfare against an invader. Although Portugal's wider metropolitan policy meandered during this period, the Australians were greatly assisted by its officials on the spot. They were also given effective help by the local people, who provided them with a large-scale intelligence screen.

The company had been specially trained for commando-style, stay-behind operations and, as the title implied, was intended to operate on its own (see INDEPENDENT COMPANIES). It had its own engineers and signallers, but it did not have heavy weapons, nor many vehicles. It was to make great use of Timor ponies and native carriers. For much of its time it operated by platoons over an arc of 100 kilometres in the rough country south of Dili and the approaches to the town. By skilful patrolling, tracking and ambushing, it maintained a moral — and for a time — a physical ascendancy over the Japanese. The remnants of Veale's force were concentrated near Mape in the southwest.

From the end of April, the company had improvised a radio link with Darwin using 'Winnie the War Winner' (q.v.), which enabled it to report its situation, receive supplies, and also receive directions as to its future. When, eventually, it was also able to evacuate its wounded and sick by small ship, its resilience was increased. By June 1942, General Thomas Blamey (q.v.), now General MacArthur's Land Forces Commander, recommended one of two courses: either recapture Timor with an overseas expedition, in which case Sparrow Force would be the principal pathfinder, or withdraw the force. MacArthur chose a middle course: leave the present force there, sustain it, but be prepared to withdraw it if the Japanese threatened to destroy it. This was the course followed.

Beginning in July, the Japanese struck in a number of columns with the intention of destroying the Australians, and also a Dutch stay-behind group in the south-west round Maucator. They also used bombing raids but, most serious of all, they used a screen of natives from Dutch Timor to convince the local people to disown the Australians. This phase ceased abruptly when the main Japanese force was withdrawn to Rabaul. But in succeeding weeks the Japanese built up their occupation force from about 1500 troops to 12 000, and their efforts to cause strife between the local people and the Portuguese administration, worked to the detriment of the Australians. In the meantime, the 2/2nd was augmented by the arrival of the 2/4th Independent Company from Darwin on 23 September 1942 on board the 'V' Class destroyer (q.v.) HMAS *Voyager*, which unfortunately went aground off Betano and had to be scuttled by its crew who were then taken off in small ships.

As Japanese efforts to wear down the Australian force and separate it from its native support became more enterprising, it was a matter of time before Sparrow Force (renamed 'Lancer Force' from November 1942) found its operations becoming too unpromising to continue. In addition, the excellent intelligence of Japanese movements, which until then had been faithfully received from the local people and in turn transmitted to Australia, now began to dry up.

From early December 1942, operations on Timor were progressively wound down. The 2/2nd Independent Company was evacuated between 10–16 December 1942. There was a flurry of uncertainty in the same month, when the Advisory War Council (q.v.) debated whether to continue operations in Timor. It was firmly dissuaded from proceeding by General Blamey, who recognised that there were not sufficient resources to continue the operation and also fight the costly bridgehead battles then in progress in the Buna area in the New Guinea Campaign. The 2/4th Independent Company was withdrawn on the Tribal Class destroyer (q.v.)

HMAS *Arunta* on 9 January 1943. Finally, a small remaining group, together with Special Operations Australia (q.v.) which had been operating independently at the eastern end of the island, was withdrawn to Fremantle by a US submarine on 10 February 1943.

The Australian operations on Timor concluded just short of year after they had begun. While in strategic terms they did not have a profound effect on the course of the war, they had tied down a large force of Japanese and had caused them a disproportionate level of casualties. They had been far more successful than other Australian operations in either Java, Ambon or Rabaul. They had also demonstrated that given reasonably favourable circumstances, unconventional operations, which an independent company was capable of pursuing, were both more versatile and cost-effective than more formal operations, for which resources, at the time, did not exist.

Peter Henning, *Doomed Battalion: Mateship and Leadership in War and Captivity: The Australian 2/40 Battalion 1940–45* (Allen & Unwin, Sydney, 1995).

JOHN COATES

TINDAL, RAAF BASE, 15 kilometres south of Katherine, is the main RAAF base in the Northern

HMAS *Tingira* in Sydney Harbour during the First World War. (AWM 301557)

Territory. Tindal covers an area of 122 square kilometres and is the largest RAAF establishment, being 10 times larger than Amberley (q.v.), the next largest base. Tindal is home to the F/A-18s (q.v.) of No. 75 Squadron. An airfield was first built on the Tindal site during the Second World War for Liberator (q.v.) bombers, but this was never used operationally. Construction of the current base began in 1982, and it was opened officially on 31 March 1989. Tindal is named for Wing-Commander A. R. Tindal, the commanding officer of No. 24 Squadron, who was killed during the bombing of Darwin (q.v.) on 19 February 1942.

TINGIRA, HMAS, originally known as *Sobraon*, was built in 1866, purchased by the Commonwealth government in 1911, renamed *Tingira* (allegedly an Aboriginal word meaning 'open sea') and commissioned on 25 April 1912 as a naval training ship for boys aged 14.5–16. Permanently moored at Rose Bay, Sydney Harbour, HMAS *Tingira* could train up to 250 boys at a time. From the first intake in June 1912 to its decommissioning on 27 June 1927, 3168 boys did their initial training there.

TOBRUK, SIEGE OF Tobruk is a port city on the Libyan coast, and in 1940 was part of the Italian empire. Its capture was important for any army moving in either direction between Egypt and Cyrenaica because of the role it could play in shortening an army's logistic lines. Well fortified by the Italians and defended by a garrison of about 25 000 men, it was taken in a brazen two-day assault by the Australian 6th Division on 21–22 January 1941, with the capture of the entire garrison and enormous quantities of stores and equipment. Left in the wake of the Allied advance, which culminated in the capture of Benghazi and the destruction of the Italian Tenth Army in February, Tobruk was chosen as a defensible position by General Sir Archibald Wavell (q.v.), and the 9th Australian Division was withdrawn into it, in the face of the first of Rommel's advances to the Egyptian border in March–April 1941. The Allied garrison, largely Australian, consisted of the 9th Division (20th, 24th and 26th Brigades), the 18th Brigade of the 7th Division, four regiments of British artillery, and a motley collection of other British units gathered up in the retreat. Lieutenant-General J. D. Lavarack was given Cyrenaica Command in place of the hapless General Neame, and he was responsible for the initial deployments for the defence of Tobruk. From 14 April and thus for most of the siege, however, the commander of Tobruk was Major-General L. J. Morshead ('Ming the merciless' to his troops, 'Ali Baba Morshead and

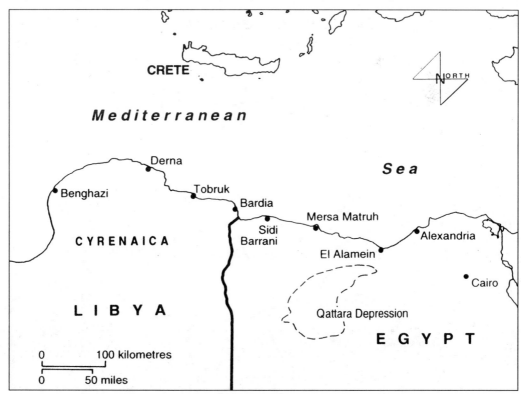

Map 26a General map of the Western Desert, 1941–42.

his 40 000 thieves' to the German propagandist, Lord Haw-Haw). Wavell's instructions for the conduct of the defence were admirably succinct: 'Your defence will be as mobile as possible and you will take any opportunity of hindering the enemy's concentration by offensive action.' The town was to be held for two months before relief from Egypt would be possible; a relieving force in fact only fought its way through in early December. Morshead was an experienced and knowledgable soldier who had commanded a battalion in France in the First World War, but this was the first time he had seen action in this war. His own division, the 9th, was in a state of some disarray regarding both training and equipment, and in fact the semi-static role adopted for the defence of Tobruk was well suited to its capabilities and experience. The defences available were essentially those created by the Italians the previous year, and Morshead conceived the defence based on a strong perimeter supported by lines of secondary defence. The Australians mounted a program of nightly patrols in strength, a technique developed during the last year of the war on the Western Front in 1918 and designed to keep the enemy off-balance and unable to build up a detailed picture of the defences through compre-

hensive reconnaissance patrols of their own. The Germans cut off the town on its eastward (i.e. Egyptian) side on 11 April, and began to test the defences. The Germans attempted combined infantry-armour assaults, but the German tanks suffered mounting casualties at the hands of the Australian anti-tank gunners (in some cases using captured Italian artillery pieces, dubbed the 'bush artillery'), and their resultant retreat left their own infantry exposed and highly vulnerable, with many being killed or captured. A number of assaults were made, one of which, at the end of April, broke through the perimeter before being held at the minefields in front of the secondary line (the Blue Line). This, the Battle of the Salient, involved the Australians in a number of counter-attacks which resulted in heavy casualties among the infantry, and even heavier ones among the German attackers, especially in armoured vehicles which they could ill-afford to lose. One German regiment went in on 1 May with 81 tanks of various marks, and had only 35 serviceable vehicles left by the following evening. Lieutenant Schorm of that regiment recorded in his diary for 6 May:

> Our opponents are Englishmen and Australians. Not trained attacking troops, but men with nerves and

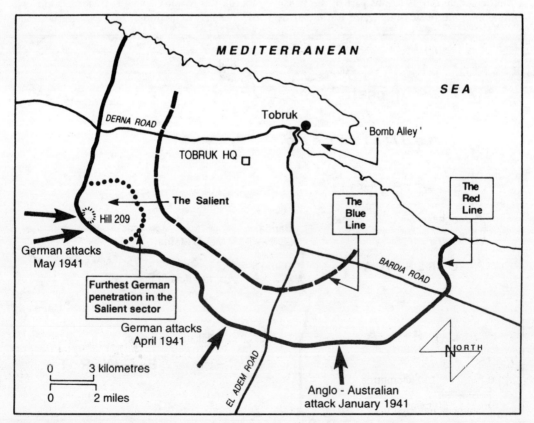

Map 26b   Tobruk, 1941, depicting the main defensive lines and principal axes of German attack.

Men of the 2/20th Battalion and the British Royal Horse Artillery use a captured Italian 110 mm gun during the siege of Tobruk, Libya, 10 August 1941. (AWM 020648)

toughness, tireless, taking punishment with obstinacy, wonderful in defence. Ah well, the Greeks also spent ten years before Troy.

Attempts by the British forces in Egypt to break through failed (codenamed Operation BATTLE-AXE, these attempts had taken place on 15 May and again on 16 June), which led to Wavell's relief and his replacement by General Sir Claude Auchinleck. The Tobruk garrison engaged the enemy beyond the perimeter on occasions, but was under constant strain from artillery fire and air bombardment. The 'Tobruk ferry service' kept the town supplied and, to some extent, reinforced with regular nightly runs. These were, however, expensive in ships and transports lost or damaged along 'Bomb Alley', the approach to Tobruk harbour. The Australian destroyer *Waterhen* and the sloop *Parramatta* were both sunk in this manner. The air effort against Tobruk was also costly for the *Luftwaffe*, however, which quickly abandoned daylight dive-bombing attacks thanks to the anti-aircraft artillery, and which during the siege lost around 150 aircraft and, more critically, dozens of experienced aircrew. Casualties from all causes mounted among the defenders, however, and in July the Australian government determined that the AIF units should be relieved; one Australian division, the 6th, had been rendered unfit for further action that year thanks to the defeat in Greece, and the

government was not prepared to lose a second so quickly. The British government and military authorities in Cairo were disinclined to accept the Australian decision, which prompted one of the more memorable exchanges between General Sir Thomas Blamey (q.v.) and his superior, Auchinleck. When Blamey chided Auchinleck by pointing out that the latter would not argue had the troops been foreign, Auchinleck responded, unwisely, that of course they were not. Blamey pointed out that his command in fact formed part of the army of a sovereign State, and that the requests of his government and its senior commanders in the field were to be accorded the respect due to them. Auchinleck had little option but to comply. The first portion of the AIF was relieved in August, the second over September–October. Only one unit, the 2/13th Battalion, saw the siege through from beginning to end when its transport was forced back before reaching the port. Ground communications were re-established on 7 December as part of Operation CRUSADER. Australian casualties were around 3000, with a further 941 taken prisoner.

Tobruk was again besieged in January 1942, but this time the garrison failed to hold in the face of a rapid assault, and some 35 000 British and South African soldiers were taken prisoner. The town changed hands again for the final time in November 1942 after the Allied victory at El Alamein (q.v.). The protracted defence of Tobruk had two important effects: it prevented Rommel from shortening his lines of communication through the use of the port while leaving a substantial Allied presence in his rear which diverted troops and resources he could ill-spare from his attempted invasion of Egypt, and it provided an enormous fillip to Allied morale in what was otherwise a very bad year for the Allied cause in Europe.

Barton Maughan, *Tobruk and Alamein* (Australian War Memorial, Canberra, 1967); Chester Wilmot, *Tobruk 1941: Capture–Siege–Relief* (Angus & Robertson, Sydney, 1944).

TOBRUK (I), HMAS see **BATTLE CLASS DESTROYERS**

TOBRUK (II), HMAS see **SMALL SHIPS SQUADRON**

TOC H was the morse signallers' name for Talbot House, a rest-house for troops established at Poperinghe, Belgium, by British Army chaplains Neville Talbot and Philip Clayton. In 1920 Clayton founded the Toc H movement as an interdenominational Christian community service organisation, and the first Australian branch was formed in 1925.

TON CLASS MINESWEEPERS (*Curlew, Gull, Hawk, Ibis, Snipe* and *Teal*). Displacement 375 tons; length 152 feet; beam 29 feet; speed 14 knots; armament 2 × 40 mm guns.

These ships were purchased from the RN in 1961 and modified in Britain to Australian requirements. They served as the 1st Mine Countermeasures Squadron in Malaysian waters during the Confrontation (q.v.). *Curlew* and *Snipe* were converted to minehunters in the late 1960s.

TOOWOOMBA (I), HMAS see **BATHURST CLASS MINESWEEPERS (CORVETTES)**

TOOWOOMBA (II), HMAS see **ANZAC CLASS FRIGATES**

TORPEDOES The name derives from a species of electric fish and was originally applied to mines. Propelled torpedoes were first demonstrated in 1866 by Robert Whitehead in the United Kingdom. His underwater missile could deliver an 18-pound explosive charge to a range of 700 yards. Initially these weapons were fitted to fast, light craft specially built to mount them, called torpedo boats. Since about 1900, however, the torpedo has developed as the main weapon of the submarine. In Australia fixed torpedoes or mines were in use in the early nineteenth century. In the mid-nineteenth century locomotive torpedoes were intro-

duced, and by the late 1880s the Whitehead Mark VIII torpedo, which could carry a warhead of 200 pounds for 1000 yards, was the main weapon of the colonial navies' (q.v.) torpedo boats. The Brennan torpedo invented by Melbourne engineer Louis Brennan (q.v.) was fired from the shore and connected by wires to a steam-driven winch which propelled the torpedo and allowed it to be steered towards the target and even retrieved if it missed. It was successfully tested in 1879 before a Victorian committee headed by Colonel Peter Scratchley (q.v.), and Brennan went to the United Kingdom where the War Department built a factory to manufacture the torpedo. They were used from the 1880s to the early 1900s to defend all the major British ports and strategic ports in the Empire such as Gibraltar. Brennan torpedoes were not used in Australia; in fact the British War Department refused an order from the Victorian government on the grounds of scarcity of supply. During the First World War Australian 'E' Class (q.v.) submarines used the 18-inch torpedo. For a brief period during the Second World War Australia manufactured torpedoes. Plans were drawn up in 1941 to make the 18-inch Mark XV torpedo, which could be dropped from the air by the Bristol Beaufort (q.v.), as well as used in submarines. Land was acquired in Sydney and in 1943 the Royal Australian Naval Torpedo Factory came into existence. The first tests of these torpedoes took place in 1944 and they passed their undersea firing and air dropping tests with great success. By then, however, the war had turned decisively in the Allies' favour, so the torpedoes were not needed. None were ever fired in anger, and the RAN Torpedo Factory was turned over to the maintenance and repair of torpedoes for the RAN and Allied navies. The RAN's *Collins* and *Oberon* Class submarines (q.v.) now use Mark 48 heavyweight torpedoes, while *Charles F. Adams* Class destroyers, Westland Sea King and Sikorsky Seahawk (qq.v.) helicopters are equipped with Mark 46 lightweight torpedoes. *Oliver Hazard Perry* Class frigates, *Anzac* Class frigates, River Class destroyer escorts and the RAAF's Lockheed Orions (qq.v.) maritime patrol aircraft also carry torpedoes.

TORRENS (I), HMAS see **RIVER CLASS TORPEDO BOAT DESTROYERS**

TORRENS (II), HMAS see **RIVER CLASS ANTI-SUBMARINE FRIGATES**

TORRES STRAIT LIGHT INFANTRY BATTALION see **ABORIGINES AND TORRES STRAIT ISLANDERS IN THE ARMED FORCES**

TOWN CLASS LIGHT CRUISERS (*Sydney, Melbourne* and *Brisbane*). Laid down 1911–13, launched 1912–15, paid off 1928–29; displacement 5400 tons; length 457 feet; beam 50 feet; speed 25.5 knots; armament 8 × 6-inch guns, 1 × 3-inch anti-aircraft gun, 4 × 3-pounder guns, 10 machine-guns, 2 × 21-inch torpedo tubes.

HMA Ships *Sydney* and *Melbourne* were built in Britain, and HMAS *Brisbane* was built at Cockatoo Island (q.v.). The *Sydney* and *Melbourne* participated in the capture of Germany's Pacific colonies, then were part of the escort for the 1st AIF in November 1914. Off the Cocos Islands HMAS *Sydney* was diverted to hunt for the German light cruiser *Emden*, which it crippled on 9 November 1914. The *Sydney* and *Melbourne* also served in the West Indies Station and with the Grand Fleet at Scapa Flow, where both ships were present at the German surrender. HMAS *Brisbane* served in the Mediterranean in 1917, then the Indian Ocean and finally in Australian waters and the Pacific. HMAS *Sydney* and HMAS *Melbourne* were scrapped in 1929, HMAS *Brisbane* in 1936.

TOWNLEY, Athol Gordon (3 October 1907–24 December 1963) was educated at the University of Tasmania and practised as a pharmacist. He served in the Royal Australian Navy Reserve between 1939 and 1945, reaching the rank of commander and serving in both Europe and the South-West Pacific Area. He had gone into local government and was a member of the State Repatriation Board before entering federal parliament as a Liberal for the seat of Denison in 1949. He entered the ministry early, in 1951, as minister for Social Services, and thereafter served in all Menzies' governments. Townley was appointed minister for Defence and Supply in 1958, and was an active and interventionist minister. A licensed pilot, he enjoyed trying out RAAF aircraft including the newly acquired Mirage fighters. Townley also held fairly forward-looking views on Australia's place in south-east Asia. Described as 'shrewd, a down-to-earth talker, sports-loving and athletic-looking' (he had played cricket at State level), his health declined and in 1963 he resigned from parliament to take up appointment as ambassador to the United States. He died six days later.

TOWNSVILLE (I), HMAS see *BATHURST* CLASS MINESWEEPERS (CORVETTES)

TOWNSVILLE (II), HMAS see *FREMANTLE* CLASS PATROL BOATS

TRACKER, GRUMMAN (4-crew carrier-borne anti-submarine aircraft). Wingspan 72 feet 7 inches; length 43 feet 6 inches; armament 2 × depth bombs or torpedoes, 6 × 250-pound bombs or 5-inch rockets; maximum speed 265 m.p.h.; range 1100 miles; power 2 × Wright Cyclone 1525 h.p. engines.

The Tracker was ordered for the Fleet Air Arm (q.v.) in 1965 at a time of tension with Indonesia (see CONFRONTATION) and replaced the Fairey Gannet (q.v.) from 1967. They were designed to hunt out and destroy enemy submarines, but could also be used for photographic surveillance, search and rescue, reconnaissance and logistical support. No. 816 Squadron flew Trackers from the aircraft-carrier HMAS *Melbourne* (q.v.). In 1976 eleven were destroyed by a deliberately lit fire at the Naval Air Station, Nowra. Trackers were withdrawn from service in 1984 after the decommissioning of the *Melbourne* in 1982.

TRANSPORT CORPS see **ROYAL AUSTRALIAN CORPS OF TRANSPORT**

TREASURY, DEFENCE DIVISION OF, was created in December 1941, following Japan's entry into the war, as part of the reorganisation of machinery for the higher direction of the war. It took over the Department of Defence Co-ordination's (q.v.) functions of financial coordination and review and the coordination of civil defence staffs other than the Public Service. It also provided financial advice to the services and was responsible for most of the executive Treasury work connected with defence, war industry and war administration. After the war it handled such matters as breakdown of war production, disposal of war assets, liquidation of canteen assets, war gratuity administration, surveillance of the Department of Post-War Reconstruction's (q.v.) activities relating to the rehabilitation of service personnel, review of demobilisation (q.v.) activities and pay and service conditions of the armed forces. It survived as a division of Treasury until 1976–77.

TREATIES AND DEFENCE AGREEMENTS see **ABCA AGREEMENT; ANGLO-AUSTRALIAN NAVAL AGREEMENTS; ANZUS; AUSTRALIA–NEW ZEALAND AGREEMENT; JOINT DEFENCE FACILITIES; RADFORD–COLLINS AGREEMENT; SOUTH-EAST ASIA TREATY ORGANIZATION**

TRELOAR, John Linton (10 December 1894–28 January 1952). First appointed to the Department of Defence as a clerk in 1911, Treloar enlisted in the AIF in August 1914 and served on Gallipoli as a staff sergeant until evacuated with enteric fever in

September 1915. He was commissioned and served as a lieutenant with No. 1 Squadron, Australian Flying Corps (q.v.) in Egypt before going to France to work as confidential clerk to the chief of staff of I Anzac Corps, Brigadier-General C. B. B. White (q.v.). From May 1917 Treloar ran the Australian War Records Section, a body set up along the lines of its Canadian counterpart to gather material for the official histories which would be written after the war. The records which he gathered and organised came to form the basis of the extensive archival holdings of the Australian War Memorial (q.v.).

From 1920 until his death in 1952 Treloar was director of the Australian War Memorial, an organisation which for much of his life existed in a very parlous state indeed. A tireless raiser of funds and advocate of a permanent building for the vast collections of records and relics which he and the official historian C. E. W. Bean (q.v.) had collected, Treloar worked six days a week, often sleeping in his office in order to get an even earlier start. In the interwar years he worked in Melbourne while his collections were housed in Sydney, and he planned a museum building to be built in Canberra. Even so, he found time to support Bean's official histories, which suffered financially in the 1930s through poor sales. The War Memorial took over publication and distribution, and Treloar devised an innovative subscription scheme whereby public servants could buy the volumes in instalments.

Treloar was made head of the Department of Information in 1939 at the beginning of the war, and from October 1941 was back in uniform as officer-in-charge of the military history section at Army Headquarters, where he repeated much of the old work in a new war. At the end of the war he returned to the War Memorial and to his obsessive work patterns.

There is little doubt that the Australian War Memorial would have foundered had it not been for Treloar's tireless and selfless labours, which almost certainly shortened his life. He had a tremendous capacity for work and an astonishing attention to detail, but was considered by some remote and formal, cold and unbending in his attitudes. His children, however, remember an affectionate and humorous man who followed South Melbourne Football Club each Saturday. The large equipment storage and display annexe of the War Memorial is named in his honour, and a commemorative plaque by the sculptor Leslie Bowles is placed outside the entrance to the memorial's archival research centre.

Michael McKernan, *Here is Their Spirit: A History of the Australian War Memorial 1917–1990* (University of Queensland Press, Brisbane, 1991).

TRENCH NEWSPAPERS see SERVICE NEWSPAPERS

TRIBAL CLASS DESTROYERS (*Arunta, Bataan* and *Warramunga*). Laid down 1939–42, completed 1942–45; length 377 feet 6 inches; beam 36 feet 6 inches; armament 6 × 4.7-inch guns, 2 × 4-inch guns, 4 × 21-inch torpedo tubes, 2 × depth-charge throwers; speed 36 knots.

The three ships of this class were built during the Second World War at Cockatoo Island (q.v.) in Sydney. *Bataan* was originally laid down as *Kurnai*, but was renamed in honour of General Douglas MacArthur (q.v.) and was launched by Mrs MacArthur in January 1944. *Arunta* and *Warramunga* served from 1942 in Australian waters on anti-submarine patrols and escort duties and supported Allied landings during New Guinea campaign (q.v.). In 1944 both ships were involved in the American landings in the Philippines and came under kamikaze attacks in Leyte Gulf. HMAS *Warramunga* and the newly completed *Bataan* were at the Japanese surrender in Tokyo Bay in 1945. HMAS *Bataan* was stationed as part of the British Commonwealth Occupation Force (q.v.) in Japan in 1950 when the Korean War began. HMAS *Bataan* served in Korean waters from July 1950 to June 1951 and on 1 August 1950 came under fire from North Korean shore positions. HMAS *Warramunga* also served in Korean waters from August 1950 to June 1951. HMAS *Bataan* was placed in reserve in 1954 and HMAS *Warramunga* in 1959. HMAS *Arunta* had been converted to an anti-submarine destroyer between 1950 and 1953 and was placed in reserve in 1957. All three ships were sold for scrap, the *Arunta* sinking off Sydney on her way to Taiwan in 1969.

TRUMBLE, Thomas (9 April 1872–2 July 1954). A member of a famous cricketing family, Trumble entered the Victorian Public Service in 1888, and from 1891 worked in the Victorian Defence Department, including a period as private secretary to the minister. After Federation he transferred to the Commonwealth Department of Defence, and from June 1910 held the position of chief clerk. He was acting secretary at the outbreak of war in August 1914 owing to the absence overseas of the Secretary, Samuel Pethebridge (q.v.), an arrangement which was confirmed in November. The Department of Defence was very small in 1914, and although it expanded in the course of the war

Tribal Class Destroyer HMAS *Bataan* (right) and HMS *Belfast* in pack ice off Korea during the Korean War. HMS *Belfast* is now moored as a museum in London on the Thames River. (AWM 042341)

the demands on experienced men like Trumble were intense. A Royal Commission into the department in 1917 revealed a number of short-comings in its administration, but no criticism was levelled at Trumble, who was generally highly regarded. Ernest Scott (q.v.), the official historian, wrote that few perhaps realised the extent to which the smooth workings of wartime defence administration were due to Trumble. He became Secretary in his own right in February 1918, a position he then held until June 1927. He was the first career civil servant so appointed. For his services during the war he was made CBE in 1918, followed by the award of CMG in 1923. In 1927 he became official secretary to the High Commission in London, and transferred to the position of defence liaison officer in 1931, which he held until his retirement in December 1932. During the Second World War he was Director of Voluntary Services in the Department of Defence Coordination from July 1940, and retired again in September 1943.

TRUSCOTT, Squadron Leader Keith William 'Bluey' (17 May 1916–28 March 1943) was the second highest-scoring Australian ace of the Second World War. Truscott worked as a school teacher and clerk, and played first-grade football for the Melbourne club before the war. He enlisted in the RAAF in July 1940, was selected for pilot training and commissioned in February 1941. Posted to No. 452 Squadron, RAF, in May that year, flying Spitfires, he shot down 16 German aircraft between August 1941 and March 1942, and was awarded the DFC and bar. He returned to Australia in May 1942 and was sent to New Guinea, flying Kittyhawks with No. 76 Squadron, RAAF. He took part in the victorious aero-naval battle of Milne Bay, for which he was mentioned in despatches for, among other things, repeatedly strafing Japanese transports and escorts at very close range in order to provide cover for attacking torpedo planes. In early 1943 he began flying duties from Darwin, and recorded his 17th confirmed victory in January that year. He was killed in a flying accident in the Exmouth Gulf area of Western Australia in March, and was buried at Karrakatta Cemetery, Perth. His unofficial tally of enemy aircraft was $20^{1}/_{2}$.

TUNNELLING COMPANIES see **MINING OPERATIONS**

28TH COMMONWEALTH INFANTRY BRIGADE Raised in April 1951 in Korea to replace the British 27th Infantry Brigade, the 28th Brigade contained infantry battalions from Australia and Britain, artillery batteries from Britain and New Zealand, a British armoured squadron and an Indian field ambulance. It formed part of the 1st Commonwealth Division, activated in July 1951, and fought right through the Korean War and remained into the cease-fire period. It was commanded by a succession of first British and then Australian officers, and was disbanded in the draw-down of United Nations forces in 1954. It was re-formed in Malaya in 1955 to form part of the British Commonwealth Far East Strategic Reserve for use in SEATO (q.v.) operations, and took part in the Malayan Emergency (q.v.), which was not in fact its primary task. It now consisted of a British, an Australian and a New Zealand infantry battalion with supporting arms and services. Elements of the brigade continued to harry remaining communist terrorists along the border with Thailand after the official end of the Emergency in July 1960. Between 1964 and 1966 its units served in southern Malaya and Borneo on operations against the Indonesians during Confrontation. It was redesignated 28 (ANZUK) Brigade in November 1971 under the Five Power Defence Arrangements (q.v.) which emerged from Britain's decision finally to withdraw from 'east of Suez' and the termination of the Anglo–Malayan Defence Agreement. It was finally disestablished in February 1975 when Australia withdrew its battalion, although the New Zealand unit remained in Singapore until the 1980s.

25-POUNDER GUN Range 13 400 yards (10 200 yards 'short' version); weight of shell 25 pounds; rate of fire 5 rounds per minute (3–4 short 25-pounder); total weight 4000 pounds (2900 pounds short 25-pounder).

The 25-pounder was the standard field gun in the British Army in the Second World War. It could fire high explosive, armour-piercing and smoke shell. It was first introduced by the British just before the Second World War, and adopted by the Australian Army in 1940. During the war over 1500 of these guns were manufactured in Australia, partly by General Motors Holden. About 500 were also purchased from the United Kingdom. They were used to support infantry attacks and in an anti-tank role in all theatres in which Australian troops fought in the Second World War. A 'short' more easily manoeuvrable version was adapted by the Australian Army for use in jungle fighting in New Guinea. The gun proved its versatility by remaining in service until 1975, though it began to be replaced by 105 mm M2A2 howitzers in 1959.

U

ULTRA

ULTRA was the code-word used during the Second World War for information obtained through the interception and deciphering of coded enemy messages. In the South-West Pacific Area (q.v.) Japanese messages, in encoded morse code, were picked up by wireless interception units of the Australian Army, RAAF or US Army, and sent to Central Bureau, a unit in General Douglas MacArthur's (q.v.) Headquarters made up of Australian, British and American personnel. Here skilled cryptanalysts deciphered the message and translated it from Japanese to English, sometimes with the assistance of Allied Translator and Interpreter Service (q.v.) personnel. ULTRA gave information of Japanese units and their movements, such as the Japanese convoy sailing from Rabaul to Lae in March 1943, which was attacked by RAAF and US aircraft in the Battle of the Bismarck Sea (q.v.). From November 1944 Australian forces also had access to ULTRA material through the Special Liaison Units (q.v.). Similar to ULTRA were MAGIC (q.v.) intercepts of Japanese diplomatic messages.

Edward J. Drea, *MacArthur's ULTRA; Codebreaking and the War against Japan, 1942–1945* (University Press of Kansas, Lawrence, 1992).

UNIFORMS The gradual but constant change in military uniforms during the nineteenth and twentieth centuries has seen the transition from bold colours to mottled camouflage. Developments in weapons and tactics, and changing theatres of war, have required that soldiers adopt suitable clothing and equipment. National characteristics, civilian fashion and military traditions and alliances also influence new uniform designs, colours and styles.

In the mid-nineteenth century the Volunteer units formed in the Australian colonies (see COLONIAL MILITARY FORCES) adopted uniforms similar to the Volunteers in England. These local Volunteers were a reserve to the British regiments serving here. However they did not imitate the British regulars, but instead modelled themselves on the British part-time soldiers. Generally the regulars' red uniforms were rejected and the greens and greys of the Volunteers in Britain adopted instead. On the whole, however, uniforms tended to follow the fashion dictated by the dominant European military power of the day; after the Crimean War (1854–56) British and colonial uniforms showed a French influence, while after the Franco-Prussian War (1870–71) they adopted certain German features.

During the 1870s there was a change in the appearance of the local corps. Red tunics such as those worn by regular infantry regiments gained popularity, while in the following years some units copied the uniforms of several distinguished British regiments. In Victoria a waist-length red or blue jacket called 'the Garibaldi jacket' was popular. For many years the infantry in New South Wales paraded in their red tunics; although khaki was increasingly being used, red remained the traditional colour for the army. Colonel C. F. Stokes, who had reported on the New South Wales military forces in 1892, recalled: 'The men like it. There is a feeling amongst them that it is associated with actual soldiering.' The New South Wales expedition to the Sudan in 1885 departed in the red of the infantry and the blue of the artillery, complete with white sun-helmets and equipment. Once in the war theatre, however, the colourful uniforms were replaced by drab khaki.

Scottish and other 'national' corps were popular in some colonies. In the former, original members were usually of Scottish birth or descent, and the units wore highland dress. Queensland and New South Wales also formed Irish Corps. The latter wore a green uniform and shamrocks featured on their badges. 'English' regiments clung to the red tunics.

In the 1890s military uniforms began to reflect local fashion and influences. Uniforms more suited to the Australian climate were being adopted. These were usually drab (i.e. khaki) with coloured regimental facings, and were worn with a wide-brimmed hat. Regimental badges even reflected this trend, with Australian flora and fauna and other national symbols being incorporated into the designs. Major-General Bevan Edwards (q.v.) recommended the wider adoption of these uniforms. In 1889 he reported:

> Such a dress is now worn by the Mounted Infantry in all the colonies, and by the Victorian Rangers, and is not only smart and soldierlike, but it is a distinctive National Dress.

The famous slouch hat (q.v.) began to be used at this time. It was first worn by the Victorian Mounted Rifles in 1885, and shortly afterwards by the New South Wales Reserve Rifle Companies, and by infantry regiments for 'shooting purposes'. It was a sensible hat which was practical as well as popular, and its use spread. The Victorians and Tasmanians wore it turned up on the right side, but after Federation all units adopted the left side.

During the South African War colourful clothing on active service virtually disappeared. Some early Australian units left wearing regimental dress, however, khaki clothing, in a variety of simple styles, was worn out on the veldt. Prior images of splendidly dressed regiments riding to battle were soon put to rest. One soldier was to write: 'Where is all the "pomp and circumstance of war"?' The slouch hat was popular among some Australian units in South Africa, although other British and irregular units also wore it. The Queenslanders wore emu plumes in their hats. It was a practice first noticed in 1891 during the shearers' strikes back home. Significantly, the Rising Sun badge (q.v.) made its appearance at this time, being adopted in two similar patterns by the battalions of Australian Commonwealth Horse (q.v.), which were sent to the war in 1902. These designs would later form the basis for the badge introduced in 1904 which, in 1911, became the general service badge, and later the distinguishing symbol of the AIF.

The creation of the Australian Army, following Federation, meant reorganisation. The Commonwealth-pattern uniform was adopted:

> The Commonwealth Uniform will consist of a General Service Dress, which will, by the addition of aiguillettes, breast-lines, and girdles, be convertible into Full Dress. By this means a single uniform coat will be provided which shall meet the requirements of a Fatigue or Service Jacket, and by the addition of lace attachments, &c., be readily converted into Ceremonial or Full Dress. Great economy ... will be insured by this means.

The jacket was khaki serge, and was worn with the slouch hat and breeches. A variety of cloth collar patches, with coloured piping on the jacket, rosettes or plumes on the hat, and attractive regimental badges brightened up the appearance. In addition, there were still some older regiments that clung to their colourful full-dress uniforms, but permission to retain these depended upon the regiments meeting the costs themselves.

The RAN resisted any suggestions to adopt a uniform other than that worn by the RN, with only minor inclusions to identify its members. The navy has made the least changes during its history, and its uniform today bears most of the characteristics evident when the service was established. The designation 'HMAS' on sailors' tally ribbons, and a distinctive button on jackets, were among the few distinguishing features on the uniforms.

The introduction of compulsory military training (see CONSCRIPTION) caused the army to expand considerably. The need for serviceable, cheap, and readily available clothing resulted in new orders for dress coming into effect in 1912. They provided for a uniform which was little more than a hat, a heavy khaki woollen shirt, cord breeches, and boots. Units and corps were distinguished by a two-inch wide coloured cloth band, and a metal numeral, worn on the hats. Officers retained their jackets and, shortly afterwards, the stepped open collar, worn with shirt and tie, was approved. These uniforms were in wide use at the outbreak of war in 1914.

The war meant a new force had to be equipped for overseas service. The slouch hat and Rising Sun badge were already available and were taken into use by the AIF. The compulsory military training uniform formed the basis of that adopted by the AIF, with the very important difference that the heavy khaki shirt was replaced with a loose-fitting, four-pocket woollen service dress jacket which had been designed just before the war. The basic AIF uniform consisted of the jacket, breeches, puttees, tan ankle-boots, and hat. The appearance was quite distinctive and meant that the Australians, unlike most Commonwealth armies, could easily be distinguished from British troops. The slouch hat, in particular, virtually became a national symbol. Some light horse regiments, especially the Queenslanders, adorned their hats with emu plumes.

All units in the AIF wore identical hat and collar badges. To distinguish the different regiments and battalions, small metal numerals and letters were worn on the shoulder straps (officers wore theirs on the collars). In 1915 a new scheme of unit identification was adopted, consisting of cloth colour patches (q.v.) worn on the upper arms of a soldier's jacket. The shape of these patches denoted the division (eg. a rectangle for the 1st Division) and the colours the battalion to which a soldier belonged. The colour-patch scheme was eventually adopted throughout the AIF and soldiers displayed a great attachment to these coloured unit-patches.

In many situations the Australians simply copied some items of dress and insignia from the wartime British Army. Although rank insignia was sometimes worn on officers' cuffs by the army at home, the AIF officers wore their rank badges on the shoulder. During the war approval was given to the wearing of a strip of two-inches of gold Russia braid to denote each wound received in action. The brass letter 'A' was authorised to be worn on the colour patch of every man and nurse who served in the 1915 Gallipoli campaign, and blue chevrons denoting each year of overseas service were adopted; a red chevron was approved to represent the first year of the war.

The equipment carried by soldiers was of two basic types: webbing equipment for the foot soldier and the leather equipment of the mounted man. (The term 'webbing' comes from the Webb system of infantry harness, introduced in 1908, that was designed to spread the load carried.) The 1908-pattern webbing had been adopted shortly before the war and was in such short supply initially that a leather copy was produced at the Commonwealth Government Harness Factory in Melbourne as an interim measure. In France, where winters were severe, a soldier's load was increased by his need for warm clothing and bedding as well as his weapons.

Following the war the army retained the AIF-pattern uniform until 1930. Meanwhile, the recently created RAAF introduced a new uniform. It was based on the RAF pattern, but made of a dark-blue woollen material. Badges had Australian motifs such as the Southern Cross and wreaths of wattle. Modifications in 1937 brought the uniform more in line with the RAF, although the dark-blue jacket, trousers, and headwear, were retained until 1972.

New and more attractive uniforms were introduced by the army in 1930 to encourage voluntary recruiting for the militia. Most of these new-style uniforms were blue or khaki with distinctive coloured collars and piping to indicate various corps. The colour-patch unit identification system was retained, to which was added regimental and corps badges. The various Scottish regiments' uniforms are probably the best remembered from this more colourful decade.

The Second World War did not bring about much immediate change. The woollen service dress worn in the First World War was resurrected for use by the 2nd AIF, except that trousers replaced breeches, and the despised puttees were abolished to be replaced by cloth anklets. The old 1908-pattern webbing equipment was gradually superseded by a lighter 1937-pattern. Khaki drill-cotton clothing was issued for summer and hot climates. Again the colour-patch system was used; the 2nd AIF units being denoted by a grey border to their patches.

The militia gradually replaced their prewar uniforms with the khaki service dress, and regimental badges were withdrawn by 1942, being replaced by the Rising Sun which became universally worn. Colour patches provided the most obvious distinction between AIF and AMF units. By mid-war the army service dress jacket underwent some minor changes which made few differences to the appearance but reduced the amount of cloth and simplified the manufacture. Cloth anklets were replaced by webbing gaiters.

Jungle warfare brought its own special demands. Troops landed at Port Moresby in September 1942 equipped with the new jungle green shirt and trousers, although for much of the campaign they found that the dye ran in the heavy tropical rain. Steel helmets were unpopular in jungle conditions, and some items of American equipment, especially their long canvas gaiters, were sought after. The need to provide equipment suitable for jungle warfare required the army to seek new designs. Clothing had to provide camouflage, and protection from mosquitoes, fungoid growth and mud. Stud-ded boots, new canvas gaiters, canvas ammunition pouches, lightweight mess tins, special water-bottle carriers, and cloth berets were among items designed for the new fighting theatres, although some of these were only available in the closing stages of the war.

The RAAF wartime uniform was worn with black badges and buttons, and an increasing range of embroidered insignia to denote specialist roles, while a drab uniform was worn in summer and tropical areas. There was a need for a range of special clothing suited to flying duty, particularly at high altitude. The British battle-dress style uniform was adopted by the RAAF (dark blue), but the army (khaki) did not take it up until after the war.

The introduction of the various women's services brought a new range of distinctive uniforms, which generally echoed the colours and styles of the equivalent male service. The Australian Army Nursing Service, however, retained the grey serge that nurses had worn since before the First World War.

Since the Second World War the trend has been towards increasing drabness with an emphasis on camouflage in battle dress and a general reduction in the range and number of uniforms. In the Korean War the army fought largely in the recently introduced British-pattern battle dress with some variations of American origin, for example in cold-weather gear ranging from boots to thermal underwear. In Malaya, Borneo and Vietnam the army wore jungle greens, although units in specialised roles such as the SAS, or individuals serving with the South Vietnamese forces such as the Australian Army Training Team Vietnam (q.v.), adopted various forms of camouflage uniforms sometimes known colloquially as 'tiger stripes'. In the late 1980s the army gradually introduced Disruptive Pattern Camouflage Uniforms to replace jungle greens. The RAAF, too, abandoned its traditional dark-blue service dress in favour of a uniform of a lighter colour and of a distinctively American cut. In the interests of utility and cost, the distinctions between ceremonial and service dress have gradually disappeared, for example through the discontinuation of officers' 'patrol blues'.

PETER BURNESS

UNIT DIARIES Known also as war diaries or commanders' diaries, these are the basic record of a unit's or formation's day-to-day activities while on active service, and are thus an important source for the study of Australian operations in any campaign. The naval equivalent is the ship's report of proceedings. The RAAF designates these records as operations records books, or unit history sheets, or simply as A50s, after the stationery number of the form. In the army, every command level from battalion up (that is, battalion, brigade, division, corps) maintains a war diary, and these become larger and more varied as the organisation becomes more complex. At unit level, a war diary consists of a narrative summary of the unit's daily activities, on a monthly basis, supplemented by various appendices which may give the unit's strength returns, officer postings, standing orders, contact or after-action reports, orders received from higher headquarters, and monthly reports from the Regimental Medical Officer, the Quartermaster, and so on. Depending on the unit, the personality of the officer (usually the intelligence officer) and other ranks charged with keeping it, and the intensity of activity, these may be anything from voluminous to scant or even non-existent. The keeping of war diaries was first prescribed by Field Service Regulations in 1909, which were amended at various times thereafter. In the First World War, units sent their war diaries to the ANZAC Section of 3rd Echelon (see CENTRAL ARMY RECORDS OFFICE) to be passed on to the War Office in London, where they were housed. From July 1917, with the creation of the Australian War Records Section under Major John Treloar (q.v.), only the duplicates were sent to the War Office, the originals henceforth being kept by the Australian authorities. Mindful of the likely demands of the official historian after the war, Treloar prodded units to keep their diaries properly and fully and to forward them regularly for safe keeping. The result was a collection of material which, when bound, came to occupy about 300 shelf feet of storage space in the Australian War Memorial. Treloar oversaw the administration of war records during the Second World War as well, and about 2000 linear feet of storage space housed these records after that war's end. Diaries for the postwar conflicts occupy a correspondingly smaller area.

Generally there are two problems with using war diaries as historical sources. The first is that, although they are meant to be kept daily, with all relevant attachments, in reality they are maintained as the exigencies of the moment permit. The war diary of the 3rd Light Horse Brigade on Gallipoli, for example, is very detailed and well kept; it was clearly written up in Egypt from notes by the brigade major, Lieutenant-Colonel J. M. Antill, after the survivors were evacuated from Gallipoli in December 1915. The diaries for 3RAR (q.v.) in the Korean War vary with each intelligence officer of the battalion; during the chaotic war of movement between September 1950 and July 1951, paradoxically, they are quite full and informative, but in certain periods of the static war in 1952–53 they become very terse and brief. The most famil-

iar, and for the historian most irritating, abbreviation in any war diary is NTR: nothing to report. The second problem is caused by the fact that Australian units and formations have frequently served as part of higher British formations, with the result that the war diaries for these formations, which provide the context for Australian operations, are not held in Australia. Thus for the Korean War, the diaries of the 27th and 28th Commonwealth (q.v.) Brigades and the 1st Commonwealth Division are in the Public Record Office, Kew, London; the same applies to pertinent records for the Malayan Emergency and Confrontation, and for both the world wars. The Canadian government systematically microfilmed such records for placement in the National Archives of Canada; Australian authorities have failed to take up the Canadian lead, ironically since the idea of a war records section in the First World War was a Canadian one to begin with.

UNIT HISTORIES were produced by unit associations after each of the world wars, and small numbers continue to be published. Units began publishing accounts of their activities, partly as souvenirs for members in the process of demobilisation and dispersal, almost as soon as the First World War ended, and histories of and by AIF units continued to appear into the late 1940s. After the Second World War 28 infantry battalions published histories in the period up to 1980; between 1980 and 1990 a further 20 appeared. In recent years the tendency has been for even sub-units to publish accounts of their wartime activities. Postwar conflicts are generally much less well served in this regard, although most of the battalions of the Royal Australian Regiment (q.v.) produced 'tour books' after each deployment to Vietnam. Such histories are rarely published commercially, and for many years after the Second World War the publishing costs were defrayed through a fund administered by the Australian Army provided through profits left over when the canteens funds (q.v.) were wound up at the end of the war, or through unit funds in the case of the Vietnam battalions. Unit histories provide a wealth of useful information often no longer extant from other sources, and a very few of them qualify as good military history also.

UNITED NATIONS PEACE-KEEPING see **PEACE-KEEPING, AUSTRALIANS AND MULTI-NATIONAL**

UNITED SERVICES INSTITUTE (USI) is an independent body dedicated to promoting the study of strategy, national defence and military affairs within Australia. It is made up of constituent bodies in each of the States and territories, and was inspired originally by the creation of the Royal United Services Institute at the direction of the Duke of Wellington in London in 1831. The first Australian institute was founded in Sydney in 1888, that in Victoria followed in 1890, and others were set up in Queensland in 1892, Western Australia in 1903, South Australia in 1919, and Tasmania in 1924. Although the ACT had been the home of the Department of Defence since the 1960s, an institute was only founded there in 1971, while that in the Northern Territory dates from 1988. In 1974 the various State institutes agreed to form the United Services Institute of Australia, with each State institute a constituent but autonomous part, in order to maximise its influence. The Queen granted the prefix 'Royal' in 1979. The activities of the RUSI of Australia, principally the production of a journal and the maintenance of significant libraries of defence-related holdings, are supported by an annual grant from the Department of Defence. For many years, the USI provided the only forum for Australian service officers to exchange views and maintain currency in the literature of their profession. Monash, for example, lectured regularly as a young officer at the Victorian USI, and joined its council in 1894 from which position he established a program of lectures on recent developments in weaponry that attracted considerable attention and favourable comment. The membership of the RUSI today draws to a considerable extent on former members of the armed forces, although those currently serving are a significant element, especially within the ACT branch.

UNIVERSITY REGIMENTS have had a distinctive history and role within the Australian Army. The first university company was formed in Melbourne in 1884, but disbanded two years later. In 1895 a University Corps of Officers was set up as a militia corps open to graduates and undergraduates of the University of Melbourne, with the aim of producing lieutenants who could then join other militia or volunteer corps. The unit was disbanded after Federation, but was replaced by the Melbourne University Rifles in 1910. At the University of Sydney, a University Volunteer Rifle Corps established in 1900 changed its name to the Sydney University Scouts in 1903. With the introduction of compulsory military training (see CONSCRIPTION) in 1911, the Scouts became a militia battalion and began recruiting senior students from the Great Public Schools (GPS) as well as undergraduates. During the First World War 61 per cent of men serving in the Scouts joined the AIF, as did many former Scouts. After the war GPS students could no longer join the unit, which was renamed the Sydney University Regiment in 1927. A detachment of a militia unit was also established at the University of Queensland in 1933. It seems likely that at this time, when universities were still largely the province of the well-to-do, there was a strong class element in the formation of university-based units. By providing military training to undergraduates, these units prepared men of the 'better classes' for their proper role as officers, a role many of them did indeed take up during the Second World War. Sydney University Regiment, for example, trained such prominent officers as J. R. Broadbent, Frederick Chilton and Ivan Dougherty (qq.v.).

The university-based units were disbanded in 1942, but as part of the postwar reorganisation of the Australian Army a Military Board (q.v.) directive in 1947 laid the basis for units to be raised at universities once again. The units were to be designated as regiments, to be organised along the lines of CMF infantry battalions, and to be open to all male students of universities and teachers' colleges who met the normal requirements for enlistment in the CMF. A regiment was to be raised at each Australian university; where there were insufficient numbers a sub-unit could be raised. Following this decision the Queensland, Sydney, Adelaide and Melbourne University Regiments were raised in 1948 and the Western Australian University Regiment (originally Perth University Regiment) in 1949. The New South Wales University of Technology Regiment was raised in 1952 (it was redesignated the University of New South Wales Regiment in 1959), and the Monash University Company was raised to regimental status in 1970. In addition, companies of the Sydney University Regiment were formed at the University of New England and the Australian National University; companies of the University of New South Wales Regiment were established at the University of Newcastle and the University of Wollongong; and a company was formed at Deakin University.

University regiments have been the subject of controversy and conflict in two areas. First, there has been some conflict with the wider university communities, particularly during the Vietnam War. At the height of student anti-war activism there were instances of vandalism of regimental property, the Queensland University Regiment's depot was occupied by protesters, and the Monash University Regiment was banned from recruiting on campus. Since that time objections to the presence of university regiments have been raised on some campuses, but the regiments have generally carried out recruiting and other activities freely.

The second controversy involving university regiments has been one within the army itself over the role of these regiments. When university regiments were raised after the Second World War their focus was on performing as infantry battalions and thereby providing university students with military training. Officer training did occur within university regiments, but it was not until the creation of Officer Cadet Training Units (OCTU) in the mid-1960s that the officer-producing role was formalised. The role of OCTUs was to train officers for the CMF, and in effect university regiments became a specialised type of OCTU, with the primary role of training university students as CMF officers. At the same time, however, they retained their battalion structures and their secondary role of providing general military training to students. As a result of this dual role, the Millar Committee (q.v.) found that the university regiments' primary role was challenged by many within the army 'who see university regiments as a very special type of infantry battalion'. The committee, however, rejected this view, seeing university regiments primarily as a source of intelligent and socially influential CMF officers. This confusion of roles was finally cleared up by the Force Structure Review of 1991, which decided to confine university regiments to their primary role of producing officers. The result of this decision has been a major reorganisation of university regiments, which has seen them change from infantry battalion structures to organisation along OCTU lines. All existing university regiments have been retained, but the university companies have been disbanded, retitled and transferred, or amalgamated.

UNKNOWN AUSTRALIAN SOLDIER The First World War left more men missing or unidentified than any previous war. Huge numbers of bodies on all sides either could not be found or were found but could not be identified. This situation was not entirely unprecedented, and unknown soldiers had been commemorated after the American Civil War, but the vastly increased number of unknown bodies produced by the First World War seemed to demand a novel response. Such a response came quickly in both Britain and France, where it was decided to entomb ceremonially a single unidentified body whose nationality could be determined from items found with the body and who would represent all the nation's war dead. On 11 November 1920, exactly two years after the signing of the Armistice, Britain's Unknown Warrior was reburied in Westminster Abbey, London. The French Unknown Soldier was entombed at the same time, and throughout the 1920s other nations which had participated in the war followed suit.

Until 1993, however, no other part of the British Empire had such a tomb, since the body in London was supposed to represent all the Empire's 'Million Dead'. Nearly half of Australia's 60 000 First World War dead were listed as missing, but they were honoured separately only on monuments to the missing, built by the Imperial War Graves Commission (see WAR GRAVES), in the countries where they died. Proposals to entomb an Unknown Australian Soldier were put forward by the RSL and others, first in the early 1920s and periodically thereafter, but the argument that the Unknown Warrior in London represented unknown Australians too remained conclusive.

It was, therefore, a sign of increasing Australian nationalism that the Australian War Memorial's (q.v.) decision to entomb an Unknown Australian Soldier on the 75th anniversary of the end of the First World War seems to have gained widespread community acceptance. The RSL initially opposed the entombment, but changed its mind in time to allow RSL participation in the burial ceremony; some RSL members, however, including outspoken Victorian RSL President Bruce Ruxton, continued to argue that the soldier's body should have remained with those of his dead comrades in France. There was also some controversy about the naming of the Unknown, with navy veterans lobbying for the title Unknown Warrior rather than Soldier. However, the War Memorial decided that Soldier was the most appropriate title, since the body was undoubtedly that of a soldier but may have been an unarmed stretcher-bearer or chaplain rather than a 'warrior'. Despite these minor controversies the Memorial secured the cooperation of ex-service organisations, the Department of Veterans' Affairs and the Commonwealth War Graves Commission, as well as bipartisan political support for the project.

In November 1993 the remains of an unidentified Australian soldier were disinterred from the Adelaide Cemetery, Villers-Bretonneux, France. (France was chosen because more Australians were killed on the Western Front [q.v.] than in any other single theatre of war in which Australians have fought.) The remains were then placed in a coffin and flown to Australia. On 11 November the body was entombed in the Australian War Memorial's Hall of Memory in a ceremony which the Memorial called the funeral service of the Unknown Australian Soldier. About 20 000 people watched the funeral parade, which featured representatives of 500 ex-service organisations and an official party that included Governor-General Bill Hayden as chief mourner and Prime Minister Paul Keating as chief pall-bearer. While the ceremony once again enshrined war as central to Australian national identity, it also reflected in some ways a society which had become more diverse and tolerant and less British since the First World War. Both officials and media commentators were at pains to emphasise that it was not about the rights or wrongs of particular wars but about honouring the spirit of service and sacrifice, while the prayers which accompanied the ceremony were intended to be inclusive and Australian in character. There was talk in the media of the futility of war, yet the high diction of terms such as 'sacrifice' and 'service' and phrases such as 'gave his life' and 'a nation's hero' which were routinely used in media reports undermined the possibility of looking critically at war. The eulogy given by Prime Minister Keating sought to resolve the apparent contradictions between honouring the Unknown Soldier and criticising the war in which he fought. Though the First World War 'was a mad, brutal, awful struggle … [in which] the waste of human life was so terrible that some said victory was scarcely discernible from defeat', Keating proclaimed, the Unknown Soldier did not die in vain because the war proved the nobility and heroism of ordinary people.

To coincide with the entombment of the Unknown Soldier Raymond Ewers' giant soldier statue was replaced at the rear of the Hall of Memory by four pillars representing earth, air, fire and water, echoing the symbolism in the Hall's mosaic. The Tomb of the Unknown Australian Soldier itself is set into the floor in the centre of the Hall. It is covered by a red marble slab bearing the words 'AN UNKNOWN AUSTRALIAN SOLDIER KILLED IN THE WAR OF 1914–1918' and surrounded by a sloping marble edge inscribed: 'HE SYMBOLISES ALL AUSTRALIANS WHO HAVE DIED IN WAR.' The Memorial encourages visitors to commemorate Australia's war dead by laying a single flower on the tomb.

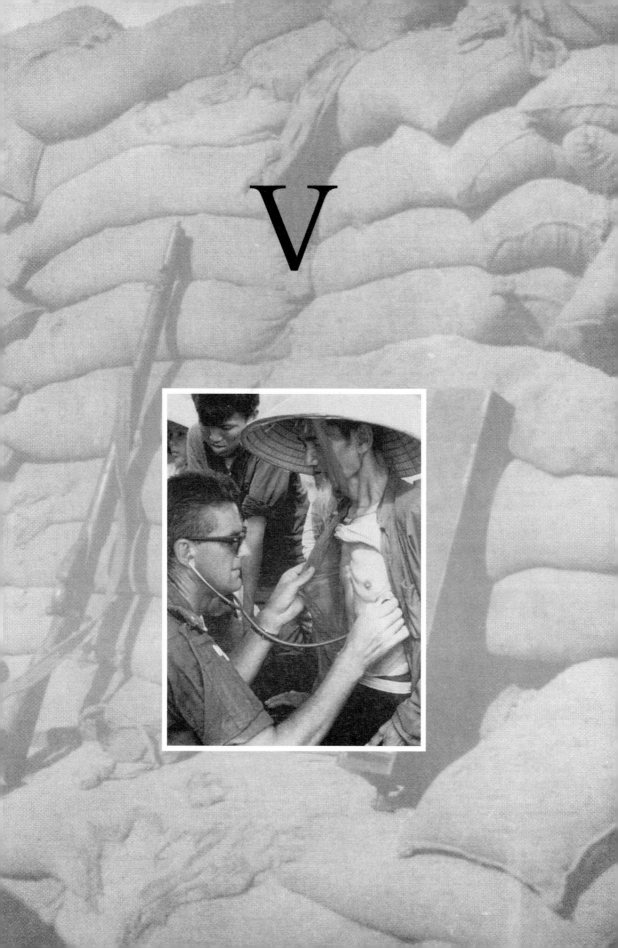

V

'V' AND 'W' CLASS DESTROYERS Laid down 1916–17, launched 1917–18; length 312 feet; beam 29 feet; displacement 1100 tons; armament 4 × 4-inch guns, five small guns, 4 × 21-inch torpedo tubes; speed 34 knots.

The four ships of this class (*Vampire, Vendetta, Voyager, Waterhen*) served with the RN during the First World War. In 1933 they were transferred to the RAN. On the outbreak of the Second World War it was decided to send the four ships to the Mediterranean. They fought against the Italian fleet at the Battle of Calabria on 9 July 1940 and the Battle of Matapan on 28 March 1941 and assisted in the evacuation of Allied troops from Greece and Crete. Their main task came in 1941 when they were used to run supplies into the besieged port of Tobruk, this duty earning them the name from the Germans of the 'Scrap Iron Flotilla' (q.v.). HMAS *Waterhen* was sunk by German bombers in this service on 30 June 1941, becoming the first Australian ship to be lost to enemy action. After the entry of Japan into the war, the remaining ships were sent to Asian waters. On 9 April 1942 HMAS *Vampire* was sunk with the aircraft-carrier HMS *Hermes* by Japanese dive-bombers off Colombo. HMAS *Voyager* did not survive much longer. On 23 September it grounded on a reef at Timor during an operation to land commandos. It was unable to be refloated and was blown up by the crew. HMAS *Vendetta* survived the war and was sunk off Sydney Heads in 1948.

VAMPIRE, DE HAVILLAND AUSTRALIA (Single-seat fighter [FB 31]). Wingspan 38 feet; length 31 feet; armament 2 × 20 mm cannon, 8 × 60-pound rockets, 2000 pounds bombs; maximum speed 548 m.p.h.; range 1220 miles; power 1 × CAC Nene 5000-pound thrust turbojet .

The Vampire was the RAAF's first operational jet and entered service in 1949. Apart from three British-made evaluation aircraft, all RAAF Vampires were built by De Havilland Australia and powered by Rolls Royce Nene engines made under licence by the Commonwealth Aircraft Corporation (q.v.). Cold War tensions led to the deployment of No. 75 and 76 Squadrons with Vampires to Malta from 1952 to 1954. During this time they also flew in the 1953 Coronation Review Flypast in the United Kingdom and in NATO exercises in Germany. Although the fighter models were quickly superseded by Meteors and Sabres (qq.v.), Vampire trainers remained in service with the RAAF until 1969 and with the RAN until 1971. One Vampire is preserved by the RAAF Historic Flight.

VAMPIRE (I), HMAS see **'V' AND 'W' CLASS DESTROYERS**

VAMPIRE (II), HMAS see *DARING* **CLASS DESTROYERS**

VASEY, Major-General George Alan (29 March 1895–5 March 1945). A graduate of RMC Duntroon (q.v.) in the class of 1915, Vasey served with the AIF in France as adjutant of the 22nd Field Artillery Brigade and then, from August 1917 until February 1919, as brigade major of the 11th Infantry Brigade. He served temporarily on the staff of the 3rd Division during the demobilisation of the AIF, and returned to Australia in September 1919.

The interwar period was one of profound disillusion for Vasey. He was forced to revert to a lieutenant's rank in the reductions which followed the war, and was not promoted substantively to major until 1935. Like all of his generation of regular officers, he spent the years between the world wars in a succession of staff and regimental postings, the latter with militia (see CITIZEN MILITARY FORCES) units and formations. In 1928–29 he attended the Staff College, Quetta, British India (now in Pakistan), and from 1934 to 1937 was an exchange officer with the British Army in India, including a stint as brigade major to an infantry brigade, which enabled him to see limited active service and to operate on a wider canvas than was possible in the tiny Australian army of the day. At the beginning of the Second World War he was GSO1 (Training) at Army Headquarters, and was only promoted substantively to lieutenant-colonel in November 1939.

The Second World War was the making of Vasey's reputation, and he shone in a variety of positions, principally command ones. At first he was Assistant Adjutant and Quartermaster-General of the 6th Division, the principal administrative staff officer on the headquarters, and in this capacity and then as GSO1, the principal operational staff officer, he oversaw first the personnel and logistic preparations of the division for its first campaign, against the Italians in Libya, and then undertook the operational planning for the same campaign. In March 1941 he inherited command of the 19th Brigade from Brigadier H. C. H. Robertson (q.v.), who was invalided sick.

Vasey led the brigade through the tough fighting in Greece and Crete. The 19th Brigade was the first formation to take the brunt of the German assault, and it fought a rearguard action at Vevi during the withdrawal to the Aliakmon line. A fortnight later Vasey's troops held the Brallos pass during the withdrawal to the Thermopylae position. On this occasion Vasey issued an instruction to his commanders that they were to hold their positions, in typically Vasey terms: 'Here you bloody well are and here you bloody well stay. And if any bloody German gets between your post and the next, turn your bloody Bren around and shoot him up the arse.' On Crete he commanded all the Australian forces there, the majority of the 6th Division having been evacuated straight back to Egypt. In desperate fighting he again commanded the rearguard as the Allies withdrew over the island to the beaches in the south, to be evacuated once again, and his tenacious defensive action enabled large numbers of men to escape German capture. Vasey came away from the disasters in Greece and Crete convinced that Australian training was deficient and that the AIF would need to take more casualties before it became a fully proficient army.

Returning to Australia at the beginning of 1942, Vasey was promoted to major-general and initially served as Lieutenant-General Iven Mackay's (q.v.) chief of staff. When General Sir Thomas Blamey (q.v.) returned Vasey was made deputy CGS, a post which he held for about five months before being given command of the 6th Division, then fighting the Japanese in Papua. A month later, in late October, he was sent forward to replace Major-General A. S. Allen (q.v.), sacked from his command of the 7th Division.

Vasey was to command the 7th Division for the rest of the war until his death. His style of command was very much in the 'digger general' tradition. He insisted on wearing his badges of rank and scarlet-banded cap, despite the danger from Japanese snipers: 'I want them [the soldiers] to know me', he said, 'good for morale'. He was frequently around the forward positions, talking to the soldiers, assessing their circumstances for himself. He was a master of the quick appreciation, able to read a tactical situation and formulate a plan with great rapidity, as he demonstrated in the 7th Division's drive towards Kumusi in November 1942. Under Vasey's leadership the 7th Division fought through the battles for Gona and Sanananda, which were mostly hard slogging and frontal assaults, and then, after a period of rest and retraining, in the fighting to capture Nadzab, in which Vasey flew the 25th Brigade in by air preparatory to the advance on Lae. In the fighting in the Ramu Valley Vasey kept his forces mobile and flexible; his use of air transport to resupply troops in forward positions enabled the 7th Division to keep up the pressure on an increasingly embattled enemy. In late December the 21st Brigade mounted the successful assault on Shaggy Ridge, and shortly after the division was withdrawn from operations.

His health impaired by the privations of the campaign and the insistent manner in which he drove himself, Vasey was evacuated sick to Australia. He became seriously ill with polyneuritis in June 1944, and was convalescent for the rest of the year. Late in the year he, Robertson, who was also convalescent, and the Commandant of RMC, Brigadier Combes, were assigned to report on the future direction of RMC Duntroon. Their report emphasised, among other things, the need for extensive professional education and degree-level studies, something which was not to come about for more than 20 years. In early 1945 he was given command of the 6th Division following the return of Major-General Jack Stevens to civilian employ. Flying north to assume command, his aircraft crashed into the sea off Cairns, killing all on board.

Vasey was probably the best divisional commander Australia produced in the Second World War, and his loss was keenly felt throughout the army. The official historian, Gavin Long (q.v.), described him as a man 'highly strung, thrustful, hard working, [who] concealed a deeply emotional even sentimental nature behind a mask of laconic and blunt speech'. General George Kenney, commander of the US 5th Air Force, wrote of him that he was 'one of my favourite generals of anyone's army, and one of the ablest leaders I have ever come in contact with'. A serious student of war and an attractive and engaging personality, his greatest legacy was undoubtedly the creation of the War Widows' Guild, pioneered by his wife Jessie after his death: before flying north for the last time, he had enjoined her to 'look after the widows, because the bloody government won't'.

David Horner, *General Vasey's War* (Melbourne University Press, Melbourne, 1992).

VENDETTA (I), HMAS see **'V' AND 'W' CLASS DESTROYERS**

VENDETTA (II), HMAS see **DARING CLASS DESTROYERS**

VENEREAL DISEASE Sexually transmitted disease has presented three problems in the military environment: discipline; personnel wastage; and the medical challenge of prevention and cure. It has also caused problems on occasion for senior officers when the rate of infection has become the source of public comment and disapproval, and hence a political issue, albeit a minor one. Venereal disease as a political issue is occasioned by what the medical historian of the First World War, A. G. Butler (q.v.), characterised as 'a conflict, in medical sociology, between reason and fetish which, in Australia . . . kept this terrible group of diseases almost wholly outside the scope of the new science of preventive medicine and thereby involved civilised man in unnecessary physical degradation and his women-folk and children in much needless misery and suffering'. Before 1914 venereal diseases were classified as notifiable diseases in most States. The official correspondent, C. E. W. Bean, was one of those who suffered from an excess of moral disapproval where the subject of venereal disease was concerned, and one of his early despatches from Egypt, commenting approvingly on the forced return to Australia of soldiers who had contracted the disease, earned him notoriety among and some hostility from members of the AIF. Egypt was notorious for venereal infection, but the problem was just as pronounced in France and Britain, brought on, as Butler noted and the Inter-Allied Sanitary Conferences of 1918–19 discussed at length, by 'the rise to prominence of the "amateur" prostitute'. In Egypt the military authorities' only recourse was to make venereal infection a punishable offence, akin to self-inflicted wounding. Morally uplifting lectures from the chaplains did little to dissuade men from sampling the delights of the bar and brothel districts of Cairo, while the idea of officially sponsored brothels, with regular medical checks on workers and clients, was anathema to Victorian moral virtue.

Once the AIF got to France, something more imaginative was clearly required, and the Director of Medical Services, Major-General Neville Howse (q.v.), developed a system of education, prophylaxis and treatment that was ahead of its time. (A broadly analogous effort was mounted on behalf of the New Zealand Expeditionary Force by a committed private individual, Ettie Rout, who suffered considerable stigmatisation for her efforts to supply New Zealand soldiers on leave with means of protection, despite the fact that Lieutenant-General Sir Alexander Godley [q.v.] offered his support for her work.) Howse's achievement was all the more remarkable in that he himself held strong moral views on the subjects of both VD and alcohol (he was a strict teetotaller). Much of the educational effort was left to the chaplains and YMCA representatives, but Howse set up two centres for the distribution of prophylactics — at Administrative Headquarters, Horseferry Road, and at each of the Australian depots — and followed this up with the establishment of early treatment centres (known as 'Blue Light Depots' for the colour of the lamp kept alight outside them after hours) at each command depot and training centre, and special VD hospitals for more sustained treatment, especially for gonorrhoea cases.

The focus of the effort was in England, for Howse recognised that VD was usually acquired on leave: the AIF's rate of infection in France was about the same as the British Army's, whereas in Britain it was four times as high. Nonetheless, it was made clear that preventive and curative measures did not imply approval, and while hospitalised for treatment of VD officers and men suffered stoppage of pay, with officers losing their field allowance also. Only if the disease was certified as a pre-existing condition (i.e. prior to enlistment) or as hereditary was forfeiture waived. Much emphasis was placed on the alleged responsibility of women in the process of infecting soldiers, an attitude present in Australia before the war in the provisions of various contagious diseases acts. Butler estimated that in no army on the Western Front did the rate of infection drop below 100 cases per 1000 soldiers, and that the rate among Australians was the highest of any British Empire army (the statistics, as always, are notoriously incomplete, but the meeting of senior medical officers at the end of the war was in general agreement on them). During the last two years of the war, when Howse's scheme was in operation, over 235 000 soldiers went on leave from AIF depots: over 142 000 prophylactic kits (i.e. condoms) were issued, nearly 170 000 attended for prophylactic (i.e. preventive) treatment, while a further 12 000, who had acquired symptoms of the disease, were admitted for abortive treatment.

In France until July 1917 Australian hospital cases were treated at No. 39 British General (VD) Hospital at Havre: 46 per cent of the admissions were Australians, and the strain on bed places was so great that General Sir Douglas Haig suggested to General Sir William Birdwood (q.v.) that an Australian VD hospital be set up. Butler notes also that VD was such a problem among military offenders sent to No. 1 Australian Dermatological Hospital at Bulford on Salisbury Plain that a special Australian VD ward was instituted for them at the Lewes Detention Barracks in Sussex. Throughout the war, the rate of VD infection per thousand soldiers of the AIF in Britain, based on recorded admissions and therefore possibly slightly on the low side, was as follows: 1915—134.05; 1916—148.1; 1917—129.2; 1918—137.12. In France, venereal cases accounted for 6.19 per cent of all non-battle casualties. For the AIF as a whole, in all theatres overseas and for the entire period of the war, the rate of infection as measured by admissions was 70.87 per thousand. When the repatriation and demobilisation of the AIF was planned at the end of the war, extra measures were taken to ensure that men who had undertaken treatment were indeed cured before they returned to Australia, while special arrangements were made for the shipping of infected men back to Australia, both during the war and as part of the repatriation program.

In discussing the problem during the Second World War, the official medical historian, A. S. Walker, was able to write that the social taboos which had inhibited discussion and treatment of VD had largely dissipated by 1939, noting that the approach within the military had swung from the largely moral to the predominantly medical. Such enlightenment was not uniform. Walker records that the isolation wing of the 107th Australian General Hospital at Puckapunyal, Victoria, was a compound under armed guard. In the Middle East a special treatment centre was set up within the 2/1st Australian General Hospital, and at least overseas the army made a genuine effort to encourage prevention through prophylactic centres in each camp. Infection rates in the Middle East nonetheless remained high, with a rate of 35.21 per thousand in 1940 and 48.46 per thousand in 1941. These rates were lower than those in the First World War, but were probably understated because of the different diagnostic categories available to medical officers, especially that of 'non-specific urethritis', which was not classified as venereal in origin. In 1942, with only one Australian division remaining in the Middle East theatre, the rate was still 47.22 per thousand, and even during the battle on Crete some men were found to need medical attention for VD. The total number of cases in the Middle East alone was at least 11 000 up to October 1942, with 'amateurs' once again being listed as a major source of infection. The process of venereal infection was reflected in soldier culture; one song from North Africa, sung to the tune of 'There is a Green Hill Far Away', described the process:

> There is a street in Cairo, full of sin and shame
> Sari Wag El Burka is the bastard's name.
>
> (Chorus) Russian, Greek and French hints, all
>   around I see
> Shouting out: 'You stupid prick, abide with me'.
>
> Two or three weeks later, when I see my dick
> Swiftly pack my small kit, and fall in with the sick.
>
> (Chorus)
>
> Five or six months later, free from sin and shame,
> Back to the El Burka, just for fun and games.

In Malaya before the outbreak of the Pacific War the rate of admission was 31 cases per week from a force of 7700, although this rose to 58 per week with the arrival of a second brigade. Not surprisingly, there were few recorded cases of venereal infection among prisoners of the Japanese, although some symptoms did present themselves. The fighting in the South-West Pacific Area posed fewer problems than the Middle East, again because of a relative absence of sources of infection. The rates per thousand in the army across the theatre (excluding Australia) were: 1942—2.33; 1943—1.06; 1944—0.36; 1945—7.69. In Australia the rates soared again, and once more the explanation, at least in part, was that VD was a 'leave disease'. In the army the rates were: 1942—18.93; 1943—16.85; 1944—13.18; 1945—17.08. Once again, the principal source of infection was 'amateurs', making control at the source more difficult. As an aspect of public health, recommendations were made during 1942 for provisions under the National Emergency Act to enable civil authorities to detain and examine persons suspected of carrying venereal infection. Such provisions were used at times as a means of social control, especially in Queensland. The biggest single advance in the treatment of venereal diseases during the Second World War was, however, the discovery of penicillin and its widespread use in Australia from 1944. Moral concerns at the spread of VD often hid other social and political agendas, such as a desire to prevent Black US servicemen from dating White women. A similar view lay behind a press campaign in 1946 over the incidence of VD among occupation troops in Japan. The highest proportion of men presenting with venereal infections among troops from four Commonwealth nations were the Australians, who accounted for nearly 65 per cent of admissions in June and July 1946. Lack of control at the source of infection was again the problem, as organised brothels had been bombed out by the end of the war, and many women who had not previously engaged in prostitution did so in a devastated economy in order to survive. No. 130 Australian General Hospital, with a capacity of 200 beds, was set up to cater for Australian VD cases in Japan, and the first C-in-C of the British Commonwealth Occupation Force (q.v.), Lieutenant-General John Northcott (q.v.), issued an order that any officer diagnosed with the disease was to be sent home to Australia. Two things drove the prurient interest in the issue on the part of the Australian press: the imminence of a federal election, and the issue of fraternisation with Japanese women which, so soon after the end of the war, many Australians professed to abhor.

In the postwar conflicts in Asia venereal diseases were a much less serious factor in military medicine than they had been previously. Soldiers fighting in Korea were most at risk while on leave in Japan. The infection rate among all Commonwealth troops was 376 per thousand in 1952; the

A venereal disease prevention sign above a bar in Nui Dat, South Vietnam, 1971. (AWM P0966/97/42)

Australian rate was 386, ahead of the British (264) but well behind the New Zealanders (410) and the Canadians (616). In March 1953 the rates showed some increase while the Commonwealth forces were out of the line for a two-month spell. This was explained partly by increased taking of leave, but also by the close proximity of the main camp and training facilities to a Korean shanty town, which proved a ready source of venereal infection. When Australian troops went to Malaya during the Emergency, they generally came under British arrangements for the control of venereal infections. In the first months, the rate among Australians soared to more than 400 cases per thousand, and in the first half of 1956 it reached 462.83 per thousand. Thereafter it declined, partly because units began operational duties well away from leave centres. Moral suasion, however, clearly had little impact, and soldiers were required to attend prophylactic centres as soon as possible after potential exposure. The rate of infection was brought under control through educational and disciplinary measures, but the concern was mainly political and moral, as the advent of antibiotic drugs meant that VD cases now had little impact on casualty and wastage rates. The incidence of venereal disease among soldiers on active service in Borneo during

Confrontation was low, not least through lack of opportunity.

In Vietnam VD again became something of a problem, with regular leave and close proximity to unsupervised brothels. Attempts were made to control infection among prostitutes and bar girls, but the Vietnamese government insisted that bars should be approved for the use of the troops if only 30 per cent of the girls who worked there were certified free of infection. This proportion was gradually increased, and Australian authorities made valiant attempts to extend treatment to civilian women at risk, but the circumstances of a traditional society under enormous strain through the exigencies of a major war and very large numbers of foreign troops 'in country' made prevention very difficult. The figures for VD rates in Vietnam are incomplete, but the total number treated for venereal infection between July 1965 and June 1972 was over 11 000. A further problem for control was the availability of 'R&R', rest and recreation leave taken outside Vietnam, often in countries like Thailand. Soldiers seem to have ignored, or forgotten, the educational messages as soon as they went on leave, and there was little or nothing the military authorities could do about this. In the era of AIDS, additional sexual hazards now face service personnel on leave or deployment in

Asia or on duty in, for example, Somalia, and very stringent controls are placed on soldiers deployed to countries like Malaysia for training, with many potential leave areas in the region banned to Australian servicemen.

A. G. Butler, *The Australian Army Medical Services in the War of 1914–1918*, vol. 3, *Problems and Services* (Australian War Memorial, Canberra, 1943); A. S. Walker, *Clinical Problems of War* (Australian War Memorial, Canberra, 1952); Darryl MacIntyre, 'Australian Army Medical Services in Korea' in Robert O'Neill, *Australia in the Korean War 1950–53*, vol. 2, *Combat Operations* (Australian War Memorial, Canberra, 1985); Brendan O'Keefe and F. B. Smith, *Medicine at War: Medical Aspects of Australia's Involvement in Southeast Asian Conflicts 1950–1972* (Allen & Unwin, Sydney, 1994).

VENGEANCE, HMAS (*Colossus* Class light fleet aircraft-carrier). Laid down 1942, launched 1944, completed 1945; displacement 13 190 tons; length 695 feet; beam 80 feet; armament 12 × 40 mm guns, 32 × 20 mm guns, aircraft 30–40 fighters; speed 24 knots.

HMAS *Vengeance* was lent to the RAN by the RN in 1952. In 1954 she was reduced to a training ship, in which service she transported a squadron of Meteor Jets (q.v.) from Korea in 1954. HMAS *Vengeance* was sold to Brazil in 1955.

VENGEANCE, VULTEE (2-seat dive-bomber). Wingspan 48 feet; length 39 feet 9 inches; armament 6 × 0.5–inch machine-guns, 2000 pounds bombs; speed 280 m.p.h.; range 600 miles; power 1 × Wright 1700 h.p. engine.

The Vengeance, which entered production in 1940 after the success of the German Stuka dive-bomber at the beginning of the Second World War, was a dive-bomber pressed into service by the RAAF in the middle of 1942 at a time when it was desperate for aircraft. It was, however, an inferior aircraft hampered by short range, light armament, and vulnerability to fighter attack. It was being withdrawn from service by the US Navy just as it was entering service with the RAAF. Vengeances were flown by No. 12, 21, 23, 24 and 25 Squadrons during the New Guinea campaign but opinion is divided on their performance. They were withdawn from operational service in March 1944 and the squadrons re-equipped with Liberators (q.v.).

VENTURA, LOCKHEED (4–5-crew bomber [Mark II]). Wingspan 65 feet 6 inches; length 51 feet 9 inches; armament 2 × 0.5–inch machine-guns, 6 × 0.303-inch machine-guns, 3000 pounds bombs; maximum speed 312 m.p.h.; range 1660 miles, power 2 × Pratt & Whitney 2000 h.p. engines.

The Ventura was a military version of the Lockheed Model 18 Lodestar passenger aircraft, which also served with the RAAF, and was similar to the Hudson (q.v.), which was a military version of the earlier Lockheed Model 14 airliner. No. 464 Squadron flew Venturas over Europe on light bombing raids from its formation in September 1942 until their replacement with Mosquitos (q.v.) in August 1943. No. 459 Squadron replaced Hudsons with Venturas in January 1944 and used them in shipping raids in the Aegean Sea until July 1944 when they in turn were replaced by Baltimores (q.v.). The personnel of these two squadrons had trained through the Empire Air Training Scheme (q.v.). No. 13 Squadron in the South-West Pacific Area (q.v.) replaced Hudsons with Venturas in 1944 and used them in bombing raids in support of the American landings in the Philippines and, as part of the 1st Tactical Air Force (q.v.), the Australian landings in Borneo. A Ventura is preserved by the RAAF Museum.

VETERINARY CORPS see **AUSTRALIAN ARMY VETERINARY CORPS**

VICKERS MEDIUM MACHINE-GUN (MMG) was a recoil gas-assisted operation, water-cooled, belt-fed automatic-fire weapon that was fired from a tripod. It was in Australian military service from 1915 until its official replacement by the GPMG M60 in 1960. However Vickers MMGs were still being used by the army for battle inoculation and at fire-power demonstrations through the 1960s to the late 1980s. The Vickers MMG saw active service with Australian soldiers during the two world wars, Korea and the Malayan Emergency.

The Vickers MMG was developed by Vickers from the Maxim Machine-Gun and was adopted by the British Army in November 1912. It is known that Australian soldiers had very early model Vickers MMGs on Gallipoli in 1915. Australia's early Vickers MMGs came from British sources and eventually the first Australian-made Vickers was issued to the army from the Small Arms Factory at Lithgow in 1929. The Vickers had an excellent reputation for reliability and production of accurate, sustained, direct and indirect firepower. A Vickers could be sustained in action by a highly trained crew of two men, but required a team of at least four men to carry the gun, tripod, sights, ammunition and water. The Vickers fired 0.303-inch calibre ammunition from a 250-round fabric belt at a cyclic rate (i.e. the rate that could theoretically be achieved given a continuous supply of ammunition) of around 450 to 500 rounds per minute out to a maximum range of 4500 yards. Its most effective range was between 800 to 1200 yards and its rapid rate of fire was 250 rounds per minute. The Vickers ready for action with its 7-pint capacity water jacket filled and mounted on its tripod weighed 90 pounds and each 250 round box of belted ammunition weighed 22 pounds.

IAN KURING

VICTORIA (I), HMVS see **VICTORIAN NAVY SHIPS**

VICTORIA (II), HMVS see **VICTORIAN NAVY SHIPS**

VICTORIA CROSS (VC), the highest British award for bravery shown in the face of an enemy by a member of the military forces, was instituted in 1856 by Queen Victoria but was made retrospective to the autumn of 1854 to cover the period of the Crimean War (1854–56). Between 1858 and 1881, eligibility for the VC was extended to cover acts of bravery not in the face of the enemy, for example, saving lives at sea. After 1881, the Albert Medal (established in 1866), and subsequently the British Empire Medal (1917) and the George Cross and George Medal (1940) were used for these sorts of actions, and the VC reverted to an award for serving personnel in the presence of the enemy. Members of British colonial military forces became eligible from 1867. The first VC to be awarded to an Australian was won by Captain Neville Howse (q.v.) for his action at Vredefort in the Boer War, on 24 July 1900. Since then Australians have won a further 95 VCs: 5 in the Boer War, 64 in the First World War, 2 in the Russian intervention (1919), 20 in the Second World War, and 4 in the Vietnam War. In January 1991, as part of the move away from imperial honours, a new award was introduced — the Victoria Cross for Australia. None has yet been awarded. The VC was designed in the form of a Maltese cross cast in bronze derived, so it is claimed, from Russian cannon captured at Sebastopol in the Crimean War, with the motto, 'For Valour', chosen by the Queen. The date of the act of bravery is inscribed on the reverse of the cross, with the rank, name and unit of the recipient on the back of the clasp. Originally there were two colours for the ribbon: red for the army and blue for the navy, but with the creation of the Royal Air Force in 1918, red was adopted as the colour for all three services. The largest single collection of VCs in the world, 54 medals, is held by the Australian War Memorial (q.v.), where the medals, along with photographs and details of the winners, are displayed in the Hall of Valour.

VICTORIAN NAVY SHIPS In 1854 fears of war with Russia (see COLONIAL WAR SCARES) prompted the Victorian government to order the steam sloop *Victoria*, a fast and graceful ship with a raking bow and funnel which arrived in Melbourne in May 1856. At first *Victoria* operated as an armed dispatch and survey vessel, then in 1860 it went to take part in the New Zealand Wars, returning to Australia the following year. Later in 1861 it was sent on an unsuccessful voyage to the Gulf of Carpentaria to find the missing explorers Burke and Wills. *Victoria* was paid off in 1864, resumed service as a survey vessel in 1865–69 and 1873–77, then was refitted as a warship following renewed Russian war scares in 1878, but was decommissioned six months later. After the Colonial Naval Defence Act (q.v.) of 1865 established for the first time the conditions under which the colonies could form their own naval forces, the Victorian government succeeded in obtaining British assistance to build up its fleet. In 1867 the wooden line-of-battle ship *Nelson* was transferred on permanent loan to Victoria, and it arrived at Williamstown in February 1868. Launched in 1814 as a three-decker, *Nelson* had been reduced to two decks and converted to a steam-screw ship between 1854 and 1860. It was an impressive-looking ship, with tall masts and a double line of chequered gunports, but the development of ironclads had made it obsolete by the time it was transferred to Victoria. Though it was reduced to one deck and reclassified as a warship during a Russian scare in 1877, *Nelson* functioned mainly as a training ship until it was sold in 1898. The other ship acquired with British help was the iron turret ship *Cerberus*, which was built in Britain with 80 per cent of the cost paid by the Admiralty and which reached Melbourne in April 1871. It had a very low freeboard, but armoured breastwork raised the funnel base, air shafts and the two gun turrets above the water. Covered in armour between 8 and 10 inches thick, *Cerberus* was invulnerable to all naval guns then in existence. It could lower itself further into the water by letting water into ballast tanks, thus leaving only the breastwork and turrets as targets, and its lack of sails meant that its guns had a wide arc of fire. Designed specifically for harbour defence, *Cerberus* had a generally uneventful career, being used mainly for exercises before it was integrated into the RAN and used as a port guard ship at Williamstown during the First World War. Renamed *Platypus* in 1921, it became the RAN's submarine depot ship, but was sold in 1924. Victoria's next naval vessels were bought during the flurry of naval purchases by Australian colonies in the early 1880s. The steel third-class gunboats *Albert* and *Victoria* (II) and the steel first-class torpedo boat *Childers* were ordered from Britain in 1883 and arrived together in Melbourne in June 1884. They had stopped off at the Red Sea port of Suakin along the way to assist with the war in the Sudan (q.v.), but found they were not needed. *Albert*, which was almost identical in design to Queensland Marine Defence Force ships (q.v.) *Gayundah* and *Paluma*, spent most of its time in Port Phillip and, along with the similar but larger *Victoria*, was regularly used for exercises. *Victoria* was sold in 1896 and *Albert* in 1897. Neither *Childers* nor the steel second-class torpedo boats *Lonsdale* and *Nepean*, which arrived in Melbourne in July 1884, had notable careers. *Childers* was integrated into the RAN but paid off in 1916 and was beached at Swan Island near *Lonsdale* and *Nepean*, which had been left there after an unsuccessful attempt to sell them in 1914. The wooden turn-about torpedo launch *Gordon*, which arrived from Britain in February 1886, is noteworthy mainly for its misfortune in being accidentally torpedoed by *Childers* in 1886 and then, as a RAN vessel in 1914, being rammed and sunk by *Cerberus*'s picket boat. The last warship built for an Australian colonial government was the steel first-class torpedo boat *Countess of Hopetoun*, a sleek and powerful vessel which reached Melbourne in May 1891 and served with the RAN from 1911 until it was sold in 1924. All the Victorian Navy's torpedo boats were used in regular exercises, as were some of its 16 auxiliary vessels, most of which were armed. The Victorian Navy was not only the largest but also the best organised and best trained Australian colonial fleet, and its ships, which bore the prefix HMVS, saw many more hours of active service than those of other colonial navies.

Victoria: displacement 580 tons; length 166 feet; speed 9.5 knots under steam, 12 under sail, 14.5 under steam and sail; armament originally 1 × 56 cwt 32-pounder, 6 × 25 cwt 32-pounders; 1859 — 4 × 25 cwt guns removed; for New Zealand Wars carried 4 × 56 cwt 32-pounders, 4 × 25 cwt 32-pounders; from 1864 carried no guns until 1878, when it carried 1 × 64-pounder, 4 × 12-pounders.

Nelson: displacement 2617 tons; length 222 feet (overall); beam 44 feet, draught 25 feet (aft); speed 13 knots; armament 1867 — 2 × 7-inch guns, 20 × 64-pounders, 20 × 32-pounders, 4 × 12-pounders; 1878 — 2 × 7-inch guns, 20 × 64-pounders, 6 × 12-pounders; 1881 — 2 × 7-inch guns, 2–4 × 64-pounders, 2 × 12.5-pounders, 2 × 10-pounders, 2 × 9-pounders, 2 × 6-pounders, 2 × 4.7-inch guns; 1887 — 2 × 7-inch guns, 14 × 64-pounders.

Cerberus: displacement 3340 tons; length 225 feet (overall); beam 45 feet; draught 15.5 feet; speed 9 knots (maximum), 6 knots (economical); armament 4 × 10-inch guns, 2 × 6 pounders, 8 × machine-guns.

Albert: displacement 370 tons; length 120 feet (overall); beam 25 feet; draught 9.5 feet; speed 10 knots; armament 1 × 8-inch gun, 1 × 6-inch gun, 2 × 9-pounders, 2 × 1.5-inch guns; 1.5-inch guns replaced by 6-pounders in 1888.

Victoria (II): displacement 511 tons; length 145 feet (overall); beam 27 feet; draught 11.5 feet; speed 12 knots; armament 1 × 10-inch gun, 2 × 12-pounders, 2 × machine-guns; 1887 — 1 × 8-inch gun, 1 × 6-inch gun, 2 × 13-pounders, Nordenfelt guns; 1888 — Nordenfelts replaced by 6-pounders.

Childers: displacement 63 tons; length 118 feet (overall); beam 12 feet; draught 5.667 feet; speed 19 knots; armament 2 × 1-pounder, 2 × 15-inch torpedo tubes; 1885 — 4 × 14-inch side-dropping torpedo gear added, number of torpedo tubes reduced to one.

Lonsdale and *Nepean*: displacement 12.5 tons; length 67 feet (overall); beam 7.5 feet; draught 3 feet (aft); speed 17 knots (maximum), 10 knots (economical); armament 2 × torpedo-dropping gear.

Gordon: displacement 12 tons; length 56 feet; beam 10 feet; draught 5 feet; speed 14 knots; armament 2 × 14-inch torpedo-dropping gear, 1 × machine-gun.

Countess of Hopetoun: displacement 75 tons; length 130 feet (overall); beam 13.5 feet; draught 7.333 feet; speed 24 knots; armament 3 × 14-inch torpedo tubes, 4 × torpedo-dropping gear, 2 × machine-guns.

(See also COLONIAL NAVIES.)

VIETNAM CAMPAIGN MEDAL Service in the Vietnam War was recognised by three campaign awards: the General Service Medal 1962 with clasp 'South Vietnam', the Vietnam Medal and the Vietnam Logistic and Support Medal. The first was awarded for service between 24 December 1962 and 29 May 1964 and was authorised by royal warrant in June 1968 for issue to Australian servicemen only. Sixty-eight were awarded to members of the Australian Army Training Team Vietnam (q.v.). Qualifying service for the Vietnam Medal was between 29 May 1964 and 27 January 1973 and was awarded to personnel who spent either 28 days on ships in inland waters or off the coast of Vietnam, one or more days on the posted strength of a unit, one operational sortie by air crew, or official visits aggregating 30 days or more. This excluded, in particular, naval personnel who served on board HMAS *Sydney* (III) (q.v.), the 'Vung Tau ferry', and after a considerable public campaign a separate medal, the Vietnam Logistic Support Medal, was instituted in 1993 for service personnel who operated in ships or aircraft in support of Australian forces, were attached to a unit in support of of Australian forces, or who served with either the Australian or Allied forces as an observer. In addition, in 1990 a small number of RAAF personnel were awarded the Australian Service Medal with clasp 'Vietnam 1975' for service in support of UNICEF evacuations of Vietnamese civilians between 19 March and 29 April 1975 before the fall of Saigon. All personnel who served inside the territory of South Vietnam were issued the Campaign Medal, with its distinctive green–and–white striped ribbon, by the South Vietnamese government.

VIETNAM LOGISTIC SUPPORT GROUP The 1st Australian Logistic Support Group was established at Vung Tau in April 1966 as part of the deployment of the 1st Australian Task Force. At first somewhat rudimentary in organisation, since the base area had to be constructed from scratch in a stretch of sand–dunes on the Vung Tau peninsula known as the Back Beach, it grew rapidly and by the end of 1967 was able to handle all the support requirements of the Task Force, which itself had grown. The difficulties in the supply system were at first severe, however, since although it had been intended that the Task Force should be self–supporting, the number of men detailed for the task was insufficient to the demands of the force, and shortages soon developed. Its development may be measured by the increase in its establishment, from 910 in July 1966 to 1317 a year later, to 1356 by December 1967. At its height, by 1969, it had under command an engineer construction squadron, Royal Australian Electrical and Mechanical Engineers (q.v.) base workshop, ordnance depot, field hospital, provost unit and amenities and welfare unit. The logistic support group was withdrawn with the Task Force at the end of 1971. Logistic support of the forces was probably the most important contribution made by the RAN to the Australian commitment in Vietnam. The aircraft carrier HMAS *Sydney* (III), converted to a fast troop transport, made 22 voyages between Australia and the port of Vung Tau ferrying troops and equipment for the 1st Australian Task Force. The first voyage was undertaken in May 1965 when it conveyed 1RAR, the last in March 1972 during the withdrawal of Australian forces. Additional lift capability was provided by two ships of the Australian National Line, MV *Jeparit* and MV *Boonaroo*. The *Jeparit* made 42 voyages to Vietnam from June 1966 to March 1972, 17 as a naval vessel, while *Boonaroo* made just two. In the course of the war the three ships carried 186 000 tons of cargo and 2400 vehicles, including Centurion (q.v.) tanks, while the *Sydney* also transported 15 600 soldiers. The effort was supplemented by the army ships AS *John Monash* and AV *Vernon Sturdee*, medium landing

ships from the army's 32 Small Ships Squadron (q.v.), and by air lift from RAAF C130 Hercules (q.v.) aircraft and commercial aircraft under charter. There was considerable ill feeling among many servicemen involved in the naval logistic support effort because of their lack of entitlement to the Vietnam Campaign Medal (q.v.), and a new medal, the Vietnam Logistic Support Medal, was struck in 1993 to recognise their service.

VIETNAM VETERANS' ASSOCIATION (VVA) Formed in 1979–80 as the Vietnam Veterans' Association of Australia (VVAA), the VVA was a product of dissatisfaction on the part of a minority of Vietnam veterans at official responses to concerns raised about the effects of toxic chemicals used for defoliation during the Vietnam War. Its early efforts were aimed at obtaining an inquiry into the issue, but the VVA also functioned as a welfare group providing counselling and other forms of support and short-term crisis intervention. From early in its existence it quarrelled publicly and often bitterly with both the Department of Veterans' Affairs and the RSL over what it saw as the neglect of veterans' problems and their betrayal by longer-established and more conservative bodies. It lobbied hard and gained a limited inquiry into veterans' health issues, resulting in the publication of the *Birth Defects Study* (1983) and the *Mortality Report* (1984), but then disowned the process when the findings contradicted its own position. It was instrumental in the convening of a Royal Commission into the Use and Effects of Chemical Agents on Australian Personnel, under Mr Justice Phillip Evatt, in 1983, which the RSL opposed. The Royal Commission's findings, which essentially cleared Agent Orange (q.v.) and similar agents of responsibility for Australian veterans' health problems, were bitterly rejected by the VVA, who proceeded to attack both Evatt and the process of the Commission on which they themselves had insisted. (In 1994, after a new medical investigation, the government accepted that chemicals used in Vietnam had caused and were continuing to cause health problems for veterans, thus overturning the findings of the Evatt Royal Commission.) With greater effect, its lobbying activities resulted in the establishment of Vietnam Veterans' Counselling Centres by the federal government in 1982, and these have been a considerable success in providing assistance to ill or troubled veterans and their families. The membership of the VVA has probably never exceeded 5000 at its height, but precision is difficult since the association will not provide detailed figures. Many Vietnam veterans dislike the dominant image of the sick, disturbed veteran with

The Australian Vietnam Veterans' National Memorial, Anzac Parade, Canberra. (Photograph by Deborah Jenkin)

a grievance which VVA publicity often presents, and point out that while such cases undoubtedly exist, the reverse is far more often true. Since the Royal Commission's findings were handed down in 1985 the VVA has lost some its former prominence, and its membership has probably declined. Any account of the experience and treatment of Vietnam veterans, however, must give a prominent place to its activities.

VIETNAM VETERANS' NATIONAL MEMORIAL

The Australian Vietnam Veterans' National Memorial is situated in the middle of the western side of Anzac Parade, Canberra. It was dedicated on 3 October 1992 in the presence of the Governor-General, Bill Hayden, the Prime Minister, Paul Keating, and about 30 000 veterans, friends and families. The memorial was designed by internationally renowned artist Ken Unsworth, AM, working with the Sydney-based architectural firm of Tonkin, Zulaikha and Harford. It consists of a number of elements making it more of a building — a shrine-like space — than a free standing sculpture or group, more typical of the other memorials along Anzac Parade. Central to the memorial are three large blocks or stelae of pre-stressed concrete, each rising approximately 10 metres from a shallow moat surrounding a triangular plinth. Suspended from the stelae is an annulus (ring) of polished black granite, hanging about 6–7

metres from the floor of the plinth. Sealed in one of the annular segments, marked with a gold cross, is an engraved steel scroll listing the names of the more than 500 dead Australian personnel.

The surrounding grounds consist of a central ramp leading up to the plinth; to the south, a ceremonial space with three flag poles, a wreath wall topped with large free-standing letters spelling VIETNAM, and inscribed below 'For All Those Who Served Suffered and Died Vietnam 1962–1973'; and on the northern side of the ramp another gathering space and a low rise of steps leading up to the sides and back of the plinth. In the surrounding grassed areas are three low concrete seats each inscribed with two names belonging to those designated as missing in action.

Each of the stela tapers towards its top and leans inwards to make an enclosed middle space. Each stela is also twisted along its vertical axis effecting a helical movement to the whole monument, a movement which draws the visitor into the internal space and then directs the eyes upwards to the inner walls and to the suspended ring. While not specifically religious, it is certainly spiritual — uplifting — a sense enhanced by the dappling of the light broken by the surrounding eucalypt canopy, the annulus, and the reflecting polished surface of the western wall. At night hidden spotlights create a softly lit environment. Both are reminiscent (intentionally) of the mottled light of the Vietnamese jungle, and of the mystical spaces of romanesque and gothic churches.

The inner back wall (west) is photo-engraved polished black granite — the image showing Australian troops on active service, adapted from a well-known photograph by Mike Colleridge. The northern wall is covered with a selection of well-known, or typical, phrases, comments or terms from the Vietnam period fixed in steel lettering. The 34 entries provide both a taste of the linguistic environment of the period and an education for future visitors — many of the inscriptions relate to events of political and military importance within the Australian involvement — others provide colloquial usages unique to Australian personnel.

The southern wall is blank, its function as a contemplative site enhanced by a large granite altar standing beneath it. This segment shaped stone echoes the annulus above it, and features service badges and lettering indicating the function of the annulus. It has become, together with the wreath wall, a site for the deposit of Vietnam memorabilia as well as floral tributes.

At a length of 3500 metres and 150 metres wide, Anzac Parade is the largest 'public space' in Australia; along its length are a number of memorials to specific services (army, navy and air force all have memorials along Anzac Parade) or to specific 'events' (such as Tobruk or the Greek campaign of the Second World War). As a legible built environment the space from the War Memorial to Parliament House along Anzac Parade may be said to map the specific histories of Australia in war. It was for this reason that the Australian Vietnam Veterans' National Memorial Committee wanted a site on Anzac Parade.

The committee was formed in 1987 just after the 'Welcome Home March' of 3 October in Sydney. By early 1988, the site had been chosen and approved. Following the selection of a Project Committee the competition for the design was announced in July 1989, but the winning entry was not announced until 3 October 1990 — the anniversary of the march now assuming its own powerful status. The interval between events indicates the care taken in choosing the winner from an initial field of over 80 entries. The first sod was turned on the site on 6 September 1991 and the final dedication happened just over a year later on 3 October 1992.

As the Australian march followed the spirit if not the precise model of the US Welcome Home Marches of the mid-1980s, the memorial committee adapted some of the concepts of the United States' National Memorial in Washington DC.

However the Australian committee was concerned to emphasise that the Australian memorial was not a monument for the dead alone; rather they stipulated that the memorial had to be 'For all those who served suffered and died'. This is why the scroll of the dead is sealed. It is not for reading in this world. At the same time the committee insisted on a strongly representational aspect, unlike Washington's stark symbolism. Equally the Australian committee was concerned that the final design had the majority approval of both the veteran and wider public communities, avoiding the disruptions which had ensued in the United States. Models of the winning design and a brochure explaining it were exhibited widely throughout Australia. The Project or Design Committee, consisting of a mixture of veterans, bureaucrats and artists, also maintained a tight control over the competition, choice and construction. The result was almost universal satisfaction with the finished memorial, proven not least by the enormous turnout for the dedication, and its subsequent evolution as one the most attended sites in Canberra.

JEFF DOYLE

VIETNAM WAR The Vietnam War was the longest and arguably the most divisive Australian military involvement in our history. It is also the least understood, and the most misrepresented.

Australia had recognised the French-created government of Bao Dai in February 1950, on the same day as the United States, and in the later 1950s had declared support for the regime of Ngo Dinh Diem, at American behest. Support for South Vietnam, and involvement in the Laotian crisis of 1961, were part of Australia's broader commitment to the policy of containment of communism by the West in both Europe and Asia, as was membership of SEATO (q.v.). Although unequivocal in their support in public, privately Australian officials were critical of the authoritarian nature of Diem's rule and recorded their distaste for his methods during, for example, the Buddhist crisis, but most felt that the wider issues of regional security were of such importance that they could not be undermined by public disapproval of the Saigon government. Following Diem's murder in 1963 and his replacement by a succession of 'revolving-door' governments, the insurgency mounted by the National Liberation Front and its North Viet-

Map 27 Phuoc Tuy province, 1966–72, showing principal areas of Australian operations.
Inset Location of Phuoc Tuy province.

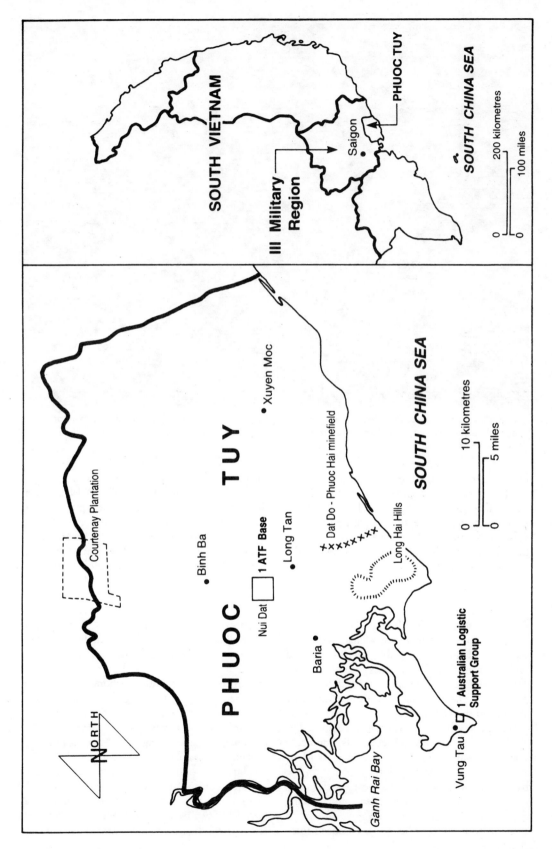

Civil Aid Programs were used during the Vietman War to gain the support of the local population. Here an Australian doctor examines a Vietnamese patient during a Medical Civil Aid Program, Binh Gia, South Vietnam, 1967. (AWM P1598.005)

namese backers gained ground rapidly, until by 1964 the Saigon government was in crisis. Diem had requested non-specific aid in 1961 and 1962, intended to improve the security situation and help strike at the roots of the revolution by improving the living standards of the rural population, and the Australian government had responded in 1962 with the dispatch of the Australian Army Training Team Vietnam (q.v.).

In December 1964 the South Vietnamese Prime Minister, Tran Van Huong, asked the Australian government for increased military assistance, which was followed a few days later by renewed pressure from the United States for an increased commitment to the defence of South Vietnam. Following further discussions with the Americans and within the Defence Committee (q.v.) and the government, a battalion of troops was offered to Washington for service in Vietnam. For domestic political consumption, Prime Minister R. G. Menzies (q.v.) tried to extract a formal request for the battalion from the South Vietnamese, which in due course was received, but the sequence and wording demonstrated clearly that the government of Prime Minister Phan Huy Quat had been pres-

sured into sending the request, and much was to be made of this in subsequent political debate both inside and outside the parliament. On the other hand, several requests for aid had been made by Diem's successors in 1963–64, and the Australian military presence in South Vietnam took place with the specific agreement of the South Vietnamese government.

In June 1965 1RAR and supporting units were dispatched to South Vietnam to serve alongside the American 173rd Airborne Brigade (Separate) in Bien Hoa. The battalion was entirely regular in composition. It operated throughout the III Corps area and was maintained and supplied largely by the Americans, seeing heavy fighting and suffering 23 killed in action. There were considerable disagreements over doctrine and tactics, and the Australian battalion was poorly served by the equipment it had brought from Australia, much of it being old, worn, or otherwise unsuitable for the conditions encountered. In March 1966 the government announced that 1RAR would be replaced by a Task Force (another name for a brigade) of two battalions and supporting arms and services, and that it would be assigned its own area of operations, Phuoc Tuy province (see Map 27). A self-sufficient task force would be capable of operating independently of the Americans, and would thus avoid some of the conflicts which had marked 1RAR's first tour, while the province concerned was chosen because it was away from the Cambodian border, could be resupplied, and evacuated, by sea, and enabled the Australians to concentrate their efforts in a single area. This would provide a test for the different tactics and techniques which the Australians brought to countering insurgency, and would provide a focus for public attention in the Australian government's efforts to sell the war to the Australian people.

The 1st Australian Task Force arrived between April and June 1966 and constructed a base at Nui Dat, while the 1st Australian Logistic Support Group was established gradually at the port of Vung Tau. In August a company of 6RAR was involved in a heavy action near Xa Long Tan, in which 18 Australians were killed and 24 wounded while inflicting 245 dead on an enemy force of at least regimental strength. Although a relatively minor engagement in comparison with other wars, or even with the big battles fought in the northern provinces, Long Tan was significant in asserting the Task Force's dominance on the province; although there were many bigger and more sustained operations against Viet Cong and North Vietnamese

units in the ensuing years, the Australians' presence in Phuoc Tuy was not again fundamentally challenged. In late 1967 a third battalion was added and armour was deployed in support. Although Phuoc Tuy was the primary area of Australian operations, units of the Task Force were available for deployment anywhere in the III Corps area, and during 1968 the Australian battalions spent a large part of their time on operations in neighbouring provinces against the enemy's Tet and subsequent offensives. It was during this period that 1RAR (on its second tour) and 3RAR with artillery and tank support fought off large-scale enemy attacks in the battle of Fire Bases Coral and Balmoral in May 1968. In June 1969 5RAR (also on its second tour) fought the other well-known major engagement of the Australians' war, at the village of Binh Ba. These spectacular, large-scale actions, however, were generally atypical of the fighting in Phuoc Tuy.

In keeping with Australian doctrine, companies spent much of their time on patrol, or in cordon and search operations, designed to put pressure on enemy units and deprive them of their links with the people. To that end, operations like Cung Chung, conducted over a period of months between June 1970 and February 1971 in four phases, and placing constant pressure on the Viet Cong infrastructure, were far more significant in terms of restoring government control in the

After a month in the field on patrol, men of 5RAR wait by the airstrip at Xuyen Moc, South Vietnam. The Caribou in the background, of No. 35 Squadron, is landing to transport them to Vung Tau, October 1969. (AWM BEL/69/741/VN)

province. By 1971 the provincial enemy forces had been pushed out of Phuoc Tuy, and relied on reinforcements from North Vietnam to make up their numbers, while Highway 15, the main route between Saigon and Vung Tau, was open to unescorted traffic. Of the nine battalions in the Royal Australian Regiment (q.v.), which carried the brunt of both the fighting and the casualties, seven served twice in Phuoc Tuy, while 8RAR and 9RAR, raised later than the others, served once each.

The RAAF sent No. 9 Squadron, equipped with Iroquois helicopters (q.v.), No. 2 Squadron, which flew Canberra bombers (q.v.) out of Phan Rang, and No. 35 Squadron, which provided logistic support with its Caribou (q.v.) aircraft, and was known accordingly as 'Wallaby airlines'. Air force personnel took part as well in logistic support flights from Australia using Hercules (q.v.) aircraft, aeromedical evacuation flights, and on attachment to US Air Force squadrons as forward air controllers and in fighter squadrons, in which latter capacity a number took part in operations over North Vietnam. The RAN sent a destroyer to provide naval gunfire support and assist in interdiction of seaborne traffic as part of the US 7th Fleet, a clearance diving team for port security and explosive ordnance demolition duties, and a helicopter flight for service with an American army assault aviation company, while a small number of navy pilots also flew on attachment to the RAAF's No. 9 Squadron.

About 50 000 Australians served in Vietnam between 1962 and 1973; the total number killed was 519, with total casualties broken down by service as follows: RAN killed in action 4, non–battle deaths 4, wounded 20; RAAF killed in action 6; non–battle deaths 11, wounded 30; Australian Army killed in action 415, non–battle deaths 79, wounded 2348; civilian deaths 7.

The commitment to Vietnam was the subject of the greatest level of domestic political dissent seen in Australia since the conscription referendums of the First World War, although Vietnam was not the only cause of such political activity. Initially with the commitment of 1RAR in mid-1965, and for some years thereafter, public opinion was strongly in support of government policy in Vietnam, and when the leader of the opposition, Arthur Calwell, declared that the 1966 federal election was to be fought specifically on the issue of Vietnam, the Labor Party was handed its biggest electoral defeat in decades. Conscription (q.v.) had been introduced in November 1964 not, as is sometimes claimed, for service in Vietnam, since at that stage no regular units had been committed, but in response to the worsening strategic situation in the region as a whole. At that stage the Australian government was far more concerned with the heightened tensions between Indonesia and Malaysia over Confrontation (q.v.), following the landings by Indonesian regulars in peninsular Malaya in August and September, and with the fear that the Indonesians might extend the fighting to the border with New Guinea. The expansion of the army to cope with a number of commitments simultaneously, and the decision to do this with national servicemen rather than improving pay and conditions in the army in order to attract more volunteers, caused the government to reintroduce conscription, this time for overseas service. A very small number of national servicemen went to Borneo with 4RAR in 1966, at which time the first conscripts were dispatched for service with the Task Force in South Vietnam. The first national serviceman, Private Errol Noack, was killed on operations with 5RAR in May 1966, and 202 conscripts were to die in Vietnam from all causes (as against 235 regulars). As a general policy, battalions were maintained at a strict 50–50 mix of regulars and national servicemen, and some national servicemen were commissioned after graduating from the Officer Training Unit at Scheyville (q.v.).

Apart from the historical basis of opposition to conscription for overseas service in Australia (and which, in the case of Arthur Calwell, for example, was direct and personal and led straight back to 1916–17), the main aspect of the scheme operating in the 1960s which made it unpopular was the inequality of sacrifice it imposed. Of the roughly 800 000 men of military age who registered for National Service between 1965–72, only about 63 000 were actually called up into the army and only 17 424 actually served in Vietnam. The system of balloting for national service quickly led to its being dubbed a 'blood ballot' or the 'lottery of death', an impression not helped by the foolish decision to televise the early draws like some sort of bizarre, and very grim, Lotto game. Indeed, it is difficult to avoid the impression that successive Liberal governments in the 1960s handled the public debate over Vietnam with increasing ineptitude. Especially after the 1969 federal election, which Labor lost again but by a reduced margin, the debate over the Vietnam War was increasingly dominated by those who were in opposition to government policy. By 1970–71, with successive, highly visible moratorium demonstrations in all capital cities and the increasing radicalism of many of the protestors, drawn increasingly but not entirely from the ranks of university staff and students, the government had lost the argument, at least on the streets, while its efforts to apprehend the small number of young men who defied their registration or call-up notices degenerated into farce. While impressive in the numbers of ordinary Australians they were able to mobilise, the moratoriums were a reflection of the shift in electoral opinion rather than a prompt to it; by 1970 many Australians, like many Americans, had concluded that despite such local successes as their troops achieved, Vietnam was an unwinnable war. For many even of those whose opposition took them on to the streets, conscription rather than the conduct of the war was the main issue, and when the announcement was made that the Task Force would be withdrawn in 1971, a great number of these people dropped out of the anti-Vietnam movements. Elsewhere in the Western world, in countries such as Britain and Germany which had no involvement in Vietnam, the same type of anti-war, anti-American, anti-establishment protests were going on, and it seems unlikely that Vietnam, by itself, was the decisive factor which led to the election of the Labor government of E. G. Whitlam in December 1972, a result which in any case was much less decisive than memory now suggests.

Like most unpopular or lost wars, Vietnam has left its own legacy of bitterness and mistrust, especially among that proportion of the veteran community who feel aggrieved at their treatment on their return from war service. This disaffection is often bound up in the twin issues of Agent Orange (q.v.) and post-traumatic stress disorders, and has been articulated most consistently through the Vietnam Veterans' Association of Australia (q.v.), whose membership at its height was probably no more than 5000. Recent signs of popular acceptance, such as the 'Welcome Home' March in 1987 and the dedication of the Vietnam Veterans' National Memorial (q.v.) in Canberra in 1992, together with the network of veterans' counselling centres which has been established, have gone some way towards incorporating Vietnam and its veterans into the national pantheon of Anzac (q.v.), but some veterans are unlikely ever to be reconciled fully to the national community which they feel betrayed them. At a wider level, the Vietnam War marked a stage in the gradual lessening of Australia's reliance on the American security connection, while the boost which participation in the anti-war movements gave to many of their young participants strongly influenced the growth and shape of the women's, environment and Aboriginal political movements of the 1970s and 1980s. The failure of the Western involvement in Vietnam in the 1960s marked the end of a period of military and diplomatic involvement in Cold War policies in Asia, the full consequences of which are still playing themselves out in the region.

VOLUNTARY ORGANISATIONS see **AUSTRALIAN COMFORTS FUND; AUSTRALIAN RED CROSS; LEGACY**

VOLUNTEER DEFENCE CORPS (VDC) was formed by the RSL in July 1940 as a home defence organisation to be composed of ex-servicemen and modelled on the British Home Guard. In May 1941 it came under the control of the Military Board (q.v.) and was given the role of training for guerrilla warfare, providing static defence of each unit's local area, protecting key points, and providing local intelligence. In February 1942 the government, seeking simultaneously to curb RSL control of the VDC and to head off growing public pressure for the arming of citizens in a people's army, expanded the VDC's strength, constituted it as a corps of the CMF, and incorporated it into the National Security Act (q.v.) Membership was open to men between the ages of 18 and 60; with the enlistment of young men working in reserved occupations (q.v.), the average age of VDC members fell to 35 and the Corps reached a peak strength of almost 100 000. There was a small number of full-time members, but most trained for an average of six hours a week at night or on weekends. The VDC's primary role gradually changed from providing static defence to operating anti-aircraft batteries, coast defences and searchlights in emergencies. In May 1944, with the threat of invasion seemingly over, members of inland VDC units were freed from their responsibility to attend regular parades so that they could concentrate on rural production. On 24 August 1945 the VDC was officially disbanded.

VOYAGER (I), HMAS see **'V' AND 'W' CLASS DESTROYERS**

VOYAGER (II), HMAS The first of three *Daring* Class destroyers (q.v.) to be laid down was HMAS *Voyager* in 1949. Built at Cockatoo Island (q.v.) Dockyard in Sydney, *Voyager* was launched on 1 May 1952 by Mrs Pattie Menzies, the wife of the Prime Minister R. G. Menzies (q.v.). After a long period of trials and evaluations as the first-of-class, *Voyager* was commissioned on 12 February 1957 under the command of Captain G. J. Crabb. *Voyager* then spent the next five years either serving in the Commonwealth Far East Strategic Reserve (q.v.), or preparing for her return there under a succession of commanding officers. On 31 January 1963, *Voyager* sailed from Sydney for her sixth deployment to the Strategic Reserve, which was to last until July of that year. She was commanded by Captain Duncan Stevens who had been given four previous seagoing commands. On her return to Australia, *Voyager* proceeded to Williamstown Naval Dockyard (q.v.) to begin a refit that continued until the beginning of 1964.

Although a number of technical defects from the refit remained, *Voyager* sailed for the East Australia Naval Exercise Area off Jervis Bay on 6 February in company with the aircraft carrier and RAN Flagship, HMAS *Melbourne* (II) (q.v.). The carrier, commanded by Captain John Robertson, had also recently completed a long refit at Garden Island (q.v.) Dockyard. After spending part of the weekend at anchor in Jervis Bay, both ships proceeded to sea on the morning of Monday 10 February and conducted independent exercises until dusk when a night-flying exercise was scheduled. *Melbourne's* air group would fly from the Naval Air Station at Nowra to conduct 'touch and goes' on the flight-deck of the carrier while *Voyager* acted as planeguard destroyer. Her task was to recover the crew from any aircraft that might ditch into the sea during the exercise.

At 1830 hours, *Voyager* and *Melbourne* were 20 miles to the south-east of Jervis Bay in very deep water. There was a low easterly swell, smooth seas and light variable winds. There was no moon. The night was clear with visibility estimated at 20 miles. Both ships were 'darkened' for the exercise with only operational lighting visible to the other ship. The two side lights in the carrier were dimmed so that their visibility range was around one nautical mile with the unassisted eye. As captain of the carrier, Robertson was the Officer in Tactical Command of the exercise.

Sunset was at 1945 hours. At 1950 hours the ships turned together to the flying course and *Voyager* assumed her station astern of the carrier. As the winds were light and variable, Robertson needed to alter *Melbourne's* course and speed so as to get the maximum amount of headwind across the flight-deck of the carrier. Darkness soon fell and the silhouettes of the ships were nearly impossible to see. All that could be seen with confidence from the carrier was the destroyer's navigation lights.

After comparing the wind on a number of different courses, Robertson ordered both ships to turn together to a course of 020 degrees from 060 degrees. This command was executed by voice-circuit radio. While the turn back to 020 degrees was in progress, Robertson signalled to *Voyager* that the carrier would operate aircraft on that course. This signal was acknowledged by the destroyer. On receipt of this signal, *Voyager* (which was then ahead of *Melbourne*) was to take up planeguard station and

manoeuvre astern of the carrier. The destroyer would then proceed with the carrier on the flying course of 020 degrees. The manoeuvre for *Voyager* to take her new station was a familiar one: the destroyer would turn away from the carrier which would pass ahead before the destroyer fell in 1000 yards astern of *Melbourne*. As *Voyager* was fine on *Melbourne's* starboard bow, Robertson expected *Voyager* to alter course to starboard before assuming her station.

The destroyer initially did as expected and turned to starboard. For the next 20 seconds, *Voyager* appeared to be slowly turning away from *Melbourne* at a distance of 1300 yards. She then altered course to port which led Robertson, who observed all of *Voyager's* movements from the carrier's starboard bridge-wing with the aid of binoculars, and the carrier's bridge staff, to assume that the destroyer was doing a manoeuvre that was later referred to as a 'fishtail'. This manoeuvre, which was rather less precise, would allow *Voyager* to reduce her speed through the water while waiting for the carrier to overtake. As the destroyer was more than 1000 yards from *Melbourne*, broad of the starboard bow and had appeared to return to the new course of 020 degrees, there was no danger and Robertson had little reason to be concerned. For more than one minute Robertson constantly watched *Voyager* and continued to believe that she was conducting a fishtail — with the expectation that the destroyer would alter course away from *Melbourne* at any time in that period.

While their ship altered course to port towards the carrier, the captain and the navigator of *Voyager* were conferring at the chart-table, probably discussing the interpretation of signals sent from *Melbourne* using the tactical manuals. The Officer of the Watch, an RN officer on exchange service in Australia, Lieutenant David Price, was responsible for keeping a close watch on *Melbourne*. Two minutes after altering course to port, *Voyager* was still turning towards the carrier. Her range from *Melbourne* was only 1000 yards. The next 15 seconds were absolutely crucial as the ships continued to close at more than 30 knots. If the bridge staff in the destroyer were to avoid a collision, action had to be taken in that short period. *Voyager* persisted with her turn. As neither ship had taken avoiding action, the two ships were now 800 yards from each other and a collision was inevitable.

Thirty seconds after 2055 hours, Commander Jim Kelly, *Melbourne's* navigator, looked towards *Voyager* and exclaimed 'Christ! What is *Voyager* doing.' He took a radar range of the destroyer and found she was 600 yards distant. Kelly ordered 'half astern both engines'. This order to the engine room was increased by Robertson who ordered 'full astern both engines' several seconds later.

The possibility that *Voyager* was not aware of her position relative to *Melbourne* did not occur to Robertson. Thus, no sound signals or messages were sent to warn *Voyager* that she was heading into danger. The carrier had little chance of altering either her course or speed in the distance available. When *Voyager* was just 80 yards away, *Melbourne* had perhaps lost a single knot of speed. Tactical Operator Everett on the carrier's bridge voiced what was in the mind of everyone, 'We are going to hit her.' They all braced themselves for the impact.

On the port bridge-wing of *Voyager*, the side of the ship nearest to *Melbourne*, the lookout was Ordinary Seaman Brian Sumpter. This was his first time at sea. He did not have any experience on which to make a judgment, but it was apparent even to him that there was something very wrong. He shouted 'Bridge!' and turned around to see Lieutenant Price looking through binoculars. Price then dropped the binoculars and stared at the approaching carrier, as if mesmerised. The shout from Sumpter brought Captain Stevens back on to the bridge from the charthouse. It took Stevens a moment to regain some night vision and establish the position of his ship relative to the carrier. At approximately 30 seconds before the collision he ordered 'Full ahead both engines. Hard a-starboard.' He then turned and said 'Quartermaster, this is an emergency. Pipe Collision Stations.' Price, who had the bridge microphone in his hand, echoed his captain's orders to the wheelhouse on the deck below in an attempt to have *Voyager* pass ahead of *Melbourne*. It was all far too late. When the order was given in *Voyager*, there was only 250 yards between their bows. A collision was imminent.

The huge bulk of the carrier, which had loomed overhead in the dark, crashed through the destroyer at right angles. The point of impact in *Voyager* was the after end of the bridge and the operations room, which was one deck below. On *Voyager's* bridge, nearly everyone was killed instantly. Following the collision, *Melbourne* slowed and stopped dead in the water as the two sections of the severed destroyer slid down the carrier's sides. The bow section, which had scraped *Melbourne's* port side as it rolled over, floated on its starboard side for five minutes. Destabilised by the weight of two large gun turrets, the bow section then rolled completely over and floated keel up for a further five minutes before it sank. The after section defiantly remained afloat for another three hours.

Rescue operations began immediately. For the next four hours the men of *Melbourne*, and some of the survivors of *Voyager*, worked to recover those in the water. Two hundred and thirty-two officers and men were recovered from the water and the after section of the ship. The bow section had contained 225 of the ship's company of 314. Eighty-two men had been killed. At 2100 hours, Robertson sent an emergency signal from *Melbourne*: 'Have been in collision with *Voyager.*' At 2114 hours, the Damage Control Officer in *Melbourne*, Lieutenant-Commander George Halley, informed Robertson that the carrier had suffered major damage but was in no danger of sinking. Fourteen minutes later, Robertson sent his second signal: '*Voyager* has lost her bows but is still floating. Am rescuing survivors. Sea calm.' The third signal sent at 2140 hours, nearly three quarters of an hour after the collision, read: '*Voyager* has settled by the bows. Waterline at forward end of torpedo tubes. She still has lights onboard. Further information will be signalled when available.' HMAS *Kimbla* was instructed to 'Proceed at best available speed and prepare to tow *Voyager.*'

Shortly after midnight, Captain Robertson was forced to send another signal. It was short and horrifying: '*Voyager* has sunk.' Preparations for towing *Voyager* back to Sydney were cancelled. She had gone. With *Voyager* sunk and the survivors recovered, *Melbourne* made her way slowly back to Sydney. It would take one and a half days.

Following the collision, Prime Minister Menzies convened a Royal Commission under Sir John Spicer, Chief Justice of the Industrial Court. Robertson was relieved of command to attend the inquiry. After 50 days of hearings, Spicer concluded that the collision was caused by the failure of *Voyager's* bridge to maintain an effective lookout although the reasons for this were inexplicable. While he did not apportion blame to *Melbourne's* officers, Spicer was critical of the performance of Robertson, Kelly and the carrier's officer-of-the-watch, Sub-Lieutenant Alex Bate. The Naval Board (q.v.) agreed that *Voyager* was to blame and that ultimate responsibility rested with Stevens. Robertson was not, however, reappointed in command and resigned in October 1964. He was subsequently contacted by Peter Cabban, *Voyager's* executive officer during her 1963 Far East deployment, who claimed that Stevens had a severe drinking problem which may have contributed to the collision. Cabban left *Voyager* and the navy several weeks before the collision. Robertson conveyed the substance of Cabban's allegations to John Jess, a government backbencher in the federal parliament. Jess lobbied for another inquiry into the collision. Mounting public pressure led to Prime Minister Holt convening a second Royal Commission in 1967. After hearing evidence for 85 days, this inquiry concluded that the sole cause for the collision lay with

Voyager and that criticism of Robertson was unfounded. The commissioners also found that Stevens's drinking was unwise given that he suffered from an active duodenal ulcer and that he was unfit to command *Voyager* by virtue of his medical condition. They were adamant that Stevens's drinking had not contributed to the collision.

The loss of HMAS *Voyager* was a key event in the development of the RAN, not as a direct result of the collision or its causes, but as a consequence of its long and controversial aftermath when the professional culture and ethos of the navy was subjected to unprecedented public scrutiny.

T. R. FRAME

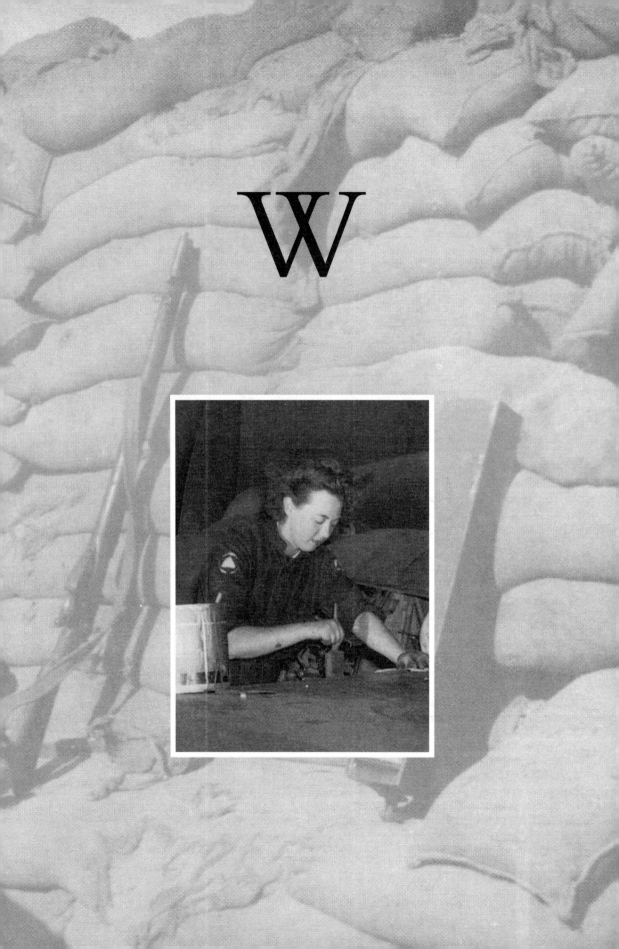

W

'W' CLASS DESTROYERS see **'V' AND 'W' CLASS DESTROYERS**

WACKETT, CAC (2-seat intermediate trainer). Wingspan 37 feet; length 26 feet; maximum speed 110 m.p.h.; range 425 miles; power 1 × Warner 175 h.p. engine.

The Wackett was designed by the Commonwealth Aircraft Corporation (q.v.) as an intermediate trainer flown by pilots between the Tiger Moth and the Wirraway (qq.v.). The first Wackett entered service with the RAAF in 1940 and 200 were used by flying training schools until 1946. The Wackett was a successful aircraft and the experience gained in designing a trainer was valuable in the creation of the Winjeel (q.v.).

WACKETT, Wing Commander Lawrence James (2 January 1896–18 March 1982). Born in Townsville, Queensland, Wackett served with the Australian Flying Corps from 1915, and transferred to the RAAF after the war, remaining with the service until 1930. His involvement with aeronautical engineering began during the First World War, when he repaired and reconstructed damaged aircraft and pioneered aerial supply dropping to Australian troops. After completing a science degree he set up the RAAF Experimental Section, which he commanded from 1924 to 1930 and where he designed, built and flew the Widgeon amphibious and Warrigal trainer aircraft, though these were not used widely. The twin-engined Wackett Gannet, produced by Wackett in small numbers after he had left the RAAF, was more successful, but his attempts to single-handedly create an Australian aircraft company required further capital investment to succeed. This came in 1936 when the Chairman of BHP, Essington Lewis (q.v.), and other business figures, concerned at the threat from Japan, formed the Commonwealth Aircraft Corporation (CAC) (q.v.) to mass-produce military aircraft in Australia and invited Wackett to become manager. Wackett then visited the United Kingdom, Europe and the United States on behalf of CAC to find a suitable aircraft design for production. After experiencing arrogant treatment at the hands of British manufacturers, who demanded CAC make a binding agreement to produce only British designs, Wackett found a more businesslike attitude and a more suitable aircraft design in the United States. On Wackett's reports, CAC chose the North American NA33 design and manufactured it under licence as the Wirraway (q.v.). During the Second World War Wackett was involved in the design of the Wackett trainer (q.v.) and the Boomerang (q.v.), a fighter design conceived and constructed while Australia was under the threat of Japanese invasion. Wackett was knighted in 1954, and retired as managing director of CAC in 1960. An accident in 1970 rendered him paraplegic, and he turned his inventive skills to the development of equipment for the disabled, becoming an adviser to the spinal injury unit at the Royal North Shore Hospital in Sydney.

WAGGA, HMAS see *BATHURST* **CLASS MINESWEEPERS (CORVETTES)**

WAKE, Nancy Grace Augusta (30 August 1912–). Born in Wellington, New Zealand, Wake was raised in Sydney, which she left in 1932 to travel to Europe. She married a Frenchman, and from 1940 was an active member of the French resistance, for which she was highly decorated. She returned to Australia after the war, and in 1994, reiterating her disappointment that she had not been awarded an Australian medal — 'I think the lack of recognition stinks and shows petty-mindedness' — she sold her medals for a large sum; they are now displayed at the Australian War Memorial.

WALERS was a term coined in India in the 1840s to describe horses from New South Wales, and was later applied more generally to Australian horses abroad. Australian horses were sold to the Indian Army from 1834 until just before the Second World War. The first Walers to be used in war by Australian troops were the 224 horses which went from Australia to the Sudan. The British and Australians used 37 245 Walers in the Boer War, but they compared poorly with the Walers used by the Indian Army. Their poor performance seems to have been due partly to the low prices paid for horses by British officials and partly to the lack of time to acclimatise them when they reached South Africa. The 135 926 Walers used by various armies in the First World War were more effective. Their very successful use by the Light Horse in the Near East (see PALESTINE CAMPAIGN) depended heavily on the work of the Australian Remount Unit (q.v.), who prepared Walers for local conditions. None of the Walers used in these three wars was returned to Australia.

(See also ANIMALS.)

A. T. Yarwood, *Walers: Australian Horses Abroad* (Melbourne University Press, Melbourne, 1989).

WALKER, Allan Seymour (19 June 1887–8 January 1958) was educated at the University of Sydney and set up medical practice at Parkes, New South Wales, in 1913. He served during the First World War in Australia as a captain in the Australian Army Medical Corps. He continued his involvement with the CMF after the war, being appointed honorary major in 1937. After undertaking postgraduate work in pathology, Walker practised in Sydney before proceeding to England in 1926 for further study, and worked in various capacities at the Royal Prince Alfred Hospital between 1920 and 1958, teaching simultaneously for much of that time at the University of Sydney. His major interests were in pathology and neurology, and this was reflected in his publications. During the Second World War he served as a lieutenant-colonel in the Middle East with the 2/1st Australian General Hospital, and then as a consultant physician on Land Headquarters with the rank of colonel. He was twice mentioned in despatches, and was discharged in June 1944. The official historian, Gavin Long, invited Walker to write the medical series of the official history in 1944, but Walker declined, considering himself obliged to return to teaching. Long's second choice, R. M. Downes, died in March 1945 and Walker was then persuaded to take on the job. He wrote three volumes of the series himself, *Clinical Problems of War* (1952), *Middle East and Far East* (1953), and *The Island Campaigns* (1957), despite periods of indifferent health and financial loss occasioned by the task. He had completed much of the work on the fourth volume on the RAN and RAAF when ill health forced him to resign in 1956. The volume was brought to publication by others. During the course of his involvement in the official history he was a central figure on the Official Medical Historians' Liaison Committee, comprising his opposite numbers from the United States, British and Commonwealth official historical teams, and in 1949 chaired the meeting of the committee which convened in Canberra. Long wrote of him that he was 'not only a medical scientist, whose learning was both wide and deep, but a man of rare literary ability, widely read, imaginative, and master of a fastidious and fluent literary style'. His volumes are well written and clearly organised, and contain much valuable military detail not found in the other Australian series. Like his predecessor, A. G. Butler (q.v.), Walker wrote with future practitioners of military medicine in mind, but his volumes have a wider appeal, evidenced by successive reprintings of the early volumes in the years immediately after publication.

WALKER, Lieutenant-General Harold Bridgwood (26 April 1862–5 November 1934). Commissioned in the Duke of Cornwall's Light Infantry in May 1884 after a period at Cambridge, Walker served in the Sudan and India before going to South Africa during the Boer War. A period in Ireland followed, and in 1914 he was appointed BGGS in India. Not long after this, his Indian and South African connections led to his appointment as chief of staff to Lieutenant-General Birdwood (q.v.) in his new role commanding the Anzac Corps. Walker opposed the Anzac landings, believing that they had little chance of success, but threw himself into the tactical command and training of troops, at which he excelled. He was the first officer of the headquarters staff to land at Anzac Cove, and by mid-morning had assumed command of the New Zealand Brigade following the indisposition of its commander. Walker's Ridge, on the left of the Anzac position, was named for him. Within a few days he had moved to command of the 1st Australian Infantry Brigade and, following Major-General W. T. Bridges's (q.v.) fatal wounding in May, took temporary command of the Australian Division. Known as 'Hooky', he alternated between brigade and divisional command until Lieutenant-General Legge's (q.v.) departure from Anzac in late July. He led the Australians of the 1st Division at Lone Pine and thereafter, and although wounded slightly in September, and more severely later the same month, he was characteristically always well forward among his men.

Returning to the division in Egypt in March 1916, he was confirmed in command in the rank of major-general and led the 1st Division through the bloody fighting at Pozières and the battle for the Hindenburg Line in the first half of 1917. He relinquished command of the division in July 1918 in favour of Major-General Sir Thomas Glasgow, as part of the 'Australianisation' of the Australian Corps. He went to a divisional command in the strategic backwater of the Italian front in July 1918, and soon after the war's end commanded all British forces in that theatre. He held a number of commands in England and India after the war, and retired in 1928. One of a small number of regular officers available to the AIF early in the war, Walker proved an important influence both in command and administration. Reluctant to squander the lives of the men under his command, he was more mindful than many British officers of the particular national obligations which command of dominion soldiers carried, and was widely respected by the men he led, his departure being particularly regretted within the 1st Division.

WALLAROO, HMAS see **BATHURST CLASS MINE-SWEEPERS (CORVETTES)**

WALLAROO, HMS see **AUXILIARY SQUADRON, AUSTRALIA STATION**

WALLER, Captain Hector MacDonald Laws (4 April 1900–1 March 1942) graduated from RANC Jervis Bay in 1917 and served with the Grand Fleet in the last year of the First World War aboard HMS *Agincourt*. He specialised as a signals officer and served with the RN in the Mediterranean during the Spanish Civil War. Between 1939 and 1941 he commanded the destroyer HMAS *Stuart* and was Captain (Destroyers) of the 10th Destroyer Flotilla in the Mediterranean. He saw action at the battles of Calabria and Cape Matapan, provided ship-to-shore support for the army during the First Libyan campaign, and took part in naval action during the evacuation of the ground forces from Greece. For his services in the Mediterranean Waller was awarded the DSO, a bar for Matapan, and two mentions in despatches. In October 1941 he was given command of the cruiser HMAS *Perth*. In company with the USS *Houston*, in early 1942, Waller attacked a Japanese convoy in the Sunda Strait off Java, engaging enemy forces considerably larger than themselves. The Japanese lost four transports and a minelayer and suffered damage to three cruisers and nine destroyers, but both the Allied ships were lost. Waller was last seen on the bridge of the *Perth*. He was awarded a posthumous mention in despatches. Highly regarded within the Australian service, his death in action was regarded as a significant loss to the still small Australian naval officer corps and one of the *Collins* Class Submarines (q.v.) has been named for him.

WALLER, HMAS see *COLLINS* CLASS SUBMARINES

WALYER see **ABORIGINAL ARMED RESISTANCE TO WHITE INVASION**

WAR ART Art on the theme of war has a history almost as long as picture-making. War scenes were painted in the paleolithic period and images of warriors can be found in Assyrian, Egyptian, Greek and Roman art. Likewise, images of fighting occur in Aboriginal art created before White settlement. Much early colonial art includes reference to the activities of the British forces in Australia and to the pre-Federation armed forces. War art extends to contemporary work that reflects on past wars, and to that which comments on the effect of past and present wars on people today. It includes work that makes a statement about war in general.

Artists have depicted war for a variety of reasons, none of which are mutually exclusive. Many made drawings in trenches or POW camps to while away the time and distract their minds from the conditions around them. Some, such as Horace Moore-Jones at Gallipoli, portrayed the landscape to provide useful information about the terrain to assist their compatriots. Others, such as George Lambert in the First World War and Murray Griffin and Louis Kahan in the Second, made drawings of fellow soldiers as gifts for them to send home to relatives — and as a means of achieving a small measure of immortality. Still others created visual jokes about the situation, drawings to entertain their mates and to help them momentarily forget the horrors of war. Some made paintings because visual ideas about the war preoccupied them: Lambert remarked:

> I must go on painting Palestine subjects, not because there is a readier market for them than my pre-Palestine work, but because they are fixed in my brain by the impressions received out there.

Some wanted to record the events that were taking place so that those back home, and future generations, would know what had happened.

The Australian official war art scheme began during the First World War. It had its conceptual origins in *The Anzac Book* (1916), a book containing stories, poems and illustrations contributed by the men on Gallipoli, and edited by the official war correspondent, and later historian, C. E. W. Bean (q.v.). The illustrations were drawn in the trenches, with the contributors, David Barker, Frank Crozier, Otho Hewitt and Cyril Leyshon White, working with whatever material was available. These artists were not official war artists, but serving soldiers taken off their usual duties to work on the project. This set a precedent for the secondment of artists to depict the activities of Australian troops during the war. In 1917 Bean established a War Records Section which collected papers, photographs and battlefield relics. This section also commissioned artists to portray the war. Bean believed that artists who had participated in the fighting would provide the best portrayals of events, and consequently sought applications from serving soldiers. The five servicemen selected by Bean — George Benson, Frank Crozier, Will Longstaff, Louis McCubbin and James Scott — received training as camoufleurs (i.e. camouflage painters) and became the Australian War Records Section artists. They were provided with sketch pads and depicted events around them when time permitted. J. S. McDonald was appointed and trained as a camouflage artist, but he became ill and never took up his position. Daryl Lindsay was a medical artist at the Queen Mary Hospital at Sidcup, Kent, in England.

Official War Artist James Quinn at Mont St Quentin, France, September 1918. (AWM E03326)

While Bean was establishing his scheme, the Australian expatriate artists Will Dyson and Arthur Streeton asked for permission to go to the front to work as war artists. Streeton lobbied for a program that would employ a range of artists, to be run along the lines of those established by the British (in July 1916) and the Canadians (in November 1916). His list of names of potential war artists included that of one woman, Dora Meeson. The official war art scheme, run by the Australian High Commission with Bean as adviser, developed out of this artist-based campaign.

Will Dyson had a well-established reputation as a war cartoonist when, in December 1916, he received permission to go to France to make sketches of the Australian troops. He did so in a civilian capacity, without pay until May 1917. When the Australian war art scheme began, he became the first official war artist. Dyson's terms of employment were vague, but he regarded his output as government property, and handed most of the drawings he made at the front to the Australian government at the end of his tour of duty. In addition to Dyson, nine of Australia's best male expatriate artists living in Britain, with a range of specialist interests and abilities, received appointments: George Bell, Charles Bryant, A. Henry Fullwood, George Lambert, Fred Leist, John Longstaff, H. Septimus Power, James Quinn and Arthur Streeton. No artist was appointed in, or from, Australia. Generally, the artists were expected to produce at least 25 works during the course of three months. They were free to draw or paint whatever subject, in whatever medium, on whatever size paper or canvas, in whatever style suited their talent and interest best. But they were expected to work within their specialist interests: Streeton, for instance, to paint landscapes, and Longstaff and Quinn to paint portraits. They were appointed as honorary lieutenants and initially were paid £1 a day for the duration of their contract, but by December 1917 they received £2 a day. This was more than the pay received by officers of an equivalent rank (lieutenant) in the AIF. All, except Lambert, served on the Western Front where they were attached to a division of the AIF. Lambert served in Palestine in 1918, and accompanied Bean on the Australian Historical Mission to Gallipoli in 1919, at which time he was granted the honorary rank of captain.

Will Dyson's drawings from the Western Front express compassion for the serviceman; they vividly portray the soldiers' exhaustion as well as their wry humour. Many artists came from a landscape tradition and were inspired by a love of nature. Moreover, the terrain was a major influence on events in this war: the topography of Gallipoli was as much the enemy as were the Turks; the unending flatness of the area around Ypres had a decisive impact on the action on the Western Front; and the vastness and intense heat of Palestine provided a constant challenge to those serving there. But the ways in which the artists responded to the land and the messages they conveyed about it were diverse: many like Fullwood and Leist portrayed the horrific scarring of the landscape, others, like Streeton, showed how the beloved landscape had been littered with the refuse of war, and yet others, such as Lambert, depicted the irony of war taking place in scenes of beauty.

Women were excluded from participating in this war in any capacity apart from nursing and related services, and on the home front. As a result women were not appointed as war artists, and Streeton's recommendation to appoint Dora Meeson was not taken up. However, several women living and working overseas during the war depicted the events as they saw them. Iso Rae was in France during the First World War, employed in the YMCA (q.v.) camp at Etaples, and portrayed aspects of military life around her. Evelyn Chapman visited the battlefields near Villers-Bretonneux in 1919–20 with her father, Francis Chapman, who was attached to the New Zealand War Graves Commission; while she was there Chapman painted a series of works depicting battlefield ruins. Vida Lahey lived in London during the war, nursing her brothers, and at the end of the war she captured the varying emotions of those who survived the war in *Rejoicing and Remembrance* (1918).

In June 1918, the first exhibition of Australian war art, 'Australian Official War Pictures and Photographs', was shown at the Grafton Galleries, London. It included works by Bryant, Crozier, Lambert, Leist, Power and Quinn. Later, in November 1918, the Royal British Colonial Society of Artists organised another exhibition of war art, 'War and Peace', at the Royal Academy, London. All the Australian war artists had works in this show, as did other Australian artists who lived in London: George Coates, Tom Roberts, Thea Proctor and Bertram Mackennal. Paintings and sculpture by British, Canadian and New Zealand artists were also included, but the Australians had the widest representation. P. G. Konody, the *Observer*'s critic, noted that the Australian artists had a 'straightforward sturdiness and honesty of purpose' and that their general aim was a faithful representation of the thing seen, 'stated with vigour and directness, without affectation, without deliberate striving for stylishness, and even with a certain disregard of the means employed'.

Dyson and Streeton also held individual war art exhibitions: Dyson's 'Australia at War' was held at the

Leicester Galleries, London, in January 1918; and Streeton's 'With the Australians on the Somme' was shown at the Alpine Club Gallery in June 1919. As a condition of their commissions, artists were asked to produce at least one large work depicting a major event or incident in the war. The size of these works was based on that selected for the paintings commissioned for a proposed British Hall of Remembrance. After the war, George Coates, W. B. McInnes, Florence Rodway and Charles Wheeler were also commissioned to paint portraits and large battle pictures. In 1927 Will Longstaff painted *Menin Gate at Midnight* (q.v.), an image of a ghostly army rising from a field of poppies in the battlefields of Ypres. This became one of the most popular pictures on the theme of war, providing comfort for those who had lost friends and relations at the front, and who had religious faith or spiritualist beliefs. When it toured Australia in 1928–29 it was viewed by record crowds.

The Australian official war artists intended their work to be displayed. However none of the officially commissioned art was reproduced on postcards, calendars or in government publications at the time. Nor was it used in Australian newspapers or magazines, as was much of the British war art. Apart from Will Dyson's *Australia at War*, published after the close of the war, there was no publication on Australian war art until 1933, when the Australian War Memorial (q.v.) published *Australian Chivalry*. This book did not include any of the images created at the front, but reproduced the large pictures portraying events of historical significance, and portraits. Its stated purpose was to provide an inspiration for living Australians.

Bean first proposed the Australian government develop a national war museum in November 1916, and its establishment was approved in August 1917. Bean envisaged the war artists' work being displayed in this museum to convey what the men had done, and how they had felt and thought during the war. He saw this museum as being a national one, located in Canberra, and, like the cultural institutions already established in most Australian States, as having three parts: a picture gallery, an historical museum and a library. The AWM opened to the public in November 1941, with over four hundred First World War paintings on show.

The 1930s and 1940s were a period of considerable controversy in Australian art. Vehement debates raged over the creation of an Australian Academy of Art and the judges' decision in awarding the 1943 Archibald Prize to William Dobell was questioned in court. Several articles appeared in the press criticising the Second World War artist appointments and maintaining (as Streeton had

done in the First World War) that the best of the Australian artists had not received appointments. The journal *Art in Australia* also discussed the war art scheme, and compared what the Australians were doing with the British program. In the end, the war brought together a variety of artists working in a range of styles. Many, who in other circumstances would not have crossed paths, mixed together and learnt from each other.

Like the First World War art scheme, the Australian Second World War program took time to become established. Again, several schemes developed. The GOC of the AIF, Major-General Sir Thomas Blamey (q.v.), wanted to have an artist in the field to record events; and in January 1941 he selected Ivor Hele, whom he regarded as 'one of the outstanding younger men in the art world', from among the ranks in the intelligence section of the AIF in North Africa to work as a war artist. Blamey promoted Hele to lieutenant, and provided him with transport and artists' materials. Unknown to Blamey, at about the same time the War Cabinet also approved Hele's appointment as an official war artist, following his nomination by the Commonwealth Art Advisory Board.

The first official war artists in the Second World War, Harold Herbert and Frank Norton, were appointed in January 1941. When Herbert declined an extension of his appointment in August 1941, William Dargie was appointed in his place as an artist in the Middle East. In November 1941, Murray Griffin joined the ranks of the official war artists and was sent to Malaya. Artists as varied as Dennis Adams, George Allen, Colin Colahan, Roy Hodgkinson, Alan Moore and Arthur Murch subsequently received appointments. In 1945, Sybil Craig, Robert Emerson Curtis, James Flett, Donald Friend, John Goodchild, Sali Herman and Max Ragless joined their ranks. Three women were commissioned: Nora Heysen, Stella Bowen and the aforementioned Sybil Craig. Classified as professionals, all women artists received similar pay to the men but only Nora Heysen served in the operational areas, recording the activities of the nurses at the Casualty Clearing Stations in New Guinea.

Initially, the war artist appointments were the responsibility of the Department of Information, who administered the program until October 1941. At this time, control of the program passed to the trustees and director of the AWM (under Colonel John Treloar [q.v.], seconded to the Military History Section of the AIF), with funds provided by the army for that purpose. Treloar monitored the appointment of war artists, and in 1943 was instrumental in changing the status of war artists from that of war correspondents (paid two guineas a day) to that of enlisted members of the AIF (with pay equivalent to that of service personnel). The Commonwealth Art Advisory Board and the Australian Academy of Art advised him on the selection of the artists until late 1944. Artists also applied directly to Treloar. In 1944 a committee of four took over this role: Daphne Mayo, Frank Medworth, Percy Meldrum and Will Rowell. Despite the fears of modernist artists that this committee would favour conservative artists, the appointments that followed included a number of artists with a more contemporary approach.

Artists such as Dargie and Norton were commissioned for an indefinite period. Murray Griffin was taken prisoner of war after the fall of Singapore on 15 February 1942, and remained in Changi (q.v.) for three and a half years. But towards the end of the war artists were appointed for shorter terms, for periods of three to six months, in order to ensure that a wider range of artists was appointed and a greater variety of responses obtained. Nonetheless, some artists found it difficult to get transport back to Australia and so remained at the front for a longer period than initially intended. From 1943, the official war artists were granted an honorary commission, with the rank of lieutenant or captain, in order to provide ease of access to the front. From this time they were required to submit all their work to the AWM. As with the First World War scheme artists were able to draw or paint in whatever manner they chose, within limits. Treloar had some expectations and dispatched artists to what he regarded as appropriate theatres of war or arenas of activity. Herbert and Dargie worked in the Middle East, and the expatriate artists, Bowen and Koala, worked in Europe, as did Moore. Many served in the South-West Pacific Area. Murch served in Darwin. Craig was specifically appointed to document the civilian activities of the women munition workers in the Commonwealth Explosives Factory in Maribyrnong (see INDUSTRY). Curtis and Ragless also worked on the home front.

Treloar was also interested in the artists' subjects. He directed Nora Heysen to paint portraits of the heads of the women's services, and was critical when she chose on another occasion to portray a recreation scene, rather than a more serious subject while in New Guinea. He anticipated that Friend might depict the heavy construction equipment and activities at Labuan. Nonetheless when the artist informed Treloar that he was not interested in the subject, Treloar tactfully commented:

> I realise that, fortunately, not all artists will find interest in the same sort of subjects. If large scale construction with the use of machinery does not appeal to you, it would, I agree, be a mistake for you to take it on.

In addition to the official war art scheme, servicemen were seconded to portray the war, as they had been during the First World War. Hele was the first such army artist. In 1941 the sculptor, Lyndon Dadswell, was wounded in action while serving in Syria, and later that year he was seconded to war artist duties. In 1943, approval was given for several other artists who were serving in the forces to be attached to the Military History Section as artists. These included Harold Abbott, Charles Bush, Ray Ewers, Henry Hanke, Geoffrey Mainwaring, Douglas Watson and Malcolm Warner. Their terms and conditions were much the same as those of the official war artists, but they were paid by the army and not the AWM. As serving soldiers they were appointed for the duration of the war, rather than for a fixed term.

Most Australian artists were keen to participate in the Second World War art scheme in the manner in which they were best trained — as artists, not as administrators or camoufleurs. In addition to the official war artists, Treloar also recruited artists from the services to work in the Military History Section to provide pictorial material to be reproduced as illustrations, and to work on the layout and design of the service annuals, *As You Were, HMAS I–IV, These Eagles, Victory Roll, RAAF Saga, RAAF Log, Khaki and Green, Soldiering On* and *Jungle Warfare*. These artists, who included Walter Beaumont, George Browning, John Dowie, Frank Hodgkinson, Tony Rafty and Ralph Walker, remained members of the AIF and received their usual army pay. Again, most of these artists worked in the South-West Pacific Area, except for Warner. After a brief period in New Guinea, Warner worked in the United States and Canada attached to the RAAF.

The official war art scheme did not end at the close of the war. Ernest Buckmaster was sent to Singapore, and Reginald Rowed and George Colville went to Japan with the British Commonwealth Occupation Force. Some artists, such as Murray Griffin, were retained as war artists to finish their work, or to do additional work. The Second World War art scheme was broad in scope, covering the activities of all the services as well as that of men and women on the home front. However the RAAF believed their involvement in the war was not being adequately covered by the official war art scheme, and as a result the RAAF War History Section appointed their own artists from within the ranks of RAAF personnel: Harold Freedman, Max Newton and Eric Thake. In fact, official war artists, Bowen, Koala and Moore, were employed to depict the activities of the RAAF in Europe, as was Warner in America, and Adams in

Australia. Adams also spent time depicting naval activities, as did Norton.

In February 1942, the Allied Works Council was established by the Australian government to organise domestic defence works, such as the camouflaging of aerodromes and the construction of roads and docks, and they too developed a war art scheme. In 1943, they appointed two artists, William Dobell and Herbert McClintock, to portray the men of the Civil Constructional Corps (q.v.) engaged in constructing roads and docks. Dobell and McClintock were paid £10 a week. They had to supply their own materials, which they found difficult to obtain. As a consequence Dobell worked on masonite, and McClintock used gouache on paper. They worked at various sites in Western Australia and New South Wales.

As in the First World War, exhibitions of war art were held during and after the Second World War. The 'All Australian Exhibition of Art by Australians in the Services', including work by army artists and serving soldiers, was shown in Sydney and Melbourne in 1943, and the 'Exhibition of Paintings, Drawings and Sculpture by Australian Official War Artists', which toured Australian galleries from 1943 to 1946, included paintings by Adams, Dargie, Dobell, Hele, Herbert, Hodgkinson, Murch and Norton, and sculpture by Dadswell. Critics of both modern and conservative orientation were generally unenthusiastic about the show. They maintained that the works were mainly topographical and factual records, lavish displays of technique, which did not provide a picture of the deeper horror and tragedy of war. They maintained that these artists did not indicate, as had Dyson in the last war, how the soldier felt. They noted that there was no record of the factories and of women's part in the war. They preferred Dobell's images, maintaining that they were the only ones that showed any variation in subject matter and treatment. And they pointed to artists who could have done better, such as Friend, Herman and Watson (who later did receive appointments), as well as Russell Drysdale, Francis Lymburner and James Gleeson (who did not).

These comments were made in 1944, about works created during the early stages of the war. They relate to eight of the 40 artists appointed during this war. There can be no doubt that these remarks, and those of the public debates about war art at this time, had an impact on the war art scheme and on which artists were subsequently selected. By the end of the war the range of artists employed and the subjects treated had expanded considerably. More modern artists such as Friend and Herman were appointed, and the topics treated included the factories and the role of women.

The 'Allied Works Council Exhibition of Paintings and Photographs of Civil Construction in Australia 1939–1944' was shown in the Myer Mural Hall, Melbourne, in 1944, and an expanded exhibition toured Australian galleries in 1944–45. It showed work by Dobell and McClintock, but also that by camouflage artists George Duncan, Raymond Lindsay, Max Ragless and Dudley Wood. The 'RAAF War Paintings Exhibition', including works by the three RAAF artists, toured Australian State galleries in 1946. From 1946–48 an exhibition to show the variety of work by a wide range of artists working in various branches of the services and on the home front (not just official artists), 'Australia at War', toured Australian galleries. In addition, exhibitions of the war work of individual artists such as James Cook, Murray Griffin, Ivor Hele, Harold Herbert and Vernon Jones were held. As well as displays of war works by Australian artists, exhibitions of war art by British and American war artists also toured Australia during the war.

By the end of the war the scope of the war art was considerably broader than that achieved in the First World War. In this war portraiture went beyond the tonal realism of Longstaff and McInnes, to the powerful characterisation and rendering of flesh of Dobell's *Billy Boy* (1943) and the symbolic and decorative arrangement of Bowen's *Bomber Crew* (1944). There is also a wide variety in the portrayal of people. The depiction of men and women during the war includes images of suffering and resourcefulness shown in Griffin's prisoner-of-war drawings, the weariness of constant travel visible in Friend's drawings, and the exhaustion of men in Hele's *Australian Troops Disembarking at Alexandria after the Evacuation of Greece 1941* (1942). But there are also images of the lighter side of war, of dreaming, having fun and seeking love. In many of these works there is an almost manic quality, as in Friend's *Crowds Dancing at King's Cross, Sydney* (1945), in which the swirling dancing figures seem almost to float within the surrounding space. Elsewhere people were portrayed with an energetic purpose and liveliness, as in Craig's images of the munition workers. As in the First World War, there was considerable diversity in the way landscape was portrayed. What pervades the images from the Second World War is, however, a sense of devastation and desolation. One of the most marked differences between the art of the Second World War and that of the First is the focus on the technology of war, and the portrayal of it in an almost surreal perspective, as in Thake's *Liberator's Face* (1945). All subjects were treated in many different ways. Some artists, such as Hele, Herbert and Norton, worked in a traditional manner with realistic detail, others like

Dobell and Friend explored the possibilities of expressionism, while yet others ventured towards surrealism.

War art is not restricted to the work of commissioned artists, it has also been produced by men and women who enlisted. Some of this art was created by artists of talent and experience, who spent their leisure time depicting their experiences to while away the time and to take their minds off the war. George Benson and Horace Moore-Jones captured aspects of Gallipoli and Penleigh Boyd and L. H. Howie depicted scenes and events on the Western Front during the First World War. Louis Kahan portrayed soldiers in North Africa; Douglas Annand, Frank Hinder, Kenneth Jack, Frank McNamara and Guy Warren recorded activities in the South-West Pacific Area; while John Brack, Donald Friend, Sali Herman, Francis Lymburner, Hal Missingham, Sidney Nolan, Oliffe Richmond and Elsa Russell depicted the army in Australia during the Second World War. But experienced artists who served as soldiers did not limit themselves to depicting military activities and soldiers at rest. Some, such as Arthur Boyd, portrayed the impact of war on civilians. Artists who were placed in internment camps in Australia, such as Erwin Fabien and Ludwig Hirschfeld-Mack, also portrayed their wartime experiences. War art was also produced by untrained amateurs, including POWs, who made images as a release from their ordeals.

Albert Tucker defies neat classification: he observed the war as a civilian before he joined up in works such as *The Futile City* (1940); he portrayed the war as a soldier in images like *Death of an Aviator* (1942); and after he was invalided out he viewed it again as a civilian in his images of modern evil, and in the painting *Victory girls* (1943). Later, he went to Japan with the British Commonwealth Occupation Force as a war correspondent, and created works such as *Osaka* (1947).

Other artists depicted war without actually being a part of the fighting forces, and portrayed the effects of war on the home front. These artists include Grace Cossington Smith (*The Sock Knitter*, 1915) and Hilda Rix Nicholas (*A Mother of France*, 1914) during the First World War; Dorrit Black (*The Wool Quilt Makers*, 1940–41), Ethel Carrick Fox (*Camouflage Net Workers*, 1942), Russell Drysdale (*Medical Examination*, 1941 and *Albury Station*, 1943) and Margaret Preston (*Tank Traps*, 1943) during the Second. Artists have also created powerful images of war from description rather than experience; some examples are Noel Counihan's *The New Order* (1942), John Perceval's *Exodus from a Bombed City* (1942) and Peter Purves Smith's *The Nazis, Nuremberg* (1938).

During the First World War and the years immediately following, there was a demand for sculptors and monumental masons to make war memorials (q.v.) to remember the dead and to discourage future wars. A substantial number of these commissions went to monumental masons, who provided stone statues for towns all around the country, but a number went to artists. These sculptures are a form of war art. Wallace Anderson, Margaret Baskerville, Charles Web Gilbert, May Butler George, Rayner Hoff, and Charles Douglas Richardson worked on such projects during the 1920s. Some of these commissions were private, for churches and for organisations such as the Chamber of Manufactures and the Commercial Travellers' Association. Lyndon Dadswell and Paul Montford worked on the Shrine of Remembrance (q.v.) in Melbourne and Bertram Mackennal on the Cenotaph (q.v.) in Sydney. Other war sculptures, such as Anderson's *Evacuation*, visualising the proud tradition of the Anzac hero, were commissioned for display in the AWM. In addition sculptors such as Anderson, Leslie Bowles and Web Gilbert were employed by the AWM to make sculptures for their First World War dioramas, and Ray Ewers and Ralph Walker worked on those for the Second World War. At the end of the Second World War town memorials were not sought after as they had been which had taken place at the close of the First. Lists of names were added to existing memorials in some towns, but many people did not want the memorials of this war to take the form they had in the last. Instead, in the words of the English *Country Life* magazine in 1944, they wanted 'a memorial which would be useful or give pleasure to those who outlive the war'. Nonetheless, sculptors continued to be commissioned to make commemorative works. George Allen, Wallace Anderson, Paul Beadle, Ray Ewers, Lyndon Dadswell and Daphne Mayo created war memorial sculptures during the 1940s and 1950s. Allen was also commissioned to make the Second World War sculpture for the Shrine of Remembrance.

War art also includes Napier Waller's major contribution, the stained-glass windows and mosaic for the AWM's Hall of Memory. The windows (1951) pay tribute to the Australians who served in the First World War, and the mosaic (1959) commemorates the service and sacrifice of Australian men and women during the Second World War. Waller created a spiritual haven, a place that stimulates reflection and remembrance, and that honours human endeavour through symbols and the emotional power of shapes and colour. Ray Ewers sculpted the figure of an Australian serviceman for this hall in 1959, and the figures of the sailor and airman which were placed in niches either side of the hall in 1964. In 1993 the *Australian Serviceman* sculpture was removed from the Hall of Memory during the construction of the Tomb of the Unknown Australian Soldier (q.v.).

Official war artists were also appointed by the AWM to cover Australian involvement in both Korea and Vietnam. Two veterans from the Second World War, Ivor Hele and Frank Norton, were selected for the Korean conflict. The AWM first recommended that an artist be appointed to go to Vietnam in July 1965, soon after the Australian government announced the commitment of troops there, but it was not until March 1967 that the first artist left for Vietnam. The selection of artists was managed by the Memorial, with advice from the Commonwealth Art Advisory Board through William Dargie. In order to ensure the widest representation, former war artists were not considered for appointment. Women artists were also overlooked. In letters to prospective artists the AWM's director, W. R. Lancaster, suggested that they could paint in their own style and express their individual personality, but he insisted that the 'subject matter [be] recognisable by the average human being', thereby suggesting that he would not consider non-figurative work. The artists finally appointed, Bruce Fletcher and Ken McFayden, were proficient realistic illustrators. They were able to make speedy and accurate sketches on the spot, but were not well known at the time. They were also required to be sufficiently fit to undergo jungle training. The army did not want the cost of the war artists to be included in its budget nor their presence in Vietnam counted against army staffing. The AWM therefore accepted responsibility for the artists' pay and allowances. In September 1967, the AWM's board noted that the list of available artists was not inspiring and that Vietnam did not appear to be 'in' with the artistic world, particularly the younger artists 'who seemed unable to appreciate the importance their role might be in capturing the scene and its social effects, whether or not they agreed with the military operations taking place there'. In February 1968 the board maintained that Fletcher and McFadyen had provided more sketches than could be displayed at any one time (overlooking the fact that these works were inevitably limited in topic and approach). Despite the mounting list of applicants, the board resolved that no further artists would be appointed.

Servicemen in Vietnam did not receive any official encouragement to depict the war, although those who were motivated found outlets for their work. Jim Kenna's cartoons were published in the 6RAR newspaper, the *Sunday Times*, but serving soldiers were not invited to become war artists or to illustrate service annuals as they were in the First and Second World Wars. A few Vietnam veterans, such as Ray Beattie and Peter Moore, painted their responses to the war soon after it was over. Artist observers such as Marcus Beilby, David Boyd, Richard Larter and Clifton Pugh also portrayed their responses to the war. Their work reflects the place the Vietnam War held in Australian society at the time. It shows that for most Australians this was a war in which everyday life went on as before.

The AWM decided not to appoint an official war artist during the Gulf War and opted instead to encourage a variety of Australian artists to depict aspects of this conflict. The Australian forces were not fighting on the ground or in the air and the Australian involvement in the Gulf was on a small scale. But television has made all wars world wars, and has involved everyone in such events as passive observers. The Gulf War was not something the Australian public could overlook. This war provoked some artists, such as Gay Hawkes and Andrew Sibley, to make a general outcry against war. Others, such as Roslyn Evans, viewed it in terms of past wars, and these wars' impact on their family. Others, like Enid Ratnam-Keese, considered the suffering of refugees. A number of artists, like Merilyn Fairskye and Jan Senbergs, referred to the way in which the media represented the Gulf War. Kevin Connor differed from most other Australian artists who portrayed this war because, in June 1991, he visited several important centres in Iraq affected by the coalition bombs and the postwar rebellion. He made drawings and paintings that showed the impact of the war on buildings and landscape as well as on people. These works report on the destructiveness of war and the suffering it brings.

War art can also be said to include that produced by artists sponsored by the Australian Army to depict their participation in peace-keeping (q.v.) missions, such as George Gittoes's work in Somalia, Cambodia and Western Sahara. Gittoes' work focuses on the human victims of these conflicts.

War art, emotional and interpretive responses to war, and reflections on past and present wars, continue to be made by contemporary artists. In the 1950s Sidney Nolan turned to creating images of war because he believed Gallipoli was the great modern Australian legend. Stimulated by his childhood memory of a poem from *The Anzac Book*, Nolan saw the Australian soldiers as resembling the heroes of the Trojan war. The landscape, the men and the events gradually took grasp of Nolan's imagination and he explored the theme over six years from 1956, returning briefly to this subject in the late 1970s. More recently, Barbara Hanrahan had looked back on women's roles in war in her screenprint, *Poppy Day* (1982). Paddy Wainburranga has reflected on the Second World War from an Aboriginal perspective, and demonstrated the important contribution of indigenous Australians to the war effort in his bark painting, *World War II Supply Ships, Darwin* (1991). Trevor Lyons has studied the severe facial injuries he received from a bomb attack in Vietnam in his series of etchings, *Journeys in my Head* (1987), which trace the progression of his disfigurement and mental stress and that point to the way in which war can damage the mind and body.

War art also includes poster art. Australian war posters include those by Norman Lindsay as well as ones by noted commercial artists such as Harry J. Weston, J. S. Watkins and James Northfield in the First World War, and Percy Trompf and Walter Jardine in the Second. These posters provide information on the development of commercial and political art in Australia, as well as being valuable records of social history. In the First World War posters were used mostly to recruit men for the AIF and they show the numerous ways in which men were persuaded to join the AIF. They also illustrate the bushman ideal and the sporting ethos of Australian society at the time. As the First World War progressed, and recruitment became more difficult, posters tended to appeal to women and to foster hatred of the enemy through grotesque portrayals of the Germans. In the Second World War posters were as much concerned with civilian behaviour as with recruitment; they were directed towards people's daily lives and encouraged health and safety. These posters sometimes used humour to achieve their end. During the First and Second World Wars posters were largely produced and promoted by government departments and authorities, but during the Vietnam War there were few government posters. At this time posters were used primarily to protest against the Australian and American involvement in this war. It was a relatively accessible medium, cheap and easy to produce and readily displayed. Some of the most powerful posters from the Vietnam war are those which dramatically portray the suffering Vietnamese. More contemporary artists, such as Ann Newmarch, Toni Robertson and Colin Russell, have made posters commenting about the threat of nuclear war.

A popular misconception about war art is that it consists of battle pictures. In fact, the majority of art on the theme of war is not the blood and thunder of tremendous charges. Many of the most moving images capture deep and abiding human emotions: they show the horror of death, grief at the loss of a friend, boredom, exhaustion and the release of energy on hearing good news. People who are lucky enough never to have lived through a war can share the experiences depicted in these works because they are universal.

While visual accuracy is an important factor in war art, it would be wrong to assume that it was the only or the prime aim of the artists. For many official war artists the arrangement of colour and tone, the handling of paint, and the personal interpretation of a scene were significant to their work; in short, the creation of a work of art was their primary concern. A number of war works have an artistic quality that would make them welcome additions to any art collection. The wartime work of George Lambert, Arthur Streeton, Donald Friend, Ivor Hele and Eric Thake has an importance that extends beyond their existence as historical records.

As well as providing a personal artistic response to the war, artists provided a valuable record of what took place. War art shows the atmosphere and general appearance of scenes of action, and the suffering and tragedy of the young men and women who fought. It portrays the equipment used, the situations under which men and women lived, and the enemy they had gone to fight. Paradoxically, the destructiveness of war generates creativity in individuals and leads to changes in society's values and concerns. The urgency of the situation, and the tremendous energy set free, gives an added intensity to life. War has an impact on most artists, even those not directly involved, and as a consequence of war new approaches to art have developed. The Australian government's involvement in the commissioning of war artists has resulted in one of the most substantial public commissioning programs in Australian art history. It is frequently thought, rightly or wrongly, that public art is propaganda and that a commissioned artist has to do what he or she is told to do. However, rather than depicting prevailing official views, most war artists have expressed their personal responses. Treloar wrote to Sali Herman, 'it will be interesting to see the different treatment which different artists will give to these subjects'. And again, in connection with Donald Friend's work, he remarked:

> although it has pleased our critics to inveigh against bureaucratic control of artists, this in fact does not and has never existed. No-one — except the critics — has ever told the artists what they should paint or how they should do it ... knowing that [artists] will handle best the subjects which make the strongest appeal to them.

Favoured artists, such as George Lambert in the First World War and Ivor Hele in the Second, were virtually given *carte blanche* to paint whatever they wanted. Moreover, as to the central issue of art in the 1940s — whether an artist should paint a close imitation of reality or make a personal expression — the Military History Section remained impartial, and employed artists working in a variety of styles.

The Australian government was also tolerant of artists' political persuasions. Will Dyson was Australia's first official war artist, despite the fact that during the first two years of the war he made cartoons for the British left-wing press, which criticised militarism, conscription and the suffering of innocent victims. Herbert McClintock was appointed a war artist by the Civil Constructional Corps during the Second World War, despite the fact that he was a member of the Communist Party. In 1966, the AWM's director explicitly stated that official war artists would not be vetted for their political views before going to Vietnam, and that the AWM was not interested in propaganda. Artists with political objections to the war were among those who exhibited in the 'Artists on War' exhibition: Noel Counihan, Robert Grieve, Daniel Moynihan, Erica McGilchrist and Udo Sellbach. These artists did not, however, seek appointments, and so the director's claim was never tested.

For many Australian artists, war has provided an occasion to become adventurous, to break free from their limitations, to adopt new subjects or new styles. Streeton portrayed machinery during the First World War, long before any other Australian artist had considered this a suitable subject for art, and depicted mechanical marching men in a futurist style when many in Australia had not even heard about futurism. During the Second World War, while employed by the Allied Works Council, Dobell painted *Billy Boy*, an image of a Civil Constructional Corps worker, which is as novel an essay in portraiture as his notorious *Joshua Smith*. Thake created some of his most haunting surrealist images while employed as an RAAF artist.

Surprising as it may be, wars have been beneficial to artists: official war art schemes have provided artists with a means of living when it was hard to sell art on any other subject. They were developed before there were government grants for artists, and before there was a highly sophisticated commercial art market. During the First World War, Dyson had a break from drawing cartoons and an opportunity to make lithographs. Waller's commission for the AWM's Hall of Memory stained-glass windows and mosaic kept him in employment for several years and resulted in one of Australia's largest and most accessible mosaics. In the Second World War, many artists at the beginning of their careers had the opportunity to learn from others whom they might not otherwise have met. The Melbourne-born artist Kenneth Jack, while serving as a survey draughtsman and cartographer with the RAAF at Labuan, met the Sydney-based artist Donald Friend, and benefited from studying his drawing technique. And although Friend, Lymburner, Richmond and Tucker hated army life, this did not stop them working. It could be argued that their anger and frustration put fire into their art.

ANNA GRAY

WAR ARTISTS The Australian official war artists were men and women specifically commissioned to depict and interpret Australia's involvement in war and warlike activity. Many significant Australian artists received these commissions. In the First World War, George Bell, Charles Bryant, Will Dyson, A. Henry Fullwood, George Lambert, Fred Leist, John Longstaff, H. Septimus Power, James Quinn and Arthur Streeton were appointed official war artists, and George Benson, Frank Crozier, Will Longstaff, Louis McCubbin and James Scott were selected to become Australian War Records Section artists. J. S. McDonald was appointed and trained as a camouflage artist but never took up a position. Daryl Lindsay was seconded to become a medical artist at the Queen Mary Hospital at Sidcup, Kent, in England. Other artists, such as George Coates, W. B. McInnes, Florence Rodway and Charles Wheeler, were commissioned to paint portraits and large battle pictures after the war. Will Longstaff, who was an Australian War Records Section artist during the war, made a name for himself later as a war artist through his painting, *Menin Gate at Midnight* (q.v.), which became one of the most popular pictures on the theme of war. It portrayed ghostly steel-helmeted soldiers rising from the cornfields of Flanders, and it provided comfort to those who welcomed its spiritualist message. When it toured Australia in 1928–29 it was viewed by record crowds.

During the Second World War several war art schemes operated: the official war artists, appointed by the Australian War Memorial (q.v.); the army artists, servicemen seconded to portray the war; and the Military History Section artists, servicemen appointed to work on the Christmas annuals. Dennis Adams, George Allen, Colin Colahan, William Dargie, Murray Griffin, Harold Herbert, Roy Hodgkinson, Alan Moore, Arthur Murch and Frank Norton received official war artist appointments. In March 1945, Sybil Craig, Robert Emerson Curtis, James Flett, Donald Friend, John Goodchild, Sali Herman and Max Ragless were added to their ranks. Three women, Stella Bowen, Sybil Craig and Nora Heysen, were appointed official war artists. Harold Abbott, Charles Bush, Lyndon Dadswell,

Ray Ewers, Henry Hanke, Ivor Hele, Geoffrey Mainwaring, Douglas Watson and Malcolm Warner enlisted in the AIF and were seconded as army artists. Richard Ashton, Walter Beaumont, George Browning, John Dowie, Frank Hodgkinson, Tony Rafty and Ralph Walker joined the AIF and were seconded as Military History Section artists. The official war art scheme did not end at the end of the war; Ernest Buckmaster was sent to Singapore, and Reginald Rowed and George Colville went to Japan with the British Commonwealth Occupation Forces (q.v.). Albert Tucker was in Japan with BCOF as a war correspondent.

Harold Freedman, Max Newton and Eric Thake were RAAF War History Section artists. William Dobell and Herbert McClintock were Allied Works Council artists, employed by the Civil Constructional Corps (q.v.). Douglas Annand, George Duncan, and Frank Hinder worked as camouflage artists. Walter Pidgeon was a war correspondent.

Sculptors such as George Allen, Wallace Anderson, Margaret Baskerville, Paul Beadle, Lyndon Dadswell, Ray Ewers, Charles Web Gilbert, May Butler George, Rayner Hoff, Bertram Mackennal, Paul Montford, Daphne Mayo and Charles Douglas Richardson worked on war memorials (q.v.) for towns all around Australia, following the First and Second World Wars. Napier Waller achieved substantial recognition as a war artist through his stained-glass windows and mosaic for the Australian War Memorial's (q.v.) Hall of Memory.

Ivor Hele and Frank Norton were war artists during the Korean conflict and Bruce Fletcher and Ken McFayden during the Vietnam War. Recently, George Gittoes was sponsored by the Australian Army to depict their participation in peace-keeping (q.v.) missions in Somalia, Cambodia and Western Sahara.

Although not specifically commissioned, some who were trained as artists enlisted in the services and produced work in their spare time. These artists might be considered to be war artists. They include Penleigh Boyd and Horace Moore-Jones in the First World War and John Brack, Kenneth Jack, Louis Kahan, Frank McNamara, Francis Lymburner, Hal Missingham, Oliffe Richmond, Elsa Russell and Albert Tucker in the Second. Wilfred McCulloch, while serving in the AIF as a stretcher-bearer, was killed while working in a volunteer relief party during the Malayan campaign.

Artists have also depicted war and aspects of military life of their own volition. They include civilians like Grace Cossington Smith and Hilda Rix Nicholas in the First World War; Ethel Carrick Fox, Noel Counihan, Russell Drysdale, Margaret Preston and John Perceval in the Second; and again Noel Counihan, as well as David Boyd, Richard Larter and Clifton Pugh during the Vietnam War. However these artists are not generally thought of as war artists, but as artists who painted war subjects. Many contemporary artists who comment about or reflect on war in their work are likewise not regarded as war artists, but are considered to be artists who turn to war as one of a number of topical subjects.

ANNA GRAY

WAR CABINET see **AUSTRALIAN WAR CABINET**

WAR COMMITMENTS COMMITTEE was established after a cabinet decision of 22 September 1942 to advise the minister for Defence on manpower requirements, availability and use. It was to oversee the manpower program as a whole and to recommend ways in which more labour could be made available for the war effort. The committee consisted of the three Chiefs of Staff, the Directors-General of War Organisation of Industry, Allied Works, Manpower, and Munitions and Aircraft Production, and the chairman of the Allied Supply Council's Standing Committee.

WAR CORRESPONDENTS The tradition of Australian war correspondence is as old as the tradition of Australian service in overseas wars. Howard Willoughby of the (Melbourne) *Argus* accompanied Australian colonial contingents to the Waikato War in the 1860s, while W. J. Lambie went to the Sudan in 1885 for the *Sydney Morning Herald*, and wrote a lengthy despatch about his own (minor) wounding. Like the expeditionary forces they covered, these accounts were precursors to the main tradition of war reporting, which in the Australian case entered the mainstream with the Boer War. Also developed from this time was the coincidence between war reportage and literary distinction, with the appointment of the poet A. B. 'Banjo' Paterson (q.v.) as a correspondent for the *Sydney Morning Herald* in mid-1901. Paterson seems to have revelled in exposure to military life on campaign; undoubtedly he did a good job in reporting the course of General Sir John French's slow and deliberate campaign to capture the capitals of the Boer republics, being retained by the major international agency, Reuters, as their second principal correspondent at the front. Something of his South African experiences was reflected later in the book, *Happy Dispatches*, published in 1934.

War correspondent Dorothy Drain of the *Australian Women's Weekly* talks to a member of the RAAF Transport Flight Vietnam, Vung Tau, South Vietnam, c. 1962. (AWM 044130)

The principal figure in the establishment of Australian war reporting, however, must be C. E. W. Bean (q.v.), official correspondent with the AIF during the First World War. Bean was an established prewar journalist, elected by his peers to the single correspondent's post attached to the first contingent of the AIF when it sailed in November 1914. He beat Keith Murdoch (q.v.), a better newspaperman but much the lesser writer, for the position. Bean's despatches struck some editors as overly wordy and yet lacking in the overblown heroic rhetoric in which the Edwardian age recorded martial endeavour. His first despatch on the landing at Gallipoli was held over, and the British correspondent Ellis Ashmead-Bartlett was the first to alert Australian readers to the achievements and sacrifices of their men in the Dardanelles through the pages of the morning papers. Bean's activities as a correspondent overlapped and complemented his wider literary concerns, not only as official historian of Australia's war effort, but through the production of the Anzac books, begun on Gallipoli and carried on in France throughout each Christmas of the war, and in his activities with the Australian War Records Section. In time he came to be assisted by F. M. Cutlack (q.v.), and by a small group of artists and photographers such as Frank Hurley (q.v.). Bean's major impact, however, was to be as an historian and as founder of the Australian War Memorial.

It was during the Second World War that Australian journalists made their most significant impression as correspondents. John Hetherington, subsequent biographer of Blamey, sailed with the 6th Division in January 1940, and by early 1941 had been joined by the poet and writer Kenneth Slessor (q.v.). Noel Monks, an expatriate working for the London *Daily Mail*, covered the disastrous campaign in France in 1940. Alan Moorehead (q.v.), later the author of a series of popular and authoritative histories of Gallipoli, the discovery of the Nile, and the European exploration of the Pacific, in addition to three volumes on the war itself, wrote for the London *Daily Express* on the fighting in the Mediterranean before moving to North-West Europe after D-Day. Gavin Long (q.v.), destined to be appointed as Australia's official historian of the Second World War in 1943, accompanied the British Expeditionary Force to France and then reported the war in New Guinea. Chester Wilmot (q.v.), with a reputation in broadcasting acquired with the ABC, worked in the Department of Information's Field Broadcasting Unit, accompanied the 6th Division through the early and triumphant campaign against the Italians in Libya and reported on the German siege of the 9th Division

in Tobruk, before writing one of the best early accounts of Australia's contribution in the Mediterranean theatre, *Tobruk*, published in 1944. After covering the fighting in New Guinea, he returned to North-West Europe to report for the BBC. Both Wilmot and Slessor ran foul of the Australian C-in-C, General Sir Thomas Blamey (q.v.), and both lost their accreditation as war correspondents in New Guinea.

In the postwar conflicts Australian correspondents continued to make their mark, often internationally. Denis Warner and Pat Burgess (q.v.) reported from South-east Asia for decades. Frank Clune and George Odgers (q.v.) reported the Korean War, and subsequently wrote books on different aspects of Australian involvement there. They were joined by experienced journalists like Harry Gordon, Ronald McKie, Michael Ramsden, Lawson Glassop and Roy Macartney, most of whom had been correspondents during the Second World War. Australian coverage of the Malayan Emergency, Confrontation and Vietnam was relatively sparse, certainly, in the last case, by comparison with American journalistic efforts, but this is explained by a number of factors: Australian media organisations kept very few correspondents based in Southeast Asia, preferring to fly them up when occasion demanded; the nature of the conflict, especially in Malaya/Malaysia militated against traditional forms of press coverage; and sensitivities regarding the conduct of operations in Vietnam led to heavy, and sometimes heavy-handed, attempts by the army and the Department of Defence to control the activities of journalists in the Australian area of operations. The most famous Australian journalist of the war, the cameraman Neil Davis, never covered Australian troops, finding it much easier, in common with many of his colleagues, to work in American or South Vietnamese areas, where restrictions were much lighter. Warner, Burgess, Keith Willey, the American Gerald Stone, and a number of young reporters mostly working for the ABC, nonetheless kept Australians informed of the activities of the Australian Task Force, most of the time.

The best known, and certainly most notorious, Australian correspondent during this period was Wilfred Burchett. Widely reviled as a traitor, Burchett came to prominence as the first Western reporter to visit Hiroshima after the atomic bombing of the city, although his journalistic career had commenced before the beginning of the Pacific War and he had reported from China and Burma. He covered the Korean War from the Chinese and North Korean side, and was instrumental in spreading allegations that the United States was waging germ warfare against North Korea; he also developed a strong line in disinformation when it came to reporting the treatment of captured UN soldiers by their Chinese and North Korean enemy. Burchett went on to champion various communist, third world and liberation causes around the world, and reported extensively during the Vietnam War from Hanoi and behind the Viet Cong lines. He remains a divisive figure after his death in 1983, not least for his role in assisting the Chinese in the interrogation of American and Commonwealth POWs in North Korea, and of American aviators shot down over North Vietnam.

In reaction to some of the perceived lessons of Army-media relations during the Vietnam War, and following on strong British and American attempts to manage the media during the Falklands War and the Grenada operation, the ADF has instituted a system of accrediting correspondents in advance of any requirement for them to cover ADF operations, and to familiarise them to some extent with the organisation and workings of the armed forces. This has extended to allocating a role to media representatives during the major Kangaroo series of exercises held in northern Australia.

WAR CRIMES TRIALS Australians' experience of trials of war criminals began after the Second World War. Originally a response to Nazi atrocities in Europe, it was Japanese atrocities against which Australian actions were directed. In practice there were three categories of criminal behaviour under which Japanese defendants were arraigned: crimes against peace, war crimes, and crimes against humanity; and three classes of criminal: A class, reserved for alleged perpetrators of major war crimes such as those covered by the International Military Tribunal for the Far East (IMTFE) in Tokyo, and B and C class, which covered lesser crimes and a decreased level of responsibility. The

majority of Japanese and Koreans tried by Australia fell into these last two classes, and included those bearing direct responsibility for the brutalisation and murder of POWs.

The major series of Australian trials was held under the War Crimes Act of 1945. Typical crimes prosecuted under this act were ill treatment of POWs, massacres of soldiers, execution of captured aircrew and ill treatment of local inhabitants. The courts set up under the act consisted of at least three service officers chosen by a convening officer. The accused were allowed proper representation by counsel, either Australian or from any other country including Japan. The courts departed from common law procedure in a number of ways. First the C-in-C could confirm a sentence without reference to any civilian authority. Second, evidence could be taken from witnesses who were not produced in court and therefore not open to cross-examination. Third, evidence could be introduced which although irrelevant to the particular case in question was prejudicial to the accused.

Crimes and charges varied widely in the Australian trials. One unique charge was that of cannibalism or mutilation of a dead body. At a trial in Wewak a Japanese officer was sentenced to death for eating part of an Australian POW body. Australia also held mass trials. In December 1945 at Labuan 70 Japanese were jointly charged with the ill treatment of civilians in Sarawak. At Amboina 93 guards were charged with calculated cruelty to Allied POWs. A summary of Australian trials, their location and verdicts is given in Table 1.

Australian war crimes trials have come in for severe criticism, generally because of legal procedure rather than partiality or unfairness. Of the 644 prisoners convicted (69.5 per cent of the total), 148 (23 per cent) were sentenced to death and executed, and 496 (77 per cent) were imprisoned.

Table 1 Summary of Australian war crimes trials

Place	Trials	Defendants	Acquitted	Convicted
Singapore	23	62	11	51
Morotai	25	148	67	81
Labuan	16	145	17	128
Wewak	2	2	1	1
Rabaul	188	390	124	266
Darwin	3	22	12	10
Hong Kong	13	42	4	38
Manus	26	113	44	69
Total	296	924	280	644

Thirty-nine prisoners were given life sentences; two were sentenced to 25 years; 152 to 11–24 years; 82 to 10 years; and 22 to less than 10 years.

Australia also participated in the IMTFE. It was composed of members from 11 nations (Australia, Canada, China, France, India, the Netherlands, New Zealand, Philippines, Soviet Union, Britain and the United States). Sir William Webb, who had originally headed the Australian War Crimes Commission, was the president of IMTFE. The task of the tribunal was to try the major Japanese war criminals. The first meeting of IMTFE was held in Tokyo on 29 April 1946, the defendants were arraigned on 3 May. Sentences were handed down in November 1948. Of the 28 defendants charged, two died during the course of the trial, one was declared unfit to stand, seven were sentenced to be executed, 16 to life imprisonment and two to lesser terms of imprisonment. The only other Australian to preside over a SCAP (Supreme Commander for the Allied Powers) tribunal was Brigadier J. W. A. O'Brien (q.v.), whose judgment in September 1949 that the last Chief of the Naval Staff of the Imperial Japanese Navy, Admiral Toyoda, was not guilty of crimes committed by naval personnel under his command seemed to overturn the doctrine of command responsibility on which a number of prominent Japanese commanders had been hanged, and greatly irritated General Douglas MacArthur (q.v.).

After the Second World War the issue of war crimes was viewed from afar in Australia as a continuing series of trials was held in West Germany, Israel and elsewhere. However, in 1986 there were widespread allegations that persons who had committed war crimes in Europe had entered Australia in the rush of postwar migration and were still living here. As a result a Special Investigations Unit (SIU) was set up in 1987 under Mr R. F. Greenwood QC (succeeded by Mr G. T. Blewett in 1991), to gather evidence on such occurrences. On 26 January 1990 officers from the SIU charged Ivan Polyukovich with the murder of 24 people and with being connected with a massacre of 850 Jews near the village of Sernike in the Ukraine in 1941. Later Mikolay Berezowsky and Heinrich Wagner were also charged with war crimes in the Ukraine.

In 1988 the War Crimes Act had been amended to take account of likely prosecutions. The amendments allowed for trials to be held in Australia for crimes committed elsewhere. Further, the crimes for which persons could be charged were carefully delineated. To be prosecuted under the act crimes had to be committed in pursuance of the policy of a government with which Australia had been at war. Thus a person could not be prosecuted for acts of murder based on personal impulse. The validity of these amendments were challenged in the High Court in 1990 but without success, and they were declared valid by the full bench of the court in 1991.

In October 1991 the committal hearings of the first of the accused, Polyukovich, commenced in Adelaide, with Greg James QC as chief prosecutor. Polyukovich was defended by well-known Adelaide QC, Michael David, and proceedings were characterised by great thoroughness. Exhumations of mass graves in the Ukraine alleged to be connected to actions of the three defendants were carried out; witnesses from the Ukraine were flown to Adelaide; experts on war crimes were engaged. In June 1992 the presiding magistrate decided that Polyukovich had a case to answer and was ordered to stand trial. The trial began in March 1993 and largely turned on the question of whether eyewitnesses could identify the accused. In May, after brief deliberations, a jury found Polyukovich not guilty on all charges. In the meantime a magistrate had dismissed the charges against B. L. Berezowsky. Later in 1993 the Director of Public Prosecutions decided that it would not proceed with the Wagner case on the grounds of the ill health of the accused. The $30 million trials therefore came to an end without a single conviction.

WAR ECONOMY As the first duty of a government is to defend the state against insurrection from within and aggression from without, the nation at war has but one aim to which everything else is largely subordinated. Wars are powerful agents of change, and the economy, the *sine qua non* of a nation's well-being, necessarily undergoes many stresses and changes in wartime. At the outset, business rarely continues as usual and the state is forced to intervene by setting priorities, allocating or rationing resources and controlling a greater proportion of the national product than in peacetime to meet the changes in demand for the goods and services that war brings. Wars can stimulate an economy and encourage innovation and efficiency, but they are also expensive undertakings — usually financed by taxation and loans from the public and the banking system, the economic consequences of which are long-lasting. Furthermore, military strategy itself is also an economic problem, because the manpower or forces at the disposal of a state at war are limited and their efficient deployment in war involves a rational choice between different plans and strategies to produce the optimum result.

By 1914 Australia was a sizeable primary producing nation depending on the export of wool, wheat and to a lesser extent frozen meat, dairy products, sugar, fruit and minerals. As Australia's prosperity depended on overseas capital to develop its resources and external market commodity prices and shipping facilities for its exports, it was not long before the war made an impact. At first, fearing the worst, employers and other businessmen panicked. The stock exchanges in Melbourne, Sydney and Adelaide closed, the price of imported goods rose sharply, and unemployment increased markedly. To make matters worse Australia was in the grip of a severe drought which affected all States and severely reduced agricultural output until the end of 1915. However, while prices, charges and unemployment continued to rise, the stock exchanges reopened at the end of September and a few weeks later the government passed the first Trading with the Enemy Act and the War Precautions Act (q.v.), the latter becoming the basis for many far-reaching regulations of an economic nature. In July 1915 another major measure was taken when an embargo was placed on the export of gold; the accumulation of gold reserves led to a rapid increase in the note issue which in turn contributed to wartime price inflation. The war, however, was to be financed by loans and taxation.

Since Federation in 1901 the Commonwealth's chief source of revenue had been customs and excise duties, a large part of which had to be returned to the States. In addition to this revenue the States depended heavily on overseas loan money for their extensive public works programs. By June 1914, as the States had incurred a public debt of £317.5 million against the Commonwealth's £19 million, there was much truth in the latter's claim that 'Australia, as a whole, was never in a stronger economic position'. At first the government expected to pay for the war from revenue and loans raised in London, but it soon became evident that revenue was insufficient and there were difficulties with London. As a result the Labor government, though naturally inclined to finance the war by high taxation rather than by extensive borrowing or high indirect taxation, committed the country to financing the war largely on loan money. These funds were raised both overseas and locally.

The States, while fully committed to the war effort, at first showed little restraint in their demand for development funds; in fact State indebtedness increased by about £100 million during the war. While the Commonwealth and British governments had reservations about raising loans in wartime for such purposes, the Commonwealth bowed to the States' pressure; however, by agreement with all States except New South Wales, the Commonwealth took control of all overseas borrowing from November 1915. In 1918 New South Wales joined in the arrangement by which time (30 June) Australia's total borrowing abroad for war purposes amounted to £92 million. In addition to this amount, all raised in the United Kingdom, seven war loans and three peace loans raised £250 million. These internal loans were floated and managed by the Commonwealth Bank (established in 1911), which during the war assumed a key position in the banking structure.

While the war was mainly financed by loans, the Commonwealth increased the rates of existing taxes, such as the land tax, and imposed new ones. By the end of the war it had built up an extensive system of direct taxation which included death duties, wartime profits, an entertainments tax and, most significantly, the first federal income tax, introduced in 1915–16. Before the war, Australia's exports were handled by wool-broking firms, grain merchants and the like, but war interfered with demand and the established means of finance and placed great stress on shipping facilities. After some initial confusion and dislocation to trade, the government decided to pool commodities, arrange finance for producers, and control exports. In general trade policy during the war Australia looked to Britain for guidance, and invariably followed it. As a result, an elaborate embargo system was set up — wheat, flour and meat being early examples of commodities where trade was controlled. Until 1916 the wool clip continued to be sold by auction, but in November the British government announced that it would take control of the whole clip at prices 55 per cent above those of 1913–14; this purchase scheme, controlled by the Central Wool Committee, continued until the 1919–20 wool-selling season. To overcome the shortage of shipping the government purchased a number of ships and established the Commonwealth Government Line.

The shortage of shipping in general forced the growth of manufacturing to replace imports, and by 1919 several hundred products were made locally which had not been made previously. But the fundamental development in secondary industry was the opening of the BHP steelworks in Newcastle in 1915; this advance was to be of far-reaching importance in the postwar period. The war also allowed Australia to wrest control from German firms in the metals industry.

A rapid rise in prices, inflation, and a marked decline in the standard of living were matters of acute concern to wage-earners, trade unions and governments and led to much social and industrial unrest. After ineffectual State attempts to control prices, the Commonwealth moved into the field in

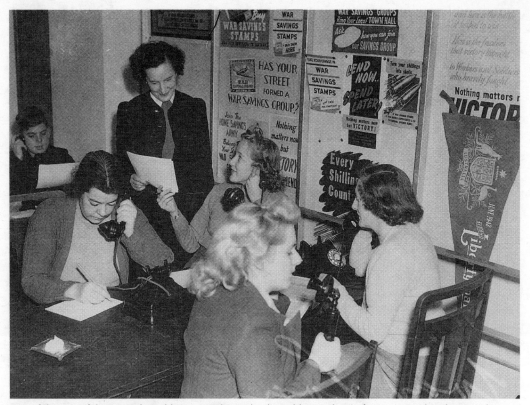

Part of the cost of the Second World War was borne by the public purchase of government bonds. Here the Commonwealth Loan Office operators are standing by to take applications for the Third Liberty Loan, Melbourne, 16 April 1943. (AWM 138661)

1916 with wide powers to fix prices by regulation. The labour movement was weak and divided after the Labor Party lost office in November 1916 and this led the more radical unions to turn from political to direct action; in 1916–17 over 6 million working days were lost through strikes, the most serious of which were in the mining and transport industries. In January 1917, in an effort to conserve energy, daylight saving was introduced, but it proved highly unpopular and was soon abandoned.

The cost of the war to Australia between 1914–15 and 1919–20, including interest and sinking funds (but not war gratuities), was £377 million. By the outbreak of the Second World War, however, including expenditure on repatriation and pensions, the total cost was approaching £900 million.

The Second World War was a much greater challenge to Australia's economic capacity than the first. Not only were there important changes in the technology of warfare, but Britain, Australia's traditional ally, was weakened and embattled, and Australia was directly threatened with invasion. Australia's participation in the war can be divided into three phases: the distant war when military campaigning took place far away in the Mediter-

ranean, the defence from Japanese invasion, and thirdly the offensive campaigns in New Guinea and the islands. Each phase produced its own characteristic demands on the economy.

The first two years of the war from September 1939 to December 1941 were in many ways a replica of Australia's experience in the First World War. In addition to a Trading with the Enemy Act the Commonwealth quickly passed the National Security Act (q.v.). This resembled the War Precautions Act of 1914 and in effect empowered the government to make regulations and orders generally on 'all matters which are necessary or convenient to be prescribed for the more effectual prosecution of the war'. The act provided the basis for most of Australia's wartime economic reorganisation. Price control and import licences were quickly introduced, a wheat board set up to control all facets of the industry and, as wool was regarded as an economic weapon, an agreement was entered into with the United Kingdom government for it to purchase the entire Australian wool clip at an agreed price for the duration of the war and for one year afterwards. Plans were announced for the reorganisation of the munitions industry, and a

Manpower Committee began work. But while the government imposed certain basic controls, took measures to protect its main export industries and gradually tooled up the war effort, its attempt to set up an 'economic cabinet' to maximise the war effort was half-hearted and ineffectual. In June 1941, however, it set up a new Department of War Organisation of Industry, with special emphasis on economic matters, and began to reallocate many prewar government functions among departments. During this first phase, while prices inevitably rose as the result of wartime shortages and a certain amount of profiteering, the effects were somewhat offset by wage increases and falling unemployment, and life on the home front was relatively normal and prosperous. This all changed when Japan entered the war.

In December 1941 the new Labor government announced 'a complete revision of the whole Australian economic, domestic and industrial life' and the War Cabinet (see AUSTRALIAN WAR CABINET) called up married and unmarried men in a wide range of occupations for full-time army service for home defence. The following month the Manpower Committee (later Directorate) was reactivated and began the full mobilisation of the nation's people and resources, drawing on hitherto untapped labour to find men and women for both the services and greatly expanded war production. Important industries and occupations were respectively declared 'essential' or 'reserved' (q.v.); industries of low priority were starved of labour and some phased out for the duration of the war. The special labour needs of the vast construction program that organisation for 'total war' demanded was the task of the Civil Constructional Corps (q.v.), independent of the Manpower Directorate but coordinated with it. The construction of military facilities such as airfields and camps outside operational areas for Australian and American forces — who had begun to arrive in Australia in December 1941 — was the work of the Allied Works Council. Between January 1942 and March 1943 the number of men and boys engaged in the armed forces or direct war work grew from 0.55 million to 1.27 million. By June 1943 Australia's manpower was fully mobilised: almost 90 per cent of all males over 14 years of age and 29 per cent of females were employed in the armed services, direct war work or 'essential' civilian jobs.

From 1 January 1942 daylight saving was introduced in all States, with the exception of Western Australia, and on 10 February the War Cabinet adopted a National Economic Plan which, among other things, stressed that the incomes of all sections of the community were to be equitably controlled.

Regulations issued the same day to give effect to the plan limited business profits to 4 per cent of capital invested, severely restricted the transfer of property, pegged wages at current award or actual rates and invested the Commonwealth Bank with control over interest rates. Rationing, which had been confined to petrol, was extended to tea, sugar and clothing. The increasing impact of the war on the economy is shown in the national accounts. War expenditure rose from £170 million in 1940–41 to £537 million in 1942–43, or from 15.3 to 36.8 per cent of gross national expenditure. Private investment fell from 15.4 per cent of gross national product in 1939–40 to less than 3 per cent in the years 1942–43 to 1944–45.

Phase three, which began after the immediate threat of invasion had passed, was directed towards a rebalancing of the war effort. The overriding problem was to balance competing economic and political interests and rationalise Australia's commitments to a level more in keeping with her resources and taking into account postwar considerations. This was largely the task of the Civil Requirements Board set up in July 1943, which consisted of representatives of the departments of War Organisation of Industry and Post-War Reconstruction (q.v.) together with the organisations dealing with food control, rationing, health, supply and shipping. Australia's allies had reasoned that Australia could contribute most to the war effort by substantially reducing the military involvement and directing the released manpower to indirect war production. This shift in the war economy began in October 1943 when the War Cabinet directed that 40 000 men be released from the armed forces and munitions production for food production (there were then 178 000 American servicemen based in Australia) and other priorities in the indirect war sector of the economy. This initial reallocation of resources was belated and small-scale and achieved little. A more determined effort was made in 1944, and while it reduced war expenditure to £385 million in 1944–45 its effect was somewhat offset by an increase in net exports, mainly to the United Kingdom. In fact the national accounts showed no rise in personal consumption in the last three years of the war; in January 1944 meat rationing was introduced and living standards were at their lowest in 1944–45.

As in the First, the Second World War was financed by loans and increased taxes. Apart from a temporary loan from the United Kingdom of £12 million sterling to help finance defence equipment, the Australian war effort was achieved without resort to a public loan overseas. Thirteen new loan campaigns — variously called liberty, austerity, special or victory loans — raised over £900 million within Australia. Among additional sources of loan funds were government bonds and war savings certificates and stamps. In addition to federal income tax, the States at the beginning of the war imposed no fewer than 11 separate taxes on income at widely varying rates. This severely restricted the Commonwealth's freedom to raise revenue for war purposes and pointed to the need for a simplification of the income tax situation. In 1942 the Commonwealth took over the whole responsibility for the collection of income tax from the States, which were compensated with annual grants. The benefits of uniform taxation were further enhanced when from July 1944 a pay-as-you-earn system replaced payment by instalments, thus giving the government immediate access to a large reserve. Total taxation as a percentage of gross national product increased steadily from 15.6 per cent in 1939 to 24.5 per cent in 1945.

Australia's war economy was greatly assisted by the Lend Lease scheme (inaugurated in March 1941) by which the United States lent or leased war materials, capital equipment and other goods and services to the Allies, payment for which was not required until the end of the war. After Pearl Harbor Australia began to receive direct allocations of aid, and in September 1942 signed a reciprocal Lend Lease agreement with the United States which took the form of provided services, foodstuffs, military stores and construction works, valued at the end of the war at over £280 million. In 1946 the Australian government paid the United States $US27 million in final settlement of the mutual aid agreement.

The Second World War resulted in great and often lasting changes to Australia's economic life and administrative arrangements. Not only had uniform taxation with a pay-as-you-earn system of deductions been introduced, but the Commonwealth Bank assumed permanent control over the trading banks, a large amount of Australia's sizeable overseas debt had been redeemed, Keynesian economics was adopted by the Treasury, and the Commonwealth public service had grown in both experience and stature. Above all perhaps, wartime experience had shown that continuous employment for almost all the workforce was possible. Not surprisingly, Australia, like many other nations, adopted full employment as its guiding postwar economic policy.

While both world wars demanded far-reaching adjustments to Australian economic life, Australia's more advanced economy by the early 1940s was able to make a far greater contribution to the requirements of total war. During the First World War productivity initially fell and then stagnated. A fall in gross domestic product was caused by generally poor seasons and the fact that the high rate of enlistment in the AIF seriously depleted the industrial work force. However, any real participation of Australia's manufacturing sector was limited by its remoteness from the conflict, and high unemployment continued throughout the war. By contrast, the Second World War saw considerable economic innovation and expansion in Australia and a steady increase in gross domestic product. This largely resulted from Australia's stronger industrial and manufacturing base, which facilitated a quick switch over to wartime production, and to the fact that Australia was in a state of full or near full employment.

GERALD WALSH

WAR GRAVES, AUSTRALIAN Australian dead from the two world wars, the Korean War, the Malayan Emergency, the Vietnam War and from the British Commonwealth Occupation Force are buried in 78 countries around the world and in 76 war cemeteries within Australia. There are also individual war graves in 900 Australian civil cemeteries. The largest number of Australian war graves are located in France (25 986), Belgium (8587) and Papua New Guinea (6164). At the other end of the scale are countries such as Portugal, Madagascar, and Romania with one each. There are 12 050 war graves in Australia. The graves are maintained by the office of Australian War Graves (established as a division of the Department of Veterans' Affairs in 1974) and by Australian participation in the Commonwealth War Graves Commission. This commission was established in Britain in 1917 and its charter established the principles under which most Australian war cemeteries are maintained. The charter required that each of the dead should be commemorated individually by name either on the headstone on the grave or by an inscription on a memorial, the headstones and memorials should be permanent, the headstones should be uniform and there should be no distinction made on the basis of military or civil rank. The commission's major war cemeteries contain two standard memorials: a cross of sacrifice in the smaller cemeteries and a stone of remembrance in cemeteries with more than 1000 graves. Engraved on the stones are the words 'Their Name Liveth for Evermore' (adapted from *Ecclesiastes* 4.14 by Rudyard Kipling). The major war cemeteries in Australia are located at Rookwood, New South Wales, with 734 graves; Springvale, Victoria 611; Perth, Western Australia 493; Lutwyche, Queensland 347; and Centennial Park, South Australia 198. Those eligible to be buried in a war grave are any Australian serviceman or woman whose death was determined to be a result of war service, VC winners, those in receipt of a TPI (i.e. totally and permanently incapacitated) pension or extreme disability allowance, multiple amputees on maximum pension rates and ex-POWs. Each headstone contains the following information: name, rank, symbol of faith, regimental badge, number, unit, date of death. At the base of the stone is a short inscription chosen by the family or next of kin of no more than 66 letters. (Families had to pay for this inscription at a rate, in 1919, of $3^{1}/_{2}$ pence per letter.) The inscriptions usually contain conventional religious references but there is the occasional exception such as 'Gone and the light of all our life gone with him.' Those war dead who have no known grave are listed on memorials to the missing. The largest of these overseas are the Australian National War Memorial at Villers-Bretonneux (q.v.) in France and the Menin Gate Memorial (q.v.) at Ypres in Belgium.

WAR MEMORIALS in Australia are essentially a twentieth-century phenomenon, but their construction reflects trends in the commemoration of war dead which had been apparent in Europe and America since the late eighteenth century, when democratic revolutions led to the creation of citizen armies of volunteers. The citizen soldier could be seen as the embodiment of the nation, and was therefore worthy of civic commemoration. In Britain, though, the practice of honouring ordinary soldiers rather than generals did not take off until the 1850s, when more sympathetic reporting, combined with official attempts to improve conditions of service, increased the status of the rank and file. These trends did not become apparent in Australia for some time; since Australians were relatively uninvolved in war in the nineteenth century they built few war memorials before the Boer War. A British regiment stationed in Hobart erected a column in 1850 to members killed in the New Zealand Wars (q.v.) and a plaque in Sydney commemorated Australians who died in the Sudan (q.v.).

Until recently no one thought of building war memorials commemorating Aboriginal resistance to White invasion (q.v.). At the time when this conflict was going on it was noted mainly in inscriptions carved on gravestones or on trees near the sites of Black–White violence, and in place names such as Murdering Creek and Slaughterhouse Gully. One gravestone in Rockhampton cemetery, Queensland, marks the death of a White man 'From the effects of wounds received in pioneer warfare'. In general, though, the few monu-

ments built up to the 1960s which relate to Aboriginal people do not put their subjects in the context of war. The eschewal of military language stands in contrast to New Zealand, where the more obviously military character of the war with the indigenous people led to the construction of war memorials to Whites who 'fell' (rather than simply being 'murdered') in the New Zealand Wars. This contrast may help to explain the fact that Australians, unlike New Zealanders, began building Boer War memorials before that war was even over. With no major memorials to earlier wars, Australians may have felt more keenly the need to show that they had entered history through involvement in war. It may also be that greater public opposition to the war in Australia provoked pro-war forces to display Australian loyalty in stone. Over 100 Boer War memorials were built in Australia (about one for every five dead), of which perhaps more than a quarter had been unveiled by war's end.

It is possible that Australia has proportionately more First World War memorials than any other country. About 1500 Australian First World War memorials have been counted, but the total number may be around 2000. If the latter figure is correct, Australia has one memorial for every 30 dead. By comparison, France, with an estimated 30 000 First World War memorials, has one for every 45 dead. This is particularly notable considering that in France local war memorial construction was subsidised by the central government, whereas in Australia it was a voluntary and municipal affair.

The large number of Australian war memorials is probably the result of three differences between Australia and most other countries which participated in the First World War. First, Australia was the only self-governing nation, apart from South Africa, whose army was composed entirely of volunteers (see CONSCRIPTION). Australians had perhaps a greater incentive, therefore, to recognise those who quite literally *gave* their lives. This characteristic of the 1st AIF may also explain the fact that about half of the Australian war memorials with names record not only those who died but also those who served and returned. Such a policy seems rare or almost unknown in France, Britain and the United States, and even in New Zealand about 85 per cent of memorials list only the dead. Without any means of compelling men to serve, psychological pressure was particularly important in sending Australian men to war. Along with the strong feeling that men should volunteer went the belief that they should be rewarded for doing so, and listing the names of all returned servicemen on war memorials was one such reward. It was also a subtle rebuke to those who did not volunteer.

The second difference between Australia and most other countries is that Australia's dead were all buried on the other side of the world (see WAR GRAVES), so far away that most Australian mourners could not consider visiting their graves. Many of the European dead were buried close to home, so that visiting their graves at least seemed like a real possibility to their relatives. In the United States the bodies of about half the dead were shipped home at the request of their families. But in Australia war memorials stood in place of the distant graves, and Anzac Day ceremonies around them each year provided a surrogate funeral. The word cenotaph, meaning an empty tomb, was often used to describe these memorials. The third difference is that in 1914 Australia had few war memorials, and indeed the period of White settlement had been so short that there were few monuments of any kind. Therefore Australian war memorials were much more prominent in the landscape than in other countries. The relative absence of monuments in Australia may also have made the building of war memorials more important as an expression of communal and national identity.

Local committees began collecting money and deliberating about suitable designs for war memorials even before the war was over; indeed, at least 60 memorials had been completed by the end of 1918. The planning of memorials almost inevitably provoked controversy, particularly over the choice of design. Debate often centred on the question of whether or not memorials should have some other use (that is, whether they should take such forms as halls or clocks as opposed to obelisks or statues). As well as trying to deal with conflicting local opinions, committees had to take account of State bodies established to advise communities on artistic taste and standards. This was particularly true in New South Wales, where all designs had to be submitted to a Public Monuments Advisory Board set up in 1919. Once war memorials had been built communities usually put disputes over the design behind them, each memorial becoming a generally accepted focus for local grief and pride. At unveiling ceremonies political, religious and military leaders made speeches which usually stressed a conservative message of social unity. Memorials were said to be an inspiration to local, national and imperial pride and loyalty, and young people were urged to emulate the virtues with which those honoured on the memorial were invested.

Though most Australian First World War memorials are purely monumental, over 40 per cent have some functional aspect. The hall and the obelisk were almost equally popular (about 20 per cent each of all First World War memorials), followed by

soldier figures and then other forms such as arches and gates, cenotaphs and columns. Many soldier figures were made in Italy, but most war memorials were produced by local tradesmen, usually stonemasons, rather than sculptors or other artists. Whatever their form, war memorials were erected in prominent and accessible sites where they would either stand out (as at intersections) or be in the midst of public activity (as in parks or school grounds or outside halls or railway stations). They were usually surrounded by open space so that they could be used for Anzac Day (q.v.) ceremonies, and were often placed on mounds or stepped bases which raised them well above eye level.

Though not the most common form, soldier figures are perhaps the most striking. Carved from pale stone or marble which gives a somewhat ghostly appearance, the digger (q.v.) in these statues is typically a private who stands at ease or with head bowed and rifle reversed. Unlike in some European countries there are few figures of dying soldiers, nor are there many aggressive ones. The figure seen by English writer D. H. Lawrence on the war memorial at Thirroul, south of Sydney, and described in his 1923 novel *Kangaroo*, is typical of many around Australia: a 'stiff, pallid, delicate fawn-coloured soldier standing forever stiff and pathetic'. There are few female figures on Australian war memorials to counterbalance the masculine figure of the digger. Australia does not have an equivalent of Britannia or France's Marianne, feminine symbols of the nation (nor is there a male symbol of the nation, though perhaps for a time the digger himself played this role).

Because war memorials were meant to act as commemorative focal points for the whole community their messages had to be as uncontroversial as possible. Christian symbolism is rare; very few memorials take the form of crosses, unlike in England where there is an Established Church. Australia's substantial Catholic minority were sensitive about any perceived attempt to undermine the separation of Church and State. Therefore, even biblical texts such as 'Greater love hath no man' and 'Their name liveth for evermore', seemingly innocuous in themselves, are less commonly inscribed on war memorials in Australia than in New Zealand. The texts inscribed on Australian war memorials are generally derivative, and usually exalt the dead rather than the war in which they fought. The most important inscriptions, though, were the names of local men who had served in the war: these were what grieving relatives came to see. They are also a valuable historical resource, since in most places there is no other listing of servicemen by locality.

In addition to the numerous local war memorials there are much larger memorials in the capital cities of each State. These were built later than most local memorials, partly because of their size but also because conflicts over design were more intense and involved more people. The first four State memorials to be completed were Hobart's obelisk (unveiled 1925), Perth's pylon and loggia on a bluff in King's Park perceived as resembling Gallipoli (1929), Brisbane's Doric Circle and eternal flame in Anzac Square (1930), and Adelaide's arch on North Terrace (1931). The greatest controversy occurred in New South Wales and Victoria, whose memorials were not completed until 1934. These two memorials seem almost deliberately designed to confirm stereotypes about the respective 'characters' of the two cities: Sydney's Anzac Memorial in Hyde Park is modernist, while Melbourne's Shrine of Remembrance on a hill in the Domain is a neoclassical pyramidal structure. Melbourne's shrine is possibly the largest purely monumental First World War memorial in the world, and is also notable for being designed so that at 11 a.m. on Armistice Day a ray of sunlight illuminates the word 'love' in the inscription 'Greater love hath no man'. The Australian War Memorial (q.v.) in Canberra, which combines the functions of national memorial, museum and archive, was completed in 1941. Its memorial function is fulfilled most clearly by the Stone of Remembrance, the Pool of Reflection, the Hall of Memory (which since 1993 has housed the Tomb of the Unknown Australian Soldier [q.v.]) and the Roll of Honour, which lists all Australian service personnel who died in war.

After the Second World War there were fewer dead to honour, and their names could simply be added to existing war memorials (as could those of the dead from later wars in Korea, Malaya/Malaysia and Vietnam). The new memorials built after the Second World War reflected the increasing preference for utilitarian memorials throughout the English-speaking world. This trend was encouraged in Australia by the federal government, which provided tax relief for war memorial halls, hospitals, libraries, schools, swimming pools, community centres and other projects. The only memorial which specifically commemorates Australians who fought in post-1945 conflicts, the Australian Vietnam Veterans' National Memorial (q.v.) on Anzac Parade in Canberra, was dedicated in 1992. Plans for a Korean War Memorial on Anzac Parade were announced in October 1994.

Those honoured on Australian war memorials are mostly White Australian male soldiers. This is hardly surprising, since they have made up the vast majority of Australia's armed forces and of its war casualties. The names of nurses appear on some memorials, and there is a War Nurses' Memorial Centre in Melbourne. A few municipal plaques honour women voluntary war workers, and in 1990 a memorial to Second World War servicewomen from New South Wales was unveiled at Gateway Park, Sydney. Since Aboriginal people did not enlist in separate battalions, unlike the Maori in New Zealand, they are not commemorated separately except on a small plaque in bushland behind the Australian War Memorial. However, the 1980s, and particularly the 'Bicentennial' year 1988, did see an increasing number of Aboriginal communities constructing memorials to Aboriginal people killed during the White invasion. Perhaps the most original of these is the Aboriginal Memorial now housed in the National Gallery of Australia, Canberra, which consists of 200 log coffins from Ramingining in Arnhem Land. As yet, though, no one has taken up historian Henry Reynolds's suggestion that 'we make room for the Aboriginal dead on our memorials, cenotaphs, boards of honour', nor is there more than a cursory mention of Aboriginal-White warfare in the Australian War Memorial.

In recent years a number of memorials have been constructed commemorating specific groups or events felt to be inadequately represented on existing memorials. This trend towards 'specialist' memorials is well illustrated along Anzac Parade in Canberra, which runs from the Australian War Memorial towards Parliament House. Eight war memorials stand on either side of Anzac Parade, six of them built in the period 1983–92. There is one memorial to each of the armed services, and two memorials which jointly commemorate Australians and their allies (New Zealanders and Greeks). Not far from Anzac Parade stands another memorial to Australia's allies, the Australian–American Memorial, which commemorates American assistance to Australia during the Second World War. Dedicated in 1954, it takes the form of an eagle with outstretched wings atop a tall metal column and is known locally as 'Bugs Bunny'. Also on Anzac Parade is the Ataturk Memorial (dedicated 1985), named after the Turkish commander at Gallipoli who became Turkey's first president. A rare example of Australian commemoration of former enemies, it honours both Australian and Turkish troops who fought in the Gallipoli campaign.

The way in which war has linked Australia to other countries is also apparent in the various Australian war memorials overseas. Some of these were erected by local communities grateful for Australian wartime assistance. Others, such as the Australian National War Memorial at Villers-Bretonneux and the Lone Pine Memorial at Gallipoli (qq.v.), were constructed by the Imperial War Graves Commission (see WAR GRAVES) to commemorate Australians whose graves could not be found.

For many years war memorials were almost unchallenged as centres of public commemoration, and only those who supported what the memorials stood for went to see them. It was not until the 1960s that they became sites for anti-war, feminist and Aboriginal protest. Even now, protests are limited to painting graffiti on memorials and disruption of or unauthorised participation in Anzac Day ceremonies. No Australian war memorial has been knocked down or blown up as has happened elsewhere in the world. In fact war memorials may be more threatened by inappropriate attempts to restore them than by people hostile to them. A number of soldier figures, for example, have been painted, destroying the intended ghostly effect. Other changes, such as relocation of memorials which had become traffic hazards, have been more justified. Fortunately there seems to be an increasing awareness of the social and historical significance of these monuments, and of the need for sympathetic efforts to preserve them.

WAR PRECAUTIONS ACT, passed 29 October 1914, gave the government extensive powers to pass regulations believed to be necessary for wartime administration. These regulations enabled the government to impose censorship (q.v.), to restrict the civil liberties of aliens (q.v.), and to establish price-fixing machinery, among other things. The act would have allowed the government to impose conscription (q.v.) had it been politically expedient to do so. The passing of this act also brought about an unprecedented centralisation of power, as everything which affected the conduct of the war came under Commonwealth control. The government's powers were so far-reaching that, when a New South Wales politician asked, 'Would it be an offence under the War Precautions Act — ?', Commonwealth Solicitor-General Robert Garran is said to have replied 'Yes' without waiting for him to finish. There were 3442 prosecutions under the act, mostly launched by the Commonwealth attorney-general or the commandants of military districts, and almost all resulted in convictions. Penalties ranged from cautionary fines of a few shillings to punitive fines of £50 or £100, or imprisonment for three or six months. The act continued in force until the passing of the War Precautions Act Repeal Act of 1920–21.

WAR SERVICES LAND SETTLEMENT SCHEME see **SOLDIER SETTLEMENT**

WARRAMANGA, HMAS see **TRIBAL CLASS DESTROYERS**

WARREGO (I), HMAS see **RIVER CLASS TORPEDO BOAT DESTROYERS**

WARREGO (II), HMAS see **GRIMSBY CLASS SLOOPS**

WARRNAMBOOL (I), HMAS see **BATHURST CLASS MINESWEEPERS (CORVETTES)**

WARRNAMBOOL (II), HMAS see **FREMANTLE CLASS PATROL BOATS**

WARUMUNGU, HMAS see **ANZAC CLASS FRIGATES**

WASHINGTON DISARMAMENT CONFERENCE, 1922 see **DISARMAMENT; JAPANESE THREAT**

WASP aeroplane engines were manufactured by the Commonwealth Aircraft Corporation (q.v.) under licence from the Pratt and Whitney Aircraft Division of the United Aircraft Corporation. Six hundred horsepower single-row Wasps were made for Wirraways (q.v.) at the CAC plant at Fishermens Bend, Melbourne. When the Beaufort (q.v.) was redesigned to use Wasps after the unsuitability of the planned Bristol Taurus engines, CAC began manufacturing 1200 h.p. twin-row Wasps at its Lidcombe factory in Sydney. Twin Wasps were also used in the Boomerang (q.v.) fighter. The output of Wasps was never enough to meet the aircraft industry's needs, and though nearly 1000 Twin Wasps were made at Lidcombe a much larger number had to be imported from the United States.

WATERHEN, HMAS see **'V' AND 'W' CLASS DESTROYERS**

WATERHEN, HMAS, at Waverton, Sydney Harbour, was commissioned on 5 December 1962 as a support base for the newly acquired *Ton* Class Minesweepers (q.v.). In March 1969 it became the base for the *Attack* Class Patrol Boats (q.v.), and its commanding officer became the Commander Australian Mine Warfare and Patrol Boat Forces. Though the ships originally based there have been superseded, *Waterhen* is still home to patrol boats and mine counter-measures vessels, as well as to RAN Clearance Diving Team (q.v.) One.

WATERS, Warrant Officer Leonard see **ABORIGINES AND TORRES STRAIT ISLANDERS IN THE ARMED FORCES**

WATSON, HMAS, whose location at South Head, Port Jackson, has long been used for defence purposes, is named after Robert Watson, quartermaster of the First Fleet ship HMS *Sirius*. South Head became the site for Australia's first radar installation in 1942, and on 14 March 1945, *Watson* was commissioned as a naval radar training establishment. Since then its role has expanded to include training in surface warfare, anti-submarine warfare, electronic warfare, navigation and tactics. It also provides support to fleet units and is the host establishment for RAN university students in the Sydney area.

WATSON, John Christian (9 April 1867–18 November 1941) was born in Valparaiso, Chile, but grew up in New Zealand. In 1886 he migrated to Sydney where he worked as a printer-compositor, became active in the labour movement and was involved in the formation of the Labor Party. He held the seat of Young for the Labor Party in the New South Wales Legislative Assembly between 1894 and 1901, and moved to the federal arena at Federation as member for the seat of Bland in 1901. He held the seat until its abolition in December 1906, after which he represented South Sydney until the election in April 1910, which he did not contest. Between 1901 and 1907 he was leader of the Federal Parliamentary Labor Party. Although it was only during his brief term as prime minister (from April to August 1904) that he exercised any direct control over defence, his involvement in debates on defence policy illustrated the changing attitudes towards defence within the ALP. In the immediate aftermath of Federation Watson argued for reduced defence expenditure, but by 1903 he was supporting compulsory military training (see CONSCRIPTION) and the creation of a 'purely Australian navy' (a position he had previously opposed). Like many within the party, his desertion of Labor's anti-militarist stance was motivated mainly by fear of the Japanese, whom he described as 'clever and warlike and ... not governed by altruistic motives', and of 'the sleeping giant — China'. In answer to the socialist charge that workers had no business defending capitalists' property he pointed out that the working class had an interest in defending Australia's social reforms. The defence of the nation was the responsibility of the entire population, since all benefited from the State's protection: 'Universal pensions in age; universal service in youth.' Though his advocacy of compulsory military training for home defence found widespread support within the ALP, Watson was among those expelled from

the party in 1916 over the issue of conscription for overseas service. He fully supported the war effort, joining the Universal Service League to assist with recruiting in 1915 and in that same year becoming an honorary organiser of the scheme to provide returned soldiers with employment. In the latter position he was an early promoter of soldier settlement (q.v.).

WATSON, Major William Thornton (10 November 1887–9 September 1961). Born in New Zealand, Watson came to Australia as a young man aged 25, and before the outbreak of the First World War played representative rugby for New South Wales. He enlisted in the Australian Naval and Military Expeditionary Force in August 1914, which may be said to have begun his association with Australia's New Guinea territories, and fought in the brief campaign that took New Britain from the Germans. He joined the AIF in March the following year as a gunner, and fought on Gallipoli from August until the evacuation. He went to France in April 1916, was promoted to sergeant, and fought on the Somme and through the winter of 1916–17. He won the DCM for rescuing wounded men under fire, and was commissioned in September 1917. He won the MC at Foucaucourt in August 1918 while directing artillery fire as a forward observation officer from exposed positions, and a bar to the award in October. In the period during which the AIF awaited repatriation, Watson captained the AIF Rugby XV which toured England and also played against a number of American university sides.

He continued his rugby career, captaining New South Wales against the traditional enemy, New Zealand, in 1920; then went to New Guinea, where he remained until 1925 and to which he returned between 1932 and 1939 as a copra planter and gold-miner. He was believed to have made a large amount of money in the latter pursuit, spending some time in the United States after 1935, having married an American in 1929. He returned from the United States on the outbreak of war in 1939 and enlisted in the 2nd Garrison Battalion before being promoted to captain and posted back to New Guinea to serve with the newly raised Papuan Infantry Battalion (PIB), precursor of the Pacific Islands Regiment (q.v.). By the time it was called into action in early 1942 he had become the unit's commanding officer, leading it through the early, gruelling fighting against the Japanese forces that landed at Buna and Gona in July, and throughout the Kokoda (q.v.) campaign. He was awarded the DSO after taking command of Maroubra

Force when its commander, Lieutenant-Colonel Owen, was mortally wounded at Kokoda during the first retreat. Thereafter, the PIB under Watson's leadership operated on the flanks of the advancing Australian and American forces and on occasions behind Japanese lines. Watson relinquished command in March 1944. After the war he returned to the United States and became vice-consul in New York between 1945 and 1952. He died in the United States. The official historian, Gavin Long, described him as 'bluff, outspoken, quick on the uptake and quick in speech … with the highest standard of personal courage'. He also regarded him as 'a great captain of a great football team [in 1919]', and noted that, in taking over from a commanding officer who had 'failed' in the position, 'his officers say that he made the PIB'.

WAVELL, Archibald Percival Field Marshal Earl (5 May 1883–24 May 1950) was the British C-in-C, Middle East Command, from July 1939 to July 1941; C-in-C in India, 1941–43; C-in-C, ABDA Command (q.v.), 15 January–25 February 1942, and Viceroy of India, 1943–47. His period in the Middle East was marked by sharp contrasts: he won important victories against Italian forces in Libya and in Ethiopia in 1940–41, and against the Vichy French in Syria in 1941; but in Greece and Crete, where operations were mounted at Churchill's insistence against his strong objections, the outcome was disastrous, and the commitment of such a large proportion of his troops, including Australian forces, enabled the Germans in North Africa to retake all that had been lost the year before, and to lay siege to the fortress of Tobruk, where significant numbers of Australians were involved. Having lost Churchill's confidence, Wavell was sent to India.

When war broke out against Japan he was appointed C-in-C of the hastily formed American–British–Dutch–Australian (ABDA) Command to coordinate resistance to Japan's southward military thrusts. So swift was the Japanese advance and so ill-prepared were the Allies that little was achieved, and Singapore fell to the Japanese on 15 February 1942, despite Churchill's orders to Wavell that it be held to the last man. Returning to India following the dissolution of ABDA Command on 25 February, he attempted to pursue the campaign in Burma, but with very limited resources had little success. On being appointed Viceroy in 1943 his attention turned to political matters, where he was barely more successful, again largely through no fault of his own. Despite his very mixed record during the war, Wavell continued to enjoy the confidence of his troops, including the Australians who fought under his command, especially in

North Africa and Greece. In 1992 a highly classified wartime report by Wavell surfaced, an appendix of which (not written by Wavell) was extremely critical of the behaviour of Australian troops in the last weeks of the Malaya–Singapore campaign.

WAZZA RIOTS took place in Cairo's brothel district on 2 April and 31 July 1915. They seem to have started when Australian and New Zealand troops sacked brothels because of grievances against the prostitutes. It was the intervention of the hated military police, however, which turned these disputes into full-scale riots involving 2000–3000 soldiers. Official inquiries held after both riots failed to reveal exactly who was responsible. The riots were probably caused by a complex mixture of hatred of Egypt and its people, frustration built up over months of enforced idleness, and contempt for military authority (see also EGYPT, AUSTRALIANS IN).

Kevin Fewster, 'The Wazza Riots, 1915', *Journal of the Australian War Memorial*, no. 4 (1984), pp. 47–53.

WELLINGTON, VICKERS (4–5 crew twin-engine bomber [Mark X]). Wingspan 86 feet 2 inches; length 64 feet 7 inches; armament 8 × 0.303-inch machine-guns, 4000 pounds bombs; maximum speed 255 m.p.h.; range 1885 miles; power 2 × Bristol Hercules 1675 h.p. engines.

The Wellington equipped No. 458, 460 and 466 Squadrons RAAF during the Second World War. These squadrons were formed through the Empire Air Training Scheme (q.v.) and served in Europe with Bomber Command and in the Mediterranean. The Wellington had entered service in 1937 and was the RAF's main bomber when the Second World War began. After bearing the brunt of the night bombing campaign against Germany, their place in Bomber Command was taken by larger and faster four-engine bombers. Wellingtons were replaced with Lancasters (q.v.) in No. 460 Squadron in 1942 and by Halifaxes (q.v.) in No. 466 Squadron in 1943. No. 458 Squadron continued to use Wellingtons in the Mediterranean against shipping until January 1945.

WESSEX, WESTLAND (4-crew anti-submarine helicopter). Rotor diameter 56 feet; length 48 feet 4.5 inches; armament 2 × torpedoes or depth charges; maximum speed 125 m.p.h.; range 200 miles; power 1 × Napier Gazelle 1600 shaft h.p. engine.

The Wessex was a British licence-built version of the American Sikorsky S-58, which entered service with the RAN in 1962. They were used in an anti-submarine role from Nowra and aboard HMAS *Melbourne* (q.v.). Four Wessexes also accompanied the troop transport HMAS *Sydney* in 1967 to South Vietnam during the Vietnam War. In 1968 all Wessexes were modified to incorporate the latest sonar equipment and more sophisticated navigational aids. The Wessex was replaced in the anti-submarine role by the Westland Sea Kings (q.v.) in 1975, and were then used as utility transports and for search and rescue. They were flown by the support ship HMAS *Stalwart* (q.v.) and remained in service until 1991.

WESTERN FRONT was the major theatre of operations in the First World War. Its development followed the implementation of the Schlieffen Plan by the Germans in August and September 1914. That plan provided for an enormous wheel through northern France and Belgium by the bulk of the German armies. They were then to envelop Paris, drive the French from the capital and pin them against their own frontier defences in Alsace and Lorraine. The plan, which was in any case beyond the capacity of the Germans to achieve, was thwarted by the French counter-strike at the Marne. A race for the open flank to the north was then instituted by both sides. Neither won it and trench lines began to appear from the Channel

coast to the border of neutral Switzerland. The last German attempt for victory in 1914 occurred at Ypres when the newly raised German formations attempted to break the British Regular Army. They failed and a stalemate ensued.

In 1915, with trench lines growing in depth and sophistication, the Allies took the initiative in launching major attacks to break the deadlock. The British launched offensives at Neuve Chapelle in March, Aubers Ridge in April and May and Loos in September. The French attacked on a much larger scale in Artois and Champagne. None of these offensives gained significant ground; all cost heavy casualties. At the beginning of 1916 the initiative lay with the Germans. They determined to bleed the French Army white by using massive artillery resources against an objective of strategic and emotional significance for the French — an objective from which they would not feel able to withdraw. Von Falkenhayn, the German commander, chose Verdun. The attack was a colossal failure. The Germans found that firing off enormous amounts of artillery was not enough. Troops had to be committed to occupy ground. But when their troops went forward they were subjected to French artillery retaliation from unattacked sections of the

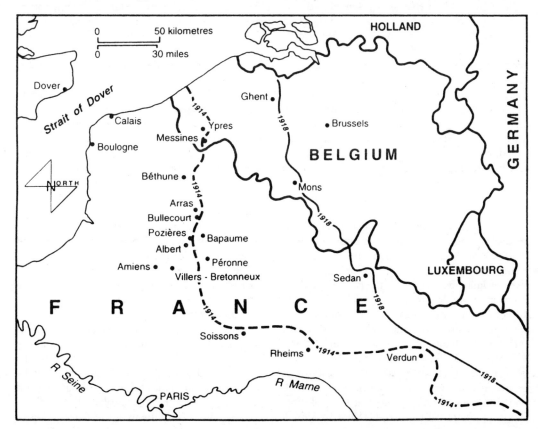

Map 28a   Western Front, 1914–18, showing the movement of the Allied line during the course of the war.

front. By June the casualties on each side were approaching 250 000. A breakthrough or the collapse of the French was nowhere in sight.

During this period the armies of the British Empire on the Western Front were rapidly increasing. By early 1916 about 1 000 000 men had been assembled. It became imperative that these armies be committed to reduce pressure on the hard-pressed French. In February 1916 the British C-in-C, Sir Douglas Haig, and the French commander, General Joffre, discussed a plan for attacking to the north of the River Somme. Initially the offensive was to be conducted mainly by the French, with flank support provided by the British. Gradually, however, as French reserves were drained away for Verdun, the main burden fell upon the British. Haig determined to attack along a 20-mile front to the north of the River Somme. The French role was now reduced to providing flank protection for the British. In the first instance the high ground facing the British — the Thiepval–Pozières–Ginchy Ridge — would be seized. Then the cavalry would sweep through the broken German ranks, advance on Bapaume and roll up the German line from south to north. A great strategic victory would have been achieved.

This plan had no chance of being realised. The cavalry were obsolete as a weapon on the Western Front. Had they managed to deploy, they would have been destroyed by rearward German machine-gunners. The fact was, however, that Haig did not possess the artillery resources to crack the German defences on a 20-mile front. The first day (1 July 1916) saw a casualty list unparalleled in British military history. About 60 000 casualties were suffered, approximately a third of which were killed. No ground at all had been gained in the north (the Thiepval–Pozières sector), and only a derisory amount in the south, where Fricourt and some woods beyond were captured. In the next two weeks operations were concentrated on the southern section of the British line. Ground was gradually gained towards the German second position around Longueval. Then in an innovative night operation on 14 July, accompanied by an overwhelming bombardment, the British seized the German second line from Longueval to just south of Pozières. Meanwhile the British command had been reorganised. General Rawlinson's Fourth Army took charge of the southern operations. General Gough's Reserve (later Fifth) Army was to concentrate on the formidable Thiepval-Pozières sector. At this point I Anzac Corps (1st, 2nd, 4th and 5th Divisions) under Lieutenant-General Birdwood (q.v.), with Brigadier-General Brudenell White (q.v.) as chief of staff, was transferred to

Gough and notified that it would shortly be committed to the battle.

At the point that the Australians were to enter battle on the Western Front the problem of the attack in trench warfare had been reduced to a matter of artillery resources. The two weapons that had prevented the infantry from gaining significant ground in all previous offensives were the machine-gun and the enemy's artillery. Machine-guns, firing about 600 rounds per minute, could destroy any formation of infantry once it had left the safety of its trenches and ventured into no man's land. Such a slow-moving body could also be hit from long range by enemy artillery which could quite easily drop the shells into the relatively large space that no man's land usually represented. The task for the attackers, therefore, was to annihilate these weapons or at least to suppress their fire for the period of time that its own infantry was exposed in no man's land. As the foot-soldiers could not carry with them the implements to overcome the barbed-wire protected and entrenched machine-gunners (let alone the distant enemy guns) these tasks fell to their own artillery. In late July 1916, however, artillery was not present in such volume to accomplish these tasks. Nor did it have sufficient accuracy to destroy the small, distant targets presented by the enemy batteries even if its numbers had been adequate.

As it happened the first operation to be carried out by Australian troops on the Western Front was not to take place on the Somme but at Fromelles, in the area around Aubers Ridge, the site of so many barren British offensives in 1915. On 5 July Haig had detected, for reasons that remain obscure, a weakening of German morale. He instructed his other army commanders to be ready to take any opportunities to attack that might present themselves. In the First Army sector held by the 5th Australian Division and 61st British Division an operation was put in place to seize the German defensive positions in front of the Aubers Ridge. The result was a fiasco. The troops had been committed to an advance across water-logged ground under the eyes of the enemy on the ridge, from where all preparations for the attack were carefully observed. The preliminary bombardment (such as it was) missed the German defences completely. The British troops were shot down from the beginning and made no progress. The Australians managed to break into the German defences, but their line was not continuous and they were vulnerable to counter-attack from either flank. They were withdrawn under the protection of a barrage, but not before they had lost 5300 casualties. The Battle of Fromelles was one of the most misconcieved operations mounted by the British on the Western

Front. An abiding memory of it is of Brigadier-General H. E. Elliott (q.v.), whose 15th Brigade had suffered particularly heavy casualties, weeping while he shook the hands of the survivors as they made their way back from the German lines.

On the Somme the first task given to the Australians was the capture of the ruins of Pozières village. This was a particularly difficult objective as it involved an attack up a steep slope with the high ground occupied by the enemy. The Australians were unfortunate to enter the battle at a time when the prevailing doctrine was to attempt to get forward by small-scale, narrow-front attacks. These tactics allowed the Germans to concentrate all their artillery fire on the area under attack and to bring enfilading artillery fire to bear from either flank. Nor were the Australians favoured by their position in the line. They were close to the boundary with the Fourth Army. During this period Gough and Rawlinson hardly ever coordinated their attacks.

The attack by 1st Australian Division on Pozières began on the night of 23 July 1916. By the 25th the village had been captured. This was significant progress in Western Front terms. The line had advanced nearly 1000 yards, and an important objective seized. However the Australians were also to find that such successes were not won without cost. The German artillery, which had not been dominated, constantly threw a rain of shells at the troops. Machine-guns, especially on the right flank, also exacted a toll. When 1st Australian Division was relieved by the 2nd, it had suffered 5285 casualties.

The objective for the Australians during the next six weeks was the high ground between Pozières village and Mouquet Farm near Thiepval. From these positions Gough intended to seize a vital section of the ridge that had been the British objective for the first day of the campaign. Between the end of July and early September, when the corps was relieved, they launched nine separate attacks. These carried the line forward to the fringe of Mouquet Farm, a distance of about 2000 yards. Most attacks gained little ground, and were extremely costly because the guns supporting the Australians could not dominate the German artillery; nor could they eliminate sufficient German machine-guns to allow the infantry to progress. Consequently the troops, not being bullet-proof, could only inch forward. The battles, units involved, objectives, and casualties are listed in Table 1.

After 3 September the Australian divisions were taken out of the line for a well-earned rest. Around Pozières and Mouquet Farm they had lost approximately 23 000 officers and men in six weeks — about equal to the casualties suffered on Gallipoli in eight months. The operations they were involved in were ill-conceived in the extreme. The fronts of attack were too narrow; and the inability of the artillery to perform tasks essential to any advance was never taken into account by the command. The ground captured proved to be of little importance to the overall campaign, although the Germans were impressed by the tenacity of the Australian attacks. Elsewhere on the Somme much the same story was repeated. Lest it be thought that the British command was especially extravagant in the use of colonial divisions, the casualty figures for British formations engaged at the same time are instructive. To give just three examples, the Guards, 1st Division and 3rd Divisions, which were operating to the right of the Australians in July and August, suffered 23 500 casualties, about the same number as the 1st, 2nd and 4th Australian Divisions.

For the remainder of September and October the Australian divisions were sent to a 'quiet' sector of the front between Ypres and Armentières. Here they rested, received reinforcements and carried

Table 1 Battles involving Australian divisions around Pozières and Mouquet Farm, 1916

Date (1916)	Division	Objective	Result	Casualties
28/29 July	2nd	Trenches NE Pozières	Failed	6848
4/5 August	2nd	Trenches NE Pozières	Captured	
9 August	4th	Trenches towards Mouquet Farm	Captured	4649
12/13 August	4th	Trenches towards Mouquet Farm	Captured	
16 August	1st	Mouquet Farm	Failed	2650
21 August	1st	Mouquet Farm	Small advance	
26 August	2nd	Mouquet Farm	Failed	900
29/30 August	4th	Mouquet Farm	Failed	1800
3 September	4th	Mouquet Farm	Small gain	

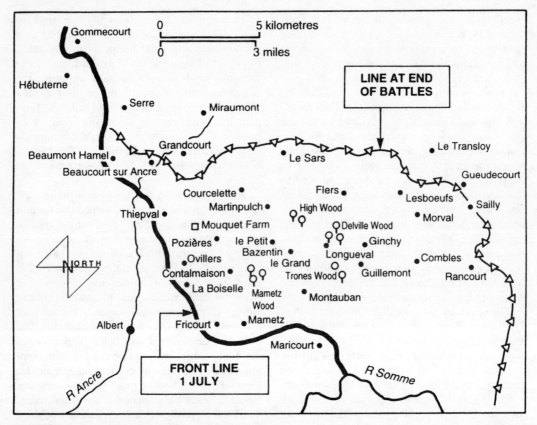

Gommecourt

0   5 kilometres

0   3 miles

LINE AT END OF BATTLES

Hébuterne

Serre

Miraumont

Le Transloy

Grandcourt

Le Sars

Gueudecourt

Beaumont Hamel

Beaucourt sur Ancre

Courcelette

Flers

Lesboeufs

Sailly

Thiepval

Martinpulch

High Wood

Morval

Mouquet Farm

Delville Wood

Ginchy

Pozières

le Petit Bazentin

Longueval

Combles

NORTH

Ovillers

le Grand

Guillemont

Rancourt

Contalmaison

Trones Wood

La Boiselle

Mametz Wood

Montauban

Albert

Fricourt

Mametz

Maricourt

FRONT LINE 1 JULY

R Ancre

R Somme

Map 28b Area of operations, Somme campaign, 1916.

out many minor operations and trench raids. They had not yet, however, finished with the Somme. While the Australians were resting the Battle of the Somme ground on. Rawlinson's Fourth Army on the right of the British front bore the main burden. Rawlinson had been employing the same narrow-front, small-scale attack policy as Gough with similar results. The front inched forward at great cost. In September, however, Haig forced a change on the reluctant Rawlinson. There would now be fewer attacks — but they would be larger in scale and on wider fronts.

The first of these operations took place near Flers on the 15th. Tanks were used for the first time in warfare in this battle and provided useful adjuncts to the infantry but little more. Nevertheless a German trench system was captured. A larger success was secured 10 days later (without tanks) when a crushing artillery bombardment allowed the British to overrun the German defensive system around Guedecourt on a 10 000 yard front. The British were, however, now faced with a further group of German defences called the Transloy Line. Rawlinson commenced operations against this position on 7 October. At this point the weather intervened. Rain turned the battlefield

into a quagmire, making aerial spotting and ground observation for the artillery impossible. As a consequence most of the bombardment missed the German lines and the infantry were unable to get forward. He tried again in similar conditions on 12, 18, 23, 28 and 29 October. The results were the same — no gains at considerable loss. After the last failure Rawlinson's exhausted divisions were rested and replaced by I Anzac Corps.

Despite the unpropitious circumstances and ignoring all previous experience, Rawlinson determined to assault the Transloy Line again. Conditions were almost indescribable. Troops had to assist each other out of their trenches. Many got bogged in no man's land. The artillery, without the assistance of observation of any kind, fired blind. Therefore, when 2nd Australian Division attacked (in conjunction with three other British divisions), there was little prospect of success. At the end of the day a miniscule amount of ground had been gained around a German position known as the Maze. The cost had been 820 casualties.

The Somme battle was at an end but the Australians were kept in the line in one of the bleakest winters of the war. The weather conditions were appalling, with frequent snow and freezing sleet creating slush; trench-foot and frostbite cases increased. Warm clothing, including 65 000 sheepskin jackets from Australia, helped reduce the misery somewhat and preventive medical procedures were stepped up. The Somme winter, however, was the most uncomfortable that the Australians would spend on the Western Front.

So far the folly of the British tactics during the Somme campaign has been stressed, but it must be noted that those employed by the Germans were hardly more sensible. Throughout most of the campaign the Germans launched counter-attacks to regain every yard of ground lost, whether it was of tactical significance or not. These tactics were almost always unprofitable and very costly. Furthermore the Germans had been shocked by the materiel superiority of the British. For every shell they fired they received many more in return and although many of these shells did not hit the target aimed at, they often hit a trench, a dugout or headquarters, which all added to the German casualties. Thus by the end of the campaign the Germans on the Somme were worn out. Their commanders, facing a repetition of the campaign in the spring, made plans to retire to a stronger, shorter line to the rear, cutting across the Noyon salient. The retreat to what became known as the Hindenburg Line began in late February 1917.

The 2nd, 1st and 5th Divisions were holding the line when the German retreat commenced. Along with British divisions to their right and left, they began slowly to follow-up the retreating Germans. Getting to grips with the enemy proved difficult. The Germans were devastating the ground over which they were retreating, causing severe supply and communications problems. Furthermore, they left strong rear-guard units, of which machine-gunners were the key, to harry and delay troops who were now advancing in the open.

For the next four weeks the Australians pressed forward against the German rear-guards, fighting occasionally for possession of a trench or shattered village. On 17 March it fell to the Australians of the 8th Brigade (5th Division) to occupy Bapaume — the objective Haig had set for the cavalry long ago on 1 July 1916. From Bapaume operations became more intense. The British were now closing on the Hindenburg line, which the Germans had protected by fortifying a number of outpost villages. Before any attempt could be made to assault the main line, these villages would have to be captured. From 2 April to the 9th the Australian forces assaulted and captured many of these outpost villages: Doignies and Noreuil on the 2nd, and Boursies, Demicourt and Hermies between the 3rd and the 8th. In the whole period of open warfare the

Australian infantry showed that, given the appropriate amount of fire support, they could undertake sophisticated manoeuvres, outflanking and enveloping the villages and then mounting converging attacks on them. By 9 April they were within striking distance of the Hindenburg Line. But by then the main action had switched to their north.

The great Allied campaign for 1917 had been decided by the British and French governments at a conference in December 1916. In that month Joffre had been replaced by General Nivelle, who was able to convince his government and then the British Prime Minister David Lloyd George of the efficacy of his new plan. Under this scheme, the main attack would be made by the French in Champagne. To assist, the British First, Third and Fifth Armies would attack at Arras on 9 April. Their contribution would be to seize the important high ground of Vimy Ridge to attract German reserves away from the French, and generally to wear down the enemy. When this had been accomplished Nivelle would strike.

The Australians were to play no part in the main operations of this campaign. However Gough's Fifth Army in which they were again located lay just to the south of General Allenby's Third Army, the main British strike force for Arras. It was considered that if the Fifth Army attacked the Hindenburg Line in this area it could threaten the German flank and thus materially assist the main operation. Gough therefore developed a plan to penetrate the Hindenburg Line around the fortified village of Bullecourt. The 4th Australian Division would attack to the right of the village and then turn left along the Hindenburg trench system and advance on it from the flank. This would be the signal for the 62nd Division to move forward and capture the village. Both divisions would then link up behind Bullecourt and push through to the rear of the German defences.

The result was a fiasco. Tanks were to be used in the Fifth Army for the first time. When it was discovered by patrols that the formidable wire protecting the Hindenburg Line had survived an artillery bombardment, Gough decided to allow tanks to cut their way through it. To enable the tanks to operate in safety the artillery barrage supporting the troops was dispensed with. Then most of the tanks failed to arrive. The result was that many troops advanced to the attack unprotected either by tanks or artillery. Amazingly, in the circumstances, some units of the division managed to break into the Hindenburg Line. But there, unsupported on either flank and cut off from reinforcements by the German barrage, the attack withered.

Over 1000 Australians were captured, the largest number in a single action for the war.

First Bullecourt caused much bitterness. Birdwood and White had warned Gough that he was not allowing sufficient time for the wire-cutting operations. Nor were they convinced that tanks were a suitable substitute for artillery in wire-cutting operations. Gough had overruled them and was proved wrong. There was also some anger at the supposed passivity of the 62nd Division, although the preconditions for its commitment had not been fulfilled and to send it against a thoroughly aroused German defence would have accomplished nothing.

Despite this failure and the cost to the 4th Division of over 3000 casualties Gough determined to try again. This time the Anzac staff insisted on much closer control over the preparations. The 3rd of May was fixed for the new attack. Bullecourt was to be enveloped by the 62nd Division to the left and 2nd Australian Division to the right. Given the failure of the Arras offensive, the original purpose of the operation had disappeared. Its only rationale, in which the Australian staff seemed to acquiesce, was to redeem the failure of the first operation.

On the first day of operations the Australians, but not the British, broke into the Hindenburg Line. This time the wire had been well cut by the artillery. The next two weeks saw some of the most intense trench fighting of the war. The Australian troops continually tried to advance down the Hindenburg Line on Bullecourt; the Germans repeatedly counter-attacked. On 8 May the 5th Division replaced the 2nd but with hardly better results. Gradually German resistance in the village crumbled and on the 17th the British on the left occupied it. A small, tactically useless village had been captured. It had cost 7000 casualties for no purpose. Serious questions had been raised about Gough's fitness for high command.

For the British, the Arras/Bullecourt offensive was only a prelude to the main offensive for 1917. Since 1916 Haig had wanted to launch a large offensive from the Ypres salient with the object of occupying the Belgian coast. The failure of Nivelle's offensive in May and the stalling of his own attack at Arras gave him the chance. The problem for Haig of launching an offensive from Ypres was that the Germans held the surrounding high ground. In particular they held the Messines Ridge which gave them perfect observation over any preparations in the salient. Haig, therefore, determined to capture Messines in a preliminary operation.

There were good reasons for selecting Messines. Since 1915 General Plumer's Second Army had been tunnelling under the Ridge and placing huge mines in the excavations. In this activity they had received considerable assistance from the 1st Australian Tunnelling Company (see MINING OPERATIONS, FIRST WORLD WAR). The AIF was also to play a role in the battle. During 1916 a new Australian division (the 3rd) had been raised in Australia and placed under the command of General Monash. It formed part of II Anzac Corps which launched the southern section of the assault. (The other formation in the corps was the New Zealand Division; the 4th Australian Division was in reserve.)

Plumer determined to subdue the German defences on the ridge and their protecting batteries behind it with a crushing artillery bombardment. This opened on 21 May and was aided by good weather and the superiority of the RFC in the air. On the day of battle (7 June) the enemy artillery had been dominated — but not eliminated — and most German defences on the ridge reduced to ruins. The attack was followed by the detonation of the mines — almost 1 million pounds of TNT which produced the largest man-made explosion in history to that date. The Germans on the ridge were either buried or stunned. As the British troops stormed up the ridge they met little resistance.

The only set back came in the area of II Anzac Corps, attacking on the southern flank. The approach of 3rd Australian Division happened to coincide with a severe gas shelling by the Germans. Some battalions lost 10 per cent of their strength before they reached the front. Altogether between 500 and 1000 men were gassed. Even so, enemy resistance was so weak that the depleted Australians took their first objective with ease.

Further north the New Zealanders occupied the ruins of Messines village while the British divisions captured the remainder of the ridge. So far the operation had been a stunning success. It was at this moment that the follow-up forces (which included the 4th Australian Division) began their advance down the ridge to capture an intermediate German defensive position, the Oostaverne Line. Communication between the front-line troops and the artillery was always a problem in the First World War. Now, as the troops began the advance, the artillery took them for enemy counter-attack forces and opened fire. When the message was passed back that these were in fact friendly troops the artillery still considered that a counter-attack was underway and merely shortened their range. They were then firing on the original attack formations digging in on the lee of the ridge. So both the 4th and the 3rd Australian Divisions had many casualties inflicted by what would now be called friendly fire. To make matters worse, the German artillery that had sur-

Trench warfare on the Western Front. Stretcher bearers of the 45th Battalion in a trench near Ypres, Belgium, 28 September 1917. (AWM E00839)

vived the bombardment opened fire on the crowded troops on the ridge. Despite this hostile fire the British prevailed. The German infantry in the vicinity of the Oostaverne Line was too stunned to put up much resistance and the reserves were too distant to intervene. Gradually, therefore, along with troops to their north, II Anzac Corps made ground and occupied their final objective. Messines had been a great success, but as was usually the case on the Western Front, success did not come cheaply. The 3rd Australian Division lost 4100 casualties and the 4th Division 2700.

With the Messines Ridge safely in his hands Haig could begin preparations for his great sweep to the Belgian coast. In planning his next stroke he made two surprising decisions. He allowed a seven week interval between Messines and the main attack, thus giving the Germans ample time to strengthen their defences in the salient. And, notwithstanding Bullecourt, he placed Gough in charge of the campaign.

The main offensive, which some units of the Australian artillery supported, but in which no Australian infantry took part, began on 31 July. On the first day Gough scored a minor success. In the north the Pilckem Ridge fell to the British and

gains of over 3000 yards were made. Against the crucial Gheluvelt Plateau, however (crucial because from its modest heights the Germans could observe all British movement to the north and behind which there lay an enormous concentration of German guns that could enfilade the northern British Corps), he made little progress.

On 1 August, before the attack could be resumed, rain started to fall. Before long the entire battlefield became a quagmire. Artillery observation was hampered by driving rain and low cloud. Forward movement was almost impossible. Logic now dictated that all operations cease until the ground dried but Gough determined to press on. A renewed attack on the Gheluvelt Plateau was mounted on the 10th. It failed completely. Then a broader front attack on the 16th (the Battle of Langemarck) came to grief in the mud. To explain the failure Gough (himself an Ulsterman) pointed to the lack of fortitude of his Irish Divisions. He might have been better served examining his own command. Despite the conditions more attacks were mounted on 22, 23 and 25 August. No gains were made.

Finally Haig intervened. He placed all troops facing the plateau under Plumer's Second Army and relegated the Fifth Army to providing flank protection to the north. Plumer wisely insisted that the operations halt until the weather improved. He also asked for fresh troops. Part of that contingent was I Anzac Corps. The Australians were about to enter the Third Battle of Ypres. Their first operation, the Battle of the Menin Road, was scheduled for 20 September

It was Plumer's intent to capture the Gheluvelt Plateau by a series of strictly limited operations or steps. Each step would be about 1500 yards and be preceded by an enormous artillery bombardment. When the final objective had been reached a barrage of shells some 700 to 1000 yards in depth would be placed in front of the troops while they consolidated their position. This measure was designed to protect the troops from the attentions of German counter-attack formations. These units were held well-back, and when committed against disorganised attacking troops had proved able to regain much of the ground taken in the initial assault.

The Menin Road battle commenced at dawn on 20 September after an artillery bombardment four times the concentration employed by Gough on 31 July. I Anzac Corps (2nd and 1st Australian Divisions) were located on the left of the Second Army area. To the south three more divisions from Plumer's force attacked; to the north Gough employed three divisions of the Fifth Army.

At the end of the day success was reported all along the front. 1st and 2nd Australian Divisions fought their way forward 1500 yards and, with the help of the standing barrage, beat off all counter-attacks. There is a general belief that these Plumer battles were relatively easy affairs — a matter of following a barrage of shells until the objectives had been reached. In truth they were difficult battles with much hard fighting. So formidable were the German defences in this area — they consisted of fortified zones, rather than lines, across which were scattered concrete pillboxes, all supported by massive artillery concentrations — that even a severe bombardment was unlikely to subdue them all. What the bombardment did do was to suppress or eliminate just enough of the zone defences and the enemy artillery to allow an infantry skilled in small unit tactics to fight its way through those that remained without intolerable loss. That is, enough troops would remain at the end of the day to consolidate and hold a line already well protected by artillery. This was what happened at the Menin Road. The Australian troops, on their way to victory, were hit by the German artillery, then fired on from a series of strong points in the belt of woods on the Plateau and numerous machine-gun rests in pillboxes, and finally came under the fire of their

own guns. In all 1st Division suffered 2500 casualties, 2nd Division 1250.

The next two battles took place at Polygon Wood on 26 September and Broodseinde on 4 October. The Australian units taking part in all three 'Plumer battles' are listed in Table 2. Each step in these battles carried the troops a further 1500 yards across the Gheluvelt Plateau after heavy fighting and moderately high casualties.

It is worth noting in regard to Broodseinde that observers at the time and historians since felt that Plumer's forces were on the verge of dealing the German armies in the Ypres salient a fatal blow. Only rain after the 4th, so the argument goes, thwarted them. There is little to recommend this view. It derived in part from the larger than usual number of enemy dead littering the battlefield. But this occurred because of a singular circumstance. The Germans too had been about to launch an attack on the 4th. They had massed troops forward to carry it out and these forces had been caught by the British bombardment.

Furthermore, Broodseinde indicated that Plumer's method could not be repeated at short intervals indefinitely. The barrage supporting the troops on the 4th had been weak because it was proving difficult to get sufficient guns forward expeditiously in the ploughed up morass of the salient. Moreover each of Plumer's steps was so short that no enemy guns were being captured. This, taken together with the fact that artillery reinforcements were being trickled forward to the German command, meant that each attack would meet with an increased barrage from the enemy. Finally, constant shelling had destroyed the drainage system of the small streams that traversed the salient. Even without rain these water courses had become bogs or swamps that were proving difficult obstacles for the attacking troops. This problem was hardly likely to improve in the future, especially as October was a notoriously wet month in Belgium and the Second and Fifth Armies were approaching an area of exceptionally low-lying ground. Taken together all these circumstances indicated it was unlikely that Broodseinde would be able to be repeated quickly with any hope of success.

Yet this is what Plumer and Gough and Haig planned. In the aftermath of Broodseinde optimism ran high. The rain, which had started on the 4th and become torrential on the following days, was discounted. German morale was sure to crack if another blow was delivered. This blow was the Battle of Poelcappelle which commenced on 9 October. Australian participation was provided by 2nd Division in II Anzac Corps. The attack was a failure, mired in mud from the beginning. The 2nd

Table 2 Australian divisions involved in the 'Plumer battles', 1917

Date (1917)	Battle	Australian divisions involved	Casualties
20 September	Menin Road	1st	1250
		2nd	2500
26 September	Polygon Wood	4th	1500
4 October	Broodseinde	1st	2500
		2nd	2200
		3rd	1800

Division was one of the few units to make any ground but units attacking on either side of it were hardly capable of movement through the sodden terrain. Eventually ferocious fire on both flanks drove the Australians back. This battle resulted in a loss of 1250 casualties for no purpose.

Amazingly Poelcappelle did not demonstrate even to Plumer the necessity of halting the attack. On the contrary, he hastened his next effort despite the fact that rain was continuing to fall. On the 12th the 3rd Australian Division, as part of II Anzac Corps, was sent against Passchendaele. There was never any chance of success. Artillery protection was practically non-existent, and in any case the mud was so deep that troops could not have made use of it. Indeed it was almost impossible to stand. Some Australian troops with the New Zealanders on their left did struggle forward a few hundred yards. But they were not in sufficient strength to hold the ground captured. Nor were they supported. At the end of the 12th (which is dignified by the name of the First Battle of Passchendaele), they were back in their own lines. So concluded major Australian participation in the Third Ypres campaign. It was left to the Canadian Corps to capture a section of the Passchendaele Ridge in early November in conditions even worse than those experienced by the Australians.

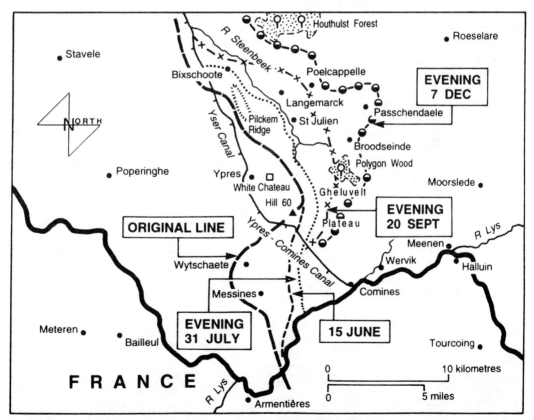

Map 28c Area of operations, Third Battle of Ypres (Passchendaele), 1917, showing the movement of the British line.

Much has been made of the three Plumer victories in which the Australians participated. They should be placed, however, in the context of the overall aims of the campaign. Plumer advanced the British line about 4500 yards. He captured a significant section of the Gheluvelt Plateau as well as a section of the Passchendaele Ridge. These achievements could at least be compared favourably with Gough's futile efforts in August. However, the Passchendaele Ridge had been the objective for the second day of the campaign. The Belgian coast lay 50 miles beyond. Thus there was never any chance that the larger aims of the campaign could be accomplished by Plumer's or any other method. German morale held, despite Haig's constant predictions that it was about to collapse. Enemy counter-attacks were as ferocious at the end of the campaign as they had been at the beginning. Nor were the Germans running out of troops or artillery. Hence the three operations in which the Australians participated were small local successes but nothing more.

After the disastrous end of the Passchendaele campaign all Australian divisions were taken out of the line to rest. The corps was transferred to the relatively quiet Messines front for the winter. All divisions took their turn in the line, the main activity being the strengthening of fortifications and raiding. It was during this period that a major reorganisation of the Australian forces took place. There had been a movement for some time for all Australian divisions to be unified in a corps under one command. On 1 November 1917 Haig agreed and the Australian Corps came into being.

It was clear during the winter of 1917–18 that the strategic situation on the Western Front was changing. The last British offensive at Cambrai on 20 November, after enjoying an initial success, had proved barren and on 30 November German counter-attack forces had driven their opponents back to their original front line. Depleted and exhausted after Passchendaele, the remainder of the British Army had lost all offensive capability for the moment.

This was not the case on the German side. In the course of 1917 the Russian front had disintegrated as that country lapsed into anarchy and revolution. By the end of 1917 the Germans were bringing back a continual stream of divisions to the west. It was obvious that they would be used offensively as soon as the ground dried in the spring. This prospect left Haig with a dilemma. In October he had agreed to an extension of the British front to the south to allow the French to rest some of their mutinous forces. Yet the Third Ypres and Cambrai campaigns had cost the British the equivalent of about 15 divisions in casualties. The result was that when the extension of the front took place the British forces were stretched very thinly along it. Haig's dilemma was that with this depleted force, he could not hope to be strong everywhere. Sensibly, he chose to concentrate his forces and strengthen his defences in the north — so safeguarding the Channel ports and his lines of communication with England. But this meant that the Fifth Army which held the St Quentin sector in the south had few divisions, hardly any reserves and only rudimentary defences. On 21 March this was the sector against which Ludendorff struck.

In the course of the next few days the battered remnants of Gough's force were relentlessly driven back. The line of the Somme around Péronne was lost, then Bapaume, then much of the old Somme battlefield. Clearly reserves were needed to shore-up the line. The Australian Corps were occupying a quiet sector — its divisions were therefore ordered to the Somme.

During the next month units from the 4th, 3rd, 5th and 2nd Australian Divisions helped stem the German attack in the area of the Somme from Hebuterne in the north to Hangard Wood in the south. The fighting was often confused, with units from various divisions detached to particular areas as the exigencies of the situation required. The areas around which the major actions were fought are listed in Table 3.

In addition, the 1st Division, between 12 and 24 April, helped to stem Ludendorff's second great offensive against the British on the River Lys. Its main area of operation was around Strazeele and Meteren.

The only major attack launched by the Austra-

Table 3 Locations of major actions on the Somme, March–April 1918

Date (1918)	Division	Location
26 March–5 April	4th Brigade (4th Div.)	Hebuterne
26 March–5 April	12th and 13th Brigades (4th Div.)	Dernancourt
27 March	3rd Division	Morlancourt
30 March–4 April	9th Brigade (3rd Div.)	Villers-Bretonneux
4 April	5th Division	River Somme–Villers-Bretonneux
4 April	2nd Division	In reserve Somme area

lians in this period took place at Villers-Breton-neux on the night of the 24–25 April. This village was an important step towards the rail junction of Amiens. The Australians had defended it in early April, successfully driving the Germans back. These troops had then been replaced by an inexperienced British unit, which lost the town to a German attack on 23 April. A counter-attack was quickly organised. It was carried out at night, the 15th Brigade (Elliott) attacking to the north of Villers-Bretonneux and 13th Brigade (Glasgow) attacking to the south. The brigades met at the eastern edge of the village and the Germans occupying it were surrounded. The attack went in on the night of the 24–25th. It was carried through with extreme ferocity, few prisoners being taken. It was a complete success. This was the scene of the first tank-to-tank battle in the history of war. The Germans were pushed back from Villers-Bretonneux. Never again did they venture down the road to Amiens.

The Australian divisions had played a notable part in halting the German offensive. However, exaggerated accounts (both then and since) have intimated that the Australians stopped the Germans almost single-handed and saved the Channel ports. This was hardly the case. Most units from the battered Fifth Army were replaced with French divisions as they were withdrawn from the line. Moreover, by the time the Australian troops arrived on the Somme, the German offensive was, in the main, running out of steam. The German troops were tired, depleted in number due to the high casualties exacted by the Fifth Army, and outrunning their artillery and logistic support. The irresistible tide of 21 March was certainly on the ebb by the time the Australians arrived.

These factors should also be taken into account when considering some of the comparisons made between the retreating British forces and the Australian troops who relieved them. The British had borne the full weight of the German attack and were exhausted. The Australians, fresh from months in a quiet sector, met the Germans when the attackers were exhausted. This was a quite different equation. Indeed it was a fortunate circumstance for Australia that its divisions were not once more in the front line with the Fifth Army. In that case half or more of the AIF might have been lost in a matter of days.

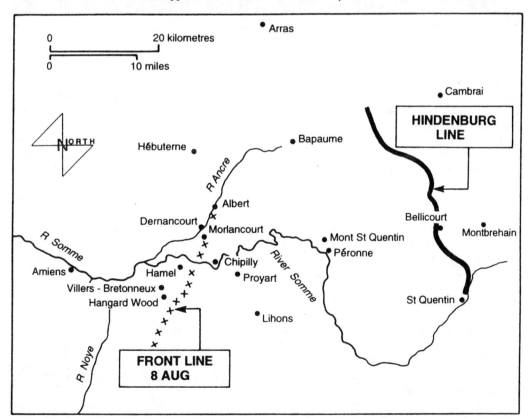

Map 28d Western Front, 1918, showing the position of the front line at the beginning of the British offensive on 8 August, and the Hindenburg defences.

On the Western Front in 1918, trench warfare ended and fighting became more open. Here troops of the 5th Brigade wait with artillery and British tanks begin an attack near Wanfusee-Abancourt, France, 8 August 1918. (AWM E03883)

As the German thrusts against the British were held and subsequent enemy offensives against the French were contained, Allied thoughts turned towards counter-attack. As they did so it is necessary to grasp what new developments in warfare were available to assist them when they once more took up the offensive. The first factor was one of quantity. In the second half of 1918, mainly due to the prodigious output of the British Ministry of Munitions, the soldiers would have many more machine-guns, Lewis guns (q.v.), rifle grenades and trench mortars per division than ever before. The second factor was one of quality. The Mark IV tank, used at Cambrai in November 1917, was a much more reliable machine than the Mark I used on the Somme. More reliable still was the Mark V tank which would come into general use from June 1918.

It was artillery, however, that witnessed the greatest qualitative improvement. For some time the British had been working on a system of sound detection to locate enemy batteries. That work had now come to fruition. Each time an enemy battery fired its guns the exact location of the battery could be established. British guns could be silently ranged

onto this location before a battle. Then when battle commenced the British artillery could suddenly spring into life with every likelihood that their shells would on most occasions find their targets. Surprise could thus be reintroduced into battle. This method, called 'sound-ranging', together with improved methods of aerial photography and sophisticated methods of correcting British guns for wear, meant that artillery could at last tilt the balance of trench warfare in favour of the attackers.

The first opportunity to test the new techniques came at the Battle of Hamel on 4 July 1918. The action was a small affair — a two brigade attack to straighten a section of the line. The infantry would be accompanied by the new Mark V tank. On the insistence of the 4th Division Commander, Major-General E. W. Sinclair-Maclagan (q.v.), a creeping barrage (i.e. a curtain of shells that advanced immediately in front of the infantry) was placed in front of the infantry and tanks. Six-hundred guns were allotted for the bombardment. The operation was a great success. The Germans were either killed by the bombardment or the tanks, or rounded up by the infantry. German supporting artillery had been completely neutralised and did not interfere with the attack. The line was straightened at a cost of 850 casualties. On the German side 1000 were taken prisoner, with 1000 killed or wounded. It was one of the rare occasions when attack had proved cheaper than defence.

Australians patrolling around Morlancourt before Hamel and on subsequent occasions after the battle, indicated that the Germans facing them were of low morale and might be vulnerable to a much larger stroke. (In preparation for the next phase the Australians, by a series of trench raids known as 'peaceful penetration', gradually pushed back the Germans until a desired start line had been obtained.)

The Battle of Amiens, as the forthcoming encounter was later known, was planned by General Rawlinson and his staff at Fourth Army with considerable contributions from Monash and Currie, the Canadian Corps Commander. The main front of attack ran from the Amiens–Roye Road to the River Somme and was shared by the Canadians and Australians. To the north of the river the British III Corps provided flank protection. The entire Tank Corps of 552 tanks (most of them Mark V) participated. The final objective lay between five and eight miles distant — the furthest distance that could be covered by the artillery. Unlike other battles there was no strategic aim. Rawlinson intended merely to push the Germans away from the Amiens rail junction.

The preparations for Amiens marked a new level of professionalism in the British Army. The Canadian Corps and the tanks were moved into position with great secrecy. The guns with the aid of sound-ranging were silently registered on the German batteries. The battle opened on 8 August 1918 in a thick fog. The Australian Corps had 3rd and 2nd Division in the line, with 4th and 5th Divisions just in rear of them ready to leapfrog through and advance on the final objective. By mid-afternoon all main objectives were in British hands. The cost had been relatively low. Overall casualties were 9000, of which about 3000 were Australian. What factors had led to this remarkable success?

Most credit must be given to the artillery. It was noted earlier that for infantry actions to be successful, the artillery was required to eliminate sufficient of the enemy machine-gunners and batteries to allow the infantry to advance with tolerable loss. This had at last been achieved. At zero hour the British gunners blanketed the enemy batteries with a deluge of fire, eliminating most of them as a factor from the outset. Then the troops advanced on the rudimentary enemy defences close behind a moving curtain of artillery fire (the creeping barrage). This enabled the foot-soldiers to storm the

enemy machine-gunners as they attempted to man their weapons. (In this the assaulting troops were also aided by the thick mist.) Once the first series of defences were overrun the tanks could be unleashed. They lumbered towards the next defensive line unhampered by enemy artillery fire and were able to create chaos among the rearward German units. Finally the second wave of infantry pushed on beyond the final tank objective and were able to consolidate their positions behind a curtain of protective artillery fire.

Not all had gone well. To the north of the Somme the understrength divisions of III Corps, operating in difficult country, were unable to keep pace with the Australian advance. The 4th Australian Division in particular suffered from severe flanking fire from the spurs which overlooked the Somme in this area. They solved this problem eventually by crossing the river and themselves dealing with the hostile machine-guns. So ended the Battle of Amiens. The Germans suffered 27 000 casualties, a large percentage of whom had been caught by the rapid advance. The Allies also captured 450 guns intact.

The next few days did not live up to the expectations of the initial results. Gradually the troops moved beyond the protection of much of their artillery. There were fewer and fewer tanks to support them — largely due to mechanical failure and crew exhaustion rather than enemy action. Attacks also proved difficult to coordinate across the whole front as commanders were unsure of the exact position of their forward troops. Gradually too German resistance stiffened. Fresh divisions were brought up, artillery replaced, new defences improvised. On 9, 10 and 11 August smaller and smaller advances were made at higher and higher cost. The Australians had some particularly difficult battles — the 5th Division at Lihons, and the 3rd Division at Proyart and north of the Somme at Chipilly and Etinehem. Casualties for the period 7–14 August were: 1st Division 2000, 2nd Division 1300, 3rd Division 1100, 4th Division 800, 5th Division 900. After the 11th a pause was instituted to enable reorganisation for a new battle. This in itself marked a new realism on the part of the high command. In this interval the Canadian Corps left the line, the front now being held by the Australian Corps and III Corps.

Haig had opened a new front to the north by ordering the Third Army into battle on 21 August. On 23 August they were followed by the Fourth Army with 3rd and 1st Australian Divisions playing the major role. From 23 August to 3 September the Fourth Army's battle front was in continual motion as the Germans were pressed back to the line of the

Somme. Rawlinson had in fact issued orders to Monash on the 25th to slacken the offensive and await events in the north. Monash interpreted these orders liberally. He considered the Germans to be in disarray and that his own corps had ample fighting power to deal with them. The pursuit continued.

The crowning achievement of this period of fighting was the capture of Mont St Quentin (north of Péronne) by 2nd Australian Division. Mont St Quentin was the key to the German defence of the line of the Somme. If it was lost the whole German position would be turned and a further retreat would become inevitable.

The 2nd Division approached the bare slopes of the Mont on 2 September. Accompanied by a formidable artillery bombardment and skilfully using rifle grenades and trench-mortars to outflank German outpost positions, the troops stormed up the slopes and ejected the German Guards Division from the defences on the summit. With Mont St Quentin secured the 5th Division was able to assault Péronne to the south. The Germans had no choice now but to retreat to their last defensive position on this sector of the front, the Hindenburg Line.

By 1918 the Germans had converted the Hindenburg Line into a defensive zone 6000 yards deep. Within this zone there were six lines, all supplied with machine-gun posts and protected by wide belts of barbed wire. In the south of the Fourth Army Sector the major obstacle of the St Quentin canal had been incorporated into the defence. Rawlinson proposed to Haig that these defences should be seized in two stages. First, the outpost line should be taken. This would allow good observation of the main line which would be assaulted in a separate operation about 10 days later. The Australian Corps was given the major task of breaking the outpost defences in the centre. It was supported to the north by III Corps and in the south by IX Corps.

Monash placed the 1st and 4th Australian Divisions in the line for the attack which took place in drizzling rain and mist on 18 September. The Australian troops, supported by a formidable artillery barrage broke right through the German positions and by the end of the day were overlooking the main Hindenburg defensive position. Casualties, numbering 1260, were light. The Australians also took 4300 prisoners and captured 76 guns. On the flanks much less progress was made. The III Corps were not blessed with a competent commander and their artillery support left much to be desired. Consequently their advance faltered and many of the outpost positions in their area remained in German hands. In the south IX Corps made better

progress but their inexperience left them lagging well behind the Australians.

After the 18th planning began at once for the main attack. Monash wished to confine operations to a very narrow front across the section of the St Quentin Canal that ran through a tunnel. Rawlinson overruled him insisting on broadening the attack to the south where IX Corps would assault across the canal. In the centre the initial attack was to be made by the 27th and 30th American Divisions lent to Rawlinson for the purpose. Once they had captured the initial objective 3rd and 5th Australian Divisions would leap-frog through them to the final objective on the far side of the Hindenburg defences.

The bombardment of the Hindenburg Line began on 26 September. From then until zero hour on the 29th 750 000 shells were fired at the German defences. Despite this ferocious bombardment all did not go well with the American/Australian attack. The bombardment had in fact missed some important German strong points which proceeded to take a great toll on the initial American assault. The 5th Australian Division that was supposed to leap-frog through the 27th Division to more distant objectives, found themselves caught up in the fighting for the first. By this time all supporting tanks had been knocked out. The battle in the centre degenerated into a series of small, confused, uncoordinated attempts to take the strong points. Further south the 30th American Division had achieved greater success thanks to the accuracy of the bombardment. The 5th Australian Division then moved through and captured Bellicourt. They were then held up by flanking fire to the north. A stalemate appeared imminent.

These partial successes were, however, redeemed by events further south. There the 46th (North Midland) Division had been carried by a bombardment of unparalleled ferocity across the St Quentin canal and through the Hindenburg defences to a distance of 4000 yards. This movement effectively outflanked the Germans holding up the Australians to the north. Gradually the enemy troops fell back — relentlessly pursued by the Australians. One line after another was captured. On 2 October the 2nd Australian Division replaced the 3rd and 5th. On 5 October the last Australian infantry action of the war took place when the 2nd Division assaulted Montbrehain. With its capture the Hindenburg Line was broken. The Australian Corps was withdrawn to rest. It was scheduled to return in November. By then it was not needed. All five British armies, the French and the Americans launched a series of offensives in October against which the Germans could not stand. On 11 November an Armistice (which really amounted to surrender on terms imposed by the Allies) was concluded. At last all was quiet on the Western Front.

From 1916 to 1918 Australian forces, for the first time in their history, engaged the main army of the main enemy in the main theatre of war. In this respect the Western Front constitutes the major episode in Australian military history. In its timing this commitment was unfortunate. Until 1918 armies of all nations were squandered in ill-conceived, poorly executed operations which gained little ground at high cost. The Australian infantry paid heavily for this tactical and strategic vacuum.

By common agreement this infantry was rated highly by friend and foe. It was the misfortune of the Australians to exhibit bravery and skill at a time when those two attributes were not enough. The First World War was a war of machines — in particular a war of artillery. Without adequate artillery support the most skilful infantry was doomed to failure — as Australian operations on the Somme demonstrated. With proper fire-support Australian soldiers (as well as British and other dominion troops) could perform well, as the Plumer battles of 1917 and the final campaigns of 1918 showed. There is no evidence, however, that the Australians were used by an unfeeling British command as shock troops in these later campaigns. It just happened that when Ludendorff attacked in March 1918, the Australian Corps (as well as most other dominion divisions) was out of the line. These rested divisions therefore represented the best reserves then available to Haig. In these circumstances it was only natural that they should play a major role in the Allied counter-offensives, along with those British units that had been resting or those brought from other theatres. Nevertheless, Australian troops played a notable part in the final victory and in freeing Western Europe from the overbearing weight of German military autocracy.

WESTRALIA (I), HMAS see **ARMED MERCHANT CRUISERS**

WESTRALIA (II), HMAS see **GULF WAR**

WHITE, Bruce (3 November 1916-19 August 1984). White joined the Commonwealth Public Service in 1934 after graduating from the University of Melbourne. During the Second World War he served as a navigator in the RAAF, enlisting in February 1942. He was commissioned in April 1943, and ended the war as a flight lieutenant, having flown with No. 13 and No. 2 Squadrons in northern Australia. He held a number of senior positions with the Overseas Telecommunications

Commission and the Public Service Board in Canberra before appointment as Secretary of the Department of the Army in April 1958.

White has been described as 'a civilised, witty, slightly old-fashioned senior public servant'. The service ministries were junior portfolios, and the secretaries of these departments were responsible for financial and administrative oversight, not for the formulation of defence or military policy. Because the Army portfolio was usually a minister's first appointment, the Secretary's job required him to compensate for the inexperience of his minister. The quality of Army ministers varied substantially during White's tenure.

White came to public attention, unwelcomed, when a speech he had given on 31 October 1966 to the Canberra Branch of the Royal Aeronautical Society was reported prominently in the press, despite his having been assured that it would not be. In it he criticised the conduct of operations in Vietnam and the bombing policy of the Americans. The speech had not been cleared with his minister, Malcolm Fraser (q.v.), in advance, but Fraser stood by White publicly and headed off any possibility of dismissal or other reprimand. White had been unwise, while Fraser's loyalty to his permanent head did him credit. White's position was saved by the fact that he had no role in policy advice or formulation with regard to Vietnam.

In February 1973 White moved from the Department of the Army to become permanent head assisting the Secretary of the Department of Defence, Sir Arthur Tange (q.v.). He headed the Reorganisation Study Group, which was dominated by civilian officials and was charged with initial research and consultation leading ultimately to the reorganisation of the defence group of departments in line with the recommendations contained in the Tange Report. He retired from the Public Service in 1975.

WHITE, General Cyril Brudenell Bingham (23 September 1876–13 August 1940). Born at St Arnaud, Victoria, and educated at Brisbane Central Boys' School and Eton Preparatory School, Nundah, White was provisionally commissioned into the 2nd Queensland Regiment in 1896 before transferring to the Permanent Military Forces in 1898, when he was commissioned in the Queensland Regiment of the Royal Australian Artillery. He saw service briefly in the Boer War in 1902, and returned to Australia to resume regimental duties. In 1904 he was appointed aide-de-camp to Major-General E. T. H. Hutton (q.v.), GOC Australian Military Forces. Through this position he travelled widely within Australia, familiarising himself with the new army, and so impressing Hutton that the latter arranged for White to attend the British Army Staff College in 1906, the first AMF officer to do so. Despite having limited regimental experience, White completed the course with creditable results, and with a growing conviction of the importance of first-rate staff work and of the necessity of an imperial approach to defence.

Upon his return to Australia in early 1908 he was promoted to captain and attached to the staff of the chief of intelligence, Colonel W. T. Bridges (q.v.). He served there only a few months before being posted back to London as a GSO3 at the War Office. This enabled him to develop a detailed knowledge of the structure of the British Army and it also deepened his belief in the imperial model of defence. He was promoted to major in 1911, and returned to Australia the following year to become Director of Military Operations. Under instructions from the Minister of Defence, Senator George Pearce (q.v.), who had approved consultations with New Zealand, White secretly drew up plans that would enable the dispatch overseas of a combined force of 18 000 (of which 12 000 would come from Australia) in time of war.

White was acting chief of staff in July 1914 when the war crisis developed, and on the outbreak of war the following month he was able to assure the government that it would be possible to raise and train a force of 20 000, though not necessarily to send it overseas within six weeks, which was the period envisaged in his prewar plan. In August Bridges was appointed to command the AIF, and he chose White to be his chief of staff. Using White's plan, they raised, equipped and trained the AIF, the first contingents of which left Australia in November 1914, disembarking in Egypt when it was found that the promised camps would not be ready to receive them in England. He assisted Bridges in planning the Australian landing on the Gallipoli peninsula, although they disagreed on the details, and accompanied Bridges on a tour of the chaotic front line on the first day of the attack. This was the first of frequent inspections of the situation in dangerous circumstances that eventually, on 15 May, resulted in Bridges suffering a mortal wound from a sniper; White was with him when he was struck. He was awarded the DSO in June, and at the beginning of October became Brigadier-General, General Staff, Australian and New Zealand Army Corps, under its new commander, Lieutenant-General Sir William Birdwood (q.v.). White's major achievement in this position was to plan and supervise the evacuation from Gallipoli, which was concluded on the night of 19–20 December without any casualties and without the Turkish defenders becoming aware of it. With the

return of the Gallipoli force to Egypt, White oversaw the expansion of the AIF to four divisions, and at the end of March 1916, having been appointed CB, left for France as chief of staff of I Anzac Corps under Birdwood.

White played a critical role in the performance of the Anzac Corps on the Western Front. While Birdwood exercised command and insisted on close and regular contact with his troops, it was White who undertook the essential planning and administration, not only for the AIF's operations in the great offensives of 1916–17 but also for its essential maintenance in France. Although committed to the ideal of imperial cooperation, White was always conscious of the need to protect Australian interests, especially where they were threatened by poor staff work. He never hesitated to reject unfair criticism of the Australian performance, or to speak plainly to senior officers when he disagreed with their proposals. On 1 January 1917 he was promoted to temporary major-general, and maintained an efficient staff system in the advance to the Hindenburg Line, planning and writing the orders for all the operations carried out by I Anzac Corps. In July he was offered command of the corps by the British C-in-C, Field Marshal Sir Douglas Haig, but he rejected the offer, preferring to maintain the working relationship with Birdwood. In December he was appointed CMG, and in May 1918 found himself involved, unwillingly, in another attempt to appoint him as corps commander in place of Birdwood who had moved to take over command of the Fifth Army after its collapse in the March German offensive. The official war correspondent, C. E. W. Bean, and the Australian journalist Keith Murdoch (qq.v.) tried to block the appointment of Major-General Sir John Monash as corps commander, urging instead that Monash be promoted to general and given command of the AIF in England, thus leaving the way for White to take command of the corps. White refused to become party to their scheming, and Monash was duly appointed to the corps command. White refused to place his own advancement before the well-being of the AIF, and when Birdwood assumed his new position, White as Major-General, General Staff, Fifth Army, retained administrative command of the AIF. When the war ended in November 1918, White served briefly in London to supervise the demobilisation and repatriation program, having been promoted to temporary lieutenant-general and made chief of staff, AIF. He was appointed KCMG in the 1919 New Year's Honours, and for his work on the Western Front he received five foreign decorations as well as being mentioned in despatches five times.

White returned to Australia in July 1919 and was appointed to a committee to consider the post-war future of the army. The committee recommended a modified compulsory training scheme to support a citizen force of 180 000 in six infantry and two mounted divisions. It fell to White, appointed CGS on 1 June 1920, to implement the scheme, which the government reduced to 130 000 in September 1920, but even this proved impossible under the swingeing defence cuts that were imposed in 1922.

He retired as CGS in 1923 and became chairman of the Commonwealth Public Service Board, a position he held until 1928 when, further honoured for his work in organising royal tours (KCVO, 1920; KCB, 1927), he left public life. Some claimed that he was active in the right-wing, secret armies (q.v.) that sprang up in the 1930s, but the evidence for these claims is flimsy, to say the least. He resumed his military career on 15 March 1940 when, newly promoted to general, he became CGS following the death in office of the incumbent, Lieutenant-General E. K. Squires. White's tenure as CGS was barely five months: he was killed in a plane crash near Canberra Airport in August 1940.

White was a central figure in the story of the 1st AIF. An outstanding staff officer, planner and administrator at the highest level, he provided the essential support that made the ultimate victories of the AIF possible. Although convinced that Australia's future lay in close cooperation with the imperial ideal, he remained fiercely protective of Australian interests, neither hesitating to defend the AIF's reputation against unwarranted criticism nor seeking to advance his own career at the expense of the AIF he had done so much to create and sustain. Bean, the chronicler and celebrator of the deeds of the AIF, wrote of White's 'perfect sense of proportion in seeing his own place and that of others in whatever situation confronted him', and concluded that 'He was the greatest man I have known.'

C. E. W. Bean, *Two Men I Knew* (Angus & Robertson, Sydney, 1957).

WHITE ARMY see **SECRET ARMIES**

WHITEHEAD, Brigadier David Adie 'Torpy' (30 September 1896–23 October 1992) was born in Scotland. He graduated from RMC Duntroon in April 1916 and was seconded immediately to the AIF in France. He commanded the 23rd Machine Gun Company in 1917–18 and was adjutant of the 3rd Machine Gun Battalion in the final weeks of the Australians' involvement in the fighting.

Between January 1919 and his repatriation in September that year Whitehead served as GSO3 on the headquarters of the 3rd Division. He was awarded the MC in September 1918 for gallantry while leading a battery of eight machine-guns under heavy fire in support of an attack by the 33rd Battalion near Bray, on the Somme, in late August 1918 during which his company suffered 23 casualties. He left the army after his return to Australia, but maintained his interest in soldiering through the CMF, and held a variety of regimental appointments between the wars culminating in his promotion to lieutenant-colonel and command of the 1st Machine Gun Regiment in October 1937.

Like many former regulars who had left after the First World War, he returned to active service in May 1940 in command of the 2/2nd Machine Gun Battalion in the 2nd AIF, as part of the 7th Division. In February 1942 he was given command of the 2/32nd Battalion in the 9th Division's 24th Brigade, and at El Alamein in October commanded the 26th Brigade which saw heavy fighting on the right of the 9th Division's position. He returned to Australia with his brigade and commanded it for the rest of the war, in the division's landing at Lae and the fighting in the Ramu and Markham valleys, and in the amphibious assault at Tarakan in June 1945 in which, heavily reinforced, the 26th Brigade was closer to a division than a brigade in size. Following the war, Whitehead returned to citizen soldiering, commanding the 2nd Armoured Brigade of the CMF from November 1947. He retired from the army in 1954. In civil life he worked as a manager for the Shell Company between 1946 and 1956, and from 1956 until his retirement in 1961 was a conciliator with the Concilitaion and Arbitration Commission. His nickname, short for 'torpedo', was a play on the Whitehead torpedo (q.v.).

WHYALLA (I), HMAS see **BATHURST CLASS MINE-SWEEPERS (CORVETTES)**

WHYALLA (II), HMAS see **FREMANTLE CLASS PATROL BOATS**

WIGMORE, Lionel Gage (14 March 1899–8 November 1989) was born in England and migrated to Australia in 1919 after working as a journalist in New Zealand. He worked on several Sydney newspapers in the early 1920s before becoming a publicist for the oil industry in 1927. Wigmore served as a citizen soldier for a few years in the 1930s, being commissioned as a lieutenant in the Australian Army Service Corps in November 1938, but by January 1940 he had been transferred to the Reserve of Officers. When the war began in 1939 he was made Deputy Director of the Australian Department of Information, and later became its representative in Singapore. He made a very fortunate escape as the Japanese assault on the city began in February 1942, returning to Australia via Tjilatjap in the Netherlands East Indies. For the rest of the war he worked for the Department of Information, including two periods as press attaché in New Delhi during and after the war. The official historian, Gavin Long (q.v.), invited Wigmore to write a volume in the army series dealing with the campaigns in Malaya and Singapore after the writer originally nominated, T. W. Mitchell, a former POW of the Japanese (and son-in-law of Chauvel [q.v.]), was elected to the Victorian parliament in 1947 and dropped out of the project. Given his experiences in Singapore Wigmore was an excellent choice, but the subject matter of his volume was highly contentious. The British official historian for the Far East, Major-General S. Woodburn Kirby, disagreed strongly with a number of Wigmore's conclusions and took the unusual step of travelling to Australia in 1950 in order to examine the Australian evidence for himself. More serious were attempts on the part of the Minister for the Interior, Colonel Sir Wilfrid Kent Hughes (another former 8th Division POW), to interfere with Wigmore's critical treatment of the 8th Division's commander, Lieutenant-General Henry Gordon Bennett (q.v.). Bennett, too, made largely ineffectual attempts to shape the history's treatment of his command. The volume, entitled *The Japanese Thrust*, was published in 1957 after being delayed as the result of a British government request that it, and its British equivalent volume, not appear before the grant of independence to Malaya had been completed successfully in August 1957. The volume included a lengthy account of the treatment of Australian POWs in the Far East by A. J. Sweeting, a member of Long's staff. Given the prominent attention that the sufferings of Japanese prisoners attracted, it is curious that POWs did not receive a separate volume. Wigmore published several local histories of Canberra and the Snowy Mountains region during the 1960s and, with Bruce Harding, a book devoted to Australian Victoria Cross and George Cross winners entitled *They Dared Mightily* (1963).

WILLIAMS, Air Marshal Richard (3 August 1890–7 February 1980). Born at Moonta Mines, South Australia, and educated at Moonta Public School, Williams was commissioned in the South Australian Infantry Regiment, AMF, in 1911, and at the end of the following year was appointed to the Permanent Military Forces and promoted to lieutenant in the Administrative and Instructional Staff. In August 1914 he attended (as one of a class of only four) the first three-month war flying course at Point Cook, Victoria, returning in July 1915 for a two-month advanced course. He expected to be sent to India, whose government had requested air assistance in Mesopotamia, but instead the Australian government decided to raise complete air units for service with the RFC. Williams was appointed a captain in the AIF, and posted to Egypt as flight commander to No. 1 Squadron, Australian Flying Corps, which from December 1916 operated in support of the ground advance into Palestine. He was awarded the DSO in 1917 for several acts of gallantry, and in May 1917 assumed command of the squadron. He was mentioned in despatches twice in early 1918, and in June, having been promoted to temporary lieutenant-colonel, was seconded to command the 40th (Army) Wing, Palestine Brigade, RFC. He never saw action again. He was awarded the OBE in 1919, and after some months in London where he liaised with the RAF over the requirements for a projected Australian air force, he returned to Australia to serve as the army's aviation expert on a joint army–navy board charged with establishing the air force.

When the Air Board was created in November 1920, Williams was appointed First Air Member with the rank of wing commander. The Second Air Member, also with the rank of wing commander, was S. J. Goble (q.v.), who as a former member of the Royal Naval Air Service and subsequently a member of the joint army–navy board had argued the navy's case against an independent air force. Thus began a bitter, long-running personal and professional dispute between the senior officers of the air force which poisoned relations for the next 20 years and set a most unfortunate precedent in RAAF circles for personal rivalries to parade as professional disagreements. Although Williams's advocacy of an independent air force seemed to bear fruit with the establishment of the RAAF on 31 March 1921, the other two services exercised considerable control over the new service through the Air Council, set up in 1920, which came between the Air Board and the minister for Defence, and which, although designed to ensure that neither service gained control of the air force, in fact enabled them both to inhibit the growth of the third service. Williams became CAS (a position Goble, with RAN support, coveted) in 1922, and the following year, to give him more formal training in staff work, he attended the British Army Staff College at Camberley and the RAF Staff College at Andover. He returned to Australia and was promoted to group captain on 1 July 1925, only to find that in his absence Goble had agreed to a navy plan to establish a fleet air arm, which Williams regarded as a serious threat to the continued existence of an independent air force. He managed to prevent the implementation of the plan, thus further deepening the rift with Goble. He was appointed CBE in 1927 for undertaking a long-range flight along the east coast of Australia to New Guinea, New Britain and the Solomon Islands to form a defence appreciation of the area, and was promoted to air commodore on 1 July.

In 1928 a report by Air Marshal Sir John Salmond, formerly CAS, RAF, had been highly critical of the state of the RAAF, which Salmond found to be 'totally unfit to undertake war operations', but he did confirm Williams's leadership of the RAAF. For the next several years, as the Depression deepened and defence cuts became more severe, Williams successfully fought off attempts by the other services to abolish the RAAF and absorb its functions into their own. After attending the Imperial Defence College (q.v.) in 1933, he was promoted to air vice-marshal in 1935 (an important step because it made him, for the first time, equal in rank to the chiefs of the other two services) and appointed CB. He resisted attempts by the Minister for Defence, Archdale Parkhill (q.v.), to man three new RAAF squadrons with members of the Citizen Air Force, which Williams rightly saw as yet another attempt to diminish the standing and professionalism of the RAAF, but he supported Parkhill's efforts to develop a local aircraft industry, thereby lessening Australia's dependence on British production, and the Commonwealth Aircraft Corporation was established in 1936.

By 1938 some progress had been made towards repairing the deficiencies identified in the Salmond report of 1928, but at a cost to operational safety standards. A further report by another former CAS, RAF, Marshal of the RAF Sir Edward Ellington (q.v.), was critical of training procedures, and as they had been part of Williams's responsibility since 1934 (which Goble ensured became known to the government), he bore the brunt of the report's adverse findings, and was sent on detachment to the RAF for two years. He was Air Officer in Command of Administration for most of 1939, and returned to Australia in early 1940 to be Air Member for Organisation and equipment, hoping, with the backing of the Minister for Air, J.V. Fairbairn (q.v.), to be made CAS. Prime Minister Menzies, however, was intent on bringing in a senior RAF officer to the position. Sir Charles Burnett was appointed CAS with the rank of air chief marshal, and Williams was given the temporary rank of air

marshal. The establishment of the Empire Air Training Scheme (q.v.) in December 1939 effectively subordinated the RAAF to the RAF, and was the antithesis of everything Williams had worked for for the previous 20 years. Williams was posted to London in October 1941 to establish and command RAAF Overseas Headquarters, which was responsible for the increasing numbers of EATS-produced RAAF personnel serving overseas. He returned to Australia shortly after the outbreak of war in the Pacific, but was not, as he had hoped, appointed CAS but instead was posted almost immediately to Washington as RAAF representative.

He retired from the RAAF on 14 September 1946 (much against his will), having already become Director-General of Civil Aviation in June. In that position he was in charge of the expansion of civil aviation in Australia, both domestic and international. He was made KBE in 1954 and retired the following year.

Williams is rightly regarded as the 'Father of the RAAF'. Had it not been for his unswerving dedication to the cause of an independent air force from 1919, and for his skill in promoting the cause in the face of fierce opposition from the other two services and their political allies, there is little doubt that the RAAF would not have been established or that it would not have survived. For all his undoubted success on this level, however, it cannot be said that Williams's legacy to the RAAF was without blemish. His protracted feud with Goble divided the senior command of the RAAF when it could least afford this display of antagonism within its own ranks. More important, Williams' understandable preoccupation with the question of survival, primarily for the RAAF but also for himself, distracted him from the important task of articulating a doctrine of air power that was clearly suited to Australian circumstances. Although in the mid-1920s he had written at length and with considerable perception on the role of the RAAF in the defence of Australia, his contributions to the debate on the Singapore strategy (q.v.) in the 1930s was surprisingly low key and failed to demonstrate the validity of his claims for the central role of air power.

C. D. Coulthard-Clark, *The Third Brother: The Royal Australian Air Force 1921–39* (Allen & Unwin, Sydney, 1991).

WILLIAMSTOWN DOCKYARD in Melbourne is the specialist yard for construction, modernisation and conversion of RAN destroyer-type ships, but it also builds and refits other ships up to destroyer size. A patent slipway was constructed at Williamstown in 1858, and this was followed in 1873 by the com-

The Alfred Graving Dock at Williamstown is used for the first time by Her Majesty's Victorian Ship *Nelson*, March 1874. (AWM 302440)

pletion of the Alfred Graving Dock, which is still in use in the 1990s. The graving dock was made part of the Victorian State Dockyard in 1911, and shipbuilding commenced in 1913. The yard was used during the First World War to convert cargo ships to transports, then was bought by the Commonwealth in 1918 to build cargo ships to offset war losses. Sold to the Melbourne Harbour Board in 1924, it began building corvettes when the Second World War broke out. It was bought by the Department of the Navy on 28 October 1942 and became HMA Naval Dockyard, Williamstown. In December 1987, Defence Minister Kim Beazley announced that the dockyard would be sold for $100 million to the Australian Marine Engineering Corporation.

WILLIAMTOWN, RAAF BASE, lies 30 kilometres north of Newcastle and is currently the RAAF's main fighter base. Williamtown is home to the F/A-18s (q.v.) of No. 3 and 77 Squadrons of No. 81 Wing. The site of 300 hectares was purchased on 8 March 1940 and the base, which became operational as a training establishment on 15 February 1941, was named for the nearby town of Williamtown. No. 450, 454 and 458 Squadrons were formed at Williamtown in early 1941 as part of the Empire Air Training Scheme (q.v.) and were

sent to the Mediterranean. The United States Army Air Force ran Williamtown during the Second World War from 16 May 1942.

WILMOT, Chester Reginald William Winchester (21 June 1911–10 January 1954). Wilmot was born in Brighton, Victoria, and was educated at Melbourne Grammar and the University of Melbourne (where he was influenced by Sir Ernest Scott [q.v]). He graduated BA in 1935 and LLB in 1936. He spent the next few years touring the world, debating in Japan and the USA and (briefly) broadcasting cricket for the Australian Broadcasting Commission (ABC) in England. On the outbreak of war he returned to Australia where he enlisted but was claimed by the ABC in February 1940 as their Middle East correspondent. He served in the Middle East from 1940 to 1942. He covered the early successes and then reverses of the British forces there (see LIBYAN CAMPAIGN) and was besieged with the 9th Division at Tobruk in 1941. Late in that year he was wounded in the leg by friendly fire and was hospitalised in Cairo. After the entry of Japan into the war he returned to Australia, from where the ABC sent him to cover the New Guinea campaign. His broadcasts from Port Moresby, Milne Bay and the Kokoda Track made him a popular figure in Australia. But he soon fell foul of the C-in-C, General Sir Thomas Blamey (q.v.). While in Australia Wilmot had been free in his criticism of Blamey, and in New Guinea became a supporter of Lieutenant-General S. F. Rowell (q.v.). When Blamey sacked Rowell Wilmot made his feelings clear and was subsequently denied accreditation as a war correspondent by the C-in-C. Wilmot returned to Australia under a cloud and despite the success of his story of Tobruk written in 1943, his career seemed to be languishing. He was rescued by the British Broadcasting Corporation, who picked him as part of their team to cover D-Day (q.v.) and subsequent operations. He accompanied the 6th Airborne Division on the day of invasion and his broadcasts from the field brought him much fame in England. He accompanied the British forces throughout the entire campaign in western Europe to the German surrender at Luneberg Heath. (There is a famous photograph of Montgomery signing the surrender with Wilmot hovering in the background.) After the war he continued to broadcast and write. He was selected as the author of one of the volumes of the Australian Official History but was killed in the crash of a Comet jet liner in 1954.

Wilmot's fame rests on his writing. His first book, *Tobruk*, was a meticulously researched work, checked by participants and eye-witnesses where possible. About 15 000 copies were sold when it appeared in 1944. His principal work, however, was *The Struggle For Europe* (1952). Although the main focus of the book is on the 1944–45 campaign in western Europe, which Wilmot witnessed, it ranges more widely than that and is better thought of as a history of the European aspect of the war — especially from the British perspective. The research on which the book is based is immensely detailed. Wilmot interviewed many of the major participants on both sides. He was among the first historians to make extensive use of captured German documents. His papers in King's College, London, are a revelation and a testimony to his great historical skills. The battle pieces in the book are models of their kind — clear, incisive and stylish. The political focus of the book now seems old-fashioned in its attempt to show how Allied (usually American) blunders left the Soviet Union predominant in Europe. Nevertheless, its pages are infused with the values of liberal democracy and a hatred of Nazism and all forms of totalitarian rule.

WILTON, General John Gordon Noel (22 November 1910–10 May 1981). Wilton graduated from RMC Duntroon into what was probably the leanest period in the Army's history, in December 1930. From a class of 12, four each went to the RAAF and the Australian Army; Wilton and three others joined the British Army, in Wilton's case serving largely in India and even gaining some minor active service experience in Burma. He returned to the Australian Army, as a captain, in May 1939.

After a stint in coastal artillery, Wilton was seconded to the 2nd AIF and departed for the Middle East with a field battery of the 7th Division artillery. In the course of the next two years he became brigade major of the divisional artillery and spent a period as a staff officer on the headquarters of the 1st Australian Corps. By the time he returned to Australia he was a lieutenant-colonel, and after commanding the 2/4th Field Regiment was appointed GSO1 to the 3rd Division, a militia formation commanded by Major-General S. G. Savige (q.v.), in June 1942. He remained in the job until August 1943, forgoing the command of the 7th Division's artillery in the process. Savige thought Wilton 'an extraordinarily able man . . . [who] produces excellent results'. Savige was not one of the army's more incisive or cerebral generals, and some observers thought him very dependent on his principal operational staff officer, ironically so given Savige's views on Duntroon graduates.

From September 1943 until the last months of the war, Wilton served in Washington with the Australian Military Mission as the senior staff officer, first to Lieutenant-General Sir Vernon Sturdee and then to Lieutenant-General Sir John Lavarack (qq.v.). He spent two months observing developments in the north-west European theatre at the end of 1944, and on return to Australia in March 1945 was posted to Advanced Land Headquarters at Morotai in a succession of senior staff positions.

In the period between 1946 and 1953, Wilton was Director of Military Operations and Plans for five years before attending the Imperial Defence College (q.v.) in 1952. He was promoted to brigadier, and given command of the 28th Commonwealth Brigade (q.v.) in Korea in succession to another Australian, Brigadier T. J. Daly (q.v.). The Korean War ended in July 1953, and Wilton's period as a brigade commander was necessarily brief, although the level of operational activity in the last months of the war was high. In 1954 he returned to Australia, and between 1955 and 1957 was, among other duties, heavily involved in Australian planning and exercises under SEATO (q.v.) arrangements. Promoted to major-general in March 1957, he served three years as commandant at RMC Duntroon, before returning to SEATO concerns as Chief of the Military Planning Office in Bangkok. From there, he became CGS in January 1963 and was appointed KBE the following year.

Wilton came to leadership of the army at a time of considerable organisational upheaval. Australian units had been on overseas service continuously since 1946 in various parts of Asia, but the army was not well equipped to meet the demands which the 1960s would provide. One of his most important decisions, in December 1964, was to abandon the Pentropic (q.v.) organisation, adopted in 1960, and return to a more traditional establishment for infantry battalions. During his period as CGS the army became increasingly heavily involved in Confrontation (q.v.) against Indonesia in 1965–66, and in the growing war in Vietnam (q.v.), to which Australia dispatched the first regular battalion in 1965. In May 1966 he was promoted to full general and made Chairman of the Chiefs of Staff Committee. His new appointment gave him scope to attempt the reform of the senior command and administration of the services, which he believed were inadequate as a consequence of ill-defined relationships between the minister, permanent head and service chief of each of the armed forces, and because of the lack of a joint service approach to operations, in Vietnam in particular.

During his time at the head of the services, Wilton made a number of moves towards integration and 'jointery': he was a strong supporter of a tri-service academy for the education of officer cadets, one moreover which would provide a tertiary education, an issue which he had first raised while commandant of RMC; the idea of a Joint Services Staff College was revived at his direction in 1967 and opened in 1969; the Joint Intelligence Organisation was established, combining the old Joint Intelligence Bureau and the overseas functions of the intelligence directorates of the three services. He also pushed reform of the defence group of departments, last proposed by the Morshead Committee (q.v.) in 1957, but in this area had little apparent impact. On the other hand, when the Labor government took office in December 1972 Wilton, by now retired, was consulted on defence reorganisation by the new Minister for Defence, Lance Barnard, who proceeded with many of the recommendations for restructuring which Morshead, Wilton and others had long advocated. After his retirement from the army in October 1970 Wilton served as Consul-General in New York between 1973 and 1975.

Wilton impressed throughout his career with the clarity and breadth of his intellect, his capacity for work, his commitment to the army, and the ability to think through a problem from all sides. He performed with equal distinction in command and on the staff at all levels. He did not suffer fools gladly, and was thought by some to lack warmth, having little patience with small talk for its own sake (his nickname was 'Happy Jack'). On the other hand he inspired great affection among those who worked for him, and went out of his way to take an interest in and foster the careers of able younger officers. His was a style well suited to the demands of a difficult period in the army's postwar development.

WINDEYER, Major-General (William John) Victor (28 July 1900–23 November 1987). From a family with long-standing connections with Sydney University and the legal profession, Windeyer was too young to serve in the First World War, but joined the Sydney University Scouts (forerunner of the Sydney University Regiment) in 1919 after entering university to read arts and law. As lieutenant-colonel, Windeyer was appointed to command that regiment in July 1937.

Seconded to the 2nd AIF in May 1940 he took a drop in rank to major. He was posted to the staff of the 7th Division before commanding the 2/48th Battalion in the 9th Division. In January 1942 he took over command of the 20th Brigade in that same division, and led it through the rest of the war in campaigns in North Africa, New Guinea and Borneo. He returned to civilian life and citizen soldiering after 1945, and in July 1950 was appointed CMF Member on the Military Board (q.v.) and commander of the 2nd Division of the CMF. He relinquished the divisional command after two years, but remained on the Military Board until July 1953. In December 1956 he was made honorary colonel of the Sydney University Regiment and was appointed KBE in 1958. His civil career was even more distinguished, as a justice of the High Court, author and company director. He was for many years a member of the Royal Historical Society of Australia.

WINJEEL, CAC (2–3-seat basic trainer). Wingspan 38 feet 8 inches; length 28 feet 1 inch; maximum speed 186 m.p.h.; range 550 miles; power 1 × Pratt & Whitney 450 h.p. engine.

The Winjeel was designed in Australia by the Commonwealth Aircraft Corporation (q.v.) to replace both the Tiger Moth (q.v.) basic trainer and the Wirraway (q.v.) advanced trainer. The name Winjeel means 'young eagle' in an Aboriginal language. The prototype was such a stable aircraft that it was almost impossible to spin, and as putting an aircraft into a spin is a necessary part of pilot training, the tail had to be redesigned to make spinning possible. The Winjeel was the RAAF's basic trainer from 1955 to 1975. It was to have been replaced by the Macchi (q.v.) in 1968 for basic training, but the failure of the 'all through' jet training concept meant the Winjeel was retained as a trainer until the arrival of the CT-4 Airtrainer. Winjeels were used by No. 4 Flight in the Forward Air Control role armed with smoke bombs for target-marking until they were replaced by the Pilatus PC-9s (q.v.) in 1994.

'WINNIE THE WAR WINNER' was a transmitter built from salvaged materials by members of Sparrow Force (q.v.). According to 2/2nd Independent Company officer Bernard Callinan, it 'occupied a room about ten feet square, and there were bits and pieces spread around on benches and joined by wires trailing across the floor'. On 18 April 1942 its weak signal was picked up in Darwin, and the next night an answer was received from Darwin before the transmitter's batteries ran out. Finally, on 20 April Sparrow Force was able to re-establish radio contact with Darwin for the first time since the Japanese landings. 'Winnie' is now held at the Australian War Memorial (q.v.).

WIRRAWAY, CAC (2-seat trainer aircraft). Wingspan 43 feet; length 27 feet 10 inches; armament 3 × 0.303-inch machine-guns; maximum speed 182 m.p.h.; range 720 miles; power 1 × Pratt & Whitney Wasp (q.v.) 600 h.p. engine.

The Wirraway, which means 'challenge' in an Aboriginal language, was a modified version of the North American Aircraft Company NA33 aircraft produced under licence by the Commonwealth Aircraft Corporation (q.v.). In all 755 Wirraways were completed at Commonwealth Aircraft Corporation's plant at Fishermens Bend, Victoria. Although the aircraft was a trainer design, Wirraways were classed 'general purpose' aircraft by the RAAF and fitted with machine-guns and bomb racks. The first Wirraway entered service in July 1939. During the Pacific War Wirraways were used in combat roles with varying degrees of success. As a fighter over Rabaul in January 1942 it was outclassed by Japanese fighters and was never used in this role again. As a dive-bomber during the Malayan campaign (q.v.) it was slightly better. The Wirraway was most successful as an army cooperation aircraft during the New Guinea campaign (q.v.), and it was in New Guinea that a Wirraway scored its only victory over a Japanese Zero, when on 26 December 1942, Pilot Officer John Archer of No. 4 Squadron shot down

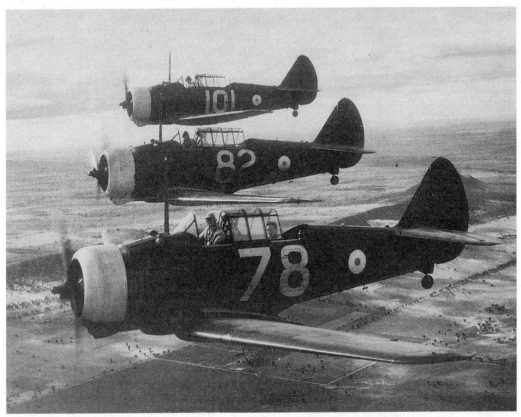

Wirraways of No. 2 Service Flying Training School, Wagga, practise formation flying, June 1941. (AWM P1254/162/006)

a Mitsubishi A6M2 near Gona. The Wirraway was also extensively used as a trainer by the Empire Air Training Scheme (q.v.) and remained in use with the RAAF until 1958 when it was replaced by the Winjeel (q.v.). The RAN used Wirraways for training from 1948 to 1957 when they were replaced by Vampire (q.v.) trainers.

WOLLONGONG (I), HMAS see **BATHURST CLASS MINESWEEPERS (CORVETTES)**

WOLLONGONG (II), HMAS see **FREMANTLE CLASS PATROL BOATS**

WOLVERENE, HMCS see **NEW SOUTH WALES NAVAL FORCES SHIPS**

WOMEN'S AIR TRAINING CORPS see **WOMEN'S AUSTRALIAN AUXILIARY AIR FORCE**

WOMEN'S AUSTRALIAN AUXILIARY AIR FORCE (WAAAF) was the first and largest of the three women's services formed during the the the Second World War. The WAAAF was established in March 1941 and at its peak in October 1944 consisted of 18664 women, or 12 per cent of RAAF personnel. The formation of the WAAAF followed the representations of women, some of whom had already joined voluntary organisations such as the Women's Air Training Corps (WATC) and the Women's Emergency Signalling Corps, and were eager to enlist, and the demands of the CAS Air Chief Marshal Charles Burnett (q.v.) who called for a women's service based on the British Women's Auxiliary Air Force. This would release men for other duties in the rapidly expanding wartime RAAF. The initial temporary commander of the WAAAF was Flight Officer Mary Bell, a pilot and leader of the WATC, however, on 21 May 1941 Squadron Officer Clare Stevenson (q.v.) was appointed first Director of the WAAAF. Air Vice-Marshal H. N. Wrigley (q.v.), then Air Member for Personnel, selected Stevenson because she had practical management experience and was not a 'socialite'. Bell had expected to be made director, and when she was not, resigned from the WAAAF in a fit of pique. As legal opinion within the RAAF in 1941 considered that women could not be enlisted under the terms of the Air Force Act, they were instead enrolled as auxiliaries for renewable

periods of 12 months. In 1943, this opinion was revised and the WAAAF was legally constituted as a part of the RAAF and its members were made liable to RAAF discipline and enlisted for the duration of the war plus a period of 12 months. WAAAF officers could not in most circumstances give orders to men, and had a separate rank structure with different titles, for example 'Group Officer' rather than 'Group Captain'. All members of the WAAAF were paid about two-thirds the salary of their male RAAF equivalents, but as this was normal procedure in the rest of the Australian workplace it was accepted by most women. Initially women enrolled in the WAAAF as wireless telegraphists, but after war broke out in the Pacific they took on other roles until, by 1945, 77 per cent of RAAF musterings, or positions, were open to women. In July 1944, 31.5 per cent of the WAAAF was serving as aircraft ground staff, 30 per cent in administration, 22 per cent in radar and signals, 20.3 per cent in stores and maintenance, and 16.5 per cent in kitchens and messes. No WAAAF personnel were stationed outside Australia, despite the fact women were serving overseas as nurses. Even a request by General Douglas MacArthur (q.v.) in 1944 for WAAAF personnel from wireless units (who intercepted and decoded Japanese messages) to join the US invasion of the Philippines was refused, and the male RAAF personnel of these units served on Leyte without them. The WAAAF was disbanded in December 1947 but during its existence it began the move towards integrating women into previously male only units that has been continued in the postwar services.

Joyce Thomson, *The WAAAF in Wartime Australia* (Melbourne University Press, Melbourne, 1991).

WOMEN'S EMERGENCY SIGNALLING CORPS see **WOMEN'S AUSTRALIAN AUXILIARY AIR FORCE**

WOMEN'S ROYAL AUSTRALIAN AIR FORCE (WRAAF) was formed after cabinet approval was given to the re-formation of all three women's services in July 1950. Unlike its wartime predecessor, it was no longer to be regarded as an auxiliary service, and this was reflected in its title; the 'Royal' prefix was granted in November 1950. Limited initially to an establishment of 30 officers and 832 other ranks, more than 2000 women applied to enlist. The first recruit courses commenced at Laverton, Victoria, and Richmond, New South Wales, in January 1951. Training for the WRAAF was subsequently centralised at Point Cook in 1954, and from 1965 moved to RAAF Edinburgh in South Australia. Originally 21 musterings (i.e. military occupation categories) were open to

WAAAF personnel took over many aspects of aircraft maintenance and repairs. Here Aircraftwomen Norma Carlton and Marion Peel apply dope and new fabric to the wing of an RAAF Avro Anson, 12 November 1942. (AWM 013531)

women in the air force, mostly in traditional employment categories as cooks, stewards, clerks, drivers, orderlies and telephone/teleprinter operators. Other ranks enlisted for four years, shortened to two in 1956 but increased to six years from 1959. Officers were given short service commissions, and permanent commissions were not offered until 1965. The first Director WRAAF was Wing Officer D. J. Carter, appointed in April 1951, who held the position until 1960. As with the other women's services, in the selection of officers in its early years the WRAAF drew heavily on women who had served in the wartime service, the WAAAF (q.v.). The RAAF persisted with different titles for female officers: squadron officer rather than squadron leader, group officer rather than group captain (the Director WRAAF was raised to Group Officer rank in 1968). An early personnel policy had attempted to post female personnel as close to their home towns as possible. This proved unworkable, and in any case was counterproductive as many women recruits, like many men, enlisted precisely to get away from their home areas. WRAAF members were not permitted to serve overseas until 1967. Again, as in the other women's services, pay and conditions were less than those accorded to airmen. In 1958 the Allison Committee determined WRAAF pay rates at 75 per cent of the civilian male basic wage, but only two-thirds of the male rank equivalent in the air force. Allowances and other conditions were identical. Officers were paid between 66 and 68 per cent of the male rank equivalent. Equal pay was introduced gradually following the National Wage Case decision of 1969, but in 1970 officers were still receiving between 76 and 78 per cent of the male wage, while NCOs received 80 per cent. Pay was not equalised until 1972. Before 1969 marriage meant discharge. Until 1960 WRAAF personnel were paid a gratuity on completion of four years service, set at £60; thereafter they became eligible for Defence Force Retirement Benefits. By 1965 the WRAAF establishment had been raised to 1050 all ranks. The broadening of opportunities for women in the air force came slowly, however, and really had to wait until full integration in the early 1980s. Subsequently there is virtually no mustering in the RAAF closed to women.

WOMEN'S ROYAL AUSTRALIAN ARMY CORPS (WRAAC)

The Australian women's services were disbanded immediately after the end of the Second World War, but the pressing need for personnel occasioned by the Korean War, the reintroduction of national service training, and full employment led to the re-formation of limited women's services, beginning in 1950 with the establishment of the Women's Australian Army Corps in December (the 'Royal' prefix was granted in June 1951). Enlistment for the corps began in April 1951 with an initial authorisation of 251 personnel. The first director was Lieutenant-Colonel K. A. L. Best (q.v.). Recruit training was centralised at Lonsdale Bight, in Victoria, with the formation of a recruit Training Company, and a WRAAC school was set up at Mildura to cater for officer and NCO courses. The first officer cadet course began in June 1952, with 22 students. In May 1957 WRAAC training was centralised, with the WRAAC School moving to George's Heights in Sydney and also incorporating a recruit training wing; the Training Company accordingly was disbanded. Permission was also given in December 1951 to raise a WRAAC CMF, for which enlistment commenced in mid-1953.

Best organised the corps closely on the model provided by the wartime Australian Women's Army Service (q.v.), although her own wartime background had been in the nursing service and the Australian Army Medical Women's Service (q.v.). The establishment of the WRAAC owed more to her than to any other individual, but her early death in November 1957 meant that it was left to her successor, Colonel Dawn Jackson, to preside over the according of permanent status to the corps in March 1959. Jackson, who held the senior post for an unprecedented 15 years, saw the purpose of the WRAAC as being to replace potential fighting men, much as it had been during the Second World War. For a long time women in the army were on a differential pay scale — officers received about two-thirds of the rates accorded their male counterparts — and for much of the existence of the WRAAC it was assumed that marriage was a barrier to a full career; all but one of the directors of the corps remained unmarried. The minimum age for enlistment was 17, originally for a period of four years but this was altered later to either three or six years, and it was assumed that few women joining the WRAAC intended to continue to retirement age.

By the final decade of the corps' existence in the 1970s, it had begun to integrate more fully with the mainstream of the regular army: WRAAC officers sat the same promotion exams, members could be posted overseas, marriage was no longer an automatic reason for discharge, and equal pay was approved in 1978. By the early 1970s there were 45 different occupation categories open to the WRAAC, a long way from the early emphasis on clerical and other traditional employment categories that had been thought suitable for women soldiers. In 1977 other ranks were 'corps streamed', that is, allotted to a non-combat corps of the regular army, and officers were given the option to transfer as part of the gradual wind-down of the WRAAC. The responsibilities of those officers who remained in the WRAAC diminished thereafter, as the rest of the army took responsibility for the administration of its women members. Training at recruit and officer cadet level continued to be separated, however, although integration of recruit training was planned when new facilities were opened at Bonegilla in 1981, and integrated officer cadet education awaited the opening of the Australian Defence Force Academy (q.v.) in 1986 and consequent changes to the regime at RMC Duntroon. The WRAAC finally disappeared as a corps of the Australian Army at the beginning of 1984. Approximately 10500 women served in the WRAAC during its existence.

Lorna Ollif, *Colonel Best and Her Soldiers* (Ollif Publishing Company, Sydney, 1985).

WOMEN'S ROYAL AUSTRALIAN NAVAL SERVICE (WRANS)

Like the other women's services, WRANS had its origins in the increasing manpower shortages experienced in 1940–41, especially in technical areas. Before 1939 a small group of women had qualified as telegraphists and formed the Women's Emergency Signalling Corps, but attempts to have them enlisted as telegraphists in the navy were not successful until April 1941, when the RAN reluctantly accepted that it could not meet its needs in this category from any other source. Even then, only 14 women were employed initially (12 telegraphists and two stewards) at HMAS *Harman*, and instructions were given that no publicity was to attend the formation of the women's service. The size of the service increased, especially once the war in the Pacific began, and by 1942 the navy was drawing the existence of the WRANS to the attention of the Women's National Register in each capital city. In July 1942 Navy Office set out formal conditions of service, predicated on an assumption that the RAN could absorb no more than about 600 women; by the end of the year that estimate had grown to 1000. Naval women mostly filled 'traditional' employment categories — drivers, typists, clerks, stewards, cooks and orderlies — although there was some diversification by the war's end, with women working in technical areas such as degaussing ranges (i.e. protecting ships from magnetic mines) and in some intelligence and cryptanalytic positions. With the creation of a service came the need for officers, and between May 1943 and September 1945 seventeen courses for WRANS officers were

conducted at Flinders Naval Depot. Chief Officer Sheilah McClemans (q.v.) was appointed the first Director, WRANS, between 1944 and 1947, and in the course of the war more than 2000 Australian women served in the RAN.

With the end of the war, the service was disbanded. In common with the other services, however, the navy experienced great difficulty meeting its postwar manpower targets, and the WRANS were reconstituted in 1951, on a temporary basis. As in 1941, however, there was considerable reluctance on the part of the RAN to enlist women into the service, and every possible alternative was examined before approval was given. The three women's services became part of the permanent establishment in December 1959, although women's conditions were still much less favourable than those of their male counterparts and many more limitations were imposed on them. Femininity was emphasised in recruiting literature, and the initial establishment was set at just 300, which it took some years to reach. The marriage ban continued to mean that most women did not pursue the navy as a career. Recruit training was conducted at HMAS *Cerberus* followed by specialist training according to occupation category. The number of occupations open to women gradually broadened, as did the range of postings which came to include a small detachment of WRANS serving with the Naval Communications Detachment in Singapore. By the early 1970s the establishment had grown to 683, with women posted to all nine major shore establishments. The ban on women serving at sea, except for very brief familiarisation periods, was maintained, however, throughout the life of the service, and was only broken down with the abolition of the WRANS and the integration of women into the mainstream of the navy in the early 1980s.

M. Curtis-Otter, *W.R.A.N.S.: The Women's Royal Australian Naval Service* (Naval Historical Society of Australia, Sydney, 1975).

WOOMERA see LONG RANGE WEAPONS PROJECT, WOOMERA

WOOTTEN, Major-General George Frederick

(1 May 1893–30 March 1970). A graduate of the first class at RMC Duntroon in August 1914, Wootten, originally appointed machine-gun officer, became adjutant of the 1st Battalion the day after the landing at Gallipoli, on 26 April 1915, and stayed with his unit until early November as a company commander. He spent most of the rest of that war in staff jobs, as brigade major at various times in France to the 8th, 11th and 9th Brigades, as GSO3 of the 3rd Division under Monash's command, and

finally as GSO2 on Haig's Headquarters between October 1918 and March 1919. Mentioned in despatches five times and awarded the DSO, he was selected to attend the first postwar course at the Staff College, Camberley, as a member of one of the most brilliant and experienced staff college courses in that institution's history, but in the postwar reductions initiated by Hughes's (q.v.) government, resigned his commission in April 1923 and left the army to practise law as a solicitor. He maintained his contact with military life through service in the CMF, becoming commanding officer of the 21st Light Horse Regiment in July 1937.

At the outbreak of the Second World War he was given command of the 2/2nd Battalion, although Major-General Blamey (q.v.) raised questions about his physical fitness because of his excessive weight. His competence and reputation won out, and he commanded the battalion until November 1940 when he was posted to set up the AIF Reinforcements Depot in Egypt, a job which he did not hold long enough, however, to make any impact at all. After a further spell elsewhere in the Middle East as a camp commandant he was given the 18th Brigade in February 1941, which he reorganised and trained before leading it through the siege of Tobruk. He returned with the 7th Division to Australia at the end of the year, and led the 18th Brigade through the difficult and demanding fighting at Buna and Sanananda which finally cleared the Japanese from Papua. He returned to Australia and succeeded Lieutenant-General Leslie Morshead (q.v.) in command of the 9th Division (newly returned from service at Alamein), and oversaw their conversion to jungle warfare through intensive training before commanding them in the fighting for Lae, Finschhafen and Satelberg and, at the war's end, at Tarakan in Borneo.

Following the war he remained in the CMF and became the first CMF Member of the Military Board and GOC of the 3rd Division, before finally retiring from the army in 1950. Between 1945 and 1958 he was Chairman of the Repatriation Commission, and helped to ease the transition of the newest generation of 'returned men' back into civilian life, for which services he was knighted in 1958. Widely respected by those he commanded, he was judged 'competent, clear-headed and knowledgeable' by his commanding general in the 7th Division, Major-General J. E. S. Stevens (q.v.);

Major-General G. F. Wootten (second from left), while GOC 9th Division, at Scarlet Beach, near Finschhafen, New Guinea, with (left to right) Colonel A. R. Garrett, Brigadier B. Evans and Lieutenant-General Sir Leslie Morshead, GOC II Corps, 25 October 1943. (AWM 059141)

the official historian, Gavin Long (q.v.), encountered him at his headquarters in New Guinea and found him 'fat, perspiring, bespectacled [and] sitting in front of his map in a large wooden armchair'. Had he remained a regular it seems probable that he would have gone on to the highest positions in the postwar army.

WOUNDED AND MISSING INQUIRY BUREAU
see **AUSTRALIAN RED CROSS; DEAKIN, VERA**

WRIGLEY, Air Marshal Henry Neilson (21 April 1892–14 September 1987). Born in Collingwood, Melbourne and trained as a school teacher, Wrigley enlisted in the Australian Flying Corps in 1916 and served with No. 3 Squadron in France, winning the DFC. In 1919, Wrigley, with Sergeant A. W. Murphy, made the first flight across Australia, flying a BE2e aircraft from Point Cook in Victoria to Darwin and was awarded the AFC. An original member of the RAAF in 1921, Wrigley attended the RAF College, and was promoted steadily throughout the interwar period. Following the outbreak of the Second World War he commanded No. 1 Group and Southern Area before being appointed to the Air Board (q.v.) as Air Member for Personnel in 1940. In this position he was involved in the formation of the Women's Australian Auxiliary Air Force (q.v.) in 1941 and selected Group Officer Clare Stevenson (q.v.) as WAAAF Director. In 1942 Wrigley went to London as AOC RAAF Overseas Headquarters, where he remained until 1946. Overseas Headquarters operated from the Australian High Commission and administered the thousands of RAAF personnel from the Empire Air Training Scheme (q.v.) who were scattered in units throughout the United Kingdom. Wrigley was, however, prevented from exerting any authority over these men by the British. Appointed CBE in 1941, he retired from the RAAF in 1946 and settled in Melbourne. Wrigley deserves recognition as an author as well as a career RAAF officer. He produced a history of his AFC Squadron entitled *The Battle Below* in 1935 and a history of the Victorian branch of the United Services Institution (q.v.) in 1980. His notebooks, including lectures he gave as RAAF Headquarters Training Officer in during the 1920s, were printed in book form as proof that, although the RAAF's *Air Power Manual* was not published until 1990, air doctrine had been taught and discussed informally within the RAAF since its formation.

H. N. Wrigley (edited by Alan Stephens and Brendan O'Loghlin) *The Decisive Factor: Air Power Doctrine* (AGPS, Canberra, 1990).

WYNTER, Lieutenant-General Henry Douglas
(5 June 1886–7 February 1945). Commissioned
into the Wide Bay Regiment as a lieutenant in
February 1907, Wynter took the examination for
transfer to the regular army in December 1910.
Passing with very high marks, he was posted to the
Administrative and Instructional Corps as a proba-
tionary lieutenant, although he had been promoted
to captain in the militia two years previously. He
held minor staff posts in Queensland and Victoria,
and was seconded to the AIF in April 1916, following the threefold expansion of that force, as
brigade major to the new 11th Infantry Brigade.
He stayed with troops only until October, and for
the rest of the war filled staff jobs on the headquarters of the 4th Division and at AIF Headquarters.
He returned to Australia only in February 1920.

After a short stint in Australia he was sent to the
Staff College, Camberley, in 1921-22. In 1925 he
published an article on the command of Empire
forces in war in the British *Army Quarterly* (q.v.),
there being no professional military journals in
Australia at that time. In 1925 he became Director
of Mobilisation, a new appointment which he held
for four and a half years and to which he returned
in 1931 after attendance at the Imperial Defence
College (q.v.) in London in 1930. In February
1935 he became Director of Military Training at
Army Headquarters, during which time his career
was to come under a cloud.

Wynter had published another paper in the
Army Quarterly in 1927, dealing with the strategic
relationships between the three Australian services.
It was critical of the emphasis placed on the navy
in Australian defence policy, with its corresponding
implications for budget share, and argued that 'the
defence of Australia is not in itself the first object of
Imperial naval policy'. This was true, but implied
criticism of prevailing government policy. He reit-
erated many of these arguments in lectures to the
Sydney and Brisbane branches of the United Service Institution in August 1935, following Sir Maurice Hankey's (q.v.) visit to Australia in late 1934.
The Minister for Defence, Sir Archdale Parkhill
(q.v.), requested and was sent a copy of Wynter's
lecture. So too was Senator H. C. Brand, a veteran
of the AIF and retired militia major-general, who
appears to have copied the paper and circulated it
among members of parliament, among them the
leader of the Opposition, John Curtin (q.v.). In the
debate on the defence estimates in late 1936

Curtin made a strong attack on government policy
and on the minister, Parkhill, based largely on Wynter's paper. Parkhill, a small-minded man and one
often out of his depth in the Defence portfolio, had
Wynter reassigned as GSO1 of the 11th Mixed
Brigade in Queensland, a clear demotion. The
announcement was made after the minister left the
country to attend the coronation in London.

Parkhill lost his seat in the 1937 election, and in
July 1938 Wynter's career was revived when he was
appointed commandant of the new Command and
Staff School in Sydney. When war came he was
posted initially as GOC of Northern Command,
based on Queensland, but in April 1940 was seconded to the AIF as Deputy Adjutant and Quartermaster-General on Blamey's 1st Australian Corps
Headquarters. Diverted with a convoy to Britain in
June, he became for a while commander of the AIF
in Britain and, in October, first GOC of the 9th
Division, which was formed from two brigades
based in the United Kingdom. The new division
moved to the Middle East in late 1940, but Wynter's health broke down and he was returned to
Australia in February the next year.

Wynter's health precluded his service in operational areas for the remainder of the war. With the
outbreak of the Pacific War and the rearrangement
of the army's command structure in April 1942, he
was appointed Lieutenant-General-in-Charge of
Administration at Allied Land Headquarters. The
responsibilities of the position included the control
of administrative policy and general direction of
the principal staff branches of the army, administrative planning, and coordination of policy and
administration between branches. He died in office
at the 115th Military Hospital in Heidelberg, Victoria, six months before the war's end.

Gavin Long (q.v.), the official historian, regarded
Wynter as 'perhaps the clearest and most profound
thinker the Australian Army of his generation had
produced'. He was one of a very small group of
Australian officers in the interwar period who tried
to generate debate on the strategic issues facing
Australia, and who offered a clear and persistent
critique of government policy, in Wynter's case at
some personal cost. Ill health denied the opportunity of command in the field which few observers
doubted he would have discharged with his usual
clear-eyed ability. His premature death was a loss to
the regular army as it adjusted to the difficult postwar period.

YZ

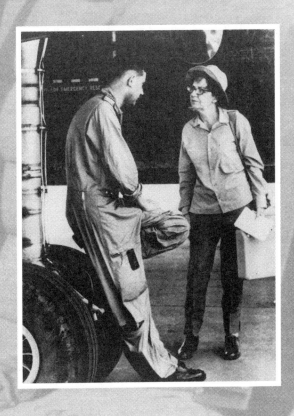

YARRA (I), HMAS see **RIVER CLASS TORPEDO BOAT DESTROYERS**

YARRA (II), HMAS see *GRIMSBY* **CLASS SLOOPS**

YARRA (III), HMAS see **RIVER CLASS ANTI-SUBMARINE FRIGATES**

YOUNG MEN'S CHRISTIAN ASSOCIATION (YMCA), an international non-sectarian Christian social organisation, began working with Australian soldiers during the Boer War. Its services to the armed forces have included providing reading, writing and recreational materials; organising social, recreational and religious activities; and setting up facilities such as clubs, cinemas and canteens. In both world wars its coffee stalls, like those of the Salvation Army (q.v.), have operated on the front lines and been very popular with troops. During the First World War, as one of three voluntary national organisations officially appointed to accompany the AIF overseas (the others were the Australian Comforts Fund and the Australian Red Cross [qq.v.]), the YMCA sent 211 men overseas as official uniformed representatives. It spent a total of £791 000 abroad and about £225 000 in Australia on First World War work. During the Second World War the Australian Comforts Fund was responsible for raising funds for the war work of the YMCA, as well as for the Salvation Army and the Young Women's Christian Association (which established clubs and hostels for servicewomen). Between September 1939 and December 1946 the YMCA spent £716 655 on services to Australian troops overseas. It had 302 official uniformed representatives attached to the Army and RAAF, 158 of them serving overseas. The Australian YMCA also supplied religious and recreational materials to Australian POWs during the Second World War and arranged for citizens of neutral countries to visit POW camps as its representatives. It had only limited success in assisting prisoners of the Japanese, however. The YMCA provided similar services for enemy POWs and alien internees (see ALIENS, WARTIME TREATMENT OF) in Australia.

Z SPECIAL UNIT was the holding unit established during the Second World War for AIF personnel serving with Special Operations Australia (q.v.), also known as the Inter-Allied Services Department or Services Reconnaissance Department. Z Special Unit was formed in June 1942 and unit members served with RAAF, RAN, British, Dutch and American personnel on secret operations in Japanese-occupied territory, the most famous of which are the JAYWICK/RIMAU raids (q.v.) on Singapore in 1943 and 1944.

APPENDIX

Chiefs of Staff of the armed forces

Chairmen, Chiefs of Staff Committee

Lieutenant-General Sir Henry Wells	23 March 1958	22 March 1959
Vice-Admiral Roy Dowling (q.v.)	23 March 1959	27 May 1961
Air Chief Marshal Sir Frederick Scherger (q.v.)	28 May 1961	18 May 1966
General Sir John Wilton (q.v.)	19 May 1966	22 November 1970
Admiral Sir Victor Smith (q.v.)	23 November 1970	22 November 1975
General Sir Francis Hassett (q.v.)	24 November 1975	8 February 1976

Chiefs of the Defence Force Staff

General Sir Francis Hassett (q.v.)	9 February 1976	20 April 1977
General Sir Arthur MacDonald (q.v.)	21 April 1977	20 April 1979
Admiral Sir Anthony Synnot (q.v.)	21 April 1979	20 April 1982
Air Chief Marshal Sir Neville McNamara (q.v.)	21 April 1982	12 April 1984
General Sir Phillip Bennett	13 April 1984	25 October 1986

Chiefs of the Defence Force

General Sir Phillip Bennett	26 October 1986	12 April 1987
General P. C. Gration	13 April 1987	16 April 1993
Admiral A. L. Beaumont	17 April 1993	6 July 1995
General J. S. Baker	7 July 1995	

Royal Australian Navy

Chiefs of Naval Staff

Rear-Admiral Sir William Creswell (q.v.)	1 March 1911	9 June 1919
Rear-Admiral Sir Edmund Grant (q.v.)	10 June 1919	14 February 1921
Commodore C. T. Hardy (acting)	15 February 1921	23 November 1921
Vice-Admiral Sir Allan Everett	24 November 1921	29 August 1923
Commodore G. F. Hyde (acting) (q.v.)	30 August 1923	24 February 1924
Rear-Admiral P. N. Hall-Thompson	25 February 1924	4 February 1925
Commodore H. P. Cayley	5 February 1925	25 April 1925
Rear-Admiral P. N. Hall-Thompson	26 April 1925	28 June 1926
Rear-Admiral W. R. Napier	29 June 1926	11 June 1929
Commodore J. B. Stevenson (acting)	12 June 1929	29 October 1929
Vice-Admiral W. M. Kerr	21 October 1929	19 October 1931
Admiral Sir George Hyde (q.v.)	20 October 1931	28 July 1937
Commodore G. P. Thompson (acting)	29 July 1937	31 October 1937
Admiral Sir Ragnar Colvin (q.v.)	1 November 1937	3 March 1941
Commodore J. W. Durnford (acting)	4 March 1941	17 July 1941
Admiral Sir Guy Royle (q.v.)	18 July 1941	28 June 1945
Vice-Admiral Sir Louis Hamilton (q.v.)	29 June 1945	1 August 1945
Commodore G. D. Moore (acting)	2 August 1945	20 September 1945
Vice-Admiral Sir Louis Hamilton (q.v.)	21 September 1945	23 February 1948
Vice-Admiral Sir John Collins (q.v.)	24 February 1948	23 February 1955
Vice-Admiral Sir Roy Dowling (q.v.)	24 February 1955	23 February 1959
Vice-Admiral Sir Henry Burrell (q.v.)	24 February 1959	23 February 1962
Vice-Admiral Sir Hastings Harrington (q.v.)	24 February 1962	23 February 1965
Vice-Admiral Sir Alan McNicoll (q.v.)	24 February 1965	2 April 1968
Vice-Admiral Sir Victor Smith (q.v.)	3 April 1968	22 November 1970
Vice-Admiral Sir Richard Peek	23 November 1970	22 November 1973
Vice-Admiral H. D. Stevenson	23 November 1973	22 November 1976
Vice-Admiral Sir Anthony Synnot (q.v.)	23 November 1976	20 April 1979
Vice-Admiral Sir James Willis	21 April 1979	20 April 1982
Vice-Admiral D. W. Leach	21 April 1982	20 April 1985
Vice-Admiral M. W. Hudson	21 April 1985	8 March 1991
Vice-Admiral I. D. G. MacDougall	9 March 1991	9 March 1994
Vice-Admiral R. G. Taylor	10 April 1994	

cont. next page

Chiefs of Staff of the armed forces *cont.*

Australian Army

Chiefs of the General Staff

Colonel William Throsby Bridges (q.v.)	1 January 1909	25 April 1909
Major-General J. C. Hoad (q.v.)	26 May 1909	30 May 1911
Lieutenant-Colonel A. F. Wilson (acting)	1 June 1911	10 May 1912
Brigadier-General J. M. Gordon (q.v.)	11 May 1912	31 July 1914
Colonel J. G. Legge (q.v.)	1 August 1914	19 May 1915
Colonel G. G. H. Irving (temporary)	24 May 1915	1 December 1915
Colonel H. J. Foster (temporary) (q.v.)	1 January 1916	30 September 1917
Major-General J. G. Legge (q.v.)	1 October 1917	31 May 1920
Major-General Sir Brudenell White (q.v.)	1 June 1920	10 June 1923
General Sir Harry Chauvel (q.v.)	11 June 1923	15 April 1930
Major-General W. A. Coxen (q.v.)	1 May 1930	30 September 1931
Major-General J. H. Bruche (q.v.)	1 October 1931	20 April 1935
Major-General J. D. Lavarack (q.v.)	21 April 1935	12 October 1939
Lieutenant-General E. K. Squires (q.v.)	13 October 1939	26 January 1940
Major-General J. Northcott (acting) (q.v.)	27 January 1940	17 March 1940
General Sir Brudenell White (q.v.)	18 March 1940	13 August 1940
Lieutenant-General V. A. H. Sturdee (q.v.)	30 August 1940	9 September 1942
Lieutenant-General J. Northcott (q.v.)	10 September 1942	30 November 1945
Lieutenant-General V. A. H. Sturdee (q.v.)	1 March 1946	16 April 1950
Lieutenant-General Sir Sydney Rowell (q.v.)	17 April 1950	15 December 1954
Lieutenant-General Sir Henry Wells	16 December 1954	22 March 1958
Lieutenant-General Sir Ragnar Garrett	23 March 1958	30 June 1960
Lieutenant-General Sir Reginald Pollard	1 July 1960	20 January 1963
Lieutenant-General Sir John Wilton (q.v.)	21 January 1963	18 May 1966
Lieutenant-General Sir Thomas Daly (q.v.)	19 May 1966	18 May 1971
Lieutenant-General Sir Mervyn Brogan	19 May 1971	19 November 1973
Lieutenant-General F. G. Hassett (q.v.)	20 November 1973	23 November 1975
Lieutenant-General A. L. MacDonald (q.v.)	24 November 1975	20 April 1977
Lieutenant-General Sir Donald Dunstan (q.v.)	21 April 1977	14 February 1982
Lieutenant-General P. H. Bennett	15 February 1982	12 February 1984
Lieutenant-General P. C. Gration	13 April 1984	12 April 1987
Lieutenant-General L. G. O'Donnell	13 April 1987	15 April 1990
Lieutenant-General H. J. Coates	16 April 1990	30 April 1992
Lieutenant-General J. C. Grey	1 May 1992	7 July 1995
Lieutenant-General J. M. Sanderson (q.v.)	8 July 1995	

Royal Australian Air Force

Chiefs of the Air Staff

Australian Air Corps	Brevet Major R. Williams (q.v.)	12 November 1920	30 March 1921
Australian Air Force	Wing-Commander R. Williams (q.v.)	31 March 1921	12 August 1921
Royal Australian Air Force	Wing-Commander R. Williams (q.v.)	13 August 1921	14 December 1922
	Wing-Commander S. J. Goble (acting) (q.v.)	15 December 1922	9 February 1925
	Air Commodore R. Williams (q.v.)	10 February 1925	6 December 1932
	Air Commodore S. J. Goble (acting) (q.v.)	7 December 1932	12 June 1934
	Air Vice-Marshal R. Williams (q.v.)	13 June 1934	27 February 1939
	Air Commodore S. J. Goble (acting) (q.v.)	28 February 1939	8 January 1940
	Air Commodore W. H. Anderson (acting)	9 January 1940	10 February 1940
	Air Chief Marshal Sir Charles Burnett (q.v.)	11 February 1940	4 May 1942
	Air Marshal G. Jones (q.v.)	5 May 1942	13 January 1952
	Air Marshal Sir Donald Hardman (q.v.)	14 January 1952	17 January 1954
	Air Marshal Sir John McCauley (q.v.)	18 January 1954	18 March 1957
	Air Marshal Sir Frederick Scherger (q.v.)	19 March 1957	28 May 1961
	Air Marshal Sir Valston Hancock (q.v.)	29 May 1961	31 May 1965
	Air Marshal Sir Alister Murdoch (q.v.)	1 June 1965	31 December 1969
	Air Marshal Sir Colin Hannah (q.v.)	1 January 1970	20 March 1972
	Air Marshal C. F. Read	21 March 1972	20 March 1975
	Air Marshal Sir James Rowland (q.v.)	21 March 1975	20 March 1979
	Air Marshal Sir Neville McNamara (q.v.)	21 March 1979	20 April 1982
	Air Marshal S. D. Evans	21 April 1982	20 May 1985
	Air Marshal J. W. Newham	21 May 1985	2 July 1987
	Air Marshal R. G. Funnell	3 July 1987	1 October 1992
	Air Marshal I. B. Gration	2 October 1992	29 November 1994
	Air Marshal I. B. Fisher	30 November 1994	

Ranks and titles given are those held during the period of office cited.

Prime ministers

Sir Edmund Barton	1 January 1901	24 September 1903
Alfred Deakin	24 September 1903	27 April 1904
John Christian Watson (q.v.)	27 April 1904	17 August 1904
George Reid	18 August 1904	5 July 1905
Alfred Deakin	5 July 1905	13 November 1908
Andrew Fisher (q.v.)	13 November 1908	1 June 1909
Alfred Deakin	2 June 1909	29 April 1910
Andrew Fisher (q.v.)	29 April 1910	24 June 1913
Joseph Cook	24 June 1913	17 September 1914
Andrew Fisher (q.v.)	17 September 1914	27 October 1915
William Morris Hughes (q.v.)	27 October 1915	9 February 1923
Stanley Melbourne Bruce	9 February 1923	22 October 1929
James Henry Scullin	22 October 1929	6 January 1932
Joseph Aloysius Lyons	6 January 1932	7 April 1939
Earle Page	7 April 1939	26 April 1939
Robert Gordon Menzies (q.v.)	26 April 1939	29 August 1941
Arthur Fadden	29 August 1941	7 October 1941
John Curtin (q.v.)	7 October 1941	5 July 1945
Francis Michael Forde (q.v.)	6 July 1945	13 July 1945
Joseph Benedict Chifley	13 July 1945	19 December 1949
Sir Robert Gordon Menzies (q.v.)	19 December 1949	26 January 1966
Harold Edward Holt	26 January 1966	19 December 1967
John McEwen	19 December 1967	10 January 1968
John Grey Gorton (q.v.)	10 January 1968	10 March 1971
William McMahon	10 March 1971	5 December 1972
Edward Gough Whitlam	5 December 1972	11 November 1975
John Malcolm Fraser (q.v.)	11 November 1975	5 March 1983
Robert James Hawke	5 March 1983	20 December 1991
Paul Keating	20 December 1991	

Titles given are those held during the period of office cited.

Ministers of defence-related departments

Air

Ministers for Air			
	James Fairbairn (q.v.)	13 November 1939	13 August 1940
	Arthur Fadden	14 August 1940	28 October 1940
	John McEwen	28 October 1940	7 October 1941
	Arthur Drakeford	7 October 1941	19 December 1949
	Thomas White	19 December 1940	11 May 1951
	Philip McBride (q.v.)	11 May 1951	17 July 1951
	William McMahon	17 July 1951	9 July 1954
	Athol Townley (q.v.)	9 July 1954	24 October 1956
	Frederick Osborne	24 October 1956	29 December 1960
	Senator Harrie Wade	29 December 1960	22 December 1961
	Leslie Bury	22 December 1961	27 July 1962
	David Fairbairn	4 August 1962	10 June 1964
	Peter Howson	10 June 1964	28 February 1968
	Gordon Freeth	28 February 1968	13 February 1969
	Dudley Erwin	13 February 1969	12 November 1969
	Senator Thomas Drake-Brockman	12 November 1969	5 December 1972
	Lance Barnard (q.v.)	5 December 1972	30 November 1973

Army

Ministers for the Army			
	Geoffrey Street (q.v.)	13 November 1939	13 August 1940
	Senator Sir Philip McBride (q.v.)	14 August 1940	28 October 1940
	Percy Spender (q.v.)	28 October 1940	7 October 1941
	Francis Michael Forde (q.v.)	7 October 1941	1 November 1946
	Cyril Chambers (q.v.)	1 November 1946	19 December 1949
	Josiah Francis	19 December 1949	7 November 1955
	Sir Eric Harrison (q.v.)	7 November 1955	28 February 1956
	John Cramer	28 February 1956	18 December 1963
	Alexander Forbes (q.v.)	18 December 1963	26 January 1966
	John Malcolm Fraser (q.v.)	26 January 1966	28 February 1968
	Phillip Lynch	28 February 1968	12 November 1969

cont. next page

Ministers of defence-related departments *cont.*

Army *cont.*

Ministers for the Army *cont.*			
	Andrew Peacock	12 November 1969	2 February 1972
	Robert Katter Snr	2 February 1972	5 December 1972
	Lance Barnard (q.v.)	5 December 1972	30 November 1973

Defence

Ministers for Defence (I)			
	Sir James Dickson	1 January 1901	10 January 1901
	Sir John Forrest (q.v.)	17 January 1901	10 August 1903
	Senator James Drake	10 August 1903	24 September 1903
	Austin Chapman	24 September 1903	27 April 1904
	Senator Anderson Dawson	27 April 1904	17 August 1904
	James McCay (q.v.)	18 August 1904	5 July 1905
	Senator Thomas Playford	5 July 1905	24 January 1907
	Sir Thomas Ewing	24 January 1907	13 November 1908
	Senator George Pearce (q.v.)	13 November 1908	1 June 1909
	Joseph Cook	2 June 1909	29 April 1910
	Senator George Pearce (q.v.)	29 April 1910	24 June 1913
	Senator Edward Millen (q.v.)	24 June 1913	17 September 1914
	Senator George Pearce (q.v.)	17 September 1914	21 December 1921
	Walter Massy-Greene	21 December 1921	5 February 1923
	Eric Bowden	9 February 1923	16 January 1925
	Sir Neville Howse (q.v.)	16 January 1925	2 April 1927
	Sir Thomas Glasgow (q.v.)	2 April 1927	22 October 1929
	Albert Green	22 October 1929	4 February 1931
	Senator John Daly	4 February 1931	3 March 1931
	Joseph Benedict Chifley	3 March 1931	6 January 1932
	Senator Sir George Pearce (q.v.)	6 January 1932	12 October 1934
	Sir Archdale Parkhill (q.v.)	12 October 1934	20 November 1937
	Joseph Aloysius Lyons	20 November 1937	29 November 1937
	Harold Thorby	29 November 1937	7 November 1938
	Geoffrey Street (q.v)	7 November 1938	13 November 1939
Assistant Ministers for Defence (I)	Granville de Laune Ryrie (q.v.)	4 February 1920	21 December 1921
	Josiah Francis	6 January 1932	12 October 1934

Ministers without Portfolio assisting the Minister for Defence (I)

	Harold Thorby	7 November 1938	24 November 1938
	James Fairbairn (q.v.)	26 April 1939	14 March 1940

Ministers for Defence Co-ordination

	Robert Gordon Menzies (q.v.)	13 November 1939	7 October 1941
	John Curtin (q.v.)	7 October 1941	14 April 1942

Ministers for Defence (II)			
	John Curtin (q.v.)	14 April 1942	5 July 1945
	John Beasley	6 July 1945	15 August 1946
	Francis Michael Forde (q.v.)	15 August 1946	1 November 1946
	John Dedman (q.v.)	1 November 1946	19 December 1949
	Eric Harrison (q.v.)	19 December 1949	24 October 1950
	Sir Philip McBride (q.v.)	24 October 1950	10 December 1958
	Athol Townley (q.v.)	10 December 1958	18 December 1963
	Paul Hasluck (q.v.)	18 December 1963	24 April 1964
	Senator Shane Paltridge (q.v.)	24 April 1964	19 January 1966
	Allen Fairhall (q.v.)	26 January 1966	12 November 1969
	John Malcolm Fraser (q.v.)	12 November 1969	8 March 1971
	John Grey Gorton (q.v.)	10 March 1971	13 August 1971
	David Fairbairn	13 August 1971	5 December 1972
	Lance Barnard (q.v.)	5 December 1972	6 June 1975
	William Morrison	6 June 1975	11 November 1975
	Denis James Killen (q.v.)	11 November 1975	7 May 1982
	Ian Sinclair	7 May 1982	11 March 1983
	Gordon Scholes	11 March 1983	13 December 1984
	Kim Beazley (q.v.)	13 December 1984	4 April 1990
	Senator Robert Ray	4 April 1990	

Ministers of defence-related departments *cont.*

Defence *cont.*

Ministers assisting the Minister for Defence (II)

Senator Reginald Bishop	19 December 1972	12 June 1974
William Morrison	12 June 1974	6 June 1975
Senator Reginald Bishop	6 June 1975	11 November 1975
John McLeay	22 December 1975	3 November 1980
Kevin Newman	3 November 1980	7 May 1982
Ian Viner	7 May 1982	11 March 1983
Kim Beazley (q.v.)	11 March 1983	13 December 1984
John Brown	13 December 1984	24 July 1987
Michael Duffy	13 December 1984	24 July 1987

Parliamentary Secretary for Defence Science and Personnel

Ros Kelly	24 July 1987	18 September 1987

Parliamentary Secretaries to the Minister for Defence

Roger Price	27 December 1991	24 March 1993
Gary Punch	24 March 1993	25 March 1993
Arch Bevis	25 March 1993	

Defence Science and Personnel

Ministers for Defence Science and Personnel

Ros Kelly	18 September 1987	6 April 1989
David Simmons	6 April 1989	4 April 1990
Gordon Bilney	4 April 1990	24 March 1993
Senator John Faulkner	24 March 1993	25 March 1994
Gary Punch	25 March 1994	

Defence Support

Ministers for Defence Support

Ian Viner	7 May 1982	11 March 1983
Brian Howe	11 March 1983	13 December 1984

Munitions

Ministers for Munitions

Robert Gordon Menzies (q.v.)	11 June 1940	28 October 1940
Senator Philip McBride (q.v.)	28 October 1940	7 October 1941
Norman Makin	7 October 1941	15 August 1946
John Dedman (q.v.)	15 August 1946	1 November 1946
Senator J. I. Armstrong	1 November 1946	6 April 1948

Ministers assisting the Minister for Munitions

Senator D. Cameron	7 October 1941	21 February 1942
E. J. Holloway	21 February 1942	21 September 1943

Navy

Ministers for the Navy (I)

Jens Jensen	12 July 1915	17 February 1917
Sir Joseph Cook	17 February 1917	28 July 1920
William Smith	28 July 1920	21 December 1921

Ministers for the Navy (II)

Sir Frederick Stewart	13 November 1939	14 March 1940
Archie Cameron	14 March 1940	28 October 1940
William Morris Hughes (q.v.)	28 October 1940	7 October 1941
Norman Makin	7 October 1941	15 August 1946
Arthur Drakeford	15 August 1946	1 November 1946
William Riordan	1 November 1946	19 December 1949
Josiah Francis	19 December 1949	11 May 1951
Philip McBride (q.v.)	11 May 1951	17 July 1951
William McMahon	17 July 1951	9 July 1954
Josiah Francis	9 July 1954	7 November 1955
Sir Eric Harrison (q.v.)	7 November 1955	11 January 1956
Senator Neil O'Sullivan	11 January 1956	24 October 1956
Charles Davidson	24 October 1956	10 December 1958
Senator John Grey Gorton (q.v.)	10 December 1958	18 December 1963
Alexander Forbes (q.v.)	18 December 1963	4 March 1964
Frederick Chaney	4 March 1964	14 December 1966
Donald Chipp	14 December 1966	28 February 1968

cont. next page

Ministers of defence-related departments *cont.*

Navy *cont.*

Ministers for the Navy (II) *cont.*

Charles Kelly	28 February 1968	12 November 1969
Denis James Killen (q.v.)	12 November 1969	22 March 1971
Malcolm Mackay	22 March 1971	5 December 1972
Lance Barnard (q.v.)	5 December 1972	30 November 1973

Repatriation

Ministers for Repatriation (I)

Senator Edward Millen (q.v.)	28 September 1917	9 February 1923
Frank Anstey	22 October 1929	3 March 1931
John McNeill	3 March 1931	6 January 1932
Charles Hawker	6 January 1932	12 April 1932
Charles Marr	12 April 1932	12 October 1934
William Morris Hughes (q.v.)	12 October 1934	6 October 1935
Joseph Aloysius Lyons	8 November 1935	6 February 1936
William Morris Hughes (q.v.)	6 February 1936	29 November 1937
Senator Hattil Foll	29 November 1937	26 April 1939
Eric Harrison (q.v.)	26 April 1939	14 March 1940
Geoffrey Street (q.v.)	14 March 1940	13 August 1940
Senator Philip McBride (q.v.)	14 August 1940	28 October 1940
Senator George McLeay	28 October 1940	26 June 1941
Senator Herbert Collett	26 June 1941	7 October 1941
Charles Frost	7 October 1941	1 November 1946
Herbert Barnard	1 November 1946	19 December 1949
Senator Walter Cooper	19 December 1949	29 December 1960
Frederick Osborne	29 December 1960	22 December 1961
Reginald Swartz	22 December 1961	22 December 1964
Senator Gerald McKellar	22 December 1964	12 November 1969
Rendele Holten	12 November 1969	5 December 1972
Lance Barnard (q.v.)	5 December 1972	19 December 1972
Senator Reginald Bishop	19 December 1972	12 June 1974

Assistant Ministers for Repatriation

Arthur Rodgers	28 July 1920	21 December 1921
Hector Lamond	21 December 1921	9 February 1923

Ministers without Portfolio assisting the Minister for Repatriation

Josiah Francis	12 October 1934	9 November 1934
Harold Thorby	9 November 1934	1 September 1935
James Hunter	1 September 1935	29 November 1937
Senator Herbert Collett	14 March 1940	26 June 1941

Ministers for Repatriation and Compensation

Senator John Wheeldon	12 July 1974	11 November 1975
Donald Chipp	11 November 1975	22 December 1975

Ministers for Repatriation (II)

Kevin Newman	22 December 1975	8 July 1976
Senator Peter Durack	8 July 1976	5 October 1976

Supply

Ministers for Supply

Richard Casey	26 April 1939	26 January 1940
Sir Frederick Stewart	26 January 1940	28 October 1940
Senator Philip McBride (q.v.)	28 October 1940	26 June 1941
Senator George McLeay	26 June 1941	7 October 1941
J. A. Beasley	7 October 1941	2 February 1945
Senator William Ashley	2 February 1945	6 April 1948
Senator John Armstrong	6 April 1948	19 December 1949
Richard Casey	19 December 1949	17 March 1950
Howard Beale	17 March 1950	10 February 1958
Athol Townley (q.v.)	10 February 1958	10 December 1958
Alan Hulme	10 December 1958	22 December 1961
Allen Fairhall (q.v.)	22 December 1961	26 January 1966
Senator Norman Henty	26 January 1966	28 February 1968
Senator Sir Kenneth Anderson	28 February 1968	2 August 1971
Ramsley Victor Garland	2 August 1971	5 December 1972
Lance Barnard (q.v.)	5 December 1972	9 October 1973
Keppel Enderby	9 October 1973	12 June 1974

Ministers of defence-related departments *cont.*

Veterans' Affairs

Ministers for Veterans' Affairs			
	Senator Peter Durack	5 October 1976	6 September 1977
	Ramsley Victor Garland	6 September 1977	4 July 1978
	Albert Adermann	4 July 1978	3 November 1980
	Senator Tony Messner	3 November 1980	11 March 1983
	Senator Arthur Gietzelt	11 March 1983	24 July 1987
	Benjamin Humphreys	24 July 1987	24 March 1993
	Senator John Faulkner	24 March 1993	25 March 1994
	Con Sciacca	25 March 1994	

Titles given are those held during the period of office cited.

Public servant heads of defence-related departments

Air

Secretaries of the Department of Air			
	M. C. Langslow	15 November 1939	20 December 1951
	Sir Edwin Hicks (q.v.)	22 December 1951	28 October 1956
	A. B. McFarlane	28 October 1956	31 May 1968
	F. J. Green	1 June 1968	30 November 1973

Army

Secretaries of the Department of the Army			
	J. T. Fitzgerald	13 November 1939	11 August 1941
	F. R. Sinclair	21 August 1941	29 January 1955
	A. D. McKnight	30 January 1955	29 April 1958
	B. White (q.v.)	30 April 1958	5 February 1973
	J. B. R. Livermore	5 February 1973	30 November 1973

Defence

Secretaries of the Department of Defence (I)			
	R. H. M. Collins (q.v.)	1 July 1901	1 March 1910
	Sir Samuel Pethebridge (q.v.)	1 March 1910	1 February 1918
	T. Trumble (q.v.)	1 February 1918	15 July 1927
	M. L. Shepherd (q.v.)	15 July 1927	17 November 1937
	F. G. Shedden (q.v.)	17 November 1937	13 November 1939

Department of Defence Co-ordination			
	F. G. Shedden (q.v.)	13 November 1939	14 April 1942

Secretaries of the Department of Defence (II)			
	Sir Frederick Shedden (q.v.)	14 April 1942	29 October 1956
	Sir Edwin Hicks (q.v.)	29 October 1956	8 January 1968
	Sir Henry Bland (q.v.)	8 January 1968	2 March 1970
	Sir Arthur Tange (q.v.)	2 March 1970	16 August 1979
	W. B. Pritchett	17 August 1979	3 February 1984
	R. W. Cole	4 February 1984	15 October 1986
	A. J. Wood	17 November 1986	31 July 1988
	A. J. Ayers	1 August 1988	

Munitions

Secretaries of the Department of Munitions			
	J. B. Brigden	1940	1941
	J. K. Jensen	1942	1948

Navy

Secretary of the Department of the Navy (I)			
	G. L. Macandie (q.v.)	12 July 1915	21 December 1921

Secretaries of the Department of the Navy (II)			
	A. R. Nankervis	30 November 1939	9 March 1950
	T. J. Hawkin	9 March 1950	4 November 1963
	S. Landau (q.v.)	4 November 1963	30 November 1973

cont. next page

Public servant heads of defence-related departments *cont.*

Repatriation

Comptrollers of the Department of Repatriation (I)

N. C. Lockyer	1917	1918
D. J. Gilbert	1918	1920

Chairmen of the Repatriation Commission

J. M. Semmens	1920	1935
N. R. Mighell	1935	1941
J. Webster	1941	1945
G. F. Wootten (q.v.)	1945	1958
Sir Frederick Chilton (q.v.)	1958	1970
R. Kingsland	1970	1974

Secretary of the Department of Repatriation and Compensation

R. Kingsland	1974	1975

Secretary of the Department of Repatriation (II)

R. Kingsland	1975	1976

Supply

Secretaries of the Department of Supply

Sir Daniel McVey	25 May 1939	30 November 1939
J. B. Bridgon	1 January 1940	30 June 1941
A. S. V. Smith	1 July 1941	11 April 1945
Sir Giles Chippindall	12 April 1945	15 September 1946
F. A. O'Connor	16 September 1946	5 April 1948
Sir John Jenson	6 April 1948	31 July 1949
H. P. Breen	1 August 1949	10 May 1951
Sir Edwin Stevens	11 May 1951	15 April 1953
F. A. O'Connor	16 April 1953	12 October 1959
Sir John Knott	13 October 1959	12 March 1965
A. S. Cooley	12 March 1965	1 November 1971
N. S. Currie	1 November 1971	12 June 1974

Veterans' Affairs

Secretaries of the Department of Veterans' Affairs

R. Kingsland	1976	1981
D. Volker	1981	1986
N. J. Tanzer	1986	1989
L. B. Woodward	1989	1994
A. Hawke	1994	

Titles given are those held during the period of office cited.